Tradeoffs and Optimization in Analog CMOS Design

Tradeoffs and Optimization in Analog CMOS Design

David M. Binkley

University of North Carolina at Charlotte, USA

A John Wiley & Sons, Ltd., Publication

Other Wiley Editorial Offices

John Wiley & Sons Inc., 111 River Street, Hoboken, NJ 07030, USA

Jossey-Bass, 989 Market Street, San Francisco, CA 94103-1741, USA

Wiley-VCH Verlag GmbH, Boschstr. 12, D-69469 Weinheim, Germany

John Wiley & Sons Australia Ltd, 42 McDougall Street, Milton, Queensland 4064, Australia

John Wiley & Sons (Asia) Pte Ltd, 2 Clementi Loop #02-01, Jin Xing Distripark, Singapore 129809

John Wiley & Sons Ltd, 6045 Freemont Blvd, Mississauga, ONT, L5R 4J3

Wiley also publishes its books in a variety of electronic formats. Some content that appears in print may not be
available in electronic books.

Library of Congress Cataloging in Publication Data

Binkley, David M.
   Tradeoffs and optimization in analog CMOS design / David M. Binkley.
      p.   cm.
   Includes bibliographical references and index.
   ISBN 978-0-470-03136-0
   1. Metal oxide semiconductors, Complementary—Design and construction.
   I. Title.
   TK7871.99.M44.B56 2007
   621.3815′2—dc22

                                              2007037318

British Library Cataloguing in Publication Data

A catalogue record for this book is available from the British Library

ISBN 978-0-470-03136-0

Typeset in 9/11pt Times by Integra Software Services Pvt. Ltd, Pondicherry, India

To my wife Jacqueline, children Anna and Christopher, and father Jerry Binkley.

To the memory of my mother Carol Binkley, her parents William and Lucie Dexter, and my father's parents Robert and Oneda Binkley.

To the memory of my teachers Charles Eason, Harry Kroll, John Abel, and T. Vaughn Blalock.

Contents

Foreword

Analog circuit design is often considered an art, sometimes even a "mystical" art with a taste of "black magic." The art aspect probably comes from the creativity that is required to find, modify, or eventually invent a circuit that implements the function to be realized, and from the skills that are necessary to understand and master the right tradeoffs between all the different constraints to achieve the target specifications. It is also an engineering science, since it definitively requires a clear methodology in order to successfully complete the design in good time. An important step in analog circuit design is the search for the appropriate circuit that can achieve the desired function. Once the circuit is defined, the designer has to find the right tradeoffs and then size the different components, and specifically choose the bias current, width, and length of MOS transistors to achieve the desired specifications. It is very often, at this early step of the design process, that simple device models that allow for hand calculation, such as the EKV MOS transistor model, are required in order to predict the performance and do a first-order circuit optimization. Once all the devices are sized, the circuit can then be simulated in more detail with a circuit simulator, fine-tuned, and finally verified over process, voltage, and temperature variations.

Most of the time the sizing methodology uses simplistic square-law transistor models that are often inaccurate in strong inversion and totally wrong in weak inversion. With today's aggressive downscaling of CMOS technologies, the operating points of transistors used in analog circuits are often chosen in between weak and strong inversion, in the so-called moderate inversion region.

The EKV MOS transistor model introduced the powerful concept of inversion coefficient (IC) – also sometimes called inversion factor – as the main transistor design parameter that is more general and replaces the long-used overdrive voltage, distinctive of strong inversion models. It allows for covering all the regions of operation of the MOS transistor in a continuous way. This approach allows for expressing all the important parameters of a single transistor, such as small-signal parameters, including transconductances, capacitances, and noise parameters, simply in terms of the inversion coefficient. Sweeping the inversion coefficient therefore allows exploration of all the design space for a single transistor to find the appropriate bias point.

For more than 25 years now, Eric Vittoz and I have been teaching analog circuit design based on the use of this concept of inversion coefficient. In 1996, I created a first sizing tool, actually called Analog Designer, which runs on a Mac [1]. The graphical user interface is shown in Figure 1. It allowed calculation of all the important design parameters from the basic inversion coefficient, corresponding to the top axis in the figure. Then all the other parameters were updated simultaneously when sweeping the inversion coefficient by moving the vertical bar. The designer could then optimize the operating point most appropriate for the particular task of the single transistor to be sized.

I remembered the first time I met David Binkley at a workshop in California in 2000. We had breakfast together and quickly entered into a discussion about analog design methodology using the inversion coefficient. I was surprised to discover that a professor in the USA was aware of this approach; one that we had been teaching for many years, but which had never really broken through into the designer community in the USA, which was still using the simplistic square law MOS transistor model. We had such a passionate discussion that I forgot about my invited talk and had to be reminded to

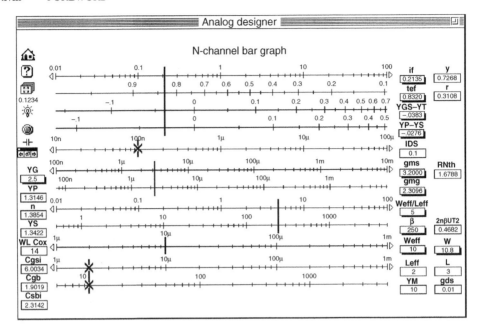

Figure 1 The Analog Designer tool illustrating the basics of the design methodology based on the inversion coefficient [1]

quickly run and give my presentation on the EKV MOS transistor model and the inversion coefficient to the waiting audience! At that time David didn't really talk about the project of writing a book, but I could already see that, with the strong practical experience he had accumulated in applying this methodology to industrial designs, he would be the right person to write a book introducing and highlighting the power of the inversion coefficient design methodology.

David's book is the first to introduce this powerful design methodology. It even extends it far beyond what was initially created by Eric Vittoz and myself, to look at all the different design cases that may be faced. For example, it extends design hand expressions to include important effects such as velocity saturation and mobility reduction due to the vertical field (VFMR), allowing for a more accurate design for short-channel devices at high IC. Drain induced barrier lowering (DIBL) is also included in the calculation of the voltage gain, since it often dominates the channel length modulation (CLM) effect at short transistor length and low IC. All these effects, including flicker-noise increases with inversion, are carefully validated by extensive normalized data measured on a 0.18 μm CMOS process. All these concepts are very nicely illustrated by several design examples, including operational transconductance amplifiers, optimized for various tradeoffs in DC and AC performance, and micropower, low-noise preamplifiers, optimized for minimum thermal and flicker noise. These examples have been fabricated and clearly demonstrate the strength and accuracy of this methodology. Finally, David has also developed a very helpful spreadsheet that not only facilitates the design and optimization of devices and circuits, but also remains as very useful documentation describing how the different transistors of a given circuit have been sized to target the specifications. All these extensions and improvements make David's work unique and his book indispensable to any analog designer.

Due to the vast design space existing for the optimization of analog circuits, and even for the sizing of a single transistor, writing a book on this topic was a real challenge. In this book, David has brought together all the important bricks required to build up a consistent and practical design methodology that can now be applied efficiently to the design of analog CMOS circuits. I'm delighted that what I have always believed to be a powerful methodology is now brought to a wider audience. This book will certainly become an extremely useful reference for any analog circuit designer. I will finally have the book I need for teaching my analog design classes!

PROFESSOR CHRISTIAN C. ENZ
Swiss Center for Electronics and Microtechnology
Swiss Federal Institute of Technology, Lausanne

REFERENCE

[1] C. C. Enz and E. A. Vittoz, "CMOS Low-power Analog Circuit Design," in *Emerging Technologies, Designing Low-power Digital Systems, Tutorial for 1996 International Symposium on Circuits and Systems*, Eds. R. Cavin and W. Liu, pp. 79–133. IEEE Service Center, Piscataway, 1996.

Preface

During the late 1990s, my colleagues and I were designing a large CMOS integrated circuit that is now installed in commercial positron emission tomography (PET) medical imaging systems. During the course of the design project, Scott Puckett asked me, "How do you select the channel width and length for MOS transistors in analog circuits?" We then went to the board in our conference room where I attempted to answer Scott's question. Because there are so many measures of analog performance and the performance depends on operation in weak, moderate, or strong inversion, the board was quickly filled with ideas and expressions. During that one afternoon, I could only begin to consider the important question of how to select sizing as well as drain current for MOS transistors in analog circuits. Now, almost 10 years later, this book is an attempt to answer the drain current and sizing question. This involves managing the many tradeoffs of performance to obtain a design that optimally meets the application requirements.

As I thought more about the drain current and sizing question, I realized that the trial-and-error practice of selecting drain current and MOS transistor sizing through iterative circuit simulations consumes considerable design time and provides little design intuition. It seemed clear that preselecting the drain current, inversion level in weak, moderate, or strong inversion, and channel length was the most physically based and time-efficient way to size MOS transistors in analog circuits. With this design approach, the usual selection of channel width is not a design choice, but width is easily found for expressions, simulations, and layout from the selected drain current, inversion level, and channel length. The use of the inversion coefficient[1] as a numerical level of MOS inversion and simple expressions of MOS performance valid in weak, moderate, and strong inversion enabled by the EKV MOS model seemed ideal for this design approach. In fact, this design approach is similar to approaches developed by Christian Enz, Eric Vittoz, and others, but these are not widely known or used at present. I hope to have introduced some new material here and to have detailed design methods that were useful to me as an industry designer.

The design methods described in this book are an extension of the work my colleagues and I were doing at Concorde Microsystems. This included an early spreadsheet design tool that Jim Rochelle referred to as "preSPICE." That name captures the intent of the material contained in this book: to provide design intuition and guidance leading the designer towards optimum analog CMOS design *prior* to launching the computer simulations required to validate a design. Chapter 1 introduces the material contained in this book further and provides an overview of the topics.

It is my hope that this book will be helpful to students and designers as they strive to design and optimize the performance of analog circuits using the remarkably interesting and complex MOS transistor.

The companion website for this book is www.wiley.com/go/binkley_tradeoffs

David M. Binkley
Charlotte, North Carolina

---

[1] References for the inversion coefficient and EKV MOS model follow in Chapter 2, along with discussions of weak, moderate, and strong inversion.

Acknowledgments

This book would not have been possible without the EKV MOS model that provides predictions of MOS performance in weak, moderate, and strong inversion with the clarity and simplicity needed to guide the designer. I am indebted to Christian Enz, Eric Vittoz, Francois Krummenacher, Matthias Bucher, Wladek Grabinski, and everyone involved in the development of the EKV model, along with Daniel Foty who first introduced me to this model. Additionally, I am indebted to Christian Enz and Eric Vittoz for introducing me to the MOS inversion factor or inversion coefficient, which is fundamental to the material contained in this book.

The field of MOS modeling and analog circuit design rests on an extensive and rich history. Although I have included several hundred references to the work of others, mostly in Chapter 3, I was unable to cite all the previous work of others. If I missed citing a key work, please feel free to let me know and I will include the citation in a possible second edition of this book.

I am grateful for the personal encouragement and ideas of Christian Enz and Yannis Tsividis. I am also grateful to Helmut Graeb and Ulf Schlichtmann for hosting me during the final months of this work at the Institute for Electronic Design Automation at the Technical University of Munich. Through a variety of discussions, I have gained an appreciation for the connections between MOS modeling, analog circuit design, and electronic design automation.

I will always be grateful to Ron Nutt, Terry Douglass, Kelly Milam, and Mike Crabtree who led the commercial development of PET medical imaging at CTI PET Systems. After over 20 years of dedicated work, PET did become the clinical reality envisioned by these remarkable individuals and is now used for the biological imaging of cancer, heart, brain, and other disorders around the world. It was the complex front-end electronic requirements in PET medical imaging that led Jim Rochelle, Brian Swann, Mike Paulus, Scott Puckett, and me to become engaged in analog CMOS design. Many thanks to Ron Nutt for taking a chance on a small design group at a small company that ventured into analog CMOS design. This was only possible because of the depth of knowledge and inspiration the design group received earlier from T. Vaughn Blalock, Ed Kennedy, and Jim Rochelle at the University of Tennessee in Knoxville. The design group was also supported by Clif Moyers and Mike Casey at CTI PET Systems and by an extended community of designers that included Chuck Britton, Mike Simpson, Nance Ericson, Alan Wintenberg, and Lloyd Clonts at the nearby Oak Ridge National Laboratory. I am also grateful to Robert Nutt, Rhonda Goble, and Stefan Siegel who led the commercial development of small-animal PET imaging at Concorde Microsystems. Both CTI PET Systems and Concorde Microsystems are now part of Siemens, which continues to advance the life-saving mission of PET medical imaging.

I will always remember with fondness the years that Jim Rochelle and I worked side by side at CTI PET Systems and Concorde Microsystems, and I was fortunate to receive so much thoughtful mentoring from Jim. I know industry designers reading this know of the countless design hours, all-night work, and excitement that Jim and I experienced releasing integrated circuits containing analog CMOS circuits. Jim also introduced analog CMOS design to me and others earlier at the University of Tennessee and was my PhD advisor.

I am grateful to Don Bouldin and Chuck Stroud for guidance and encouragement during my mid-career transition to the university. I am also grateful to my dean, Bob Johnson, and department chairs, Farid Tranjan and Lee Casperson, at the University of North Carolina at Charlotte for their encouragement as I wrote this book. Thanks also to students Norbert Ulshoefer, David Ihme, Clark Hopper, Brian Moss (it is true that his middle initial is "C"), Steve Tucker, Srikanth Mohan, Jeremy Yager, and Nikhil Verma, and university colleagues Arun Ravindran, Tom Weldon, and Steve Bobbio for help with design ideas, measurements, computer programming, integrated circuit layout, and photomicrographs. Thanks also to Mohammad Mojarradi, Ben Blalock, Harry Lee, Alan Mantooth, and Bill Kuhn for enjoyable research collaborations, and to Ben Blalock, Nance Ericson, and Andries Scholten for measurements contributed to this book.

I am grateful to John Wiley & Sons Ltd for publishing this book during its 201st year as a publisher. I very much appreciate the encouragement, patience, and professionalism of my senior editor, Simone Taylor, and production manager, Mary Lawrence, during this long project. Thanks also to Sarah Kiddle for content editing and Laura Bell for help with the cover design. I also appreciate the thoughtful comments from Wiley's reviewers and the guidance and encouragement from experienced authors Chuck Stroud and Jim Conrad.

Finally, I am especially grateful for the encouragement of my wife Jacqueline and children Anna and Christopher during the continuous and seemingly endless five years it took to write this book. I am really enjoying gaining back our time together. I am also grateful for the encouragement of my father Jerry Binkley.

List of Symbols and Abbreviations

Common symbols and abbreviations are organized below by meaning. Some symbols and abbreviations defined and used locally are not included.

Table 0.1 Symbols related to MOS operating parameters

Symbol	Description	Reference (first or primary)
	Geometry	
A_D	Drain area	3.9.4
A_S	Source area	3.9.4
l_p	Depletion-region width between drain and pinched-off channel	Figure 3.9
L	Effective channel length,[a] $L = L_{drawn} - DL$	Figure 3.9
L_{drawn}	Drawn channel length	3.3.2
m	Interdigitation multiplier (number of gate stripes)	3.9.4
P_D	Drain perimeter	3.9.4
P_S	Source perimeter	3.9.4
S	Shape factor or aspect ratio, W/L	3.6.1
S	Distance between devices (the context clarifies this from S above)	3.11.2.1
t_{dep}	Depletion-region thickness in p/n junction or below MOS channel	Figure 3.9
W	Effective channel length,[a] $W = W_{drawn} - DW$	Figure 3.9
W_{drawn}	Drawn channel width	3.3.2
WL	Effective gate area	3.6.3
	Voltages and potentials	
$\delta V_T / \delta V_{DS}$	Change in threshold voltage with drain–source voltage caused by DIBL	3.8.4.2
ϕ_s	Silicon surface potential[b]	3.4.1
LE_{CRIT}	Velocity saturation voltage, $L \cdot E_{CRIT}$	3.7.1.5
$(LE_{CRIT})'$	Equivalent velocity saturation voltage including VFMR effect, $L \cdot E_{CRIT} \parallel 1/\theta$	3.7.1.5
V_{CB}	Channel–body voltage[b]	Table 3.1
V_{DB}	Drain–body voltage[b]	Figure 3.9
V_{DS}	Drain–source voltage[b]	Figure 3.9
$V_{DS,sat}$	Drain–source saturation voltage[b]	3.7.3
V_{EFF}	Effective gate–source voltage,[c] $V_{GS} - V_T$	3.7.2
$V_{EFF,L}$	Large-geometry, effective gate–source voltage,[c] $V_{GS} - V_T$, without increases due to velocity saturation and VFMR	3.7.2.2

Table 0.1   *(continued)*

Symbol	Description	Reference (first or primary)
V_{GS}	Gate–source voltage[b]	Figure 3.9
V_{INV}	Voltage related to inversion charge	3.12.1.1
V_{SB}	Source–body voltage[b]	Figure 3.9
V_T	Threshold voltage[b] (U_T is used for the thermal voltage)	3.7.1.7
V_{TO}	Threshold voltage[b] for zero V_{SB}	3.7.1.7
V_{ox}	Voltage across the gate oxide	3.12.1.1
Electric fields		
E_x	Horizontal electric field (in direction of drain to source)	3.7.1.2
E_y	Vertical electric field (in direction of gate to channel)	3.7.1.3
Currents and current densities		
I_D	Total drain current,[b] $I_D = I_{DS} + I_{DB}$, equal to I_{DS} when I_{DB} is negligible as usual	Figure 3.9
$I_{D,L}$	Large-geometry, total drain current[b] without decreases due to velocity saturation and VFMR	3.7.3.3
I_{DB}	Drain–body current[b]	3.8.4.3, 3.12.3
I_{DS}	Drain–source current[b]	3.8.4.3
I_G	Total gate current, $I_G = I_{GS} + I_{GD}$	3.12.1.1
I_{GD}	Gate–drain current	3.12.1.1
I_{GS}	Gate–source current	3.12.1.1
I_{SB}	Source–body current[b]	3.12.3
J_{DB}	Drain–body leakage current density	3.12.3
J_{GS}	Gate–source leakage current density	3.12.1.1
J_{SB}	Source–body leakage current density	3.12.3
J_s	Diode saturation current density	3.12.3
Parameters related to drain current		
k	Operating value of transconductance factor, $\mu C'_{OX}$	3.3.1
μ	Operating value of carrier mobility	3.3.1
$\mu C'_{OX}$	Operating value of transconductance factor, k	3.3.1
n	Operating value of substrate factor	3.4.1
S	Weak-inversion (subthreshold) swing voltage, $2.3nU_T$ (the context clarifies this from $S = W/L$)	2.4.1
VSF	Velocity saturation factor, $V_{EFF}/(LE_{CRIT})$	3.7.1.5
VSF'	Velocity saturation factor including VFMR, $V_{EFF}/(LE_{CRIT})'$	3.7.1.5
Charges		
Q_{inv}	Total inversion layer charge	3.10.2.1
Q'_{inv}	Inversion layer charge per unit area	3.12.1.1
Resistances		
R_B	Body series resistance including contact resistance	3.10.4
R_D	Drain series resistance including source–drain extension and contact resistance	3.10.4
R_G	Gate series resistance including contact resistance	3.10.4
R_S	Source series resistance including source–drain extension and contact resistance	3.10.4
Total capacitances		
C_{DB}	Total drain–body capacitance[d]	Figure 3.21
C_{DBO}	Total drain–body capacitance,[d] zero bias ($V_{DB} = 0\,\text{V}$)	3.9.4
C_{GB}	Total gate–body capacitance, $C_{gbi} + C_{GBO}$	Figure 3.21

C_{GD}	Total gate–drain capacitance, $C_{gdi} + C_{GDO}$	Figure 3.21
C_{GS}	Total gate–source capacitance, $C_{gsi} + C_{GSO}$	Figure 3.21
C_{SB}	Total source–body capacitance[d]	Figure 3.21
C_{SBO}	Total source–body capacitance,[d] zero bias ($V_{SB} = 0\,\mathrm{V}$)	3.9.4

Intrinsic capacitances

C_{dbi}	Intrinsic drain–body capacitance	3.9.5
C_{gbi}	Intrinsic gate–body capacitance	3.9.2
C_{gdi}	Intrinsic gate–drain capacitance	3.9.2
C_{gsi}	Intrinsic gate–source capacitance	3.9.2
C_{sbi}	Intrinsic source–body capacitance	3.9.5
$\hat{C}_{gbi}$	Normalized intrinsic gate–body capacitance, C_{gbi}/C_{GOX}	3.9.2
$\hat{C}_{gdi}$	Normalized intrinsic gate–drain capacitance, C_{gdi}/C_{GOX}	3.9.2
$\hat{C}_{gsi}$	Normalized intrinsic gate–source capacitance, C_{gsi}/C_{GOX}	3.9.2

Extrinsic overlap capacitances

C_{GBO}	Extrinsic gate–body overlap capacitance	3.9.3
C_{GDO}	Extrinsic gate–drain overlap capacitance	3.9.3
C_{GSO}	Extrinsic gate–source overlap capacitance	3.9.3

Other capacitances

C_{GOX}	(Total) gate-oxide capacitance, $C'_{OX} WL$	3.9.1
C'_{DEP}	Depletion capacitance below channel per unit area	3.4.1
C'_{INT}	Interface-state capacitance per unit area	3.4.1
$\hat{C}_J$	Drain, source area capacitance reduction factor due to non-zero V_{DB} or V_{SB}	3.9.4
$\hat{C}_{JSW}$	Drain, source perimeter capacitance reduction factor due to non-zero V_{DB} or V_{SB}	3.9.4

Small-signal parameters

η	Body-effect transconductance ratio, g_{mb}/g_m, where $g_{mb} = \eta \cdot g_m$	3.8.3
g_{ds}	Drain–source conductance ($1/r_{ds}$)	3.8.4
g_{ds}/I_D	Drain–source conductance efficiency, where $g_{ds} = (g_{ds}/I_D) \cdot I_D = I_D/V_A{}^e$	3.8.4
g_{gs}	Gate–source conductance associated with gate leakage	3.12.1.2
g_{gs}/I_{GS}	Gate–source conductance efficiency, where $g_{gs} = (g_{gs}/I_{GS}) \cdot I_{GS}$	3.12.1.2
g_m	Transconductance, operated by gate	3.8.2
g_m/I_D	Transconductance efficiency, where $g_m = (g_m/I_D) \cdot I_D$	3.8.2
g_{mb}	Body-effect transconductance, operated by body	3.8.3
g_{ms}	Total source transconductance, $g_{ms} = g_m + g_{mb} + g_{ds}$	3.8.1
r_{ds}	Drain–source resistance ($1/g_{ds}$)	3.8.4
V_A	Early voltage, used as normalized drain–source conductance, where $g_{ds} = 1/r_{ds} = I_D/V_A{}^e$	3.8.4
V_{AL}	Early voltage factor, V_A/L, where $V_A = V_{AL} \cdot L$	3.8.4.1
V_{gds}	Drain–source conductance effective voltage, $(g_{ds}/I_D)^{-1}$, where $g_{ds} = 1/r_{ds} = I_D/V_{gds}$; $V_{gds} = V_A{}^e$	3.8.4
V_{gm}	Transconductance effective voltage, $(g_m/I_D)^{-1}$, where $g_m = I_D/V_{gm}$	3.8.2.1

Intrinsic gain measures

A_{ii}	Intrinsic gate-to-drain current gain, g_m/g_{gs}, for short-circuit drain load	3.12.2.2
A_{Vi}	Intrinsic gate-to-drain voltage gain, $g_m/g_{ds} = g_m \cdot r_{ds}$, for open-circuit drain load	3.8.5

Frequencies related to bandwidth

f_{diode}	Diode-connected bandwidth including extrinsic gate overlap and drain–body capacitances, $g_m/[2\pi(C_{gsi} + C_{GSO} + C_{gbi} + C_{GBO} + C_{DB})]$	3.9.7

Table 0.1 (*continued*)

Symbol	Description	Reference (first or primary)
f_{gate}	Minimum frequency of operation before gate–source conductance affects operation	3.12.2.1
f_{Ti}	Intrinsic bandwidth, $g_m/[2\pi(C_{gsi}+C_{gbi})]$	3.9.6
f'_{Ti}	Intrinsic bandwidth including C_{gsi} only, $g_m/(2\pi C_{gsi})$	3.10.6
f_T	Extrinsic bandwidth including extrinsic gate overlap capacitances, $g_m/[2\pi(C_{gsi}+C_{GSO}+C_{gbi}+C_{GBO})]$	3.9.7
	Frequencies related to noise	
χ	Dimensionless parameter associated with relating f_c to f_{Ti}, $(\hat{C}_{gsi}+\hat{C}_{gbi})/(n\Gamma)$	3.10.3.8
f_c	Corner frequency, usually for flicker noise	3.10.3.8
$f_{flicker}$	Frequency for flicker-noise evaluation	A.2.1.3
f_m	Measurement frequency for noise measurements	3.10.3.8
	Noise	
$\Delta Q'_t$	Fluctuating flicker-noise charge per unit area near the gate and gate-oxide interface	3.10.3.1
ΔV_{fb}	Fluctuating flicker-noise, flat-band voltage due to $\Delta Q'_t$	3.10.3.1
ΔV_g	Fluctuating flicker-noise, gate voltage due to $\Delta Q'_t$	3.10.3.1
Γ	Thermal-noise factor relative to g_m for operation in saturation	3.10.2.1
K_F	Operating value of flicker-noise factor or normalized flicker noise where S_{VG} flicker $(f) = K_F/(C'^2_{OX}WLf^{AF})$; in units of C^2/cm^2	3.10.3.4
K'_F	Same as K_F but in hand units of $(nV)^2 \cdot \mu m^2$	3.10.3.4
K_{FSPICE}	Same as K_F but in SPICE units of $V^2 \cdot F$	3.10.3.4
N	Total number of carriers or fluctuators	3.10.3.2
N'_t	Effective trap density near the gate and gate-oxide interface	3.10.3.1
$N_t(E)$	Number of traps near the gate and gate-oxide interface per unit volume and energy	3.10.3.1
R_N	Equivalent thermal-noise resistance	3.10.2.3
S_I	General noise current PSD[f]	3.10.1
$S_I^{1/2}$	General noise current density[f]	3.10.1
S_{ID}	Drain-referred noise current PSD[f]	Figure 3.60, 3.10.2.3, 3.10.3.7
$S_{ID}^{1/2}$	Drain-referred noise current density[f]	Figure 3.60, 3.10.2.3, 3.10.3.7
S_{IG}	Gate noise current PSD[f]	Figure 3.60, 3.10.6, 3.10.7
$S_{IG}^{1/2}$	Gate noise current density[f]	Figure 3.60, 3.10.6, 3.10.7
S_V	General noise voltage PSD[f]	3.10.1
$S_V^{1/2}$	General noise voltage density[f]	3.10.1
S_{Vfb}	Flat-band, flicker-noise voltage PSD	3.10.3.1
S_{VG}	Gate-referred noise voltage PSD[f]	Figure 3.60, 3.10.2.3, 3.10.3.7
$S_{VG}^{1/2}$	Gate-referred noise voltage density[f]	Figure 3.60, 3.10.2.3, 3.10.3.7
	Mismatch	
A_{VGS}	Operating value of local-area, gate–source voltage mismatch factor, including non-zero-V_{SB} effect and transconductance factor mismatch	3.11.1.4

A_{VT}	Operating value of local-area, threshold-voltage mismatch factor, including non-zero-V_{SB} effect	3.11.1.1
$\Delta\gamma$	Body-effect factor mismatch	3.11.1.1
Δg_{ds}	Drain–source conductance mismatch	3.11.4.2
$\Delta g_{ds}/g_{ds}$	Relative drain–source conductance mismatch	3.11.4.2
Δg_m	Transconductance mismatch	3.11.4.1
$\Delta g_m/g_m$	Relative transconductance mismatch	3.11.4.1
ΔI_D	Drain current mismatch	3.11.1.4
$\Delta I_D/I_D$	Relative drain current mismatch	3.11.1.4
ΔI_{GS}	Gate–source current mismatch	3.12.2.5
ΔK_P	Transconductance factor mismatch	3.11.1.1
$\Delta K_P/K_P$	Relative transconductance factor mismatch	3.11.1.1
Δn_{dop}	Mismatch in number of active dopant atoms in the body below the gate	3.11.1.1
$\Delta Q'_B$	Mismatch in depletion charge density in the body below the gate	3.11.1.1
ΔV_{GS}	Gate–source voltage mismatch	3.11.1.4
ΔV_{TO}	Threshold-voltage mismatch for zero V_{SB}	3.11.1.1
ΔV_T	Threshold-voltage mismatch for non-zero V_{SB}	3.11.1.1
n_{dop}	Average number of active dopant atoms in the body below the gate	3.11.1.1
Q'_B	Average depletion charge density in the body below the gate	3.11.1.1
S_{VT}	Operating value of distance, threshold-voltage mismatch factor, including non-zero-V_{SB} effect	3.11.2.1

Leakage and distortion

P_{TUN}	Probability of tunneling through the gate oxide	3.12.1.1
$V_{INDIF1dB}$	Input, 1 dB compression voltage for a differential pair	3.8.2.6

Inversion coefficient

IC	Inversion coefficient, fixed normalized, where $IC = I_D/[2n_0\mu_0 C'_{OX} U_T^2(W/L)] = I_D/[I_0(W/L)]$ is linearly linked to drain current and device geometry; velocity saturation and VFMR reductions of μ are considered outside the IC definition	3.4.2.2
IC'	Inversion coefficient, traditional, where $IC' = I_D/[2n\mu C'_{OX} U_T^2(W/L)]$	3.4.2.1
IC_{CRIT}	Critical inversion coefficient where g_m/I_D is reduced to approximately 70.7 % of its value without velocity saturation and VFMR effects	3.8.2.2

---

[a] Sometimes, especially in schematic diagrams and measurements, L is used interchangeably with L_{drawn} and W is used interchangeably with W_{drawn}.

[b] These voltages and currents are taken positively for nMOS devices. In this book, they are also taken positively for pMOS devices as shown in Figure 3.9.

[c] $V_{EFF} = V_{GS} - V_T$ is negative for nMOS devices for V_{GS} less than V_T (as in weak inversion) and positive for V_{GS} greater than V_T (as in strong inversion). In this book, these signs are also used for pMOS devices as shown in Figure 3.9.

[d] The drain–body and source–body capacitances consist of the diode junction capacitances where the intrinsic drain–body and source–body capacitances, C_{dbi} and C_{sbi}, are neglected as described in Section 3.9.5.

[e] As customary, this definition of V_A assumes V_{DS} is included in the value of V_A such that $g_{ds} = 1/r_{ds} = I_D/V_A$.

[f] The noise PSD or density is qualified directly or by context as thermal, flicker, shot, or total noise. Noise voltage PSD and density are frequently given in design units of $(nV)^2/Hz$ and $nV/Hz^{1/2}$, respectively. Noise current PSD and density are frequently given in design units of $(pA)^2/Hz$ and $pA/Hz^{1/2}$, respectively.

Table 0.2   Symbols related to MOS process parameters

Symbol	Description	Reference (first or primary)
	Geometry	
DL	Lateral diffusion for length where $L = L_{drawn} - DL$	3.3.2
DW	Lateral diffusion for width where $W = W_{drawn} - DW$	3.3.2
L_{min}	Minimum channel length in the process	3.3.2
t_{ox}	Gate-oxide thickness, effective electrical value	Figure 3.9
W_{DIF}	Drain (source) stripe width used in calculation of A_D, A_S, P_D, and P_S	3.9.4
x_j	Drain junction depth	3.8.4.1
	Doping concentration	
N_A	Acceptor (p-type) doping concentration	3.3.1
N_B	Average body doping concentration for the region of interest	3.3.1
N_D	Donor (n-type) doping concentration	3.3.1
N_{DIF}	Doping concentration in drain (source) regions	3.12.3
n_i	Intrinsic carrier concentration of silicon, $1.5 \times 10^{10}/\text{cm}^3$ at $T = 300\,\text{K}$	3.5.1
	Drain current	
α	Velocity saturation transition exponent for drain current	3.7.1.2
β	Velocity saturation transition exponent for V_{EFF}	3.7.2.2
$DVTDIBL$	Threshold-voltage change with V_{DS}, $\delta V_T/\delta V_{DS}$, due to DIBL for L_{min}	3.8.4.2
$DVTDIBLEX$	Channel length exponent describing decrease in magnitude of $\delta V_T/\delta V_{DS}$ with increasing L	3.8.4.2
E_{CRIT}	Critical horizontal, drain–source electric field for velocity saturation	3.7.1.2
E_{CRIT0}	Critical horizontal, drain–source electric field for velocity saturation for low-field mobility	3.7.1.2
γ	Body-effect factor	3.3.1
I_0	Technology current, $2n_0\mu_0 C'_{OX} U_T^2$, used in IC definition	3.4.2.2
k_0	Transconductance factor, low field, $\mu_0 C'_{OX}$	3.3.2
λ_C	Dimensionless fitting parameter associated with CLM	3.8.4.1
λ_{DIBL}	Dimensionless fitting parameter associated with DIBL	3.8.4.2
L_C	Characteristic length associated with CLM	3.8.4.1
L_{DIBL}	Characteristic length associated with DIBL	3.8.4.2
μ_0	Carrier mobility, low field	3.3.2
$\mu_0 C'_{OX}$	Transconductance factor, k_0, low field	3.3.2
n_0	Substrate factor, average value in moderate inversion	3.4.2.2
ϕ_0	Approximate silicon surface potential[a] in strong inversion, taken at approximate pinned value of $2\phi_F$ $(PHI) + 4U_T$	3.4.1
ϕ_D	Built-in potential[a] of drain and pinched-off channel or voltage related to the electric field at the pinched-off channel	3.8.4.1
ϕ_F	Fermi potential[a] associated with the voltage required to convert channel doping concentration to intrinsic concentration	3.3.1
PHI	Process Fermi potential,[a] twice the Fermi potential or $2\phi_F$	3.3.1
θ	Vertical field mobility reduction factor	3.7.1.3
V_E	Voltage related to V_A dependence with V_{DS}	3.8.4.1
v_{sat}	Carrier saturation velocity	3.7.1.2
V_{TO} (large)	Large-geometry threshold voltage for zero V_{SB}	3.3.2, 3.7.1.7
	Resistance	
ρ_G	Gate material resistivity	3.10.4
R_{GC}	Gate contact resistance (each finger, including gate extension)	3.10.4

Capacitance

$CGBO$	Gate–body overlap capacitance per unit gate length	3.9.3
$CGDO$	Gate–drain overlap capacitance per unit gate width	3.9.3
$CGSO$	Gate–source overlap capacitance per unit gate width	3.9.3
C'_{OX}	Gate-oxide capacitance per unit area	3.3.1
C_J	Drain–(source–)body, area, junction diode capacitance per unit area at zero bias (V_{DB}, $V_{SB} = 0\,\mathrm{V}$)	3.9.4
C_{JSW}	Drain–(source–)body, perimeter, junction diode capacitance per unit perimeter at zero bias (V_{DB}, $V_{SB} = 0\,\mathrm{V}$)	3.9.4
M_J	Drain–(source–)body, area, junction diode capacitance voltage exponent	3.9.4
M_{JSW}	Drain–(source–)body, perimeter, junction diode capacitance voltage exponent	3.9.4
P_B	Drain–(source–)body, area, junction diode built-in potential	3.9.4
P_{BSW}	Drain–(source–)body, perimeter, junction diode built-in potential	3.9.4

Noise

α_H	Hooge's bulk mobility parameter	3.10.3.2
α_{SC}	Coulomb scattering parameter	3.10.3.3
AF	Flicker-noise PSD frequency slope	3.10.3.4
A_{GR}	Amplitude factor for Lorentzian power spectrum associated with generation-recombination noise	3.10.3.1
γ_t	McWhorter's tunneling constant for traps near the gate and gate-oxide interface	3.10.3.1
K	Flicker-noise factor for Hooge mobility fluctuation model ($\mathrm{V^2 \cdot F}$)	3.10.3.2
K_{F0}	Flicker-noise factor value for weak inversion or the weak-inversion side of moderate inversion in units of $\mathrm{C^2/cm^2}$	3.10.3.4
K'_{F0}	Same as K_{F0} but in hand units of $\mathrm{(nV)^2 \cdot \mu m^2}$	3.10.3.4
$K_{FSPICE0}$	Same as K_{F0} but in SPICE units of $\mathrm{V^2 \cdot F}$	3.10.3.4
τ_t	Time constant associated with traps near the gate and gate-oxide interface	3.10.3.1
V_{KF}	Voltage associated with inversion level or V_{EFF} increase in gate-referred flicker-noise voltage	3.10.3.4

Mismatch

A_{GAMMA}	Local-area, body-effect factor mismatch factor	3.11.1.1
A_{IGS}	Local-area, gate–source current mismatch factor	3.12.2.5
A_{KP}	Local-area, transconductance factor mismatch factor	3.11.1.1
A_{VTO}	Local-area, threshold-voltage mismatch factor for $V_{SB} = 0\,\mathrm{V}$	3.11.1.1
L_L	Length mismatch factor for edge effects	3.11.1.3
S_{GAMMA}	Distance, body-effect factor mismatch factor	3.11.2.1
S_{KP}	Distance, transconductance factor mismatch factor	3.11.2.1
S_{VTO}	Distance, threshold-voltage mismatch factor for $V_{SB} = 0\,\mathrm{V}$	3.11.2.1
W_W	Width mismatch factor for edge effects	3.11.1.3

Leakage and impact ionization

D_n	Diffusion constant for electrons	3.12.3
D_p	Diffusion constant for holes	3.12.3
E_B	Characteristic electric field for gate-oxide tunneling	3.12.1.1
η	Diode ideality factor (the context clarifies this from $\eta = g_{mb}/g_m$)	3.12.3
K_{GA}	Constant associated with gate–source tunneling leakage current	3.12.1.1
K_{GB}	Constant related to exponential gate–source tunneling leakage current dependence with V_{GS}	3.12.1.1
K_i	Constant related to impact ionization current	3.8.4.3
L_n	Diffusion length for electrons	3.12.3
L_p	Diffusion length for holes	3.12.3
ϕ_{bi}	Built-in p/n junction potential	3.12.3
τ_{gen}	Carrier generation time constant in the depletion region	3.12.3

Table 0.2 (*continued*)

Symbol	Description	Reference (first or primary)
V_i	Voltage related to exponential impact ionization current dependence with V_{DS}	3.8.4.3
X_B	Oxide voltage barrier for gate-oxide tunneling	3.12.1.1
	Temperature	
BEX	Mobility temperature exponent	3.5.2
TCV	Threshold-voltage temperature coefficient	3.5.4
UCEX	Velocity saturation, critical electric field temperature exponent	3.3.5

[a] ϕ_0, ϕ_D, ϕ_F, and *PHI* are taken positively for nMOS devices. These are also taken positively for pMOS devices.

Table 0.3 Symbols related to MOS circuit operation

Symbol	Description	Reference (first or primary)
$A_{CC,AC}$	Accuracy, AC	4.3.6.6
$A_{CC,DC}$	Accuracy, DC	3.11.3.1, 3.11.3.2
A_V	Voltage gain	5.3.3
A_{vbd}	Body-to-drain voltage gain	3.8.1
A_{vbs}	Body-to-source voltage gain	3.8.1
A_{vbso}	Open-circuit (Thévenin) body-to-source voltage gain	3.8.1
A_{vgd}	Gate-to-drain voltage gain	3.8.1
A_{vgs}	Gate-to-source voltage gain	3.8.1
A_{vgso}	Open-circuit (Thévenin) gate-to-source voltage gain	3.8.1
A_{vsd}	Source-to-drain voltage gain	3.8.1
BW	Bandwidth	3.11.3.1, 3.11.3.2
C_{INDIF}	Differential input capacitance	5.3.9
C_{OUT}	Output capacitance	5.3.9
$E_{matching}$	Matching energy, $C'_{OX} \cdot A_{VT}^2$	3.11.3.1, 3.11.3.2
$E_{thermal}$	Thermal energy, kT	4.3.6.6
f_{-3db}	$-3\,$dB bandwidth	5.3.4
g_d	External small-signal conductance connected to drain, $1/r_d$	3.8.1
g_s	External small-signal conductance connected to source, $1/r_s$	3.8.1
G_M	Effective MOS, short-circuit (Norton) transconductance including resistive source degeneration; also circuit short-circuit (Norton) transconductance	3.8.1, 5.3.1
G_{MB}	Effective MOS, short-circuit (Norton) body-effect transconductance including resistive source degeneration	3.8.1
H_{sd}	Short-circuit (Norton) source-to-drain current gain, including current-splitting effect of external source resistance	3.8.1
i_d	Small-signal drain current	3.8.1
i_{ins}	Small-signal current connected to source	3.8.1
I_{OS}	General offset current	3.11.3.1
NAF	Noise area factor	6.3.2
NEF	Noise efficiency factor	6.3.1
NEF'	Thermal-noise efficiency factor	6.3.1
P	Power consumption	3.11.3.1, 3.11.3.2

r_b	External small-signal resistance connected to body	3.8.1
r_d	External small-signal resistance connected to drain	3.8.1
r_g	External small-signal resistance connected to gate	3.8.1
r_s	External small-signal resistance connected to source	3.8.1
r_{inb}	Small-signal input resistance at body	3.8.1
r_{ing}	Small-signal input resistance at gate	3.8.1
$r_{ins} = r_{outs}$	Small-signal input and output resistance at source	3.8.1
r_{outd}	Small-signal output resistance at drain	3.8.1
RMP	Relative mismatch power compared to input device(s)	5.3.7
RNP	Relative noise power compared to input device(s)	5.3.5, 5.3.6
R_{OUT}	Output resistance	5.3.2
R_{sd}	MOS drain transresistance for input current at source, including current-splitting effect of external source resistance and loading of external drain resistance	3.8.1
SR	Slew rate	5.3.10
S_{VIN}	Input-referred noise voltage PSD[a]	5.3.5, 5.3.6
$S_{VIN}^{1/2}$	Input-referred noise voltage density[a]	5.3.5, 5.3.6
v_b	Small-signal voltage at body	3.8.1
v_d	Small-signal voltage at drain	3.8.1
v_g	Small-signal voltage at gain	3.8.1
v_s	Small-signal voltage at source	3.8.1
V_{DD}	Positive supply voltage	3.11.3.1, 3.11.3.2, 5.2
V_{INCM+}	Maximum common-mode input voltage	5.3.11
V_{INCM-}	Minimum common-mode input voltage	5.3.11
$V_{INDIF1dB}$	Input, 1 dB compression voltage (the context clarifies this from $V_{INDIF1dB}$ given in Table 0.1 for a differential pair)	5.3.12
$V_{INOFFSET}$	Input-referred offset voltage (1σ value)	5.3.7
$V_{INOFFSET}^2$	Input-referred offset voltage variance (square of 1σ value)	5.3.7
V_{OS}	General offset voltage	3.11.3.2
V_{OUT+}	Maximum output voltage	5.3.11
V_{OUT-}	Minimum output voltage	5.3.11
V_{SS}	Negative supply voltage	5.2

[a] The noise PSD or density is qualified directly or by context as thermal, flicker, or total noise. Noise voltage PSD and density are frequently given in design units of $(nV)^2/Hz$ and $nV/Hz^{1/2}$, respectively.

Table 0.4 Common symbols and symbols related to physical constants

Symbol	Description
C	Capacitance (F)
ε_0	Permittivity of free space, 8.85×10^{-3} fF/μm
ε_{Si}	Permittivity of silicon, $11.8 \cdot 8.85 \times 10^{-3}$ fF/μm
ε_{SiO_2}	Permittivity of silicon dioxide, $3.9 \cdot 8.85 \times 10^{-3}$ fF/μm
f	Frequency (Hz)
G, g	Conductance, usually upper case for system or DC and lower case for small signal or AC (S)
I, i	Current, usually upper case for DC and lower case for small signal or AC (A)
k	Boltzmann's constant, 1.3806×10^{-23} J/K
kT	Thermal-noise energy, 4.142×10^{-21} VA/Hz (VA$\cdot$s or J) at $T = 300$ K
$4kT$	Thermal-noise energy, 1.657×10^{-20} VA/Hz (VA$\cdot$s or J) at $T = 300$ K

Table 0.4 (*continued*)

Symbol	Description
P	Power (W)
q	Elemental unit of charge, 1.602×10^{-19} C
ρ	Electrical resistivity ($\Omega \cdot$ cm)
R, r	Resistance, usually upper case for system or DC and lower case for small signal or AC (Ω)
σ	Electrical conductivity (S/cm)
σ	Statistical standard deviation (the context clarifies this from σ above)
T	Temperature (K or °C)
U_T	Thermal voltage, kT/q, 25.85 mV at $T = 300$ K (U_T is used to differentiate from MOS threshold voltage, V_T)
V, v	Voltage, usually upper case for DC and lower case for small signal or AC (V)
ω	Angular frequency, $2\pi f$ (radians/s)

Table 0.5 Abbreviations related to MOS operation

Abbreviation	Description	Reference (first or primary)
CLM	Channel length modulation	3.8.4.1
CMOS	Complementary metal–oxide semiconductor	1.1
DIBL	Drain-induced barrier lowering	3.8.4.2
FD	Fully depleted	3.8.2.5
GIDL	Gate-induced drain leakage	3.12.3
GR	Generation, recombination	3.10.3.1
LDD	Lightly doped drain, usually through source and drain extensions	3.8.4.3, 3.10.3.3
MDD	Moderately doped drain	3.10.3.3
MI	Moderate inversion	2.4.4, 3.4.2.4, 4.3.2
MOS	Metal–oxide semiconductor[a]	1.1
MOSFET	Metal–oxide semiconductor field-effect transistor[a]	1.1
nMOS	n-type metal–oxide semiconductor	Figure 3.9
pMOS	p-type metal–oxide semiconductor	Figure 3.9
PD	Partially depleted	3.8.3.3
RTS	Random telegraph signal	3.10.3.1
SOI	Silicon-on-insulator	3.8.3.3
SI	Strong inversion	2.4.2, 3.4.2.4, 4.3.2
VFMR	Vertical field mobility reduction	3.7.1.3
WI	Weak inversion	2.4.1, 3.4.2.4, 4.3.2

[a] The abbreviations for MOS and MOSFET are often used interchangeably, although MOS can include a broader definition beyond MOSFET devices alone.

Table 0.6   Common abbreviations and units

Abbreviation	Description
A	Ampere, unit of current
AC	Alternating current
BSIM	Berkeley short-channel, insulated gate, field-effect transistor model
°C	Centigrade, unit of temperature
C	Coulomb, unit of charge
CAD	Computer-aided design
DC	Direct current
EDA	Electronic design automation
EKV	Enz–Krummenacher–Vittoz MOS model
F	Farad, unit of capacitance
J	Joule, unit of energy
K	Kelvin, (absolute) unit of temperature
Hz	Hertz, unit of frequency
OTA	Operational transconductance amplifier
PSD	Power spectral density
rms	Root mean square
s	Second, unit of time
S	Siemens, unit of conductance
V	Volt, unit of voltage
W	Watt, unit of power
Ω	Ohm, unit of resistance

1

Introduction

1.1 IMPORTANCE OF TRADEOFFS AND OPTIMIZATION IN ANALOG CMOS DESIGN

Wireless, wire-line, and optical communications, along with entertainment, multimedia, biomedical, and many other applications, require analog circuits for interfacing with the physical world. The proliferation of these applications has resulted in a continual increase in the quantity and complexity of analog circuits fabricated in the prevalent, complementary metal–oxide semiconductor (CMOS), integrated circuit technology.

Analog CMOS design requires the design of system specifications and architectures, followed by the design of circuits containing various topologies of interconnected metal–oxide semiconductor, field-effect transistors (MOSFETs). Following this, the designer must select a drain current, channel width, and channel length for every MOSFET in a circuit. This book considers these MOSFET design selections.

In this book, the inversion coefficient[1] replaces channel width as a design choice because the inversion coefficient provides a numerical measure of metal–oxide semiconductor (MOS)[2] inversion where values below 0.1 correspond to weak inversion, values between 0.1 and 10 correspond to moderate inversion, and values above 10 correspond to strong inversion. Channel width, required for layout, is easily calculated from the selected drain current, inversion coefficient, and channel length and is implicitly considered in predictions of MOS performance.

The three independent degrees of MOS design freedom, whether drain current, inversion coefficient, and channel length used in this book, or the traditional choices of drain current, channel width, and channel length, influence all measures of MOS performance,[3] including physical size, gate–source bias and drain–source saturation voltages, small-signal parameters, signal distortion, intrinsic voltage gain, capacitances, intrinsic and extrinsic bandwidths, thermal noise, flicker noise, mismatch, and leakage. Performance such as capacitances, intrinsic voltage gain, and intrinsic bandwidth can vary over several

---

[1] References for the inversion coefficient follow in Chapter 2, along with discussions of weak, moderate, and strong inversion.

[2] The abbreviations MOSFET and MOS are often used interchangeably, although MOS can include a broader definition beyond MOSFET devices alone.

[3] Chapter 3 describes measures of MOS performance in detail.

Tradeoffs and Optimization in Analog CMOS Design   David M. Binkley
© 2008 John Wiley & Sons, Ltd

decades, resulting in a wide range of available circuit performance for a given circuit topology. Although three degrees of design freedom for each MOSFET in a circuit greatly complicate analog CMOS design, this affords significant opportunities to manage performance tradeoffs and optimize designs.

The subject of this book is guiding the designer in the selection of MOS drain current, inversion coefficient, and channel length for desired tradeoffs in performance leading towards optimum design. This is done through hand and spreadsheet expressions and graphical presentations of MOS performance that are valid in all regions of operation. A key aspect of this book is the detailed treatment of moderate inversion, which is increasingly important in low-voltage, low-power design. Operation in moderate inversion offers the advantages of low MOS drain–source saturation voltage, high transconductance efficiency (the ratio of transconductance to the drain bias current), moderate intrinsic bandwidth, and good immunity to velocity saturation effects that otherwise could deteriorate performance. Design in moderate inversion, with hand expressions recently facilitated by developments like the EKV MOS model,[4] is rarely covered in existing books and has traditionally required iterative computer simulations. This book complements existing books by specifically addressing tradeoffs and optimization of analog CMOS circuits in weak, moderate, and strong inversion, over the full range of process channel length. In addition to addressing common measures of performance like intrinsic voltage gain and bandwidth, this book contains considerable material addressing the tradeoffs and optimization of thermal noise, flicker noise, and mismatch.

The methods and design examples presented in this book are intended to help the designer manage performance tradeoffs and rapidly create optimum or near-optimum designs before launching the computer simulations required to verify a design for production. In the author's experience, this minimizes trial-and-error simulations, saves design time, provides a cross-check with production MOS models, builds design intuition, and enhances the enjoyment of design.

1.2 INDUSTRY DESIGNERS AND UNIVERSITY STUDENTS AS READERS

This book is written for the industry designer or university student familiar with the analysis and design of analog CMOS circuits using traditional, strong-inversion, square-law MOS modeling. The book hopes to extend this knowledge and enable design and design optimization freely in weak, moderate, and strong inversion, inclusive of small-geometry effects and other advanced effects like thermal noise, flicker noise, and mismatch. It is assumed that the reader is familiar with operational transconductance amplifiers, operational amplifiers, and other analog CMOS circuits along with their DC bias and small-signal analysis. While full details are provided for thermal- and flicker-noise analysis, it is also assumed that the reader is familiar with the basic concepts of noise analysis. These core analog CMOS design topics are covered in detail in a number of excellent books.

Most of the material contained in this book was used by the author at Concorde Microsystems[5] in the late 1990s for the design of front-end, CMOS integrated circuits for positron emission tomography (PET) medical imaging systems manufactured by Siemens. Some of the material has also been successfully taught several times in a course, *Advanced Analog CMOS Design*, at the University of North Carolina at Charlotte. Both in industry and at the university, the material was successful in minimizing time-consuming, trial-and-error circuit simulations by building design intuition and guiding the designer towards optimum design.

---

[4] References for the EKV MOS model follow in Chapter 2.
[5] In 2005, Concorde Microsystems, Inc., became part of Siemens Medical Solutions.

1.3 ORGANIZATION AND OVERVIEW OF BOOK

This book begins by introducing the complexity of analog CMOS design resulting from the MOS design choices of drain current, inversion coefficient, and channel length. This is followed by an in-depth study of MOS performance and tradeoffs resulting from these design choices. The book then presents design examples of CMOS operational transconductance amplifiers optimized for DC, balanced, and AC performance, and micropower, low-noise, CMOS preamplifiers optimized for low thermal and flicker noise. Finally, the book concludes with a discussion on how the design methods can be extended to smaller-geometry CMOS processes as well as emerging non-CMOS processes.

The book is organized into two parts having emphasis on devices and circuits, respectively.

Part I, MOS Device Performance, Tradeoffs and Optimization for Analog CMOS Design, contains Chapters 2, 3, and 4 and is overviewed below.

Chapter 2, MOS Design from Weak through Strong Inversion, briefly introduces MOS operation in all regions, emphasizing decreasing transconductance efficiency at increasing inversion levels, including the additional decrease associated with small-geometry effects like velocity saturation. This chapter also introduces normalized MOS drain–source resistance as an Early voltage, which has a strong dependence on the channel length. The complexities associated with selecting MOS drain current, inversion coefficient, and channel length are compared to design using bipolar transistors where generally only the collector current must be selected. This emphasizes the importance of optimization methods that permit design freely in all regions of MOS operation, over the full range of channel length. The chapter concludes by presenting previously reported optimization methods, including both electronic design automation (EDA) and hand methods.

Chapter 3, MOS Performance versus Drain Current, Inversion Coefficient, and Channel Length, is an extensive "book within a book" chapter that is the basis for understanding MOS performance tradeoffs that lead the designer towards optimum design. The chapter begins by introducing the inversion coefficient as a primary design choice governing the region and degree of MOS inversion. The chapter describes the advantages of selecting MOS drain current, inversion coefficient, and channel length for design optimization, with channel width found from these design selections. Additionally, the chapter introduces the *MOSFET Operating Plane*, which illustrates tradeoffs in performance. For example, operation at low inversion coefficients in weak or moderate inversion at long channel lengths optimally maximizes voltage gain and minimizes gate-referred thermal- and flicker-noise voltage, mismatch, and the drain–source saturation voltage. This is referred to as a DC or low-frequency optimization. Conversely, operation at high inversion coefficients in strong inversion at short channel lengths maximizes bandwidth and minimizes transconductance distortion. This is referred to as an AC optimization. Operation at intermediate inversion coefficients and channel lengths provides a balance of DC and AC performance.

The majority of Chapter 3 presents expressions and design graphs of MOS performance versus the design choices of drain current, inversion coefficient, and channel length. Performance includes channel width and gate area, effective gate–source voltage, drain–source saturation voltage, transconductance efficiency, transconductance distortion, body-effect transconductance ratio, normalized drain–source resistance or conductance, intrinsic voltage gain, intrinsic and extrinsic capacitances, intrinsic and extrinsic bandwidths, thermal noise, flicker noise, mismatch, and leakage. Much of this material is based on or extended from the EKV MOS model and includes hand expressions, often with experimental validations. Simple hand model extensions are included for small-geometry effects like velocity saturation, vertical field mobility reduction, drain-induced barrier lowering, and increases in gate-referred flicker noise voltage with the inversion level. Although performance can be predicted and measured for any process, most validating experimental measurements are for a typical $0.18\,\mu m$ CMOS processes. Finally, the chapter contains predictions of gate leakage current and describes its effect on circuit performance. This can be significant for gate-oxide thickness less than around 2 nm.

Chapter 4, Tradeoffs in MOS Performance, and Design of Differential Pairs and Current Mirrors, presents tradeoffs of MOS performance versus the design choices of drain current, inversion coefficient, and channel length. Expressions and design graphs show tradeoffs in device channel width and gate area, effective gate–source bias and drain–source saturation voltages, intrinsic voltage gain, intrinsic bandwidth, capacitances, gate-referred thermal- and flicker-noise voltage, drain-referred thermal- and flicker-noise current, and gate–source voltage and drain current mismatch. Predicted and measured tradeoffs are given for a typical 0.18 μm CMOS process. Tradeoffs, for example in MOS intrinsic voltage gain and bandwidth, show that gain decreases while bandwidth increases as the inversion coefficient increases in strong inversion. These tradeoffs, however, are even more significant as channel length increases where voltage gain increases, but bandwidth decreases more rapidly. The chapter also shows tradeoffs in thermal noise efficiency, bandwidth-power-accuracy, and other figures of merit that combine multiple aspects of performance.

Chapter 4 concludes by illustrating performance tradeoffs through the design of differential pairs and current mirrors. This includes designs for separate design choices of the inversion coefficient, channel length, and drain current, and designs optimized for various tradeoffs in DC and AC performance at both millipower (100 μA) and micropower (1 μA) levels of drain current. The designs use the *Analog CMOS Design, Tradeoffs and Optimization* spreadsheet to show MOS performance for the selected drain current, inversion coefficient, and channel length. This spreadsheet is available at the web site for this book and is described in the Appendix.

Part II, Circuit Design Examples Illustrating Optimization for Analog CMOS Design, contains Chapters 5 and 6 and is overviewed below.

Chapter 5, Design of CMOS Operational Transconductance Amplifiers Optimized for DC, Balanced, and AC Performance, illustrates design optimization using two different operational transconductance amplifier (OTA) topologies in two different CMOS processes. Simple, 0.5 μm CMOS OTAs are described first where devices in a given version operate at equal drain currents, inversion coefficients, and channel lengths. This is typical of general-purpose designs where input devices do not dominate noise or mismatch and provide a simple design example for introducing the design methods. The OTAs are optimized for DC, balanced, and AC performance by operating devices at low, moderate, and high inversion coefficients with long, moderate, and short channel lengths. The DC-optimized OTA has high transconductance, output resistance, voltage gain, and input and output voltage ranges, combined with small input-referred thermal-noise voltage, flicker-noise voltage, and offset voltage due to local-area mismatch and systematic offset. The AC-optimized OTA has high transconductance bandwidth, combined with small transconductance, transconductance distortion, input and output capacitances, and layout area. The balanced optimized OTA has a balance of DC and AC performance.

Detailed circuit analysis and performance trends are developed, and device performance resulting from drain current, inversion coefficient, and channel length selections is mapped into OTA circuit performance using the *Analog CMOS Design, Tradeoffs and Optimization* spreadsheet. Measured voltage gain is 326, 110, and 16.8 V/V, transconductance bandwidth is 5, 51, and 350 MHz, input-referred flicker-noise voltage density at 100 Hz is 80, 450, and 2000 nV/Hz$^{1/2}$, and input-referred offset voltage due to mismatch is 1.1, 2.2, and 10.2 mV (1σ) for the DC-, balanced, and AC-optimized OTAs, respectively. Measured transconductance is 912, 647, and 383 μS, input-referred thermal-noise voltage density is 11.2, 14.4, and 19.4 nV/Hz$^{1/2}$, and the input, 1 dB compression voltage is 78, 115, and 218 mV. This illustrates the wide range of available performance tradeoffs. The OTAs operate at equal core (excludes bias references) supply currents of 200 μA and supply voltages of ±1.25 V, illustrating performance comparisons at equal power consumptions of 500 μW.

Chapter 5 also illustrates design optimization through the design of three, cascoded, 0.18 μm CMOS OTAs where devices in a given version operate at different drain currents and inversion coefficients so input devices dominate the thermal noise. The OTAs are optimized for DC, balanced, and AC performance by operating devices at fixed inversion coefficients with long, moderate, and short channel lengths. Again, detailed circuit analysis and performance trends are developed, and device performance is mapped into OTA circuit performance using the *Analog CMOS Design, Tradeoffs and Optimization* spreadsheet. Measured voltage gain is 19 100, 4400, and 490 V/V, transconductance bandwidth is 75, 285, and 850 MHz, input-referred flicker-noise voltage density at 100 Hz is 96, 420, and 1700 nV/Hz$^{1/2}$, and input-referred offset voltage due to mismatch is 0.24, 1.1, and 3.2 mV (1σ) for the DC-, balanced, and AC-optimized OTAs, respectively. Operating input devices near the center of moderate inversion results in a nearly constant transconductance of 1900 µS, input-referred thermal-noise voltage density of 5 nV/Hz$^{1/2}$, and input, 1 dB compression voltage of 55 mV. Although non-input devices do not have sufficient device area to ensure negligible flicker-noise and mismatch contributions, the measured input-referred offset voltage for the DC-optimized OTA is very low at 0.72 mV (3σ) and is more typical of that found in bipolar transistor circuits. The OTAs operate at equal core (excludes bias references) supply currents of 300 µA and supply voltages of ±0.9 V, illustrating performance comparisons at equal power consumptions of 540 µW.

Chapter 6, Design of Micropower CMOS Preamplifiers Optimized for Low Thermal and Flicker Noise, illustrates design optimization through the design of 0.35 µm, silicon-on-insulator (SOI), CMOS micropower, low-noise preamplifiers having differential and single-ended inputs. This chapter begins by describing performance measures useful for minimizing input-referred thermal-noise voltage for a given level of power consumption, and the preamplifiers presented here are compared to others reported in the literature. The chapter then presents methods for minimizing input-referred thermal- and flicker-noise voltage, including operating input devices in weak or moderate inversion for high transconductance and low input-referred thermal-noise voltage at minimum current consumption. Operating non-input devices well into strong inversion for low transconductance and low drain-referred thermal- and flicker-noise current is described to minimize non-input device noise contributions. The advantage of resistive degeneration of non-input device flicker noise is also presented, and noise is optimized against the constraints of available bias compliance voltage. This shows that managing non-input device noise becomes increasingly difficult at low supply voltages where bias compliance voltage is reduced.

After detailed circuit analysis for the differential and single-ended input preamplifiers, device performance resulting from device drain current, inversion coefficient, and channel length selections is again mapped into overall circuit performance using the *Analog CMOS Design, Tradeoffs and Optimization* spreadsheet. The measured input-referred thermal-noise voltage density is 63.8 and 35.3 nV/Hz$^{1/2}$ and flicker noise at 1 Hz is 240 and 160 nV/Hz$^{1/2}$, giving flicker-noise corner frequencies of 12 and 19 Hz for the differential and single-ended input preamplifiers, respectively. The preamplifiers operate at equal core supply currents of 2 µA, supply voltages of 3.3 V, and power consumptions of 6.6 µW, excluding bias references and optional output buffers.

Chapter 7, Extending Optimization Methods to Smaller-Geometry Processes and Future Technologies, concludes this book by discussing the extension of optimization methods to smaller-geometry CMOS processes and even emerging non-CMOS technologies like organic, thin-film and carbon nanotube, field-effect transistor technologies. Evaluating measures of performance like transconductance efficiency, normalized drain–source resistance, intrinsic voltage gain, capacitances, intrinsic bandwidth, noise, mismatch, and leakage in terms of device bias current, inversion level, and channel length permits ready extension of the methods. The technology normalization inherent in the inversion coefficient facilitates this extension.

1.4 FULL OR SELECTIVE READING OF BOOK

Recognizing that few industry designers or university students will have the time to read this book fully, it was written to be read either fully or selectively. The extensive use of design tables facilitates selective reading by summarizing predictions of MOS device and circuit performance, often with trends listed for the design choices of device drain current, inversion coefficient, and channel length. Additionally, figures facilitate selective reading by showing predicted and measured performance where trends and tradeoffs are readily observed. The figures contain narrative captions that, in addition to titling the figures, summarize key information. The tables and figures, with their captions, are intended to rapidly "tell the story" with the book text providing detailed, supportive information.

In addition to material provided in tables and figures, the *Analog CMOS Design, Tradeoffs and Optimization* spreadsheet permits the reader to rapidly estimate the performance of individual MOS devices and complete circuits as design choices are explored. This spreadsheet is used in the design examples presented in this book and can be extended to other designs by the reader. The spreadsheet, available from the book's web site, listed on the cover, is summarized in the Appendix.

A brief reintroduction of topics, symbols, abbreviations, and meanings within separate chapters also permits selective reading where individual chapters can be read with basic comprehension without requiring frequent reference to material contained in other chapters. Reference is made to previous material where the reader can obtain detailed explanations, derivations, measured data, and literature citations.

Finally, the table of contents facilitates selective reading where topics can be identified rapidly, while hopefully avoiding less efficient page searches using the *Index*. While many symbols and abbreviations are redefined in new chapters, these can also be found in the *List of Symbols and Abbreviations*.

1.5 EXAMPLE TECHNOLOGIES AND TECHNOLOGY EXTENSIONS

Design examples presented in this book are in 0.5, 0.35, and 0.18 μm CMOS processes, with most experimental measurements provided for a typical 0.18 μm CMOS process. At publication time, some designers in large companies are working in 0.09 and 0.065 μm processes, but many designers in smaller companies or those designing dedicated analog or mixed-signal integrated circuits are working in 0.18 μm or even larger feature size processes. Although not available for system-on-chips fabricated in the highest-density digital CMOS processes, except through thick gate-oxide options, many analog designers have a preference for the thicker gate oxide and higher supply voltages of 0.18 μm and larger processes.

Regardless of the CMOS process used, the performance trends and tradeoffs presented here guide the designer towards optimum design. For example, as described in Chapter 3, transconductance efficiency versus the inversion coefficient follows the universal behavior of being maximum in weak inversion, beginning to decrease in moderate inversion, and decreasing continually in strong inversion, decreasing faster for short channel lengths due primarily to velocity saturation. While this is illustrated in 0.5 and 0.18 μm processes here, the behavior is similar for smaller-geometry processes with transconductance efficiency decreasing faster for channel lengths below 0.18 μm. As discussed in Chapter 7, the inversion coefficient with its inherent technology normalization facilitates extending the methods presented here to smaller-geometry processes.

The hand expressions of MOS performance presented here can be extended for improved accuracy in smaller-geometry processes as long as these expressions are still simple enough to permit design guidance and intuition. Additionally, the *Analog CMOS Design, Tradeoffs and Optimization* spreadsheet is a "work in progress" that in the future could be coupled to the EKV MOS model or other simulation MOS models. This would permit MOS device and circuit performance prediction with improved accuracy in smaller-geometry processes. The reader is invited to check the book's web site for updates to the spreadsheet.

1.6 LIMITATIONS OF THE METHODS

This book presents approximate hand or spreadsheet expressions of MOS performance and resulting circuit performance as the designer explores device drain current, inversion coefficient, and channel length selections and makes initial, informed design choices. The expressions are necessarily simplified to show performance trends and tradeoffs useful to guide the designer towards an optimum or near-optimum design before launching computer circuit simulations. Although predictions of MOS transconductance, thermal noise, flicker noise, and mismatch are often within measured values by 10 %, errors in the prediction of drain–source resistance and resulting open-loop, circuit voltage gain can be greater. This is because of complex dependencies of drain–source resistance on the inversion level, channel length, and drain–source voltage. Fortunately, closed-loop circuit configurations are commonly used to desensitize circuit performance to variations in open-loop gain. The methods and expressions presented provide useful design guidance, especially since performance measures, like voltage gain and bandwidth, often vary over several decades with inversion coefficient and channel length selections.

The methods and expressions presented here involve quasi-static MOS modeling where the frequency of operation is sufficiently low such that device charges immediately track terminal voltages. The methods provide guidance up to frequencies of approximately 25 % of the intrinsic MOS bandwidth, which is developed in Chapter 3. Non-quasi-static extensions, for example where resistances are placed in series with intrinsic capacitances, can be envisioned, but are not developed here. Although the methods and expressions presented here are quasi-static and approximate, they do consider important small-geometry effects such as velocity saturation, vertical field mobility reduction, drain-induced barrier lowering, and increases in gate-referred flicker noise voltage with increasing inversion level. Additionally, the methods permit the inclusion of excess thermal noise through the use of modeling parameters.

For gate-oxide thickness around 2 nm and below, gate leakage current and the resulting gate–source conductance, gate shot- and flicker noise current, and increase in mismatch can become significant. As described in Chapter 3, smaller channel length and gate area may be required to balance the traditional improvements in gain, flicker noise, and local-area mismatch at increasing channel length and gate area with the deterioration of these associated with increasing gate leakage current caused by increasing gate area. The *Analog CMOS Design, Tradeoffs and Optimization* spreadsheet does not include gate leakage effects, which are negligible for the device and design examples contained in this book. These effects, however, could be included in later versions of the spreadsheet.

The methods and expressions presented in this book do not replace production MOS models used in computer simulations that are critical to verify the performance of candidate designs. Such simulation models can include non-quasi-static effects, especially for radio frequency (RF) applications, and gate leakage current effects. As mentioned, the *Analog CMOS Design, Tradeoffs and Optimization* spreadsheet might later be linked to the EKV MOS model or other simulation MOS models. This could then provide prediction accuracies approaching that of computer simulations.

As mentioned at various places in this book and well known to experienced designers, all candidate designs must be thoroughly verified by computer simulations using MOS models with parameters appropriate for the production process. This normally involves extensive simulations, inclusive of layout parasitics, over nominal and corner process conditions, temperature, and supply voltage. Portions of the disclaimer below appear at various places in this book and in the *Analog CMOS Design, Tradeoffs and Optimization* spreadsheet as a reminder.

1.7 DISCLAIMER

The design tradeoff and optimization methods, predictions, examples, and measurements given in this book or its associated spreadsheet software are intended for design guidance only, not for actual design, and do not correspond to any particular CMOS fabrication process. The designer must independently validate designs using MOS models and parameters appropriate for the actual fabrication process used.

Use of information in this book or its associated spreadsheet software expressly indicates the assumption of risk that this information should only be used for design guidance and should not be used for actual design or the validation of actual design. Additionally, use of information in this book or its associated spreadsheet software expressly indicates acknowledgment of responsibility for independently validating designs using MOS models and parameters appropriate for the actual fabrication process used.

The information in this book and associated spreadsheet software is provided without express or implied warranties that the information is accurate or reliable, and there are no warranties as to fitness for any particular purpose. Neither the author nor John Wiley & Sons, Ltd. accept any responsibility or liability for loss or damage occasioned to any person or property through using the material, instructions, methods, or ideas contained herein, or acting or refraining from acting as a result of such use. The author and publisher expressly disclaim all implied warranties, including merchantability of fitness for any particular purpose. There will be no duty on the author or publisher to correct any errors or defects in this book or its associated spreadsheet software.

Part I
MOS Device Performance, Tradeoffs and Optimization for Analog CMOS Design

Part I presents MOS device performance, tradeoffs and optimization leading to the optimum design of analog CMOS circuits. The contents of Part I were described earlier in Chapter 1 and are reviewed briefly here.

Chapter 2 introduces MOS design from weak through strong inversion, emphasizing the declining transconductance efficiency at increasing inversion levels and the increasing normalized drain–source resistance or Early voltage at increasing channel lengths. The chapter then describes the importance of designing freely in weak, moderate, and strong inversion and reviews hand and electronic design automation methods that support design continuously in these regions.

Chapter 3 presents MOS device performance in terms of the selected drain current, inversion coefficient, which numerically represents the level of MOS inversion, and channel length. The advantages of using these degrees of design freedom for optimum circuit design are developed where channel width is implicitly considered and easily found for layout. The chapter then presents tabular hand expressions and graphs for the MOS effective gate–source voltage, drain–source saturation voltage, transconductance efficiency, body-effect transconductance ratio, Early voltage, intrinsic voltage gain, intrinsic and extrinsic capacitances, intrinsic and extrinsic bandwidths, thermal noise, flicker noise, DC and small-signal parameter mismatch, and leakage current. These are developed in considerable detail, interpreted for circuit design, and often compared to measured data from a $0.18\,\mu m$ CMOS process to provide a comprehensive understanding of MOS performance. The expressions and graphs include small-geometry effects like velocity saturation, vertical field mobility reduction (VFMR), and drain-induced barrier lowering (DIBL), along with increases in gate-referred flicker-noise voltage at increasing inversion levels. Because of the process normalization present when using the inversion coefficient, the expressions and graphs may be extended to different processes, including smaller-geometry processes.

Chapter 4 describes tradeoffs in MOS device performance, again for the selected drain current, inversion coefficient, and channel length. The chapter presents tradeoffs for single aspects of performance as well as tradeoffs for figures of merit combining multiple aspects of performance. Performance tradeoffs are illustrated by the design of differential pair and current mirror devices using the *Analog CMOS Design, Tradeoffs and Optimization* spreadsheet.

The methods developed in Part I for the design of MOS devices are applied to the design of complete circuits in the design examples contained in Part II.

Tradeoffs and Optimization in Analog CMOS Design   David M. Binkley
© 2008 John Wiley & Sons, Ltd

2

MOS Design from Weak through Strong Inversion

2.1 INTRODUCTION

This chapter begins by comparing the complexity of design using MOS transistors with design using bipolar transistors. Increased complexity in MOS design is a result of multiple regions of MOS inversion giving different drain current and transconductance characteristics, combined with the choice of channel length giving different drain–source resistance characteristics. The complexity of MOS design is first illustrated by reviewing collector current and transconductance for the bipolar transistor, which has one usual region of operation where collector current is proportional to the exponential of base–emitter voltage. Following this, MOS drain current and transconductance in weak inversion is reviewed where, similar to the bipolar transistor, drain current is proportional to the exponential of gate–source voltage. MOS operation in strong inversion is then reviewed where drain current is proportional to the square of gate–source voltage, and this is extended to include the reduction of drain current and transconductance caused by velocity saturation. The increasingly important moderate inversion region that spans nearly two decades of drain current between weak and strong inversion is also reviewed.

MOS operation in weak, moderate, and strong inversion is summarized to illustrate the declining drain current control or transconductance efficiency with increasing inversion level compared to the constant collector current control or transconductance efficiency present in the bipolar transistor. This shows that MOS inversion level is a primary design parameter not present in bipolar transistor design. The chapter also reviews MOS drain–source resistance to illustrate that channel length, which strongly influences drain–source resistance, is a second MOS design parameter not present in bipolar transistor design where collector–emitter resistance is usually fixed by fixed base width in the process. The additional degrees of MOS design freedom, inversion level and channel length, in addition to the selection of bias current common to both MOS and bipolar transistor design, illustrate the need for efficient analog CMOS design methods. The chapter concludes by reviewing electronic design automation (EDA) tools and design methods valid from weak through strong inversion, including previous applications of the methods described in this book.

Tradeoffs and Optimization in Analog CMOS Design   David M. Binkley
© 2008 John Wiley & Sons, Ltd

2.2 MOS DESIGN COMPLEXITY COMPARED TO BIPOLAR DESIGN

As just introduced, and detailed in the following discussions, MOS design includes two added degrees of design freedom – inversion level and channel length – beyond the choice of bias current common to both MOS and bipolar transistor design. The choices of inversion level and channel length greatly complicate analog CMOS design as these separately or jointly influence gate–source bias and drain–source saturation voltages, small-signal parameters, distortion, intrinsic voltage gain, capacitances, intrinsic bandwidth, noise, DC and small-signal matching, and all aspects of circuit performance. To understand the importance of MOS inversion level in analog design, MOS drain current and transconductance are compared to bipolar transistor collector current and transconductance in the following discussions.

2.3 BIPOLAR TRANSISTOR COLLECTOR CURRENT AND TRANSCONDUCTANCE

The bipolar transistor collector current, base–emitter voltage relationship is given by [1, p. 24]

$$I_C = I_S e^{\frac{V_{BE}}{U_T}} \tag{2.1}$$

where I_S is the collector saturation current, which is dependent on the process and emitter area, V_{BE} is the base–emitter voltage, and $U_T = kT/q$ is the thermal voltage.[1] Collector current is given for the active region where the collector–emitter voltage exceeds its saturation value ($V_{CE} > V_{CESAT}$). Collector–emitter resistance effects are not considered here where collector current increases slightly with increasing collector–emitter voltage. Bipolar transistor collector current follows Equation 2.1 over as many as 10 decades of current before falling off due to high-level injection and other high-current effects.

Transconductance, g_m, is the small-signal parameter relating the change in short-circuit output current resulting from a change in the primary controlling voltage. Bipolar transistor transconductance, found from the collector current given in Equation 2.1, is given by

$$g_m = \frac{\partial I_C}{\partial V_{BE}} = \frac{I_C}{U_T} \tag{2.2}$$

Bipolar transistor transconductance efficiency, g_m/I_C, is the ratio of transconductance to collector bias current. Transconductance efficiency can be thought of as a quality factor describing how much transconductance is produced for a given unit of bias current. Transconductance can be found by multiplying transconductance efficiency and bias current as given by

$$g_m = \left(\frac{g_m}{I_C}\right) \cdot I_C \tag{2.3}$$

Bipolar transistor transconductance efficiency, found from Equations 2.1 and 2.2, is given by

$$\frac{g_m}{I_C} = \frac{\left(\dfrac{I_C}{U_T}\right)}{I_C} = \frac{1}{U_T} \tag{2.4}$$

---

[1] In this book, U_T represents the thermal voltage to differentiate it from V_T, which is used for the MOS threshold voltage.

which, unlike transconductance efficiency for the MOSFET, is maintained at a constant high value over multiple decades of collector current. This, again, assumes high-level injection and other high-current effects are avoided by properly scaling the emitter area for the operating range of collector current. For a thermal voltage $U_T = 25.9$ mV at room temperature (300 K or 27 °C), bipolar transistor transconductance efficiency is 38.6/V or 38.6 μS/μA. This means a transconductance of 38.6 μS or 38.6 mS is produced for a collector bias current of 1 μA or 1 mA, respectively, found by multiplying the transconductance efficiency by the collector bias current. Unfortunately, MOSFET transconductance efficiency is always less than that of the bipolar transistor and drops significantly for operation in strong inversion.

2.4 MOS DRAIN CURRENT AND TRANSCONDUCTANCE

2.4.1 In Weak Inversion

Weak inversion occurs for MOSFETs operating at sufficiently low effective gate–source voltages ($V_{EFF} = V_{GS} - V_T < -72$ mV) where the gate–source voltage, V_{GS}, is below the threshold voltage, V_T, by at least 72 mV for a typical bulk CMOS process at room temperature.[2] In this region, the channel is weakly inverted and drain diffusion current dominates. MOS drain current in weak inversion is proportional to the exponential of the effective gate–source voltage. Weak inversion drain current is approximated from the EKV MOS model [2, 3], using the source as the reference voltage, by

$$I_D \, (\text{WI}) = 2n\mu \, C'_{OX} U_T^2 \left(\frac{W}{L} \right) \left(e^{\frac{V_{GS} - V_T}{nU_T}} \right) \tag{2.5}$$

W and L are the effective channel width and length,[3] μ is the channel carrier mobility, and C'_{OX} is the gate-oxide capacitance per unit area. Drain current, I_D,[4] is given for MOS saturation where the drain–source voltage, V_{DS}, exceeds its saturation value, $V_{DS,sat}$.[5] In weak inversion, the drain–source saturation voltage is approximately $4U_T$, which is 104 mV at room temperature. Drain–source resistance is not considered here where drain current increases with increasing drain–source voltage.

The substrate factor, n, in Equation 2.5 appears in the EKV [2, 3] and other MOS models. Conceptually, it represents a loss of coupling efficiency between the gate and channel caused by the substrate or body, which acts as a back gate. In weak inversion, n is related to the capacitive voltage division between the gate voltage and silicon surface potential resulting from the gate-oxide, depletion, and interface state capacitances. In weak inversion, n is expressed by

$$n \, (\text{WI}) = \frac{C'_{OX} + C'_{DEP} + C'_{INT}}{C'_{OX}} = 1 + \frac{C'_{DEP}}{C'_{OX}} + \frac{C'_{INT}}{C'_{OX}} \approx 1 + \frac{C'_{DEP}}{C'_{OX}} \tag{2.6}$$

where C'_{DEP} and C'_{INT} are the depletion and interface state capacitances per unit area, respectively. Usually, as indicated in the rightmost term of Equation 2.6, the interface state capacitance is negligible and is not included in the prediction of n. In weak inversion, n is approximately 1.4–1.5 for typical bulk CMOS processes, but can be as low as 1.1 [4] for fully depleted (FD) silicon-on-insulator (SOI)

---

[2] As described in Section 3.7.2.4, this assumes $V_{EFF} < -2nU_T$, which corresponds to an inversion coefficient less than 0.1 for weak inversion. V_{EFF} is evaluated for $n = 1.4$ and a thermal voltage $U_T = 25.9$ mV at room temperature.

[3] Effective W and L include the decrease in width and length due to mask biases, lateral diffusion, and other effects.

[4] Drain current, I_D, drain–source voltage, V_{DS}, gate–source voltage, V_{GS}, and threshold voltage, V_T, are positive for nMOS devices. In this book, these are also taken positively for pMOS devices as shown in Figure 3.9.

[5] MOS operation in saturation is assumed where $V_{DS} > V_{DS,sat}$. Here, the device acts a current source under control of the gate–source voltage.

CMOS processes where there is little substrate effect. n drops slightly with increasing inversion level (increasing $V_{EFF} = V_{GS} - V_T$) and is approximately 1.35 in moderate inversion and 1.3 in strong inversion for typical bulk CMOS processes.

As discussed in Section 3.7.2.4, interface state capacitance may be significant at temperatures below 200 K ($-73\,°$C) where n can double in value at temperatures near 77 K ($-196\,°$C). Operation at such cold temperatures, which is well below even the military temperature range of -55 to $+125\,°$C, is not addressed in the methods and design examples in this book. However, the micropower, differential-input, SOI CMOS preamplifier described in Section 6.7 is analyzed and evaluated at temperatures down to 77 K ($-196\,°$C) in a cited paper.

In weak inversion, n is the weak inversion slope factor that describes the degradation of exponential MOS drain current compared to bipolar transistor collector current where the equivalent value of n, the emitter injection efficiency, is very close to unity. Often the MOS weak inversion slope factor is expressed by the weak inversion or subthreshold swing given by

$$S = \ln(10)\, nU_T = 2.303 \cdot nU_T \text{ (mV/decade)} \tag{2.7}$$

This is the required increase in gate–source voltage for a factor-of-10 increase in drain current. The weak inversion swing is approximately 90 mV/decade for bulk CMOS processes at room temperature, assuming $n = 1.5$ and $U_T = 25.9$ mV.

MOS transconductance, $g_m = \delta I_D / \delta V_{GS}$, in weak inversion found from Equation 2.5 is given by

$$g_m \text{ (WI)} = \frac{I_D}{nU_T} \tag{2.8}$$

Transconductance efficiency, g_m/I_D, in weak inversion is then given by

$$\frac{g_m}{I_D} \text{ (WI)} = \frac{1}{nU_T} \tag{2.9}$$

MOS transconductance, like bipolar transistor transconductance given by Equation 2.3, is equal to the product of transconductance efficiency and drain current, as given by

$$g_m = \left(\frac{g_m}{I_D}\right) \cdot I_D \tag{2.10}$$

Interestingly, g_m and g_m/I_D in weak inversion given by Equations 2.8 and 2.9 are independent of MOS sizing and process parameters, except for dependence on n, which varies little for bulk CMOS processes. For a given temperature and value of n, g_m depends only on the DC bias current, I_D.

For a thermal voltage $U_T = 25.9$ mV at room temperature and substrate factor $n = 1.5$, MOS weak inversion transconductance efficiency is 25.7 µS/µA, which is approximately 67 % of the bipolar transistor transconductance efficiency of 38.6 µS/µA where n, again, is effectively unity. The exponential current–voltage relationship in weak inversion results in optimally high transconductance efficiency. Unfortunately, weak inversion operation requires a large MOS shape factor ($S = W/L$). This results in large gate area (WL), high gate capacitances, and relatively poor circuit bandwidth.

2.4.2 In Strong Inversion without Velocity Saturation Effects

Strong inversion occurs for MOSFETs operating at sufficiently high effective gate–source voltages ($V_{EFF} = V_{GS} - V_T > 225$ mV) where the gate–source voltage is above the threshold voltage by at least 225 mV for a typical bulk CMOS process at room temperature.[6] Here, the channel is strongly inverted

---

[6] As described in Section 3.7.2.4, this assumes $V_{EFF} > 6.24 \cdot nU_T$, which corresponds to an inversion coefficient greater than 10 for strong inversion. V_{EFF} is evaluated for $n = 1.4$ and $U_T = 25.9$ mV at room temperature.

and drain drift current dominates. Strong inversion drain current, excluding small-geometry effects like velocity saturation and vertical field mobility reduction, is proportional to the square of the effective gate–source voltage. It is approximated from the EKV MOS model [2, 3, 5, p. 177], using the source as the reference voltage, by

$$I_D \text{ (SI, no vel. sat.)} = \frac{1}{2}\left(\frac{\mu\, C'_{OX}}{n}\right)\left(\frac{W}{L}\right)(V_{GS} - V_T)^2 \qquad (2.11)$$

Drain current, again, is given for MOS saturation where the drain–source voltage exceeds its saturation value $(V_{DS} > V_{DS,sat})$. In strong inversion, the drain–source saturation voltage is frequently taken as $V_{DS,sat} \approx (V_{GS} - V_T)$, while $V_{DS,sat} \approx (V_{GS} - V_T)/n$ may be a better choice since body effect along the channel raises the local threshold voltage and lowers the drain–source voltage required for inversion charge pinch-off. Drain–source resistance effects, again, are not considered where drain current increases with increasing drain–source voltage.

MOS transconductance $(g_m = \delta I_D/\delta V_{GS})$ in strong inversion found from Equation 2.11 is given by

$$g_m \text{ (SI, no vel. sat.)} = \sqrt{2I_D\left(\frac{\mu\, C'_{OX}}{n}\right)\left(\frac{W}{L}\right)} = \left(\frac{\mu\, C'_{OX}}{n}\right)\left(\frac{W}{L}\right)(V_{GS} - V_T)$$

$$= \frac{2I_D}{V_{GS} - V_T} = \frac{2I_D}{V_{EFF}} \qquad (2.12)$$

Transconductance efficiency in strong inversion found from Equations 2.11 and 2.12 is then given by

$$\frac{g_m}{I_D} \text{ (SI, no vel. sat.)} = \sqrt{\frac{2}{I_D}\left(\frac{\mu\, C'_{OX}}{n}\right)\left(\frac{W}{L}\right)} = \frac{2}{V_{GS} - V_T} = \frac{2}{V_{EFF}} \qquad (2.13)$$

Finally, MOS effective gate–source voltage in strong inversion found from Equation 2.11 is given by

$$V_{EFF} \text{ (SI, no vel. sat.)} = V_{GS} - V_T = \sqrt{\frac{2I_D}{\frac{\mu\, C'_{OX}}{n}\left(\frac{W}{L}\right)}} \qquad (2.14)$$

Interestingly, g_m in strong inversion given by the last expression in Equation 2.12 is independent of MOS sizing and process parameters, depending only on the DC bias conditions, I_D and $V_{EFF} = V_{GS} - V_T$. Similarly, g_m/I_D given by the last expression in Equation 2.13 depends only on V_{EFF}. Technology- and sizing-independent expressions of transconductance and thermal noise, valid from weak through strong inversion, will be linked to the drain–source saturation voltage later in Section 6.5. These expressions will be used to minimize MOS thermal noise for a given level of supply voltage and current.

For bulk CMOS processes at room temperature, transconductance efficiency drops from approximately $25.7\,\mu\text{S}/\mu\text{A}$ $(1/(nU_T), \; n = 1.5, \; U_T = 25.9\,\text{mV})$ in weak inversion to $8.9\,\mu\text{S}/\mu\text{A}$ $(2/V_{EFF}, \; V_{EFF} = 225\,\text{mV})$ at the onset of strong inversion, which is 35 % of the weak inversion value. Transconductance efficiency continues to drop further as MOS inversion (V_{EFF}) increases. Decreasing transconductance efficiency and increasing drain–source saturation voltage are disadvantages of MOS operation in strong inversion. However, strong inversion operation requires smaller MOS shape factors $(S = W/L)$ compared to weak inversion. This results in smaller gate area (WL), lower gate capacitances, and relatively good circuit bandwidth since the decrease in capacitances exceeds the decrease in transconductance.

Strong inversion operation is often assumed in texts and papers describing analog MOS design. Equations 2.11–2.14 are usually used, although n is frequently excluded. The assumption of MOS strong inversion operation, however, should always be carefully stated and verified by ensuring sufficient effective gate–source voltage. Additionally, as described next, it is important to consider small-geometry, high-field effects, including velocity saturation. These effects cause departure from the ideal, square-law, drain current behavior of strong inversion.

2.4.3 In Strong Inversion with Velocity Saturation Effects

MOS drain current and transconductance can drop significantly from the ideal square-law relationships in strong inversion due to high-field, small-geometry effects. A primary effect is velocity saturation associated with the loss of effective carrier mobility caused by a high tangential or horizontal electric field between the pinched-off drain (again, saturation operation is considered) and source. This electric field is approximately equal to $E_x \approx (V_{GS} - V_T)/L$, which is the drain–source saturation or effective gate–source voltage appearing between the pinched-off drain and source divided by the channel length. As the effective gate–source voltage increases for small-channel-length devices, the horizontal electric field becomes so high that the average channel carrier velocity begins to drop from its normal value that linearly tracks the electric field. This causes the drain current to drop from its expected strong-inversion, square-law value. Ultimately, the field can become so high that the carrier velocity saturates, resulting in drain current that ceases to increase with increasing effective gate–source voltage.

While velocity saturation usually dominates in saturation for high effective gate–source voltages at short channel lengths, MOS source resistance also contributes to the falloff in drain current and transconductance. The reduction of drain current caused by velocity saturation effects can itself be modeled by introducing a degeneration resistance in the source of the MOSFET [1, p. 63]. This illustrates that velocity saturation effects and source degeneration resistance work collectively to reduce drain current and transconductance.

Another effect lowering drain current and transconductance is referred to here as vertical field mobility reduction (VFMR). This describes the loss of effective carrier mobility due to a high normal or vertical electric field, E_y, between the gate and the inverted MOS channel. As the effective gate–source voltage increases, mobile carriers are increasingly attracted closer to the Si–SiO$_2$ interface where their mobility reduces due to interface states and other imperfections at this interface. The loss of drain current due to VFMR is dominant for devices operating as resistors in the deep linear or deep ohmic region where the effective gate–source voltage is high and the drain–source voltage is low, well below the drain–source saturation voltage. Here, the average carrier velocity is low, resulting in small velocity saturation effects, but the carrier mobility can be significantly reduced due to VFMR associated with the high vertical field. Additionally, VFMR effects often dominate for pMOS devices and long-channel nMOS devices operating at high effective gate–source voltages. Here, carrier velocities are lower and less likely to saturate compared to carriers in short-channel nMOS devices, but carrier mobility is reduced by the high vertical field causing a falloff in drain current from ideal strong-inversion, square-law behavior. VFMR effects are not predicted here, but are considered with velocity saturation effects later in Sections 3.7.1.3–3.7.1.5.

The decrease in drain current due to velocity saturation effects is often estimated by dividing the strong-inversion, square-law drain current by the term $1 + (V_{GS} - V_T)/LE_{CRIT}$, which corresponds to $1 + E_x/E_{CRIT}$ [1, p. 62]. Dividing Equation 2.11 by this term gives

$$I_D \text{ (SI, with vel. sat.)} = \frac{\dfrac{1}{2}\left(\dfrac{\mu\, C'_{OX}}{n}\right)\left(\dfrac{W}{L}\right)(V_{GS} - V_T)^2}{1 + \left(\dfrac{V_{GS} - V_T}{L\, E_{CRIT}}\right)} \qquad (2.15)$$

$E_{CRIT} \approx v_{sat}/\mu$ is the critical, velocity saturation, electric field where average carrier velocity is reduced to approximately one-half the expected value without velocity saturation. v_{sat} is the carrier saturation velocity, and μ is the mobility for low, horizontal, drain–source electric fields. E_{CRIT} is typically 4 and 12 V/μm, assuming mobility equal to its low-field values, $\mu = \mu_0$, of 400 and 133 cm^2/V · s for nMOS and pMOS devices, respectively, and saturated carrier velocities of 1.6×10^7 cm/s. Actual values of E_{CRIT} depend not only on the operating mobility, but also on the velocity saturation model used as described in Sections 3.7.1.2 and 3.7.1.4. pMOS devices are less prone to velocity saturation since their hole mobility and, correspondingly, carrier velocity are less for a given horizontal, drain–source electric field. Equation 2.15 provides an estimate of drain current reduction due to velocity saturation. Other estimates are described in Sections 3.7.1.2, 3.7.1.4, and 3.7.1.5.

The degree of velocity saturation can be estimated by comparing $(V_{GS} - V_T)/LE_{CRIT}$ in the dominator of Equation 2.15 to unity. This can be done by defining a velocity saturation factor, *VSF*, equal to

$$VSF = \frac{V_{GS} - V_T}{L\,E_{CRIT}} = \frac{V_{EFF}}{L\,E_{CRIT}} \qquad (2.16)$$

Velocity saturation factors well below unity correspond to negligible velocity saturation, factors approaching unity correspond to significant velocity saturation, and factors well above unity correspond to nearly full velocity saturation.

The velocity saturation factor can be found by first evaluating LE_{CRIT}, which gives a voltage that can be compared to the effective gate–source voltage. The value of LE_{CRIT} for nMOS devices having $E_{CRIT} = 4$ V/µm is 16, 2, and 0.6 V for channel lengths of 4, 0.5, and 0.15 µm, respectively. For an effective gate–source voltage of 0.225 V (approximately the onset of strong inversion for bulk CMOS processes at room temperature), the corresponding velocity saturation factors $(VSF = V_{EFF}/(LE_{CRIT}))$ are 0.014, 0.11, and 0.38. This indicates that velocity saturation is negligible for the channel length of 4 µm, small but observable for the channel length of 0.5 µm, and significant for the channel length of 0.15 µm. For an effective gate–source voltage a factor-of-four higher at 0.9 V, the velocity saturation factors are also a factor-of-four higher, where velocity saturation is now significant for the channel length of 0.5 µm and nearly full velocity saturation occurs for the channel length of 0.15 µm.

Significant or full velocity saturation is common in modern nMOS devices having short channel lengths and high effective gate–source voltages. Full velocity saturation occurs when $VSF = (V_{GS} - V_T)/LE_{CRIT}$ is well above unity. Here, the drain current given by Equation 2.15 is approximated by

$$I_D \text{ (SI, full vel. sat.)} = \frac{1}{2}\left(\frac{\mu\,C'_{OX}}{n}\right)\left(\frac{W}{L}\right)(L\,E_{CRIT})(V_{GS} - V_T)$$

$$= \frac{1}{2}\left(\frac{\mu\,C'_{OX}}{n}\right)(W)\,E_{CRIT}(V_{GS} - V_T) \qquad (2.17)$$

In full velocity saturation, the drain current is linearly proportional to the effective gate–source voltage and is substantially below the strong-inversion, square-law value predicted by Equation 2.11.

The first expression in Equation 2.17 provides some insight into velocity saturation operation. Here, instead of squaring the effective gate–source voltage as in strong-inversion, square-law operation, the effective gate–source voltage is multiplied by a fixed voltage given by LE_{CRIT} to estimate the drain current. This illustrates that drain current increases linearly with effective gate–source voltage. The second expression in Equation 2.17 provides a different insight. Here, it is observed that drain current ceases to be a function of channel length once full velocity saturation is reached.

MOS transconductance $(g_m = \delta I_D/\delta V_{GS})$ for strong-inversion, full velocity saturation operation is approximated from Equation 2.17 by

$$g_m \text{ (SI, full vel. sat.)} = \frac{\mu\,C'_{OX}W\,E_{CRIT}}{2n} = \frac{I_D}{V_{GS} - V_T} = \frac{I_D}{V_{EFF}} \qquad (2.18)$$

Transconductance efficiency is then given by

$$\frac{g_m}{I_D}\text{(SI, full vel. sat.)} = \frac{\left(\dfrac{\mu\,C'_{OX}W\,E_{CRIT}}{2n}\right)}{I_D} = \frac{1}{V_{GS} - V_T} = \frac{1}{V_{EFF}} \qquad (2.19)$$

Like the last expression given in Equation 2.12 for strong-inversion operation without velocity saturation effects, g_m given by the last expression in Equation 2.18 is independent of MOS sizing and process parameters, depending only on the DC bias conditions, I_D and $V_{EFF} = V_{GS} - V_T$. Similarly, g_m/I_D given by the last expression in Equation 2.19 depends only on V_{EFF}.

Transconductance efficiency halves from its normal strong-inversion, square-law value given in Equation 2.13 if full velocity saturation is present. For typical bulk CMOS processes at room temperature, g_m/I_D drops from approximately $25.7\,\mu\text{S}/\mu\text{A}$ ($1/(nU_T)$, $n = 1.5$, $U_T = 25.9\,\text{mV}$) in weak inversion to $8.9\,\mu\text{S}/\mu\text{A}$ ($2/V_{EFF}$, $V_{EFF} = 225\,\text{mV}$) at the onset of strong inversion, to $2\,\mu\text{S}/\mu\text{A}$ ($2/V_{EFF}$) for $V_{EFF} = V_{GS} - V_T = 1\,\text{V}$ if no velocity saturation is present. If full velocity saturation is present for $V_{EFF} = V_{GS} - V_T = 1\,\text{V}$, g_m/I_D drops to $1\,\mu\text{S}/\mu\text{A}$ ($1/V_{EFF}$), which is one-half its value without velocity saturation effects and only $3.9\,\%$ of its weak-inversion value. The drop in strong-inversion g_m/I_D is significant without velocity saturation effects, but is even more significant when velocity saturation effects are present. Velocity saturation is a disadvantage of operating short-channel MOSFETs at high levels of inversion. However, for operation here, MOSFET shape factors ($S = W/L$) and gate areas (WL) are minimal, giving minimum gate capacitances and optimal circuit bandwidth since the decrease in capacitances generally exceeds the decrease in transconductance.

2.4.4 In Moderate Inversion and All Regions of Operation

As described above, the MOSFET has two distinct physical regions of operation: weak and strong inversion. In weak inversion, diffusion current dominates and drain current is proportional to the exponential of effective gate–source voltage. In strong inversion, drift current dominates and drain current is proportional to the square of effective gate–source voltage. Velocity saturation and other small-geometry, high-field effects reduce drain current below the strong-inversion, square-law value, especially for short-channel devices operating at high effective gate–source voltages.

Between weak and strong inversion, there is a transition region known as moderate inversion [6] where both diffusion and drift current are significant. As anticipated over 25 years ago [6], moderate inversion is an increasingly important region for modern analog MOS design. Moderate inversion offers higher transconductance efficiency and lower drain–source saturation voltage compared to strong inversion, combined with smaller gate area and capacitances and higher bandwidth compared to weak inversion.

Equation 2.5 estimates drain current in weak inversion where the effective gate–source voltage ($V_{EFF} = V_{GS} - V_T$) is below $-72\,\text{mV}$, and Equation 2.11 estimates drain current in strong inversion where the effective gate–source voltage is above $225\,\text{mV}$. These transition points assume typical bulk CMOS processes at room temperature with an average value of $n = 1.4$ at the onset of weak and strong inversion and $U_T = 25.9\,\text{mV}$. Moderate inversion occurs between the effective gate–source voltages of -72 and $+225\,\text{mV}$, and neither the weak- nor the strong-inversion drain current expressions can be used. Instead, interpolation expressions are used as described later in Section 3.7.1.

Figure 2.1, motivated by [7, p. 67], shows the measured drain current versus gate–source voltage[7] for an nMOS device in a $0.18\,\mu\text{m}$ CMOS process having a long channel length of $L = 4\,\mu\text{m}$. For low gate–source voltages in weak inversion, the measured drain current increases exponentially with gate–source voltage and follows the weak-inversion prediction of Equation 2.5 shown in the figure for a low-field transconductance factor, $\mu_0 C'_{OX}$, of $355\,\mu\text{A}/\text{V}^2$ ($\mu_0 = 422\,\text{cm}^2/\text{V} \cdot \text{s}$; $C'_{OX} = 8.41\,\text{fF}/\mu\text{m}^2$) and an

---

[7] MOSFET DC and small-signal parameter measurements given in this book were made on a semiconductor parameter analyzer developed by the author and colleagues at CTI PET Systems and Concorde Microsystems in the late 1990s. This system includes custom voltage force units having full-scale current measurement from 10 pA to 100 mA in decade ranges. Data is acquired using commercial 16-bit analog input/output boards, and small-signal parameters are computed using custom software.

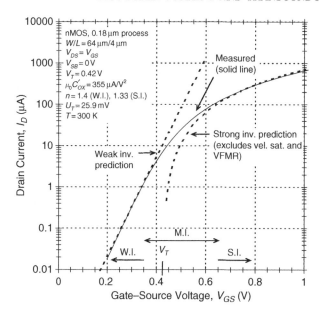

Figure 2.1 Measured drain current versus gate–source voltage for an $L = 4\,\mu m$, nMOS device in a $0.18\,\mu m$ CMOS process in weak, moderate, and strong inversion. The drain current follows the predicted values in weak and strong inversion but deviates significantly from these values in moderate inversion

average weak-inversion substrate factor of $n = 1.4$. Weak-inversion drain current appears as a straight line since a log scale is used for drain current with a linear scale used for the gate–source voltage. As the gate–source voltage increases and the device moves into moderate inversion, the measured drain current drops from its predicted weak-inversion value. As the gate–source voltage increases further and the device moves into strong inversion, the measured drain current follows the strong-inversion, square-law (without velocity saturation and VFMR effects) prediction of Equation 2.11, shown in the figure for $\mu_0 C'_{OX} = 355\,\mu A/V^2$ and an average strong-inversion substrate factor of $n = 1.33$. As observed in the figure, the measured drain current deviates from both its predicted weak- and strong-inversion values in moderate inversion, which spans nearly two decades of drain current from approximately 1 to $100\,\mu A$ for the device shown. Since drain current in the moderate-inversion region obeys neither weak- nor strong-inversion expressions, and this region spans nearly two decades of drain current, interpolation expressions are required to predict drain current and transconductance.

Figure 2.2 shows the measured drain current versus gate–source voltage for nMOS devices in a $0.18\,\mu m$ CMOS process having channel lengths of $L = 4$, 0.48, and $0.18\,\mu m$ and constant shape factors of $S = W/L = 16$. As the gate–source voltage increases, the drain current first follows the weak-inversion prediction of Equation 2.5 shown in the figure, drops from this in moderate inversion, and begins following the strong-inversion, square-law (again without velocity saturation and VFMR effects) prediction of Equation 2.11, also shown in the figure. However, as the gate–source voltage continues increasing, the drain current drops from the predicted strong-inversion, square-law value. The decrease, due primarily to velocity saturation, is much greater for the short-channel, $L = 0.18\,\mu m$ device. At high gate–source voltages, the drain current for this device begins following the prediction of Equation 2.15 shown in the figure, which includes velocity saturation effects. This prediction uses the previous values of $\mu_0 C'_{OX} = 355\,\mu A/V^2$ and average strong-inversion substrate factor of $n = 1.33$,

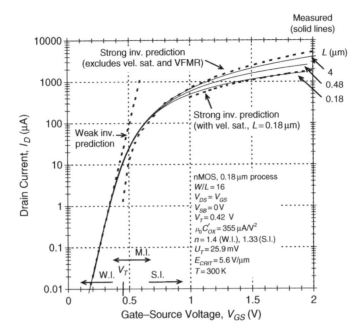

Figure 2.2 Measured drain current versus gate–source voltage for $L = 4$, 0.48, and 0.18 μm, nMOS devices in a 0.18 μm CMOS process in weak, moderate, and strong version. At the onset of strong inversion, drain current follows the predicted strong-inversion value without velocity saturation and VFMR effects but drops significantly at high gate–source voltages (also high effective gate–source voltages) for the $L = 0.48$ and 0.18 μm devices due primarily to velocity saturation. Drain current for the long-channel, $L = 4$ μm device drops modestly due primarily to VFMR

but includes a critical, horizontal electric field of $E_{CRIT} = 5.6\,V/\mu m$.[8] While drain current drops significantly for the short-channel, $L = 0.18\,\mu m$ and medium-channel, $L = 0.48\,\mu m$ devices, due primarily to velocity saturation, drain current for the long-channel, $L = 4\,\mu m$ device drops modestly due primarily to VFMR. Drain current falloff from ideal strong-inversion, square-law behavior illustrates the need for drain current expressions that include small-geometry, high-field effects like velocity saturation and VFMR.

Figure 2.3 shows the measured drain current of Figure 2.2 using a square-root scale for drain current compared to the log scale used earlier. The predicted square root of drain current shown in the figure for strong-inversion, square-law operation (again without velocity saturation and VFMR effects) increases linearly with the effective gate–source voltage ($V_{EFF} = V_{GS} - V_T$) and, correspondingly, increases linearly with the gate–source voltage, V_{GS}. This is because the strong-inversion, square-law drain current predicted in Equation 2.11 increases as the square of the effective gate–source voltage. As the gate–source voltage increases in strong inversion, the measured square root of drain current for the long-channel, $L = 4\,\mu m$ device follows the linear, straight-line prediction shown in the figure until the drain current begins dropping primarily due to VFMR. As observed, the long-channel, $L = 4\,\mu m$ device has a relatively wide region of strong-inversion, square-law operation. However, the short-channel, $L = 0.18\,\mu m$ device has little region of strong-inversion, square-law operation as its drain current

---

[8] As discussed in Sections 3.7.1.2 and 3.7.1.4, this value of E_{CRIT} is likely increased from the value associated with the low-field mobility, $\mu_0 = 422\,cm^2/V\cdot s$, due to the decrease in mobility caused by VFMR. As mobility decreases due to VFMR, E_{CRIT} increases if carrier saturation velocity, v_{sat}, remains nearly constant.

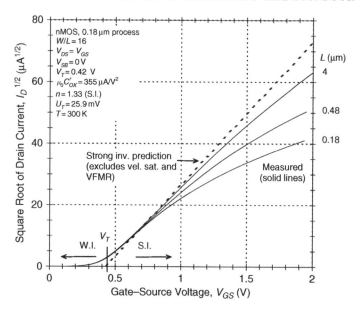

Figure 2.3 Measured drain current versus gate–source voltage for $L = 4$, 0.48, and 0.18 μm, nMOS devices in a 0.18 μm CMOS process using a square-root scale for drain current. At the onset of strong inversion, drain current follows the predicted, straight-line, square-law value without velocity saturation and VFMR effects but drops significantly at high gate–source voltages (also high effective gate–source voltages) for the $L = 0.48$ and 0.18 μm devices due primarily to velocity saturation. Drain current drops modestly for the long-channel, $L = 4$ μm device due primarily to VFMR. As observed, for short channel lengths, there is little or no region of strong-inversion, square-law operation

begins almost immediately dropping from the predicted straight-line, square-law value. As mentioned, this decrease is due primarily to velocity saturation for this short-channel device.

While Figures 2.1, 2.2, and 2.3 show drain current behavior, transconductance behavior is related to the drain current and is important for small-signal circuit analysis and design. Figure 2.4 shows the measured transconductance efficiency (g_m/I_D) versus drain current for five nMOS devices in a 0.18 μm CMOS process having channel lengths of $L = 4$, 1, 0.48, 0.28, and 0.18 μm. The devices have equal shape factors of $S = W/L = 16$ so the transitions from weak, moderate, and strong inversion occur at the same drain currents.

In Figure 2.4, devices operating at drain currents below 1 μA are in weak inversion where g_m/I_D is maximum and constant, tracking the predicted weak-inversion value shown in the figure from Equation 2.9. For drain currents above 100 μA, the devices are in strong inversion where g_m/I_D for the long-channel, $L = 4$ μm device nearly tracks the predicted strong-inversion, square-law value (without velocity saturation and VFMR effects) shown in the figure from Equation 2.13. For long-channel devices, g_m/I_D drops inversely with the square root of drain current as seen in Equation 2.13 and shown in the figure, while for short-channel devices experiencing full velocity saturation, g_m/I_D drops more rapidly as the inverse of drain current as seen in Equation 2.19. The devices shown in the figure do not reach full velocity saturation, except for the $L = 0.18$ μm device operating in deep strong inversion at a drain current of 1000 μA.

In Figure 2.4, devices operating at drain currents from 1 to 100 μA are in moderate inversion where g_m/I_D lies between the predicted weak-inversion and strong-inversion, square-law values shown in the figure. The center of moderate inversion occurs for drain currents near 10 μA where the predicted weak- and strong-inversion values of g_m/I_D intercept. Like the drain current shown in Figures 2.1 and 2.2,

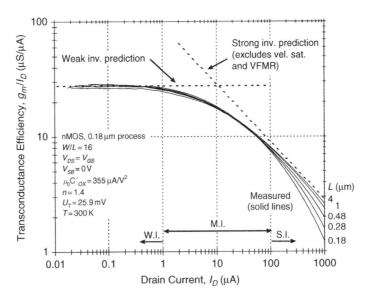

Figure 2.4   Measured transconductance efficiency versus drain current for $L = 4$, 1, 0.48, 0.28, and 0.18 μm, nMOS devices in a 0.18 μm CMOS process. Device shape factors ($S = W/L = 16$) are equal so transconductance efficiencies can be compared at equal currents. Transconductance efficiency is maximum in weak inversion, begins dropping in moderate inversion, and drops as the inverse square root of drain current in strong inversion, dropping more rapidly for short-channel devices due primarily to velocity saturation. The region of moderate inversion spans nearly two decades of drain current, emphasizing the importance of predicting performance in this region

g_m/I_D deviates considerably from either its weak- or strong-inversion values in moderate inversion, again illustrating the importance of expressions that are continuously valid from weak through strong inversion. Interestingly, g_m/I_D for devices having a long-channel length of $L = 4$ μm down to a short-channel length of $L = 0.18$ μm nearly overlay in moderate inversion, indicating small velocity saturation effects. High g_m/I_D and good immunity from velocity saturation are major advantages of operation in moderate inversion.

The measured g_m/I_D in Figure 2.4 for the long-channel, $L = 4$ μm device drops from approximately 27.6 μS/μA in weak inversion ($I_D < 1$ μA), to 18.3 μS/μA in the center of moderate inversion ($I_D = 10$ μA), to 8.1 μS/μA at the onset of strong inversion ($I_D = 100$ μA), to 2.6 μS/μA in deep strong inversion ($I_D = 1000$ μA). In deep strong inversion, g_m/I_D for the short-channel, $L = 0.18$ μm device drops further to approximately 1.3 μS/μA, which is one-half the long-channel value of 2.6 μS/μA, indicating the short-channel device is nearly fully velocity saturated. g_m/I_D for the long-channel device drops over a factor of 10 from weak inversion to deep strong inversion, and the drop is even greater for short-channel devices that experience velocity saturation. In contrast, the bipolar transistor normally maintains a constant, high transconductance efficiency at 38.6 μS/μA ($1/U_T$ with $U_T = 25.9$ mV at room temperature).

The drain currents associated with the transitions from weak, moderate, and strong inversion in Figure 2.4 depend on the shape factor (W/L), device type (nMOS or pMOS), and process characteristics. These transitions were conveniently aligned for the devices shown by maintaining constant, nominal shape factors of $S = W/L = 16$ and by making calibration adjustments in drain current resulting from effective channel lengths that are 0.028 μm below the drawn channel lengths shown. However, for generally sized devices, drain current is not conveniently aligned to the level of MOS inversion. As a result, in Chapter 3, drain current will be normalized using the inversion coefficient, IC [8, p. 62], where values below 0.1 correspond to weak inversion, values between 0.1

and 10 correspond to moderate inversion, and values above 10 correspond to strong inversion. The inversion coefficient provides a convenient, technology-independent measure of MOS inversion. It will be used throughout this book to evaluate MOS performance tradeoffs and provide guidance leading towards optimum design.

As described in this section, MOS drain current follows three regions of mathematical operation: exponential operation in weak inversion, square-law operation in strong inversion when small-geometry, high-field effects are negligible, and linear-law operation in strong inversion when full velocity saturation is present. Additionally, there are two transitional regions: moderate inversion between weak and strong inversion, and a region between strong-inversion, square-law and velocity-saturated, linear-law operation. The presence of effectively five regions or subregions of drain current operation, which also controls transconductance, illustrates the complexity of analog CMOS design compared to bipolar transistor design, where a single region of exponential operation exists over as many as 10 decades of collector current. The changing drain current behavior and declining transconductance efficiency with increasing inversion level emphasizes the importance of selecting the level of MOS inversion in analog CMOS design.

2.5 MOS DRAIN–SOURCE CONDUCTANCE

As described in the previous section, the choice of weak, moderate, or strong inversion controls MOS transconductance efficiency, which can decrease by a factor of 10 in strong inversion from its maximum weak-inversion value. While the inversion level primarily controls the transconductance efficiency, the channel length primarily controls the MOS drain–source resistance or its reciprocal, the drain–source conductance. In the following discussion, the MOS drain–source resistance and conductance are interchanged as convenient for the context.

MOS drain–source resistance causes an increase in drain current as the drain–source voltage increases for constant gate–source and gate–body voltages. As for the previous drain current and transconductance discussions, operation in saturation is assumed where the drain–source voltage exceeds the drain–source saturation voltage ($V_{DS} > V_{DS,sat}$). Over a limited signal range around a fixed bias point, the increase in drain current with drain–source voltage is modeled as a fixed, small-signal, drain–source resistance, r_{ds}, or conductance, $g_{ds} = 1/r_{ds}$, appearing between the drain and source. This resistance has a pronounced effect on open-loop voltage gains in analog CMOS circuits since signal currents typically flow into circuit resistances, developing voltage gain.

MOS drain–source resistance can be modeled, at least for drain voltages and currents near the bias point, using the Early voltage concept originally used for bipolar transistors [9]. Here, the increase in MOS drain current with drain–source voltage is modeled by the term $1 + V_{DS}/V_A$ multiplied by the saturation drain current having no drain–source voltage dependency, as given by

$$I_D = (I_D \text{without } V_{DS} \text{ dependency}) \cdot \left(1 + \frac{V_{DS}}{V_A}\right)$$

$$= (I_D \text{without } V_{DS} \text{ dependency}) \cdot (1 + \lambda V_{DS}) \tag{2.20}$$

V_A is the Early voltage, which is also the reciprocal of the channel length modulation (CLM) parameter, $\lambda = 1/V_A$. V_A, however, is used in this book to include more than just CLM components of drain–source resistance as described later in Section 3.8.4.

Graphically, the Early voltage corresponds to the intersection of a tangent line touching the bias point on the drain current, drain–source voltage transistor curve with the drain–source voltage corresponding to zero drain current. As detailed later in Section 3.8.4, this tangent line normally intersects zero drain current at a negative drain–source voltage outside the actual operating range of the device. Here, as is customary, the Early voltage is taken as a positive quantity for both nMOS and pMOS devices.

The slope of the tangent line touching the bias point on the drain current, drain–source voltage transistor curve is the drain–source conductance at the bias point. This slope or drain–source conductance is given by

$$g_{ds} = \frac{1}{r_{ds}} = \frac{I_D}{V_{DS} + V_A} \approx \frac{I_D}{V_A} \qquad (2.21)$$

where the effect of the V_{DS} bias voltage is often neglected. MOS drain–source resistance is the reciprocal of drain–source conductance and is given by

$$r_{ds} = \frac{1}{g_{ds}} = \frac{V_{DS} + V_A}{I_D} \approx \frac{V_A}{I_D} \qquad (2.22)$$

Unlike bipolar transistors where the Early voltage may be around 100 V for vertical transistor structures, MOS Early voltages can be much lower, especially for short-channel devices, as shown in measurements later in Section 3.8.4.5. Since the MOS Early voltage often does not exceed the drain–source voltage significantly by a factor of 10 or more, the approximations given in Equations 2.21 and 2.22, which neglect the drain–source voltage, provide a conservatively higher estimate of drain–source conductance or a conservatively lower estimate of drain–source resistance. However, often in Equations 2.21 and 2.22 it is customary to include the value of V_{DS} in the value of V_A such that $g_{ds} = I_D/V_A$ and $r_{ds} = V_A/I_D$. This convention is used in this book where the device operating conditions, including the value of V_{DS}, are specified for the value of V_A.

MOS Early voltage (V_A) is used in this book as a normalized drain–source resistance parameter, giving $r_{ds} = V_A/I_D$, in a similar way as transconductance efficiency (g_m/I_D) is used as a normalized transconductance parameter, giving $g_m = (g_m/I_D) \cdot I_D$. Both are quality factors where a high Early voltage corresponds to a high drain–source resistance or low drain–source conductance for a given drain current, while a high transconductance efficiency corresponds to a high transconductance for a given drain current.

The Early voltage and transconductance efficiency are not fixed quantities but complex functions of MOS operating conditions and geometry. Transconductance efficiency is primarily a function of MOS inversion level, while having a dependency on channel length for short-channel devices where velocity saturation deteriorates transconductance efficiency. MOS Early voltage, however, is primarily a function of MOS channel length where the Early voltage increases with channel length because of the lower effect of CLM, which raises the drain–source resistance. Additionally, the Early voltage associated with drain-induced barrier lowering (DIBL) increases with channel length because the decrease in threshold voltage with increasing drain–source voltage is less, which also raises the drain–source resistance. Finally, the MOS Early voltage depends strongly on the drain–source voltage, increasing as the drain–source voltage increases above the drain–source saturation voltage. These MOS Early voltage and corresponding drain–source resistance trends are discussed in detail later in Section 3.8.4.

Measured data shown later in Figure 3.47 in Section 3.8.4.5 confirms that the MOS Early voltage is a strong function of channel length, typically ranging from 1 to over 50 V for drain–source voltage sufficiently above the drain–source saturation voltage. This is in contrast to bipolar transistor Early voltage, which is usually fixed by a fixed base–collector width in the process.

MOS inversion level, which primarily controls transconductance efficiency, and MOS channel length, which primarily controls drain–source resistance, are degrees of design freedom normally not present in bipolar transistor design. As a result, this book uses MOS inversion coefficient as a measure of inversion and channel length, along with the drain current, as design choices for obtaining desired performance tradeoffs leading towards optimum analog CMOS design. Channel width, required for layout, is easily found from the selected drain current, inversion coefficient, and channel length, and is implicitly considered in predictions of performance.

2.6 ANALOG CMOS ELECTRONIC DESIGN AUTOMATION TOOLS AND DESIGN METHODS

Transistor-level, analog CMOS design involves the selection of a circuit topology followed by the selection of drain currents, channel widths, and channel lengths for MOSFETs in the circuit. The circuit topology will usually be selected to implement a particular function, for example an operational transconductance amplifier (OTA) as part of a data converter in a larger mixed-signal system containing both analog and digital circuits. The selection of a circuit topology will, of course, have a profound impact on circuit performance.

Once a candidate circuit topology is selected, the time-consuming process of selecting MOSFET drain currents, channel widths, and channel lengths begins. Often, iterative, trial-and-error computer simulations are done where different drain currents and sizing (channel width and length selections) are tried. In addition to being time consuming, this approach provides little design intuition and often excludes important considerations such as DC mismatch, not included in the widely used BSIM3 and BSIM4 MOS models [10].

2.6.1 Electronic Design Automation Tools

The significant time required for analog CMOS design combined with the shortage of experienced designers have prompted considerable research in electronic design automation (EDA) tools [11–13]. Before ceasing business operations in 2005, Barcelona Design, Inc., provided analog synthesis tools for the design of operational amplifier, data converter, and other circuits. These tools used monotonic, posynomial equations to predict circuit performance with MOS drain bias current and sizing found by rapid, geometric solution to these equations [14]. Cadence Design Systems, Inc., currently offers its Virtuoso NeoCircuit circuit sizing and optimization tool, which finds MOS bias current and sizing through iterative simulations of a user-defined circuit topology [15]. In addition to commercial EDA tools, many university methods and tools have been reported [13], including tools for MOS sizing design that consider variations in process parameters and operating conditions (temperature and supply voltage) [16] and local-area, DC mismatch [17]. The EDA tools introduced in [16, 17] are now commercially offered for manufacturability and yield enhancement by MunEDA, GmbH.

Unlike automatic sizing EDA tools, there are tools that do not automatically select device biasing and sizing, but rather guide the designer in these selections. In one tool, an operational amplifier is first simulated at various bias currents and sizing [18]. Following this, the designer explores bias current and sizing selections while observing performance like gain and phase margin. This provides valuable design insight and guides the designer towards optimum selections of bias current and MOS sizing.

Another EDA tool, developed by the author and students at the University of North Carolina at Charlotte (UNC Charlotte), provides design guidance by displaying MOS effective gate–source and drain–source saturation voltages, small-signal parameters including transconductance efficiency and Early voltage, intrinsic voltage gain and bandwidth, DC drain current and gate–source voltage mismatch, and thermal and flicker noise as the designer explores drain current, inversion coefficient, and channel length [19]. The intrinsic voltage gain is the gate-to-drain voltage gain of a single, grounded-source device, while the intrinsic bandwidth is the bandwidth associated with a single device, excluding extrinsic overlap capacitances. The tool utilizes the modeling equations and parameter set of the EKV MOS model [2, 3] and provides both graphical and numerical views of MOS performance. In the graphical view, the designer observes MOS performance displayed by vertical bargraphs. In addition to observing performance trends and tradeoffs, the designer can also assign numerical performance goals and explore design choices that meet the assigned goals. In the numerical view, the designer obtains a spreadsheet-like report documenting design selections and resulting MOS performance for multiple MOSFETs in a circuit.

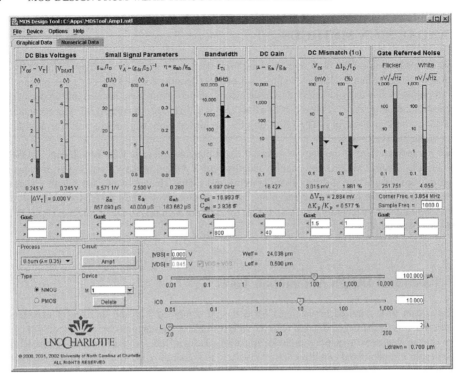

Figure 2.5 Graphical view of the UNC Charlotte, analog CMOS optimization tool for a drain current of 100 μA, an inversion coefficient of 10 (the onset of strong inversion), and a minimum channel length of 0.5 μm for an nMOS device in a 0.5 μm CMOS process. At this short channel length, bandwidth is successfully above the goal, but voltage gain is unsuccessfully below the goal and DC mismatch is unsuccessfully above the goals. The bargraphs appear darker (green in the actual tool) when goals are successfully met. Reproduced by permission of the Institute of Electrical and Electronic Engineers © 2003, from [19]

Figures 2.5, 2.6, and 2.7 show the graphical views of the UNC Charlotte EDA tool as channel length is explored for an nMOS device in 0.5 μm CMOS process operating at a drain current of 100 μA and inversion coefficient of 10, the onset of strong inversion. In all three figures, the following performance goals are selected: intrinsic voltage gain greater than 40 V/V, intrinsic bandwidth greater than 800 MHz, DC drain current mismatch for a pair of devices less than 1 % (1σ), and DC gate–source voltage mismatch for a pair of devices less than 1.5 mV (1σ). At the minimum channel length of 0.5 μm shown in Figure 2.5, bandwidth is successfully above the goal, but voltage gain is unsuccessfully below the goal, and DC mismatch is unsuccessfully above the goals. At a moderate channel length of 1.2 μm shown in Figure 2.6, voltage gain and bandwidth are successfully above the goals, and DC mismatch is successfully below the goals. Finally, at a long channel length of 2.6 μm shown in Figure 2.7, voltage gain is successfully above the goal and DC mismatch successfully below the goals, but bandwidth is unsuccessfully below the goal. The figures show that as channel length increases, gain increases, bandwidth decreases, and DC mismatch decreases. Gain, bandwidth, and DC mismatch goals are simultaneously met only for the moderate channel length selection of 1.2 μm shown in Figure 2.6, which provides a balance of gain, bandwidth, and DC mismatch performance. While not observed in the black-and-white figures, bargraphs of performance appear green when goals are met, yellow when nearly met, and red when not met. Bargraphs appear blue when no goals have been assigned.

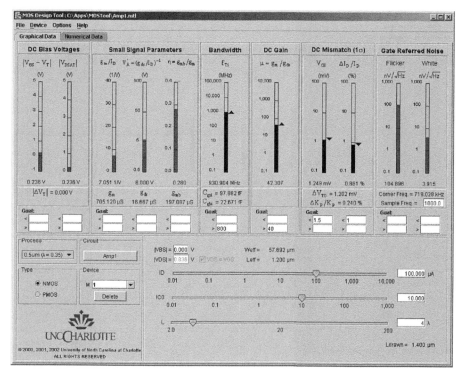

Figure 2.6 Graphical view of the UNC Charlotte, analog CMOS optimization tool for a drain current of 100 μA, an inversion coefficient of 10 (the onset of strong inversion), and a channel length of 1.2 μm for an nMOS device in a 0.5 μm CMOS process. At this moderate channel length, gain, bandwidth, and DC mismatch goals are simultaneously met where bandwidth and voltage gain are successfully above the goals, while DC mismatch is successfully below the goals. The bargraphs appear darker (green in the actual tool) when goals are successfully met. Reproduced by permission of the Institute of Electrical and Electronic Engineers © 2003, from [19]

While the graphical view of the UNC Charlotte EDA tool provides rapid, visual, MOS performance insight as the designer explores drain current, inversion coefficient, and channel length, this tool has been replaced by the *Analog CMOS Design, Tradeoffs and Optimization* spreadsheet. This spreadsheet permits operation on all computers having the popular Microsoft Excel spreadsheet program and facilitates easy user programming of circuit performance equations for user-defined circuits. The spreadsheet, used for design examples in this book, is described in the Appendix and is available at the book's web site listed on the cover.

In addition to the UNC Charlotte EDA tool and the *Analog CMOS Design, Tradeoffs and Optimization* spreadsheet, the simplicity and design intuition provided by the EKV MOS model [2, 3] has resulted in another tool, known as the Procedural Analog Design (PAD) tool [20, 21]. This tool provides numerical and graphical displays of MOS performance for selected drain current, channel width, and channel length. Unlike the UNC Charlotte tool, the inversion coefficient is not directly selected in PAD, but it is displayed as a performance parameter. Also, unlike the UNC Charlotte tool, PAD maps MOS performance into the performance of complete circuits, permitting complete design. Additionally, PAD contains a converter that converts widely used BSIM3 version 3 [10] MOS model parameters into EKV version 2.6 parameters [2, 3] required for the tool. Previously developed at the Swiss Federal Institute of Technology (EPFL) was probably the earliest tool for displaying MOS device performance as a function of the inversion coefficient or inversion factor as named in the EKV MOS model [22]. This tool displays MOS performance using a slide-rule and text-box presentation and was shown in the Foreword.

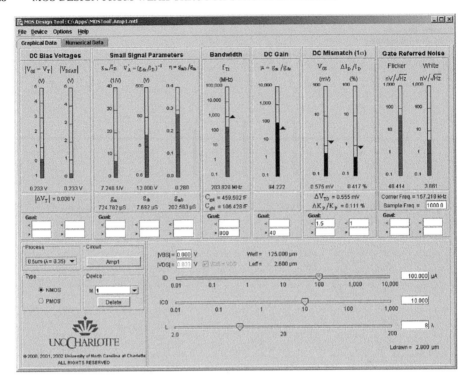

Figure 2.7   Graphical view of the UNC Charlotte, analog CMOS optimization tool for a drain current of 100 μA, an inversion coefficient of 10 (the onset of strong inversion), and a channel length of 2.6 μm for an nMOS device in a 0.5 μm CMOS process. At this long channel length, voltage gain is successfully above the goal and DC mismatch is successfully below the goals, but bandwidth is unsuccessfully below the goal. The bargraphs appear darker (green in the actual tool) when goals are successfully met. Reproduced by permission of the Institute of Electrical and Electronic Engineers © 2003, from [19]

The section just discussed gives a brief survey of EDA tools providing automation, synthesis, or design guidance for analog CMOS circuits. Many more tools are available or under development. The reader is referred to a book-length discussion on this topic [13] and the most recent literature for additional information.

2.6.2 Design Methods

Just as there is considerable interest in analog CMOS automation, synthesis, or design guidance EDA tools, there is considerable interest in manual design methods or methodologies, including the underlying methodologies employed in EDA tools. Many of these methodologies recognize the importance of transconductance efficiency, g_m/I_D, or the inversion coefficient as a numerical measure of MOS inversion.

One g_m/I_D-based methodology is illustrated through the design of a silicon-on-insulator (SOI) CMOS, micropower OTA [23]. Here, increased g_m/I_D resulting from the reduced SOI substrate effect results in less power consumption compared to that of bulk CMOS circuits. Another g_m/I_D-based design methodology is illustrated through the design of regulated cascode stages with emphasis on optimizing settling time [24].

Other design methodologies permit optimum transistor sizing for amplifiers and OTAs [25–28, 29, pp. 42–48] using a compact, current-based MOS model [29, pp. 7–42, 30] similar to the EKV MOS model [2, 3]. In these methods, a different but related numerical measure of MOS inversion is used instead of the inversion coefficient. A methodology similar to the methodology reported by the author for the optimization of individual MOSFETs [19] describes the optimization of amplifiers using the inversion coefficient as a design parameter [31]. The inversion coefficient is also used in the design of micropower CMOS preamplifiers for neural and electroencephalogram (EEG) [32] and nerve [33] applications. These and other CMOS preamplifiers are reviewed in Chapter 6, which describes the design of two micropower CMOS preamplifiers optimized for low thermal and flicker noise.

The increasing use of the inversion coefficient in reported designs, in addition to designs reported by the author in the next section that use the inversion coefficient, suggest that inversion coefficient design methods are gaining in popularity. As designers increasingly utilize moderate inversion for its high transconductance efficiency, low drain–source saturation voltage, immunity to velocity saturation and VFMR effects, and moderate bandwidth, inversion-coefficient-based design methods are expected to continue gaining in popularity.

The above is a brief survey of analog CMOS design methodologies. Many other methodologies have been reported or are under development, and a book is dedicated to the topic of tradeoffs in analog circuit design [34]. This book includes discussions of noise, gain, and bandwidth [pp. 227–256], frequency, dynamic range, power [pp. 283–313], and tradeoffs in the design of CMOS comparators [pp. 407–441]. Additionally, there are unpublished or proprietary design methodologies used in industry [35] and at universities [36].

2.6.3 Previous Application of Design Methods Presented in this Book

The analog CMOS design methods presented in this book were introduced in [19, 37–45] and implemented in an early Microsoft Excel spreadsheet tool at Concorde Microsystems, Inc., in the late 1990s. The methods were used to design large portions of a CMOS front-end chip [46] for positron emission tomography (PET) medical imaging scanners. This chip included a high count-rate, time pick-off circuit [47] and time-to-digital converter having 100 ps resolution [48] for determining detector event timing. Elements of the design methods were also used for a low-noise CMOS preamplifier having an input-referred noise voltage of $0.65\,\mathrm{nV/Hz}^{1/2}$ at a supply current of 3 mA [49] and for a micropower, CMOS, direct-conversion, very low-frequency (VLF) single-chip receiver operating at a total supply current of $80\,\mu\mathrm{A}$ [50]. In these industry designs, the design methods were used for the optimization of gain, bandwidth, transconductance linearity, thermal noise, flicker noise, and DC mismatch as needed for the specific circuits.

In addition to their use at Concorde Microsystems, Inc., the design methods presented in this book have been used for designs done at UNC Charlotte for the Jet Propulsion Laboratory (JPL). These designs include micropower, low-noise CMOS preamplifiers for neural implants [51], and preamplifiers in an SOI CMOS process for the extreme temperature and radiation environment of deep space [52, 53]. In these designs, the design methods were used to minimize thermal and flicker noise at minimum power consumption. This topic is discussed in detail in Chapter 6.

The design methods and early design spreadsheet, which are expanded in this book, allow the designer to rapidly select MOS drain current, inversion coefficient, and channel length while observing the effective gate–source voltage, drain–source saturation voltage, small-signal parameters including transconductance efficiency and Early voltage, voltage gain, capacitances, bandwidth, distortion, thermal noise, flicker noise, and DC mismatch prior to launching circuit simulations. Channel width is calculated for layout and is implicitly considered in the predictions of MOS performance. The methods reduce design time by minimizing iterative, trial-and-error sizing simulations, while providing design intuition. Additionally, the methods directly include DC mismatch, not included in the commonly used

BSIM3 and BSIM4 MOS models [10]. Finally, the methods permit technology-independent design and provide a cross-check with simulation MOS models. The advantages of these methods, where MOS drain current, inversion coefficient, and channel length are independent degrees of design freedom, are discussed further in Section 3.2.

REFERENCES

[1] P. R. Gray, P. J. Hurst, S. H. Lewis, and R. G. Meyer, *Analysis and Design of Analog Integrated Circuits*, fourth edition, John Wiley and Sons, Ltd, 2001.

[2] C. Enz, F. Krummenacher, and E. A. Vittoz, "An analytical MOS transistor model valid in all regions of operation and dedicated to low-voltage and low-current applications," *Analog Integrated Circuits and Signal Processing*, vol. 8, pp. 83–114, July 1995.

[3] M. Bucher, C. Lallement, C. Enz, F. Théodoloz, and F. Krummenacher, "The EPFL-EKV MOSFET model equations for simulation, version 2.6," Technical Report, EPFL, July 1998, Revision II, available on-line at http://legwww.epfl.ch/ekv/.

[4] J. P. Colinge, "Fully-depleted SOI CMOS for analog applications," *IEEE Transactions on Electron Devices*, vol. 45, pp. 1010–1016, May 1998.

[5] Y. Tsividis, *Operation and Modeling of the MOS Transistor*, second edition, McGraw-Hill, 1999.

[6] Y. Tsividis, "Moderate inversion in MOS devices," *Solid-State Electronics*, vol. 25, pp. 1099–1104, Nov. 1982.

[7] Y. Tsividis, *Mixed Analog-Digital VLSI Devices and Technology: An Introduction*, McGraw-Hill, 1996.

[8] E. A. Vittoz, "Micropower techniques," in *Design of MOS VLSI Circuits for Telecommunications*, ed. J. Franca and Y. Tsividis, Prentice Hall, 1994.

[9] J. M. Early, "Effects of space-charge layer widening in junction transistors," *Proceedings of the IRE*, vol. 40, pp. 1401–1406, Nov. 1952.

[10] University of California at Berkeley BSIM3 and BSIM4 MOS models, available on-line at http://www-device.eecs.Berkeley.edu/bsim3.

[11] G. E. Gielen and R. A. Rutenbar, "Computer-aided design of analog and mixed-signal integrated circuits," *Proceedings of the IEEE*, vol. 88, pp. 1825–1854, Dec. 2000.

[12] R. Martin, "Automation comes to analog," *IEEE Spectrum*, pp. 70–75, June 2001.

[13] R. A. Rutenbar, G. G. E. Gielen, and B. A. Antao, eds, *Computer-Aided Design of Analog Integrated Circuits and Systems*, Wiley – IEEE Press, 2002.

[14] M. del Mar Hershenson, S. P. Boyd, and T. H. Lee, "Optimal design of a CMOS op-amp via geometric programming," *IEEE Transactions on Computer-Aided Design of Integrated Circuits and Systems*, vol. 20, pp. 1–21, Jan. 2001.

[15] Cadence Design Systems, Inc., Virtuoso NeoCircuit CAD tool, available on-line at http://www.cadence.com.

[16] R. Schwencker, F. Schenkel, M. Pronath, and H. Graeb, "Analog circuit sizing using adaptive worst-case parameters sets," *Proceedings of the 2002 Design and Test in Europe Conference (DATE)*, pp. 581–585, Mar. 2002.

[17] K. Antreich, J. Eckmueller, H. Graeb, M. Pronath, F. Schenkel, and S. Zizala, "Wicked: analog circuit synthesis incorporating mismatch," *Proceedings of the IEEE 2000 Custom Integrated Circuits Conference (CICC)*, pp. 511–514, May 2000.

[18] R. Spence, "The facilitation of insight for analog design," *IEEE Transactions on Circuits and Systems – II*, vol. 46, pp. 540–548, May 1999.

[19] D. M. Binkley, C. E. Hopper, S. D. Tucker, B. C. Moss, J. M. Rochelle, and D. P. Foty, "A CAD methodology for optimizing transistor current and sizing in analog CMOS design," *IEEE Transactions on Computer-Aided Design of Integrated Circuits and Systems*, vol. 22, pp. 225–237, Feb. 2003.

[20] D. Stefanovic, M. Kayal, and M. Pastre, "PAD: a new interactive knowledge-based analog design approach," *Analog Integrated Circuits and Signal Processing*, vol. 42, pp. 291–299, Mar. 2005.

[21] Procedural Analog Design (PAD) CAD tool, Electronics Laboratories (LEG) of the Swiss Federal Institute of Technology (EPFL), available on-line at http://legwww.epfl.ch/CSL.

[22] C. C. Enz and E. A. Vittoz, "CMOS low-power analog circuit design," in *Emerging Technologies, Designing Low-Power Digital Systems*, tutorial for *1996 International Symposium on Circuits and Systems (ISCAS)*, eds. R. Cavin and W. Liu, IEEE Service Center, Piscataway, pp. 79–133, 1996.

[23] F. Silveira, D. Flandre, and P. G. A. Jespers, "A g_m/I_D based methodology for the design of CMOS analog circuits and its application to the synthesis of a silicon-on-insulator micropower OTA," *IEEE Journal of Solid-State Circuits*, vol. 31, pp. 1314–1319, Sept. 1996.

[24] D. Flandre, A. Viviani, J.-P. Eggermont, B. Gentinne, and P. G. A. Jespers, "Improved synthesis of gain-boosted regulated-cascode CMOS stages using symbolic analysis and g_m/I_D methodology," *IEEE Journal of Solid-State Circuits*, vol. 32, pp. 1006–1012, July 1997.

[25] R. L. Pinto, A. I. A. Cunha, M. C. Schneider, and C. Galup-Montoro, "An amplifier design methodology derived from a MOSFET current-based model," *Proceedings of the 1998 International Symposium on Circuits and Systems (ISCAS)*, pp. I-301–I-304, May 1998.

[26] X. Xie, M. C. Schneider, S. H. K. Embabi, and E. Sanchez-Sinencio, "Optimal design of low power nested gm-C compensation amplifiers using a current-based MOS transistor model," *Proceedings of the 1998 International Symposium on Circuits and Systems (ISCAS)*, pp. II-29–II-32, May 1998.

[27] R. L. Pinto, M. C. Schneider, and C. Galup-Montoro, "Sizing of MOS transistors for amplifier design," *Proceedings of the 2000 International Symposium on Circuits and Systems (ISCAS)*, pp. IV-185–IV-188, May 2000.

[28] H. C. M. Santos and A. I. A. Cunha, "Application of ACM model to the design of CMOS OTA through a graphical approach," *Proceedings of the 2003 International Symposium on Circuits and Systems (ISCAS)*, pp. I-305–I-308, May 2003.

[29] E. Sanchez-Sinencio and A. G. Andreou, eds, *Low-Voltage/Low-Power Integrated Circuits and Systems: Low-Voltage, Mixed-Signal Circuits*, IEEE Press, 1999.

[30] A. I. A. Cunha, M. C. Schneider, and C. Galup-Montoro, "An MOS transistor model for analog circuit design," *IEEE Journal of Solid-State Circuits*, vol. 33, pp. 1510–1519, Oct. 1998.

[31] T. M. Hollis, D. J. Comer, and D. T. Comer, "Optimization of MOS amplifier performance through channel length and inversion level selection," *IEEE Transactions on Circuits and Systems – II: Express Briefs*, vol. 52, pp. 545–549, Sept. 2005.

[32] R. R. Harrison and C. Charles, "A low-power low-noise CMOS amplifier for neural recording applications," *IEEE Journal of Solid-State Circuits*, vol. 38, pp. 958–965, June 2003.

[33] J. H. Nielsen and E. Bruun, "A CMOS low-noise instrumentation amplifier using chopper modulation," *Analog Integrated Circuits and Signal Processing*, vol. 42, pp. 65–76, Jan. 2005.

[34] C. Toumazou, G. Moschytz, and B. Gilbert, eds, *Trade offs in Analog Circuit Design*, Kluwer Academic, 2002.

[35] M. Bell, "Amplifiers for driving LCDs," National Semiconductor, unpublished notes, 2002.

[36] J. R. Brews, "Excel and PSPICE in analog design," Arizona State University, unpublished notes, 2005.

[37] D. M. Binkley, "A methodology for transistor level analog CMOS design," Conference tutorial with D. P. Foty, "MOSFET modeling and circuit design: Re-establishing a lost connection," *37th Annual Design Automation Conference (DAC)*, Los Angeles, June 2000.

[38] D. M. Binkley, "A methodology for analog CMOS design based on the EKV MOS model," *2000 Fabless Semiconductor Association Design Modeling Workshop: SPICE Modeling*, Santa Clara, Oct. 2000.

[39] D. M. Binkley, "A methodology for analog CMOS design based on the EKV MOS model," Conference tutorial with D. P. Foty, "Re-connecting MOS modeling and circuit design: new methods for design quality," *2001 IEEE 2nd International Symposium on Quality Electronic Design (ISQED)*, San Jose, Mar. 2001.

[40] D. M. Binkley, "A methodology for analog CMOS design based on the EKV MOS model," Conference tutorial with D. P. Foty, "MOS modeling as a basis for design methodologies: new techniques for next-generation analog circuit design," *2001 15th European Conference on Circuit Theory and Design (ECCTD)*, Helsinki, Aug. 2001.

[41] D. M. Binkley, "A methodology for analog CMOS design based on the EKV MOS model," Conference tutorial with D. P. Foty, "MOS modeling as a basis for design methodologies: new techniques for modern analog design," *2002 IEEE International Symposium on Circuits and Systems (ISCAS)*, Scottsdale, May 2002.

[42] D. M. Binkley, M. Bucher, and D. Kazazis, "Modern analog CMOS design from weak through strong inversion," *Proceedings of the European Conference on Circuit Theory and Design (ECCTD)*, pp. I-8–I-13, Krakow, Aug. 2003.

[43] D. M. Binkley, "Optimizing analog CMOS design from weak through strong inversion," Seminar talk, Columbia University Integrated Systems Laboratory, New York, November 21, 2003.

[44] D. M. Binkley, "Optimizing analog CMOS design from weak through strong inversion," Seminar talk, North Carolina State University Electrical and Computer Engineering Department, Raleigh, NC, January 19, 2005.

[45] D. M. Binkley, B. J. Blalock, and J. M. Rochelle, "Optimizing drain current, inversion level, and channel length in analog CMOS design," *Analog Integrated Circuits and Signal Processing*, vol. 47, pp. 137–163, May 2006.

[46] B. K. Swann, J. M. Rochelle, D. M. Binkley, B. S. Puckett, B. J. Blalock, S. C. Terry, J. W. Young, M. S. Musrock, J. E. Breeding, and K. M. Baldwin, "A custom mixed signal CMOS integrated circuit for high performance PET tomograph front-end applications," *IEEE Transactions on Nuclear Science*, vol. 50, pp. 909–914, Aug. 2003.

[47] D. M. Binkley, B. S. Puckett, B. K. Swann, J. M. Rochelle, M. S. Musrock, and M. E. Casey, "A 10-Mcps, 0.5-μm CMOS constant-fraction discriminator having built-in pulse tail cancellation," *IEEE Transactions on Nuclear Science*, vol. 49, pp. 1130–1140, June 2002.

[48] B. K. Swann, B. J. Blalock, L. G. Clonts, D. M. Binkley, J. M. Rochelle, E. Breeding, and K. M. Baldwin, "A 100-ps time-resolution CMOS time-to-digital converter for positron emission tomography imaging applications," *IEEE Journal of Solid-State Circuits*, vol. 39, pp. 1839–1853, Nov. 2004.

[49] D. M. Binkley, M. E. Casey, B. S. Puckett, R. Lecomte, and A. Saoudi, "A power efficient, low noise, wideband, integrated CMOS preamplifier for LSO/APD PET systems," *IEEE Transactions on Nuclear Science*, vol. 47, pp. 810–817, June 2000.

[50] D. M. Binkley, J. M. Rochelle, B. K. Swann, L. G. Clonts, and R. N. Goble, "A micropower CMOS, direct-conversion, VLF receiver chip for magnetic-field wireless applications," *IEEE Journal of Solid-State Circuits*, vol. 33, pp. 344–358, Mar. 1998.

[51] M. Mojarradi, D. M. Binkley, B. J. Blalock, R. Andersen, N. Ulshoefer, T. Johnson, and L. Del Castillo, "A miniaturized, neuroprosthesis suitable for implantation into the brain," *IEEE Transactions on Neural Systems and Rehabilitation Engineering*, vol. 11, pp. 38–42, Mar. 2003.

[52] D. M. Binkley, C. E. Hopper, B. J. Blalock, M. M. Mojarradi, J. D. Cressler, and L. K. Yong, "Noise performance of 0.35-μm SOI CMOS devices and micropower preamplifier from 77 to 400 K," *Proceedings of the 2004 IEEE Aerospace Conference*, vol. 4, pp. 2495–2506, Mar. 2004.

[53] D. M. Binkley, C. E. Hopper, J. D. Cressler, M. M. Mojarradi, and B. J. Blalock, "Noise performance of 0.35-μm SOI CMOS devices and micropower preamplifier following 63-MeV, 1-Mrad (Si) proton irradiation," *IEEE Transactions on Nuclear Science*, vol. 51, pp. 3788–3794, Dec. 2004.

3

MOS Performance versus Drain Current, Inversion Coefficient, and Channel Length

3.1 INTRODUCTION

Chapter 2 introduced MOS design from weak through strong inversion, emphasizing the complexity of design resulting from the design choices of inversion level and channel length not normally present in bipolar transistor design. The chapter described MOS transconductance efficiency, g_m/I_D, and reviewed drain current and transconductance efficiency expressions in weak and strong inversion. Strong-inversion expressions were then extended to include velocity saturation effects. Transconductance efficiency is constant and maximum in weak inversion, begins dropping in moderate inversion, and continues dropping in strong inversion, dropping even faster for devices experiencing velocity saturation and vertical field mobility reduction effects. Measured drain current and transconductance efficiency showed that the moderate-inversion region spans nearly two decades of drain current between weak and strong inversion where neither weak- nor strong-inversion expressions apply. This illustrated the need for interpolative expressions, continuously valid in all regions of operation. The inversion coefficient was then mentioned as a convenient numerical measure of MOS inversion.

In addition to transconductance efficiency, Chapter 2 introduced MOS drain–source resistance, r_{ds}, using the Early voltage, V_A, as a normalized measure of r_{ds}, where $r_{ds} = V_A/I_D$. Transconductance efficiency and the Early voltage are quality factors describing the desired production of trans-conductance and drain–source resistance at a given drain current. MOS Early voltage is not a fixed value but increases nearly directly with the channel length, increases with the drain–source voltage, and increases slightly with the inversion level. The chapter then suggested MOS inversion level, which primarily controls transconductance efficiency, and channel length, which primarily controls the Early voltage, as primary degrees of analog MOS design freedom. These are not degrees of design freedom in bipolar transistor design because transconductance efficiency and Early voltage are usually constant and fixed by the process. The chapter concluded by reviewing analog CMOS design EDA tools and design methodologies with emphasis on methods that save design time while providing design intuition. This review included the previous industry application of the design methods presented in this book.

Tradeoffs and Optimization in Analog CMOS Design   David M. Binkley
© 2008 John Wiley & Sons, Ltd

This chapter begins by describing the advantages of selecting MOS drain current, inversion coefficient, and channel length to obtain desired tradeoffs in geometry, bias conditions, small-signal parameters, intrinsic voltage gain, capacitances, intrinsic and extrinsic bandwidths, thermal noise, flicker noise, mismatch, and leakage current. The *MOSFET Operating Plane* is introduced to illustrate these tradeoffs over the design choices of inversion coefficient and channel length. The chapter then lists process parameters for example CMOS processes used for predictions of device performance in the chapter and for design examples in later chapters. Following this, the chapter develops expressions for the substrate factor and inversion coefficient. Process temperature effects are then reviewed, including changes in the inversion coefficient resulting from changes in temperature.

The majority of this chapter describes MOS performance as a function of the drain current, inversion coefficient, and channel length. MOS channel width and gate area, effective gate–source bias and drain–source saturation voltages, transconductance efficiency, transconductance distortion, body-effect transconductance ratio, drain–source resistance normalized as an Early voltage, intrinsic voltage gain, intrinsic and extrinsic capacitances, intrinsic and extrinsic bandwidths, thermal and flicker noise, DC voltage and current mismatch, small-signal parameter mismatch, and leakage current are discussed in considerable detail. Hand or spreadsheet predictions of MOS performance are developed and summarized in design tables and figures that illustrate trends resulting from the selected drain current, inversion coefficient, and channel length. Small-geometry effects like velocity saturation, vertical field mobility reduction, drain-induced barrier lowering, increases in gate-referred flicker-noise voltage with increasing inversion, and gate leakage current are included. Additionally, tables of reported flicker noise and DC mismatch factors illustrate technology trends. Finally, measured data, usually for a typical 0.18 μm CMOS process, is included to validate performance predictions and trends. The study of MOS performance in this chapter, presented from a design perspective, permits the study of analog performance tradeoffs in Chapter 4, followed by the optimization of designs described in Chapters 5 and 6.

3.2 ADVANTAGES OF SELECTING DRAIN CURRENT, INVERSION COEFFICIENT, AND CHANNEL LENGTH IN ANALOG CMOS DESIGN

As described in Chapter 2, analog CMOS design is greatly complicated by the additional degrees of design freedom – inversion level and channel length – not usually present in bipolar transistor design. The level of MOS inversion will be considered here using the inversion coefficient, IC, a numerical measure of MOS inversion where unity corresponds to the center of moderate inversion [1, p. 62]. The inversion coefficient can be found by the drain current, I_D, divided by the product of the shape factor, $S = W/L$, and a technology current, I_0, for the process at the operating temperature. This gives $IC = I_D/[I_0(W/L)]$, which is developed further in Section 3.4.2.2. Inversion coefficient values less than 0.1 correspond to weak inversion, values between 0.1 and 10 correspond to moderate inversion, and values above 10 correspond to strong inversion. The inversion coefficient is used throughout this chapter and the remainder of this book.

The choice of MOS inversion coefficient strongly influences transconductance efficiency, g_m/I_D, whereas channel length strongly influences the Early voltage, V_A, or normalized drain–source resistance, where $r_{ds} = V_A/I_D$. Additionally, the inversion coefficient and channel length strongly influence intrinsic voltage gain, capacitances, intrinsic bandwidth, thermal and flicker noise, DC voltage and current mismatch, and other aspects of analog circuit performance described later in this chapter.

Selecting drain current, inversion coefficient, and channel length as the three independent degrees of design freedom represents a conceptual shift from the traditional selections of drain current, channel width, and channel length. Selecting the inversion coefficient permits a conscious choice of the region and level of MOS inversion, which is not directly known when making the traditional selections of drain current, channel width, and channel length. While drain current, inversion coefficient, and channel length are the degrees of design freedom used throughout this book, *channel width is easily found for layout and is implicitly considered throughout* in predictions of capacitance, flicker noise, DC mismatch, and other performance measures where channel width and resulting gate area affect performance.

EDA tools for guiding the designer towards optimum analog CMOS design were mentioned earlier in Section 2.6.1. This included tools developed by the author and others that utilize the inversion coefficient as a measure of MOS inversion. Design methods, including those utilizing the inversion coefficient or related measures of MOS inversion, were mentioned earlier in Section 2.6.2. Finally, designs by the author using the methods described in this book were mentioned earlier in Section 2.6.3. The increase in reported EDA tools and design methods that utilize the inversion coefficient suggests that inversion coefficient design methods are gaining popularity. This increase is expected to continue, especially as designers increasingly utilize moderate inversion for high transconductance efficiency, low drain–source saturation voltage, minimal velocity saturation effects, and moderate bandwidth. The following discussions describe the advantages of selecting drain current, inversion coefficient, and channel length in analog CMOS design.

3.2.1 Optimizing Drain Current, Inversion Coefficient, and Channel Length Separately

When selecting drain current, inversion coefficient, and channel length as the independent degrees of MOS design freedom, the question arises, "What happens if drain current is changed for a device having a fixed channel width and length?" As observed from $IC = I_D/[I_0(W/L)]$ described later in Section 3.4.2.2, if drain current is increased for a device having fixed channel width and length, *both* the drain current and inversion coefficient increase simultaneously. As a result, g_m/I_D decreases due to the increase in inversion level, represented by the inversion coefficient, while overall $g_m = (g_m/I_D) \cdot I_D$ usually increases because the increase in drain current usually exceeds the decrease in g_m/I_D. This obscures optimization choices compared to separately exploring drain current, inversion coefficient, and channel length, which are independent, non-interacting degrees of design freedom. When drain current, inversion coefficient, and channel length are explored separately, MOS channel width must be adjusted to maintain the selected inversion coefficient at the selected drain current and channel length. This is illustrated in the MOS channel width relationships described later in Section 3.6.2.

When exploring drain current, inversion coefficient, and channel length separately, selecting an initial drain current permits a separate optimization of inversion coefficient and channel length. Holding drain current constant also permits a "fair" optimization where power consumption remains constant. As a result, in this book drain current will normally be selected first and held constant, followed by exploration of the inversion coefficient and channel length. Of course, drain current can also be explored, especially for thermal noise and other parameters that depend strongly on the drain current.

3.2.2 Design in Moderate Inversion

Using the inversion coefficient permits design freely in all regions of MOS operation, including moderate inversion. Moderate inversion [2], a transitional region spanning nearly two decades of drain current between weak and strong inversion, has become increasingly important in modern design. This is because it offers a compromise of high transconductance (high transconductance efficiency), low drain–source saturation voltage, minimal velocity saturation degradation of transconductance, and moderate bandwidth necessary for power-efficient, low-voltage design. However, as described in Section 2.4.4, drain current, transconductance, and transconductance efficiency cannot be modeled in moderate inversion by the exponential, weak-inversion equations (Equations 2.5–2.10) or the traditional, square-law, strong-inversion equations (Equations 2.11–2.13). Instead, interpolative expressions, utilizing the inversion coefficient, can be used that are continuously valid in all regions of MOS operation. Predictions and measurements of MOS performance will be given later in this chapter, with considerable emphasis on the moderate inversion region. Additionally, design examples in Chapters 5 and 6 will utilize input devices operating in moderate inversion for high transconductance, giving low input-referred thermal-noise voltage.

3.2.3 Design Inclusive of Velocity Saturation Effects

As described in Section 2.4.4, MOSFETs having short channel lengths can operate in a transitional region between strong-inversion, square-law operation described by Equations 2.11–2.13 and strong-inversion, velocity-saturated, linear-law operation described by Equations 2.17–2.19. Both this region and the moderate inversion transitional region complicate analog design as devices operating in these regions have neither exponential, square-law, nor linear drain current versus gate–source voltage characteristics. As a result, traditional hand analysis expressions, like the widely used strong-inversion, square-law equations, are not valid to provide performance prediction and design intuition. As mentioned for moderate inversion, interpolative expressions, utilizing the inversion coefficient, can be used that are continuously valid in all regions of MOS operation. These expressions can then be extended to include the reductions in drain current and transconductance associated with velocity saturation and vertical field mobility reduction (VFMR) effects. Predictions and measurements of MOS performance, including performance between square-law and linear-law operation caused by velocity saturation and VFMR effects, will be included later in this chapter.

3.2.4 Design with Technology Independence

As described later in Section 3.4.2, the inversion coefficient is a normalized measure of MOS inversion independent of technology parameters such as mobility, gate-oxide capacitance, and the threshold voltage. As a result, key performance measures like the effective gate–source voltage, $V_{EFF} = V_{GS} - V_T$, and transconductance efficiency, g_m/I_D, are dependent primarily on the inversion coefficient, the substrate factor n (typically 1.3–1.4 for bulk CMOS processes operating in moderate and strong inversion), and the thermal voltage. Velocity saturation and VFMR effects require consideration of the velocity saturation critical field, E_{CRIT}, and the mobility reduction factor, θ, but still few process parameters are embedded in design expressions developed from the inversion coefficient. This greatly simplifies performance predictions, design optimization, and the transfer of candidate designs to different MOS processes. The final chapter of this book, Chapter 7, discusses the extension of performance predictions and optimization to increasing smaller-geometry CMOS processes and emerging, non-CMOS processes.

3.2.5 Simple Predictions of Performance and Trends

For the three independent degrees of MOS design freedom – drain current, inversion coefficient, and channel length – used here, the effective gate–source voltage, $V_{EFF} = V_{GS} - V_T$, could replace the inversion coefficient as the design choice governing the level of MOS inversion. However, MOS drain current, channel width, and gate area are non-linearly linked to the effective gate–source voltage, compared to the simple linkage provided by the inversion coefficient as observed in the expression $IC = I_D/[I_0(W/L)]$. The linkage of selected drain current, inversion coefficient, and channel length to MOS channel width and gate area is described in detail in Section 3.6. This simple linkage permits the development of performance predictions that implicitly consider the impact of channel width and gate area on capacitances, intrinsic bandwidth, flicker noise, and DC mismatch. Such expressions appear throughout this chapter, in the performance tradeoffs described in Chapter 4, and in the design examples of Chapters 5 and 6. These expressions provide simple predictions and trends of MOS device and circuit performance, continuously valid from weak through strong inversion.

3.2.6 Minimizing Iterative Computer Simulations – "PreSPICE" Guidance

During more than 20 years of experience as an industry designer, the author observed the common practice of iterative, trial-and-error computer simulations for optimizing drain current, channel width, and channel length selection in analog CMOS design. Such design methods were arguably necessary when designing in moderate inversion prior to the recent availability of hand analysis for this region of operation. However, iterative computer simulations are time consuming, provide little design intuition, and rarely lead to an optimized design. The design methods presented here minimize iterative simulations by guiding the designer towards initial optimal or near-optimal selections of drain current, inversion coefficient, and channel length. Additionally, DC mismatch effects, excluded in the widely used BSIM3 and BSIM4 MOS models [3], are directly considered in the design process.

Since performance predictions presented here are necessarily simplified to provide design intuition during the initial design process, a candidate design must be computer simulated using production MOS models to verify performance over process and temperature. However, when using an early version of the *Analog CMOS Design, Tradeoffs and Optimization* spreadsheet at Concorde Microsystems, Inc., in the late 1990s, the author frequently obtained final designs prior to launching computer simulations. While some design iteration may be required, the design methods presented here have the potential to significantly reduce design iterations and design time. The name "preSPICE," suggested by Jim Rochelle at Concorde Microsystems, Inc., captures the intent of the design methods and the *Analog CMOS Design, Tradeoffs and Optimization* spreadsheet. This spreadsheet lists MOS device performance as the designer explores drain current, inversion coefficient, and channel length, and maps device performance into complete circuit performance. It is used for differential-pair and current-mirror designs in Chapter 4 and for circuit design examples in Chapters 5 and 6. The spreadsheet is described further in the Appendix and is available at the book's web site listed on the cover.

3.2.7 Observing Performance Tradeoffs – The *MOSFET Operating Plane*

In addition to permitting the prediction of device and circuit performance for all regions of MOS operation, the selection of drain current, inversion coefficient, and channel length permits the observation of performance tradeoffs. Performance tradeoffs are available through the selection of drain current, inversion coefficient, and channel length for individual MOSFETs or related MOSFETs, for example differential pairs and current mirrors, in an analog circuit.

Figure 3.1 introduces the *MOSFET Operating Plane*, which illustrates performance tradeoffs versus the selected inversion coefficient and channel length for a fixed drain current [4–11]. Each MOSFET or related group of MOSFETs operates at a selected inversion coefficient and channel length, corresponding to a point on the operating plane. Operation at low inversion coefficients in weak or moderate inversion (the left side of the plane) maximizes transconductance efficiency and transconductance, minimizes gate-referred thermal-noise voltage, maximizes drain-referred thermal-noise current, maximizes intrinsic voltage gain, and minimizes the effective gate–source bias and drain–source saturation voltages. Since shape factors ($S = W/L$), channel widths, gate areas, and capacitances are large at low inversion coefficients, intrinsic bandwidth, gate-referred flicker-noise voltage, drain-referred flicker-noise current, and DC gate–source voltage and drain current mismatch are minimized. Conversely, operation at high inversion coefficients in strong inversion (the right side of the plane) minimizes transconductance efficiency and transconductance, maximizes gate-referred

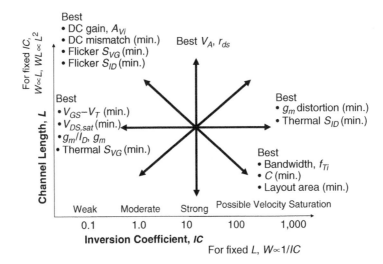

Figure 3.1 *MOSFET Operating Plane* illustrating MOS analog performance tradeoffs versus selected inversion coefficient and channel length for a fixed drain current. Each MOSFET in an analog circuit operates at some inversion coefficient and channel length defining a point on the operating plane

thermal-noise voltage, minimizes drain-referred thermal-noise current, minimizes intrinsic voltage gain, and maximizes the effective gate–source bias and drain–source saturation voltages. Since shape factors, channel widths, gate areas, and capacitances are small at high inversion coefficients, intrinsic bandwidth, gate-referred flicker-noise voltage, drain-referred flicker-noise current, and DC gate–source voltage and drain current mismatch are maximized.

Operation at short channel lengths (the bottom side of the plane) minimizes drain–source resistance, Early voltage, and intrinsic voltage gain. Since channel width, gate area, and capacitances must decrease with length to maintain the required shape factor, these are small at short channel lengths, maximizing intrinsic bandwidth, gate-referred flicker-noise voltage, drain-referred flicker-noise current, and DC gate–source voltage and drain current mismatch. Conversely, operation at long channel lengths (the top side of the plane) maximizes drain–source resistance, Early voltage, and intrinsic voltage gain. Since channel width, gate area, and capacitances must increase with length to maintain the required shape factor, these are large at long channel lengths, minimizing intrinsic bandwidth, gate-referred flicker-noise voltage, drain-referred flicker-noise current, and DC gate–source voltage and drain current mismatch.

The combination of low inversion coefficients and long channel lengths (the upper left side of the plane) further maximizes intrinsic voltage gain, maximizes gate area and capacitances, minimizes intrinsic bandwidth, and minimizes gate-referred flicker-noise voltage, drain-referred flicker-noise current, and DC gate–source voltage and drain current mismatch. Conversely, the combination of high inversion coefficients and short channel lengths (the lower right side of the plane) further minimizes intrinsic voltage gain, minimizes gate area and capacitances, maximizes intrinsic bandwidth, and maximizes gate-referred flicker-noise voltage, drain-referred flicker-noise current, and DC gate–source voltage and drain current mismatch. These and other performance tradeoffs will become clear during the course of this chapter. Additionally, performance tradeoffs are discussed in detail in Chapter 4.

Modern, low-voltage analog CMOS design requires lower levels of inversion to keep the drain–source saturation and effective gate–source voltages to sufficiently low values. Thus, design choices are increasing required in weak ($IC < 0.1$), moderate ($0.1 < IC < 10$), and at the onset of strong inversion ($IC = 10$). As a result, design is increasingly shifting to the left side of the *MOSFET Operating Plane* where inversion coefficient values are less than 10. Since operation in weak inversion at $IC < 0.1$ provides little increase in transconductance efficiency but requires larger shape factors, channel widths, and gate areas giving high capacitances and low intrinsic bandwidth, there is little reason to operate in weak inversion compared to the weak inversion side of moderate inversion. This further illustrates the importance of the moderate-inversion region.

Different transistors or transistor groups may be optimized differently, as denoted by different locations on the *MOSFET Operating Plane*. For example, transistors in the signal path may be optimized for tradeoffs in gain, bandwidth, and noise while transistors in DC bias and control paths may be optimized solely for low DC mismatch. The optimization of different transistors in an analog circuit is illustrated in Chapter 5 for cascoded, operational transconductance amplifiers (OTAs) optimized for low thermal noise. Here, input devices are operated at low inversion coefficients for high transconductance efficiency and high transconductance, giving low gate-referred thermal-noise voltage. Conversely, non-input devices are operated at high inversion coefficients for low transconductance efficiency and transconductance, giving low drain-referred thermal-noise current contributions such that input devices dominate the thermal noise. In addition to different inversion coefficient choices, different channel lengths are selected for input and non-input devices for the micropower, low-noise preamplifier design examples of Chapter 6. Here, input devices dominate both thermal and flicker noise. In all cases, visualizing or even marking the location of MOSFET inversion coefficient and channel length selections on the *MOSFET Operating Plane* provides design guidance and intuition that can lead towards optimum design.

The performance tradeoffs shown in the *MOSFET Operating Plane* of Figure 3.1 assume negligible gate leakage current. However, for gate-oxide thickness around 2 nm and below, gate–source conductance, gate shot- and flicker-noise current, and mismatch due to gate leakage current can affect performance tradeoffs. As a result, operating at arbitrarily long channel lengths where gate area is large may not maximize intrinsic voltage gain and minimize flicker noise and DC mismatch because the large gate area increases gate leakage current and deteriorates performance. As described later in Section 3.12.2, a smaller channel length and resulting gate area may be required to balance the traditional improvements in intrinsic voltage gain, flicker noise, and DC mismatch with the deterioration of these associated with gate leakage current.

3.2.8 Cross-Checking with Computer Simulation MOS Models

An analog CMOS design is vulnerable to simulation model errors, especially when the designer is not aware of potential errors. These errors can include structural errors where, for example, transconductance may not be accurately modeled in moderate inversion regardless of the choice of model parameters. Additionally, serious errors can result from inaccurate model parameters, especially when devices in a design are operated at inversion levels and geometries considerably different than those used in the model extractions. Significant errors in MOS drain–source resistance are most common, with simulated resistance sometimes as much as a factor of 10 below or above the actual measured value. In these cases, a simple, rule-of-thumb, Early voltage prediction of $r_{ds} = V_A/I_D$ with the Early voltage approximated as 10 V/μm multiplied by the channel length ($V_A = 10\,\text{V}/\mu\text{m} \cdot L$) would have provided a better prediction of drain–source resistance. As discussed later in Section 3.8.4.5, the measured drain–source resistance and Early voltage can still be as much as a factor of three below or above these rule-of-thumb predictions.

The design methods presented here, while not replacing the need for computer simulations for a candidate design, do provide an independent cross-check on the simulation MOS models used. As seen later in Section 3.8.2.5, MOS transconductance efficiency behavior in weak, moderate, and strong inversion is fundamental and independent of process technology, excepting the onset of velocity saturation for short-channel devices that depends on the value of the critical, velocity saturation electric field. Also, measured transconductance efficiency and Early voltage presented later in Sections 3.8.2.3 and 3.8.4.5 provide a reality check and rigorous benchmark for analog MOS models operating from weak through strong inversion over a wide range of channel lengths. As a result, simulation modeling errors in transconductance or, more likely, drain–source resistance, may be identified from the hand expressions and measured performance data presented here. Additionally, MOS threshold-voltage mismatch parameters reported later in Section 3.11.1.2, which follow clear technology tends with gate-oxide thickness, provide a cross-check on mismatch model parameters, or a starting point when these parameters are unavailable from the foundry. The MOS performance predictions, trends, and measured data presented here have the potential to help designers recognize and avoid design problems resulting from errors in simulation models.

3.3 PROCESS PARAMETERS FOR EXAMPLE PROCESSES

This chapter describes MOS performance with example predictions and measurements usually given for a $0.18\,\mu$m CMOS process, with some predictions and measurements given for 0.5 and $0.35\,\mu$m CMOS processes to illustrate process trends. Chapter 4 describes performance tradeoffs with example predictions and measurements given for a $0.18\,\mu$m process. Finally, Chapter 5 describes design examples in 0.5 and $0.18\,\mu$m processes, while Chapter 6 describes design examples in a $0.35\,\mu$m process.

Parameters for the 0.5, 0.35, and $0.18\,\mu$m example processes are grouped into tables below for those modeling drain current, small-signal parameters, and intrinsic gate capacitances, those modeling flicker noise and local-area DC mismatch, those modeling gate-overlap and drain–body capacitances, and those modeling temperature effects. Process parameters may be compared across the different processes to show technology trends. Process parameters will be discussed in detail in this chapter as they relate to specific areas of MOS performance.

3.3.1 Calculation of Composite Process Parameters

Table 3.1 gives expressions for the gate-oxide capacitance, C'_{OX}, and transconductance factor, $k = \mu C'_{OX}$, as functions of the electrical gate-oxide thickness, t_{ox}, and mobility, μ. Following this, the table gives an expression for the technology current, I_0, which will be used for calculating the MOS inversion coefficient later in Section 3.4.2.2. The table also gives expressions for the body-effect factor, γ, and the Fermi potential,[1] $PHI = 2\phi_F$, as functions of the doping concentration in the substrate or body, N_B. The body-effect factor and Fermi potential are often given for a process, but expressions are included to illustrate process trends. Finally, the table gives expressions for the thermal voltage, U_T, built-in p/n junction potential, ϕ_{bi}, and depletion region thickness, t_{dep}, present for a p/n junction or below the MOS channel. For the p/n junction, N_A and N_D are used for acceptor (p-type) and donor (n-type) doping concentrations, respectively. The table includes notes describing the use of ϕ_{bi} and reverse bias voltage, V_R, for calculating p/n junction t_{dep}. Additionally, the table includes notes describing the use of MOS surface potential, ϕ_s, N_B, and channel-to-body (channel-to-substrate) reverse bias voltage, V_{CB}, for calculating MOS t_{dep} below the channel.

Table 3.1 also includes convenient expressions[2] with built-in unit conversions to permit rapid evaluation of process characteristics. This enables, for example, the immediate calculation of MOS $t_{dep} = 0.361\,\mu$m for a body doping concentration of $N_B = 10^{16}$/cm³, assuming a strong-inversion surface potential of $\phi_s = 1$ V. Increasing the doping concentration to $N_B = 10^{18}$/cm³ quickly reveals that t_{dep} is reduced by a factor of 10 to $t_{dep} = 0.0361\,\mu$m. As noted in the table, transconductance factor (through the mobility), technology current, Fermi potential, thermal voltage, and built-in potential are strong functions of temperature and must be evaluated for the temperature of interest. t_{dep} is a slight function of temperature through either ϕ_{bi} or ϕ_s.

---

[1] $PHI = 2\phi_F$ is actually twice the Fermi potential. One unit of Fermi potential is required to convert the doped silicon channel to intrinsic silicon. The second unit of Fermi potential inverts the channel silicon to a carrier concentration equal to its initial doped value.

[2] Convenient expressions were motivated by [12, p. 51] and are used throughout this book.

Table 3.1 Expressions for calculating composite process parameters and depletion region width

Expression	Convenient expression[a]
Gate-oxide capacitance[b] $$C'_{OX} = \frac{\varepsilon_{SiO_2}}{t_{ox}}$$	$$C'_{OX} = 34.5\,\frac{\text{fF}}{\mu\text{m}^2}\left(\frac{1\,\text{nm}}{t_{ox}}\right)$$ $$1\,\text{nm} = 10\,\text{Å}$$
Transconductance factor[c] $$k = \mu C'_{OX}$$	$$k = 10\,\frac{\mu\text{A}}{\text{V}^2}\left(\frac{\mu}{100\,\frac{\text{cm}^2}{\text{V}\cdot\text{s}}}\right)\left(\frac{C'_{OX}}{\frac{\text{fF}}{\mu\text{m}^2}}\right)$$
Technology current[c] $$I_0 = 2n_0\mu_0 C'_{OX} U_T^2 = 2n_0 k_0 U_T^2$$	$$I_0 = 0.1871\,\mu\text{A}\left(\frac{n_0}{1.4}\right)\left(\frac{k_0}{100\,\frac{\mu\text{A}}{\text{V}^2}}\right)\left(\frac{T}{300\,\text{K}}\right)^2$$
Body-effect factor[d] $$\gamma = \frac{\sqrt{2q\varepsilon_{Si}N_B}}{C'_{OX}}$$	$$\gamma = 0.578\sqrt{\text{V}}\left(\sqrt{\frac{N_B}{10^{16}/\text{cm}^3}}\right)\left(\frac{\frac{\text{fF}}{\mu\text{m}^2}}{C'_{OX}}\right)$$
Fermi potential[c,d] $$PHI = 2\phi_F = 2\,U_T\ln\left(\frac{N_B}{n_i}\right)$$	$$PHI = 0.693\,\text{V} + 0.0517\,\text{V}\cdot\ln\left(\frac{N_B}{10^{16}/\text{cm}^3}\right)$$ For $T = 300\,\text{K}$ (see note[e])
Thermal voltage[c] $$U_T = \frac{kT}{q}$$	$$U_T = 25.85\,\text{mV}\left(\frac{T}{300\,\text{K}}\right)$$ $$T(\text{K}) = T(°\text{C}) + 273$$
Built-in potential[c] $$\phi_{bi} = U_T\ln\left(\frac{N_A N_D}{n_i^2}\right)$$ N_A and N_D are acceptor (p-type) and donor (n-type) doping concentrations respectively	$$\phi_{bi} = 0.693\,\text{V} + 0.0259\,\text{V}\cdot\ln\left(\frac{N_A}{10^{16}/\text{cm}^3}\cdot\frac{N_D}{10^{16}/\text{cm}^3}\right)$$ For $T = 300\,\text{K}$ (see note[e])
Depletion-region width $$t_{dep} = \sqrt{\frac{2\varepsilon_{Si}}{qN_A}}\sqrt{\phi + V_R}$$	$$t_{dep} = 0.361\,\mu\text{m}\cdot\sqrt{\frac{10^{16}/\text{cm}^3}{N_A}}\cdot\sqrt{\frac{\phi + V_R}{1\,\text{V}}}$$

For p/n junction t_{dep}, N_A is the doping concentration in the lightly doped side, $\phi = \phi_{bi}$, and V_R is the applied reverse bias voltage taken positively. For MOSFET t_{dep} below the channel, $N_A = N_B$ (described in note[d]), $\phi = \phi_s$ is the surface potential ($\phi_s = \phi_0 \approx 2\phi_F + 4U_T$ in SI), and $V_R = V_{CB}$ is the reverse bias, channel to body voltage

[a] The use of convenient expressions with built-in unit conversions was motivated by [12, p. 51]. Convenient expressions are for $\varepsilon_{SiO_2} = 3.9\cdot 8.85 \times 10^{-3}\,\text{fF}/\mu\text{m}$, $\varepsilon_{Si} = 11.8\cdot 8.85 \times 10^{-3}\,\text{fF}/\mu\text{m}$, $q = 1.602 \times 10^{-19}\,\text{C}$, $k = 1.3806 \times 10^{-23}\,\text{J/K}$, $n_i = 1.5 \times 10^{10}/\text{cm}^3$ ($T = 300\,\text{K}$), and $U_T = 25.85\,\text{mV}$ ($T = 300\,\text{K}$).
[b] Assumes t_{ox} is the effective, electrical gate-oxide thickness, which includes the increase from physical thickness associated with inversion charge position and polysilicon gate depletion.
[c] This is a strong function of temperature and must be evaluated at the temperature of interest.
[d] N_B is the average substrate or body doping concentration for the region of interest.
[e] U_T and especially n_i increase with temperature while PHI and ϕ_{bi} decrease with temperature.

3.3.2 DC, Small-Signal, and Intrinsic Gate Capacitance Parameters

Table 3.2 lists drain current, small-signal parameter, and intrinsic gate capacitance process parameter values for the example 0.5, 0.35, and 0.18 μm CMOS processes. As noted, the 0.35 μm process is a partially depleted (PD), silicon-on-insulator (SOI) process. The electrical gate-oxide thickness, t_{ox}, gate-oxide capacitance, C'_{OX}, low-field mobility, μ_0, low-field transconductance factor, $k_0 = \mu_0 C'_{OX}$, body-effect factor, γ, and Fermi potential, $PHI = 2\phi_F$, model the drain current, transconductance, body-effect transconductance, and intrinsic gate capacitances. The velocity saturation, critical electric field, E_{CRIT}, and mobility reduction factor, θ, model the reduction in drain current and transconductance due to horizontal

Table 3.2 Drain current, small-signal parameter, and intrinsic gate capacitance parameters for 0.5, 0.35, and 0.18 μm CMOS processes ($T = 300$ K)

Parameter	Description	nMOS	pMOS	Units
t_{ox}	Gate-oxide thickness[a]			
	0.5 μm	13.5	13.5	nm
	0.35 μm, PD SOI	8	8	
	0.18 μm	4.1	4.1	
C'_{OX}	Gate-oxide capacitance			
	0.5 μm	2.56	2.56	fF/μm^2
	0.35 μm, PD SOI	4.31	4.31	
	0.18 μm	8.41	8.41	
μ_0	Mobility (low field)[b]			
	0.5 μm	438	152	cm^2/V · s
	0.35 μm, PD SOI	372	135	
	0.18 μm	422	89.2	
$k_0 = \mu_0 C'_{OX}$	Transconductance factor (low field)[b]			
	0.5 μm	112	39	μA/V^2
	0.35 μm, PD SOI	160	60	
	0.18 μm	355	75	
γ	Body-effect factor			
	0.5 μm	0.72	0.59	V$^{1/2}$
	0.35 μm, PD SOI	0.68	0.55	
	0.18 μm	0.56	0.61	
$PHI = 2\phi_F$	Fermi potential[b]			
	0.5 μm	0.8	0.8	V
	0.35 μm, PD SOI	0.7	0.7	
	0.18 μm	0.85	0.85	
E_{CRIT}	Velocity saturation critical electric field (horizontal field)[b]			
	0.5 μm	4	10.5	V/μm
	0.35 μm, PD SOI	4	10.5	
	0.18 μm	5.6	14	
α	Velocity saturation transition exponent (enhanced drain current model)			
	0.5 μm	1.5	1.5	
	0.35 μm, PD SOI	1.5	1.5	
	0.18 μm	1.3	1.3	
θ	Mobility reduction factor (vertical field)			
	0.5 μm	0.14	0.17	1/V
	0.35 μm, PD SOI	0.2	0.24	
	0.18 μm	0.28	0.35	

n_0	Substrate factor (average moderate inversion value)[c]			
	0.5 μm	1.4	1.35	
	0.35 μm, PD SOI	1.4	1.35	
	0.18 μm	1.35	1.35	
$I_0 = 2n_0\mu_0 C'_{OX} U_T^2$	Technology current[b,c]			
	0.5 μm	0.21	0.07	μA
	0.35 μm, PD SOI	0.30	0.105	
	0.18 μm	0.64	0.135	
V_{AL}	Early voltage factor	Typically		
	0.5 μm	3–40, depending on		V/μm
	0.35 μm, PD SOI	inversion level, L, and V_{DS}		
	0.18 μm	(see Figure 3.48)		
V_{TO}	Threshold voltage (large)[b]			
	0.5 μm	0.7	(−)0.95	V
	0.35 μm, PD SOI	0.65	(−)0.9	
	0.18 μm	0.42	(−)0.42	
$DVTDIBL$	Threshold voltage change with V_{DS} (due to DIBL) for min. L			
	0.5 μm	−12	(+)15	mV/V
	0.35 μm, PD SOI			
	0.18 μm	−8	(+)10	
$DVTDIBLEXP$	Exponent describing reduction in magnitude of $DVTDIBL$ for L above min. L			
	All	3	3	
DW, DL	Lateral diffusion			
	0.5 μm	0.1	0.1	μm
	0.35 μm, PD SOI	0.05	0.05	
	0.18 μm ($DW = 0.0$ μm)	0.028	0.051	

[a] Effective, electrical gate-oxide thickness, including the increase from physical thickness associated with inversion charge position and polysilicon gate depletion.

[b] Values are for $T = 300$ K. Mobility, transconductance factor, Fermi potential, velocity saturation critical electric field, technology current, threshold voltage, and thermal voltage are strong functions of temperature and must be evaluated at the temperature of interest.

[c] n_0 and I_0 are for $V_{SB} = 0$ V. These drop approximately 6 % in moderate inversion for $V_{SB} = 1$ V for $PHI = 2\phi_F = 0.85$ V and $\gamma = 0.6$ $V^{1/2}$.

field velocity saturation and VFMR. The velocity saturation transition exponent, α, models the transition from little velocity saturation to full velocity saturation. The technology current, I_0, and moderate-inversion substrate factor value, n_0, will be used for inversion coefficient calculations, again described later in Section 3.4.2.2. The Early voltage factor, V_{AL}, models small-signal, drain–source resistance.

The "large-geometry," zero-V_{SB} (no body effect) threshold voltage, V_{TO}, given in Table 3.2 models the threshold voltage prior to geometry corrections required for small width and length dimensions. In the design methods presented here, we will rarely consider the threshold voltage, which varies widely over device geometry and process. Instead, we will utilize current-biased designs and use the inversion coefficient, transconductance efficiency, and other performance measures that are independent of the threshold voltage. The threshold voltage, however, is needed to find the operating gate–source voltage that consists of the sum of the effective gate–source and threshold voltages. The drop in threshold voltage with increasing drain–source voltage due to drain-induced barrier lowering (DIBL) is modeled by the parameter $DVTDIBL$. This parameter, specified for the minimum channel length in the process, is used to evaluate small-signal, drain–source resistance due to DIBL. The parameter $DVTDIBLEXP$ is an exponent describing the rapid decrease in the magnitude of DIBL drop in threshold voltage as channel length increases from the process minimum. Finally, DW and DL model the reduction of effective channel width and length, respectively, due to lateral diffusion effects. The effective width is $W = W_{drawn} - DW$, where W_{drawn} is the drawn width. Similarly, the effective length is $L = L_{drawn} - DL$, where L_{drawn} is the drawn length. Process parameters listed in Table 3.2 are described further in the DC drain current and bias voltage, small-signal parameter, and intrinsic gate capacitance discussions of Sections 3.7, 3.8, and 3.9.2.

3.3.3 Flicker Noise and Local-Area DC Mismatch Parameters

Table 3.3 lists flicker noise and local-area DC mismatch parameter values for the example 0.5, 0.35, and 0.18 µm CMOS processes. The flicker-noise factor, given in semiconductor (K_{F0}), common SPICE MOS model ($K_{FSPICE0}$), and hand calculation (K'_{F0}) units, models the gate-referred flicker-noise voltage, power spectral density at a frequency of 1 Hz for operation in weak inversion or the weak-inversion side of moderate inversion. Increasing flicker noise at increasing inversion levels or increasing $V_{EFF} = V_{GS} - V_T$ is modeled by the voltage, V_{KF}. The flicker-noise slope, AF, models the frequency slope of the flicker-noise power spectral density. Finally, the threshold-voltage mismatch parameter, A_{VTO}, and transconductance mismatch factor, A_{KP}, model local-area DC mismatch, with the increase in threshold-voltage mismatch due to non-zero source–body voltage, V_{SB}, modeled separately as noted in the table. Process parameters listed in Table 3.3 are described further in the flicker noise and local-area DC mismatch discussions of Sections 3.10.3 and 3.11.1.

Table 3.3 Flicker noise and local-area DC mismatch parameters for 0.5, 0.35, and 0.18 µm CMOS processes ($T = 300$ K)

Parameter	Description	nMOS	pMOS	Units
K_{F0}	Flicker-noise factor, weak inversion (semiconductor units)			
	0.5 µm	1.23×10^{-31}	1.57×10^{-32}	C^2/cm^2
	0.35 µm, PD SOI	2.68×10^{-31}	6.04×10^{-32}	
	0.18 µm	3.18×10^{-31}	2.38×10^{-31}	
$K_{FSPICE0}$	Flicker-noise factor, weak inversion (SPICE units)			
	0.5 µm	4.80×10^{-25}	6.15×10^{-26}	V^2F
	0.35 µm, PD SOI	6.21×10^{-25}	1.40×10^{-25}	
	0.18 µm	3.78×10^{-25}	2.83×10^{-25}	
K'_{F0}	Flicker-noise factor, weak inversion (hand calculation units)			
	0.5 µm	$(13700)^2$	$(4900)^2$	$(nV)^2 \cdot µm^2$
	0.35 µm, PD SOI	$(12000)^2$	$(5700)^2$	
	0.18 µm	$(6700)^2$	$(5800)^2$	
V_{KF}	Flicker-noise factor voltage to model increase with inversion level			
	0.5 µm	note[a]	note[a]	V
	0.35 µm, PD SOI	note[a]	0.55	
	0.18 µm	1	0.25	
AF	Flicker-noise slope			
	0.5 µm	0.82	1.0	
	0.35 µm, PD SOI	1	1	
	0.18 µm	0.85	1.05	
A_{VTO} (1σ)	Threshold-voltage mismatch factor[b]			
	0.5 µm	14	14	$mV \cdot µm$
	0.35 µm, PD SOI	9	9	
	0.18 µm	5	5	
A_{KP} (1σ)	Transconductance mismatch factor			
	All	0.02	0.02	µm

[a] Flicker noise factor increase with inversion level is modest and not modeled
[b] Threshold-voltage mismatch increase for non-zero V_{SB} due to body-effect mismatch is modeled by A_{VT} (1σ) given by Equation 3.147 as described in Section 3.11.1.4.

3.3.4 Gate-Overlap and Drain–Body Capacitance Parameters

Table 3.4 lists gate-overlap and drain–body, diode junction capacitance parameter values for the example 0.5, 0.35, and 0.18 μm CMOS processes. While gate-overlap and drain–body capacitances are not part of intrinsic MOS operation, these can significantly lower frequency response.

Table 3.4 Gate-overlap and drain–body, (source–body), diode junction capacitance parameters for 0.5, 0.35, and 0.18 μm CMOS processes ($T = 300$ K)

Parameter	Description	nMOS	pMOS	Units
$CGDO$, $CGSO^a$	Gate–drain, gate–source overlap capacitances (average or worst case values since overlap capacitances are voltage dependent)			
	0.5 μm	0.21	0.28	fF/μm
	0.35 μm, PD SOI	0	0	
	0.18 μm	0.94	0.64	
$CGBO^a$	Gate–body overlap capacitance			
	0.5 μm	1	1	fF/μm
	0.35 μm, PD SOI	0	0	
	0.18 μm	0	0	
C_J	Drain, source area capacitance			
	0.5 μm	0.43	0.72	fF/μm²
	0.35 μm, PD SOI	0.1	0.1	
	0.18 μm	0.96	1.2	
P_B	Drain, source area capacitance built-in potential			
	0.5 μm	0.94	0.98	V
	0.35 μm, PD SOI	1	1	
	0.18 μm	0.8	0.85	
M_J	Drain, source area capacitance grading coefficient			
	0.5 μm	0.44	0.5	
	0.35 μm, PD SOI	0.5	0.5	
	0.18 μm	0.38	0.41	
C_{JSW}	Drain, source perimeter capacitance			
	0.5 μm	0.3	0.27	fF/μm
	0.35 μm, PD SOI	0.5	0.3	
	0.18 μm	0.27	0.24	
P_{BSW}	Drain, source perimeter capacitance built-in potential			
	0.5 μm	0.8	1	V
	0.35 μm, PD SOI	0.8	0.8	
	0.18 μm	0.8	0.82	
M_{JSW}	Drain, source perimeter capacitance grading coefficient			
	0.5 μm	0.18	0.28	
	0.35 μm, PD SOI	0.4	0.4	
	0.18 μm	0.15	0.33	
W_{DIFF}	Drain, source finger width^b			
	0.5 μm	1.8	1.8	μm
	0.35 μm, PD SOI	1.2	1.2	
	0.18 μm	0.6	0.6	

^{a} Process parameters, $CGDO$, $CGSO$, and $CGBO$, are not subscripted. This distinguishes them from the related MOS overlap capacitances, C_{GDO}, C_{GSO}, and C_{GBO}.

^{b} Exterior finger width may be less since gate overlap is required on one side only.

Table 3.5 Temperature parameters for 0.5, 0.35, and 0.18 μm CMOS processes

Parameter	Description	nMOS	pMOS	Units
TCV	Threshold voltage temperature coefficient[a]			
	All	-1×10^{-3}	-1×10^{-3}	V/C
BEX	Mobility temperature exponent			
	0.5 μm	-1.5	-1.5	
	0.35 μm, PD SOI	-1.40	-1.02	
	0.18 μm	-1.5	-1.5	
UCEX	Velocity saturation, critical electric field, temperature exponent			
	All	0.8	0.8	

[a] The threshold-voltage temperature coefficient refers to the change in the absolute value of threshold voltage for both nMOS and pMOS devices.

The gate–drain, $CGDO$, gate–source, $CGSO$, and gate–body, $CGBO$, overlap capacitance parameters model extrinsic capacitances between the gate and drain, gate and source, and gate and body, respectively. The overlap capacitance parameter names are not subscripted to differentiate them from the MOS gate–drain, C_{GDO}, gate–source, C_{GSO}, and gate–body, C_{GBO}, overlap capacitances, respectively.

The area capacitance parameter, C_J, models the capacitance of the area or "bottom" of the diode junction associated with the drain and the substrate or body. The perimeter capacitance, C_{JSW}, models the capacitance of the perimeter or "side wall" of this diode junction. P_B and P_{BSW} are the built-in potentials for the area and perimeter parts of the diode junction, and M_J and M_{JSW} are grading coefficients governing the falloff of capacitance with increasing reverse bias for the area and perimeter parts of the diode junction. Finally, W_{DIFF} is the width of a drain stripe needed for calculating the area and perimeter required for calculating the drain–body diode junction capacitance. While described for the drain–body capacitance, these parameters apply equally for the source–body capacitance associated with the source–body diode junction. Process parameters listed in Table 3.4 are described further in the gate-overlap and drain–body capacitance discussions of Sections 3.9.3 and 3.9.4.

3.3.5 Temperature Parameters

Table 3.5 lists temperature parameter values for the example 0.5, 0.35, and 0.18 μm CMOS processes. These parameters include the threshold-voltage temperature coefficient, TCV, the mobility temperature exponent, BEX, and the velocity saturation, critical electric field, temperature exponent, $UCEX$. Threshold voltage, V_T, mobility, μ, Fermi potential, PHI, substrate factor, n, transconductance factor, k, and technology current, I_0, temperature effects are discussed later in Section 3.5.

3.4 SUBSTRATE FACTOR AND INVERSION COEFFICIENT

In the following discussions, MOS drain current, introduced in Section 2.4, will be briefly reviewed to describe the substrate factor. Following this, the substrate factor and inversion coefficient will be described. The substrate factor is important since it describes substrate action and appears in many MOS expressions. The inversion coefficient is important since it allows the designer to explicitly select the region and level of MOS inversion.

3.4.1 Substrate Factor

Table 3.6 gives MOS drain current expressions derived from the EKV MOS model[3] [12, p. 136, 13, 14, 15, p. 178] with the source used as the reference instead of the substrate.[4] The first expression in

Table 3.6 MOS drain current, substrate factor, and inversion coefficient expressions

Parameter	Expression
Drain current[a,b]	$I_D(\text{WI}) = 2n\mu C'_{OX} U_T^2 \left(\dfrac{W}{L}\right)\left(e^{\left(\frac{V_{GS}-V_T}{nU_T}\right)}\right)$ $I_D(\text{SI}) = \dfrac{1}{2}\left(\dfrac{\mu C'_{OX}}{n}\right)\left(\dfrac{W}{L}\right)(V_{GS}-V_T)^2$ $I_D(\text{WI–SI}) = 2n\mu C'_{OX} U_T^2 \left(\dfrac{W}{L}\right)\left[\ln\left(1+e^{\frac{V_{GS}-V_T}{2nU_T}}\right)\right]^2$
Substrate factor[c]	$n = 1 + \dfrac{\gamma}{2\sqrt{(V_{GS}-V_{TO}+V_{SB})/n+\phi_0}}$ $= 1 + \dfrac{\gamma}{2\sqrt{(V_{GS}-V_T)/n+V_{SB}+\phi_0}}$ $\phi_0 \approx 2\phi_F + 4U_T$
Inversion coefficient (traditional)	$IC' = \dfrac{I_D}{2n\mu C'_{OX}U_T^2\left(\dfrac{W}{L}\right)}$
Fixed–normalized inversion coefficient (used in this book)	$IC = \dfrac{I_D}{2n_0\mu_0 C'_{OX}U_T^2\left(\dfrac{W}{L}\right)} = \dfrac{I_D}{I_0\left(\dfrac{W}{L}\right)}$ $IC < 0.1$, weak inversion $0.1 < IC < 10$, moderate inversion $IC > 10$, strong inversion $IC \approx IC'$ for $IC \lesssim 10$, until μ decreases significantly from μ_0 VFMR and velocity saturation reduction of μ is considered outside the IC definition as discussed in Section 3.4.2.3
Technology current	$I_0 = 2n_0\mu_0 C'_{OX}U_T^2 = 2n_0 k_0 U_T^2$, where $k_0 = \mu_0 C'_{OX}$

Expressions are for saturation where the drain–source voltage exceeds its saturation value ($V_{DS} > V_{DS.sat}$).

[a] The decrease in drain current due to velocity saturation and VFMR is not included here. This is included later in Section 3.7.1.

[b] The increase in drain current with increasing drain–source voltage is not included here. This is included later in Section 3.8.4.

[c] Expression is actually for strong inversion but provides an estimate in all regions because negative values of $V_{GS} - V_T$ in moderate and weak inversion reduce the surface potential from its strong-inversion estimate of $\phi_0 \approx 2\phi_F + 4U_T$. This is described in Section 3.8.3.1.

[3] While the initial EKV MOS model ideas are referenced here to the original EKV MOS model papers [13, 14], [16] contains a recent book-length discussion of the model.

[4] Drain current, I_D, drain–source voltage, V_{DS}, gate–source voltage, V_{GS}, threshold voltage, V_T, and zero-V_{SB} threshold voltage, V_{TO}, are positive for nMOS devices. The source–body voltage, V_{SB}, is zero or positive for nMOS devices representing zero or reverse bias for the source–body diode. Throughout this book, these quantities will also be taken positively for pMOS devices as shown later in Figure 3.9. W and L are effective channel width and length.

the table is the weak-inversion drain current where I_D is exponentially proportional to $V_{GS} - V_T$. Weak-inversion operation occurs at sufficiently low effective gate–source voltages, $V_{EFF} = V_{GS} - V_T$, where $V_{EFF} < -72\,\text{mV}$ ($-2 \cdot nU_T$). The second expression in the table is the strong-inversion drain current where I_D is proportional to the square of $V_{GS} - V_T$. Strong-inversion operation occurs at sufficiently high effective gate–source voltages where $V_{EFF} > 225\,\text{mV}$ ($6.24 \cdot nU_T$). The boundaries between weak and strong inversion, discussed further in Section 3.7.2.4, assume operation at inversion coefficients of 0.1 and 10, respectively, a substrate factor $n = 1.4$, and a thermal voltage $U_T = 25.9\,\text{mV}$ for a typical bulk CMOS process at room temperature ($T = 300\,\text{K}$). The third expression in the table is a continuous expression for drain current interpolated from weak through strong inversion. The reduction in drain current due to velocity saturation and VFMR, introduced in Section 2.4.3, is not included in the drain current expressions but will be considered later in Section 3.7.1.

The substrate factor, n, appears in the weak, strong, and continuous weak through strong-inversion expressions of drain current in Table 3.6, illustrating its importance in all regions of MOS operation. As introduced in Section 2.4.1, it represents a loss of coupling efficiency between the gate and channel caused by the substrate or body, which acts as a back gate.

In weak inversion, n is related to the capacitive voltage division between the gate voltage and silicon surface potential resulting from the gate-oxide, C'_{OX}, depletion, C'_{DEP}, and interface-state, C'_{INT}, capacitances as introduced in Equation 2.6. In weak inversion, the surface potential tracks the gate voltage with a slope of $1/n$ [15, p.75]. Normally, the interface-state capacitance is negligible such that n is governed by substrate effects. However, as described later in Section 3.5.1 and in Figure 3.4, n can increase significantly at low temperatures below $200\,\text{K}$ ($-73\,°\text{C}$) due, most likely, to significant increases in interface-state capacitance. In weak inversion, n is commonly recognized as the weak-inversion slope factor appearing in the weak-inversion drain current in Table 3.6 and in the subthreshold or weak-inversion swing voltage, $S = 2.303 \cdot nU_T$, introduced in Equation 2.7. The weak-inversion swing voltage, which is the required gate–source voltage increase for a factor-of-10 increase in drain current, is approximately $90\,\text{mV/decade}$ for $n = 1.5$ and $U_T = 25.9\,\text{mV}$ at $T = 300\,\text{K}$.

While most easily recognized in the expression of weak-inversion drain current, n causes a reduction in strong-inversion drain current through division by n as illustrated in the strong-inversion drain current expression of Table 3.6. As described for strong-inversion operation in the EKV MOS model, the slope of the channel, inversion-charge, pinch-off voltage, V_P, with increasing gate voltage is equal to $1/n$ [13]. An n value above unity corresponds to body or substrate effect along the channel that raises the local threshold voltage, reducing the source-referenced, inversion-charge, pinch-off voltage to $V_P = (V_{GS} - V_T)/n$ compared to $V_P = V_{GS} - V_T$ present if there were no substrate effects. As a result, the strong-inversion drain current is then divided by n.

n is approximated in Table 3.6 from the EKV MOS model for operation in strong inversion where a source-referenced, inversion-charge, pinch-off voltage of $V_P = (V_{GS} - V_T)/n$ and constant, "pinned," strong-inversion, silicon surface potential of $\phi_0 = 2\phi_F + 4U_T$ are assumed [13, 14]. As described later in Section 3.8.3.1, the approximations for n will be used in all regions of operation because the silicon surface potential is effectively reduced for negative values of $V_{GS} - V_T$ appearing in weak and moderate inversion. Like the continuous drain current expression given in the table, n is modified using the source as the reference terminal.

As seen in Table 3.6, n is a function of the gate–source, V_{GS}, source–body, V_{SB}, and zero-V_{SB} threshold, V_{TO}, voltages, where the body-referenced $V_P = (V_{GS} - V_{TO} + V_{SB})/n$. Alternatively, as shown in the table, $V_P = (V_{GS} - V_T)/n + V_{SB}$ since the threshold voltage, $V_T \approx V_{TO} + (n - 1) \cdot V_{SB}$, approximates the threshold-voltage increase from V_{TO} due to non-zero V_{SB}. n is also a function of the body-effect factor, γ, the Fermi potential, $PHI = 2\phi_F$,[5] and the thermal voltage, U_T. n increases with increasing γ, corresponding to increased substrate doping concentration and increased body or substrate effect. As a result, n can be as low as 1.1 for fully depleted (PD), silicon-on-insulator (SOI)

---

[5] Like the MOS drain current and terminal voltages described in Footnote 4, the Fermi potential, $PHI = 2\phi_F$, is also taken positively for both nMOS and pMOS devices.

CMOS processes [17], where there is little substrate effect, compared to typical values of $n = 1.3$–1.4 for bulk CMOS processes. As shown later, n decreases slightly with increasing V_{GS} and V_{SB}. While n given in the table is an implicit expression since it appears inside the expression itself, it converges rapidly. A value of $n = 1.4$ can be assumed initially with the results of that calculation used for the next calculation. Two calculations usually render a value for n within 1 % of its converged value.

Figure 3.2 shows the calculated n for nMOS devices in a $0.18\,\mu$m CMOS process having parameters of $\gamma = 0.56\,\mathrm{V}^{1/2}$, $PHI = 2\phi_F = 0.85\,\mathrm{V}$, $V_{TO} = 0.42\,\mathrm{V}$, and $U_T = 25.9\,\mathrm{mV}$ ($T = 300\,\mathrm{K}$) listed in Table 3.2. n is calculated for $V_{SB} = 0\,\mathrm{V}$ and 1 V using the expressions in Table 3.6. For $V_{SB} = 0\,\mathrm{V}$, n is 1.3 near the onset of weak inversion where V_{GS} is 72 mV below the threshold voltage. It decreases slightly to 1.28 in the center of moderate inversion where V_{GS} is 40 mV above the threshold voltage. n continues dropping slightly to 1.26 near the onset of strong inversion where V_{GS} is approximately 225 mV above the threshold voltage, and it continues dropping to 1.24 and 1.21 deeper in strong inversion where V_{GS} is 0.5 and 1 V above the threshold voltage. n is within $\pm1.6\%$ of its moderate inversion value from the onset of weak inversion to the onset of strong inversion. Since n is a slight function of the gate–source voltage, and correspondingly the inversion level, we will often approximate it as a constant at its moderate inversion value. This approximation improves for low-voltage processes where V_{GS} and the level of MOS inversion are necessarily limited.

As shown in Figure 3.2, n decreases approximately by 6 % in moderate inversion for $V_{SB} = 1\,\mathrm{V}$. This drop, which is greater in weak inversion, is a result of reverse, source–body bias that reduces the depletion capacitance. Since V_{SB} is necessarily limited in low-voltage design, n can be expected to vary only slightly with V_{SB}.

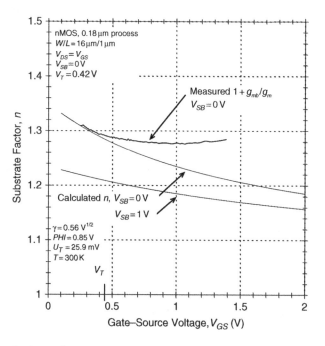

Figure 3.2 Predicted substrate factor, n, versus gate–source voltage with overlay of measured $1 + g_{mb}/g_m$ for nMOS devices in a $0.18\,\mu$m CMOS process. The measured $1 + g_{mb}/g_m$ overlay tracks n in weak inversion before leveling off in strong inversion. n decreases by 3 % from the onset of weak inversion to the onset of strong inversion and decreases by approximately 6 % for a source–body voltage of $V_{SB} = 1\,\mathrm{V}$. n is slightly higher for pMOS devices due to an increase in process γ

Figure 3.2 also shows the measured body-effect transconductance ratio added to unity, $1 + g_{mb}/g_m$, for a $W/L = 16\,\mu\text{m}/1\,\mu\text{m}$ nMOS device in the $0.18\,\mu\text{m}$ CMOS process. This closely tracks calculated n in weak inversion, providing a cross-check for the process body-effect factor, γ, where the relationships between n and γ are described later in Section 3.8.3. When V_{GS} increases for operation in strong inversion, the measured $1 + g_{mb}/g_m$ levels off before increasing slightly. While not shown here, this increase is greater for pMOS devices compared to nMOS devices in the 0.18 and $0.5\,\mu\text{m}$ CMOS processes described in Table 3.2.

The calculated n with measured $1 + g_{mb}/g_m$ overlay is similar for pMOS devices in the $0.18\,\mu\text{m}$ CMOS process. n for pMOS devices is higher by about 2.3 % or the addition of 0.03 for an average value of 1.3 because γ is higher at $0.61\,\text{V}^{1/2}$ (listed in Table 3.2) compared to $0.56\,\text{V}^{1/2}$ for nMOS devices in the process. Similar results are also obtained for the 0.35 and $0.5\,\mu\text{m}$ CMOS processes described in Table 3.2, with n increasing or decreasing slightly with the process value of γ. Results are similar for the $0.35\,\mu\text{m}$ process, even though this is a partially depleted (PD) SOI process. This process, which has local body ties permitting independent body connection for both nMOS and pMOS devices, has similar n and γ as the bulk processes since the bodies are partially depleted giving considerable substrate effect. As mentioned earlier, n for fully depleted (FD) SOI processes can be as low as 1.1 due to the reduced substrate effect [17]. This is well below values of 1.3–1.4 typical for bulk CMOS processes.

As described later in Section 3.8.3.1, n and, correspondingly, the body-effect transconductance ratio, g_{mb}/g_m, can decrease for minimum channel length devices in a process because the apparent γ can decrease due to charge sharing with the nearby source and drain. This charge-sharing effect, the separation of n into both voltage and charge components [16, pp. 150–154], and the effect of polysilicon gate depletion [16, pp. 150–154] are not considered in the simple predictions of n described here. These complex, small-geometry effects are not considered to enable design guidance through simple predictions of MOS performance. These effects, however, are included in many MOS models, including the EKV MOS model [13, 14], used for computer simulations and verification of candidate designs.

3.4.2 Inversion Coefficient

In the design methods described in this book, the designer selects drain current, inversion coefficient, and channel length to optimize analog MOS performance. The inversion coefficient, IC, is used here as a numerical representation of MOS inversion level for operation anywhere in weak through strong inversion [1, p. 62]. The derivation below shows that an inversion coefficient of unity corresponds to the center of moderate inversion where the asymptotic values of weak- and strong-inversion transconductance are equal.

3.4.2.1 Traditional inversion coefficient

Equating the weak-inversion transconductance, Equation 2.8, with the strong-inversion transconductance, Equation 2.12, gives

$$g_m\,(\text{WI}) = \frac{I_D}{nU_T} = g_m\,(\text{SI}) = \sqrt{2I_D\left(\frac{\mu\,C'_{OX}}{n}\right)\left(\frac{W}{L}\right)} \tag{3.1}$$

Solving for the drain current $I_D = I_D$ (moderate) that gives equal weak- and strong-inversion transconductance gives

$$I_D\,(\text{moderate}) = 2n\mu\,C'_{OX}U_T^2\left(\frac{W}{L}\right) \tag{3.2}$$

This is the drain current for a device operating in the center of moderate inversion where the predicted weak- and strong-inversion transconductances are equal. Actual transconductance, however, is approximately 37 % below the predicted weak- and strong-inversion values (shown later in Section 3.8.2.1), illustrating the need for MOS modeling valid in moderate inversion.

The moderate-inversion current of Equation 3.2 is the specific or normalizing current for the MOS inversion coefficient, IC', given here and in Table 3.6 as [1, p. 62]

$$IC' \text{ (traditional)} = \frac{I_D}{I_D \text{ (moderate)}} = \frac{I_D}{2n\mu\, C'_{OX} U_T^2 \left(\dfrac{W}{L}\right)} \tag{3.3}$$

IC' is used in this book to denote the traditional inversion coefficient,[6] where the normalizing current varies with the inversion level due to variations in n and mobility, μ. IC' is also equal to the normalized, forward drain current (dominant in saturation), i_f, in the EKV MOS model [13, 14]. Another normalized drain current is reported in a similar MOS model where the normalization current, the denominator of Equation 3.3, is four times smaller [18]. A sheet current or current-per-square normalization, $I_D/(W/L)$, has also been reported in a g_m/I_D-based design methodology [19]. This normalization, however, does not include the technology parameters n, μ, and C'_{OX}, and thus does not provide a dimensionless, technology-independent normalization. The inversion coefficient is convenient for design since a value of unity corresponds to the center of moderate inversion regardless of the process technology. In Section 3.8.2.2, measured transconductance efficiency will be used to verify an inversion coefficient value of unity in the center of moderate inversion where the asymptotic values of weak- and strong-inversion transconductance are equal.

3.4.2.2 Fixed–normalized inversion coefficient

As shown in Figure 3.2, n decreases slightly with the increasing gate–source voltage associated with increasing levels of inversion. Additionally, μ decreases with the increasing gate–source voltage and vertical field associated with increasing inversion since carriers are increasingly attracted to interface states and imperfections at the Si–SiO$_2$ interface. As a result, the inversion coefficient, normalization current in the denominator of Equation 3.3 decreases somewhat with increasing inversion level and decreases even more if velocity saturation effects are modeled by additional decreases in μ. While this change in normalization current with inversion level is not a difficulty for MOS models used in computer circuit simulations, it complicates hand calculations of inversion coefficient and associated expressions for MOS device and circuit performance.

To permit simple design guidance and evaluation of measured data, we define and use in this book a fixed–normalized inversion coefficient, IC, which is related to, but distinct from, the traditional inversion coefficient, IC', given by Equation 3.3 [4–11]. In our definition of IC, $n = n_0$ is held fixed at its average value in moderate inversion and $\mu = \mu_0$ is held fixed at its low-field value. This permits normalization by a fixed current, independent of inversion level, that we will call the technology current [4–11]. As summarized in Table 3.6, the technology current is

$$I_0 = 2n_0\mu_0\, C'_{OX} U_T^2 \tag{3.4}$$

and the fixed–normalized inversion coefficient is

$$IC \text{ (fixed–normalized)} = \frac{I_D}{2n_0\mu_0\, C'_{OX} U_T^2 \left(\dfrac{W}{L}\right)} = \frac{I_D}{I_0 \left(\dfrac{W}{L}\right)} \tag{3.5}$$

---

[6] In [1, p. 62], the traditional inversion coefficient is denoted as IC. IC, however, is used in this book to denote the fixed–normalized inversion coefficient described in Section 3.4.2.2.

The technology current is the drain current of a unity-shape-factor ($S = W/L = 1$) device operating in the center of moderate inversion. Only two technology currents are required for a given process at a given temperature: one for nMOS devices and another for pMOS devices. These are given in Table 3.2 for the 0.5, 0.35, and 0.18 μm processes illustrated in this book. The fixed–normalized inversion coefficient provides an easy way to determine the level and region of MOS inversion by simply dividing the drain current by the product of the technology current and shape factor, W/L. The fixed–normalized inversion coefficient directly tracks drain current and inversely tracks the shape factor regardless of the inversion level.

The technology current can be verified and calibrated, if necessary, by observing the asymptotic intersection of measured weak- and strong-inversion transconductance efficiency as shown later in Section 3.8.2.2. Since mobility is a strong function of temperature and thermal voltage a lesser function of temperature, it is necessary to find the technology current for the temperature of interest, or for the nominal design temperature. As temperature changes, both the traditional and fixed–normalized inversion coefficients change for a device operating at constant current. This is discussed later in Section 3.5.5.

3.4.2.3 Using the fixed–normalized inversion coefficient in design

As described in Section 3.4.1 and shown in Figure 3.2, n for nMOS devices in a 0.18 μm CMOS process is 1.3 near the onset of weak inversion, 1.28 in the center of moderate inversion, 1.26 near the onset of strong inversion, 1.24 where V_{GS} is 0.5 V above the threshold voltage, and 1.21 in deep strong inversion where V_{GS} is 1 V above the threshold voltage. This represents a variation of $+1.6\%$ in weak inversion and -5.5% in deep strong inversion from the moderate-inversion value of n. The variation is less at $+1.6\%$ and -3.1% when operation is constrained between weak inversion and a level of strong inversion where V_{GS} is limited to 0.5 V above the threshold voltage. Restricting operation over this range, or less, is typically required to limit effective gate–source and drain–source saturation voltages for design in low-voltage processes. Assuming n is constant at its moderate inversion value in the fixed–normalized inversion coefficient, IC, results in an approximate 3 % variation from the traditional inversion coefficient, IC'. Such errors are not significant for the development of predictions of MOS performance intended to provide initial design guidance.

While n varies slightly over the level of MOS inversion allowable for low-voltage design in modern processes, mobility can potentially vary, or decrease, much more from its low-field, $\mu = \mu_0$ value. An estimation of $\mu = \mu_0/[1 + \theta(V_{GS} - V_T)]$ reveals a mobility drop of 12 % for a mobility reduction factor $\theta = 0.28/\text{V}$ (listed in Table 3.2) for nMOS devices in a 0.18 μm CMOS process operating at a level of strong inversion where V_{GS} is 0.5 V above the threshold voltage. The reduction of μ and drain current associated with this VFMR, and the more important reductions in short-channel devices due to velocity saturation, are considered in later drain current (Section 3.7.1.4), effective gate–source voltage (Section 3.7.2.2), and transconductance efficiency (Section 3.8.2.2) discussions.

Although the effects of mobility reduction significantly affect the evaluation of the traditional inversion coefficient, IC', given by Equation 3.3, expressions will be developed in this book to accept the fixed–normalized inversion coefficient, IC, given by Equation 3.5, while *separately* including VFMR and velocity saturation effects *outside* the inversion coefficient definition. This inversion coefficient definition used throughout this book is summarized in Table 3.6 and repeated below with important notations:

$$IC \text{ (used in this book)} = \frac{I_D}{2n_0\mu_0 C'_{OX} U_T^2 \left(\dfrac{W}{L}\right)} = \frac{I_D}{I_0\left(\dfrac{W}{L}\right)} \tag{3.6}$$

- $IC \approx IC'$ (traditional given by Equation 3.3) for $IC \lesssim 10$, until μ decreases significantly below μ_0.

- In the IC definition used in this book, the reduction in μ due to VFMR and velocity saturation is considered *separately* and *outside* the IC definition.

- Variations in operating temperature must be considered through the value of I_0.

The first notation mentions that the inversion coefficient used here is nearly equal to the traditional inversion coefficient, $IC \approx IC'$, until μ decreases significantly from its low-field value, μ_0. As mentioned, the small decrease in n with increasing inversion level results in a small difference between the two inversion coefficients. The two inversion coefficients are nearly equal in weak and moderate inversion, and, in many cases, up to the onset of strong inversion at $IC = 10$. As the inversion level increases further and μ decreases due to VFMR and velocity saturation, the two inversion coefficients begin to diverge, with the traditional one increasing because of the decrease in normalization current associated with decreasing μ. However, in the inversion coefficient used here, the reduction of μ is considered *separately* and *outside* the inversion coefficient definition as mentioned in the second notation and described later in Sections 3.7.2.2 and 3.8.2.2. Finally, the last notation below Equation 3.6 emphasizes that a fixed μ or the related technology current, I_0, cannot be used across temperatures. As described later in Section 3.5.5, changes in temperature affect the inversion coefficient value through changes in I_0. Thus I_0 must be evaluated for the nominal temperature of operation and then re-evaluated at temperature limits.

In addition to performance predictions, all measured data will be presented assuming the inversion coefficient definition of Equation 3.6 such that MOS drain current, shape factor $S = W/L$, and inversion coefficient, IC, are *linearly* linked. A factor-of-10 increase in IC corresponds to a drain current increase of a factor of 10 for a fixed-geometry device. Alternatively, a factor-of-10 increase in IC corresponds to a shape factor decrease of a factor of 10 for a fixed drain current. The relationships between IC, drain current, and MOS sizing are described in detail later in Section 3.6.

As noted in Table 3.2 and observed in Figure 3.2, n decreases by approximately 6% for a body-effect voltage of $V_{SB} = 1$ V, assuming values of $PHI = 2\phi_F = 0.85$ V and $\gamma = 0.6$ V$^{1/2}$ (0.56 V$^{1/2}$ in the figure). This will reduce I_0 by 6% and increase the value of IC given by Equation 3.6 by 6% compared to the value present for $V_{SB} = 0$ V. For simplicity in performance predictions, we will neglect this change in IC by defining it here for $V_{SB} = 0$ V. This approximation improves as supply voltage decreases in modern, lower-voltage processes because of smaller n and I_0 variations due to the lower allowable V_{SB}.

Approximations where n, μ, and V_{SB} are assumed fixed in the IC definition of Equation 3.6, with the effects of VFMR and velocity saturation considered separately, permit hand or spreadsheet predictions of performance where IC, drain current, and MOS sizing are linearly related. This facilitates rapid, intuitive design guidance where full accuracy is obtained later for candidate designs by computer simulations using full MOS models.

3.4.2.4 Regions and subregions of inversion

The inversion coefficient describes the region and level of MOS inversion from weak through strong inversion as illustrated in Figure 3.3. The primary regions are weak inversion for $IC < 0.1$, moderate inversion for $0.1 < IC < 10$, and strong inversion for $IC > 10$.

We will define subregions of inversion to facilitate discussion of MOS inversion-level choices. The subregions are defined in terms of the inversion coefficient and the associated effective gate–source voltage, $V_{EFF} = V_{GS} - V_T$. As mentioned in Section 3.2.5, V_{EFF} can be used to describe the level of MOS inversion, but it is non-linearly linked to drain current and the MOS shape factor, $S = W/L$. V_{EFF}, however, is included here since it provides insight and connection to traditional analog CMOS design methods.

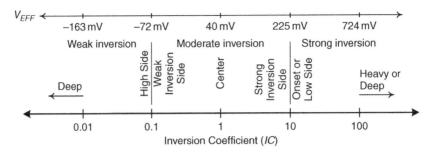

Figure 3.3 The inversion coefficient presented as a number line showing the regions and subregions of MOS inversion with corresponding effective gate–source voltage, $V_{EFF} = V_{GS} - V_T$. The effective gate–source voltage is for room temperature ($T = 300$ K) and an average substrate factor of $n = 1.4$. It is higher than values shown for short-channel devices operating in strong inversion due to velocity saturation. Moderate inversion is increasingly important in modern, low-voltage processes because of lower allowable effective gate–source and drain–source saturation voltages

The value of V_{EFF} associated with a given IC, described later in Section 3.7.2, is approximately -4.5, -2, 1.08, 6.24, and $20 \cdot nU_T$ for $IC = 0.01$, 0.1, 1, 10, and 100, respectively. Values given assume $U_T = 25.9$ mV ($T = 300$ K) and a constant value of n at 1.4. Values given exclude the increase associated with velocity saturation present for short-channel devices at high levels of inversion.

The subregions of inversion are described below in terms of IC and the associated V_{EFF}. Additionally, a few comments about MOS transconductance efficiency, g_m/I_D, effective gate–source voltage, V_{EFF}, drain–source saturation voltage, $V_{DS,sat}$, intrinsic voltage gain, and intrinsic bandwidth are included. These aspects of MOS performance are developed in Sections 3.8.2, 3.7.2, 3.7.3, 3.8.5, and 3.9.6. Additionally, performance tradeoffs associated with the level of inversion are discussed in detail in Section 4.2.2 and summarized in Figure 4.1.

The subregions of MOS inversion shown in Figure 3.3 are:

- **Deep weak inversion ($IC < 0.01$, $V_{EFF} < -163$ mV).** Because the large device shape factor, channel width, and gate area required results in high gate capacitances, very low bandwidth, and potentially high DC leakage, operation here is not desirable. There is little increase in g_m/I_D or decrease in $V_{DS,sat}$ here compared to the *high side of weak inversion*. Operation may be required here for very low drain currents.

- **High side of weak inversion ($IC = 0.1$, $V_{EFF} = -72$ mV).** This occurs at the boundary of moderate inversion. Operation here provides nearly the full g_m/I_D of *deep weak inversion*, low V_{EFF} and $V_{DS,sat}$, high gain, and low bandwidth, although improved bandwidth compared to *deep weak inversion*.

- **Weak-inversion side of moderate inversion ($0.1 < IC < 1$, -72 mV $< V_{EFF} < 40$ mV).** This occurs between the boundary of weak inversion and the center of moderate inversion, with $IC = 0.3$ corresponding to the geometric center between weak and moderate inversion. Operation here provides high g_m/I_D, low V_{EFF} and $V_{DS,sat}$, high gain, and improved bandwidth compared to the *weak-inversion side of moderate inversion*.

- **Center of moderate inversion ($IC = 1$, $V_{EFF} = 40$ mV).** Operation in the center of moderate inversion provides good g_m/I_D, low V_{EFF} and $V_{DS,sat}$, good gain, and modest bandwidth.

- **Strong-inversion side of moderate inversion ($1 < IC < 10$, 40 mV $< V_{EFF} < 225$ mV).** This occurs between the center of moderate inversion and the boundary or onset of strong inversion, with $IC = 3$ corresponding to the geometric center between moderate and strong inversion. Operation here provides modest g_m/I_D, modestly increasing V_{EFF} and $V_{DS,sat}$, modest gain, and good bandwidth.

- **Onset of strong inversion ($IC = 10$, $V_{EFF} = 225\,mV$).** Operation at the onset of strong inversion provides low g_m/I_D, high V_{EFF} and $V_{DS,sat}$, low gain, and very good bandwidth.

- **Low side of strong inversion ($10 < IC < 100$, $225\,mV < V_{EFF} < 724\,mV$).** Operation here provides low and declining g_m/I_D, high and increasing V_{EFF} and $V_{DS,sat}$, low and declining gain, and excellent bandwidth. Because of high V_{EFF} and $V_{DS,sat}$, operation here may not be feasible in low-voltage processes. Additionally, significant velocity saturation reduction in g_m/I_D and increases in V_{EFF} are likely for short-channel nMOS devices. This can also result in saturation or leveling off of bandwidth.

- **Heavy or deep strong inversion ($IC > 100$, $V_{EFF} > 724\,mV$).** Because of the small device shape factor and channel width required, it is increasingly difficult to size devices here. Operation here is generally not useful in modern, low-voltage processes because of the very high V_{EFF} and $V_{DS,sat}$. g_m/I_D and voltage gain are very low, and the velocity saturation effects described for the *low side of strong inversion* are generally severe for short-channel nMOS devices.

In this book, moderate inversion is considered in detail with selected devices operating here in the design examples described in Chapters 5 and 6. Moderate inversion is advantageous because V_{EFF} and $V_{DS,sat}$ are moderate (only slightly above their weak-inversion values), permitting low-voltage design. Additionally, g_m/I_D and voltage gain are high, bandwidth is moderate, and velocity saturation reductions in g_m/I_D and increases in V_{EFF} are generally small, even for short-channel devices.

3.5 TEMPERATURE EFFECTS

The operating temperature affects many MOS parameters and must be considered in design. The following discussion describes changes in MOS parameters resulting from changes in temperature. For operation at temperatures outside the nominal model parameter temperature, temperature-corrected parameters should be used to predict MOS performance.

Table 3.7 summarizes major MOS temperature effect expressions with process temperature parameters listed earlier in Table 3.5 for example 0.5, 0.35, and 0.18 μm CMOS processes. The thermal voltage, U_T, increases directly with absolute temperature, T. The bandgap energy, E_g, decreases slightly while the intrinsic carrier concentration, n_i, increases significantly with temperature. The Fermi potential, $PHI = 2\phi_F$, decreases with temperature because the decrease due to increasing intrinsic carrier concentration exceeds the increase due to increasing thermal voltage as described in the next section. Mobility, μ, decreases with temperature and is modeled here using the mobility temperature exponent *BEX* from the EKV MOS model [14]. The mobility temperature exponent has typical values between -1.2 and -2 [15, p. 189]. The velocity saturation, critical electric field, E_{CRIT}, increases with temperature because mobility decreases, lowering carrier velocity. It is modeled here using the temperature exponent *UCEX* from the EKV MOS model, which lists a typical *UCEX* value of 0.8 [14]. Finally, the zero-V_{SB} threshold voltage, V_{TO}, exhibits a near straight-line decrease with increasing temperature, having a temperature coefficient typically between -0.5 and $-3\,mV/C$ [15, p. 190]. It is modeled here using the EKV MOS model parameter *TCV*, although *TCV* is given a negative sign here to represent the drop in threshold voltage (including the drop in magnitude for pMOS threshold voltage) with increasing temperature.

3.5.1 Bandgap Energy, Thermal Voltage, and Substrate Factor

Table 3.8 shows the bandgap energy, thermal voltage, intrinsic carrier concentration, and Fermi potential over the -55 to 125 °C military temperature range as calculated from the expressions in Table 3.7. A room temperature ($T = 300\,K$) value of $PHI = 2\phi_F = 0.85\,V$ is used for the Fermi

Table 3.7 MOS temperature effect expressions

Parameter	Expression	
Thermal voltage	$U_T(T) = \dfrac{kT}{q} = 25.85\,\text{mV}\left(\dfrac{T}{300\,\text{K}}\right)$ From Table 3.1	(V)
Bandgap energy	$E_g(T) = 1.16 - 0.000\,702\left(\dfrac{T^2}{T+1108}\right)$ From [14]	(eV)
Intrinsic carrier concentration	$n_i(T) = 1.45 \times 10^{10}\left(\dfrac{T}{300\,\text{K}}\right)\exp\left(\dfrac{E_g(300\,\text{K})}{2U_T(300\,\text{K})} - \dfrac{E_g(T)}{2U_T(T)}\right)$ $= 1.126 \times 10^{17} \cdot T\, e^{\dfrac{-E_g}{2U_T}}$ From [14]	$(1/\text{cm}^3)$
Fermi potential	$PHI(T) = 2\phi_F = 2U_T \ln\left(\dfrac{N_B}{n_i}\right)$ $\qquad = PHI \cdot \left(\dfrac{T}{T_{NOM}}\right) - 2U_T \ln\left(\dfrac{T}{T_{NOM}}\right)$ $\qquad - E_g(T_{NOM})\left(\dfrac{T}{T_{NOM}}\right) + E_g(T)$ where PHI is the Fermi potential (actually twice the Fermi potential ϕ_F) at the model parameter temperature T_{NOM}. From Table 3.1 and [14]	(V)
Mobility	$\mu_0(T) = U_0 \cdot \left(\dfrac{T}{T_{NOM}}\right)^{BEX}$ where U_0 is the low-field mobility at the model parameter temperature T_{NOM}. From [14]	$(\text{cm}^2/\text{V}\cdot\text{s})$
Velocity saturation, critical electric field	$E_{CRIT}(T) = E_{CRIT} \cdot \left(\dfrac{T}{T_{NOM}}\right)^{UCEX}$ where E_{CRIT} is the velocity saturation, critical electric field at the model parameter temperature T_{NOM}. From [14]	$(\text{V}/\mu\text{m})$
Threshold voltage	$V_{TO}(T) = V_{TO} + TCV \cdot (T - T_{NOM})$ where V_{TO} is the large-geometry threshold voltage at the model parameter temperature T_{NOM}. From [14]	(V)

The transconductance factor, $k_0 = \mu_0 C'_{OX}$, is proportional to T^{BEX}, and the technology current, $2n_0\mu_0 C'_{OX} U_T^2$, is proportional to $T^{(2+BEX)}$ due to mobility and thermal voltage temperature dependencies.
T is in K.

potential, which from the expression in Table 3.1 corresponds to a doping concentration in the substrate or body of $N_B = 2 \times 10^{17}/\text{cm}^3$ and an intrinsic carrier concentration of $n_i = 1.5 \times 10^{10}/\text{cm}^3$. The actual value of N_B may be higher but increases PHI only slightly. Table 3.8 also shows the substrate factor as calculated from the expressions in Table 3.6. The substrate factor is calculated in moderate inversion for $V_{EFF} = V_{GS} - V_T = 40\,\text{mV}$ at $V_{SB} = 0\,\text{V}$ using a body-effect factor of $\gamma = 0.56\,\text{V}^{1/2}$. Process values of PHI and γ, listed in Table 3.2, are for nMOS devices in a $0.18\,\mu\text{m}$ CMOS process.

Table 3.8 shows a slight decrease in bandgap energy while the thermal voltage increases directly with absolute temperature. The intrinsic carrier concentration, however, increases substantially with

Table 3.8 MOS bandgap energy, thermal voltage, intrinsic carrier concentration, Fermi potential, and moderate-inversion substrate factor over the −55 to 125°C military temperature range. The Fermi potential and substrate factor are for nMOS devices in a 0.18 μm CMOS process

T (°C)	T (K)	E_g (eV)	U_T (mV)	n_i (1/cm^3)	$PHI = 2\phi_F$ (V)	n (n_0) (in MI)
−55	218	1.135	18.784	1.87×10^6	0.954	1.272
−20	253	1.127	21.800	1.69×10^8	0.911	1.276
0	273	1.122	23.524	1.35×10^9	0.885	1.279
27	300	1.115	25.850	1.45×10^{10}	0.850	1.282
50	323	1.109	27.832	8.12×10^{10}	0.819	1.286
70	343	1.103	29.555	3.04×10^{11}	0.792	1.289
100	373	1.094	32.140	1.70×10^{12}	0.750	1.294
125	398	1.086	34.294	5.94×10^{12}	0.715	1.298

Values are calculated for $n_i = 1.45 \times 10^{10}$/cm^3 ($T = 300$ K), $N_B = 2 \times 10^{17}$/cm^3, PHI ($T = 300$ K) = $2\phi_F = 0.85$ V, $\gamma = 0.56$ V$^{1/2}$, $V_{SB} = 0$ V, and $V_{EFF} = V_{GS} - V_T = 40$ mV (moderate inversion) to model nMOS devices in a 0.18 μm CMOS process.

temperature. As a result, the Fermi potential, $PHI = 2\phi_F = 2U_T \cdot \ln(N_B/n_i)$ from Table 3.7, decreases modestly with temperature since the decrease associated with increasing intrinsic carrier concentration dominates the increase associated with increasing thermal voltage. Finally, the substrate factor increases by only 2 % over the −55 to 125 °C temperature range because the approximate surface potential, taken in strong inversion at $\phi_s = \phi_0 = 2\phi_F + 4U_T$ and appearing in the denominator of the square-root term in the n expressions in Table 3.6, decreases slightly with temperature. The surface potential decreases with temperature because the decrease in the Fermi potential, $PHI = 2\phi_F$, exceeds the increase in the thermal voltage term, $4U_T$. The slight temperature sensitivity of the substrate factor is fortunate and will permit us to approximate it at a fixed, moderate-inversion value, n_0, even over wide temperature ranges.

While the substrate factor varies little over the −55 to 125 °C temperature range just described, it can increase significantly at very low temperatures. Figure 3.4 shows this where the measured substrate factor, n, is displayed for temperatures from −196 to 127 °C (77 to 400 K) for nMOS and pMOS devices in a 0.35 μm, partially depleted (PD) SOI CMOS process [20]. Here, it is observed that n increases significantly below −73 C (200 K), nearly doubling in value at −196 °C (77 K). This results in nearly a factor-of-two decrease in the slope of $\log(I_D)$ versus V_{GS} in weak inversion, which is proportional to $1/(nU_T)$, and results in nearly a factor-of-two increase in the subthreshold swing voltage, which is equal to $S = 2.3 \cdot nU_T$. Weak-inversion drain current was described earlier in Section 2.4.1 and in Figures 2.1 and 2.2.

The large increase in n at very low temperatures is likely due to increased interface capacitance, which appears in the expression for n given in Equation 2.6 [20]. Operation at such extremely cold temperatures is not considered in the methods and design examples described in this book. However, the micropower, low-noise, 0.35 μm PD SOI CMOS preamplifier described in Section 6.7 is evaluated from −196 to 127 °C (77 to 400 K) in [20]. Figure 3.4, like Table 3.8, confirms n varies little over the −55 to 125 °C military temperature range that encompasses the temperature requirements of most designs.

3.5.2 Mobility, Transconductance Factor, and Technology Current

Table 3.9 shows the substrate factor, thermal voltage, low-field mobility, and low-field transconductance factor with resulting technology current over the −55 to 125 °C military temperature range. The substrate factor is from Table 3.8, and the technology current is from the expression

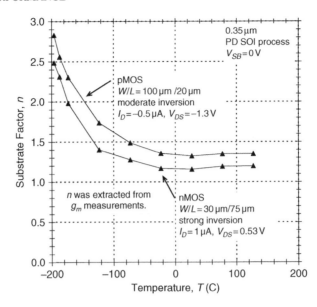

Figure 3.4   Measured substrate factor, n, from -196 to $127\,°C$ (77 to 400 K) for nMOS and pMOS devices in a 0.35 µm PD SOI CMOS process. n is nearly independent of temperature except at very cold temperatures below $-73\,°C$ (200 K). n almost doubles at $-196\,°C$ (77 K), most likely due to increasing interface capacitance. Reproduced by permission of the Institute of Electrical and Electronic Engineers © 2004, from [20]

Table 3.9   MOS substrate factor, thermal voltage, low-field mobility, and low-field transconductance factor with resulting technology current over the -55 to 125°C military temperature range for nMOS devices in a 0.18 µm CMOS process

T (°C)	T (K)	$n\ (n_0)$ (in MI)	U_T (mV)	μ_0 (cm²/V·s)	$k_0 = \mu_0 C'_{OX}$ (µA/V²)	$I_0 = 2n_0 k_0 U_T^2$ (µA)
-55	218	1.272	18.784	681.3	572.9	0.514
-20	253	1.276	21.800	544.9	458.3	0.556
0	273	1.279	23.524	486.1	408.8	0.578
27	300	1.282	25.850	422.0	354.9	0.608
50	323	1.286	27.832	377.7	317.7	0.633
70	343	1.289	29.555	345.2	290.3	0.654
100	373	1.294	32.140	304.4	256.0	0.684
125	398	1.298	34.294	276.2	232.3	0.709

Values are calculated for μ_0 $(T = 300\,K) = 422\,cm^2/V \cdot s$, $C'_{OX} = 8.41\,fF/µm^2$, and a mobility temperature exponent of -1.5 to model nMOS devices in a 0.18 µm CMOS process. The substrate factor is from Table 3.8.

in Table 3.6 using the temperature-corrected thermal voltage, mobility, and transconductance factor expressions from Table 3.7. A room temperature $(T = 300\,K)$, low-field mobility of $\mu_0 = 422\,cm^2/V \cdot s$, gate-oxide capacitance of $C'_{OX} = 8.41\,fF/µm^2$, and mobility temperature exponent of $BEX = -1.5$ are used to again model nMOS devices in a 0.18 µm CMOS process. Mobility and gate-oxide capacitance process parameters are from Table 3.2, and temperature process parameters are from Table 3.5.

Table 3.9 shows a large decrease in mobility and the transconductance factor as temperature increases. Mobility, μ_0, and correspondingly the transconductance factor, $k_0 = \mu_0 C'_{OX}$, are proportional

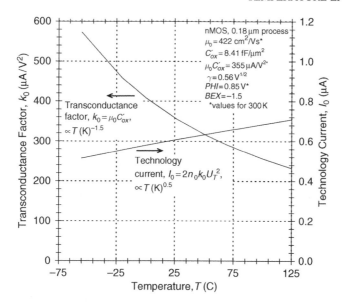

Figure 3.5   Low-field transconductance factor and technology current for the −55 to 125 °C military temperature range for nMOS devices in a 0.18 μm CMOS process. While mobility and the transconductance factor decrease significantly with temperature, the technology current used in the inversion coefficient definition increases modestly because of the increase in the thermal voltage

to T^{BEX}, the absolute temperature raised to the mobility temperature exponent, which is taken as $BEX = -1.5$ for the example process. The technology current, $I_0 = 2n_0 k_0 U_T^2 = 2n_0 \mu_0 C'_{OX} U_T^2$, is proportional to the product of T^{BEX} and T^2 resulting from temperature dependency in the mobility and squared thermal voltage, U_T^2, respectively. As a result, the technology current is proportional to $T^{(2+BEX)}$. This is an interesting result where decreasing mobility is countered by the square of increasing thermal voltage. For a high value of $BEX = -2$, the mobility and thermal voltage temperature dependencies cancel, giving a temperature-independent technology current. On the other extreme for a low value of $BEX = -1$, the technology current is directly proportional to T. For the example here with $BEX = -1.5$, the technology current is proportional to $T^{(2-1.5)}$, or $T^{0.5}$. This is a modest temperature sensitivity, well below that of the mobility or thermal voltage individually. As mentioned earlier, the moderate inversion substrate factor, $n = n_0$, given in Table 3.9, is nearly constant over the −55 to 125 °C temperature range. It contributes little to technology current temperature effects.

Figure 3.5 shows the transconductance factor and technology current over the −55 to 125 °C military temperature range from values given in Table 3.9. Over the military temperature range, the transconductance factor decreases by 60 % while the technology current increases by 38 %. However, over the 0 to 70 °C commercial temperature range, the transconductance factor decreases by 29 % while the technology current increases by only 13 %.

3.5.3 Inversion Coefficient

The increase in technology current, I_0, with temperature means that the inversion coefficient, $IC = I_D/[I_0(W/L)]$ from Table 3.6, decreases with temperature if drain current, I_D, and device shape factor, $S = W/L$, remain constant. Over the −55 to 125 °C military temperature range listed in Table 3.9, the technology current increases by 38 % resulting in an inversion coefficient decrease of 28 % if drain current and shape factor are held constant. Over the 0 to 70 °C commercial temperature range, however, the technology current increases by only 13 % resulting in an inversion coefficient decrease of less than 12 %. These temperature changes are for the mobility temperature exponent of $BEX = -1.5$ used for Table 3.9 and Figure 3.5.

As mentioned in the previous section, if $BEX = -2$, the technology current is temperature independent. This would result, correspondingly, in a temperature-independent inversion coefficient. If the inversion coefficient were temperature independent, however, this would mean only that the region and numerical level of MOS inversion would remain constant with temperature. Transconductance, for example, would still decrease with temperature due to increasing thermal voltage as described later in Section 3.8.2.4.

The usual modest drop in inversion coefficient with temperature due to increasing technology current lowers the level of inversion. This would normally be expected to increase the transconductance efficiency and transconductance as described later in Section 3.8.2.4. However, the increase in the thermal voltage is more significant, resulting in an overall decrease of transconductance efficiency and transconductance.

3.5.4 Threshold Voltage

The design methods and examples described in this book involve selecting drain current, inversion coefficient, and channel length for optimum tradeoffs in performance. This is consistent with threshold-voltage, independent design where drain bias current is held fixed or perhaps varied with temperature to hold transconductance constant. Threshold voltage, however, is important in evaluating the operating gate–source voltage, $V_{GS} = V_T + V_{EFF}$, which is equal to the sum of the threshold voltage, V_T, and the effective gate–source voltage, $V_{EFF} = V_{GS} - V_T$. V_{GS} must be maintained at sufficiently low values to ensure bias compliance, especially in low-voltage designs.

The threshold voltage change with temperature, described in Table 3.7, is modeled by the threshold-voltage temperature coefficient, TCV. Figure 3.6 shows the measured threshold voltage for nMOS devices in a 0.8 μm, PD SOI CMOS process from 20 to 300 °C (293 to 573 K) [21]. This is an extremely wide temperature range encompassing very high temperatures supported by SOI CMOS processes because of low drain–body leakage resulting from drains placed on insulating material instead of inside the normal silicon substrate. In Figure 3.6, the threshold voltage decreases almost linearly by 0.45 V over the temperature increase of 280 °C. This gives a value of $TCV = -0.45\,\text{V}/280\,°\text{C} = -1.6\,\text{mV}/°\text{C}$. The magnitude of pMOS threshold voltage decreases a similar amount with increasing temperature. While increasing temperatures usually lower transconductance, voltage gain, and bandwidth, increasing temperatures actually lower the threshold voltage. This usually favorably lowers V_{GS} since the threshold voltage usually decreases more compared to the temperature-related increase in V_{EFF}.

3.5.5 Design Considerations

Temperature effects on parameters like the thermal voltage, mobility, transconductance factor, technology current, inversion coefficient, and threshold voltage must be considered in design. In this book, we will focus on selecting drain current, inversion coefficient, and channel length for optimum tradeoffs in performance at a selected nominal temperature. This is done using performance predictions given in tables and figures, which are included in the *Analog CMOS Design, Tradeoffs and Optimization* spreadsheet. Performance can then be predicted at other temperatures. This involves first finding the temperature-corrected inversion coefficient as described in Section 3.5.3, followed by evaluating performance using the temperature-corrected thermal voltage, mobility, or other relevant process parameters. When using the spreadsheet, the temperature input cell can be changed, and multiple columns can be used for the minimum, nominal, and maximum temperature. However, device inversion coefficient inputs must be changed to hold a constant channel width since the inversion coefficient changes modestly with temperature for a constant drain current and device shape factor.

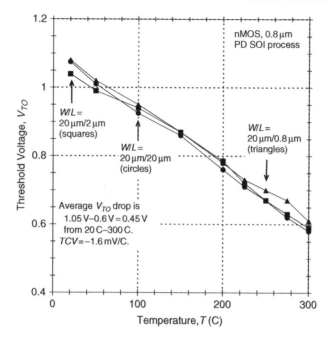

Figure 3.6   Measured threshold voltage from 20 to 300 °C (293 to 573 K) for nMOS devices in a 0.8 μm PD SOI CMOS process. The threshold voltage decreases almost linearly with temperature, having a temperature coefficient of $TCV = -1.6$ mV/C. There is a similar drop in threshold-voltage magnitude with temperature for pMOS devices. Reproduced by permission of the Institute of Electrical and Electronic Engineers © 2005, from [21]

Alternatively, a candidate design optimized at a nominal temperature can be computer simulated over temperature to assess temperature effects as required prior to production release. While design over temperature is not explicitly addressed in design examples contained in this book, [20] considers design and performance from -196 to 127 °C (77 to 400 K) for the micropower, low-noise, 0.35 μm SOI CMOS preamplifier described in Section 6.7.

3.6 SIZING RELATIONSHIPS

Before investigating tradeoffs in analog MOS performance resulting from the design choices of drain current, inversion coefficient, and channel length, it is necessary to first look at MOS sizing relationships. MOS channel width, channel length, and the resulting gate area strongly affect capacitances, bandwidth, flicker noise, local area mismatch, and other aspects of performance. Additionally, channel width must be determined for circuit layout.

Table 3.10 gives MOS sizing relationships in terms of the design choices of inversion coefficient, IC, channel length, L, and drain current, I_D. As described in Section 3.4.2.3 and considered throughout this book, a fixed–normalized inversion coefficient, IC, is used where IC tracks drain current for a fixed-geometry device. Given in the table are the shape factor, $S = W/L$, channel width, W, and gate area, WL, along with their dependencies on the inversion coefficient, channel length, and drain current. The following are observed from the MOS sizing relationships in the table [10, 11]:

Table 3.10 MOS sizing expressions and trends in terms of drain current, inversion coefficient, and channel length

Sizing parameter	$IC \uparrow$ L, I_D fixed	$L \uparrow$ IC, I_D fixed	$I_D \uparrow$ IC, L fixed
Shape factor: $S = \dfrac{W}{L} = \left(\dfrac{1}{IC}\right)\left(\dfrac{I_D}{I_0}\right)$	$\downarrow \propto \dfrac{1}{IC}$	Unchanged	$\uparrow \propto I_D$
Width: $W = \left(\dfrac{L}{IC}\right)\left(\dfrac{I_D}{I_0}\right)$	$\downarrow \propto \dfrac{1}{IC}$	$\uparrow \propto L$	$\uparrow \propto I_D$
Gate area: $WL = \left(\dfrac{L^2}{IC}\right)\left(\dfrac{I_D}{I_0}\right)$	$\downarrow \propto \dfrac{1}{IC}$	$\uparrow \propto L^2$	$\uparrow \propto I_D$

$I_0 = 2n_0\mu_0 C'_{OX} U_T^2$ is the process technology current for the nominal temperature of operation. Sizing relationships are found from the fixed–normalized inversion coefficient, given by $IC = I_D/[I_0(W/L)]$ from Table 3.6.

- *If the inversion coefficient is increased (while holding drain current and the channel length fixed),* shape factor, channel width, and gate area (and gate-oxide capacitance) all must drop inversely with increasing inversion coefficient.

- *If the channel length is increased (while holding drain current and the inversion coefficient fixed),* the channel width must increase directly with channel length to maintain the required shape factor. Since the channel width must increase directly with channel length, gate area (and gate-oxide capacitance) increases as the square of increasing channel length.

- *If the drain current is increased (while holding the inversion coefficient and channel length fixed),* shape factor, channel width, and gate area (and gate-oxide capacitance) all must increase directly with increasing drain current.

In the design methods presented here, MOS sizing, in particular channel width and gate area, is found from the selection of drain current, inversion coefficient, and channel length. MOS sizing must be found using the technology current, I_0, for the nominal temperature. This is because the technology current and inversion coefficient are functions of temperature as described in Sections 3.5.2 and 3.5.3. Often the drain current will be fixed in a design. If drain current varies for a device having fixed sizing, the inversion coefficient directly tracks the current.

3.6.1 Shape Factor

The MOS shape factor or aspect ratio, $S = W/L$, is the ratio of channel width to channel length. For the design methods described in this book, the channel width is found by multiplying the selected channel length with the required shape factor resulting from the selected drain current and inversion coefficient.

An nMOS device in a $0.18\,\mu\text{m}$ CMOS process operating at $IC = 10$ and $I_D = 100\,\mu\text{A}$ requires a shape factor, given in Table 3.10, of

$$S = \frac{W}{L} = \left(\frac{1}{IC}\right)\left(\frac{I_D}{I_0}\right) = \left(\frac{1}{10}\right)\left(\frac{100\,\mu\text{A}}{0.64103\,\mu\text{A}}\right) = 15.6 \tag{3.7}$$

To permit convenient sizing, the technology current of $I_0 = 2n_0\mu_0 C'_{OX} U_T^2 = 0.64103\,\mu\text{A}$ was adjusted slightly from the value of $0.64\,\mu\text{A}$ ($T = 300\,\text{K}$) given in the process parameters of Table 3.2.

Figure 3.7 shows the required shape factor versus inversion coefficient and drain current for nMOS devices in a $0.18\,\mu\text{m}$ CMOS process. The shape factor decreases inversely with inversion

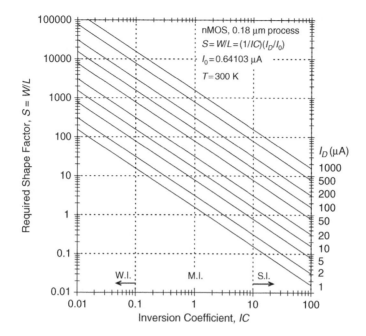

Figure 3.7 Required shape factor, $S = W/L$, versus inversion coefficient, IC, for nMOS devices in a $0.18\,\mu m$ CMOS process operating at drain currents from 1 to $1000\,\mu A$. Shape factor decreases inversely with inversion coefficient and directly tracks drain current. Shape factors for pMOS devices are nearly five times larger due to lower mobility

coefficient for a fixed drain current as listed in Table 3.10. For example, increasing the inversion coefficient by a factor of 10 from the center of moderate inversion ($IC = 1$) to the onset of strong inversion ($IC = 10$) requires a factor-of-10 decrease in shape factor and, correspondingly, the same decrease in channel width for a fixed channel length. The shape factor directly tracks drain current for a fixed inversion coefficient. Thus, increasing the drain current by a factor of 10 requires the same increase in the shape factor and, correspondingly, the same increase in channel width for a fixed channel length.

As seen in Figure 3.7, the largest shape factors and corresponding channel widths are required for devices operating at low inversion coefficients in weak inversion at high drain currents. Conversely, the smallest shape factors and corresponding channel widths are required for devices operating at high inversion coefficients in strong inversion at low drain currents. As a result, micropower design for drain currents below $1\,\mu A$ complicates MOS sizing since the required shape factors often have values much less than unity.

The required shape factor and corresponding channel width are inversely proportional to the transconductance factor, $k_0 = \mu_0 C'_{OX}$, and technology current, $I_0 = 2n_0\mu_0 C'_{OX}U_T^2$, which track the mobility, μ_0. As a result, the required shape factor and corresponding channel width are nearly five times larger for pMOS devices compared to the values shown in Figure 3.7 for nMOS devices. This is because the mobility, transconductance factor, and technology current are nearly a factor of five lower for pMOS devices compared to nMOS devices as shown in the $0.18\,\mu m$ CMOS process parameters of Table 3.2.

Since the shape factor shown in Figure 3.7 and the corresponding channel width are inversely proportional to the transconductance factor and technology current, the shape factor and channel width decrease in smaller-geometry processes for a given drain current, inversion coefficient, and channel length. This is because the decrease in gate-oxide thickness, t_{ox}, results in an increase in gate-oxide capacitance, C'_{OX}, giving increased transconductance factor and technology current.

3.6.2 Channel Width

Unlike traditional analog CMOS design, channel width is not a degree of design freedom in the design methods described here. Drain current, inversion coefficient, and channel length are selected to optimize tradeoffs in analog circuit performance. Channel width is implicitly considered in the design methods and then calculated as needed for layout.

An nMOS device in a $0.18\,\mu m$ CMOS process operating at $I_D = 100\,\mu A$, $IC = 10$, and $L = 0.18\,\mu m$ requires a channel width, given in Table 3.10, of

$$W = S \cdot L = \left(\frac{L}{IC}\right)\left(\frac{I_D}{I_0}\right) = \left(\frac{0.18\,\mu m}{10}\right)\left(\frac{100\,\mu A}{0.64103\,\mu A}\right) = 2.8\,\mu m \qquad (3.8)$$

This is equal to the shape factor given by Equation 3.7 multiplied by the channel length for the technology current $I_0 = 0.64103\,\mu A$ mentioned in the previous section.

Table 3.11 lists shape factors and channel widths required for nMOS devices in a $0.18\,\mu m$ CMOS process. This is listed for selected inversion coefficients and channel lengths for a "millipower" drain current of $100\,\mu A$. MOS channel width is given with channel length as W/L to reflect common sizing practice. Inversion coefficients from $IC = 0.1$ (weak inversion) to 100 (deep strong inversion), and channel lengths of $L = 0.18$, 0.36, 0.72, and $1.44\,\mu m$ are considered.

In Table 3.11, the shape factor and channel width decrease inversely with increasing inversion coefficient as listed in Table 3.10. Channel width increases directly with channel length as required to maintain the shape factor. Gate area, observed by multiplying channel width and length, decreases inversely with inversion coefficient. Gate area, however, increases as the square of channel length because an increase in channel length must be countered by an increase in channel width to maintain the shape factor. As seen in Table 3.11, long-channel devices in weak and moderate inversion require large channel widths and gate areas while short-channel devices in strong inversion require small channel widths and gate areas.

Table 3.12 lists shape factors and channel widths required for nMOS devices in a $0.18\,\mu m$ CMOS process, this time for a "micropower" drain current of $1\,\mu A$. Inversion coefficients from $IC = 0.1$

Table 3.11 Required shape factor and width/length sizing versus inversion coefficient and channel length for nMOS devices in a $0.18\,\mu m$ CMOS process operating at a "millipower" drain current of $100\,\mu A$. Bold entries denote selected design choices. pMOS shape factors and widths are approximately five times larger due to lower mobility. Shape factor and width directly track drain current

I_D (μA)	IC	Inv.	$S = W/L$	$W(\mu m)/L(\mu m)$			
				$L = 0.18\,\mu m$	$L = 0.36\,\mu m$	$L = 0.72\,\mu m$	$L = 1.44\,\mu m$
100	**0.1**	**WI**	1560	280.8/0.18	561.6/0.36	1123/0.72	2246/1.44
100	**0.2**	**MI**	780	140.4/0.18	280.8/0.36	561.6/0.72	1123/1.44
100	**0.5**	**MI**	312	56.2/0.18	112.3/0.36	224.6/0.72	449.3/1.44
100	**1**	**MI**	156	28.1/0.18	56.2/0.36	112.3/0.72	224.6/1.44
100	**2**	**MI**	78	14.0/0.18	28.1/0.36	56.2/0.72	112.3/1.44
100	**5**	**MI**	31.2	5.62/0.18	11.2/0.36	22.5/0.72	44.9/1.44
100	**10**	**SI**	15.6	2.81/0.18	5.62/0.36	11.2/0.72	22.5/1.44
100	**20**	**SI**	7.8	1.40/0.18	2.81/0.36	5.62/0.72	11.2/1.44
100	**50**	**SI**	3.12	0.56/0.18	1.12/0.36	2.25/0.72	4.49/1.44
100	**100**	**SI**	1.56	0.28/0.18	0.56/0.36	1.12/0.72	2.25/1.44

Sizing is for $I_0 = 2n_0\mu_0 C'_{OX} U_T^2 = 0.64103\,\mu A$ adjusted from $0.64\,\mu A$ value given in Table 3.2 to permit convenient sizing. Sizing is for room temperature ($T = 300\,K$).

Table 3.12 Required shape factor and width/length sizing versus inversion coefficient and channel length for nMOS devices in a 0.18 μm CMOS process operating at a "micropower" drain current of 1 μA. Bold entries denote selected design choices. pMOS shape factors and widths are approximately five times larger due to lower mobility. Shape factor and width directly track drain current

I_D (μA)	IC	Inv.	$S = W/L$	$W(\mu m)/L(\mu m)$			
				$L = 0.72\,\mu m$	$L = 1.44\,\mu m$	$L = 2.88\,\mu m$	$L = 5.76\,\mu m$
1	0.1	**WI**	15.6	11.2/0.72	22.5/1.44	44.9/2.88	89.9/5.76
1	0.2	**MI**	7.8	5.62/0.72	11.2/1.44	22.5/2.88	44.9/5.76
1	0.5	**MI**	3.12	2.25/0.72	4.49/1.44	8.99/2.88	18.0/5.76
1	1	**MI**	1.56	1.12/0.72	2.25/1.44	4.49/2.88	8.99/5.76
1	2	**MI**	0.78	0.56/0.72	1.12/1.44	2.25/2.88	4.49/5.76
1	5	**MI**	0.312	NA	0.45/1.44	0.90/2.88	1.80/5.76
1	10	**SI**	0.156	NA	NA	0.45/2.88	0.90/5.76

Sizing is for $I_0 = 2n_0\mu_0 C'_{OX} U_T^2 = 0.64103\,\mu A$ adjusted from 0.64 μA value given in Table 3.2 to permit convenient sizing. Sizing is for room temperature ($T = 300\,K$).

NA: Device width is below minimum required for layout.

(weak inversion) to 10 (the onset of strong inversion) and longer channel lengths of $L = 0.72$, 1.44, 2.88, and 5.76 μm are considered. The shape factors and channel widths (for equal inversion coefficients and channel lengths) are 100 times less than those in Table 3.11 where drain current is 100 μA. For the "micropower" drain current of Table 3.12, the shape factors required near strong inversion are much less than unity. Typical of micropower design, the channel length must be increased to obtain a viable channel width and gate area. This comes at the expense of intrinsic bandwidth, which is inversely proportional to the square of channel length (unless velocity saturation is significant) as discussed in Section 3.9.6.

Like the shape factor mentioned in the previous section, channel width and gate area are inversely proportional to the transconductance factor and technology current, which track the mobility. As noted for the shape factor shown in Figure 3.7 and the channel width and length sizing of Tables 3.11 and 3.12, pMOS shape factors, channel widths, and gate areas for the 0.18 μm CMOS process are approximately five times those given for nMOS devices. This is because of a nearly factor-of-five reduction in pMOS mobility as listed in Table 3.2.

Like the shape factor mentioned in the previous section, channel width and gate area decrease in smaller-geometry processes for a given drain current, inversion coefficient, and channel length. This is because of the increased transconductance factor and technology current resulting from increased gate-oxide capacitance associated with thinner gate oxide.

In Tables 3.11 and 3.12, effective MOS channel width and length are given inclusive of mask biasing and lateral diffusion effects. It is important to understand how drawn dimensions associated with EDA layout relate to the effective dimensions that actually govern MOS performance. This is especially important for short channel lengths or widths where the difference between drawn and effective dimensions can be significant. For nMOS devices in the 0.18 μm process, a drawn channel length of 0.18 μm corresponds to an effective channel length of 0.152 μm for $DL = 0.028$ μm given in Table 3.2. For simplicity, rounded effective channel lengths are given in Tables 3.11 and 3.12.

3.6.3 Gate Area and Silicon Cost

The selection of MOS drain current, inversion coefficient, and channel length results in a required channel width for layout. The channel width and length combination then results in a gate area. Gate area is important since it controls area-related parameters like intrinsic gate capacitances (Section 3.9.2),

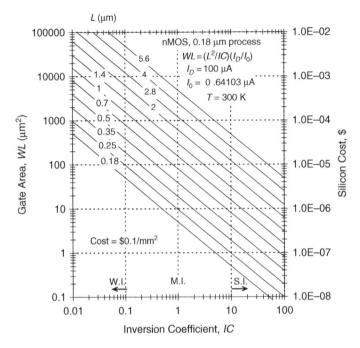

Figure 3.8   Gate area, *WL*, versus inversion coefficient, *IC*, for $L = 0.18$, 0.25, 0.35, 0.5, 0.7, 1, 1.4, 2, 2.8, 4, and 5.6 μm, nMOS devices in a 0.18 μm CMOS process operating at 100 μA. Cost is also estimated assuming a \$0.1/mm² silicon cost. Gate area and cost for pMOS devices are approximately five times larger due to lower mobility. Gate area and cost scale directly with drain current

intrinsic bandwidth (Section 3.9.6), low-frequency flicker noise (Section 3.10.3), and local-area DC mismatch (Section 3.11.1). Additionally, gate area influences die area and silicon production cost.

Figure 3.8 shows gate area versus selected inversion coefficient and channel length for nMOS devices in a 0.18 μm CMOS process operating at a "millipower" drain current of 100 μA. Gate area is calculated from the expression in Table 3.10, again using the technology current of $I_0 = 0.64103$ μA rounded slightly from the CMOS process parameters of Table 3.2. The figure also shows silicon cost associated with gate area assuming a production cost of \$0.1/mm².

Figure 3.8 shows that gate area and device cost scale inversely with the inversion coefficient and as the square of channel length as listed in Table 3.10. The rapid increase of gate area with increasing channel length results in desirable reductions in flicker noise and local-area DC mismatch at the expense of substantial increases in intrinsic capacitances and reductions in intrinsic bandwidth. Gate area ranges from less than 1 μm² for short-channel devices operating at high inversion coefficients in strong inversion to over 100 000 μm² (0.1 mm²) for long-channel devices operating at low inversion coefficients in weak inversion. A 100 000 μm² gate area corresponds to an approximate cost of \$0.01 per device, exclusive of the additional area required for drain and source implants, substrate connections, and metal interconnects. However, the silicon cost figure of \$0.1/mm² is probably higher than that available in volume production, somewhat compensating for these extra costs. Device cost is especially important for circuits that are replicated in quantity, like comparators in flash analog-to-digital converters.

Gate area and device cost, like the shape factor and channel width, scale directly with drain current and are thus a factor of 100 less for a "micropower" drain current of 1 μA compared to values shown in Figure 3.8 for a "millipower" drain current of 100 μA. For micropower circuits, however, overall die area is significantly above the small gate area of devices since drain and source implants, substrate connections, and metal interconnects require proportionally larger area.

Gate area and device cost, like the shape factor and channel width, are inversely proportional to the transconductance factor and technology current. As a result, pMOS gate area and device cost are approximately five times the values given in Figure 3.8 for nMOS devices in the 0.18 μm CMOS process. This, again, is because of the nearly factor-of-five reduction in pMOS mobility listed in Table 3.2. The factor-of-five increase in pMOS channel width and gate area, compared to nMOS devices operating at equal drain current, inversion coefficient, and channel length, immediately gives rise to a factor-of-five increase in intrinsic gate capacitances. This results in nearly a factor-of-five reduction in intrinsic bandwidth.

Gate area, like the shape factor and channel width, decreases in smaller-geometry processes for a given drain current, inversion coefficient, and channel length. This, again, is because of the increased transconductance factor and technology current resulting from increased gate-oxide capacitance associated with thinner gate oxide.

3.7 DRAIN CURRENT AND BIAS VOLTAGES

The following discussions describe MOS drain current, the effective gate–source voltage, and the drain–source saturation voltage. Drain current governs device transconductance and is related to the effective gate–source voltage, which controls the operating gate–source voltage. The drain–source saturation voltage sets the minimum operating drain–source voltage for usual operation in saturation. The management of gate–source and drain–source bias voltages is increasingly important as supply voltage decreases in increasingly smaller-geometry processes.

Figure 3.9 shows the location and polarity of MOS drain current, I_D, and drain–source, V_{DS}, gate–source, V_{GS}, source–body, V_{SB}, and drain–body, V_{DB}, voltages, which are positive for nMOS devices. In this book, as is commonly done, these are also taken positively for pMOS devices as shown in the figure. The drain–source saturation, $V_{DS,sat}$, threshold, V_T, and zero-V_{SB} threshold, V_{TO}, voltages are also taken positively for both nMOS and pMOS devices. The effective gate–source voltage, $V_{EFF} = V_{GS} - V_T$, is positive for nMOS devices for V_{GS} greater than V_T (as in strong inversion and most of moderate inversion) and is negative for V_{GS} less than V_T (as in weak inversion). In this book, the same polarity for V_{EFF} is also used for pMOS devices where a positive value corresponds to a magnitude of V_{GS} above the magnitude of V_T.

As observed in Figure 3.9, gate and drain voltages are source referenced as V_{GS} and V_{DS}, respectively. Although substrate referencing is advantageous for MOS models like the EKV model [13, 14], for example to avoid polarity reversal in voltages like V_{DS}, this book uses source referencing as is traditional for design [15, pp. 179–180]. Source referencing de-emphasizes the substrate, which ideally would not be present. The substrate effect, however, is considered through V_{SB} and its influence on V_T, the body-effect transconductance, g_{mb}, and other MOS parameters.

In addition to showing MOS current and voltage polarities, Figure 3.9 also shows a physical cross-section of MOS operation. Here, the source, gate, drain, and body or substrate regions are shown along with physical dimensions. These dimensions include the gate-oxide thickness, t_{ox}, channel length, L, channel pinch-off length, l_p, and depletion region width, t_{dep}, below the channel. The channel width, W, appears perpendicular to the cross-section. The channel width and length are effective values inclusive of mask bias and lateral diffusion effects.

3.7.1 Drain Current

The design methods presented in this book involve selecting drain current, inversion coefficient, and channel length for optimum tradeoffs in analog performance. Current biasing, typical for analog CMOS design, is used since it removes the bias current variations caused by threshold-voltage variation with device geometry, voltage bias (V_{SB} and, for short-channel devices, V_{DS}), temperature, and process.

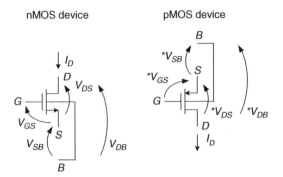

(a)

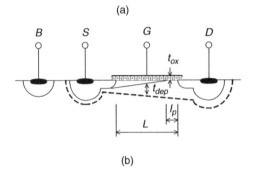

(b)

Figure 3.9   MOSFET (a) schematic diagram symbol showing current and voltage definitions, and (b) cross-section showing physical regions and dimensions.

The current and voltage definitions in (a) are the drain current, I_D, and drain–source, V_{DS}, gate–source, V_{GS}, source–body, V_{SB}, and drain–body, V_{DB}, voltages. In this book, positive values associated with nMOS devices are also used for pMOS devices. The drain–source saturation, $V_{DS,sat}$, threshold, V_T, and zero-V_{SB} threshold, V_{TO}, voltages are also taken positively for both nMOS and pMOS devices. Source referencing for V_{GS}, V_{DS}, and $V_{DS,sat}$ is traditional for design and de-emphasizes the substrate, which ideally would not be present.

The cross-section of (b) shows the source (S), gate (G), drain (D), and body (B) regions, where the body region extends across the other regions. The physical dimensions are the gate-oxide thickness, t_{ox}, channel length, L, channel pinch-off length, l_p, and depletion-region width, t_{dep}. The channel width, W, is perpendicular to the cross-section

Although drain current is selected for design, the drain current, gate–source voltage characteristic is important to evaluate the effective gate–source voltage, $V_{EFF} = V_{GS} - V_T$, and operating gate–source voltage, $V_{GS} = V_{EFF} + V_T$. Additionally, the drain current characteristic is needed to predict device transconductance.

3.7.1.1 Without small-geometry effects

Table 3.13 includes expressions for drain current in weak, moderate, and strong inversion with mobility held fixed at its low-field value, $\mu = \mu_0$. Correction factors included in the table separately model the reduction of drain current due to velocity saturation and VFMR effects. Initially, drain current will be considered without these small-geometry effects. As throughout this book, operation in saturation is assumed where $V_{DS} > V_{DS,sat}$.

Table 3.13   MOS drain current expressions

Parameter	Expression
Drain current (weak inversion)	$I_D(\text{WI}) = 2n\,\mu_0 C'_{OX} U_T^2 \left(\dfrac{W}{L}\right)\left(e^{\frac{V_{GS}-V_T}{nU_T}}\right)$
Drain current (strong inversion, no small-geometry effects)	$I_D(\text{SI}) = \dfrac{1}{2}\left(\dfrac{\mu_0 C'_{OX}}{n}\right)\left(\dfrac{W}{L}\right)(V_{GS}-V_T)^2$
Drain current (weak through strong inversion, no small-geometry effects)	$I_D(\text{WI–SI}) = 2n\,\mu_0 C'_{OX} U_T^2 \left(\dfrac{W}{L}\right)\left[\ln\left(1+e^{\frac{V_{GS}-V_T}{2nU_T}}\right)\right]^2$
Drain current correction factor[a] (simple)	$\dfrac{1}{1+\left(\theta+\dfrac{1}{L\,E_{CRIT}}\right)(V_{GS}-V_T)} = \dfrac{1}{1+\dfrac{V_{GS}-V_T}{(L\,E_{CRIT})'}} = \dfrac{1}{1+VSF'}$ where the velocity saturation factor including VFMR is $VSF' = \dfrac{V_{GS}-V_T}{(L\,E_{CRIT})'} = \left(\theta+\dfrac{1}{L\,E_{CRIT}}\right)(V_{GS}-V_T)$ and the equivalent velocity saturation voltage is $(L\,E_{CRIT})' = \dfrac{1}{\theta+\dfrac{1}{L\,E_{CRIT}}} = \dfrac{1}{\theta}\;\|\;L\,E_{CRIT}$
Drain current correction factor[a] (enhanced)	$\dfrac{1}{[1+\theta(V_{GS}-V_T)]\left[1+\left(\dfrac{V_{GS}-V_T}{L\,E_{CRIT}}\right)^\alpha\right]^{\left(\frac{1}{\alpha}\right)}}$ $\alpha = 1\text{–}2\ (1.3\ \text{used here})$
Drain current correction factor[a] (complex)	$\dfrac{1}{[1+\theta(V_{GS}-V_T)]\left[1+\left(\dfrac{V_{GS}-V_T}{L\,E_{CRIT0}[1+(V_{GS}-V_T)\theta]}\right)^\alpha\right]^{\left(\frac{1}{\alpha}\right)}}$ $\alpha = 1\text{–}2\ (1.7\ \text{used here})$
Drain current (strong inversion, full velocity saturation)	$I_D\,(\text{SI, full vel. sat.}) = \dfrac{1}{2}\left(\dfrac{\mu_0 C'_{OX}}{n}\right)\left(\dfrac{W}{L}\right)(L\,E_{CRIT})'\,(V_{GS}-V_T)$
$(VSF' = (V_{GS}-V_T)/$ $(L\,E_{CRIT})' \gg 1)$	$\approx \dfrac{1}{2}\left(\dfrac{\mu_0 C'_{OX}}{n}\right)(W)\,E_{CRIT}\,(V_{GS}-V_T)$

Drain current is for saturation where $V_{DS} > V_{DS,sat}$. The increase in drain current with increasing drain–source voltage is not included here. This is included separately through drain–source resistance expressions.

[a] Approximates the drop in drain current due to velocity saturation and VFMR. $V_{GS} - V_T$ in the drain current correction factors can be replaced with $V_{GS} - V_T = (nU_T/2)\ln\left[1+e^{2(V_{GS}-V_T)/(nU_T)}\right]$ to smoothly "clamp" it to zero when it is negative. When $V_{GS} - V_T$ is negative or zero, the correction factors have value of unity.

The first expression in Table 3.13 models drain current in weak inversion, which was introduced in Section 2.4.1. The second expression models drain current in strong inversion, which was introduced in Section 2.4.2. These expressions and the third expression for modeling drain current continuously from weak through strong inversion were included in Table 3.6 for the purpose of developing the substrate factor, n, and inversion coefficient, IC. Drain current expressions, again, were derived from the EKV MOS model with the source as the reference terminal instead of the substrate [12, p. 136, 13, 14, 15, p. 178].

The continuous expression for drain current in Table 3.13 is a non-physically based interpolation expression predicting drain current in weak, moderate, and strong inversion. The discussion below shows that this interpolation expression converges to physically based, weak- and strong-inversion expressions.

In weak inversion[7] for $IC < 0.1$, $V_{EFF} = (V_{GS} - V_T) < -2nU_T$ as described later in Section 3.7.2.4. This corresponds to V_{EFF} below -72 mV assuming $n = 1.4$ and $U_T = 25.9$ mV ($T = 300$ K). In weak inversion, the exponential argument $(V_{GS} - V_T)/2nU_T$ in the continuous drain current expression is equal to -1 or lower yielding an exponential term that is well below unity. Since $\ln(1 + x) \approx x$ for $x \ll 1$, the following is observed:

$$\left[\ln\left(1 + e^{\frac{V_{GS} - V_T}{2nU_T}} \right) \right]^2 \approx \left(e^{\frac{V_{GS} - V_T}{2nU_T}} \right)^2 \approx e^{\frac{V_{GS} - V_T}{nU_T}} , \quad e^{\frac{V_{GS} - V_T}{2nU_T}} \ll 1 \quad \text{in WI} \quad (3.9)$$

The rightmost approximation in Equation 3.9 multiplied by the continuous drain current, prefix term of $2n\mu_0 C'_{OX} U_T^2 (W/L)$ converges to the weak-inversion, drain current expression.

In strong inversion for $IC > 10$, $V_{EFF} = (V_{GS} - V_T) > 6.24nU_T$, which is, again, described later in Section 3.7.2.4. This corresponds to V_{EFF} above 225 mV assuming $n = 1.4$ and $U_T = 25.9$ mV ($T = 300$ K). In strong inversion, the exponential argument $(V_{GS} - V_T)/2nU_T$ in the continuous drain current expression is equal to 3.12 or higher yielding an exponential term that is much greater than unity. Since $\ln(1 + x) \approx \ln(x)$ for $x \gg 1$, the following is observed:

$$\left[\ln\left(1 + e^{\frac{V_{GS} - V_T}{2nU_T}} \right) \right]^2 \approx \left[\ln\left(e^{\frac{V_{GS} - V_T}{2nU_T}} \right) \right]^2 \approx \left(\frac{V_{GS} - V_T}{2nU_T} \right)^2 , \quad e^{\frac{V_{GS} - V_T}{2nU_T}} \gg 1 \quad \text{in SI} \quad (3.10)$$

The rightmost approximation in Equation 3.10 multiplied by the continuous drain current, prefix term of $2n\mu_0 C'_{OX} U_T^2 (W/L)$ converges to the strong-inversion, drain current expression.

As illustrated later in measured data, the continuous drain current expression provides a smooth and accurate interpolation in moderate inversion where drain current cannot be modeled by weak-inversion, exponential or strong-inversion, square-law expressions.

3.7.1.2 With velocity saturation effects

Table 3.13 includes simple, enhanced, and complex correction factors that are multiplied by the continuous drain current expression to approximate the loss of drain current due to velocity saturation and VFMR effects. As introduced in Section 2.4.3, velocity saturation is the primary effect for short-channel devices operating in saturation ($V_{DS} > V_{DS,sat}$) at high levels of inversion (high $V_{EFF} =$

---

[7] Sometimes the weak-inversion region is called the subthreshold region. However, the subthreshold region includes operation at gate–source voltages just below the threshold voltage, which is actually in the moderate-inversion region. For clarity, this book will define MOS operation using weak-, moderate-, and strong-inversion regions.

$V_{GS} - V_T$) in strong inversion. Here, average carrier velocity and the resulting drain current drop below expected low-field values due to the high horizontal, drain–source, electric field. If the field is sufficiently high, carrier velocity and drain current level off at saturated values.

The reduction of carrier velocity and drain current associated with high values of the horizontal electric field, E_x, can be modeled by a loss of mobility given by [22–25]

$$\mu(E_x) = \frac{\mu}{\left[1 + \left(\dfrac{E_x}{E_{CRIT}}\right)^\alpha\right]^{\frac{1}{\alpha}}} \approx \frac{\mu}{\left[1 + \left(\dfrac{V_{GS} - V_T}{L\,E_{CRIT}}\right)^\alpha\right]^{\frac{1}{\alpha}}}$$

$$\approx \frac{\mu}{1 + \left(\dfrac{V_{GS} - V_T}{L\,E_{CRIT}}\right)}, \text{ if } \alpha = 1 \qquad (3.11)$$

Here, the horizontal electric field in strong inversion is roughly equal to the voltage across the pinched-off channel and source, $V_{DS,sat} \approx V_{GS} - V_T$, divided by the channel length, or $E_x \approx (V_{GS} - V_T)/L$. μ is taken as a fixed, low-field mobility associated with low values of E_x. μ, however, may be reduced from its specified low-field value of μ_0 by VFMR effects as described later. Equation 3.11 appears in the drain current correction factors of Table 3.13 to model velocity saturation effects.

Average carrier velocity is equal to the mobility given in Equation 3.11 multiplied by E_x, or $v = \mu(E_x) \cdot E_x$. For values of $E_x \ll E_{CRIT}$, corresponding to $V_{GS} - V_T \ll L E_{CRIT}$, velocity saturation effects are negligible and average carrier velocity is equal to the product of low-field mobility and E_x, or $v = \mu \cdot E_x$. When E_x reaches the critical, velocity saturation, electric field, E_{CRIT}, the average carrier velocity decreases to approximately one-half the low-field value predicted by $v = \mu \cdot E_x$. Operation at $E_x = E_{CRIT}$, corresponding to $V_{GS} - V_T = L E_{CRIT}$, results in a mobility reduction of approximately 50 % as seen in the bottom term of Equation 3.11, giving an approximate 50 % loss of average carrier velocity and drain current from the expected low-field values. The actual loss of mobility, average carrier velocity, and drain current depends on the value of the velocity saturation, transition exponent, α, which is described below.

Operation at $E_x \gg E_{CRIT}$, corresponding to $V_{GS} - V_T \gg L E_{CRIT}$, results in full velocity saturation at a saturated carrier velocity and saturated drain current. The saturated carrier velocity is $v = \mu(E_x) \cdot E_x = \mu \cdot E_{CRIT} = v_{sat}$, which is found by evaluating $\mu(E_x)$ in Equation 3.11 for $E_x \gg E_{CRIT}$. $E_{CRIT} = v_{sat}/\mu$ has typical values of 4 and 12 V/μm for nMOS and pMOS devices, respectively, assuming v_{sat} of 1.6×10^7 cm/s and $\mu = \mu_0$ of 400 and 133 cm^2/V $\cdot$ s, respectively. The value of E_{CRIT}, however, may be increased because of the decrease in the operating value of μ from its μ_0 value due to VFMR. Additionally, the interpretation and value of E_{CRIT} depends on the velocity saturation model used. In one model, a 2 appears multiplied by E_{CRIT} in the denominator of the model, where carrier velocity is assumed to be fully saturated at $E_x = 2E_{CRIT}$ [16, pp. 169–171]. Here, a point can be defined near the drain end of the channel where $E_x = 2E_{CRIT}$. Non-velocity-saturated and velocity-saturated regions are defined on the source and drain sides of this point.

Carrier velocity overshoot where carriers can exceed the saturation velocity over short times or short distances [15, p. 280] is not included in Equation 3.11 or the drain current correction terms of Table 3.13. Such advanced effects can be considered during computer simulations of a candidate design.

The velocity saturation, transition exponent α in Equation 3.11 models the roll-off or transition of carrier velocity from its linear increase with low values of E_x to its saturated value at high values of E_x [23–25]. α has typical values between 1 and 2 [23–25], with possibly different values for electrons (nMOS devices), holes (pMOS) devices, and different temperatures. α equal to unity gives the familiar velocity saturation denominator term of $1 + (V_{GS} - V_T)/(L E_{CRIT})$ shown in the bottom term of Equation 3.11. Regardless of the exponent selected, the denominator terms of Equation 3.11 approach $(V_{GS} - V_T)/(L E_{CRIT})$ in full velocity saturation when this term, introduced in Section 2.4.3 as the velocity saturation factor, $VSF = (V_{GS} - V_T)/(LE_{CRIT})$, considerably exceeds unity. Drain current in full velocity saturation was described in Section 2.4.3 and is repeated at the bottom of Table 3.13 with a modified velocity saturation factor, $VSF' = (V_{GS} - V_T)/(LE_{CRIT})'$. This considers the additional reduction of drain current due to VFMR described below in Section 3.7.1.5. In full velocity saturation, drain current is linearly dependent on $V_{GS} - V_T$ and is nearly independent of channel length. Drain current is independent of channel length if VFMR effects are negligible compared to velocity saturation effects. Under normal strong-inversion, square-law operation, drain current is dependent on the square of $V_{GS} - V_T$ and is inversely proportional to channel length.

Velocity saturation is significant for short-channel devices operating at high $V_{EFF} = V_{GS} - V_T$ because $E_x \approx (V_{GS} - V_T)/L$ is large due to a high voltage appearing across a short channel length. Velocity saturation is especially significant for short-channel nMOS devices because of lower E_{CRIT} associated with higher mobility and the resulting higher carrier velocity.

3.7.1.3 With VFMR effects

As introduced in Section 2.4.3, another loss of drain current occurs for devices operating at high $V_{EFF} = V_{GS} - V_T$, including long-channel devices having little velocity saturation. At high $V_{EFF} = V_{GS} - V_T$, carriers are attracted to interface states and other imperfections at the Si–SiO$_2$ interface due to the high, vertical, electric field between the gate and channel. This causes a reduction in carrier mobility that is referred to here as vertical field mobility reduction or VFMR to differentiate it from the loss of mobility associated with velocity saturation in the high, horizontal, drain–source electric field. VFMR becomes more severe as the vertical electric field increases, which is roughly equal to the effective gate–source voltage divided by the gate-oxide thickness, or $E_y \approx (V_{GS} - V_T)/t_{ox}$.

VFMR is modeled by denominator terms of $1 + \theta(V_{GS} - V_T)$ in the drain current correction factors of Table 3.13. θ is the mobility reduction factor, which has values of 0.28 and 0.35 per volt for nMOS and pMOS devices, respectively, for $t_{ox} = 4.1\,\text{nm}$ in the 0.18 µm CMOS process described in Table 3.2. θ generally increases as t_{ox} decreases in smaller-geometry processes, but the overall VFMR effect is somewhat compensated for by the lower values of $V_{GS} - V_T$ required for lower supply voltages. θ is often higher for pMOS devices.

Non-zero V_{SB} (in the normal reverse bias, source–body direction) reduces mobility slightly, which can be modeled by an additional mobility reduction coefficient, θ_{SB}, multiplied by V_{SB}. Including this compensates for the otherwise false increase in mobility associated with increasing V_T when V_{SB} is present [12, p. 72, 15, p. 187]. Here, we exclude the V_{SB} effect to permit simple expressions for design intuition and optimization. This effect is reduced in low-voltage processes where V_{SB} must necessarily be small and would be included in computer simulations of a candidate design.

VFMR dominates for devices, even short-channel devices, in the deep ohmic, linear, or triode region ($V_{DS} \ll V_{DS,sat}$) at high $V_{EFF} = V_{GS} - V_T$. Here, VFMR can be isolated from velocity saturation since carrier velocity is low in the low, horizontal, drain–source electric field associated with low drain–source voltage. VFMR effects also dominate for long-channel devices in saturation ($V_{DS} > V_{DS,sat}$) where carrier velocity is low in the low electric field associated with a long channel length. VFMR effects are more noticeable for pMOS devices because of higher θ and less velocity saturation effects due to higher E_{CRIT} associated with lower mobility and lower carrier velocity.

3.7.1.4 With velocity saturation and VFMR effects

Drain current reduction due to velocity saturation and VFMR is predicted by the simple, enhanced, and complex drain current correction factors listed in Table 3.13. These will be discussed next and then compared using measured data.

The complex drain current correction factor of Table 3.13 may make the most physical sense of the correction factors. Here, the drop in mobility and drain current due to VFMR is modeled by the multiplication of $1 + \theta(V_{GS} - V_T)$ in the denominator. The drop in mobility and drain current due to velocity saturation is modeled separately by the multiplication of $\{1 + [(V_{GS} - V_T)/(LE_{CRIT})]^\alpha\}^{1/\alpha}$ from Equation 3.11 in the denominator, with E_{CRIT} replaced by $E_{CRIT0}[1 + \theta(V_{GS} - V_T)]$. Here, E_{CRIT} is increased from its low-field value, E_{CRIT0}, associated with low-field mobility, μ_0, due to the VFMR reduction of mobility. As mobility decreases due to VFMR, carrier velocity would be expected to drop, increasing the value of E_{CRIT}. The complex correction factor predicts drain current within 4 % of measured $L = 0.18 \mu m$, nMOS data for a velocity saturation, transition exponent of $\alpha = 1.7$.

The enhanced drain current correction factor of Table 3.13 is identical to the complex correction factor except the velocity saturation term $\{1 + [(V_{GS} - V_T)/(LE_{CRIT})]^\alpha\}^{1/\alpha}$ appears directly in the denominator. When $\alpha = 1$, the enhanced correction factor denominator is recognized as $[1 + \theta(V_{GS} - V_T)] \cdot [1 + (V_{GS} - V_T)/(LE_{CRIT})]$, which is the product of the VFMR correction term and a commonly used velocity saturation correction term [26, pp. 62 and 65]. The value of E_{CRIT} is likely to be selected above the value associated with μ_0 to consider an average critical field increase caused by the reduction of mobility and carrier velocity due to VFMR. The enhanced correction factor predicts drain current within 4 % of measured $L = 0.18 \mu m$, nMOS data for a velocity saturation, transition exponent of $\alpha = 1.3$.

Unlike the complex and enhanced factors, the simple drain current correction factor of Table 3.13 does not consider the multiplication of separate VFMR and velocity saturation correction terms. In the simple correction factor, $\alpha = 1$ is assumed, and VFMR and velocity saturation corrections are added in the denominator term giving $1 + [\theta + 1/(LE_{CRIT})] \cdot (V_{GS} - V_T)$. The mathematical addition of these effects is recognized in other predictions of drain current [27, p. 589]. As for the enhanced correction factor, the value of E_{CRIT} is likely to be selected above the value associated with μ_0 to consider an average critical field increase caused by the reduction of mobility and carrier velocity due to VFMR. The simple correction factor predicts drain current within 4 % of measured $L = 0.18 \mu m$, nMOS data, except at the onset of velocity saturation effects where it under-predicts drain current by 13 %.

Figure 3.10 compares drain current predicted by the simple, enhanced, and complex correction factors listed in Table 3.13 with measured data for a $W = 3.2 \mu m$, $L = 0.18 \mu m$ nMOS device in the $0.18 \mu m$ CMOS process described in Table 3.2. Drain current without velocity saturation and VFMR effects is predicted using the continuous drain current expression in the table with process parameters described in the figure and later in Section 3.7.1.6. All three correction factors use $\theta = 0.28/V$ to model VFMR. The simple and enhanced correction factors use $E_{CRIT} = 5.6 V/\mu m$ to model velocity saturation, while the complex correction factor uses $E_{CRIT0} = 3.6 V/\mu m$. A lower level of critical field is used because the complex correction factor increases the critical field with the degree of VFMR present as mentioned earlier. Finally, the enhanced and complex correction factors use $\alpha = 1.3$ and 1.7, respectively, to model the transition into velocity saturation. The lower fitted value of α for the enhanced correction factor may result from the use of a fixed, average E_{CRIT} compared to an E_{CRIT} that increases with VFMR in the complex correction factor. The simple correction factor uses $\alpha = 1$, which cannot be adjusted.

As mentioned, all three drain current correction factors give predicted drain current within 4 % of measured values, except for the simple correction factor that under-predicts drain current by 13 % at the transition into velocity saturation. This is a result of the velocity saturation transition exponent of $\alpha = 1$, which gives a faster roll-off of carrier velocity and drain current compared to higher values used in the enhanced and complex correction factors. All correction factors provide a good estimation of velocity saturation and VFMR effects that result in a significant 70 % drop of drain current (corresponding to correction factor values of 30 %) at the maximum gate–source voltage. This is primarily the result of velocity saturation for the short-channel nMOS device considered. The substantial drop of drain current seen in Figure 3.10 illustrates the importance of considering velocity saturation and VFMR effects in the predictions of drain current and the corresponding effective gate–source voltage (Section 3.7.2.2) and transconductance efficiency (Section 3.8.2.2).

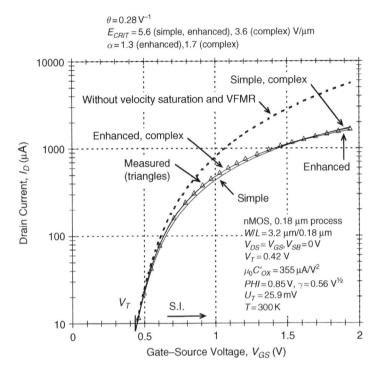

Figure 3.10 Comparison of measured and predicted drain current using simple, enhanced, and complex drain current, correction factors for velocity saturation and VFMR. For the $L = 0.18\,\mu m$, nMOS device shown from a $0.18\,\mu m$ CMOS process, predicted drain current is within 4% of measured values except at the transition into velocity saturation effects where the simple correction factor under-predicts drain current by 13%. At the maximum gate–source voltage, drain current is down 70% from the value predicted without velocity saturation and VFMR

While the simple drain current correction factor does not predict drain current as accurately as the enhanced and complex correction factors, it does permit the development of simple hand analysis expressions inclusive of velocity saturation and VFMR effects. Since the simple correction factor under-predicts drain current at the onset of velocity saturation effects, the resulting hand analysis expressions conservatively over-predict the effective gate–source voltage and under-predict the transconductance efficiency. Here, accuracy is traded off to permit usable circuit prediction and optimization expressions. Full accuracy is obtained during computer simulations using production MOS models as required to validate a candidate design.

For many applications, the simple correction factor, where VFMR and velocity saturation effects are mathematically added, is nearly equal to the enhanced correction factor, where these effects are mathematically multiplied. This is easily seen when the velocity saturation, transition exponent α is set to unity. For α of unity and the substitution $V_{EFF} = V_{GS} - V_T$, the enhanced correction factor, located on the left and recognized in practice [26, pp. 62 and 65], can be compared to the simple correction factor, located on the right and recognized in practice [27, p. 589], as

$$\frac{1}{[1+\theta V_{EFF}]\left[1+\left(\dfrac{V_{EFF}}{L\,E_{CRIT}}\right)\right]} \lesssim \frac{1}{1+\left[\theta+\dfrac{1}{L\,E_{CRIT}}\right]V_{EFF}} \tag{3.12}$$

Mathematical expansion of the enhanced correction factor on the left gives

$$
\cfrac{1}{1+\left[\theta+\cfrac{1}{L\,E_{CRIT}}\right]V_{EFF}+\left(\cfrac{\theta}{L\,E_{CRIT}}\right)V_{EFF}^2} \lesssim \cfrac{1}{1+\left[\theta+\cfrac{1}{L\,E_{CRIT}}\right]V_{EFF}}
\tag{3.13}
$$

The enhanced correction factor (left side of Equations 3.12 and 3.13) is nearly equal to the simple correction factor (right side of the equations) if

$$
\left(\frac{\theta}{L\,E_{CRIT}}\right)V_{EFF}^2 \ll 1+\left[\theta+\frac{1}{L\,E_{CRIT}}\right]V_{EFF}
\tag{3.14}
$$

This condition is met for most short- and long-channel devices, especially if $V_{EFF} < 1$ V as required for most low-voltage applications.

For example, for $E_{CRIT} = 5.6$ V/μm, $\theta = 0.28$/V, $L = 0.18\,\mu$m, and $V_{EFF} = 0.25$ V (near the onset of strong inversion) for nMOS devices in the 0.18 μm CMOS process described in Table 3.2, the enhanced correction factor (with $\alpha = 1$) on the left side of Equations 3.12 and 3.13 is equal to 0.749, while the simple correction factor on the right side is 1.3 % higher at 0.759. The difference between these factors becomes even less as channel length increases. As a result, the simple correction factor that adds velocity saturation and VFMR effects often provides an accurate replacement for the enhanced correction factor that multiplies these effects.

Multiplication of velocity saturation and VFMR effects likely makes more physical sense compared to adding these effects if mobility used in Equation 3.11 is lowered due to VFMR in the vertical electric field prior to lowering due to velocity saturation in the horizontal electric field. While both velocity saturation and VFMR effects are present for devices in saturation, these effects cannot be rigorously isolated and added or multiplied in simple mathematical terms as is common practice for design guidance [26, pp. 62 and 65, 27, p. 598] and done here. The drain current correction factors in Table 3.13 provide an estimation of velocity saturation and VFMR effects and are not intended to replace production MOS models used in computer simulation or represent actual physical MOS behavior. The drain current correction factors contain a velocity saturation component that is channel length dependent, as expected, while the VFMR component is independent of length. Source resistive degeneration, caused by bulk source and contact resistance, can additionally be implicitly included in the velocity saturation correction since velocity saturation itself can be modeled by the addition of source resistance [26, p. 63].

3.7.1.5 The equivalent velocity saturation voltage

It is useful to combine velocity saturation and VFMR effects in the simple drain current correction factor of Table 3.13 using a single, equivalent, velocity saturation voltage as detailed in the table. This voltage, $(LE_{CRIT})'$, simultaneously models velocity saturation and VFMR effects and is equal to the velocity saturation voltage, LE_{CRIT}, in parallel with the mobility reduction voltage, $1/\theta$. As seen in the table, multiplying V_{EFF} by $[\theta+1/(LE_{CRIT})]$ gives the same value as dividing V_{EFF} by $(LE_{CRIT})'$. This simplifies drain current correction to division by $1+VSF' = 1+V_{EFF}/(LE_{CRIT})'$, or division by unity plus the ratio of voltages V_{EFF} and $(LE_{CRIT})'$. This ratio of voltages or $VSF' = V_{EFF}/(LE_{CRIT})'$ is the velocity saturation factor including VFMR effects, which is extended from the velocity saturation factor, $VSF = V_{EFF}/(LE_{CRIT})$, given in Section 2.4.3 for velocity saturation effects alone.

The equivalent velocity saturation voltage, in addition to permitting easy prediction of drain current reduction, permits a rapid assessment of the relative effects of velocity saturation and VFMR. Since the equivalent velocity saturation voltage is the parallel combination of the velocity saturation and mobility reduction voltages, the lower voltage dominates. For example, an $L = 0.18\,\mu$m, nMOS device in the 0.18 μm CMOS process described in Table 3.2 has a velocity saturation voltage of $LE_{CRIT} = 0.18\,\mu$m $\cdot$ 5.6 V/μm $= 1$ V and a mobility reduction voltage of $1/\theta = 1/(0.28$/V$) = 3.6$ V.

The velocity saturation voltage dominates giving a parallel combination of $(LE_{CRIT})' = 0.783\,\text{V}$. The drain current reduction for $V_{EFF} = 0.25\,\text{V}$ (near the onset of strong inversion), as an example, is then found by dividing by $1 + VSF' = 1 + V_{EFF}/(LE_{CRIT})'$ or $1 + 0.25\,\text{V}/0.783\,\text{V}$, giving a drain current reduction by the division of 1.32. In contrast, an $L = 4\,\mu\text{m}$, pMOS device in the $0.18\,\mu\text{m}$ CMOS process has a typical velocity saturation voltage of $LE_{CRIT} = 4\,\mu\text{m} \cdot 14\,\text{V}/\mu\text{m} = 56\,\text{V}$ and a mobility reduction voltage of $1/\theta = 1/(0.35/\text{V}) = 2.86\,\text{V}$. Here, the mobility reduction voltage dominates giving a parallel combination of $(LE_{CRIT})' = 2.72\,\text{V}$. The drain current reduction for $V_{EFF} = 0.25\,\text{V}$ is found by dividing by $1 + VSF' = 1 + V_{EFF}/(LE_{CRIT})'$ or $1 + 0.25\,\text{V}/2.72\,\text{V}$, giving a drain current reduction by the division of 1.09.

While velocity saturation clearly dominates for short-channel devices, especially for nMOS devices having low E_{CRIT}, the VFMR component is important to include. This is especially true for long-channel devices where VFMR is more dominant than velocity saturation. VFMR causes a falloff of drain current, even for long-channel devices, and can be thought of as a small-geometry (thin gate oxide), high-field effect versus a short-channel effect.

Minimizing the ratio of V_{EFF} to the velocity saturation voltage, $V_{EFF}/(LE_{CRIT})$, minimizes the velocity saturation reduction of drain current. This can be done by limiting the value of V_{EFF}, increasing L, or using pMOS devices that have higher E_{CRIT}. Minimizing the ratio of V_{EFF} to the mobility reduction voltage, $V_{EFF}/(1/\theta)$, or the product of $V_{EFF} \cdot \theta$ minimizes VFMR of drain current. This can be done only by limiting the value of V_{EFF}, although using nMOS devices reduces VFMR if θ is meaningfully smaller. Channel length can be increased to increase the velocity saturation voltage and reduce velocity saturation effects, but the mobility reduction voltage and VFMR effects are fixed by the process.

3.7.1.6 Predicted and measured values

Figure 3.11 shows the predicted and measured drain current for $L = 0.18$ and $4\,\mu\text{m}$, nMOS devices having shape factors of $S = W/L = 16$ in the $0.18\,\mu\text{m}$ CMOS process described in Table 3.2. Drain current is predicted by the continuous drain current expression in Table 3.13 multiplied by the enhanced correction factor given in the table. The substrate factor, n, is found from the expressions in Table 3.6. Drain current is predicted using $\mu_0 = 422\,\text{cm}^2/\text{V} \cdot \text{s}$, $C'_{OX} = 8.41\,\text{fF}/\mu\text{m}^2$, $k_0 = 355\,\mu\text{A}/\text{V}^2$, $\gamma = 0.56\,\text{V}^{1/2}$, $PHI = 2\phi_F = 0.85\,\text{V}$, $E_{CRIT} = 5.6\,\text{V}/\mu\text{m}$, $\theta = 0.28/\text{V}$, $V_{TO} = 0.42\,\text{V}$, and $\alpha = 1.3$ from the process parameters in Table 3.2. Threshold voltage is adjusted slightly to match that of the test devices. Channel length is adjusted from the drawn values given by subtracting $DL = 0.028\,\mu\text{m}$, which is also listed in the table.

At high V_{GS} values, drain current for the short-channel, $L = 0.18\,\mu\text{m}$ device decreases well below that of the long-channel, $L = 4\,\mu\text{m}$ device due to velocity saturation. Drain current for the long-channel, $L = 4\,\mu\text{m}$ device, however, decreases below that predicted without velocity saturation and VFMR, due primarily to VFMR. The predicted and measured that are within 4 % for the short-channel, $L = 0.18\,\mu\text{m}$ device for a V_{GS} range of 0.2 V to the process maximum of 1.8 V. Predicted and measured values are within 6 % for the long-channel, $L = 4\,\mu\text{m}$ device except in deep weak inversion at $V_{GS} < 0.25\,\text{V}$ or $V_{EFF} < -170\,\text{mV}$ $(V_{GS} - V_T$, with $V_T = 0.42\,\text{V})$. Predicted and measured values are within 12 % for an intermediate channel-length, $L = 0.48\,\mu\text{m}$ device having measured values shown on the figure.

Figure 3.12 shows predicted and measured drain current for $L = 0.18$ and $4\,\mu\text{m}$, pMOS devices having shape factors of $S = W/L = 76$ in the $0.18\,\mu\text{m}$ CMOS process described in Table 3.2. Drain current is predicted using $\mu_0 = 89.2\,\text{cm}^2/\text{V} \cdot \text{s}$, $C'_{OX} = 8.41\,\text{fF}/\mu\text{m}^2$, $k_0 = 75\,\mu\text{A}/\text{V}^2$, $\gamma = 0.61\,\text{V}^{1/2}$, $PHI = 2\phi_F = 0.85\,\text{V}$, $E_{CRIT} = 14\,\text{V}/\mu\text{m}$, $\theta = 0.35/\text{V}$, $V_{TO} = 0.42\,\text{V}$, and $\alpha = 1.3$ from the process parameters in the table. Threshold voltage is, again, adjusted slightly to match that of the test devices. Channel length is adjusted from the drawn values given by subtracting $DL = 0.051\,\mu\text{m}$, which is also listed in the table.

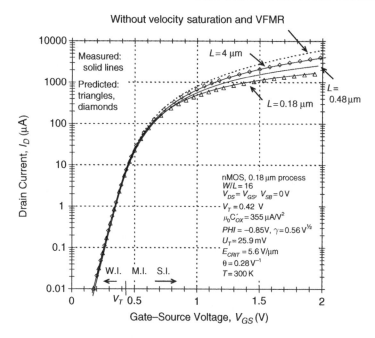

Figure 3.11 Predicted and measured drain current versus gate–source voltage for $L = 0.18$ and $4\,\mu m$, nMOS devices in a $0.18\,\mu m$ CMOS process. Measured drain current is also shown for an $L = 0.48\,\mu m$ device. At high gate–source voltages (also high $V_{EFF} = V_{GS} - V_T$), drain current is reduced significantly for the short-channel, $L = 0.18\,\mu m$ device primarily due to velocity saturation. Drain current is reduced for the long-channel, $L = 4\,\mu m$ device primarily due to VFMR

At high V_{GS} values, drain current for the short-channel, $L = 0.18\,\mu m$ device decreases below that of the long-channel, $L = 4\,\mu m$ device, but the decrease is considerably less than the decrease for the $L = 0.18\,\mu m$, nMOS device shown in Figure 3.11. This is because velocity saturation effects are lower for pMOS devices due to their higher E_{CRIT} associated with lower mobility and lower carrier velocities. Drain current for the long-channel, $L = 4\,\mu m$ device, like the $L = 4\,\mu m$, nMOS device shown in Figure 3.11, decreases below drain current predicted without velocity saturation and VFMR, primarily due to VFMR. Predicted and measured drain currents are within 3 % for the short-channel, $L = 0.18\,\mu m$ device for a V_{GS} range of 0.2 V to the process maximum of 1.8 V. Predicted and measured values are within 8 % for the long-channel, $L = 4\,\mu m$ device except in deep weak inversion at $V_{GS} < 0.31$ V or $V_{EFF} < -110\,mV$ ($V_{GS} - V_T$, with $V_T = 0.42$ V). Predicted and measured values are also within 8 % over the same V_{GS} range for an intermediate channel-length, $L = 0.48\,\mu m$ device not shown on the figure. As shown in Figure 3.9, V_{GS}, V_{EFF}, and V_T values are treated positively (except for V_{EFF} when V_{GS} is below V_T) for pMOS devices as they are for nMOS devices.

The predicted drain current without velocity saturation and VFMR effects is nearly identical for both nMOS and pMOS devices shown in Figures 3.11 and 3.12. This is because the product of nMOS device shape factor and transconductance factor ($16 \cdot 355\,\mu A/V^2 = 5680\,\mu A/V^2$) is nearly equal to that of pMOS devices ($76 \cdot 75\,\mu A/V^2 = 5700\,\mu A/V^2$) and threshold voltages are nearly equal at $V_T = 0.42$ V. As a result, the nMOS and pMOS drain currents of Figures 3.11 and 3.12 can be overlaid to assess relative velocity saturation and VFMR effects. As observed, drain current reduction for the $L = 0.18\,\mu m$, nMOS device is considerably greater than that of the $L = 0.18\,\mu m$, pMOS device because of greater nMOS velocity saturation. Drain current reduction for the $L = 4\,\mu m$, pMOS device,

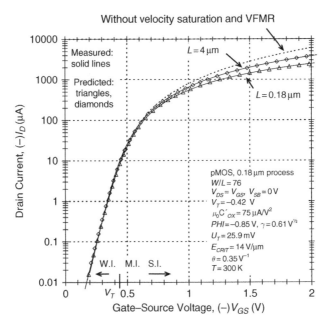

Figure 3.12 Predicted and measured drain current versus gate–source voltage for $L = 0.18$ and $4 \mu m$, pMOS devices in a $0.18 \mu m$ CMOS process. Velocity saturation reduction of drain current for these devices is less than the nMOS devices shown in Figure 3.11. Drain current reduction due to VFMR, however, is significant for both the short- and long-channel pMOS devices

however, is slightly greater than that of the $L = 4 \mu m$, nMOS device because of increased pMOS VFMR associated with a slightly higher value of θ.

Interestingly, the reduction of pMOS drain current due to velocity saturation is nearly equal to that of nMOS drain current if pMOS channel length is decreased by a factor of 2.5 $((14 \, V/\mu m)/(5.6 \, V/\mu m))$, the ratio of pMOS and nMOS E_{CRIT}. This can be seen where the $L = 0.18 \mu m$, pMOS current in Figure 3.12 is nearly equal to the $L = 0.48 \mu m$, nMOS current in Figure 3.11.

Predicted and measured drain current were also compared for nMOS and pMOS devices in the $0.5 \mu m$ CMOS process described in Table 3.2. Agreement between predicted and measured drain current is similar to that of the $0.18 \mu m$ CMOS process devices, even up to a much higher maximum process V_{GS} of 5 V. Interestingly, the $L = 0.5 \mu m$, nMOS device in the $0.5 \mu m$ process, which has three times the channel length of the $L = 0.18 \mu m$ device in the $0.18 \mu m$ process, experiences comparable velocity saturation as the $L = 0.18 \mu m$ device. This is because the maximum $V_{EFF} = 5 - 0.65 = 4.35 \, V$ for the $0.5 \mu m$ process is nearly three times greater than $V_{EFF} = 1.8 - 0.42 = 1.38 \, V$ for the $0.18 \mu m$ process. Because of the much higher V_{EFF}, devices in the $0.5 \mu m$ process actually experience more VFMR than devices in the $0.18 \mu m$ process, even though θ for the $0.18 \mu m$ process is nearly double that of the $0.5 \mu m$ process. If the maximum or usable V_{EFF} in analog circuits decreases with supply voltage in smaller-geometry processes as minimum channel length decreases and the mobility reduction factor increases, the relative degree of velocity saturation and VFMR remains comparable.

The agreement, generally within 8 % and in many cases much better, between predicted and measured drain current in the 0.18 and $0.5 \mu m$ processes is surprising for operation from weak through strong inversion over five or more decades of drain current. This is especially surprising since only a few, fixed (e.g., not binned for channel length) process parameters are used for drain current prediction over the wide range of channel lengths. Additionally, a fixed threshold voltage is assumed (even in weak inversion), no channel length modulation is considered, and no source bulk or contact resistance is considered. The threshold voltage magnitude change with increasing drain–source voltage is estimated at -8 and $-10\,\text{mV/V}$ (Table 3.2) for minimum channel length nMOS and pMOS devices in the $0.18\,\mu\text{m}$ process due to drain-induced barrier lowering (DIBL). Since the drain–source voltage increased significantly during the measurements (it was set equal to the gate–source voltage to ensure uniform saturation operation), significant increases in drain current were expected due to the drop in threshold voltage. However, neither DIBL (Section 3.8.4.2) nor channel length modulation (Section 3.8.4.1) increases in drain current were considered in the simple drain current modeling.

While drain current was predicted using the enhanced correction factor of Table 3.13, similar accuracy is obtained using the simple correction factor, which is used later for the development of V_{EFF} (Section 3.7.2.2) and g_m/I_D (Section 3.8.2.2) expressions. As shown in Figure 3.10, the simple correction factor under-predicts drain current by about 13 % at the transition into velocity saturation. This is primarily observed for short-channel nMOS devices, which experience the most velocity saturation.

3.7.1.7 The extrapolated threshold voltage

As mentioned at the beginning of Section 3.7.1, current biasing is used throughout this book to support MOS operation at a selected drain current, inversion coefficient, and channel length without regard to the threshold voltage, V_T, which varies widely across device geometry, voltage bias (V_{SB} and, for short-channel devices, V_{DS}), temperature, and process. However, the threshold voltage appears in the drain current expressions of Tables 3.6 and 3.13 and is required for the prediction of the gate–source voltage, $V_{GS} = V_{EFF} + V_T$, from the effective gate–source voltage, $V_{EFF} = V_{GS} - V_T$, which is developed in the next section.

As described in [15, pp. 196–197], it is important to clarify the meaning of the threshold voltage used in this book. Here, as in [15, pp. 196–197], the threshold voltage is taken to mean the extrapolated threshold voltage corresponding to the fitting of the drain current, gate–source voltage characteristics. The threshold voltage is the extrapolated gate–source voltage for zero drain current from the strong-inversion, square-law drain current, which excludes small-geometry effects like velocity saturation and VFMR. The strong-inversion drain current without small-geometry effects is given in Tables 3.6 and 3.13, and Figure 2.3 shows an example of the extrapolated threshold voltage from the strong-inversion, square-law drain current. Additionally, the threshold voltage can be extracted from the drain current, gate–source voltage characteristic in weak inversion or the continuous drain current expression, both of which are also given in Tables 3.6 and 3.13. Substituting $V_{GS} = V_T$ in the continuous drain current expression reveals that $I_D = 2n\mu C_{OX}' U_T^2(W/L) \cdot \ln^2(2) = 2n\mu C_{OX}' U_T^2(W/L) \cdot 0.48$. From the traditional inversion coefficient definition listed in Table 3.6, this corresponds to an inversion coefficient of 0.48. This is also essentially the value for the fixed–normalized inversion coefficient definition used in this book because n and μ have changed little near the center of moderate inversion. Thus, operation at the threshold voltage corresponds to operation at a drain current of approximately one-half the specific value, I_D (moderate) $= 2n\mu C_{OX}' U_T^2(W/L)$ given by Equation 3.2, associated with operation in the center of moderate inversion where the inversion coefficient is unity. It is important to note that operation at the threshold voltage corresponds to operation neither in weak nor strong inversion, but operation in moderate inversion.

For experimental extraction, the threshold voltage can be found from the value of the gate–source voltage giving $I_D = 2n\mu C_{OX}' U_T^2(W/L) \cdot 0.48$. Because of the threshold voltage increase with increasing V_{SB} (Figure 3.70) and decrease with temperature (Figure 3.6), it is important to specify V_{SB} and the temperature for threshold-voltage extraction. Additionally, because of the decrease in threshold voltage with increasing V_{DS} for small channel lengths associated with DIBL (Section 3.8.4.2), and the threshold-voltage sensitivity to small values of channel width and length, it is important to specify V_{DS} and geometry for small-geometry devices. Table 3.2 gives values of the zero-V_{SB} threshold voltage, V_{TO}, for large-geometry devices in 0.5, 0.35, and 0.18 μm CMOS processes. As observed, the threshold voltage decreases in smaller-geometry processes, but it does not decrease as rapidly as the process scaling.

3.7.2 Effective Gate–Source Voltage

The effective gate–source voltage, $V_{EFF} = V_{GS} - V_T$, represents how much the operating gate–source voltage is above, if V_{EFF} is positive, or below, if V_{EFF} is negative, the MOS threshold voltage. V_{EFF} is important since it controls the operating gate–source voltage, $V_{GS} = V_{EFF} + V_T$, which must be known to ensure sufficient bias compliance, especially in low-voltage designs. In this book, the polarity of V_{EFF} is taken to be the same for both nMOS and pMOS devices, since as mentioned earlier and described in Figure 3.9, voltage polarities for both nMOS and pMOS devices are defined positively.

3.7.2.1 Without small-geometry effects

V_{EFF} from weak through strong inversion can be derived from the continuous drain current expression. This expression, excluding reductions in mobility, μ, due to VFMR and velocity saturation, is repeated from Table 3.13 as

$$I_D = 2n\,\mu_0\,C'_{OX}\,U_T^2\left(\frac{W}{L}\right)\left[\ln\left(1+e^{\frac{V_{GS}-V_T}{2nU_T}}\right)\right]^2 \tag{3.15}$$

Solving for V_{EFF} in terms of the inversion coefficient requires expressing drain current in terms of the inversion coefficient. From Equation 3.6, which is also summarized in Table 3.6, drain current can be expressed as

$$I_D = 2n_0\,\mu_0\,C'_{OX}\,U_T^2\left(\frac{W}{L}\right)\cdot IC \tag{3.16}$$

where in this book, the fixed–normalized inversion coefficient, described earlier in Sections 3.4.2.2 and 3.4.2.3, is used and referred to as *IC*. *IC* and drain current track each other for a fixed-geometry device since the substrate factor, n, and μ are assumed fixed at moderate inversion and low-field values, respectively. In the next section, the increase in V_{EFF} due to mobility reduction associated with VFMR and velocity saturation is evaluated outside the *IC* definition, which does not include these effects.

Substituting the drain current given by Equation 3.16 into Equation 3.15 assuming n constant at its moderate inversion value of n_0 and solving for *IC* gives

$$IC = \left[\ln\left(1+e^{\frac{V_{GS}-V_T}{2nU_T}}\right)\right]^2 \tag{3.17}$$

This equation shows the linkage between *IC* and $V_{EFF} = V_{GS} - V_T$, again, excluding mobility reduction associated with VFMR and velocity saturation. Interestingly, if variable n and μ are used in Equations 3.15 and 3.16, corresponding to the traditional inversion coefficient, *IC'*, and in drain current expressions of Table 3.6, the variable n and μ divide out in Equation 3.17. This gives a fixed relationship between *IC'* and $V_{EFF} = V_{GS} - V_T$, regardless of mobility reduction due to VFMR and velocity saturation. However, *IC* is more convenient to use since it is linearly linked to drain current and sizing, and independent of variations in n and μ. When mobility is close to its low-field value, μ_0, $IC \approx IC'$ as described in Section 3.4.2.3 with only a slight difference due to small variations in n.

Solving Equation 3.17 for V_{EFF} gives

$$V_{EFF} = 2nU_T \ln \left(e^{\sqrt{IC}} - 1 \right) \tag{3.18}$$

Table 3.14 lists expressions for V_{EFF} along with dependencies on the design choices of inversion coefficient and channel length. V_{EFF} excluding velocity saturation and VFMR effects is included as the first expression from Equation 3.18. Since $e^{\sqrt{IC}} \approx 1 + \sqrt{IC}$ for $\sqrt{IC} \ll 1$, V_{EFF} approaches $2nU_T \ln \sqrt{IC}$ or $nU_T \ln(IC)$ in Equation 3.18 in weak inversion as noted in the table. This is also the value of V_{EFF} found from the weak-inversion drain current of Table 3.13 with drain current expressed from the inversion coefficient in Equation 3.16. $V_{EFF} = nU_T \ln(IC)$ is consistent with weak-inversion, exponential drain current operation (introduced in Section 2.4.1) where V_{EFF} increases as the logarithm of drain current or inversion coefficient. V_{EFF} and V_{GS} increase by the subthreshold swing, $S = \ln(10)nU_T = 2.303nU_T$, for each decade increase in drain current or inversion coefficient. $S = 90 \, \text{mV/decade}$ at room temperature ($T = 300 \, \text{K}$) for devices in a typical CMOS process having $n = 1.5$ in weak inversion.

Since $e^{\sqrt{IC}} \gg 1$ for $\sqrt{IC} \gg 1$, V_{EFF} approaches $2nU_T\sqrt{IC}$ in Equation 3.18 in strong inversion as noted in the first expression of Table 3.14. This is also the value of V_{EFF} found from the strong-inversion drain current (excluding small-geometry effects) of Table 3.13 with drain current expressed from the inversion coefficient in Equation 3.16. $V_{EFF} = 2nU_T\sqrt{IC}$ is consistent with strong-inversion,

Table 3.14 MOS effective gate–source voltage, $V_{EFF} = V_{GS} - V_T$, expressions and trends in terms of inversion coefficient and channel length. When evaluated in terms of the inversion coefficient, V_{EFF} is independent of the drain current

Effective gate–source voltage, $V_{EFF} = V_{GS} - V_T$	$IC \uparrow$ L fixed	$L \uparrow$ IC fixed
All regions, without small-geometry effects: $V_{EFF} = 2nU_T \ln \left(e^{\sqrt{IC}-1} \right)$ $\approx nU_T \ln(IC)$ in WI $\approx 2nU_T\sqrt{IC}$ in SI	$\uparrow \propto \ln(IC)$ in WI	Unchanged
$V_{EFF} = -4.5, -2, 0, 1.08, 6.24, 20,$ and $63.2 \cdot nU_T = -163, -72, 0, 40, 225,$ $724,$ and $2290 \, \text{mV}$ for $IC = 0.01, 0.1, 0.48, 1, 10, 100,$ and $1000,$ respectively, assuming $n = 1.4,$ and $U_T = 25.9 \, \text{mV}$ ($T = 300 \, \text{K}$)	$\uparrow \propto \sqrt{IC}$ in SI	
All regions, with velocity saturation and VFMR effects: Replace IC in above expression with $IC \left[1 + \left(\dfrac{IC}{4IC_{CRIT}} \right)^{\beta} \right]^{\frac{1}{\beta}} \approx IC \left[1 + \left(\dfrac{IC}{4IC_{CRIT}} \right) \right]$ where $\beta = 0.7 - 1$ (0.8 for nMOS and 0.9 for pMOS) and $IC_{CRIT} = \left(\dfrac{(LE_{CRIT})'}{4nU_T} \right)^2,$ and $(LE_{CRIT})' = \dfrac{1}{\theta} \| L \, E_{CRIT}$		
Strong inversion with full velocity saturation ($VSF' = V_{EFF}/(LE_{CRIT})' \gg 1,$ or $IC \gg IC_{CRIT}$): $V_{EFF} = \dfrac{(2nU_T)^2 \, IC}{(LE_{CRIT})'} = \dfrac{(LE_{CRIT})' \, IC}{4IC_{CRIT}}$	$\uparrow \propto IC$	$\downarrow \propto 1/L$

V_{EFF} is for saturation where $V_{DS} > V_{DS,sat}$.

square-law, drain current operation (introduced in Section 2.4.2) where V_{EFF} increases as the square root of drain current or inversion coefficient. As noted in the table, V_{EFF} is independent of channel length when velocity saturation effects are excluded. Unfortunately, at high IC, velocity saturation significantly increases V_{EFF} for short-channel devices, while VFMR moderately increases V_{EFF} for devices of any channel length. These effects are discussed in the next section.

As noted in Table 3.14, operation at the threshold voltage ($V_{GS} = V_T$) where $V_{EFF} = 0\,\text{V}$ corresponds to $IC = 0.48$, which is approximately one half $IC = 1$ associated with the center of moderate inversion. Operation at the threshold voltage was described in Section 3.7.1.7.

3.7.2.2 With velocity saturation and VFMR effects

The reduction of drain current associated with velocity saturation and VFMR was discussed in Sections 3.7.1.2–3.7.1.6. This reduction requires an increase in V_{EFF} to maintain the selected drain bias current. We will initially consider the increase in V_{EFF} in strong inversion where the effects of velocity saturation and VFMR are most prevalent.

The increase in V_{EFF} due to velocity saturation and VFMR effects can be estimated by equating drain current with and without these effects. This gives

$$I_D = \frac{\frac{1}{2}\left(\frac{\mu_0 \, C'_{OX}}{n}\right)\left(\frac{W}{L}\right)V_{EFF}^2}{1 + \frac{V_{EFF}}{(LE_{CRIT})'}} = \frac{1}{2}\left(\frac{\mu_0 \, C'_{OX}}{n}\right)\left(\frac{W}{L}\right)V_{EFF,L}^2 \tag{3.19}$$

The left side of the equation gives drain current with velocity saturation and VFMR effects where the effective gate–source voltage present with these effects is denoted in the usual way, V_{EFF}. The right side of the equation gives drain current without these effects where the effective gate–source voltage present without these effects is denoted as $V_{EFF,L}$. $V_{EFF,L}$ refers to the large-geometry value of V_{EFF} where long channel length eliminates velocity saturation effects (equivalent to $LE_{CRIT} \gg V_{EFF}$) and thick gate-oxide thickness eliminates VFMR effects (equivalent to $\theta = 0$). The strong-inversion drain current is from the expression in Table 3.13 using the simple correction factor to model velocity saturation and VFMR effects. As described earlier in Section 3.7.1.5, $(LE_{CRIT})'$ is the equivalent velocity saturation voltage, which is equal to the velocity saturation voltage, LE_{CRIT}, in parallel with the mobility reduction voltage, $1/\theta$.

Solving Equation 3.19 gives

$$V_{EFF} = V_{EFF,L}\sqrt{1 + \frac{V_{EFF}}{(LE_{CRIT})'}} = V_{EFF,L}\sqrt{1 + VSF'} \tag{3.20}$$

where, from Section 3.7.1.5, $VSF' = V_{EFF}/(LE_{CRIT})'$ is the velocity saturation factor including VFMR effects. Equation 3.20 can be solved explicitly in terms of $V_{EFF,L}$ giving

$$V_{EFF} = V_{EFF,L}\left[\sqrt{1 + \left(\frac{1}{2}\right)^2\left(\frac{V_{EFF,L}}{(LE_{CRIT})'}\right)^2} + \left(\frac{1}{2}\right)\frac{V_{EFF,L}}{(LE_{CRIT})'}\right] \tag{3.21}$$

$V_{EFF,L}$ from the previous section and summarized in Table 3.14 is given by

$$V_{EFF,L} = V_{EFF} \text{ (SI, no vel. sat., VFMR)} = 2nU_T\sqrt{IC} \tag{3.22}$$

Substituting this for the inside terms of Equation 3.21 gives

$$V_{EFF} = V_{EFF,L} \left[\sqrt{1 + \left(\frac{1}{2}\right)^2 \frac{IC}{4IC_{CRIT}}} + \left(\frac{1}{2}\right) \sqrt{\frac{IC}{4IC_{CRIT}}} \right] \qquad (3.23)$$

where

$$IC_{CRIT} = \left(\frac{(LE_{CRIT})'}{4nU_T} \right)^2 \qquad (3.24)$$

As developed later in Section 3.8.2.2, IC_{CRIT} is the critical value of the inversion coefficient where transconductance efficiency, g_m/I_D, transitions between values associated with strong-inversion, square-law and velocity-saturated, linear-law drain current.

When $V_{EFF,L} \ll (LE_{CRIT})'$ or $IC \ll 4IC_{CRIT}$, velocity saturation and VFMR increases in V_{EFF} are negligible. Here, Equations 3.21 and 3.23 give $V_{EFF} = V_{EFF,L}$ directly as the bracketed terms have values of unity. When $V_{EFF,L} \gg (LE_{CRIT})'$ or $IC \gg 4IC_{CRIT}$, devices are in full velocity saturation and V_{EFF} is increased considerably from $V_{EFF,L}$ that would be present without velocity saturation and VFMR effects. Under full velocity saturation, Equations 3.22 and 3.23 give

$$V_{EFF} \text{ (SI, full vel. sat.)} = \frac{(2nU_T)^2}{(LE_{CRIT})'} IC = (LE_{CRIT})' \frac{IC}{4IC_{CRIT}} \qquad (3.25)$$

This is also the value of V_{EFF} found by solving the strong-inversion, velocity-saturated drain current of Table 3.13 with drain current expressed from the inversion coefficient in Equation 3.16. Under full velocity saturation, V_{EFF} increases directly with drain current or IC. This is summarized with the V_{EFF} expressions given in Table 3.14 and is consistent with strong-inversion, velocity-saturated drain current introduced in Section 2.4.3. Under full velocity saturation, V_{EFF} increases inversely with declining equivalent, velocity saturation voltage, $(LE_{CRIT})'$, as seen by the middle term of Equation 3.25. While VFMR decreases drain current for all channel lengths, requiring a modest increase in V_{EFF}, velocity saturation dominates for short-channel devices. Under these conditions, V_{EFF} increases inversely with the product of declining channel length and E_{CRIT}, and IC must decrease as channel length decreases to maintain a selected V_{EFF}. Again, the inversion coefficient and drain current track each other for a fixed-geometry device since n and mobility are assumed fixed in the inversion coefficient definition shown in Equation 3.16 and mentioned in Sections 3.4.2.2 and 3.4.2.3. Mobility reduction associated with VFMR and velocity saturation effects is considered outside the inversion coefficient definition.

V_{EFF} given in Equations 3.21 and 3.23 was derived from drain current using the simple correction factor for velocity saturation and VFMR. As a result, these equations over-predict V_{EFF} at the onset of velocity saturation effects because drain current is under-predicted here as shown in Figure 3.10. A better fit with measured data from 0.18 and 0.5 μm CMOS processes is obtained by a modification of Equation 3.23 given by

$$V_{EFF} = V_{EFF,L} \left[\sqrt{1 + \left(\frac{4}{5}\right)^2 \frac{IC}{4IC_{CRIT}}} + \left(\frac{1}{5}\right) \sqrt{\frac{IC}{4IC_{CRIT}}} \right] \qquad (3.26)$$

This gives less initial increase in V_{EFF} over $V_{EFF,L}$ while still approaching the same limit in full velocity saturation.

Equations 3.21, 3.23, and 3.26 permit the estimation of V_{EFF} inclusive of the increase associated with velocity saturation and VFMR effects. V_{EFF} is found by first finding $V_{EFF,L}$, the value present without these effects given by Equation 3.18. $V_{EFF,L}$ is then multiplied by the rightmost term in Equations 3.21,

3.23, or 3.26 to obtain V_{EFF}. While the increase in V_{EFF} due to velocity saturation and VFMR was derived assuming strong-inversion operation, the equations can also be used to approximate the small increase present in moderate inversion.

Predicted V_{EFF} from Equations 3.18 and 3.26 was compared to measured values for nMOS and pMOS devices from the 0.18 and 0.5 μm CMOS processes described in Table 3.2. V_{EFF} was considered from deep weak inversion to deep strong inversion for IC exceeding 100 and 1000 for the 0.18 and 0.5 μm processes, respectively. While predicted values were generally within measured values by 50 mV and were often within 20 mV, the prediction is cumbersome, requiring the prediction of $V_{EFF,L}$ from Equation 3.18 followed by the prediction of V_{EFF} resulting from the multiplication of $V_{EFF,L}$ and Equation 3.26.

A simpler estimate of V_{EFF} can be obtained by using Equation 3.18 with a modified value of IC to include velocity saturation and VFMR effects. This estimate of V_{EFF}, which involves a non-physically based interpolation from strong-inversion, square-law and fully velocity-saturated, linear-law, drain current operation, is described below.

An expression for V_{EFF}, inclusive of velocity saturation and VFMR effects, must provide an interpolation between the values given by Equation 3.22 when these effects are negligible and the values given in Equation 3.25 when full velocity saturation is present. Replacing IC in these equations with

$$IC \text{ (replacement for } V_{EFF}) = IC\left[1 + \left(\frac{IC}{4IC_{CRIT}}\right)^{\beta}\right]^{\frac{1}{\beta}} \approx IC\left[1 + \left(\frac{IC}{4IC_{CRIT}}\right)\right] \qquad (3.27)$$

provides an interpolation that meets these conditions. When $IC \ll 4IC_{CRIT}$, velocity saturation and VFMR effects are negligible and IC is nearly unchanged as it appears in the continuous (Equation 3.18) or strong-inversion (Equation 3.22) equations for V_{EFF}. As a result, these equations give the expected value of V_{EFF} excluding velocity saturation and VFMR effects. However, under full velocity saturation when $IC \gg 4IC_{CRIT}$, IC is replaced with $IC \cdot (IC/(4IC_{CRIT}))$ in Equation 3.18 or 3.22. As a result, these equations give the value of V_{EFF} given in Equation 3.25 when full velocity saturation is present. In Equation 3.27, β models the transition of V_{EFF} without velocity saturation and VFMR effects to the value with these effects. Values of β near unity provide a good fit with measured data.

Like Equations 3.21, 3.23, and 3.26 derived to predict V_{EFF} inclusive of velocity saturation and VFMR effects, the prediction using the modified value of IC in Equation 3.27 was also was derived assuming strong-inversion operation. As for the previous equations, Equation 3.27 can also be used to approximate the small increase in V_{EFF} present in moderate inversion. This will be shown below in comparisons of predicted and measured data.

Figure 3.13 illustrates the use of Equation 3.27 for the prediction of V_{EFF} inclusive of velocity saturation and VFMR effects. Predicted V_{EFF} in strong inversion is overlaid with measured values for an $L = 0.18$ μm, nMOS device in the 0.18 μm CMOS process described in Table 3.2. V_{EFF} is predicted using the continuous, weak through strong-inversion expression given by Equation 3.18, summarized at the top of Table 3.14. To consider velocity saturation and VFMR effects, the value of IC is modified by Equation 3.27, which is also summarized in the table, using a transition exponent of $\beta = 0.8$. A fixed value of $n = 1.33$, thermal voltage of $U_T = 25.9$ mV ($T = 300$ K), $E_{CRIT} = 5.6$ V/μm, and $\theta = 0.28$/V are used from the process values of Table 3.2. Channel length is adjusted from the $L = 0.18$ μm drawn value given by subtracting $DL = 0.028$ μm, which is also listed in the table. The inversion coefficient for measured data is found by $IC = I_D/[I_0(W/L)]$ from Table 3.6, using a technology current of $I_0 = 0.64$ μA from Table 3.2.

In Figure 3.13, V_{EFF} initially increases as the square root of IC, tracking the expected strong-inversion $V_{EFF} = 2nU_T\sqrt{IC}$ from Equation 3.22 for negligible velocity saturation and VFMR effects. As IC increases further, V_{EFF} increases from this value and begins to approach the expected $V_{EFF} = (2nU_T)^2IC/(LE_{CRIT})'$ from Equation 3.25 for full velocity saturation. As seen in the figure, the interception of these expected values occurs at a critical or corner value of IC. This critical value denotes the transition of V_{EFF} associated with square-law and velocity-saturated, linear-law, drain

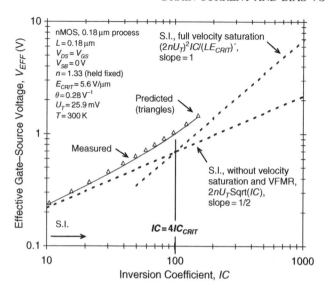

Figure 3.13 Predicted and measured effective gate–source voltage, $V_{EFF} = V_{GS} - V_T$, with trends versus inversion coefficient, IC, for an $L = 0.18\,\mu m$, nMOS device in a $0.18\,\mu m$ CMOS process operating in strong inversion. V_{EFF} initially increases as the square root of IC when velocity saturation and VFMR effects are small. At high levels of IC where velocity saturation is significant for this short-channel device, V_{EFF} increases significantly and is nearly proportional to IC. Operation at $IC = 4IC_{CRIT}$ corresponds to the V_{EFF} transition associated with square-law and velocity-saturated, linear-law, drain current

current operation. This critical value of IC is found by equating Equations 3.22 and 3.25 and solving for IC. This gives

$$IC \ (V_{EFF} \text{ critical value, vel. sat.}) = \left(\frac{(LE_{CRIT})'}{2nU_T}\right)^2 = 4IC_{CRIT} \tag{3.28}$$

As seen in Figure 3.13, the critical value of IC for V_{EFF} occurs at $IC = 4IC_{CRIT} = 100$, which is a factor of four above IC_{CRIT}, the critical value of IC where g_m/I_D transitions between values associated with strong-inversion, square-law and fully velocity-saturated, linear-law drain current as discussed later in Section 3.8.2.2. At $IC = 4IC_{CRIT}$, V_{EFF} is approximately 41 % higher than the value present without velocity saturation and VFMR effects. This is because $IC \approx 2IC$ from Equation 3.27 is placed in Equations 3.18 or 3.22 for V_{EFF}. This gives $V_{EFF} = 2nU_T(2IC)^{1/2}$ compared to $V_{EFF} = 2nU_TIC^{1/2}$ present without velocity saturation and VFMR effects.

In Figure 3.13, V_{EFF} is over-predicted by a maximum of 58 mV up to the maximum available $IC = 152$ where measured $V_{EFF} = 1.42\,V$ results in V_{GS} above the 1.8 V supply voltage limit. Selecting the transition exponent of $\beta = 0.9$ in Equation 3.27 nearly halves the over-prediction, but a value of $\beta = 0.8$ is used as a best fit for all channel lengths. For $IC < 100$, the V_{EFF} prediction generally improves. Here, V_{EFF} is under-predicted by a maximum of 32 mV for $\beta = 1$, which permits simple hand predictions where IC is replaced with $IC \cdot [1 + IC/(4IC_{CRIT})]$. For all considered transition exponents, the prediction of V_{EFF} is adequate to provide design guidance where $V_{EFF} = 1.42\,V$ at the maximum $IC = 152$ is 67 % above $V_{EFF} = 0.848\,V$ predicted without velocity saturation and VFMR effects. This illustrates the importance of considering these effects.

The V_{EFF} versus inversion coefficient behavior shown in Figure 3.13 is also observed for devices having different channel lengths and for devices in different CMOS processes. Measurements for an $L = 0.5\,\mu m$, nMOS device in the $0.5\,\mu m$ CMOS process described in Table 3.2 show a measured critical IC for V_{EFF} at $IC = 4IC_{CRIT} = 500$, which is a factor of five higher than that shown in Figure 3.13 for an $L = 0.18\,\mu$m device. This critical IC is found from Equation 3.28 for a fixed value of $n = 1.35$, thermal voltage of $U_T = 25.9\,$mV ($T = 300\,$K), $E_{CRIT} = 4\,$V/μm, and $\theta = 0.14$/V from Table 3.2 and is higher primarily as a result of longer channel length.

As seen in Equation 3.28, the critical value of IC for V_{EFF} is proportional to the square of the equivalent velocity saturation voltage, $(LE_{CRIT})'$. For short-channel devices where velocity saturation dominates over VFMR, the critical value of IC is nearly proportional to the square of the velocity saturation voltage, LE_{CRIT}, or the square of channel length and E_{CRIT}. As a result, the critical value of IC decreases significantly for short-channel nMOS devices, which have lower E_{CRIT} compared to pMOS devices. This suggests that the critical value of IC might be approximately 13 % (($0.065\,\mu$m/$0.18\,\mu$m$)^2$) of the value of 100 shown in Figure 3.13 for an $L = 0.18\,\mu$m, nMOS device, or 13 for an $L = 0.065\,\mu$m, nMOS device. Such a critical value of IC is only slightly into strong inversion, suggesting a modified interpolation in Equation 3.27 may be required as velocity saturation effects begin to encroach closer to moderate inversion for very short-channel devices.

3.7.2.3 Predicted and measured values

Figure 3.14 shows predicted and measured V_{EFF} for $L = 0.18$, 0.28, 0.48, and $4\,\mu$m, nMOS devices in the $0.18\,\mu$m CMOS process described in Table 3.2. V_{EFF} is predicted as described for Figure 3.13, using the process parameters previously mentioned. V_{EFF} is presented as a function of the inversion coefficient from $IC = 0.01$ (deep weak inversion) to 100 (deep strong inversion) over a wide range of channel lengths, providing a full characterization of the process for analog design.

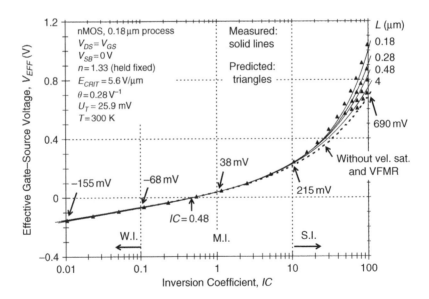

Figure 3.14 Predicted and measured effective gate–source voltage, $V_{EFF} = V_{GS} - V_T$, versus inversion coefficient, IC, for $L = 0.18$, 0.28, 0.48, and $4\,\mu$m, nMOS devices in a $0.18\,\mu$m CMOS process. In weak inversion, V_{EFF} increases as the logarithm of IC before increasing as the square root of IC in strong inversion. For short-channel devices at high IC, V_{EFF} increases significantly due to velocity saturation and is nearly proportional to IC. At high IC, V_{EFF} increases modestly for all channel lengths due to VFMR effects. V_{EFF} is similar for all CMOS processes

V_{EFF} increases linearly in weak inversion for the logarithmic presentation of inversion coefficient. V_{EFF} is -68 mV at the boundary of weak inversion ($IC = 0.1$), increases to zero ($V_{GS} = V_T$) at $IC = 0.48$ in moderate inversion, and increases to 38 mV at the center of moderate inversion ($IC = 1$). V_{EFF} increases proportionally with $\sqrt{IC}$ in strong inversion, reaching 215 mV at the onset of strong inversion ($IC = 10$) before reaching 690 mV at the boundary of deep strong inversion ($IC = 100$). These values are for room temperature ($T = 300$ K), assuming no velocity saturation or VFMR effects, and are slightly less than values listed in Table 3.14 because an average value of $n = 1.33$ is used compared to 1.4 in the table. As shown in Figure 3.2, n decreases slightly from weak through strong inversion. While giving some loss of accuracy, the prediction of V_{EFF} is greatly simplified by assuming a fixed value of n.

Figure 3.14 also shows the increase in V_{EFF} due to velocity saturation and VFMR effects. For the $L = 0.18\,\mu$m device, V_{EFF} increases by 45 % from 0.69 to 1.00 V at $IC = 100$, primarily due to velocity saturation effects. The increases are progressively less as channel length increases, but some increase is always present due to VFMR effects. No velocity saturation increase in V_{EFF} is observed in weak inversion and little increase is observed in moderate inversion. Small velocity saturation effects in moderate inversion, even for short-channel devices, are a major advantage of operation here. The increase of V_{EFF} in strong inversion for short channel lengths, however, is significant and must be considered since this raises the operating gate–source voltage, $V_{GS} = V_{EFF} + V_T$. This must be kept within voltage compliance limits imposed by the supply voltage and circuit topology.

Figure 3.15 shows predicted and measured V_{EFF} for $L = 0.18$ and $4\,\mu$m, pMOS devices in the 0.18 μm CMOS process described in Table 3.2. V_{EFF} is predicted using a fixed value of $n = 1.33$, thermal voltage of $U_T = 25.9$ mV ($T = 300$ K), $E_{CRIT} = 14$ V/μm, and $\theta = 0.35$/V from the process values of Table 3.2. A transition exponent of $\beta = 0.9$ in Equation 3.27 is used compared to 0.8 for the nMOS devices shown in Figures 3.13 and 3.14. Channel length is adjusted

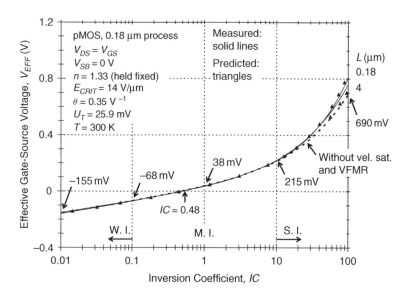

Figure 3.15 Predicted and measured effective gate–source voltage, $V_{EFF} = V_{GS} - V_T$, versus inversion coefficient, IC, for $L = 0.18$ and $4\,\mu$m, pMOS devices in a 0.18 μm CMOS process. V_{EFF} increases less at high IC for short-channel devices compared to the nMOS values shown in Figure 3.14 because of lower pMOS device velocity saturation. At high IC, V_{EFF} increases modestly for all channel lengths due to VFMR effects. V_{EFF} is similar for all CMOS processes

from the drawn values given by subtracting $DL = 0.051\,\mu m$, which is also listed in the table. The inversion coefficient for measured data is found using a technology current $I_0 = 0.135\,\mu A$ from the table.

In weak and moderate inversion ($IC < 10$), pMOS V_{EFF} in Figure 3.15 is nearly identical to that of nMOS devices shown in Figure 3.14. However, the V_{EFF} increase in deep strong inversion ($IC = 100$) is less for short-channel pMOS devices because of less velocity saturation resulting from higher pMOS E_{CRIT}. As a result, pMOS V_{EFF} is similar for both short- and long-channel devices, while increasing modestly for all channel lengths due to VFMR. This is a small-geometry (thin gate oxide), high-field effect present even for long-channel devices. Like the drain current shown in Figures 3.11 and 3.12, pMOS V_{EFF} is nearly equal to that of nMOS devices if pMOS channel length is decreased by a factor of 2.5 (($14\,V/\mu m)/(5.6\,V/\mu m)$), the ratio of pMOS and nMOS E_{CRIT}. This can be seen where the $L = 0.18\,\mu m$, pMOS V_{EFF} in Figure 3.15 is nearly equal to the $L = 0.48\,\mu m$, nMOS value shown in Figure 3.14.

Predicted nMOS V_{EFF} values in Figure 3.14 are within measured values by 26 mV, except for the $L = 0.18\,\mu m$ device operating at $IC = 100$ where V_{EFF} is over-predicted by 58 mV compared to a measured value of 1.00 V. Predicted pMOS values in Figure 3.15 are within measured values by 41 mV. Similar prediction accuracy was also obtained for nMOS and pMOS devices in the 0.5 µm CMOS process described in Table 3.2. As mentioned for Figure 3.13, a transition exponent of $\beta = 1$ permits simple hand analysis. Here, there is less over-prediction of V_{EFF} for short channel lengths but more under-prediction for long channel lengths.

The V_{EFF} prediction described is clearly adequate for design guidance, especially since it is only required, again, to ensure the operating gate–source voltage, $V_{GS} = V_{EFF} + V_T$, is within voltage compliance limits. The prediction accuracy is encouraging since a fixed value of n and only a few, fixed (e.g., not binned for channel length) process parameters are used for prediction over the wide range of inversion level and channel lengths. The simple V_{EFF} prediction described here permits hand or spreadsheet calculation and is valid from weak through strong inversion, inclusive of velocity saturation and VFMR effects.

Predicted and measured V_{EFF} values, like those shown in Figure 3.13, were taken for IC up to values of 150 to 300 depending on the channel length where the supply voltage limit of $V_{GS} = 1.8\,V$ was reached. Data shown in Figures 3.14 and 3.15, however, was limited to IC values below 100 as operation above this is unusually high for a supply voltage of 1.8 V. V_{EFF} was measured for $V_{DS} = V_{GS}$, which corresponds to the bias conditions for a "diode-connected" MOSFET that ensures consistent saturation operation regardless of the level of inversion. Actual operation would likely be for V_{DS} somewhat below V_{GS}, especially for low-voltage design. However, V_{EFF} increases only slightly as V_{DS} decreases as required to support the drain current, providing operation in saturation is maintained. Variations in V_{EFF} with V_{DS} are considered by the required computer simulations of a candidate design using full MOS models.

3.7.2.4 Summary of trends

As mentioned and listed in Table 3.14, $V_{EFF} = nU_T \ln(IC)$ increases logarithmically with IC in weak inversion. It increases modestly in moderate inversion before increasing as the square root of IC in strong inversion, approaching $V_{EFF} = 2nU_T\sqrt{IC}$. V_{EFF} increases modestly above this due to VFMR effects, but increases significantly for short-channel devices experiencing velocity saturation. Under full velocity saturation, V_{EFF} increases directly with IC, approaching $V_{EFF} = (2nU_T)^2 IC/(LE_{CRIT})'$. When velocity saturation effects are negligible, as for long-channel devices, V_{EFF} is independent of the channel length. However, under conditions of full velocity saturation, V_{EFF} increases nearly inversely with channel length. To maintain a given value of V_{EFF} under full velocity saturation, IC must decrease as channel length decreases as seen in the expression for velocity-saturated V_{EFF} above. Finally, V_{EFF} is independent of the selected drain current when evaluated in terms of the inversion coefficient.

In addition to expressions for V_{EFF} and trends in terms of the design choices of inversion coefficient and channel length, Table 3.14 lists predicted values without small-geometry velocity saturation and

VFMR effects. $V_{EFF} = -4.5, -2, 0, 1.08, 6.24, 20,$ and $63.2 \cdot nU_T$ for $IC = 0.01, 0.1, 0.48, 1, 10,$ 100, and 1000, respectively. This corresponds to $V_{EFF} = -163, -72, 0, 40, 225, 724,$ and 2290 mV for $U_T = 25.9$ mV $(T = 300$ K) and a constant value of $n = 1.4$. The increase of V_{EFF} with IC is actually slightly less than shown. This is because, as shown earlier in Figure 3.2 and later in Figure 3.30, n decreases slightly with increasing gate–source voltage (and V_{EFF}) associated with increasing inversion level. Additionally, as shown in Figure 3.2, n decreases by about 6 % in the presence of $V_{SB} = 1$ V, which is a large value of V_{SB} for low-voltage designs. When n decreases with the inversion level or V_{SB}, V_{EFF} decreases slightly because it tracks the value of n. Finally, V_{EFF} increases with temperature because it tracks the thermal voltage.

When V_{EFF}, excluding small-geometry effects like velocity saturation and VFMR, is evaluated in terms of the inversion coefficient, it is independent of the MOS process except for the substrate factor, n, which has typical values of 1.4, 1.35, and 1.3 in weak, moderate, and strong inversion for typical bulk CMOS processes. As mentioned in Section 3.2.4, the development of process-independent design expressions is one of the advantages of using the inversion coefficient in analog CMOS design. When small-geometry effects are excluded, V_{EFF} is also independent of channel length. The increase in V_{EFF} associated with velocity saturation effects depends on the process through the critical, velocity saturation, electric field, E_{CRIT}, and the channel length. The smaller increase associated with VFMR depends on the process through the mobility reduction factor, θ.

3.7.3 Drain–Source Saturation Voltage

As described in the previous section, the effective gate–source voltage, V_{EFF}, must be known to ensure sufficient circuit bias compliance for the operating gate–source voltage, $V_{GS} = V_{EFF} + V_T$. The drain–source saturation voltage, $V_{DS,sat}$,[8] must also be known to ensure sufficient circuit bias compliance for the operating drain–source voltage, V_{DS}.

Usually operation in saturation is desired where $V_{DS} > V_{DS,sat}$. Here, the MOS device acts as a voltage-controlled current source having optimally high transconductance, g_m, drain–source resistance, r_{ds}, and intrinsic gate-to-drain voltage gain, $A_{Vi} = g_m \cdot r_{ds}$. Operation near the boundary of saturation where $V_{DS} = V_{DS,sat}$ results in a slight loss of transconductance, but a larger loss of drain–source resistance and voltage gain. Operation in the deep ohmic, triode, or linear region where $V_{DS} \ll V_{DS,sat}$ results in a collapse of transconductance, drain–source resistance, and voltage gain. Here, the device acts as a voltage-controlled resistance. The design methods throughout this book consider optimization of analog MOS circuits in saturation. Thus, it is necessary to evaluate $V_{DS,sat}$ to ensure V_{DS} is sufficiently high for operation in saturation.

3.7.3.1 Physical versus circuit definition

Unlike drain current or gate–source voltage, which can be explicitly measured, $V_{DS,sat}$ is an inferred quantity based on either physical or circuit operation. In strong inversion, the traditional value is $V_{DS,sat} = V_{GS} - V_T = V_{EFF}$. This corresponds approximately to the value of V_{DS} where channel inversion charge pinches off at the drain side. However, the body effect along the channel raises the local threshold voltage at the drain side causing inversion charge pinch-off at a lower V_{DS}. $V_{DS,sat}$ is then estimated by $V_{DS,sat} = (V_{GS} - V_T)/n = V_{EFF}/n$, where the substrate factor, n, is approximately 1.3 for bulk CMOS processes in strong inversion. This is the strong-inversion value of $V_{DS,sat}$ predicted

---

[8] $V_{DS,sat}$ is source referenced as is V_{DS}.

by continuous weak through strong-inversion expressions given in the EKV [14] and similar [18, 28, p. 20] MOS models.

Operation at V_{DS} at the previously mentioned physical definitions for $V_{DS,sat}$ does not ensure optimal circuit performance. This is because MOS drain current versus drain–source voltage, transistor curves do not abruptly flatten in saturation giving optimally low drain–source conductance, g_{ds} (the slope of I_D versus V_{DS}), or high $r_{ds} = 1/g_{ds}$ when V_{DS} is near $V_{DS,sat}$. Instead, the curves are rounded at the $V_{DS,sat}$ transition giving g_{ds} significantly above its value available at higher V_{DS} deeper in saturation. This is shown later in Figures 3.37 and 3.38 in Section 3.8.4.1 where drain current is down by only 3.6 % at $V_{DS} = V_{DS,sat}$, but g_{ds} is a factor of five above its value available deeper in saturation. This suggests a $V_{DS,sat}$ definition that is circuit based rather than physically based. One such definition might be the operating V_{DS} where g_{ds} is twice its average value in saturation, as this would ensure no worse than a factor-of-two increase in g_{ds} when operating at the boundary of saturation. Another definition used for measurements later is the operating V_{DS} where drain current is down by 5 % from the value extrapolated by a line tangent to the drain current in saturation. This lower definition of $V_{DS,sat}$ is close to the physical definition of $V_{DS,sat} = (V_{GS} - V_T)/n = V_{EFF}/n$ in strong inversion, but results in g_{ds} well above that available deeper in saturation. Traditional, physically based definitions are widely used for $V_{DS,sat}$ predictions and are used here. This requires the designer to carefully select V_{DS} sufficiently above $V_{DS,sat}$ for applications where g_{ds} is critical for small-signal resistance and related signal gain.

3.7.3.2 Without small-geometry effects

Table 3.15 gives an expression for MOS $V_{DS,sat}$ along with dependencies on the design choices of inversion coefficient and channel length. The expression, developed from those given in [18, 28, p. 20], gives $V_{DS,sat}$ without velocity saturation and VFMR effects. This expression is similar to $V_{DS,sat} = 2U_T\sqrt{IC} + 4U_T$ developed from the EKV MOS model [14], but over-predicts $V_{DS,sat}$ less at the onset of strong inversion ($IC = 10$) compared to $V_{DS,sat} = V_{EFF}/n$ expressions.

Figure 3.16 shows $V_{EFF} = V_{GS} - V_T$ and $V_{DS,sat}$ predicted by the first expression in Table 3.14 and the expression in Table 3.15, respectively, assuming a fixed value of $n = 1.33$ for V_{EFF}, and

Table 3.15   MOS drain–source saturation voltage, $V_{DS,sat}$, expression and trends in terms of inversion coefficient and channel length. When evaluated in terms of the inversion coefficient, $V_{DS,sat}$ is independent of the drain current

Drain–source saturation voltage, $V_{DS,sat}$	$IC \uparrow$ L fixed	$L \uparrow$ IC fixed
All regions, with or without small-geometry effects:[a] $V_{DS,sat} = 2U_T\sqrt{IC + 0.25} + 3U_T$ $\approx 4U_T$ in WI $\approx 2U_T\sqrt{IC} = V_{EFF}/n$ in SI, where V_{EFF} is the effective gate–source voltage present without small-geometry effects	Unchanged in WI	Unchanged
$V_{DS,sat} = 4.02, 4.18, 5.24, 9.40, 23,$ and $66.3 \cdot U_T = 104, 108, 135, 243, 595,$ and 1710 mV for $IC = 0.01, 0.1, 1, 10, 100,$ and 1000, respectively, assuming $U_T = 25.9$ mV ($T = 300$ K)	$\uparrow \propto \sqrt{IC}$ in SI	

As shown later in Figures 3.36 and 3.38, operation at $V_{DS} = V_{DS,sat}$ can result in a significant increase in drain–source conductance, g_{ds}, above that available at higher V_{DS} values in saturation.
[a]The drop in $V_{DS,sat}$ relative to V_{EFF} caused by velocity saturation is largely compensated for by the increase in V_{EFF} as discussed in Sections 3.7.3.3 and 3.7.3.4. As a result, the expression provides a good estimate even when velocity saturation is significant.

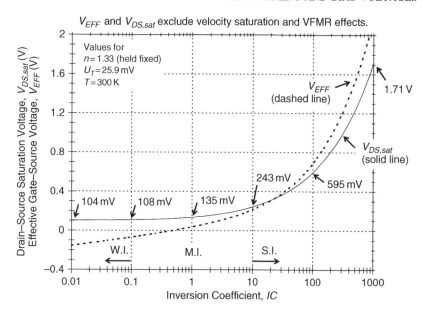

Figure 3.16   Predicted drain–source saturation voltage, $V_{DS,sat}$, and effective gate–source voltage, $V_{EFF} = V_{GS} - V_T$, versus the inversion coefficient, IC, exclusive of velocity saturation and VFMR effects. $V_{DS,sat}$ reaches a minimum, constant value of $4U_T$ in weak inversion and increases modestly in moderate inversion. It increases as the square root of IC in strong inversion and approaches V_{EFF}/n. Although V_{EFF} increases with velocity saturation and VFMR effects, $V_{DS,sat}$ is nearly unaffected by these effects as described in Section 3.7.3.3. $V_{DS,sat}$ is similar for all CMOS processes

$U_T = 25.9\,\text{mV}$ ($T = 300\,\text{K}$) for both V_{EFF} and $V_{DS,sat}$. Both V_{EFF}, shown earlier by the dashed lines in Figures 3.14 and 3.15, and $V_{DS,sat}$ exclude velocity saturation and VFMR effects.

In weak inversion ($IC < 0.1$), V_{EFF} is negative, indicating that the gate–source voltage is less than the threshold voltage. $V_{DS,sat}$, of course, is positive and reaches a minimum value of four times the thermal voltage, $4U_T$, or 104 mV at room temperature ($T = 300\,\text{K}$). In the center of moderate inversion ($IC = 1$), $V_{EFF} = 38\,\text{mV}$ while $V_{DS,sat}$ has risen slightly to 135 mV. This illustrates that there is little $V_{DS,sat}$ penalty in moderate inversion compared to weak inversion. At the onset of strong inversion ($IC = 10$), $V_{DS,sat}$ has risen to 243 mV and is close to the value of V_{EFF}. Here, $V_{DS,sat}$ is over-predicted compared to the expected $V_{DS,sat} = V_{EFF}/n$ value in strong inversion. Deeper in strong inversion, $V_{DS,sat}$ does drop below V_{EFF} and approach $V_{DS,sat} = V_{EFF}/n$. Both $V_{DS,sat}$ and V_{EFF} increase as the square root of inversion coefficient in strong inversion.

Figure 3.17 shows $V_{DS,sat}$ predicted in Figure 3.16 overlaid with measured drain current versus drain–source voltage curves. This illustrates the leveling or saturation of drain current for $V_{DS} > V_{DS,sat}$. The curves are for a $W/L = 64\,\mu\text{m}/4\,\mu\text{m}$, nMOS device in the 0.18 µm CMOS process described in Table 3.2 operating over nearly five decades of current from weak through strong inversion. Instead of being labeled with the usual gate–source voltage, each curve is labeled with the inversion coefficient associated with the drain current in saturation. This is found from $IC = I_D/[I_0(W/L)]$ from Table 3.6, using a technology current of $I_0 = 0.64\,\mu\text{A}$ from Table 3.2.

The figure illustrates constant $V_{DS,sat} = 4U_T$ (again, 104 mV at $T = 300\,\text{K}$) in weak inversion for IC values less than 0.1 and the slight increase in moderate inversion for IC values of 0.21, 0.52, 0.95, and 2.14. $V_{DS,sat}$ begins to increase noticeably at the strong-inversion side of moderate inversion for $IC = 5.07$ and in strong inversion for IC values greater than 10. This shows how the increase of $V_{DS,sat}$ in strong inversion works against low-voltage analog design. Low $V_{DS,sat}$ along with high transconductance efficiency, moderate bandwidth, and minimal velocity saturation and VFMR effects a compelling reason to consider moderate inversion in low-voltage designs.

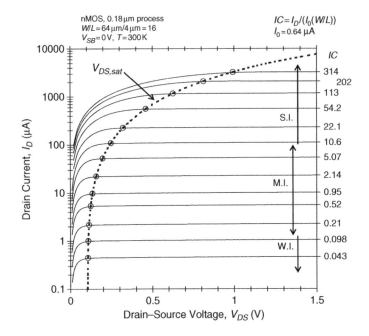

Figure 3.17 Drain current versus drain–source voltage curves for a $W/L = 64\,\mu m/4\,\mu m$, nMOS device in a 0.18 μm CMOS process with estimated drain–source saturation voltages, $V_{DS,sat}$. Each curve is identified by the inversion coefficient associated with drain current in saturation. Drain current is nearly leveled off for $V_{DS} > V_{DS,sat}$. $V_{DS,sat}$ is constant at $4U_T$ in weak inversion, increases modestly in moderate inversion, and increases as the square root of drain current or inversion coefficient in strong inversion

3.7.3.3 With velocity saturation effects

Sections 3.7.2.2 and 3.7.2.3 described the increase in $V_{EFF} = V_{GS} - V_T$ when velocity saturation and VFMR effects are present. These effects cause a reduction of drain current, requiring an increase in V_{EFF} to support the selected drain current or inversion coefficient. The inversion coefficient and drain current track each other for a fixed-geometry device since n and mobility are assumed fixed in the inversion coefficient definition used in this book as shown in Equation 3.16 and mentioned in Section 3.4.2.3.

This section shows that $V_{DS,sat}$ decreases from its usual value of V_{EFF} or V_{EFF}/n in strong inversion when velocity saturation effects are present. However, when evaluated in terms of the drain current or inversion coefficient, $V_{DS,sat}$ is relatively unchanged with velocity saturation effects since the increase in V_{EFF} is largely countered by the decrease in $V_{DS,sat}$ relative to V_{EFF}.

Figure 3.18 illustrates the decrease in drain current and $V_{DS,sat}$ associated with velocity saturation. Here, measured drain current versus drain–source voltage curves are overlaid for $L = 0.18$ and $4\,\mu m$, nMOS devices in the 0.18 μm CMOS process described in Table 3.2. Both devices have nearly equal threshold voltages of 0.42 V and nearly equal shape factors of 16. Without velocity saturation effects, this should result in similar saturation drain currents for equal V_{GS} values.

For V_{GS} near 0.7 V (a slightly higher V_{GS} was used for the $L = 0.18\,\mu m$ device because its threshold voltage was slightly higher) where $V_{EFF} = V_{GS} - V_T = 0.7\,V - 0.42\,V = 0.28\,V$, the drain current near $V_{DS} = V_{DS,sat}$ is modestly lower for the $L = 0.18\,\mu m$ device because of some velocity saturation, even here near the onset of strong inversion. However, the measured $V_{DS,sat}$, determined from an enlarged view of the curves, is at the expected value of $V_{DS,sat} = V_{EFF}/n = 0.28\,V/1.33 = 0.21\,V$ for both devices assuming $n = 1.33$. As denoted in the figure, the measured

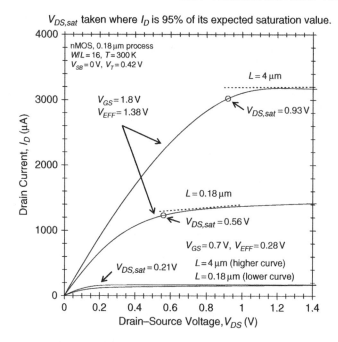

V_{DS,sat} taken where I_D is 95% of its expected saturation value.

Figure 3.18   Illustration of drain current and drain–source saturation voltage, $V_{DS,sat}$, reduction resulting from velocity saturation. $L = 0.18$ and $4\,\mu m$, nMOS devices having equal shape factors are operated at equal $V_{EFF} = V_{GS} - V_T$ values. At the lower V_{EFF} associated with $V_{GS} = 0.7\,V$, drain current and $V_{DS,sat}$ are nearly equal for both devices. However, at the higher V_{EFF} associated with $V_{GS} = 1.8\,V$, the drain current and $V_{DS,sat}$ are considerably lower for the $L = 0.18\,\mu m$ device because of velocity saturation. The reduction of drain current and $V_{DS,sat}$ is less for pMOS devices because of less velocity saturation

$V_{DS,sat}$ was taken where drain current is 5 % down from the value extrapolated by a line tangent to the drain current in saturation.

For $V_{GS} = 1.8\,V$ where $V_{EFF} = V_{GS} - V_T = 1.8\,V - 0.42\,V = 1.38\,V$, the drain current for the $L = 0.18\,\mu m$ device is less than half that of the $L = 4\,\mu m$ device because of significant velocity saturation. This results in a measured $V_{DS,sat} = 0.56\,V$ for the $L = 0.18\,\mu m$ device, well below $V_{DS,sat} = 0.93\,V$ measured for the $L = 4\,\mu m$ device that experiences little velocity saturation. $V_{DS,sat} = 0.56\,V$ for the $L = 0.18\,\mu m$ device is well below $V_{DS,sat} = V_{EFF}/n = 1.38\,V/1.33 = 1.04\,V$ predicted in strong inversion without velocity saturation, while $V_{DS,sat} = 0.93\,V$ for the $L = 4\,\mu m$ device is only slightly below this value. Measured and predicted $V_{DS,sat}$ values are not directly, numerically comparable since measured $V_{DS,sat}$ is found from a circuit definition where drain current is down by 5 %. However, $V_{DS,sat}$ is clearly significantly lower for the short-channel, $L = 0.18\,\mu m$ device compared to the long-channel, $L = 4\,\mu m$ device, even though both devices are operated at equal V_{EFF}.

Before estimating the value of $V_{DS,sat}$ including velocity saturation effects, it is necessary to estimate V_{EFF} including these effects. Dividing V_{EFF} found earlier in Equation 3.20 by $V_{EFF,L}$ shows the increase in V_{EFF} relative to $V_{EFF,L}$, the large-geometry value present without velocity saturation and VFMR effects. This gives

$$\frac{V_{EFF}}{V_{EFF,L}} = \sqrt{1 + \frac{V_{EFF}}{(LE_{CRIT})'}} = \sqrt{1 + VSF'} \approx \sqrt{1 + \frac{V_{EFF}}{LE_{CRIT}}} = \sqrt{1 + VSF} \qquad (3.29)$$

The two rightmost terms neglect the relatively small increase in V_{EFF} (shown earlier in Figures 3.14 and 3.15) due to VFMR at all channel lengths compared to the large increase at short channel lengths due to velocity saturation. VFMR effects are neglected by setting $\theta = 0$ in the equivalent velocity saturation voltage, $(LE_{CRIT})'$, and velocity saturation factor, VSF', terms, which converts these to LE_{CRIT} and VSF, respectively. $VSF = V_{EFF}/(LE_{CRIT})$ is the regular velocity saturation factor from Sections 2.4.3 and 3.7.1.5, which excludes VFMR effects. $V_{EFF}/V_{EFF,L}$ is unity when velocity saturation effects are not present ($VSF \ll 1$), but increases well above this when these effects become significant. For example, at $VSF = V_{EFF}/(LE_{CRIT}) = 1$, Equation 3.29 predicts $V_{EFF}/V_{EFF,L} = 1.41$.

In [26, p. 60], $V_{DS,sat}$ inclusive of velocity saturation effects is estimated by differentiating MOS drain current in strong inversion with respect to the drain–source voltage to find the $V_{DS} = V_{DS,sat}$ transition point between the ohmic and saturation regions. Strong-inversion drain current in the ohmic region ($V_{DS} < V_{DS,sat}$) is approximated by

$$
I_D = \frac{\mu\, C_{OX}' \left(\dfrac{W}{L}\right) \left((V_{GS} - V_T)\,V_{DS} - \dfrac{V_{DS}^2}{2}\right)}{1 + \dfrac{V_{DS}}{LE_{CRIT}}}
\tag{3.30}
$$

where μ may include the loss of mobility due to VFMR but excludes the mobility loss due to velocity saturation. The denominator term $1 + V_{DS}/(LE_{CRIT})$ models the reduction in drain current due to velocity saturation reduction of mobility and carrier velocity where $E_x = V_{DS}/L$ approximates the horizontal, drain–source electric field across the non-pinched-off channel. This is analogous to the velocity saturation reduction term $1 + V_{EFF}/(LE_{CRIT})$ described earlier in Section 3.7.1.2 for operation in saturation where the horizontal electric field across the pinched-off channel is approximately $E_x = V_{EFF}/L$.

The derivative of drain current in Equation 3.30 with respect to drain–source voltage is zero at $V_{DS} = V_{DS,sat}$, which corresponds to the transition point between the ohmic and saturation regions. This occurs at [26, p. 61]

$$
V_{DS,sat} = LE_{CRIT}\left(\sqrt{1 + \frac{2V_{EFF}}{LE_{CRIT}}} - 1\right) = LE_{CRIT}\left(\sqrt{1 + 2VSF} - 1\right)
\tag{3.31}
$$

where $V_{GS} - V_T$ is replaced with V_{EFF}. When $VSF = V_{EFF}/(LE_{CRIT}) \ll 1$, $\sqrt{1 + 2VSF} \approx 1 + VSF$, giving $V_{DS,sat} = LE_{CRIT} \cdot VSF = V_{EFF}$. This is the expected, traditional, strong-inversion value of $V_{DS,sat} = V_{EFF}$ when velocity saturation effects are not present. $V_{DS,sat}$, however, decreases below V_{EFF} when velocity saturation effects are present.

Interestingly, a reduction of drain current due to VFMR (e.g., mobility in Equation 3.30 divided by $1 + V_{EFF} \cdot \theta$) or the substrate factor (e.g., Equation 3.30 divided by n) does not change the value of $V_{DS,sat}$ predicted by Equation 3.31. This is a result of differentiating Equation 3.30 with respect to the drain–source voltage assuming VFMR and substrate effects are constant with the drain–source voltage.

Dividing $V_{DS,sat}$ in Equation 3.31 by V_{EFF} shows the reduction in $V_{DS,sat}$ relative to V_{EFF} resulting from velocity saturation. This gives

$$
\frac{V_{DS,sat}}{V_{EFF}} = \frac{LE_{CRIT}}{V_{EFF}}\left[\sqrt{1 + \frac{2V_{EFF}}{LE_{CRIT}}} - 1\right] = \frac{\sqrt{1 + 2VSF} - 1}{VSF}
\tag{3.32}
$$

$V_{DS,sat}/V_{EFF}$ is unity when velocity saturation effects are not present ($VSF \ll 1$), but decreases well below this when these effects become significant. For example, at $VSF = V_{EFF}/(LE_{CRIT}) = 1$, Equation 3.32 predicts $V_{DS,sat}/V_{EFF} = 0.732$.

Multiplying Equations 3.32 and 3.29 gives $V_{DS,sat}$ relative to $V_{EFF,L}$ as

$$\frac{V_{DS,sat}}{V_{EFF,L}} = \frac{V_{DS,sat}}{V_{EFF}} \cdot \frac{V_{EFF}}{V_{EFF,L}} \approx \left(\frac{\sqrt{1+2VSF}-1}{VSF} \right) \sqrt{1+VSF} \qquad (3.33)$$

where the approximation results from neglecting the normally small increase in V_{EFF} in Equation 3.29 due to VFMR. $V_{DS,sat}/V_{EFF,L}$ is unity when velocity saturation effects are not present ($VSF \ll 1$), but interestingly changes little when these effects are significant. This is because the decrease in $V_{DS,sat}$ relative to V_{EFF} is largely countered by the increase of V_{EFF} relative to $V_{EFF,L}$. For example, at $VSF = V_{EFF}/(LE_{CRIT}) = 1$, Equation 3.33 predicts $V_{DS,sat}/V_{EFF,L} = 1.035$, which indicates that $V_{DS,sat}$ is essentially unchanged from the large-geometry value of $V_{EFF,L}$.

Figure 3.19 shows Equations 3.29, 3.32, and 3.33 plotted against $VSF = V_{EFF}/(LE_{CRIT})$, which represents the degree of velocity saturation. Additionally, the reduction of drain current, $I_D/I_{D,L} = 1/(1+VSF)$, is plotted from the simple drain current correction factor of Table 3.13, where $I_{D,L}$ is the large-geometry drain current when velocity saturation and VFMR effects are not present. Here, as in Equation 3.29, the velocity saturation factor VSF is used instead of VSF' to exclude VFMR effects.

The $I_D/I_{D,L}$ curve shows that as VSF increases, I_D decreases from $I_{D,L}$. This requires an increase in V_{EFF} relative to $V_{EFF,L}$ as shown by the $V_{EFF}/V_{EFF,L}$ curve (Equation 3.29) to support the selected drain current. While V_{EFF} increases relative to $V_{EFF,L}$ with increasing VSF, the $V_{DS,sat}/V_{EFF}$ curve (Equation 3.32) shows that $V_{DS,sat}$ decreases relative to V_{EFF}. As a result, the $V_{DS,sat}/V_{EFF,L}$ curve (Equation 3.33), which is the product of the $V_{EFF}/V_{EFF,L}$ and $V_{DS,sat}/V_{EFF}$ curves, shows that $V_{DS,sat}$ remains nearly unchanged from $V_{EFF,L}$ as VSF increases. This shows that as the degree of velocity

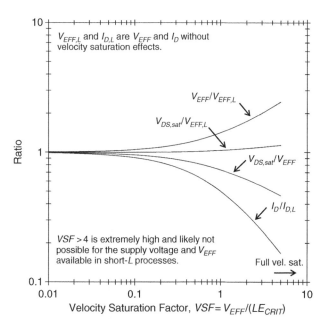

Figure 3.19 Ratio of drain current, I_D, effective gate–source voltage, $V_{EFF} = V_{GS} - V_T$, and drain–source saturation voltage, $V_{DS,sat}$, to values present without velocity saturation effects. I_D decreases with increasing velocity saturation while V_{EFF} increases to support the selected I_D. $V_{DS,sat}$ is nearly unaffected with increasing velocity saturation since its decrease relative to V_{EFF} is countered by the increase in V_{EFF}

saturation increases, the increase in V_{EFF} is countered by a decrease in $V_{DS,sat}$ relative to V_{EFF} leaving $V_{DS,sat}$ nearly equal to $V_{EFF,L}$, its expected value for no velocity saturation effects. This is a fortunate result that will permit us to predict $V_{DS,sat}$ from the expression in Table 3.15, where $V_{DS,sat}$ increases as the square root of IC in strong inversion, with or without velocity saturation effects.

$V_{DS,sat}/V_{EFF,L}$ (Equation 3.33) increases slightly above unity for VSF values shown ranging up to four. VSF is likely to be limited to these values by the maximum supply voltage and V_{EFF}. If the maximum V_{EFF} and minimum channel length decrease nearly the same in increasingly smaller-geometry processes, the maximum $VSF = V_{EFF}/(LE_{CRIT})$ will remain nearly unchanged.

The ratios of $V_{EFF}/V_{EFF,L}$, $V_{DS,sat}/V_{EFF}$, and $V_{DS,sat}/V_{EFF,L}$ given by Equations 3.29, 3.32, and 3.33, respectively, and shown in Figure 3.19 are developed from drain current estimates where the velocity saturation, transition exponent, α, described in Section 3.7.1.2, is assumed equal to unity. As discussed in Section 3.7.1.4, this gives the simple drain current correction factor of Table 3.13. This results in the division of drain current by $1 + VSF$ or $1 + V_{EFF}/(LE_{CRIT})$ when velocity saturation effects are considered without VFMR effects. As shown earlier in Figure 3.10, the simple correction factor under-predicts drain current by 13 % at the transition into velocity saturation. As a result, predictions of V_{EFF} and $V_{DS,sat}$ are approximate, especially near the transition into velocity saturation. While approximate, the preceding analysis shows that $V_{DS,sat}$ is nearly unaffected by velocity saturation effects because it tracks V_{EFF} present without these effects ($V_{EFF,L}$) as drain current or inversion coefficient increases. This is verified by measured data in the following discussion.

3.7.3.4 Predicted and measured values

Figure 3.20 shows predicted and measured V_{EFF} and $V_{DS,sat}$ for $L = 0.18$ and $10\,\mu m$, nMOS devices in the $0.18\,\mu m$ CMOS process described in Table 3.2. Predicted V_{EFF} is from Figure 3.14 and includes values both with and without velocity saturation and VFMR effects. Predicted $V_{DS,sat}$ is from Figure 3.16, which excludes velocity saturation and VFMR effects. Channel length is adjusted from the $0.18\,\mu m$ value given by subtracting $DL = 0.028\,\mu m$, which is listed in Table 3.2. Since measured V_{EFF} and $V_{DS,sat}$ are evaluated in terms of the inversion coefficient, this is calculated as done for the drain current curves of Figure 3.17. As described earlier for Figure 3.18, measured $V_{DS,sat}$ is taken using a circuit definition where drain current is 5 % down from the value extrapolated by a line tangent to the drain current in saturation.

In Figure 3.20, the dashed heavier line shows predicted $V_{EFF,L}$ without velocity saturation and VFMR effects while the upper dashed lines show predicted V_{EFF} for the $L = 0.18$ and $10\,\mu m$ devices with these effects. As expected, V_{EFF} increases slightly above $V_{EFF,L}$ for the long-channel, $L = 10\,\mu m$ device due to VFMR effects, but increases significantly above this for the short-channel, $L = 0.18\,\mu m$ device due to velocity saturation. Measured values given by the diamonds and triangles for the $L = 0.18$ and $10\,\mu m$ devices, respectively, closely align with the predicted values. The solid heavier line shows predicted $V_{DS,sat}$ without velocity saturation and VFMR effects. Interestingly, measured $V_{DS,sat}$ given by the lower solid lines lies below the predicted values for both the $L = 0.18$ and $10\,\mu m$ devices, even though the $L = 0.18\,\mu m$ device experiences considerable velocity saturation at high inversion coefficients. As observed, as V_{EFF} increases significantly for the $L = 0.18\,\mu m$ device due to velocity saturation effects, $V_{DS,sat}$ decreases relative to V_{EFF}, giving a $V_{DS,sat}$ value that is nearly independent of velocity saturation effects. Measured $V_{DS,sat}$ for both devices tracks near or below predicted $V_{DS,sat}$, which is nearly equal to $V_{EFF,L}/n$ in strong inversion and independent of velocity saturation effects. Measured $V_{DS,sat}$ values are slightly higher for the $L = 0.18\,\mu m$ device, which could be a result of the much steeper drain current versus drain–source voltage characteristic where drain current changes more rapidly near $V_{DS,sat}$. Again, measured and predicted $V_{DS,sat}$ values cannot be directly compared numerically as a circuit definition is used for measured values.

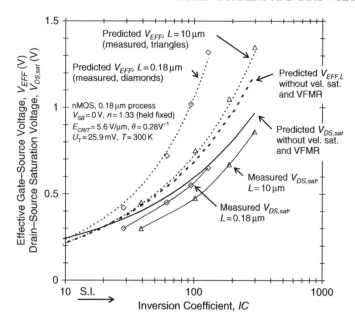

Figure 3.20   Predicted and measured drain–source saturation voltage, $V_{DS,sat}$, and effective gate–source voltage, $V_{EFF} = V_{GS} - V_T$, versus the inversion coefficient, IC, for $L = 0.18$ and $10\,\mu m$, nMOS devices in a $0.18\,\mu m$ CMOS process operating in strong inversion. For the short-channel, $L = 0.18\,\mu m$ device, V_{EFF} increases significantly from $V_{EFF,L}$, the value present without velocity saturation and VFMR effects, while $V_{DS,sat}$ decreases well below its expected value of V_{EFF}/n. Because the decrease in $V_{DS,sat}$ relative to V_{EFF} is largely countered by the increase in V_{EFF}, $V_{DS,sat}$ remains near its value predicted without velocity saturation effects, even when velocity saturation is significant. $V_{DS,sat}$ was measured where drain current is 5 % below its expected value

The measured $V_{DS,sat}$ data of Figure 3.20 confirms the predicted $V_{DS,sat}/V_{EFF,L}$ curve of Figure 3.19 (Equation 3.33) where $V_{DS,sat}$ changes little from the strong inversion value of V_{EFF} or V_{EFF}/n predicted without velocity saturation and VFMR effects, even if velocity saturation is significant.

3.7.3.5 Summary of trends

Since $V_{DS,sat}$ evaluated from the inversion coefficient is nearly independent of velocity saturation effects, we will use the expression in Table 3.15, which was plotted in Figure 3.16 and used for predicted $V_{DS,sat}$ in Figure 3.20, for predicting $V_{DS,sat}$ for all applications. As mentioned and listed in the table, $V_{DS,sat} = 4U_T$ is constant at its minimum value in weak inversion. It increases modestly in moderate inversion before increasing as the square root of IC in strong inversion, approaching $V_{DS,sat} = 2U_T\sqrt{IC} = V_{EFF}/n$, where V_{EFF} is the effective gate–source voltage ($V_{EFF,L}$) described earlier without velocity saturation or VFMR effects. $V_{DS,sat}$ is independent of the channel length as a result of being nearly independent of velocity saturation effects. Finally, $V_{DS,sat}$, like V_{EFF} summarized in Table 3.14, is independent of the selected drain current when evaluated in terms of the inversion coefficient.

In addition to the expression for $V_{DS,sat}$ and trends with the inversion coefficient and channel length, Table 3.15 lists predicted values. $V_{DS,sat} = 4.02, 4.18, 5.24, 9.40, 23,$ and $66.3 \cdot U_T$ for $IC = 0.01,$ $0.1, 1, 10, 100,$ and 1000, respectively. This corresponds to $V_{DS,sat} = 104, 108, 135, 243, 595,$ and

1710 mV for $U_T = 25.9$ mV $(T = 300$ K). $V_{DS,sat}$ increases with temperature through the thermal voltage.

Like V_{EFF} summarized in Table 3.14, $V_{DS,sat}$ is largely independent of the MOS process when evaluated in terms of the inversion coefficient. In fact, $V_{DS,sat}$ predicted from the expression in Table 3.15 is independent even of n, depending only on the inversion coefficient and thermal voltage. As mentioned in Section 3.2.4, the development of process-independent design expressions is one of the advantages of using the inversion coefficient in analog CMOS design.

As mentioned in Section 3.7.3.1 and described later in Section 3.8.4.1, DC drain current may be down only slightly for operation at $V_{DS} = V_{DS,sat}$, but drain–source conductance can be up significantly compared to values available at higher V_{DS}. As described later in Section 3.8.4.4, operating near $V_{DS,sat}$ can cause significant reductions in small-signal resistance and gain, especially for cascode or regulated cascode circuits where small-signal resistance has nearly a squared or cubed dependency on drain–source conductances. As a result, the operating V_{DS} should be set somewhat above $V_{DS,sat}$ if possible, which becomes increasingly difficult in low-voltage designs. Additionally, the designer should be aware of the $V_{DS,sat}$ prediction used, either the lower $V_{DS,sat} = V_{EFF}/n$ prediction used here, or the higher $V_{DS,sat} = V_{EFF}$ traditional prediction, both given for strong inversion.

3.8 SMALL-SIGNAL PARAMETERS AND INTRINSIC VOLTAGE GAIN

The following discussions describe MOS small-signal parameters and intrinsic voltage gain. First, the MOS small-signal model is described along with its application in circuits. This is important to permit analysis of small-signal resistance, signal gain, and other aspects of analog circuit performance. Then, small-signal transconductance, body-effect transconductance, and drain–source resistance parameters are described in their respective normalized forms of transconductance efficiency, body-effect transconductance ratio, and the Early voltage. The prediction of these parameters is important since they influence the transconductance, resistance, signal gain, bandwidth, and thermal noise of analog circuits. As part of the transconductance discussion, transconductance distortion is described in terms of the input, 1 dB compression voltage for a MOS differential pair where the output current is 1 dB down from its expected, linear value. The prediction of transconductance distortion is important since this often influences the maximum allowable input signal for analog circuits. Finally, MOS intrinsic voltage gain, the product of transconductance and drain–source resistance, is described. The prediction of intrinsic voltage gain is important since this is the maximum voltage gain available for a single device and is a quality factor appearing in many small-signal resistance and gain calculations.

3.8.1 Small-Signal Model and its Application

Figure 3.21 shows the MOS small-signal model used in this book. The model includes the small-signal transconductance, g_m, body-effect transconductance, g_{mb}, drain–source resistance, r_{ds} (or the drain–source conductance, $g_{ds} = 1/r_{ds}$), and drain–body resistance, r_{db} (or the drain–body conductance, $g_{db} = 1/r_{db}$), which are discussed in detail later in Sections 3.8.2–3.8.4. The small-signal drain current is controlled by the two voltage-controlled current sources, which have values of $g_m \cdot v_{gs}$ and $g_{mb} \cdot v_{bs}$. These are controlled by the small-signal, gate–source and body–source voltages, v_{gs} and v_{bs}. The drain is loaded to the source by r_{ds} and is loaded to the body by r_{db}. The model also includes intrinsic gate–source, C_{gsi}, gate–body, C_{gbi}, and gate–drain, C_{gdi}, capacitances, along with extrinsic overlap, gate–source, C_{GSO}, gate–body, C_{GBO}, and gate–drain, C_{GDO}, capacitances,

MOS small-signal voltage and current definitions

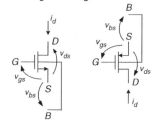

MOS small-signal model

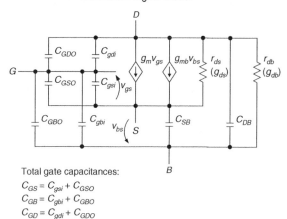

Total gate capacitances:

$$C_{GS} = C_{gsi} + C_{GSO}$$
$$C_{GB} = C_{gbi} + C_{GBO}$$
$$C_{GD} = C_{gdi} + C_{GDO}$$

Figure 3.21 MOS small-signal model including small-signal parameters and capacitances. The model neglects gate, drain, and source series or access resistance associated with contacts and bulk resistance. The model is quasi-static and is useful for general analog purposes at operating frequencies up to approximately 25 % of the device intrinsic bandwidth, f_{Ti}. Non-quasi-static enhancements can include, for example in strong inversion, the addition of a resistance having value of $1/(5g_m)$ placed in series with the intrinsic gate–source capacitance, C_{gsi}, combined with low-passed sampling of the gate–source voltage used to control g_m. When gate leakage current is significant for gate-oxide thickness below approximately 2 nm, a gate–source conductance, g_{gs}, appears between the gate and source

and drain–body, C_{DB}, and source–body, C_{SB}, capacitances. Model capacitances are discussed later in Sections 3.9.2–3.9.4.

The model excludes gate, drain, and source series or access resistance associated with contacts and bulk material resistance. Such resistance is often made negligible for general-purpose analog applications by careful layout. Source resistance, which degenerates transconductance, however, can impact performance for short-channel devices operating at high levels of inversion where current density is high.

When gate leakage current is significant for gate-oxide thickness below approximately 2 nm, a small-signal, gate–source conductance, g_{gs}, should be connected between the gate and source in the model shown in Figure 3.21. The prediction of gate leakage current and the resulting gate–source conductance is discussed later in Sections 3.12.1.1–3.12.1.2.

The model shown in Figure 3.21 is a simple, quasi-static [15, pp. 315–316] model where the frequency of operation is sufficiently low such that device charges immediately track terminal voltages. The model is useful for general analog applications up to frequencies of approximately 25 % of the MOS intrinsic bandwidth, $f_{Ti} = g_m/[2\pi(C_{gsi} + C_{gbi})]$, which is described later in Section 3.9.6. Interestingly, since f_{Ti} is inversely proportional to the square of channel length (or inversely proportional to channel length when velocity saturation effects are significant), the allowable frequencies for quasi-static models can be low for long-channel devices.

Possible non-quasi-static extensions for operation in strong inversion in saturation include, for example, the addition of resistance having a value of $1/(5g_m)$ placed in series with C_{gsi} [15, p. 489]. This can also be combined with a low-pass roll-off of g_m at a frequency above f_{Ti}, which can be implemented by a low-pass sampling of the gate–source voltage used for controlling g_m. Such model extensions attempt to model resistive gate loading and the delay in inversion charge at high frequencies. Non-quasi-static MOS model extensions involving the use of y parameters, the addition of series elements, or the use of smaller, distributed MOS devices are discussed in [15, pp. 467–504]. Additionally, a first-order, non-quasi-static enhancement for the EKV MOS model is discussed in [13] along with more recent enhancements discussed in [16, pp. 240–259]. The development of non-quasi-static models, which are especially important for radio frequency (RF) circuits, is a subject of current interest and research.

Predictions of MOS device and resulting circuit performance must be necessarily simple to provide design guidance, requiring a simple, small-signal model like that shown in Figure 3.21. However, as mentioned, non-quasi-static extensions are possible to provide design guidance for RF applications. In all cases, the predictions and design guidance suggested by this book must be followed by computer simulations using full MOS models to validate a candidate design. Depending on the operating frequency and application, such models may need to include non-quasi-static effects.

Figure 3.22 shows the MOS small-signal model of Figure 3.21 applied in an external circuit for low frequencies (capacitances removed) with resistance, voltage gain, current gain, transconductance, and transresistance expressions listed in Table 3.16. As listed in the table, the gate input resistance, r_{ing}, and body input resistance, r_{inb}, are infinite. The drain output resistance, r_{outd}, is equal to r_{ds}, raised by the multiplication of $1 + g_{ms}r_s$ as a result of external source, degeneration resistance, r_s. The source output or input resistance, $r_{outs} = r_{ins}$, is equal to $1/g_{ms}$, raised by the multiplication of $1 + r_d/r_{ds}$ as a result of external drain resistance, r_d. The total source transconductance, $g_{ms} = g_m + g_{mb} + g_{ds}$, appears in many of the expressions and is also listed in the table.

The lower portion of Figure 3.22 provides design intuition by representing the drain as a Norton equivalent circuit and the source as a Thévenin equivalent circuit. At the drain, three dependent current sources under control of the gate voltage, v_g, body voltage, v_b, and source input current, i_{ins}, provide current having values of $v_g \cdot G_M$, $v_b \cdot G_{MB}$, and $i_{ins} \cdot H_{sd}$, respectively. As listed in Table 3.16, G_M and G_{MB} are the source-degenerated transconductances associated with g_m and g_{mb}, respectively, which are reduced by the division of $1 + g_{ms}r_s$. This factor, $1 + g_{ms}r_s$, can be defined as the source degeneration factor where native transconductance is divided and reduced by this factor, while native drain–source resistance, r_{ds}, is multiplied and increased by this same factor to give the drain output resistance, r_{outd}, mentioned earlier. H_{sd} represents the current division of i_{ins} into the MOS source for a short-circuit drain, which is given by $g_{ms}/(g_{ms} + g_s)$ where $g_s = 1/r_s$ is the external source conductance. To complete the Norton equivalent circuit at the drain, the drain output resistance, r_{outd}, appears in parallel with the dependent current sources.

At the source, two dependent voltage sources under control of v_g and v_b provide voltage having values of $v_g \cdot A_{vgso}$ and $v_b \cdot A_{vbso}$, respectively. As listed in Table 3.16, A_{vgso} and A_{vbso} are the open-circuit, gate-to-source and body-to-source voltage gains, having values of g_m/g_{ms} and g_{mb}/g_{ms}, respectively. To complete the Thévenin equivalent circuit at the source, the source output resistance, r_{outs}, appears in series with the dependent voltage sources.

When the gate is driven by a voltage, the $v_g \cdot G_M$ current source in Figure 3.22 is active at the drain providing signal current into the parallel combination of r_{outd} and r_d, giving potential voltage gain. At the same time, the $v_g \cdot A_{vgso}$ voltage source is active at the source providing slightly less than unity voltage gain into a relatively low output resistance of r_{outs}. If the body is connected and driven by the source, these sources are still active at the drain and source, but g_{mb} is set to zero

MOSFET applied in circuit with signal and resistance definitions

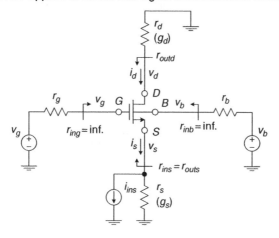

Equivalent Norton circuit at drain and Thévenin circuit at source

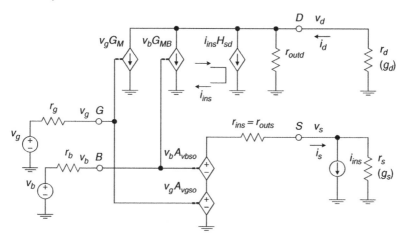

Figure 3.22 MOS small-signal model of Figure 3.21 applied in an external circuit with resistance, voltage gain, current gain, and transconductance quantities given in Table 3.16. A Norton model represents drain current components and drain output resistance, while a Thévenin model represents source voltage components and source output resistance. The model permits rapid small-signal analysis for gate, body, or source input signals for any level of resistive loading

giving $g_{ms} = g_m + g_{ds}$ compared to $g_{ms} = g_m + g_{mb} + g_{ds}$ as noted in Table 3.16. The lower value of g_{ms} increases G_M and decreases r_{outd} slightly through the division and multiplication, respectively, by a lower value of $1 + g_{ms}r_s$. Additionally, r_{outs} increases slightly through division by a lower value of g_{ms}. Of greater importance, however, is the increase in unloaded, gate-to-source voltage gain, $A_{vgso} = g_m/g_{ms} = g_m/(g_m + g_{ds})$ compared to $A_{vgso} = g_m/g_{ms} = g_m/(g_m + g_{mb} + g_{ds})$ present when the body is not driven by the source. This voltage gain is nearly unity when the body is driven by the source, permitting a nearly unity-gain source follower circuit when the source is lightly loaded by a high value of r_s.

Table 3.16 Small-signal quantities and expressions for the circuit model shown in Figure 3.22 using the MOS small-signal parameters from Figure 3.21. The expressions permit rapid small-signal analysis for gate, body, or source input signals for any level of resistive loading

---

Small-signal quantity and expression

Small-signal resistances:

$$r_{ing} = \infty \qquad\qquad\qquad r_{inb} = \infty$$

$$r_{outd} = r_{ds}\left(1 + g_{ms}r_s\right) \qquad r_{outs} = r_{ins} = \frac{1 + r_d/r_{ds}}{g_{ms}}$$

Short-circuit (Norton) transconductances and current gain at drain:[a]

$$G_M = \left.\frac{i_d}{v_g}\right|_{r_d=0} = \frac{g_m}{1 + g_{ms}r_s} \qquad G_{MB} = \left.\frac{i_d}{v_b}\right|_{r_d=0} = \frac{g_{mb}}{1 + g_{ms}r_s}$$

$$H_{sd} = \left.\frac{i_d}{i_{ins}}\right|_{r_d=0} = \frac{g_{ms}}{g_{ms} + g_s}$$

Open-circuit (Thévenin) voltage gains at source:[a]

$$A_{vgso} = \left.\frac{v_s}{v_g}\right|_{r_s=\infty} = \frac{g_m}{g_{ms}} \qquad A_{vbso} = \left.\frac{v_s}{v_b}\right|_{r_s=\infty} = \frac{g_{mb}}{g_{ms}}$$

Terminal voltage gains:[a]

$$A_{vgd} = \frac{v_d}{v_g} = -G_M\left(r_{outd}\,\|\,r_d\right) = \frac{-g_m}{g_{ds} + g_d\left(1 + g_{ms}r_s\right)} = \frac{-g_m r_d}{1 + g_{ms}r_s + r_d/r_{ds}}$$

$$A_{vbd} = \frac{v_d}{v_b} = -G_{MB}\left(r_{outd}\,\|\,r_d\right) = \frac{-g_{mb}}{g_{ds} + g_d\left(1 + g_{ms}r_s\right)} = \frac{-g_{mb} r_d}{1 + g_{ms}r_s + r_d/r_{ds}}$$

$$A_{vgs} = \frac{v_s}{v_g} = A_{vgso}\left(\frac{r_s}{r_s + r_{outs}}\right) = \frac{g_m}{g_{ms} + g_s\left(1 + r_d/r_{ds}\right)}$$

$$A_{vbs} = \frac{v_s}{v_b} = A_{vbso}\left(\frac{r_s}{r_s + r_{outs}}\right) = \frac{g_{mb}}{g_{ms} + g_s\left(1 + r_d/r_{ds}\right)}$$

$$A_{vsd} = \frac{v_d}{v_s} = g_{ms}\left(r_{ds}\,\|\,r_d\right) = \frac{g_{ms}}{g_{ds} + g_d}$$

Drain transresistance for current input at source:[a]

$$R_{sd} = \frac{v_d}{i_{ins}} = -H_{sd}\left(r_{outd}\,\|\,r_d\right) = \frac{-g_{ms}r_s}{g_{ds} + g_d\left(1 + g_{ms}r_s\right)} = \frac{-g_{ms}r_d}{g_{ms} + g_s\left(1 + r_d/r_{ds}\right)}$$

$g_{ms} = g_m + g_{mb} + g_{ds}$ is the source transconductance.
The MOS drain–body resistance, r_{db}, is not included, but can be included separately.
If the body is driven by the source, A_{vbso}, A_{vbs}, A_{vbd}, and G_{MB} no longer apply, and g_{mb}
  is set to zero where $g_{ms} = g_m + g_{ds}$.
[a] Only the single, relevant input signal is applied for these expressions.

When the body is driven by a voltage, the $v_b \cdot G_{MB}$ current source is active at the drain, and the $v_b \cdot A_{vbso}$ voltage source is active at the source. The current and voltage behavior and resistance levels at the drain and source are unchanged, but, as seen in Table 3.16, G_{MB} and A_{vbso} are considerably lower than G_M and A_{vgso} described earlier when the gate was driven because g_{mb} is considerably lower than g_m. The lower value of G_{MB} permits the design of low-transconductance, body-driven circuits. Additionally, analysis for body input signals permits noise analysis for body (substrate) noise.

Finally, when the source is driven by a current, i_{ins}, as for the top or non-rail device in a cascode circuit, the $i_{ins} \cdot H_{sd}$ current source is active at the drain. This current source replicates source input current not lost through current division into the external source resistance, r_s, and presents it into the parallel combination of r_{outd} and r_d at the drain for current-to-voltage or transresistance conversion.

H_{sd}, mentioned earlier, represents the current division of i_{ins} into the MOS source for a short-circuit drain, which is appropriate for a Norton equivalent circuit at the drain. If the body is connected and driven by the source, again, g_{mb} is set to zero, giving a lower value of g_{ms}. This results in a slight increase of the source input resistance, r_{ins}, through division by the lower value of g_{ms}. Regardless of the body connection, however, r_{ins} is strongly affected by the value of external drain resistance, r_d, since r_{ins} is increased through the multiplication by $1 + r_d/r_{ds}$ as mentioned earlier. If the drain is connected to an equal value of r_{ds} ($r_d = r_{ds}$), r_{ins} is doubled compared to its value present if the drain is connected to a short-circuit load. If the drain is connected to a high-resistance load associated with a cascoded circuit, r_{ins} increases significantly and can even approach or exceed the value of r_{ds}.

In addition to small-signal resistances and values associated with the Norton and Thévenin models at the drain and source, Table 3.16 also lists terminal, gate-to-drain, body-to-drain, gate-to-source, body-to-source, and source-to-drain voltage gains, A_{vgd}, A_{vbd}, A_{vgs}, A_{vbs}, and A_{vsd}, respectively, derived from the circuit model of Figure 3.22. Additionally, the table lists the source-to-drain transresistance, R_{sd}, which describes the drain voltage produced for a given level of source input current. The expressions for voltage gains at the drain are organized to show the effective transconductance, G_M or G_{MB}, multiplied by the total drain resistance given by the parallel combination of r_{outd} and the external drain resistance, r_d. The expressions for voltage gains at the source are organized to show the open-circuit voltage gains, A_{vgso} or A_{vgbo}, multiplied by voltage division terms associated with r_{outs} and the external source resistance, r_s.

The general circuit model of Figure 3.22, with expressions given in Table 3.16, permits rapid analysis of small-signal resistances and signal gains for gate voltage, body voltage, or source current input signals. Additionally, the model permits analysis for any level of external resistive loading, including resistive source degeneration. Resistive loading also includes short-circuit or open-circuit limits where, for example, r_s can be assigned a value of zero to give a grounded-source amplifier configuration, or assigned a value approaching infinity to give an unloaded, source follower amplifier configuration. Finally, the model permits an understanding of the role MOS small-signal parameters g_m, g_{mb}, and g_{ds} play on overall circuit performance.

3.8.2 Transconductance

Section 2.4 introduced MOS transconductance, g_m, the small-signal parameter describing the change in short-circuit drain current resulting from a change in controlling gate–source voltage, V_{GS}. It is given by the partial derivative of drain current with respect to V_{GS}, $g_m = \delta I_D / \delta V_{GS}$, with the drain–source and body–source voltages held constant. g_m is the gain factor for the gate–source voltage-controlled, dependent current source appearing between the drain and source in the MOS small-signal model shown earlier in Figure 3.21. g_m appears frequently in small-signal resistance, voltage gain, current gain, transconductance, and transresistance expressions listed in Table 3.16 for a MOSFET connected to the general circuit shown in Figure 3.22. As shown later in this chapter, intrinsic voltage gain (Section 3.8.5) and bandwidth (Section 3.9.6) are favorably increased while gate-referred thermal-noise voltage density (Section 3.10.2) is favorably decreased as g_m increases.

Section 2.4 also introduced MOS transconductance efficiency, g_m/I_D, the transconductance normalized to the drain bias current, I_D. Transconductance efficiency is a quality factor describing the production of desired transconductance for a given level of drain bias current. It has units of 1/V, or offering more intuition, units of $\mu S/\mu A$, which describes the transconductance in μS ($\mu A/V$) produced per unit of drain current in μA. As introduced in Section 2.4, MOS transconductance efficiency is maximum in weak inversion, decreases modestly in moderate inversion, and continues dropping in strong inversion, dropping even faster if velocity saturation effects are present.

3.8.2.1 Without small-geometry effects

Transconductance efficiency can be derived from the continuous, weak through strong-inversion, drain current expression given in Table 3.13 by differentiating the drain current with respect to the gate–source voltage followed by dividing by the drain current. This gives [13]

$$\frac{g_m}{I_D} = \frac{1 - e^{-\sqrt{IC}}}{nU_T\sqrt{IC}} \tag{3.34}$$

Equation 3.34 offers prediction accuracy to within 6 % for a body-effect factor $\gamma = 1\,V^{1/2}$ [13], but the combined exponential and square-root terms do not permit simple circuit optimization expressions.

A more accurate g_m/I_D expression is given by [13]

$$\frac{g_m}{I_D} = \frac{1}{nU_T\left(\sqrt{IC + 0.5\sqrt{IC} + 1}\right)} \tag{3.35}$$

This expression gives accuracy to within 3 % for a body-effect factor between $\gamma = 0.5$ and $2\,V^{1/2}$ [13], but involves a double square-root term that complicates simple circuit optimization expressions.

A simpler g_m/I_D expression is given by [14, 16, pp. 62–63]

$$\frac{g_m}{I_D} = \frac{1}{nU_T\left(\sqrt{IC + 0.25} + 0.5\right)} \tag{3.36}$$

This expression, while more accurate than the expression in Equation 3.34 but not as accurate as the expression in Equation 3.35, facilitates the derivation of simple circuit optimization expressions since it has only a single square-root term.

Figure 3.23 compares g_m/I_D predicted by the exponential expression of Equation 3.34, the double square-root expression of Equation 3.35, and the single square-root expression of Equation 3.36. The

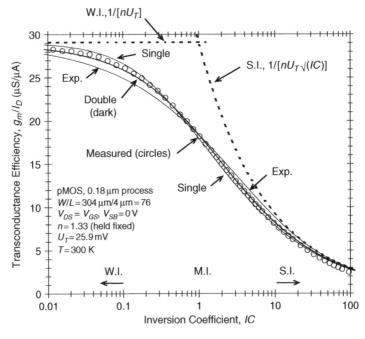

Figure 3.23 Comparison of different transconductance efficiency, g_m/I_D, predictions versus the inversion coefficient, IC, with measured overlay for an $L = 4\,\mu m$, pMOS device in a $0.18\,\mu m$ CMOS process. g_m/I_D predictions are from the exponential, double square-root, and single square-root expressions of Equations 3.34, 3.35, and 3.36, respectively. Predicted g_m/I_D is within measured values by 2.3 % and 3.7 % for the double and single square-root expressions, respectively, for $IC < 10$. For $IC > 10$, g_m/I_D is over-predicted since velocity saturation and VFMR effects are not included

figure compares predicted g_m/I_D with measured values for a $W/L = 304\,\mu\text{m}/4\,\mu\text{m}$, pMOS device in the $0.18\,\mu\text{m}$ CMOS process described in Table 3.2. A long-channel pMOS device was selected to minimize velocity saturation effects since the predictions do not include these effects. The inversion coefficient for measured data was found by $IC = I_D/[I_0(W/L)]$ from Table 3.6 using a technology current of $I_0 = 0.135\,\mu\text{A}$ from Table 3.2.

The exponential expression of Equation 3.34 under-predicts g_m/I_D in weak inversion by 5 % and over-predicts it by 9 % at the transition into strong inversion. The double square-root expression of Equation 3.35 under-predicts g_m/I_D in weak inversion by 1.3 % and over-predicts it by 2.3 % at the transition into strong inversion. Finally, the single square-root expression of Equation 3.36 over-predicts g_m/I_D in weak inversion by 1.8 % and under-predicts it by 3.7 % at the transition into strong inversion. As seen in Figure 3.23, the double square-root expression provides the most accurate prediction, but the single square-root expression provides a good prediction, especially given its simplicity. Predictions are compared to measured values for operation below the onset of strong inversion $(IC = 10)$ because all expressions over-predict g_m/I_D above this because, again, velocity saturation and VFMR effects are not included in the predictions. For $IC > 10$, VFMR effects lower measured g_m/I_D for the $L = 4\,\mu\text{m}$, pMOS device used for comparing the predictions.

All g_m/I_D predictions assume a fixed, average value of $n = 1.33$. Prediction accuracy can be improved if n is decreased slightly as the gate–source voltage (and effective gate–source voltage) increases with increasing inversion coefficient, as shown earlier in Figure 3.2 and later in Figure 3.30. However, computing a variable n greatly increases the complexity of the g_m/I_D predictions. In this book, we trade off some accuracy in order to derive simple expressions that permit design intuition and guidance.

Figure 3.23 shows that g_m/I_D in moderate inversion is significantly below values predicted using either the weak-inversion, $1/(nU_T)$, or strong-inversion, $1/(nU_T\sqrt{IC})$, expression listed on the figure. These expressions result from evaluating any of the g_m/I_D predictions at $IC < 0.1$ or $IC > 10$, respectively. At the center of moderate inversion where $IC = 1$, predicted g_m/I_D using the weak- and strong-inversion expressions is equal, which is consistent with the inversion coefficient definition described earlier in Section 3.4.2.1. However, measured g_m/I_D here is only 63 % of the value predicted by the weak- and strong-inversion expressions. The prediction error gets significantly worse if weak-inversion expressions are further misapplied in strong inversion, or if strong-inversion expressions are further misapplied in weak inversion. This emphasizes the importance of using a g_m/I_D expression that is continuously valid in weak, moderate, and strong inversion.

Table 3.17 lists expressions for MOS g_m/I_D along with dependencies on the design choices of inversion coefficient and channel length. The first expression is the single square-root expression from Equation 3.36 that will be used for predictions of performance in this book because of its simplicity and good accuracy. The second expression is the double square-root expression from Equation 3.35 that provides improved accuracy. Both expressions link g_m/I_D to the transconductance effective voltage, $V_{gm} = (g_m/I_D)^{-1}$, which is the reciprocal of g_m/I_D [4–11]. V_{gm} permits prediction of transconductance by $g_m = I_D/V_{gm}$, which is analogous to predicting drain–source conductance, $g_{ds} = I_D/V_A$, using the Early voltage, V_A. The first expression in the table additionally includes a linkage to $V_{DS,sat}$, which is developed later in Section 6.5.1. The g_m/I_D, V_{gm}, and $V_{DS,sat}$ linkages show that g_m/I_D and g_m decrease as V_{gm} and $V_{DS,sat}$ increase with increasing inversion coefficient. As described later in Sections 6.5.4 and 6.5.5, this decreases drain-referred thermal-noise current for non-input devices in low-noise circuits, but comes at the expense of requiring higher voltages across these devices to maintain operation in saturation.

For weak inversion at $IC < 0.1$, both expressions in Table 3.17 and the exponential expression of Equation 3.34 show that g_m/I_D approaches $1/(nU_T)$, which is the value introduced earlier in Equation 2.9 for weak-inversion, exponential, drain current operation. This is also the value found from the weak-inversion drain current of Table 3.13. For strong inversion at $IC > 10$, the g_m/I_D expressions

Table 3.17 MOS transconductance efficiency, g_m/I_D, expressions and trends in terms of inversion coefficient and channel length. When evaluated in terms of the inversion coefficient, g_m/I_D is independent of the drain current

Transconductance efficiency, g_m/I_D	$IC \uparrow$ L fixed	$L \uparrow$ IC fixed
All regions without small-geometry effects: $$\frac{g_m}{I_D} = \frac{1}{V_{gm}} = \frac{1}{nU_T\left(\sqrt{IC+0.25}+0.5\right)} = \frac{2}{n\left(V_{DS,sat}-2U_T\right)}$$ where $V_{DS,sat}$ is from Table 3.15	Unchanged in WI	
$$\frac{g_m}{I_D} = \frac{1}{V_{gm}} = \frac{1}{nU_T\left(\sqrt{IC+0.5\sqrt{IC}+1}\right)}$$ (above gives improved accuracy)		Unchanged
$\approx 1/(nU_T)$ in WI	$\downarrow\propto \dfrac{1}{\sqrt{IC}}$ in SI	
$\approx 1/(nU_T\sqrt{IC}) = 2/V_{EFF}$ in SI See Table 3.18 for representative g_m/I_D values		
All regions, with velocity saturation and VFMR effects: Replace IC in above expressions with $IC\left(1+\dfrac{IC}{IC_{CRIT}}\right)$ where $IC_{CRIT} = \left(\dfrac{(LE_{CRIT})'}{4nU_T}\right)^2$ and $(LE_{CRIT})' = \dfrac{1}{\theta} \parallel LE_{CRIT}$		
Strong inversion with full velocity saturation $(VSF' = V_{EFF}/(LE_{CRIT})' \gg 1$, or $IC \gg IC_{CRIT})$: $$\frac{g_m}{I_D} = \frac{1}{V_{gm}} = \frac{(LE_{CRIT})'}{(2nU_T)^2\,IC} = \frac{4IC_{CRIT}}{(LE_{CRIT})'\,IC} = \frac{1}{V_{EFF}}$$	$\downarrow\propto \dfrac{1}{IC}$	$\uparrow\propto L$

g_m/I_D is for saturation where $V_{DS} > V_{DS,sat}$.

show that g_m/I_D approaches $1/(nU_T\sqrt{IC})$. This is consistent with $g_m/I_D = 2/V_{EFF}$ introduced earlier in Equation 2.13 for strong-inversion, square-law, drain current operation using $V_{EFF} = 2nU_T\sqrt{IC}$ from Equation 3.22. This is also the g_m/I_D value found from the strong-inversion drain current of Table 3.13 with drain current expressed from the inversion coefficient in Equation 3.16.

As noted in Table 3.17, g_m/I_D is constant and maximum in weak inversion. It decreases modestly in moderate inversion before decreasing as $1/\sqrt{IC}$ in strong inversion. As noted, g_m/I_D is independent of channel length when velocity saturation effects are excluded. Unfortunately, at high IC, velocity saturation significantly decreases g_m/I_D for short-channel devices, while VFMR moderately decreases g_m/I_D for devices of any channel length. These effects are discussed in the next section.

3.8.2.2 With velocity saturation and VFMR effects

The reduction of drain current associated with velocity saturation and VFMR was discussed in Sections 3.7.1.2–3.7.1.6. This reduction results in a reduction of g_m/I_D as well. We will initially consider the decrease in g_m/I_D in strong inversion where the effects of velocity saturation and VFMR are most prevalent.

MOS strong-inversion drain current using the simple correction factor to model velocity saturation and VFMR effects is repeated from Table 3.13 as

$$I_D = \frac{\frac{1}{2}\left(\frac{\mu_0 C'_{OX}}{n}\right)\left(\frac{W}{L}\right)V_{EFF}^2}{1 + \frac{V_{EFF}}{(LE_{CRIT})'}} \tag{3.37}$$

As described earlier in Section 3.7.1.5, $(LE_{CRIT})'$ is the equivalent velocity saturation voltage, which is equal to the velocity saturation voltage, LE_{CRIT}, in parallel with the mobility reduction voltage, $1/\theta$.

Differentiating Equation 3.37 with respect to the gate–source voltage or V_{EFF} gives the transconductance. Dividing this by the drain current then gives the transconductance efficiency, which is equal to

$$\frac{g_m}{I_D} = \frac{2}{V_{EFF}}\left(\frac{1 + VSF'/2}{1 + VSF'}\right) \tag{3.38}$$

$VSF' = V_{EFF}/(LE_{CRIT})'$ is the velocity saturation factor including VFMR effects from Section 3.7.1.5.

When $VSF' = V_{EFF}/(LE_{CRIT})' \ll 1$, velocity saturation and VFMR effects are negligible. Here, Equation 3.38 predicts $g_m/I_D = 2/V_{EFF}$, which is consistent with g_m/I_D introduced earlier in Equation 2.13 for strong-inversion, square-law drain current where drain current increases as the square of V_{EFF}. Under these conditions, g_m/I_D is given by

$$\frac{g_m}{I_D}\text{ (SI, no vel. sat., VFMR)} = \frac{2}{V_{EFF}\text{ (SI, no vel. sat., VFMR)}} = \frac{1}{nU_T\sqrt{IC}} \tag{3.39}$$

where V_{EFF} excluding velocity saturation and VFMR effects is from Equation 3.22. Equation 3.39 is also equal to the strong-inversion limits of the g_m/I_D expressions that exclude velocity saturation and VFMR effects listed in Table 3.17. When these effects are negligible, g_m/I_D is inversely proportional to the square root of inversion coefficient or drain current for a fixed-geometry device. As throughout this book, the inversion coefficient and drain current track each other for a fixed-geometry device since n and mobility are assumed fixed in the inversion coefficient definition given in Equation 3.6 and mentioned in Section 3.4.2.3. Mobility reduction associated with VFMR and velocity saturation effects is considered outside the inversion coefficient definition.

When $VSF' = V_{EFF}/(LE_{CRIT})' \gg 1$, devices are in full velocity saturation. Here, Equation 3.38 shows that g_m/I_D is reduced to one-half its square-law value or $1/V_{EFF}$. This is consistent with $g_m/I_D = 1/V_{EFF}$ introduced earlier in Equation 2.19 for strong-inversion, linear-law drain current where drain current increases linearly with V_{EFF}. Under these conditions, g_m/I_D is given by

$$\frac{g_m}{I_D}\text{ (SI, full vel. sat.)} = \frac{1}{V_{EFF}\text{ (SI, full vel. sat.)}} = \frac{(LE_{CRIT})'}{(2nU_T)^2 IC} \tag{3.40}$$

where velocity-saturated V_{EFF} is from Equation 3.25. Equation 3.40 is also the value of g_m/I_D found by solving the strong-inversion, velocity-saturated drain current of Table 3.13 with drain current expressed from the inversion coefficient in Equation 3.16. Under full velocity saturation, g_m/I_D decreases inversely with the inversion coefficient or drain current for a fixed-geometry device.

Rather than attempt to estimate g_m/I_D from Equation 3.38, we will replace IC that appears in the g_m/I_D expressions of Table 3.17 (Equations 3.35 and 3.36) with a modified value to estimate velocity saturation and VFMR effects. This estimate of g_m/I_D involves a non-physically based interpolation from strong-inversion, square-law and fully velocity-saturated, linear-law, drain current operation and is analogous to the interpolation described earlier in Section 3.7.2.2 for V_{EFF}.

The g_m/I_D interpolation expression must smoothly connect the strong-inversion, square-law, asymptotic value exclusive of velocity saturation and VFMR effects given by Equation 3.39 with the fully velocity-saturated, linear-law, asymptotic value given by Equation 3.40. These asymptotic values are equal at a critical inversion coefficient given by

$$IC_{CRIT} = \left[\frac{(LE_{CRIT})'}{4nU_T} \right]^2 \tag{3.41}$$

Just as $IC = 1$ denotes the transition point between weak- and strong-inversion g_m/I_D without velocity saturation and VFMR effects, $IC = IC_{CRIT}$ denotes the transition point between strong-inversion g_m/I_D without velocity saturation and VFMR effects and the velocity-saturated value.

Similar to the substitution of IC described for V_{EFF} earlier in Equation 3.27, we consider replacing IC in the g_m/I_D expressions of Table 3.17 (Equations 3.35 and 3.36) with

$$IC \text{ (replacement for } g_m/I_D) = IC \left(1 + \frac{IC}{IC_{CRIT}} \right) \tag{3.42}$$

to include velocity saturation and VFMR effects. When $IC \ll IC_{CRIT}$, velocity saturation and VFMR effects are negligible and IC is nearly unchanged as it appears in the g_m/I_D expressions in the table. As a result, these equations are unchanged, giving g_m/I_D exclusive of velocity saturation and VFMR effects. However, under full velocity saturation when $IC \gg IC_{CRIT}$, IC is replaced with $IC \cdot (IC/IC_{CRIT})$ in the g_m/I_D expressions. As a result, these equations give the velocity-saturated value of g_m/I_D given in Equation 3.40.

In addition to converging to the values present without and with full velocity saturation, it is also necessary that the g_m/I_D interpolation between negligible and full velocity saturation using Equation 3.42 be sufficiently accurate. This will be first shown by comparing predicted and measured g_m/I_D for an $L = 0.5\,\mu m$, nMOS device in a $0.5\,\mu m$ CMOS process that has a large region of operation (high IC_{CRIT}) before velocity saturation effects become significant. This will then be followed by comparing predicted and measured g_m/I_D for an $L = 0.18\,\mu m$, nMOS device in a $0.18\,\mu m$ CMOS process that has a small region of operation (low IC_{CRIT}) before these effects become significant. The comparison of measured and predicted g_m/I_D will also show that the g_m/I_D interpolation, which was derived for operation in strong inversion, can be used to predict the small velocity saturation decrease in g_m/I_D present in moderate inversion, at least for channel lengths down to $0.18\,\mu m$.

Figure 3.24 shows predicted g_m/I_D from weak through strong inversion overlaid with measured values for an $L = 0.5\,\mu m$, nMOS device in the $0.5\,\mu m$ CMOS process described in Table 3.2. g_m/I_D is predicted using the double square-root expression in Table 3.17 (Equation 3.35) using the value of IC modified by Equation 3.42, also listed in the table, to consider velocity saturation and VFMR effects. A fixed value of $n = 1.35$, $U_T = 25.9\,mV$ $(T = 300\,K)$, $E_{CRIT} = 4\,V/\mu m$, and $\theta = 0.14/V$ are used from the process values of Table 3.2. The inversion coefficient for measured data is found by $IC = I_D/[I_0(W/L)]$ from Table 3.6 using a technology current of $I_0 = 0.21\,\mu A$ from Table 3.2. Finally, the channel length of $L = 0.5\,\mu m$ includes the reduction by $DL = 0.1\,\mu m$ (also listed in Table 3.2) for a drawn length of $0.6\,\mu m$.

In Figure 3.24, g_m/I_D is maximum and constant in weak inversion, tracking the weak-inversion value of $g_m/I_D = 1/(nU_T)$. It decreases modestly (by about 37 % at $IC = 1$) in moderate inversion before dropping as the inverse square root of IC, tracking the strong-inversion value of $g_m/I_D = 1/(nU_T\sqrt{IC})$ from Equation 3.39 for negligible velocity saturation and VFMR effects. g_m/I_D then decreases further and approaches the velocity-saturated value of $g_m/I_D = (LE_{CRIT})'/[(2nU_T)^2 IC]$ from Equation 3.40 where it decreases inversely with IC.

Figure 3.24 reveals two critical values for the inversion coefficient. One is the center of moderate inversion at $IC = 1$ where the predicted, asymptotic values of weak- and strong-inversion values of g_m/I_D are equal, excluding velocity saturation and VFMR effects. Measured g_m/I_D data can be used to verify and, if needed, calibrate the technology current, $I_0 = 2n_0\mu_0 C'_{OX}U_T^2$, used in the inversion coefficient calculation, $IC = I_D/[I_0(W/L)]$, described in Table 3.6. When properly calibrated, the asymptotes of weak- and strong-inversion g_m/I_D intercept at $IC = 1$ as shown in the figure. This,

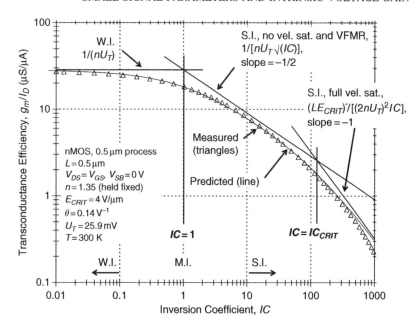

Figure 3.24   Predicted and measured transconductance efficiency, g_m/I_D, with trends versus inversion coefficient, IC, for an $L = 0.5\,\mu m$, nMOS device in a $0.5\,\mu m$ CMOS process. g_m/I_D is maximum in weak inversion, decreases modestly in moderate inversion, and decreases as $1/\sqrt{IC}$ in strong inversion when velocity saturation and VFMR effects are small. At high levels of IC where velocity saturation is significant for this short-channel device, g_m/I_D decreases significantly and is nearly proportional to $1/IC$. Operation at $IC = IC_{CRIT}$ corresponds to the g_m/I_D transition between values associated with strong-inversion, square-law, and fully velocity-saturated, linear-law, drain current. The device shown can operate at unusually high IC because of high allowable supply voltage and V_{EFF} in the $0.5\,\mu m$ process

however, may only be clearly observed for sufficiently long channel length devices in the process that have a sufficient region of strong-inversion g_m/I_D operation without significant velocity saturation effects.

The second critical value of the inversion coefficient occurs at $IC = IC_{CRIT} = [(LE_{CRIT})'/(4nU_T)]^2 = 125$ (expression given in Table 3.17 and in Equation 3.41) where the predicted strong-inversion g_m/I_D, excluding velocity saturation and VFMR effects, is equal to the predicted velocity saturated value. Here, g_m/I_D is down to approximately 70.7 % of the value present without velocity saturation and VFMR effects. This is because $IC = 2IC$ from Equation 3.42 is placed in the continuous, weak-through strong-inversion expressions for g_m/I_D given in Table 3.17 (Equations 3.35 and 3.36). In strong inversion at $IC > 10$, this gives $g_m/I_D = 1/[nU_T(2IC)^{1/2}]$ compared to $g_m/I_D = 1/[nU_T(IC)^{1/2}]$ present without velocity saturation and VFMR effects.

Figure 3.25 shows predicted g_m/I_D overlaid with measured values for an $L = 0.18\,\mu m$, nMOS device in the $0.18\,\mu m$ CMOS process described in Table 3.2. g_m/I_D is predicted using a fixed value of $n = 1.33$, thermal voltage of $U_T = 25.9\,mV$ ($T = 300\,K$), $E_{CRIT} = 5.6\,V/\mu m$, and $\theta = 0.28/V$ from the process values in the table. Channel length is adjusted from the $L = 0.18\,\mu m$ drawn value given by subtracting $DL = 0.028\,\mu m$, which is also listed in the table. The inversion coefficient for measured data is found using a technology current of $I_0 = 0.64\,\mu A$ from the table.

As seen in Figure 3.24 for the longer $L = 0.5\,\mu m$ device, g_m/I_D for the $L = 0.18\,\mu m$ device in Figure 3.25 decreases in strong inversion from its normal, strong-inversion, square-law value due to velocity saturation and VFMR effects. However, the decrease occurs at much lower inversion coefficients for the $L = 0.18\,\mu m$ device where $IC_{CRIT} = [(LE_{CRIT})'/(4nU_T)]^2 = 25$ compared to

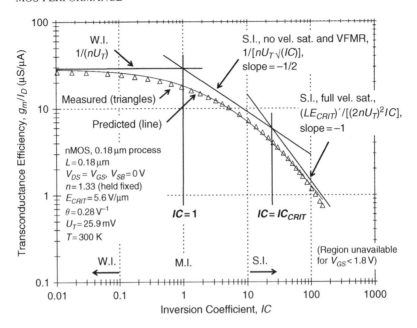

Figure 3.25 Predicted and measured transconductance efficiency, g_m/I_D, with trends versus inversion coefficient, IC, for an $L = 0.18\,\mu m$, nMOS device in a $0.18\,\mu m$ CMOS process. The shape of the g_m/I_D versus IC curve is identical to that shown in Figure 3.24 for an $L = 0.5\,\mu m$, nMOS device in a $0.5\,\mu m$ CMOS process. However, IC_{CRIT}, corresponding to the g_m/I_D transition associated with strong-inversion, square-law, and velocity-saturated, linear-law, drain current is considerably lower for this shorter-channel, $L = 0.18\,\mu m$ device. As a result, g_m/I_D never reaches the strong-inversion, square-law value predicted without velocity saturation and VFMR effects

125 found for the $L = 0.5\,\mu m$ device. As a result, g_m/I_D decreases significantly in strong inversion for the $L = 0.18\,\mu m$, nMOS device and never reaches the predicted strong-inversion value without velocity saturation and VFMR effects shown on the figure. As seen later in Sections 3.8.5 and 3.9.6, this decrease in g_m/I_D lowers intrinsic voltage gain and bandwidth.

For the $L = 0.5\,\mu m$ device shown in Figure 3.24, predicted g_m/I_D is within 5 % of measured values for $IC < 160$ and is over-predicted by 10 % at the high value of $IC = 400$. For the $L = 0.18\,\mu m$ device shown in Figure 3.25, predicted g_m/I_D is within 7 % of measured values for $IC < 100$ and is over-predicted by 10 % at the high value of $IC = 120$. Since both velocity saturation and VFMR effects are merged into a single, equivalent, velocity saturation voltage, $(LE_{CRIT})'$, in the simple drain current correction factor of Table 3.13 used for predicting g_m/I_D, predicted g_m/I_D is proportional to $1/IC$ under full velocity saturation. At very high IC, however, g_m/I_D potentially decreases more than this due to separate velocity saturation and VFMR effects as seen in the enhanced and complex correction factors in the table. Additionally, effects like source contact and bulk resistance degenerate g_m and lower the value of g_m/I_D further. All of this likely results in some g_m/I_D over-prediction at very high IC, but such high levels of inversion are not commonly used in design because of high V_{EFF} and $V_{DS,sat}$.

Predicted g_m/I_D is adequate for design guidance, even when g_m/I_D is decreased significantly due to significant velocity saturation effects. For the $L = 0.5\,\mu m$ device operating at $IC = 400$, or the $L = 0.18\,\mu m$ device operating at $IC = 100$, measured g_m/I_D is down by 57 % from the value predicted without velocity saturation and VFMR effects. This illustrates the importance of considering these effects.

As observed in Equation 3.41 (listed in Table 3.17), IC_{CRIT} for g_m/I_D is proportional to the square of $(LE_{CRIT})'$. For short-channel devices where velocity saturation dominates over VFMR, IC_{CRIT} is nearly proportional to the square of the velocity saturation voltage, LE_{CRIT}, or the square of channel length and E_{CRIT}. This suggests that IC_{CRIT} might be approximately 13 % ($(0.065\,\mu m/0.18\,\mu m)^2$) of the value of 25 shown in Figure 3.25 for an $L = 0.18\,\mu m$, nMOS device, or 3.3 for an $L = 0.065\,\mu m$, nMOS device. Such a critical value of IC_{CRIT} is well into moderate inversion and is likely too low since IC_{CRIT} was derived for operation in strong inversion. A modified interpolation in Equation 3.42 may be required as velocity saturation effects begin to encroach closer to moderate inversion for very short-channel devices. The designer can minimize velocity saturation reduction of g_m/I_D by increasing IC_{CRIT} by increasing channel length and using pMOS devices that have higher E_{CRIT}. Additionally, operation can be confined to lower IC values in moderate inversion.

3.8.2.3 Predicted and measured values

Figure 3.26 shows measured and predicted g_m/I_D for $L = 0.18, 0.28, 0.48, 1$, and $4\,\mu m$, nMOS devices in the $0.18\,\mu m$ process described in Table 3.2. Measurements for $L = 2\,\mu m$ devices are not shown but lie essentially on the $L = 4\,\mu m$ values. g_m/I_D is predicted as described for Figure 3.24, using the process parameters previously mentioned for Figure 3.25. g_m/I_D is presented as a function of the inversion coefficient from $IC = 0.01$ (deep weak inversion) to 100 (deep strong inversion) over a wide range of channel lengths, providing a full characterization of the process for analog design.

In Figure 3.26, g_m/I_D is constant and maximum in weak inversion,[9] before decreasing modestly in moderate inversion. In strong inversion, g_m/I_D decreases as $1/\sqrt{IC}$ for the long-channel devices that

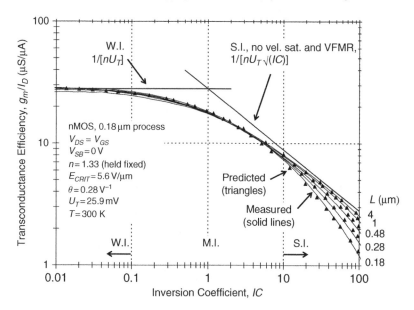

Figure 3.26 Predicted and measured transconductance efficiency, g_m/I_D, versus inversion coefficient, IC, for $L = 0.18, 0.28, 0.48, 1$, and $4\,\mu m$, nMOS devices in a $0.18\,\mu m$ CMOS process. g_m/I_D is maximum in weak inversion, decreases modestly in moderate inversion, and decreases as $1/\sqrt{IC}$ in strong inversion. For short-channel devices at high IC, g_m/I_D decreases significantly due to velocity saturation and is nearly proportional to $1/IC$. At high IC, g_m/I_D decreases modestly for all channel lengths due to VFMR effects. g_m/I_D is similar for all CMOS processes

---

[9] g_m/I_D actually peaks very slightly in weak inversion before decreasing slightly for operation deeper into weak inversion because of the slight increase in n.

experience little velocity saturation. However, at high levels of IC, g_m/I_D decreases nearly as $1/IC$ for the short-channel devices that experience significant velocity saturation. g_m/I_D decreases more for short channel lengths due to increased velocity saturation effects, while it decreases modestly for all channel lengths due to VFMR effects. No velocity saturation decrease in g_m/I_D is observed in weak inversion and little decrease is observed in moderate inversion. Small velocity saturation effects in moderate inversion, even for short-channel devices, are a major advantage of operation here. The decrease in g_m/I_D in strong inversion for short channel lengths, however, is significant and must be considered in design.

Figure 3.27 shows measured and predicted g_m/I_D for $L = 0.18$, 0.28, and $1\,\mu m$, pMOS devices in the $0.18\,\mu m$ process described in Table 3.2. Measurements for $L = 0.48$, 2, and $4\,\mu m$ devices are not shown but lie essentially on the $L = 1\,\mu m$ values. g_m/I_D is predicted using a fixed value of $n = 1.33$, thermal voltage of $U_T = 25.9\,mV$ ($T = 300\,K$), $E_{CRIT} = 14\,V/\mu m$, and $\theta = 0.35/V$ from Table 3.2. Channel length is adjusted from the drawn values given by subtracting $DL = 0.051\,\mu m$, which is also listed in the table. The inversion coefficient for measured data is found using a technology current of $I_0 = 0.135\,\mu A$ from the table.

In weak and moderate inversion ($IC < 10$), pMOS g_m/I_D shown in Figure 3.27 is nearly identical to that of nMOS devices shown in Figure 3.26. However, the g_m/I_D decrease in deep strong inversion ($IC = 100$) at short channel lengths is significantly less for pMOS devices because of less velocity saturation resulting from a higher $E_{CRIT} = 14\,V/\mu m$ compared to $5.6\,V/\mu m$ for nMOS devices. While the g_m/I_D decrease at short channel lengths is greater for nMOS devices due to velocity saturation, the g_m/I_D decrease at long channel lengths is greater for pMOS devices because of higher VFMR effects resulting from a higher mobility reduction factor of $\theta = 0.35/V$ compared to $0.28/V$ for nMOS devices. Like the drain current shown in Figures 3.11 and 3.12, and V_{EFF} shown in Figures 3.14 and 3.15, pMOS g_m/I_D is nearly equal to that of nMOS devices shown if pMOS channel length is decreased

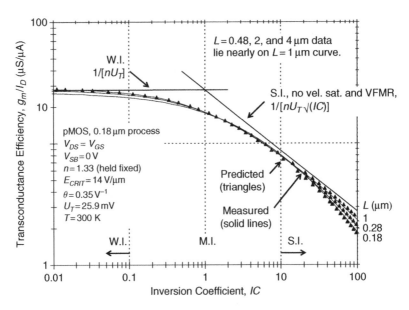

Figure 3.27 Predicted and measured transconductance efficiency, g_m/I_D, versus inversion coefficient, IC, for $L = 0.18$, 0.28, and $1\,\mu m$, pMOS devices in a $0.18\,\mu m$ CMOS process. g_m/I_D decreases less at high IC for short-channel devices compared to the nMOS values shown in Figure 3.26 because of lower pMOS velocity saturation. At high IC, g_m/I_D decreases modestly for all channel lengths due to VFMR effects. Measurements not shown for $L = 0.48$, 2, and $4\,\mu m$ devices lie nearly on the measured, $L = 1\,\mu m$ curve. g_m/I_D is similar for all CMOS processes

by a factor of 2.5 $((14\,\text{V}/\mu\text{m})/(5.6\,\text{V}/\mu\text{m}))$, the ratio of pMOS and nMOS E_{CRIT}. This can be seen where $L = 0.18\,\mu\text{m}$, pMOS g_m/I_D shown in Figure 3.27 is nearly equal to $L = 0.48\,\mu\text{m}$, nMOS g_m/I_D shown in Figure 3.26. As mentioned and observed in the figures, there is some additional pMOS g_m/I_D decrease due to higher VFMR effects.

Predicted nMOS g_m/I_D in Figure 3.26 is within measured values by 5 % for $IC < 65$. For $IC >$ 65, g_m/I_D is over-predicted by a maximum of 10 % for some channel lengths. This over-prediction, as mentioned earlier for Figures 3.24 and 3.25, is likely a result of merging velocity saturation and VFMR effects and excluding source resistance. Predicted pMOS g_m/I_D in Figure 3.27 is within measured values by 5 % for all IC. Similar prediction accuracy was also obtained for nMOS and pMOS devices in the 0.5 μm CMOS process described in Table 3.2.

g_m/I_D prediction accuracy is clearly adequate for design guidance and is encouraging since a fixed value of n and only a few, fixed (e.g., not binned for channel length) process parameters are used for prediction over the wide range of inversion level and channel lengths. Accuracy can be improved for a particular region of inversion by slightly adjusting the value of n, since n decreases slightly as inversion increases, or by binning process parameters for different channel lengths. However, the intention here is to develop simple expressions for design intuition and optimization, recognizing that computer simulations using complete MOS models are required to validate a candidate design. The simple g_m/I_D prediction described here permits hand or spreadsheet calculation and is valid from weak through strong inversion, inclusive of velocity saturation and VFMR effects.

While g_m/I_D was predicted using the double square-root expression of Table 3.17 (Equation 3.35), similar accuracy is obtained using the single square-root expression in the table (Equation 3.36), which is used in design expressions in this book because of its simplicity. As mentioned earlier for Figure 3.23, the single square-root expression under-predicts g_m/I_D by 3.7 % at the transition into strong inversion compared to an over-prediction of 2.3 % by the double square-root expression.

Predicted and measured g_m/I_D values, like those shown in Figure 3.25, were taken for IC up to values of 150 to 300 depending on the channel length where the supply voltage limit of $V_{GS} = 1.8\,\text{V}$ was reached. Data shown in Figures 3.26 and 3.27, however, was limited to IC values below 100 as operation above this is unusually high for a supply voltage of 1.8 V. Like measured V_{EFF} shown in Figures 3.14 and 3.15, g_m/I_D was measured for $V_{DS} = V_{GS}$, which corresponds to the bias conditions for a "diode-connected" MOSFET that ensures consistent saturation operation regardless of the level of inversion. Actual operation would likely be for V_{DS} somewhat below V_{GS}, especially for low-voltage design. However, g_m/I_D decreases only slightly as V_{DS} decreases, providing operation in saturation is maintained. Variations in g_m/I_D with V_{DS} are considered by the required computer simulations of a candidate design using full MOS models.

3.8.2.4 Summary of trends

As mentioned and listed in Table 3.17, $g_m/I_D = 1/(nU_T)$ is maximum and constant in weak inversion. It decreases modestly in moderate inversion before decreasing as $1/\sqrt{IC}$ in strong inversion, approaching $g_m/I_D = 1/(nU_T\sqrt{IC})$. g_m/I_D decreases modestly below this due to VFMR effects, but decreases significantly for short-channel devices experiencing velocity saturation. Under full velocity saturation, g_m/I_D decreases as $1/IC$, approaching $g_m/I_D = (LE_{CRIT})'/[(2nU_T)^2 IC]$. When velocity saturation effects are negligible, as for long-channel devices, g_m/I_D is independent of the channel length. However, under conditions of full velocity saturation, g_m/I_D increases nearly directly with channel length. To maintain a given value of g_m/I_D under full velocity saturation, IC must decrease as channel length decreases as seen in the expression for velocity-saturated g_m/I_D above. Finally, like V_{EFF} and $V_{DS,sat}$ given in Tables 3.14 and 3.15, g_m/I_D is independent of the selected drain current when evaluated in terms of the inversion coefficient.

Table 3.18 lists predicted g_m/I_D from Figure 3.26 for $L = 0.18$, 0.28, 0.48, 1, and 4 μm, nMOS devices in the 0.18 μm process described in Table 3.2. g_m/I_D is listed as a percentage of its maximum

Table 3.18   Transconductance efficiency, g_m/I_D, expressed as a percentage of its maximum weak inversion value for $L = 0.18$, 0.28, 0.48, 1, and 4 μm, nMOS devices in a 0.18 μm CMOS process. In strong inversion, g_m/I_D decreases from its large-geometry value for short-channel devices due primarily to velocity saturation effects. The decrease is less for short-channel pMOS devices since the velocity saturation, critical electric field is higher

IC	Region	g_m/I_D as a percentage of its weak-inversion value[a]					
		$L = $ Large[b] $IC_{CRIT} = \to \infty$	4 μm 500	1 μm 242	0.48 μm 116	0.28 μm 54.1	0.18 μm 25
0.01	WI	97.1	97.1	97.1	97.1	97.1	97.1
0.02	WI	95.8	95.8	95.8	95.8	95.8	95.8
0.05	WI	92.8	92.8	92.8	92.8	92.8	92.8
0.1	WI	89.2	89.2	89.2	89.2	89.2	89.2
0.2	MI	83.8	83.8	83.8	83.8	83.8	83.7
0.5	MI	73.5	73.4	73.4	73.4	73.3	73.2
1	MI	63.2	63.2	63.2	63.1	63.0	62.6
2	MI	51.9	51.9	51.8	51.7	51.3	50.7
5	MI	37.5	37.3	37.2	36.9	36.2	34.9
10	SI	28.2	28.0	27.7	27.2	26.2	24.3
20	SI	20.7	20.4	20.0	19.3	18.0	15.8
50	SI	13.5	12.9	12.4	11.4	9.9	8.0
100	SI	9.7	8.9	8.2	7.2	5.8	4.4

Values are for saturation where $V_{DS} > V_{DS,sat}$.
g_m/I_D and IC_{CRIT} are found for $n = 1.33$, $U_T = 25.9$ mV $(T = 300$ K$)$, $E_{CRIT} = 5.6$ V/μm, $\theta = 0.28$/V, and $DL = 0.028$ μm.
[a] The maximum value of g_m/I_D in weak inversion is 29.1 μS/μA for n and U_T given.
[b] Large denotes large geometry with long channel length and thick gate oxide resulting in no velocity saturation or VFMR effects.

weak inversion value, $1/(nU_T)$, which permits an understanding of the normal decrease with increasing inversion coefficient and the additional decrease associated with velocity saturation and VFMR effects.

g_m/I_D is near its maximum weak-inversion value in weak inversion and is 89.2 % of this value at the onset of moderate inversion at $IC = 0.1$. At the center of moderate inversion at $IC = 1$, g_m/I_D is down modestly to 63.2 % of its weak-inversion value. However, at the onset of strong inversion at $IC = 10$, g_m/I_D is down significantly to 28 % of its weak-inversion value for the $L = 4$ μm device and is down further at 24.3 % for the $L = 0.18$ μm device because of some velocity saturation effects. When IC increases to deep strong inversion at $IC = 100$, g_m/I_D is 8.9 % and 4.4 % of its weak-inversion value for $L = 4$ and 0.18 μm devices, respectively. For the $L = 0.18$ μm device, this is one-half the already low value for the long-channel, $L = 4$ μm device due primarily to velocity saturation. VFMR effects drop g_m/I_D modestly at $IC = 100$ as seen by comparing the $L = 4$ μm device that has negligible velocity saturation effects with the device denoted as large geometry that has no velocity saturation or VFMR effects. Interestingly, operation even up to the strong inversion side of moderate inversion at $IC = 5$ results in little velocity saturation loss of g_m/I_D for the $L = 0.18$ μm device. This, again, illustrates an advantage of operation in moderate inversion where velocity saturation effects are small.

Also noted in the table, IC_{CRIT} is 500, 242, 116, 54.1, and 25 for $L = 4$, 1, 0.48, 0.28, and 0.18 μm devices, respectively. As mentioned, this denotes the IC breakpoint where g_m/I_D is down to approximately 70.7 % of the value present without velocity saturation and VFMR effects. Operation at IC well below IC_{CRIT} ensures negligible velocity saturation and VFMR loss of g_m/I_D.

The decrease of g_m/I_D with IC given in Table 3.18 is slightly less than shown, at least for the device denoted as large geometry that has no velocity saturation or VFMR effects. This is because, as shown earlier in Figure 3.2 and later in Figure 3.30, n decreases slightly with increasing gate–source voltage (and the effective gate–source voltage, $V_{EFF} = V_{GS} - V_T$) associated with increasing inversion level. Additionally, as shown in Figure 3.2, n decreases by about 6 % in the presence of $V_{SB} = 1$ V, which is a large value of V_{SB} for low-voltage designs. When n decreases with the inversion level or V_{SB}, g_m/I_D increases slightly because is it inversely proportional to n. Finally, g_m/I_D decreases with temperature because it is inversely proportional to the thermal voltage.

3.8.2.5 Universal g_m/I_D characteristic in CMOS technologies

Like V_{EFF} and $V_{DS,sat}$ summarized in Tables 3.14 and 3.15, g_m/I_D (summarized in Table 3.17) is largely independent of the MOS process when evaluated in terms of the inversion coefficient. Excluding small-geometry effects like velocity saturation and VFMR, g_m/I_D is independent of the process except for the substrate factor, n, which has typical values of 1.4, 1.35, and 1.3 in weak, moderate, and strong inversion for typical bulk CMOS processes. As mentioned in Section 3.2.4, the development of process-independent design expressions is one of the advantages of using the inversion coefficient in analog CMOS design. Like V_{EFF}, g_m/I_D is also independent of channel length when small-geometry effects are excluded. The decrease in g_m/I_D associated with velocity saturation effects depends on the process through the critical, velocity saturation, electric field, E_{CRIT}, and the channel length. The smaller decrease associated with VFMR depends on the process through the mobility reduction factor, θ.

The characteristic shape of the g_m/I_D versus inversion coefficient curve is *universal* for all CMOS technologies as seen in Figures 3.24 and 3.25, where $IC = 1$ is associated with the asymptotic intersection of weak- and strong-inversion g_m/I_D, excluding velocity saturation and VFMR effects. Additionally, $IC = IC_{CRIT}$ is associated with the asymptotic intersection of strong-inversion g_m/I_D, again, excluding velocity saturation and VFMR effects, and velocity-saturated g_m/I_D. The universal g_m/I_D characteristic is also observed by comparing g_m/I_D for an $L = 0.5\,\mu$m device in a 0.5 μm process shown in Figure 3.24 with an almost identical length, $L = 0.48\,\mu$m device in a 0.18 μm process shown in Figure 3.26. While process parameters are considerably different for these different processes as described in Table 3.2, g_m/I_D is nearly identical for these nearly equal-length, nMOS devices, although IC_{CRIT} will vary with the process E_{CRIT} and θ.

Additionally, measured g_m/I_D for a 0.5 μm, fully depleted (FD), silicon-on-insulator (SOI), CMOS process has an identical g_m/I_D characteristic as the bulk processes characterized here even though this process does not have a body or substrate connection [29]. The weak inversion value of g_m/I_D, however, is higher at 32 μS/μA ($T = 300$ K) compared to approximately 27.6 μS/μA for the bulk processes. This is because of less substrate effect for the FD SOI process where weak-inversion n is approximately 1.2 compared to approximately 1.4 for the bulk processes.

In addition to permitting the prediction of device and circuit performance in weak, moderate, and strong inversion with a considerable degree of technology independence, the inversion coefficient is useful to characterize processes for analog design as done here and in [21, 29–32]. This permits experimental characterization of small-signal parameters like g_m/I_D and $V_A = I_D/g_{ds}$ (described later in Section 3.8.4) from weak through strong inversion over the full range of available channel length. Additionally, inversion coefficient evaluation of g_m/I_D and V_A provides a rigorous benchmark for computer simulation MOS models [30, 31].

3.8.2.6 Distortion

In the small-signal model shown earlier in Figure 3.21, the small-signal, gate–source voltage, v_{gs}, and drain current, i_d, represent sufficiently small changes from the bias conditions such that $i_d = v_{gs} \cdot g_m$ linearly tracks v_{gs} through the transconductance, g_m. However, as v_{gs} increases above the small-signal limit, i_d ceases to linearly track v_{gs}, resulting in signal distortion. Distortion sets an upper limit for input signals in an analog circuit causing corrupted representations of the signals having undesired signal harmonics or harmonic distortion. Additionally, distortion causes the production of distortion products, or intermodulation distortion, when multiple inputs are present.

Distortion will be studied here by evaluating the transconductance distortion of a MOS differential pair, which is a widely used transconductor providing a differential output current in response to a differential input voltage. An example of a MOS differential pair is shown by devices M_1 and M_2 in Figure 5.1 for the simple operational transconductance amplifier (OTA) described later in Chapter 5. Here, a differential input voltage is applied across the two gates with a differential output current appearing as the difference of the two drain currents.

Distortion will be quantified by the input, differential, 1 dB compression voltage, $V_{INDIF1dB}$, which corresponds to the differential input voltage, applied either positively or negatively, where the differential output current decreases by 1 dB or -10.9% from its expected linear value. Operation at inputs sufficiently below $V_{INDIF1dB}$ results in linear operation, while operation at inputs above this results in the compression of expected output current and signal distortion. The input, 1 dB compression point is a widely used distortion measure for RF circuits and is an indicator of intermodulation distortion measures like the third-order intercept point, which is traditionally around 10 to 16 dB (a factor of three or four in voltage) above the 1 dB compression point. Harmonic distortion, intermodulation distortion, and spurious free dynamic range (SFDR), which is the dynamic range where signals are sufficiently free of circuit noise and intermodulation distortion products, are discussed in [33, pp. 636–643, 34, pp. 14–22, and 35, pp. 295–304].

The compression of differential output current for large input signals is observed in Figure 5.8 for the differential pair shown in the simple OTA of Figure 5.1. This OTA receives the differential output current from the M_1 and M_2 differential pair and passes it directly to the output with unity current gain. Figure 5.8 shows the measured output current versus input voltage for three versions of the OTA and is labeled with the corresponding values of $V_{INDIF1dB}$. The output current linearly tracks the input voltage until it approaches $V_{INDIF1dB}$ and begins compressing for input voltages above this. As seen in the figure, the three OTA versions have different values of $V_{INDIF1dB}$. This results from the selection of different inversion coefficients for differential-pair devices in the OTA versions.

To predict $V_{INDIF1dB}$, we will initially consider the MOS differential pair operating in weak inversion and follow this by considering operation in strong inversion. The differential output current, I_{OUTDIF}, for operation in weak inversion in saturation is given by

$$I_{OUTDIF} \text{ (WI)} = I_{BIAS} \tanh\left(\frac{V_{INDIF}}{2nU_T}\right) \tag{3.43}$$

where I_{BIAS} is the bias current at the common source connection for the two devices, and V_{INDIF} is the differential input voltage. The differential output current is found from the MOS weak-inversion, exponential drain current introduced in Equation 2.5 (repeated in Table 3.13) and is identical to that found for a bipolar transistor differential pair [26, pp. 215–216, 36, pp. 372–373] except n is removed in the expression for the bipolar pair. This is a result of n not appearing in the bipolar transistor, exponential collector current given earlier in Equation 2.1.

The ideal, linear value of I_{OUTDIF} present for small values of V_{INDIF} is given by

$$I_{OUTDIF} \text{ (WI, ideal)} = I_{BIAS}\left(\frac{V_{INDIF}}{2nU_T}\right) = V_{INDIF}\left(\frac{\frac{1}{2}I_{BIAS}}{nU_T}\right) = V_{INDIF} \cdot g_m \tag{3.44}$$

where $g_m = \frac{1}{2}I_{BIAS}/(nU_T)$ is the bias point, differential-pair transconductance and the weak-inversion transconductance of the individual pair devices, $g_m = I_D/(nU_T)$, each operating at $I_D = \frac{1}{2}I_{BIAS}$. Equation 3.44 is found by recognizing that $\tanh(x) \approx x$, for $x \ll 1$, where $x = V_{INDIF}/(2nU_T)$. This corresponds to operation for $V_{INDIF} \ll 2nU_T$ and is consistent with small-signal analysis using device, weak-inversion transconductance given by Equation 2.8 and summarized in terms of g_m/I_D in Table 3.17.

When evaluated at $V_{INDIF} = V_{INDIF1dB}$, the actual output current given by Equation 3.43 is down by 1 dB from the ideal current given by Equation 3.44 as expressed by

$$I_{OUTDIF} \text{ (WI)} = I_{BIAS} \tanh\left(\frac{V_{INDIF1dB}}{2nU_T}\right) = 0.89125 \cdot I_{BIAS}\left(\frac{V_{INDIF1dB}}{2nU_T}\right) \tag{3.45}$$

Solving Equation 3.45 for $V_{INDIF1dB}$ gives

$$V_{INDIF1dB} \text{ (WI)} = 1.22 \cdot (nU_T) = 1.22 \cdot V_{gm} = 1.22 \cdot \left(\frac{g_m}{I_D}\right)^{-1} \tag{3.46}$$

where the transconductance effective voltage is given by $V_{gm} = (g_m/I_D)^{-1}$, and is recognized as the reciprocal of weak-inversion $g_m/I_D = 1/(nU_T)$ given by Equation 2.9 and summarized in Table 3.17. V_{gm} and g_m/I_D are taken at the differential-pair bias point. $V_{INDIF1dB}$ given by Equation 3.46 will be discussed after its value is found for operation in strong inversion.

To predict $V_{INDIF1dB}$ in strong inversion, we now consider the differential output current for a differential pair operating in strong inversion in saturation. This output current can be expressed from derivations given in [26, pp. 218–221, 27, pp. 107–109, 36, pp. 358–360, and 37, pp. 182–183] as

$$I_{OUTDIF} \text{ (SI)} = V_{INDIF} \cdot \left(\frac{I_{BIAS}}{V_{EFF,BIAS}}\right)\sqrt{1 - \frac{1}{4}\left(\frac{V_{INDIF}}{V_{EFF,BIAS}}\right)^2}$$

$$= V_{INDIF} \cdot g_m \sqrt{1 - \frac{1}{4}\left(\frac{V_{INDIF}}{V_{EFF,BIAS}}\right)^2} \tag{3.47}$$

where $V_{EFF,BIAS}$ is the effective gate–source voltage for the individual pair devices at the bias point. $g_m = I_{BIAS}/V_{EFF,BIAS}$ is the bias point, differential-pair transconductance and the strong-inversion transconductance of the individual pair devices, $g_m = 2I_D/V_{EFF}$, each operating at $I_D = \frac{1}{2}I_{BIAS}$ and $V_{EFF} = V_{EFF,BIAS}$. This is also the strong-inversion value of device transconductance given by Equation 2.12 and summarized in terms of g_m/I_D in Table 3.17. Equation 3.47 is factored here where the square-root term shows signal loss or compression as V_{INDIF} increases. The output current is derived assuming strong-inversion, square-law drain current, excluding small-geometry effects like velocity saturation and VFMR. The derivation also assumes both pair devices remain in strong inversion for an input of V_{INDIF}.

The ideal, linear value of I_{OUTDIF} present for small values of V_{INDIF} is given by

$$I_{OUTDIF} \text{ (SI, ideal)} = V_{INDIF}\left(\frac{I_{BIAS}}{V_{EFF,BIAS}}\right) = V_{INDIF} \cdot g_m \tag{3.48}$$

which is found by recognizing that the square-root term in Equation 3.47 is nearly unity for $V_{INDIF} \ll V_{EFF,BIAS}$ and is consistent with small-signal analysis using the previously mentioned, strong-inversion value of transconductance for the individual devices. When evaluated at $V_{INDIF} = V_{INDIF1dB}$, the actual output current given by Equation 3.47 is down by 1 dB from the ideal current given by Equation 3.48 as expressed by

$$I_{OUTDIF} \text{ (SI)} = V_{INDIF1dB} \cdot g_m \sqrt{1 - \frac{1}{4}\left(\frac{V_{INDIF1dB}}{V_{EFF,BIAS}}\right)^2} = 0.89125 \cdot V_{INDIF1dB} \cdot g_m \tag{3.49}$$

Solving Equation 3.49 for $V_{INDIF1dB}$ gives

$$V_{INDIF1dB} \text{ (SI)} = 0.907 \cdot V_{EFF,BIAS} = 1.81 \cdot V_{gm} = 1.81 \cdot \left(\frac{g_m}{I_D}\right)^{-1} = 1.81 \cdot nU_T\sqrt{IC} \tag{3.50}$$

Table 3.19   MOS differential-pair, input, 1 dB compression voltage, $V_{INDIF1dB}$, expressions and trends in terms of inversion coefficient and channel length. When evaluated in terms of the inversion coefficient, $V_{INDIF1dB}$ is independent of the drain current

Input, 1 dB compression voltage, $V_{INDIF1dB}$	$IC \uparrow$ L fixed	$L \uparrow$ IC fixed
Weak inversion: $$V_{INDIF1dB} = 1.22 \cdot V_{gm} = 1.22 \cdot \left(\frac{g_m}{I_D}\right)^{-1} = 1.22 \cdot nU_T$$ where $V_{gm} = (g_m/I_D)^{-1} = nU_T$ from Table 3.17	Unchanged	Unchanged
Strong inversion, without small-geometry effects: $$V_{INDIF1dB} = 1.81 \cdot V_{gm} = 1.81 \cdot \left(\frac{g_m}{I_D}\right)^{-1} = 1.81 \cdot nU_T\sqrt{IC}$$ where $V_{gm} = (g_m/I_D)^{-1} = nU_T\sqrt{IC}$ from Table 3.17	$\uparrow \propto \sqrt{IC}$	Unchanged
Strong inversion, with velocity saturation and VFMR effects: $$V_{INDIF1dB} > 1.81 \cdot V_{gm} = 1.81 \cdot \left(\frac{g_m}{I_D}\right)^{-1} = 1.81 \cdot nU_T\sqrt{IC}$$ where IC is replaced in expression with $IC\left(1 + \dfrac{IC}{IC_{CRIT}}\right)$ with IC_{CRIT} from Table 3.17	$\uparrow \propto \approx IC$ as vel. sat. becomes significant	$\downarrow \propto \approx 1/L$ as vel. sat. becomes less significant

$V_{INDIF1dB}$ is for saturation where $V_{DS} > V_{DS,sat}$.

where $V_{gm} = \frac{1}{2}V_{EFF,BIAS}$ is the reciprocal of strong-inversion $g_m/I_D = 2/V_{EFF}$ given by Equation 2.13, with $g_m/I_D = 1/(nU_T\sqrt{IC})$ as summarized in Table 3.17. Like $V_{EFF,BIAS}$, V_{gm}, g_m/I_D, and IC are taken at the differential-pair bias point.

Table 3.19 summarizes expressions for $V_{INDIF1dB}$ from Equations 3.46 and 3.50 along with dependencies on the design choices of inversion coefficient and channel length. $V_{INDIF1dB}$ is directly proportional to V_{gm} and inversely proportional to g_m/I_D with proportionality factors of 1.22 and 1.81 for operation in weak and strong inversion, respectively. In weak inversion, $V_{INDIF1dB}$ is minimum because V_{gm} is minimum and g_m/I_D is maximum here, requiring a small input voltage to steer differential-pair output current to the 1 dB compression point. In strong inversion, $V_{INDIF1dB}$ increases as $\sqrt{IC}$ with V_{gm}, as g_m/I_D decreases as $1/\sqrt{IC}$, requiring increasing input voltage to steer the output current to the 1 dB compression point.

As listed in Table 3.19, $V_{INDIF1dB}$ increases when velocity saturation and VFMR effects are present. If the strong-inversion proportionality factor of 1.81 remained constant, $V_{INDIF1dB}$ would increase nearly as IC when velocity saturation is significant because V_{gm} increases nearly as IC, as g_m/I_D decreases nearly as $1/IC$, under these conditions. However, in addition to the increase associated with increasing V_{gm} and decreasing g_m/I_D, $V_{INDIF1dB}$ increases because the proportionality factor increases as a result of the linearization of drain current versus V_{EFF} when velocity saturation effects are present. The linearization of drain current was introduced in Equation 2.17 where drain current increases linearly with V_{EFF} under conditions of full velocity saturation.

No attempt will be made here to develop continuous, weak through strong-inversion expressions of $V_{INDIF1dB}$, beyond the relationships with g_m/I_D shown in Table 3.19. Instead, simulated values of $V_{INDIF1dB}$ are used in the *Analog CMOS Design, Tradeoffs and Optimization* spreadsheet used in the differential-pair and current-mirror designs of Chapter 4 and in the circuit designs of Chapters 5 and 6.

Figure 3.28 shows predicted $V_{INDIF1dB}$ from weak through strong inversion, inclusive of velocity saturation and VFMR effects, for $L = 0.18$, 0.28, 0.48, 1, and $4\,\mu m$, nMOS devices in the $0.18\,\mu m$ process described in Table 3.2. $V_{INDIF1dB}$ is predicted by iterative solution of differential-pair,

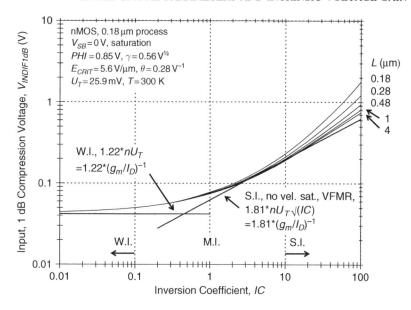

Figure 3.28 Predicted input, 1 dB compression voltage, $V_{INDIF1dB}$, for a MOS differential pair versus inversion coefficient, IC, at the bias point for $L = 0.18, 0.28, 0.48, 1$, and $4\,\mu m$, nMOS devices in a $0.18\,\mu m$ CMOS process. $V_{INDIF1dB}$, which is inversely proportional to g_m/I_D and directly proportional to the transconductance effective voltage, V_{gm}, is minimum in weak inversion, increases modestly in moderate inversion, and increases as $\sqrt{IC}$ in strong inversion. At high IC, $V_{INDIF1dB}$ increases further because of decreasing g_m/I_D and improved device linearity due to velocity saturation for short-channel lengths and VFMR for all channel lengths. $V_{INDIF1dB}$ is similar for all CMOS processes

differential output current using the continuous drain current expression in Table 3.13 multiplied by the enhanced correction factor given in the table. The substrate factor, n, is found from the expressions in Table 3.6. Drain current, like that shown earlier in Figure 3.11, is predicted using $\mu_0 = 422\,cm^2/V \cdot s$, $C'_{OX} = 8.41\,fF/\mu m^2$, $k_0 = 355\,\mu A/V^2$, $\gamma = 0.56\,V^{1/2}$, $PHI = 2\phi_F = 0.85\,V$, $E_{CRIT} = 5.6\,V/\mu m$, $\theta = 0.28/V$, and $\alpha = 1.3$ from the process parameters in Table 3.2. Channel length is adjusted from the drawn values given by subtracting $DL = 0.028\,\mu m$, which is also listed in the table.

To predict $V_{INDIF1dB}$ in terms of the inversion coefficient, device geometry and drain current at the bias point are directly related to the inversion coefficient since n and mobility are assumed fixed in the inversion coefficient definition shown in Equation 3.6 and mentioned in Section 3.4.2.3. As mentioned for V_{EFF} (Section 3.7.2.2), $V_{DS,sat}$ (Section 3.7.3.3), and g_m/I_D (Section 3.8.2.2), mobility reduction associated with velocity saturation and VFMR effects is considered outside the inversion coefficient definition. Here, these effects are considered through the enhanced drain current correction factor mentioned from Table 3.13.

In Figure 3.28, $V_{INDIF1dB}$ is observed to be proportional to $V_{gm} = (g_m/I_D)^{-1}$ and inversely proportional to g_m/I_D, which can be seen by flipping the g_m/I_D curve of Figure 3.26 vertically. $V_{INDIF1dB}$ is constant and minimum in weak inversion, approaching the value predicted by Equation 3.46 before increasing modestly in moderate inversion. In strong inversion, $V_{INDIF1dB}$ increases nearly as $\sqrt{IC}$, approaching the value predicted by Equation 3.50 for the long-channel devices that experience little velocity saturation. However, for short-channel devices operating at high IC, $V_{INDIF1dB}$ increases significantly above this due to the velocity saturation decrease in g_m/I_D and the increase in proportionality factor between $V_{INDIF1dB}$ and V_{gm} resulting from improved device linearity. At high IC, there is a modest $V_{INDIF1dB}$ increase at all channel lengths due to the g_m/I_D decrease and improved device linearity resulting from VFMR effects. Like g_m/I_D shown in Figures 3.26 and 3.27, there is little change in $V_{INDIF1dB}$ in moderate inversion between long- and short-channel devices. This, again, is because velocity saturation and VFMR effects are nearly negligible in moderate inversion. As seen, operating short-channel devices at high IC maximizes $V_{INDIF1dB}$ and, correspondingly, minimizes transconductance distortion. Unfortunately, as described later in Section 3.10.2.3, gate-referred thermal-noise voltage is maximum for these operating conditions.

Figure 3.29 shows the proportionality factor between $V_{INDIF1dB}$ shown in Figure 3.28 and $V_{gm} = (g_m/I_D)^{-1}$, where g_m/I_D is found by the numerical derivative of device drain current at the bias point. The proportionality factor, the $V_{INDIF1dB}/V_{gm}$ ratio or $V_{INDIF1dB} \cdot (g_m/I_D)$ product, is nearly 1.22 predicted by Equation 3.46 in weak inversion and rises to nearly 1.81 predicted by Equation 3.50 in strong inversion when velocity saturation and VFMR effects are excluded. As seen in the figure, the proportionality factor increases significantly for short-channel devices operating at high IC. This, as mentioned, is expected as a result of improved device linearity resulting from velocity saturation effects. For long-channel devices, the proportionality factor increases modestly due to some improvement in device linearity caused by VFMR.

Like the drain current shown in Figures 3.11 and 3.12, V_{EFF} shown in Figures 3.14 and 3.15, and g_m/I_D shown in Figures 3.26 and 3.27, pMOS $V_{INDIF1dB}$ and its proportionality factor are nearly equal to that of the nMOS devices shown in Figures 3.28 and 3.29 if pMOS channel length is decreased by a factor of 2.5 $((14\,\text{V}/\mu\text{m})/(5.6\,\text{V}/\mu\text{m}))$, the ratio of pMOS and nMOS critical, velocity saturation, electric field, E_{CRIT}. This is a result of nearly equal g_m/I_D for $L = 0.18\,\mu\text{m}$, pMOS devices shown in Figure 3.27 compared to $L = 0.48\,\mu\text{m}$, nMOS devices shown in Figure 3.26. There is some additional

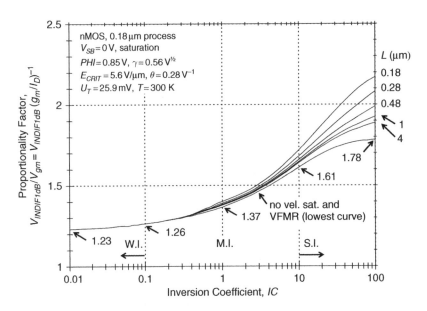

Figure 3.29 Predicted proportionality factor between input, 1 dB compression voltage, $V_{INDIF1dB}$, for a MOS differential pair and device transconductance effective voltage, $V_{gm} = (g_m/I_D)^{-1}$, at the bias point. The proportionality factor is shown versus inversion coefficient, IC, for $L = 0.18$, 0.28, 0.48, 1, and 4 μm, nMOS devices in a 0.18 μm CMOS process. The proportionality factor, $V_{INDIF1dB}/V_{gm} = V_{INDIF1dB} \cdot (g_m/I_D)$, approaches its expected minimum value of 1.22 in weak inversion, increases modestly in moderate inversion, and approaches its expected maximum value of 1.81 in strong inversion. At high IC, the proportionality factor increases because of improved device linearity due to velocity saturation for short-channel lengths and VFMR for all channel lengths. The proportionality factor is similar for all CMOS processes

pMOS $V_{INDIF1dB}$ increase due to higher VFMR effects where $\theta = 0.35/V$ compared to $0.28/V$ for nMOS devices.

Like V_{EFF}, $V_{DS,sat}$, and g_m/I_D summarized in Tables 3.14, 3.15, and 3.17, $V_{INDIF1dB}$ (summarized in Table 3.19) is largely independent of the MOS process when evaluated in terms of the inversion coefficient. Excluding small-geometry effects like velocity saturation and VFMR, $V_{INDIF1dB}$ is independent of the process except for the substrate factor, n, which has typical values of 1.4, 1.35, and 1.3 in weak, moderate, and strong inversion for typical bulk CMOS processes. As mentioned in Section 3.2.4, the development of process-independent design expressions is one of the advantages of using the inversion coefficient in analog CMOS design. Like V_{EFF} and g_m/I_D, $V_{INDIF1dB}$ is also independent of channel length when small-geometry effects are excluded. The increase in $V_{INDIF1dB}$ associated with velocity saturation effects depends on the process through E_{CRIT} and the channel length. The smaller increase associated with VFMR depends on the process through the mobility reduction factor, θ. Like V_{EFF}, $V_{DS,sat}$, and g_m/I_D, $V_{INDIF1dB}$ is independent of the selected drain current (in this case, the drain current at the differential-pair bias point) when evaluated in terms of the inversion coefficient. Finally, $V_{INDIF1dB}$ depends on the operating temperature, increasing as the thermal voltage increases with temperature.

3.8.3 Body-Effect Transconductance and Relationship to Substrate Factor

MOS body-effect transconductance, g_{mb}, is the small-signal parameter that describes the change in short-circuit drain current resulting from a change in body–source voltage, V_{BS}. It is given by the partial derivative of drain current with respect to V_{BS}, $g_{mb} = \delta I_D/\delta V_{BS}$, with the drain–source and gate–source voltages held constant. g_{mb} is the gain factor for the V_{BS}-controlled, dependent current source appearing between the drain and source in the MOS small-signal model shown earlier in Figure 3.21.

An increase in V_{BS} corresponds to a decrease in V_{SB}, which lowers the threshold voltage, V_T, due to the body effect. This, then, increases $V_{EFF} = V_{GS} - V_T$ for V_{GS} held constant, resulting in an increase in drain current. So, when the body voltage is source referenced as V_{BS} just as the gate voltage is source referenced as V_{GS}, an increase in V_{BS} creates an increase in drain current through g_{mb} just as an increase in V_{GS} creates an increase in drain current through g_m. As a result, the body or substrate is often considered a "back gate".

Body-effect transconductance is often expressed by the body-effect transconductance ratio given as a fraction of the gate transconductance as

$$\eta = \frac{g_{mb}}{g_m} \tag{3.51}$$

η is typically 20–40 % for bulk CMOS and partially depleted SOI CMOS processes, increasing with increasing substrate doping concentration and decreasing in the presence of source–body voltage in the normal, reverse-biased direction. η is lower for fully depleted SOI CMOS processes since the substrate has little control over the channel.

Body-effect transconductance both unfavorably and favorably affects analog MOS performance. It is best known for lowering below unity the gate-to-source voltage gain of source followers, where the unloaded voltage gain given earlier in Table 3.16 is $A_{vgso} = g_m/(g_m + g_{mb} + g_{ds}) \approx g_m/(g_m + g_{mb}) = 1/(1 + \eta)$. For $\eta = 0.3$, the unloaded voltage gain is 0.77 V/V, neglecting the normally slight decrease associated with the drain–source conductance, g_{ds}. The loss of voltage gain from unity can be avoided if the body is driven by the source, where, as noted in Table 3.16, the g_{mb} term is set to zero, giving nearly unity voltage gain. In addition to lowering source follower voltage gain, body-effect transconductance also unfavorably increases MOS thermal noise. This is considered later in Section 3.10.2 using the substrate factor, n.

Body-effect transconductance has a small, favorable effect on small-signal source and drain resistances given earlier in Table 3.16 and illustrated in Figure 3.22. It raises the total source conductance,

$g_{ms} = g_m + g_{mb} + g_{ds}$. This, then, favorably lowers the source output or input resistance for source follower and grounded-gate stages. The increase in g_{ms} also favorably increases the drain output resistance when external, source degeneration resistance is present.

For all circuit applications except where the body is used as a signal input, g_{mb} appears added to the much larger gate transconductance (g_m) in $g_{ms} = g_m + g_{mb} + g_{ds}$ terms. As a result, g_{mb} has a second-order effect of circuit performance. For example, for $\eta = 0.3$, g_{mb} is 30 % of g_m. As a result, a 10 % error in the prediction of g_{mb} results in only a 2.3 % error (1.33/1.3) in g_{ms} and related circuit performance measures again, neglecting the normally small effect of g_{ds}. This is a fortunate result that permits a simple prediction of $g_{mb} = (n-1) \cdot g_m$ for most applications as described later. When the body is driven as a signal input, g_{mb} directly controls the circuit body transconductance as summarized in Table 3.16 and Figure 3.22. In this case, errors in g_{mb} directly appear as errors in circuit performance.

3.8.3.1 Substrate factor

When predicting g_{mb}, it is useful to relate it to the substrate factor, n, since both parameters result from body or substrate effect. Additionally, it is useful to observe the small variation in n from weak through strong inversion since n is assumed fixed in predictions of g_m/I_D (Section 3.8.2.3), V_{EFF} (Section 3.7.2.3), and simple predictions of g_{mb} described later. n was introduced earlier in Sections 2.4.1 and 3.4.1, but is described in additional detail here and predicted in terms of the inversion coefficient. Table 3.20 summarizes expressions for n along with dependencies on the design choices of inversion coefficient and channel length. The table also summarizes expressions for $\eta = g_{mb}/g_m$ that will be discussed later.

The first expression in Table 3.20 gives n in weak inversion [15, p. 379], where the silicon surface potential is assumed fixed at $PHI = 2\phi_F$, which is its value at the boundary of weak and moderate inversion [15, p. 87]. In weak inversion, n is related to the capacitive voltage division between the gate voltage and silicon surface potential where the surface potential increases with a slope of $1/n$ [15, p. 75] with increasing gate voltage. As a result, n is linked to $\eta = g_{mb}/g_m$ and the depletion, C'_{DEP}, and gate-oxide, C'_{OX}, capacitances through the expression $n = 1 + \eta = 1 + g_{mb}/g_m = 1 + C'_{DEP}/C'_{OX}$ [15, p. 379]. As mentioned earlier for Equation 2.6, this assumes that the interface state capacitance, C'_{INT}, is negligible. This is usually the case, except possibly for operation at very low temperatures below $-73\,°C$ (200 K) where n shown earlier in Figure 3.4 increases most likely as a result of increasing interface capacitance. In weak inversion, n is commonly recognized as the weak-inversion slope factor appearing in the weak-inversion drain current in Table 3.13 and in the subthreshold or weak-inversion swing voltage, $S = 2.303 \cdot nU_T$, introduced earlier in Equation 2.7.

The second expression in Table 3.20 is from the EKV MOS model for operation in strong inversion [13, 14], but provides an approximation of n in all regions as described below. The expression assumes a body-referenced inversion charge, pinch-off voltage of $V_P = (V_{GS} - V_T)/n + V_{SB}$ and a constant, "pinned," strong-inversion, silicon surface potential of $\phi_0 = 2\phi_F + 4U_T$. As mentioned earlier in Section 3.4.1, $V_P = (V_{GS} - V_{TO} + V_{SB})/n$ in the EKV MOS model is replaced with $V_P = (V_{GS} - V_T)/n + V_{SB}$ since $V_T \approx V_{TO} + (n-1) \cdot V_{SB}$ approximates the threshold voltage increase with V_{SB} where V_{TO} is the zero-V_{SB} value. In strong inversion, the derivative of V_P with respect to the gate voltage is $1/n$ [13]. The body or substrate effect, indicated by n greater than unity, raises the local threshold voltage along the channel and reduces the source-referenced $V_P = (V_{GS} - V_T)/n$ from a value of $V_P = V_{GS} - V_T$ if there were no substrate effect. This results in the reduction of strong-inversion drain current by the division of n as seen in the expressions in Table 3.13.

As mentioned in [13], V_P applies to operation in strong inversion where there is significant inversion charge that is pinched off for operation in saturation. However, V_P can also be used to estimate the silicon surface potential for operation in weak inversion [13]. This can be seen by observing that negative values of source-referenced $V_P = (V_{GS} - V_T)/n = V_{EFF}/n$ in weak inversion subtract from the strong-inversion, silicon surface potential of $\phi_0 = 2\phi_F + 4U_T$ that appears in the second

Table 3.20   MOS substrate factor, n, and body-effect transconductance ratio, $\eta = g_{mb}/g_m$, expressions and trends in terms of inversion coefficient and channel length. When evaluated in terms of the inversion coefficient, n and η are independent of the drain current

Expression	$IC \uparrow$ L fixed	$L \uparrow$ IC fixed
Substrate factor, n Weak inversion: $$n = 1 + \frac{\gamma}{2\sqrt{V_{SB} + 2\phi_F}} = 1 + \frac{g_{mb}}{g_m} = 1 + \frac{C'_{DEP}}{C'_{OX}}$$ All regions:[a] $$n = 1 + \frac{\gamma}{2\sqrt{V_{EFF}/n + V_{SB} + \phi_0}}$$ For $IC > 0.5$ (note[b]): $$n \approx 1 + \frac{\gamma}{2\sqrt{2U_T\sqrt{IC} + V_{SB} + \phi_0}}$$	$\downarrow$ slightly, but decreases more rapidly if vel. sat. is present where $V_{EFF} \propto IC$	Unchanged, but increases slightly if V_{EFF} decreases due to less vel. sat.
Body-effect transconductance ratio, $\eta = g_{mb}/g_m$ Weak inversion: $$\eta = \frac{g_{mb}}{g_m} = \frac{\gamma}{2\sqrt{V_{SB} + 2\phi_F}} = n - 1 = \frac{C'_{DEP}}{C'_{OX}} =$$ $$\frac{\partial V_T}{\partial V_{SB}}$$ where $V_T = V_{TO} + \gamma\left[\sqrt{V_{SB} + 2\phi_F} - \sqrt{2\phi_F}\right]$ or $V_T \approx V_{TO} + (n-1)V_{SB}$ All regions:[a] $$\eta = \frac{g_{mb}}{g_m} = \frac{\gamma}{\sqrt{V_{EFF}/n + V_{SB} + \phi_0} + \sqrt{V_{SB} + \phi_0}}$$ For $IC > 0.5$ (note[b]): $$\eta = \frac{g_{mb}}{g_m} \approx \frac{\gamma}{\sqrt{2U_T\sqrt{IC} + V_{SB} + \phi_0} + \sqrt{V_{SB} + \phi_0}}$$ Estimate, all regions: $$\eta = \frac{g_{mb}}{g_m} \approx n - 1$$ where n may be taken as n_0, its value at $IC = 1$ in moderate inversion	$\downarrow$ slightly, but decreases more rapidly if vel. sat. is present where $V_{EFF} \propto IC$	Unchanged, but increases slightly if V_{EFF} decreases due to less vel. sat.

n and $\eta = g_{mb}/g_m$ are for saturation where $V_{DS} > V_{DS,sat}$.

[a] Expressions are actually for strong inversion but provide an estimate in all regions because negative values of V_{EFF} in moderate and weak inversion reduce the surface potential from its strong-inversion estimate of $\phi_0 \approx 2\phi_F + 4U_T$. $V_{EFF} = V_{GS} - V_T$ is from Table 3.14.

[b] V_{EFF} is approximated as $V_{EFF} \approx 2nU_T\sqrt{IC}$ for $IC > 0.5$, excluding velocity saturation and VFMR increases. These increases may be included by replacing IC with $IC(1 + IC/(4IC_{CRIT}))$ in the $\sqrt{IC}$ term as described in Table 3.14.

n expression in Table 3.20. Additionally, decreasing V_{EFF} in weak inversion decreases the silicon surface potential by the expected slope of $1/n$ through the V_{EFF}/n term.

Usually, using an expression derived in strong inversion for operation in weak and moderate inversion leads to significant errors. This can be seen, for example, in Figure 3.23 for g_m/I_D predicted using weak- and strong-inversion expressions. However, the second expression in Table 3.20 provides a usable estimate for n in all regions because of the compensating effects of negative V_{EFF} values in weak inversion. For example, at the boundary of weak and moderate inversion at $IC = 0.1$, $V_{EFF}/n = -2U_T$ from Table 3.14. Using this in the expression for n gives a silicon surface potential, $V_{EFF}/n + \phi_0$, that is $2U_T$ (52 mV) above the usual value of $2\phi_F$ (850 mV). This results in an under prediction of n by only 0.7 % for the process parameters described below for Figure 3.30.

The second expression for n in Table 3.20 also appeared in Table 3.6 earlier and was used to show the small variation in n versus the gate–source voltage in Figure 3.2. n can now be related to the inversion coefficient by using the V_{EFF} expressions from Table 3.14. The third expression for n in Table 3.20 provides a simple estimate in terms of the inversion coefficient for $IC > 0.5$ where V_{EFF} is positive. Here, V_{EFF} is estimated by its strong-inversion value of $V_{EFF} = 2nU_T\sqrt{IC}$. There is, of course, error in this V_{EFF} prediction in moderate inversion, but the small value of V_{EFF} present in moderate inversion has little effect on the value of n. As observed in the table, n decreases slightly with increasing IC because of the increase in V_{EFF}. n is independent of channel length, as V_{EFF} is, unless velocity saturation is present and V_{EFF} increases as described in the table. n, however, may decrease for short-channel devices because of a reduction in the apparent value of γ due to charge sharing. This can occur when the source and drain depletion regions cut off some of the normal, substrate, depletion-region control below the channel [15, pp. 259–263]. This reduction of substrate control can be significant for minimum channel length devices since the source and drain regions are in close proximity and their depletion regions may extend under much of the channel. The effect of this charge sharing is an electrical or apparent γ having a value below that set by the normal substrate doping concentration, and a reduction in the threshold voltage. Charge sharing in small-geometry processes is reduced by reducing the depletion region thickness below the channel and at the source and drain. This is done by retrograde doping, which increases the doping vertically below the channel surface, and by halo doping, which increases the doping laterally near the source and drain.

Figure 3.30 shows predicted n as a function of the inversion coefficient from deep weak inversion ($IC = 0.01$) to deep strong inversion ($IC = 100$) for nMOS devices in the 0.18 µm CMOS process having process values listed in Table 3.2. n is predicted using the second expression in Table 3.20 for a body-effect factor, $\gamma = 0.56$ V$^{1/2}$, Fermi potential, $PHI = 2\phi_F = 0.85$ V, and thermal voltage, $U_T = 25.9$ mV ($T = 300$ K). n is predicted using V_{EFF} from Table 3.14 using a critical, velocity saturation, electric field of $E_{CRIT} = 5.6$ V/µm and a mobility reduction factor of $\theta = 0.28$/V to model the increase in V_{EFF} due to velocity saturation and VFMR effects. A fixed value of $n = 1.3$ is used for V_{EFF} predictions, but this is divided out in the V_{EFF}/n term used for the predicted value of n.

Figure 3.30 shows that n decreases slightly with increasing inversion coefficient. n is 1.295 at the boundary of weak inversion ($IC = 0.1$), 1.283 at the center of moderate inversion ($IC = 1$), 1.265 at the onset of strong inversion ($IC = 10$), and 1.231 in deep strong inversion ($IC = 100$) for long-channel devices. This corresponds to a variation of less than ± 3 % from an average value of 1.270. For $L = 0.18$ µm devices at $IC = 100$, n is reduced to 1.212 or 5 % below the average value of 1.270 because of velocity saturation increases in V_{EFF}. However, it is unlikely that short-channel devices would be operated at such a high level of inversion since the gate–source voltage is near the supply voltage limit. Here, V_{EFF} is approximately 1 V for $IC = 100$ and $L = 0.18$ µm from Figure 3.14, giving $V_{GS} = V_{EFF} + V_T = 1.45$ V for a threshold voltage of 0.45 V. If IC is constrained to values less than 20 in low-voltage designs, n varies by less than ± 3 % from a fixed, average value for all channel lengths, assuming charge-sharing effects do not change n significantly. As mentioned, assuming a constant value of n greatly simplifies estimates for V_{EFF}, g_m/I_D, and g_{mb}/g_m (described in the next section) while providing sufficient accuracy for design optimization.

The variation in n with inversion coefficient for pMOS devices is nearly equal to that shown in Figure 3.30 for nMOS devices. However, there is little additional n decrease for short-channel pMOS devices because of smaller velocity saturation increases in V_{EFF}. n is approximately 3 % or an additive factor of 0.04 higher for pMOS devices because of a higher process value of $\gamma = 0.61$ V$^{1/2}$ compared to $\gamma = 0.56$ V$^{1/2}$ for nMOS devices. These process values are listed in Table 3.2 for the 0.18 µm CMOS process considered.

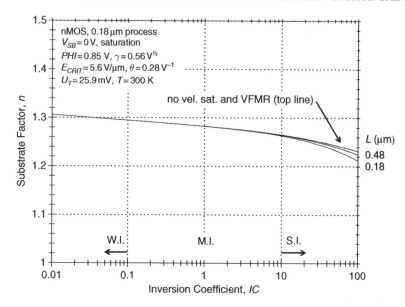

Figure 3.30 Predicted substrate factor, n, versus inversion coefficient, IC, for $L = 0.18$ and $0.48\,\mu m$, nMOS devices in a $0.18\,\mu m$ CMOS process. n decreases slightly with increasing inversion coefficient and is within $\pm 3\%$ of an average value of 1.270, except for short-channel devices where n decreases further because of the increase in gate–source voltage due to velocity saturation effects. n decreases by approximately 6% for a source–body voltage of $V_{SB} = 1\,V$ and is slightly higher for pMOS devices due to an increase in process γ. Figure 3.2 showed n versus the gate–source voltage. n is similar for all CMOS processes, increasing with γ.

3.8.3.2 Body-effect transconductance

In addition to the predictions of n described above, Table 3.20 lists predictions for the body-effect transconductance ratio, $\eta = g_{mb}/g_m$. The first expression gives η for operation in weak inversion [15, p. 379] where the silicon surface potential is, again, assumed fixed at $2\phi_F$, its value at the boundary of weak and moderate inversion [15, p. 87]. In weak inversion, η, like n given in the table, is related to the capacitive voltage division between the gate voltage and silicon surface potential. As a result, η is linked to n through $\eta = g_{mb}/g_m = n - 1 = C'_{DEP}/C'_{OX}$, where $\eta = C'_{DEP}/C'_{OX} \approx (\varepsilon_{Si}/\varepsilon_{SiO_2})(t_{ox}/t_{dep}) \approx 3t_{ox}/t_{dep}$ [15, p. 379]. η is approximately equal to the ratio of silicon, ε_{Si}, and silicon dioxide, ε_{SiO_2}, dielectric constants multiplied by the ratio of gate-oxide thickness, t_{ox}, and depletion region thickness, t_{dep} (expression given in Table 3.1), located below the channel under formation. This represents the body or substrate control over the channel relative to that of the gate. For a weak-inversion value of $n = 1.33$ ($V_{SB} = 0\,V$), $\eta = 0.33$ implies $t_{dep} \approx 9 \cdot t_{ox}$ at the source end of the channel or for low values of V_{DS}. While t_{dep} is nearly a factor of nine larger than t_{ox}, the body exerts a larger than expected control over the channel because of the higher silicon permittivity or dielectric constant compared to that of the gate oxide. In addition to its linkage with n, the weak-inversion expression for η given in Table 3.20 is recognized as $\delta V_T/\delta V_{SB}$ using the common prediction of V_T given in the table for the presence of non-zero V_{SB}. Finally, like the expressions of n given in the table and discussed above, the expression for η assumes that interface state capacitance is negligible.

The second expression for η in Table 3.20 is adapted from [15, p. 369] where $\phi_0 \approx 2\phi_F + 4U_T$ approximates the silicon surface potential and V_{EFF}/n approximates $V_{DS,sat}$ for operation in strong inversion. While this expression was derived in strong inversion, it provides a rough approximation of η in all regions since the V_{EFF}/n term goes negative in weak inversion and lowers the silicon surface potential as mentioned earlier for the second prediction of n given in the table. The surface potential in the rightmost, denominator, square-root term of η remains fixed and is not lowered in weak inversion, but its presence in a square-root term reduces its effect.

As for the expression of n versus the inversion coefficient given in Table 3.20, η can be related to the inversion coefficient using the V_{EFF} expressions from Table 3.14. The third expression for η in Table 3.20 provides a simple estimate in terms of the inversion coefficient for $IC > 0.5$ where V_{EFF} is positive. Here, again, V_{EFF} is estimated by its strong-inversion value of $V_{EFF} = 2nU_T\sqrt{IC}$ where the error in moderate inversion has little effect because V_{EFF} is small here. As observed in the table, η decreases slightly with IC, dropping less than $IC^{-0.25}$. Like n, η is independent of channel length unless velocity saturation is present and V_{EFF} increases as described in the table. However, like n, η may decrease for short-channel devices because of a decrease in the apparent value of γ due to charge sharing.

3.8.3.3 Predicted and measured values

Figure 3.31 shows measured and predicted $\eta = g_{mb}/g_m$ for $L = 1.2\,\mu\text{m}$, nMOS devices in the $0.35\,\mu\text{m}$, partially depleted (PD), SOI CMOS process having process values listed in Table 3.2. η is presented as a function of the inversion coefficient from $IC = 0.01$ (deep weak inversion) to 100 (deep strong inversion) with predicted values obtained from the second η expression in Table 3.20 for a body-effect factor, $\gamma = 0.68\,\text{V}^{1/2}$, Fermi potential, $PHI = 2\phi_F = 0.7\,\text{V}$, and thermal voltage, $U_T = 25.9\,\text{mV}$ ($T = 300\,\text{K}$). η is predicted using V_{EFF} from Table 3.14 using a critical, velocity saturation, electric field of $E_{CRIT} = 4\,\text{V}/\mu\text{m}$ and a mobility reduction factor of $\theta = 0.2/\text{V}$ to model the increase in V_{EFF} due

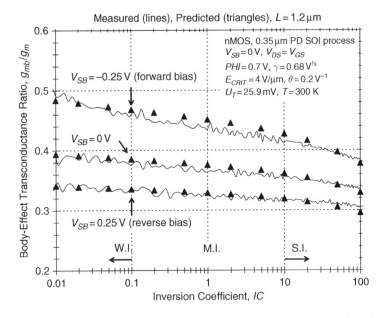

Figure 3.31 Predicted and measured body-effect transconductance ratio, $\eta = g_{mb}/g_m$, versus inversion coefficient, IC, for $L = 1.2\,\mu\text{m}$, nMOS devices in a $0.35\,\mu\text{m}$ PD SOI CMOS process. Unlike data from a $0.18\,\mu\text{m}$, bulk CMOS process shown later in Figures 3.32 and 3.33, measured η decreases with increasing inversion coefficient in strong inversion and closely matches the predicted values. η decreases under normal, reverse bias, source–body voltage and increases if this voltage is forward biased

to velocity saturation and VFMR effects. The increase in V_{EFF}, however, is small for the moderate, channel length devices considered. A fixed value of $n = 1.3$ is used for V_{EFF} predictions, but this is divided out in the V_{EFF}/n term used for the prediction of η. η is evaluated for $V_{SB} = 0.25$, 0, and -0.25 V corresponding to reverse, zero, and forward source–body bias, permitting a study of the effect of V_{SB}. In the PD SOI CMOS process, individual MOS bodies are isolated and connected to body ties. This permits the application of independent body voltages for both nMOS and pMOS devices. The inversion coefficient for measured data is found by $IC = I_D/[I_0(W/L)]$ from Table 3.6 using a technology current of $I_0 = 0.30 \mu$A from Table 3.2.

η is accurately predicted in Figure 3.31, being well within 5 % of measured values for $V_{SB} = 0.25$, 0, and -0.25 V from deep weak to deep strong inversion. Like n shown earlier in Figure 3.30, η decreases slightly with increasing inversion coefficient as a result of increasing V_{EFF}. η decreases when the MOS source–body voltage is reverse biased ($V_{SB} = 0.25$ V in the figure) and increases when this voltage is forward biased ($V_{SB} = -0.25$ V in the figure). Such forward bias is not usual and must be limited to avoid turning on the source–body diode junction. Forward bias, however, illustrates the potential of bipolar signal drive at the body terminal where the device transconductance is governed by g_{mb}. When the source–body voltage is reverse biased, η is at its minimum value and decreases less with increasing inversion coefficient and V_{EFF}. This is observed in the second η expression in Table 3.20 for positive V_{SB} and is a result of reduced depletion capacitance.

Figure 3.32 shows measured and predicted η for $L = 1 \mu$m, nMOS devices in the 0.18 μm CMOS process having process parameters listed in Table 3.2. Measurements for $L = 0.18, 0.28, 0.38, 0.48, 2$, and 4 μm devices were also taken, but lie nearly on the $L = 1 \mu$m measurements shown. Again, η is presented as a function of the inversion coefficient from $IC = 0.01$ to 100 with predicted values obtained from the second η expression in Table 3.20. Additionally, predicted values of $\eta = n - 1$ are shown, using the second n expression in the table. Process values of $\gamma = 0.56$ V$^{1/2}$, $PHI = 2\phi_F = 0.85$ V,

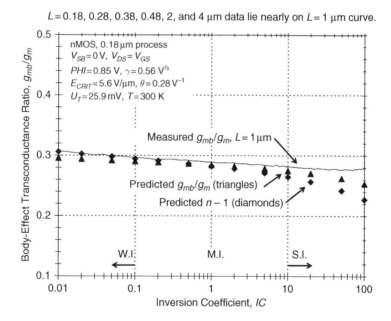

Figure 3.32 Predicted and measured body-effect transconductance ratio, $\eta = g_{mb}/g_m$, versus inversion coefficient, IC, for $L = 1 \mu$m, nMOS devices in a 0.18 μm CMOS process. Predicted η decreases slightly with increasing inversion coefficient because the gate–source voltage increases, but measured η flattens in strong inversion. As a result, a better prediction for η is obtained by $\eta = n - 1$, assuming a fixed value of $n = 1.28$

$U_T = 25.9\,\text{mV}$ $(T = 300\,\text{K})$, $E_{CRIT} = 5.6\,\text{V}/\mu\text{m}$, and $\theta = 0.28/\text{V}$ are used from Table 3.2. The inversion coefficient for measured data is found using a technology current of $I_0 = 0.64\,\mu\text{A}$ from Table 3.2.

Below the onset of strong inversion at $IC = 10$, η predicted by the second expression in Table 3.20 is within 3 % of measured values, while η predicted by $\eta = n - 1$ is within 6 % of measured values. η, however, is increasingly under-predicted in strong inversion with errors of 9 % and 18 %, respectively, in deep strong inversion at $IC = 100$. This is because measured η flattens in strong inversion compared to declining as shown earlier in Figure 3.31 for the 0.35 μm, PD SOI CMOS process. Because η flattens in strong inversion, a better prediction is obtained using the estimate $\eta = n - 1$, listed in Table 3.20, using a fixed, moderate-inversion value of $n_0 = 1.28$ from Figure 3.30. Predicted η using this simple estimate is within measured values by 5 % for $IC > 0.1$. In weak inversion, η is under-predicted by more than this amount because measured η increases.

Figure 3.33 shows measured and predicted η for $L = 1\,\mu\text{m}$, pMOS devices in the 0.18 μm CMOS process, again having process parameters listed in Table 3.2. Measurements for $L = 0.18$, 0.28, 0.38, 0.48, 2, and 4 μm devices were also taken, but lie nearly on the $L = 1\,\mu\text{m}$ measurements shown, except for $L = 0.18\,\mu\text{m}$ measurements where η flattens at high values of IC. η is predicted as described for Figure 3.32 using $\gamma = 0.61\,\text{V}^{1/2}$, $PHI = 2\phi_F = 0.85\,\text{V}$, $U_T = 25.9\,\text{mV}$ $(T = 300\,\text{K})$, $E_{CRIT} = 14\,\text{V}/\mu\text{m}$, and $\theta = 0.35/\text{V}$ from Table 3.2. The inversion coefficient for measured data is found using a technology current of $I_0 = 0.135\,\mu\text{A}$ from Table 3.2.

Below the onset of strong inversion at $IC = 10$, η predicted by the second expression in Table 3.20 is within 9 % of measured values, while η predicted by $\eta = n - 1$ is within 13 % of measured values. η, again, is increasingly under-predicted in strong inversion with errors now of 22 % and 30 %, respectively, in deep strong inversion at $IC = 100$. The increased under-prediction of pMOS η compared to that shown for nMOS devices in Figure 3.32 results from an increase in measured pMOS η in strong inversion. As for the nMOS devices shown in Figure 3.32, a better prediction is obtained using the estimate $\eta = n - 1$, listed in Table 3.20, using a fixed, moderate-inversion value of $n_0 = 1.32$. This

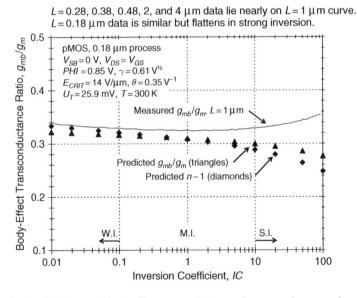

Figure 3.33 Predicted and measured body-effect transconductance ratio, $\eta = g_{mb}/g_m$, versus inversion coefficient, IC, for $L = 1\,\mu\text{m}$, pMOS devices in a 0.18 μm CMOS process. Like data shown for nMOS devices in Figure 3.32, predicted η decreases slightly with increasing inversion coefficient. However, measured η for pMOS devices shown here increases slightly in strong inversion. As a result, a better prediction for η is obtained by $\eta = n - 1$, assuming a fixed value of $n = 1.32$

value of n_0 is slightly higher than $n_0 = 1.28$ mentioned for nMOS devices because of a higher value of pMOS process γ. Predicted pMOS η using the simple estimate is within measured values by 5 % for $IC < 30$. Deeper into strong inversion, η is under-predicted by more than this amount because measured η increases.

3.8.3.4 Summary of trends

The slight decrease in η with increasing inversion level predicted by the second expression in Table 3.20 is in good agreement (within 5 %) with measured data shown in Figure 3.31 for nMOS devices in a 0.35 μm, PD SOI CMOS process. However, measured η shown in Figures 3.32 and 3.33 for nMOS and pMOS devices in a 0.18 μm, bulk CMOS process shows a better prediction is obtained from the simple prediction of $\eta = n - 1$ listed in Table 3.20, using a fixed, moderate-inversion value of n. This constant prediction of η is more accurate because measured η flattens in strong inversion for nMOS devices (Figure 3.32), while it actually increases for pMOS devices (Figure 3.33). Measured η (not shown) for the 0.5 μm, bulk CMOS process described in Table 3.2 decreases in strong inversion for nMOS devices, while it flattens for pMOS devices.

Unlike the technology-independent characteristic of g_m/I_D described in Section 3.8.2.5, no clear trend of η with increasing inversion level is observed here. However, measured η, whether decreasing, flattening, or increasing in strong inversion, changes only modestly from its moderate-inversion value. The change is even less when the inversion coefficient is limited to values below 20 as usually required for low-voltage design. As a result, we will use the prediction $\eta = n - 1$ in this book, assuming a fixed, average value of n. This provides adequate accuracy for design guidance, especially, since as mentioned at the beginning of Section 3.8.3, errors in g_{mb} have a limited effect on most aspects of analog performance where g_{mb} is added to the much larger g_m.

As noted earlier in the n versus V_{GS} prediction of Figure 3.2, n decreases by about 6 % in the presence of $V_{SB} = 1$ V, which is a large and probably unallowable value for low-voltage designs. The decrease of n and, correspondingly, $\eta = n - 1$ with V_{SB} is seen in the expressions given in Table 3.20 for positive V_{SB} (normal, reverse-biased, body–source diode junction) and is due to the decrease in depletion capacitance. As mentioned earlier in Section 3.4.2.3, we normally assume $V_{SB} = 0$ V or neglect the small change in n and η with V_{SB}. This permits simple expressions for design guidance where n and η are assumed constant from weak through strong inversion and in the presence of non-zero V_{SB}. As always, a candidate design must be validated using computer simulations with full MOS models, and such simulations would normally include variations in n and η. Finally, as shown in Figure 3.4, n and, correspondingly, $\eta = n - 1$ are nearly independent of temperature, except for temperatures below approximately $-73\,°C$ (200 K) where n can increase significantly due to possible increases in the interface-state capacitance.

As described in Section 3.8.3.1, charge sharing, especially in short-channel devices, can alter the value of n and η through a change in the apparent, electrical value of γ. Additionally, as mentioned earlier for n in Section 3.4.1, the separation of n into voltage and charge components and the effects of polysilicon gate depletion [16, pp. 150–154] are not considered in the design guidance expressions listed in Table 3.20. Measured η for minimum, channel length, $L = 0.18$ μm devices in the 0.18 μm CMOS process described in Table 3.2 is nearly identical to that of $L = 1$ μm, nMOS and pMOS devices shown in Figures 3.32 and 3.33, excepting a flattening of pMOS η in strong inversion. This indicates good control of short-channel, charge-sharing effects. Interestingly, these effects are much more significant for minimum, channel length, $L = 0.5$ μm devices in the older 0.5 μm CMOS process described in Table 3.2. Here, measured η indicates that the apparent γ is reduced by approximately 10 % for short-channel devices. Increasing channel length only slightly from the process minimum to $L = 0.85$ μm nearly eliminates the drop in apparent γ. As discussed later in Section 3.8.4.2, increasing channel length only slightly above the process minimum can also significantly reduce short-channel, drain, charge-sharing effects leading to drain-induced barrier lowering (DIBL). DIBL reduces the threshold voltage for increasing drain–source voltage, increasing drain current and increasing the value of small-signal, drain–source conductance.

3.8.4 Drain Conductance

The drain–source conductance, g_{ds}, and drain–body conductance, g_{db}, shown in the small-signal model of Figure 3.21 influence small-signal resistances and voltage gains resulting from signal currents flowing into circuit resistances. This section discusses the usually dominant g_{ds} along with g_{db}, which is present when hot-electron effects or leakage create substrate current.

Section 2.5 introduced g_{ds}, the small-signal parameter describing the change in drain current resulting from a change in drain–source voltage, V_{DS}. It is given by the partial derivative of drain current with respect to V_{DS}, $g_{ds} = \delta I_D / \delta V_{DS}$, with the gate–source and body–source voltages held constant. An increase in V_{DS} creates an increase in drain current through g_{ds}.[10]

Often, designers describe MOS drain–source conductance using its reciprocal form, the drain–source resistance, $r_{ds} = 1/g_{ds}$. r_{ds} is convenient to use in design optimization since MOS intrinsic voltage gain, $A_{Vi} = g_m \cdot r_{ds}$ described later in Section 3.8.5, increases as r_{ds} increases, just as it increases as g_m increases. In the following discussions, we interchange freely between conductance and resistance forms, g_{ds} and r_{ds}, using the most convenient form for the predictions or measurements described. g_{ds} or r_{ds} appears frequently in small-signal resistance, voltage gain, and transresistance expressions listed in Table 3.16 for a MOSFET connected to the general circuit shown in Figure 3.22.

Section 2.5 also introduced the MOS Early voltage, V_A, as a measure of normalized drain–source conductance or resistance, where $g_{ds} = I_D / (V_A + V_{DS})$ or $r_{ds} = (V_A + V_{DS}) / I_D$. The Early voltage,[11] named after bipolar–transistor pioneer James Early, was originally used to model the small-signal, collector–emitter resistance of bipolar transistors [38]. For MOSFETs, V_A is a quality factor describing the production of g_{ds} or r_{ds} for a given level of drain bias current. Analogous to the normalized transconductance or transconductance efficiency, g_m / I_D, which should be maximized to maximize g_m for a given level of drain current, V_A should be maximized to minimize g_{ds} or maximize r_{ds}.

V_A is used here as a normalized measure of g_{ds} or r_{ds} and is not assumed constant for the process, as frequently done for bipolar transistors. V_A increases as channel length increases and also increases as V_{DS} increases above $V_{DS,sat}$, assuming channel length modulation effects (Section 3.8.4.1) dominate DIBL (Section 3.8.4.2) and hot-electron effects (Section 3.8.4.3). V_A is also dependent on the level of inversion, generally increasing slightly in strong inversion for fixed V_{DS} provided V_{DS} is sufficiently above the operating value of $V_{DS,sat}$.

Figure 3.34 shows the calculation of $g_{ds} = 1/r_{ds}$ using V_A. The I_D versus V_{DS} curve shown is equal to $I_D = (I_D$ without V_{DS} dependency$) \cdot (1 + V_{DS}/V_A)$ described earlier in Equation 2.20 for a limited range of V_{DS} in saturation where I_D increases nearly linearly with V_{DS}. In Figure 3.34, a tangent line is drawn touching the I_D versus V_{DS} curve at the selected bias point. This tangent line intersects the V_{DS} axis ($I_D = 0\,\mu\text{A}$) at V_A.[12] g_{ds} is the drain current slope at the bias point, which is the slope of the tangent line. This slope is the drain bias current, I_D, divided by the sum of the Early voltage, V_A, and the drain–source bias voltage, V_{DS}, as given by

$$g_{ds} = \frac{1}{r_{ds}} = \frac{I_D}{V_A + V_{DS}} \approx \frac{I_D}{V_A}$$

$$= \frac{I_D}{V_A\ (V_{DS}\ \text{included})} \tag{3.52}$$

---

[10] As mentioned in Footnote 4 and shown earlier in Figure 3.9, I_D and V_{DS} are taken positively for both nMOS and pMOS devices.

[11] When bipolar transistor, collector–emitter voltage increases, the collector–base depletion region widens, reducing base width and increasing collector current. This effect is known as the Early effect.

[12] For nMOS devices, the V_A intersection occurs at a negative V_{DS} value. However, like V_{DS}, V_A is taken positively in this book for both nMOS and pMOS devices.

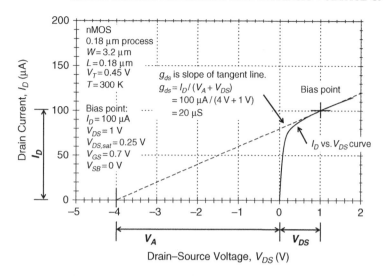

Figure 3.34 Illustration of MOS drain–source conductance calculation using the Early voltage. A tangent line is drawn touching the I_D versus V_{DS} curve at the bias point, and this line intersects the V_{DS} axis at the Early voltage. The drain–source conductance, g_{ds}, is the slope of the tangent line given by the drain bias current, I_D, divided by the sum of the Early voltage, V_A, and drain–source bias voltage, V_{DS}. Frequently, as throughout this book, V_{DS} is included in V_A such that $V_A = V_{gds} = (g_{ds}/I_D)^{-1}$, where $g_{ds} = I_D/V_A$. V_A is not a constant value for the process, but depends upon the channel length, inversion level, and V_{DS}

For the $W/L = 3.2\,\mu m/0.18\,\mu m$, nMOS device in a $0.18\,\mu m$ CMOS process shown in Figure 3.34, $V_A = 4\,V$ for the bias point of $I_D = 100\,\mu A$, $V_{GS} = 0.7\,V$, and $V_{DS} = 1\,V$. Throughout this book, we express the MOS bias point in terms of the inversion coefficient, which is near the onset of strong inversion at $IC = I_D/[I_0(W/L)] = 7.4$ from Table 3.6. This is for a shape factor of $S = W/L = 3.2\,\mu m/0.152\,\mu m$ using a technology current of $I_0 = 0.64\,\mu A$ and lateral diffusion, length reduction of $DL = 0.028\,\mu m$ listed in the process parameters of Table 3.2. At the bias point shown, the value of g_{ds} is $g_{ds} = I_D/(V_A + V_{DS}) = 100\,\mu A/(4\,V + 1\,V) = 20\,\mu S$.

Since V_A is usually somewhat greater than V_{DS}, it is customary in Equation 3.52 to neglect V_{DS} and approximate g_{ds} as the ratio of I_D to V_A ($g_{ds} = I_D/V_A$) or r_{ds} as the ratio of V_A to I_D ($r_{ds} = V_A/I_D$). Neglecting V_{DS} simplifies the estimate of g_{ds} or r_{ds}, while providing a conservatively larger or lower estimate, respectively, for design optimization.

Often, including throughout this book, the operating V_{DS} is effectively included in the value of V_A as noted in the second line of Equation 3.52. In this case, V_A corresponds to the full horizontal voltage seen in Figure 3.34, which includes V_{DS} such that g_{ds} directly equals I_D/V_A or r_{ds} directly equals V_A/I_D. When V_{DS} is included in the definition of V_A, V_A is actually the drain–source conductance, effective voltage, $V_{gds} = (g_{ds}/I_D)^{-1}$, which is the reciprocal of the drain–source conductance efficiency, g_{ds}/I_D. $g_{ds} = I_D/V_{gds}$ in an analogous way to $g_m = I_D/V_{gm}$, where $V_{gm} = (g_m/I_D)^{-1}$ is the transconductance effective voltage described earlier in Section 3.8.2.1. In applications where V_A includes the operating V_{DS} (V_A actually being V_{gds}), it is important to specify V_{DS}. Additionally, regardless of the definition of V_A, it is important to specify V_{DS} since g_{ds} or r_{ds} depends strongly on V_{DS} as discussed in the next section.

3.8.4.1 Due to channel length modulation

The primary contributor to MOS drain–source conductance for typical operating conditions is channel length modulation, commonly abbreviated as CLM. Unlike DIBL and hot-electron contributions described in Sections 3.8.4.2 and 3.8.4.3 that can be made nearly negligible through channel length and bias selections, CLM contributions to drain–source conductance are always present.

In strong-inversion in saturation ($V_{DS} > V_{DS,sat}$), the channel inversion charge is nearly zero or pinched off at the drain side of the channel located near the drain depletion region as shown in the MOS cross-section of Figure 3.9. As V_{DS} increases, the drain depletion-region width widens due to increasing reverse bias between the drain and drain side of the channel. This reduces the electrical channel length below the physical spacing between the source and drain junctions. As a result, drain current increases as expected for a reduction in channel length. The increase in drain current with increasing V_{DS} then creates an effective conductance between the drain and source.

We will estimate CLM effects in strong inversion starting with drain current, excluding CLM effects, repeated from Table 3.6 as

$$I_D \text{ (without CLM)} = \frac{1}{2} \left(\frac{\mu\, C'_{OX}}{n} \right) \left(\frac{W}{L} \right) (V_{GS} - V_T)^2 \tag{3.53}$$

Here, mobility, μ, and the substrate factor, n, correspond to their values at the bias point, which do not need to be known in the CLM analysis. Drain current, including CLM effects, can by estimated by substituting the electrical value of channel length, $L - l_p$, into Equation 3.53. l_p is the pinch-off length, the distance from the pinched-off channel to the drain [15, p. 251], as shown in the MOS cross-section of Figure 3.9. Substituting $L - l_p$ for the channel length gives

$$I_D \text{ (with CLM)} = \frac{1}{2} \left(\frac{\mu\, C'_{OX}}{n} \right) \left(\frac{W}{L - l_p} \right) (V_{GS} - V_T)^2 \tag{3.54}$$

As observed in the equation, an increasing value of l_p lowers the electrical channel length and increases the drain current.

The drain–source conductance is found by the partial derivative of the drain current, including CLM effects, with respect to the drain–source voltage. This gives [15, p. 372]

$$g_{ds} = \frac{\partial}{\partial V_{DS}} (I_D \text{ with CLM}) = \frac{\partial}{\partial l_p} (I_D \text{ with CLM}) \cdot \frac{\partial l_p}{\partial V_{DS}}$$

$$= \frac{1}{2} \left(\frac{\mu\, C'_{OX}}{n} \right) \left(\frac{W}{(L - l_p)^2} \right) (V_{GS} - V_T)^2 \cdot \frac{\partial l_p}{\partial V_{DS}} \tag{3.55}$$

Equation 3.55 can be expressed in terms of the drain current without CLM effects given by Equation 3.53. This gives

$$g_{ds} = \frac{\partial}{\partial V_{DS}} (I_D \text{ with CLM}) = (I_D \text{ without CLM}) \cdot \frac{1}{L \left(1 - \dfrac{l_p}{L} \right)^2} \cdot \frac{\partial l_p}{\partial V_{DS}} \tag{3.56}$$

Assuming the pinch-off length is much shorter than the channel length, $l_p \ll L$, g_{ds} is approximated by

$$g_{ds} \approx (I_D \text{ without CLM}) \frac{1}{L} \cdot \frac{\partial l_p}{\partial V_{DS}} \tag{3.57}$$

Finally, assuming the drain bias current with CLM effects is not significantly above the current if these effects were not present (this is consistent with $l_p \ll L$ assumed above), Equation 3.57 can be related to the bias current, I_D, by

$$g_{ds} \approx I_D \frac{1}{L} \cdot \frac{\partial l_p}{\partial V_{DS}} \tag{3.58}$$

The Early voltage, $V_A = I_D/g_{ds}$ from Equation 3.52, is found from Equation 3.58 as

$$V_A = \frac{I_D}{g_{ds}} \approx L \left(\frac{\partial l_p}{\partial V_{DS}} \right)^{-1} \tag{3.59}$$

As mentioned for Equation 3.52 near the beginning of Section 3.8.4, excluding the value of V_{DS} in the g_{ds} or r_{ds} relationship with V_A effectively includes the value of V_{DS} in the value of V_A as is common practice. Dividing Equation 3.59 by channel length gives the Early voltage factor, V_{AL}, as

$$V_{AL} = \frac{V_A}{L} \approx \left(\frac{\partial l_p}{\partial V_{DS}} \right)^{-1} \tag{3.60}$$

V_{AL} is a quality factor, having units of volts per micron of channel length (V/μm), that describes the production of V_A for a given value of channel length. Since V_A is itself a quality factor describing the production of $r_{ds} = V_A/I_D$, V_{AL} is effectively the combination of two quality factors.

Using V_{AL}, V_A is expressed by

$$V_A = V_{AL} \cdot L \tag{3.61}$$

As seen from Equation 3.59, V_A and the corresponding drain–source resistance, $r_{ds} = V_A/I_D$, increase favorably as channel length increases. This is because the relative degree of CLM decreases as channel length increases. As seen from Equations 3.59 and 3.60, V_A and V_{AL} increase favorably as $\delta l_p/\delta V_{DS}$ decreases, corresponding to a smaller pinch-off length increase with increasing drain–source voltage.

Estimating V_A and V_{AL} from Equations 3.59 and 3.60 appears simple, but requires estimating $\delta l_p/\delta V_{DS}$, which is very complex. We will initially consider a simple, historical prediction of $\delta l_p/\delta V_{DS}$ assuming a one-dimensional analysis where electric field lines extend only horizontally, parallel to the source and drain axis. This assumes that vertical components of electric field from the gate and drain are negligible near the pinched-off channel. The prediction assumes the depletion-region width associated with a simple, reverse-biased, abrupt junction between the drain and pinched-off channel where the channel is much more lightly doped than the drain (or the extended drain in the usual case of lightly doped drain fabrication). These assumptions are invalid given the thin gate-oxide, small drain junction depth, and high channel doping concentration present in modern, small-geometry processes, but provide a historical starting point.

The pinch-off length or depletion-region width between the drain and pinched-off channel is given by [33, p. 25]

$$l_p \text{ (1-D analysis)} = \sqrt{\frac{2\varepsilon_{Si}}{qN_B}} \sqrt{(V_{DS} - V_{DS,sat}) + \phi_D} \tag{3.62}$$

where ε_{Si} is the permittivity of silicon, q is the electronic charge, and N_B is the doping concentration in the substrate or body. ϕ_D is the built-in potential (expression given in Table 3.1) associated with the junction between the pinched-off drain side of the channel and the drain. The $V_{DS} - V_{DS,sat}$ term represents the reverse bias between the pinched-off channel and drain, where channel pinch-off (relative to the source voltage) occurs at $V_{DS,sat} = V_{GS} - V_T$ (as mentioned earlier in Section 3.7.3.1, the body effect along the channel may actually lower $V_{DS,sat}$ closer to $V_{DS,sat} = (V_{GS} - V_T)/n$). $V_{DS} - V_{DS,sat}$ can be considered as the excess drain–source voltage above $V_{DS,sat}$.

An alternative estimate of pinch-off length is given by [15, p. 251]

$$l_p \text{ (1-D analysis, alternative)} = \sqrt{\frac{2\varepsilon_{Si}}{qN_B}} \left(\sqrt{(V_{DS} - V_{DS,sat}) + \phi_D} - \sqrt{\phi_D} \right) \tag{3.63}$$

where $\phi_D = (\varepsilon_{Si} E_x^2)/(2qN_B)$ considers the horizontal electric field, E_x, where the inversion charge pinches off.

Finding $\delta l_p / \delta V_{DS}$ from either Equation 3.62 or 3.63 (assuming ϕ_D is nearly constant with V_{DS}) gives

$$\frac{\partial l_p}{\partial V_{DS}} \text{ (1-D analysis)} = \frac{\sqrt{\dfrac{2\varepsilon_{Si}}{qN_B}}}{2\sqrt{(V_{DS} - V_{DS,sat}) + \phi_D}} \tag{3.64}$$

Substituting Equation 3.64 into the Early voltage given by Equation 3.59 gives

$$V_A \text{ (1-D analysis)} \approx L \left(\frac{\partial l_p}{\partial V_{DS}} \right)^{-1} = 2L \sqrt{\frac{qN_B}{2\varepsilon_{Si}}} \sqrt{(V_{DS} - V_{DS,sat}) + \phi_D} \tag{3.65}$$

where V_{AL} is then given from Equation 3.60 by

$$V_{AL} \text{ (1-D analysis)} = \frac{V_A}{L} \approx \left(\frac{\partial l_p}{\partial V_{DS}} \right)^{-1} = 2 \sqrt{\frac{qN_B}{2\varepsilon_{Si}}} \sqrt{(V_{DS} - V_{DS,sat}) + \phi_D} \tag{3.66}$$

In Equations 3.64–3.66, ϕ_D can be taken as either the built-in potential used in Equation 3.62 or $\phi_D = (\varepsilon_{Si} E_x^2)/(2qN_B)$ used in Equation 3.63. Alternatively, from a different derivation, l_p is related to where the field is high enough near the drain to cause velocity saturation [15, p. 252]. In this case, ϕ_D can be taken as a fixed value associated with a value of E_x above the critical, velocity saturation, electric field, E_{CRIT}. Velocity saturation effects on l_p are considered in the two-dimensional analysis that follows.

Equation 3.65 predicts that V_A and, correspondingly, r_{ds}, favorably increase directly with channel length, L, increase with the square root of channel doping, N_B, and increase with the square root of the excess V_{DS} above $V_{DS,sat}$, $V_{DS} - V_{DS,sat}$. However, Equation 3.65 significantly under-predicts the measured decrease in $g_{ds}(1/r_{ds})$ observed with increasing V_{DS} shown later in Figures 3.36 and 3.38.

While the preceding analysis provides some historical guidance on the value of V_A associated with CLM in large-geometry processes, it is not valid for modern, small-geometry processes because it excludes important, vertical, electric field components coming from the drain and gate [15, p. 372]. In modern processes, the gradual, inversion charge, channel approximation fails and current flow may flow below the surface in the pinch-off region. Additionally, the high vertical electric field from the gate may effectively "grab" the inversion charge and reduce the horizontal CLM influence from the drain. The importance of two-dimensional effects on drain–source conductance was recognized over 30 years ago in the work reported in [39].

A common approximation for l_p that considers two-dimensional effects is derived from [39] and has the form [14, 15, p. 254, 28, p. 39]

$$l_p \text{ (2-D analysis)} = \lambda_C L_C \ln \left(1 + \frac{V_{DS} - V_{DS,sat}}{V_E} \right) \tag{3.67}$$

where L_C is a characteristic length depending on vertical dimensions given by [14, 15, p. 254]

$$L_C = \sqrt{\frac{\varepsilon_{Si}}{C'_{OX}}} x_j = \sqrt{\frac{\varepsilon_{Si}}{\varepsilon_{SiO_2}}} t_{ox} x_j \approx \sqrt{3 t_{ox} x_j} \tag{3.68}$$

ε_{Si} is the permittivity of silicon, ε_{SiO_2} is the permittivity of silicon dioxide, t_{ox} is the gate-oxide thickness, and x_j is the junction depth of the drain. λ_C is a dimensionless fitting parameter having values ranging from approximately 0.5 to 3 [14]. V_E is a voltage, best found by measurements [15, p. 254], having extracted values of approximately 0.15 and 0.45 V, respectively, for the 0.18 and

$0.5\,\mu m$ CMOS processes described in Table 3.2. V_E is taken as $V_E = L_C \cdot E_{CRIT}$ in l_p given by [14, 28, p. 39], but its definition as an independent parameter enables fitting with measured data. Additionally, measured data shown later in this section shows that V_E is nearly equal for both nMOS and pMOS devices, while $V_E = L_C \cdot E_{CRIT}$ predicts a significantly higher value of V_E for pMOS devices because of the higher value of E_{CRIT}. The $V_{DS} - V_{DS,sat}$ term of Equation 3.67 appears in l_p given by [15, p. 254, 28, p. 39], while a more complex, related term appears in [14]. $V_{DS} - V_{DS,sat}$ terms appear directly and squared within a square-root term from the original work in [39] and in [16, p. 188], but Equation 3.67 is used here for its simplicity and use of λ_C and V_E fitting parameters.

Although l_p appears logarithmically compressed in Equation 3.67, it changes rapidly for V_{DS} near $V_{DS,sat}$. l_p can be readily differentiated with respect to V_{DS}, giving

$$\frac{\partial l_p}{\partial V_{DS}} \text{ (2-D analysis)} = \frac{\lambda_C L_C}{V_E}\left(1 + \frac{V_{DS} - V_{DS,sat}}{V_E}\right)^{-1} \tag{3.69}$$

Substituting Equation 3.69 into the Early voltage given by Equation 3.59 gives

$$V_A \text{ (2-D analysis)} \approx L\left(\frac{\partial l_p}{\partial V_{DS}}\right)^{-1} = L \cdot \frac{V_E}{\lambda_C L_C}\left(1 + \frac{V_{DS} - V_{DS,sat}}{V_E}\right) \tag{3.70}$$

V_A given by Equation 3.70 is of the same form as that given in [15, p. 373, 28, p. 41], resulting from the use of l_p having the form given by Equation 3.67. The Early voltage factor, V_{AL}, is given from Equation 3.60 by

$$V_{AL} \text{ (2-D analysis)} = \frac{V_A}{L} \approx \left(\frac{\partial l_p}{\partial V_{DS}}\right)^{-1} = \frac{V_E}{\lambda_C L_C}\left(1 + \frac{V_{DS} - V_{DS,sat}}{V_E}\right) \tag{3.71}$$

Equation 3.70, which considers two-dimensional effects, predicts that V_A and, correspondingly, r_{ds}, favorably increase directly with L as found earlier by Equation 3.65, which considers only single-dimensional effects. However, Equation 3.70 predicts that V_A and r_{ds} increase directly with the excess V_{DS} above saturation, $V_{DS} - V_{DS,sat}$, with a slope given by $1/V_E$ compared to a much lower square-root increase predicted by Equation 3.65. Additionally, Equation 3.70 predicts that V_A and r_{ds} increase inversely with L_C as this becomes smaller in smaller-geometry processes since L_C is proportional to the square root of the product of vertical dimensions t_{ox} and x_j. Equation 3.70 also predicts that V_A is independent of channel doping, N_B. In contrast, Equation 3.65 predicts V_A and r_{ds} increase with the square root of N_B as this becomes higher in smaller-geometry processes, while being independent of t_{ox} and x_j. As mentioned, Equation 3.65 is not valid for modern, small-geometry processes because it ignores important, vertical, electric field components from the gate and drain. Finally, both Equations 3.70 and 3.65 incorrectly predict that V_A increases linearly with L, when the actual increase due to CLM is sublinear as described below.

Equation 3.70 can be expressed in terms of the Early voltage CLM factor, $V_{AL}(V_{DS,sat})$, found at $V_{DS} = V_{DS,sat}$, giving

$$V_A \text{ (2-D analysis)} \approx V_{AL}\left(V_{DS,sat}\right) \cdot L \cdot \left(1 + \frac{V_{DS} - V_{DS,sat}}{V_E}\right) \tag{3.72}$$

where

$$V_{AL}\left(V_{DS,sat}\right) \approx \frac{V_E}{\lambda_C L_C} \tag{3.73}$$

Equations 3.72 and 3.73 provide a simple prediction of V_A that includes the significant increase with V_{DS} shown later through the decrease in g_{ds} with V_{DS} in Figures 3.36 and 3.38. However, the equations predict a fixed value of $V_{AL}(V_{DS,sat})$ across channel lengths while Figure 3.48 in Section 3.8.4.5 shows significant variation in V_{AL} across channel lengths. This is a result of V_A increasing faster than L at short channel lengths (actually due primarily to DIBL discussed in Section 3.8.4.2) and sublinearly with L at longer channel lengths as shown in Figure 3.47. Additionally, the equations exclude the increase in V_A with increasing inversion in strong inversion described in Section 3.8.4.5 for constant V_{DS} (Figures 3.43 and 3.44) and the larger increase in V_A for $V_{DS} = V_{GS}$ (Figures 3.49 and 3.50) where the excess V_{DS}, $V_{DS} - V_{DS,sat}$, remains nearly constant. Under the condition of constant excess V_{DS}, the equations incorrectly predict constant V_A by excluding inversion-level effects, which were not considered in the derivation.

Equation 3.72 can be extended to consider $V_{AL}(V_{DS,sat})$, which specifies the Early voltage CLM factor at $V_{DS} = V_{DS,sat}$, and V_E, which specifies how rapidly V_A (or V_{AL}) increases with V_{DS}, as independent fitting parameters. This is an interesting result and is repeated below with notations about its use:

$$V_A \text{ (CLM only)} \approx V_{AL}\left(V_{DS,sat}\right) \cdot L \cdot \left(1 + \frac{V_{DS} - V_{DS,sat}}{V_E}\right) \tag{3.74}$$

- $V_{AL}(V_{DS,sat})$ is the Early voltage CLM factor (V_A/L having units of V/μm) extracted at $V_{DS} = V_{DS,sat}$ near the level of inversion (IC) used since V_{AL} varies with the level of inversion.

- $V_{AL}(V_{DS,sat})$ is extracted from measurements near the L used since V_{AL} varies significantly with L.

- V_A increases with V_{DS} with a slope governed by $1/V_E$. Measurements indicate $V_E \approx 0.15$ and 0.45 V for the 0.18 and 0.5 μm CMOS processes described in Table 3.2.

- V_A decreases rapidly as V_{DS} decreases close to $V_{DS,sat}$. Considerable prediction errors are likely here.

Equation 3.74 provides design insight and correctly considers the increase in V_A with V_{DS} shown later through the decrease in g_{ds} with V_{DS} in Figures 3.36 and 3.38. However, Equation 3.74 is difficult to use in practice because it involves the extraction of $V_{AL}(V_{DS,sat})$ at $V_{DS} = V_{DS,sat}$ where there is uncertainty in the value of $V_{DS,sat}$ and where V_A changes rapidly. Additionally, $V_{AL}(V_{DS,sat})$ requires extraction near the IC and L used.

Instead of attempting to fit variable values of $V_{AL}(V_{DS,sat})$ and potentially variable values of V_E for Equation 3.74, we will use a prediction of V_A based on V_{AL} values extracted at fixed values of V_{DS} over a limited range of IC and L. This prediction is given below with notations about its use:

$$V_A \approx V_{AL}(\text{near operating } IC, L, V_{DS}) \cdot L \tag{3.75}$$

- V_{AL} is extracted near the V_{DS} used since it increases significantly with V_{DS} as shown in Figure 3.48 in Section 3.8.4.5.

- V_{AL} is extracted near the L used since it varies significantly with L as shown in Figure 3.48. This corresponds to V_A not directly tracking L as shown in Figure 3.47.

- V_{AL} is extracted over a range of IC at fixed values of V_{DS} where V_A remains nearly constant as shown in Figures 3.43–3.46.

- Since V_{AL} is extracted from measured data, it implicitly includes DIBL effects discussed in Section 3.8.4.2 and can be extended to include second-order effects like non-zero V_{SB}.

Equation 3.75 is not an analytical prediction of V_A, but is a connection to measured data and trends described later in Section 3.8.4.5. At present, no simple, reasonably accurate, hand prediction for V_A is available because of the complexities associated with dependencies on IC, L, and V_{DS} and dependencies on process engineering. Instead, Equation 3.75 can be used with extracted V_{AL} values, like those shown in Figure 3.48 for the 0.18 μm CMOS process described in Table 3.2. Here, V_{AL} ranges from 3 to 40 V/μm from weak through strong inversion for $L = 0.18 - 4$ μm and $V_{DS} = 0.25 - 1$ V. These values include DIBL effects described in Section 3.8.4.2 and can be extended to include second-order effects like non-zero V_{SB}.

Although the simple g_m/I_D predictions shown in Section 3.8.2.3 are generally accurate to within 10 % or less, inclusive of velocity saturation and VFMR effects, no attempt will be made here to specify the accuracy of Equations 3.74 and 3.75 for V_A across the design choices of IC, L, and V_{DS}. However, measured data below shows that Equation 3.74 provides useful design guidance, while measured data in Section 3.8.4.5 shows that Equation 3.75 provides accurate predictions when V_{AL} is extracted in the vicinity of the IC, L, and V_{DS} used.

Figure 3.35 illustrates CLM by showing drain current curves for pMOS devices having channel lengths of $L = 0.18, 0.28, 0.48$, and 4 μm. The devices are from the 0.18 μm CMOS process described in Table 3.2 and are operated in strong inversion at $V_{EFF} = V_{GS} - V_T = -0.35$ V. Bias currents of $I_D = 300$ μA (for V_{DS} near $V_{DS,sat}$) correspond to $IC = I_D/[I_0(W/L)] \approx 30$ from Table 3.6 for device shape factors of $S = W/L = 76$ and a technology current of $I_0 = 0.135$ μA from Table 3.2. The devices have widths of $W = 15.2, 22.8, 38$, and 304 μm, respectively, giving nearly equal shape factors of 76. This permits their curves to be overlaid since their drain currents are nearly equal. Drain currents for the short-channel devices were scaled down slightly to compensate for channel length reductions due to lateral diffusion ($L = L_{DRAWN} - DL$, where $DL = 0.051$ μm from Table 3.2) and permit a direct overlay. In order to study CLM effects, it is necessary to minimize other g_{ds} effects. The low g_m/I_D in strong inversion minimizes DIBL increases in g_{ds}, discussed later in Section 3.8.4.2, which is especially a concern for the minimum channel length device. pMOS devices are selected as these do not experience hot-electron increases in g_{ds}, discussed later in Section 3.8.4.3, that are likely to be present in nMOS devices at high values of V_{DS}.

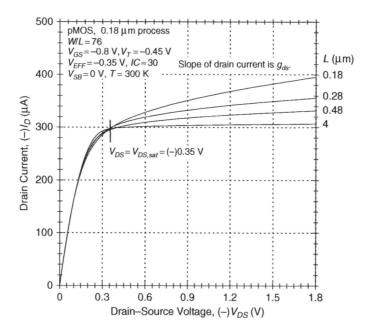

Figure 3.35 Drain current curves illustrating channel length modulation for $L = 0.18, 0.28, 0.48$, and 4 μm, pMOS devices operating in strong inversion at $V_{EFF} = V_{GS} - V_T = -0.35$ V. In saturation, the slope of the curves, g_{ds}, is progressively less as channel length increases due to less CLM. The slope or g_{ds} decreases with increasing V_{DS} above $V_{DS,sat}$

The slope of the drain current curves in Figure 3.35, which is g_{ds}, is noticeably highest in saturation for the $L = 0.18\,\mu\mathrm{m}$ device and is progressively less for the $L = 0.28, 0.48$, and $4\,\mu\mathrm{m}$ devices. CLM effects are reduced as channel length increases, resulting in lower g_{ds}, higher r_{ds}, and higher V_A. A careful examination of the curves reveals a noticeable slope or g_{ds} decrease as V_{DS} increases above $V_{DS,sat}$, which is taken here at the traditional strong-inversion value of $V_{DS,sat} = V_{EFF} = -0.35\,\mathrm{V}$. The slope decrease is easily seen for the shorter-channel devices that have large slopes, but cannot be reliably, visually observed for longer-channel devices where the curves become nearly horizontal. Measured g_{ds} versus V_{DS} curves should be used for assessing g_{ds} as visual slope observations can result in significant errors.

Figure 3.36 shows measured g_{ds} versus V_{DS} for the drain current curves of Figure 3.35 with an overlay of g_{ds} predicted from $g_{ds} = I_D/V_A$ using the V_A prediction of Equation 3.74. V_A is predicted using V_{AL} $(V_{DS,sat} = -0.35\,\mathrm{V}) = 8.5, 9, 7$, and $3\,\mathrm{V}/\mu\mathrm{m}$ for the $L = 0.18, 0.28, 0.48$, and $4\,\mu\mathrm{m}$ devices, respectively. These V_{AL} values are very similar to those extracted at the lower $V_{DS} = (-)0.25\,\mathrm{V}$ value shown later in Figure 3.48 in Section 3.8.4.5, which include a reduction in V_{AL} for the $L = 0.18\mu\mathrm{m}$ device due to DIBL effects. V_E is $0.15\,\mathrm{V}$, which results in V_A increasing more rapidly than V_{DS}. Channel length is adjusted to effective values by the subtraction of $DL = 0.051\,\mu\mathrm{m}$. As always, predictions of pMOS performance are made assuming positive nMOS polarities of $V_{DS}, V_{DS,sat}, V_E$, and other relevant variables.

Predicted and measured g_{ds} track each other for V_{DS} sufficiently above $V_{DS,sat}$, but g_{ds} is over-predicted at $V_{DS,sat}$ for the short-channel, $L = 0.18\,\mu\mathrm{m}$ device and under-predicted for the long-channel, $L = 4\,\mu\mathrm{m}$ device. Measured g_{ds} decreases by approximately a factor of four from its value at $V_{DS} = V_{DS,sat} = -0.35\,\mathrm{V}$ to its value deeper in saturation at $V_{DS} = V_{GS} = -0.8\,\mathrm{V}$, with the decrease less for the $L = 0.18\,\mu\mathrm{m}$ device and more for the $L = 4\,\mu\mathrm{m}$ device. This illustrates the importance of considering V_{DS} in the V_A prediction where V_A increases by the same approximate factor of four, predicted by

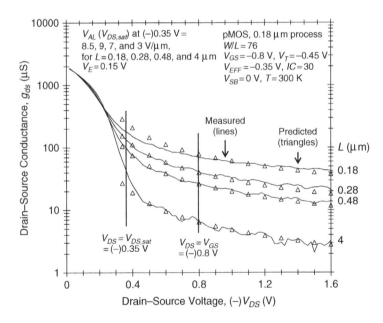

Figure 3.36 Measured and predicted drain–source conductance, g_{ds}, versus drain–source voltage, V_{DS}, for the devices shown in Figure 3.35. Measured g_{ds}, found from the slope of the drain current curves in Figure 3.35, decreases as channel length increases and decreases with increasing V_{DS}. Predicted and measured g_{ds} track each other for V_{DS} above $V_{DS,sat}$, but deviate for some devices as V_{DS} approaches $V_{DS,sat}$. g_{ds} decreases by nearly a factor of four from $V_{DS} = V_{DS,sat} = -0.35\,\mathrm{V}$ to $V_{DS} = V_{GS} = -0.8\,\mathrm{V}$ deeper into saturation, showing a strong V_{DS} dependency

Equation 3.74 with the term $1 + (V_{DS} - V_{DS,sat})/V_E = 1 + (0.8\,V - 0.35\,V)/0.15\,V = 4$ for $V_{DS} = -0.8\,V$ compared to a value of unity for $V_{DS} = -0.35\,V$. The rapid increase of V_A with V_{DS}, corresponding to a rapid decrease in g_{ds}, shows that Equation 3.65 cannot be used for predicting V_A. Equation 3.65, derived assuming a simple, depletion-region approximation of l_p exclusive of two-dimensional effects, predicts that V_A increases very slowly with V_{DS}. This is a result of incorrectly predicted l_p resulting from neglecting vertical electric fields from the gate and drain.

The g_{ds} versus V_{DS} behavior shown for pMOS devices in Figure 3.36 is also observed for nMOS devices where g_{ds} increases significantly for V_{DS} near $V_{DS,sat}$ compared to the lower values available for V_{DS} well above $V_{DS,sat}$. Figure 3.37 illustrates this with measured drain current curves for a $W/L = 8\,\mu m/0.48\,\mu m$, nMOS device from the 0.18 μm CMOS process described in Table 3.2. The top two curves are for operation in strong inversion and are marked with two estimates for $V_{DS,sat}$: $V_{DS,sat} = (V_{GS} - V_T)/n$ (n assumed equal to 1.3 in strong inversion) to the left and the traditional estimate, $V_{DS,sat} = V_{GS} - V_T$, to the right. These estimates of $V_{DS,sat}$ were discussed earlier in Section 3.7.3.1 where the lower left estimate considers the local threshold-voltage increase along the channel that lowers the pinch-off voltage. This $V_{DS,sat}$ estimate closely matches the continuous weak- through strong-inversion prediction given in Table 3.15.

As seen in Figure 3.37, operation at the lower left estimate of $V_{DS,sat}$ results in a small loss of drain current, but with a substantial increase in g_{ds} (again, observed as the slope of I_D versus V_{DS}) above that available at higher V_{DS}. Operation at the higher, right estimate of $V_{DS,sat}$ results in less g_{ds} increase, but g_{ds} is still well above that available at higher V_{DS}. As shown in the figure, drain current does not transition abruptly from the ohmic, linear, or triode region into the saturation region at $V_{DS,sat}$, but makes a smooth, rounded transition. This prevents an abrupt discontinuity in g_{ds} near $V_{DS,sat}$, but substantially increases g_{ds} near $V_{DS,sat}$.

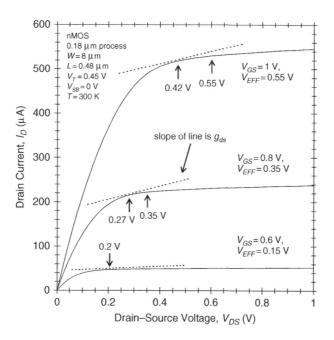

Figure 3.37 Drain current curves for a $W/L = 8\,\mu m/0.48\,\mu m$, nMOS device in a 0.18 μm CMOS process showing significant increases in drain–source conductance, g_{ds}, for operation near the drain–source saturation voltage, $V_{DS,sat}$. While operation at any of the labeled $V_{DS,sat}$ estimates (discussed in the text) results in only a small loss of drain current, g_{ds} increases significantly from the value available further in saturation at higher V_{DS}. This is detailed in Figure 3.38

The bottom curve in Figure 3.37 is for operation in moderate inversion at $V_{EFF} = V_{GS} - V_T = 0.15\,V$ and $IC = 5$ ($IC = I_D/[I_0(W/L)]$ from Table 3.6, with $I_0 = 0.64\,\mu A$ from Table 3.2). This curve is marked with $V_{DS,sat} = 0.2\,V$ found from the continuous, weak- through strong-inversion expression for $V_{DS,sat}$ given in Table 3.15. The traditional, strong-inversion $V_{DS,sat} = V_{GS} - V_T$ prediction does not apply in weak or moderate inversion and is not marked on the curve.

Figure 3.38 details the rounded transition from the ohmic, linear, or triode region into saturation for the bottom curve in Figure 3.37 by showing measured drain current and g_{ds} versus V_{DS}. Unlike Figure 3.36 shown earlier that used the traditional log scale for g_{ds}, Figure 3.38 presents g_{ds} on an expanded, linear scale. The figure shows an average $g_{ds} = 5\,\mu S$ in saturation, while g_{ds} continues to decrease as V_{DS} increases. However, at $V_{DS} = 0.2\,V = V_{DS,sat}$, $g_{ds} = 25\,\mu S$ or $500\,\%$ of its average value in saturation, while drain current is down by only 3.6 % from its expected value. Operating at $V_{DS} = 0.3\,V = 1.5 \times V_{DS,sat}$ results in $g_{ds} = 10\,\mu S$, or $200\,\%$ of its average value in saturation. Finally, operating at $V_{DS} = 0.4\,V = 2 \times V_{DS,sat}$ results in $g_{ds} = 7.5\,\mu S$, or $150\,\%$ of its average value in saturation. These increases in g_{ds} show that it is necessary to operate at V_{DS} values well above $V_{DS,sat}$ to obtain the optimally low g_{ds} (or high r_{ds}) available in saturation. As described later in Section 3.8.4.4, the increase in g_{ds} near $V_{DS,sat}$ can cause a significant loss of small-signal resistance in circuits, resulting in a significant loss of signal gain. Additionally, the variation of g_{ds} with V_{DS} can cause signal non-linearity and distortion.

Figure 3.38 also includes predicted $g_{ds} = I_D/V_A$ using V_A predicted by Equation 3.74. V_A is predicted using $V_{AL}(V_{DS,sat}) = 6.5\,V/\mu m$, extracted at $V_{DS} = 0.25\,V$, which is slightly above the moderate inversion value of $V_{DS,sat} = 0.2\,V$. V_E is taken at 0.15 V, the same value used for predicting pMOS g_{ds} in Figure 3.36. Predicted and measured g_{ds} track each other with increased variation for V_{DS} near $V_{DS,sat}$. The significant decrease of g_{ds} with increasing V_{DS} shown in Figure 3.38, like that shown

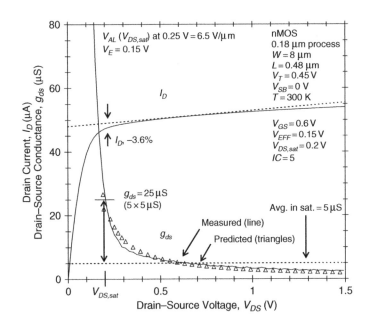

Figure 3.38 Drain current and drain–source conductance, g_{ds}, versus drain–source voltage, V_{DS}, corresponding to the $V_{GS} = 0.6\,V$ curve in Figure 3.37. While drain current decreases by only 3.6 % at $V_{DS,sat}$, g_{ds} increases significantly to five times its average value in saturation. The increase in g_{ds} near $V_{DS,sat}$ can cause a significant loss of small-signal gain while the variation in g_{ds} can cause signal non-linearity and distortion. The figure also includes predicted g_{ds}, which tracks measured g_{ds} as V_{DS} increases

in Figure 3.36, again illustrates the importance of including the V_A increase with V_{DS} predicted by Equation 3.74.

The g_{ds} versus V_{DS} behavior seen in Figures 3.36 and 3.38 for pMOS and nMOS devices in the example 0.18 μm CMOS process is similar to that observed for devices in the much larger example 0.5 μm process (both processes are described in Table 3.2). However, comparisons of predicted and measured g_{ds} indicate that V_E must be increased to 0.45 V for the 0.5 μm process from 0.15 V for the 0.18 μm process. This results in an effective scaling of V_{DS} where a 3 V increase in V_{DS} in the 0.5 μm process results in a similar decrease of g_{ds} as a 1 V increase in V_{DS} for the 0.18 μm process. In effect, the decrease of g_{ds} over the full supply voltage range of 5 V and 1.8 V, respectively, is similar for the two processes.

The larger value of V_E for the 0.5 μm process is consistent with a larger value of characteristic length, L_C, where V_E is proportional to L_C in the predictions of g_{ds} given in [14, 28, p. 39]. From Equation 3.68, L_C has predicted values of 27 and 82 nm for the 0.18 and 0.5 μm processes, assuming $t_{ox} = 4.1$ and 13.5 nm, and $x_j = 60$ and 165 nm. If V_E is taken as $V_E = L_C E_{CRIT}$ from [14, 28, p. 39], it then has predicted values of 0.15 and 0.33 V for nMOS devices in the 0.18 and 0.5 μm processes, assuming $E_{CRIT} = 5.6$ and 4 V/μm. Process values are from Table 3.2 except for the drain junction depth, x_j, which is estimated at one-third the minimum channel length. The predicted value of $V_E = 0.15$ V is equal to the extracted value for the 0.18 μm process, while the predicted value of $V_E = 0.33$ V is below the extracted value of 0.45 V for the 0.5 μm process. The values cannot be directly compared since the actual values of x_j are unknown, but do show the trends where L_C and, correspondingly, V_E, decrease in smaller-geometry processes. As mentioned, measurements here indicate that V_E is equal for nMOS and pMOS devices in a given process, whereas $V_E = L_C \cdot E_{CRIT}$ predicts higher values for pMOS values because of higher E_{CRIT}. This suggests that V_E is best found by fitting to measured data.

3.8.4.2 Due to DIBL

While CLM is the primary contributor to MOS drain–source conductance for typical operating conditions, drain-induced barrier lowering (DIBL) can be a significant contributor, especially for short-channel devices operating at low levels of inversion. When V_{DS} increases, the threshold voltage, V_T, decreases, especially for short-channel devices. This effect is commonly referred to as DIBL. As V_T decreases with increasing V_{DS}, $V_{EFF} = V_{GS} - V_T$ increases, increasing the drain current. This increase of drain current with increasing V_{DS} gives rise to a component of drain–source conductance.

The decrease of V_T with increasing V_{DS} can be explained by either the barrier-lowering or charge-sharing concept. In the barrier-lowering concept [15, pp. 257–259], decreasing the channel length places the drain and source closer together resulting in a deeper depletion region under the channel because of the nearby source and drain depletion regions. The channel depletion region is deepened further with increasing V_{DS} as the depletion region around the drain increases. The deeper channel depletion region results in less substrate control, lower depletion capacitance, and increased silicon surface potential. Increasing V_{DS} then lowers the potential barrier, attracts more carriers to the channel, and, correspondingly, increases the drain current.

In the charge-sharing concept [15, p. 259], the source and drain influence channel operation in addition to the normal influence of the gate and substrate. This is especially true for short-channel devices where the drain is close to the channel and acts as an undesired, secondary gate generating field lines terminating on the channel. Increasing V_{DS} then enhances the channel beyond its regular gate and substrate control and, correspondingly, increases the drain current.

Both the charge-sharing and barrier-lowering concepts describe increasing channel enhancement and drain current with increasing V_{DS} for a fixed value of V_{GS}. Additionally, this effect is sometimes referred to as static feedback between the drain and gate. As is common, the effects will be collectively referred to as DIBL here and described by a decreasing value of V_T (from the unaffected, large-geometry value) with increasing V_{DS}.

As shown later in measured $V_A = I_D/g_{ds} = I_D \cdot r_{ds}$ data in Section 3.8.4.5, DIBL effects significantly increase g_{ds}, or decrease r_{ds} and V_A, for minimum channel length devices operating in weak inversion. DIBL effects must be carefully considered in all processes, especially for devices having channel lengths near the process minimum.

The change in drain current, ΔI_D, resulting from a change in V_T, ΔV_T, due to DIBL is given by

$$\Delta I_D \text{ (DIBL)} = g_m (-\Delta V_T) \tag{3.76}$$

ΔV_T acts as a negative, small-signal, gate–source voltage since its increase lowers $V_{EFF} = V_{GS} - V_T$, lowering the drain current. Dividing Equation 3.76 by ΔV_{DS} that causes ΔV_T and ΔI_D gives

$$g_{ds} \text{ (DIBL)} = \frac{\Delta I_D \text{ (DIBL)}}{\Delta V_{DS}} = \frac{g_m (-\Delta V_T)}{\Delta V_{DS}} = g_m \left(\frac{-\partial V_T}{\partial V_{DS}} \right) \tag{3.77}$$

where $\delta V_T/\delta V_{DS}$ is the change in threshold voltage with respect to V_{DS}. Since $\delta V_T/\delta V_{DS}$ is negative (V_T decreases as V_{DS} increases), g_{ds} due to DIBL is positive, consistent with the increase in drain current caused by increasing V_{DS}. The Early voltage associated with DIBL is then estimated as

$$V_A \text{ (DIBL)} = \frac{I_D}{g_{ds} \text{ (DIBL)}} = \frac{I_D}{g_m \left(\dfrac{-\partial V_T}{\partial V_{DS}} \right)} = \frac{1}{\left(\dfrac{g_m}{I_D} \right) \dfrac{-\partial V_T}{\partial V_{DS}}} = \frac{V_{gm}}{\dfrac{-\partial V_T}{\partial V_{DS}}} \tag{3.78}$$

where, in the rightmost term, g_m/I_D is alternatively expressed from the transconductance effective voltage, $V_{gm} = (g_m/I_D)^{-1}$, described earlier in Section 3.8.2.1. As mentioned for Equation 3.52 near the beginning of Section 3.8.4, excluding the value of V_{DS} in the g_{ds} or r_{ds} relationship with V_A effectively includes the value of V_{DS} in the value of V_A.

Predicting V_A and the corresponding g_{ds} or r_{ds} due to DIBL appears simple using Equation 3.78 above. However, the prediction is complex because of the difficulty in predicting the change in threshold voltage with drain–source voltage, $\delta V_T/\delta V_{DS}$. This is analogous to predicting V_A due to CLM in Equation 3.59, which also appears simple. In the case of CLM, though, the prediction is complex because of the difficulty in predicting the change in pinch-off length with drain–source voltage, $\delta l_p/\delta V_{DS}$.

One prediction of $\delta V_T/\delta V_{DS}$ is based on a pseudo, two-dimensional solution of Poisson's equation giving [40]

$$\Delta V_T \text{ (DIBL)} \approx - \left[3 \left(\phi_D - 2\phi_F \right) + V_{DS} \right] e^{-L/L_{DIBL}}$$
$$- 2\sqrt{\left(\phi_D - 2\phi_F \right) \left(\phi_D - 2\phi_F + V_{DS} \right)} e^{-L/(2L_{DIBL})} \tag{3.79}$$

ΔV_T is the threshold voltage change for a given value of V_{DS}, ϕ_D (expression given in Table 3.1) is the built-in potential for the drain and channel junction, and $2\phi_F$ is the silicon surface potential taken at the threshold of surface inversion (the boundary of weak and moderate inversion [15, p. 87]); $2\phi_F$ is equal to the process parameter PHI that describes twice the Fermi potential. L_{DIBL} is a characteristic length for DIBL given by

$$L_{DIBL} = \lambda_{DIBL} \sqrt{\frac{\varepsilon_{Si}}{C'_{OX}} t_{dep}} = \lambda_{DIBL} \sqrt{\frac{\varepsilon_{Si}}{\varepsilon_{SiO_2}} t_{ox} t_{dep}} \approx \lambda_{DIBL} \sqrt{3 t_{ox} t_{dep}} \tag{3.80}$$

which depends on vertical dimensions and is identical in form to the characteristic length for CLM, L_C, given by Equation 3.68 except that the depletion-region width below the channel, t_{dep} (expression given in Table 3.1), is used in place of the drain junction depth, x_j, and the fitting parameter λ_{DIBL} is included. λ_{DIBL} has values in the vicinity of unity.

Equation 3.79 predicts that ΔV_T depends negatively on V_{DS} and the square root of V_{DS}, with the square-root relationship dominating for small values of V_{DS} and channel lengths somewhat above L_{DIBL} [40]. The equation also predicts that ΔV_T is proportional to the exponential of $-L/L_{DIBL}$ and the exponential of $-L/(2L_{DIBL})$, as seen in the first term and the second, square-root term. While making the prediction of ΔV_T very dependent on the value of L_{DIBL}, this shows that the magnitude of ΔV_T decreases very rapidly as L increases. ΔV_T is roughly proportional to $1/L^2$ in the MOS model described in [28, p. 36] and proportional to $1/L^3$ in the SPICE Level 3 MOS model [36, pp. 57–58]. All models predict a rapid decrease of ΔV_T with increasing channel length.

A rapid decrease in the magnitude of ΔV_T with increasing channel length is consistent with measured data where the DIBL increase in g_{ds} and the corresponding decrease in V_A are nearly eliminated by increasing channel length modestly from the process minimum. This is shown in Figures 3.43–3.47 and 3.49 in Section 3.8.4.5 where measured V_A for $L = 0.28\,\mu m$ increases very rapidly towards normal expected values compared to reduced values for the process minimum (L_{min}) of $L = 0.18\,\mu m$ where DIBL effects are severe. The rapid decrease of ΔV_T is also consistent with increased, simulated, channel surface potential due to DIBL effects that decreases rapidly as channel length increases from the process minimum [40].

In addition to a rapid decrease in the magnitude of ΔV_T with increasing channel length, Equation 3.79 also predicts decreased ΔV_T for high channel doping concentration where t_{dep} and, correspondingly, L_{DIBL} are smaller. The equation also predicts increased ΔV_T for non-zero V_{SB} where t_{dep} and L_{DIBL} are larger. Both trends are observed in practice [15, p. 267]. The equation, however, does not predict the decrease in ΔV_T resulting from decreased drain junction depth, x_j, as this does not appear in the expression [15, p. 267]. In smaller-geometry processes, the depletion region thickness below the channel, t_{dep}, is reduced by retrograde doping where doping is increased vertically below the channel surface. Additionally, the depletion region thickness at the source and drain is reduced by halo doping where doping is increased laterally near the source and drain. This reduces DIBL effects, especially increased doping near the drain. The reduction of L_{DIBL} predicted by Equation 3.80 with decreasing t_{ox} and t_{dep} (t_{dep} decreases because of higher channel doping) and the reduction of x_j suggest that ΔV_T could remain at comparable levels as L_{min} decreases in smaller-geometry processes.

L_{DIBL}, unlike the characteristic length L_C given in Equation 3.68 for CLM predictions, is a function of the inversion level (through the silicon surface potential) and V_{SB} since these influence t_{dep}. As mentioned in Section 3.8.3.2, $t_{dep} \approx 9t_{ox}$ in weak inversion for $n = 1.33$ ($V_{SB} = 0\,V$), where $n - 1 = \eta = 0.33 \approx 3t_{ox}/t_{dep}$. For the $0.18\,\mu m$ CMOS process described in Table 3.2 with $t_{ox} = 4.1\,nm$, this gives $t_{dep} \approx 37$ nm. Using $t_{dep} \approx 37$ nm, $t_{ox} = 4.1\,nm$, and $\lambda_{DIBL} = 1$, Equation 3.80 predicts $L_{DIBL} \approx 21$ nm $\approx L_{min}/8$ for $L_{min} = 0.18\,\mu m$ (this neglects the reduction associated with DL, which lowers L_{min} somewhat). When using such a small value of L_{DIBL}, Equation 3.79 predicts that the magnitude of ΔV_T decreases very, very rapidly as channel length increases from L_{min}. While ΔV_T does decrease rapidly as channel length increases from L_{min}, it is unknown if it decreases as rapidly as predicted by $L_{DIBL} \approx L_{min}/8$ using Equation 3.79. Measured data would be required to verify this.

While Equation 3.79 shows that the magnitude of ΔV_T and, correspondingly, $\delta V_T/\delta V_{DS}$ decrease exponentially with increasing channel length with a length constant related to L_{DIBL}, we will use the simple relationship mentioned earlier for some MOS models where $\delta V_T/\delta V_{DS}$ is proportional to $(1/L)$ raised to some power. This avoids the complexity of estimating L_{DIBL} while providing some ability to select fitting parameters that correspond to measured values of $\delta V_T/\delta V_{DS}$. $\delta V_T/\delta V_{DS}$ is then estimated by

$$\frac{\partial V_T}{\partial V_{DS}} \approx DVTDIBL \cdot \left(\frac{L_{min}}{L}\right)^{DVTDIBLEXP} \qquad (3.81)$$

where $DVTDIBL = \delta V_T/\delta V_{DS}$ for the minimum channel length, L_{min}, in the process, and $DVTDIBLEXP$ is the exponent describing the decrease of $\delta V_T/\delta V_{DS}$ with increasing channel length. Estimated $DVTDIBL$ is -12 and $-8\,\text{mV/V}$, respectively, for nMOS devices in the example 0.5 and 0.18 μm CMOS processes listed in Table 3.2. As listed, the magnitude of pMOS values is slightly higher. Estimated $DVTDIBLEXP$ is three, corresponding to a $1/L^3$ decrease in $\delta V_T/\delta V_{DS}$ with increasing channel length. While there is uncertainty about the value of $\delta V_T/\delta V_{DS}$ for channel lengths above L_{min}, Equation 3.81 provides a good estimate for the minimum channel length because it uses the extracted value of $\delta V_T/\delta V_{DS}$ here. If non-zero V_{SB} is present, $DVTDIBL$ should be higher because of higher t_{dep} as mentioned earlier.

Substituting $\delta V_T/\delta V_{DS}$ from Equation 3.81 into Equation 3.78 gives V_A due to DIBL as

$$V_A \text{ (DIBL)} = \frac{1}{\left(\dfrac{g_m}{I_D}\right)\dfrac{-\partial V_T}{\partial V_{DS}}} = \frac{V_{gm}}{\dfrac{-\partial V_T}{\partial V_{DS}}} \approx \frac{1}{\left(\dfrac{g_m}{I_D}\right)\left(-DVTDIBL \cdot \left(\dfrac{L_{min}}{L}\right)^{DVTDIBLEXP}\right)} \quad (3.82)$$

Substituting g_m/I_D, excluding velocity saturation and VFMR effects, from Table 3.17 then gives

$$V_A \text{ (DIBL)} = \frac{1}{\left(\dfrac{g_m}{I_D}\right)\dfrac{-\partial V_T}{\partial V_{DS}}} = \frac{V_{gm}}{\dfrac{-\partial V_T}{\partial V_{DS}}} \approx \frac{nU_T\left(\sqrt{IC+0.25}+0.5\right)}{-DVTDIBL \cdot \left(\dfrac{L_{min}}{L}\right)^{DVTDIBLEXP}} \quad (3.83)$$

This predicts that V_A due to DIBL is equal to the reciprocal of the product of g_m/I_D and $-\delta V_T/\delta V_{DS}$ or V_{gm} divided by $-\delta V_T/\delta V_{DS}$. As mentioned for Equation 3.78, $\delta V_T/\delta V_{DS}$ and, correspondingly, $DVTDIBL$ are negative giving the expected positive value for V_A.

Equations 3.82 and 3.83 show that increasing L, where the magnitude of $\delta V_T/\delta V_{DS}$ decreases rapidly, favorably maximizes V_A, minimizing g_{ds}, due to DIBL. Also, operating in strong inversion ($IC > 10$) increases V_A through the reduction in g_m/I_D, where the additional reduction in g_m/I_D caused by velocity saturation and VFMR effects can be included in Equation 3.83 by the IC modification described in Table 3.17. Operating short-channel devices in weak inversion gives the worst case DIBL reduction in V_A and increase in $g_{ds} = I_D/V_A$ because of maximum $\delta V_T/\delta V_{DS}$ and g_m/I_D.

Figure 3.39 illustrates g_{ds} increases caused by DIBL. The figure shows drain current curves for a $W/L = 25.2\,\mu\text{m}/0.5\,\mu\text{m}$, nMOS device in the 0.5 μm CMOS process described in Table 3.2. The curves are for operation in weak inversion or the weak-inversion side of moderate inversion where DIBL effects are severe for the minimum channel length device shown. As seen in the figure, the curves turn up exponentially with increasing V_{DS}. This is a result of weak-inversion drain current that is exponentially proportional to $V_{EFF} = V_{GS} - V_T$, where V_{EFF} increases by 12 mV per volt of increasing V_{DS}. This is caused by a 12 mV/V decrease in V_T, given by $DVTDIBL = \delta V_T/\delta V_{DS} = -12\,\text{mV/V}$ listed in Table 3.2 for the minimum channel length device in the process.

The I_D versus V_{DS} characteristics in Figure 3.39 are very steep, giving unattractively high g_{ds} for analog design. Additionally, the increasing slope (g_{ds}) at increasing V_{DS} indicates that DIBL or hot-electron effects (described in Section 3.8.4.3) are present as CLM effects result in decreasing slope and g_{ds} at increasing V_{DS} as seen earlier in Figures 3.35–3.38. Interestingly, a tangent line touching any of the curves in Figure 3.39 at high V_{DS} intercepts the V_{DS} axis at $I_D = 0\,\mu\text{A}$ for positive values of V_{DS}. This corresponds to a negative value of the Early voltage, V_A, for the definition shown in Figure 3.34 and indicates an unusually high value of g_{ds}. DIBL effects are very easy to observe in Figure 3.39 because the large V_{DS} allowed in the 0.5 μm process results in a significant drop in V_T, resulting in a significant increase in drain current.

In Figure 3.39, the identified bias point is at $I_D = 1\,\mu\text{A}$, corresponding to $IC = 0.094$ in weak inversion. This is found from $IC = I_D/[I_0(W/L)]$ from Table 3.6 with a technology current $I_0 = 0.21\,\mu\text{A}$ from Table 3.2. The channel length of $L = 0.5\,\mu\text{m}$ includes the reduction by $DL = 0.1\,\mu\text{m}$ (also listed

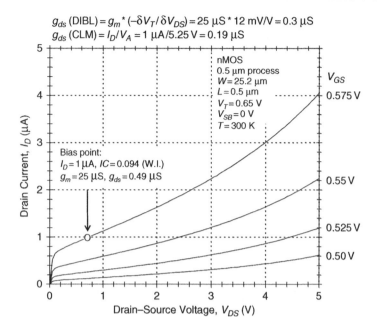

$$g_{ds}\,(\text{DIBL}) = g_m * (-\delta V_T/\delta V_{DS}) = 25\,\mu\text{S} * 12\,\text{mV/V} = 0.3\,\mu\text{S}$$
$$g_{ds}\,(\text{CLM}) = I_D/V_A = 1\,\mu\text{A}/5.25\,\text{V} = 0.19\,\mu\text{S}$$

nMOS
0.5 μm process
W = 25.2 μm
L = 0.5 μm
V_T = 0.65 V
V_{SB} = 0 V
T = 300 K

Bias point:
I_D = 1 μA, IC = 0.094 (W.I.)
g_m = 25 μS, g_{ds} = 0.49 μS

V_{GS}
0.575 V
0.55 V
0.525 V
0.50 V

Drain Current, I_D (μA)

Drain–Source Voltage, V_{DS} (V)

Figure 3.39   Drain current curves for a $W/L = 25.2\,\mu\text{m}/0.5\,\mu\text{m}$, nMOS device in a 0.5 μm CMOS process illustrating increased drain current slope, g_{ds}, at high V_{DS} due to DIBL. The threshold voltage decreases by approximately 12 mV per volt with increasing V_{DS}, causing increased drain current and turn-up in the curves. Increasing channel length from the process minimum significantly reduces DIBL effects, while increasing the inversion level from the weak-inversion operation shown also reduces these effects by lowering g_m/I_D. pMOS DIBL effects are similar

in Table 3.2) from a drawn length of 0.6 μm. At the bias point, measured g_m is approximately 25 μS, corresponding to $g_m/I_D = 25\,\mu\text{S}/\mu\text{A}$ (or per volt). Measured g_{ds}, found from the slope of the tangent line at the bias point, is 0.49 μS.

For the bias point shown in Figure 3.39, evaluating Equation 3.77 for g_{ds} due to DIBL gives $g_{ds} = g_m \cdot (-\delta V_T/\delta V_{DS}) = 25\,\mu\text{S} \cdot (12\,\text{mV/V}) = 0.3\,\mu\text{S}$. The combined DIBL and CLM effects give a total $g_{ds} = 0.3 + 0.19 = 0.49\,\mu\text{S}$, where g_{ds} due to CLM is estimated at 0.19 μS as summarized at the top of the figure. As expected by the shape of the drain current curves, DIBL clearly dominates the overall value of g_{ds} and raises it significantly from the value associated with CLM alone.

Figure 3.40 shows drain current curves for $L = 0.18, 0.28, 0.48$, and 4 μm, nMOS devices in the 0.18 μm CMOS process described in Table 3.2. The devices have widths of $W = 3.2, 4.8, 8$, and 64 μm, respectively, giving nearly equal shape factors of $S = W/L = 16$. This permits their curves to be overlaid since their drain currents are nearly equal. The curves are for operation in weak inversion where DIBL effects are severe for the minimum channel length, $L = 0.18\,\mu\text{m}$ device. Unlike the curves shown in Figure 3.39, the $L = 0.18\,\mu\text{m}$ curve does not show an exponential turn-up of drain current with increasing V_{DS} because of smaller $\delta V_T/\delta V_{DS} = -8\,\text{mV/V}$, compared to $-12\,\text{mV/V}$, and reduced V_{DS}. Instead, nearly equal contributions of DIBL and CLM result in nearly constant drain current slope and g_{ds} as V_{DS} increases compared to the typical decrease observed in Figure 3.35 when CLM effects dominate. Figure 3.40 also shows that DIBL increases in slope and g_{ds} are significantly reduced by increasing channel length modestly from the process minimum of $L = 0.18\,\mu\text{m}$ to $L = 0.28\,\mu\text{m}$. For the $L = 0.28\,\mu\text{m}$ and longer-channel devices, CLM effects dominate, giving the typical reduction of slope and g_{ds} as V_{DS} increases. For the bias point identified at $I_D = 1\,\mu\text{A}$ for the $L = 0.18\,\mu\text{m}$ device, estimated g_{ds} due to DIBL and CLM is 0.2 μS and 0.23 μS, respectively, for a total g_{ds} of 0.43 μS

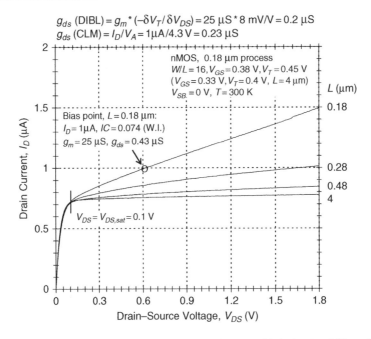

Figure 3.40   Drain current curves for $L = 0.18, 0.28, 0.48,$ and $4\,\mu m$, nMOS devices in a $0.18\,\mu m$ CMOS process illustrating increased drain current slope, g_{ds}, due to DIBL for the $L = 0.18\,\mu m$ device. The lower threshold-voltage reduction of approximately $8\,mV$ per volt with increasing drain–source voltage and limited drain–source voltage reduce DIBL effects compared to those shown in Figure 3.39. Increasing channel length modestly from the process minimum of $L = 0.18\,\mu m$ to $L = 0.28\,\mu m$ significantly reduces DIBL contributions to g_{ds} while also lowering CLM contributions. pMOS DIBL effects are similar

as summarized at the top of the figure. The bias point corresponds to operation at $IC = 0.074$ in weak inversion found using the technology current $I_0 = 0.64\,\mu A$ and the channel length reduction of $DL = 0.028\,\mu m$ from Table 3.2.

While the examples shown are for minimum channel length, nMOS devices, DIBL effects are also significant and even slightly higher for minimum channel length, pMOS devices in the example processes. This can be seen by the slightly higher values of $DVTDIBL = \delta V_T / \delta V_{DS}$ (at the minimum channel length) listed in Table 3.2.

For the processes illustrated here, DIBL effects are nearly non-existent for an $L = 0.5\,\mu m$ device in the $0.18\,\mu m$ process compared to the same channel length, now the minimum channel length device, in the $0.5\,\mu m$ process. This is typical of processes where short-channel, charge-sharing effects are substantially reduced for channel lengths modestly above the process minimum. In contrast, velocity saturation effects on V_{EFF} (Section 3.7.2.2) and g_m/I_D (Section 3.8.2.2) are similar across processes for equal channel length devices, varying only slightly through the value of E_{CRIT}.

3.8.4.3 Due to hot-electron effects

Hot-electron or impact ionization effects cause an increase in drain current with increasing V_{DS} for nMOS devices, giving rise to another component of MOS drain conductance. When the electric field across the depletion region between the pinched-off channel and drain is sufficiently high, above E_{CRIT} associated with velocity saturation, electrons acquire kinetic energy from lattice collisions that limit their velocity to near the velocity-saturated value. Some of these electrons will acquire enough energy and become "hot carriers" creating impact ionization or weak avalanching in silicon lattice atoms [15, pp. 286–288]. When this occurs, hole–electron pairs are created with the holes swept into the substrate and electrons swept into the drain. This gives rise to a drain–substrate current, I_{DB}, which adds to the total drain current taken here as $I_D = I_{DS} + I_{DB}$, where I_{DS} is the drain–source current. The increase in I_{DB} with increasing V_{DS} can be modeled as a conductance between the drain and substrate. This conductance is g_{db} shown earlier in the small-signal model of Figure 3.21.

Since g_{db} appears between the drain and substrate (body) it directly affects the drain output conductance for the usual case when the substrate is at small-signal ground. This is different from g_{ds} appearing between the drain and source where the presence of source degeneration resistance (as for the top, non-rail device in a cascode circuit) significantly lowers the drain output conductance from the g_{ds} value. Drain output conductance with source degeneration resistance is described in Figure 3.22 and Table 3.16 given earlier. Because g_{db} is not favorably reduced by circuit cascoding, it can be especially significant for cascode circuits having low, drain output conductance.

Hot-electron substrate current also generates a voltage drop across the substrate resistance that lowers the operating value of the source–substrate (source–body) voltage, V_{SB} [15, p. 376]. This lowers the threshold voltage, V_T, increases the effective gate–source voltage, $V_{EFF} = V_{GS} - V_T$, and, correspondingly, increases the drain–source current. This increase in drain–source current with increasing V_{DS} can be modeled as a component of g_{ds} added to those already associated with CLM and DIBL. Sufficiently low substrate resistance and low values of I_{DB} can ensure that the hot-electron component of g_{ds} is negligible compared to the normal CLM and DIBL components.

The presence of substrate current caused by hot-electron effects does not necessarily permanently damage the device. However, a small number of electrons created by impact ionization may create damage at the Si–SiO$_2$ interface causing an increase in interface-state density [15, pp. 287–288]. Additionally, some of the electrons may become trapped in the oxide, changing the oxide charge. Over a sufficiently long period of time, these effects can permanently increase the threshold voltage and lower device transconductance. This is known as hot-electron device degradation or damage. While hot-electron effects can significantly affect analog circuit operation, operation at supply voltages below the process maximum often prevents significant hot-electron device degradation as a result of process design.

Hot-electron effects are managed in some processes by utilizing a lightly doped drain (LDD) or extended drain between the normal, highly doped drain and the channel [15, pp. 289–290]. Lighter doping in the LDD region increases the depletion-region width at the drain near the channel, resulting in a lower electric field between the pinched-off channel and drain. This reduces electron velocity and energy, lowering impact ionization and the resulting substrate current. Coupling or extending the LDD region to the normal, highly doped drain helps minimize the drain resistance, or, more importantly, the source resistance when the drain and source layout is interchangeable. Source resistance creates an undesired degeneration of drain current and transconductance. Additionally, an extended LDD region (shown in Figure 3.9) has a smaller junction depth, x_j, reducing charge-sharing effects that contribute to a drop in the apparent value of the body-effect factor, γ (Section 3.8.3.1) and DIBL decreases in the threshold voltage with V_{DS} (previous section). These effects are especially severe for the minimum channel length device where the source and drain are in close proximity to the channel.

The prediction of substrate current due to hot-electron effects is useful for design to help minimize this current and the resulting production of g_{db}. A widely used estimate for substrate current in saturation ($V_{DS} > V_{DS,sat}$) is given by the form [14, 15, pp. 288–289]

$$I_{DB} \approx I_{DS} K_i \left(V_{DS} - V_{DS,sat} \right) \cdot e^{\left(\dfrac{-V_i}{V_{DS} - V_{DS,sat}} \right)} \tag{3.84}$$

where I_{DS}, again, is the drain–source current, $V_{DS} - V_{DS,sat}$ is the excess V_{DS} above $V_{DS,sat}$, K_i is a fitting parameter having values in the vicinity of 1 to 3/V, and V_i is a fitting voltage having values in the vicinity of $3 - 20$ V. In the EKV MOS model [14], K_i is equal to I_{BA}/I_{BB} and V_i is equal to

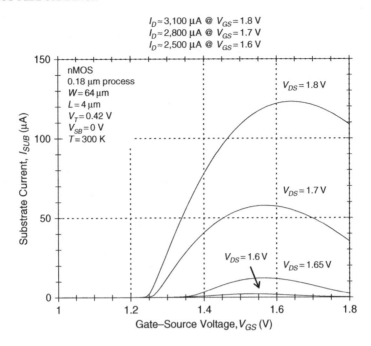

Figure 3.41 Substrate current, I_{DB}, curves for a $W/L = 64\,\mu\text{m}/4\,\mu\text{m}$, nMOS device in a $0.18\,\mu\text{m}$ CMOS process showing I_{DB} associated with hot-electron effects. I_{DB} increases at high values of V_{DS} and peaks for V_{GS} somewhat below V_{DS}. For the curves shown, maximum I_{DB} is less than $125\,\mu\text{A}$ or 4% of the total drain current

$I_{BB} \cdot L_C$, where I_{BA} and I_{BB} are impact ionization process parameters and L_C is the characteristic length for CLM given earlier by Equation 3.68. Equation 3.84 shows that I_{DB} increases exponentially with the excess V_{DS} above $V_{DS,sat}$, $V_{DS} - V_{DS,sat}$. Operating V_{DS} only modestly above $V_{DS,sat}$, which is often a requirement for modern, low-voltage processes, significantly reduces I_{DB}.

Figure 3.41 shows substrate current curves for a $W/L = 64\,\mu\text{m}/4\,\mu\text{m}$, nMOS device in the $0.18\,\mu\text{m}$ CMOS process described in Table 3.2. Unlike drain current curves where V_{DS} is swept for fixed values of V_{GS}, V_{GS} is swept for fixed values of V_{DS} in substrate current curves.

I_{DB} is not observed on the scale shown in Figure 3.41 for V_{DS} less than 1.6 V, but increases rapidly for voltages above this. This is consistent with the exponential increase in V_{DS} predicted by Equation 3.84. For a given V_{DS}, I_{DB} reaches a peak for a value of V_{GS} somewhat below V_{DS} where both I_{DS} and $V_{DS} - V_{DS,sat}$ are high. Here, I_{DB} is maximum because of the large number of electrons present in the drain region and the high electric field that increases the probability of impact ionization. Operation at low V_{GS} gives high $V_{DS} - V_{DS,sat}$ because of low $V_{DS,sat}$, but I_{DS} is low giving low I_{DB}. Conversely, operation at high V_{GS} gives high I_{DS}, but $V_{DS} - V_{DS,sat}$ is low because of high $V_{DS,sat}$ giving reduced I_{DB}. As noted at the top of the figure, $I_D = I_{DS} + I_{DB}$ ranges from 2500 to $3100\,\mu\text{A}$ for the substrate current curves shown. I_{DB} reaches a peak value of $I_{DB} = 125\,\mu\text{A}$ or 4% of I_D and is well below this for most points on the substrate current curves.

While I_{DB} may be a small fraction of I_D or I_{DS} ($I_D \approx I_{DS}$ when $I_{DB} \ll I_D$), it can significantly increase the drain conductance as illustrated by the drain current curves in Figure 3.42 for a $W/L = 64\,\mu\text{m}/4\,\mu\text{m}$, nMOS device in the $0.18\,\mu\text{m}$ CMOS process described in Table 3.2. For $V_{DS} = 1.8$ V on the $V_{GS} = 1.6$ V curve, I_{DB} is approximately $50\,\mu\text{A}$ or 2.1% of $I_D = 2400\,\mu\text{A}$. However, this small amount of substrate current significantly increases the drain current slope or drain conductance, illustrating the presence of a significant value of g_{db} compared to the normal value of g_{ds}. Hot-electron effects result in a turn-up of drain current and increased drain conductance (the sum of normal g_{ds} and I_{DB}-induced g_{db}) at high values of V_{DS}. This is unlike CLM effects where drain current turns down slightly giving lower drain conductance (the sum of normal g_{ds} and negligible g_{db}) at high values of V_{DS} as shown earlier in Figures 3.35–3.38. As described below, hot-electron increases in drain conductance can be more significant for long-channel devices where g_{ds} due to CLM is low and g_{ds} due to DIBL is even lower.

While predicting I_{DB} provides some design insight, the increased drain current slope shown in Figure 3.42 shows that g_{db} can significantly exceed the normal g_{ds} even for small values of I_{DB}. Additional insight can be gained by predicting g_{db}, especially since it contributes directly to drain conductance.

g_{db} is given by $\delta I_{DB}/\delta V_{DB}$, which, in the partial derivative definition, assumes that V_{GB} and V_{SB} are held fixed. Fixed V_{SB} gives $V_{DB} = V_{DS}$, such that g_{db} is also equal to $\delta I_{DB}/\delta V_{DS}$. Evaluating this from Equation 3.84 gives

$$g_{db} = \frac{\partial I_{DB}}{\partial V_{DB}} \approx I_{DB} \cdot \frac{V_i + (V_{DS} - V_{DS,sat})}{(V_{DS} - V_{DS,sat})^2} \tag{3.85}$$

Rather than assessing g_{db} separately, comparing it to the normal g_{ds} allows an assessment of its significance. The ratio of g_{db} given by Equation 3.85 and g_{ds} given by $g_{ds} = I_{DS}/V_A$ (without hot-electron effects) gives

$$\frac{g_{db}}{g_{ds}} \approx \frac{I_{DB}}{I_{DS}} \cdot \frac{V_i + (V_{DS} - V_{DS,sat})}{(V_{DS} - V_{DS,sat})^2} \cdot V_A \text{ (without hot-electron effects)} \tag{3.86}$$

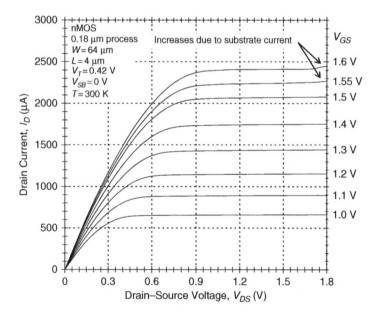

Figure 3.42   Drain current curves for a $W/L = 64\,\mu\text{m}/4\,\mu\text{m}$, nMOS device in a 0.18 μm CMOS process illustrating drain current and drain conductance increases associated with hot-electron effects. Increased drain current shown in the $V_{GS} = 1.55$ and 1.6 V curves at $V_{DS} = 1.8$ V comes from the substrate, causing increased drain current slope and drain conductance. Hot-electron effects can occur for nMOS devices operating at high V_{DS} and can be especially noticeable for long-channel devices where drain–source conductance due to CLM and DIBL is small

As expected, Equation 3.86 shows that g_{db} relative to g_{ds}, or g_{db}/g_{ds}, is primarily minimized by minimizing I_{DB} relative to I_{DS}, or I_{DB}/I_{DS}. However, the equation also shows that g_{db} is more significant (g_{db}/g_{ds} larger) for long-channel devices where V_A (without hot-electron effects) is large and the corresponding g_{ds} is small. Hot-electron effects are present even for long-channel devices as a result of the narrow depletion region between the pinched-off channel and drain. Hot-electron-induced I_{DB} and g_{db} can be easily masked by the normally steep, drain current slope and high g_{ds} present for short-channel devices, whereas the shallow drain current slope and low g_{ds} present for long-channel devices make the presence of I_{DB} and g_{db} resulting from hot-electron effects much more significant [15, p. 378]. It is important to include g_{db} (shown earlier in Figure 3.21) in small-signal analysis or ensure that it is small enough to have negligible effects on circuit performance.

Using Equation 3.84 to express I_{DB}, Equation 3.86 can be expressed as

$$\frac{g_{db}}{g_{ds}} \approx K_i \cdot \left(\frac{V_i + (V_{DS} - V_{DS,sat})}{V_{DS} - V_{DS,sat}} \cdot V_A \text{ (without hot-electron effects)} \right) \cdot e^{\left(\frac{-V_i}{V_{DS} - V_{DS,sat}} \right)} \quad (3.87)$$

The exponential term shows that g_{db}/g_{ds} is primarily minimized, as I_{DB}/I_{DS} is, by minimizing the excess V_{DS} above $V_{DS,sat}$, $V_{DS} - V_{DS,sat}$. Again, operating V_{DS} only modestly above $V_{DS,sat}$, which is often a requirement for modern, low-voltage processes, substantially reduces I_{DB}/I_{DS} and g_{db}/g_{ds}.

The final consideration discussed here is determining the maximum allowable value of g_{db}/g_{ds} such that hot-electron increases in drain conductance are negligible. This is easy to determine for grounded-source configurations where drain conductance equals the sum of g_{db} and the normal g_{ds} as seen in the small-signal model shown earlier in Figure 3.21. For grounded-source configurations, g_{db}/g_{ds} need only be in the vicinity of 0.1 to ensure negligible drain conductance increases. However, the presence of resistive source degeneration decreases the appearance of the normal g_{ds} while g_{db} remains largely unaffected. Drain output conductance, found from the expression summarized earlier in Table 3.16, is given by $g_{outd} = g_{ds}/(1 + g_{ms} \cdot r_s) \approx g_{ds}/(1 + g_m \cdot r_s)$, where $g_{ms} = g_m + g_{mb} + g_{ds}$ is the source transconductance and r_s is the external, source degeneration resistance. For configurations having resistive source degeneration, g_{db} then needs to be compared to g_{outd}, which can be factor of 20 to 100 lower than g_{ds} depending primarily on the product of g_m and r_s. Thus, cascode applications place demanding requirements on g_{db}/g_{ds}, which may need to be in the vicinity of $0.001 - 0.005$ to ensure negligible hot-electron increases in drain output conductance.

Fortunately, the presence of the excess V_{DS} above $V_{DS,sat}$, $V_{DS} - V_{DS,sat}$, in the exponential term of Equation 3.87 shows that g_{db}/g_{ds} is substantially reduced for low values of $V_{DS} - V_{DS,sat}$. Additionally, this ensures low I_{DB} such that the voltage drop across the substrate resistance is low, minimizing the related component of g_{ds} mentioned earlier. As mentioned, V_{DS} must often be constrained to values only modestly above $V_{DS,sat}$, especially for stacked, cascoded devices. While being aware of potentially significant increases in drain conductance at high V_{DS}, we will assume sufficiently low V_{DS} values where these increases are negligible. This is also recommended for design because of considerable temperature and process uncertainties in the prediction of hot-electron effects.

3.8.4.4 Impact of increase near $V_{DS,sat}$

Figures 3.36 and 3.38 showed that g_{ds} increases significantly for operation at V_{DS} near $V_{DS,sat}$ compared to operation at higher V_{DS} deeper into saturation. In addition to increasing the prediction uncertainty near $V_{DS,sat}$, the increased value of g_{ds} can significantly impact analog circuit performance.

Increases in g_{ds} and the corresponding decreases in r_{ds} can significantly lower small-signal voltage (V/V) and transresistance (V/I) gains. This is because the Ohm's law product of signal currents and MOS drain output resistances generates signal voltages. While cascode circuits greatly increase small-signal resistances and related signal gains compared to non-cascode circuits, cascode circuits are more affected by MOS operation near $V_{DS,sat}$. This is because cascode output resistance depends on g_{ds} or r_{ds} of both the rail and non-rail cascode devices.

Cascode, small-signal output resistance, found from the drain output resistance expression summarized earlier in Table 3.16, is given by

$$r_{out}(\text{cascode}) = r_{ds1}\left(1 + g_{ms1}r_{ds2}\right)$$

$$\approx g_{ms1}r_{ds1}r_{ds2} \approx (nA_{Vi})_1 \cdot r_{ds2} \tag{3.88}$$

where r_{ds1} $(1/g_{ds1})$ is the drain–source resistance and $g_{ms1} = g_{m1} + g_{mb1} + g_{ds1}$ is the source transconductance of the non-rail cascode device. r_{ds2} $(1/g_{ds2})$ is the drain–source resistance of the rail device that acts as the external, source degeneration resistance for the cascode device. The first expression shows that cascode output resistance is equal to drain–source resistance of the cascode device multiplied by the source degeneration factor given in the brackets. While the cascode output resistance is raised from the native drain–source resistance, r_{ds1}, by multiplication by the degeneration factor, the effective transconductance is lowered from the native device transconductance, g_{m1}, by division by this same factor. The low effective transconductance results in normally negligible noise and DC mismatch contributions from cascode devices.

When $g_{ms1} \cdot r_{ds2} \gg 1$, which is usually the case for cascode circuits, the second expression in Equation 3.88 is valid where the cascode output resistance depends on the product of drain–source resistances for both the cascode and rail devices. The rightmost expression shows that the output resistance is also equal to the product of the substrate factor and intrinsic voltage gain of the cascode device, $(nA_{Vi})_1$, multiplied by the small-signal resistance appearing at its source, r_{ds2}. Here, the intrinsic voltage gain is equal to $A_{Vi} = g_m \cdot r_{ds}$ as described later in Section 3.8.5 and $g_{ms} \approx g_m + g_{mb} \approx n \cdot g_m$, where $g_{mb} \approx (n-1)g_m$ from Table 3.20. Regardless of the expression used, the cascode output resistance depends again on the product of the drain–source resistances for both the cascode and rail devices. Since operation near $V_{DS,sat}$ can easily result in r_{ds} at one-half its expected value available deeper in saturation (corresponding to g_{ds} at twice its expected value available deeper in saturation as seen in Figures 3.36 and 3.38), cascode output resistance can be a factor of four or more lower than expected. This can then result in a factor of four or more loss of expected small-signal gain.

Decreases in drain–source resistances in low-voltage circuits operating near $V_{DS,sat}$ have a squaring effect on decreased cascode output resistance and the resulting loss of signal gain. It is easy to extend this analysis to regulated cascode circuits where output resistance depends on the product of drain–source resistances for cascode and rail devices and additionally the product of drain–source resistance associated with the grounded-source device that provides inverting voltage gain for regulated cascode action. Here, decreases in drain–source resistances have a cubed effect on decreased cascode output resistance and the resulting loss of signal gain.

In addition to the significant increase in g_{ds} or loss of r_{ds}, operation near $V_{DS,sat}$ results in significant variation in these with signal voltage variation at the drain. This is seen in Figures 3.36 and 3.38 where g_{ds} decreases by over a factor of four as V_{DS} increases from $V_{DS,sat}$. Changes in g_{ds} with signal voltage can give rise to gain non-linearity and signal distortion [41], further emphasizing the importance of selecting V_{DS} well above $V_{DS,sat}$ if possible.

If supply and signal compliance voltages are sufficiently high, V_{DS} can be selected sufficiently above $V_{DS,sat}$ to avoid significant losses of small-signal resistance and signal gain as well as increased signal distortion. However, this may not be possible in low-voltage designs where operating V_{DS} may necessarily be close to $V_{DS,sat}$, especially for stacked devices. Additionally, the rapid increase in g_{ds} near $V_{DS,sat}$ places demanding requirements on the prediction of g_{ds}, including the prediction in production-release simulations using full MOS models. Loss and uncertainties in small-signal resistances and signal gains are expected to worsen because of required operation closer to $V_{DS,sat}$ as supply voltages decrease. Fortunately, uncertainties in signal gains are often mitigated by closed-loop, negative feedback gain regulation, which requires only a sufficient level of open-loop gain to regulate the closed-loop gain.

3.8.4.5 Measured values

Unlike transconductance efficiency, g_m/I_D (Sections 3.8.2.1–3.8.2.5), which has a universal characteristic for CMOS technologies, and body-effect transconductance ratio, $\eta = g_{mb}/g_m$ (Sections 3.8.3.2–3.8.3.4), which is nearly constant, $V_A = I_D/g_{ds}$ is very complex to predict. Assuming operation in saturation ($V_{DS} > V_{DS,sat}$) is maintained, both g_m/I_D and η vary slightly with V_{DS}, requiring little consideration of V_{DS} for design guidance predictions. As a result, measured values of g_m/I_D and η were given in Sections 3.8.2.3 and 3.8.3.3 for $V_{DS} = V_{GS}$. This ensures operation consistently into saturation even for $IC = 100$ deep into strong inversion.

In contrast, V_A is a strong function of V_{DS} as shown by g_{ds} in Figures 3.36 and 3.38 and described in Section 3.8.4.1. It is also a function of the level of inversion, like g_m/I_D, but, unlike g_m/I_D, is a dominant function of the channel length at all levels of inversion. The prediction of V_A is further complicated by the possibility of DIBL and hot-electron components in addition to the always present CLM component. For these reasons, MOS drain–source conductance, represented here in normalized form using V_A, is the most difficult of the small-signal parameters to predict, having a long history of modeling errors, well known to experienced designers.

Other than the simple predictions provided by Equations 3.74 or 3.75, which require measured extraction of V_A over some range of inversion coefficients, channel lengths, and V_{DS}, no reasonably accurate, hand prediction is presently available for V_A. Instead, a set of measured values will be presented for the $0.18\,\mu m$ CMOS process described in Table 3.2. These measurements will then be displayed as average values of V_A and the Early voltage factor, $V_{AL} = V_A/L$, as a function of channel length and V_{DS} over ranges of inversion where V_A and V_{AL} are nearly constant.

While not universal across CMOS technologies, V_A measurements over a wide range of inversion coefficients, channel lengths, and V_{DS} provide insight and trends useful for design intuition and optimization. Additionally, the measurements provide a rigorous benchmark and "reality check" for drain–source conductance modeling provided by MOS models used in computer simulation. Similar benchmarking has been reported for a $0.75\,\mu m$ CMOS process at fixed V_{DS} [28, pp. 41, 43; 42], 0.5 and $0.25\,\mu m$ processes at $V_{DS} = V_{GS}$ [30, 31], and 0.18 and $0.13\,\mu m$ processes at $V_{DS} = V_{GS}$ [43]. The measurement methodology described here, however, considers multiple, fixed values of V_{DS} that correspond more closely to the required design choices for low-voltage processes. This measurement methodology can be extended to any process of interest, permitting the designer to estimate V_A and V_{AL} near the level of inversion, channel length, and V_{DS} actually used. Additionally, second-order effects like the inclusion of non-zero V_{SB} can be included. The increased V_A prediction difficulties associated with non-uniform lateral, halo channel doping make effective measurement methodologies even more important for smaller-geometry processes.

V_A measurements, like the I_D, V_{EFF}, g_m/I_D, and $\eta = g_{mb}/g_m$ measurements presented earlier in Sections 3.7.1.6, 3.7.2.3, 3.8.2.3 and 3.8.3.3, were made on a custom semiconductor parameter analyzer. This analyzer was developed by the author and colleagues and was described earlier in Footnote 7 in Section 2.4.4. Sixty-four repeated g_{ds} measurements (g_{db} was also included in the measurements but is believed negligible for the measurement conditions) were averaged for each bias point before calculating the corresponding value of V_A. In addition to the need for statistical averaging, especially for long-channel devices where g_{ds} is low, a major concern in g_{ds} measurements is device self-heating. This is because a stepped V_{DS} level required for assessing g_{ds} results in increased power dissipation and heating in the device. This can result in decreased mobility giving lower than expected stepped drain current and a correspondingly lower value of g_{ds}. The fact that nMOS and pMOS V_A measurements are very similar suggests that self-heating was well managed. This is because pMOS devices in the $0.18\,\mu m$ CMOS process evaluated here have nearly a factor-of-five larger channel width and gate area, compared to nMOS devices operating at the same inversion coefficient and drain current. The larger device geometry lowers pMOS device susceptibility to self-heating.

V_A measurements presented here include the operating V_{DS} in the V_A value, such that V_A is actually equal to $V_A = V_{gds} = (g_{ds}/I_D)^{-1}$ where $g_{ds} = I_D/V_A = I_D/V_{gds}$. This is a common definition for V_A, but as mentioned for Equation 3.52 near the beginning of Section 3.8.4, requires specifying V_{DS} used for the measurement. Regardless of the V_A definition, V_{DS} must always be specified for V_A measurements because of the strong V_{DS} dependency.

Figure 3.43 shows measured V_A for $L = 0.18, 0.28, 0.38, 0.48, 1, 2,$ and $4\,\mu$m, nMOS devices in the $0.18\,\mu$m process described in Table 3.2. V_A is presented as a function of the inversion coefficient from $IC = 0.01$ (deep weak inversion) to 100 (deep strong inversion) over a wide range of channel lengths, providing a full characterization of the process for analog design. The inversion coefficient for measured data is found by $IC = I_D/[I_0(W/L)]$ from Table 3.6 using a technology current of $I_0 = 0.64\,\mu$A from Table 3.2. The channel length given is reduced by $DL = 0.028\,\mu$m (also listed in the table) to consider lateral diffusion effects. V_A is first evaluated for $V_{DS} = 1\,$V.

In Figure 3.43 for channel lengths above the process minimum of $L = 0.18\,\mu$m, V_A is nearly constant from weak inversion to the onset of strong inversion at $IC = 10$ before increasing slightly in strong inversion. V_A then levels off at $IC = 100$ deep in strong inversion as $V_{DS,sat} = 0.6\,$V approaches V_{DS}, which is held fixed at $1\,$V. $V_{DS,sat}$ is found from Table 3.15 for $T = 300\,$K (room temperature), while the traditional, $V_{DS,sat} = V_{GS} - V_T = V_{EFF}$ value is higher at $0.67\,$V from $V_{EFF} \approx 2nU_T\sqrt{IC}$ from Table 3.14 for a strong-inversion value of $n = 1.3$. The decrease, or leveling of V_A if it is increasing, is expected as V_{DS} nears $V_{DS,sat}$, corresponding to the increase in g_{ds} shown in Figures 3.36 and 3.38. As observed, V_A is not entirely constant from weak inversion to the onset of strong inversion. It decreases slightly for shorter channel lengths, again, excluding the minimum channel length, while increasing slightly for longer channel lengths. Nearly constant V_A in weak inversion with an increase in moderate and strong inversion is also reported in [28, pp. 41, 43; 42] for a constant and sufficiently high V_{DS}.

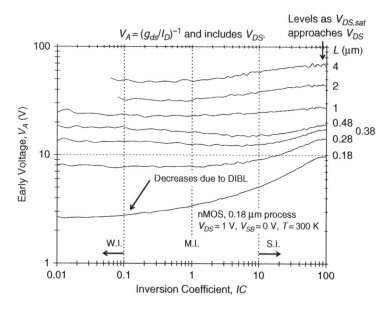

Figure 3.43 Measured Early voltage, V_A, versus inversion coefficient, IC, for $L = 0.18, 0.28, 0.38, 0.48, 1, 2,$ and $4\,\mu$m, nMOS devices in a $0.18\,\mu$m CMOS process operating at $V_{DS} = 1\,$V. For channel lengths above the process minimum, V_A is nearly constant in weak and moderate inversion, increasing slightly in strong inversion before leveling off at $IC = 100$ deep in strong inversion as $V_{DS,sat}$ approaches V_{DS}. DIBL effects significantly reduce V_A for the minimum channel length, especially in weak inversion where g_m/I_D is high. V_A is a strong function of channel length, increasing with increasing channel length. V_A is also a strong function of V_{DS}, increasing with increasing V_{DS}

Interestingly, in additional measurements taken, V_A is more constant for a different set of identical test devices showing some variation with processing.

In Figure 3.43 for the minimum channel length of $L = 0.18 \, \mu m$, V_A decreases significantly due to DIBL effects, especially for operation in weak inversion where high g_m/I_D increases drain current and g_{ds} more as threshold voltage decreases with increasing V_{DS}. As described in Section 3.8.4.2, only a moderate increase in channel length from the process minimum, corresponding to $L = 0.28 \, \mu m$ in the figure, significantly reduces the decrease in V_A associated with DIBL. Finally, as expected, V_A is primarily controlled by L, increasing significantly with L.

Figure 3.44 shows measured V_A for pMOS devices in the $0.18 \, \mu m$ process. The inversion coefficient for measured data is found using a technology current of $I_0 = 0.135 \, \mu A$ from Table 3.2, and channel length given is reduced by $DL = 0.051 \, \mu m$ (also listed in the table) to consider lateral diffusion effects. V_A is evaluated for the same channel lengths as nMOS devices shown in Figure 3.43 at the same $V_{DS} = (-)1 \, V$. V_A shown in Figure 3.44 for pMOS devices behaves almost identically to that shown in Figure 3.43 for nMOS devices, excepting a slight decrease for pMOS devices at $IC = 100$ deep in strong inversion as $V_{DS,sat}$ approaches V_{DS}. DIBL effects are significant for both nMOS and pMOS devices at the minimum channel length of $L = 0.18 \, \mu m$ as seen by a lower value of V_A. These effects, however, appear nearly negligible at the moderately increased channel length of $L = 0.28 \, \mu m$ for both device types.

Figure 3.45 extends the nMOS V_A measurement survey to a lower $V_{DS} = 0.5 \, V$, compared to $V_{DS} = 1 \, V$ considered in Figure 3.43. The behavior of V_A remains essentially unchanged except for the significant decrease around $IC = 30$ where $V_{DS,sat} = 0.36 \, V$ (from Table 3.15) nears the lower value of $V_{DS} = 0.5 \, V$. The other notable difference is the lower value of V_A observed in Figure 3.45 for $V_{DS} = 0.5 \, V$, which is approximately one-half the value observed in Figure 3.43 for $V_{DS} = 1 \, V$. The V_A decrease is less at shorter channel lengths but is significant at all channel lengths, illustrating a strong

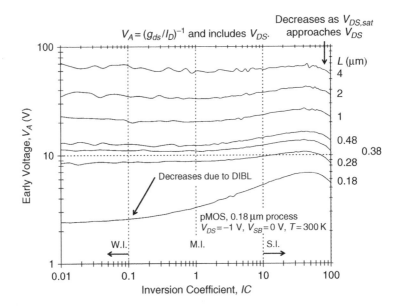

Figure 3.44 Measured Early voltage, V_A, versus inversion coefficient, IC, for $L = 0.18, 0.28, 0.38, 0.48, 1, 2,$ and $4 \, \mu m$, pMOS devices in a $0.18 \, \mu m$ CMOS process operating at $V_{DS} = -1 \, V$. V_A closely follows the trends described in Figure 3.43 for nMOS devices except that it decreases slightly at $IC = 100$ deep in strong inversion as $V_{DS,sat}$ approaches V_{DS}

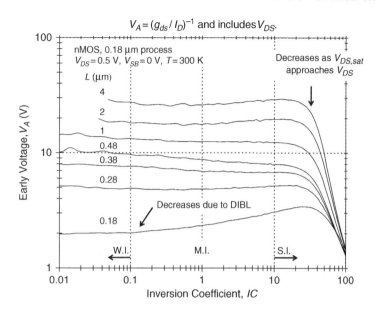

Figure 3.45   Measured Early voltage, V_A, versus inversion coefficient, IC, for $L = 0.18$, 0.28, 0.38, 0.48, 1, 2, and 4 μm, nMOS devices in a 0.18 μm CMOS process operating at $V_{DS} = 0.5$ V. V_A closely follows the trends described in Figure 3.43 for $V_{DS} = 1$ V, except that it decreases at $IC = 30$ in strong inversion as $V_{DS,sat}$ approaches the lower value of $V_{DS} = 0.5$ V. Additionally, V_A is lower because of lower V_{DS}. pMOS V_A is similar

V_{DS} dependency. Measured pMOS V_A for $V_{DS} = (-)0.5$ V is very similar to that shown in Figure 3.45 for nMOS devices.

Finally, Figure 3.46 extends the nMOS V_A measurement survey to an even lower $V_{DS} = 0.25$ V, which is more representative of allowable values for stacked devices in modern, low-voltage processes. Again, the behavior of V_A remains essentially unchanged from that associated with higher values of V_{DS}, except that V_A now decreases slightly from weak inversion into moderate inversion before decreasing significantly around $IC = 5$ where $V_{DS,sat} = 0.2$ V (again, from Table 3.15) nears the lower value of $V_{DS} = 0.25$ V. Again, the other notable difference is the lower value of V_A for the lower value of V_{DS}. V_A for $V_{DS} = 0.25$ V shown in Figure 3.46 is nearly one-half that for $V_{DS} = 0.5$ V shown in Figure 3.45 and nearly one-fourth that for $V_{DS} = 1$ V shown in Figure 3.43. Again, the V_A decrease is less at shorter channel lengths, but is significant for all channel lengths. Measured pMOS V_A for $V_{DS} = (-)0.25$ V is very similar to that shown in Figure 3.46 for nMOS devices.

Since V_A for channel lengths above the process minimum is nearly constant with inversion level for constant V_{DS} up to the point where $V_{DS,sat}$ begins to approach V_{DS}, an average value can be found that corresponds to the nearly constant region. Here, an average value of V_A is found from $IC = 0.1 - 10$ for $V_{DS} = 0.5$ and 1 V and from $IC = 0.1 - 3$ for $V_{DS} = 0.25$ V, corresponding to the nearly constant regions observed in Figures 3.43–3.46. Although V_A decreases slightly with IC for the low value of $V_{DS} = 0.25$ V shown in Figure 3.46 and increases with IC for the minimum channel length at higher V_{DS} shown in Figures 3.43–3.45 because of decreasing DIBL effects, an average value represents typical operation for $V_{DS,sat}$ sufficiently below V_{DS}.

Figure 3.47 shows the average value of V_A versus channel length for $V_{DS} = 0.25$, 0.5, and 1 V found from the nMOS V_A data of Figures 3.46, 3.45, and 3.43, and the pMOS V_A data of Figure 3.44 for $V_{DS} = (-)1$ V. Additional pMOS data was also collected for $V_{DS} = (-)0.25$ and $(-)0.5$ V to permit a direct comparison with the nMOS data.

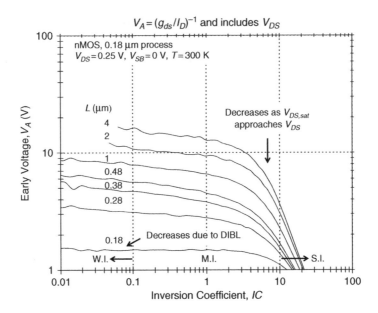

Figure 3.46 Measured Early voltage, V_A, versus inversion coefficient, IC, for $L = 0.18, 0.28, 0.38, 0.48, 1, 2,$ and $4\,\mu m$, nMOS devices in a $0.18\,\mu m$ CMOS process operating at $V_{DS} = 0.25\,V$. V_A follows the trends described in Figure 3.43 for $V_{DS} = 1\,V$, except that it decreases slightly with increasing inversion coefficient before decreasing rapidly at $IC = 5$ near strong inversion as $V_{DS,sat}$ approaches the lower value of $V_{DS} = 0.25\,V$. Additionally, V_A is lower because of lower V_{DS}. pMOS V_A is similar

In Figure 3.47, the trends of V_A are similar for both nMOS and pMOS devices. V_A increases rapidly as L initially increases from the process minimum of $L_{min} = 0.18\,\mu m$ ($L \approx 0.15\,\mu m$ shown in the figure because of the DL reduction), corresponding to a rapid decrease in DIBL effects. For the data shown, V_A increases approximately as L^2 for L near L_{min} for $V_{DS} = 1\,V$, but it is not possible to accurately assess the increase due to an insufficient number of channel lengths near L_{min}. Additionally, V_A shown are average values versus the minimum values found in weak inversion alone where DIBL effects are most severe. The actual V_A increase is likely to be higher at approximately L^3 in weak inversion, dropping from this rate in strong inversion, as observed in similar processes where multiple channel lengths were studied near L_{min} [43]. This increase would be consistent with a reduction in DIBL $\delta V_T/\delta V_{DS}$ proportional to $1/L^3$ assumed for Equation 3.81, when applied in Equation 3.82 for operation in weak inversion where g_m/I_D is high and DIBL effects dominate for L near L_{min}.

For $L > 3L_{min}$ ($L \gtrsim 0.45\,\mu m$), CLM dominates and V_A increases at a much slower rate with almost constant power-law behavior observed by a straight-line increase in the log–log presentation in Figure 3.47. Extracted power-law exponentials are included in the figure with V_A increasing as $L^{0.49}$, $L^{0.53}$, and $L^{0.54}$ for nMOS devices and $L^{0.62}$, $L^{0.67}$, and $L^{0.7}$ for pMOS devices at $V_{DS} = 0.25, 0.5,$ and $1\,V$, respectively. The V_A increase is slightly less, or more sublinear, at lower values of V_{DS} and is less for nMOS devices compared to pMOS devices. A smaller V_A increase is also observed in [43] for nMOS devices where V_A increases approximately as $L^{0.6}$ compared to $L^{0.75}$ for pMOS devices. This data is not directly comparable as V_{DS} was set equal to V_{GS} compared to constant V_{DS} values used here. Additionally, a different $0.18\,\mu m$ process and a $0.13\,\mu m$ process were considered in [43].

Finally, in addition to the V_A increase with L, Figure 3.47 shows the strong V_A increase with V_{DS}. For $L > 3L_{min}$, V_A nearly doubles from $V_{DS} = 0.25$ to $0.5\,V$, and nearly doubles again from $V_{DS} = 0.5$ to $1\,V$. The V_{DS} increase is less for $L < 3L_{min}$, but is still significant.

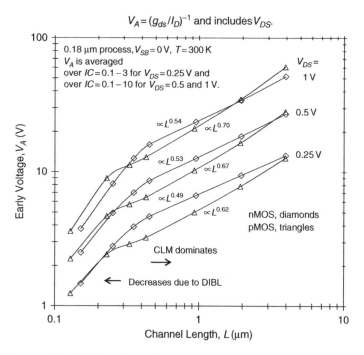

Figure 3.47 Average value of Early voltage, V_A, versus channel length, L, for nMOS and pMOS devices in a 0.18 μm CMOS process operating at $V_{DS} = 0.25$, 0.5, and 1 V. V_A is found from the measurements given in Figures 3.43–3.46 and additional pMOS measurements not shown. As L increases from the process minimum, V_A increases rapidly as DIBL effects are significantly reduced. For $L \geq 0.48$ μm, where CLM dominates, V_A increases less rapidly and is proportional to L raised to the power listed. V_A is also a strong function of V_{DS}, increasing approximately by a factor of two and factor of four, respectively, as V_{DS} increases from 0.25 to 0.5 V and 0.25 to 1 V. The data shows that V_A cannot be assumed to increase linearly with L

The data in Figure 3.47 shows that large errors result in using a "rule-of-thumb" prediction where V_A is assumed linearly proportional to L. As observed, V_A increases rapidly as L initially increases from L_{min}, but increases sublinearly with L for $L > 3L_{min}$. Here, doubling the choice of L does not double the value of V_A. The data also shows the importance of V_{DS} on the prediction of V_A. V_A ranges from nearly 1 to 70 V for $L = 0.18 - 4$ μm and $V_{DS} = 0.25 - 1$ V, showing the large design range available to the designer. As mentioned, the V_A values shown are average values taken over the wide ranges of inversion mentioned where V_A is nearly constant for the fixed values of V_{DS} listed.

Figure 3.48 presents the V_A data of Figure 3.47 in terms of the Early voltage factor, $V_{AL} = V_A/L$. V_{AL} represents an Early voltage efficiency factor describing the desired production of V_A for a given channel length, where V_A is found by $V_A = V_{AL} \cdot L$. V_{AL} is also used in Equation 3.75, which describes the prediction of V_A from extracted V_{AL} values, illustrated here, near the operating IC, L, and V_{DS}.

For both nMOS and pMOS devices, V_{AL} reaches a peak shown in Figure 3.48 for $L \approx 2L_{min}$ ($L \approx 0.3$ μm), which corresponds to the rapid decrease in DIBL effects. The V_{AL} peak occurs at slightly shorter L for pMOS devices, suggesting a more rapid decrease in DIBL $\delta V_T/\delta V_{DS}$ with increasing L compared to nMOS devices in the process. V_{AL} decreases as L increases beyond the value associated with the peak, corresponding to the sublinear increase in V_A with L observed in Figure 3.47. Finally, V_{AL} increases with V_{DS} consistent with the increase in V_A observed in Figure 3.47. While not giving the maximum value of V_A, which is obtained by making L arbitrarily long, operation at L near peak values of V_{AL} gives an efficiently high V_A at a moderate L. This permits considerably improved

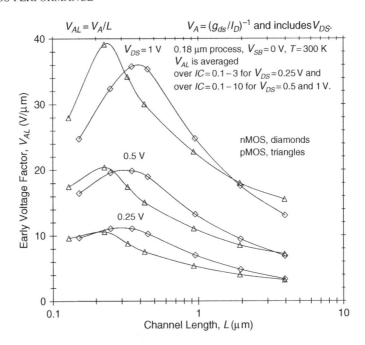

Figure 3.48 Average value of Early voltage factor, $V_{AL} = V_A/L$, versus channel length, L, for nMOS and pMOS devices in a 0.18 μm CMOS process operating at $V_{DS} = 0.25$, 0.5, and 1 V. V_{AL} is found from the data in Figure 3.47. As L increases from the process minimum, V_{AL} increases rapidly as DIBL effects are significantly reduced. As L increases further, V_{AL} decreases, corresponding to a sublinear increase in V_A with L. V_{AL}, like V_A shown in Figure 3.47, is also a strong function of V_{DS}, increasing by approximately a factor of two and factor of four, respectively, as V_{DS} increases from 0.25 to 0.5 V and 0.25 to 1 V. The data shows that a single value of V_{AL} cannot be assumed for a process

bandwidth compared to operation at longer L. Section 3.9.6 discusses MOS intrinsic bandwidth, which is proportional to $1/L^2$, assuming velocity saturation effects are negligible. As mentioned for V_A, measurements taken for additional channel lengths near L_{min} will better define the rapidly changing V_{AL} behavior and peaking for L near L_{min}.

Figure 3.48 shows that V_{AL} ranges from 3 to 40 V/μm as L ranges from 0.18 to 4 μm and V_{DS} ranges from 0.25 to 1 V. Again, except for channel length near the process minimum where DIBL effects are severe, V_A and, correspondingly, V_{AL} are nearly constant with IC for fixed values of V_{DS} until the operating $V_{DS,sat}$ nears V_{DS}. The widely varying V_{AL} shows that assuming a constant, "rule-of-thumb" $V_{AL} = 10$ V/μm, which assumes V_A increases linearly with L, results in as much as a factor of three or more under- and over-prediction of V_A and the corresponding g_{ds} as L and V_{DS} vary. However, V_{AL} extracted over a limited range of IC, L, and V_{DS} can provide a prediction of V_A useful for design intuition and optimization as described in the notes below Equation 3.75.

The preceding V_A data taken at constant and sufficiently high V_{DS} for channel length above the process minimum showed that V_A (Figures 3.43–3.45) is nearly constant with IC before increasing slightly in strong inversion, provided operation is constrained to sufficiently low IC where the operating $V_{DS,sat}$ is sufficiently below V_{DS}. This is an interesting result since Equation 3.72 predicts that V_A should actually decrease as IC increases since $V_{DS,sat}$ increases, lowering the excess V_{DS} or $V_{DS} - V_{DS,sat}$. The question then arises of what happens if V_A is evaluated for constant excess V_{DS}, for example at $V_{DS} = V_{GS}$. Under this condition, the excess V_{DS} in strong inversion is equal to $V_{DS} - V_{DS,sat} = V_{GS} - (V_{GS} - V_T) = V_T$, the threshold voltage, assuming the traditional strong-inversion estimate of

$V_{DS,sat} = V_{GS} - V_T = V_{EFF}$. Operation at $V_{DS} = V_{GS}$ corresponds to the constant and high excess V_{DS} present for diode-connected devices.

Figure 3.49 shows measured V_A for nMOS devices in the 0.18 μm process considered for $V_{DS} = V_{GS}$. Unlike the data shown earlier for constant V_{DS}, V_A increases significantly in moderate and especially strong inversion as IC and the corresponding $V_{DS} = V_{GS}$ increase. Similar V_A increases with IC were observed for $V_{DS} = V_{GS}$ in [30] for the 0.5 μm process described in Table 3.2 (shown later in Figure 3.50), in [31] for a 0.25 μm process, and in [43] for 0.18 and 0.13 μm processes. Since the excess V_{DS} in strong inversion is nearly constant for $V_{DS} = V_{GS}$, this suggests that the increasing vertical gate field associated with increasing inversion and V_{GS} "grabs" inversion charge in the channel and reduces the expected level of CLM. Such behavior is not predicted in the simple estimate of Equation 3.72 that predicts constant V_A for constant excess V_{DS}. As mentioned, inversion-level effects were not considered in the derivation of Equation 3.72.

In Figure 3.49, nMOS V_A at $IC = 10$ at the onset of strong inversion is 3.9, 7, 9.3, 10.8, 16.7, 25.5, and 38 V for $L = 0.18, 0.28, 0.38, 0.48, 1, 2,$ and 4 μm, respectively. The corresponding V_{AL} is 25.7, 27.8, 26.4, 23.9, 17.5, 13.1, and 9.6 V/μm. V_{AL} follows the trends shown in Figure 3.48 where it peaks for L near $2L_{min}$ before decreasing as L increases further, corresponding to a sublinear increase in V_A with L. pMOS V_A values for the 0.18 μm process are close to nMOS values, except for $L = 0.18$ μm (process minimum) below strong inversion where increased DIBL effects lower pMOS V_A slightly. A nearly identical increase in nMOS and pMOS V_A at $IC = 100$ in deep strong inversion for all channel lengths suggests that the measured data is representative and not affected by possible self-heating, which, as mentioned earlier, is significantly less for larger-area pMOS devices.

In order to observe behavior in a different process, Figure 3.50 extends the V_A study for $V_{DS} = V_{GS}$ to $L = 0.5, 0.85, 1.2, 1.9, 4, 8.2, 16.6,$ and 33.4 μm, nMOS devices in the 0.5 μm process described in Table 3.2. The inversion coefficient for measured data is found by $IC = I_D / [I_0(W/L)]$ from Table 3.6

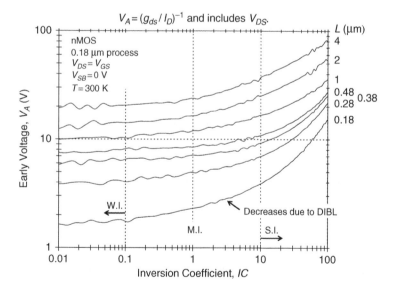

Figure 3.49 Measured Early voltage, V_A, versus inversion coefficient, IC, for $L = 0.18, 0.28, 0.38, 0.48, 1, 2,$ and 4 μm, nMOS devices in a 0.18 μm CMOS process operating at $V_{DS} = V_{GS}$. Operation at $V_{DS} = V_{GS}$ results in nearly constant excess V_{DS} ($V_{DS} - V_{DS,sat}$) in strong inversion. For channel lengths above the process minimum, V_A increases consistently with increasing IC and the corresponding increasing $V_{DS} = V_{GS}$. As seen in Figures 3.43–3.46, DIBL effects significantly reduce V_A for the minimum channel length device, especially in weak inversion where g_m/I_D is high. As seen in Figures 3.43–3.46, V_A is a strong function of channel length, increasing with increasing channel length. pMOS V_A is similar

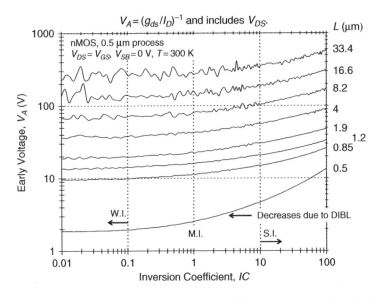

Figure 3.50 Measured Early voltage, V_A, versus inversion coefficient, IC, for $L = 0.5$, 0.85, 1.2, 1.9, 4, 8.2, 16.6, and 33.4 μm, nMOS devices in a 0.5 μm CMOS process operating at $V_{DS} = V_{GS}$. V_A closely follows the trends described in Figure 3.49 for a 0.18 μm CMOS process, but increases less rapidly with increasing IC and the corresponding increasing $V_{DS} = V_{GS}$. pMOS V_A is similar

using a technology current of $I_0 = 0.21\,\mu A$ from Table 3.2. Channel lengths given are the effective values since larger drawn lengths are used for this process. Again, V_A increases in moderate and especially in strong inversion, but the increase is less for devices in the 0.5 μm process compared to those in the 0.18 μm process shown in Figure 3.49. The lower V_A increase with increasing $V_{DS} = V_{GS}$ associated with increasing IC is consistent with a higher value of $V_E \approx 0.45\,V$ for the 0.5 μm process compared to $V_E \approx 0.15\,V$ for the 0.18 μm process. The increase of V_A with V_{DS} is observed in Equations 3.72 and 3.74, with estimated values of V_E discussed in Section 3.8.4.1.

In Figure 3.50, nMOS V_A at $IC = 10$ at the onset of strong inversion is 4.7, 15, 20.8, 30, 57, 105, 185, and 350 V for $L = 0.5$, 0.85, 1.2, 1.9, 4, 8.2, 16.6, and 33.4 μm, respectively. The unusually long channel lengths result in V_A values comparable to those traditionally found for vertical bipolar transistors. The corresponding V_{AL} is 9.4, 17.6, 17.3, 15.8, 14.3, 12.8, 11.1, and 10.5 V/μm. This, even for a considerably different process, follows the trends shown in Figure 3.48 where V_{AL} peaks for L near $2L_{min}$ before decreasing as L increases further, corresponding to a sublinear increase in V_A with L. However, V_{AL} is lower compared to values given earlier for the 0.18 μm process, except for long channel lengths where values are comparable. pMOS V_A values for the 0.5 μm process are close to nMOS values, except for $L = 0.5$ μm (process minimum) below strong inversion where increased DIBL effects lower pMOS V_A slightly.

The increasing V_A in moderate and strong inversion observed in Figures 3.49 and 3.50 for constant excess V_{DS} when $V_{DS} = V_{GS}$ is a fortunate result. This is because the V_A increase with increasing inversion may counter the expected V_A decrease with decreasing excess V_{DS} resulting from increasing $V_{DS,sat}$ and a fixed value of V_{DS}. This may explain the nearly constant V_A, except for channel length near the process minimum where DIBL effects are severe, with slight increase in strong inversion when V_{DS} is held fixed as shown in Figures 3.43–3.45. As mentioned, V_A decreases slightly with increasing inversion in Figure 3.46 for the low value of $V_{DS} = 0.25\,V$ because the operating $V_{DS,sat}$ is closer to V_{DS}.

As mentioned near the beginning of this section, g_{ds} behavior, evaluated here as $V_A = V_{gds} = (g_{ds}/I_D)^{-1}$, cannot be considered fundamental across CMOS processes. However, it is interesting that the structure of V_A when evaluated in terms of the inversion coefficient is very similar for devices in $0.18\,\mu\text{m}$ and $0.5\,\mu\text{m}$ processes as observed in Figures 3.49 and 3.50. For both processes, V_A increases in moderate and especially strong inversion for all channel lengths for $V_{DS} = V_{GS}$. Additionally, V_A increases rapidly as L increases from near the process minimum, but increases sublinearly as L increases further. V_A is significantly lower for the minimum channel length due to significant DIBL effects.

Data taken at $V_{DS} = V_{GS}$ for 0.18 and $0.13\,\mu\text{m}$ processes having pocket implants, giving halo or non-uniform lateral channel doping with higher doping near the source and drain, showed anomalous V_A behavior for nMOS devices at intermediate channel lengths [43]. Here, nMOS V_A increased initially with L but failed to continue increasing at intermediate channel lengths before resuming the expected increase at longer channel lengths. This emphasizes the importance of using measurement methodologies that provide a thorough characterization of processes for analog design and for the validation of MOS models. As done for most of the measurements in this section, evaluation of V_A for fixed values of V_{DS} is recommended to reflect the limited V_{DS} values available for low-voltage processes. Evaluations at $V_{DS} = V_{GS}$ were included here to consider performance at constant excess V_{DS}.

3.8.4.6 Summary of trends

Table 3.21 summarizes expressions and trends for MOS Early voltage, V_A, in terms of the design choices of inversion coefficient, channel length, and drain–source voltage. The first expression summarizes the calculation of $g_{ds} = 1/r_{ds}$ where $g_{ds} = I_D/(V_A + V_{DS})$, or $g_{ds} = I_D/V_A$ when V_{DS} is included in the value of V_A as is common practice and done here. In this case, $V_A = V_{gds} = (g_{ds}/I_D)^{-1}$ as discussed at the beginning of Section 3.8.4. As shown in the table, a composite V_A, inclusive of both CLM and DIBL effects, can be obtained by paralleling V_A due to CLM with a separate V_A due to DIBL. This is equivalent to paralleling (adding) the separate g_{ds} components.

The table then summarizes V_A associated with CLM, including the prediction given by Equation 3.74 from Section 3.8.4.1. While not predicted by the equation, V_A increases slightly in strong inversion for constant V_{DS} assuming the operating $V_{DS,sat}$ is sufficiently below V_{DS}. While the equation predicts a linear increase with L, V_A increases sublinearly with L with an increase ranging from $L^{0.4}$ to $L^{0.8}$, which is expanded from the increases shown in Figure 3.47 to include additional measured data like that given in [43]. Finally, as predicted by the equation, V_A increases with V_{DS}. These trends were observed when CLM effects were dominant in the measured V_A data in Section 3.8.4.5 as summarized in Figure 3.47. CLM effects are generally dominant for L a factor of three or more above the process minimum ($L > 3L_{min}$). V_A associated with CLM is maximized by operating long-channel devices at V_{DS} well above $V_{DS,sat}$.

Table 3.21 then summarizes V_A associated with DIBL, including the prediction given by Equation 3.83 from Section 3.8.4.2. As predicted by the equation, V_A increases with increasing IC because of decreasing g_m/I_D. This lowers the conversion of threshold voltage decrease with V_{DS} ($\delta V_T/\delta V_{DS}$) into increased drain current and the resulting component of g_{ds}. V_A, however, increases more rapidly by increasing L where the V_A increase is approximately L^3, corresponding to the rapid decrease in $\delta V_T/\delta V_{DS}$ set by $DVTDIBLEXP$. Finally, V_A can decrease with V_{DS} if drain current turns up as shown in Figure 3.39 in Section 3.8.4.2. V_A due to DIBL is low and usually dominant (lower than V_A due to CLM) when minimum channel length devices are operated at low levels of inversion. Selecting $L > 3L_{min}$ usually makes DIBL effects negligible compared to CLM effects.

Finally, Table 3.21 summarizes the composite V_A due to both DIBL and CLM, using Equation 3.75 from Section 3.8.4.1. This equation uses extracted values of V_{AL} near the operating IC, L, and V_{DS} to provide a prediction of V_A since no reasonably accurate, hand expression is available. V_{AL} values

Table 3.21 MOS Early voltage, V_A, expressions and trends in terms of inversion coefficient, channel length, and drain–source voltage. When evaluated in terms of the inversion coefficient, V_A is independent of the drain current

Early Voltage, V_A, where $$g_{ds} = \frac{1}{r_{ds}} = \frac{I_D}{V_A + V_{DS}} = \frac{I_D}{V_A(\text{incl.}V_{DS})}$$ $$V_A = V_A(\text{CLM}) \parallel V_A(\text{DIBL})$$	$IC \uparrow$ L, V_{DS} fixed	$L \uparrow$ IC, V_{DS} fixed	$V_{DS} \uparrow$ IC, L fixed
Due to channel length modulation (CLM): $$V_A(\text{CLM}) \approx \frac{L}{\frac{\partial l_p}{\partial V_{DS}}}$$ $$\approx V_{AL}\left(V_{DS,sat}\right) \cdot L \cdot \left(1 + \frac{V_{DS} - V_{DS,sat}}{V_E}\right)$$ where V_{AL} (V/μm) $= V_A/L$ is the Early voltage factor extracted at $V_{DS} = V_{DS,sat}$	$\uparrow$ slightly in SI (inversion dependency not included in expression)	$\uparrow \propto L^{0.4-0.8}$ (sublinear increase not included in expression)	$\uparrow$ especially as V_{DS} increases from near $V_{DS,sat}$
Due to drain-induced barrier lowering (DIBL): $$V_A(\text{DIBL}) = \frac{1}{\left(\frac{g_m}{I_D}\right)\frac{-\partial V_T}{\partial V_{DS}}}$$ $$\approx \frac{nU_T\left(\sqrt{IC+0.25}+0.5\right)}{-DVTDIBL \cdot \left(\frac{L_{min}}{L}\right)^{DVTDIBLEXP}}$$ where $DVTDIBL = \delta V_T/\delta V_{DS}$ for the minimum process channel length, L_{min}, and $DVTDIBLEXP = 2\text{--}4$ (3 used here)	$\uparrow$ as g_m/I_D decreases	$\uparrow \propto L^3$ as $\delta V_T/\delta V_{DS}$ decreases rapidly	$\downarrow$ if drain current turns up as V_{DS} increases
Due to CLM and DIBL from extracted measurements: $$V_A \approx V_{AL}(IC, L, V_{DS}) \cdot L$$ where V_{AL} (V/μm) $= V_A/L$ is the Early voltage factor extracted near the operating IC, L, and V_{DS}. See measured data in Figure 3.48 where V_{AL} ranges from 3 to 40 V/μm for a 0.18 μm CMOS process	$\uparrow$ slightly in SI (increases more from WI if DIBL is significant)	$\uparrow \propto L^3$ near L_{min} in WI due to DIBL; increase less in SI $\uparrow \propto L^{0.4-0.8}$ for $L > 3L_{min}$ due to CLM	$\uparrow$ especially as V_{DS} increases from near $V_{DS,sat}$

Expressions assume saturation operation where V_{DS} is sufficiently above $V_{DS,sat}$. As discussed in Section 3.8.4.1, V_A decreases significantly (g_{ds} increases significantly) for V_{DS} near $V_{DS,sat}$.

Hot-electron, substrate current leading to the presence of g_{db} is not included. As discussed in Section 3.8.4.3, this is made negligible by operating nMOS devices at sufficiently low values of V_{DS}.

were shown in Figure 3.48 in Section 3.8.4.5 for the 0.18 μm CMOS process described in Table 3.2. Here, V_{AL} ranges from 3 to 40 V/μm as L ranges from 0.18 to 4 μm and V_{DS} ranges from 0.25 to 1 V. As mentioned, except for channel length near the process minimum where DIBL effects are severe, V_A and, correspondingly, V_{AL} are nearly constant with IC for constant V_{DS} before increasing slightly in strong inversion. This is true if operation is constrained to sufficiently low IC where the operating $V_{DS,sat}$ is sufficiently below V_{DS}. For L near L_{min} where DIBL effects dominate at low levels of inversion, V_A increases rapidly with L at approximately L^3. For $L > 3L_{min}$ where CLM effects usually dominate, V_A increases sublinearly with increases ranging from approximately $L^{0.4}$ to $L^{0.8}$. Finally, when CLM effects dominate, V_A increases as V_{DS} increases above $V_{DS,sat}$. Each of the trends was, again, described in Section 3.8.4.5 and summarized in Figure 3.47. Selecting $L \approx 2L_{min}$ or slightly higher in the vicinity of the V_{AL} peak shown in Figure 3.48 provides an efficiently high value of V_A

for the given L. Increasing L further increases V_A at a slower rate while significantly reducing intrinsic bandwidth as described in Section 3.9.6.

As noted below Table 3.21 and shown in Figures 3.36 and 3.38 in Section 3.8.4.1, V_A decreases and, correspondingly, g_{ds} increases very rapidly as V_{DS} decreases to near $V_{DS,sat}$. The impact of this on analog design was discussed in Section 3.8.4.4 and emphasizes the importance of selecting V_{DS} sufficiently above $V_{DS,sat}$ if possible. The high g_m/I_D and low $V_{DS,sat}$ in moderate inversion are compelling reasons to consider operation here, especially for power-efficient, low-voltage circuits.

Finally, as noted below the table, hot-electron effects that introduce g_{db} through the presence of substrate current are not included in the summary and trends. These effects were discussed in Section 3.8.4.3 and are minimized by avoiding the operation of nMOS devices at high values of V_{DS}. V_{DS} is normally limited for stacked transistors in low-voltage designs. Non-zero V_{SB} effects are also not included in the expressions and trends, but can be considered using the measurement methodology described in Section 3.8.4.5.

As mentioned at the end of Section 3.8.4.5, g_{ds} behavior, evaluated here in normalized form as $V_A = V_{gds} = (g_{ds}/I_D)^{-1}$, cannot be considered fundamental across CMOS processes. However, as mentioned, V_A measurements taken over a wide range of IC, L, and V_{DS} provide an extensive characterization of a process for analog design and provide a rigorous benchmark for MOS models. This is especially true for V_A and the related g_{ds}, which are very complex to model and have a long history of modeling errors. Although no simple, reasonably accurate hand prediction exists for V_A, one might be anticipated that includes a DIBL component from Equation 3.82 and a CLM component from Equation 3.70. In addition to the V_A increase with V_{DS} included in Equation 3.70, the sublinear V_A increase with L shown in Figure 3.47 could be included by an exponent for L. Finally, the V_A increase with the level of inversion for constant excess V_{DS}, $V_{DS} = V_{DS,sat}$, shown in Figures 3.49 and 3.50, could be included.

While measured V_A from a 0.18 μm CMOS process taken over the multiple design choices of IC, L, and V_{DS} provides design insight, V_A will need to be continually re-evaluated for smaller-geometry processes. As always, computer simulations using accurate MOS models are required to validate a candidate design. Fortunately, the uncertainty in signal gain resulting from variations in g_{ds} is often mitigated by the use of negative feedback.

Lower V_{DS} required for lower supply voltages in smaller-geometry processes can be expected to lower V_A and raise g_{ds}. It is also likely that channel lengths somewhat above the process minimum, or maybe even further above the process minimum, will continue to be required to obtain desirable levels of V_A and g_{ds} for many analog circuits.

3.8.5 Intrinsic Voltage Gain

MOS intrinsic voltage gain, A_{Vi}, is the low-frequency, small-signal, gate-to-drain voltage gain of a grounded-source MOSFET with its drain connected to the infinite, small-signal resistance of an ideal current source load. A_{Vi} represents the maximum voltage gain achievable for a single device and often appears in cascode small-signal resistance and related signal gain calculations. A_{Vi} is a quality factor related to low-frequency voltage gain.

As derived from the MOS small-signal model shown earlier in Figure 3.21 in Section 3.8.1, A_{Vi} is equal to the ratio of transconductance, g_m, and drain–source conductance, g_{ds}, or the corresponding product of g_m and drain–source resistance, r_{ds}. This gives

$$A_{Vi} = \frac{g_m}{g_{ds}} = g_m \cdot r_{ds} \tag{3.89}$$

Equation 3.89 assumes that the small-signal, drain–body resistance, r_{db}, due to drain–substrate leakage current or hot-electron effects (described in Section 3.8.4.3) is much higher than r_{ds} and is thus negligible. While the actual gate-to-drain voltage gain is inverting or negative, it is customary to express A_{Vi} as a positive value.

Expressing g_m and r_{ds} in terms of the transconductance efficiency, g_m/I_D, and Early voltage, V_A, gives A_{Vi} as

$$A_{Vi} = g_m \cdot r_{ds} = \left(\frac{g_m}{I_D} \cdot I_D\right) \cdot \left(\frac{V_A + V_{DS}}{I_D}\right) = \frac{g_m}{I_D} \cdot (V_A + V_{DS}) \approx \frac{g_m}{I_D} \cdot V_A$$

$$= \frac{g_m}{I_D} \cdot V_A \ (V_{DS} \text{ included}) \tag{3.90}$$

The rightmost approximation results from neglecting the operating V_{DS} since V_A is usually somewhat greater than V_{DS}. However, as mentioned for Equation 3.52 near the beginning of Section 3.8.4, V_A is often defined where it includes V_{DS} such that r_{ds} directly equals V_A/I_D or g_{ds} directly equals I_D/V_A. This convention is used throughout this book.

Equation 3.90 shows that A_{Vi} is equal to the product of the transconductance efficiency and Early voltage. While it is sometimes believed that maximum voltage gain is achieved at minimum drain current, A_{Vi} is independent of the drain current when evaluated in terms of the inversion coefficient and channel length. A given voltage gain can be realized at any drain current providing devices can be physically sized for the selected drain current, inversion coefficient, and channel length. For example, if a voltage gain of 200 V/V is obtained in a MOS circuit at bias currents of $1\,\mu A$, this same voltage gain can be obtained at bias currents of $10\,\mu A$ by simply scaling all widths up by a factor of 10 to maintain the selected inversion coefficients and channel lengths. This is equivalent to paralleling 10 identical devices everywhere giving a factor-of-10 increase in drain current, a factor-of-10 increase in transconductance, and a factor-of-10 increase in drain–source conductance. This gives no change in voltage gain because transconductance and drain–source conductance increase equally. For voltage gain alone, there would obviously be no advantage in selecting higher drain currents. However, gate-referred noise and DC mismatch voltages are lower at higher drain current because of higher input-device g_m, lowering thermal noise, and larger device areas, lowering flicker noise and local-area DC mismatch. These topics will be summarized later in Sections 3.10.2.3, 3.10.3.7, and 3.11.1.7.

As discussed in Section 3.8.4.2, DIBL effects can significantly contribute to g_{ds}, especially for short-channel devices operating at low levels of inversion. As a result, DIBL effects can significantly affect A_{Vi}. It is useful then to separate out the CLM component of A_{Vi} that is usually dominant and the DIBL component that may be dominant. Separating A_{Vi} using CLM and DIBL components of g_{ds} gives

$$A_{Vi} = \frac{g_m}{g_{ds}(\text{CLM}) + g_{ds}(\text{DIBL})} = g_m \left(r_{ds}(\text{CLM}) \parallel r_{ds}(\text{DIBL})\right)$$

$$= A_{Vi}(\text{CLM}) \parallel A_{Vi}(\text{DIBL}) \tag{3.91}$$

where

$$A_{Vi}(\text{CLM}) = \frac{g_m}{g_{ds}(\text{CLM})} = g_m \cdot r_{ds}(\text{CLM}) = \frac{g_m}{I_D} \cdot V_A(\text{CLM}) \tag{3.92}$$

and

$$A_{Vi}(\text{DIBL}) = \frac{g_m}{g_{ds}(\text{DIBL})} = g_m \cdot r_{ds}(\text{DIBL}) = \frac{g_m}{I_D} \cdot V_A(\text{DIBL}) \tag{3.93}$$

Sections 3.8.4.1 and 3.8.4.2 described V_A (CLM) and V_A (DIBL) as normalized measures of g_{ds} (CLM) and g_{ds} (DIBL). Equation 3.91 shows that A_{Vi} can be expressed as the parallel combination of CLM and DIBL components.

While A_{Vi} (CLM) is best expressed as the product of g_m/I_D and V_A (CLM), A_{Vi} (DIBL) can be best expressed using g_m and g_{ds} (DIBL) from Equation 3.77. This gives

$$A_{Vi}(DIBL) = \frac{g_m}{g_{ds}(DIBL)} = \frac{g_m}{g_m \dfrac{-\partial V_T}{\partial V_{DS}}} = \frac{1}{\dfrac{-\partial V_T}{\partial V_{DS}}}$$

$$= \frac{1}{-DVTDIBL \cdot \left(\dfrac{L_{min}}{L}\right)^{DVTDIBLEXP}} \quad (3.94)$$

where from Equation 3.81 $DVTDIBL = \delta V_T/\delta V_{DS}$ for the minimum channel length, L_{min}, and $DVTDIBLEXP$ is the exponent describing how rapidly the magnitude of $\delta V_T/\delta V_{DS}$ decreases with increasing channel length.

Equation 3.94 is an interesting result showing that A_{Vi} (DIBL) is independent of g_m/I_D and the level of inversion, depending only on the channel length and process characteristics. For $DVTDIBL = -8\,mV/V$ for nMOS devices in the 0.18 μm CMOS process described in Table 3.2, A_{Vi} (DIBL) = 125 V/V for the minimum channel length, $L = 0.18\,μm$. Because A_{Vi} (DIBL) is paralleled with A_{Vi} (CLM), A_{Vi} is limited to a maximum of $A_{Vi} = 125$ V/V for the minimum channel length regardless of the level of inversion. The DIBL effect, however, is greater in weak or moderate inversion because high g_m/I_D results in high A_{Vi} (CLM) that usually exceeds A_{Vi} (DIBL). In this case, A_{Vi} is dominated by the lower, DIBL component. In contrast, in deep strong inversion where g_m/I_D is low, A_{Vi} (CLM) is usually lower than A_{Vi} (DIBL) and thus dominates A_{Vi}. As channel length increases, A_{Vi} (DIBL) increases significantly such that A_{Vi} (CLM) quickly dominates A_{Vi} for all levels of inversion.

Table 3.22 summarizes predicted A_{Vi} and trends in terms of the design choices of inversion coefficient, channel length, and V_{DS}. Expressions and trends for A_{Vi} result from the product of g_m/I_D, described in Table 3.17 and Section 3.8.2.4, and V_A, described in Table 3.21 and Section 3.8.4.6. As observed, the independent behaviors of g_m/I_D and V_A complicate the expressions and trends for A_{Vi} listed in Table 3.22. As in Table 3.21, V_A is expressed as $V_A = V_{AL} \cdot L$ where V_{AL} is the Early voltage factor having units of V/μm. V_{AL} is not constant as sometimes assumed but varies significantly with the selected IC, L, and V_{DS} as illustrated in Figure 3.48 in Section 3.8.4.5 (except for the minimum channel length, the IC variation is slight for constant V_{DS} as discussed in the section).

In Table 3.22, the first expression describes A_{Vi} excluding small-geometry effects. Without these effects, g_m/I_D is not reduced by velocity saturation and VFMR effects and V_A is not reduced by DIBL effects. A_{Vi} is maximum and constant in weak inversion where g_m/I_D is maximum before decreasing as $1/\sqrt{IC}$ in strong inversion corresponding to the decrease in g_m/I_D. A_{Vi} increases sublinearly with L as $L^{0.4}$ to $L^{0.8}$, corresponding to the increase of V_A. Finally, A_{Vi} increases with V_{DS} corresponding to the increase in V_A. As noted at the bottom of the table, A_{Vi} decreases rapidly as V_{DS} decreases to near $V_{DS,sat}$, emphasizing the importance of selecting V_{DS} sufficiently above $V_{DS,sat}$.

In addition to IC, the table expresses A_{Vi} in terms of $V_{DS,sat}$ through its relationship with g_m/I_D. A_{Vi} is maximized for long-channel devices, where V_A is high, for operation in weak inversion, where g_m/I_D is high and $V_{DS,sat}$ low. It is also maximized for high values of V_{DS} where V_A is increased, assuming, as noted at the bottom of the table, that hot-electron effects are negligible. The maximum, weak-inversion value of A_{Vi} is equal to $A_{Vi} = V_{AL} \cdot L/(nU_T)$, which is the ratio of the voltage $V_A = V_{AL} \cdot L$ to the voltage nU_T. However, A_{Vi} is down only modestly in moderate inversion because of the modest loss of g_m/I_D. Moderate inversion is generally a better choice compared to operation in weak inversion because of smaller device shape factor, smaller gate area, and, correspondingly, smaller capacitances giving improved bandwidth.

In Table 3.22, the second expression describes the inclusion of velocity saturation and VFMR effects for A_{Vi} due to the reduction in g_m/I_D. As described in Section 3.8.2.2, these effects are included by replacing IC in the g_m/I_D component of the expressions with $IC = IC(1 + IC/IC_{CRIT})$. IC_{CRIT} is the critical or corner inversion coefficient where velocity saturation and VFMR effects reduce g_m/I_D to

Table 3.22 MOS intrinsic voltage gain, A_{Vi}, expressions and trends in terms of inversion coefficient, channel length, and drain–source voltage. When evaluated in terms of the inversion coefficient, A_{Vi} is independent of the drain current

Intrinsic voltage gain $A_{Vi} = g_m/g_{ds} = g_m \cdot r_{ds} \approx (g_m/I_D) \cdot V_A$ $= (g_m/I_D) \cdot V_A$ (when V_A includes V_{DS})	$IC \uparrow$ L, V_{DS} fixed	$L \uparrow$ IC, V_{DS} fixed	$V_{DS} \uparrow$ IC, L fixed
Without small-geometry effects: $A_{Vi} = \dfrac{V_{AL} \cdot L}{nU_T\left(\sqrt{IC+0.25}+0.5\right)} = \dfrac{2V_{AL} \cdot L}{n\left(V_{DS,sat}-2U_T\right)}$ $\approx V_{AL} \cdot L/(nU_T)$ in WI $\approx V_{AL} \cdot L/(nU_T\sqrt{IC}) = 2V_{AL} \cdot L/(nV_{DS,sat}) =$ $2V_{AL} \cdot L/V_{EFF}$ in SI	Unchanged in WI $\downarrow \propto \dfrac{1}{\sqrt{IC}}$ in SI	$\uparrow \propto L^{0.4-0.8}$ (due to $V_{AL} \cdot L$)	$\uparrow$ especially as V_{DS} increases from near $V_{DS,sat}$ (due to $V_{AL} \cdot L$)
With velocity saturation and VFMR effects: Replace IC in above expression with $IC\left(1+\dfrac{IC}{IC_{CRIT}}\right)$			
Strong inversion with full velocity saturation ($VSF' = V_{EFF}/(LE_{CRIT})' \gg 1$, or $IC \gg IC_{CRIT}$): $A_{Vi} = \left(\dfrac{(LE_{CRIT})'}{(2nU_T)^2 IC}\right)(V_{AL} \cdot L)$ $= \left(\dfrac{4IC_{CRIT}}{(LE_{CRIT})' IC}\right)(V_{AL} \cdot L) = \dfrac{V_{AL} \cdot L}{V_{EFF}}$	$\downarrow \propto \dfrac{1}{IC}$	$\uparrow \propto L \cdot L^{0.4-0.8}$ (due to g_m/I_D and $V_{AL} \cdot L$)	$\uparrow$ especially as V_{DS} increases from near $V_{DS,sat}$ (due to $V_{AL} \cdot L$)
DIBL component of gain: $A_{Vi}(\text{DIBL}) = \dfrac{1}{\dfrac{-\partial V_T}{\partial V_{DS}}}$ $= \dfrac{1}{-DVTDIBL \cdot \left(\dfrac{L_{min}}{L}\right)^{DVTDIBLEXP}}$ DIBL effects may be included by paralleling the above gain with the gains due to CLM above or by including DIBL effects directly in V_{AL}	Unchanged	$\uparrow \propto\approx L^3$	$\downarrow$ if drain current turns up as V_{DS} increases

Expressions assume saturation operation where V_{DS} is sufficiently above $V_{DS,sat}$. As discussed in Section 3.8.4.4, voltage gain decreases significantly for V_{DS} near $V_{DS,sat}$.

V_{EFF} is from Table 3.14, $V_{DS,sat}$ is from Table 3.15, IC_{CRIT} and $(LE_{CRIT})'$ are from Table 3.17, $DVTDIBL$ and $DVTDIBLEXP$ are from Table 3.21, and V_{AL} (V/μm) $= V_A/L$ is the Early voltage factor extracted near the operating IC, L, and V_{DS} from Table 3.21 (see Figure 3.48 for measured V_{AL}).

Hot-electron, substrate current leading to the presence of r_{db} or g_{db} is not included. As discussed in Section 3.8.4.3, this is made negligible by operating nMOS devices at sufficiently low values of V_{DS}.

approximately 70.7 % its value without these effects. When velocity saturation effects are significant for short-channel devices in strong inversion, A_{Vi} decreases more rapidly than $1/\sqrt{IC}$, decreasing as $1/IC$ when full velocity saturation is present ($IC \gg IC_{CRIT}$). Also when full velocity saturation is present, A_{Vi} increases more rapidly with L because of the improvement in g_m/I_D as velocity saturation effects decrease, along with the usual improvement in V_A as L increases.

Finally, the last expression in the table considers the DIBL component of A_{Vi}, A_{Vi} (DIBL). This is independent of the level of inversion, but increases as approximately L^3 as the magnitude of $\delta V_T/\delta V_{DS}$ decreases rapidly. A_{Vi} (DIBL) is considered by paralleling it with the CLM component given by the other expressions in the table, or by including DIBL effects directly in the value of V_{AL} as shown in

Figure 3.48. As mentioned, DIBL effects are especially significant for short-channel devices operating at low levels of inversion where A_{Vi}(DIBL) is often less than A_{Vi}(CLM), reducing the overall A_{Vi}. Increasing L only modestly from the process minimum significantly reduces the DIBL loss of A_{Vi} as shown in measured data that follows.

Figure 3.51 shows measured A_{Vi} for $L = 0.18$, 0.28, 0.38, 0.48, 1, 2, and 4 μm, nMOS devices in the 0.18 μm process described in Table 3.2 operating at IC ranging from 0.01 (deep weak inversion) to 100 (deep strong inversion) at $V_{DS} = 1$ V. A_{Vi} was found by multiplying measured g_m/I_D for $V_{DS} = 1$ V, which is nearly identical to that shown in Figure 3.26 for $V_{DS} = V_{GS}$, with measured V_A for $V_{DS} = 1$ V shown in Figure 3.43.

A_{Vi} closely tracks g_m/I_D, being constant and maximum in weak inversion, decreasing modestly in moderate inversion, and then decreasing nearly as $1/\sqrt{IC}$ in strong inversion. The decrease in strong inversion is actually less because of a slight increase in V_A shown in Figure 3.43, but is greater for short channel lengths because of the velocity saturation decrease in g_m/I_D. A_{Vi} increases sublinearly with L for L somewhat above the minimum channel length, corresponding to the increase in V_A shown in Figure 3.47. As noted for the minimum channel length, A_{Vi} is decreased in weak and moderate inversion because of DIBL effects that reduce V_A. As expected, increasing L only modestly from $L = 0.18$ μm to $L = 0.28$ μm nearly eliminates the DIBL effects.

Measured A_{Vi} in Figure 3.51 extends over two orders of magnitude from 11 to 1200 V/V illustrating the wide range available to the designer through the choice of IC and L. Even higher A_{Vi} is available using channel lengths above the maximum of 4 μm evaluated here. While not shown, A_{Vi} decreases by nearly a factor of two for $V_{DS} = 0.5$ V and a factor of four for $V_{DS} = 0.25$ V compared to the value shown for $V_{DS} = 1$ V, with an additional decrease occurring when the operating $V_{DS,sat}$ nears V_{DS}. This is consistent with the decrease in V_A observed in Figure 3.45 ($V_{DS} = 0.5$ V) and Figure 3.46 ($V_{DS} = 0.25$ V), compared to Figure 3.43 ($V_{DS} = 1$ V).

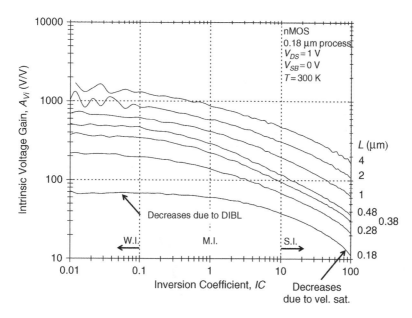

Figure 3.51 Measured intrinsic voltage gain, A_{Vi}, versus inversion coefficient, IC, for $L = 0.18$, 0.28, 0.38, 0.48, 1, 2, and 4 μm, nMOS devices in a 0.18 μm CMOS process operating at $V_{DS} = 1$ V. A_{Vi} decreases as the inversion coefficient increases, tracking g_m/I_D, and it increases with channel length, tracking V_A. A_{Vi} for the minimum channel length is reduced at low levels of inversion because of DIBL reduction in V_A and is reduced at high levels of inversion because of velocity saturation reduction of g_m/I_D. Measured A_{Vi} is lower for lower V_{DS} (not shown) because of lower V_A. A_{Vi} for pMOS devices is similar, but with less velocity saturation decrease

Measured pMOS A_{Vi} is very similar to that shown in Figure 3.51 for the same $V_{DS} = (-)1$ V. As for nMOS devices, pMOS A_{Vi} is significantly reduced for the minimum channel length at low levels of inversion because of DIBL effects. pMOS A_{Vi}, however, is reduced less for short-channel devices at high levels of inversion because of less velocity saturation reduction of g_m/I_D. This is a result of higher $E_{CRIT} = 14$ V/μm for pMOS devices compared to 5.6 V/μm for nMOS devices in the 0.18 μm process.

Figure 3.52 presents a three-dimensional display of the measured A_{Vi} shown in Figure 3.51. Looking along the IC axis reveals the A_{Vi} decrease with increasing IC due primarily to decreasing g_m/I_D (as seen in Figure 3.43, V_A increases slightly in strong inversion, which also affects A_{Vi}). Looking along the L axis reveals the A_{Vi} increase with increasing L due primarily to increasing V_A (as seen in Figure 3.26, g_m/I_D increases with L when velocity saturation is significant, which also affects A_{Vi}). The figure also shows the problematic, low values of A_{Vi} present for short-channel devices: one at low IC due to the DIBL reduction of V_A, and the other at high IC due to the velocity saturation reduction of g_m/I_D. The figure shows A_{Vi} over the design choices of IC and L where A_{Vi} is maximized at low IC and long L, and is minimized at high IC and short L. Intrinsic bandwidth, shown later in Figure 3.58 in Section 3.9.6, behaves oppositely. It is maximized at high IC and short L, and minimized at low IC and long L.

As discussed in Section 3.8.2.5, g_m/I_D when evaluated in terms of the inversion coefficient has a fundamental characteristic for CMOS processes. The V_A component of A_{Vi}, however, does not have a fundamental characteristic and is subject to considerable change with processes as discussed at the end of Section 3.8.4.5. A_{Vi} can be expected to decrease in smaller-geometry processes as channel length and allowable V_{DS} decrease.

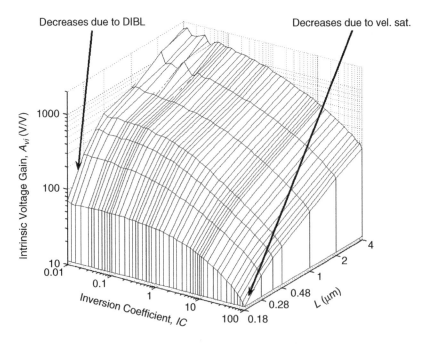

Figure 3.52 Measured intrinsic voltage gain, A_{Vi}, versus inversion coefficient, IC, and channel length, L, displayed in three dimensions from the data in Figure 3.51. A_{Vi} follows the trends described in Figure 3.51, being maximum at low IC and long L, and minimum at high IC and short L. Comparing A_{Vi} with the intrinsic bandwidth, f_{Ti}, shown later in Figure 3.58, shows the opposing tradeoffs of MOS gain and bandwidth

3.9 CAPACITANCES AND BANDWIDTH

The prediction and trends of MOS capacitances are important since these, in combination with small-signal parameters, control operating signal bandwidth. The following discussions describe MOS gate-oxide capacitance, intrinsic gate capacitances, extrinsic gate-overlap capacitances, and intrinsic and extrinsic drain–body and source–body capacitances. To illustrate the undesirable loading of extrinsic capacitances, these are compared to intrinsic capacitances that are fundamental to MOS operation. Finally, MOS bandwidth measures are described. These include intrinsic bandwidth, extrinsic bandwidth that includes overlap capacitances, and diode-connected bandwidth that includes overlap capacitances and the drain–body capacitance. For short-channel devices having small gate area and small intrinsic capacitances, the extrinsic and diode-connected bandwidths are significantly below the intrinsic bandwidth. The extrinsic and diode-connected bandwidth measures are useful as these are better linked to the actual operating bandwidth.

3.9.1 Gate-Oxide Capacitance

MOS gate-oxide capacitance, C_{GOX}, controls intrinsic gate capacitances and is the capacitance associated with the gate separated by the gate oxide from the body or substrate as seen in the MOS cross-section of Figure 3.9. The first entry in Table 3.23 lists expressions for C_{GOX} along with dependencies on the design choices of inversion coefficient, channel length, and drain current. Since C_{GOX} is equal to the product of gate area, WL, and the process gate-oxide capacitance, C'_{OX}, it increases directly with gate area, which is found from the sizing relationships of Table 3.10. C_{GOX} and gate area decrease as $1/IC$ with increasing IC, corresponding to the decrease in shape factor, $S = W/L$, and W required to maintain the selected L and I_D. C_{GOX} and gate area increase as the square of L since an increase in L must be accompanied by an equal increase in W to maintain the selected IC and I_D. Finally, C_{GOX} and gate area increase directly with I_D, corresponding to the increase in S and W required to maintain the selected IC and L.

Interestingly, when C_{GOX} is evaluated in terms of the inversion coefficient, it is independent of the process gate-oxide capacitance. This is observed in the second line of the first expression in Table 3.23 where C'_{OX} in the numerator cancels the denominator C'_{OX} term in the technology current, $I_0 = 2n_0\mu_0 C'_{OX} U_T^2$ from Table 3.6. As C'_{OX} and, correspondingly, I_0 increase in smaller-geometry processes, smaller shape factors, widths, and gate areas are required to maintain the selected IC, L, and I_D. C_{GOX}, however, remains unchanged since the smaller gate area is multiplied by the increased value of C'_{OX}. While C_{GOX} is independent of C'_{OX}, it depends inversely on process mobility, μ_0. pMOS devices, having lower mobility compared to nMOS devices, require larger shape factors, widths, and gate areas, resulting in higher C_{GOX}.

Figure 3.53 shows C_{GOX} as a function of IC and L for nMOS devices in the 0.18 μm CMOS process described in Table 3.2 operating at $I_D = 100\,\mu A$. C_{GOX} is shown for $n_0 = 1.35$, $\mu_0 = 422\,cm^2/V \cdot s$, and $U_T = 25.9\,mV$ ($T = 300\,K$). As observed, C_{GOX} decreases as $1/IC$ and increases as the square of L. C_{GOX} ranges from less than 1 fF for short-channel devices operating at high IC in strong inversion to over 100000 fF (100 pF) for long-channel devices operating at low IC in weak inversion. This is a wide range of C_{GOX}, especially for a fixed drain current. As mentioned, C_{GOX} scales directly with drain current, so values are lower for drain currents below 100 μA and higher for drain currents above this. C_{GOX} shown is similar for nMOS devices in all processes having similar mobility and substrate factors. C_{GOX} is nearly a factor of five larger for pMOS devices in the 0.18 μm process because mobility is nearly a factor of five lower.

When evaluated in terms of the inversion coefficient, C_{GOX} is independent of the MOS process except for the mobility and n, which varies only slightly from approximately $n = 1.3$ to 1.4 for bulk processes. C_{GOX} directly influences intrinsic gate–source and gate–body capacitances, which, in turn, influence intrinsic bandwidth. As described in Section 3.9.6 later, the rapid increase in C_{GOX} with increasing L results in a rapid decrease in intrinsic bandwidth.

Table 3.23   MOS gate-oxide capacitance and intrinsic gate capacitance expressions and trends in terms of the inversion coefficient, channel length, and drain current

Expression	$IC \uparrow$ L, I_D fixed	$L \uparrow$ IC, I_D fixed	$I_D \uparrow$ IC, L fixed
Gate-oxide capacitance: $$C_{GOX} = WLC'_{OX} = \left(\frac{L^2}{IC}\right)\left(\frac{I_D}{I_0}\right)C'_{OX}$$ $$= \left(\frac{L^2}{IC}\right)\left(\frac{I_D}{2n_0\mu_0 U_T^2}\right)$$	$\downarrow \propto \dfrac{1}{IC}$	$\uparrow \propto L^2$	$\uparrow \propto I_D$
Intrinsic gate capacitances:[a] $$C_{gsi} = \hat{C}_{gsi}C_{GOX}$$ $$C_{gbi} = \hat{C}_{gbi}C_{GOX}$$ $$C_{gdi} \approx 0 \cdot C_{GOX} \text{ (in saturation)}$$	$\downarrow \propto \dfrac{1}{IC}$ in SI	$\uparrow \propto L^2$	$\uparrow \propto I_D$
Normalized intrinsic gate capacitances:[a,b] $$\hat{C}_{gsi} = \frac{C_{gsi}}{C_{GOX}} = \frac{2-x}{3}$$ $$= \frac{0}{3}, \frac{1}{3}, \frac{2}{3} \text{ in WI, MI, and SI}$$	$\hat{C}_{gsi} \uparrow$, leveling off in SI	Unchanged	Unchanged
$$\hat{C}_{gbi} = \frac{C_{gbi}}{C_{GOX}} = \left(1 - \hat{C}_{gsi}\right)\frac{n-1}{n} = \frac{1+x}{3}\left(\frac{n-1}{n}\right)$$ $$= \left(\frac{3}{3}, \frac{2}{3}, \frac{1}{3}\right)\frac{n-1}{n} \text{ in WI, MI, and SI}$$	$\hat{C}_{gbi} \downarrow$, leveling off in SI		
$$\hat{C}_{gsi} + \hat{C}_{gbi} = \frac{n-(1+x)/3}{n}$$ $$= \frac{n-1}{n}, \frac{n-2/3}{n}, \frac{n-1/3}{n} \text{ in WI, MI, and SI}$$ $$x = \frac{\left(\sqrt{IC+0.25}+0.5\right)+1}{\left(\sqrt{IC+0.25}+0.5\right)^2}$$ $$= 2, 1, 0 \text{ for WI, MI, and SI}$$	$\hat{C}_{gsi} + \hat{C}_{gbi} \uparrow$, leveling off in SI		

[a] For intrinsic gate capacitances, saturation is assumed where $V_{DS} > V_{DS,sat}$. n is assumed fixed at an average value. The more rapid increase in $\hat{C}_{gsi}$ and C_{gsi}, and decrease in $\hat{C}_{gbi}$ and C_{gbi}, due to increased $V_{EFF} = V_{GS} - V_T$ resulting from velocity saturation and VFMR effects is neglected. Polysilicon depletion effects that reduce intrinsic capacitances are also neglected.
[b] Values given for moderate inversion (MI) are for the center of moderate inversion at $IC = 1$.

3.9.2 Intrinsic Gate Capacitances

MOS intrinsic gate capacitances appear between the gate and source, gate and body (substrate), and gate and drain, and are denoted as C_{gsi}, C_{gbi}, and C_{gdi}, respectively. Unlike extrinsic overlap, junction, and layout capacitances, intrinsic capacitances are part of physical MOS operation, consisting of portions of the gate-oxide capacitance. Intrinsic gate capacitances are included in the small-signal model shown earlier in Figure 3.21 in Section 3.8.1.

Intrinsic gate capacitances are conveniently normalized to the gate-oxide capacitance where $\hat{C}_{gsi} = C_{gsi}/C_{GOX}$, $\hat{C}_{gbi} = C_{gbi}/C_{GOX}$, and $\hat{C}_{gdi} = C_{gdi}/C_{GOX}$. This provides convenient expressions linked to C_{GOX} where $C_{gsi} = \hat{C}_{gsi}C_{GOX}$, $C_{gbi} = \hat{C}_{gbi}C_{GOX}$, and $C_{gdi} = \hat{C}_{gdi}C_{GOX}$. Since devices are assumed to be

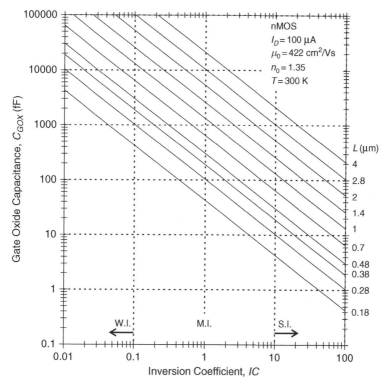

Figure 3.53 Gate-oxide capacitance, C_{GOX}, versus inversion coefficient, IC, for $L = 0.18$, 0.28, 0.38, 0.48, 0.7, 1, 1.4, 2, 2.8, and 4 μm, nMOS devices in a 0.18 μm CMOS process operating at $I_D = 100$ μA. C_{GOX} decreases as $1/IC$, increases as L^2, and scales directly with I_D. C_{GOX}, which is independent of C'_{OX} when evaluated in terms of IC, is inversely proportional to the mobility and is similar for all nMOS processes having similar mobility. C_{GOX} is nearly a factor of five higher for pMOS devices in the 0.18 μm process because of lower mobility

in saturation ($V_{DS} > V_{DS,sat}$), the intrinsic gate–drain capacitance, $\hat{C}_{gdi}$ or C_{gdi}, is nearly zero because the channel is pinched off at the drain side. The extrinsic gate–drain overlap capacitance, however, is non-zero and will be considered later.

Table 3.23 lists intrinsic gate capacitance expressions derived from the EKV MOS model [14] along with dependencies on the inversion coefficient, channel length, and drain current. Expressions for the normalized intrinsic capacitances are somewhat complex and are expressed here using an intermediate variable, x. x equals two for $IC \ll 0.1$ deep in weak inversion, unity for $IC = 1$ in the center of moderate inversion, and zero for $IC \gg 10$ deep in strong inversion. In strong inversion where $\hat{C}_{gsi}$ and $\hat{C}_{gbi}$ are at constant maximum and minimum values, respectively, C_{gsi} and C_{gbi} directly track C_{GOX}, decreasing as $1/IC$. In all regions, C_{gsi} and C_{gbi} increase as L^2 and increase directly with the drain current, directly tracking C_{GOX} because $\hat{C}_{gsi}$ and $\hat{C}_{gbi}$ are independent of L and the drain current.

Figure 3.54 shows the normalized intrinsic gate capacitances as a function of the inversion coefficient. $\hat{C}_{gsi}$ increases from nearly zero deep in weak inversion, where there is negligible inversion charge, to $^1/_3$ in the center of moderate inversion, corresponding to $C_{gsi} = ^1/_3 C_{GOX}$. Deep in strong inversion, $\hat{C}_{gsi}$ reaches its maximum value of $^2/_3$, corresponding to $C_{gsi} = ^2/_3 C_{GOX}$. These values are commonly assumed in strong inversion, but are sometimes incorrectly assumed for weak or moderate inversion where the values are considerably smaller. While $\hat{C}_{gsi}$ increases from weak to strong inversion, $\hat{C}_{gbi}$ decreases from its maximum value of $(n-1)/n$ deep in weak inversion to $^2/_3(n-1)/n$ at the center of moderate inversion, to $^1/_3(n-1)/n$ deep in strong inversion. For $n = 1.4$ assumed in the figure, $\hat{C}_{gbi}$ is

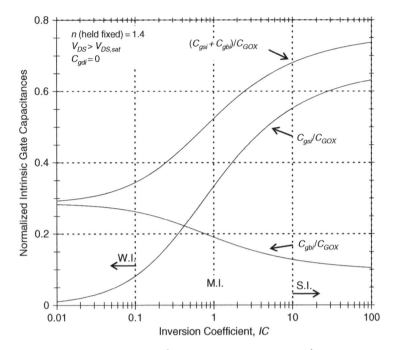

Figure 3.54 Normalized intrinsic gate–source, $\hat{C}_{gsi} = C_{gsi}/C_{GOX}$, and gate–body, $\hat{C}_{gbi} = C_{gbi}/C_{GOX}$, capacitances versus the inversion coefficient, IC. As IC increases from weak to strong inversion, $\hat{C}_{gsi}$ increases from nearly zero to a maximum of $^2/_3$, while $\hat{C}_{gbi}$ decreases from a maximum of $(n-1)/n$ to a minimum of $^1/_3(n-1)/n$. A fixed value of $n = 1.4$ is assumed, and decreases in capacitance associated with polysilicon depletion are neglected. Operation in saturation ($V_{DS} > V_{DS,sat}$) is assumed where the normalized intrinsic gate–drain capacitance, $\hat{C}_{gdi} = C_{gdi}/C_{GOX}$, is nearly zero

0.29, 0.19, and 0.095 in deep weak, moderate, and deep strong inversion, respectively, corresponding to $C_{gbi} = 0.29C_{GOX}$, $0.19C_{GOX}$, and $0.095C_{GOX}$. $\hat{C}_{gbi}$ and C_{gbi} are often excluded in design, but are dominant in weak inversion and important in moderate inversion where $\hat{C}_{gsi}$ and C_{gsi} are reduced from their maximum strong-inversion values.

Finally, Figure 3.54 shows the sum of $\hat{C}_{gsi}$ and $\hat{C}_{gbi}$, representing the total, normalized intrinsic capacitance connected to the gate (again, $\hat{C}_{gdi}$ is nearly zero in saturation). The sum of $\hat{C}_{gsi}$ and $\hat{C}_{gbi}$ is 0.29, 0.52, and 0.76 for operation in deep weak, moderate, and deep strong inversion, respectively, corresponding to a sum of C_{gsi} and C_{gbi} equal to $0.29C_{GOX}$, $0.52C_{GOX}$, and $0.76C_{GOX}$. As observed in the figure, there is over a factor-of-two increase in the sum of $\hat{C}_{gsi}$ and $\hat{C}_{gbi}$ from weak to strong inversion. In weak and moderate inversion where C_{GOX} is large because of large gate areas, the reduction in the sum of $\hat{C}_{gsi}$ and $\hat{C}_{gbi}$ is helpful in reducing the intrinsic gate capacitance.

$\hat{C}_{gsi}$ and $\hat{C}_{gbi}$ given by Table 3.23 and Figure 3.54 are approximate and assume a constant value of n, while n actually decreases slightly with increasing inversion as described in Section 3.8.3.1. The expressions are also based on the traditional inversion coefficient, taken in this book as IC' as described in Section 3.4.2.1, where $V_{EFF} = V_{GS} - V_{T}$ is linked to IC' by Equation 3.18 given in Section 3.7.2.1, using IC' instead of IC in the equation. $\hat{C}_{gsi}$ and $\hat{C}_{gbi}$, however, increase and decrease, respectively, more rapidly towards their constant strong-inversion values when evaluated in terms of the fixed–normalized inversion coefficient, IC, used here and described in Section 3.4.2.2. This is a result of the additional increase in V_{EFF} when evaluated in terms of IC, resulting primarily from velocity saturation for short-channel devices with a slight increase due to VFMR for all channel lengths as described in Section 3.7.2.2. Since the increase in V_{EFF} occurs primarily in strong inversion where

$\hat{C}_{gsi}$ and $\hat{C}_{gbi}$ are near constant values, little effect is expected. Using *IC* without separately including velocity saturation increases in V_{EFF} permits expressions for $\hat{C}_{gsi}$ and $\hat{C}_{gbi}$ that are conveniently independent of the channel length. Finally, the expressions given exclude the reduction in capacitances associated with polysilicon depletion [44]. While not included here in simple expressions for design guidance, intrinsic gate capacitance effects due to velocity saturation increases in V_{EFF} and polysilicon depletion can be included in simulations of a candidate design using full MOS models.

The expressions given by Table 3.23 and Figure 3.54, again, assume operation in saturation ($V_{DS} > V_{DS,sat}$) where $\hat{C}_{gdi}$ is nearly zero. As V_{DS} decreases below $V_{DS,sat}$, $\hat{C}_{gdi}$ increases from zero while $\hat{C}_{gsi}$ decreases. When V_{DS} decreases well below $V_{DS,sat}$ in the deep ohmic region, $\hat{C}_{gsi} = \hat{C}_{gdi} = \frac{1}{2}$, corresponding to $C_{gsi} = C_{gdi} = \frac{1}{2}C_{GOX}$. Here, the gate-oxide capacitance is divided equally between the drain and source.

When evaluated in terms of the inversion coefficient, $\hat{C}_{gsi}$ is independent of the MOS process, while $\hat{C}_{gbi}$ depends on n, which varies only slightly from approximately $n = 1.3$ to 1.4 for bulk processes. The complete intrinsic capacitances, C_{gsi} and C_{gbi}, additionally depend directly on C_{GOX}, which is independent of the process except for mobility and n as described in the previous section.

3.9.3 Extrinsic Gate-Overlap Capacitances

As described in the previous section, MOS intrinsic gate capacitances arise from intrinsic MOS operation and are strong functions of the inversion level. Extrinsic gate-overlap capacitances, however, result from the physical overlap of the gate over the drain, source, and body as seen in the MOS cross-section of Figure 3.9. The gate–drain, gate–source, and gate–body overlap capacitances are denoted as C_{GDO}, C_{GSO}, and C_{GBO}, respectively, and are included in the small-signal model shown earlier in Figure 3.21 in Section 3.8.1. As observed in the figure, the total gate–drain, gate–source, and gate–body capacitances are the sums of intrinsic and overlap capacitances given by $C_{GD} = C_{gdi} + C_{GDO}$, $C_{GS} = C_{gsi} + C_{GSO}$, and $C_{GB} = C_{gbi} + C_{GBO}$, respectively. Overlap capacitances can be significant contributors to overall gate capacitances, especially for short-channel devices that have small gate areas and small related intrinsic gate capacitances. C_{GDO} has an additional importance since it usually dominates C_{GD} since C_{gdi} is usually negligible in saturation.

Table 3.24 gives expressions for gate-overlap capacitances along with dependencies on the design choices of inversion coefficient, channel length, and drain current. Overlap capacitance is simply calculated as the overlap dimension multiplied by a process overlap capacitance for that overlap dimension. The gate–drain, gate–source, and gate–body process overlap capacitance parameters are denoted without subscripts as *CGDO*, *CGSO*, and *CGBO*, respectively, which distinguishes them from the related overlap capacitances C_{GDO}, C_{GSO}, and C_{GBO}. Table 3.4 gives process overlap capacitances, which have units of fF/μm, for the example 0.5, 0.35, and 0.18 μm CMOS processes.

To permit simple hand expressions useful for design guidance, bias voltage dependency on overlap capacitances is neglected. Here, average or worst-case levels of overlap capacitance are assumed through the choice of *CGDO*, *CGSO*, and *CGBO* process overlap capacitances. Bias voltage dependency can be considered during computer simulations of a candidate design using BSIM3, 4 [3], MOS Model 11 [45], SP (surface potential) [46], or other MOS models. The gate–drain and gate–source overlap capacitances are especially bias voltage dependent because the drain or source extension that is overlapped by the gate (Figure 3.9) can be accumulated, depleted, or inverted depending upon the gate, drain, and source voltages [16, pp. 220–222].

As seen in Table 3.24, C_{GDO} and C_{GSO} increase directly with channel width since width is the overlap dimension for these capacitances. Like channel width summarized in Table 3.10, C_{GDO} and C_{GSO} decrease as $1/IC$, increase directly with L, and increase directly with I_D. C_{GBO}, however, depends and increases with L only since this is the overlap dimension for this capacitance.

It is convenient to compare C_{GDO} and C_{GSO} to the gate-oxide capacitance using the ratios of C_{GDO}/C_{GOX} and C_{GSO}/C_{GOX} shown in Table 3.24. These ratios show the relative importance of

Table 3.24 MOS extrinsic gate-overlap capacitance expressions and trends in terms of the inversion coefficient, channel length, and drain current

Expression	$IC \uparrow$ L, I_D fixed	$L \uparrow$ IC, I_D fixed	$I_D \uparrow$ IC, L fixed
Gate–drain (gate–source) overlap capacitance: $C_{GDO} = W \cdot CGDO = \left(\dfrac{L}{IC}\right)\left(\dfrac{I_D}{I_0}\right)CGDO$ C_{GSO} expression uses $CGSO$	$\downarrow \propto \dfrac{1}{IC}$	$\uparrow \propto L$	$\uparrow \propto I_D$
Ratio of gate–drain (gate–source) overlap to gate-oxide capacitance: $\dfrac{C_{GDO}}{C_{GOX}} = \dfrac{W \cdot CGDO}{WLC'_{OX}} = \dfrac{1}{L}\left(\dfrac{CGDO}{C'_{OX}}\right)$ C_{GSO} expression uses $CGSO$	Unchanged	$\downarrow \propto \dfrac{1}{L}$	Unchanged
Gate–body overlap capacitance: $C_{GBO} = L \cdot CGBO$	Unchanged	$\uparrow \propto L$	Unchanged
Ratio of gate–body overlap to gate-oxide capacitance: $\dfrac{C_{GBO}}{C_{GOX}} = \dfrac{L \cdot CGBO}{WLC'_{OX}} = \dfrac{1}{W}\left(\dfrac{CGBO}{C'_{OX}}\right)$ $= \left(\dfrac{IC}{L}\right)\left(\dfrac{I_0}{I_D}\right)\left(\dfrac{CGBO}{C'_{OX}}\right)$	$\uparrow \propto IC$	$\downarrow \propto \dfrac{1}{L}$	$\downarrow \propto \dfrac{1}{I_D}$

Expressions exclude bias voltage dependencies on overlap capacitances and assume average or worst-case capacitances through the choice of process overlap capacitance parameters.

$CGDO$ and $CGSO$ are the process gate–drain and gate–source overlap capacitances per unit of gate width. $CGBO$ is the process gate–body overlap capacitance per unit of gate length.

overlap capacitances compared to the gate-oxide capacitance and will be used in bandwidth calculations later in Section 3.9.7. C_{GDO}/C_{GOX} and C_{GSO}/C_{GOX} are equal to $(1/L) \cdot (CGDO/C'_{OX})$ and $(1/L) \cdot (CGSO/C'_{OX})$, respectively, and are independent of channel width since the overlap capacitances, C_{GDO} and C_{GSO}, and C_{GOX} depend directly on channel width. Interestingly, because of the channel width independence, the capacitance ratios are independent of drain current and inversion level. The ratios, however, depend inversely on channel length, indicating that C_{GDO} and C_{GSO} are most significant for short-channel devices where gate area, gate-oxide capacitance, and the related intrinsic gate capacitances are small.

Figure 3.55 shows the overlap capacitance ratios, C_{GDO}/C_{GOX} and C_{GSO}/C_{GOX}, as a function of channel length for nMOS devices in the $0.18\,\mu m$ CMOS process described in Tables 3.2 and 3.4. The ratios are shown for $CGDO = CGSO = 0.94\,fF/\mu m$ (Table 3.4) and $C'_{OX} = 8.41\,fF/\mu m^2$ (Table 3.2). The figure shows that C_{GDO}/C_{GOX} and C_{GSO}/C_{GOX} are equal to 0.62 for the minimum channel length in the process, corresponding to $C_{GDO} = C_{GSO} = 0.62 C_{GOX}$. This illustrates the importance of including C_{GDO} and C_{GSO} since these can be appreciable components of the total gate–drain, $C_{GD} = C_{gdi} + C_{GDO}$, and gate–source, $C_{GS} = C_{gsi} + C_{GSO}$, capacitances. For the example given where $C_{GDO} = C_{GSO} = 0.62 C_{GOX}$, C_{GSO} exceeds C_{gsi} in weak and moderate inversion and is nearly equal to $C_{gsi} \approx {}^2/_3 C_{GOX}$ in strong inversion. C_{GDO} is important for all applications since it dominates C_{GD} as C_{gdi} is nearly zero in saturation. C_{GDO} gives rise to a Miller-multiplied component of gate input capacitance equal to $C_{Miller} = C_{GDO} \cdot (1 - A_{Vgd})$, where A_{Vgd} is the voltage gain, which is negative or inverting, between the gate and drain. Additionally, C_{GDO} can give rise to a right-half-plane zero in the gate-to-drain frequency response. As observed in the figure, increasing channel length reduces the

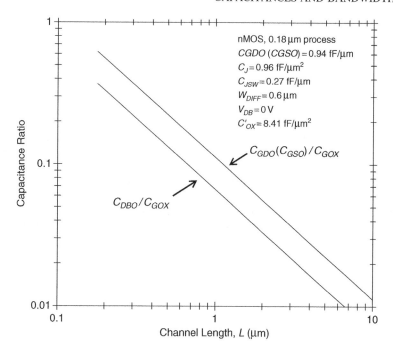

nMOS, 0.18 μm process
CGDO (CGSO) = 0.94 fF/μm
C_J = 0.96 fF/μm²
C_{JSW} = 0.27 fF/μm
W_{DIFF} = 0.6 μm
V_{DB} = 0 V
C'_{OX} = 8.41 fF/μm²

$C_{GDO}(C_{GSO})/C_{GOX}$

C_{DBO}/C_{GOX}

Figure 3.55 Ratio of gate–drain and gate–source overlap capacitances to the gate-oxide capacitance, C_{GDO}/C_{GOX} and C_{GSO}/C_{GOX}, for nMOS devices in a 0.18 μm CMOS process. Additionally, the figure shows the ratio of zero-bias ($V_{DB} = 0$ V), drain–body, junction capacitance to the gate-oxide capacitance, C_{DBO}/C_{GOX}, for devices having an even number of interdigitations with the drain junction minimized. At short channel lengths where gate area and C_{GOX} are small, C_{GDO}, C_{GSO}, and C_{DBO} are significant compared to C_{GOX} and the related intrinsic gate capacitances. C_{GDO} is always significant since the related intrinsic gate–drain capacitance is nearly zero in saturation

C_{GDO}/C_{GOX} and C_{GSO}/C_{GOX} ratios, corresponding to overlap capacitances that are a smaller portion of the gate-oxide capacitance and related intrinsic gate capacitances. The figure also shows the ratio of $V_{DB} = 0$ V, drain–body junction capacitance, C_{DBO}, to the gate-oxide capacitance. This will be described in the next section.

While C_{GDO} and C_{GSO} depend on channel width, C_{GBO} depends on channel length since, as mentioned, this is the gate–body overlap dimension. As done with C_{GDO} and C_{GSO}, it is convenient to compare C_{GBO} to the gate-oxide capacitance. As listed in Table 3.24, the ratio of C_{GBO}/C_{GOX} is equal to $(1/W) \cdot (CGBO/C'_{OX})$. Unlike the C_{GDO}/C_{GOX} and C_{GSO}/C_{GOX} ratios that depend inversely on channel length, C_{GBO}/C_{GOX} depends inversely on channel width. As a result, C_{GBO}/C_{GOX} increases directly with IC, decreases as $1/L$, and decreases as $1/I_D$. C_{GBO} is most significant for short-channel devices operating at high IC and low I_D. These devices have small gate area and short channel width where the channel length and resulting value of C_{GBO} are more significant. Often C_{GBO} is neglected during initial design, but is included in simulations of a candidate design using full MOS models.

As mentioned in the previous section, when evaluated in terms of the inversion coefficient, the intrinsic MOS capacitances depend only on process mobility and n, which varies only slightly from approximately $n = 1.3$ to 1.4 for bulk processes. The dependence on process mobility results from the dependence on C_{GOX}. In contrast, overlap capacitances vary considerably over processes as observed in the process overlap capacitance parameters listed in Table 3.4 for the example CMOS processes. As overlap capacitances can contribute significantly to overall device capacitances, it is important to carefully consider these for the process of interest. This includes considering the strong bias voltage dependence mentioned earlier for simulations of a candidate design.

3.9.4 Drain–Body and Source–Body Junction Capacitances

Drain–body and source–body junction capacitances result from the drain–body and source–body diode junctions as seen in the MOS cross-section of Figure 3.9. These capacitances are referred to here as drain–body, C_{DB}, and source–body, C_{SB}, capacitances, where the intrinsic MOS drain–body, C_{dbi}, and source–body, C_{sbi}, capacitances described in the next section are neglected. C_{DB} and C_{SB} are included in the small-signal model shown earlier in Figure 3.21 in Section 3.8.1. C_{DB} is described here since it often loads the signal path. C_{SB} is found nearly identically as C_{DB}.

Table 3.25 lists expressions for C_{DB} along with dependencies on the design choices of inversion coefficient, channel length, and drain current. A_D and P_D are the drain junction area and perimeter. C_J and C_{JSW} are process capacitance parameters for the drain area and perimeter, having units of fF/μm^2 and fF/μm. $\hat{C}_J$ and $\hat{C}_{JSW}$ are factors describing the reduction of area and perimeter capacitances resulting from non-zero, drain–body voltage, V_{DB}.[13] As V_{DB} increases from zero, $\hat{C}_J$ and $\hat{C}_{JSW}$ decrease from unity reflecting the reduction in diode depletion capacitance associated with increasing reverse bias. Table 3.4 gives process parameters for drain–body and source–body junction capacitances for the example 0.5, 0.35, and 0.18 μm CMOS processes.

If a MOS device is interdigitated with drain sharing, it is possible to halve the drain area and perimeter, halving C_{DB}, compared to a non-interdigitated device. This is commonly done using an even number of gate fingers, $m = 2, 4, 6, 8$, etc., corresponding to the paralleling of smaller devices having individual channel widths of W/m. For this layout method, Table 3.25 shows that $A_D = {}^1/_2 W \cdot W_{DIF}$ and $P_D = W + m \cdot W_{DIF}$, where W_{DIF} is the process width of individual drain fingers. Since the length of individual drain fingers is usually somewhat greater than the process width of these fingers ($W/m \gg W_{DIF}$), this implies $W \gg m \cdot W_{DIF}$, resulting in $P_D \approx W$ as shown in the table. This is a convenient result since both A_D and P_D can now be expressed directly in terms of channel width.

The first expression in Table 3.25 gives C_{DB} resulting from A_D, P_D, and process parameters. Since C_{DB} depends nearly directly on channel width through A_D and P_D, C_{DB} decreases as $1/IC$, increases directly with L, and increases directly with I_D from the channel width relationships given in Table 3.10. These trends are identical to those of the width-dependent overlap capacitances, C_{GDO} and C_{GSO}, listed earlier in Table 3.24. As noted at the bottom of Table 3.25, if interdigitation is not used, corresponding to $m = 1$, C_{DB} is equal to twice the value listed.

As also noted in Table 3.25, the expressions for C_{DB} and, when evaluated accordingly, C_{SB} exclude capacitance associated with the source–drain extensions shown in the MOS cross-section of Figure 3.9. This component of capacitance can be included by removing the perimeter associated with the extension from P_D and adding $W \cdot C_{JSWG}$ to C_{DB} and C_{SB}, where C_{JSWG} is the capacitance per unit of channel width associated with the source–drain extensions [16, pp. 225–226]. This component of capacitance has a voltage dependence similar to that of the area and perimeter capacitances. The capacitance associated with source–drain extensions is partially compensated for here by using the full value of P_D, but can be considered during computer simulations of a candidate design using full, simulation MOS models.

As was done with the overlap capacitances, it is convenient to compare C_{DB} to the gate-oxide capacitance. Here, we compare the maximum C_{DB} occurring when $V_{DB} = 0$ V, denoted as C_{DBO}, to the gate-oxide capacitance. This capacitance ratio of C_{DBO}/C_{GOX} shows the relative importance of the drain–body capacitance compared to the gate-oxide capacitance and will be used in bandwidth calculations later in Section 3.9.7.

Table 3.25 gives expressions for C_{DBO}/C_{GOX}, which is independent of channel width since both C_{DBO} and C_{GOX} increase directly with channel width. Like the overlap capacitance ratios, C_{GDO}/C_{GOX} and C_{GSO}/C_{GOX}, listed earlier in Table 3.24 that are also independent of channel width, C_{DBO}/C_{GOX} depends inversely on L and is independent of IC and I_D. Like the overlap capacitances, C_{DBO} or C_{DB}

---

[13] V_{DB} is zero or positive for nMOS devices reflecting zero or reverse bias drain–body voltage. Here, it is also taken positively for pMOS devices as shown in Figure 3.9.

Table 3.25 MOS drain–body junction capacitance expressions and trends in terms of the inversion coefficient, channel length, and drain current. Source–body junction capacitance is calculated similarly using the source area, A_S, source perimeter, P_S, and source–body voltage, V_{SB}

Expression	$IC \uparrow$ L, I_D fixed	$L \uparrow$ IC, I_D fixed	$I_D \uparrow$ IC, L fixed
Drain–body capacitance:[a] $C_{DB} = A_D C_J \hat{C}_J + P_D C_{JSW} \hat{C}_{JSW}$ $\approx W \left(\frac{1}{2} W_{DIF} C_J \hat{C}_J + C_{JSW} \hat{C}_{JSW} \right)$ $\approx \left(\frac{L}{IC} \right) \left(\frac{I_D}{I_0} \right) \cdot \left(\frac{1}{2} W_{DIF} C_J \hat{C}_J + C_{JSW} \hat{C}_{JSW} \right)$ $A_D = \frac{m}{2} \left(\frac{W}{m} \cdot W_{DIF} \right) = \frac{1}{2} W \cdot W_{DIF}$ $P_D = \frac{m}{2} \left[2 \left(\frac{W}{m} + W_{DIF} \right) \right] = (W + m \cdot W_{DIF}) \approx W$	$\downarrow \propto \frac{1}{IC}$	$\uparrow \propto L$	$\uparrow \propto I_D$
Ratio of drain–body ($V_{DB} = 0\,\text{V}$) to gate-oxide capacitance:[a,b] $\frac{C_{DBO}}{C_{GOX}} \approx \frac{W \left(\frac{1}{2} W_{DIF} C_J + C_{JSW} \right)}{WLC'_{OX}}$ $\approx \frac{1}{L} \left(\frac{\frac{1}{2} W_{DIF} C_J + C_{JSW}}{C'_{OX}} \right)$	Unchanged	$\downarrow \propto \frac{1}{L}$	Unchanged
Drain–body capacitance reduction factors for V_{DB}:[c] $\hat{C}_J = \dfrac{1}{\left(1 + \dfrac{V_{DB}}{P_B} \right)^{M_J}}$ $\hat{C}_{JSW} = \dfrac{1}{\left(1 + \dfrac{V_{DB}}{P_{BSW}} \right)^{M_{JSW}}}$	Unchanged	Unchanged	Unchanged

A_D and P_D are the area and perimeter of the drain junction, m is the number of gate fingers, and W_{DIF} is the process width of drain fingers.

C_J, P_B, and M_J are process area capacitance and related built-in voltage and V_{DB} exponent parameters. C_{JSW}, P_{BSW}, and M_{JSW} are process perimeter (side-wall) capacitance and related built-in voltage and V_{DB} exponent parameters.

Expressions neglect the component of C_{DB} (or C_{SB}) associated with source–drain extensions. This can be included by removing the perimeter associated with the extension from P_D and adding a component of capacitance equal to $W \cdot C_{JSWG}$, where C_{JSWG} is the capacitance per unit of channel width associated with the source–drain extensions. This component of capacitance has voltage dependence similar to the area and perimeter capacitances.

[a] m is assumed even at 2, 4, 6, etc., with the drain junction minimized. For small devices with $m = 1$, A_D, P_D, C_{DB}, and C_{DBO}/C_{GOX} are doubled from the expressions given.

[b] $V_{DB} = 0\,\text{V}$, giving worst case $C_{DB} = C_{DBO}$.

[c] V_{DB} is taken positively for normal reverse bias.

is most significant for short-channel devices where gate area, gate-oxide capacitance, and the related intrinsic gate capacitances are small. As noted at the bottom of Table 3.25, if interdigitation is not used, C_{GDO}/C_{GOX} is equal to twice the value listed.

Figure 3.55, which showed C_{GDO}/C_{GOX} and C_{GSO}/C_{GOX} in the previous section, also shows C_{DBO}/C_{GOX} as a function of channel length for nMOS devices in the 0.18 μm CMOS process described in Tables 3.2 and 3.4. C_{DBO}/C_{GOX} is shown for $C_J = 0.96$ fF/μm², $C_{JSW} = 0.27$ fF/μm, $W_{DIF} = 0.6$ μm from Table 3.4, and $C'_{OX} = 8.41$ fF/μm² from Table 3.2. The figure shows that $C_{DBO}/C_{GOX} = 0.37$ for the minimum channel length in the process, corresponding to $C_{DBO} = 0.37 C_{GOX}$. While not as large as the overlap capacitance ratios of $C_{GDO}/C_{GOX} = C_{GSO}/C_{GOX} = 0.62$, corresponding to $C_{GDO} = C_{GSO} = 0.62 C_{GOX}$, C_{DBO}/C_{GOX} is still significant, indicating a significant value of C_{DBO} or C_{DB}. If interdigitation is not used, C_{GDO}/C_{GOX} is doubled to 0.74, corresponding to $C_{DBO} = 0.74 C_{GOX}$, which now exceeds C_{GDO}, C_{GSO}, and the maximum strong-inversion value of $C_{gsi} = {}^2/_3 C_{GOX}$. As observed in the figure, increasing channel length reduces the C_{DBO}/C_{GOX} ratio as it does the C_{GDO}/C_{GOX} and C_{GSO}/C_{GOX} ratios. This corresponds to drain–body and overlap capacitances that are a smaller portion of the gate-oxide capacitance and related intrinsic gate capacitances.

As mentioned, C_{DBO} is the maximum zero-bias ($V_{DB} = 0$ V) value of C_{DB}. As summarized in Table 3.25, the area component of C_{DB} is equal to the area component of C_{DBO} multiplied by the area capacitance reduction factor, $\hat{C}_J$, associated with non-zero V_{DB}. Similarly, the perimeter component of C_{DB} is equal to the perimeter component of C_{DBO} multiplied by the perimeter capacitance reduction factor, $\hat{C}_{JSW}$. For the area capacitance reduction factor, P_B is the built-in potential, and M_J is the reverse bias voltage exponent, theoretically 0.5 for abrupt junctions. For the perimeter or side-wall capacitance reduction factor, P_{BSW} is the built-in potential, and M_{JSW} is the reverse bias voltage exponent. Table 3.4 gives these process parameters for the example CMOS processes.

Figure 3.56 shows the C_{DB} area capacitance reduction factor, $\hat{C}_J$, and perimeter capacitance reduction factor, $\hat{C}_{JSW}$, resulting from non-zero V_{DB} for nMOS devices in the 0.18 μm CMOS process described in Table 3.4. This is shown for $P_B = 0.8$ V, $M_J = 0.38$, $P_{BSW} = 0.8$ V, and $M_{JSW} = 0.15$. For $V_{DB} = 0.5$ V, C_{DB} is between 83 and 93 % of C_{DBO}, corresponding to area and perimeter capacitances at

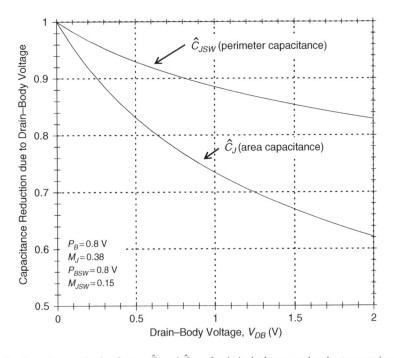

Figure 3.56 Capacitance reduction factors, $\hat{C}_J$ and $\hat{C}_{JSW}$, for drain–body area and perimeter capacitances versus the drain–body voltage, V_{DB}, for nMOS devices in a 0.18 μm CMOS process. The reduction in total drain–body capacitance, C_{DB}, from its zero-bias ($V_{DB} = 0$ V), C_{DBO} value lies between the two curves since C_{DB} consists of both area and perimeter components

83 and 93% of their zero-bias values ($\hat{C}_J = 0.83$ and $\hat{C}_{JSW} = 0.93$). For $V_{DB} = 1\,\text{V}$, C_{DB} is lower, between 73 and 89% of C_{DBO}, corresponding to area and perimeter capacitances at 73 and 89% of their zero-bias values ($\hat{C}_J = 0.73$ and $\hat{C}_{JSW} = 0.89$). The actual value of C_{DB} versus V_{DB} depends on the relative amount of area and perimeter capacitances. Interestingly, for many devices these capacitance components are similar, resulting from drain junction area, A_D, in units of μm^2, that is similar to the junction perimeter, P_D, in units of μm. For low-voltage design where V_{DB} is necessarily small, perhaps between 0.25 and 0.5 V, the figure shows that C_{DB} is only modestly reduced from C_{DBO}. Predicting C_{DBO} is useful for design guidance where C_{DBO} is conservatively larger than C_{DB} when V_{DB} is non-zero.

The source–body, junction capacitance, C_{SB}, is found identically as C_{DB} using the area, A_S, and perimeter, P_S, associated with the source. Here, the source–body voltage, V_{SB}, governs the drop in capacitance associated with non-zero V_{SB}. If an interdigitation factor of $m = 2$ is used, corresponding to two gate fingers, either the source or drain junction will have the full area and perimeter of a non-interdigitized device. Often, the full area and perimeter will be assigned to the source, especially if the source is connected to a supply rail where its capacitance does not impact circuit performance. For m of four or greater, both the source and drain junctions enjoy nearly the full interdigitation advantage where areas, perimeters, and related junction capacitances are nearly half those of a non-interdigitated device. Since Table 3.25 and Figure 3.55 assume the full interdigitation advantage in calculating C_{DB} and C_{DB}/C_{GOX}, or calculating C_{SB} and C_{SB}/C_{GOX}, these values must be doubled if either the drain or source does not enjoy the interdigitation decrease in junction area and perimeter. As mentioned, m is often selected as an even number using drain area and perimeter minimization with interior drain fingers. This minimizes drain junction capacitance and ensures that each interior drain finger receives channel current symmetrically from two adjacent source fingers. Symmetrical current flow in identically laid-out fingers improves matching across different devices.

The drain–body and source–body capacitances, like the overlap capacitances described in the previous section, vary considerably across processes. This is observed in the process drain–body (source–body) junction capacitance parameters listed in Table 3.4 for the example CMOS processes. The area capacitance parameter for nMOS devices in the 0.18 μm, bulk process is $C_J = 0.96\,\text{fF}/\mu\text{m}^2$, compared to only $C_J = 0.1\,\text{fF}/\mu\text{m}^2$ for the 0.35 μm, PD (partially depleted) SOI (silicon-on-insulator) process. The low area capacitance results from drain and source elements residing on top of silicon dioxide in the 0.35 μm, PD SOI process compared to residing inside the silicon substrate in the 0.18 μm bulk process. As drain–body and source–body capacitances can contribute significantly to overall device capacitances, it is important to carefully consider these for the process of interest.

3.9.5 Intrinsic Drain–Body and Source–Body Capacitances

As mentioned earlier in Section 3.9.2, intrinsic gate capacitances result from physical MOS operation. MOS operation, however, also results in intrinsic drain–body and source–body capacitances, denoted as C_{dbi} and C_{sbi}. C_{dbi} and C_{sbi} are equal to $(n-1)$ multiplied by intrinsic gate–drain and gate–source capacitances, C_{gdi} and C_{gsi}, respectively [13]. C_{dbi} and C_{sbi} model the coupling of intrinsic gate capacitances into the body or substrate.

Since C_{gdi} is assumed negligibly small for operation in saturation, C_{dbi} will also be neglected since it is even smaller at 40% of C_{gdi} for $n = 1.4$. C_{sbi} at 40% of C_{gsi} may not be negligible compared to the source–body junction capacitance, C_{SB}, described in the previous section. However, for design guidance described here, C_{sbi} will also be neglected for simplicity. C_{sbi} can be included in the value of C_{SB} and can be included in simulations of candidate designs using full MOS models.

3.9.6 Intrinsic Bandwidth

Section 3.8.5 described MOS intrinsic voltage gain, A_{Vi}, which is the gate-to-drain voltage gain for a grounded-source device with its drain connected to the infinite, small-signal resistance of an ideal current source load. This represents the maximum voltage gain available for a single device. Intrinsic bandwidth, f_{Ti}, described in this section, is the frequency where gate-to-drain current gain is unity for a grounded-source, intrinsic device with its drain connected to a small-signal ground.

The circuit for assessing intrinsic bandwidth can be visualized from the MOS small-signal model shown earlier in Figure 3.21 in Section 3.8.1. Here, the source, body, and drain are grounded and a signal current is applied to the gate. The intrinsic gate–source, C_{gsi}, and gate–body, C_{gbi}, capacitances are included, while the intrinsic gate–drain capacitance, C_{gdi}, is not included, although it is nearly zero for operation in saturation. The extrinsic gate–source, C_{GSO}, gate–body, C_{GBO}, and gate–drain, C_{GDO}, overlap capacitances are excluded when evaluating the intrinsic bandwidth. Additionally, the drain–body capacitance, C_{DB}, is excluded, although it and the drain–source conductance, g_{ds}, have no effect on the evaluation of intrinsic bandwidth since the drain, source, and body are grounded. Finally, the body-effect transconductance, g_{mb}, has no effect because the small-signal, body–source voltage, V_{bs}, is zero since both the source and body are grounded. Intrinsic bandwidth is then dependent only on C_{gsi} and C_{gbi}, and the transconductance, g_m. Intrinsic bandwidth – the frequency where current gain from the gate input to the short-circuit, drain output is equal to unity – is then given by

$$f_{Ti} = \frac{g_m}{2\pi\left(C_{gsi} + C_{gbi}\right)} \tag{3.95}$$

Sometimes C_{gbi} is excluded in f_{Ti} expressions, but it must be included in expressions valid in weak and moderate inversion. As seen earlier in Figure 3.54 in Section 3.9.2, C_{gbi} (presented as normalized to the gate-oxide capacitance) exceeds C_{gsi} in weak inversion and the weak-inversion side of moderate inversion, where it is clearly not negligible.

Another interpretation of f_{Ti} is the pole frequency associated with a circuit output conductance of g_m connected to the input, C_{gsi} and C_{gbi} capacitances of a device in a succeeding stage. An output conductance of g_m can be generated, for example, by unity feedback around a transconductance stage or an operational transconductance amplifier having transconductance of g_m.

Yet another interpretation of f_{Ti} is the pole frequency associated with a diode-connected (gate connected to the drain) device with the source, again, grounded. Again, the extrinsic gate-overlap and drain–body capacitances are excluded and g_{mb} has no effect because of zero V_{bs}. For a diode-connected device, a small-signal conductance level of g_m (g_{ds} is neglected since it is usually much smaller than g_m) appears in parallel with C_{gsi} and C_{gbi}, giving a pole frequency equal to f_{Ti}. In the next section, extrinsic gate-overlap and the drain–body capacitances are included in bandwidth measures since these can significantly lower the bandwidth of MOS circuits.

In the following discussion, velocity saturation and VFMR effects that reduce g_m are initially neglected in the evaluation of f_{Ti}. From Equation 3.95, f_{Ti} is given by

$$f_{Ti} = \frac{g_m}{2\pi\left(C_{gsi} + C_{gbi}\right)} = \frac{g_m}{2\pi\left(\hat{C}_{gsi} + \hat{C}_{gbi}\right)C_{GOX}} = \frac{g_m}{2\pi\left(\hat{C}_{gsi} + \hat{C}_{gbi}\right)WLC'_{OX}} \tag{3.96}$$

where, from Section 3.9.2, $\hat{C}_{gsi} = C_{gsi}/C_{GOX}$ and $\hat{C}_{gbi} = C_{gbi}/C_{GOX}$ are the intrinsic gate–source and gate–body capacitances normalized to the gate-oxide capacitance, $C_{GOX} = WLC'_{OX}$. Substituting g_m, given by I_D multiplied by the first g_m/I_D expression in Table 3.17, and gate area, WL, given by the third expression in Table 3.10, gives

$$f_{Ti} = \frac{\dfrac{I_D}{nU_T\left(\sqrt{IC + 0.25} + 0.5\right)}}{2\pi\left(\hat{C}_{gsi} + \hat{C}_{gbi}\right)\left(\dfrac{L^2}{IC}\dfrac{I_D}{I_0}\right)C'_{OX}}$$

$$= \left(\frac{IC}{\sqrt{IC + 0.25} + 0.5}\right)\left(\frac{I_0/nU_T}{2\pi\left(\hat{C}_{gsi} + \hat{C}_{gbi}\right)L^2 C'_{OX}}\right) \tag{3.97}$$

Substituting the technology current, $I_0 = 2n_0\mu_0 C'_{OX} U_T^2$, from Table 3.6, and assuming n constant at its moderate inversion value, n_0, gives

$$f_{Ti} = \left(\frac{IC}{\sqrt{IC + 0.25} + 0.5} \right) \left(\frac{\mu_0 U_T}{\pi \left(\hat{C}_{gsi} + \hat{C}_{gbi} \right) L^2} \right) \qquad (3.98)$$

Equation 3.98 shows that f_{Ti}, like A_{Vi} described in Section 3.8.5, is independent of the drain current. However, circuits operating at low drain currents are more likely to have significant extrinsic layout capacitances because of small device capacitances present for small devices used at low drain currents. As a result, operating bandwidth is typically lower for circuits operating at low drain currents compared to circuits operating at the same inversion coefficients and channel lengths at higher drain currents. Equation 3.98 shows that f_{Ti} is independent of C'_{OX}, while directly tracking the low-field mobility, μ_0 (again, velocity saturation and VFMR effects are excluded). As a result, f_{Ti} is nearly a factor of five lower for pMOS devices compared to nMOS devices in the example 0.18 μm CMOS process described in Table 3.2, because of the near factor of five lower pMOS mobility.

Table 3.26 lists expressions for f_{Ti} along with dependencies on the design choices of inversion coefficient and channel length. The first expression summarizes f_{Ti}, excluding velocity saturation and

Table 3.26 MOS intrinsic bandwidth, f_{Ti}, expressions and trends in terms of inversion coefficient and channel length. When evaluated in terms of the inversion coefficient, f_{Ti} is independent of the drain current

Intrinsic bandwidth	$IC \uparrow$ L fixed	$L \uparrow$ IC fixed
$f_{Ti} = \dfrac{g_m}{2\pi \left(C_{gsi} + C_{gbi} \right)}$		
All regions, without small-geometry effects:		
$f_{Ti} = \left(\dfrac{IC}{\sqrt{IC + 0.25} + 0.5} \right) \left(\dfrac{\mu_0 U_T}{\pi \left(\hat{C}_{gsi} + \hat{C}_{gbi} \right) L^2} \right)$		
$= \dfrac{IC \mu_0 U_T}{\pi L^2} \left(\dfrac{n}{n-1} \right)$ in WI	$\uparrow \propto IC$ in WI	$\downarrow \propto \dfrac{1}{L^2}$
$= \dfrac{\sqrt{IC} \mu_0 U_T}{\pi L^2} \left(\dfrac{n}{n - \frac{1}{3}} \right)$ in SI	$\uparrow \propto \sqrt{IC}$ in SI	
With velocity saturation and VFMR effects:		
Replace IC in above $\sqrt{IC + 0.25}$ expression with $IC \left(1 + \dfrac{IC}{IC_{CRIT}} \right)$ where IC_{CRIT} is given in Table 3.17		
Strong inversion with full velocity saturation $(VSF' = V_{EFF}/(LE_{CRIT})' \gg 1$, or $IC \gg IC_{CRIT})$: $f_{Ti} = \dfrac{E_{CRIT} \mu_0}{4\pi \left(n - \frac{1}{3} \right) L}$	Unchanged	$\downarrow \propto \dfrac{1}{L}$

f_{Ti} is for saturation where $V_{DS} > V_{DS,sat}$.
C_{gsi}, C_{gbi}, $\hat{C}_{gsi}$, and $\hat{C}_{gbi}$ are from Table 3.23.
Expressions assume n is fixed at an average value and neglect polysilicon depletion effects that can reduce intrinsic capacitances and increase f_{Ti}.

VFMR effects, from Equation 3.98. Additionally, the table gives f_{Ti} in weak and strong inversion using $\hat{C}_{gsi}$ and $\hat{C}_{gbi}$ values from Table 3.23.

In weak inversion, f_{Ti} increases directly with IC showing that bandwidth increases significantly as the inversion level increases towards moderate inversion. This is because there is little decrease in g_m/I_D and g_m in weak inversion with increasing inversion, while gate area, gate-oxide capacitance, and the associated intrinsic gate capacitances decrease as $1/IC$. Interestingly, in weak inversion, f_{Ti} is inversely proportional to $(n-1)$, showing that fully depleted (FD) SOI CMOS processes having n equal to 1.2 [29] or lower [17] offer potentially more bandwidth compared to bulk processes where n is typically near 1.4 (in weak inversion).

In strong inversion, f_{Ti} increases as $\sqrt{IC}$ showing that maximum bandwidth is achieved at high levels of inversion. This is because g_m/I_D and g_m decrease as $1/\sqrt{IC}$ in strong inversion (again, velocity saturation and VFMR are excluded), while gate area, gate-oxide capacitance, and the associated intrinsic gate capacitances decrease faster as $1/IC$.

In all regions of operation, f_{Ti} decreases as $1/L^2$, corresponding to a rapid decrease in bandwidth with increasing L. This is because gate area, gate-oxide capacitance, and the associated intrinsic gate capacitances increase as L^2 since channel width must be increased with channel length to maintain the selected inversion coefficient and drain current. The rapid decrease in f_{Ti} with increasing L is unfortunate, since increasing L increases voltage gain through the increase in V_A described earlier in Section 3.8.4.5, while also decreasing low-frequency flicker noise and DC mismatch through the increase in gate area as described later in Sections 3.10.3.7 and 3.11.1.7. f_{Ti} decreases as $1/L^2$ until velocity saturation effects on g_m/I_D and g_m become significant where f_{Ti} then decreases as $1/L$.

Since velocity saturation and VFMR effects reduce g_m/I_D and g_m, f_{Ti} is reduced proportionally. As described in Table 3.26 and earlier in Section 3.8.2.2, these effects are included by replacing IC in the g_m/I_D component of expressions with $IC = IC(1+IC/IC_{CRIT})$. IC_{CRIT} is the critical or corner inversion coefficient where velocity saturation and VFMR effects reduce g_m/I_D to approximately 70.7 % its value without these effects.

Substituting the velocity-saturated value of g_m in the numerator of the first expression in Equation 3.97 gives the velocity-saturated value of f_{Ti}. Substituting g_m, found by I_D multiplied by the velocity-saturated g_m/I_D expression in Table 3.17, gives

$$f_{Ti}(\text{full vel. sat.}) = \frac{\dfrac{(LE_{CRIT})' I_D}{(2nU_T)^2 IC}}{2\pi \left(\hat{C}_{gsi} + \hat{C}_{gbi}\right) \left(\dfrac{L^2}{IC} \dfrac{I_D}{I_0}\right) C'_{OX}}$$

$$= \frac{\dfrac{E_{CRIT}I_0}{(2nU_T)^2}}{2\pi \left(\hat{C}_{gsi} + \hat{C}_{gbi}\right) LC'_{OX}} = \frac{E_{CRIT}\mu_0}{4\pi \left(n - 1/3\right) L} \tag{3.99}$$

Equation 3.99 gives f_{Ti} in strong inversion under full velocity saturation where $IC \gg IC_{CRIT}$. As done for Equation 3.98, $I_0 = 2n_0\mu_0 C'_{OX}U_T^2$ and $n = n_0$ are substituted to simplify the final expression. Additionally, the strong inversion values of $\hat{C}_{gsi} = {}^2/_3$ and $\hat{C}_{gbi} = {}^1/_3(n-1)/n$ are substituted from Table 3.23. Finally, the equivalent velocity saturation voltage, inclusive of VFMR effects, $(LE_{CRIT})'$, is replaced with the velocity saturation voltage, LE_{CRIT}, as velocity saturation effects dominate under full velocity saturation. These voltages were described with the g_m/I_D expressions in Table 3.17 and were derived earlier in Section 3.7.1.5.

When velocity saturation effects are significant for short-channel devices in strong inversion, f_{Ti} increases as less than $\sqrt{IC}$, reaching a constant value under full velocity saturation as seen by Equation 3.99. In addition to the increase in V_{EFF} (Section 3.7.2.2) and decrease in g_m/I_D (Section 3.8.2.2), the leveling or flattening of f_{Ti} is a penalty of operation when velocity saturation is significant. Under full velocity saturation, f_{Ti} decreases as $1/L$, compared to decreasing as $1/L^2$

when velocity saturation effects are not significant. Under full velocity saturation, f_{Ti} is proportional to the product of low-field mobility, μ_0, and the critical, velocity saturation, electric field, E_{CRIT}. Under full velocity saturation, f_{Ti} is also proportional to the carrier saturation velocity, v_{sat}, which itself is proportional to the product of μ_0 and E_{CRIT}. Table 3.26 summarizes Equation 3.99 and trends for velocity-saturated f_{Ti} along with the expressions and trends mentioned earlier that exclude velocity saturation and VFMR effects.

Figure 3.57 shows predicted f_{Ti} for $L = 0.18, 0.28, 0.38, 0.48, 1, 1.4, 2, 2.8$, and $4\,\mu$m, nMOS devices in the $0.18\,\mu$m process described in Table 3.2 operating at IC ranging from 0.01 (deep weak inversion) to 100 (deep strong inversion). f_{Ti} is found from the first expression in Table 3.26 using the modification described in the table to include velocity saturation and VFMR effects. $U_T = 25.9\,$mV $(T = 300\,$K$)$ and $n = 1.33$, which is held fixed and equal to the value used for g_m/I_D shown in Figure 3.26, are used along with process values of $\mu_0 = 422\,$cm^2/V $\cdot$ s, $E_{CRIT} = 5.6\,$V/μm, and $\theta = 0.28$/V from Table 3.2. Channel length is reduced by $DL = 0.028\,\mu$m (also given in Table 3.2) from the values given to include lateral diffusion effects.

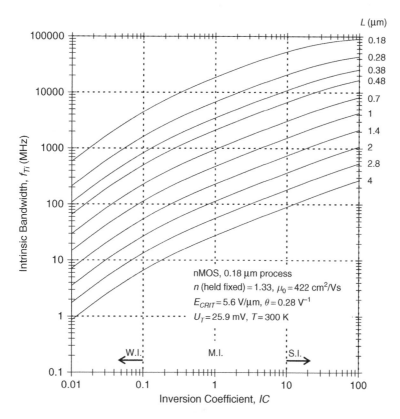

Figure 3.57 Intrinsic bandwidth, f_{Ti}, versus inversion coefficient, IC, for $L = 0.18, 0.28, 0.38, 0.48, 0.7, 1, 1.4, 2, 2.8$, and $4\,\mu$m, nMOS devices in a $0.18\,\mu$m CMOS process. In weak inversion, f_{Ti} increases directly with IC before increasing as $\sqrt{IC}$ in strong inversion and leveling off for short channel lengths at high IC due to velocity saturation. f_{Ti} decreases as $1/L^2$, except for short channel lengths at high IC where it decreases as $1/L$ due to velocity saturation. f_{Ti} is proportional to the mobility and is similar for all nMOS processes having similar mobility. f_{Ti} is approximately a factor-of-five lower for pMOS devices in the $0.18\,\mu$m process because of lower mobility, but is reduced less for short channel lengths at high IC because of less pMOS velocity saturation

As described for the expressions given earlier, f_{Ti} increases directly with IC in weak inversion and increases as $\sqrt{IC}$ in strong inversion before leveling off for short channel lengths at high IC due to velocity saturation. Under significant velocity saturation, f_{Ti} can actually roll down slightly with increasing IC because of combined velocity saturation, VFMR, and source resistance degeneration effects. f_{Ti} increases significantly as L decreases in all regions of operation, with less increase present for short channel lengths at high IC due, again, to velocity saturation effects. f_{Ti} shown in Figure 3.57 extends over five orders of magnitude from less than 1 MHz to nearly 100000 MHz illustrating the wide range available to the designer through the choice of IC and L. As discussed in the next section, extrinsic gate-overlap and the drain–body capacitances significantly reduce operating bandwidth from f_{Ti}, especially for short-channel devices that have small gate area, small gate-oxide capacitance, and, correspondingly, small intrinsic gate capacitances.

Predicted f_{Ti} for pMOS devices in the 0.18 μm process is approximately a factor-of-five less than that shown in Figure 3.57 for nMOS devices. As mentioned, this is due to the approximate factor-of-five lower pMOS mobility. pMOS f_{Ti}, however, is reduced less for short channel lengths at high IC because of less velocity saturation reduction of g_m/I_D and g_m. This is a result of higher $E_{CRIT} = 14$ V/μm for pMOS devices compared to 5.6 V/μm for nMOS devices.

Figure 3.58 presents a three-dimensional display of f_{Ti} from the values shown in Figure 3.57. Looking along the IC axis reveals that f_{Ti} increases directly with IC in weak inversion before increasing as $\sqrt{IC}$ in strong inversion, and then leveling off for short channel lengths due to velocity saturation. Looking along the reversed L axis reveals the significant increase in f_{Ti} as L decreases, where f_{Ti} is

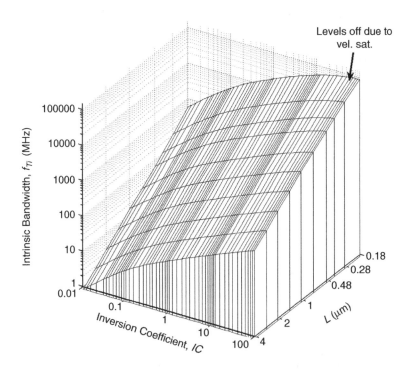

Figure 3.58 Intrinsic bandwidth, f_{Ti}, versus inversion coefficient, IC, and channel length, L, displayed in three dimensions from the data in Figure 3.57. f_{Ti} follows the trends described in Figure 3.57, being maximum at high IC and short L, and minimum at low IC and long L. Comparing f_{Ti} with the intrinsic voltage gain, A_{Vi}, shown in Figure 3.52 shows the opposing tradeoffs of MOS gain and bandwidth

proportional to $1/L^2$. The f_{Ti} increase is smaller for short channel lengths at high IC, where f_{Ti} is proportional to $1/L$ under full velocity saturation.

Figure 3.58 shows f_{Ti} over the design choices of IC and L where f_{Ti} is maximized at high IC and short L, and is minimized at low IC and long L. Intrinsic voltage gain, A_{Vi}, shown earlier in Figure 3.52 in Section 3.8.5, behaves oppositely. It is maximized at low IC and long L, and minimized at high IC and short L. Comparing f_{Ti} and A_{Vi}, presented in three dimensions in Figures 3.58 and 3.52 or presented in two dimensions for specific channel lengths in Figures 3.57 and 3.51, shows opposing gain and bandwidth tradeoffs available through the selection of IC and L. Operating at a channel length somewhat above the process minimum in moderate inversion provides one potential tradeoff of gain and bandwidth. Tradeoffs in MOS performance are discussed later in Chapter 4.

f_{Ti} listed in Table 3.26, when evaluated in terms of the inversion coefficient, has a universal characteristic across processes. This is expected since the intrinsic gate capacitances, described in Section 3.9.2, and g_m/I_D, described in Section 3.8.2.5, have universal characteristics. As mentioned in Section 3.2.4, the development of process-independent design expressions is one of the advantages of using the inversion coefficient in analog CMOS design. As mentioned, f_{Ti} is independent of C'_{OX}, while depending on the low-field mobility, μ_0, and substrate factor, n. Additionally, when velocity saturation effects are significant, the decrease in f_{Ti} depends on the critical, velocity saturation, electric field, E_{CRIT}, and the channel length. The smaller decrease associated with VFMR depends on the process through the mobility reduction factor, θ. f_{Ti} shown in Figures 3.57 and 3.58 is similar for nMOS devices in all processes having similar μ_0, n, and E_{CRIT}, and is lower for pMOS devices, proportional to the decrease in μ_0. The reduction of pMOS f_{Ti} for short-channel devices at high levels of inversion, however, is less because of less velocity saturation reduction of g_m/I_D associated with higher E_{CRIT}. Since f_{Ti} increases significantly as channel length decreases, continued increases in f_{Ti} can be expected in smaller-geometry processes.

3.9.7 Extrinsic and Diode-Connected Bandwidths

While intrinsic bandwidth, f_{Ti}, discussed in the previous section is a traditional and useful measure of MOS bandwidth, it considers capacitive loading from the intrinsic gate–source, C_{gsi}, and gate–body, C_{gbi}, capacitances, while excluding loading from the extrinsic gate–source, C_{GSO}, and gate–body, C_{GBO}, overlap capacitances. These overlap capacitances are included in the MOS small-signal model shown earlier in Figure 3.21 in Section 3.8.1 and can significantly lower the bandwidth measure, which, again, is the frequency where gate-to-drain current gain is unity for a grounded-source device with its drain connected to a small-signal ground.

Here, we define an extrinsic bandwidth, f_T, that includes the extrinsic C_{GSO} and C_{GBO} gate-overlap capacitances along with the intrinsic C_{gsi} and C_{gbi} gate capacitances considered in f_{Ti}. Additionally, we define a diode-connected bandwidth, f_{diode}, which is the pole frequency associated with a diode-connected (gate connected to the drain) device with the source grounded. For a diode-connected device, a small-signal resistance level of $1/g_m$(g_{ds} is neglected since it is usually much smaller than g_m) appears in parallel with capacitances connected between the common gate and drain connection and ground: C_{gsi}, C_{gbi}, C_{GSO}, C_{GBO}, and C_{DB} [11]. Since it considers the full capacitive loading of the gate and drain, f_{diode} corresponds more closely to the $-3\,\text{dB}$ bandwidth associated with a diode-connected device, which is similar to bandwidth at the source of a source follower or grounded-gate device. f_T and f_{diode} can be significantly less than f_{Ti}, especially for short-channel devices where small gate area results in small gate-oxide capacitance and, correspondingly, small intrinsic gate capacitances.

Table 3.27 gives expressions for f_T and f_{diode}, which are equal to Equation 3.95 for f_{Ti}, with the addition of C_{GSO} and C_{GBO} for f_T, and the addition of C_{GSO}, C_{GBO}, and C_{DB} for f_{diode}. The table also gives f_T/f_{Ti} and f_{diode}/f_{Ti}, which are useful to assess the loss of extrinsic and diode-connected bandwidth compared to the intrinsic bandwidth. Since the g_m term is equal in all bandwidth expressions, f_T/f_{Ti} and f_{diode}/f_{Ti} are simply found by the inverse ratio of loading capacitances associated with

Table 3.27 MOS extrinsic bandwidth, f_T, and diode-connected bandwidth, f_{diode}, expressions and trends in terms of inversion coefficient and channel length. When evaluated in terms of the inversion coefficient, f_T and f_{diode} are independent of the drain current

Expression	$IC \uparrow$ L fixed	$L \uparrow$ IC fixed
Extrinsic bandwidth: $$f_T = \frac{g_m}{2\pi \left(C_{gsi} + C_{gbi} + C_{GSO} + C_{GBO} \right)}$$	Similar to f_{Ti} trends described in Table 3.26	
Ratio of f_T to f_{Ti} (notes[a,c]): $$\frac{f_T}{f_{Ti}} = \frac{C_{gsi} + C_{gbi}}{C_{gsi} + C_{gbi} + C_{GSO} + C_{GBO}}$$ $$\approx \frac{\hat{C}_{gsi} + \hat{C}_{gbi}}{\hat{C}_{gsi} + \hat{C}_{gbi} + \dfrac{C_{GSO}}{C_{GOX}}}$$	$\uparrow$ approaching unity as L increases; increases slightly with IC; see Figure 3.59	
Diode-connected bandwidth: $$f_{diode} = \frac{g_m}{2\pi \left(C_{gsi} + C_{gbi} + C_{GSO} + C_{GBO} + C_{DB} \right)}$$	Similar to f_{Ti} trends described in Table 3.26	
Ratio of f_{diode} to f_{Ti} (notes[b,c]): $$\frac{f_{diode}}{f_{Ti}} = \frac{C_{gsi} + C_{gbi}}{C_{gsi} + C_{gbi} + C_{GSO} + C_{GBO} + C_{DB}}$$ $$\approx \frac{\hat{C}_{gsi} + \hat{C}_{gbi}}{\hat{C}_{gsi} + \hat{C}_{gbi} + \dfrac{C_{GSO}}{C_{GOX}} + \dfrac{C_{DBO}}{C_{GOX}}}$$	$\uparrow$ approaching unity as L increases; increases slightly with IC; see Figure 3.59	

C_{gsi}, C_{gbi}, $\hat{C}_{gsi}$, and $\hat{C}_{gbi}$ are from Table 3.23, C_{GSO} and C_{GSO}/C_{GOX} are from Table 3.24, and C_{DB}, C_{DBO}, and C_{DBO}/C_{GOX} are from Table 3.25.

[a] Approximation neglects C_{GBO}.

[b] Approximation neglects C_{GBO} and assumes C_{DB} at its maximum value of C_{DBO}, corresponding to $V_{DB} = 0\,\text{V}$.

[c] Approximation neglects bias voltage effects on C_{GSO} if this is found from Table 3.24, which assumes average or worst-case capacitance through the choice of process overlap capacitance parameter, $CGSO$.

each bandwidth measure. The table also gives trends for the design choices of inversion coefficient and channel length.

Each capacitance term in the f_T/f_{Ti} and f_{diode}/f_{Ti} expressions can be normalized to the gate-oxide capacitance, C_{GOX}, by dividing by this. This gives normalized, intrinsic gate–source and gate–body capacitance terms, $\hat{C}_{gsi}$ and $\hat{C}_{gbi}$, and C_{GSO}/C_{GOX}, C_{GBO}/C_{GOX}, and C_{DB}/C_{GOX}. As discussed in Section 3.9.2, $\hat{C}_{gsi}$ and $\hat{C}_{gbi}$ depend on IC and are independent of the drain current. As discussed in Section 3.9.3, C_{GSO}/C_{GOX} depends on L and is also independent of the drain current. However, C_{GBO}/C_{GOX} depends on W and thus depends on the drain current. Here, we neglect C_{GBO}/C_{GOX} to permit f_T/f_{Ti} and f_{diode}/f_{Ti} expressions that are independent of the drain current. Finally, as discussed in Section 3.9.4, C_{DB}/C_{GOX} depends on L and is independent of the drain current. To eliminate V_{DB} dependency in the C_{DB}/C_{GOX} term, we assume $V_{DB} = 0\,\text{V}$, giving the maximum value of $C_{DB}/C_{GOX} = C_{DBO}/C_{GOX}$.

Using the normalized $\hat{C}_{gsi}$, $\hat{C}_{gbi}$, C_{GSO}/C_{GOX}, and C_{DBO}/C_{GOX} capacitance terms mentioned above permits the approximate expressions for f_T/f_{Ti} and f_{diode}/f_{Ti} given in Table 3.27 that, like the expressions for f_{Ti} in Table 3.26, depend on IC and L, while being independent of drain current. As mentioned, drain current independence is possible by assuming that C_{GBO} is negligible. Additionally, V_{DB} independence is possible by assuming that C_{DB} is at its maximum value of C_{DBO} for $V_{DB} = 0$ V. These assumptions permit initial design guidance, while the effects of C_{GBO} and non-zero V_{DB} would be included in simulations of a candidate design using complete MOS models.

Figure 3.59 shows predicted f_T/f_{Ti} and f_{diode}/f_{Ti} as a function of channel length for $IC = 0.1$ (boundary of weak inversion), $IC = 1$ (center of moderate inversion), and $IC = 10$ (onset of strong inversion) for nMOS devices in the 0.18 μm CMOS process described in Tables 3.2 and 3.4. These bandwidth reductions are calculated from the expressions given in Table 3.27, using the normalized, intrinsic gate capacitances, $\hat{C}_{gsi}$ and $\hat{C}_{gbi}$, given in Figure 3.54, and the normalized gate–source overlap, C_{GSO}/C_{GOX}, and drain–body, C_{DBO}/C_{GOX}, capacitances given in Figure 3.55.

As shown in Figure 3.59, f_T/f_{Ti} and f_{diode}/f_{Ti} decrease well below unity for the minimum channel length where C_{GSO}/C_{GOX} and C_{DBO}/C_{GOX} shown in Figure 3.55 are maximum, corresponding to maximum C_{GSO} and G_{DBO} contributions. At long channel lengths, f_T/f_{Ti} and f_{diode}/f_{Ti} approach unity corresponding to minimum C_{GSO} and G_{DBO} contributions. While the channel length affects f_T/f_{Ti} and f_{diode}/f_{Ti} more, these are lower in weak inversion where the sum of $\hat{C}_{gsi} + \hat{C}_{gbi}$ shown in Figure 3.54 is smaller due to smaller $\hat{C}_{gsi}$. For the minimum channel length at $IC = 1$, f_T/f_{Ti} and f_{diode}/f_{Ti} are 45 % and 34 %, respectively, corresponding to f_T and f_{diode} equal to 45 % and 34 %, respectively, of f_{Ti} shown in Figure 3.57. As expected, the bandwidth reduction in f_{diode}, which includes C_{DBO} and C_{GSO}, is greater than the reduction in f_T, which includes C_{GSO}. f_{diode}/f_{Ti}, like C_{DBO}/C_{GOX} shown in

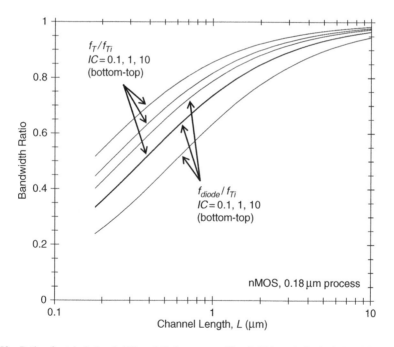

Figure 3.59   Ratio of extrinsic bandwidth and diode-connected bandwidth to the intrinsic bandwidth, f_T/f_{Ti} and f_{diode}/f_{Ti}, versus channel length for nMOS devices in a 0.18 μm CMOS process. For short-channel devices, f_T and f_{diode} are well below f_{Ti} because extrinsic gate–source overlap and drain–body capacitances are significant compared to small intrinsic gate capacitances associated with small gate area. f_T and f_{diode} are reduced further below f_{Ti} in weak inversion where intrinsic gate–source capacitance is small. f_{diode}/f_{Ti} shown assumes a device having even interdigitation with the drain area and perimeter minimized

Figure 3.55, assumes an evenly interdigitated device ($m = 2, 4, 6, 8$, etc.) where the drain perimeter and area are minimized. If interdigitation is not used ($m = 1$), C_{DBO} is twice the value considered, reducing f_{diode}/f_{Ti} and the associated value of f_{diode}.

Figure 3.59 shows the bandwidth reduction from f_{Ti} present in a single device as a result of the extrinsic capacitances, C_{GSO} and C_{DBO}. The bandwidth reduction is greater for multiple devices, for example for a unity-gain current mirror where two units of C_{gsi}, C_{gbi}, C_{GSO}, and C_{GBO} (again, excluded in Figure 3.59) load the circuit from the gate of input and output devices along with one unit of C_{DB} from the drain of the input device alone. As a rough estimate, the $-3\,\text{dB}$ frequency associated with the input impedance of a unity-gain current mirror is $f_{diode}/2$, which assumes an additional unit of C_{DB} not present at the input while neglecting C_{GBO} from both input and output devices [11]. Layout capacitances, of course, reduce operating bandwidth further.

As mentioned in the previous section, f_{Ti} has a universal characteristic across CMOS processes. f_T and f_{diode}, however, depend on f_{Ti} and the extrinsic, gate-overlap and drain–body capacitances described in Sections 3.9.3 and 3.9.4. These depend strongly on the process, giving a strong process dependence on f_T and f_{diode}. Extrinsic device and layout capacitances can be expected to continue to lower bandwidth from the intrinsic bandwidth in smaller-geometry processes.

3.10 NOISE

The prediction and trends of MOS noise are important since noise determines the minimum AC signal that can be processed by an analog circuit. This section describes MOS thermal noise in both the ohmic and saturation regions. Following this, low-frequency flicker noise is described, including a review of reported flicker-noise factors. The section then describes channel avalanche noise and thermal noise from the gate, substrate, and source resistances. Finally, the section concludes by describing induced gate noise current caused by channel thermal noise, and gate noise current caused by gate resistance noise and DC leakage current.

Figure 3.60 shows MOS drain-referred noise current and gate-referred noise voltage sources that are added to the small-signal model shown earlier in Figure 3.21 in Section 3.8.1. As described in this section, the sources model thermal and low-frequency flicker noise where the drain-referred or gate-referred sources are used separately for circuit analysis. Figure 3.60 describes the conversion

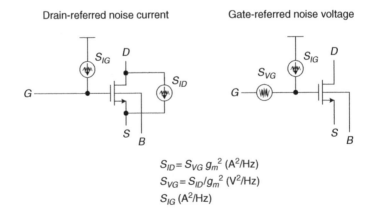

$$S_{ID} = S_{VG}\, g_m{}^2 \; (\text{A}^2/\text{Hz})$$
$$S_{VG} = S_{ID}/g_m{}^2 \; (\text{V}^2/\text{Hz})$$
$$S_{IG} \; (\text{A}^2/\text{Hz})$$

Figure 3.60 MOS noise model showing drain-referred noise current and gate-referred noise voltage sources along with a gate noise current source. The drain-referred and gate-referred noise sources are used separately and are added to the small-signal model shown in Figure 3.21 along with the gate noise current source that is always present. Expressions are given showing the PSD conversion between drain-referred and gate-referred noise sources

between drain-referred and gate-referred noise sources and also shows a gate noise current source that is always connected to the gate. As standard for noise sources, the noise sources are characterized using the power spectral density (PSD) or noise power in a 1Hz bandwidth.

3.10.1 Thermal Noise in the Ohmic Region

Since MOS devices operating in the linear, triode, or ohmic region ($V_{DS} < V_{DS,sat}$) are often used as resistors, resistor noise will be reviewed first. Einstein predicted that Brownian motion of charge carriers would lead to a fluctuation of voltage across a resistance in thermal equilibrium [47]. Resistor thermal noise is caused by random thermal motion of carriers and is sometimes called Johnson noise, since Johnson first observed and reported it [48]. Thermal noise is also sometimes called white noise because of its constant or flat PSD with frequency. However, thermal noise is a better descriptor because it describes the physical cause of the noise and differentiates it from, for example, shot noise that is also white noise.

Nyquist [49] originally analyzed resistor thermal noise where the noise current PSD is given by

$$S_I(R) = \frac{4kT}{R} \tag{3.100}$$

k is Boltzmann's constant, T is the absolute temperature (K), and R is the resistance value. The noise current PSD has units of A^2/Hz, which is the mean square current appearing in a 1 Hz bandwidth. This noise current appears in parallel with a noiseless resistance having the value of R, creating a Norton equivalent circuit.

Resistor thermal noise can also be expressed by a Thévenin equivalent circuit as a noise voltage in series with a noiseless resistor, again, having the value of R. Here, the noise voltage PSD is given by

$$S_V(R) = 4kT \cdot R \tag{3.101}$$

The noise voltage PSD has units of V^2/Hz, which is the mean square voltage appearing in a 1 Hz bandwidth.

Table 3.28 summarizes resistor thermal-noise expressions, including convenient expressions motivated by [12, p. 51]. While the noise of uncorrelated noise sources is conveniently added using PSD expressions, designers usually specify noise using the square root of PSD. Taking the square root of the convenient PSD expressions in the table reveals that the noise voltage density of a $1\,k\Omega$ resistor at $T = 300\,K$ (room temperature) is $S_V^{1/2} = 4.07\,nV/Hz^{1/2}$ and the noise current density is $S_I^{1/2} = 4.07\,pA/Hz^{1/2}$. Noise voltage PSD increases directly with R while noise voltage density increases as $\sqrt{R}$. Noise current PSD decreases as $1/R$ while noise current density decreases as $1/\sqrt{R}$. Finally, both voltage and current noise PSD increase directly with T, while both voltage and current noise density increase as $\sqrt{T}$.

As noted at the bottom of Table 3.28, the drain-referred thermal-noise current PSD, S_{ID} shown in Figure 3.60, for a deep ohmic ($V_{DS} \ll V_{DS,sat}$) MOS device acting as a resistor between the drain and source can be found from resistor thermal-noise expressions. Here, the deep ohmic, drain–source resistance, $r_{ds} = 1/g_{ds}$, is assigned to R in the expressions. MOS flicker noise, described later, appears as an added noise current at the drain having PSD given by $S_{VG}(\text{flicker}) \cdot g_m^2$, where $S_{VG}(\text{flicker})$ is the gate-referred flicker-noise voltage PSD and g_m is the device transconductance. Often, the flicker noise is negligible compared to the thermal noise because of the collapsed g_m of a deep ohmic device. This is described later in Section 6.5.5.3 for the design of micropower, low-noise CMOS preamplifiers.

The collapse of g_m for devices operating in the ohmic region resulted in serious noise modeling errors in some early SPICE MOS models where the thermal noise was erroneously linked to g_m. This resulted in the appearance of seemingly noiseless resistors in designs where devices operating in the deep ohmic region were used for resistive noise degeneration at the sources of devices operating in

Table 3.28   Resistor and ohmic MOSFET thermal-noise expressions. Voltage density $(nV/Hz^{1/2})$ and current density $(pA/Hz^{1/2})$ expressions are found by taking the square root of the convenient PSD expressions

Expression	Convenient expression
Thermal-noise voltage PSD appearing in series with resistor:	
$S_V = 4kT \cdot R$	$S_V = \dfrac{(4.07\,\text{nV})^2}{\text{Hz}} \left(\dfrac{T}{300\ \text{K}}\right)\left(\dfrac{R}{1\,\text{k}\Omega}\right)$
Thermal-noise current PSD appearing across resistor:	
$S_I = \dfrac{4kT}{R}$	$S_I = \dfrac{(4.07\,\text{pA})^2}{\text{Hz}} \left(\dfrac{T}{300\ \text{K}}\right)\left(\dfrac{1\,\text{k}\Omega}{R}\right)$

For ohmic MOS resistors $(V_{DS} < V_{DS,sat})$:

$$\frac{1}{R} = \frac{1}{r_{ds}} = g_{ds} = \frac{\mu_0 C'_{OX}}{1 + \theta V_{EFF}}\left(\frac{W}{L}\right)(V_{EFF} - V_{DS}),\ \text{where } V_{EFF} = V_{GS} - V_T$$

Near-ideal resistor behavior occurs in the deep ohmic region where $V_{DS} \ll V_{DS,sat}$. Here, MOS drain-referred thermal-noise current PSD, S_{ID}, is equal to S_I above with $R = r_{ds} = 1/g_{ds}$. Flicker noise is considered separately by multiplying gate-referred flicker-noise voltage PSD, S_{VG} (flicker), by g_m^2. Flicker noise is often negligible for deep ohmic MOS resistors because g_m is collapsed to a low value. Convenient expressions assume $k = 1.3806 \times 10^{-23}\,\text{J/K}$.

saturation. This modeling error is avoided by using expressions similar to Equation 3.104 described later where noise is linked to the total source conductance, $g_{ms} = g_m + g_{mb} + g_{ds}$, for both the ohmic and saturation regions. MOS noise modeling is complicated by the dual dimensions of inversion level, related to the inversion coefficient or $V_{EFF} = V_{GS} - V_T$, and operation in the ohmic or saturation regions, related to V_{DS} and $V_{DS,sat}$. Noise models can be verified by simulating test devices from weak through strong inversion in both the ohmic and saturation regions.

3.10.2 Thermal Noise in the Saturation Region

3.10.2.1 Without small-geometry effects

Thermal noise has been studied in field-effect transistors since the early 1960s [50–57], and research has continued in recent years [58–80]. MOS thermal noise is analyzed by integrating the noise associated with a differential resistive element over the full channel length [13, 15, pp. 414–418, 54, 58, p. 75, 80]. This leads to the expression for MOS drain-referred thermal-noise PSD, S_{ID} shown in Figure 3.60, given by

$$S_{ID}(\text{thermal}) = 4kT \cdot \frac{\mu_{eff}}{L^2} Q_{inv} \qquad (3.102)$$

Here, μ is assumed constant at μ_{eff}, carrier temperature is assumed constant at T, L is the physical, effective channel length, and Q_{inv} is the total inversion layer charge, which is taken positively for both nMOS and pMOS devices. For operation in saturation $(V_{DS} > V_{DS,sat})$, Equation 3.102 can be written similar to how it is given in [13] as

$$S_{ID}(\text{thermal}) = 4kT \cdot \Gamma (g_m + g_{mb}) = 4kT \cdot (n\,\Gamma)\,g_m \qquad (3.103)$$

where n is taken as $(g_m + g_{mb})/g_m$ or $1 + g_{mb}/g_m$ (this approximation is consistent with $g_{mb}/g_m = n - 1$ from Table 3.20) and Γ is the thermal-noise factor for operation in saturation. Γ is used here to distinguish it from the body-effect factor, γ. Neglecting small-geometry effects including velocity saturation, carrier heating, VFMR, and CLM described in the next section, Γ is one-half in weak inversion, rising to two-thirds in strong inversion.

It is important to distinguish our use of Γ, which relates MOS thermal noise in saturation to operating g_m, from other noise factors that relate the noise to g_{do}, the value of g_{ds} at $V_{DS} = 0\,\mathrm{V}$ [58, p. 76]. As described in [79], MOS thermal noise in saturation is best related to operating g_m since g_{do} is not relevant for operation in saturation (g_{do} is relevant for operation in the deep ohmic region where $V_{DS} \ll V_{DS,sat}$). In [79], thermal noise is related to g_m by γ_{sat}, which is equal to $n \cdot \Gamma$ used here.

MOS drain-referred thermal-noise current PSD can be expressed in both the saturation and ohmic regions by [13]

$$S_{ID}(\text{thermal, saturation and ohmic}) = 4kT \cdot \Gamma \left(g_m + g_{mb} + g_{ds} \right) = 4kT \cdot \Gamma g_{ms} \qquad (3.104)$$

where the total source conductance, g_{ms}, is given by $g_{ms} = g_m + g_{mb} + g_{ds}$. In saturation, g_{ds} is negligible and Equation 3.104 simplifies to the thermal noise given by Equation 3.103. In the deep ohmic region where $V_{DS} \ll V_{DS,sat}$, g_{ds} is dominant and Equation 3.104 with $\Gamma = 1$ simplifies to the thermal noise of a resistor given by Equation 3.100 where $R = 1/g_{ds}$. Equation 3.104 predicts thermal noise from weak through strong inversion, as Equation 3.103 does for saturation, but additionally predicts thermal noise continuously across the ohmic and saturation regions. This requires an interpolation of Γ from weak through strong inversion across both the ohmic and saturation regions, where Γ approaches unity in the deep ohmic region [13]. Here, we normally consider operation in saturation and use Equation 3.103. When devices are used in the deep ohmic region as resistors, we use Equation 3.100.

Substituting the transconductance in weak inversion, $g_m = I_D/(nU_T) = I_D/(nkT/q)$, and the thermal noise factor in weak inversion, $\Gamma = 0.5$, into Equation 3.103 yields the interesting result

$$S_{ID} \,(\text{thermal, WI}) = 4kT \cdot n \tfrac{1}{2} g_m = 4kT \cdot n \tfrac{1}{2} \frac{I_D}{n\,(kT/q)} = 2qI_D \qquad (3.105)$$

Equation 3.105 shows that the weak-inversion, drain thermal-noise current is equal to the full shot noise associated with the DC drain current [13, 15, p. 421, 62]. Shot noise, having a constant or white-noise current PSD of $2qI$, results from the random arrival of discrete charges across a junction where I is the average current. As inversion increases from weak inversion, drain thermal-noise current falls progressively below that predicted by drain current shot noise.

Γ, again neglecting small-geometry effects, increases smoothly from one-half in weak inversion to two-thirds in strong inversion for operation in saturation. As required for g_m/I_D, an interpolation expression is required for Γ to permit design freely in all regions of MOS operation. Several expressions have been reported [13, 61, 67], with the simplest given by [13]

$$\Gamma = \frac{1}{1+IC} \left(\tfrac{1}{2} + \tfrac{2}{3}IC \right) \qquad (3.106)$$

Another expression is given by [61]

$$\Gamma = \tfrac{1}{2} + \tfrac{1}{6} \frac{IC}{\left(\sqrt{IC} + 0.5\sqrt{IC} + 1 \right)^2} \qquad (3.107)$$

and another is given by [67]

$$\Gamma = \tfrac{2}{3} \left(1 + \frac{1 - \sqrt{4IC+1}}{8IC} \right) \qquad (3.108)$$

Γ given by Equations 3.106–3.108 is based on the traditional inversion coefficient, taken in this book as IC' and described in Section 3.4.2.1, where $V_{EFF} = V_{GS} - V_T$ is linked to IC' by Equation 3.18 given in Section 3.7.2.1. Γ may increase more rapidly towards its constant strong-inversion value of two-thirds when evaluated in terms of the fixed–normalized inversion coefficient, IC, used here and described in Section 3.4.2.2. This is a result of the additional increase in V_{EFF} when evaluated in terms of IC, resulting primarily from velocity saturation for short-channel devices with a slight increase due to VFMR for all channel lengths as described in Section 3.7.2.2. However, the V_{EFF} increase occurs primarily in strong inversion where Γ is near its constant value of two-thirds and is expected to have little effect. More important are separate increases in Γ due to small-geometry effects that are described in the next section.

Figure 3.61 shows Γ in saturation predicted by Equations 3.106–3.108. All predictions show that Γ increases smoothly from one-half in weak inversion to two-thirds in strong inversion. Equation 3.106 is, again, the simplest prediction, while also giving the largest Γ increase in moderate inversion. Equation 3.107 involves the square of the improved accuracy g_m/I_D interpolation factor appearing in Equation 3.35, which suggests another prediction of Γ is available using the simple g_m/I_D interpolation factor from Equation 3.36. Equation 3.108 gives the most gradual Γ transition from weak to strong inversion.

In this book, we use Γ predicted by Equation 3.106 shown in Figure 3.61 because the more rapid increase in moderate inversion provides a conservatively higher estimate of thermal noise. Using this estimate, Γ is equal to 0.515, 0.583, and 0.652 for $IC = 0.1$, 1, and 10, respectively, corresponding to the onset of weak inversion, the center of moderate inversion, and the onset of strong inversion. Interestingly, at $IC = 1$ where the weak-inversion and strong-inversion values of transconductance are

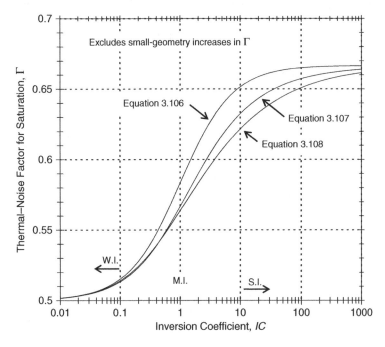

Figure 3.61 Predicted thermal-noise factor, Γ, for operation in saturation versus the inversion coefficient, IC. Three predictions of Γ are shown, with the one given by Equation 3.106 rising most rapidly in moderate inversion. Γ is one-half in weak inversion, increasing to two-thirds in strong inversion. The predictions exclude increases associated with small-geometry effects described in Section 3.10.2.2

equal as described in Sections 3.4.2.1 and 3.8.2.1, $\Gamma = 0.583$ given by Equation 3.106 is the average of the weak-inversion value of one-half and the strong-inversion value of two-thirds. Comparisons of measured thermal noise with that predicted using Γ from Equations 3.106–3.108 would be required to suggest which equation provides the best prediction. Equation 3.106 provides an adequate prediction for design guidance where, as always, simulations of a candidate design are required using full MOS models.

3.10.2.2 With small-geometry effects

Increases in thermal noise above the expected large-geometry values, also referred to as excess thermal noise, have been discussed in the literature, often with experimental measurements [15, p. 424, 16, pp. 198–212, 58, p. 79, 59, 60, 63–80]. Early measurements indicate that thermal-noise PSD increases by as much as a factor of 10 above the expected large-geometry value for a $0.7\,\mu m$ process [60]. Later measurements for 0.7 [63, 69], 0.65 [66], 0.35 [71], 0.25 [70, 72], and $0.18\,\mu m$ [70] processes also indicate significant thermal noise increases, generally a factor of two or greater. Recent measurements, however, for a $0.18\,\mu m$ process indicate modest thermal-noise increases, generally less than a factor of two [74, 76, 77, 78]. All measurements suggest little thermal-noise increase for long-channel devices. Increases in thermal noise were originally believed due to carrier heating associated with high carrier velocity for short-channel devices operating at high levels of inversion. However, recent research discussed below describes important additional components of noise increase and suggests that little noise is generated in the region where carriers are velocity saturated.

Possible reasons for differences in excess thermal-noise measurements could include the high-flicker-noise corner frequency (discussed later in Section 3.10.3.8) present for short-channel nMOS devices operating at high levels of inversion. This requires high measurement bandwidth to ensure thermal noise alone is measured. Additionally, thermal noise from gate and substrate resistances and channel avalanche noise can significantly contribute to the measured noise. Finally, differing definitions of thermal-noise factors complicate the interpretation of research results [79]. Unfortunately, most reported data relates noise in saturation to g_{do} instead of operating g_m. As mentioned, for operation in saturation, thermal noise is best related to the operating g_m.

Recent research suggests that MOS channel thermal noise should be analyzed distinctly in two regions: a gradual channel region near the source and a velocity-saturated region near the drain [16, pp. 198–212, 63, 69, 70, 74, 76, 77, 79, 80]. In [74], it is postulated that carriers traveling at the saturation velocity will not respond to changes in the local electric field caused by noise voltage fluctuation. As a result, little noise contribution might be expected in the velocity-saturated region. Noise in this region is neglected in much of the recent research including in [16, pp. 198–212, 79, 80].

CLM effects are believed to increase thermal noise for short-channel devices [74, 76, 77, 80]. This can be seen by replacing L in Equation 3.102 with $L = L_{elec}$, the electrical channel length including the reduction associated with the CLM effect. Additionally, Q_{inv} in Equation 3.102 is replaced with Q_{inv} present in the gradual channel region. A given inversion charge appearing across a smaller electrical channel length would increase thermal noise as validated experimentally in [74, 76, 77]. CLM effects suggest a larger increase in thermal noise for short-channel devices since L_{elec} is a smaller fraction of L for these devices.

In recent research, MOS thermal noise is analyzed where noise in the velocity-saturated region is neglected, but the interactions of velocity saturation, carrier heating, VFMR, and CLM are carefully considered in the gradual channel region [16, pp. 198–212, 79]. This results in some interesting findings. When thermal noise is evaluated relative to g_m, the noise of short-channel devices increases significantly due to the combination of velocity saturation, carrier heating, and CLM. However, VFMR lowers the noise to only modestly above its large-geometry value. This is because VFMR lowers the mobility and, correspondingly, lowers carrier velocity and heating. Additionally, the increased value of critical, velocity saturation, electric field, E_{CRIT}, associated with lower mobility lowers CLM as seen by a decreased value of $\delta l_p / \delta V_{DS}$ in Equation 3.69 when V_E is proportional to E_{CRIT}.

Measured data from nMOS and pMOS, $L = 0.18\,\mu$m devices in a $0.18\,\mu$m process indicates that thermal-noise PSD is up by about 20 % from its large-geometry value for high $V_{GS} = 1.7$ V values associated with deep strong inversion [77]. The increase, however, approaches 100 % for V_{GS} values associated with moderate inversion. The noise increase is much less for $L = 0.25\,\mu$m devices with little increase observed for channel lengths above this. Similar measured results for other $0.18\,\mu$m processes are also reported in [74, 76]. While these modest thermal-noise increases are encouraging, noise measurements were unfortunately evaluated relative to g_{do} instead of g_m, which is relevant for operation in saturation. As discussed in [79], thermal-noise increases can be much greater when evaluated relative to g_m when velocity saturation reduces g_m and the related thermal noise, as expected for $V_{GS} = 1.7$ V mentioned above. When thermal noise is evaluated relative to g_m including velocity saturation, carrier heating, VFMR, and CLM, it increases by about 100 % for $L = 0.18\,\mu$m devices in a $0.18\,\mu$m process over a wide range of inversion [79]. This corresponds to a factor-of-two increase in Γ or γ_{sat} ($\gamma_{sat} = n \cdot \Gamma$) given in [79].

At the time of this writing, small-geometry increases in MOS thermal noise remain an area of emerging research, although consensus is building that noise can be neglected in the velocity-saturated region. However, as discussed in [16, pp. 198–212, 79], velocity saturation, carrier heating, VFMR, and CLM must be considered simultaneously in the gradual channel region with noise compared to the operating g_m. Although we will use the large-geometry value given in Equation 3.106 and Figure 3.61 for graphs of predicted thermal noise, Γ will appear in all expressions and can be readily modified to consider small-geometry noise increases. As channel length decreases in smaller-geometry processes, thermal noise is expected to continue increasing above the large-geometry prediction. The reader is encouraged to consult the contemporary literature for the latest reported results and experimentally validate thermal noise for short-channel devices used in critical applications.

3.10.2.3 Summary of drain-referred and gate-referred thermal noise

Table 3.29 summarizes expressions for drain-referred thermal-noise current PSD, S_{ID}, shown in the MOS noise model of Figure 3.60. S_{ID} is expressed in terms of device g_m/I_D and g_m in the first expression from Equation 3.103 where the equivalent noise resistance, R_N, is inversely proportional to g_m. R_N corresponds to the resistance value giving thermal-noise current equal to that appearing at the MOSFET drain. S_{ID} decreases with increasing inversion level, tracking the decrease in g_m/I_D and g_m. It increases directly with drain current, tracking the increase in g_m. The table also gives a convenient expression for S_{ID} depending on g_m, n, Γ, and T.

Table 3.29 also gives expressions and trends for S_{ID} in terms of the design choices of IC, L, and I_D. Interestingly, when evaluated in terms of the inversion coefficient, S_{ID} is independent of temperature, assuming the inversion coefficient is held constant. This is best seen in weak inversion where $4kT$ increases directly with T while $g_m = I_D/(nU_T)$ decreases inversely with T through the increase in $U_T = kT/q$. S_{ID}, when evaluated in terms of the inversion coefficient, is also independent of n since n in the noise expression cancels with n in the denominator of the expression for g_m. The table also gives weak inversion, strong inversion without velocity saturation, and strong-inversion, velocity-saturated values of S_{ID}, factored in the usual $4kT$ noise resistance form. S_{ID} can also be factored with $2q \cdot I_D$ appearing at the beginning of the expressions. This shows that S_{ID} decreases from the full shot-noise value of $2q \cdot I_D$ in weak inversion to progressively smaller values in strong inversion.

As noted in Table 3.29, S_{ID} is unchanged at its maximum value in weak inversion, decreases as $1/\sqrt{IC}$ in strong inversion, and decreases as $1/IC$ under conditions of full velocity saturation. It is independent of L, unless short channel length results in velocity saturation reduction of g_m/I_D and g_m, which then decreases S_{ID}. As mentioned, S_{ID} increases directly with drain current. Each of these trends corresponds to the trends for g_m since S_{ID} directly tracks g_m. The noise expressions, which are based on g_m found by I_D multiplied by the g_m/I_D expressions given in Table 3.17, include velocity saturation and VFMR effects through the replacement of IC with $IC = IC(1 + IC/IC_{CRIT})$. As described earlier in Section 3.8.2.2, IC_{CRIT} is the critical or corner inversion coefficient where velocity saturation and

Table 3.29 MOS drain-referred thermal-noise current PSD expressions and trends in terms of the inversion coefficient, channel length, and drain current. Current density ($pA/Hz^{1/2}$) is found by taking the square root of the convenient PSD expressions

Drain-referred thermal-noise current PSD	$IC \uparrow$ L, I_D fixed	$L \uparrow$ IC, I_D fixed	$I_D \uparrow$ IC, L fixed
From g_m: $S_{ID} = 4kT \cdot$ $\quad n\Gamma g_m = 4kT\left(n\Gamma \cdot \dfrac{g_m}{I_D} \cdot I_D\right) = \dfrac{4kT}{R_N}$ $R_N = \dfrac{1}{n\Gamma g_m}$	$\downarrow$ as g_m/I_D decreases from maximum value in WI	Unchanged, but increases if less vel. sat. reduction of g_m/I_D	$\uparrow \propto I_D$

Convenient expression:
$$S_{ID} = \frac{(4.07\,pA)^2}{Hz}\left(\frac{T}{300K}\right) n\Gamma\left(\frac{g_m}{1000\,\mu S}\right)$$

From IC, L, I_D: $S_{ID} = 4kT\left(\dfrac{\Gamma I_D}{U_T\sqrt{IC + 0.25} + 0.5}\right)$ $\quad = 4kT\,[(1/2)I_D/U_T] = 2qI_D$ in WI $\quad = 4kT\,[(2/3)I_D/(U_T\sqrt{IC})]$ in SI without vel. sat. $\quad = 4kT\{[(LE_{CRIT})'/(6nU_T)]I_D/(U_T IC)\}$ in SI with full vel. sat.	Unchanged in WI $\downarrow \propto \dfrac{1}{\sqrt{IC}}$ in SI with no vel. sat. $\downarrow \propto \dfrac{1}{IC}$ in SI with full vel. sat.	Unchanged, but increases if less vel. sat. reduction of g_m/I_D $\uparrow \propto L$ in SI with full vel. sat.	$\uparrow \propto I_D$

Convenient expression:
$$S_{ID} = \frac{(8\,pA)^2}{Hz} \cdot \Gamma \frac{\left(\dfrac{I_D}{100\,\mu A}\right)}{\sqrt{IC + 0.25} + 0.5}$$

To include velocity saturation and VFMR decreases in g_m, replace IC with $IC(1 + IC/IC_{CRIT})$ as described in Table 3.17

S_{ID} is for saturation where $V_{DS} > V_{DS,sat}$.
Velocity saturation refers to both velocity saturation and VFMR effects.
Γ, given by Equation 3.106, increases from one-half in weak inversion to two-thirds in strong inversion, excluding small-geometry increases described in Section 3.10.2.2. IC and L trends exclude small-geometry increases in Γ. Convenient expressions assume $k = 1.3806 \times 10^{-23}$ J/K.

VFMR effects reduce g_m/I_D to approximately 70.7 % its value without these effects. Finally, the table includes a convenient expression for S_{ID} in terms of IC, L, I_D, and Γ. The square root of this expression gives drain-referred thermal-noise current density, $S_{ID}^{1/2}$, in its usual design units of $pA/Hz^{1/2}$.

Figure 3.62 shows predicted $S_{ID}^{1/2}$ in $pA/Hz^{1/2}$ for $L = 0.18$, 0.28, 0.48, and 1 μm, nMOS devices in the 0.18 μm CMOS process described in Table 3.2 operating at IC ranging from 0.01 (deep weak inversion) to 100 (deep strong inversion). Unlike the normalized small-signal parameters, g_m/I_D, $\eta = g_{mb}/g_m$, and $V_A = I_D/g_{ds}$, and intrinsic voltage gain and bandwidth, A_{Vi} and f_{Ti}, $S_{ID}^{1/2}$ is not independent of the drain current when evaluated in terms of the inversion coefficient. Here, a drain current of 100 μA is assumed, where the increase in $S_{ID}^{1/2}$ with increasing drain current resulting from increasing g_m is easily found by multiplying the values shown in the figure by $(I_D/100\,\mu A)^{1/2}$.

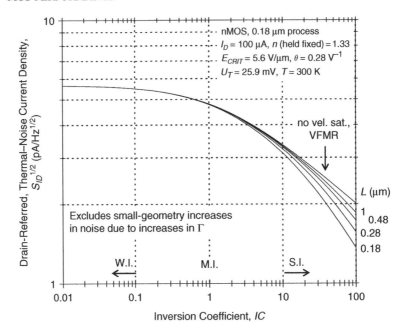

Figure 3.62 Predicted drain-referred thermal-noise current density, $S_{ID}^{1/2}$, versus inversion coefficient, IC, for $L = 0.18, 0.28, 0.48$, and $1\,\mu m$, nMOS devices in a $0.18\,\mu m$ CMOS process operating at a drain current of $100\,\mu A$. $S_{ID}^{1/2}$, which tracks the square root of g_m/I_D for constant drain current, decreases in strong inversion, decreasing more for short-channel devices because of velocity saturation. $S_{ID}^{1/2}$ is similar for all CMOS processes because of the fundamental g_m/I_D characteristic and is found for different drain currents by multiplying the values shown by $(I_D/100\,\mu A)^{1/2}$. The prediction shown excludes possible noise increases associated with small-geometry effects described in Section 3.10.2.2

$S_{ID}^{1/2}$ is predicted from the last expression in Table 3.29 using the modification described in the table to include velocity saturation and VFMR effects that decrease g_m and $S_{ID}^{1/2}$. $U_T = 25.9\,mV$ ($T = 300\,K$) and $n = 1.33$, which is held fixed and equal to the value used for g_m/I_D shown in Figure 3.26, are used along with process values of $E_{CRIT} = 5.6\,V/\mu m$ and $\theta = 0.28/V$ from Table 3.2. Channel length is reduced by $DL = 0.028\,\mu m$ (also given in Table 3.2) from the values given to include lateral diffusion effects. In addition to the channel lengths shown, $S_{ID}^{1/2}$ is also shown assuming velocity saturation and VFMR reductions of g_m/I_D and g_m are not present.

In Figure 3.62, $S_{ID}^{1/2}$, which tracks the square root of g_m/I_D shown in Figure 3.26 for constant drain current, decreases from its maximum weak-inversion, full shot-noise value as the inversion coefficient increases. The decrease is greater in strong inversion for short-channel devices because of velocity saturation decreases in g_m/I_D and g_m. The noise decrease in strong inversion is advantageous for non-input devices where drain noise current is divided by input device transconductance to find the contribution to input-referred noise voltage for amplifiers receiving voltage inputs. For these applications, non-input devices should be operated at high IC well into strong inversion, consistent with allowable $V_{DS,sat}$, and at minimum drain current. This is illustrated for $0.18\,\mu m$ CMOS operational transconductance amplifiers and micropower, low-noise, $0.35\,\mu m$, SOI CMOS preamplifiers later in Chapters 5 and 6, respectively.

The lowest values of $S_{ID}^{1/2}$ shown in Figure 3.62 for short-channel devices in deep strong inversion may not be attainable for $I_D = 100\,\mu A$ shown because resulting widths described in Table 3.10 may be below values possible for layout. However, when verified for practical device geometries and scaled for the drain current used, the figure provides an estimate for $S_{ID}^{1/2}$ useful for design guidance. Possible small-geometry noise increases described in Section 3.10.2.2 are excluded in the figure.

The gate-referred thermal-noise voltage PSD, S_{VG}, shown in the MOS noise model of Figure 3.60, is important for many applications, including when voltage signals appear at the gate of transconductor input stages. S_{VG} is found by dividing the drain-referred thermal-noise current PSD, S_{ID}, by g_m^2. This refers drain noise current PSD to a gate noise voltage PSD.

Table 3.30 summarizes expressions for S_{VG}. In the first expression, S_{VG} is expressed in terms of device g_m/I_D and g_m where the equivalent noise resistance, R_N, is inversely proportional to g_m as it was for S_{ID}. R_N corresponds to the resistance value giving thermal-noise voltage equal to that effectively

Table 3.30 MOS gate-referred thermal-noise voltage PSD expressions and trends in terms of the inversion coefficient, channel length, and drain current. Voltage density ($nV/Hz^{1/2}$) is found by taking the square root of the convenient PSD expressions

Gate-referred thermal-noise voltage PSD	$IC \uparrow$ L, I_D fixed	$L \uparrow$ IC, I_D fixed	$I_D \uparrow$ IC, L fixed
From g_m: $S_{VG} = 4kT \cdot \dfrac{n\Gamma}{g_m} = 4kT \left(\dfrac{n\Gamma}{\frac{g_m}{I_D} \cdot I_D} \right) = 4kT \cdot R_N$ $R_N = \dfrac{n\Gamma}{g_m}$	$\uparrow$ as g_m/I_D decreases from maximum value in WI	Unchanged, but decreases if less vel. sat. reduction of g_m/I_D	$\downarrow \propto \dfrac{1}{I_D}$
Convenient expression: $S_{VG} = \dfrac{(4.07\,nV)^2}{Hz} \left(\dfrac{T}{300K} \right) n\Gamma \left(\dfrac{1000\,\mu S}{g_m} \right)$			

From IC, L, I_D:
 	
$S_{VG} = 4kT \left(n\Gamma \left(\sqrt{IC+0.25} + 0.5 \right) \left(\dfrac{nU_T}{I_D} \right) \right)$
$= 4kT \left[(1/2)n^2 U_T/I_D \right]$ in WI
$= 4kT \left[(2/3)n^2 U_T \sqrt{IC}/I_D \right]$ in SI without
   vel. sat.
$= 4kT \left[(8/3)n^3 U_T (U_T/(LE_{CRIT})')IC/I_D \right]$ in
   SI with full vel. sat.
	Constant in WI
	
	$\uparrow \propto \sqrt{IC}$ in SI if no vel. sat.
	
	$\uparrow \propto IC$ in SI with full vel. sat.
		Unchanged, but decreases if less vel. sat. reduction of g_m/I_D
		
		$\downarrow \propto \dfrac{1}{L}$ in SI with full vel. sat.
			$\downarrow \propto \dfrac{1}{I_D}$

Convenient expression: $S_{VG} = \dfrac{(2.069\,nV)^2}{Hz} \left(\dfrac{T}{300K} \right)^2$ $\cdot n^2 \Gamma \left(\sqrt{IC+0.25} + 0.5 \right) \left(\dfrac{100\,\mu A}{I_D} \right)$			
To include velocity saturation and VFMR decreases in g_m, replace IC with $IC(1 + IC/IC_{CRIT})$ as described in Table 3.17			

S_{VG} is for saturation where $V_{DS} > V_{DS,sat}$.
Velocity saturation refers to both velocity saturation and VFMR effects.
Γ, given by Equation 3.106, increases from one-half in weak inversion to two-thirds in strong inversion, excluding small-geometry increases described in Section 3.10.2.2. IC and L trends exclude small-geometry increases in Γ.
Convenient expressions assume $k = 1.3806 \times 10^{-23}$ J/K.

appearing in series with the MOSFET gate. Opposite to the S_{ID} trends, S_{VG} increases with increasing inversion level, inversely tracking the decrease in g_m/I_D and g_m. It decreases with increasing drain current, inversely tracking the increase in g_m. The table also gives a convenient expression for S_{VG} depending on g_m, n, Γ, and T.

Table 3.30 also gives expressions and trends for S_{VG} in terms of the design choices of IC, L, and I_D. While S_{ID} is independent of temperature for constant inversion coefficient, S_{VG} tracks the square of absolute temperature, T. This is best seen in weak inversion where, again, $4kT$ increases directly with T while $g_m = I_D/(nU_T)$ decreases inversely with T through the increase in $U_T = kT/q$. The resultant S_{VG}, which depends inversely on g_m, is then proportional to T^2. This suggests that extremely low S_{VG} is available in weak inversion at cryogenic temperatures where $T = 77\,\text{K}$. Here, $4kT$ is small combined with a high value of $g_m = I_D/(nU_T)$ since U_T is also small. However, as shown earlier in Figure 3.4, n nearly doubles at $T = 77\,\text{K}$, likely due to increased interface capacitance. As a result, g_m is not increased as much as expected, giving a higher than expected value of S_{VG}. Interestingly, while n increases at very low temperatures and degrades g_m, the value of n used in the $n\Gamma$ terms of noise expressions appears to remain nearly unchanged. This is discussed in [20] for the micropower, low-noise, $0.35\,\mu\text{m}$, SOI CMOS preamplifier example described later in Section 6.7. Unlike S_{ID}, which is independent of n when evaluated in terms of the inversion coefficient, S_{VG} depends on n^2. This is because n appears in the noise expression and in the denominator of the g_m expression, which is inverted in the noise expression. Except for the very low temperatures mentioned, n varies little with temperature as discussed earlier in Section 3.5.1 and it decreases only slightly with increasing inversion level as shown earlier in Figure 3.30 in Section 3.8.3.1. Table 3.30 also gives weak inversion, strong inversion without velocity saturation, and strong-inversion, velocity-saturated values of S_{VG}, factored in the usual $4kT$ noise resistance form.

As noted in Table 3.30, S_{VG} is unchanged at its minimum value in weak inversion, increases as $\sqrt{IC}$ in strong inversion, and increases directly with IC under conditions of full velocity saturation. It is independent of L, unless short channel length results in velocity saturation reduction of g_m/I_D and g_m, which then raises S_{VG}. As mentioned, S_{VG} decreases inversely with increasing drain current. Each of these trends corresponds to the trends for the inverse of g_m since S_{VG} inversely tracks g_m. The noise expressions, like those given in Table 3.29 for S_{ID}, are based on g_m found by I_D multiplied by the g_m/I_D expressions given in Table 3.17. Again, velocity saturation and VFMR effects are included through the replacement of IC with $IC = IC(1 + IC/IC_{CRIT})$. Finally, the table includes a convenient expression for S_{VG} in terms of IC, L, I_D, n, Γ, and T. The square root of this expression gives gate-referred thermal-noise voltage density, $S_{VG}^{1/2}$, in its usual design units of $\text{nV}/\text{Hz}^{1/2}$.

Figure 3.63 shows predicted $S_{VG}^{1/2}$ in $\text{nV}/\text{Hz}^{1/2}$ for $L = 0.18$, 0.28, 0.48, and $1\,\mu\text{m}$, nMOS devices in the $0.18\,\mu\text{m}$ CMOS process described in Table 3.2 operating at IC ranging from 0.01 to 100. Like $S_{ID}^{1/2}$, $S_{VG}^{1/2}$ is a function of drain current through g_m. Here, a drain current of $100\,\mu\text{A}$ is assumed, where the decrease in $S_{VG}^{1/2}$ with increasing drain current resulting from increasing g_m is easily found by multiplying the values shown in the figure by $(100\,\mu\text{A}/I_D)^{1/2}$. $S_{VG}^{1/2}$ is found from the last expression in Table 3.30, using the process values described for prediction of $S_{ID}^{1/2}$ shown in Figure 3.62. In addition to the channel lengths shown, $S_{VG}^{1/2}$ is also shown assuming velocity saturation and VFMR reductions of g_m/I_D and g_m are not present.

In Figure 3.63, $S_{VG}^{1/2}$, which tracks the inverse square root of g_m/I_D shown in Figure 3.26 for constant drain current, increases from its minimum weak-inversion value as the inversion coefficient increases. The increase is greater in strong inversion for short-channel devices because of velocity saturation decreases in g_m/I_D and g_m. Increasing $S_{VG}^{1/2}$ with increasing inversion level is undesirable for input transconductor devices receiving voltage inputs. For example, $S_{VG}^{1/2}$ increases from $2.67\,\text{nV}/\text{Hz}^{1/2}$ at the center of moderate inversion ($IC = 1$) to at least $4.29\,\text{nV}/\text{Hz}^{1/2}$ (as shown, it increases further for short-channel devices) at the onset of strong inversion ($IC = 10$) for the drain current of $100\,\mu\text{A}$ shown. Operation at the onset of strong inversion requires a drain current increase of at least $(4.29/2.67)^2 = 2.58$ or an I_D of at least $258\,\mu\text{A}$ to achieve $2.67\,\text{nV}/\text{Hz}^{1/2}$ available at the center of moderate inversion at $I_D = 100\,\mu\text{A}$. This emphasizes the significant advantage of moderate-inversion

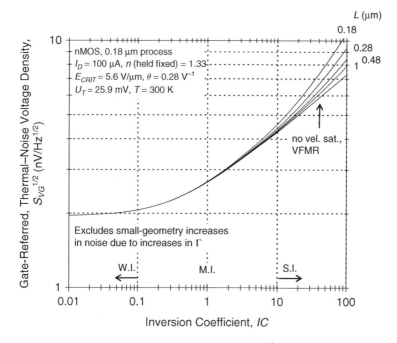

Figure 3.63 Predicted gate-referred thermal-noise voltage density, $S_{VG}^{1/2}$, versus inversion coefficient, IC, for $L = 0.18, 0.28, 0.48$, and $1\,\mu m$, nMOS devices in a $0.18\,\mu m$ CMOS process operating at a drain current of $100\,\mu A$. $S_{VG}^{1/2}$, which tracks the inverse square root of g_m/I_D for constant drain current, increases in strong inversion, increasing more for short-channel devices because of velocity saturation. $S_{VG}^{1/2}$ is similar for all CMOS processes because of the fundamental g_m/I_D characteristic and is found for different drain currents by multiplying the values shown by $(100\,\mu A/I_D)^{1/2}$. The prediction shown excludes possible noise increases associated with small-geometry effects described in Section 3.10.2.2

operation for power-efficient, low, input-referred thermal-noise voltage. Weak-inversion operation gives slightly lower input-referred thermal-noise voltage due to the slightly higher g_m/I_D and g_m, but at a considerable reduction in intrinsic and operating bandwidth as discussed earlier in Sections 3.9.6 and 3.9.7. Attaining low, input-referred thermal noise efficiently at low power consumption is discussed later in Section 6.3.1 using a thermal-noise efficiency figure of merit. While input transconductor devices should be operated at low inversion coefficients and high drain currents to maximize g_m and minimize input-referred thermal-noise voltage, non-input devices should be operated at high inversion coefficients and low drain currents to minimize g_m and their drain thermal-noise current contributions as mentioned. The choice of input and non-input device inversion coefficients and drain currents for minimizing input-referred thermal-noise voltage is discussed for $0.18\,\mu m$ CMOS operational transconductance amplifiers and micropower, low-noise, $0.35\,\mu m$, SOI CMOS preamplifiers later in Chapters 5 and 6.

Like the minimum values of $S_{ID}^{1/2}$ shown in Figure 3.62, the highest values of $S_{VG}^{1/2}$ shown in Figure 3.63 for short-channel devices in deep strong inversion may not be attainable for $I_D = 100\,\mu A$ considered because resulting widths may be below values possible for layout. However, when verified for practical device geometries and scaled for the drain current used, Figure 3.63 provides an estimate of $S_{VG}^{1/2}$ useful for design guidance. Possible small-geometry noise increases described in Section 3.10.2.2 are excluded in the figure.

Since thermal $S_{ID}^{1/2}$ and $S_{VG}^{1/2}$ shown in Figures 3.62 and 3.63 are linked to g_m, the noise of pMOS devices is nearly equal to that of nMOS devices shown if pMOS channel length is decreased by a factor of 2.5 $((14\,V/\mu m)/(5.6\,V/\mu m))$, the ratio of pMOS and nMOS E_{CRIT}. This is a result of nearly equal g_m/I_D for

$L = 0.18\,\mu m$, pMOS devices shown in Figure 3.27 compared to $L = 0.48\,\mu m$, nMOS devices shown in Figure 3.26. Smaller reductions in g_m/I_D and the resulting smaller changes in thermal noise present for all channel lengths due to VFMR are similar for both nMOS and pMOS devices because of similar values of the mobility reduction factor, θ.

Since thermal $S_{ID}^{1/2}$ and $S_{VG}^{1/2}$ are linked to g_m, the noise is largely independent of process technology when evaluated in terms of the inversion coefficient. This is because of the technology independence of g_m/I_D as described earlier in Section 3.8.2.5. As mentioned in Section 3.2.4, the development of process-independent design expressions is one of the advantages of using the inversion coefficient in analog CMOS design. Like g_m/I_D, thermal noise is independent of channel length when small-geometry effects are excluded. The decrease in g_m/I_D associated with velocity saturation effects depends on the process through E_{CRIT} and the channel length. This decrease causes the decrease in $S_{ID}^{1/2}$ and increase in $S_{VG}^{1/2}$ observed for short-channel devices operating in deep strong inversion. Since S_{VG} depends on n^2, $S_{VG}^{1/2}$ varies directly as n varies with the process technology. Such variations, however, are small for typical, bulk CMOS processes.

Thermal $S_{ID}^{1/2}$ and $S_{VG}^{1/2}$ shown in Figures 3.62 and 3.63 along with the trends for S_{ID} and S_{VG} listed in Tables 3.29 and 3.30 do not include possible small-geometry noise increases discussed in Section 3.10.2.2. Selecting channel length somewhat above the process minimum reduces possible noise increases, which, as mentioned, can be considered by increasing the value of Γ. Additionally, avoiding operation at the extremes of weak inversion or deep strong inversion may minimize CLM and carrier heating increases, respectively. Avoiding minimum channel length devices and extremes of inversion are general rules for general-purpose analog design. Unfortunately, RF designs may require short channel lengths and high levels of inversion. Experimental measurements on test devices are again recommended for critical noise applications.

3.10.3 Flicker Noise

Johnson observed low-frequency flicker noise in vacuum tubes having coated filaments [81]. Here, additional low-frequency noise was observed beyond the normal shot noise that has a white frequency spectrum. MOS devices also have flicker noise, which is dominant at low frequencies and often exceeds thermal noise at frequencies even above 1 MHz. Sometimes called "$1/f$ noise," flicker noise is a better description since the noise PSD is actually proportional to $1/f^{AF}$, where the flicker-noise exponent, AF, ranges between 0.7 and 1.2.

MOS flicker noise has been extensively studied over the last several decades [15, p. 422, 16, pp. 96–105, 58, p. 162, 72, 76, 82–103, 104, p. 12, 105–165], and the reader is referred to [15, pp. 432–437] and the current literature for additional references. Interest in MOS flicker noise continues, especially since previous trends, like pMOS devices having lower flicker noise compared to nMOS devices, do not hold in modern, small-geometry processes. Unlike the normalized small-signal parameters, g_m/I_D, $\eta = g_{mb}/g_m$, and $V_A = I_D/g_{ds}$, intrinsic voltage gain and bandwidth, A_{Vi} and f_{Ti}, and drain-referred and gate-referred thermal-noise PSD, S_{ID} and S_{VG}, that have a universal characteristic when evaluated in terms of the inversion coefficient, flicker noise varies significantly across processes and process generations.

MOS flicker noise results from carrier trapping and de-trapping near the Si–SiO$_2$ interface. Radiation effects experiments show that gate-referred flicker-noise voltage and trapped oxide charge both increase with total ionizing dose [102, 109, 116]. Additionally, processes having low-flicker-noise sustain less trapped oxide charge damage following irradiation [103, 109]. As a result, flicker noise serves as a process diagnostic for the Si–SiO$_2$ interface and gate oxide. Processing steps to minimize flicker noise include optimizing gate-oxide growth conditions, reducing post-oxidation temperatures, minimizing hydrogen in post-oxidation anneals, and avoiding implant or other process-induced oxide damage [109]. Unfortunately, as described later in Section 3.10.3.5, reported flicker noise is higher in processes having minimum gate lengths of 0.18 μm and smaller, most likely due to increased interface and oxide damage during processing.

MOS flicker noise is described using the carrier density fluctuation model, the mobility fluctuation model, or the unified, carrier density, correlated mobility fluctuation model. These models are described in the following discussions. The expressions given are intended to link flicker noise to likely physical causes where equalities should be replaced with proportionalities in many cases. Final noise expressions utilize extracted noise parameters.

3.10.3.1 Carrier density fluctuation model

The carrier density fluctuation model is also known as the McWhorter model [82], the number density model, the Δn model, and the charge trapping and de-trapping model [15, p. 422, 58, p. 148, 83, 100, 104, p. 15, 108, 112, 129, 147, 152]. In this model, carriers are trapped and released near the Si–SiO$_2$ interface resulting in a modulation of the flat-band voltage. This results in a modulation of the channel carrier density and drain current.

Charge trapping and de-trapping result in the appearance of a voltage fluctuation, ΔV_g, in series with the gate equal to the fluctuation in the flat-band voltage, ΔV_{fb}, given by

$$\Delta V_g = \Delta V_{fb} = \frac{\Delta Q'_t}{C'_{OX}} \tag{3.109}$$

Here, $\Delta Q'_t$ is the fluctuating charge per unit area resulting from charge trapping and de-trapping near the Si–SiO$_2$ interface. This includes trapped charge at various depths in the gate oxide near the interface and other trapped charge near the interface that gives rise to flicker noise [109]. Squaring Equation 3.109 gives

$$\Delta V_g^2 = \Delta V_{fb}^2 = \left(\frac{\Delta Q'_t}{C'_{OX}}\right)^2 \tag{3.110}$$

which shows that the mean square, gate-referred flicker-noise voltage is inversely proportional to the square of the gate-oxide capacitance.

Trapping and de-trapping events associated with a single trap give rise to random telegraph signals (RTS), which look like random square waves having discrete levels [93, 100, 118, 129, 136, 147, 161, 162]. RTS are observed in small-area MOS devices and have a Lorentzian frequency spectrum typical of generation–recombination (GR) noise. The Lorentzian power spectrum is given by

$$F(f) = A_{GR}\frac{\tau_t}{1+(2\pi f\tau_t)^2} \tag{3.111}$$

where A_{GR} is the amplitude factor, and τ_t is the tunneling time constant associated with trapping [129]. This is a low-pass spectrum having constant content at low frequencies before rolling off at a frequency that is inversely proportional to the tunneling time constant. The roll-off frequency is higher for shorter time constants associated with shallower traps and lower for longer time constants associated with deeper traps. The summation of multiple spectra from RTS signals having varying roll-off frequencies associated with a uniform distribution of tunneling time constants gives rise to noise having a $1/f$ power spectrum.

MOS gate-referred flicker-noise voltage PSD, S_{VG} shown in the MOS noise model of Figure 3.60, is given from the carrier density fluctuation model by the general form of

$$S_{VG}\,(\text{flicker},\Delta n) = \frac{K_F}{C'^2_{OX}WLf^{AF}} \tag{3.112}$$

with one derivation given in [104, pp. 15–21]. Equation 3.112 is valid for both the ohmic, linear, or triode and the saturation regions [122, 139] and is widely used in both regions. Some treatments, however, include a constant factor in the denominator that can be effectively included in the value of K_F. K_F is the flicker-noise factor commonly having units of C^2/cm^2, and AF is the frequency exponent, which has been added to consider the deviation from the ideal $1/f$ PSD slope.

K_F is related to the effective trap density, N_t', by

$$K_F = q^2 N_t' \tag{3.113}$$

Often, N_t' is expressed as [83, 108, 139]

$$N_t' = \frac{kT \cdot N_t(E)}{\gamma_t} \tag{3.114}$$

where $N_t(E)$ is the number of traps per unit volume and energy, having units per $eV \cdot cm^3$, kT is the product of Boltzmann's constant and absolute temperature expressed in eV, and γ_t is McWhorter's tunneling constant, having units per cm (in some treatments, γ_t appears in the numerator of the expression and has units of cm). K_F, N_t', and $N_t(E)$ vary widely with process technologies, having typical values ranging from 10^{-33} to $10^{-29} C^2/cm^2$, 3.9×10^4 to $3.9 \times 10^8/cm^2$, and 1.5×10^{14} to $1.5 \times 10^{18}/eV \cdot cm^3$, respectively. N_t' is linked to K_F assuming $q = 1.602 \times 10^{-19}$ C, and $N_t(E)$ is linked to N_t' assuming $kT = 0.026\,eV$ ($T = 300$ K) and $\gamma_t = 1/(10^{-8}\,cm)$. As discussed later for reported K_F values in Section 3.10.3.5, K_F, N_t', and $N_t(E)$ are nearer the lower range of values for pMOS devices in older, large-geometry processes, but are unfortunately nearer the higher range of values for nMOS and pMOS devices in modern, small-geometry processes.

Equation 3.112 reveals the following trends for S_{VG} predicted by the carrier density fluctuation model. S_{VG} directly tracks K_F and measures of interface trap density, N_t' or $N_t(E)$, indicating that flicker noise is lower for cleaner processes having less oxide and interface damage. S_{VG} inversely tracks gate area and is lower for large-area devices. S_{VG} inversely tracks the square of C_{OX}' because noise voltage resulting from trapped and de-trapped charge is lower in the presence of larger capacitance. This indicates that S_{VG} decreases as t_{ox} decreases and C_{OX}' increases in smaller-geometry processes for equal gate area and K_F. Unfortunately, as described later in Section 3.10.3.5, K_F usually increases in smaller-geometry processes largely countering the noise advantage associated with higher C_{OX}'. S_{VG} is proportional to $1/f^{AF}$, where AF typically ranges from 0.7 to 1.2. Finally, S_{VG} is *independent* of the MOS inversion level ($V_{EFF} = V_{GS} - V_T$) or inversion coefficient. Since S_{VG} observed for nMOS devices is often nearly constant with inversion level, the McWhorter, carrier density fluctuation model was historically associated with nMOS devices [108, 111, 112]. There may be a consensus developing that the McWhorter model applies to both nMOS and pMOS devices operating at low levels of inversion, for example weak inversion, before inversion-level, flicker-noise increases are observed due possibly to correlated mobility fluctuation as described in Section 3.10.3.3.

Gate-referred flicker-noise voltage PSD (S_{VG}) is easy to predict from the McWhorter, carrier density fluctuation model because it is independent of the inversion level, depending only on the gate area, frequency, and process values of K_F and AF. This is a compelling reason to refer flicker noise to the gate as a noise voltage since drain flicker-noise current depends additionally on the transconductance of the device, where the drain-referred flicker-noise current PSD is given by $S_{ID} = S_{VG} \cdot g_m^2$. Additionally, referring the noise to the gate is appropriate since flicker noise results from carrier trapping and de-trapping near the Si–SiO$_2$ interface below the gate and is a gate-noise process.

Often, including in this book, gate-referred flicker-noise voltage PSD is evaluated using Equation 3.112 even when measured data given later in Section 3.10.3.6 shows that the noise increases with increasing inversion level, especially for pMOS devices. Inversion-level noise increases are then handled by increasing the value of K_F, including the K_F increase derived from the unified or carrier density, correlated mobility fluctuation model in Section 3.10.3.3. With values of K_F appropriate for the level of inversion, Equation 3.112 provides a pragmatic and useful method for characterizing and predicting MOS flicker noise for analog design.

3.10.3.2 Carrier mobility fluctuation model

The carrier mobility fluctuation model is also known as the Hooge model [85, 88, 111] or the $\Delta\mu$ model [15, p. 423, 58, p. 151, 100, 104, p. 13, 108, 112, 129, 147, 152]. This model predicts mobility fluctuations resulting from carrier lattice or phonon scattering in the bulk semiconductor below the Si–SiO$_2$ interface. Carrier mobility fluctuations then give rise to drain current fluctuations.

The Hooge model was originally experimentally verified for homogeneous materials by normalizing the current noise PSD, S_I, to the DC bias current squared, I^2, giving

$$\frac{S_I}{I^2} = \frac{\alpha_H}{Nf} \tag{3.115}$$

Here, N is the total number of carriers or "fluctuators," and α_H is the Hooge parameter. Originally, this dimensionless parameter was reported at approximately 2×10^{-3} for homogeneous materials [85, 88, 111].

For MOS devices operating in strong inversion in the deep ohmic, linear, or triode region ($V_{DS} \ll V_{DS,sat}$), the total inversion charge is $C'_{OX}WL(V_{GS} - V_T)$ and the corresponding total number of carriers is equal to $N = C'_{OX}WL(V_{GS} - V_T)/q$. Substituting this into Equation 3.115 gives

$$\frac{S_{ID}}{I_D^2} = \frac{q\alpha_H}{C'_{OX}WL(V_{GS} - V_T)f} \tag{3.116}$$

Since $S_{VG} = S_{ID}/g_m^2$, Equation 3.116 is converted to the gate-referred flicker-noise voltage PSD through division by $(g_m/I_D)^2$. For the deep ohmic region where I_D is linearly proportional to $V_{GS} - V_T$, $g_m/I_D = 1/(V_{GS} - V_T)$, giving gate-referred flicker-noise voltage PSD predicted by the mobility fluctuation model of [108, 129, 147]

$$S_{VG} \text{ (flicker, } \Delta\mu) = \frac{q\alpha_H(V_{GS} - V_T)}{C'_{OX}WLf^{AF}} \tag{3.117}$$

AF is the frequency exponent, which, again, has been added to consider the deviation from the ideal $1/f$ PSD slope. Equation 3.117 is a strong-inversion, deep ohmic region approximation that assumes constant channel mobility. For operation in saturation, the equation is divided by two [104, pp. 13–15, 112], or the value of α_H can be reduced by a factor of two.

As mentioned, α_H was originally reported at approximately 2×10^{-3} for homogeneous materials. However, this value includes only lattice or phonon scattering and excludes other scattering components in MOS devices that reduce mobility from the lattice mobility, resulting in a reduced value of α_H [100, 111, 112, 129, 152]. Experimental values of α_H range from 4×10^{-7} to 8.1×10^{-5} for pMOS devices in large-geometry processes having $t_{ox} \geq 11$ nm [108]. Values are higher at 7×10^{-4} [156] and 3.6×10^{-4} [154] for processes having minimum channel lengths of 0.13 and 0.09 μm and t_{ox} of 2 and 1.5 nm, respectively. As described later in Section 3.10.3.5, the increase in α_H in smaller-geometry processes is likely the result of surface channel operation compared to buried channel operation present for pMOS devices in large-geometry processes.

Equation 3.117 reveals the following trends for S_{VG} predicted by the carrier mobility fluctuation model. S_{VG} directly tracks the Hooge parameter, α_H, indicating that flicker noise is lower for processes having improved bulk purity and reduced overall mobility compared to lattice scattering mobility alone. S_{VG} inversely tracks gate area, like flicker noise predicted by the carrier density fluctuation model given by Equation 3.112. S_{VG} inversely tracks the gate-oxide capacitance, compared to inversely tracking the square of gate-oxide capacitance in the carrier density fluctuation model. Again, S_{VG} is proportional to $1/f^{AF}$, where AF typically ranges from 0.7 to 1.2. Finally, S_{VG} is *dependent* on the MOS inversion level, increasing with increasing inversion level ($V_{EFF} = V_{GS} - V_T$) or inversion coefficient. This is in contrast to the carrier density fluctuation model, which is inversion-level independent. Since S_{VG} observed for pMOS devices often increases with the inversion level, the Hooge, carrier mobility fluctuation model was historically associated with pMOS devices [108, 111, 112].

There is a growing consensus that the Hooge, carrier mobility fluctuation model does not describe most of the flicker noise observed in MOS devices because it does not consider carrier trapping and de-trapping near the Si–SiO$_2$ interface as the basic noise process. As shown in [76], small-area pMOS devices, like small-area nMOS devices, show flicker-noise spectral humps consistent with discrete Lorentzian spectra associated with trapping and de-trapping RTS from discrete traps. This suggests that the noise is clearly associated with carrier trapping and de-trapping compared to mobility fluctuations in the bulk semiconductor. The Hooge, carrier mobility fluctuation model, however, is described here for historical context where Equation 3.117 is still sometimes used as a flicker-noise extraction method. Additionally, carrier mobility fluctuation may be a significant contributor to MOS flicker noise for operation deep in weak inversion [16, p. 105].

As mentioned earlier, Equation 3.112 from the McWhorter, carrier density fluctuation model provides a pragmatic and useful way to characterize and predict MOS flicker noise where inversion-level noise increases are considered separately through increases in the value of K_F. These increases will be derived from the unified or carrier density, correlated mobility fluctuation model described in the next section.

3.10.3.3 Unified, carrier density, correlated mobility fluctuation model

The relatively inversion-level-independent, gate-referred flicker-noise voltage of nMOS devices combined with the more inversion-level-dependent noise of pMOS devices suggest combining both the McWhorter carrier density and Hooge mobility fluctuation models. These models have been compared in the literature [108, 111, 112] and added without correlation [91].

The unified, flicker-noise model [98, 100, 101] used in the BSIM3, BSIM4 [3], and MOS 9, MOS 11 [45] MOS models is not, as mentioned in [76], the direct addition of the McWhorter carrier density and Hooge mobility fluctuation models. In the unified model, which we will call the carrier density, correlated mobility fluctuation model, charge trapping and de-trapping near the Si–SiO$_2$ interface is the basic noise process. However, in addition to the resulting carrier density fluctuations associated with the McWhorter model, the carrier mobility also fluctuates due to Coulomb scattering caused by carrier trapping and de-trapping. This mobility fluctuation is different from that described by the Hooge mobility fluctuation model, which, at least in its original form, considers bulk mobility fluctuations independent of carrier trapping and de-trapping at an interface.

The presence of correlated mobility fluctuations resulting from charge trapping and de-trapping near the Si–SiO$_2$ interface is supported by measurements on small-area MOS devices where single trapping and de-trapping RTS are observed [93]. Here, channel resistance fluctuations are found to be above or below values predicted by the missing trapped carrier. This suggests that both carrier density and mobility fluctuate as a result of charge trapping and de-trapping.

Drain current fluctuation predicted by the carrier density, correlated mobility fluctuation model can be described by the general form [100, 129, 147]

$$\Delta I_d = g_m \Delta V_{fb} + I_D \frac{\alpha_{SC} \mu}{q} \Delta Q'_t \tag{3.118}$$

From Section 3.10.3.1, $\Delta Q'_t$ is the fluctuating trapped and de-trapped charge per unit area and ΔV_{fb} is the resulting fluctuating flat-band voltage. α_{SC} is the Coulomb scattering coefficient, which has approximate values of 1.6×10^{-15} and 1.6×10^{-14} V·s for electrons and holes, respectively, or 10^4 and 10^5 V·s/C if division by q is included in the scattering coefficient [147]. I_D, g_m, and μ are the usual MOS drain current, transconductance, and channel mobility, respectively. The first term in Equation 3.118 describes fluctuations predicted by the carrier density fluctuation model, where ΔV_{fb} provides fluctuating drain current through the device transconductance. The second term describes drain current fluctuations resulting from correlated mobility fluctuations. These mobility fluctuations result from the same carrier trapping and de-trapping that cause the carrier density fluctuations and usually add, as shown by the positive sign of the second term [100].

Substituting $\Delta Q'_t = C'_{OX} \Delta V_{fb}$ found from Equation 3.109 and dividing Equation 3.118 by g_m gives a gate-referred voltage fluctuation of

$$\Delta V_g = \Delta V_{fb}\left[1 + \frac{\alpha_{SC}\,\mu\,C'_{OX}}{q}\left(\frac{I_D}{g_m}\right)\right] \tag{3.119}$$

Equation 3.119, while usually given for the ohmic, linear, or triode region, also provides an estimate for operation in saturation [147]. Substituting $g_m/I_D = 2/(V_{GS} - V_T)$ from Equation 2.13 for strong inversion, saturation operation and squaring gives

$$\Delta V_g^2 = \Delta V_{fb}^2\left(1 + \frac{\alpha_{SC}\,\mu\,C'_{OX}}{2q}(V_{GS} - V_T)\right)^2 \tag{3.120}$$

The noise PSD associated with ΔV_g^2 is S_{VG} and the noise PSD associated with ΔV_{fb}^2 is S_{Vfb}, giving

$$S_{VG}\ \text{(flicker, unified)} = S_{Vfb}\left(1 + \frac{\alpha_{SC}\,\mu\,C'_{OX}}{2q}(V_{GS} - V_T)\right)^2$$

$$= \frac{K_F}{C'^2_{OX}WLf^{AF}}\left(1 + \frac{\alpha_{SC}\,\mu\,C'_{OX}}{2q}(V_{GS} - V_T)\right)^2 \tag{3.121}$$

The leftmost terms in Equation 3.121 are the flat-band noise voltage PSD, S_{Vfb}, which is equal to the gate-referred noise voltage PSD predicted in Equation 3.112 by the carrier density fluctuation model. The rightmost terms describe the increase in flicker noise associated with correlated mobility fluctuations. Equation 3.121, estimated for strong-inversion, saturation operation, is equal to expressions given in [100, 129, 147] derived for strong-inversion, deep ohmic operation, except for the "2" appearing in the denominator of the $V_{GS} - V_T$ term. The "2" appears because of the substitution in Equation 3.119 of $g_m/I_D = 2/(V_{GS} - V_T)$ from Equation 2.13 for operation in saturation compared to $g_m/I_D = 1/(V_{GS} - V_T)$ from Equation 6.13 for operation in the deep ohmic region where V_{DS} is nearly zero. The expressions for g_m/I_D are for strong inversion excluding velocity saturation and VFMR effects. The presence or absence of the "2" in Equation 3.121 is not significant since the form of the expression will be used to predict flicker noise using extractions from measured noise.

At low levels of inversion where $V_{GS} - V_T$ is small, Equation 3.121 converges to the inversion-level-independent, McWhorter, carrier density fluctuation model given by Equation 3.112. However, the correlated mobility fluctuation component increases with increasing inversion level or $V_{GS} - V_T$, giving increased flicker noise as predicted by the Hooge, carrier mobility fluctuation model. Interestingly, Equation 3.121 predicts that gate-referred flicker-noise voltage PSD increases as $(V_{GS} - V_T)^2$ compared to an increase of $V_{GS} - V_T$ predicted in Equation 3.117 by the Hooge model. Additionally, the higher Coulomb scattering coefficient for holes, estimated at approximately $\alpha_{SC} = 1.6 \times 10^{-14}$ V·s versus 1.6×10^{-15} V·s for electrons [147], predicts larger flicker-noise increases with increasing inversion for pMOS devices compared to nMOS devices. Both the $(V_{GS} - V_T)^2$ noise increase and greater inversion-level noise increase for pMOS devices will be shown later in measured data in Section 3.10.3.6.

Equation 3.121 requires modification for operation in weak and moderate inversion, and the BSIM3, BSIM4 [3], and MOS 9, MOS 11 [45] MOS models include additional complexity for predicting flicker noise [101]. Equation 3.121, however, is modified here to provide a simple prediction of flicker noise in weak, moderate, and strong inversion. Here, the gate-referred flicker-noise voltage PSD is given by

$$S_{VG} \text{ (flicker, unified)} = \frac{K_{F0}}{C_{OX}'^2 WLf^{AF}} \left(1 + \frac{V_{GS} - V_T, \text{taken as zero when negative}}{V_{KF}}\right)^2$$

$$\approx \frac{K_{F0}}{C_{OX}'^2 WLf^{AF}} \left(1 + \frac{2nU_T\sqrt{IC}}{V_{KF}}\right)^2 \tag{3.122}$$

which can also be expressed from the McWhorter, carrier density fluctuation model using a modified value of K_F, giving

$$S_{VG} \text{ (flicker, unified)} = \frac{K_F}{C_{OX}'^2 WLf^{AF}}$$

where

$$K_F = K_{F0} \left(1 + \frac{V_{GS} - V_T, \text{taken as zero when negative}}{V_{KF}}\right)^2 \approx K_{F0} \left(1 + \frac{2nU_T\sqrt{IC}}{V_{KF}}\right)^2 \tag{3.123}$$

In Equations 3.122 and 3.123, K_{F0} is the flicker-noise factor for operation in weak or the weak-inversion side of moderate inversion where the gate-referred flicker-noise voltage PSD is usually minimum and constant. V_{KF} is a voltage describing the increase in K_F and flicker noise with increasing $V_{EFF} = V_{GS} - V_T$, where V_{EFF} is clamped to zero when it is negative in weak or the weak-inversion side of moderate inversion. In the approximations in the equations, $V_{EFF} = 2\,nU_T\sqrt{IC}$ is expressed in terms of the inversion coefficient from Table 3.14 for operation in strong inversion, exclusive of velocity saturation and VFMR increases. This expression for V_{EFF} clamps the value to nearly zero when it is negative for $IC < 0.5$ while providing a rough estimate for $IC > 1$ in moderate inversion, before providing a good estimate in strong inversion. Since flicker noise usually increases slightly in weak and moderate inversion, this estimate of V_{EFF} is adequate for inclusion in Equations 3.122 and 3.123. As noted in Table 3.14, IC can be replaced with $IC(1 + IC/4IC_{CRIT})$ to include the increase in V_{EFF} due to velocity saturation and VFMR effects. IC_{CRIT}, given in the table, is the critical inversion coefficient where g_m/I_D is down to approximately 70.7 % of its value present without velocity saturation and VFMR effects.

Comparing Equation 3.122 with Equation 3.121 where K_F is replaced with K_{F0} because noise increases are considered by the product term shows that V_{KF} is equal to

$$V_{KF} = \frac{2q}{\alpha_{SC}\,\mu\,C_{OX}'} \tag{3.124}$$

Rather than predict the value of α_{SC} required in Equations 3.122 and 3.123, we will extract V_{KF} from measured data illustrated later in Section 3.10.3.6. Following this, we will infer values of α_{SC}.

Following several decades of controversy over the physical origin and modeling of MOS flicker noise, there is general consensus that the noise is driven by charge trapping and de-trapping near the Si–SiO$_2$ interface [76, 129]. However, open questions remain with regard to correlated mobility fluctuations considered in the unified or carrier density, correlated mobility fluctuation model. These questions include arguments in [134] that correlated mobility fluctuations are negligible due to weak Coulomb scattering. Additionally, it has been recently reported that inversion layer quantization and increases in trap cross-section may explain the increase in flicker noise with increasing inversion level [148, 161].

Regardless of the physical basis, the unified or carrier density, correlated mobility fluctuation model provides good agreement with measured data, at least for $0.18\,\mu m$ and larger processes, including data from a $0.18\,\mu m$ CMOS process described in [76]. Here, pMOS gate-referred flicker-noise voltage PSD increases by about a factor of 44 (factor of 6.6 in voltage density) from moderate through deep strong inversion, while nMOS noise PSD increases by a factor of four (factor of two in voltage density). Equations 3.122 and 3.123, derived from the unified or carrier density, correlated mobility fluctuation model, provide a simple flicker-noise prediction useful for design guidance that includes the increase with increasing inversion. Comparisons of measured and predicted flicker noise in Section 3.10.3.6 show that flicker noise predicted by Equations 3.122 and 3.123 agrees well with measured data from [76] and additional measured data described in Section 3.10.3.6 from a separate $0.18\,\mu m$ CMOS process.

Recent measurements in a $0.13\,\mu m$ CMOS process having $t_{ox} = 2$ nm indicate that gate-referred flicker-noise voltage PSD may increase more rapidly than the $(V_{GS} - V_T)^2$ increase predicted by the unified or carrier density, correlated mobility fluctuation model [161]. This may be the result of increased flicker noise due to inversion layer quantization [148, 161]. Additionally, flicker noise in the source and drain channel access resistances, R_S and R_D, may increase flicker noise at high levels of inversion (high $V_{GS} - V_T$), especially for devices having lightly doped drain (LDD) or moderately doped drain (MDD) drain and source extensions [16, pp. 102–105, 106, 107, 133, 138, 154, 156]. However, $V_{DS,sat}$ and $V_{EFF} = V_{GS} - V_T$ must often be limited for low-voltage designs in modern processes, restricting operation to near or below the onset of strong inversion where excessive increases in flicker noise are likely avoided. In addition to possible noise increases at very high levels of inversion that are above that predicted by Equations 3.122 and 3.123, flicker noise may increase from its expected constant value at very low levels of inversion at $IC < 0.001$ due to mobility fluctuations predicted by the Hooge model [16, pp. 100–105]. However, even though the physical basis for Equations 3.122 and 3.123 may be unclear, these or modifications of these equations can be used to structurally model flicker noise, including its increase with inversion, for typical levels of inversion encountered in design. Still, the reader is encouraged to experimentally measure flicker noise for the inversion levels and geometries of interest, especially given likely increases in flicker noise in smaller-geometry processes, including increased sensitivity to the inversion level.

3.10.3.4 Flicker-noise prediction from flicker-noise factors

Equations 3.112 and 3.123 provide a useful way to characterize and predict MOS flicker noise, while considering inversion-level increases in the noise through increases in the value of K_F. While this characterization and prediction of MOS flicker noise is common in practice, it is complicated by multiple unit systems for K_F.

Table 3.31 summarizes MOS gate-referred flicker-noise voltage PSD, S_{VG}, shown in the MOS noise model of Figure 3.60. S_{VG} is found from Equations 3.112 and 3.123 using three different unit systems for the flicker-noise factor. K_F, which was used for the flicker-noise models described in Sections 3.10.3.1 and 3.10.3.3, has semiconductor units of C^2/cm^2. This is the preferred unit system for comparing processes as seen in Table 3.33 in the next section because K_F is linked to the trap density through Equations 3.113 and 3.114. S_{VG} is found by dividing K_F by the process gate-oxide capacitance squared and the gate area, $C'^2_{OX}WL$, and the frequency, f^{AF}. K_{FSPICE}, which is commonly used in SPICE MOS models, has units of $V^2 \cdot F$. S_{VG} is found by dividing K_{FSPICE} by the gate-oxide capacitance, WLC'_{OX}, and the frequency. Finally, K'_F, which is used for hand analysis, has units of $(nV)^2 \cdot \mu m^2$. S_{VG} in design units of $(nV)^2/Hz$ is easily found by dividing K'_F by the gate area, WL, in μm^2 and the frequency in Hz. In addition to permitting simple hand analysis, K'_F permits a direct comparison of flicker noise across processes for equal gate area.

Each of the flicker-noise expressions in Table 3.31 includes a convenient expression motivated by [12, p. 51] where S_{VG} is found in units of $(nV)^2/Hz$ from C'_{OX} in units of $fF/\mu m^2$, W and L in units of μm, and frequency in Hz. This permits easy hand prediction of flicker noise without tedious unit

Table 3.31 MOS gate-referred flicker-noise voltage PSD expressions from flicker-noise factors having different unit systems. Voltage density in the design units of $nV/Hz^{1/2}$ is found by taking the square root of the convenient PSD expressions

---

Gate-referred flicker-noise voltage PSD with convenient expressions

---

From K_F in semiconductor units (C^2/cm^2):

$$S_{VG}(f) = \frac{K_F}{C_{OX}'^2 WL f^{AF}}$$

$$S_{VG}(f) = \frac{(nV)^2}{Hz} \left(\frac{K_F}{10^{-40} C^2/cm^2}\right) \left(\frac{fF/\mu m^2}{C_{OX}'}\right)^2 \left(\frac{\mu m^2}{WL}\right) \left(\frac{Hz}{f}\right)^{AF}$$

From K_{FSPICE} in SPICE units ($V^2 \cdot F$):

$$S_{VG}(f) = \frac{K_{FSPICE}}{C_{OX}' WL f^{AF}}$$

$$S_{VG}(f) = \frac{(nV)^2}{Hz} \left(\frac{K_{FSPICE}}{10^{-33} V^2 \cdot F}\right) \left(\frac{fF/\mu m^2}{C_{OX}'}\right) \left(\frac{\mu m^2}{WL}\right) \left(\frac{Hz}{f}\right)^{AF}$$

From K_F' in hand calculation units ($(nV)^2 \cdot \mu m^2$):

$$S_{VG}(f) = \frac{K_F'}{WL f^{AF}}$$

$$S_{VG}(f) = \frac{(nV)^2}{Hz} \left(\frac{K_F'}{(nV)^2 \cdot \mu m^2}\right) \left(\frac{\mu m^2}{WL}\right) \left(\frac{Hz}{f}\right)^{AF}$$

Inversion-level increase in flicker-noise factors from weak-inversion values:

$$K_F = K_{F0}\left(1 + \frac{V_{GS} - V_T, \text{ taken as zero when negative}}{V_{KF}}\right)^2 \approx K_{F0}\left(1 + \frac{2nU_T\sqrt{IC}}{V_{KF}}\right)^2$$

As described for Equation 3.123, K_{F0} is the flicker-noise factor in weak or the weak-inversion side of moderate inversion, V_{KF} is a voltage describing the inversion-level dependency, and $V_{EFF} = V_{GS} - V_T$ is clamped to zero when negative in weak or the weak-inversion side of moderate inversion. To include velocity saturation and VFMR increases in V_{EFF}, replace IC in $\sqrt{IC}$ term with $IC(1 + IC/4IC_{CRIT})$ as described in Table 3.14. Less V_{EFF} over-prediction in moderate inversion is obtained by replacing $V_{EFF} = 2nU_T\sqrt{IC}$ with V_{EFF} given in Table 3.14. Flicker-noise factors in all unit systems, K_F, K_{FSPICE}, and K_F', are increased as shown. Alternatively, flicker-noise factors can be extracted directly for the level of inversion

---

$C_{OX}' = 34.5\,fF/\mu m^2\,(1\,nm/t_{ox})$, where t_{ox} is the electrical gate-oxide thickness described in Table 3.1.

---

conversions. Taking the square root of S_{VG} gives gate-referred flicker-noise voltage density, $S_{VG}^{1/2}$, in the common design units of $nV/Hz^{1/2}$. The table also includes an expression from Equations 3.122 and 3.123 describing the increase in flicker noise with increasing inversion. Here, K_{F0}, $K_{FSPICE0}$, and K_{F0}' are the flicker-noise factors in weak or the weak-inversion side of moderate inversion, and V_{KF} is a voltage describing the flicker-noise factor increase with increasing effective gate–source voltage, $V_{EFF} = V_{GS} - V_T$. Flicker-noise increases with increasing inversion are modeled by increasing the flicker-noise factors from their weak-inversion values by the factor $(1 + V_{EFF}/V_{KF})^2$, where V_{EFF} is clamped to zero when it is negative in weak or the weak-inversion side of moderate inversion.

Table 3.32 gives expressions for converting between MOS flicker-noise factors, K_F (C^2/cm^2), K_{FSPICE}($V^2 \cdot F$), and K_F'($(nV)^2 \cdot \mu m^2$). As in Table 3.31, convenient expressions are given that eliminate tedious unit conversions. Table 3.32 also shows the extraction of K_F' from measured flicker noise. Here, $S_{VG}(f)$ is the measured gate-referred flicker-noise voltage PSD, f is the measurement frequency, AF is the flicker-noise PSD slope, and WL is the device gate area. Caution is required when extrapolating

Table 3.32 Expressions for MOS flicker-noise factor conversions and extraction from measured data

Expression	Convenient expression
Finding K_F in semiconductor units (C^2/cm^2):	
$K_F = K_{FSPICE} C'_{OX}$	$K_F = 10^{-7} \dfrac{C^2}{cm^2} \left(\dfrac{K_{FSPICE}}{V^2 \cdot F} \right) \left(\dfrac{C'_{OX}}{fF/\mu m^2} \right)$
$K_F = K'_F C'^2_{OX}$	$K_F = 10^{-40} \dfrac{C^2}{cm^2} \left(\dfrac{K'_F}{(nV)^2 \cdot \mu m^2} \right) \left(\dfrac{C'_{OX}}{fF/\mu m^2} \right)^2$
Finding K_{FSPICE} in SPICE units ($V^2 \cdot F$):	
$K_{FSPICE} = \dfrac{K_F}{C'_{OX}}$	$K_{FSPICE} = 10^7 \ V^2 \cdot F \left(\dfrac{K_F}{C^2/cm^2} \right) \left(\dfrac{fF/\mu m^2}{C'_{OX}} \right)$
$K_{FSPICE} = K'_F C'_{OX}$	$K_{FSPICE} = 10^{-33} \ V^2 \cdot F \left(\dfrac{K'_F}{(nV)^2 \cdot \mu m^2} \right) \left(\dfrac{C'_{OX}}{fF/\mu m^2} \right)$
Finding K'_F in hand calculation units ($(nV)^2 \cdot \mu m^2$):	
$K'_F = \dfrac{K_F}{C'^2_{OX}}$	$K'_F = (nV)^2 \cdot \mu m^2 \left(\dfrac{K_F}{10^{-40} \ C^2/cm^2} \right) \left(\dfrac{fF/\mu m^2}{C'_{OX}} \right)^2$
$K'_F = \dfrac{K_{FSPICE}}{C'_{OX}}$	$K'_F = (nV)^2 \cdot \mu m^2 \left(\dfrac{K_{FSPICE}}{10^{-33} \ V^2 \cdot F} \right) \left(\dfrac{fF/\mu m^2}{C'_{OX}} \right)$
Finding K'_F from measured data:	
$K'_F = WL \cdot S_{VG}(f) \cdot f^{AF}$	$K'_F = (nV)^2 \cdot \mu m^2 \left(\dfrac{WL}{\mu m^2} \right) \left(\dfrac{S_{VG}(f)}{\dfrac{(nV)^2}{Hz}} \right) \left(\dfrac{f}{Hz} \right)^{AF}$

$C'_{OX} = 34.5 \ fF/\mu m^2 \ (1 \ nm/t_{ox})$, where t_{ox} is the electrical gate-oxide thickness described in Table 3.1.

from frequencies well above 1 Hz because significant errors can occur if AF is not carefully evaluated. For example, extrapolating from 100 Hz assuming $AF = 1$ would give K'_F a factor of 2.5 higher than the value if AF were actually 0.8. Including 1 Hz in the measurement spectrum helps ensure proper noise extraction to 1 Hz. Once K'_F is extracted from experimental measurements, it is easily converted to K_F or K_{FSPICE} using expressions given in the table. K'_F should be extracted from measurements at different inversion levels to consider inversion-level dependency. K'_{F0} could be used to denote the value in weak or the weak-inversion side of moderate inversion before the increase with inversion level described in Table 3.31.

3.10.3.5 Reported flicker-noise factors and trends

Table 3.33 summarizes MOS flicker-noise factors, K_F, and flicker-noise slopes, AF, reported directly from the literature or extracted from reported measurements. The use of K_F appears to assume the inversion-level-independent, McWhorter, carrier density fluctuation model described by Equation 3.112, but, again inversion-level increases in flicker noise are considered through increased values of K_F as summarized in Table 3.31. As a result, K_F is used as a normalized flicker-noise measure inclusive of all effects.

Table 3.33 presents a limited selection of K_F values. Often drain-referred flicker-noise current PSD, S_{ID}, or S_{ID} normalized to the drain current squared, S_{ID}/I_D^2, is reported without mention of device trans-conductance, making it impossible to refer flicker noise to the gate and extract K_F. Additionally, full bias conditions and device geometries are often omitted, making it difficult to assess the region of operation.

Table 3.33 Reported MOS flicker-noise factors and slopes for submicron CMOS processes. Flicker-noise factors are reported or calculated from measured data

t_{ox}(nm)	L_{min} (μm)	K_F ($\times 10^{-31}$ C^2/cm^2)		AF		Notes, reference
		nMOS	pMOS	nMOS	pMOS	
17.5	0.8	15.1	0.71			[141]
—	0.7	0.66	0.06	0.7	0.86	MI [117]
13.5	0.5	1.23	0.16	0.82	1	SI [11]
11.5	0.5	0.06–0.1	0.01–0.07			SI, $V_{GS} = 1$–2.5 V, low-noise process [139][a]
8	0.35	2.67	0.67–1.86	1	1	nMOS, SI; pMOS, MI–SI, $IC = 0.95$–38; PD SOI, author measured
7.2	0.35	13	2			MI [143]
		4.8	1.92	1	1.2	MI [163][b]
		4.8	4.32	0.9	1.1	MI [163][b]
5.5	0.25	1.1–2.1	0.3–0.6	0.9–1	0.8–0.9	WI–SI, $L > 0.8$ μm [72]
		3.4-3.8	0.6–0.8	0.9–1	0.8–0.9	WI–SI, $L = 0.35$ μm [72][a,b]
		21	0.8			MI [143]
		4.48	3.84	0.85	1.1	MI [163][b]
		5.12	2.56	0.85	1.15	MI [163][b]
4.1	0.18	2.75–7.95	2.38–27.2	0.85	1.05	MI–SI; nMOS, $IC = 1$–100; pMOS, $IC = 0.9$–90, author measured[a]
4	0.18	32	3.3			MI [143]
		8.8	4.4	0.9	1	MI [163][b]
		13.2	7.04	0.9	1.1	MI [163][b]
—	0.18	4.3–17.1	5.1–223			MI–SI, $V_{GS} = 0.5$–1.8 V, $L = 10$ μm [76][a]
2.8	0.13	55				Deep SI [159]
	0.18		616		1.15	Deep SI [155]
2.4	0.13	22.2	44.4–96.2	0.85	1.2	[165][a,b]
2.2	—	79	314			SI, $V_{EFF} = 0.5$ V [128]
1.5	0.09		52.4, 524			WI, SI at $V_{EFF} = 0.25$ V [154]

[a]Noise increases slightly or modestly for nMOS devices while increasing significantly for pMOS devices as inversion ($V_{EFF} = V_{GS} - V_T$ or IC) increases.
[b]Noise increases for shorter L devices.

K_F values in Table 3.33 should not be used for actual design as flicker noise must be evaluated for the specific process used. However, the K_F values listed are useful to observe technology trends. As mentioned in the previous section, K_F(C^2/cm^2) is related to the carrier trap density near the Si–SiO$_2$ interface by Equations 3.113 and 3.114. This is especially true at low levels of inversion where the inversion-level-independent, McWhorter, carrier density fluctuation model described in Section 3.10.3.1 likely applies.

As seen in Table 3.33, K_F is usually significantly lower in 0.25 µm and larger-geometry processes for pMOS devices compared to nMOS devices, making pMOS devices preferable for low-flicker-noise applications in these processes [72, 97, 105, 108, 113, 117, 122, 139, 141, 143]. The higher, 4.7 eV barrier for injection of holes into the gate oxide for pMOS devices compared to a lower, 3.2 eV barrier for injection of electrons for nMOS devices [166, p. 400] was sometimes cited as a possible reason for lower pMOS flicker noise. However, the general consensus is that flicker noise is lower for pMOS devices in older, larger-geometry processes because of the use of n+ polysilicon gates that results in buried channel operation. This keeps carriers further from the Si–SiO$_2$ interface and likely lowers the flicker noise. Additionally, buried channel operation is often cited as a reason for using the Hooge, mobility fluctuation model for pMOS devices, since the original Hooge model applies to carrier scattering in bulk materials. In [129], which utilizes data from [114], the Hooge parameter, α_H, is reported at approximately 1.6×10^{-7} to 7×10^{-7} for n+ polysilicon, buried channel pMOS devices and is nearly two orders of magnitude higher at 1.2×10^{-5} to 3×10^{-5} for p+ polysilicon, surface-mode devices. For equal inversion level and gate area, this suggests an increase of nearly two orders of magnitude in gate-referred flicker-noise voltage PSD or one order of magnitude in noise voltage density for surface-mode devices compared to buried channel devices.

The use of p+ polysilicon pMOS devices in 0.18 µm and smaller-geometry processes gives surface channel operation like that present for n+ polysilicon nMOS devices traditionally used in all processes. This likely explains comparable levels of nMOS and pMOS K_F observed in Table 3.33 for 0.18 µm and smaller-geometry processes. K_F values for pMOS devices in strong inversion, however, are often higher than values for nMOS devices because of significant pMOS increases in flicker noise with increasing inversion level. K_F values can best be compared at their weak-inversion K_{F0} values prior to increases with inversion level. Then, inversion-level increases can be evaluated separately.

Flicker-noise trends with inversion level, including those given by Footnote a in Table 3.33, most often show that gate-referred flicker-noise voltage increases slightly to modestly for nMOS devices and modestly to significantly for pMOS devices as inversion or $V_{EFF} = V_{GS} - V_T$ increases [72, 76, 108, 110, 118, 122, 125, 137, 139, 147, 153, 165]. This is especially well shown in the 0.18 µm process measurements of [76] shown in the table where, as mentioned in Section 3.10.3.3, pMOS K_F and the corresponding gate-referred flicker-noise voltage PSD increase by about a factor of 44 (factor of 6.6 for voltage density) from moderate through deep strong inversion, while nMOS noise PSD increases by a factor of four (factor of two for voltage density). Measurements in [137] show similar noise increases for processes having $t_{ox} = 20$, 7.5, 5, and 2 nm, with inversion-level dependency increasing for thinner gate oxides. As mentioned in Section 3.10.3.3, a higher Coulomb scattering coefficient for holes may explain increased inversion-level dependency on pMOS flicker noise as seen by a lower value of V_{KF} given by Equation 3.124. Additionally, a higher value of C'_{OX} associated with thinner t_{ox} corresponds to a lower value of V_{KF}, again suggesting increased inversion-level dependency on flicker noise as seen by Equations 3.122 and 3.123. While gate-referred flicker-noise voltage generally increases with the inversion level, especially for pMOS devices, noise increases have been occasionally reported in weak inversion [108, 167]. Nearly constant, gate-referred flicker-noise voltage from weak to the onset of strong inversion will be shown in measurements later in this section. This suggests operating in this range of inversion to minimize flicker noise, while also minimizing $V_{DS,sat}$ as required in modern, low-voltage design.

Predictions of gate-referred flicker-noise voltage, summarized in Table 3.31, depend geometrically only on total gate area, WL. However, flicker noise often increases for short-channel devices [72, 126, 157, 163–165], including for devices denoted by Footnote b in Table 3.33. The noise increase may be caused by increased trap density near the Si–SiO$_2$ interface due to the close proximity of the source and drain implants to the channel [126, 164]. Avoiding short-channel devices, which also increases the gate area and reduces flicker noise, minimizes the potential flicker-noise increase.

The flicker-noise slope, AF, is often lower for nMOS devices compared to pMOS devices [11, 105, 117, 163, 165]. This is generally observed in Table 3.33 where nMOS AF is typically below unity while pMOS AF is typically close to unity. While predictions of gate-referred flicker-noise voltage assume AF is constant with the inversion level, AF often increases with increasing inversion [95, 98]. This might be attributed to deeper oxide traps resulting from the increasing vertical electric field under the gate caused by increasing $V_{EFF} = V_{GS} - V_T$. Deeper oxide traps would have longer trapping time constants giving increased lower-frequency noise components reflected by a higher value of AF. The author's measurements indicate that pMOS devices often have greater AF changes, especially a reduction of AF or shallowing of the flicker-noise slope in weak inversion.

Finally, the K_F summary in Table 3.33 reveals technology trends. The most notable trend is the unfortunate increase in K_F for $0.18\,\mu m$ and smaller-geometry processes, especially for pMOS devices in strong inversion. The increase in K_F is only partially countered by the reduction of gate-referred flicker-noise voltage, summarized in Table 3.31, resulting from higher C'_{OX}. Rapid thermal oxidation, chemical vapor deposition, and the use of nitrided oxides in smaller-geometry processes generally increase K_F [129]. Nitrided oxides, while providing resistance to hot-carrier interface-state generation and the diffusion of dopants into the oxide, are associated with higher trap density near the Si–SiO$_2$ interface giving increased K_F [129, 137, 155, 159, 161]. Nitrided oxides having reoxidization generally reduce the high flicker noise associated with regular nitrided oxides by reducing the density of traps, moving these further from the Si–SiO$_2$ interface, and increasing the trap energy [129]. In addition to higher values of K_F, K_F increases more with increasing inversion in smaller-geometry processes, consistent with a lower value of V_{KF} mentioned for higher C'_{OX}. Another technology trend observed for smaller-geometry processes is increased dispersion or variability of flicker noise [129, 147]. Also, as discussed later in Section 3.10.7, flicker noise associated with gate leakage current, significant for gate-oxide thickness below approximately 2 nm, is another unfortunate technology trend. Flicker-noise technology trends are discussed further in [129, 137, 147, 155, 159, 161] with current research directed towards gate-oxide processing that minimizes flicker noise. Additionally, gate processing using fine-grained, n+ polysilicon for nMOS devices may reduce flicker noise [156].

3.10.3.6 Measured and predicted flicker noise

As mentioned and included in Table 3.33, flicker noise was reported in [76] from moderate through deep strong inversion for a $0.18\,\mu m$ CMOS process. This thorough noise study was done for both nMOS and pMOS devices having sizes of $W/L = 10/10$, $10/5$, $10/2.5$, $10/1$, $10/0.28$, and $10/0.18\,\mu m/\mu m$. Devices were evaluated for V_{GS} equal to 0.5 V up to the process maximum of 1.8 V for operation in saturation at $V_{DS} = 1.8$ V. With permission, this noise data is reformatted in Figure 3.64 in terms of the square root of the hand analysis flicker-noise factor, $K'_F((nV)^2 \cdot \mu m^2)$, or $K'^{1/2}_F(nV \cdot \mu m)$. As observed in the flicker-noise expression in Table 3.31, $K'^{1/2}_F$ corresponds to the gate-referred flicker-noise voltage density $(nV/Hz^{1/2})$ at 1 Hz for a device having a gate area of $1\,\mu m^2$. Noise for other frequencies and device areas is easily found by multiplying $K'^{1/2}_F$ shown in Figure 3.64 by $1/(WLf^{AF})^{1/2}$, where W and L are MOS channel width and length in μm, f is the frequency in Hz, and AF is the flicker-noise PSD slope. The figure shows measured $K'^{1/2}_F$ for devices having $W/L = 10/10$ (squares) and $10/0.26$ (triangles) $\mu m/\mu m$. $K'^{1/2}_F$ for the other device geometries is similar.

In addition to measured values, Figure 3.64 includes a fitted overlay (solid lines) of predicted $K'^{1/2}_F$. $K'^{1/2}_F$ is predicted from the square root of $K'_F = K'_{F0}[1 + (V_{GS} - V_T)/V_{KF}]^2$ or $K'^{1/2}_F = K'^{1/2}_{F0}[1 + (V_{GS} - V_T)/V_{KF}]$ from Equation 3.123 as summarized in Table 3.31. $K'^{1/2}_{F0}$ is the square root of the flicker-noise factor, K'_{F0}, in weak or the weak-inversion side of moderate inversion before flicker noise increases, and V_{KF} is a voltage describing the increase in flicker noise with increasing inversion or increasing $V_{EFF} = V_{GS} - V_T$. V_{EFF} is clamped to zero when it is negative in weak or the weak-inversion side of moderate inversion.

Figure 3.64 reveals several things about flicker noise for the $0.18\,\mu m$ process evaluated. First, for operation near weak inversion, $K'^{1/2}_F = K'^{1/2}_{F0}$ is comparable for both nMOS and pMOS devices, having values of approximately 9000 and 7000 nV $\cdot \mu m$, respectively. Second, the figure shows

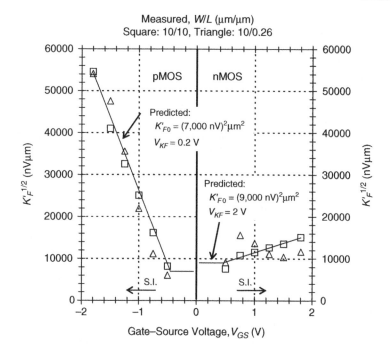

Measured, W/L (μm/μm)
Square: 10/10, Triangle: 10/0.26

Gate–Source Voltage, V_{GS} (V)

Figure 3.64 Measured and predicted flicker noise for a $0.18\,\mu m$ CMOS process from data in [76] with noise presented as the square root of the hand calculation, flicker-noise factor, $K_F'^{1/2}$. Gate-referred flicker-noise voltage density, $S_{VG}^{1/2}$, in units of $nV/Hz^{1/2}$, is found by multiplying the data shown by $1/(WLf^{AF})^{1/2}$, where W and L are MOS width and length in μm, f is the frequency in Hz, and AF is the flicker-noise PSD slope. nMOS devices show modest flicker-noise increase with inversion level, while pMOS devices show significant increase. Flicker noise is comparable for nMOS and pMOS devices for operation below the onset of strong inversion

significant inversion-level ($V_{EFF} = V_{GS} - V_T$) dependency for pMOS devices and only a slight or modest dependency for nMOS devices. For pMOS devices, $K_F'^{1/2}$ increases from 7000 to 56000 $nV \cdot \mu m$ as V_{GS} ranges from near or below 0.5 V to the process maximum of 1.8 V. This corresponds to a gate-referred flicker-noise voltage density, $S_{VG}^{1/2}$, at 1 Hz for a gate area of $1\,\mu m^2$ that increases from 7000 to 56000 $nV/Hz^{1/2}$, a factor-of-eight increase in noise density or factor-of-64 increase in noise PSD. For nMOS devices, $K_F'^{1/2}$ increases from 9000 to 14000 $nV \cdot \mu m$, which corresponds to a 56 % increase in noise density or 142 % increase in noise PSD. The noise increases are modestly different than values reported for [76] in Table 3.33 because the noise is fitted across different channel lengths.

As observed in Figure 3.64, $K_F'^{1/2}$ and the associated gate-referred flicker-noise voltage density increase linearly with V_{GS} or $V_{EFF} = V_{GS} - V_T$ for operation in moderate and strong inversion. This corresponds to an increase in K_F' and the associated noise PSD with $(V_{EFF} = V_{GS} - V_T)^2$ as predicted by the unified or carrier density, correlated mobility fluctuation model described in Section 3.10.3.3. This model was used for the development of Equation 3.123 (summarized in Table 3.31) used for the $K_F'^{1/2}$ prediction shown in the figure. In contrast, the Hooge, carrier mobility fluctuation model described in Section 3.10.3.2 predicts a noise PSD increase linearly with $V_{EFF} = V_{GS} - V_T$, while the McWhorter, carrier density fluctuation model described in Section 3.10.3.1 predicts constant noise PSD in its original form having a fixed value of K_F'. The significant increase in flicker noise for pMOS devices is well predicted using $V_{KF} = 0.2$ V, while the modest flicker-noise increase for nMOS devices is well predicted using a factor-of-10 higher value of $V_{KF} = 2$ V.

While Figure 3.64 shows a significant increase in gate-referred flicker-noise for pMOS devices, the increase is much less if operation is constrained between weak inversion and the onset of strong inversion ($IC = 10$). This is often required to keep $V_{EFF} = V_{GS} - V_T$ and $V_{DS,sat}$ sufficiently low for design in low-voltage processes. When approximating $V_{EFF} = 0.25\,\text{V}$ (Table 3.14) for maximum operation at $IC = 10$, $K_F'^{1/2}$ ranges from 7000 to 15800 nV $\cdot$ μm for pMOS devices and 9000 to 10100 nV $\cdot$ μm for nMOS devices. Here, gate-referred flicker-noise voltage density ($S_{VG}^{1/2}$) increases by 125 % for pMOS devices and 12.5 % for nMOS devices.

Figure 3.64 shows that the simple, inversion-dependent, flicker-noise prediction given by Equation 3.123 (summarized in Table 3.31) provides useful design guidance. This is, at least, likely true for 0.18 μm and larger-geometry processes. As mentioned at the end of Section 3.10.3.3, increased inversion-level dependency on gate-referred flicker noise is possible in smaller-geometry processes. In this case, Figure 3.64 provides a methodology for interpreting flicker noise and verifying its prediction for analog design.

Flicker noise for the 0.18 μm CMOS process described in Tables 3.2 and 3.3 and used for differential pair and current mirror, and OTA design examples in Chapters 4 and 5 was measured by the author. Measurements on this process, a different process from that considered in Figure 3.64 [76], show similarities in the flicker noise and its inversion-level dependency. Measurements were made using an author-designed noise test set that includes a transimpedance amplifier to convert MOS drain noise current to a sufficiently large noise voltage for spectral analysis. Measurements were made from moderate to deep strong inversion in saturation, as appropriate for most analog applications. Noise from the transimpedance amplifier feedback resistance, R_F, input noise current, and input noise voltage was analyzed in a spreadsheet and compared to the MOS noise to ensure negligible noise components from the measurement system. The noise analysis included the multiplication of amplifier input noise voltage by $1 + R_F/r_{ds}$ at the amplifier output, where r_{ds} is the MOS drain–source resistance. Finally, the amplifier included variable feedback capacitance around the feedback resistance to minimize noise peaking and manage the measurement bandwidth.

Multiple, decade-wide, fast Fourier transform (FFT) spectra were collected using 10000 spectral averages for each spectrum. Voltage gain from the MOS gate to the spectrum analyzer input was measured, and measured noise was referred to the gate as a noise voltage as recommended for flicker-noise measurements. MOS channel width and length, measured bias conditions, I_D, V_{DS}, V_{GS}, and V_{SB}, and measured transconductance, g_m, and drain–source conductance, $g_{ds} = 1/r_{ds}$, were recorded along with the calculated inversion coefficient. The inversion coefficient was calculated by $IC = I_D/[I_0(W/L)]$ from Table 3.6 using a technology current of $I_0 = 0.64$ and 0.135 μA for nMOS and pMOS devices, respectively, from Table 3.2. Channel length given was reduced by $DL = 0.028$ and 0.051 μm (also listed in the table) for nMOS and pMOS devices, respectively, to consider lateral diffusion effects.

The predicted noise given by Equation 3.123 and summarized in Table 3.31 is overlaid with the measured gate-referred flicker-noise voltage density. Noise is predicted using K_{F0}', which corresponds to the minimum value of K_F' in weak inversion, equal to $(5800\,\text{nV})^2 \cdot \text{μm}^2$ and $(5000\,\text{nV})^2 \cdot \text{μm}^2$ for nMOS and pMOS devices, respectively. V_{KF} is equal to 1 and 0.25 V for nMOS and pMOS devices, respectively. Finally, the flicker-noise PSD slope AF is equal to 0.85 and 1.05 for nMOS and pMOS devices, respectively. These extracted values of V_{KF} and AF are used in Table 3.3 for the example 0.18 μm CMOS process. The K_{F0}' values used in the table, however, are approximately 32 % higher, corresponding to 15 % higher gate-referred flicker-noise voltage density, than the values extracted in the measurements described here. The values are increased in the table to partially compensate for K_{F0}' increases observed for devices having channel lengths below 0.48 and 1 μm considered in measurements here.

Figure 3.65 shows measured and predicted gate-referred flicker-noise voltage density, $S_{VG}^{1/2}$, in design units of nV/Hz$^{1/2}$ for a $W/L = 16\,\text{μm}/1\,\text{μm}$, nMOS device in the 0.18 μm CMOS process described in Tables 3.2 and 3.3. For operation in moderate inversion at $IC = 1$, measured noise lies slightly below the predicted noise shown by the bottom dashed line. For operation at the onset of strong inversion at $IC = 10$, measured noise increases slightly and closely overlays predicted noise shown by the middle dashed line. Finally, for operation in deep strong inversion at $IC = 100$, measured noise increases modestly and averages about the predicted noise shown by the top dashed line. The predicted noise at 1 Hz, which is near the measured noise, is 1550, 1830, and 2620 nV/Hz$^{1/2}$ for $IC = 1$, 10, and 100, respectively. The noise voltage density increases by 18.4 % at $IC = 10$ from $IC = 1$, but increases by 69 % at $IC = 100$. As mentioned earlier for Figure 3.64, constraining operation below the onset of strong inversion ($IC = 10$) minimizes the increase in flicker noise.

Figure 3.66 shows measured and predicted $S_{VG}^{1/2}$ for a $W/L = 38 \mu m/0.48 \mu m$, pMOS device in the 0.18 μm CMOS process. For operation in moderate inversion at $IC = 0.9$, measured noise averages about the predicted noise shown by the bottom dashed line. For operation at $IC = 9$, 27, and 90, measured noise increases significantly and closely overlays predicted noise shown by the higher dashed lines. The predicted noise at 1 Hz, which is near the measured noise, is 1440, 2320, 3150, and 4860 nV/Hz$^{1/2}$ for $IC = 0.9$, 9, 27, and 90, respectively. The noise voltage density increases by 61 % at $IC = 9$ from $IC = 0.9$, but increases by 238 % at $IC = 90$. While constraining operation below the onset of strong inversion ($IC = 10$) manages the flicker-noise increase for the nMOS device shown in Figure 3.65, pMOS flicker-noise increases significantly even near the onset of strong inversion. Clearly, this increase must be considered in design.

Like the flicker-noise data of Figure 3.64 shown for a different 0.18 μm CMOS process, the gate-referred flicker-noise voltage density increases modestly with the inversion level for the nMOS device shown in Figure 3.65, while increasing significantly for the pMOS device shown in Figure 3.66. Additionally, as shown in Figure 3.64, flicker noise in moderate inversion is comparable for the nMOS and pMOS devices shown in Figures 3.65 and Figure 3.66. The noise can be directly compared because

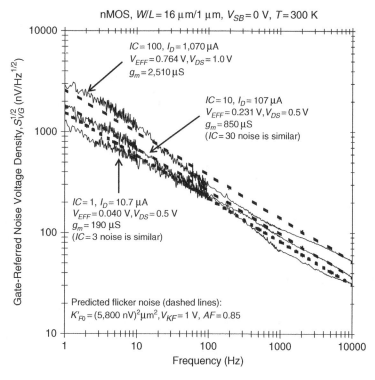

Figure 3.65 Measured and predicted gate-referred flicker-noise voltage density, $S_{VG}^{1/2}$, for a $W/L = 16 \mu m/1 \mu m$, nMOS device in a 0.18 μm CMOS process evaluated from moderate inversion to deep strong inversion. The noise increases modestly in strong inversion as predicted. Measured spectral humps that average around the straight-line noise predictions are a result of discrete traps in the small-area device

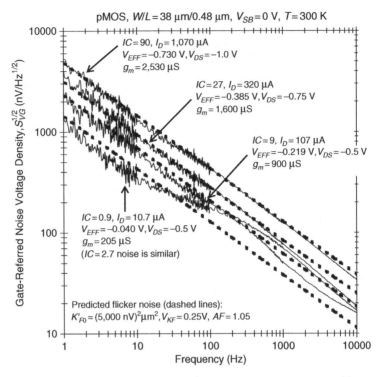

Figure 3.66 Measured and predicted gate-referred flicker-noise voltage density, $S_{VG}^{1/2}$, for a $W/L = 38\,\mu m/0.48\,\mu m$, pMOS device in a $0.18\,\mu m$ CMOS process evaluated from moderate inversion to deep strong inversion. The noise increases significantly in strong inversion as predicted. For operation in moderate inversion, the pMOS noise shown is comparable to the nMOS noise shown in Figure 3.65, with both devices having comparable gate area. However, pMOS noise increases much more in strong inversion compared to the nMOS noise

the devices have gate areas that are nearly equal. Finally, consistent with most reported values in Table 3.33, $AF = 0.85$ for the nMOS device is lower than $AF = 1.05$ for the pMOS device.

The comparison of measured and predicted flicker noise for the nMOS and pMOS devices shown in Figures 3.65 and 3.66 is complicated by spectral humps appearing in the measured noise, especially for the nMOS device. These spectral humps are a result of discrete traps in the devices that have small gate areas, WL, of approximately $16\,\mu m^2$. As mentioned in Section 3.10.3.1, discrete traps give rise to discrete RTS having discrete Lorentzian or generation–recombination noise spectra. Spectral humps are minimized by using large-area devices where the flicker-noise spectrum approaches the expected straight-line spectrum. This is well shown in [76] where the addition of multiple flicker-noise PSD spectra having spectral humps from small-area devices results in a composite spectrum equal to that of a large-area device having the expected straight-line spectrum. Large-area devices could have been used to obtain nearly straight-line noise spectra, but the devices evaluated illustrate the difficulty in predicting the flicker-noise spectra of small-area devices. The spectral humps are even more significant for devices having gate areas below $16\,\mu m^2$. For low-flicker-noise designs, like the micropower, low-noise CMOS preamplifiers described later in Sections 6.7 and 6.8, large gate areas required for minimizing flicker noise also minimize the presence of spectral humps.

As mentioned for Equation 3.123 and summarized in Table 3.31, the voltage V_{KF} models the increase in gate-referred flicker-noise voltage with increasing inversion (increasing $V_{EFF} = V_{GS} - V_T$). V_{KF}

was extracted at 1 and 0.25 V for the nMOS and pMOS devices shown in Figures 3.65 and 3.66. Equation 3.124 relates V_{KF} to α_{SC}, the Coulomb scattering coefficient. Assuming a low-field value of $\mu C'_{OX}$ of 355 and 75 μA/V^2 for nMOS and pMOS devices in the 0.18 μm process from Table 3.2 and $q = 1.602 \times 10^{-19}$ C, α_{SC} is extracted at 9×10^{-16} and 1.7×10^{-14} V $\cdot$ s for nMOS and pMOS devices, respectively. This is in the range of approximate values of 1.6×10^{-15} and 1.6×10^{-14} V $\cdot$ s for electrons and holes [147] and, again, shows increased Coulomb scattering that might be attributed to holes in pMOS devices. As mentioned in Section 3.10.3.3, this increased scattering accounts for the higher increase in pMOS flicker noise with inversion level compared to nMOS devices. This assumes that the physical basis of the unified or carrier density, correlated mobility fluctuation flicker-noise model is, in fact, valid.

3.10.3.7 Summary of gate-referred and drain-referred flicker noise

Table 3.34 summarizes MOS flicker-noise expressions and trends in terms of the design choices of IC, L, and I_D. The first entry gives the gate-referred flicker-noise voltage PSD, S_{VG}, shown in the MOS noise model of Figure 3.60. S_{VG} expressions are taken from Table 3.31 as derived from Equation 3.123. Inversion-level increases in the noise are found by increases in the operating flicker-noise factor, $K'_F = K'_{F0}[1 + (V_{GS} - V_T)/V_{KF}]^2$, where K'_{F0} is the flicker-noise factor in weak or the weak-inversion side of moderate inversion and V_{KF} is a voltage describing the flicker-noise increase with increasing effective gate–source voltage, $V_{EFF} = V_{GS} - V_T$. V_{EFF} is clamped to zero when it is negative in weak or the weak-inversion side of moderate inversion.

In order to express S_{VG} in Table 3.34 in terms of IC, L, and I_D, MOS gate area is expressed as $WL = (L^2/IC)(I_D/I_0)$ from Table 3.10 where $I_0 = 2n_0\mu_0 C'_{OX}U_T^2$ is the technology current described in Table 3.6. Additionally, $V_{EFF} = 2nU_T\sqrt{IC}$ is expressed in terms of the inversion coefficient from Table 3.14 for operation in strong inversion, exclusive of velocity saturation and VFMR increases. As mentioned for Equation 3.123, this clamps V_{EFF} to nearly zero for $IC < 0.5$ while providing a rough estimate in moderate inversion, before providing a good estimate in strong inversion. Velocity saturation and VFMR increases in V_{EFF}, which can increase S_{VG}, can be approximated by the inversion coefficient modification noted at the bottom of Table 3.34. The approximate expression for S_{VG} in the table gives units of (nV)^2/Hz for K'_{F0} in hand calculation units of (nV)$^2 \cdot \mu$m^2, L in μm, I_D and I_0 in μA, and f in Hz. Taking the square root of this gives gate-referred flicker-noise voltage density, $S_{VG}^{1/2}$, in the common design units of nV/Hz$^{1/2}$.

The trends listed in Table 3.34 result primarily from S_{VG} being inversely proportional to the gate area. When inversion-level or V_{EFF} increases in flicker noise are not significant, S_{VG} increases as IC, decreases as $1/L^2$, and decreases as $1/I_D$. While minimizing IC lowers S_{VG}, increasing L lowers the noise more because of the larger increase in required gate area. This is because an increase in L requires an equal increase in W, giving a squared increase in gate area for a given IC and I_D.

As noted in Table 3.34, when inversion-level or V_{EFF} increases in flicker noise are significant, S_{VG} increases as IC^2, with an increase with IC due to the decrease of required gate area and another increase with IC due to the noise increase with V_{EFF}. This assumes that $V_{EFF} = 2nU_T\sqrt{IC}$ in strong inversion. As noted in the table, the noise increase is even greater with IC if velocity saturation and VFMR effects increase V_{EFF}, where the noise decreases more rapidly than $1/L^2$ since increasing L lowers V_{EFF}.

S_{VG} is minimized at low IC, long L, and high I_D, which maximizes gate area. Additionally, operating at low IC helps ensure that S_{VG} does not increase with the level of inversion or V_{EFF}. The IC, L, and I_D trends for gate-referred flicker-noise voltage density, $S_{VG}^{1/2}$, and gate–source voltage mismatch, described later in Table 3.41, are the same when inversion-level increases in S_{VG} are not present and when mismatch is dominated by threshold-voltage mismatch. In this case, both $S_{VG}^{1/2}$ and gate–source voltage mismatch are inversely proportional to the square root of gate area. Finally, S_{VG} tracks the flicker-noise factor giving lower flicker noise for cleaner processes having lower trap density near

Table 3.34 MOS gate-referred flicker-noise voltage and drain-referred flicker-noise current PSD expressions and trends in terms of the inversion coefficient, channel length, and drain current. Voltage density ($nV/Hz^{1/2}$) and current density ($pA/Hz^{1/2}$) are found by taking the square-root of the approximate PSD expressions

Expression	$IC \uparrow$ L, I_D fixed	$L \uparrow$ IC, I_D fixed	$I_D \uparrow$ IC, L fixed
Gate-referred flicker-noise voltage PSD: $$S_{VG}(f) = \frac{K'_F}{WLf^{AF}} = \frac{K'_{F0}(1+V_{EFF}/V_{KF})^2}{WLf^{AF}}$$ $$\approx \left(\frac{IC}{L^2}\right)\left(\frac{I_0}{I_D}\right)\frac{K'_{F0}}{f^{AF}}\left(1+\frac{2nU_T\sqrt{IC}}{V_{KF}}\right)^2$$ Last expression gives units of $(nV)^2/Hz$ for L in µm, I_D and I_0 in µA, f in Hz, and K'_{F0} in $(nV)^2 \cdot \mu m^2$.	$\uparrow \propto IC$ $\uparrow \propto IC^2$ in SI when K'_F increase is significant, but increases more rapidly if V_{EFF} increases due to vel. sat.	$\downarrow \propto \dfrac{1}{L^2}$, decreasing more rapidly when K'_F increase is significant and V_{EFF} decreases due to less vel. sat.	$\downarrow \propto \dfrac{1}{I_D}$
Drain-referred flicker-noise current PSD: $$S_{ID}(f) = \frac{K'_F \cdot g_m^2}{WLf^{AF}} = \frac{K'_{F0}(1+V_{EFF}/V_{KF})^2 \cdot g_m^2}{WLf^{AF}}$$ $$\approx 2\left(\frac{IC}{L^2}\right)\left(\frac{I_D}{(\sqrt{IC+0.25}+0.5)^2}\right)\left(\frac{k_0}{n}\right)$$ $$\cdot\frac{K'_{F0}}{f^{AF}}\left(1+\frac{2nU_T\sqrt{IC}}{V_{KF}}\right)^2$$ Last expression gives units of $(fA)^2/Hz$ for L in µm, I_D in µA, k_0/n in $\mu A/V^2$, f in Hz, and K'_{F0} in $(nV)^2 \cdot \mu m^2$. Multiplying by 10^{-6} converts result to $(pA)^2/Hz$	$\uparrow \propto IC$ in WI Unchanged in SI when K'_F increase is not significant, but decreases if g_m decreases due to vel. sat. $\uparrow \propto IC$ in SI when K'_F increase is significant[a]	$\downarrow \propto \dfrac{1}{L^2}$, with more complex behavior when K'_F increase is significant and V_{EFF} and g_m are affected by vel. sat.	$\uparrow \propto I_D$

S_{VG} and S_{ID} are for saturation where $V_{DS} > V_{DS,sat}$.
Velocity saturation refers to both velocity saturation and VFMR effects.
To include velocity saturation and VFMR increases in V_{EFF}, replace IC in $\sqrt{IC}$ terms with $IC(1+IC/4IC_{CRIT})$ as described in Table 3.14. Less V_{EFF} over-prediction in moderate inversion is obtained by replacing $V_{EFF} = 2nU_T\sqrt{IC}$ with V_{EFF} given in Table 3.14. Negative values of V_{EFF} in weak or the weak-inversion side of moderate inversion are clamped to zero.
To include velocity saturation and VFMR decreases in g_m, replace IC in $\sqrt{IC+0.25}$ terms with $IC(1+IC/IC_{CRIT})$ as described in Table 3.17.
[a] When velocity saturation is significant, the decrease in g_m somewhat counters the increase in gate-referred noise voltage due to increasing V_{EFF}. Flicker noise, however, continues to increase with IC because of reduced gate area.

the Si–SiO$_2$ interface. The S_{VG} increase with inversion level or V_{EFF} is separately minimized through maximum process values of V_{KF}. As shown in measured data in Section 3.10.3.6, V_{KF} is considerably lower for pMOS devices in the 0.18 µm CMOS processes considered where the noise of these devices increases considerably more than that of nMOS devices as the level of inversion or V_{EFF} increases.

Figure 3.67 shows predicted $S_{VG}^{1/2}$ in $nV/Hz^{1/2}$ at 1 Hz for $L = 0.18$, 0.28, 0.48, 1, 2, and 4 µm pMOS devices in the 0.18 µm CMOS process described in Tables 3.2 and 3.3 operating at IC ranging from 0.01 (deep weak inversion) to 100 (deep strong inversion). Noise is found for other frequencies by multiplying the values given by $(1/f^{AF})^{1/2}$, where f is in Hz. Unlike the normalized small-signal parameters, g_m/I_D, $\eta = g_{mb}/g_m$, and $V_A = I_D/g_{ds}$, and intrinsic voltage gain and bandwidth, A_{Vi} and f_{Ti}, $S_{VG}^{1/2}$ is not independent of the drain current when evaluated in terms of the inversion coefficient.

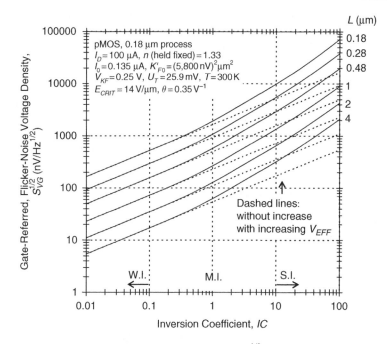

Figure 3.67 Predicted gate-referred flicker-noise voltage density, $S_{VG}^{1/2}$, at 1 Hz versus inversion coefficient, IC, for $L = 0.18$, 0.28, 0.48, 1, 2, and 4 μm, pMOS devices in a 0.18 μm CMOS process operating at a drain current of 100 μA. $S_{VG}^{1/2}$ increases as the square root of IC and decreases inversely with L, tracking the inverse square root of gate area, until it increases further due to inversion level or $V_{EFF} = V_{GS} - V_T$ increases in the noise. $S_{VG}^{1/2}$ for different drain currents is found by multiplying the values shown by $(100 \, \mu A / I_D)^{1/2}$. For comparable flicker-noise factors, nMOS $S_{VG}^{1/2}$ is higher in weak and moderate inversion because of lower gate area, but increases less with increasing inversion

Here, a drain current of 100 μA is assumed. The decrease in $S_{VG}^{1/2}$ with increasing drain current resulting from the required increase in gate area is easily found by multiplying the values shown in the figure by $(100 \, \mu A / I_D)^{1/2}$. pMOS devices are considered because of their greater inversion level or V_{EFF} increases in flicker noise compared to nMOS devices in the process.

$S_{VG}^{1/2}$ is predicted from the expression in Table 3.34 assuming $I_0 = 0.135 \, \mu A$ from Table 3.2, and $K_{F0}' = (5800 \, nV)^2 \cdot \mu m^2$ and $V_{KF} = 0.25 \, V$ from Table 3.3. K_{F0}' values given in Table 3.3 for the example 0.18 μm CMOS process are $(6700 \, nV)^2 \cdot \mu m^2$ and $(5800 \, nV)^2 \cdot \mu m^2$ for nMOS and pMOS devices, respectively, which are increased slightly from values of $(5800 \, nV)^2 \cdot \mu m^2$ and $(5000 \, nV)^2 \cdot \mu m^2$ extracted from the measurements shown in Figures 3.65 and 3.66. The values of K_{F0}' are increased to provide a better fit with other noise measurements in the process, including increased noise observed for channel lengths below $L = 1 \, \mu m$ (nMOS) and $L = 0.48 \, \mu m$ (pMOS) considered in Figures 3.65 and 3.66. As mentioned earlier in Section 3.10.3.5, flicker noise for short-channel devices may increase due to increased trap density near the Si–SiO$_2$ interface caused by the close proximity of the source and drain implants to the channel. While K_{F0}' values are increased from the values extracted in Figures 3.65 and 3.66, the values of V_{KF} and AF listed in Table 3.3 are equal to those extracted in the figures.

$S_{VG}^{1/2}$ is predicted, as noted at the bottom of Table 3.34, using V_{EFF} given by the expression in Table 3.14. This expression gives less over-prediction of V_{EFF} in moderate inversion and potentially less over-prediction of $S_{VG}^{1/2}$ since V_{EFF} controls the inversion-level increase of $S_{VG}^{1/2}$. $U_T = 25.9 \, mV$ ($T = 300 \, K$) and $n = 1.33$, which is held fixed and equal to the value used for g_m/I_D shown in Figure 3.27, are used for the V_{EFF} prediction, along with process values of $E_{CRIT} = 14 \, V/\mu m$ and $\theta = 0.35/V$ from Table 3.2 that model the increase associated with velocity saturation and VFMR effects. Finally, channel length is reduced by $DL = 0.051\,\mu m$ (also given in Table 3.2) from the values given in the $S_{VG}^{1/2}$ prediction of Figure 3.67 to include lateral diffusion effects.

As observed in Figure 3.67, when inversion-level or V_{EFF} increases in flicker noise are not significant, $S_{VG}^{1/2}$ tracks the inverse square root of gate area, increasing as $\sqrt{IC}$ and decreasing as $1/L$. This is shown by the dashed lines in the figure that overlay the solid lines for $IC < 1$. When inversion-level or V_{EFF} increases in flicker noise are significant, $S_{VG}^{1/2}$ increases closer to IC in strong inversion because of the noise increase with V_{EFF}. This is seen in the figure when the solid lines increase above the dashed lines that exclude the increase in noise with V_{EFF}.

As observed in Figure 3.67, $S_{VG}^{1/2}$ spans nearly four decades, being minimum for long-channel devices in deep weak inversion and maximum for short-channel devices in deep strong inversion. The highest values of $S_{VG}^{1/2}$ shown for short-channel devices in deep strong inversion may not be attainable for $I_D = 100\,\mu A$ shown because resulting widths described in Table 3.10 may be below values possible for layout. However, when verified for practical device geometries and scaled for the drain current used, the figure provides an estimate for $S_{VG}^{1/2}$ useful for design guidance.

Since $S_{VG}^{1/2}$ scales inversely with the square root of I_D, again tracking the inverse square root of gate area, it is a factor of 10 higher for $I_D = 1\,\mu A$ compared to values shown in Figure 3.67 for $I_D = 100\,\mu A$. This illustrates the flicker-noise problem associated with micropower design where channel length is often increased to obtain attainable widths for layout and manage flicker noise through increased gate area at the expense of intrinsic bandwidth described in Section 3.9.6.

For the comparable values of K_{F0}' listed in Table 3.3, nMOS $S_{VG}^{1/2}$ in weak or moderate inversion is higher than the pMOS values shown in Figure 3.67 because the required nMOS channel width and gate area are smaller by a factor of 4.7 for equal IC, L, and I_D. This is seen in the $WL = (L^2/IC)(I_D/I_0)$ expression in Table 3.10 for $I_0 = 0.64\,\mu A$ and $0.135\,\mu A$ listed in Table 3.2 for nMOS and pMOS devices, respectively. However, comparing pMOS $S_{VG}^{1/2}$ shown in Figure 3.67 with nMOS $S_{VG}^{1/2}$ shown later in Figure 4.14 shows that pMOS noise is initially smaller in weak and moderate inversion, but then approaches or exceeds the nMOS noise in strong inversion. This is because pMOS noise increases more with V_{EFF} as characterized by $V_{KF} = 0.25\,V$ compared to $V_{KF} = 1\,V$ for nMOS devices, with the V_{KF} values extracted from the measurements shown in Figures 3.65 and 3.66. Additionally, a larger pMOS noise increase is shown in Figure 3.64 for a different but similar $0.18\,\mu m$ CMOS process. As observed for the reported flicker-noise factors in Section 3.10.3.5 and the measurements mentioned, pMOS devices may no longer be preferred devices for low flicker noise in small-geometry processes, especially given their generally greater noise increase with inversion level or V_{EFF}.

While MOS flicker noise is a gate noise process that should be referred to the gate for prediction, it is sometimes useful to refer the noise to the drain. This is especially true for non-input devices in circuits where voltage inputs are converted to currents by input transconductors. In this case, drain noise current of non-input devices is divided by the input transconductance to refer the noise current to a noise voltage at the circuit input.

Drain-referred flicker-noise current PSD, S_{ID}, shown in the MOS noise model of Figure 3.60 is found from the gate-referred flicker-noise voltage PSD, S_{VG}, multiplied by the square of device transconductance, which refers the noise to the drain. Multiplying S_{VG} given by the first entry in Table 3.34 with the square of transconductance given by the first entry in Table 3.17 gives

$$S_{ID}(\text{flicker}) = S_{VG}(\text{flicker}) \cdot g_m^2 = \frac{K_{F0}'\left(1 + V_{EFF}/V_{KF}\right)^2}{WL f^{AF}} \cdot g_m^2$$

$$= \left(\frac{IC}{L^2}\right)\left(\frac{I_0}{I_D}\right)\frac{K_{F0}'}{f^{AF}}\left(1 + \frac{2nU_T\sqrt{IC}}{V_{KF}}\right)^2 \cdot \left(\frac{I_D}{nU_T\left(\sqrt{IC + 0.25} + 0.5\right)}\right)^2$$

$$= 2\left(\frac{IC}{L^2}\right)\left(\frac{I_D}{\left(\sqrt{IC + 0.25} + 0.5\right)^2}\right)\left(\frac{k_0}{n}\right)\cdot\frac{K_{F0}'}{f^{AF}}\left(1 + \frac{2nU_T\sqrt{IC}}{V_{KF}}\right)^2 \qquad (3.125)$$

In Equation 3.125, the substrate factor, n, which varies only slightly with the inversion level as described in Section 3.8.3.1, is assumed fixed where one unit of n in the g_m expression cancels with its average moderate inversion value, n_0, appearing in the technology current, $I_0 = 2n_0\mu_0 C'_{OX} U_T^2$. As a result, the flicker noise is expressed in terms of the low-field transconductance factor divided by a remaining unit of n, or $k_0/n = \mu_0 C'_{OX}/n$. This is the same transconductance factor that appears in expressions for strong-inversion drain current given in Tables 3.6 and 3.13.

As for S_{VG} described earlier, Table 3.34 summarizes expressions and trends for S_{ID} in terms of IC, L, and I_D. Because S_{ID} is given by the product of S_{VG} and g_m^2, S_{ID} expressions and trends are more complex than those for S_{VG}. The second expression for S_{ID}, repeated from Equation 3.125, excludes velocity saturation and VMFR increases in V_{EFF} and decreases in g_m, but these can be considered by the inversion coefficient modifications noted at the bottom of the table. In this expression, S_{ID} is given in $(fA)^2/Hz$ for L in μm, I_D in μA, k_0/n in $\mu A/V^2$, f in Hz, and K'_{F0} in hand calculation units of $(nV)^2 \cdot \mu m^2$. As noted in the table, multiplying by 10^{-6} converts the noise PSD to $(pA)^2/Hz$. Taking the square root of this then gives drain-referred flicker-noise current density, $S_{ID}^{1/2}$, in the common design units of $pA/Hz^{1/2}$.

As noted in Table 3.34, in weak inversion, S_{ID} increases directly with IC since S_{VG} increases as gate area decreases, while g_m/I_D and g_m are unchanged at their maximum values. Assuming inversion-level or V_{EFF} increases in S_{VG} are not significant, S_{ID} is nearly constant with IC in strong inversion since the S_{VG} increase associated with decreasing gate area is countered by the decrease in g_m/I_D and g_m. S_{VG} increases as IC, while g_m/I_D and g_m decrease as $1/\sqrt{IC}$ or $1/IC$ when squared, giving $S_{ID} = S_{VG} \cdot g_m^2$ constant with IC. If velocity saturation and VFMR effects reduce g_m/I_D and g_m further, S_{ID} actually rolls over and begins decreasing with increasing IC. However, when inversion-level or V_{EFF} increases in S_{VG} are significant, S_{ID} does not level off in strong inversion, but increases as the increase in S_{VG} due to decreasing gate area and increasing V_{EFF} exceeds the decrease in g_m^2.

S_{ID} is minimized at low IC, long L, and low I_D. While g_m/I_D and g_m are maximum at low IC in weak inversion, the decreased value of S_{VG} due to the required increase in gate area reduces S_{ID}. Additionally, operation at low IC helps ensure little inversion-level or V_{EFF} increases in S_{VG}. In all regions of operation, unless potentially if velocity saturation and VFMR effects are significant as noted in Table 3.34, S_{ID} decreases as $1/L^2$ through the large increase in gate area, just as S_{VG} does. This, again, emphasizes the reduction in flicker noise available when using longer-channel-length devices. The IC and L trends for drain-referred flicker-noise current density, $S_{ID}^{1/2}$, and relative drain current mismatch, described later in Table 3.41, are the same when inversion-level increases in S_{VG} are not present and when mismatch is dominated by threshold-voltage mismatch. In this case, both $S_{ID}^{1/2}$ and relative drain current mismatch are inversely proportional to the square root of gate area and proportional to g_m/I_D. However, $S_{ID}^{1/2}$ depends differently on the drain current and is minimized at low drain currents as described below, while relative drain current mismatch is minimized at high drain currents.

S_{ID} increases as I_D since the $1/I_D$ decrease in S_{VG} associated with the required increase in gate area is exceeded by the I_D^2 increase associated with the increase in g_m^2. Both drain-referred thermal-noise and flicker-noise current are minimized by reducing I_D as commonly done for non-input devices in circuits where input transconductors receive voltage inputs. Finally, like S_{VG}, S_{ID} tracks the flicker-noise factor, again indicating that processes having lower trap density near the Si–SiO$_2$ interface have lower flicker noise. Like the S_{VG} increase with inversion level or V_{EFF}, the associated S_{ID} increase is separately minimized through maximum process values of V_{KF}.

Figure 3.68 shows predicted $S_{ID}^{1/2}$ in $pA/Hz^{1/2}$ at 1 Hz for $L = 0.18, 0.28, 0.48, 1, 2$, and $4\,\mu m$ pMOS devices in the $0.18\,\mu m$ CMOS process described in Tables 3.2 and 3.3 operating at IC ranging from 0.01 to 100. Like $S_{VG}^{1/2}$ shown in Figure 3.67, noise is found for other frequencies by multiplying the values given by $(1/f^{AF})^{1/2}$, where f is in Hz. $S_{ID}^{1/2}$ is a function of drain current through the required gate area, like $S_{VG}^{1/2}$, but is also a function of drain current through g_m. Here, a drain current of $100\,\mu A$ is assumed, where the increase in $S_{ID}^{1/2}$ with increasing drain current is easily found by multiplying the values shown in the figure by $(I_D/100\,\mu A)^{1/2}$. $S_{ID}^{1/2}$ is predicted from the expression in Table 3.34 assuming that $k_0/n = 75\,\mu A/V^2/1.33 = 56.4\,\mu A/V^2$ with k_0 given in Table 3.2. As for $S_{VG}^{1/2}$ predicted

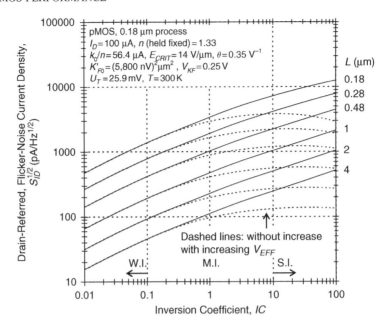

Figure 3.68 Predicted drain-referred flicker-noise current density, $S_{ID}^{1/2}$, at 1 Hz versus inversion coefficient, IC, for $L = 0.18, 0.28, 0.48, 1, 2,$ and 4 μm, pMOS devices in a 0.18 μm CMOS process operating at a drain current of 100 μA. $S_{ID}^{1/2}$ increases as the square root of IC in weak inversion and levels off in strong inversion, until it increases due to inversion level or $V_{EFF} = V_{GS} - V_T$ increases in the gate-referred flicker-noise voltage density, $S_{VG}^{1/2}$, shown in Figure 3.67. $S_{ID}^{1/2}$ for different drain currents is found by multiplying the values shown by $(I_D/100\,\mu A)^{1/2}$. For comparable flicker-noise factors, nMOS $S_{ID}^{1/2}$ is higher in weak and moderate inversion because of lower gate area, but increases less with increasing inversion

in Figure 3.67, V_{EFF} is predicted with less over-prediction in moderate inversion using the expression in Table 3.14 and the previously mentioned process parameter values.

As observed in Figure 3.68, in weak inversion where g_m/I_D and g_m are constant and maximum, $S_{ID}^{1/2}$ increases as $\sqrt{IC}$, tracking $S_{VG}^{1/2}$, which tracks the inverse square root of gate area. When inversion-level or V_{EFF} increases in $S_{VG}^{1/2}$ are not significant, which is shown by the dashed lines in the figure, $S_{ID}^{1/2}$ levels off in strong inversion. This is because, as mentioned, the increase in $S_{VG}^{1/2}$ due to the decrease in gate area is countered by the decrease in g_m/I_D and g_m. Once velocity saturation and VFMR effects are significant, $S_{ID}^{1/2}$ begins decreasing because of the decrease in g_m. However, when inversion-level or V_{EFF} increases in $S_{VG}^{1/2}$ are significant, $S_{ID}^{1/2}$ does not level off in strong inversion but continues increasing as shown in the figure when the solid lines increase above the dashed lines. $S_{ID}^{1/2}$ decreases as $1/L$, again tracking the inverse square root of gate area, with a potentially more complex behavior if velocity saturation effects increase V_{EFF} and decrease g_m/I_D.

As observed in Figure 3.68, $S_{ID}^{1/2}$ spans nearly three decades, being minimum for long-channel devices in weak inversion and maximum for short-channel devices in strong inversion. In weak inversion, the low $S_{VG}^{1/2}$ resulting from large required gate area exceeds the effect of maximum g_m that maximizes the conversion of $S_{VG}^{1/2}$ to $S_{ID}^{1/2}$. In strong inversion, the increase in $S_{VG}^{1/2}$ due to decreasing gate area and increasing V_{EFF}, when inversion-level noise increases are significant, is somewhat countered by the decrease in g_m/I_D and g_m giving reduced conversion of $S_{VG}^{1/2}$ to $S_{ID}^{1/2}$. This accounts for the smaller three-decade range of $S_{ID}^{1/2}$ shown in Figure 3.68 compared to the four-decade range of $S_{VG}^{1/2}$ shown in Figure 3.67.

Like $S_{VG}^{1/2}$ shown in Figure 3.67, the highest values of $S_{ID}^{1/2}$ shown in Figure 3.68 for short-channel devices in deep strong inversion may not be attainable for $I_D = 100\,\mu A$ considered because resulting widths may be below values possible for layout. However, when verified for practical device geometries and scaled for the drain current used, Figure 3.68 provides an estimate of $S_{ID}^{1/2}$ useful for design guidance.

For the comparable values of K_{F0}' listed in Table 3.3, nMOS $S_{ID}^{1/2}$ in weak or moderate inversion is higher than the pMOS values shown in Figure 3.68. This is because, as mentioned, nMOS $S_{VG}^{1/2}$ is higher because nMOS channel width and gate area are smaller by a factor of 4.7 for equal IC, L, and I_D. However, comparing pMOS $S_{ID}^{1/2}$ shown in Figure 3.68 with nMOS $S_{ID}^{1/2}$ shown later in Figure 4.18 shows that pMOS noise is initially smaller in weak and moderate inversion, but then approaches or exceeds the nMOS noise in strong inversion. This, again, is because pMOS $S_{VG}^{1/2}$ increases considerably more with increasing inversion or V_{EFF} compared to nMOS devices.

Flicker-noise $S_{VG}^{1/2}$ and $S_{ID}^{1/2}$ are generally inversely proportional to L due to the L^2 increase in gate area, where flicker noise is most effectively minimized by increasing L. Unfortunately, as described in Section 3.9.6, increasing L decreases intrinsic bandwidth by the inverse square of L, unless full velocity saturation is present where bandwidth is reduced by the inverse of L. The favorable decrease in flicker noise and unfavorable decrease in bandwidth with increasing L illustrate some of the important analog CMOS design tradeoffs. Chapter 4 discusses performance tradeoffs, and these tradeoffs are discussed further for specific design examples in Chapter 5.

This summary of MOS flicker is concluded with the following design suggestions:

- In processes having minimum channel length around $0.25\,\mu\mathrm{m}$ and below, re-evaluate the historical preference for pMOS devices for low flicker noise. nMOS devices may have comparable flicker noise, but with much smaller increases in gate-referred flicker-noise voltage at increasing inversion levels.

- Maximize gate area as this lowers flicker noise, except for a possible increase resulting from flicker noise in gate leakage current described below. In addition to lower flicker noise, large-area devices have less significant spectral humps associated with discrete traps near the Si–SiO$_2$ interface.

- Avoid short-channel devices to minimize possible flicker-noise increases associated with higher trap density caused potentially by nearby source and drain implants. Avoiding short-channel devices also helps increase gate area.

- Avoid extremes of operation in deep weak inversion where flicker noise may increase and flicker-noise slope decrease. Moderate inversion, giving nearly as low input-referred thermal-noise voltage due to nearly as high g_m/I_D and g_m, may be a better choice.

- Avoid extremes of operation in deep strong inversion where flicker noise may increase significantly, especially for pMOS devices where flicker noise can increase significantly even at the onset of strong inversion. Additionally, operation in deep strong inversion may increase flicker noise through noise contributions from source and drain access resistances.

- Use partially depleted (PD) or fully depleted (FD) SOI processes with caution as generation–recombination noise may exceed flicker noise because of high defect density at the buried oxide and back-gate interface [113, 121]. Generation–recombination noise can be shifted to lower frequencies by body and back-gate bias voltage choices.

- Consider flicker noise of non-input devices carefully to ensure negligible noise contributions for low-noise applications. This is discussed in Sections 6.7 and 6.8 for micropower, low-noise, CMOS preamplifier design examples.

- Measure gate-referred flicker-noise voltage for critical flicker-noise applications using relevant device geometries and various levels of inversion. This permits an assessment of the noise in weak or moderate inversion and its increase with increasing inversion.

- Finally, in processes having gate-oxide thickness around $2\,\mathrm{nm}$ and below, evaluate gate leakage current and its associated flicker-noise current. This could be a significant contributor to flicker noise in smaller-geometry processes as discussed later in Section 3.10.7. Unfortunately, gate leakage current and its associated flicker-noise current increase with gate area, working against the usual improvement in flicker noise as described in Sections 3.12.2.4 and 3.12.2.6. Noise contributions from gate leakage current can be minimized by selecting thicker gate-oxide options when these are available.

3.10.3.8 Flicker-noise corner frequency

The final discussion of MOS flicker noise considers the flicker-noise corner frequency, f_c. This is the frequency where thermal and flicker noise are equal. At frequencies below f_c, flicker noise dominates, and at frequencies above this, thermal noise dominates.

f_c can be observed directly from the noise spectrum or found experimentally from separate flicker- and thermal-noise measurements. When using separate measurements, flicker noise must be measured at sufficiently low frequencies, $f \ll f_c$, where the thermal noise is negligible. Correspondingly, thermal noise must be measured at sufficiently high frequencies, $f \gg f_c$, where the flicker noise is negligible. f_c is then found where the thermal and flicker noise are equal, which occurs when

$$S_{VG}(\text{thermal}) = S_{VG}(\text{flicker}, f_c) = \left(\frac{f_m}{f_c}\right)^{AF} S_{VG}(\text{flicker}, f_m) \tag{3.126}$$

S_{VG} (thermal) is the gate-referred thermal-noise voltage PSD and S_{VG} (flicker, f) is the gate-referred flicker-noise voltage PSD evaluated at either f_c or a general measurement frequency, f_m. The rightmost expression describes the measurement of flicker noise at f_m, where the measurement is extrapolated to f_c using the noise PSD slope, AF, described earlier in the flicker-noise expressions given in Tables 3.31 and 3.34. Measurement extrapolation requires knowing AF (typically 0.7–1.2) from the noise spectrum.

Solving Equation 3.126 gives

$$f_c = f_m \left[\frac{S_{VG}(\text{flicker}, f_m)}{S_{VG}(\text{thermal})}\right]^{1/AF} \tag{3.127}$$

For $AF = 1$, f_c is equal to the flicker-noise PSD at a measurement frequency of $f_m = 1\,\text{Hz}$ divided by the thermal-noise PSD. f_c increases with increasing flicker noise, but decreases with increasing thermal noise. While Equations 3.126 and 3.127 are illustrated for the gate-referred thermal- and flicker-noise voltage PSD, the drain-referred thermal- and flicker-noise current PSD can also be used as these are linked to the gate-referred noise voltage PSD by the square of device transconductance.

f_c can also be predicted from MOS parameters. Equating MOS gate-referred thermal-noise voltage PSD given in Table 3.30 and gate-referred flicker-noise voltage PSD given in Tables 3.31 and 3.34 evaluated at $f = f_c$ gives

$$\frac{4kT n \Gamma}{g_m} = \frac{K_F}{C_{OX}^{\prime 2} WL f_c^{AF}} \tag{3.128}$$

Solving for f_c where thermal and flicker noise are equal gives

$$f_c = \left(\frac{g_m}{4kT n \Gamma} \cdot \frac{K_F}{C_{OX}^{\prime 2} WL}\right)^{1/AF} \tag{3.129}$$

f_c is conveniently linked to the MOS intrinsic bandwidth, f_{Ti}, described earlier in Section 3.9.6. From Table 3.26, MOS intrinsic bandwidth is given by

$$f_{Ti} = \frac{g_m}{2\pi \left(\hat{C}_{gsi} + \hat{C}_{gbi}\right) WL C_{OX}^{\prime}} \tag{3.130}$$

where $\hat{C}_{gsi} = C_{gsi}/C_{GOX}$ and $\hat{C}_{gbi} = C_{gbi}/C_{GOX}$ are the intrinsic gate–source and gate–body capacitances normalized to the gate-oxide capacitance, C_{GOX}, as given in Table 3.23. Solving Equation 3.129 using f_{Ti} given by Equation 3.130 gives

$$f_c = \left(\frac{2\pi K_F}{4kT\, C'_{OX}} \cdot \chi \cdot f_{Ti}\right)^{1/AF} = \left[\frac{2\pi K_{F0}\,(1 + V_{EFF}/V_{KF})^2}{4kT\, C'_{OX}} \cdot \chi \cdot f_{Ti}\right]^{1/AF} \qquad (3.131)$$

where the dimensionless parameter, χ, is given by

$$\chi = \frac{\hat{C}_{gsi} + \hat{C}_{gbi}}{n\,\Gamma} \qquad (3.132)$$

In the right side of Equation 3.131, inversion-level or $V_{EFF} = V_{GS} - V_T$ increases in flicker noise are included by increases in the operating flicker-noise factor, $K_F = K_{F0}(1 + V_{EFF}/V_{KF})^2$, where K_{F0} is the flicker-noise factor in weak or the weak-inversion side of moderate inversion and V_{KF} is a voltage describing the flicker-noise increase with increasing V_{EFF}. This flicker-noise increase is described in Equation 3.123 and Table 3.31, where V_{EFF} is clamped to zero in weak or the weak-inversion side of moderate inversion when it is negative. The left side of Equation 3.131 can also be used where K_F is the flicker-noise factor for the actual level of inversion or V_{EFF} used.

Table 3.35 summarizes expressions and trends for f_c in terms of the design choices of inversion coefficient and channel length. The first expression gives f_c in terms of the measured thermal and flicker noise from Equation 3.127, extending the equation to show the use of either gate-referred noise voltage or drain-referred noise current PSD. The second expression gives f_c in terms of MOS parameters from Equation 3.131 using the value of χ given by Equation 3.132. As noted in the table, χ has values for n assumed fixed at $n = 1.33$ of $\chi = 0.45$, 0.64, and 0.76 for $IC = 0.1$, 1, and 10, respectively, corresponding to the onset of weak inversion, the center of moderate inversion, and the onset of strong inversion. These χ values correspond to $\hat{C}_{gsi} + \hat{C}_{gbi} = 0.309$, 0.499, and 0.663 from the expressions given in Table 3.23, and $\Gamma = 0.515$, 0.583, and 0.652 from Equation 3.106. The remaining expressions in Table 3.35 give convenient expressions for f_c in terms of flicker-noise factors, K_F in semiconductor units of C^2/cm^2, K_{FSPICE} in SPICE units of $V^2 \cdot F$, and K'_F in hand units of $(nV)^2 \cdot \mu m^2$. As discussed in Section 3.10.3.4, different flicker-noise factors result in tedious unit conversions for the calculation of flicker noise. The convenient expressions also involve T, C'_{OX}, χ, and f_{Ti}. T is expressed in K, C'_{OX} is expressed in convenient units of $fF/\mu m^2$, and f_{Ti} is expressed in Hz.

Assuming $K'_F = (5000\,nV)^2 \cdot \mu m^2$, $C'_{OX} = 8.41\,fF/\mu m^2$, and $AF = 1$, which are similar to values given for the $0.18\,\mu m$ CMOS process summarized in Tables 3.2 and 3.3 except for AF, gives $f_c \approx 0.005\% \cdot f_{Ti}$ for $\chi = 0.64$ at $IC = 1$. It is important to note that assuming AF equal to unity when it is not unity can result in significant errors in the prediction of f_c for the usual case where f_c is much greater than 1 Hz. pMOS devices have lower f_{Ti} and, thus, lower f_c due to their larger required widths and gate areas. In older, large-geometry processes where pMOS flicker-noise factors are significantly lower than those of nMOS devices, both f_c as a fraction of f_{Ti} and f_{Ti} are lower compared to values for nMOS devices.

Finally, Table 3.35 gives trends for f_c in terms of IC and L. When inversion-level or V_{EFF} increases in flicker noise are small, f_c tracks f_{Ti}, increasing directly with IC in weak inversion, as the square root of IC in strong inversion, and remaining nearly fixed in strong inversion under full velocity saturation. Like f_{Ti}, f_c decreases as $1/L^2$ or $1/L$ under full velocity saturation. When inversion-level or V_{EFF} increases in flicker noise are significant, f_c increases with IC more rapidly than f_{Ti}, increasing as $IC^{3/2}$ in strong inversion and as IC^2 under full velocity saturation. The more rapid increase in f_c corresponds to K_F and gate-referred flicker-noise voltage PSD increasing as V_{EFF}^2 or IC in strong inversion and IC^2 under full velocity saturation as found from Equation 3.131 and Table 3.14. If small-geometry effects described in Section 3.10.2.2 increase the thermal-noise factor, Γ, this increases the thermal noise and lowers f_c.

Table 3.35 MOS flicker-noise corner frequency, f_c, expressions and trends in terms of the inversion coefficient and channel length. When evaluated in terms of the inversion coefficient, f_c is independent of the drain current

Flicker-noise corner frequency	$IC \uparrow$ L fixed	$L \uparrow$ IC fixed
From measured data: $$f_c = f_m \left[\frac{S_{VG}(\text{flicker}, f_m)}{S_{VG}(\text{thermal})} \right]^{1/AF} = f_m \left[\frac{S_{ID}(\text{flicker}, f_m)}{S_{ID}(\text{thermal})} \right]^{1/AF}$$ where f_m is the measurement frequency in Hz		
From MOS parameters: $$f_c = \left[\frac{2\pi K_F}{4kT\, C'_{OX}} \cdot \chi \cdot f_{Ti} \right]^{1/AF}$$ where $\chi = \dfrac{\hat{C}_{gsi} + \hat{C}_{gbi}}{n\Gamma} = 0.45, 0.64, 0.76$ for $IC = 0.1, 1, 10$ and $n = 1.33$ Convenient expressions: $$f_c = \left[3.79 \times 10^{27} \left(\frac{300\,\text{K}}{T} \right) \left(\frac{K_F}{\text{C}^2/\text{cm}^2} \right) \left(\frac{\text{fF}/\mu\text{m}^2}{C'_{OX}} \right) \cdot \chi \cdot f_{Ti} \right]^{1/AF}$$ $$= \left[3.79 \times 10^{20} \left(\frac{300\,\text{K}}{T} \right) \left(\frac{K_{FSPICE}}{\text{V}^2 \cdot \text{F}} \right) \cdot \chi \cdot f_{Ti} \right]^{1/AF}$$ $$= \left[3.79 \times 10^{-13} \left(\frac{300\,\text{K}}{T} \right) \left(\frac{K'_F}{(\text{nV})^2 \cdot \mu\text{m}^2} \right) \left(\frac{C'_{OX}}{\text{fF}/\mu\text{m}^2} \right) \cdot \chi \cdot f_{Ti} \right]^{1/AF}$$	Generally trends are similar to those for f_{Ti} described in Table 3.26 $\uparrow \propto IC$ in WI $\uparrow \propto \sqrt{IC}$ in SI if no vel. sat.[a] Unchanged if full vel. sat.[a]	Generally trends are similar to those for f_{Ti} described in Table 3.26 $\downarrow \propto \dfrac{1}{L^2}$ if no velocity saturation[b] $\downarrow \propto \dfrac{1}{L}$ if full vel. sat.[b]

f_c is for saturation where $V_{DS} > V_{DS,sat}$.
Velocity saturation refers to both velocity saturation and VFMR effects.
$\hat{C}_{gsi} = C_{gsi}/C_{GOX}$ and $\hat{C}_{gbi} = C_{gbi}/C_{GOX}$ are given in Table 3.23.
Γ, given by Equation 3.106, increases from one-half in weak inversion to two-thirds in strong inversion, excluding small-geometry increases described in Section 3.10.2.2. IC and L trends exclude small-geometry increases in Γ.
$C'_{OX} = 34.5\,\text{fF}/\mu\text{m}^2$ ($1\,\text{nm}/t_{ox}$), where t_{ox} is the electrical gate-oxide thickness described in Table 3.1.
Inversion-level or V_{EFF} increases in flicker noise can be included by increasing flicker-noise factors from their values in weak or the weak-inversion side of moderate inversion, e.g., $K_F = K_{F0}(1 + V_{EFF}/V_{KF})^2$, as described in Tables 3.31 and 3.34. V_{EFF} is clamped to zero when it is negative in weak or the weak-inversion side of moderate inversion.
Convenient expressions assume $k = 1.3806 \times 10^{-23}\,\text{J/K}$.
[a] f_c increases more rapidly when inversion-level or V_{EFF} increases in flicker noise are significant.
[b] Behavior is more complex when inversion-level or V_{EFF} increases in flicker noise are significant.

f_c generally follows the f_{Ti} trends shown in Figure 3.57, spanning nearly five decades over the design choices shown for inversion coefficient and channel length. When inversion-level or V_{EFF} increases in flicker noise are significant, f_c spans more than the five decades shown for f_{Ti} because of additional increases in strong inversion. Like f_{Ti}, f_c is minimized for long-channel devices in weak inversion and maximized for short-channel devices in strong inversion. Also like f_{Ti}, f_c is independent of drain current when evaluated in terms of the inversion coefficient. Finally, the f_c fraction of f_{Ti} is governed primarily by the flicker-noise factor K_F divided by C'_{OX}, K_{FSPICE} directly, or K'_F multiplied by C'_{OX}. As a result, cleaner processes having lower trap density near the Si–SiO$_2$ interface and, correspondingly, lower flicker-noise factors, have lower values of f_c. The increase in f_c associated with inversion-level or V_{EFF} increases in flicker noise is minimized by the use of devices having lower flicker-noise increases. As described in Sections 3.10.3.5 and 3.10.3.6, nMOS devices often have lower flicker-noise increases compared to pMOS devices.

3.10.4 Gate, Substrate, and Source Resistance Thermal Noise

In addition to channel thermal noise described in Section 3.10.2, gate resistance resulting from the gate polysilicon material and metal contacts contributes to MOS thermal noise [15, pp. 498–500, 64, 65, 71, 72, 76, 104, pp. 22–24, 168–170]. This noise is easily modeled as a thermal-noise voltage appearing in series with the gate resistance. Additionally, as described below, this noise along with that associated with substrate and source resistance thermal noise is easily included in either the MOS gate-referred noise voltage PSD, S_{VG}, or drain-referred noise current PSD, S_{ID}, shown in the MOS noise model in Figure 3.60.

Table 3.36 gives expressions for MOS drain-referred thermal-noise current and gate-referred thermal-noise voltage PSD resulting from gate resistance, R_G. Ensuring that R_G is much less than $R_N = n\Gamma/g_m$ ensures that R_G noise contributions are negligible. R_N is the effective noise resistance associated with the gate-referred thermal-noise voltage resulting from channel thermal noise. The comparison of R_G and R_N is included in the table where R_N is described in Table 3.30.

Table 3.36 Expressions for MOS drain-referred noise current and gate-referred noise voltage PSD resulting from gate, substrate, and source resistance noise contributions

---

Expression

Noise PSD contribution from gate resistance, R_G:
$$S_{ID} = 4kT \cdot R_G g_m^2, \quad S_{VG} = 4kT \cdot R_G$$

Calculation of R_G:

Gate contacted on one end: $R_G = \dfrac{\rho_G}{3m^2}\left(\dfrac{W}{L}\right) + \dfrac{R_{GC}}{m}$

Gate contacted on both ends: $R_G = \dfrac{\rho_G}{12m^2}\left(\dfrac{W}{L}\right) + \dfrac{R_{GC}}{2m}$

R_G noise contribution is negligible if $R_G \ll R_N$
W/L is the total composite shape factor
m is the number of gate fingers
ρ_G is the gate polysilicon resistance in ohms per square
R_{GC} is the total resistance associated with each individual gate contact, including the gate extension
R_G noise contribution can be lowered by contacting gates on both ends on smaller, paralleled MOS devices corresponding to a sufficiently high number of gate fingers, m

---

Noise PSD contribution from substrate resistance, R_B:

$$S_{ID} = 4kT \cdot R_B g_{mb}^2, \quad S_{VG} = 4kT \cdot R_B \left(\frac{g_{mb}}{g_m}\right)^2$$

R_B noise contribution is negligible if $R_B \ll (g_m/g_{mb})^2 \cdot R_N$
R_B noise contribution can be lowered through the use of a low-resistivity epitaxial layer below the MOS substrate along with frequent substrate contacts
R_B noise contribution can also be lowered by increasing V_{SB}, which lowers n and g_{mb}

Noise PSD contribution from source resistance, R_S:

$$S_{ID} = 4kT \cdot \frac{1}{R_S}\left(\frac{g_{ms} R_S}{1 + g_{ms} R_S}\right)^2, \quad S_{VG} = 4kT \cdot R_S \left(\frac{g_{ms}}{g_m}\right)^2 = 4kT \cdot R_S n^2$$

R_S noise contribution is negligible if $R_S \ll (1/n)^2 \cdot R_N$

$R_N = n\Gamma/g_m$ is the equivalent noise resistance for MOS gate-referred thermal-noise voltage resulting from channel thermal noise as described in Table 3.30.
$g_{ms} = g_m + g_{mb} + g_{ds} \approx g_m + g_{mb} = n g_m$, where $g_{mb}/g_m = n - 1$ from Table 3.20.

Table 3.36 also includes expressions for estimating R_G [15, pp. 498–500, 104, pp. 23–24, 169]. Here, each gate finger has dimensions of $W/m \times L$, where W is the composite MOS channel width, L is the channel length, and m is the number of gate fingers. Each gate finger has $(W/L)/m$ squares of resistance reduced to $(W/L)/(3m)$ through a distributed analysis if the gate fingers are contacted at one end. Contacting the gate fingers on both ends reduces this to $(W/L)/(12m)$. Finally, paralleling all gate fingers together gives a total gate resistance of $(W/L)/(3m^2)$ or $(W/L)/(12m^2)$ squares for contacts at one or both ends, respectively. Multiplying the gate resistance in squares by the gate resistance per square, ρ_G, gives R_G excluding the gate extension and contact resistance. This is included by the addition of R_{GC}/m or $R_{GC}/(2m)$ for gate contacts on one or both ends, respectively. R_{GC} is the contact resistance of each individual gate contact including the gate extension, vias, and contacts. As mentioned in [76], contact resistance can be a significant contributor to R_G in silicided processes having low gate resistivity.

Contacting MOS gates on both ends and ensuring enough individual gate fingers (sufficiently high value of m) minimize R_G through the paralleling of many shorter, low-resistance gate fingers. Large-shape-factor (W/L) devices typical of low-noise input devices require careful layout to avoid long and narrow gate fingers having high resistance. Additionally, protection pad circuits must be carefully evaluated to ensure negligible noise contributions. From the expression in Table 3.28, a 50 Ω series protection resistance introduces a thermal-noise voltage of $0.91\,\mathrm{nV/Hz}^{1/2}$ at $T = 300\,\mathrm{K}$.

In addition to the noise contributed by R_G, substrate resistance, R_B, contributes thermal noise by introducing a noise voltage in series with the substrate that drives the body-effect transconductance, g_{mb} [104, pp. 24–25, 171–174]. Table 3.36 gives expressions for MOS drain-referred thermal-noise current and gate-referred thermal-noise voltage PSD resulting from R_B. Referring R_B to the gate as an effective noise resistance gives $R_B(g_{mb}/g_m)^2$. Comparing this to R_N shows that R_B must be much less than the substrate-referred effective noise resistance, $R_N(g_m/g_{mb})^2$, to ensure negligible noise contributions. This comparison is included in the table.

While estimating R_G is relatively simple, excluding perhaps contact resistance estimation, estimating R_B is complex [172] and will not be attempted here. In [71], R_B is extracted from noise measurements associated with different values of source–body voltage, V_{SB}. Here, a $W/L = 200\,\mu\mathrm{m}/0.35\,\mu\mathrm{m}$, nMOS device in a $0.35\,\mu\mathrm{m}$ CMOS process has an estimated $R_B = 901\,\Omega$, while a $W/L = 2000\,\mu\mathrm{m}/0.35\,\mu\mathrm{m}$ device has an estimated $R_B = 382\,\Omega$. Assuming $g_{mb}/g_m = 0.3$, typical for bulk CMOS processes, these values of R_B are reduced by $(g_{mb}/g_m = 0.3)^2$ when referred to the gate, giving gate-referred resistance values of 100 and 42 Ω, respectively. The associated gate-referred thermal-noise voltage then corresponds to 1.29 and $0.83\,\mathrm{nV/Hz}^{1/2}$, respectively, at $T = 300\,\mathrm{K}$ from the expression in Table 3.28. This noise can be significant since, at sufficiently high drain currents, overall gate-referred thermal-noise voltage is in the vicinity of $1\,\mathrm{nV/Hz}^{1/2}$ [71] for these large-shape-factor devices.

R_B is minimized by the use of a highly doped, low-resistivity epitaxial layer below the lighter-doped substrate combined with frequent substrate contacts. If a low-resistivity epitaxial layer is not available, frequent substrate contacts become increasingly important. This could require the use of smaller, paralleled devices with local perimeter substrate contacts. Such layout techniques, like common-centroid layout techniques, necessarily increase layout area and layout capacitance. In addition to layout techniques, the application of non-zero V_{SB} lowers n and g_{mb}, and reduces R_B noise contributions. However, modern low-voltage processes limit V_{SB} values well below 1.5 V [71] and 2 V [173] reported for reducing R_B effects.

In addition to the noise contributed by R_G and R_B, the source access resistance, R_S, associated with contacting the channel on the source side contributes thermal noise. Table 3.36 gives expressions for MOS drain-referred thermal-noise current and gate-referred thermal-noise voltage PSD resulting from R_S. Referring R_S to the gate as an effective noise resistance gives $R_S \cdot n^2$. Comparing this to R_N shows that R_S must be much less than the source-referred effective noise resistance, $R_N(1/n)^2$, to ensure negligible noise contributions. This comparison is included in the table.

R_S is often kept sufficiently low by contacts along the full length of source. This is different from the gate that is contacted, at most, at two ends and the substrate that is contacted along the perimeter of devices. R_S can be significant for operation at high levels of inversion at high current densities, but operation here is usually avoided in low-noise designs. This is because operation at high inversion levels gives low g_m/I_D (Table 3.17) resulting in inefficient production of g_m and higher gate-referred thermal-noise voltage than that available at lower levels of inversion. When R_S is significant, it not only increases the thermal noise but also degenerates MOS transconductance. The degenerated transconductance from the expression in Table 3.16 is given by $g_m/(1 + g_{ms}R_S)$, where g_m is the native transconductance and $g_{ms} = g_m + g_{mb} + g_{ds}$ is the source transconductance. The drain-referred thermal-noise current PSD due to R_S given in Table 3.36 is divided by the square of degenerated transconductance to refer the noise to the gate-referred thermal-noise voltage PSD.

Drain access resistance, R_D, noise contributions are not included in the expressions of Table 3.36. These noise contributions are often negligible since this extrinsic resistance appears in series with the high drain–source resistance, r_{ds}, associated with operation in saturation.

3.10.5 Channel Avalanche Noise

Normally, MOS channel thermal noise increases only slightly with increasing V_{DS} in saturation. However, if V_{DS} is sufficiently high, weak avalanching can occur that can significantly increase the noise [76, 175, 176]. Weak avalanching was deliberately introduced in a $0.18\,\mu m$ CMOS process for V_{DS} up to 3 V, well above 1.8 V allowed for the process [76]. This resulted in a significant increase in MOS thermal noise.

The presence of avalanche noise can be evaluated by noise measurements in saturation at various values of V_{DS}. If noise increases significantly as V_{DS} increases, avalanche noise is likely present. While avalanche noise was not observed at the maximum process V_{DS} of 1.8 V for one $0.18\,\mu m$ process [76], the presence of significant hot-electron substrate current effects in the example $0.18\,\mu m$ process considered here suggests that avalanche noise is possible. Figures 3.41 and 3.42 showed significant substrate current and drain current increases for V_{DS} equal to the maximum process specification of 1.8 V. Avalanche noise is possible because hot-electron substrate current is caused by weak avalanching [15, p. 286].

3.10.6 Induced Gate Noise Current

Capacitive coupling between the gate and thermal noise in the MOS channel results in induced gate noise current [15, pp. 492–495, 51, 55, 58, pp. 88–91, 64, 65, 67, 76]. Induced noise current is included in the gate noise current PSD, S_{IG}, shown in the MOS noise model in Figure 3.60. This noise current flows into the gate circuit impedance and can contribute significant circuit noise at high frequencies.

Table 3.37 gives an expression for induced gate noise current PSD where C_{gsi} is the intrinsic gate–source capacitance coupling the gate into the channel. While induced gate noise current results from channel thermal noise, it does not have the constant or white spectrum with frequency associated with thermal noise. Instead, induced gate noise current is "blue noise" where the PSD increases as the square of frequency and the density increases directly with frequency. This is because, as frequency increases, the gate is increasing coupled to the noisy MOS channel through C_{gsi}.

The expression given in Table 3.37 for induced gate noise current PSD excludes a 4/3 multiplier term often included for long-channel devices operating in strong inversion [58, p. 90, 64, 76]. This term is reported as $(4/3)/n$ in [16, p. 282, 67], which for $n = 1.3$, typical of bulk CMOS processes in strong inversion, gives a value of $(4/3)/n$ very close to unity. As a result of its near-unity value and the reduced accuracy required for expressions providing design guidance, the $(4/3)/n$ term is neglected here. This term can increase for short-channel devices, mostly due to the induced gate current noise from gate resistance [76], which is considered separately later.

As noted in Table 3.37, induced gate noise current is partially correlated with drain thermal-noise current because both arise from channel thermal noise. The measured correlation coefficient, ρ, varies with frequency and devices, but is generally near $\rho = j(5/32)^{1/2}$ or $\rho = j0.395$ (0.395 with a 90 degree phase relationship) [65, 76] as predicted for long-channel devices [58, p. 90].

Table 3.37 conveniently gives induced gate noise current PSD in terms of the ratio of operating frequency to the intrinsic bandwidth, f'_{Ti}, associated with C_{gsi}. Here, f'_{Ti} is substituted for $g_m/(2\pi C_{gsi})$, which allows C_{gsi} to be removed from the noise expression. At $f = f'_{Ti}$, induced gate noise current PSD rises to $4kT \cdot (g_m/5)$, which is approximately one-fifth the drain-referred thermal-noise current PSD of $4kT \cdot n\Gamma g_m$ (Table 3.29) for $n\Gamma \approx 1$. As noted in Table 3.37, using the regular and lower intrinsic bandwidth, f_{Ti}, associated with both C_{gsi} and C_{gbi} increases the prediction of induced gate

Table 3.37   MOS gate noise current PSD expressions. Current density in the design units of $\text{pA}/\text{Hz}^{1/2}$ is found by taking the square root of the convenient PSD expressions

---

Expression

Induced gate noise current PSD from channel thermal noise:

$$S_{IG} = 4kT \cdot \frac{(2\pi f)^2 C_{gsi}^2}{5g_m} = 4kT \left(\frac{f}{f'_{Ti}}\right)^2 \left(\frac{g_m}{5}\right)$$

$$= \frac{(4.07\,\text{pA})^2}{\text{Hz}} \left(\frac{T}{300\,\text{K}}\right) \left(\frac{f}{f'_{Ti}}\right)^2 \left(\frac{g_m/5}{1000\,\mu\text{S}}\right)$$

Gate noise current is partially correlated with drain thermal-noise current, with correlation coefficient $\rho = j0.395$ for long-channel devices. The multiplier of 4/3 or $(4/3)/n$ described in the text is neglected since its value is close to unity

---

Induced gate noise current PSD from gate resistance:

$$S_{IG} = 4kT \cdot \left(R_G (2\pi f)^2 C_{gsi}^2\right) = 4kT \left(\frac{f}{f'_{Ti}}\right)^2 \left(\frac{g_m}{5}\right)(5g_m R_G)$$

$$= \frac{(4.07\,\text{pA})^2}{\text{Hz}} \left(\frac{T}{300\,\text{K}}\right) \left(\frac{f}{f'_{Ti}}\right)^2 \left(\frac{g_m/5}{1000\,\mu\text{S}}\right)(5g_m R_G)$$

Gate noise current is correlated with gate resistance thermal noise with correlation coefficient $\rho = j1$

Gate noise current is negligible compared to induced gate noise from channel if $R_G \ll 1/(5g_m)$, which is also required for negligible R_G thermal-noise contributions as described in Table 3.36

Gate noise current PSD from gate leakage current:

$$S_{IG} = 2qI_G = \frac{(0.566\,\text{pA})^2}{\text{Hz}} \cdot \left(\frac{I_G}{1\,\mu\text{A}}\right)$$

I_G is gate DC leakage current

Gate noise current also has a flicker-noise component giving a low-frequency corner frequency in the vicinity of 10 kHz. Flicker noise can be included by multiplying the noise expression by $(1 + f_c/f)$ where f_c is the corner frequency

---

Corner frequency between gate leakage and induced gate noise currents:

$$f_c(\text{induced}) = f'_{Ti}\sqrt{\frac{5I_G}{2U_T g_m}} = f'_{Ti} \cdot 0.311 \sqrt{\left(\frac{300\,\text{K}}{T}\right) \left(\frac{I_G}{1\mu\text{A}}\right) \left(\frac{1000\,\mu\text{S}}{g_m}\right)}$$

Induced noise current from gate resistance can be included by replacing g_m with $g_m(1 + 5g_m R_G)$

$f'_{Ti} = g_m/(2\pi C_{gsi})$. Using $f_{Ti} = g_m/[2\pi (C_{gsi} + C_{gbi})]$ from Table 3.26 instead of f'_{Ti} conservatively over-predicts S_{IG} induced from channel thermal noise and gate resistance and under-predicts f_c (induced).

Convenient expressions assume $k = 1.3806 \times 10^{-23}\,\text{J/K}$ and $q = 1.602 \times 10^{-19}\,\text{C}$.

noise current slightly. In strong inversion where C_{gbi} is well below C_{gsi}, f'_{Ti} and f_{Ti} are nearly equal, giving nearly equal predictions of gate noise current.

Table 3.37 includes a second component of induced gate noise current, this noise current arising from the thermal noise of R_G in series with C_{gsi} [76]. This noise current has the same form and frequency dependency as the induced gate noise current arising from channel thermal noise, but is capacitively correlated to the thermal noise associated with R_G with $\rho = j1$ (1 with a 90 degree phase relationship) and is uncorrelated with channel thermal noise. As noted in the table, gate noise current resulting from R_G is negligible if R_G itself is a negligible contributor to MOS thermal noise. As mentioned in Table 3.36, this occurs when R_G is much less than $R_N = n\Gamma/g_m$.

The impact of induced gate noise current from both channel and gate resistance thermal noise is found by multiplying the noise current by the gate circuit impedance, which predicts the resulting gate noise voltage. If the gate circuit impedance is low, for example equal to or less than $1/g_m$, induced gate noise current is usually a negligible contributor to circuit noise. However, it can be significant for applications having high gate circuit impedances operating at frequencies approaching f'_{Ti} or f_{Ti}.

3.10.7 Gate Leakage Noise Current

Section 3.12.1.1 describes DC gate leakage current associated with direct tunneling when the gate-oxide thickness is less than approximately 3 nm. In addition to drawing current from circuits connected to the gate, the leakage current contains shot and flicker noise that introduce circuit noise [76, 154, 164, 177–179]. Like induced gate noise current described above, noise current associated with gate leakage current is included in the gate noise current PSD, S_{IG}, shown in the MOS noise model in Figure 3.60. This noise current flows into the gate circuit impedance and can contribute significant circuit noise for devices having significant gate leakage current.

Measured gate leakage noise current PSD closely matches $S_{IG} = 2qI_G$, the full shot noise associated with the DC gate leakage current, I_G [76, 154]. In one experiment, gate leakage noise current PSD closely tracked the expected shot noise as I_G varied from 1 to $100\,\mu A$ for a $W/L = 10\,\mu m/10\,\mu m$, nMOS device having $t_{ox} = 1.5\,nm$ (as always, this is the effective electrical value of t_{ox} as described at the bottom of Table 3.1) [76]. In another experiment, the noise current closely tracked the expected shot noise as I_G varied from 1 nA to $10\,\mu A$ for a pMOS device having this same size and value of t_{ox} [154]. Measurements were taken at mid-band frequencies to avoid gate-flicker-noise current at low frequencies and induced gate noise current at high frequencies.

Table 3.37 lists gate noise current PSD resulting from gate leakage current shot noise. As shown in the convenient expression, a gate leakage current of $1\,\mu A$ results in a gate noise current density of $S_{IG}^{1/2} = 0.566\,pA/Hz^{1/2}$. Unlike induced gate noise current given in the table that increases with frequency, gate leakage current shot noise is constant or "white" with frequency.

Solving for the frequency where gate leakage and induced gate noise current PSD expressions in Table 3.37 are equal finds a noise corner frequency, f_c (induced), for the induced noise. At frequencies sufficiently below f_c (induced), gate noise current is dominated by the leakage noise current, which is constant with frequency except at low frequencies where it increases due to flicker noise. At frequencies sufficiently above f_c (induced), gate noise current is dominated by the induced noise current, which increases with frequency. As listed in the table, f_c (induced) is equal to f'_{Ti} multiplied by $[5I_G/(2U_Tg_m)]^{1/2}$. This considers induced gate noise current from the channel, but excludes induced noise current arising from the gate resistance. As mentioned earlier, this can be made negligible through a sufficiently low value of R_G. Induced noise current from the gate resistance can be included by the expression modification listed in the table.

In addition to shot noise appearing in the gate leakage current, the leakage current also contains low-frequency flicker noise [76, 154, 164, 177–179]. For the $W/L = 10\,\mu m/10\,\mu m$, nMOS and pMOS devices having $t_{ox} = 1.5\,nm$ mentioned earlier, the flicker-noise corner frequency, f_c, associated with gate leakage current is in the vicinity of 10 kHz [76, 154]. Flicker noise in gate leakage current may be caused by some combination of trap-assisted tunneling [76, 177], DC leakage or conduction into flicker noise in the channel [76, 164], and modulation of direct tunneling by the Coulomb field associated with trapping and de-trapping near the Si–SiO$_2$ interface [76]. As mentioned in [76], it is unclear if any of these effects alone describes the level of flicker noise observed in gate leakage current.

Figure 3.69 shows gate noise current density, $S_{IG}^{1/2}$, arising from gate leakage and induced noise currents predicted by the expressions given in Table 3.37. Here, an nMOS device in a 0.09 μm process operating at $I_D = 100$ μA, $IC = 6$, $L = 1$ μm, $g_m/I_D = 10$ μS/μA, $g_m = 1000$ μS, and $f_{Ti}'(\approx f_{Ti}) = 1000$ MHz is assumed. The values of g_m/I_D and $f_{Ti}'(\approx f_{Ti})$ for $IC = 6$ and $L = 1$ μm correspond to those given in Figures 3.26 and 3.57, respectively. The values given in the figures are for nMOS devices in a 0.18 μm process but are similar for the 0.09 μm process considered because, when evaluated in terms of IC and L, g_m/I_D is largely independent of the process while f_{Ti} depends primarily on the mobility. Finally, a gate leakage current of $I_G = 1$ μA is assumed for $t_{ox} = 1.5$ nm, with a flicker-noise corner frequency of $f_c = 10$ kHz as mentioned above from reported noise experiments. The actual level of I_G may be lower for $IC = 6$ at the strong-inversion side of moderate inversion and for the gate area, and the value of f_c may be different due to different geometry and bias conditions compared to those reported. However, the values assumed serve to illustrate noise in the gate leakage current, which increases as I_G increases with increasing gate area and decreasing t_{ox}.

In Figure 3.69, the gate noise current density is 0.566 pA/Hz$^{1/2}$ over a wide range of mid-band frequencies, corresponding to shot noise associated with $I_G = 1$ μA. Below $f = f_c = 10$ kHz, the noise increases because of gate leakage flicker noise. Above $f = f_c$ (induced) $= 310$ MHz or 31 % of $f_{Ti}' = 1000$ MHz, the noise increases due to induced gate noise current. f_c (induced) agrees with f_{Ti}' multiplied by $[5I_G/(2U_T g_m)]^{1/2}$ as given in Table 3.37. At frequencies below $f = f_c$ (induced), gate leakage noise current dominates, and at frequencies above this, induced gate noise current dominates. While gate leakage noise current is negligible for processes having t_{ox} greater than 3 nm giving negligible I_G, the induced gate noise current is always significant at high frequencies.

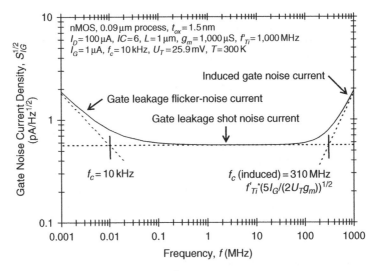

Figure 3.69 Predicted gate noise current density, $S_{IG}^{1/2}$, for an nMOS device operating at $I_D = 100$ μA, $IC = 6$, and $L = 1$ μm with a gate leakage current of 1 μA from a $t_{ox} = 1.5$ nm, 0.09 μm CMOS process. At low frequencies, noise is dominated by gate leakage shot and flicker noise. At high frequencies, noise is dominated by gate noise current induced from the channel thermal noise. Gate leakage current and the resulting gate noise current are negligible for processes having $t_{ox} > 3$ nm, while induced gate noise current is important at high frequencies in all processes

For both gate leakage noise current and high-frequency induced gate noise current shown in Figure 3.69, the gate circuit impedance must be sufficiently low to prevent significant noise contributions. Since gate leakage current and the corresponding leakage noise current increase significantly as t_{ox} decreases, management of these will become an increasing problem in smaller-geometry processes [180] as discussed later in Section 3.12.2. Using thicker t_{ox} options in these processes reduces the level of gate leakage current and leakage noise current, mitigating the deterioration of analog circuit performance.

3.11 MISMATCH

Manufacturing variations result in electrical parameter mismatches for MOSFETs having identical layout and design bias conditions. Mismatches in the threshold voltage, V_{TO}, transconductance factor, $\mu C'_{OX}(W/L)$, and body-effect factor, γ, lead to mismatches in the gate–source voltage of differential pairs and the drain current of current mirrors. MOS gate–source voltage mismatch can easily reach 10 mV, and drain current mismatch can easily reach 5 % compared to bipolar transistor base–emitter voltage and collector current mismatches that are often a factor of 10 or more lower. As a result, MOS mismatch must often be considered in design and minimized for specific applications.

The following sections discuss MOS local-area and distance DC mismatch, and provide expressions for predicting and minimizing the mismatch. Additionally, the impact of DC mismatch on the accuracy and accuracy–bandwidth–power tradeoffs of circuits is discussed. Finally, small-signal parameter and capacitance mismatch, and its impact on circuit performance, is discussed.

3.11.1 Local-Area DC Mismatch

MOS mismatch has been discussed in the literature for over two decades and remains a topic of considerable interest as technology feature sizes continue to shrink [181–236]. Local mismatch is most commonly considered, which results from local variations in channel dopants, geometries, and other effects separate from process gradients across a wafer that give rise to distance, gradient, or global mismatch. The following discussions of local mismatch, specifically local-area mismatch, assume adjacent or common-centroid layout of identically laid-out devices having the same orientation and design bias conditions such that there is no systematic mismatch. Layout techniques for optimal MOS matching are described in detail in [237, pp. 426–442].

3.11.1.1 Modeling

Research on the mismatch of MOS capacitors [181] preceded research on the mismatch of MOS devices [182, 183, 185]. In [181], MOS capacitor mismatch was attributed to edge effects, variations in oxide thickness, and variations in the oxide dielectric constant. For sufficiently large capacitors, area-dependent oxide variations exceed edge effects. In this case, absolute capacitance mismatch, ΔC, is proportional to the square root of capacitance, or $\Delta C \propto \sqrt{C}$. Relative capacitance mismatch, $\Delta C/C$, is then proportional to the inverse square root of capacitance, or $\Delta C/C \propto 1/\sqrt{C}$. Thus, relative capacitance mismatch is minimized by maximizing the capacitance or, correspondingly, maximizing capacitor area.

In [182], the mismatch of MOS current sources was attributed to edge effects, gate-oxide thickness variations, mobility variations, and depletion, surface-state and threshold-implant charge variations. Variations in charge and gate-oxide thickness give rise to threshold-voltage mismatch, while variations in geometry, gate-oxide thickness, and mobility give rise to transconductance factor mismatch.

Increasing MOS gate area reduces most variations, and experimental data confirms lower relative drain current mismatch, $\Delta I_D / I_D$, with increasing gate area.

In [183], MOS mismatch was separated into threshold-voltage, V_T, and transconductance factor, $\mu C'_{OX}(W/L)$, components. These were plotted separately against the inverse square root of gate area, $(WL)^{-1/2}$, showing a linear increase in mismatch with increasing $(WL)^{-1/2}$, which corresponds to decreasing gate area. These plots, which were developed further in [185], are commonly referred to as "Pelgrom" plots where both threshold-voltage and transconductance factor mismatch are proportional to the inverse square root of gate area, at least when area mismatch effects are dominant.

In [185], mismatch in the body-effect parameter, γ, was included and also shown to be proportional to the inverse square root of gate area through experimental "Pelgrom" plots. Variations in substrate doping, gate-oxide capacitance, and device geometry give rise to mismatch in γ. In the presence of non-zero V_{SB}, mismatch in γ results in additional threshold-voltage mismatch. This is discussed later in Section 3.11.1.5.

Threshold-voltage mismatch is usually the largest contributor to gate–source voltage and drain current mismatch as described later in Section 3.11.1.6. Threshold-voltage mismatch results from variations in threshold-voltage components, which include the gate-semiconductor work function difference, the substrate or body Fermi potential, and the depletion, fixed-oxide, surface-state, and implant charge densities divided by the gate-oxide capacitance [182, 183].

Variations in depletion charge caused by variations in the number of active dopant atoms under the gate usually dominate threshold-voltage mismatch [190, 206, 208, 227, 231, 236]. Threshold-voltage mismatch between two devices is then proportional to the mismatch in depletion charge density, $\Delta Q'_B$, divided by the gate-oxide capacitance, C'_{OX}, as expressed by

$$\Delta V_T \propto \frac{\Delta Q'_B}{C'_{OX}} \tag{3.133}$$

Threshold-voltage mismatch is then proportional to the relative mismatch in the number of active dopant atoms as expressed by

$$\Delta V_T \, (1\sigma) \propto \frac{Q'_B}{C'_{OX}} \left(\frac{\Delta n_{dop}}{n_{dop}} \right) = \frac{Q'_B}{C'_{OX}} \left(\frac{\sqrt{2}\sqrt{n_{dop}}}{n_{dop}} \right) = \frac{Q'_B}{C'_{OX}} \left(\frac{\sqrt{2}}{\sqrt{n_{dop}}} \right) \tag{3.134}$$

Q'_B is the average depletion charge density, n_{dop} is the average number of active dopant atoms under each gate, and $\Delta n_{dop} = \sqrt{2}\sqrt{n_{dop}}$ is the standard deviation or 1σ value of the number difference of dopant atoms between two MOS devices. The $\sqrt{2}$ term considers the additional standard deviation associated with two devices, where the standard deviation of the number of dopant atoms for a single device is equal to $\Delta n_{dop} = \sqrt{n_{dop}}$.

The rightmost expression in Equation 3.134 shows that threshold-voltage mismatch is inversely proportional to the square root of the average number of active dopant atoms under the gate. This makes intuitive sense as lower threshold-voltage mismatch would be expected from large devices having a large number of dopant atoms compared to small devices having a small number where the variability in the number is greater. The average number of dopant atoms under the gate is equal to the doping concentration in the substrate or body, N_B, multiplied by the depletion volume under the gate, WLt_{dep}, where t_{dep} is the depletion-region thickness under the gate. This gives

$$n_{dop} = N_B \, WL \, t_{dep} \tag{3.135}$$

Here, the body doping concentration N_B is assumed constant or average through the depletion region.

The average depletion charge density is given by the average dopant density under the gate, $N_B t_{dep}$, multiplied by the unit of electronic charge, giving

$$Q'_B = q \, N_B \, t_{dep} \tag{3.136}$$

Substituting Equations 3.135 and 3.136 into the rightmost expression in Equation 3.134 gives

$$\Delta V_T(1\sigma) \propto \frac{q}{C'_{OX}} \left(\frac{\sqrt{2N_B t_{dep}}}{\sqrt{WL}} \right) = \sqrt{2\sqrt{2}} \frac{t_{ox}}{\varepsilon_{SiO_2}} \frac{\left(q^3 \varepsilon_{Si} \phi_s N_B \right)^{1/4}}{\sqrt{WL}}$$

$$= \frac{A_{VTO}(V_{SB} = 0 \text{ V})}{\sqrt{WL}} \qquad (3.137)$$

In the top expression, $C'_{OX} = \varepsilon_{SiO_2}/t_{ox}$ and $t_{dep} = (2\varepsilon_{Si}\phi_s/(qN_B))^{1/2}$ are substituted from the expressions in Table 3.1 where t_{ox} is the gate-oxide thickness, ε_{SiO_2} is the permittivity of silicon dioxide, ε_{Si} is the permittivity of silicon, and ϕ_s is the silicon surface potential.

The top expression in Equation 3.137 is directly equal to the 1σ, threshold-voltage mismatch given in [208]. However, this expression is not directly equal to threshold-voltage mismatch given in [16, p. 129, 190, 231], but is larger by a constant factor resulting from differences in referring depletion charge mismatch to the gate. The top expression in Equation 3.137 is intended to illustrate mismatch trends with actual estimates of mismatch found from the bottom expression using the extracted threshold-voltage mismatch factor, A_{VTO}. A_{VTO}, summarized later with other local-area mismatch factors in Table 3.38, corresponds to the zero-V_{SB}, 1σ, threshold-voltage (ΔV_{TO}) mismatch between two identically laid-out devices having gate areas of $1\,\mu m^2$ each.

From Equation 3.137, A_{VTO} is proportional to process parameters as given by

$$A_{VTO}(V_{SB} = 0 \text{ V}) \propto \frac{q}{C'_{OX}} \sqrt{2N_B t_{dep}} = \sqrt{2\sqrt{2}} \frac{t_{ox}}{\varepsilon_{SiO_2}} \left(q^3 \varepsilon_{Si} \phi_s N_B \right)^{1/4} \qquad (3.138)$$

For a $0.18\,\mu m$ CMOS process having $N_B = 7 \times 10^{17}/cm^3$ ($7 \times 10^5/\mu m^3$), the depletion region thickness is $t_{dep} = 0.043\,\mu m$. This assumes $V_{SB} = 0\,V$, $PHI = 2\phi_F = 0.913\,V$ at $T = 300\,K$, and an approximate strong-inversion surface potential of $\phi_s = \phi_0 \approx 2\phi_F + 4U_T \approx 1\,V$ from the expressions given in Table 3.1. The average doping density under the gate is then equal to $N_B t_{dep} = 30000/\mu m^2$. For $t_{ox} = 4.1\,nm$, which gives $C'_{OX} = 8.41\,fF/\mu m^2$ from the expression in Table 3.1, $A_{VTO} = 4.7\,mV \cdot \mu m$ when directly evaluated from Equation 3.138. This is close to $A_{VTO} = 5\,mV \cdot \mu m$ listed in Table 3.3 for the example $0.18\,\mu m$ process, but is above A_{VTO} values estimated from threshold-voltage mismatch given in [16, p. 129, 190, 231] because of differences in constant factors mentioned earlier. As described later in Section 5.5.2.5, measured input-referred offset voltages for example $0.18\,\mu m$ CMOS operational transconductance amplifiers are well predicted using the value of $A_{VTO} = 5\,mV \cdot \mu m$.

Equations 3.137 and 3.138 suggest that threshold-voltage mismatch decreases directly as t_{ox} decreases in smaller-geometry processes. However, mismatch is proportional to $N_B^{1/4}$, which increases in smaller-geometry processes and partially counters the mismatch decrease associated with decreasing t_{ox} [208]. As a result, threshold-voltage mismatch decreases less rapidly than t_{ox} decreases in new processes as shown in reported mismatch factors in the next section. Additionally, as described in the next section and Section 3.11.1.3, gate non-uniformity [227], inversion layer quantization [229], and other effects may contribute to threshold-voltage mismatch and affect technology trends.

Threshold-voltage mismatch is usually assumed to be independent of the inversion level or $V_{EFF} = V_{GS} - V_T$. Equations 3.137 and 3.138 suggest only a small mismatch increase as t_{dep} increases slightly in strong inversion through the slight increase in ϕ_s. ϕ_s increases from approximately $\phi_s = PHI = 2\phi_F$ at the boundary of weak and moderate inversion [15, p. 87] to a nearly "pinned" value of $\phi_s = \phi_0 \approx 2\phi_F + 4U_T$ in strong inversion.

Threshold-voltage mismatch increases for non-zero V_{SB} in the usual, reverse-biased direction [185, 188, 192, 196, 209]. Equations 3.137 and 3.138 predict a mismatch increase through the increase in t_{dep} associated with adding V_{SB} to the value of ϕ_s. In the less usual case when V_{SB} is slightly forward biased, mismatch is reported to decrease as the result of gated lateral bipolar transistor action [192, 196]. The V_{SB} component of threshold-voltage mismatch is not included in the value of A_{VTO}, which, again, is defined for $V_{SB} = 0\,\text{V}$. This can be considered separately through mismatch in the body-effect factor, γ. Also, as described later in Sections 3.11.1.4 and 3.11.1.5, mismatch in the general, non-zero-V_{SB} threshold voltage, V_T, is frequently described using a threshold-voltage mismatch factor, A_{VT}, that directly considers increases associated with non-zero V_{SB}.

Equation 3.137 illustrates the similarity between threshold-voltage mismatch and gate-referred flicker-noise voltage predicted by the square root of noise voltage PSD given in Equation 3.112 from the carrier density fluctuation, number fluctuation, or McWhorter flicker-noise model. Threshold-voltage mismatch is a DC noise (Equation 3.133) resulting from fixed depletion-region charge mismatch across two devices, whereas gate-referred flicker-noise voltage is an AC noise (Equation 3.109) resulting from fluctuating charge associated with carrier trapping and de-trapping near the Si–SiO$_2$ interface. Both decrease inversely with the square root of increasing gate area and decrease inversely with increasing C'_{OX}. However, as seen in the next section, reported threshold-voltage mismatch factors, A_{VTO}, vary considerably less across similar process generations compared to reported flicker-noise factors, K_F, listed earlier in Table 3.33. This is because threshold-voltage mismatch depends primarily on mismatch in the number of active dopant atoms in the depletion region under the gate, whereas flicker noise depends on the much more variable density of traps near the Si–SiO$_2$ interface as well as on buried or surface channel operation. Gate-referred flicker-noise voltage can also increase with increasing inversion or $V_{EFF} = V_{GS} - V_T$ as described in Sections 3.10.3.3 and 3.10.3.5–3.10.3.7, whereas, as mentioned, threshold-voltage mismatch is usually considered independent of the level of inversion.

While Equations 3.137 and 3.138 consider threshold-voltage mismatch due to the mismatch in depletion charge, mismatches in the fixed-oxide, surface-state, and implant charges are also largely area dependent and are divided by the gate-oxide capacitance. As a result, these mismatch components are included structurally in the equations and in the local-area mismatch model described below where mismatch is proportional to the inverse square root of gate area.

Table 3.38 summarizes MOS local-area mismatch expressions and trends for the zero-V_{SB} threshold voltage, V_{TO}, transconductance factor, $\mu C'_{OX}(W/L)$, and body-effect factor, γ, in terms of the

Table 3.38 MOS local-area mismatch expressions and trends in terms of the inversion coefficient, channel length, and drain current

Parameter	$IC \uparrow$ $L,\ I_D$ fixed	$L \uparrow$ $IC,\ I_D$ fixed	$I_D \uparrow$ $IC,\ L$ fixed
Threshold-voltage mismatch ($V_{SB} = 0\,\text{V}$): $\Delta V_{TO} = \dfrac{A_{VTO}}{\sqrt{WL}} = \left(\dfrac{\sqrt{IC}}{L}\sqrt{\dfrac{I_0}{I_D}}\right) A_{VTO}$			
Threshold-voltage mismatch (general V_{SB} that includes mismatch in γ for non-zero V_{SB}): $\Delta V_T = \dfrac{A_{VT}}{\sqrt{WL}} = \left(\dfrac{\sqrt{IC}}{L}\sqrt{\dfrac{I_0}{I_D}}\right) A_{VT}$			
Transconductance factor ($K_P = \mu C'_{OX}(W/L)$) mismatch: $\dfrac{\Delta K_P}{K_P} = \dfrac{A_{KP}}{\sqrt{WL}} = \left(\dfrac{\sqrt{IC}}{L}\sqrt{\dfrac{I_0}{I_D}}\right) A_{KP}$	$\uparrow \propto \sqrt{IC}$	$\downarrow \dfrac{1}{L}$	$\downarrow \propto \dfrac{1}{\sqrt{I_D}}$
Body-effect factor mismatch: $\Delta \gamma = \dfrac{A_{GAMMA}}{\sqrt{WL}} = \left(\dfrac{\sqrt{IC}}{L}\sqrt{\dfrac{I_0}{I_D}}\right) A_{GAMMA}$			

design choices of inversion coefficient, IC, channel length, L, and drain current, I_D. The zero-V_{SB}, threshold-voltage mismatch factor, A_{VTO}, transconductance factor mismatch factor, A_{KP}, and body-effect mismatch factor, A_{GAMMA}, are named as those in the EKV 2.6 MOS model [13, 14]. When expressed in terms of the gate area, WL, the expressions listed in the table are commonly known as the local-area mismatch or "Pelgrom" model from [185]. Edge effects and other limitations of the local-area mismatch model are discussed later in Section 3.11.1.3.

Table 3.38 also includes an expression for local-area mismatch in the general, non-zero-V_{SB} threshold voltage, V_T. As mentioned, this is frequently found from a local-area, threshold-voltage mismatch factor, A_{VT}, that directly includes the threshold-voltage mismatch increase present for non-zero V_{SB} due to mismatch in γ. As discussed in Sections 3.11.1.4 and 3.11.1.5, $A_{VT} = A_{VTO}$ for zero V_{SB}, increasing only slightly above A_{VTO} when V_{SB} is constrained to low values as typically required in low-voltage design.

In order to express mismatch in Table 3.38 in terms of IC, L, and I_D, MOS gate area is expressed as $WL = (L^2/IC)(I_D/I_0)$ from Table 3.10 where $I_0 = 2n_0\mu_0 C'_{OX} U_T^2$ is the technology current described in Table 3.6. Threshold-voltage, transconductance factor, and body-effect mismatch increase as the square root of IC, decrease inversely with L, and decrease inversely with the square root of I_D, tracking the inverse square root of gate area. Mismatch is minimized for large gate areas associated with low levels of inversion, long channel length, and high drain currents. Unfortunately, operation at low levels of inversion and long channel lengths yields low intrinsic (Section 3.9.6) and operating bandwidths (Section 3.9.7) because of large intrinsic (Section 3.9.2) and extrinsic (Sections 3.9.3 and 3.9.4) capacitances.

3.11.1.2 Reported mismatch factors and trends

Local-area mismatch factors can be experimentally extracted using the "Pelgrom" plots mentioned in the previous section. Here, the inverse square root of gate area, $(WL)^{-1/2}$, is used as the x axis, and the measured mismatch for the parameter of interest is plotted on the y axis. A straight line is fitted through the data, and the slope of this line is the mismatch factor. Usually, 1σ values are used for mismatch measurements and extracted mismatch factors.

Table 3.39 lists reported 1σ values for the local-area, threshold-voltage mismatch factor, A_{VTO}, and transconductance factor mismatch factor, A_{KP}, for submicron CMOS processes having specified values of gate-oxide thickness, t_{ox}, and minimum channel length, L_{min}. Values are taken from reported values or extracted from reported data, often given as "Pelgrom" plots.

As predicted by Equation 3.138 and described in the previous section, A_{VTO} decreases as t_{ox} decreases and C'_{OX} increases in smaller-geometry processes. In fact, A_{VTO} can be roughly estimated using a "rule of thumb" where A_{VTO} $(1\sigma) \approx (1\,\text{mV} \cdot \mu\text{m}) \cdot (t_{ox}/\text{nm})$ [27, p. 465, 208]. However, as seen in Table 3.39, A_{VTO} ceases to decrease as rapidly as t_{ox}, due in part to the impact of higher body doping concentration, N_B, described in the previous section. In addition to decreasing less rapidly than t_{ox} in smaller-geometry processes, it is possible that A_{VTO} could saturate for small values of t_{ox}.

For the 0.5, 0.35, and 0.18 μm CMOS processes considered in design examples in this book, we assume A_{VTO} (1σ) values of 14, 9, and 5 mV $\cdot$ μm given in Table 3.3 for t_{ox} of 13.5, 8, and 4.1 nm, respectively. Lower values of A_{VTO} in smaller-geometry processes correspond to lower threshold-voltage mismatch for devices having equal gate area.

As described in the previous section, A_{VTO} varies much less across similar process generations compared to the flicker-noise factor, K_F, reported earlier in Table 3.33. This is observed in Table 3.39 where A_{VTO} varies modestly for processes having similar t_{ox}. This is fortunate because mismatch data is often unavailable to the designer and requires extensive mismatch measurements across many pairs of devices. In contrast, flicker-noise K_F extractions require noise measurements on a moderate number of devices as described earlier in Section 3.10.3.6.

While A_{VTO} is similar across similar process generations, it varies with gate fabrication type for the $t_{ox} = 3.2$ nm, $L_{min} = 0.18\,\mu$m process [227] listed in Table 3.39. Here, three different gate

Table 3.39 Reported local-area, threshold-voltage and transconductance factor mismatch factors for submicron CMOS process

Process		V_{TO} mismatch A_{VTO} (mV·μm, 1σ)		$\mu C'_{OX}(W/L)$ mismatch A_{KP} (%·μm, 1σ)		Reference
t_{ox}(nm)	L_{min}(μm)	nMOS	pMOS	nMOS	pMOS	
—	0.8		13		1.3	[205]
—	0.7	11.5	12.3	1.8	3.1	[193]
17.6	0.7	12.2			1.3	[195]
17	0.7	11	22	1.9	2.8	[199]
16	—	13				[208]
15	0.8	10	15			[208]
	0.5	19	16			[217]
12	0.6	10.3	9			[208]
—	0.5	11	13	1.8	2.3	[222]
11	—	8.5				[208]
9	0.4	8.9		2.1		[211]
—	0.35	9	9	1.9	2.25	[222]
7.5	0.35	9	7.7			[208]
		6.2	10.7	1.1	0.7	[223]
		8.75	8.5			[217]
7	0.35	8.55	7.2	1.46	0.84	[197]
		8.2		1.5		[211]
6	—	5.5				[208]
—	0.25	6	6	1.85	1.85	[222]
5.5	0.25	6.7	6.9			[217]
		7.8				[213]
5	0.28	5		1.3		[211]
	0.25	7.7	6.3			[208]
—	0.18	6 (SI) 12.5 (WI)				[232]
—	0.18	5	5.49	1.04	0.99	[214]
3.5	0.18	5.8	4	1.3	0.6	[224]
			5			[213]
3.3	0.18	4.3				[208]
3.2	0.18	6.08	11.2	Amor. silicon gate, furnace anneal		
		5.31	4.52	Amor. silicon gate, rapid thermal oxidation		
		3.46	2.85	Polysilicon gate, rapid thermal oxidation		[227]
2	0.12	3.5		1.1	1.1	[220]
—	0.12			0.99	0.71	[230]

fabrications are evaluated: amorphous silicon gate deposition with furnace anneal, amorphous silicon gate deposition followed by rapid thermal oxidation, and polysilicon gate deposition followed by rapid thermal oxidation. A notable decrease in A_{VTO} is observed for the last type of gate fabrication. Reduced gate grain size and improved gate uniformity, especially in the region near the Si–SiO$_2$ interface, are believed to be responsible for the decrease in A_{VTO} [227].

As described in the previous section, threshold voltage mismatch is usually assumed to be independent of the level of inversion. The mismatch is reported as nearly unchanged between weak and strong inversion by some researchers as summarized in [232], while increasing in weak inversion for the $t_{ox} = 4$ nm, $L_{min} = 0.18$ μm process [232] listed in Table 3.39. Here, $A_{VTO}(1\sigma) = 12.5$ mV · μm in weak inversion compared to $A_{VTO} = 6$ mV · μm in strong inversion. Reasons for the mismatch increase in weak inversion are unclear, but suggest it might be advisable to avoid operation at the extreme of weak inversion for critical matching applications.

While A_{VTO} clearly decreases with decreasing t_{ox} as seen in Table 3.39, the transconductance factor mismatch factor, A_{KP}, displays a less clear technology trend. $A_{KP}(1\sigma)$ is generally around 1–3 % · μm, decreasing some with decreasing t_{ox} as would be expected with improved geometry uniformity. A_{KP} is related to the variation in the transconductance factor, $\mu C'_{OX}(W/L)$, resulting from variations in mobility, gate-oxide capacitance, and device geometry. For the 0.5, 0.35, and 0.18 μm CMOS processes considered in design examples, we assume A_{KP} (1σ) values at 2 % · μm given in Table 3.3. As described later in Section 3.11.1.6, transconductance factor mismatch is often a negligible contributor to overall mismatch compared to threshold-voltage mismatch.

Table 3.40 lists reported 1σ values for the local-area, body-effect factor mismatch factor, A_{GAMMA}, for CMOS processes. The only values listed for submicron processes are 6.5 and 7×10^{-3} V$^{1/2}$ · μm for nMOS and pMOS devices, respectively, in a process having $t_{ox} = 17$ nm and $L_{min} = 0.7$ μm [209]. Sections 3.11.1.4 and 3.11.1.5 describe similar predictions of threshold-voltage mismatch increase with V_{SB} using both A_{GAMMA} and the silicon surface potential that does not require knowledge of A_{GAMMA}. As described in Section 3.11.1.5, γ mismatch is often a negligible contributor to threshold-voltage mismatch when V_{SB} is necessarily constrained to low values in low-voltage processes.

3.11.1.3 Edge effects and other model limitations

Gate area often dominates mismatch, especially when both width and length are made sufficiently large as usually required for critical matching applications. However, the local-area MOS mismatch model may not provide sufficient accuracy for small gate areas where edge effects can be significant.

The 1σ mismatch in width and length due to edge roughness is given by

$$\Delta W\,(1\sigma) = \frac{W_W}{\sqrt{L}}$$

$$\Delta L\,(1\sigma) = \frac{L_L}{\sqrt{W}} \tag{3.139}$$

Table 3.40 Reported local-area, body-effect factor mismatch factors for CMOS processes

Process		γ mismatch		Reference
t_{ox} (nm)	L_{min} (μm)	A_{GAMMA} (V$^{1/2}$ · μm, 1σ) nMOS	pMOS	
50	2.5	16×10^{-3}	12×10^{-3}	[185]
22.5	1.2	16×10^{-3}	6.4×10^{-3}	[209]
17	0.7	6.5×10^{-3}	7×10^{-3}	[209]

W_W and L_L are mismatch factors describing mismatch in width and length associated with the nominal width, W, and nominal length, L. Width mismatch is averaged or reduced by division by the square root of length, while length mismatch is averaged or reduced by division by the square root of width [16, pp. 126–127, 226]. This corresponds to reductions in the width and length variances, the square of 1σ mismatches, through division by length and width, respectively.

Separate width and length mismatches cause MOS mismatch through mismatch in the aspect ratio or shape factor $S = W/L$ that appears in the transconductance factor, $\mu C'_{OX}(W/L)$. Assuming small mismatches in width and length, or $\Delta W \ll W$ and $\Delta L \ll L$, the mismatch in $S = W/L$ is given by

$$\Delta(W/L) = \frac{L\Delta W - W\Delta L}{L^2} \tag{3.140}$$

The relative mismatch is then given by

$$\frac{\Delta(W/L)}{(W/L)} = \frac{L\Delta W - W\Delta L}{WL} = \frac{\Delta W}{W} - \frac{\Delta L}{L} \tag{3.141}$$

Assuming that ΔW and ΔL are uncorrelated, the variance for the relative mismatch is given by

$$\left(\frac{\Delta(W/L)}{(W/L)}\right)^2 = \left(\frac{\Delta W}{W}\right)^2 + \left(\frac{\Delta L}{L}\right)^2 \tag{3.142}$$

which from ΔW and ΔL given by Equation 3.139 gives

$$\left(\frac{\Delta(W/L, 1\sigma)}{(W/L)}\right)^2 = \frac{W_W^2}{W^2 L} + \frac{L_L^2}{WL^2} = \left(\frac{1}{WL}\right)\left(\frac{W_W^2}{W} + \frac{L_L^2}{L}\right) \tag{3.143}$$

Equation 3.143 is derived in [16, pp. 126–127] and has the same form as edge-effect mismatch terms appearing in early MOS mismatch models [182, 185]. The 1σ mismatch due to edge effects given by the square root of Equation 3.143 and local-area mismatch expressions given in Table 3.38 decrease inversely with the square root of gate area, WL. However, mismatch due to edge effects is additionally minimized by avoiding short dimensions for either channel width or length. This also minimizes the area loss associated with short channel width or length due to the lateral diffusion reduction of width or length. The importance of using effective width and length for calculating gate area in mismatch prediction is discussed in [205]. Unlike area mismatch factors reported in Table 3.39, width and length edge-effect mismatch factors are rarely reported in the literature.

In addition to excluding edge effects, the local-area mismatch model summarized in Table 3.38 is not physically based since the threshold voltage, transconductance factor, and body-effect factor are not independent process parameters, but are electrical parameters resulting from multiple, independent process parameters [16, pp. 120–122, 226]. For example, variations in gate-oxide thickness affect the gate-oxide capacitance and mismatch in the threshold voltage, transconductance factor, and body-effect factor, γ, and the related substrate factor, n, yet these are assumed to be statistically independent in the local-area mismatch model. For a CMOS process having an average value of $n = 1.4$, an average strong-inversion surface potential of $\phi_0 \approx 2\phi_F + 4U_T = 0.8\,\text{V}$, and a threshold-voltage component of $-200\,\text{mV}$ due to fixed interface charge, a 1% increase in the gate-oxide capacitance results in a $-8.4\,\text{mV}$ decrease in the threshold voltage, a 1% increase in the transconductance factor, and a -0.3% decrease in n [16, pp. 120–122]. Additionally, variations in the depletion charge in the depletion region under the gate result in variations in both the threshold voltage and γ or n. However, variations in mobility and width and length affect transconductance factor mismatch only. The correlation between threshold voltage and γ or n mismatch is considered in the next section, while the threshold voltage and transconductance factor mismatch are assumed uncorrelated as usual for the local-area mismatch model. This is a good assumption when threshold-voltage mismatch is dominated by variations in the depletion charge under the gate that has no effect on the transconductance factor. When threshold-voltage mismatch dominates transconductance factor mismatch as frequently the case as discussed later in Section 3.11.1.6, the degree of correlation between these is not important.

In [226], independent process variations are mapped into electrical mismatch using the propagation of variance. While such a model is physically based, gives good results with experimental data, and can be implemented during computer simulations of a candidate design, it is too complex to enable simple hand expressions that provide design guidance. Additionally, extensive process mismatch data is required that is often unavailable to the designer.

Measurements in 0.18 [208, 221], 0.13 [228], and 0.1 μm [204] CMOS processes indicate good agreement with the local-area mismatch model, at least for sufficiently large width and length. Additionally, comparisons of predicted and measured input-referred offset voltages for 0.5 and 0.18 μm CMOS operational transconductance amplifier design examples shown later in Sections 5.4.2.5 and 5.5.2.5 indicate good mismatch prediction accuracy considering only local-area mismatch. Measurements suggest line edge roughness does not significantly contribute to mismatch in processes having minimum channel length down to 0.09 μm [225]. However, mismatch prediction may need to include gate non-uniformity [227], and short-channel charge sharing and inversion layer quantization effects for threshold-voltage mismatch in smaller-geometry processes [229]. Additionally, it is possible that Coulomb scattering involving carriers in the inversion layer could impact transconductance factor mismatch [230] in addition to the impact of line edge roughness.

Because of adequate prediction accuracy for many applications, the unavailability of edge-effect and independent process mismatch parameters, and the complexity of including these and small-geometry mismatch effects in design expressions, we will consider only local-area mismatch as summarized in Table 3.38 for design guidance. Additional mismatch effects could be considered in computer simulations of a candidate design if available in MOS models.

3.11.1.4 Calculating gate–source voltage and drain current mismatch

The local-area, zero-V_{SB} threshold-voltage, ΔV_{TO}, transconductance factor, $\Delta K_P/K_P$, and body-effect factor, $\Delta \gamma$, mismatch described in Table 3.38 map into gate–source voltage mismatch, ΔV_{GS}, or drain current mismatch, ΔI_D, for a pair of devices having equal layout and design bias conditions. The gate–source voltage mismatch, usually associated with two devices in a differential pair, will be studied first.

Threshold-voltage mismatch maps directly into gate–source voltage mismatch. Threshold-voltage mismatch, ΔV_T, is equal to the ΔV_{TO} value given in Table 3.38 when V_{SB} is zero. However, ΔV_T increases from ΔV_{TO} for non-zero V_{SB} as a result of $\Delta \gamma$ mismatch. This can be observed from the increase in the threshold voltage, V_T, from its V_{TO} value present for zero V_{SB}. V_T from Table 3.20 is given by[14]

$$V_T = V_{TO} + \gamma \left(\sqrt{V_{SB} + 2\phi_F} - \sqrt{2\phi_F} \right) \tag{3.144}$$

where $PHI = 2\phi_F$ is twice the Fermi potential. As listed in the table, V_T is also conveniently approximated by

$$V_T = V_{TO} + (n - 1) V_{SB} \tag{3.145}$$

---

[14] As described for Figure 3.9, V_T, V_{TO}, $2\phi_F$, and V_{SB} are taken positively for both nMOS and pMOS devices. Positive V_{SB} corresponds to normal, reverse bias source–body voltage.

Equations 3.144 and 3.145 give similar values of V_T assuming V_{SB} is sufficiently small as typically required for design in low-voltage processes. Both equations show that V_T increases as V_{SB} increases, with mismatch in the V_T increase caused by mismatch in γ or n.

ΔV_T mismatch can be found from Equation 3.144 as

$$\Delta V_T(1\sigma) = \Delta V_{TO}(1\sigma) + \Delta\gamma\,(1\sigma)\left(\sqrt{V_{SB}+2\phi_F} - \sqrt{2\phi_F}\right)$$

$$= \frac{1}{\sqrt{WL}}\left[A_{VTO} + A_{GAMMA}\left(\sqrt{V_{SB}+2\phi_F} - \sqrt{2\phi_F}\right)\right]$$

$$= \frac{1}{\sqrt{WL}}[A_{VT}] \tag{3.146}$$

where the expression in the second line is obtained from the 1σ mismatch expressions for ΔV_{TO} and $\Delta\gamma$ given in Table 3.38. The bracketed expression in the second line is equal to the local-area, threshold-voltage mismatch factor, A_{VT}, given in the third line, which includes the mismatch increase associated with non-zero V_{SB}. When V_{SB} is non-zero, $A_{VT} > A_{VTO}$ where A_{VTO} is the mismatch factor for zero V_{SB}. Equation 3.146 assumes worst case, full correlation between ΔV_{TO} and $\Delta\gamma$, with mismatch dominated by mismatch in the number of active dopant atoms under the gate as described for ΔV_T given by the top line in Equation 3.137.

ΔV_T mismatch can also be conveniently expressed by [16, p. 129]

$$\Delta V_T(1\sigma) = \Delta V_{TO}(1\sigma)\left(1 + \frac{V_{SB}}{2\phi_0}\right)$$

$$= \frac{1}{\sqrt{WL}}\left[A_{VTO}\left(1 + \frac{V_{SB}}{2\phi_0}\right)\right]$$

$$= \frac{1}{\sqrt{WL}}[A_{VT}] \tag{3.147}$$

where $\phi_0 \approx 2\phi_F + 4U_T$ is taken as the approximate silicon surface potential in strong inversion. As for Equation 3.146, the bracketed expression in the second line is equal to A_{VT}, which, again, includes the mismatch increase associated with non-zero V_{SB}. Equation 3.147, like Equation 3.146, assumes worst case, full correlation between ΔV_{TO} and mismatch in γ or n, with mismatch, again, dominated by mismatch in the number of active dopant atoms under the gate.

Equations 3.146 and 3.147 show that ΔV_T is equal to ΔV_{TO} given in Table 3.38 when V_{SB} is zero and show that ΔV_T increases above ΔV_{TO} for non-zero V_{SB}. Later, Section 3.11.1.5 will show that ΔV_T predicted by Equations 3.146 and 3.147 is similar and increases only slightly from ΔV_{TO} when $V_{SB} \ll 2\phi_0$ as typically required for design in low-voltage processes. While Equation 3.147 will be used for its simplicity, Equation 3.146 is included as part of the traditional, local-area, "Pelgrom" mismatch model summarized in Table 3.38.

In addition to threshold-voltage mismatch, transconductance factor mismatch causes gate–source voltage mismatch. Transconductance factor mismatch causes a mismatch in drain current that is equal to $\Delta I_D = (\Delta K_P/K_P)I_D$. This drain current mismatch is then referred to a gate–source voltage mismatch by division by device transconductance, g_m. The resulting gate–source voltage mismatch is then given by

$$\Delta V_{GS}(\Delta K_P, 1\sigma) = \frac{\Delta I_D(\Delta K_P, 1\sigma)}{g_m} = \left(\frac{\Delta K_P(1\sigma)}{K_P}\cdot I_D\right)\frac{1}{g_m}$$

$$= \frac{1}{\sqrt{WL}}\left(\frac{A_{KP}}{g_m/I_D}\right) \tag{3.148}$$

where the expression in the second line is obtained from the 1σ mismatch expression for $\Delta K_P/K_P$ given in Table 3.38, and g_m/I_D is the device transconductance efficiency.

The total gate–source voltage mismatch is given from Equations 3.147 and 3.148 as

$$\Delta V_{GS}(1\sigma) = \sqrt{\Delta V_T^2(1\sigma) + \Delta V_{GS}^2(\Delta K_P, 1\sigma)}$$

$$= \frac{1}{\sqrt{WL}}\sqrt{\left[A_{VTO}\left(1 + \frac{V_{SB}}{2\phi_0}\right)\right]^2 + \left(\frac{A_{KP}}{g_m/I_D}\right)^2}$$

$$= \frac{1}{\sqrt{WL}}\sqrt{A_{VT}^2 + \left(\frac{A_{KP}}{g_m/I_D}\right)^2} \tag{3.149}$$

In Equation 3.149, threshold-voltage mismatch associated with variations in the number of active dopant atoms under the gate is assumed uncorrelated with transconductance factor mismatch associated with variations in mobility, gate-oxide capacitance, and geometry. When threshold-voltage mismatch dominates transconductance factor mismatch as frequently the case as discussed later in Section 3.11.1.6, the degree of correlation between these is not important.

Equation 3.149 can be conveniently expressed as

$$\Delta V_{GS}(1\sigma) = \frac{A_{VGS}}{\sqrt{WL}} \tag{3.150}$$

where

$$A_{VGS} = \sqrt{\left[A_{VTO}\left(1 + \frac{V_{SB}}{2\phi_0}\right)\right]^2 + \left(\frac{A_{KP}}{g_m/I_D}\right)^2}$$

$$= \sqrt{A_{VT}^2 + \left(\frac{A_{KP}}{g_m/I_D}\right)^2} \tag{3.151}$$

A_{VGS}, having units of V $\cdot$ μm, is a composite gate–source voltage mismatch factor depending on the mismatch factors, A_{VTO} and A_{KP}, the silicon surface potential, ϕ_0, and the operating V_{SB} and g_m/I_D. When $V_{SB} \ll 2\phi_0$ and $g_m/I_D \gg A_{KP}/A_{VTO}$, $A_{VGS} \approx A_{VTO}$ and the gate–source voltage mismatch is nearly equal to the zero-V_{SB} threshold-voltage mismatch, ΔV_{TO}, directly as given by

$$\Delta V_{GS}(1\sigma) \approx \frac{A_{VTO}}{\sqrt{WL}}, \text{ when } V_{SB} \ll 2\phi_0 \text{ and } g_m/I_D \gg A_{KP}/A_{VTO} \tag{3.152}$$

Sections 3.11.1.5 and 3.11.1.6 will show that Equation 3.152 provides a good estimate of gate–source voltage mismatch except when V_{SB} is unusually high or the inversion level is unusually high, giving unusually low g_m/I_D.

The preceding analysis considered the gate–source voltage mismatch usually associated with two devices in a differential pair having identical layout and design bias conditions. The drain current mismatch, however, is important for a pair of current-mirror devices having identical layout and design bias conditions.

Drain current mismatch, ΔI_D, is found from the threshold-voltage mismatch, ΔV_T, multiplied by the device transconductance, g_m, which refers the threshold-voltage mismatch to a drain current mismatch. The relative drain current mismatch, $\Delta I_D/I_D$, is then found by ΔV_T multiplied by the device transconductance efficiency, g_m/I_D. Additionally, $\Delta I_D/I_D$ associated with transconductance factor mismatch is found directly from the relative transconductance factor mismatch, $\Delta K_P/K_P$. The total relative, drain current mismatch is then given by

$$\frac{\Delta I_D(1\sigma)}{I_D} = \sqrt{(\Delta V_T(1\sigma) \cdot g_m/I_D)^2 + \left(\frac{\Delta K_P(1\sigma)}{K_P}\right)^2}$$

$$= \frac{1}{\sqrt{WL}} \sqrt{\left[A_{VTO}\left(1 + \frac{V_{SB}}{2\phi_0}\right) \cdot g_m/I_D\right]^2 + A_{KP}^2}$$

$$= \frac{1}{\sqrt{WL}} \sqrt{(A_{VT} \cdot g_m/I_D)^2 + A_{KP}^2} \tag{3.153}$$

where the expression in the second line is obtained from the 1σ mismatch expressions for ΔV_T given in Equation 3.147 and $\Delta K_P/K_P$ given in Table 3.38. Like Equation 3.149 for the gate–source voltage mismatch, Equation 3.153 assumes that threshold-voltage and transconductance factor mismatch are uncorrelated.

Analogous to Equation 3.150 given for the gate–source voltage mismatch, Equation 3.153 can be conveniently expressed as

$$\frac{\Delta I_D(1\sigma)}{I_D} = \frac{A_{VGS}}{\sqrt{WL}} \cdot g_m/I_D \tag{3.154}$$

Finally, when $V_{SB} \ll 2\phi_0$ and $g_m/I_D \gg A_{KP}/A_{VTO}$, $A_{VGS} \approx A_{VTO}$ and the relative drain current mismatch is nearly equal to the zero-V_{SB} threshold-voltage mismatch, ΔV_{TO}, multiplied by g_m/I_D as given by

$$\frac{\Delta I_D(1\sigma)}{I_D} \approx \frac{A_{VTO}}{\sqrt{WL}} \cdot g_m/I_D, \quad \text{when } V_{SB} \ll 2\phi_0 \text{ and } g_m/I_D \gg A_{KP}/A_{VTO} \tag{3.155}$$

As for the gate–source voltage mismatch given by Equation 3.152, Sections 3.11.1.5 and 3.11.1.6 will show that Equation 3.155 provides a good estimate of drain current mismatch except when V_{SB} is unusually high or the inversion level is unusually high, giving unusually low g_m/I_D.

Comparing gate–source voltage mismatch, ΔV_{GS}, given by Equations 3.149 and 3.150 to relative drain current mismatch, $\Delta I_D/I_D$, given by Equations 3.153 and 3.154 reveals that ΔV_{GS} is equal to $\Delta I_D/I_D$ divided by g_m/I_D, and $\Delta I_D/I_D$ is equal to ΔV_{GS} multiplied by g_m/I_D. Thus, the gate–source voltage mismatch and relative drain current mismatch are interrelated by the transconductance efficiency, g_m/I_D. As expected, both mismatches are minimized by maximizing gate area, which minimizes mismatch in the zero-V_{SB} threshold-voltage, transconductance factor, and body-effect factor. However, relative drain current mismatch has additional dependency on the level of inversion, which affects the conversion of threshold-voltage mismatch to drain current mismatch through g_m/I_D. Section 3.11.1.7 summarizes expressions and trends for gate–source voltage mismatch and relative drain current mismatch in terms of the design selections of drain current, inversion coefficient, and channel length.

3.11.1.5 Threshold-voltage mismatch increase for non-zero V_{SB}

Equations 3.144 and 3.145 show that threshold voltage, V_T, increases from its zero-V_{SB} value, V_{TO}, when non-zero V_{SB} is present. Equations 3.146 and 3.147 show that threshold-voltage mismatch, ΔV_T, increases from its zero-V_{SB} value, ΔV_{TO}, when non-zero V_{SB} is present. This mismatch increase is a result of mismatch in the body-effect factor, γ, or the related substrate factor, n.

Figure 3.70 shows the predicted threshold-voltage increase, $V_T - V_{TO}$, versus V_{SB} from Equations 3.144 and 3.145 for nMOS devices in the example 0.18 μm CMOS process described in Table 3.2. The predictions assume $T = 300$ K (room temperature), $\gamma = 0.56$ V$^{1/2}$, and $PHI = 2\phi_F = 0.85$ V for Equation 3.144, and $n = 1.33$ (held fixed at the same value used for g_m/I_D in Figure 3.26) for Equation 3.145. The threshold voltage increases by approximately 0.15 V above $V_{TO} = 0.42$ V given in

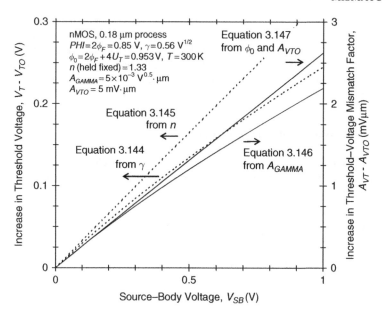

Figure 3.70   Predicted increase in threshold voltage, V_T, and threshold voltage mismatch factor, A_{VT}, versus the source–body voltage, V_{SB}, for nMOS devices in a 0.18 μm CMOS process. V_T shown by the dashed lines increases by approximately 0.15 V above $V_{TO} = 0.42$ V for $V_{SB} = 0.5$ V, while A_{VT} shown by the solid lines increases by approximately 1.25 mV · μm above $A_{VTO} = 5$ mV · μm

Table 3.2 for $V_{SB} = 0.5$ V, or increases by approximately 0.3 V per volt of applied V_{SB}. The threshold voltage increases more rapidly for the simple, linear prediction given by Equation 3.145 compared to the sublinear prediction given by Equation 3.144, although increases are similar for V_{SB} below 0.5 V. The threshold voltage predicted by Equation 3.145 is given by $V_T = V_{TO} + (n-1) \cdot V_{SB}$, indicating a negligible threshold-voltage increase if $(n-1) \cdot V_{SB} \ll V_{TO}$ or $0.33 V_{SB} \ll V_{TO}$ for $n = 1.33$.

Figure 3.70 also shows the predicted increase in threshold-voltage mismatch factor, $A_{VT} - A_{VTO}$, versus V_{SB} from Equations 3.146 and 3.147 for nMOS devices in the example 0.18 μm CMOS process. The predictions assume a body-effect mismatch parameter of $A_{GAMMA} = 5 \times 10^{-3}$ V$^{1/2}$μm for Equation 3.146 and $\phi_0 = 2\phi_F + 4U_T = 0.953$ V for Equation 3.147, which assumes that $U_T = 25.9$ mV and, again, $PHI = 2\phi_F = 0.85$ V. The threshold-voltage mismatch factor increases by approximately 1.25 mV · μm above $A_{VTO} = 5$ mV · μm given in Table 3.3 for $V_{SB} = 0.5$ V, corresponding to an increase in threshold-voltage mismatch of 25 %. The threshold-voltage mismatch factor increases more rapidly for the simple, linear prediction given by Equation 3.147 compared to the sublinear prediction given by Equation 3.146, although increases are similar for V_{SB} below 0.5 V. The threshold-voltage mismatch factor predicted by Equation 3.147 is given by $A_{VT} = A_{VTO}(1 + V_{SB}/(2\phi_0))$, indicating a negligible threshold-voltage mismatch increase if $V_{SB}/(2\phi_0) \ll 1$ or $V_{SB}/1.9$ V $\ll 1$ for $\phi_0 = 0.953$ V.

Figure 3.70 shows a modest 0.15 V increase in the threshold voltage from $V_{TO} = 0.42$ V and a modest 1.25 mV · μm increase in the threshold-voltage mismatch factor from $A_{VTO} = 5$ mV · μm when $V_{SB} = 0.5$ V. Since V_{SB} must often be constrained to 0.5 V or lower for design in low-voltage processes, the increase in threshold voltage and threshold-voltage mismatch associated with V_{SB} can likely be neglected for initial design guidance.

The gate–source voltage mismatch, ΔV_{GS}, for differential pairs is unaffected by mismatch in γ or n if the common sources are connected to their local body or substrate, which sets $V_{SB} = 0$ V. However, this is not possible for native devices, usually nMOS devices in bulk CMOS processes, that reside in the common substrate. Drain current mismatch, $\Delta I_D/I_D$, for current mirrors is often unaffected by mismatch in γ or n since the common sources are often connected to the main supply rails, which easily sets $V_{SB} = 0\,V$.

For limited values of V_{SB}, Equation 3.152 provides a good estimate for ΔV_{GS} and Equation 3.155 provides a good estimate for $\Delta I_D/I_D$ where, in both cases, the additional mismatch due to non-zero V_{SB} is neglected. Both estimates assume that threshold-voltage mismatch dominates transconductance factor mismatch. This is often the case as described in the next section.

3.11.1.6 Threshold-voltage dominance of mismatch

As seen from Equations 3.149 and 3.153, threshold-voltage mismatch, ΔV_T, and transconductance factor mismatch, $\Delta K_P/K_P$, contribute equally to gate–source voltage mismatch, ΔV_{GS}, and drain current mismatch, $\Delta I_D/I_D$, when

$$\left[A_{VTO}\left(1 + \frac{V_{SB}}{2\phi_0}\right) = A_{KP}\left(\frac{g_m}{I_D}\right)^{-1} \right] = \left[A_{VTO}\left(1 + \frac{V_{SB}}{2\phi_0}\right)\left(\frac{g_m}{I_D}\right) = A_{KP} \right] \tag{3.156}$$

This occurs at a critical value of g_m/I_D given by

$$\frac{g_m}{I_D}(\Delta V_T, \Delta K_P) = \frac{A_{KP}}{A_{VTO}\left(1 + \dfrac{V_{SB}}{2\phi_0}\right)} = \frac{A_{KP}}{A_{VT}} \tag{3.157}$$

where $A_{VT} = A_{VTO}(1 + V_{SB}/(2\phi_0))$ includes the threshold-voltage mismatch increase present for non-zero V_{SB} as given by Equation 3.147. The rightmost term in Equation 3.157 is equal to the critical g_m/I_D for equal threshold-voltage and transconductance factor mismatch contributions given in [238].

For $V_{SB} = 0\,V$, $A_{KP} = 2\% \cdot \mu m$, and $A_{VT} = A_{VTO} = 14$, 9, and $5\,mV \cdot \mu m$ for the example 0.5, 0.35, and $0.18\,\mu m$ CMOS processes listed in Table 3.3, the critical $g_m/I_D = 1.4$, 2.2, and $4\,\mu S/\mu A$ (1/V), respectively. All three values of critical g_m/I_D are low, corresponding to operation deep into strong inversion. When V_{SB} is non-zero, the critical g_m/I_D is even lower because of increased threshold-voltage mismatch. Threshold-voltage mismatch dominates for typical operation below deep strong inversion where operating g_m/I_D is above the critical value. Transconductance factor mismatch dominates in deep strong inversion where operating g_m/I_D is below the critical value.

In addition to finding the critical g_m/I_D, a critical effective gate–source voltage, $V_{EFF} = V_{GS} - V_T$, can be found where threshold-voltage and transconductance factor mismatch contribute equally to mismatch. Assuming strong-inversion operation without velocity saturation and VFMR effects, $g_m/I_D = 2/V_{EFF}$ from Table 3.17. Solving for the critical g_m/I_D in Equation 3.157 in terms of V_{EFF} gives

$$V_{EFF}(\Delta V_T, \Delta K_P, SI) \approx \frac{2A_{VTO}\left(1 + \dfrac{V_{SB}}{2\phi_0}\right)}{A_{KP}} = \frac{2A_{VT}}{A_{KP}} \tag{3.158}$$

The rightmost term in Equation 3.158 is equal to the critical V_{EFF} for equal threshold-voltage and transconductance factor mismatch contributions given in [239].

For $V_{SB} = 0\,V$, the critical $V_{EFF} = 1.4$, 0.9, and $0.5\,V$ for the example 0.5, 0.35, and $0.18\,\mu m$ processes, respectively. All three values of critical V_{EFF} are high, corresponding, again, to operation deep into strong inversion. Threshold-voltage mismatch dominates when operating V_{EFF} is below the critical value, corresponding to operation at g_m/I_D above its critical value. When velocity saturation and VFMR effects are significant, the critical V_{EFF} is below that predicted by Equation 3.158. Under full velocity saturation in strong inversion where $g_m/I_D = 1/V_{EFF}$ from Table 3.17, the critical V_{EFF} is one-half that given by Equation 3.158.

In addition to finding the critical g_m/I_D and V_{EFF}, a critical inversion coefficient can be found where threshold-voltage and transconductance factor mismatch contribute equally to mismatch. Assuming strong-inversion operation without velocity saturation and VFMR effects, $g_m/I_D = 1/(nU_T\sqrt{IC})$ from Table 3.17. Solving for the critical g_m/I_D in Equation 3.157 in terms of IC then gives

$$IC(\Delta V_T, \Delta K_P, \mathrm{SI}) \approx \left(\frac{A_{VTO}\left(1 + \dfrac{V_{SB}}{2\phi_0}\right)}{A_{KP} \cdot nU_T} \right)^2 = \left(\frac{A_{VT}}{A_{KP} \cdot nU_T} \right)^2 \tag{3.159}$$

For $V_{SB} = 0\,\mathrm{V}$, the critical $IC = 432$, 179, and 55 for the example 0.5, 0.35, and 0.18 μm processes, respectively, assuming $n = 1.3$ and $U_T = 25.9\,\mathrm{mV}$ ($T = 300\,\mathrm{K}$). All three values of critical IC are high, corresponding, again, to operation deep into strong inversion. Threshold-voltage mismatch dominates when operating IC is below the critical value, corresponding to operation at g_m/I_D above its critical value. Under full velocity saturation where g_m/I_D is given in Table 3.17, the critical IC value is reduced to the square root of the value given by Equation 3.159 multiplied by $\sqrt{IC_{CRIT}}$. IC_{CRIT}, given in Table 3.17, corresponds to the g_m/I_D transition between strong-inversion operation without velocity saturation and VFMR effects, and full velocity saturation operation.

The critical g_m/I_D given in Equation 3.157 for equal threshold-voltage and transconductance factor mismatch contributions is valid in all regions of operation, with or without velocity saturation and VFMR effects. The corresponding critical V_{EFF} and IC given in Equations 3.158 and 3.159, however, are valid only in strong inversion when velocity saturation and VFMR effects are negligible. Since these critical values often correspond to operation in strong inversion and long channel lengths are often selected for critical matching applications, Equations 3.158 and 3.159 can often be used with appropriate caution. Equation 3.157 can then be used to verify if operating g_m/I_D exceeds the critical value by at least a factor of two to ensure that threshold-voltage mismatch dominates. If threshold-voltage and transconductance factor mismatch are largely uncorrelated, operation at only a factor of two above the critical g_m/I_D results in an under-prediction of gate–source voltage and drain current mismatch by only 10.6 % when considering only threshold-voltage mismatch. When threshold-voltage mismatch dominates, Equations 3.152 and 3.155 provide simple predictions of gate–source voltage and drain current mismatch, assuming V_{SB} is sufficiently low as noted in the equations.

Technology matching trends can be observed by evaluating the critical $g_m/I_D = A_{KP}/A_{VT}$ from Equation 3.157. This can be done for the reported A_{KP} and A_{VTO} values listed in Table 3.39 as described in [238] where A_{VT} is assumed equal to A_{VTO}. The critical g_m/I_D increases as t_{ox} decreases in smaller-geometry processes because the threshold-voltage mismatch factor, A_{VTO}, decreases with t_{ox}, while the transconductance mismatch factor, A_{KP}, decreases less as described in Section 3.11.1.2. This requires operation at lower levels of inversion where g_m/I_D is higher to ensure that threshold-voltage mismatch dominates. However, inversion levels are increasingly limited to near the onset of strong inversion because V_{EFF} and $V_{DS,sat}$ must be lower to support the lower supply voltages in smaller-geometry processes. As a result, threshold-voltage mismatch may continue to dominate for most applications.

3.11.1.7 Summary of gate–source voltage and drain current mismatch

Table 3.41 summarizes MOS gate–source voltage and drain current, local-area mismatch expressions and trends in terms of the design choices of IC, L, and I_D. The first entry gives gate–source voltage mismatch, ΔV_{GS}, from Equations 3.149 and 3.150. ΔV_{GS} is found from the threshold-voltage mismatch factor, A_{VT}, which is increased from the zero-V_{SB} threshold-voltage mismatch factor, A_{VTO}, when V_{SB} is non zero as noted at the bottom of the table. ΔV_{GS} is also found from the transconductance mismatch factor, A_{KP}, which can be neglected when threshold-voltage mismatch dominates. As noted in the table and described in Section 3.11.1.6, threshold voltage often dominates mismatch and dominates ΔV_{GS} mismatch when operating g_m/I_D exceeds the normally low critical value of $g_m/I_D = A_{KP}/A_{VT}$. Finally, ΔV_{GS} is expressed using the composite gate–source voltage mismatch factor, A_{VGS}, described by Equation 3.151. A_{VGS} includes both threshold-voltage and transconductance factor mismatch effects.

Table 3.41 MOS gate–source voltage and drain current, local-area mismatch expressions and trends in terms of the inversion coefficient, channel length, and drain current

Mismatch (1σ) for standard 1σ mismatch factors	$IC \uparrow$ L, I_D fixed	$L \uparrow$ IC, I_D fixed	$I_D \uparrow$ IC, L fixed
Gate–source voltage mismatch: $$\Delta V_{GS} = \sqrt{(\Delta V_T)^2 + \left[\frac{\Delta K_P}{K_P}\left(\frac{g_m}{I_D}\right)^{-1}\right]^2} = \frac{A_{VGS}}{\sqrt{WL}}$$ $$= \left(\frac{\sqrt{IC}}{L}\sqrt{\frac{I_0}{I_D}}\right)$$ $$\cdot \sqrt{A_{VT}^2 + \left(A_{KP}\cdot nU_T\left(\sqrt{IC+0.25}+0.5\right)\right)^2}$$ When ΔV_T dominates,[a] A_{KP} term can be neglected	When ΔV_T dominates,[a] $\uparrow\propto\sqrt{IC}$ When ΔK_P dominates, $\uparrow\propto IC$ in SI, but increases more rapidly if g_m/I_D decreases due to vel. sat.	$\downarrow\propto\frac{1}{L}$, decreasing more rapidly when ΔK_P dominates and g_m/I_D increases due to less vel. sat.	$\downarrow\propto\frac{1}{\sqrt{I_D}}$
Drain current mismatch: $$\frac{\Delta I_D}{I_D} = \sqrt{\left(\Delta V_T\cdot\frac{g_m}{I_D}\right)^2 + \left(\frac{\Delta K_P}{K_P}\right)^2} = \frac{A_{VGS}\cdot\frac{g_m}{I_D}}{\sqrt{WL}}$$ $$= \left(\frac{\sqrt{IC}}{L}\sqrt{\frac{I_0}{I_D}}\right)$$ $$\cdot \sqrt{\left(\frac{A_{VT}}{nU_T\left(\sqrt{IC+0.25}+0.5\right)}\right)^2 + A_{KP}^2}$$ When ΔV_T dominates,[a] A_{KP} term can be neglected	When ΔV_T dominates,[a] $\uparrow\propto\sqrt{IC}$ in WI; almost constant in SI, but decreases if g_m/I_D decreases due to vel. sat. When ΔK_P dominates, $\uparrow\propto\sqrt{IC}$	$\downarrow\propto\frac{1}{L}$, decreasing less rapidly when ΔV_T dominates and g_m/I_D increases due to less vel. sat.	$\downarrow\propto\frac{1}{\sqrt{I_D}}$

Mismatch expressions are for saturation where $V_{DS} > V_{DS,sat}$.
Velocity saturation refers to both velocity saturation and VFMR effects.
Expressions assume ΔV_T and ΔK_P are uncorrelated. Some correlation is present through C'_{OX} variations.
$A_{VT} = A_{VTO}\left[1 + V_{SB}/(2\phi_0)\right]$. When $V_{SB} \ll 2\phi_0$, $A_{VT} \approx A_{VTO}$ as discussed in Section 3.11.1.5.
$\phi_0 \approx 2\phi_F + 4U_T$ is the approximate silicon surface potential in strong inversion.
To include velocity saturation and VFMR decreases in g_m/I_D, replace IC in $\sqrt{IC+0.25}$ terms with $IC(1 + IC/IC_{CRIT})$ as described in Table 3.17.
[a] ΔV_T often dominates mismatch. This occurs when $g_m/I_D > A_{KP}/A_{VT}$, which corresponds to $IC < (A_{VT}/(nU_T A_{KP}))^2$ in strong inversion. The critical value of IC is lower if velocity saturation and VFMR reduction effects are significant. This is discussed in Section 3.11.1.6.

In order to express ΔV_{GS} in Table 3.41 in terms of IC, L, and I_D, MOS gate area is expressed as $WL = (L^2/IC)(I_D/I_0)$ from Table 3.10 where $I_0 = 2n_0\mu_0 C'_{OX}U_T^2$ is the technology current described in Table 3.6. Additionally, g_m/I_D is expressed in terms of IC from Table 3.17. Velocity saturation and VFMR reduction effects are not included in the g_m/I_D component of ΔV_{GS}, but, as noted, can be included by replacing IC in the g_m/I_D components with $IC = IC(1 + IC/IC_{CRIT})$. IC_{CRIT}, given in Table 3.17, corresponds to the g_m/I_D transition between strong-inversion operation without velocity saturation and VFMR effects, and full velocity saturation operation.

The trends listed in Table 3.41 result primarily from threshold-voltage mismatch being inversely proportional to the square root of gate area. When threshold-voltage mismatch dominates, ΔV_{GS} increases as the square root of IC, decreases inversely with L, and decreases inversely with the square root of I_D. While minimizing IC lowers ΔV_{GS}, increasing L lowers the mismatch more because of the larger increase in gate area. This is because an increase in L requires an equal increase in W, giving a squared increase in gate area for a given IC and I_D. When threshold-voltage mismatch dominates, the trends for ΔV_{GS} are identical to those for gate-referred flicker-noise voltage density when the gate-referred noise voltage remains constant with the inversion level. This can be observed by comparing ΔV_{GS} given by Table 3.41 with the square root of gate-referred flicker-noise voltage PSD, S_{VG}, given in Table 3.34.

As noted in Table 3.41, in strong inversion at high levels of inversion where transconductance factor mismatch dominates, ΔV_{GS} increases directly with IC. This is because the transconductance factor mismatch increases with the square root of IC due to the decrease in gate area while another increase in ΔV_{GS} with the square root of IC results from division by decreasing g_m/I_D to refer relative drain current mismatch to ΔV_{GS}. The ΔV_{GS} increase is even greater when velocity saturation and VFMR effects lower g_m/I_D.

Both ΔV_{GS} and gate-referred flicker-noise voltage density (the square root of gate-referred flicker-noise voltage PSD, S_{VG}, given in Table 3.34) are minimized at low IC, long L, and high I_D, which maximizes gate area. ΔV_{GS} is also minimized by limiting the value of V_{SB} so there is little additional threshold-voltage mismatch associated with body-effect mismatch. Finally, ΔV_{GS} is minimized in processes having lower threshold-voltage mismatch factors, A_{VTO}, analogous to the gate-referred flicker-noise voltage density that is minimized in processes having lower flicker-noise factors, K_F. As shown in Table 3.39, A_{VTO} varies only modestly across similar process generations and generally decreases in smaller-geometry processes having smaller t_{ox}.

Figure 3.71 shows predicted 1σ, ΔV_{GS} for $L = 0.18, 0.28, 0.48, 1, 2$, and $4\,\mu m$, nMOS devices in the $0.18\,\mu m$ CMOS process described in Tables 3.2 and 3.3 operating at IC ranging from 0.01

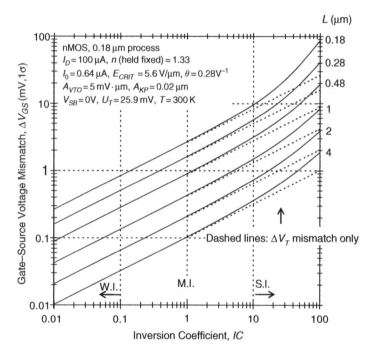

Figure 3.71   Predicted 1σ, gate–source voltage mismatch, ΔV_{GS}, versus inversion coefficient, IC, for $L = 0.18$, $0.28, 0.48, 1, 2$, and $4\,\mu m$, nMOS devices in a $0.18\,\mu m$ CMOS process operating at a drain current of $100\,\mu A$. When threshold-voltage mismatch dominates, ΔV_{GS} increases as the square root of IC and decreases inversely with L, tracking the inverse square root of gate area. At higher levels of inversion, transconductance factor mismatch increases ΔV_{GS} further. ΔV_{GS} for different drain currents is found by multiplying the values shown by $(100\,\mu A/I_D)^{1/2}$. For equal mismatch factors, pMOS ΔV_{GS} is approximately 46 % $(1/\sqrt{4.7})$ of the values shown because of required width and area increases

(deep weak inversion) to 100 (deep strong inversion). Unlike the normalized small-signal parameters, g_m/I_D, $\eta = g_{mb}/g_m$, and $V_A = I_D/g_{ds}$, and intrinsic voltage gain and bandwidth, A_{Vi} and f_{Ti}, ΔV_{GS} (like gate-referred flicker-noise voltage density) is not independent of the drain current when evaluated in terms of the inversion coefficient. Here, a drain current of 100 µA is assumed, where the decrease in ΔV_{GS} with increasing drain current resulting from the required increase in gate area is easily found by multiplying the values shown in the figure by $(100\,\mu A/I_D)^{1/2}$.

ΔV_{GS} is predicted from the expression in Table 3.41 assuming $I_0 = 0.64\,\mu A$ from Table 3.2, and $A_{VT} = A_{VTO} = 5\,\text{mV} \cdot \mu m$ (assumes $V_{SB} = 0\,\text{V}$) and $A_{KP} = 2\% \cdot \mu m$ from Table 3.3. g_m/I_D, which controls the conversion of transconductance factor mismatch to ΔV_{GS}, is predicted assuming $U_T = 25.9\,\text{mV}$ ($T = 300\,\text{K}$) and $n = 1.33$, which is held fixed and equal to the value used for g_m/I_D shown in Figure 3.26. Velocity saturation and VFMR reductions in g_m/I_D are predicted as noted at the bottom of Table 3.41 assuming $E_{CRIT} = 5.6\,\text{V}/\mu m$ and $\theta = 0.28/\text{V}$ from Table 3.2. Finally, channel length is reduced by $DL = 0.028\,\mu m$ (also given in Table 3.2) from the values given to include lateral diffusion effects.

As observed in Figure 3.71, when threshold-voltage mismatch dominates, ΔV_{GS} tracks the inverse square root of gate area, increasing as the square root of IC and decreasing inversely with increasing L. This is shown by the dashed lines that consider only threshold-voltage mismatch and overlay the solid lines for $IC < 10$. However, when transconductance factor mismatch dominates at high levels of inversion, ΔV_{GS} increases directly with IC, increasing even further when velocity saturation and VFMR effects lower g_m/I_D. This is seen in the figure by the solid lines that consider both threshold-voltage and transconductance factor mismatch and increase above the dashed lines. For IC below the critical value of $IC = 55$ found from Equation 3.159, threshold-voltage mismatch dominates. For IC above this, transconductance factor mismatch dominates. Finally, the similarity of trends for ΔV_{GS} and gate-referred flicker-noise voltage density mentioned can be observed by comparing Figures 3.71 and 3.67. At high levels of inversion, additional increases in ΔV_{GS} are caused by transconductance factor mismatch, whereas additional increases in flicker noise are caused by increasing V_{EFF}.

As observed in Figure 3.71, ΔV_{GS} spans nearly four decades being minimum for long-channel devices in deep weak inversion and maximum for short-channel devices in deep strong inversion. The lowest values of ΔV_{GS} shown for long-channel devices in deep weak inversion may not be attainable because global, gradient, or distance mismatch effects described in Section 3.11.2 may increase the mismatch for very large devices. The highest values of ΔV_{GS} shown for short-channel devices in deep strong inversion may not be attainable because resulting widths described in Table 3.10 may be below values possible for layout. However, when verified for practical device geometries and geometries below the critical distance value described in Section 3.11.2.5, and scaled for the drain current used, the figure provides an estimate of ΔV_{GS}. Typically, at least three times the 1σ mismatch shown, giving at least 3σ mismatch, would be considered for production design.

Since ΔV_{GS} scales inversely with the square root of I_D, again tracking the inverse square root of required gate area, it is a factor of 10 higher for $I_D = 1\,\mu A$ compared to values shown in Figure 3.71 for $I_D = 100\,\mu A$. This illustrates the mismatch problem associated with micropower design where channel length is often increased to obtain attainable widths for layout and manage mismatch through increased gate area at the expense of intrinsic bandwidth described in Section 3.9.6.

Predicted ΔV_{GS} for pMOS devices is approximately 46 % ($1/\sqrt{4.7}$) the nMOS values shown in Figure 3.71 for equal mismatch factors listed in Table 3.3 due to a factor-of-4.7 increase in required pMOS width and gate area. This is seen in the width and gate area expressions in Table 3.10 for $I_0 = 0.135\,\mu A$ and $0.64\,\mu A$ listed in Table 3.2 for pMOS and nMOS devices, respectively. However, at high levels of inversion where transconductance factor mismatch dominates, pMOS ΔV_{GS} increases less rapidly for short-channel devices since drain current mismatch is reflected to lower ΔV_{GS} because of division by larger g_m/I_D due to less velocity saturation effects.

As for ΔV_{GS}, Table 3.41 summarizes relative drain current mismatch, $\Delta I_D/I_D$, expressions and trends in terms of IC, L, and I_D. $\Delta I_D/I_D$ is found from Equations 3.153 and 3.154 using the process mismatch factors, gate area, and g_m/I_D described earlier for ΔV_{GS}.

Since $\Delta I_D / I_D$ is equal to ΔV_{GS} multiplied by g_m / I_D as seen by Equations 3.149 and 3.153, the trends listed in Table 3.41 are more complex for $\Delta I_D / I_D$ than those for ΔV_{GS}. As for ΔV_{GS}, threshold-voltage mismatch dominates $\Delta I_D / I_D$ mismatch when operating g_m / I_D exceeds the normally low critical value of $g_m / I_D = A_{KP} / A_{VT}$.

When threshold-voltage mismatch dominates, $\Delta I_D / I_D$ is equal to the threshold-voltage mismatch multiplied by g_m / I_D. In weak inversion where g_m / I_D is constant, $\Delta I_D / I_D$ increases as the square root of IC and decreases inversely with L, tracking threshold-voltage mismatch that is inversely proportional to the square root of gate area. However, in strong inversion, $\Delta I_D / I_D$ levels off and becomes nearly constant with IC because threshold-voltage mismatch increases as the square root of IC due to the reduction of gate area while g_m / I_D decreases as the inverse square root of IC. $\Delta I_D / I_D$ still decreases inversely with L, permitting control of the mismatch. When velocity saturation and VFMR reduce g_m / I_D, $\Delta I_D / I_D$ actually rolls over and begins to decrease with increasing IC. When threshold-voltage mismatch dominates, the IC and L trends for $\Delta I_D / I_D$ are identical to those for drain-referred flicker-noise current density when the gate-referred noise voltage remains constant with the inversion level. This can be observed by comparing $\Delta I_D / I_D$ given by Table 3.41 with the square root of drain-referred flicker-noise current PSD, S_{ID}, given in Table 3.34.

As noted in Table 3.41, in strong inversion at high levels of inversion where transconductance factor mismatch dominates, $\Delta I_D / I_D$ begins increasing as the square root of IC. Here, $\Delta I_D / I_D$ directly tracks transconductance factor mismatch, which tracks the inverse square root of gate area.

Both $\Delta I_D / I_D$ and drain-referred flicker-noise current density (the square root of drain-referred flicker-noise current PSD, S_{ID}, given in Table 3.34) are minimized at low IC and long L, which maximizes gate area. Interestingly, both the drain mismatch current and drain flicker-noise current are minimized at low IC even though g_m / I_D and g_m are maximum here. This is because the reduction of threshold-voltage mismatch and gate-referred flicker-noise voltage density due to large gate area exceeds the effect of maximum g_m / I_D and g_m. Operating at low IC also helps to ensure that threshold-voltage mismatch dominates and gate-referred flicker-noise voltage does not increase with the inversion level. As for ΔV_{GS} and gate-referred flicker-noise voltage density, increasing L, which increases gate area by the square of L, most significantly lowers $\Delta I_D / I_D$ and drain-referred flicker-noise current density.

While both $\Delta I_D / I_D$ and drain-referred flicker-noise current density are minimized at low IC and long L due to maximum gate area, $\Delta I_D / I_D$ decreases as the inverse square root of I_D while drain-referred flicker-noise current density increases as the square root of I_D. Absolute drain current mismatch, ΔI_D, and drain-referred flicker-noise current density both increase as the square root of I_D because ΔV_{GS} and gate-referred flicker-noise voltage density decrease inversely with the square root of I_D through the increase in required gate area, while g_m increases directly with I_D. This results in increased conversion of ΔV_{GS} and gate-referred flicker-noise voltage to ΔI_D and drain-referred flicker-noise current. The relative drain mismatch current, $\Delta I_D / I_D$, however, is unaffected by the g_m increase associated with increasing I_D since I_D also appears in the denominator.

As for ΔV_{GS}, $\Delta I_D / I_D$ is minimized by limiting the value of V_{SB}, but V_{SB} is often zero for current-mirror applications. Additionally, as for ΔV_{GS}, $\Delta I_D / I_D$ is minimized in processes having lower threshold-voltage mismatch factors, A_{VTO}, analogous to drain-referred flicker-noise current density that is minimized in processes having lower flicker-noise factors, K_F.

Figure 3.72 shows predicted 1σ, $\Delta I_D / I_D$ for $L = 0.18, 0.28, 0.48, 1, 2,$ and $4\,\mu\text{m}$, nMOS devices in the $0.18\,\mu\text{m}$ CMOS process described in Tables 3.2 and 3.3 operating at IC ranging from 0.01 to 100. Like ΔV_{GS} shown in Figure 3.71, $\Delta I_D / I_D$ is not independent of the drain current when evaluated in terms of the inversion coefficient. Again, a drain current of $100\,\mu\text{A}$ is assumed, where the decrease in $\Delta I_D / I_D$ with increasing drain current resulting from the required increase in gate area is easily found by multiplying the values shown in the figure by $(100\,\mu\text{A}/I_D)^{1/2}$. $\Delta I_D / I_D$ is predicted from the expression in Table 3.41 using the process parameters used for ΔV_{GS} shown in Figure 3.71.

As observed in Figure 3.72, in weak inversion where threshold-voltage mismatch is multiplied by constant g_m / I_D, $\Delta I_D / I_D$ tracks the inverse square root of gate area, increasing as the square root of IC and decreasing inversely with increasing L. However, in strong inversion, $\Delta I_D / I_D$ levels off and

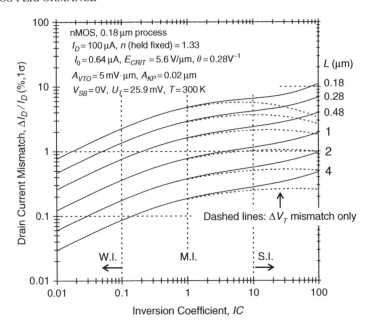

Figure 3.72 Predicted 1σ, relative drain current mismatch, $\Delta I_D/I_D$, versus inversion coefficient, IC, for $L = 0.18$, 0.28, 0.48, 1, 2, and 4 μm, nMOS devices in a 0.18 μm CMOS process operating at a drain current of 100 μA. When threshold-voltage mismatch dominates, $\Delta I_D/I_D$ increases as the square root of IC in weak inversion and levels off in strong inversion before decreasing due to velocity saturation and VFMR reductions in g_m/I_D. At higher levels of inversion where transconductance factor mismatch dominates, $\Delta I_D/I_D$ increases again as the square root of IC. $\Delta I_D/I_D$ for different drain currents is found by multiplying the values shown by $(100\,\mu A/I_D)^{1/2}$. For equal mismatch factors, pMOS $\Delta I_D/I_D$ is approximately 46 % $(1/\sqrt{4.7})$ of the values shown because of required width and area increases

is nearly constant with IC because, as mentioned, threshold-voltage mismatch increases as the square root of IC while g_m/I_D decreases as the inverse square root of IC. As seen by the dashed lines in the figure that include only threshold-voltage mismatch, $\Delta I_D/I_D$ begins to roll over and decrease at high IC when velocity saturation and VFMR reduce g_m/I_D. At high levels of inversion where transconductance factor mismatch becomes significant, $\Delta I_D/I_D$ begins increasing again as the square root of IC, tracking the inverse square root of gate area. This is seen in the figure by the solid lines that consider both threshold-voltage and transconductance factor mismatch and increase above the dashed lines. As for ΔV_{GS} shown in Figure 3.71, when IC is below the critical value of $IC = 55$, threshold-voltage mismatch dominates. Transconductance factor mismatch dominates when IC is above the critical value. Finally, the similarity of trends for $\Delta I_D/I_D$ and drain-referred flicker-noise current density mentioned can be observed by comparing Figures 3.72 and 3.68. At high levels of inversion, increases in $\Delta I_D/I_D$ are caused by transconductance factor mismatch, whereas increases in flicker noise are caused by increasing gate-referred flicker noise voltage with V_{EFF}.

As observed in Figure 3.72, $\Delta I_D/I_D$ spans nearly two and a half decades being minimum for long-channel devices in deep weak inversion and maximum for short-channel devices in deep strong inversion. In weak inversion, the low threshold-voltage mismatch resulting from large required gate area exceeds the effect of maximum g_m/I_D that maximizes the conversion of threshold-voltage mismatch to $\Delta I_D/I_D$. In strong inversion, the increase in threshold-voltage mismatch due to decreasing gate area is largely countered by the decrease in g_m/I_D giving reduced conversion of threshold-voltage mismatch to $\Delta I_D/I_D$. This accounts for the smaller range of two and a half decades of $\Delta I_D/I_D$ shown in Figure 3.72 compared to the four-decade range of ΔV_{GS} shown in Figure 3.71.

Like ΔV_{GS} shown in Figure 3.71, the lowest values of $\Delta I_D / I_D$ shown in Figure 3.72 for long-channel devices in deep weak inversion may not be attainable because global, gradient, or distance mismatch effects may increase the mismatch for very large devices. Additionally, like ΔV_{GS}, the highest values of $\Delta I_D / I_D$ shown for short-channel devices in deep strong inversion may not be attainable for $I_D = 100 \, \mu A$ considered because resulting widths may be below values possible for layout. However, when verified for practical device geometries and geometries below the critical distance value described in Section 3.11.2.5, and scaled for the drain current used, Figure 3.72 provides an estimate of $\Delta I_D / I_D$. Again, typically at least three times the 1σ mismatch shown, giving at least 3σ mismatch, would be considered for production design.

As for ΔV_{GS} shown in Figure 3.71, $\Delta I_D / I_D$ scales inversely with the square root of I_D. For $I_D = 1 \, \mu A$, predicted $\Delta I_D / I_D$ is a factor of 10 larger than values shown in Figure 3.72 for $I_D = 100 \, \mu A$. This, again, illustrates the mismatch problem for micropower design where channel length must necessarily be increased to obtain sufficient channel width and gate area. Similar to ΔV_{GS}, predicted $\Delta I_D / I_D$ for pMOS devices is approximately 46 % $(1/\sqrt{4.7})$ the nMOS values shown in Figure 3.72 due to the 4.7 times larger required width and gate area.

ΔV_{GS} and $\Delta I_D / I_D$, as well as gate-referred flicker-noise voltage density and drain-referred flicker-noise current density, are generally inversely proportional to L due to the L^2 increase in gate area. As a result, mismatch and flicker noise are most effectively minimized by increasing L. Unfortunately, as described in Section 3.9.6, increasing L decreases intrinsic bandwidth by the inverse square of L, unless full velocity saturation is present where bandwidth is reduced by the inverse of L. The favorable decrease in mismatch and flicker noise and unfavorable decrease in bandwidth with increasing L illustrate some of the important analog CMOS design tradeoffs. Chapter 4 discusses performance tradeoffs, and these tradeoffs are discussed further for specific design examples in Chapter 5.

This summary of local-area DC mismatch is concluded with the following design suggestions. Many of these suggestions are similar to those given earlier in Section 3.10.3.7 for minimizing flicker noise.

- Maximize gate area as this lowers local-area and edge-effect mismatch, except for a possible increase in mismatch resulting from gate leakage current described below.

- Avoid short-width or short-length devices to minimize loss of effective gate area, minimize edge effects, and minimize effects of depletion charge sharing [229] and other small-geometry effects.

- Avoid extremes of operation in deep weak inversion where threshold-voltage mismatch may increase [232]. Moderate inversion, giving nearly as high g_m / I_D with improved intrinsic bandwidth, may be a better choice.

- Avoid extremes of operation in deep strong inversion, where as described in Section 3.11.1.6, transconductance factor mismatch may become significant.

- Maintain a low value of V_{SB} to minimize n or γ mismatch increases in threshold-voltage mismatch. V_{SB} is usually necessarily limited to small values in low-voltage processes. V_{SB} can be set to zero if the sources and bodies of differential-pair devices are tied together. V_{SB} is normally zero for current-mirror devices connected to supply rails.

- Consider the mismatch of non-input devices carefully to ensure negligible mismatch contributions for low-mismatch applications. This is discussed in Section 5.3.7 for operational transconductance amplifier design examples.

- Lay out matched devices using identical structures and orientation. For sufficiently large devices, use common-centroid layout to help average out distance, gradient, or global mismatch. Additionally, dummy structures should be placed at the perimeter of devices to ensure an identical external environment. Layout techniques for optimal MOS matching are described in detail in [237, pp. 426–442].

- Avoid covering matched devices with metal traces or fill as this can significantly degrade matching [198, 235]. Symmetrical use of higher-level metals reduces the mismatch contribution.

- Be aware of mismatch associated with self-heating of devices, especially for SOI CMOS processes, which have nearly a factor-of-100 lower thermal conductivity into the main substrate compared to bulk processes. Thermally induced mismatch is discussed in [202].

- Finally, in processes having gate-oxide thickness around 2 nm and below, evaluate gate leakage current and its effect on circuit mismatch. Unfortunately, as described in Sections 3.12.2.5 and 3.12.2.6, gate leakage current and its mismatch increase with gate area, working against the usual improvement in local-area mismatch [180]. Mismatch contributions from gate leakage current can be minimized by selecting thicker gate-oxide options when these are available.

3.11.2 Distance DC Mismatch

The previous section considered local-area mismatch, which results from rapidly varying, local process variations that cause mismatch in the threshold voltage, transconductance factor, and body-effect factor. This then causes gate–source voltage and drain current mismatch for identically laid-out, adjacent devices.

Distance, gradient, or global mismatch results from slowly varying process variations across a wafer and also causes mismatch in the threshold voltage, transconductance factor, and body-effect factor. Unlike local-area mismatch, distance mismatch is usually not significant unless devices are unusually large or laid out with large separation distances. When large devices are used, common-centroid and other layout techniques described in [237, pp. 426–442] minimize distance mismatch, but this mismatch cannot be totally eliminated. Additionally, layout along thermal center lines having little temperature gradient minimizes the thermal component of distance mismatch. Thermally induced mismatch depends on the power dissipation, location, and thermal characteristics of the matched devices of interest as well as neighboring devices that produce heat. This is specific to a circuit and is not considered in distance mismatch modeling described in the next section. While local-area mismatch dominates for most applications, it is believed that distance mismatch dominates input-referred offset voltage for the DC optimized, simple, operational transconductance amplifier described in Sections 5.2.1 and 5.4.2.5. As described later in Section 3.11.2.5, unusually large devices are used where distance mismatch likely exceeds local-area mismatch.

3.11.2.1 Modeling

Distance mismatch was considered in early MOS mismatch papers [182, 183] and was modeled by Pelgrom [185] using distance mismatch parameters multiplied by the separation distance between devices. Table 3.42 summarizes expressions for distance mismatch in the zero-V_{SB} threshold voltage, ΔV_{TO}, the transconductance factor, $\Delta K_P/K_P$, and the body-effect factor, $\Delta\gamma$. ΔV_{TO}, $\Delta K_P/K_P$, and $\Delta\gamma$ mismatch are modeled by the distance or spacing, S, multiplied by distance mismatch factors, S_{VTO}, S_{KP}, and S_{GAMMA}, respectively. This is similar to local-area mismatch given earlier in Table 3.38, except S replaces the inverse square root of gate area, $1/(WL)^{1/2}$, and S_{VTO}, S_{KP}, and S_{GAMMA} replace local-area mismatch factors, A_{VTO}, A_{KP}, and A_{GAMMA}. Since distance mismatch in ΔV_{TO}, $\Delta K_P/K_P$, and $\Delta\gamma$ depends only on device spacing and not gate area, the mismatch is independent of operating *IC*, *L*, and I_D. Differential-pair gate–source voltage and current-mirror drain current mismatch, however, may depend on operating conditions as described later in Section 3.11.2.3.

Table 3.42 MOS distance mismatch expressions for threshold voltage, transconductance factor, and body-effect factor

Parameter
Threshold-voltage mismatch ($V_{SB} = 0\,\text{V}$): $\Delta V_{TO} = S_{VTO} \cdot S$
Transconductance factor ($K_P = \mu C'_{OX}(W/L)$) mismatch: $\dfrac{\Delta K_P}{K_P} = S_{KP} \cdot S$
Body-effect factor mismatch (for non-zero V_{SB}): $\Delta\gamma = S_{GAMMA} \cdot S$

S is the spacing between identically laid-out MOSFETs.

3.11.2.2 Reported mismatch factors and trends

Table 3.43 lists reported 1σ values for the distance mismatch parameters S_{VTO} and S_{KP} for different MOS process generations. Additionally, the table notes if mismatch measurements were performed at the wafer level or on packaged parts. This is important because mechanical strain resulting from device packaging often raises device mismatch as shown in the comparative study included in the table. Here, mismatch was measured for a $0.7\,\mu\text{m}$ CMOS process at the wafer level and for packaged parts [200]. S_{VTO} increased from less than $0.1\,\mu\text{V}/\mu\text{m}$ for wafer measurements to 0.1 and $0.6\,\mu\text{V}/\mu\text{m}$ for polyimide and eutectic solder die attachment, respectively. Similarly, S_{KP} increased from $5 \times 10^{-5}\,\%/\mu\text{m}$ for wafer measurements to 6×10^{-5} and $2 \times 10^{-4}\,\%/\mu\text{m}$ for the two die attachments, respectively. This suggests that polyimide die attachment, which has less residual strain, lowers distance mismatch compared to eutectic solder attachment. While a separate study showed little distance mismatch increase for 0.25 and $0.18\,\mu\text{m}$ processes when dies were deliberately stressed [213], an increase in circuit DC offset due to mismatch is commonly observed between die and packaged part testing.

Table 3.43 Reported MOS distance mismatch factors for threshold voltage and transconductance factor

Process		V_{TO} mismatch S_{VTO} ($\mu\text{V}/\mu\text{m}, 1\sigma$)		$\mu C'_{OX}(W/L)$ mismatch S_{KP} ($\%/\mu\text{m}, 1\sigma$)		Notes, reference
t_{ox} (nm)	L_{min} (μm)	nMOS	pMOS	nMOS	pMOS	
50	2.5	4	4	2×10^{-4}	2×10^{-4}	Note[a], [185]
22.5	1.2	0.3	0.6	3×10^{-4}	5×10^{-4}	Note[b], [191]
17	0.7	0.3	0.3	5×10^{-5}	5×10^{-5}	Note[b], [199]
17	0.7	< 0.1		5×10^{-5}		Note[a]
		0.1		6×10^{-5}		Note[c]
		0.6		2×10^{-4}		Note[d], [200]
7.5	0.35	0.8 (ΔV_{GS})				Note[a], $V_{EFF} = 0.2\,\text{V}$, $L = 0.71\,\mu\text{m}$, [217]
5.5	0.25		0.73 (ΔV_{GS})			Note[a], $V_{EFF} = 0.2\,\text{V}$, $L = 0.57\,\mu\text{m}$ [217]

[a] Wafer measurement.
[b] Packaged measurement, die attachment unknown.
[c] Packaged measurement, die attached with polyimide.
[d] Packaged measurement, die attached with eutectic solder.

Although little has been reported for distance mismatch parameters summarized in Table 3.43 compared to local-area mismatch parameters summarized earlier in Table 3.39, S_{VTO} is reported at approximately $1\,\mu\mathrm{V}/\mu\mathrm{m}$ or below for most CMOS processes. This suggests that ΔV_{TO} is approximately $100\,\mu\mathrm{V}$ (1σ) or below for a device spacing of $S = 100\,\mu\mathrm{m}$. Such spacing can occur within the layout of large, matched devices or between a bias reference and remote, related devices.

Little has also been reported for S_{KP}, but reported values range from 6×10^{-5} to $2 \times 10^{-4}\,\%/\mu\mathrm{m}$ for packaged parts in a $0.7\,\mu\mathrm{m}$ CMOS process depending on the die attachment mentioned earlier. For the higher $S_{KP} = 2 \times 10^{-4}\,\%/\mu\mathrm{m}$, $\Delta K_P/K_P$ is approximately $0.02\,\%$ (1σ) for a device spacing of $100\,\mu\mathrm{m}$. As described in Section 3.11.2.4, threshold-voltage mismatch usually dominates transconductance factor mismatch for distance parameters having values around $S_{VTO} = 1\,\mu\mathrm{V}/\mu\mathrm{m}$ and $S_{KP} = 2 \times 10^{-4}\,\%/\mu\mathrm{m}$.

Considerable variability in the distance, threshold-voltage mismatch parameter S_{VTO} (Table 3.43) compared to the local-area, threshold-voltage mismatch parameter A_{VTO} (Table 3.39) is observed and expected across processes. As mentioned in Section 3.11.1.1, A_{VTO} nearly tracks the gate-oxide thickness as a result primarily of referring local, statistical variations in the number of active dopants in the depletion region under the gate to threshold-voltage mismatch. However, S_{VTO} depends on gradients in the body doping concentration, gate-oxide capacitance, and other process parameters across a wafer. These can be expected to vary considerably across processes and, as mentioned, vary with residual strain resulting from packaging. S_{KP} can also be expected to vary widely across processes due to variations in mobility, gate-oxide capacitance, and geometry across a wafer.

3.11.2.3 Gate–source voltage and drain current mismatch

Table 3.44 summarizes MOS gate–source voltage, ΔV_{GS}, and drain current, $\Delta I_D/I_D$, distance mismatch expressions and trends in terms of the design choices of IC and L. When evaluated in terms of IC, distance mismatch, unlike local-area mismatch, is independent of I_D. This is because distance mismatch is independent of the gate area while local-area mismatch depends on the gate area that itself depends on I_D. Distance mismatch is predicted identically to local-area mismatch given earlier in Table 3.41, with $\Delta V_T = S_{VT} \cdot S$ and $\Delta K_P/K_P = S_{KP} \cdot S$, where S_{VT} is increased above S_{VTO} to consider the increase in ΔV_T from ΔV_{TO} for non-zero V_{SB} as described later. Again, threshold-voltage and transconductance factor mismatch are assumed to be uncorrelated, although as mentioned earlier in Section 3.11.1.3, some correlation can be expected through variations in the gate-oxide capacitance. When threshold-voltage mismatch dominates transconductance factor mismatch as discussed in the next section, the degree of correlation between these is not important.

Unlike expressions for local-area mismatch given in Table 3.41, the effect of non-zero V_{SB} is not easily included in the distance mismatch expressions given in Table 3.44. This is because the mismatch and correlation of threshold voltage and body-effect factor, γ, or substrate factor, n, dominated by variations in the number of active dopants in the depletion region under the gate for local-area mismatch, cannot be assumed for distance mismatch. As noted in Table 3.44, the distance, threshold-voltage mismatch factor, S_{VT}, can be increased from the zero-V_{SB} factor, S_{VTO}, to consider non-zero V_{SB}. S_{VT} can be approximated by equating the second and third lines of Equation 3.146 with the local-area mismatch factors A_{VT}, A_{VTO}, and A_{GAMMA} replaced with distance mismatch factors S_{VT}, S_{VTO}, and S_{GAMMA}. This prediction assumes worst case, full correlation between threshold-voltage and body-effect mismatch, and requires S_{GAMMA}, which is rarely reported. Fortunately, low V_{SB} required for designs in small-geometry processes or the assignment of $V_{SB} = 0\,\mathrm{V}$ by connecting sources to their local bodies minimizes or eliminates the V_{SB} effect.

When threshold-voltage mismatch dominates, ΔV_{GS} given by Table 3.44 is equal to $S_{VT} \cdot S$ and is independent of the bias conditions. However, when transconductance factor mismatch dominates,

Table 3.44 MOS gate–source voltage and drain current, distance mismatch expressions and trends in terms of the inversion coefficient and channel length. When evaluated in terms of the inversion coefficient, distance mismatch is independent of the drain current

Mismatch (1σ) for standard 1σ mismatch factors	$IC \uparrow$ L fixed	$L \uparrow$ IC fixed
Gate–source voltage mismatch: $\Delta V_{GS} = \sqrt{(\Delta V_T)^2 + \left[\dfrac{\Delta K_P}{K_P}\left(\dfrac{g_m}{I_D}\right)^{-1}\right]^2}$ $= S \cdot \sqrt{S_{VT}^2 + \left(S_{KP} \cdot nU_T \left(\sqrt{IC+0.25}+0.5\right)\right)^2}$ When ΔV_T dominates,[a] S_{KP} term can be neglected	When ΔV_T dominates,[a] unchanged When ΔK_P dominates, unchanged in WI; $\uparrow \propto \sqrt{IC}$ in SI, but increases more rapidly if g_m/I_D decreases due to vel. sat.	Unchanged, but decreases when ΔK_P dominates and g_m/I_D increases due to less vel. sat.
Drain current mismatch: $\dfrac{\Delta I_D}{I_D} = \sqrt{\left(\Delta V_T \cdot \dfrac{g_m}{I_D}\right)^2 + \left(\dfrac{\Delta K_P}{K_P}\right)^2}$ $= S \cdot \sqrt{\left(\dfrac{S_{VT}}{nU_T\left(\sqrt{IC+0.25}+0.5\right)}\right)^2 + S_{KP}^2}$ When ΔV_T dominates,[a] S_{KP} term can be neglected	When ΔV_T dominates,[a] unchanged in WI; $\downarrow \propto 1/\sqrt{IC}$ in SI, but decreases more rapidly if g_m/I_D decreases due to vel. sat. When ΔK_P dominates, unchanged	Unchanged, but increases when ΔV_T dominates and g_m/I_D increases due to less vel. sat.

Mismatch expressions are for saturation where $V_{DS} > V_{DS,sat}$.
Velocity saturation refers to both velocity saturation and VFMR effects.
Expressions assume ΔV_T and ΔK_P are uncorrelated. Some correlation is present through C'_{OX} variations.
When $V_{SB} = 0\,\text{V}$, $S_{VT} = S_{VTO}$. When $V_{SB} > 0\,\text{V}$, $S_{VT} > S_{VTO}$ as described in text.
To include velocity saturation and VFMR decreases in g_m/I_D, replace IC in $\sqrt{IC+0.25}$ terms with $IC(1+IC/IC_{CRIT})$ as described in Table 3.17.
[a]ΔV_T dominates mismatch when $g_m/I_D > S_{KP}/S_{VT}$, which corresponds to $IC < (S_{VT}/(nU_T S_{KP}))^2$ in strong inversion. The critical value of IC is lower if velocity saturation and VFMR reduction effects are significant. This is discussed in Section 3.11.2.4.

ΔV_{GS} is equal to $S_{KP} \cdot S$ divided by the operating g_m/I_D. In this case, ΔV_{GS} is constant in weak inversion and increases as the square root of IC in strong inversion as g_m/I_D decreases. ΔV_{GS} increases further if velocity saturation and VFMR effects reduce g_m/I_D. ΔV_{GS} depends only on L when transconductance factor mismatch dominates and g_m/I_D is decreased by velocity saturation effects. In this case, ΔV_{GS} decreases as g_m/I_D increases with increased L because of lower velocity saturation effects. As mentioned, ΔV_{GS} is independent of I_D, and, of course, is strongly dependent on the spacing between devices.

When threshold-voltage mismatch dominates, $\Delta I_D/I_D$ given by Table 3.44 is equal to $S_{VT} \cdot S$ multiplied by the operating g_m/I_D. $\Delta I_D/I_D$ is constant in weak inversion and decreases as the inverse square root of IC in strong inversion as g_m/I_D decreases. $\Delta I_D/I_D$ decreases further if velocity saturation and VFMR effects reduce g_m/I_D. However, when transconductance factor mismatch dominates, $\Delta I_D/I_D$ is simply equal to $S_{KP} \cdot S$ and is independent of the bias conditions. $\Delta I_D/I_D$ depends only on L when threshold-voltage mismatch dominates and g_m/I_D is decreased by velocity saturation effects. In this case, $\Delta I_D/I_D$ increases as g_m/I_D increases with increased L because of lower velocity saturation effects. As mentioned, $\Delta I_D/I_D$ is independent of I_D, and, of course, is strongly dependent on the spacing between devices.

3.11.2.4 Threshold-voltage dominance of mismatch

As discussed in Section 3.11.1.6, threshold-voltage mismatch usually dominates local-area mismatch for typical operation below deep strong inversion. Threshold-voltage mismatch dominance of distance mismatch cannot be generally assumed because few mismatch parameters (Table 3.43) have been reported. However, as for local-area mismatch, critical g_m/I_D, V_{EFF}, and IC values can be found where distance threshold-voltage and transconductance factor mismatch components are equal. These critical values are found by replacing the local-area mismatch factors, A_{VT} and A_{KP}, with their distance mismatch counterparts, S_{VT} and S_{KP}, in Equations 3.157, 3.158, and 3.159. The replacement of mismatch factors is valid because the inverse square root of gate area, which is common to both ΔV_T and $\Delta K_P/K_P$ local-area mismatch components, can be replaced with device distance, which is common to these same distance mismatch components.

For $S_{VT} = S_{VTO} = 1\,\mu V/\mu m$ and $S_{KP} = 2 \times 10^{-4}\,\%/\mu m$ from Section 3.11.2.2, the critical $g_m/I_D = S_{KP}/S_{VT} = 2/V$, $V_{EFF} = 1\,V$, and $IC = 221$ from Equations 3.157, 3.158, and 3.159, respectively, when A_{VT} and A_{KP} are replaced with S_{VT} and S_{KP}. The critical IC additionally assumes a strong-inversion value of $n = 1.3$ and $U_T = 25.9\,mV$ for $T = 300\,K$ (room temperature), and is decreased as described for Equation 3.159 when velocity saturation and VFMR effects are significant. The critical values of g_m/I_D, V_{EFF}, and IC correspond to the same operating condition in deep strong inversion. Operation at $g_m/I_D > 2/V$, $V_{EFF} < 1\,V$, or $IC < 221$ below deep strong inversion, required in small-geometry, low-voltage processes for sufficiently low V_{EFF} and $V_{DS,sat}$, ensures that threshold-voltage mismatch dominates. When threshold-voltage mismatch dominates, the calculation of ΔV_{GS} and $\Delta I_D/I_D$ summarized in Table 3.44 is simplified because the S_{KP} terms can be neglected.

3.11.2.5 Critical spacing for comparable distance and local-area mismatch

This final discussion of distance mismatch compares distance mismatch with the always present and usually dominant, local-area mismatch. Assuming threshold-voltage mismatch dominates, this can be done by equating distance and local-area mismatch as given by

$$\left[\Delta V_T(S) = S_{VT} \cdot S\right] \quad = \quad \left[\Delta V_T(WL) = \frac{A_{VT}}{\sqrt{WL}}\right] \tag{3.160}$$

Here, distance and local-area, threshold-voltage mismatch is given by the expressions in Tables 3.42 and 3.38 with S_{VT} and A_{VT} replacing S_{VTO} and A_{VTO}, respectively, to include the mismatch increase for non-zero V_{SB}. Solving Equation 3.160 where distance and local-area mismatch are equal gives a critical spacing of

$$S(S, WL) = \frac{A_{VT}}{S_{VT}\sqrt{WL}} \tag{3.161}$$

The critical spacing corresponds to the maximum allowable spacing between devices where distance mismatch becomes significant relative to local-area mismatch and significantly increases the overall mismatch. Since the critical spacing decreases with increasing gate area because of reduced local-area mismatch, the critical spacing can approach or become less than layout dimensions of large matched devices. In this case, distance mismatch can be significant even for a careful, common-centroid layout of immediately adjacent devices.

For $S_{VT} = S_{VTO} = 1\,\mu V/\mu m$ given in Section 3.11.2.2 and $A_{VT} = A_{VTO} = 5\,mV \cdot \mu m$ summarized in Table 3.3 for the example 0.18 μm CMOS process, the critical spacing is $S = 158\,\mu m$ for two devices having gate areas of $WL = 1000\,\mu m$ each. A dimension of at least 32 μm on a side is required for a square layout of each device, which is well below the critical spacing. This suggests that distance mismatch is not significant for a careful, common-centroid layout of the two devices. Of course, device layout dimensions exceed $32\,\mu m$ on a side because of interdigitated drain and sources, and body connections, but layout dimensions should still be well below the critical spacing. If the two devices are separated by a spacing of $S = 158\,\mu m$, distance mismatch would be comparable to local-area mismatch, and the total mismatch would increase appreciably.

For the DC optimized, simple, operational transconductance amplifier (OTA) described in Section 5.2.1, unusually large devices are used with pMOS devices having gate areas of $WL = 2870\,\mu m \cdot 4\,\mu m = 11500\,\mu m^2$ each. For $S_{VT} = S_{VTO} = 1\,\mu V/\mu m$ and $A_{VT} = A_{VTO} = 14\,mV \cdot \mu m$ summarized in Table 3.3 for the example $0.5\,\mu m$ CMOS process used in this design example, the critical spacing is $S = 131\,\mu m$. A dimension of at least $107\,\mu m$ on a side is required for a square layout of each device, which is comparable to the critical spacing. This suggests that distance mismatch could contribute significantly to overall mismatch. In fact, the measured input-referred offset voltage for the DC optimized OTA described in Section 5.4.2.5 is $1.1\,mV$ (1σ), which significantly exceeds $0.4\,mV$ predicted from local-area mismatch. The balanced and AC optimized versions of the OTA that have much smaller devices, however, have measured offset voltages of 2.2 and $10.2\,mV$, which compare closely with predicted values of 2.3 and $9.8\,mV$, respectively, from local-area mismatch. The large increase in measured offset voltage compared to that predicted from local-area mismatch suggests that distance mismatch is significant for the large devices used in the DC optimized OTA.

3.11.3 DC Mismatch Effects on Circuit Performance

As discussed in Sections 3.11.1.7 and 3.11.2.3, local-area and distance mismatch results in mismatch in the DC gate–source voltage or drain current for identically laid-out devices. This DC mismatch usually dominates the DC offset voltage or current of analog CMOS circuits and is analyzed for OTA design examples in Section 5.3.7. DC mismatch directly impacts the performance of DC reference and bias circuits. However, this mismatch also impacts the performance of amplifiers, filters, comparators, and other circuits, including circuits used in analog-to-digital converters [238–241]. Finally, DC mismatch even impacts the propagation delay and time dispersion of digital circuits [208].

3.11.3.1 Bandwidth, power, and accuracy tradeoffs in current-mode circuits

DC mismatch impacts the accuracy of current-mode circuits and the resulting tradeoffs in bandwidth, power, and accuracy. The DC accuracy of a 1:1 current mirror due to the usually dominant, local-area, threshold-voltage mismatch (Section 3.11.1.6) can be expressed similarly to the derivations in [238, 239] by

$$A_{CC,DC} = \frac{\left(\frac{I_{IN}}{I_D}\right) I_D}{I_{OS}(3\sigma)} = \frac{\left(\frac{I_{IN}}{I_D}\right) I_D}{3 \cdot g_m \Delta V_T(1\sigma)} = \frac{\left(\frac{I_{IN}}{I_D}\right) I_D}{3 \cdot g_m \dfrac{A_{VT}}{\sqrt{WL}}}$$

$$= \left(\frac{I_{IN}}{I_D}\right) \frac{\sqrt{WL}}{3\left(\dfrac{g_m}{I_D}\right) A_{VT}} \tag{3.162}$$

The DC accuracy is the signal-to-noise ratio or ratio of signal to offset currents. I_{IN} is the input signal current represented as a fraction of the DC drain bias current, I_D, for each current-mirror device. I_{IN} may need to be sufficiently below I_D, giving a value of I_{IN}/I_D sufficiently below unity, to manage signal distortion. The denominators in the first row of Equation 3.162 list the 3σ, DC offset current found from the threshold-voltage mismatch, ΔV_T, given by Equation 3.147 multiplied by the device transconductance, g_m. g_m/I_D is the operating transconductance efficiency, and WL is the device gate area. A_{VT} is the local-area, threshold-voltage mismatch factor, which can include the increase from the zero-V_{SB} factor, A_{VTO}, resulting from non-zero V_{SB} as summarized in Table 3.41.

For a gate area of $WL = 100\,\mu\text{m}^2$, $g_m/I_D = 10\,\mu\text{S}/\mu\text{A}$ ($1/\text{V}$) near the onset of strong inversion, and $A_{VT} = 5\,\text{mV}\cdot\mu\text{m}$ for the example $0.18\,\mu\text{m}$, CMOS process described in Tables 3.2 and 3.3, Equation 3.162 predicts an accuracy of 66.7 for $I_{IN} = I_D$ ($I_{IN}/I_D = 1$). The reciprocal of this corresponds to an error of 1.5 %, again, for a 3σ value of DC offset current. The accuracy would be reduced by one-half and the error doubled if I_{IN} were lowered to $\frac{1}{2}I_D$ ($I_{IN}/I_D = \frac{1}{2}$) to manage signal distortion.

As described in [238, 239], it is useful to also estimate the bandwidth or speed to understand tradeoffs in bandwidth, power, and accuracy. The $-3\,\text{dB}$ bandwidth of a 1:1 current mirror is roughly approximated by

$$BW = \frac{g_m}{2\pi\left(2C_{gsi}\right)} = \frac{f'_{Ti}}{2} = \frac{\left(\dfrac{g_m}{I_D}\right)I_D}{\frac{8}{3}\pi C'_{OX}WL} \tag{3.163}$$

where f'_{Ti} is the intrinsic bandwidth described in Section 3.10.6 that includes the intrinsic gate–source capacitance, C_{gsi}, but excludes the intrinsic gate–body capacitance, C_{gbi}. f'_{Ti} is higher than the intrinsic bandwidth, f_{Ti}, described in Section 3.9.6 that includes C_{gbi}, especially in weak and moderate inversion where C_{gbi} is significant. As described in Section 3.9.7, f_{Ti} and, thus, f'_{Ti} underestimate operating bandwidth because they exclude gate overlap and drain–body capacitances that can significantly lower bandwidth, especially for short-channel devices where gate area and intrinsic capacitances are small. Equation 3.163 is used here to illustrate design tradeoffs rather than accurately predict bandwidth. C_{gsi} is taken at its strong-inversion value of $(2/3)C'_{OX}WL$, although the 2/3 factor is less in weak and moderate inversion as shown in Figure 3.54, somewhat compensating for the exclusion of C_{gbi}, which is itself less significant in strong inversion.

The power consumption of a 1:1 current mirror is given by

$$P = 2I_D \cdot V_{DD} \tag{3.164}$$

where V_{DD} is the power supply voltage and I_D, again, is the bias current for each device. Dividing Equation 3.163 by Equation 3.164 gives a bandwidth–power figure of merit expressing the bandwidth for a given level of power. Dividing the equations gives [238, 239]

$$\frac{BW}{P} = \frac{\dfrac{g_m}{I_D}}{\frac{16}{3}\pi C'_{OX}WL \cdot V_{DD}} \tag{3.165}$$

It is important to note that the DC accuracy given by Equation 3.162 and the bandwidth measures given by Equations 3.163 and 3.165 are expressed in terms of g_m/I_D and the gate area, WL, which obscure the observation of performance trends for the identical current-mirror devices. This is because the gate area (Table 3.10) is proportional to I_D, inversely proportional to IC, and proportional to the square of L, while g_m/I_D (Table 3.17) is inversely proportional to the square-root of IC in strong inversion assuming negligible velocity saturation and VFMR effects. As a result, the DC accuracy, which is inversely proportional to the relative drain-current mismatch, $\Delta I_D/I_D$ (Table 3.41), is maximized at high I_D, low IC in weak inversion, and long L. The bandwidth measures, which are proportional to the intrinsic bandwidth, f_{Ti} (Table 3.26), are optimized oppositely at high IC in strong inversion and short L.

Finally, as described in [238, 239], a bandwidth–power–accuracy figure of merit can be developed by squaring the accuracy given in Equation 3.162 and multiplying this by the bandwidth–power figure of merit given by Equation 3.165. This gives

$$\frac{A_{CC,DC}^2 \cdot BW}{P} = \frac{\left(\dfrac{I_{IN}}{I_D}\right)^2}{48\pi \, C_{OX}' A_{VT}^2 \dfrac{g_m}{I_D} V_{DD}} \tag{3.166}$$

In [238], unity appears in the numerator and a factor of 384π appears in the denominator of Equation 3.166, corresponding to a fixed input signal current of $I_{IN} = I_D/(2\sqrt{2})$. This result is consistent with the result given by Equation 3.166 when evaluated for this value of I_{IN}.

The bandwidth–power–accuracy figure of merit given by Equation 3.166 is independent of the drain bias current, I_D. It is also independent of the gate area, because gate area cancels in the $A_{CC,DC}^2$ and BW terms. As a result, the figure of merit is independent of the channel length, except when velocity saturation effects cause reduced g_m/I_D for short-channel devices operating at high IC. Under these conditions, transconductance factor mismatch, neglected in the analysis, can contribute to increased DC offset and reduce the figure of merit, combined with increases otherwise resulting from the reduction of g_m/I_D.

Operation at constant IC giving constant g_m/I_D results in constant bandwidth-power-accuracy figures of merit and, correspondingly, constant bandwidth-accuracy figures of merit for constant drain bias current and power consumption. This is seen in the continuation sheet of Figure 4.21 in Section 4.4.2 where the products of accuracy squared and bandwidth are constant with L. This is because accuracy, which is inversely proportional to local-area, drain-current mismatch listed in Table 3.41, increases as L while bandwidth listed in Table 3.26 decreases as $1/L^2$, both due to gate area that increases as L^2. The product of accuracy (not squared) and bandwidth, however, is not constant with L, but decreases as $1/L$. Bandwidth may actually decrease as $1/L$, changing the trends (Section 5.5.2.11). Finally, for constant IC and g_m/I_D, bandwidth-accuracy figures of merit (products of accuracy squared and bandwidth) increase directly with the drain bias current as seen in the continuation sheet of Figure 4.22 in Section 4.4.3.

The bandwidth–power–accuracy figure of merit given by Equation 3.166 for a current mirror is maximized for the maximum allowable value of I_{IN}/I_D, operation at low g_m/I_D in strong inversion, low V_{DD}, and a small process value of $C_{OX}' A_{VT}^2$. However, operation at decreasing values of g_m/I_D in strong inversion results in higher $V_{DS,sat}$ and higher required V_{DD}. As a result, the $g_m/I_D = 1/(nU_T\sqrt{IC})$ (Table 3.17) and $V_{DD} = V_{DS,sat} = 2U_T\sqrt{IC}$ (Table 3.15) trends cancel in strong inversion assuming negligible velocity saturation, revealing that, at the design limit, the figure of merit is inversely proportional to $C_{OX}' A_{VT}^2$ alone as given by

$$\frac{A_{CC,DC}^2 \cdot BW}{P} \propto \frac{1}{C_{OX}' A_{VT}^2} = \frac{1}{E_{matching}} \tag{3.167}$$

As mentioned in [238], this result suggests there is little the designer can do to maximize the bandwidth–power–accuracy figure of merit as this is largely set by the technology through the value of $C_{OX}' A_{VT}^2$.

In [208], the technology constant $C_{OX}' A_{VT}^2$ is identified as a matching energy resulting from the threshold-voltage mismatch, ΔV_T, appearing across the device gate-oxide capacitance, C_{GOX}. The matching energy is given by

$$E_{matching} = C_{GOX} \left(\Delta V_T\right)^2 = C_{OX}' WL \left(\frac{A_{VT}}{\sqrt{WL}}\right)^2 = C_{OX}' A_{VT}^2 \tag{3.168}$$

$E_{matching}$ is minimized by minimizing the process gate-oxide capacitance and the square of the threshold-voltage mismatch factor. Fortunately, as described in Section 3.11.1.2 and observed in Table 3.39, as the gate-oxide thickness, t_{ox}, decreases in smaller-geometry processes, A_{VT} or A_{VTO} for zero V_{SB} decreases almost directly with t_{ox}. The square of a smaller value of A_{VT} then dominates the increase in C_{OX}' associated with decreasing t_{ox}. This indicates that $E_{matching}$ decreases nearly with t_{ox} as process geometry decreases. However, because it is not clear if A_{VT} will continue decreasing nearly with t_{ox} as discussed in Section 3.11.1.2, it is also not clear if $E_{matching}$ will continue decreasing favorably as process geometry continues to decrease. Additionally, increased drain–body capacitance and transconductance factor mismatch effects in smaller-geometry processes not considered here deteriorate performance [222].

As mentioned in [208, 238], the matching energy, $E_{matching}$, greatly exceeds the thermal-noise energy, kT, in MOS circuits. As a result, errors for circuits sensitive to DC mismatch are much greater than errors associated with thermal noise. This is shown later in Section 4.3.6.7 by comparing the bandwidth–power–accuracy figures of merit presented here that consider DC offset due to mismatch with the figures of merit presented in Section 4.3.6.6 that consider thermal noise. Additionally, increased errors resulting from DC mismatch compared to errors from thermal noise are shown in Figure 4.21.

While the bandwidth–power–accuracy figure of merit is largely set by the technology and not by the designer, it is important to note that many applications do not require the simultaneous optimization of bandwidth, power, and accuracy. Circuits requiring high accuracy can be readily optimized at low bandwidths, while circuits requiring low accuracy can be readily optimized at high bandwidths. These tradeoffs will be discussed later in Chapter 4 and illustrated in the OTA design examples of Chapter 5.

3.11.3.2 Bandwidth, power, and accuracy tradeoffs in voltage-mode circuits

The bandwidth–power–accuracy figure of merit described by Equations 3.166 and 3.167 applies to current-mode circuits, like the 1:1 current mirror used for the derivation. A figure of merit can also be found for voltage-mode circuits. Here, the circuit considered is an input differential pair driven by a buffer amplifier, like a comparator driven by a buffer amplifier in an analog-to-digital converter. A similar figure of merit can also be developed for a single differential pair in an operational amplifier using negative feedback [238]. The DC accuracy of a differential pair due to the usually dominant, local-area, threshold-voltage mismatch (Section 3.11.1.6) can be expressed as [238, 239]

$$A_{CC,DC} = \frac{V_{IN}}{V_{OS}(3\sigma)} = \frac{V_{IN}}{3 \cdot \Delta V_T(1\sigma)} = \frac{V_{IN}}{3 \cdot \dfrac{A_{VT}}{\sqrt{WL}}}$$

$$= \frac{V_{IN}\sqrt{WL}}{3\,A_{VT}} \tag{3.169}$$

The DC accuracy is the signal-to-noise ratio or ratio of signal to offset voltages. V_{IN} is the input voltage signal, which for linear amplification may need to be below the input, 1 dB compression point, $V_{INDIF1dB}$, for a differential pair because of signal distortion as described in Section 3.8.2.6. The denominators in the first row of Equation 3.169 list the 3σ, input-referred DC offset voltage found directly from the threshold-voltage mismatch, ΔV_T, given by Equation 3.147.

Equation 3.169 shows that DC voltage accuracy is maximized by large V_{IN}, large gate area for the differential pair devices, and a low process value of A_{VT}. Unfortunately, large gate area results in large input capacitance and reduced bandwidth at the differential pair input. For an input signal of $V_{IN} = 100\,\text{mV}$, gate area of $WL = 100\,\mu\text{m}^2$, and $A_{VT} = 5\,\text{mV} \cdot \mu\text{m}$ for the example 0.18 μm CMOS process described in Tables 3.2 and 3.3, Equation 3.169 predicts an accuracy of 66.7%. The reciprocal of this corresponds to an error of 1.5 %, again, for a 3σ value of input-referred DC offset voltage. The accuracy would be reduced by one-half and the error doubled if the signal were lowered to one-half of $V_{IN} = 100\,\text{mV}$ or $V_{IN} = 50\,\text{mV}$ to manage signal distortion.

Estimating the bandwidth at the input of the differential pair involves the differential input capacitance and the output resistance of the preceding buffer amplifier. The differential input capacitance in strong inversion is $C_{in} = (1/2)C_{gsi} = (1/2)(2/3)C'_{OX}WL$. As for the bandwidth estimation of the current mirror described in the previous section, the capacitive loading from C_{gbi}

and the gate overlap and drain–body (this at the buffer amplifier output) capacitances is neglected. As before, the bandwidth estimation is intended to illustrate circuit tradeoffs rather than accurately predict bandwidth. Assuming the output resistance of the preceding buffer amplifier is equal to $1/g_m$, as for an OTA having transconductance of g_m and unity-gain negative feedback, the bandwidth at the input of the differential pair is roughly approximated by

$$BW = \frac{g_m}{2\pi\, C_{in}} = \frac{g_m}{2\pi\left(\frac{1}{2}C_{gsi}\right)} = 2f'_{Ti}$$

$$= \frac{\left(\frac{g_m}{I_D}\right)I_D}{\frac{2}{3}\pi\, C'_{OX}WL} \tag{3.170}$$

Here, g_m/I_D corresponds to the transconductor (differential pair) devices in the OTA buffer amplifier. This bandwidth estimate is a factor of four higher than that given in Equation 3.163 for the current mirror because the capacitance is approximately equal to $(1/2)C_{gsi}$ for the series connection of the differential-pair input capacitances compared to $2C_{gsi}$ for the parallel connection of the current-mirror capacitances.

The power associated with the buffer amplifier driving the differential pair is the same as that associated with the current mirror given by Equation 3.164. For the buffer amplifier, a bias current of I_D is assumed to be flowing on each side of the differential signal path. Dividing Equation 3.170 by Equation 3.164 gives a bandwidth–power figure of merit expressing the bandwidth for a given level of power. Dividing the equations gives

$$\frac{BW}{P} = \frac{\frac{g_m}{I_D}}{\frac{4}{3}\pi\, C'_{OX}WL \cdot V_{DD}} \tag{3.171}$$

This bandwidth–power figure of merit for a buffer amplifier and differential-pair input is a factor of four larger than the one given by Equation 3.165 for a current mirror. This is a result of the lower differential-pair input capacitance and correspondingly higher bandwidth. The bandwidth and bandwidth–power figure of merit are maximized by maximizing g_m/I_D of the buffer amplifier devices in weak or moderate inversion and minimizing the gate area, WL, of the differential-pair devices. This is in conflict with maximizing the accuracy where, from Equation 3.169, gate area should be maximized. Unlike the 1:1 current mirror described in the previous section that uses identical devices, the buffer amplifier and differential-pair devices can be optimized separately.

Finally, as described in [238, 239], a bandwidth–power–accuracy figure of merit can be developed by squaring the accuracy given in Equation 3.169 and multiplying this by the bandwidth–power figure of merit given by Equation 3.171. This gives

$$\frac{A^2_{CC,DC} \cdot BW}{P} = \frac{V^2_{IN}\frac{g_m}{I_D}}{12\pi\, C'_{OX}A^2_{VT}V_{DD}} \tag{3.172}$$

Like the bandwidth–power–accuracy figure of merit given by Equation 3.166 for current-mode circuits, the figure of merit for voltage-mode circuits given by Equation 3.172 is independent of the drain bias current, I_D. As for current-mode circuits, it is also independent of the gate area, because gate area cancels in the $A^2_{CC,DC}$ and BW terms. As a result, the figure of merit is independent of the channel length, except when velocity saturation effects cause reduced g_m/I_D for short-channel devices operating at high IC in the buffer amplifier where the bandwidth and figure of merit are reduced.

As described for current-mode circuits in the previous section, operation at constant IC giving constant g_m/I_D results in constant bandwidth-power-accuracy figures of merit and, correspondingly, constant bandwidth-accuracy figures of merit for constant drain bias current and power consumption.

This, again, is seen in the continuation sheet of Figure 4.21 in Section 4.4.2 where the products of accuracy squared and bandwidth are constant with L. This is because accuracy, which is inversely proportional to local-area, gate–source voltage mismatch listed in Table 3.41, increases as L while bandwidth listed in Table 3.26 decreases as $1/L^2$, both due to gate area that increases as L^2. The product of accuracy (not squared) and bandwidth, however, is not constant with L, but decreases as $1/L$. Bandwidth may actually decrease as $1/L$, changing the trends (Section 5.5.2.11). Finally, for constant IC and g_m/I_D, bandwidth-accuracy figures of merit (products of accuracy squared and bandwidth) increase directly with the drain bias current as seen in the continuation sheet of Figure 4.22 in Section 4.4.3.

The bandwidth–power–accuracy figure of merit given by Equation 3.172 is maximized for operation at the maximum allowable V_{IN} at the differential-pair input, high g_m/I_D in weak or moderate inversion for the preceding buffer amplifier, low V_{DD}, and a small process value of $C'_{OX}A^2_{VT}$. For an OTA buffer amplifier having unity-gain negative feedback, high g_m/I_D for the transconductor (differential pair) devices gives high OTA g_m and low output resistance of $1/g_m$. This results in improved signal bandwidth at the differential-pair input. However, beyond operation at high g_m/I_D for the buffer amplifier, only V_{IN} can be selected and this is limited by the value of V_{DD}, revealing that, at the design limit, the figure of merit is inversely proportional to $C'_{OX}A^2_{VT}$ alone as given by

$$\frac{A^2_{CC,DC} \cdot BW}{P} \propto \frac{1}{C'_{OX}A^2_{VT}} = \frac{1}{E_{matching}} \tag{3.173}$$

Equation 3.173 is identical to the result obtained in Equation 3.167 for current-mode circuits. Again, this suggests there is little the designer can do to maximize the bandwidth–power–accuracy figure of merit as this is largely set by the technology through the value of $C'_{OX}A^2_{VT}$ [238]. As before, $C'_{OX}A^2_{VT}$ is the matching energy given by Equation 3.168 that depends only on the process technology.

Again, as mentioned in [208, 238], the mismatch energy, $E_{matching}$, greatly exceeds the thermal-noise energy, kT, in MOS circuits. As a result, errors for circuits sensitive to DC mismatch are much greater than errors associated with thermal noise. As mentioned, this is shown later in Section 4.3.6.7 by comparing the bandwidth–power–accuracy figures of merit presented here that consider DC offset due to mismatch with the figures of merit presented in Section 4.3.6.6 that consider thermal noise. Additionally, as mentioned, increased errors resulting from DC mismatch compared to errors from thermal noise are shown in Figure 4.21.

While the bandwidth–power–accuracy figure of merit is largely set by the technology and not by the designer, it is important to note, again, that many applications do not require the simultaneous optimization of bandwidth, power, and accuracy. This is illustrated in the DC, balanced, and AC optimized OTAs described later in Chapter 5. The DC optimized OTAs have high accuracy, as evidenced by low input-referred offset voltage, and low bandwidth, while the AC optimized OTAs have low accuracy and high bandwidth. The balanced OTAs have a tradeoff of both accuracy and bandwidth. However, as mentioned in [222, 238–240], optimizing the bandwidth–power–accuracy figure of merit is required for some analog-to-digital converter circuits where high bandwidth or speed and high accuracy are desired at minimum power consumption.

3.11.3.3 Timing skew in digital circuits

While DC offset errors caused by MOS mismatch clearly affect the performance of analog circuits, mismatch can affect even the performance of digital circuits [208]. Here, threshold-voltage mismatch changes the threshold point of digital circuits resulting in timing skew. For example, a threshold point mismatch of 20 mV, possible for small-geometry devices used in digital circuits, causes a timing skew of 4 ps (20 mV/(5 V/ns)) for a signal slope of 5 V/ns. This signal slope corresponds to a 1.8 V transition over a rise-time of 0.36 ns. The timing skew gets progressively worse as digital signals propagate through multiple digital circuits. Failure of operation can result if the timing skew in clock trees becomes an appreciable portion of the clock period.

The input-referred DC offset and wide-band noise voltages for digital circuits are both divided by the signal slope resulting in timing skew or timing jitter, respectively. For example, an input-referred noise voltage of 20 mV rms creates a timing jitter of 4 ps rms for a signal slope of 5 V/ns, just as a DC threshold error of 20 mV creates a timing skew of 4 ps. However, an input-referred noise voltage of 20 mV rms over a noise bandwidth of 10 GHz corresponds to an input-referred thermal-noise voltage density of $200 \, nV/Hz^{1/2}$, which is likely more than a factor of 10 above the actual noise. As mentioned in the previous section, errors due to DC mismatch dominate errors associated with thermal noise for circuits sensitive to DC mismatch.

3.11.4 Small-Signal Parameter and Capacitance Mismatch

Just as mismatch between identically laid-out devices results in DC gate–source voltage and drain current mismatch, small-signal parameter and capacitance mismatch are also present. Mismatch in small-signal parameters and capacitances gives rise to non-DC signal errors, including reduction of common-mode rejection and even-order distortion cancellation in differential circuits. Additionally, the mismatch results in the reduction of port isolation in balanced mixers.

3.11.4.1 Transconductance mismatch

Transconductance mismatch results, for example, for devices in a differential pair that have mismatched drain currents. The relationship between drain current and transconductance mismatch is easily analyzed for the usual case described in Section 3.11.1.6 where threshold-voltage mismatch, ΔV_T, dominates mismatch.

In weak inversion where $g_m/I_D = 1/(nU_T)$, relative drain current mismatch is found from Equation 3.153 by

$$\frac{\Delta I_D}{I_D}(\text{WI}) = \frac{g_m \Delta V_T}{I_D} = \left(\frac{g_m}{I_D}\right)\Delta V_T = \frac{\Delta V_T}{V_{gm}} = \frac{\Delta V_T}{nU_T} \tag{3.174}$$

The mismatch is equal to ΔV_T multiplied by g_m/I_D, or ΔV_T divided by the transconductance effective voltage, $V_{gm} = (g_m/I_D)^{-1} = nU_T$. V_{gm} was described in Section 3.8.2.1 and summarized in Table 3.17. Mismatch in the substrate factor, n, is neglected because threshold-voltage mismatch is assumed to dominate. Since transconductance is $g_m = I_D/(nU_T)$ and transconductance mismatch is then $\Delta g_m = \Delta I_D/(nU_T)$, relative transconductance mismatch, $\Delta g_m/g_m$, is directly equal to relative drain current mismatch, $\Delta I_D/I_D$. Combining this result with Equation 3.174 and ΔV_T given by Equation 3.147 gives

$$\frac{\Delta g_m}{g_m}(\text{WI}, 1\sigma) = \frac{\Delta I_D}{I_D}(\text{WI}, 1\sigma) = \left(\frac{g_m}{I_D}\right)\Delta V_T(1\sigma) = \frac{\Delta V_T(1\sigma)}{nU_T}$$

$$= \frac{A_{VT}}{nU_T\sqrt{WL}} = \frac{\sqrt{IC}}{L}\sqrt{\frac{I_0}{I_D}}\frac{A_{VT}}{nU_T} \tag{3.175}$$

WL is the device gate area, and A_{VT} is the local-area, threshold-voltage mismatch factor, which can include the increase from the zero-V_{SB} factor, A_{VTO}, resulting from non-zero V_{SB} as summarized in Table 3.41. In the last term of the expression, mismatch is expressed using I_D, IC, and L by replacing the gate area with $WL = (L^2/IC)(I_D/I_0)$ from Table 3.10 where $I_0 = 2n_0\mu_0 C'_{OX}U_T^2$ is the technology current described in Table 3.6. For a fixed gate area, mismatch in weak inversion is independent of the level of inversion because of constant g_m/I_D. Table 3.41 described the relative drain current (and thus relative transconductance) mismatch trends in weak inversion observed from the last term of Equation 3.175 for the design choices of I_D, IC, and L.

Relative transconductance and drain current mismatch are equal in weak inversion because transconductance directly tracks drain current since g_m/I_D is constant. For a gate area of $WL = 10\,\mu m^2$, $A_{VT} = 5\,mV \cdot \mu m$ for the example 0.18 μm CMOS process described in Tables 3.2 and 3.3, $n = 1.4$ (weak inversion), and $U_T = 25.9\,mV$ ($T = 300\,K$), Equation 3.175 predicts a 3σ, relative transconductance and drain current mismatch of 13 %. Increasing gate area reduces the mismatch at the expense of increased gate capacitance and reduced bandwidth.

Relative transconductance and drain current mismatch are not equal in strong inversion because transconductance is proportional to the square root of drain current for a fixed-geometry device. Neglecting velocity saturation and VFMR reductions of g_m/I_D, relative drain current mismatch is given by

$$\frac{\Delta I_D}{I_D}(\text{SI, no vel. sat.}) = \frac{g_m \Delta V_T}{I_D} = \left(\frac{g_m}{I_D}\right)\Delta V_T = \frac{\Delta V_T}{V_{gm}} = \frac{2\Delta V_T}{V_{EFF}} \tag{3.176}$$

Here, $g_m/I_D = 2/V_{EFF}$ from Equation 2.13 and, correspondingly, $V_{gm} = (g_m/I_D)^{-1} = V_{EFF}/2$, where $V_{EFF} = V_{GS} - V_T$ is the effective gate–source voltage.

Relative transconductance mismatch is given by

$$\frac{\Delta g_m}{g_m}(\text{SI, no vel. sat.}) = \frac{\Delta V_{EFF}(k/n)(W/L)}{V_{EFF}(k/n)(W/L)} = \frac{\Delta V_{EFF}}{V_{EFF}}$$

$$= \left(\frac{g_m}{I_D}\right)\frac{\Delta V_T}{2} = \frac{\Delta V_T}{2V_{gm}} = \frac{\Delta V_T}{V_{EFF}} \tag{3.177}$$

Here, $g_m = V_{EFF}(k/n)(W/L)$ from Equation 2.12 where $k = \mu C'_{OX}$ is the transconductance factor. Mismatch in the total transconductance factor, $k(W/L)$, and n are neglected because threshold-voltage mismatch, again, is assumed to dominate. Mismatch in V_{EFF}, ΔV_{EFF}, is expressed as mismatch in the threshold voltage, ΔV_T. Finally, g_m/I_D, V_{gm}, and V_{EFF} are interrelated as described for Equation 3.176.

Comparing Equations 3.176 and 3.177 reveals that in strong inversion relative transconductance mismatch is one-half the relative drain current mismatch. Combining the equations and ΔV_T given by Equation 3.147 gives

$$\frac{\Delta g_m}{g_m}(\text{SI, no vel. sat., } 1\sigma) = \frac{1}{2}\frac{\Delta I_D}{I_D}(\text{SI, no vel. sat., } 1\sigma) = \left(\frac{g_m}{I_D}\right)\frac{\Delta V_T(1\sigma)}{2}$$

$$= \frac{\Delta V_T(1\sigma)}{V_{EFF}} = \frac{A_{VT}}{V_{EFF}\sqrt{WL}} = \frac{\sqrt{IC}}{L}\sqrt{\frac{I_0}{I_D}}\frac{A_{VT}}{2nU_T\sqrt{IC}} \tag{3.178}$$

In the last term of the expression, mismatch is expressed using I_D, IC, and L as described for Equation 3.175 with $V_{EFF} = 2nU_T\sqrt{IC}$ from Table 3.17. For a fixed gate area, mismatch in strong inversion decreases with increasing inversion because of the decrease in g_m/I_D. Table 3.41 described the relative drain current (and thus relative transconductance) mismatch trends in strong inversion observed from the last term of Equation 3.178 for the design choices of I_D, IC, and L. For these design choices, the mismatch is independent of IC because increasing IC requires a reduction in shape factor, $S = W/L$, channel width, and gate area to maintain the selected drain current and channel length. This results in increasing ΔV_T that is countered by decreasing g_m/I_D associated with increasing IC.

For the gate area of $WL = 10\,\mu m^2$, $A_{VT} = 5\,mV \cdot \mu m$, and $U_T = 25.9\,mV$ mentioned for mismatch evaluation in weak inversion using Equation 3.175, Equation 3.178 predicts a 3σ, relative transconductance mismatch of 1.9 % and a relative drain current mismatch of 3.8 % for $V_{EFF} = 0.25\,V$

near the onset of strong inversion. For equal gate area giving equal ΔV_T, the mismatch is lower in strong inversion compared to weak inversion because of the reduction of g_m/I_D. However, the smaller required shape factor in strong inversion requires a decrease in channel width and increase in channel length, the latter resulting in a loss of intrinsic bandwidth as described in Section 3.9.6.

The relationship between relative transconductance and drain current mismatch is observed for pMOS devices in a 0.65 μm, CMOS process having a gate-oxide thickness of $t_{ox} = 15$ nm [242]. Here, both measured transconductance and drain current mismatch are inversely proportional to the square root of gate area as expected from Equations 3.175 and 3.178. For operation near the onset of strong inversion at $V_{EFF} = 0.2$ V, $\Delta g_m/g_m(1\sigma) = 9\% \cdot \mu\text{m}/(WL)^{1/2}$ while $\Delta I_D/I_D(1\sigma) = 17\% \cdot \mu\text{m}/(WL)^{1/2}$. The transconductance mismatch is nearly one-half the drain current mismatch as predicted by Equation 3.178 for operation in strong inversion. The mismatch reported, which includes the effect of g_m/I_D, is for $V_{EFF} = 0.2$ V only and corresponds to mismatch predicted by Equation 3.178 using $A_{VT} = 18$ mV · μm. This value of A_{VT} is typical for $t_{ox} = 15$ nm as discussed in Section 3.11.1.2 and seen in Table 3.39.

When threshold-voltage mismatch dominates, relative transconductance mismatch, $\Delta g_m/g_m$, is equal to relative drain current mismatch, $\Delta I_D/I_D$, in weak inversion (Equation 3.175), decreasing to 50 % of the drain current mismatch in strong inversion (Equation 3.178), assuming velocity saturation and VFMR effects are negligible. Transconductance mismatch then varies from approximately 100 % to 50 % of the drain current mismatch from the weak-inversion side of moderate inversion ($IC = 0.1$) to the strong-inversion side ($IC = 10$). Assuming threshold-voltage mismatch continues to dominate, transconductance mismatch is expected to be less than 50 % of the drain current mismatch when velocity saturation and VFMR effects lower g_m/I_D. However, as described in Section 3.11.1.6, mismatch in the transconductance factor can dominate mismatch in the threshold voltage for low values of g_m/I_D present at high levels of inversion.

Under conditions of full velocity saturation in strong inversion, the transconductance given by Equation 2.18 is $g_m = (\mu C'_{OX}/2n)WE_{CRIT}$, and drain current given by Equation 2.17 is equal to this multiplied by V_{EFF}. μ includes the VFMR effect while the reduction in mobility associated with velocity saturation is considered using the critical velocity saturation, electric field, E_{CRIT}. When mismatch in $(\mu C'_{OX}/2n)WE_{CRIT}$ dominates, relative transconductance and drain current mismatch are expected to be similar since both depend directly on this factor. This condition likely occurs when mismatch in the transconductance factor, $(\mu C'_{OX})(W/L)$, dominates over threshold-voltage mismatch and is similar to mismatch in $(\mu C'_{OX}/n)WE_{CRIT}$ through, for example, mismatches in the mobility, gate-oxide capacitance, and channel width.

3.11.4.2 Drain–source conductance mismatch

Drain–source conductance, g_{ds}, mismatch would be expected to track drain current mismatch as seen by $g_{ds} = I_D/V_A$ if the normalized g_{ds} or Early voltage, $V_A = I_D/g_{ds}$, described at the beginning of Section 3.8.4 remains constant with drain current. Figures 3.43–3.46 showed that V_A remains nearly constant with the level of inversion or drain current for a fixed-geometry device at a given V_{DS}. If g_{ds} mismatch tracks drain current mismatch that is dominated by threshold-voltage mismatch, both relative g_{ds} and drain current mismatch should be inversely proportional to the square root of gate area multiplied by g_m/I_D as shown by Equations 3.175 and 3.178 for drain current mismatch.

However, g_{ds} mismatch for pMOS devices in the 0.65 μm CMOS process described in the previous section was found to be inversely proportional to the square root of channel width alone with values of $\Delta g_{ds}/g_{ds}$ (1σ) = 27 % · μm$^{1/2}$/$(W)^{1/2}$ for $V_{EFF} = 0.2$ V, decreasing to $\Delta g_{ds}/g_{ds}$ (1σ) = 10 % · μm$^{1/2}$/$(W)^{1/2}$ for $V_{EFF} = 0.7$ V deeper into strong inversion [242]. Similar values of g_{ds} mismatch, also inversely proportional to the square root of width, were observed for nMOS devices in the 0.65 μm process and in a 0.35 μm CMOS process having $t_{ox} = 7.5$ nm [242]. The width dependence on g_{ds} mismatch might be due to g_{ds} dependency on the high, horizontal electric field between the drain and channel pinch-off point that is largely independent of channel length [242]. g_{ds} mismatch may then result from parameter variations in the volume between the drain and channel pinch-off point. This volume depends primarily on width and not the product of width and length associated with gate area in the local-area mismatch model.

Fortunately, many circuits are relatively insensitive to g_{ds} and its mismatch compared to transconductance and its mismatch. For example, the input conductance for a current mirror is equal to the transconductance added to the much smaller g_{ds}.

3.11.4.3 *Mismatch effects on circuit performance*

In [243], the harmonic distortion resulting from device mismatch is analyzed for simple class A and complementary class AB MOS current mirrors. Analysis and supporting simulations show that threshold-voltage mismatch usually dominates distortion compared to transconductance factor mismatch. As expected, increasing channel length and gate area reduces mismatch and distortion at the expense of bandwidth.

In [244], the loss of port isolation resulting from device mismatch is analyzed for high-frequency, balanced MOS mixers. If mixer differential-pair devices are perfectly matched, radio frequency (RF) and local oscillator (LO) crosstalk is zero and the port isolation is infinite. Actual port isolation is estimated from the mismatch in measured s-parameters for identical test devices. This mismatch includes both low-frequency, small-signal parameter and capacitance mismatches.

For initial design guidance, the 3σ mismatch in DC drain currents due to device mismatch can be estimated and related to transconductance mismatch as described in Section 3.11.4.1. As an initial estimate, the drain current mismatch might be directly related to g_{ds} mismatch, which assumes a nearly constant V_A across mismatched drain currents. Mismatch in transconductance, g_{ds}, and even capacitances can then be considered in the analysis and simulation of circuits.

An interesting small-signal parameter mismatch study would include not only the mismatch associated with mismatch in the DC drain current, but the mismatch present if mismatch in the DC drain current was adjusted out. In this case, the small-signal parameter mismatch should be significantly less, suggesting performance improvement when drain current mismatch is sensed and removed. Other methods of managing small-signal parameter mismatch could include the tuning of DC drain currents in differential pairs for improved cancellation of second-order distortion or the tuning of drain currents in balanced mixer circuits, which include differential pairs, to maximize the isolation between ports.

3.12 LEAKAGE CURRENT

The final section of this chapter considers leakage current and its effects on MOS performance. Gate leakage current is considered first because it can significantly affect circuit performance in processes having gate-oxide thicknesses around 2 nm and below. Following this, drain–body and source–body leakage current, and subthreshold drain leakage current for $V_{GS} = 0\,\text{V}$, are considered.

3.12.1 Gate Leakage Current and Conductance

Direct carrier tunneling through thin gate oxides results in gate leakage current. This leakage current then gives rise to small-signal gate conductance. Gate leakage current and the resulting conductance are described in the following sections.

3.12.1.1 Gate current

Gate leakage current for the example 0.18 μm CMOS process considered in this chapter, in the performance tradeoffs in Chapter 4, and in the cascoded, OTA design examples in Chapter 5 is negligible because gate-oxide thickness, t_{ox}, is 4.1 nm. Additionally, gate leakage current is negligible for the 0.5 and 0.35 μm CMOS processes considered in the design examples in Chapters 5 and 6 because t_{ox} is greater than 4.1 nm. However, when t_{ox} is less than 3 nm and especially less than 2 nm, gate leakage current can become significant. Gate leakage current is a topic of current research [16, pp. 161–166, 45, 180, 245–249] and is included in the BSIM4 [3], MOS Model 11 [45], and SP (surface potential) [46] MOS models.

Gate leakage current is proportional to MOS geometry, bias, and process characteristics as given by [16, p. 161, 45]

$$I_G \propto WL \cdot \frac{V_{ox}}{t_{ox}} \cdot Q'_{INV} \cdot P_{TUN}(V_{ox}) \tag{3.179}$$

The current is proportional to the gate area, WL, and the voltage across the gate oxide, V_{ox}, divided by the gate-oxide thickness, t_{ox}. The current is also proportional to the inversion charge density, Q'_{INV}, in the channel below the gate oxide and the gate-oxide tunneling probability, $P_{TUN}(V_{ox})$.

When t_{ox} decreases, the increase in gate leakage current results primarily from the increase in $P_{TUN}(V_{ox})$, which is given by [16, p. 161, 45]

$$P_{TUN}(V_{ox}) = \exp\left\{-\frac{E_B t_{ox}}{V_{ox}}\left[1 - \left(1 - \frac{V_{ox}}{X_B}\right)^{3/2}\right]\right\} \tag{3.180}$$

X_B is the oxide voltage barrier, which is approximately 3.1 V for electrons and 4.5 V for holes. E_B is a characteristic electric field, which is approximately 29 V/nm for electrons and 43 V/nm for holes. Equation 3.180 gives the tunneling probability for direct tunneling when $V_{ox} < X_B$ as typical for MOS values of V_{ox}. Field-induced, Fowler–Nordheim tunneling occurs for $V_{ox} > X_B$ in high-voltage applications where gates isolated by thick t_{ox} are charged or discharged in EPROM and EEPROM memories [16, p. 161].

The probability of direct tunneling given by Equation 3.180 is a function of V_{ox}, but is a much stronger function of t_{ox}. For $V_{ox} = 0.2$ V, and X_B and E_B given above for electrons in nMOS devices, the probability given by Equation 3.180 increases by six orders of magnitude as t_{ox} decreases from 3 to 2 nm, increases by yet another three orders of magnitude as t_{ox} decreases from 2 to 1.5 nm, and increases by yet another three orders of magnitude as t_{ox} decreases from 1.5 to 1 nm [16, p. 162]. These increases with decreasing t_{ox} correspond to increases in gate leakage current for equal gate area and equal operating conditions at $V_{ox} = 0.2$ V and equal Q'_{INV}, with an additional increase associated with decreasing t_{ox} as seen in Equation 3.179. The significant increase in gate leakage current for small values of t_{ox} has resulted in the research for alternative dielectrics having less tunneling current [248].

Equations 3.179 and 3.180 are difficult to use for predicting gate leakage current for design guidance because V_{ox}, Q'_{INV}, and $P_{TUN}(V_{ox})$ are not simply related to MOS bias conditions. However, a simplified approximation derived from the MOS Model 11 [45] MOS model permits design guidance [180]. An adaptation of this approximation is included in the first entry in Table 3.45 for predicting the gate–source leakage current, I_{GS}. WL is the gate area, V_{INV} is a voltage representing the effective level of inversion related to the value of Q'_{INV}, V_{GS} is the gate–source voltage, and K_{GA} and K_{GB} are constants that depend on the value of t_{ox}. V_{INV} is found from the expression given in the table and increases as $nU_T\exp[(V_{GS} - V_T)/(nU_T)]$ in weak inversion to $V_{EFF} = V_{GS} - V_T$ in strong inversion. While only I_{GS} is given, the total gate current, I_G, splits approximately into two-thirds to the source as $I_{GS} = {}^2/_3 I_G$ and one-third to the drain as $I_{GD} = {}^1/_3 I_G$ in saturation [45]. Thus, in saturation, the total gate current can be taken as 150 % of I_{GS}. In the deep ohmic, linear, or triode region where V_{DS} is well below $V_{DS,sat}$, the gate current splits nearly equally between the source and drain [45].

The gate current relationship given by Equation 3.179 and the approximation given in Table 3.45 exclude overlap tunneling current in the small overlap regions between the gate and source, and gate and drain. The overlap tunneling current is often negligible compared to the intrinsic tunneling current except possibly for short-channel, small-area devices operating at high values of V_{DS} [45].

Table 3.45 MOS gate–source leakage current and conductance, and drain–body leakage current expressions and trends in terms of the inversion coefficient, channel length, and drain current. Source–body leakage current is calculated similarly to drain–body leakage current using the source area, A_S, and source–body voltage, V_{SB}

Expression	$IC \uparrow$ $L,\ I_D$ fixed	$L \uparrow$ $IC,\ I_D$ fixed	$I_D \uparrow$ $IC,\ L$ fixed
Gate–source leakage current:[a] $$I_{GS} \approx WL \cdot K_{GA} V_{INV} V_{GS} e^{K_{GB} V_{GS}}$$ $$\approx \left(\frac{L^2}{IC}\right)\left(\frac{I_D}{I_0}\right) K_{GA} V_{INV} V_{GS} e^{K_{GB} V_{GS}}$$ $$V_{INV} = nU_T \ln\left(1 + e^{\frac{V_{EFF}}{nU_T}}\right),$$ where $V_{EFF} = V_{GS} - V_T$ is given in Table 3.14, and $V_{GS} = V_T + V_{EFF}$ $$K_{GA}(\text{nMOS}) \approx \frac{80\ e^{-14(t_{ox}/1\,\text{nm})}}{(t_{ox}/1\,\text{nm})^2}\ (\text{A}/\mu\text{m}^2 \cdot \text{V}^2)$$ $$K_{GB}(\text{nMOS}) \approx 1.13 \cdot (t_{ox}/1\,\text{nm})\ (\text{V}^{-1})$$	$\uparrow$ as effect of increasing V_{EFF} and V_{GS} exceeds the decrease in gate area	$\uparrow \propto L^2$	$\uparrow \propto I_D$
Gate–source conductance efficiency:[a] $$\frac{g_{gs}}{I_{GS}} \approx \frac{\dfrac{\partial V_{INV}}{\partial V_{GS}}}{V_{INV}} + \frac{1}{V_{GS}} + K_{GB}$$ $$\approx \frac{1}{nU_T} + \frac{1}{V_{GS}} + K_{GB}\ (\text{WI})$$ $$\approx \frac{1}{V_{EFF}} + \frac{1}{V_{GS}} + K_{GB}\ (\text{SI})$$	$\downarrow$ as V_{EFF} and V_{GS} increase	Unchanged, but increases if V_{EFF} and V_{GS} decrease due to less vel. sat.	Unchanged
Drain–body leakage current:[b] $$I_{DB} \approx A_D J_{DB}(V_{DB})$$ $$\approx W \cdot \frac{1}{2} W_{DIF} J_{DB}(V_{DB})$$ $$\approx \left(\frac{L}{IC}\right)\left(\frac{I_D}{I_0}\right) \cdot \frac{1}{2} W_{DIF} J_{DB}(V_{DB})$$ $J_{DB}(V_{DB})$ is the drain–body leakage current density for a given value of drain–body voltage, V_{DB} I_{DB} is for an even number of gate fingers, m, with gate area, A_D, minimized. Table 3.25 describes A_D, m, and W_{DIF}. If $m = 1$, A_D and I_{DB} are doubled from values given	$\downarrow \propto \dfrac{1}{IC}$	$\uparrow \propto L$	$\uparrow \propto I_D$

Expressions are for saturation where $V_{DS} > V_{DS,sat}$.

[a] Gate–source leakage current and conductance efficiency are due to direct gate tunneling current with estimations adapted from [180]. Estimations exclude gate-overlap tunneling current. The total gate leakage current (in saturation) can be taken as 150 % of the gate–source leakage current.

[b] Additional components of leakage current related to channel width can be included for GIDL, especially for negative values of gate–source voltage, and for perimeter or side-wall leakage. This can include separate or averaged geometry considerations for the source–drain extensions.

In addition to being present under normal inversion operation, gate tunneling current is also present when MOS devices are operated in accumulation, obtained for operation below the flat-band voltage when the gate is sufficiently reverse biased. Gate tunneling current is less in accumulation compared to inversion and flows mostly into the substrate [45].

Figure 3.73 shows the gate–source leakage current, I_{GS}, versus the gate–source voltage, V_{GS}, predicted by the expression in Table 3.45 for an nMOS device in saturation having $W/L = 1\,\mu\text{m}/1\,\mu\text{m}$ and $t_{ox} = 2\,\text{nm}$. I_{GS} is predicted using $K_{GA} = 1.38 \times 10^{-11}\,\text{A}/\mu\text{m}^2 \cdot \text{V}^2$ and $K_{GB} = 2.26/\text{V}$ found for the value of t_{ox}. V_{INV} required for the I_{GS} prediction is predicted assuming a threshold voltage of $V_T = 0.25\,\text{V}$, substrate factor of $n = 1.3$ (held fixed), and $U_T = 25.9\,\text{mV}$ at $T = 300\,\text{K}$ (room temperature). The drain current, I_D, is also predicted for comparison with I_{GS} using the weak- through strong-inversion expression in Table 3.13 with the enhanced velocity saturation and VFMR correction factor. I_D is predicted assuming $\mu_0 = 250\,\text{cm}^2/\text{V} \cdot \text{s}$, $C'_{OX} = 17.3\,\text{fF}/\mu\text{m}^2$ (found from the expression in Table 3.1 for $t_{ox} = 2\,\text{nm}$), and, again, $n = 1.3$ and $U_T = 25.9\,\text{mV}$. The modest reduction in I_D due to velocity saturation (modest because L is sufficiently long at $1\,\mu\text{m}$) is predicted assuming $E_{CRIT} = 5\,\text{V}/\mu\text{m}$, and the larger reduction due to VFMR is predicted assuming $\theta = 0.5/\text{V}$.

As observed in Figure 3.73, in weak inversion, I_{GS} and I_D increase rapidly with increasing V_{GS}, with I_{GS} nearly seven orders of magnitude below I_D. In strong inversion, I_{GS} and I_D increase more slowly, but I_{GS} increases more rapidly than I_D, with I_{GS} nearly five orders of magnitude below I_D at the maximum value of $V_{GS} = 1.5\,\text{V}$. The small-signal, gate–source conductance, g_{gs}, resulting from I_{GS} is also shown in the figure. This will be discussed later in the next section.

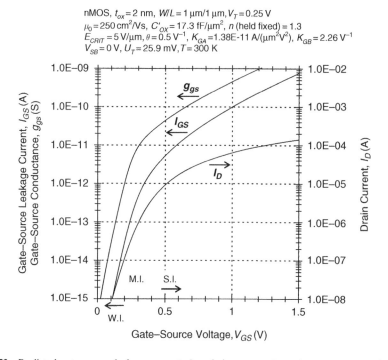

nMOS, $t_{ox} = 2$ nm, $W/L = 1\,\mu\text{m}/1\,\mu\text{m}$, $V_T = 0.25$ V
$\mu_0 = 250$ cm^2/Vs, $C'_{OX} = 17.3$ fF/μm^2, n (held fixed) $= 1.3$
$E_{CRIT} = 5$ V/μm, $\theta = 0.5$ V^{-1}, $K_{GA} = 1.38\text{E}-11$ A/(μm^{2}V^2), $K_{GB} = 2.26$ V^{-1}
$V_{SB} = 0$ V, $U_T = 25.9$ mV, $T = 300$ K

Figure 3.73 Predicted gate–source leakage current, I_{GS}, drain current, I_D, and gate–source conductance, g_{gs}, versus the gate–source voltage, V_{GS}, for a $W/L = 1\,\mu\text{m}/1\,\mu\text{m}$, nMOS device having a gate-oxide thickness of 2 nm. In weak inversion, I_{GS} is approximately seven orders of magnitude below I_D before increasing more rapidly than I_D in strong inversion. In weak inversion, the g_{gs} efficiency, g_{gs}/I_{GS}, is maximum before decreasing in strong inversion. I_{GS} and g_{gs} scale directly with the gate area and are lower for pMOS devices because of lower gate current tunneling probability. I_{GS} and g_{gs} increase significantly as gate-oxide thickness decreases

Since it is given for $W = L = 1\,\mu$m giving a gate area of $WL = 1\,\mu\text{m}^2$, I_{GS} in Figure 3.73 is the gate leakage current density per μm^2 of gate area. I_{GS} is easily found for other gate areas at a given value of V_{GS} by multiplying the value shown in the figure by the gate area in μm^2. I_{GS} shown for $t_{ox} = 2\,$nm corresponds to gate leakage current densities ranging from below 10^{-15} to nearly $10^{-9}\,$A/μm^2 as V_{GS} ranges from 0 to 1.5 V, again, for nMOS devices. Measurements on a device (believed also to be an nMOS device) having $W = 10\,\mu$m, $L = 10\,\mu$m, and $t_{ox} = 1.5\,$nm reveal gate leakage current between 1 and 100 μA as V_{GS} varies for a constant $V_{DS} = 1\,$V [76]. This corresponds to gate leakage current densities ranging from 10^{-8} to $10^{-6}\,$A/μm^2. Assuming the gate current densities can be compared at the maximum values of V_{GS}, the current density is three orders of magnitude higher for $t_{ox} = 1.5\,$nm compared to $t_{ox} = 2\,$nm. Although the values of V_{ox} and Q'_{INV} may not be exactly comparable, this increase in gate leakage current density corresponds to the three-order-of-magnitude increase in gate tunneling probability, P_{TUN} ($V_{ox} = 0.2\,$V), mentioned earlier as t_{ox} decreases from 2 to 1.5 nm.

In addition to giving an expression for I_{GS} as a function of the gate area, V_{GS}, and V_{INV}, which itself depends on $V_{EFF} = V_{GS} - V_T$, Table 3.45 expresses I_{GS} and its trends in terms of the design choices of IC, L, and I_D. Here, MOS gate area is expressed as $WL = (L^2/IC)(I_D/I_0)$ from Table 3.10 where $I_0 = 2n_0\mu_0 C'_{OX}U_T^2$ is the technology current described in Table 3.6. V_{EFF} is found from the expression given in Table 3.14 using the IC substitution listed to model the V_{EFF} increase associated with velocity saturation and VFMR effects. Finally, V_{GS} is found from $V_{GS} = V_{EFF} + V_T$.

Figure 3.74 shows predicted I_{GS} for $L = 0.09, 0.18, 0.28, 0.48, 1, 2,$ and $4\,\mu$m, nMOS devices having $t_{ox} = 2\,$nm operating at IC ranging from 0.01 (deep weak inversion) to 100 (deep strong inversion). Unlike the normalized small-signal parameters, g_m/I_D, $\eta = g_{mb}/g_m$, and $V_A = I_D/g_{ds}$, and intrinsic

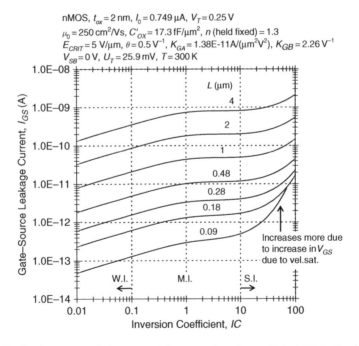

Figure 3.74 Predicted gate–source leakage current, I_{GS}, versus inversion coefficient, IC, for $L = 0.09, 0.18, 0.28,$ 0.48, 1, 2, and 4 μm, nMOS devices in a CMOS process having a gate-oxide thickness of 2 nm, operating at a drain current of 100 μA. I_{GS} increases modestly with IC because the increase in gate leakage current density associated with increasing V_{GS} exceeds the decrease in gate area. I_{GS} increases significantly with increasing L because of the significant increase in gate area. I_{GS} increases significantly as gate-oxide thickness decreases

voltage gain and bandwidth, A_{Vi} and f_{Ti}, I_{GS}, like flicker noise and local-area DC mismatch, is not independent of the drain current when evaluated in terms of the inversion coefficient. Here, a drain current of $100\,\mu A$ is assumed, where the increase in I_{GS} with increasing drain current resulting from the required increase in gate area is easily found by multiplying the values shown in the figure by $(I_D/100\,\mu A)$. I_{GS} is predicted from the expression in Table 3.45 using the process parameters described earlier for Figure 3.73 and $I_0 = 0.749\,\mu A$ found from the process parameters. Finally, channel length is reduced by $DL = 0.015\,\mu m$ from the values given to include lateral diffusion effects.

As observed in Figure 3.74, I_{GS} increases modestly with IC in weak inversion and the weak-inversion side of moderate inversion. This is because the increase in gate leakage current density associated with increasing V_{GS} observed in Figure 3.73 exceeds the decrease in leakage current associated with the reduction in required gate area as gate area, channel width, W, and shape factor, $S = W/L$, decrease inversely with IC for a given L and I_D. In the strong-inversion side of moderate inversion between $IC = 1$ and 10, I_{GS} appears to be nearly independent of IC, suggesting that the increase in gate current density is countered by the decrease in gate area. In strong inversion, I_{GS} increases modestly, except for short-channel devices where it increases significantly because V_{EFF} and V_{GS} increase significantly due to velocity saturation effects.

While Figure 3.74 suggests that I_{GS} increases usually modestly with IC, I_{GS} increases significantly with channel length, L. This is because an increase in L requires an equal increase in W, giving a squared increase in gate area for a given IC and I_D. As summarized later in Section 3.12.2.6, the increase of I_{GS} with increasing L is in opposition to the increase in drain–source resistance and circuit gain associated with increasing L. Additionally, the increase in I_{GS} with increasing L is in opposition to the reduction in flicker noise and local-area DC mismatch associated with increasing gate area caused by increasing L. Finally, as mentioned, I_{GS} directly tracks I_D through the required gate area increase for a given IC and L. As a result, the ratio of I_{GS}/I_D is independent of I_D.

I_{GS} for short-channel devices in deep strong inversion may not be attainable for $I_D = 100\,\mu A$ considered in Figure 3.74 because resulting widths may be below values possible for layout. However, when verified for practical device geometries and scaled for the drain current used, the figure provides an estimate of I_{GS} useful for design guidance.

Figures 3.73 and 3.74 show predicted I_{GS} for nMOS devices, and the prediction constants, K_{GA} and K_{GB}, given in Table 3.45 are also for nMOS devices. I_{GS} is lower for pMOS devices compared to nMOS devices having equal gate area and bias conditions. This is because of the higher oxide barrier for holes compared to electrons, which lowers the value of $P_{TUN}(V_{ox})$ given by Equation 3.180 and I_{GS} by approximately a factor of two for $V_{ox} = 0.2\,V$ and $t_{ox} = 2\,nm$. However, for a given IC, L, and I_D, pMOS gate area is larger than nMOS gate area by a factor of approximately five resulting from reduced pMOS μ_0 and I_0. This then increases pMOS gate leakage current for a given level of gate leakage current density. Finally, as mentioned, gate leakage current increases significantly as t_{ox} decreases. The increase is in the vicinity of three orders of magnitude for t_{ox} decreasing from 2 nm to 1.5 nm, at least for the bias conditions mentioned.

3.12.1.2 Gate conductance

Unfortunately, the increase in I_{GS} with V_{GS} results in a small-signal, gate–source conductance, g_{gs}, appearing between the gate and source in the MOS small-signal model shown earlier in Figure 3.21. The gate–source conductance is given by $g_{gs} = \delta I_{GS}/\delta V_{GS}$, which assumes that the other bias voltages, V_{DS} and V_{SB}, are held constant. g_{gs} is shown in Figure 3.73 as found from the slope of I_{GS} with V_{GS}.

Like the drain–source conductance, g_{ds}, g_{gs} is best evaluated by normalizing it to its associated DC current. Analogous to the drain–source conductance efficiency, $g_{ds}/I_D = 1/V_A$ described near the beginning of Section 3.8.4, the gate–source conductance efficiency, g_{gs}/I_{GS}, should be minimized. This minimizes the production of undesired g_{gs} for a given value of I_{GS}, where $g_{gs} = (g_{gs}/I_{GS}) \cdot I_{GS}$.

Along with the simplified prediction of I_{GS}, a simplified prediction of g_{gs}/I_{GS} is given in Table 3.45 as adapted from [180]. g_{gs}/I_{GS} is found from $g_{gs} = \delta I_{GS}/\delta V_{GS}$ divided by I_{GS}, which results in the cancellation of many terms. g_{gs}/I_{GS} is equal to $(\delta V_{INV}/\delta V_{GS})/V_{INV}$ added with $1/V_{GS}$ and K_{GB}. As described in the previous section, V_{INV} is a voltage that represents the effective level of inversion, and K_{GB} is a constant related to the value of t_{ox}. The table lists weak- and strong-inversion values of g_{gs}/I_{GS} where the $(\delta V_{INV}/\delta V_{GS})/V_{INV}$ term is equal to $1/(nU_T)$ and $1/V_{EFF}$, respectively.

For nMOS devices having $t_{ox} = 2\,\text{nm}$ considered in Figures 3.73 and 3.74 with $n = 1.3$, $U_T = 25.9\,\text{mV}$, and $K_{GB} = 2.26/\text{V}$, g_{gs}/I_{GS} predicted from the expression in Table 3.45 is $g_{gs}/I_{GS} = 37.6/\text{V}$ for $V_{GS} = 0.18\,\text{V}$. This corresponds to $V_{EFF} = V_{GS} - V_T = -0.07\,\text{V}$ for $V_T = 0.25\,\text{V}$, which is near the onset of weak inversion as described in Table 3.14. For $V_{GS} = 0.47\,\text{V}$ where $V_{EFF} = 0.22\,\text{V}$, which is near the onset of strong inversion, g_{gs}/I_{GS} is lower at $g_{gs}/I_{GS} = 8.9/\text{V}$. The decreasing value of g_{gs}/I_{GS} with increasing inversion or V_{GS} is observed in Figure 3.73 by visualizing the decreasing vertical distance between g_{gs} and I_{GS}. Although g_{gs}/I_{GS} decreases with increasing V_{GS}, the significant increase in I_{GS} results in a significant increase in g_{gs} as seen in the figure.

As mentioned in the previous section, when t_{ox} decreases from 2 to 1.5 nm, I_{GS} increases by approximately three orders of magnitude for equal gate area and equal $V_{ox} = 0.2\,\text{V}$ and Q'_{INV} operating conditions. Assuming an increase of three orders of magnitude in I_{GS}, g_{gs} predicted in Figure 3.73 increases from $g_{gs} = 4 \times 10^{-11}\,\text{S}$ $(r_{gs} = 1/g_{gs} = 25000\,\text{M}\Omega)$ at $V_{GS} = 0.47\,\text{V}$ near the onset of strong inversion for $t_{ox} = 2\,\text{nm}$ to approximately $g_{gs} = 4 \times 10^{-8}\,\text{S}$ $(r_{gs} = 25\,\text{M}\Omega)$ for $t_{ox} = 1.5\,\text{nm}$, with the corresponding values of I_{GS} at 5 pA and 5 nA. Since these values are for a gate area of $1\,\mu\text{m}^2$, I_{GS} and g_{gs} would increase by an order of magnitude for a gate area of $10\,\mu\text{m}^2$. This would result in $g_{gs} = 4 \times 10^{-7}\,\text{S}$ $(r_{gs} = 2.5\,\text{M}\Omega)$ for $t_{ox} = 1.5\,\text{nm}$, which could significantly lower the low-frequency signal gain due to resistive gate–source loading. This places a minimum limit on the operating frequency as described in Section 3.12.2.1 below.

3.12.2 Gate Leakage Current Effects on Circuit Performance

Gate leakage current results in a gate conductance that causes a loss of signal gain at frequencies below a minimum frequency of operation associated with the gate conductance and capacitance. Additionally, gate conductance causes a loss of low-frequency, intrinsic gate-to-drain current gain that without gate conductance is nearly infinite. Gate leakage current also increases the discharge rate or droop for capacitances connected to MOS gates. Finally, gate leakage current increases circuit noise and mismatch because the leakage current contains shot and flicker noise current along with mismatch current. These gate leakage current effects are described below.

3.12.2.1 Minimum frequency of operation

The gate–source input impedance of a MOSFET seen in the small-signal model of Figure 3.21 includes the intrinsic gate–source capacitance, C_{gsi} (Section 3.9.2), in parallel with the gate–source conductance, g_{gs}. The parallel combination of C_{gsi} and g_{gs} provides useful insight on the effects of g_{gs} on circuit performance, even though this excludes extrinsic overlap and other circuit capacitances.

A corner frequency can be defined for the C_{gsi} and g_{gs} gate–source input impedance with rule-of-thumb values in strong inversion given by [180]

$$f_{gate} = \frac{g_{gs}}{2\pi C_{gsi}}$$

$$\approx 1.5 \times 10^{16}\ (\text{Hz/V}^2) \cdot V_{GS}^2 \cdot e^{(t_{ox}/1\,\text{nm})((V_{GS}-13.6\ \text{V})/1\ \text{V})} \quad (\text{nMOS, SI})$$

$$\approx 0.5 \times 10^{16}\ (\text{Hz/V}^2) \cdot V_{GS}^2 \cdot e^{(t_{ox}/1\,\text{nm})((V_{GS}-13.6\ \text{V})/1\ \text{V})} \quad (\text{pMOS, SI}) \qquad (3.181)$$

C_{gsi} is proportional to the gate area having a strong inversion value of $C_{gsi} = {}^2/_3 C'_{OX} WL$ as summarized in Table 3.23. g_{gs} is also proportional to the gate area through the value of I_{GS} as summarized in

Table 3.45. As a result, f_{gate} is independent of the gate area [180]. As listed in Equation 3.181, f_{gate} for pMOS devices is approximately one-third that of nMOS devices because of lower pMOS I_{GS} and g_{gs} due to the higher gate-oxide barrier [180].

For frequencies sufficiently above f_{gate}, C_{gsi} dominates the gate–source input impedance (again, extrinsic overlap capacitances are not considered) as it does when I_{GS} and g_{gs} are not significant. However, for frequencies sufficiently below f_{gate}, g_{gs} dominates the input impedance. f_{gate} then corresponds to the minimum frequency of operation where signals can be processed without loss of gate–source impedance and loss of signal gain resulting from the presence of g_{gs} [180].

For nMOS devices with $V_{GS} = 0.5\,\text{V}$ and $V_{EFF} = V_{GS} - V_T = 0.25\,\text{V}$ near the onset of strong inversion, Equation 3.181 predicts f_{gate} of 32 mHz, 16 kHz, and 11 MHz for $t_{ox} = 3$, 2, and 1.5 nm, respectively. $f_{gate} = 11\,\text{MHz}$ for $t_{ox} = 1.5\,\text{nm}$ suggests potential loss of impedance and signal gain for frequencies even into the low-megahertz range. f_{gate} is favorably reduced for smaller values of V_{GS} and, especially, for increased t_{ox}. The actual impact of g_{gs} on circuit performance depends, of course, on the overall small-signal resistance and capacitance levels. For example, g_{gs} lowers small-signal resistance and low-frequency gain and raises the minimum frequency of operation (the minimum frequency giving little gain loss due to g_{gs}) when connected to the output of a cascode circuit where g_{gs} exceeds the low cascode output conductance.

3.12.2.2 Intrinsic current gain

In addition to f_{gate}, another metric useful for assessing the effects of g_{gs} on circuit performance is the low-frequency, small-signal, intrinsic, gate-to-drain current gain [180]. This is defined as the ratio of short-circuit drain current to gate current for a grounded-source MOSFET as given by

$$A_{ii} = \frac{i_d}{i_{gs}} = \frac{g_m v_{gs}}{g_{gs} v_{gs}} = \frac{g_m}{g_{gs}} \tag{3.182}$$

Expressing g_m in terms of g_m/I_D in strong inversion (neglecting velocity saturation and VFMR reductions) from Table 3.17 and g_{gs} in strong inversion in terms of g_{gs}/I_{GS} from Table 3.45 allows A_{ii} to be expressed as the ratio of DC drain current, I_D, to gate–source leakage current, I_{GS}. This gives

$$A_{ii} = \frac{g_m}{g_{gs}} = \frac{I_D \cdot \dfrac{g_m}{I_D}}{I_{GS} \cdot \dfrac{g_{gs}}{I_{GS}}} \approx \frac{I_D \cdot \left(\dfrac{2}{V_{EFF}}\right)}{I_{GS} \cdot \left(\dfrac{1}{V_{EFF}} + \dfrac{1}{V_{GS}} + K_{GB}\right)} \text{(SI)}$$

$$\approx \left(\frac{I_D}{I_{GS}}\right) \cdot \frac{2}{1 + \dfrac{V_{EFF}}{V_{GS}} + V_{EFF} K_{GB}} \text{(SI)} \tag{3.183}$$

For nMOS devices with $t_{ox} = 2\,\text{nm}$ and $K_{GB} = 2.26/\text{V}$ (found from the expression in Table 3.45) operating at $V_{GS} = 0.5\,\text{V}$ and $V_{EFF} = V_{GS} - V_T = 0.25\,\text{V}$ near the onset of strong inversion, Equation 3.183 predicts that A_{ii} is nearly equal to the ratio of I_D to I_{GS}, or $A_{ii} = g_m/g_{gs} \approx I_D/I_{GS}$. This is a result of similar transconductance efficiency, g_m/I_D, and gate–source conductance efficiency, g_{gs}/I_{GS}. As mentioned in Section 3.12.1.2 and summarized in Table 3.45, g_{gs}/I_{GS} is maximum in weak inversion before decreasing in strong inversion. g_m/I_D is also maximum in weak inversion before decreasing in strong inversion as summarized in Table 3.17. As a result, A_{ii} roughly tracks I_D/I_{GS} in all regions of operation, although g_m/I_D and g_{gs}/I_{GS} should be evaluated and replaced with the strong-inversion expressions given in Equation 3.183. When solved for g_{gs}, $A_{ii} = g_m/g_{gs} \approx I_D/I_{GS}$ gives $g_{gs} \approx g_m \cdot (I_{GS}/I_D)$. This provides insight on the value of g_{gs}, which is below g_m by roughly the ratio of I_{GS} to I_D.

As developed in [180], it is useful to estimate A_{ii} in terms of the channel length, L. Since g_m is proportional to the shape factor, $S = W/L$, for fixed V_{EFF} in strong inversion (neglecting velocity saturation and VFMR reductions from Equation 2.12) and I_{GS} and g_{gs} are proportional to the gate area, WL, $A_{ii} = g_m/g_{gs}$ is independent of the channel width, but is proportional to $1/L^2$.

A rule-of-thumb value for A_{ii} in strong inversion is given by [180]

$$A_{ii} = \frac{g_m}{g_{gs}} \approx 7 \times 10^{-6} \cdot e^{13.6\,(t_{ox}/1\,\mathrm{nm})} \frac{V_{EFF}^2}{(L/1\mu\mathrm{m})^2\, V_{GS}^2 \cdot e^{(t_{ox}/1\,\mathrm{nm})(V_{GS}/1\,\mathrm{V})}} \,(\mathrm{SI}) \tag{3.184}$$

which applies to both nMOS and pMOS devices because a reduction of nearly a factor of three in pMOS mobility affecting g_m is countered by nearly the same reduction in tunneling probability affecting I_{GS} and g_{gs} [180]. Equation 3.184 shows that A_{ii} decreases significantly as t_{ox} decreases, just as Equation 3.183 shows that A_{ii} decreases significantly as t_{ox} decreases as a result of increasing I_{GS}. Equation 3.184 additionally shows that A_{ii} decreases rapidly with increasing L.

For nMOS devices with $L = 1\,\mu\mathrm{m}$ and $t_{ox} = 1.5\,\mathrm{nm}$ operating at $V_{GS} = 0.5\,\mathrm{V}$ and $V_{EFF} = V_{GS} - V_T = 0.25\,\mathrm{V}$ near the onset of strong inversion, Equation 3.184 predicts A_{ii} at approximately 600. Increasing L to $5\,\mu\mathrm{m}$ lowers the predicted value of A_{ii} to approximately 24. An intrinsic current gain of $A_{ii} = g_m/g_{gs} = 24$ indicates that g_{gs} is only a factor of 24 below g_m. If a grounded-source MOSFET having transconductance of g_m were loaded at the drain by this value of g_{gs} from a subsequent device, the gate-to-drain voltage gain would be only 24, excluding a loss in gain associated with device g_{ds}. While the rule-of-thumb values of A_{ii} given by Equations 3.183 and 3.184 are very approximate, these show the significant reduction of A_{ii} when gate leakage current and conductance are significant for long channel length and, correspondingly, large-area devices having small t_{ox}.

As channel length increases, the increasing intrinsic, gate-to-drain voltage gain, $A_{Vi} = g_m/g_{ds}$ (Section 3.8.5), which is nearly proportional to L, is in conflict with the decreasing intrinsic gate-to-drain current gain, $A_{ii} = g_m/g_{gs}$, which is nearly proportional to $1/L^2$. Thus, when gate leakage current is significant, increasing L to increase circuit small-signal resistance and gain through the decrease in drain–source conductance, g_{ds}, is countered by the increase in gate–source conductance, g_{gs}. This suggests an optimum value of L that trades off the normal circuit loading associated with g_{ds} with the additional loading associated with g_{gs} arising from gate leakage current. Tradeoffs related to gate leakage current are summarized later in Section 3.12.2.6.

3.12.2.3 Discharge of capacitances

Gate leakage current causes a discharge path for capacitors connected to the gates of MOSFETs. It also causes self-discharge in MOS capacitors. As mentioned in Section 3.12.1.1, gate leakage current is present for normal inversion as well as accumulation when the gate is sufficiently reverse biased.

The change in voltage across a capacitor with time, or droop rate, due to total gate leakage current, $I_G = I_{GS} + I_{GD}$, is given by

$$\frac{dV_C}{dt} = -\frac{I_G}{C} \tag{3.185}$$

where C is the capacitance value. As mentioned earlier in Section 3.12.1.1, I_{GD} is approximately one-half I_{GS} saturation. This increases I_G to 150 % of I_{GS} alone. In [180], the droop rate in V/s for MOS capacitors operating in strong inversion is found to be roughly equal to f_{gate} in Hz. This suggests a droop rate of roughly 3.2×10^{-3}, 16×10^3, and 11×10^6 V/s for f_{gate} of 32 mHz, 16 kHz, and 11 MHz given in Section 3.12.2.1 for $t_{ox} = 3$, 2, and 1.5 nm, respectively. f_{gate} and the droop rate decrease for V_{GS} and V_{EFF} below the values of 0.5 V and 0.25 V considered while decreasing significantly for increased t_{ox}. The droop rate caused by gate leakage current can significantly decrease the allowable hold time for capacitors in switched circuits.

3.12.2.4 Noise

Gate leakage current contains full shot noise and also flicker noise that contributes to overall circuit noise. This noise was discussed in Section 3.10.7 and illustrated by the white shot-noise and low-frequency, flicker-noise current density shown in Figure 3.69. Unfortunately, increasing gate area to reduce normal MOS flicker noise as discussed in Section 3.10.3.7 increases the gate shot-noise and flicker-noise current through the increase in gate leakage current.

When gate leakage current is significant, gate area should be limited to a value that trades off normal flicker noise with the added shot noise and flicker noise arising from the gate current. A similar tradeoff, described in the next section, exists for DC mismatch due to normal local-area mismatch and the additional mismatch arising from gate leakage current mismatch. These tradeoffs are summarized later in Section 3.12.2.6.

3.12.2.5 Mismatch

If gate leakage currents were equal in identically laid out MOSFETs connected in identical circuit environments, there would be no increase in overall circuit mismatch. However, manufacturing variations in the gate-oxide thickness and other process characteristics result in mismatch in gate leakage current that contributes to circuit mismatch.

Mismatch in the gate–source leakage current, ΔI_{GS}, can be estimated by

$$\Delta I_{GS}(1\sigma) = \frac{A_{IGS}}{\sqrt{WL}} \cdot I_{GS}$$

$$= \frac{A_{IGS}}{\sqrt{WL}} \cdot J_{GS}(V_{GS}, V_{EFF}) \cdot WL$$

$$= A_{IGS} \cdot J_{GS}(V_{GS}, V_{EFF}) \cdot \sqrt{WL} \tag{3.186}$$

where A_{IGS} is a gate–source current, area mismatch factor estimated at roughly $A_{IGS} \approx 3\% \cdot \mu m$ from [180]. As observed in the first line of Equation 3.186, the relative leakage current mismatch $\Delta I_{GS}/I_{GS}$ is inversely proportional to the square root of the gate area, WL, as is local-area threshold voltage and transconductance factor mismatch summarized earlier in Table 3.38. However, as $\Delta I_{GS}/I_{GS}$ decreases as gate area increases, I_{GS} itself increases with the gate area as seen by the second line of the equation. Here, I_{GS} is expressed as the gate–source leakage current density, $J_{GS}(V_{GS}, V_{EFF})$, multiplied by the gate area (WL). ΔI_{GS} then increases with the square root of gate area as shown in the third line of the equation. $J_{GS}(V_{GS}, V_{EFF})$, which increases with V_{GS} and V_{EFF}, is found from the I_{GS} expression in Table 3.45 with the gate area (WL) term removed.

Drain current mismatch for a current mirror is increased across the reference and output devices by ΔI_{GS}. However, mismatch in the gate–drain current, ΔI_{GD}, increases mismatch across multiple output devices. ΔI_{GD} might be approximated as one-half ΔI_{GS}, since I_{GD} itself is approximately one-half I_{GS} in saturation as mentioned earlier in Section 3.12.1.1. However, to consider possible increases in ΔI_{GD} associated with its generation over a smaller portion of the gate area compared to ΔI_{GS}, ΔI_{GD} will be taken as approximately equal to ΔI_{GS}.

Combining drain current mismatch due to threshold-voltage mismatch from Equation 3.153 with mismatch due to $\Delta I_{GD} \approx \Delta I_{GS}$ where ΔI_{GS} is given from Equation 3.186 gives an overall relative drain current mismatch of

$$\frac{\Delta I_D(1\sigma)}{I_D} = \sqrt{(\Delta V_T(1\sigma) \cdot g_m/I_D)^2 + \left(\frac{\Delta I_{GD}(1\sigma)}{I_D}\right)^2}$$

$$= \frac{1}{\sqrt{WL}} \sqrt{\left[A_{VTO}\left(1 + \frac{V_{SB}}{2\phi_0}\right) \cdot g_m/I_D\right]^2 + \left(\frac{A_{IGS} \cdot J_{GS}(V_{GS}, V_{EFF}) \cdot WL}{I_D}\right)^2} \tag{3.187}$$

Assuming $\Delta I_{GD} \approx \Delta I_{GS}$, the expression approximates mismatch across a reference and output device or mismatch across two output devices. Equation 3.187, which is similar in form to drain current mismatch given in [180], assumes that transconductance factor mismatch is negligible compared to threshold-voltage mismatch. This is often the case as described in Section 3.11.1.6. Additionally, the equation assumes that mismatch in the threshold voltage and gate leakage current are uncorrelated. Finally, the $A_{VTO} \cdot (1 + V_{SB}/(2\phi_0))$ term represents the threshold-voltage mismatch factor, A_{VT}, inclusive of the additional mismatch associated with non-zero V_{SB} as described in Equation 3.147.

The left term inside the square-root term in the second line of Equation 3.187 shows that drain current mismatch due to threshold-voltage mismatch decreases as the inverse square root of increasing gate area. However, the right term shows that drain current mismatch due to gate leakage current mismatch increases as the square root of increasing gate area. As a result, there is an optimum gate area that trades off the threshold-voltage and gate leakage mismatch components. The optimum gate area is smaller in strong inversion because lower g_m/I_D reduces the threshold-voltage mismatch contribution for a given gate area, while increased gate leakage current density increases the gate leakage current mismatch contribution.

Drain current mismatch given by Equation 3.187 can be referred to a gate–source voltage mismatch for a differential pair through division by g_m/I_D as described for local-area mismatch in Section 3.11.1.4. However, input-referred voltage mismatch is increased from this if equal gate leakage current from the two devices flows through unequal gate circuit resistances, or if mismatch in leakage current flows through equal gate circuit resistances. Gate leakage current and leakage mismatch current, like input bias current and offset current in bipolar operational amplifier circuits, contribute to mismatch through the external circuit.

3.12.2.6 Summary of tradeoffs

The presence of significant gate leakage current for small values of t_{ox} changes some of the long-established tradeoffs in analog CMOS design. Small-signal resistances and related signal gains, normally maximized by increasing channel length to minimize g_{ds}, are now affected by increasing g_{gs} associated with increasing gate leakage current as gate area increases with channel length. As a result, there will be an optimum channel length and gate area for signal gain that trade off the decrease in g_{ds} with the increase in g_{gs}. Also, there is a minimum frequency of operation set by the gate capacitance and g_{gs} that increases as g_{gs} increases with increasing gate leakage current. Signals at frequencies sufficiently below this minimum frequency are affected by the presence of g_{gs} loading while signals at frequencies sufficiently above this experience normal capacitive loading.

In addition to small-signal resistance and gain loading, gate leakage current results in increased capacitance discharging or droop in circuits utilizing switched capacitances. The maximum allowable hold time for capacitors decreases as gate leakage current increases, requiring increased clocking frequencies.

White noise arising from thermal noise, normally unaffected by gate area, is now increased by gate shot-noise current as gate leakage current increases with gate area. Additionally, flicker noise, normally minimized by increasing gate area, is now increased by gate-flicker-noise current as gate leakage current increases with gate area. As a result, there will be an optimum channel length and gate area that trade off normal thermal and flicker noise with the increase in circuit noise caused by gate shot-noise and flicker-noise current arising from gate leakage current.

Finally, gate leakage current flowing through unequal circuit resistances and mismatch in gate leakage current result in an additional component of circuit mismatch. As a result, there will be an optimum channel length and gate area that trade off the decrease in threshold-voltage and transconductance factor mismatch as gate area increases with the increasing mismatch associated with increasing gate leakage current. Active matching techniques can help manage increased mismatch associated with gate leakage current [180].

Like the gate–source leakage current shown in Figure 3.74 predicted from the expression in Table 3.45, gate leakage current and resulting tradeoffs in circuit performance can be predicted in terms of the independent design choices of drain current, inversion coefficient, and channel length. The tradeoffs of normal signal gain, noise, and mismatch against the decreased signal gain and increased noise and mismatch due to gate leakage current can be first evaluated for a given drain current. Then, an inversion coefficient, with corresponding g_m/I_D, and channel length, with corresponding gate area, can be found for optimum performance where components of normal performance are nearly in balance with components due to gate leakage current. In many cases, the flow of gate leakage current into the circuit requires the consideration of multiple devices. While not included in the initial *Analog CMOS Design, Tradeoffs and Optimization* spreadsheet or in the design examples in Chapters 5 and 6, inclusion of gate leakage current effects is anticipated in future extensions of this work.

Designs in processes having t_{ox} of at least 2 nm may not be significantly affected by gate leakage current. However, designs in processes having t_{ox} of 1.5 nm and below likely require the management of gate leakage current and the resulting gate conductance, noise, and mismatch. When available, thicker gate-oxide options can permit the use of traditional design tradeoffs that exclude gate leakage current effects. Thicker gate oxide can be used for critical analog circuits, while less critical analog circuits and digital circuits can use thinner gate-oxide options. At the time of this writing, analog design issues related to gate leakage current are becoming increasingly important, and new design methodologies are evolving to manage these [180].

3.12.3 Drain–Body and Source–Body Leakage Current

The drain–body diode junction is reverse biased for normal, positive drain-to-body voltage, V_{DB}, in nMOS devices and negative V_{DB} in pMOS devices. Here, as shown earlier in Figure 3.9, V_{DB} is taken positively for both nMOS and pMOS devices. Drain–body leakage current, I_{DB}, results from the reverse-biased, drain–body diode junction. Like V_{DB}, I_{DB} is taken positively here for both nMOS and pMOS devices.

Traditionally, the drain–body leakage current per drain–body junction area, or the drain–body leakage current density, $J_{DB}(V_{DB})$, is found from the diode equation [16, p. 223, 37, p. 36] evaluated for reverse bias as

$$
\begin{aligned}
J_{DB}(V_{DB}, \text{traditional}) &= -J_s\left(e^{-V_{DB}/(\eta U_T)} - 1\right) \\
&= -\left[qn_i^2\left(\frac{D_p}{L_p N_D} + \frac{D_n}{L_n N_A}\right)\right]\left(e^{-V_{DB}/(\eta U_T)} - 1\right)
\end{aligned} \tag{3.188}
$$

q is the unit of electronic charge, n_i is the intrinsic carrier concentration, and η is a diode ideality factor that is usually near unity. D_p and L_p, and D_n and L_n are the diffusion constants and diffusion lengths for minority-carrier holes on the n-side of the junction and minority-carrier electrons on the p-side of the junction, respectively. N_D and N_A are the doping concentrations for the n- and p-side of the junction, respectively. For $V_{DB} > 5\eta U_T$ (130 mV at $T = 300$ K), $J_{DB}(V_{DB})$, taken positively here for reverse bias leakage current, is essentially equal to the saturation current density, J_s, given by the bracketed term on the second line of the equation. Under reverse bias, the small leakage current density is caused by a small supply of minority carriers near the edge of the depletion region.

Equation 3.188 under-predicts $J_{DB}(V_{DB})$ because it excludes the usually more dominant generation leakage current caused by electron–hole pair generation in the depletion region. $J_{DB}(V_{DB})$ due to this generation current is given by [16, pp. 223–224]

$$J_{DB}(V_{DB}, \text{ generation}) = \frac{q\,n_i\,t_{dep}}{\tau_{gen}}$$

$$= \frac{q\,n_i}{\tau_{gen}}\sqrt{\frac{2\,\varepsilon_{Si}}{qN_B}}\sqrt{\phi_{bi} + V_{DB}} \tag{3.189}$$

where t_{gen} is the carrier generation time constant in the depletion region, and t_{dep} is the depletion-region thickness. In the second line of the equation, t_{dep} is given from the expression in Table 3.1 where ε_{Si} is the permittivity of silicon, N_B is the doping concentration in the body or substrate, and ϕ_{bi} is the built-in voltage of the drain–body diode junction. The expression for t_{dep} assumes that the doping concentration, N_{DIF}, in the drain region or diffusion is much greater than N_B. ϕ_{bi} is found from the expression in Table 3.1 using the doping concentrations N_{DIF} and N_B.

I_{DB} can be calculated by the sum of J_{DB} (V_{DB}) given by Equations 3.188 and 3.189 multiplied by the drain–body junction area, A_D. For the usual case where the generation leakage current dominates, J_{DB} (V_{DB}) and I_{DB} increase with V_{DB} because of the increase in t_{dep}. Additionally, J_{DB} (V_{DB}) and I_{DB} increase significantly with temperature due primarily to the increase in n_i with temperature shown earlier in Table 3.8. J_{DB} (V_{DB}) and I_{DB} typically double for every 8 to 12 °C increase in temperature. For this reason, silicon-on-insulator (SOI) CMOS processes are often used for high-temperature applications because the drain region sits on an insulator, significantly reducing the leakage current at high temperatures.

J_{DB} (V_{DB}) is reported at approximately $0.02\,\text{fA}/\mu\text{m}^2$ for a 0.35 μm, bulk CMOS process at room temperature [250], although the value of V_{DB} is unreported. Assuming that the leakage current density doubles over nearly 10, factor-of-10 °C increases in temperature from room temperature at 27 °C (300 K) to the maximum military temperature of 125 °C (398 K), the current density increases by nearly three orders of magnitude from $0.02\,\text{fA}/\mu\text{m}^2$ at 27 °C to $0.02\,\text{pA}/\mu\text{m}^2$ at 125 °C.

While Equations 3.188 and 3.189 may adequately predict J_{DB} (V_{DB}) and I_{DB} in large-geometry processes, the equations exclude band-to-band tunneling current that can be dominant in modern, small-geometry processes [247]. In these processes, band-to-band tunneling results from the high electric field across the narrow depletion region associated with high doping concentration on both sides of the drain–body junction [247]. This results from non-uniform halo doping along the channel where doping is higher near the source and drain. Halo doping reduces the depletion-region thickness near the drain, reducing the associated charge-sharing effects. This processing technique, while helping to manage short-channel effects, increases drain–body leakage current because of band-to-band tunneling current. The band-to-band tunneling current density is in the vicinity of $1\,\text{A/cm}^2$ or $10\,\text{nA}/\mu\text{m}^2$ for $V_{DB} \approx 1$ V, an electric field of 1.7 MV/cm, and a halo body doping concentration near the drain of approximately $10^{19}/\text{cm}^3$ [251]. This leakage current density is substantially above $J_{DB}(V_{DB}) = 0.02\,\text{fA}/\mu\text{m}^2$ given for a 0.35 μm process above, although the tunneling current density is reduced some since it applies to a portion of the drain area near the channel. Band-to-band tunneling dominates $J_{DB}(V_{DB})$ in smaller-geometry process, except potentially for small or negative values of V_{GS} where gate-induced drain leakage current can dominate.

In addition to excluding band-to-band tunneling current in the overall drain–body region, Equations 3.188 and 3.189 exclude leakage current that is influenced by the gate voltage. This includes gate-induced drain leakage current (GIDL) that flows from the drain to the substrate in the drain–body region below the gate overlap [247]. GIDL is maximum for reverse bias gate voltages, for example in off-state devices, combined with high drain voltages. This is because of the high electric field in the narrow depletion region between the channel, which when accumulated has higher doping than the substrate, and the drain. The electric field is even greater when high negative gate voltage and high positive drain voltage cause depletion or even inversion in the drain region under the gate. The high electric field in the drain region under the gate increases the probability of band-to-band tunneling, trap-assisted tunneling, and avalanching causing GIDL.

Because of the complexity predicting J_{DB} (V_{DB}) and I_{DB} due to potential band-to-band tunneling and other effects, no attempt will be made here to predict J_{DB} (V_{DB}) and I_{DB} from process parameters. Instead, a process value of J_{DB} (V_{DB}) will be used that is assumed to include generation, band-to-band tunneling, GIDL, and other important components of leakage current. This is illustrated in Table 3.45 where I_{DB} is calculated from A_D multiplied by J_{DB} (V_{DB}). Additionally, A_D is expressed in terms of the channel width, W, using the process drain diffusion width, W_{DIF}, and number of gate fingers, m. A_D is expressed as $A_D = \frac{1}{2}W \cdot W_{DIF}$, which assumes that m is even and the area of the interdigitated drain is minimized. If m is one, A_D is equal to twice this value as noted in the table. The calculation of A_D was summarized earlier in Table 3.25 for the calculation of drain–body junction capacitance.

Since the expression for I_{DB} given in Table 3.45 is linearly related to W through the A_D calculation, it is possible to include GIDL, which depends or W, in the value of J_{DB} (V_{DB}) for a selected gate–source voltage, especially for negative voltages where GIDL is maximum. Additionally, since the drain perimeter is linearly related to W, as summarized earlier in Table 3.25, leakage current associated with the perimeter or side wall of the drain–body junction can also be included in J_{DB} (V_{DB}). Finally, separate or averaged drain extension geometries can be included for different components of leakage current. As always, the predictions provided in this book are intended for design guidance only. Full computer simulations of a candidate design could include complex I_{DB} dependencies on the geometry of the drain, drain extension, and gate overlap, as well as dependencies on the drain and gate voltages.

In addition to expressing I_{DB} in terms of A_D and W, Table 3.45 gives I_{DB} and its trends in terms of the design choices of IC, L, and I_D. Here, W is expressed as $W = (L/IC)(I_D/I_0)$ from Table 3.10 where $I_0 = 2n_0\mu_0 C'_{OX} U_T^2$ is the technology current described in Table 3.6. I_{DB} decreases as $1/IC$ because of the reduction of required shape factor, $S = W/L$, and W with increasing IC. I_{DB} increases directly with L because of the required increase in W. Finally, I_{DB} increases directly with I_D because of the required increase in S and W. Thus, for a fixed IC and L, I_{DB} remains a constant fraction of I_D since both track W. As mentioned, I_{DB} increases with V_{DB} and with temperature, and also increases for low or negative values of gate–source voltage due to GIDL.

Long-channel devices operating at low IC in weak inversion have the highest values of I_{DB} relative to I_D because their widths and corresponding drain areas are large for a given I_D. In contrast, short-channel devices operating at high IC in strong inversion have the lowest values of I_{DB} relative to I_D because their widths and corresponding drain areas are small. Increased drain–body leakage, along with low intrinsic and extrinsic bandwidths described earlier in Sections 3.9.6 and 3.9.7, are disadvantages of weak-inversion operation. Leakage is reduced and bandwidth improved in moderate and strong inversion because of the reduction of width and the resulting reduction of drain and gate areas.

The prediction of I_{DB} discussed here excludes drain–body or substrate current due to weak avalanching associated with hot-electron effects. This was discussed earlier in Section 3.8.4.3. Additionally, I_{DB} due to stronger avalanching or breakdown is not considered here. Hot-electron, avalanche, or breakdown effects are generally negligible when the drain voltage is kept sufficiently low. Finally, while not a high-field effect or causing current flow to the substrate, the small leakage current between the drain and source in accumulation is also not considered here. This leakage current may result from carrier diffusion under the accumulation layer [252].

The increase in I_{DB} with V_{DB} due to generation leakage current density (Equation 3.189) and other components of leakage results in a small-signal, drain–body conductance, g_{db}, appearing between the drain and body in the MOS small-signal model shown earlier in Figure 3.21. The drain–body conductance is given by $g_{db} = \delta I_{DB}/\delta V_{DB}$, which assumes that the other bias voltages, V_{GS} and V_{SB}, are held constant. The components of g_{db} due to hot-electron effects, discussed earlier in Section 3.8.4.3, avalanching, or breakdown may be negligible for low drain voltage but should be included when significant.

As noted in Table 3.45, the source–body leakage current, I_{SB}, resulting from non-zero, reverse-biased, source–body voltage, V_{SB}, can be predicted analogously to I_{DB}. This can be done by replacing the area of the drain, A_D, with the area of the source, A_S, and replacing the drain–body leakage current density, J_{DB} (V_{DB}), with the source–body leakage current density, J_{SB} (V_{SB}). Like V_{DB}, V_{SB} is taken positively here for both nMOS and pMOS devices for normal reverse bias, source–body junction voltage.

MOS circuits can be designed with very low levels of drain leakage current at room temperature, at least for large-geometry processes, and gate leakage current can also be very low for thick, typically at least 4 nm, gate oxides. The operation of MOS circuits at Femtoampere (fA $= 10^{-15}$ A) current levels is discussed in [250] for a 0.35 μm process, and the operation of an analog storage cell having 10 aA (aA $= 10^{-18}$ A) leakage is discussed in [252] for a 1.5 μm process. The analog storage cell makes use of low leakage switches. Unfortunately, gate leakage current, due to tunneling through the gate oxide, and drain–body leakage current, due to band-to-band tunneling and GIDL, increase greatly in smaller-geometry processes [247, 251].

3.12.4 Subthreshold Drain Leakage Current

MOS drain current for $V_{GS} = 0$ V, usually corresponding to the off-state condition, is commonly referred to as subthreshold leakage current in digital circuits. For analog circuits, this is the normal weak-inversion drain current evaluated at $V_{GS} = 0$ V. Subthreshold leakage current and its process and temperature dependencies are discussed below.

The subthreshold drain leakage current, referred to here as the off-state leakage current, is given from the weak-inversion drain current in Table 3.13 evaluated at $V_{GS} = 0$ V by

$$I_D(\text{off}, V_{GS} = 0 \text{ V}) = 2n\,\mu_0\,C'_{OX}\,U_T^2\left(\frac{W}{L}\right)e^{\frac{-V_T}{nU_T}} \tag{3.190}$$

Mobility, μ_0, is taken at its low-field value in weak inversion, and the threshold voltage, V_T, includes device geometry, bias voltage (V_{DS} and V_{SB}), and temperature adjustments described later.

The off-state leakage current can be normalized to the number of device squares or the shape factor, $S = W/L$, as

$$\frac{I_D(\text{off}, V_{GS} = 0 \text{ V})}{W/L} = 2n\,\mu_0\,C'_{OX}\,U_T^2 e^{\frac{-V_T}{nU_T}} \tag{3.191}$$

For nMOS devices having $t_{ox} = 2$ nm, $C'_{OX} = 17.3$ fF/μm^2, $\mu_0 = 250$ cm^2/V·s, weak inversion $n = 1.4$, $V_T = 0.25$ V, and $U_T = 25.9$ mV at $T = 300$ K (room temperature), the off-state current is 0.82 nA/square. The off-state current for a digital device in the process at room temperature is 0.82 nA for a shape factor of unity.

It is useful to compare the off-state current with the on-state current at a selected V_{GS}, often at the supply voltage, V_{DD}, for digital circuits. Evaluating the ratio of off-state, weak-inversion current at $V_{GS} = 0$ V to the on-state, strong-inversion current at $V_{GS} = V_{DD}$ using the drain current expressions in Table 3.13 gives

$$\frac{I_D(\text{off}, V_{GS} = 0 \text{ V})}{I_D(\text{on}, V_{GS} = V_{DD})} = \frac{2n\,\mu_0\,C'_{OX}\,U_T^2\left(\dfrac{W}{L}\right)e^{\frac{-V_T}{nU_T}}}{\dfrac{1}{2n}\cdot\dfrac{\mu_0\,C'_{OX}}{1+\dfrac{V_{DD}-V_T}{(L\,E_{CRIT})'}}\left(\dfrac{W}{L}\right)(V_{DD}-V_T)^2}$$

$$= \left(\frac{2nU_T}{V_{DD}-V_T}\right)^2 e^{\frac{-V_T}{nU_T}}\left(1+\frac{V_{DD}-V_T}{(L\,E_{CRIT})'}\right) \tag{3.192}$$

Here, the $1+(V_{DD}-V_T)/(LE_{CRIT})'$ term in the denominator of the on-state current in the denominator of the first line describes the loss of current for $V_{GS} = V_{DD}$ due to velocity saturation and VFMR effects. $(LE_{CRIT})'$ is the equivalent velocity saturation voltage inclusive of VFMR effects described earlier in

Section 3.7.1.5. The high on-state, effective gate–source voltage, $V_{EFF} = V_{GS} - V_T = V_{DD} - V_T$, combined with the usual minimum channel length used for digital devices, results in significant velocity saturation reduction of on-state current, especially for nMOS devices that have lower values of critical, horizontal electric field, E_{CRIT}, compared to pMOS devices. The reduction of on-state current unfavorably increases the ratio of off- to on-state current. Increased off-state current results from width increases required to maintain a given level of on-state drive current when velocity saturation is significant.

Equations 3.190–3.192 show that the off-state current and the off- to on-state current ratio increase significantly as V_T decreases because $-V_T$ appears in exponential terms. The off-state current and off- to on-state current ratio increase for a worst case combination of decreasing process V_T in smaller-geometry processes, decreasing V_T for the minimum channel length device, decreasing V_T for non-zero V_{DS}, and decreasing V_T with increasing temperature. Process V_T decreases with decreasing process V_{DD} in smaller-geometry processes to maintain adequate on-state current. V_T decreases for the usual minimum channel length device because of V_T roll-off. This is caused by charge sharing where the nearby source and drain depletion regions cut off some of the normal, substrate, depletion-region control below the channel, lowering V_T [15, pp. 259–263]. V_T decreases further for non-zero V_{DS}, especially for the minimum channel length device, due to charge sharing or DIBL described earlier in Section 3.8.4.2. As mentioned in Section 3.8.4.2, higher retrograde doping below the channel surface and higher halo doping near the source and drain reduce the depletion regions under the channel and near the source and drain, reducing charge-sharing and DIBL effects. Finally, V_T decreases with temperature as described earlier in Section 3.5.4 and shown in Figure 3.6.

The off-state current given by Equations 3.190 and 3.191 increases by nearly a factor of 50 for an operating die temperature increase of 100 °C from 300 to 400 K (27 to 127 °C). This assumes a 1 mV/C decrease in V_T from 250 mV at 300 K, a 0.086 mV/C increase in U_T from 25.9 mV at 300 K, and $n = 1.4$ for nMOS devices in the process mentioned earlier. The current increase is dominated by the exponential term in the equations as the preceding technology current term, $I_0 = 2n_0\mu_0 C'_{OX} U_T^2$, increases by only about 15 % over the 100 °C range as seen in Figure 3.5 for a different process I_0 having a same mobility temperature exponent of $BEX = -1.5$. Since temperature dependency in the off-state current dominates temperature dependency in the on-state current, the off- to on-state current ratio experiences a similar, significant increase with temperature as the off-state current. Low process V_T combined with the V_T decrease resulting from elevated die temperature significantly increases the off-state current and off- to on-state current ratio in large digital circuits.

Since off-state, subthreshold drain leakage current usually exceeds the current associated with dynamic switching in digital circuits, considerable research is underway to manage the subthreshold leakage current [247]. Techniques to manage the leakage current include the use of stacked transistors where the presence of reverse bias V_{GS} and non-zero V_{SB} for at least one transistor significantly reduces the subthreshold current [247]. The current is reduced through the negative value of V_{GS} and the increase in V_T associated with non-zero V_{SB}. Additionally, multiple V_T, dynamic V_T using switched V_{SB} levels, V_{DD} scaling, and dynamic V_{DD} techniques can be used to reduce subthreshold leakage current [247]. Although this discussion has considered subthreshold drain leakage current, the drain–body leakage current, due to GIDL and band-to-band tunneling (Section 3.12.3), and gate leakage current (Section 3.12.1.1) can be signicant contributors to digital circuit leakage current in smaller-geometry processes [247, 251].

REFERENCES

[1] E. A. Vittoz, "Micropower techniques," in *Design of MOS VLSI Circuits for Telecommunications*, ed. J. Franca and Y. Tsividis, Prentice Hall, 1994.

[2] Y. Tsividis, "Moderate inversion in MOS devices," *Solid-State Electronics*, vol. 25, pp. 1099–1104, Nov. 1982.

[3] University of California at Berkeley BSIM3 and BSIM4 MOS Models, available on-line at www.device. eecs.Berkeley.edu/bsim3.

[4] D. M. Binkley, "A methodology for transistor level analog CMOS design," Conference tutorial with D. P. Foty, "MOSFET modeling and circuit design: re-establishing a lost connection," *37th Annual Design Automation Conference (DAC)*, Los Angeles, June 2000.

[5] D. M. Binkley, "A methodology for analog CMOS design based on the EKV MOS model," *2000 Fabless Semiconductor Association Design Modeling Workshop: SPICE Modeling*, Santa Clara, Oct. 2000.

[6] D. M. Binkley, "A methodology for analog CMOS design based on the EKV MOS model," Conference tutorial with D. P. Foty, "Re-connecting MOS modeling and circuit design: new methods for design quality," *2001 IEEE 2nd International Symposium on Quality Electronic Design (ISQED)*, San Jose, Mar. 2001.

[7] D. M. Binkley, "A methodology for analog CMOS design based on the EKV MOS model," Conference tutorial with D. P. Foty, "MOS modeling as a basis for design methodologies: new techniques for next-generation analog circuit design," *2001 15th European Conference on Circuit Theory and Design (ECCTD)*, Helsinki, Aug. 2001.

[8] D. M. Binkley, "A methodology for analog CMOS design based on the EKV MOS model," Conference tutorial with D. P. Foty, "MOS modeling as a basis for design methodologies: new techniques for modern analog design," *2002 IEEE International Symposium on Circuits and Systems (ISCAS)*, Scottsdale, May 2002.

[9] D. M. Binkley, M. Bucher, and D. Kazazis, "Modern analog CMOS design from weak through strong inversion," *Proceedings of the European Conference on Circuit Theory and Design (ECCTD)*, pp. I-8–I-13, Krakow, Aug. 2003.

[10] D. M. Binkley, C. E. Hopper, S. D. Tucker, B. C. Moss, J. M. Rochelle, and D. P. Foty, "A CAD methodology for optimizing transistor current and sizing in analog CMOS design," *IEEE Transactions on Computer-Aided Design of Integrated Circuits and Systems*, vol. 22, pp. 225–237, Feb. 2003.

[11] D. M. Binkley, B. J. Blalock, and J. M. Rochelle, "Optimizing drain current, inversion level, and channel length in analog CMOS design," *Analog Integrated Circuits and Signal Processing*, vol. 47, pp. 137–163, May 2006.

[12] Y. Tsividis, *Mixed Analog-digital VLSI Devices and Technology*, McGraw-Hill, 1996.

[13] C. Enz, F. Krummenacher, and E. A. Vittoz, "An analytical MOS transistor model valid in all regions of operation and dedicated to low-voltage and low-current applications," *Analog Integrated Circuits and Signal Processing*, vol. 8, pp. 83–114, July 1995.

[14] M. Bucher, C. Lallement, C. Enz, F. Théodoloz, and F. Krummenacher, "The EPFL-EKV MOSFET model equations for simulation, version 2.6," Technical Report, EPFL, July 1998, Revision II, available on-line at http://legwww.epfl.ch/ekv/.

[15] Y. Tsividis, *Operation and Modeling of the MOS Transistor*, second edition, McGraw-Hill, 1999.

[16] C. C. Enz and E. A. Vittoz, *Charge-based MOS Transistor Modeling: The EKV Model for Low-power and RF IC Design*, John Wiley and Sons, Inc., 2006.

[17] J. P. Colinge, "Fully-depleted SOI CMOS for analog applications," *IEEE Transactions on Electron Devices*, vol. 45, pp. 1010–1016, May 1998.

[18] A. I. A. Cunha, M. C. Schneider, and C. Galup-Montoro, "An MOS transistor model for analog circuit design," *IEEE Journal of Solid-State Circuits*, vol. 33, pp. 1510–1519, Oct. 1998.

[19] F. Silveira, D. Flandre, and P. G. A. Jespers, "A g_m/I_D based methodology for the design of CMOS analog circuits and its application to the synthesis of a silicon-on-insulator micropower OTA," *IEEE Journal of Solid-State Circuits*, vol. 31, pp. 1314–1319, Sept. 1996.

[20] D. M. Binkley, C. E. Hopper, B. J. Blalock, M. M. Mojarradi, J. D. Cressler, and L. K. Yong, "Noise performance of 0.35-μm SOI CMOS devices and micropower preamplifier from 77–400 K," *Proceedings of the 2004 IEEE Aerospace Conference*, vol. 4, pp. 2495–2506, Mar. 2004.

[21] M. N. Ericson, M. Hasanuzzaman, S. C. Terry, C. L. Britton, B. Ohme, S. S. Frank, J. A. Richmond, and B. J. Blalock, "1/f noise and DC characterization of partially depleted SOI n- and p-MOSFETs from 20–250 C," *Proceedings of the 2005 IEEE Aerospace Conference*, vol. 1, pp. 1–13, Mar. 2005.

[22] F. N. Trofimenkoff, "Field-dependent mobility analysis of the field-effect transistor," *Proceedings of the IEEE*, vol. 53, pp. 1765–1766, Nov. 1965.

[23] D. M. Caughey and R. E. Thomas, "Carrier mobilities in silicon empirically related to doping and field," *Proceedings of the IEEE*, vol. 55, pp. 2192–2193, Dec. 1967.

[24] C. Canali, G. Majni, R. Minder, and G. Ottaviani, "Electron and hole drift velocity measurements in silicon and their empirical relation to electric field and temperature," *IEEE Transactions on Electron Devices*, vol. 22, pp. 1045–1047, Nov. 1975.

[25] T. Grotjohn and B. Hoefflinger, "A parametric short-channel MOS transistor model for subthreshold and strong-inversion current," *IEEE Journal of Solid-State Circuits*, vol. 19, pp. 100–112, Feb. 1984.

[26] P. R. Gray, P. J. Hurst, S. H. Lewis, and R. G. Meyer, *Analysis and Design of Analog Integrated Circuits*, fourth edition, John Wiley and Sons, Inc., 2001.

[27] B. Razavi, *Design of Analog CMOS Integrated Circuits*, McGraw-Hill, 2001.

[28] E. Sanchez-Sinencio and A. G. Andreou, eds, *Low-Voltage/Low-Power Integrated Circuits and Systems: Low-Voltage, Mixed-Signal Circuits*, IEEE Press, 1999.

[29] M. N. Ericson, C. L. Britton, A. L. Wintenberg, J. M. Rochelle, B. J. Blalock, D. M. Binkley, and B. D. Williamson, "Noise behavior of MOSFETs fabricated in 0.5-μm fully-depleted silicon-on-sapphire CMOS in weak, moderate, and strong inversion," *IEEE Transactions on Nuclear Science*, vol. 50, pp. 963–968, Aug. 2003.

[30] D. Binkley, M. Bucher, and D. Foty, "Design-oriented characterization of CMOS over the continuum of inversion level and channel length," *Proceedings of the 7th IEEE International Conference on Electronics, Circuits and Systems (ICECS)*, pp. 161–164, Dec. 2000.

[31] M. Bucher, D. Kazazis, F. Krummenacher, D. Binkley, D. Foty, and Y. Papanos, "Analysis of transconductance at all levels of inversion in deep submicron CMOS," *Proceedings of the 9th IEEE International Conference on Electronics, Circuits and Systems (ICECS)*, vol. III, pp. 1183–1186, Sept. 2002.

[32] S. C. Terry, J. M. Rochelle, D. M. Binkley, B. J. Blalock, and D. Foty, "Comparison of a BSIM3V3 and EKV MOST model for a 0.5-μm CMOS process and implications for analog circuit design," *IEEE Transactions on Nuclear Science*, vol. 50, pp. 915–920, Aug. 2003.

[33] D. A. Johns and K. W. Martin, *Analog Integrated Circuit Design*, John Wiley and Sons, Inc., 1997.

[34] B. Razavi, *RF Microelectronics*, Prentice Hall, 1998.

[35] T. Lee, *The Design of CMOS Radio-Frequency Integrated Circuits*, Cambridge University Press, 1998.

[36] K. R. Laker and W. M. C. Sansen, *Design of Analog Integrated Circuits and Systems*, McGraw-Hill, 1994.

[37] P. E. Allen and D. R. Holberg, *CMOS Analog Circuit Design*, second edition, Oxford University Press, 2002.

[38] J. M. Early, "Effects of space-charge layer widening in junction transistors," *Proceedings of the IRE*, vol. 40, pp. 1401–1406, Nov. 1952.

[39] Y. A. El-Mansy and A. R. Boothroyd, "A simple two-dimensional model for IGFET operation in the saturation region," *IEEE Transactions on Electron Devices*, vol. ED-24, pp. 254–262, Mar. 1977.

[40] Z.-H. Liu, C. Hu, J.-H. Huang, T.-Y. Chan, M.-C. Jeng, P. K. Ko, and Y. C. Cheng, "Threshold voltage model for deep-submicrometer MOSFETs," *IEEE Transactions on Electron Devices*, vol 40, pp. 86–95, Jan. 1993.

[41] R. van Langevelde and F. M. Klaassen, "Accurate drain conductance modeling for distortion analysis in MOSFETs," *Technical Digest of the 1997 IEEE International Electron Devices Meeting*, pp. 313–316, Dec. 1997.

[42] M. C. Schneider, C. Galup-Montoro, O. C. Gouveia Filho, and A. I. A. Cunha, "A single-piece charge-based model for the output conductance of MOS transistors," *Proceedings of the IEEE International Conference on Electronics, Circuits, and Systems (ICECS)*, vol. 1, pp. 545–548, Sept. 1998.

[43] M. Bucher, D. Kazazis, F. Krummenacher, D. Binkley, D. Foty, P. Bendix, T. Yoshitomi, T. Ohguro, K. Kojima, and Y. Papanos, "Anomalous scaling of output conductance in pocket implanted deep submicron CMOS," Unpublished research notes.

[44] J. M. Sallese, M. Bucher, and C. Lallement, "Improved analytical modeling of polysilicon depletion in MOSFETs for circuit simulation," *Solid-State Electronics*, vol. 44, pp. 905–912, June 2000.

[45] R. van Langevelde, A. J. Scholten, and D. B. M. Klaassen, "Physical background of MOS model 11, level 1101," Philips Electronics, NV, Technical Report NL-TN 2003/00239, available on-line at http://www.semiconductors.philips.com/Philips_Models/mos_models, Apr. 2003.

[46] G. Gildenblat, H. Wang, T.-L. Chen, X. Gu, and X. Cai, "SP: an advanced surface-potential-based compact MOSFET model," *IEEE Journal of Solid-State Circuits*, vol. 39, pp. 1394–1406, Sept. 2004.

[47] A. Einstein, "Eine neue Bestimmung der Molekuldimensionen," *Annalen der Physik*, Band 19, pp. 289–306, 1906.

[48] J.B. Johnson, "Thermal agitation of electricity in conductors," *Physical Review*, vol. 29, p. 367, Feb. 1928 (minutes of the Philadelphia meeting, December 28–30, 1926), and vol. 32, pp. 97–109, July 1928 (full article).

[49] H. Nyquist, "Thermal agitation of electric charge in conductors," *Physical Review*, vol. 32, pp. 110–113, July 1928.

[50] A. van der Ziel, "Thermal noise in field effect transistors," *Proceedings of the IRE*, pp. 1808–1812, Aug. 1962.

[51] A. van der Ziel, "Gate noise in field effect transistors at moderately high frequencies," *Proceedings of the IEEE*, pp. 460–467, Mar. 1963.

[52] A. G. Jordan and N. A. Jordan, "Theory of noise in metal oxide semiconductor devices," *IEEE Transactions on Electron Devices*, vol. ED-12, pp. 148–156, Mar. 1965.

[53] C. T. Sah, S. Y. Wu, and F. H. Hielscher, "The effects of fixed bulk charge on the thermal noise in metal-oxide-semiconductor transistors," *IEEE Transactions on Electron Devices*, vol. ED-13, pp. 410–414, Apr. 1966.

[54] F. M. Klaassen and J. Prins, "Thermal noise of MOS transistors," *Philips Research Reports*, vol. 22, pp. 505–514, Oct. 1967.

[55] F. M. Klaassen and J. Prins, "Noise in field effect transistors at very high frequencies," *IEEE Transactions on Electron Devices*, vol. ED-16, pp. 952–957, Nov. 1969.

[56] P. S. Rao, "The effect of the substrate upon the gate and drain noise parameters of MOSFETS," *Solid-State Electronics*, vol. 12, pp. 549–555, July 1969.

[57] C. Huang and A. van der Ziel, "Thermal noise in ion-implanted MOSFETS," *Solid-State Electronics*, vol. 18, pp. 509–510, June 1975.

[58] A. van der Ziel, *Noise in Solid State Devices and Circuits*, Wiley-Interscience, 1986.

[59] R. P. Jindal, "Hot-electron effects on channel thermal noise in fine-line field-effect transistors," *IEEE Transactions on Electron Devices*, vol. ED-33, pp. 1395–1397, Sept. 1986.

[60] A. Abidi, "High frequency noise measurement on FETs with small dimensions," *IEEE Transactions on Electron Devices*, vol. ED-33, pp. 1801–1805, Nov. 1986.

[61] C. Enz, "High precision CMOS micropower amplifiers," PhD thesis no. 802, Ecole Polytechnique Federal de Lausanne, 1989.

[62] R. Sarpeshkar, T. Delbruck, and C. A. Mead, "White noise in MOS transistors and resistors," *IEEE Circuits and Devices Magazine*, vol. 9, pp. 23–29, Nov. 1993.

[63] D. P. Triantis, A. N. Birbas, and D. Kondis, "Thermal noise modeling for short-channel MOSFETs," *IEEE Transactions on Electron Devices*, vol. 43, pp. 1950–1955, Nov. 1996.

[64] D. Shaeffer and T. Lee, "A 1.5-V, 1.5-GHz CMOS low noise amplifier," *IEEE Journal of Solid-State Circuits*, vol. 32, pp. 745–759, May 1997.

[65] T. Manku, "Microwave CMOS–device physics and design," *IEEE Journal of Solid-State Circuits*, vol. 34, pp. 277–285, Mar. 1999.

[66] P. Klein, "An analytical thermal noise model of deep submicron MOSFETs," *IEEE Electron Device Letters*, vol. 20, pp. 399–401, Aug. 1999.

[67] C. Enz and Y. Cheng, "MOS transistor modeling for RF IC design," *IEEE Journal of Solid-State Circuits*, vol. 25, pp. 186–201, Feb. 2000.

[68] J.-S. Goo, C.-H. Choi, F. Danneville, E. Morifuji, H. S. Momose, Z. Yu, H. Iwai, T. H. Lee, and R. W. Dutton, "An accurate and efficient high frequency noise simulation technique for deep submicron MOSFETs," *IEEE Transactions on Electron Devices*, vol. 47, pp. 2410–2419, Dec. 2000.

[69] C. H. Park and Y. J. Park, "Modeling of thermal noise in short-channel MOSFETs at saturation," *Solid-State Electronics*, vol. 44, pp. 2053–2057, Nov. 2000.

[70] G. Knoblinger, P. Klein, and M. Tiebout, "A new model for thermal channel noise of deep-submicron MOSFETS and its application in RF-CMOS design," *IEEE Journal of Solid-State Circuits*, vol. 36, pp. 831–837, May 2001.

[71] V. Re, I. Bietti, R. Castello, M. Manghisoni, V. Speziali, and F. Svelto, "Experimental study and modeling of the white noise sources in submicron P- and N- MOSFETs," *IEEE Transactions on Nuclear Science*, vol. 48, pp. 1577–1586, Aug. 2001.

[72] G. Anelli, F. Faccio, S. Florian, and P. Jarron, "Noise characterization of a 0.25-μm CMOS technology for the LHC experiments," *Nuclear Instruments and Methods in Physics Research A*, vol. 457, pp. 361–368, Jan. 2001.

[73] C. Enz, "An MOS transistor model for RF IC design valid in all regions of operation," *IEEE Transactions on Microwave Theory and Techniques*, vol. 50, pp. 342–359, Jan. 2002.

[74] C.-H. Chen and M. J. Deen, "Channel noise modeling of deep submicron MOSFETs," *IEEE Transactions on Electron Devices*, vol. 49, pp. 1484–1487, Aug. 2002.

[75] S. Spedo and C. Fiegna, "Analysis of thermal noise in scaled MOS devices and RF circuits," *Solid-State Electronics*, vol. 46, pp. 1933–1939, Nov. 2002.

[76] A. J. Scholten, L. F. Tienmeijer, R. van Langevelde, R. J. Havens, A. T. A. Zegers-van Duijnhoven, and V. C. Venezia, "Noise modeling for RF CMOS circuit simulation," *IEEE Transactions on Electron Devices*, vol. 50, pp. 618–632, Mar. 2003.

[77] K. Han, H. Shin, and K. Lee, "Drain current thermal noise modeling for deep submicron n- and p-channel MOSFETs," *Solid-State Electronics*, vol. 48, pp. 2255–2262, Dec. 2004.

[78] H. Wang and R. Zeng, "Experimental verification of the effect of carrier heating on channel noise in deep submicron NMOSFETs by substrate bias," *Proceedings of the 2004 IEEE Radio Frequency Integrated Circuits Symposium*, pp. 599–602, June 2004.

[79] A. S. Roy and C. C. Enz, "Compact modeling of thermal noise in the MOS transistor," *Proceedings of the 11th International Conference Mixed Design of Integrated Circuits and Systems (MIXDES)*, Szczecin, Poland, June 2004.

[80] A. S. Roy and C. C. Enz, "Compact modeling of thermal noise in the MOS transistor," *IEEE Transactions on Electron Devices*, vol. 52, pp. 611–614, Apr. 2005.

[81] J. B. Johnson, "The Schottky effect in low frequency ciruits," *Physical Review*, vol. 26, pp. 71–85, July 1925.

[82] A. L. McWhorter, *1/f Noise and Germanium Surface Properties*, University of Philadelphia Press, pp. 207–228, 1957.

[83] S. Christensson, I. Lundstrom, and C. Svensson, "Low frequency noise in MOS transistors–I. Theory," *Solid-State Electronics*, vol. 11, pp. 792–812, Sept. 1968.

[84] S. Christensson, I. Lundstrom, and C. Svensson, "Low frequency noise in MOS transistors–II. Experiments," *Solid-State Electronics*, vol. 11, pp. 813–820, Sept. 1968.

[85] F. N. Hooge, "1/f noise is no surface effect," *Physics Letters*, vol. 29A, pp. 139–140, 1969.

[86] F. M. Klaassen, "Characterization of low 1/f noise in MOS transistors," *IEEE Transactions on Electron Devices*, vol. ED-18, pp. 887–891, Oct. 1971.

[87] H.-S. Fu and C.-T. Sah, "Theory and experiments on surface 1/f noise," *IEEE Transactions on Electron Devices*, vol. ED-19, pp. 273–285, Feb. 1972.

[88] F. N. Hooge, "1/f noise," *Physica*, vol. 83B, pp. 14–23, 1976.

[89] M. Aoki, H. Katto, and E. Yamada, "Low frequency 1/f noise in MOSFETs at low current levels," *Journal of Applied Physics*, vol. 48, pp. 5135–5140, Dec. 1977.

[90] M. Aoki, Y. Sakai, and T. Masuhara, "Low 1/f noise design of Hi-CMOS devices," *IEEE Transactions on Electron Devices*, vol. ED-29, pp. 296–299, Feb. 1982.

[91] H. Mikoshiba, "1/f noise in n-channel, silicon-gate MOS transistors," *IEEE Transactions on Electron Devices*, vol. ED-29, pp. 965–970, June 1982.

[92] G. Reimbold, "Modified 1/f trapping noise theory and experiments in MOS transistors biased from weak to strong inversion – influence of surface states," *IEEE Transactions on Electron Devices*, vol. ED-31, pp. 1190–1198, Sept. 1984.

[93] K. S. Ralls, W. J. Skoepol, L. D. Jackel, R. E. Howard, L. A. Fetter, R. W. Epworth, and D. M. Tennant, "Discrete resistance switching in submicrometer silicon inversion layers: Individual interface traps and low frequency (1/f?) noise," *Physical Review Letters*, vol. 52, pp. 228–231, Jan. 1984.

[94] C. Surya and T. Y. Hsiang, "Theory and experiment on the $1/f^\gamma$ noise in p-channel metal-oxide-semiconductor field-effect transistors at low drain bias," *Physical Review B*, vol. 33, no. 7, pp. 4898–4905, 1986.

[95] Z. Celik-Butler and T. Y. Hsiang, "Spectral dependence of $1/f^\gamma$ noise on gate bias in N-MOSFETS," *Solid-State Electronics*, vol. 30, pp. 419–423, Apr. 1987.

[96] C. Surya and T. Y. Hsiang, "Surface mobility fluctuations in metal-oxide-semiconductor field-effect transistors," *Physical Review B*, vol. 35, pp. 6343–6347, Apr. 1987.

[97] Z. Y. Chang and W. Sansen, "Test structure for evaluation of 1/f noise in CMOS technologies," *Proceedings of the 1989 International Conference on Microelectronic Test Structures*, pp. 143–146, Mar. 1989.

[98] R. Jayaraman and C. G. Sodini, "A 1/f noise technique to extract the oxide trap density near the conduction band edge of silicon," *IEEE Transactions on Electron Devices*, vol. 36, pp. 1773–1782, Sept. 1989.

[99] J. Chang, C. R. Viswanathan, and C. Anagnostopoulos, "Flicker-noise measurements in enhancement mode and depletion mode n-MOS transistors," *Proceedings of the 1989 International Symposium on VLSI Technology*, pp. 217–221, May 1989.

[100] K. K. Hung, P. K. Ko, C. Hu, and Y. C. Cheng, "A unified model for flicker noise in metal-oxide-semiconductor field-effect transistors," *IEEE Transactions on Electron Devices*, vol. 37, pp. 654–665, Mar. 1990.

[101] K. K. Hung, P. K. Ko, C. Hu, and Y. C. Cheng, "Physics-based MOSFET noise model for circuit simulators," *IEEE Transactions on Electron Devices*, vol. 37, pp. 1323–1333, May 1990.

[102] T. L. Meisenheimer and D. M. Fleetwood, "Effect of radiation-induced charge on 1/f noise in MOS devices," *IEEE Transactions on Nuclear Science*, vol. 37, pp. 1696–1702, Dec. 1990.

[103] J. H. Scofield and D. M. Fleetwood, "Physical basis for non-destructive tests of MOS radiation hardness," *IEEE Transactions on Nuclear Science*, vol. 38, pp. 1567–1577, Dec. 1991.

[104] Z. Y. Chang and W. M. C. Sansen, *Low-noise, Wide-band Amplifiers in Bipolar and CMOS Technologies*, Kluwer Academic, 1991.

[105] S. Tedja, H. H. Williams, J. Van der Spiegel, F. M. Newcomer, and R. Van Berg, "Noise spectral density measurements of a radiation hardened CMOS process in the weak and moderate inversion," *IEEE Transactions on Nuclear Science*, vol. 39, pp. 804–808, Aug. 1992.

[106] X. Li and L. K. J. Vandamme, "1/f noise in series resistance of LDD MOSTs," *Solid-State Electronics*, vol. 35, pp. 1471–1475, Oct. 1992.

[107] X. Li and L. K. J. Vandamme, "Explanation of 1/f noise in LDD MOSFETs from the ohmic region to saturation," *Solid-State Electronics*, vol. 36, pp. 1515–1521, Nov. 1993.

[108] J. C. Chang, A. A. Abidi, and C. R. Viswananthan, "Flicker noise in CMOS transistors from subthreshold to strong inversion at various temperatures," *IEEE Transactions on Electron Devices*, vol. 41, pp. 1965–1971, Nov. 1994.

[109] D. M. Fleetwood, T. L. Meisenheimer, and J. H. Scofield, "1/f noise and radiation effects in MOS devices," *IEEE Transactions on Electron Devices*, vol. 41, pp. 1953–1964, Nov. 1994.

[110] J. H. Scofield, N. Borland, and D. M. Fleetwood, "Reconciliation of different gate-voltage dependencies of 1/f noise in n-MOS and p-MOS transistors," *IEEE Transactions on Electron Devices*, vol. 41, pp. 1953–1964, Nov. 1994.

[111] F. N. Hooge, "1/f noise sources," *IEEE Transactions on Electron Devices*, vol. 41, pp. 1926–1935, Nov. 1994.

[112] L. K. J. Vandamme, X. Li, and D. Rigaud, "1/f noise in MOS devices, mobility or number fluctuations?," *IEEE Transactions on Electron Devices*, vol. 41, pp. 1936–1945, Nov. 1994.

[113] F. Faccio, M. Bianchi, M. Fornasari, E. H. M. Heijne, P. Jarron, G. Rossi, G. Borel, and J. Redolfi, "Noise characterization of transistors in a 1.2 μm CMOS SOI technology up to a total-dose of 12 Mrad (Si)," *IEEE Transactions on Nuclear Science*, vol. 41, pp. 2310–2316, Dec. 1994.

[114] X. Li, C. Barros, E. P. Vandamme, and L. K. J. Vandamme, "Parameter extraction and 1/f noise in a surface and bulk-type, p-channel LDD MOSFET," *Solid-State Electronics*, vol. 37, pp. 1853–1862, Nov. 1994.

[115] F. Faccio, P. Aspell, E. H. M. Heijne, and P. Jarron, "Analog Performance of SOI MOSFETs up to 25 Mrad (Si)," *Third European Conference on Radiation and its Effects on Components and Systems*, Arcachon, France, pp. 137–141, Sept. 1995.

[116] T. Matsushita, C. Fukunaga, H. Ikeda, and Y. Saitoh, "Total-dose effects of gamma-ray irradiation on SOI-MOS transistors," *Nuclear Instruments and Methods in Physics Research Section A*, vol. 366, pp. 366–371, Dec. 1995.

[117] J. C. Santiard and F. Faccio, "Noise and speed characteristics of test transistors and charge amplifiers designed using a submicron CMOS technology," *Nuclear Instruments and Methods in Physics Research Section A*, vol. 380, pp. 350–352, Oct. 1996.

[118] T. Boutchacha, G. Ghibaudo, G. Guegan, and M. Haond, "Low frequency noise characterization of 0.25 μm Si CMOS transistors," *Journal of Non-Crystalline Solids*, vol. 216, pp. 192–197, Aug. 1997.

[119] B. Iniguez, M. Tambani, V. Dessard, and D. Flandre, "Unified 1/f noise SOI MOSFET modeling for circuit simulation," *Electronic Letters*, vol. 33, pp. 1781–1782, Oct. 1997.

[120] T. Boutchacha, G. Ghibaudo, G. Guégan, and T. Skotnicki, "Low frequency noise characterization of 0.18 μm Si CMOS transistors," *Microelectronics Reliability*, vol. 37, pp. 1599–1602, Oct.–Nov. 1997.

[121] J. A. Babcock, D. K. Schroder, and Y.-C. Tseng, "Low-frequency noise in nearly-fully-depleted TFSOI MOSFETs," *IEEE Electron Device Letters*, vol. 19, pp. 40–43, Feb. 1998.

[122] C. Jakobson, I. Bloom, and Y. Nemirovsky, "1/f noise in CMOS transistors for analog applications from subthreshold to saturation," *Solid-State Electronics*, vol. 42, pp. 1807–1817, Oct. 1998.

[123] S. Villa, G. De Geronimo, A. Pacelli, A. L. Lacaita, and A. Longoni, "Application of 1/f noise measurements to the characterization of near-interface oxide traps in ULSI n-MOSFETs," *Microelectronics Reliability*, vol. 38, pp. 1919–1923, Dec. 1998.

[124] C. G. Jakobson and Y. Nemirovsky, "1/f noise in ion sensitive field effect transistors from subthreshold to saturation," *IEEE Transactions on Electron Devices*, vol. 46, pp. 259–261, Jan. 1999.

[125] T. Boutchacha, G. Ghilbaudo, and B. Belmekki, "Study of low frequency noise in the 0.18 μm silicon CMOS transistors," *Proceedings of the IEEE 1999 International Conference on Microelectronic Test Structures*, vol. 12, pp. 84–88, Mar. 1999.

[126] Z. Celik-Butler and P. Vasina, "Channel length scaling on 1/f noise in 0.18-μm technology MDD n-MOSFETs," *Solid-State Electronics*, vol. 43, pp. 1695–1701, Sept. 1999.

[127] R. Brederlow, W. Weber, D. Schmitt-Landsiedel, and R. Thewes, "Fluctuations of the low frequency noise of MOS transistors and their modeling in analog and RF-circuits," *Technical Digest of the 1999 IEEE International Electron Devices Meeting*, pp. 159–162, Dec. 1999.

[128] S. D'Souza, L. Hwang, M. Matloubian, S. Martin, P. Sherman, A. Joshi, H. Wu, S. Bhattacharya, and P. Kempf, "1/f noise characterization of deep sub-micron, dual thickness, nitrided gate-oxide n- and p-MOSFETs," *Technical Digest of the 1999 IEEE International Electron Devices Meeting*, pp. 839–841, Dec. 1999.

[129] E. Simoen and C. Claeys, "On the flicker noise in submicron silicon MOSFETs," *Solid-State Electronics*, vol. 43, pp. 865–882, May 1999.

[130] J. C. Vildeuil, M. Valenza, and D. Rigaud, "Extraction of the BSIM3 1/f noise parameters in CMOS transistors," *Microelectronics Journal*, vol. 30, pp. 199–205, Feb. 1999.

[131] S. Martin, G. P. Li, H. Guan, and S. D'Souza, "A BSIM3-based flat-band voltage perturbation model for RTS and 1/f noise," *IEEE Electron Device Letters*, vol. 21, pp. 30–33, Jan. 2000.

[132] E. A. M. Klumperink, S. L. J. Gierkink, A. P. van der Wel, and B. Nauta, "Reducing MOSFET 1/f noise and power consumption by switched biasing," *IEEE Journal of Solid-State Circuits*, vol. 35, pp. 994–1001, July 2000.

[133] N. Lukyanchikova, N. Garbar, M. Petrichuk, E. Simoen, and C. Claeys, "Flicker noise in deep submicron nMOS transistors," *Solid-State Electronics*, vol. 44, pp. 1239–1245, July 2000.

[134] E. P. Vandamme and L. K. J. Vandamme, "Critical discussion on unified 1/f noise models for MOSFETs," *IEEE Transactions on Electron Devices*, vol. 47, pp. 2146–2152, Nov. 2000.

[135] D. Xie, M. Cheng, and L. Forbes, "SPICE models for flicker noise in n-MOSFETs from subthreshold to strong inversion," *IEEE Transactions on Computer-Aided Design of Integrated Circuits and Systems*, vol. 19, pp. 1293–1303, Nov. 2000.

[136] H. M. Bu, Y. Shi, X. L. Yuan, Y. D. Zheng, S. H. Gu, H. Majima, H. Ishikuro, and T. Hiramoto, "Impact of the device scaling on the low-frequency noise in n-MOSFETs," *Materials Science and Processing in Applied Physics A*, vol. 71, pp. 133–136, Aug. 2000.

[137] M. J. Knitel, P. H. Woerlee, A. J. Scholten, and A. T. A. Zegers-Van Duijnhoven, "Impact of process scaling on 1/f noise in advanced CMOS technologies," *Technical Digest of the 2000 IEEE International Electron Devices Meeting*, pp. 463–466, Dec. 2000.

[138] R. Kolarova, T. Shotnicki, and J. A. Chroboczek, "Low frequency noise in thin gate oxide MOSFETs," *Microelectronics Reliability*, vol. 41, pp. 579–585, Apr. 2001.

[139] Y. Nemirovsky, I. Brouk, and C. Jakobson, "1/f noise in CMOS transistors for analog applications," *IEEE Transactions on Electron Devices*, vol. 48, pp. 921–927, May 2001.

[140] J. Zhou, M. Cheng, and L. Forbes, "SPICE models for flicker noise in p-MOSFETs in the saturation region," *IEEE Transactions on Computer-Aided Design of Integrated Circuits and Systems*, vol. 20, pp. 763–767, June 2001.

[141] P. K. Chan, L. S. Ng, L. Siek, and K. T. Lau, "Designing CMOS folded-cascode operational amplifier with flicker noise minimization," *Microelectronics Journal*, vol. 32, pp. 69–73, Jan. 2001.

[142] V. Dessard, B. Iniguez, S. Adriaensen, and D. Flandre, "SOI n-MOSFET low-frequency noise measurements and modeling from room temperature up to 250 C," *IEEE Transactions on Electron Devices*, vol. 49, pp. 1289–1295, July 2002.

[143] M. Manghisoni, L. Ratti, V. Re, and V. Speziali, "Submicron CMOS technologies for low-noise analog front-end circuits," *IEEE Transactions on Nuclear Science*, vol. 49, pp. 1783–1790, Aug. 2002.

[144] N. Lukyanchikova, M. Petrichuk, N. Garbar, E. Simoen, A. Mercha, C. Claeys, H. van Meer, and K. De Meyer, "The $1/f^{1.7}$ noise in submicron SOI MOSFETs with 2.5 nm nitrided gate oxide," *IEEE Transactions on Electron Devices*, vol. 49, pp. 2367–2370, Dec. 2002.

[145] H. D. Xiong, D. M. Fleetwood, B. K. Choi, and A. L. Sternberg, "Temperature dependence and irradiation response of 1/f-Noise in MOSFETs," *IEEE Transactions on Nuclear Science*, vol. 49, pp. 2718–2723, Dec. 2002.

[146] Y. A. Allogo, M. de Murcia, J. C. Vildeuil, M. Valenza, P. Llinares, and D. Cottin, "1/f noise measurements in n-channel MOSFETs processed in 0.25 μm technology–extraction of BSIM3v3 noise parameters," *Solid-State Electronics*, vol. 46, pp. 361–366, July 2002.

[147] G. Ghibaudo and T. Boutchacha, "Electrical noise and RTS fluctuations in advanced CMOS devices," *Microelectronics Reliability*, vol. 42, pp. 573–582, Apr.–May 2002.

[148] A. Mercha, E. Simoen, G. Richardson, and C. Claeys, "Inversion layer quantization impact on the interpretation of 1/f noise in deep submicron CMOS transistors," *Proceedings of the 32nd European Solid-State Device Research Conference (ESSDERC)*, pp. 79–82, 2002.

[149] A. Arnaud and C. Galup-Montoro, "A compact model for flicker noise in MOS transistors for analog circuit design," *IEEE Transactions on Electron Devices*, vol. 50, pp. 1815–1818, Aug. 2003.

[150] F. Dieudonné, S. Haendler, J. Jomaah, and F. Balestra, "Low frequency noise in 0.12 μm partially and fully depleted SOI technology," *Microelectronics Reliability*, vol. 43, pp. 243–248, Feb. 2003.

[151] F. Dieudonné, S. Haendler, J. Jomaah, and F. Balestra, "Shrinking from 0.25 down to 0.12 μm SOI CMOS technology node: a contribution to low-frequency noise in partially depleted N-MOSFETs," *Solid-State Electronics*, vol. 47, pp. 1213–1218, July 2003.

[152] H. Wong, "Low-frequency noise study in electron devices: review and update," *Microelectronics Reliability*, vol. 43, pp. 585–599, Apr. 2003.

[153] J. Jomaah and F. Balestra, "Low-frequency noise in advanced CMOS/SOI devices," *IEE Proceedings: Circuits, Devices and Systems*, vol. 151, pp. 111–117, Apr. 2004.

[154] M. Valenza, A. Hoffmann, D. Sodini, A. Laigle, F. Martinez, and D. Rigaud, "Overview of the impact of downscaling technology on 1/f noise in p-MOSFETs to 90 nm," *IEE Proceedings: Circuits, Devices and Systems*, vol. 151, pp. 102–110, Apr. 2004.

[155] K. W. Chew, K. S. Yeo, and S.-F. Chu, "Effect of technology scaling on the 1/f noise of deep submicron PMOS transistors," *Solid-State Electronics*, vol. 48, pp. 1101–1109, July 2004.

[156] M. Marin, Y. A. Allogo, M. de Murcia, P. Llinares, and J. C. Vildeuil, "Low frequency noise characterization in 0.13 μm p-MOSFETs. Impact of scaled-down 0.25, 0.18 and 0.13 μm technologies on 1/f noise," *Microelectronics Reliability*, vol. 44, pp. 1077–1085, July 2004.

[157] J. W. Wu, C. C. Cheng, K. L. Chiu, J. C. Guo, W. Y. Lien, C. S. Chang, G. W. Huang, and T. Wang, "Pocket implantation effect on drain current flicker noise in analog nMOSFET devices," *IEEE Transactions on Electron Devices*, vol. 51, pp. 1262–1266, Aug. 2004.

[158] A. Arnaud and C. Galup-Montoro, "Consistent noise models for analysis and design of CMOS circuits," *IEEE Transactions on Circuits and Systems I*, vol. 51, pp. 1909–1915, Oct. 2004.

[159] K. W. Chew, K. S. Yeo, and S.-F. Chu, "Impact of technology scaling on the 1/f noise of thin and thick gate oxide deep submicron NMOS transistors," *IEE Proceedings: Circuits, Devices and Systems*, vol. 151, pp. 415–421, Oct. 2004.

[160] D. M. Binkley, C. E. Hopper, J. D. Cressler, M. M. Mojarradi, and B. J. Blalock, "Noise performance of 0.35-μm SOI CMOS devices and micropower preamplifier following 63-MeV, 1-Mrad (Si) proton irradiation," *IEEE Transactions on Nuclear Science*, vol. 51, pp. 3788–3794, Dec. 2004.

[161] C. Claeys, A. Mercha, and E. Simoen, "Low-frequency noise assessment for deep submicrometer CMOS technology nodes," *Journal of the Electrochemical Society*, vol. 151, no. 5, pp. G307–G318, 2004.

[162] F. Martinez, C. Leyris, G. Neau, M. Valenza, A. Hoffmann, J. C. Vildeuil, E. Vincent, F. Boeuf, T. Skotnicki, M. Bidaud, D. Barge, and B. Tavel, "Oxide traps characterization of 45 nm MOS transistors by gate current R.T.S. noise measurements," *Microelectronic Engineering*, vol. 80, pp. 54–57, June 2005.

[163] V. Re, M. Manghisoni, L. Ratti, V. Speziali, and G. Traversi, "Survey of noise performances and scaling effects in deep submicrometer CMOS devices from different foundries," *IEEE Transactions on Nuclear Science*, vol. 52, pp. 2733–2740, Dec. 2005.

[164] T. Contaret, K. Romanjek, T. Boutchacha, G. Ghibaudo, and F. Boeuf, "Low frequency noise characterization and modelling in ultrathin oxide MOSFETs," *Solid-State Electronics*, vol. 50, pp. 63–68, Jan. 2006.

[165] V. Re, M. Manghisoni, L. Ratti, V. Speziali, and G. Traversi, "Total ionizing dose effects on the noise performances of a 0.13 μm CMOS technology," *IEEE Transactions on Nuclear Science*, vol. 53, pp. 1599–1606, June 2006.

[166] S. M. Sze, *Physics of Semiconductor Devices*, John Wiley & Sons, Inc., 1981.

[167] D. M. Binkley, J. M. Rochelle, B. K. Swann, L. G. Clonts, and R. N. Goble, "A micropower CMOS, direct-conversion, VLF receiver chip for magnetic-field wireless applications," *IEEE Journal of Solid-State Circuits*, vol. 33, pp. 344–358, Mar. 1998.

[168] R. P. Jindal, "Noise associated with distributed resistance of MOSFET gate structures in integrated circuits," *IEEE Transactions on Electron Devices*, vol. ED-31, pp. 1501–1509, Oct. 1984.

[169] B. Razavi, R.-H. Yan, and K. F. Lee, "Impact of distributed gate resistance on the performance of MOS devices," *IEEE Transactions on Circuits and Systems I*, vol. 41, pp. 750–754, Nov. 1994.

[170] E. F. Tsakas and A. N. Birbas, "Noise associated with interdigitated gate structures in submicron MOSFETs," *IEEE Transactions on Electron Devices*, vol. 47, pp. 1745–1750, Sept. 2000.

[171] R. P. Jindal, "Distributed substrate resistance noise in fine-line NMOS field-effect transistors," *IEEE Transactions on Electron Devices*, vol. ED-32, pp. 2450–2453, Nov. 1985.

[172] L. Deferm, C. Claeys, and G. J. Declerck, "Two- and three-dimensional calculation of substrate resistance," *IEEE Transactions on Electron Devices*, vol. 35, pp. 339–352, Mar. 1988.

[173] S. Tedja, J. van der Spiegel, and H. H. Williams, "Analytical and experimental studies of thermal noise in MOSFETs," *IEEE Transactions on Electron Devices*, vol. 41, pp. 2069–2075, Nov. 1994.

[174] S. V. Kishore, G. Chang, G. Asmanis, C. Hull, and F. Stubbe, "Substrate-induced high-frequency noise in deep sub-micron MOSFETs for RF applications," *Proceedings of the IEEE 1999 Custom Integrated Circuits Conference (CICC)*, pp. 365–368, May 1999.

[175] A. van der Ziel and E. R. Chenette, "Noise in solid-state devices," *Advanced Electronic Physics*, vol. 46, pp. 313–383, 1978.

[176] J. P. Jindal, "Noise associated with substrate current in fine-line nMOS field-effect transistors," *IEEE Advanced Electronic Physics on Electron Devices*, vol. ED-32, pp. 1047–1052, June 1985.

[177] G. B. Alers, K. S. Krish, D. Monroe, B. E. Weir, and A. M. Chang, "Tunneling current noise in thin gate oxides," *Applied Physics Letters*, vol. 69, pp. 2885–2887, Nov. 1996.

[178] H. S. Momose *et al.*, "A study of flicker noise in n- and p-MOSFETs with ultrathin gate oxide in the direct-tunneling regime," *Technical Digest of the 1998 IEEE International Electron Devices Meeting*, pp. 923–926, Dec. 1998.

[179] J. Lee and G. Bosman, "Comprehensive noise performance of ultrathin oxide MOSFETs at low frequencies," *Solid-State Electronics*, vol. 48, pp. 61–71, Jan. 2004.

[180] A.-J. Annema, B. Nauta, R. van Langevelde, and H. Tuinhout, "Analog circuits in ultra-deep-submicron CMOS," *IEEE Journal of Solid-State Circuits*, vol. 40, pp. 132–143, Jan. 2005.

[181] J. B. Shyu, G. C. Temes, and K. Yao, "Random errors in MOS capacitors," *IEEE Journal of Solid-State Circuits*, vol. SC-17, pp. 1070–1076, Dec. 1982.

[182] J. B. Shyu, G. C. Temes, and F. Krummenacher, "Random error effects in matched MOS capacitors and current sources," *IEEE Journal of Solid-State Circuits*, vol. SC-19, pp. 948–955, Dec. 1984.

[183] K. R. Lakshmikumar, R. A. Hadaway, and M. A. Copeland, "Characterization and modeling of mismatch in MOS transistors for precision analog design," *IEEE Journal of Solid-State Circuits*, vol. SC-21, pp. 1057–1066, Dec. 1986.

[184] T.-W. Pan and A. A. Abidi, "A 50-dB variable gain amplifier using parasitic bipolar transistors in CMOS," *IEEE Journal of Solid-State Circuits*, vol. 24, pp. 951–961, Aug. 1989.

[185] M. Pelgrom, A. Duinmaijer, and A Welbers, "Matching properties of MOS transistors," *IEEE Journal of Solid-State Circuits*, vol. 24, pp. 1433–1440, Oct. 1989.

[186] C. Michael and M. Ismail, "Statistical modeling of device mismatch for analog MOS integrated circuits," *IEEE Journal of Solid-State Circuits*, vol. 27, pp. 154–166, Feb. 1992.

[187] Z. Wong, "Automatic V_T extractors based on an $n \times n^2$ MOS transistor array and their application," *IEEE Journal of Solid-State Circuits*, vol. 27, pp. 1277–1285, Sept. 1992.

[188] F. Forti and M. E. Wright, "Measurement of MOS current mismatch in the weak inversion region," *IEEE Journal of Solid-State Circuits*, vol. 29, pp. 138–142, Feb. 1994.

[189] M. Steyaert, J. Bastos, R. Roovers, P. Kinget, W. Sansen, B. Graindourze, A. Pergoot, and E. Janssens, "Threshold voltage mismatch in short-channel MOS transistors," *Electronics Letters*, vol. 30, pp. 1546–1548, Sept. 1994.

[190] T. Mizuno, J. Okamura, and A. Toriumi, "Experimental study of threshold voltage fluctuation due to statistical variation of channel dopant number in MOSFETs," *IEEE Transactions on Electron Devices*, vol. 41, pp. 2216–2221, Nov. 1994.

[191] J. Bastos, M. Steyaert, R. Roovers, P. Kinget, W. Sansen, B. Graindourze, A. Pergoot, and E. Janssens, "Mismatch characterization of small size MOS transistors," *Proceedings of the 1995 IEEE International Conference on Microelectronic Test Structures*, vol. 8, pp. 271–276, Mar. 1995.

[192] M.-J. Chen, J.-S. Ho, and T.-H. Huang, "Dependence on current match on back-gate bias in weakly inverted MOS transistors and its modeling," *IEEE Journal of Solid-State Circuits*, vol. 31, pp. 259–262, Feb. 1996.

[193] J. Bastos, M. Steyaert, B. Graindourze, and W. Sansen, "Matching of MOS transistors with different layout styles," *Proceedings of the 1996 IEEE International Conference on Microelectronic Test Structures*, vol. 9, pp. 17–18, Mar. 1996.

[194] H. Elzinga, "On the impact of spatial parametric variations on MOS transistor mismatch," *Proceedings of the 1996 IEEE International Conference on Microelectronic Test Structures*, vol. 9, pp. 173–177, Mar. 1996.

[195] S. J. Lovett, R. Clancy, M. Welten, A. Mathewson, and B. Mason, "Characterizing the mismatch of submicron MOS transistors," *Proceedings of the 1996 IEEE International Conference on Microelectronic Test Structures*, vol. 9, pp. 39–42, Mar. 1996.

[196] M.-J. Chen, J.-S. Ho, and D.-Y. Chang, "Optimizing the match in weakly inverted MOSFETs by gated lateral bipolar action," *IEEE Transactions on Electron Devices*, vol. 43, pp. 766–773, May 1996.

[197] O. Buisson and G. Morin, "MOSFET matching in deep submicron technology," *Proceedings of the 26th European Solid State Device Research Conference (ESSDERC)*, pp. 731–734, 1996.

[198] H. Tuinhout, M. Pelgrom, R. Penning de Vries, and M. Vertregt, "Effects of metal coverage on MOSFET matching," *Technical Digest of the 1996 IEEE International Electron Devices Meeting*, pp. 735–738, Dec. 1996.

[199] J. Bostos, M. Steyaert, A. Pergoot, and W. Sansen, "Mismatch characterization of submicron MOS transistors," *Analog Integrated Circuits and Signal Processing*, vol. 12, pp. 95–106, Feb. 1997.

[200] J. Bostos, M. Steyaert, A. Pergoot, and W. Sansen, "Influence of die attachment on MOS transistor matching," *IEEE Transactions on Semiconductor Manufacturing*, vol. 10, pp. 209–218, May 1997.

[201] S.-C. Wong, K.-H. Pan, and D. J. Ma, "A CMOS mismatch model and scaling effects," *IEEE Electron Device Letters*, vol. 18, pp. 261–263, June 1997.

[202] B. M. Tenbroek, M. S. L. Lee, W. Redman-White, C. F. Edwards, M. J. Uren, and R. J. T. Bunyan, "Drain current mismatch in SOI CMOS current mirrors and D/A converters due to localised internal and coupled heating," *Proceedings of the 1997 European Solid-State Circuits Conference (ESSCIRC)*, pp. 276–278, 1997.

[203] H. Tuinhout, A. Montree, J. Schmitz, and P. Stolk, "Effects of gate depletion and boron penetration on matching of deep submicron CMOS transistors," *Technical Digest of the 1997 IEEE International Electron Devices Meeting*, pp. 631–634, Dec. 1997.

[204] J. T. Horstmann, U. Hilleringmann, and K. F. Goser, "Matching analysis of deposition defined 50-nm MOSFETs," *IEEE Transactions on Electron Devices*, vol. 45, pp. 299–306, Jan. 1998.

[205] S. J. Lovett, M. Welten, A. Mathewson, and B. Mason, "Optimizing MOS transistor mismatch," *IEEE Journal of Solid-State Circuits*, vol. 33, pp. 147–150, Jan. 1998.

[206] P. Stolk, F. Widdershoven, and D. Klaassen, "Modeling statistical dopant fluctuations in MOS transistors," *IEEE Transactions on Electron Devices*, vol. 45, pp. 1960–1971, Sept. 1998.

[207] C. G. Linnenbank *et al.*, "What do matching results of medium area MOSFETs reveal for large area devices in typical analog applications?" *Proceedings of the 1998 European Solid-State Device Research Conference (ESSDERC)*, pp. 104–107, Sept. 1998.

[208] M. Pelgrom, H. Tuinhout, and M. Vertregt, "Transistor matching in analog CMOS applications," *Technical Digest of the 1998 IEEE International Electron Devices Meeting*, pp. 915–918, Dec. 1998.

[209] J. Bastos, "Characterization of MOS transistor mismatch for analog design," PhD thesis, Katholieke Universiteit, Leuven, 1998.

[210] A. Van den Bosch, M. Steyaert, and W. Sansen, "A high density matched hexagonal transistor structure in standard CMOS technology for high speed applications," *Proceedings of the IEEE 1999 International Conference on Microelectronic Test Structures*, pp. 212–215, Mar. 1999.

[211] H. Thibieroz and A. Duvallet, "Mismatch characterization and modelization in deep submicron CMOS transistors," *Proceedings of the SPIE Conference on Microelectronic Technology III*, vol. 3881, pp. 121–128, Sept. 1999.

[212] T. Serrano-Gotarredona and B. Linares-Barranco, "A new five-parameter MOS transistor mismatch model," *IEEE Electron Device Letters*, vol. 21, pp. 37–39, Jan. 2000.

[213] U. Schaper, C. Linnenbank, U. Kollmer, H. Mulatz, T. Mensing, R. Schmidt, R. Tilgner, and R. Thewes, "Evaluation of the impact on mechanical stress on CMOS device mismatch," *Proceedings of the 2001 IEEE International Conference on Microelectronic Test Structures*, vol. 14, pp. 311–317, Mar. 2001.

[214] T.-H. Yeh, J. Lin, S.-C. Wong, H. Huang, and J. Sun, "Mismatch characterization of 1.8 V and 3.3 V devices in 0.18 μm mixed-signal CMOS technology," *Proceedings of the 2001 IEEE International Conference on Microelectronic Test Structures*, vol. 14, pp. 77–82, Mar. 2001.

[215] H. P. Tuinhout, J. H. Klootwijk, W. C. Goeke, and L. K. Stauffer, "Impact of transistor noise on high precision parametric matching measurements," *Proceedings of the 2001 IEEE International Conference on Microelectronic Test Structures*, vol. 14, pp. 201–206, Mar. 2001.

[216] Q. Zhang, J. J. Liou, J. R. McMacken, J. Thomson, and P. Layman, "SPICE modeling and quick estimation of MOSFET mismatch based on BSIM3 model and parametric tests," *IEEE Journal of Solid-State Circuits*, vol. 36, pp. 1592–1595, Oct. 2001.

[217] U. Schaper, C. Linnenbank, and R. Thewes, "Precise characterization of long-distance mismatch of CMOS devices," *IEEE Transactions on Semiconductor Manufacturing*, vol. 14, pp. 311–317, Nov. 2001.

[218] C. Paulus, R. Brederlow, U. Kleine, and R. Thewes, "An efficient and precise design method to optimize device areas in mismatch and flicker noise sensitive analog circuits," *Proceedings of the 2001 IEEE International Conference on Electronics, Circuits, and Systems (ICECS)*, vol. 1, pp. 107–111, Sept. 2001.

[219] J. A. Croon, H. P. Tuinhout, R. Difrenza, J. Knol, A. J. Moonen, S. Decoutere, H. E. Maes, and W. Sansen, "A comparison of extraction techniques for threshold voltage mismatch," *Proceedings of the 2002 IEEE International Conference on Microelectronic Test Structures*, vol. 15, pp. 235–240, April 2002.

[220] R. Difrenza, P. Llinares, S. Taupin, R. Palla, C. Garnier, and G. Ghibaudo, "Comparison between matching parameters and fluctuations at the wafer level," *Proceedings of the 2002 IEEE International Conference on Microelectronic Test Structures*, vol. 15, pp. 241–246, Apr. 2002.

[221] S. B. Yeo, J. Bordelon, S. Chu, M. F. Li, B. A. Tranchina, M. Harward, L. H. Chan, and A. See, "A robust and production worthy addressable array architecture for deep sub-micron MOSFETs matching characterization," *Proceedings of the 2002 IEEE International Conference on Microelectronic Test Structures*, vol. 15, pp. 229–234, Apr. 2002.

[222] K. Uyttenhove, and M. Steyaert, "Speed-power-accuracy tradeoff in high-speed CMOS ADCs," *IEEE Transactions on Circuits and Systems II*, vol. 49, pp. 280–287, Apr. 2002.

[223] P. Martin, M. Bucher, and C. Enz, "MOSFET modeling and parameter extraction for low temperature analog circuit design," *5th European Workshop on Low Temperature Electronics (WOLTE-5)*, June 2002, in *Journal de Physique IV: JP*, vol. 12, no. 3, pp. Pr3/51–Pr3/56, May 2002.

[224] J. A. Croon, M. Rosmeulen, S. Decoutere, W. Sansen, and H. E. Maes, "An easy-to-use mismatch model for the MOS transistor," *IEEE Journal of Solid-State Circuits*, vol. 37, pp. 1056–1063, Aug. Dec. 2002.

[225] J. A. Croon, G. Storms, S. Winkelmeier, I. Pollentier, M. Ercken, S. Decoutere, W. Sansen, and H. E. Maes, "Line edge roughness: characterization, modeling and impact on device behavior," *Technical Digest of the 2002 IEEE International Electron Devices Meeting*, pp. 307–310, Dec. 2002.

[226] P. G. Drennan and C. C. McAndrew, "Understanding MOSFET mismatch for analog design," *IEEE Journal of Solid-State Circuits*, vol. 38, pp. 450–456, Mar. 2003.

[227] R. Difrenza, J. C. Vildeuil, P. Llinares, and G. Ghibaudo, "Impact of grain number fluctuations in the MOS transistor gate on matching performance," *Proceedings of the 2003 IEEE International Conference on Microelectronic Test Structures*, vol. 16, pp. 244–249, Mar. 2003.

[228] M. Quarantelli, S. Saxena, N. Dragone, J. A. Babcock, C. Hess, S. Minehane, S. Winters, J. Chen, H. Karbasi, and C. Guardiani, "Characterization of modeling of MOSFET mismatch of a deep submicron technology," *Proceedings of the 2003 IEEE International Conference on Microelectronic Test Structures*, vol. 16, pp. 238–243, Mar. 2003.

[229] R. Difrenza, P. Llinares, and G. Ghibaudo, "The impact of short channel and quantum effects on the MOS transistor mismatch," *Solid-State Electronics*, vol. 47, pp. 1161–1165, July 2003.

[230] R. Difrenza, P. Llinares, and G. Ghibaudo, "A new model for the current factor mismatch in the MOS transistor," *Solid-State Electronics*, vol. 47, pp. 1167–1171, July 2003.

[231] H. Yang, V. Macary, J. L. Huber, W.-G. Min, B. Baird, and J. Zuo, "Current mismatch due to local dopant fluctuations in MOSFET channel," *IEEE Transactions on Electron Devices*, vol. 50, pp. 2248–2254, Nov. 2003.

[232] J. P. de Gyvez and H. P Tuinhout, "Threshold voltage matching and intra-die leakage current in digital CMOS circuits," *IEEE Journal of Solid-State Circuits*, vol. 39, pp. 157–168, Jan. 2004.

[233] K. Rochereau, R. Difrenza, J. McGinley, O. Noblanc, C. Julien, S. Parihar, and P. Llinares, "Impact of pocket implant on MOSFET mismatch for advanced CMOS technology," *Proceedings of the 2004 International Conference on Microelectronic Test Structures*, pp. 123–126, Mar. 2004.

[234] J. A. Croon, S. Decoutere, W. Sansen, and H. E. Maes, "Physical modeling and prediction of the matching properties of MOSFETs," *Proceedings of the 34th European Solid-State Device Research Conference (ESSDERC)*, pp. 193–196, Sept. 2004.

[235] J. Fukumoto, T. Das, K. Paradis, J. Burleson, J. E. Moon, and P. R. Mukund, "Impact of asymmetric metal coverage on high performance MOSFET mismatch," *Solid-State Electronics*, vol. 48, pp. 1767–1770, Oct.–Nov. 2004.

[236] C. Galup-Montoro, M. C. Schneider, H. Klimach, and A. Arnaud, "A compact model of MOSFET mismatch for circuit design," *IEEE Journal of Solid-State Circuits*, vol. 40, pp. 1649–1657, Aug. 2005.

[237] A. R. Hastings, *The Art of Analog Layout*, Prentice Hall, 2001.

[238] P. Kinget, "Device mismatch and tradeoffs in the design of analog circuits," *IEEE Journal of Solid-State Circuits*, vol. 40, pp. 1212–1224, June 2005.

[239] P. Kinget and M. Steyaert, "Impact of transistor mismatch on the speed-accuracy-power trade-off of analog CMOS circuits," *Proceedings of the IEEE 1996 Custom Integrated Circuits Conference (CICC),* pp. 333–336, May 1996.

[240] M. Steyaert, V. Peluso, J. Bastos, P. Kinget, and W. Sansen, "Custom analog low power design: the problem of low voltage and mismatch," *Proceedings of the IEEE 1997 Custom Integrated Circuits Conference (CICC),* pp. 285–292, May 1997.

[241] M. P. Flynn, S. Park, and C. C. Lee, "Achieving analog accuracy in nanometer CMOS," *International Journal of High Speed Electronics and Systems,* vol. 15, pp. 255–275, June 2005.

[242] R. Thewes, C. Linnenbank, U. Kollmer, S. Burges, U. Schaper, R. Brederlow, and W. Weber, "Mismatch of MOSFET small signal parameters under analog operation," *IEEE Electron Device Letters,* vol. 21, pp. 552–553, Dec. 2000.

[243] E. Bruun, "Analytical expressions for harmonic distortion at low frequencies due to device mismatch in CMOS current mirrors," *IEEE Transactions on Circuits and Systems II,* vol. 46, pp. 937–941, July 1999.

[244] S. Laursen, "High frequency MOS transistor matching measurements for the determination of mixer port crosstalk," *Proceedings of the 2001 IEEE International Conference on Microelectronic Test Structures,* vol. 14, pp. 11–14, Mar. 2001.

[245] H. S. Momose, M. Ono, T. Yoshitomi, T. Ohguro, S. Nakamura, M. Saito, and H. Iwai, "1.5 nm direct-tunneling gate oxide Si MOSFETs," *IEEE Transactions on Electron Devices,* vol. 43, pp. 1233–1242, Aug. 1996.

[246] R. van Langevelde, A. J. Scholten, R. Duffy, F. N. Cubaynes, M. J. Knitel, and D. B. M. Klaasseen, "Gate current: modeling, ΔL extraction, and impact on RF performance," *Technical Digest of the 2001 IEEE International Electron Devices Meeting,* pp. 289–292, Dec. 2001.

[247] K. Roy, S. Mukhopadhyay, and H. Mahmoodi-Meimand, "Leakage current mechanisms and leakage reduction techniques in deep-submicrometer CMOS circuits," *Proceedings of the IEEE,* vol. 91, pp. 305–327, Feb. 2003.

[248] Y.-C. Yeo, T.-J. King, and C. Hu, "MOSFET gate leakage modeling and selection guide for alternative gate dielectrics based on leakage considerations," *IEEE Transactions on Electron Devices,* vol. 50, pp. 1027–1035, Apr. 2003.

[249] D. Lee, D. Blaauw, and D. Sylvester, "Gate oxide leakage current analysis and reduction for VLSI circuits," *IEEE Transactions on Very Large Scale Integration Systems,* vol. 12, pp. 155–166, Feb. 2004.

[250] B. Linares-Barranco and T. Serrano-Gotarredona, "On the design and characterization of femtoampere current-mode circuits," *IEEE Journal of Solid-State Circuits,* vol. 38, pp. 1353–1363, Aug. 2003.

[251] D. J. Frank, R. H. Dennard, E. Nowak, P. M. Solomon, Y. Taur, and H.-S. P. Wong, "Device scaling limits of Si MOSFETs and their application dependencies," *Proceedings of the IEEE,* vol. 89, pp. 259–288, Mar. 2001.

[252] M. O'Halloran and R. Sarpeshkar, "A 10-nW 12-bit accurate analog storage cell with 10-aA leakage," *IEEE Journal of Solid-State Circuits,* vol. 39, pp. 1985–1996, Nov. 2004.

4

Tradeoffs in MOS Performance, and Design of Differential Pairs and Current Mirrors

4.1 INTRODUCTION

Chapter 3 described the advantages of using the inversion coefficient, IC, [1, p. 62] as one of three, independent degrees of analog MOS design freedom: drain current, inversion coefficient, and channel length. Channel width is not used as a degree of design freedom here because it does not directly reflect the region or level of MOS inversion. Channel width, required for layout, is easily calculated from the selected drain current, inversion coefficient, and channel length as summarized earlier in Table 3.10 and is implicitly considered in estimations of MOS performance. Selecting the inversion coefficient instead of channel width permits a conscious choice of operation in weak, moderate, or strong inversion, which controls many aspects of MOS performance. Inversion coefficients below 0.1 correspond to weak inversion, values between 0.1 and 10 correspond to moderate inversion where unity corresponds to the center of moderate inversion, and values greater than 10 correspond to strong inversion. The inversion coefficient also permits technology-independent design since the substrate factor, mobility, and gate-oxide capacitance are built into the inversion coefficient definition. This enables design expressions that are largely independent of the process, excluding small-geometry effects like velocity saturation and vertical field mobility reduction (VFMR) that depend on the critical, velocity saturation electric field, E_{CRIT}, and the VFMR coefficient, θ, respectively.

Chapter 3 then presented MOS performance in detail as a function of the drain current, inversion coefficient, and channel length. Performance included width, W, gate area, WL, silicon cost, effective gate–source voltage, $V_{EFF} = V_{GS} - V_T$, drain–source saturation voltage, $V_{DS,sat}$, transconductance efficiency, g_m/I_D, substrate factor, n, body-effect transconductance ratio, $\eta = g_{mb}/g_m$, and drain–source conductance, g_{ds}, expressed in normalized form as an Early voltage, $V_A = I_D \cdot r_{ds} = I_D/g_{ds}$. Additionally, performance included intrinsic voltage gain, $A_{Vi} = g_m \cdot r_{ds} = (g_m/I_D) \cdot V_A$, intrinsic and extrinsic capacitances, intrinsic bandwidth, $f_{Ti} = g_m/[2\pi(C_{gsi} + C_{gbi})]$, extrinsic bandwidth, thermal noise, flicker noise, local-area and distance mismatch, and gate and drain leakage current. For design guidance, the chapter included approximate, hand modeling expressions in summary tables and also included figures to illustrate performance trends. In addition to being continuously valid in weak, moderate, and strong inversion, the expressions included small-geometry effects like velocity saturation, VFMR, drain-induced barrier lowering (DIBL), and flicker-noise increases with increasing inversion or

Tradeoffs and Optimization in Analog CMOS Design   David M. Binkley
© 2008 John Wiley & Sons, Ltd

V_{EFF}. To validate the expressions, overlays of measured data were often included for a typical $0.18\,\mu m$ CMOS process used for design examples in this book. Predicting MOS performance over the design choices of drain current, inversion coefficient, and channel length provides design intuition and leads towards optimum design, while minimizing time-consuming, trial-and-error computer simulations.

This chapter begins by reviewing MOS performance trends for the design choices of inversion coefficient, channel length, and drain current. Then, while Chapter 3 described predictions and trends for single aspects of MOS performance, this chapter presents graphical tradeoffs across multiple aspects of performance. Tradeoffs involving geometry, W and WL, intrinsic gate capacitance and drain–body capacitance, $C_{gsi} + C_{gbi}$ and C_{DB}, bias compliance voltages, V_{EFF} and $V_{DS,sat}$, small-signal parameters, g_m/I_D and V_A, and intrinsic voltage gain and bandwidth, A_{Vi} and f_{Ti}, are described first as these are common to all applications. Tradeoffs specific to differential pairs or single, gate input, common source devices are then described, including transconductance distortion, gate-referred thermal- and flicker-noise voltage, and gate–source voltage mismatch. Additionally, tradeoffs specific to current mirrors are described, including drain-referred thermal- and flicker-noise current, and drain current mismatch. Finally, tradeoffs in figures of merit are described that involve the combination of multiple aspects of performance.

The chapter then concludes by illustrating numerical MOS performance tradeoffs leading to the optimum design of differential pairs and current mirrors. Optimization is facilitated by the *Analog CMOS Design, Tradeoffs and Optimization* spreadsheet, which details performance for selected values of drain current, inversion coefficient, and channel length. The performance of differential pairs and current mirrors is explored over these design choices. This is followed by design examples where differential pairs and current mirrors are optimized for DC, balanced, and AC performance at millipower and micropower operation, corresponding to drain currents of 100 and $1\,\mu A$, respectively. The DC optimization provides high transconductance, high drain–source resistance, high intrinsic voltage gain, low gate-referred thermal- and flicker-noise voltage, low drain-referred flicker-noise current, and low gate–source voltage and drain current mismatch. The AC optimization provides high intrinsic bandwidth, small capacitances, low drain-referred thermal-noise current, and low transconductance distortion at the expense of the DC optimized performance. The balanced optimization provides a compromise of DC and AC performance. Finally, performance for the DC, balanced, and AC optimizations is compared for millipower and micropower operation at equal gate areas.

The *Analog CMOS Design, Tradeoffs and Optimization* spreadsheet is extended later in Chapters 5 and 6 where MOS device performance is mapped into the performance of complete operational transconductance amplifier (OTA) and micropower, low-noise preamplifier circuits. The spreadsheet is described further in the Appendix and is available on the web site for this book listed on the cover.

4.2 PERFORMANCE TRENDS

This section reviews the importance of separately exploring drain current, inversion coefficient, and channel length to assess the performance trends and tradeoffs associated with each of these independent design choices. Then, trends in MOS performance for separate increases in the inversion coefficient, channel length, and drain current are reviewed from Chapter 3. This is followed by a discussion of MOS performance tradeoffs in Section 4.3 and a discussion of the optimum design of differential-pair and current-mirror devices in Section 4.4.

4.2.1 Exploring Drain Current, Inversion Coefficient, and Channel Length Separately

Throughout this book, as described earlier in Section 3.2.1, we separately explore drain current, inversion coefficient, and channel length to evaluate trends and tradeoffs in analog MOS performance leading towards optimum circuit design. Exploring drain current, inversion coefficient, and channel length separately allows an understanding of performance trends associated with each of these independent degrees of design freedom, but requires the resizing of channel width. For example, increasing drain current, while maintaining a fixed inversion coefficient and channel length, requires scaling channel width directly with the drain current. Increasing the inversion coefficient, while maintaining a fixed drain current and channel length, requires scaling the channel width inversely with the inversion coefficient. Finally, increasing the channel length, while maintaining a fixed drain current and inversion coefficient, requires scaling channel width directly with the channel length. These sizing relationships were summarized earlier in Table 3.10.

Normally, we select drain current first and then select the inversion coefficient and channel length. The selected inversion coefficient and channel length then define a point on the *MOSFET Operating Plane* [2–9], which was shown in Figure 3.1 and described in Section 3.2.7. This plane illustrates tradeoffs in performance for the selected inversion coefficient, the *x* axis, and channel length, the *y* axis. Since devices in DC bias and control paths are often optimized differently than devices in the signal path, devices within a given circuit will often be located at different points on the *MOSFET Operating Plane*, corresponding to different inversion coefficient and channel length selections.

4.2.2 Trends as Inversion Coefficient Increases

Before presenting MOS performance trends with increasing inversion coefficient, we first review the definition of the inversion coefficient. As used throughout this book and discussed earlier in Section 3.4.2.2, we define the inversion coefficient, *IC*, as a fixed–normalized inversion coefficient where the substrate factor, $n = n_0$, is held fixed at its average value in moderate inversion and mobility, $\mu = \mu_0$, is held fixed at its low-field value. While *n* decreases slightly with the inversion level as shown earlier in Figure 3.30, at high levels of inversion, μ decreases modestly due to VFMR and decreases significantly for short channel lengths due to velocity saturation. However, as described earlier in Section 3.7.2.2 for V_{EFF} and Section 3.8.2.2 for g_m/I_D, these effects are considered *separately* and *outside* the fixed–normalized inversion coefficient definition used here.

The fixed–normalized definition of *IC* facilitates design because *IC*, drain current, I_D, shape factor, $S = W/L$, channel width, *W*, and channel length, *L*, are linearly linked as $IC = I_D/[I_0(W/L)]$, where $I_0 = 2n_0k_0U_T^2 = 2n_0\mu_0C_{OX}U_T^2$ is a fixed technology current for the device type, process, and temperature of interest. As a result of the linear linkage, *IC* increases by a factor of 10 for a factor-of-10 increase in I_D for a fixed-geometry device. Correspondingly, for a fixed selection of I_D typically considered here, increasing the value of *IC* by a factor of 10 requires a factor-of-10 decrease in *S*, or a factor-of-10 decrease in *W* for a fixed selection of *L*. As mentioned, MOS sizing relationships for the design choices of drain current, inversion coefficient, and channel length are summarized in Table 3.10.

As described earlier in Section 3.4.2.3, in weak and moderate inversion, the fixed–normalized *IC* is nearly equal to the traditional *IC*, taken as *IC'* in this book, which considers reductions in *n* and μ. When VFMR and velocity saturation effects are significant, the traditional *IC* increases beyond the fixed–normalized *IC* because of the reduction of μ. However, as mentioned, these effects are considered outside the fixed–normalized *IC* definition used here. Table 3.6 summarizes the fixed–normalized *IC* and I_0 definitions used here. Table 3.2 lists process values of $I_0 = 0.64$ and $0.135\,\mu A$ for nMOS and pMOS devices, respectively, at $T = 300\,K$ (room temperature) for the example $0.18\,\mu m$ CMOS process considered in this chapter.

If MOS inversion coefficient is increased, while holding drain current and channel length fixed, the following occur [7, 9]:

- **Shape factor, channel width, gate area, and gate-oxide capacitance decrease** inversely with increasing inversion coefficient, tracking the decrease in shape factor required for increasing inversion coefficient. Sections 3.6.1, 3.6.2, and 3.6.3 described shape factor, channel width, and gate area, respectively. Section 3.9.1 described gate-oxide capacitance.

- **Layout area and silicon cost decrease** inversely with increasing inversion coefficient, tracking the decrease in gate area. Section 3.6.3 described layout area and silicon cost.

- **Effective gate–source voltage, $V_{EFF} = V_{GS} - V_T$, increases** as inversion coefficient increases. The increase is greater for short-channel devices at high levels of inversion due to velocity saturation effects. Some additional increase also occurs for long-channel devices at high levels of inversion due to VFMR effects. Sections 3.7.2.3 and 3.7.2.4 described V_{EFF}.

- **Drain–source saturation voltage, $V_{DS,sat}$, increases** as inversion coefficient increases. While there is a significant additional increase in V_{EFF} for short-channel devices at high levels of inversion due to velocity saturation effects, $V_{DS,sat}$ increases little due to these effects when evaluated in terms of the inversion coefficient. Sections 3.7.3.4 and 3.7.3.5 described $V_{DS,sat}$.

- **Transconductance efficiency, g_m/I_D, and transconductance, g_m, decrease** as inversion coefficient increases. The decrease is greater for short-channel devices at high levels of inversion due to velocity saturation effects. Some additional decrease also occurs for long-channel devices at high levels of inversion due to VFMR effects. Sections 3.8.2.3 and 3.8.2.4 described g_m/I_D and g_m. The decrease in g_m/I_D with increasing inversion coefficient results in increasing bandwidth-accuracy and bandwidth–power–accuracy figures of merit for current-mode circuits and decreasing figures of merit for voltage-mode circuits. Sections 4.3.6.5 and 4.3.6.6 discuss these figures of merit for DC offset and thermal noise.

- **Transconductance distortion decreases** as inversion coefficient increases since g_m/I_D decreases, increasing the gate–source voltage required to modulate a given fraction of drain bias current. Distortion decreases further for short-channel devices at high levels of inversion due to reduced g_m/I_D and increased linearity resulting from velocity saturation. The input, 1 dB compression voltage, $V_{INDIF1dB}$, for a differential pair increases as inversion coefficient increases, corresponding to decreased g_m/I_D and decreased transconductance distortion as described in Section 3.8.2.6.

- **Normalized drain–source resistance, $V_A = I_D \cdot r_{ds} = I_D/g_{ds}$, and drain–source resistance, r_{ds}, increase slightly** as inversion coefficient increases. When V_A and r_{ds} are reduced for short-channel devices because of drain-induced barrier lowering (DIBL), these increase more rapidly as inversion coefficient increases. This is because decreasing transconductance reduces the DIBL effect on V_A and r_{ds}. Sections 3.8.4.5 and 3.8.4.6 described V_A and r_{ds}, including their slight dependency on inversion level, strong dependency on channel length, and strong dependency on V_{DS}.

- **Intrinsic voltage gain, $A_{Vi} = g_m \cdot r_{ds} = (g_m/I_D) \cdot V_A$, decreases** as inversion coefficient increases since g_m/I_D and transconductance decrease while V_A and r_{ds} increase slightly. The decrease in A_{Vi} is greater for short-channel devices at high levels of inversion due to velocity saturation reduction of transconductance. Section 3.8.5 described A_{Vi}, including its dependency on the inversion level, channel length, and V_{DS}.

- **Intrinsic gate–source and gate–body capacitances, C_{gsi} and C_{gbi}, decrease** as inversion coefficient increases. In strong inversion, C_{gbi} becomes a nearly negligible portion of the gate-oxide capacitance, which decreases inversely with increasing inversion coefficient, while C_{gsi} approaches two-thirds of the gate-oxide capacitance. Section 3.9.2 described C_{gsi} and C_{gbi}.

- **Extrinsic gate–source and gate–drain overlap capacitances, C_{GSO} and C_{GDO}, decrease** inversely with increasing inversion coefficient, tracking the decrease in channel width. Section 3.9.3 described C_{GSO} and C_{GDO}.

- **Extrinsic drain–body and source–body junction capacitances, C_{DB} and C_{SB}, decrease** inversely with increasing inversion coefficient, tracking the decrease in channel width. Section 3.9.4 described C_{DB} and C_{SB}.

- **Intrinsic bandwidth,** $f_{Ti} = g_m/[2\pi(C_{gsi} + C_{gbi})]$, **increases** as inversion coefficient increases since intrinsic gate capacitances decrease more rapidly, nearly tracking the gate-oxide capacitance, than the decrease in transconductance. f_{Ti} levels off for short-channel devices at high levels of inversion due to velocity saturation reduction of transconductance. Section 3.9.6 described f_{Ti}. Operating extrinsic, f_T, and diode-connected, f_{diode}, bandwidths also increase as inversion coefficient increases due to reduced intrinsic and extrinsic capacitances associated with reduced gate area and channel width. Section 3.9.7 described f_T and f_{diode}.

- **Gate-referred thermal-noise voltage PSD,** S_{VG} **thermal, increases** as inversion coefficient increases since transconductance decreases. The increase is greater for short-channel devices at high levels of inversion due to velocity saturation reduction of transconductance. Small-geometry effects can increase the noise further. Section 3.10.2.3 described S_{VG} thermal.

- **Drain-referred thermal-noise current PSD,** S_{ID} **thermal, decreases** as inversion coefficient increases since transconductance decreases. The decrease is greater for short-channel devices at high levels of inversion due to velocity saturation reduction of transconductance, although small-geometry effects can increase the noise. Section 3.10.2.3 described S_{ID} thermal.

- **Gate-referred flicker-noise voltage PSD,** S_{VG} **flicker, increases** as inversion coefficient increases since gate area decreases. The noise may increase further due to increasing inversion or V_{EFF}, especially for pMOS devices. Section 3.10.3.7 described S_{VG} flicker.

- **Drain-referred flicker-noise current PSD,** S_{ID} **flicker, increases** as inversion coefficient increases in weak inversion, since gate area decreases causing S_{VG} flicker to increase while transconductance remains nearly constant. However, in strong inversion, S_{ID} flicker levels off as the increase in S_{VG} flicker is countered by the decrease in transconductance. S_{ID} flicker can then decrease for short-channel devices at high levels of inversion due to velocity saturation reduction of transconductance. The noise, however, may increase due to increasing inversion or V_{EFF}, especially for pMOS devices. Section 3.10.3.7 described S_{ID} flicker.

- **Flicker-noise corner frequency,** f_c, **increases** as inversion coefficient increases, tracking a small fraction of f_{Ti}. Section 3.10.3.8 described f_c.

- **Local-area, gate–source voltage mismatch,** ΔV_{GS}, **increases** as inversion coefficient increases since gate area decreases causing threshold-voltage mismatch to increase. At high levels of inversion where transconductance factor mismatch becomes significant, ΔV_{GS} increases further, especially for short-channel devices where velocity saturation reduces transconductance. Section 3.11.1.7 described ΔV_{GS}.

- **Local-area, drain current mismatch,** $\Delta I_D/I_D$, **increases** as inversion coefficient increases in weak inversion, since gate area decreases causing threshold-voltage mismatch to increase while transconductance remains nearly constant. However, in strong inversion, $\Delta I_D/I_D$ levels off as the increase in threshold-voltage mismatch is countered by the decrease in transconductance. $\Delta I_D/I_D$ can then decrease for short-channel devices at high levels of inversion due to velocity saturation reduction of transconductance. At high levels of inversion where transconductance factor mismatch becomes significant, $\Delta I_D/I_D$ begins increasing again. Section 3.11.1.7 described $\Delta I_D/I_D$.

- **For gate-oxide thickness around 2 nm and below, gate leakage current, gate–source conductance, gate shot- and flicker-noise current, and mismatch in gate leakage current can increase** as inversion coefficient increases because the increase in gate leakage current density due to increasing V_{EFF} can exceed the effect of decreasing gate area. Gate leakage current can level off in portions of moderate and strong inversion if the increase in gate leakage current density is countered by the decrease in gate area, as seen earlier in Figure 3.74. Section 3.12.1 described gate leakage current and gate–source conductance, Section 3.10.7 described gate leakage noise current, and Section 3.12.2 described the resulting effects on circuit performance.

4.2.3 Trends as Channel Length Increases

The preceding section presented MOS performance trends for separate increases in the inversion coefficient. This section presents performance trends for separate increases in the channel length. References to sections in Chapter 3 describing the MOS performance listed below are given in Section 4.2.2.

If MOS channel length is increased, while holding drain current and inversion coefficient fixed, the following occur [7, 9]:

- **Channel width increases** directly with increasing channel length to maintain the shape factor associated with the selected inversion coefficient.

- **Gate area and gate-oxide capacitance increase** as the square of increasing channel length since channel width and length must increase together to maintain the selected inversion coefficient.

- **Layout area and silicon cost increase** as the square of increasing channel length, tracking the increase in gate area.

- V_{EFF} **is nearly unchanged** as channel length increases. When velocity saturation for short-channel devices operating at high levels of inversion causes increased V_{EFF}, V_{EFF} decreases as channel length increases because of reduced velocity saturation.

- $V_{DS,sat}$ **is nearly unchanged** as channel length increases. When velocity saturation is significant, $V_{DS,sat}$ remains nearly constant as channel length increases because, as mentioned in the previous section, it is nearly unaffected by velocity saturation when evaluated in terms of the inversion coefficient.

- g_m/I_D **and** g_m **are nearly unchanged** as channel length increases. When velocity saturation causes reduced g_m/I_D and g_m, these increase as channel length increases because of reduced velocity saturation. Nearly constant g_m/I_D with increasing channel length results in nearly constant bandwidth-accuracy and bandwidth–power–accuracy figures of merit for DC offset and thermal noise for both current-mode and voltage-mode circuits. Accuracy squared is proportional to channel length squared while bandwidth is proportional to the inverse square of channel length.

- **Transconductance distortion is nearly unchanged** as channel length increases since $V_{INDIF1dB}$ remains nearly constant because g_m/I_D remains nearly constant. When velocity saturation causes reduced g_m/I_D, linearization of drain current, and reduced distortion, the distortion increases and $V_{INDIF1dB}$ decreases as channel length increases because of reduced velocity saturation.

- V_A **and** r_{ds} **increase** with increasing channel length since channel length modulation (CLM) becomes less significant. When DIBL effects for short-channel devices cause V_A and r_{ds} to be reduced, these increase more rapidly as channel length increases because DIBL effects are reduced.

- A_{Vi} **increases** with increasing channel length since V_A and r_{ds} increase while g_m/I_D and transconductance remain nearly constant. When velocity saturation or DIBL effects cause A_{Vi} to be reduced, it increases more rapidly as channel length increases because of less velocity saturation reduction of transconductance or less DIBL reduction of r_{ds}.

- C_{gsi} **and** C_{gbi} **increase** as the square of increasing channel length, tracking the increase in gate area and gate-oxide capacitance.

- C_{GSO} **and** C_{GDO} **increase** directly with increasing channel length, tracking the increase in channel width.

- C_{DB} **and** C_{SB} **increase** directly with increasing channel length, tracking the increase in channel width.

- f_{Ti} **decreases** as the inverse square of increasing channel length since intrinsic gate capacitances track the increasing gate area and gate-oxide capacitance while transconductance remains nearly constant.

When velocity saturation is significant causing reduced transconductance, f_{Ti} decreases inversely with increasing channel length because of the increase in transconductance due to less velocity saturation. f_T and f_{diode} approach f_{Ti} as channel length increases because intrinsic capacitances dominate extrinsic gate overlap and drain–body capacitances.

- S_{VG} **thermal is nearly unchanged** as channel length increases since transconductance remains nearly constant. When velocity saturation is significant causing reduced transconductance and increased noise, the noise decreases as channel length increases because of less velocity saturation reduction of transconductance. The noise may decrease further as channel length increases because of lower small-geometry noise increases.

- S_{ID} **thermal is nearly unchanged** as channel length increases since transconductance remains nearly constant. When velocity saturation is significant causing reduced transconductance and reduced noise, the noise increases as channel length increases because of less velocity saturation reduction of transconductance. The noise, however, may decrease as channel length increases because of lower small-geometry increases.

- S_{VG} **flicker decreases** as the inverse square of increasing channel length since gate area increases. When velocity saturation is significant causing increased V_{EFF} and potentially increased flicker noise, the noise decreases further as channel length increases because of reduced velocity saturation increases in V_{EFF}.

- S_{ID} **flicker decreases** as the inverse square of increasing channel length since gate area increases causing S_{VG} flicker to decrease while transconductance remains nearly constant. When velocity saturation is significant causing reduced transconductance and reduced S_{ID} flicker, the noise decreases less as channel length increases because of less velocity saturation reduction of transconductance. The noise potentially decreases further as channel length increases because of reduced velocity saturation increases in V_{EFF}.

- f_c **decreases** as the inverse square of increasing channel length, tracking a small fraction of f_{Ti}. When velocity saturation is significant causing reduced f_{Ti} and increased V_{EFF}, more complex trends are possible.

- **Local-area** ΔV_{GS} **decreases** inversely with increasing channel length since gate area increases causing lower threshold-voltage mismatch. When transconductance factor mismatch dominates at high levels of inversion, ΔV_{GS} also decreases inversely with increasing channel length, tracking transconductance factor mismatch that is divided by nearly constant transconductance. When transconductance factor mismatch dominates and velocity saturation is significant causing increased ΔV_{GS} through reduced transconductance, increasing channel length decreases the mismatch further because of less velocity saturation reduction of transconductance.

- **Local-area** $\Delta I_D/I_D$ **decreases** inversely with increasing channel length since gate area increases causing lower threshold-voltage mismatch, while transconductance remains nearly constant. When velocity saturation is significant and $\Delta I_D/I_D$ due to threshold voltage mismatch is reduced through reduced transconductance, increasing channel length decreases $\Delta I_D/I_D$ less because of less velocity saturation reduction of transconductance. When transconductance factor mismatch dominates at high levels of inversion, $\Delta I_D/I_D$ also decreases inversely with channel length, tracking the transconductance factor mismatch.

- **For gate-oxide thickness around 2 nm and below, gate leakage current, gate–source conductance, gate shot- and flicker-noise current, and mismatch in gate leakage current increase** as the square of increasing channel length because of the increase in gate area. Mismatch in gate leakage current actually increases slightly less than this because increased gate area reduces the relative mismatch. When velocity saturation is significant causing increased V_{EFF} and gate leakage current density, the leakage current increases less rapidly as channel length increases because of reduced velocity saturation increases in V_{EFF}.

4.2.4 Trends as Drain Current Increases

The previous sections presented MOS performance trends for separate increases in the inversion coefficient and channel length. This section presents performance trends for separate increases in the drain current. References to sections in Chapter 3 describing the MOS performance listed below are given in Section 4.2.2.

If MOS drain current is increased, while holding inversion coefficient and channel length fixed, the following occur [7, 9]:

- **Shape factor, channel width, gate area, and gate-oxide capacitance increase** directly with increasing drain current, tracking the increase in shape factor required to maintain the selected inversion coefficient.

- **Layout area and silicon cost increase** directly with increasing drain current, tracking the increase in gate area.

- V_{EFF} **and** $V_{DS,sat}$ **are unchanged** as drain current increases since the inversion coefficient and channel length are held fixed.

- g_m/I_D **and** V_A **are unchanged** as drain current increases since the inversion coefficient and channel length are held fixed. These quality factors are normalized to the drain current. Constant g_m/I_D results in constant bandwidth–power–accuracy figures of merit for DC offset and thermal noise for both current-mode and voltage-mode circuits, while the bandwidth-accuracy figures of merit increase with the drain current.

- g_m **increases** directly with increasing drain current since it is equal to the drain current multiplied by constant g_m/I_D associated with fixed inversion coefficient and channel length.

- r_{ds} **decreases** inversely with increasing drain current since it is equal to constant V_A associated with fixed inversion coefficient and channel length divided by the drain current. The drain–source conductance, $g_{ds} = 1/r_{ds}$, increases directly with drain current since it is equal to the drain current divided by V_A.

- **Transconductance distortion is unchanged** as drain current increases. $V_{INDIF1dB}$ is unchanged since g_m/I_D is constant due to fixed inversion coefficient and channel length.

- A_{Vi} **is unchanged** as drain current increases since transconductance increases and r_{ds} decreases by an equal amount with increasing drain current, while g_m/I_D and V_A are constant due to fixed inversion coefficient and channel length.

- C_{gsi} **and** C_{gbi} **increase** directly with increasing drain current, tracking the increase in gate area.

- C_{GSO} **and** C_{GDO} **increase** directly with increasing drain current, tracking the increase in channel width.

- C_{DB} **and** C_{SB} **increase** directly with increasing drain current, tracking the increase in channel width.

- f_{Ti} **is unchanged** as drain current increases since intrinsic gate capacitances and transconductance increase equally with drain current when inversion coefficient and channel length are held fixed. f_T and f_{diode} are also unchanged since intrinsic and extrinsic MOS capacitances and transconductance increase equally with drain current. However, layout capacitances reduce operating bandwidth at low drain currents where layout capacitances are significant compared to small intrinsic and extrinsic MOS capacitances.

- S_{VG} **thermal decreases** inversely with increasing drain current since transconductance increases with drain current, while g_m/I_D is constant due to fixed inversion coefficient and channel length.

- S_{ID} **thermal increases** directly with increasing drain current since transconductance increases with drain current, while g_m/I_D is constant due to fixed inversion coefficient and channel length.

- S_{VG} **flicker decreases** inversely with increasing drain current since gate area increases.

- S_{ID} **flicker increases** directly with increasing drain current since the square of increasing transconductance, which converts S_{VG} flicker to S_{ID} flicker, exceeds the decrease in S_{VG} flicker associated with increasing gate area.

- f_c **is unchanged** as drain current increases since it tracks f_{Ti}, which is independent of the drain current.

- **Local-area** ΔV_{GS} **decreases** inversely as the square root of increasing drain current since gate area increases, reducing mismatch.

- **Local-area** $\Delta I_D/I_D$ **decreases** inversely as the square root of increasing drain current since gate area increases, reducing mismatch.

- **For gate-oxide thickness around 2 nm and below, gate leakage current, gate–source conductance, gate shot- and flicker-noise current, and mismatch in gate leakage current increase** directly with increasing drain current, tracking the increase in gate area. Mismatch in gate leakage current actually increases as the square root of drain current because increased gate area reduces the relative mismatch. For fixed inversion coefficient and channel length, gate leakage current remains a constant fraction of the drain current.

4.3 PERFORMANCE TRADEOFFS

Section 4.2 described trends of individual aspects of MOS performance for separate increases in the design choices of inversion coefficient, channel length, and drain current. This section presents tradeoffs of multiple aspects of MOS performance over these design choices, beginning with an overview using the *MOSFET Operating Plane* followed by a discussion of the region and level of inversion as expressed by the inversion coefficient. Graphical performance tradeoffs common to all applications are then presented, followed by tradeoffs specific to differential-pair and current-mirror applications. Finally, this section presents tradeoffs in figures of merit that involve the combination of multiple aspects of performance.

An understanding of MOS performance tradeoffs provides design intuition and leads towards optimum design. This is illustrated through the design of differential-pair and current-mirror devices in Section 4.4 and the design of complete circuits in Chapters 5 and 6. A discussion of general analog circuit performance tradeoffs, not specific to the selection of MOS drain current, inversion coefficient, and channel length presented here, is contained in another book-length work [10].

4.3.1 Overview – The *MOSFET Operating Plane*

The *MOSFET Operating Plane* [2–9] shown earlier in Figure 3.1 provides an overview of tradeoffs in analog MOS performance resulting from inversion coefficient and channel length selections. Operation at low inversion coefficients in weak inversion on the left side of the *Plane* results in optimally high g_m/I_D and g_m, optimally low, gate-referred thermal-noise voltage, and optimally low V_{EFF} and $V_{DS,sat}$, while requiring large shape factor devices. Operation at high inversion coefficients in strong inversion on the right side of the *Plane* results in optimally high g_m linearity and optimally low, drain-referred thermal-noise current, while requiring small shape factor devices.

Operation at low inversion coefficients and long channel lengths on the upper left of the *Plane* results in optimally high A_{Vi} since both g_m and r_{ds} are maximum here. Additionally, operation here results in optimally low, gate-referred flicker-noise voltage, optimally low, drain-referred flicker-noise current, and optimally low, local-area gate–source voltage and drain current mismatch. This is because devices have large shape factors, long channel lengths, and, correspondingly, large gate areas. Even though g_m is maximized at low inversion coefficients, drain-referred flicker-noise current and drain mismatch current are minimized here because the reduction of gate-referred flicker-noise voltage and gate–source mismatch voltage associated with large gate areas exceeds the increase in g_m, which converts gate noise and mismatch to the drain. Operation at high inversion coefficients and short channel lengths on the lower right of the *Plane* results in optimally high f_{Ti} since the decrease in intrinsic capacitances exceeds the decrease in g_m. Operation here requires small shape factors, short channel lengths, and, correspondingly, small gate areas and optimally minimal layout areas.

While A_{Vi}, flicker noise, and mismatch are optimized at low inversion coefficients and f_{Ti} is optimized at high inversion coefficients, these are even stronger functions of channel length. Increasing the channel length increases A_{Vi} because of increased r_{ds} due to reduced CLM effects. Additionally, increasing the channel length requires an equal increase in channel width to maintain the shape factor for the selected inversion coefficient. This results in a squaring of gate area and gate-oxide capacitance as channel length increases. As a result, flicker-noise density and local-area mismatch decrease as $1/L$ as channel length increases since these are proportional to the inverse square root of gate area. Unfortunately, increasing channel length decreases f_{Ti} rapidly as $1/L^2$ ($1/L$ if full velocity saturation is present) because of the large increase in gate area and, correspondingly, large increase in intrinsic gate capacitances.

If inversion coefficient and channel length are held fixed, corresponding to a fixed point on the *Plane*, and drain current is explored, V_{EFF} and $V_{DS.sat}$, g_m/I_D and V_A, and A_{Vi} and f_{Ti} are unchanged. However, g_m increases with increasing drain current, lowering gate-referred thermal-noise voltage while increasing drain-referred thermal-noise current. Additionally, channel width and gate area increase with increasing drain current as shape factor increases to maintain the selected inversion coefficient. The increase in gate area results in a decrease in gate-referred flicker-noise voltage and local-area mismatch since these, again, are inversely proportional to the square root of gate area. Drain-referred flicker-noise current, however, increases with drain current because the increase in g_m, which converts gate noise to the drain, exceeds the decrease in gate-referred flicker-noise voltage.

The tradeoffs summarized here are affected somewhat when small-geometry velocity saturation, DIBL, increases in thermal noise, or inversion-level (V_{EFF}) increases in gate-referred flicker-noise voltage are present. These effects, which were discussed in the preceding sections, do not change the general location on the *Plane* corresponding to desired performance. However, as discussed in the preceding sections, gate leakage current present for gate-oxide thickness around 2 nm and below affects performance tradeoffs and potentially the location on the *Plane* for desired performance. As described earlier in Section 3.12.2, a smaller channel length and resulting gate area may be required to balance the traditional improvements in gain, flicker noise, and local-area mismatch at increasing channel length and gate area with the deterioration of these associated with increasing gate leakage current caused by increasing gate area.

4.3.2 Region and Level of Inversion – The Inversion Coefficient as a Number Line

Examining performance tradeoffs over design choices of the inversion coefficient provides guidance on the optimal region and level of inversion for a given application. Figure 4.1 presents design choices for the inversion coefficient along a number line and summarizes values of W, WL (gate area), g_m/I_D, A_{Vi}, f_{Ti}, $V_{DS,sat}$, and V_{EFF} for n held fixed at $n = 1.4$ and $U_T = 25.9$ mV at $T = 300$ K (room temperature). Values are found from the expressions summarized earlier in Tables 3.10, 3.17, 3.22, 3.26, 3.15, and 3.14, respectively, excluding small-geometry VFMR, velocity saturation, and DIBL effects. Like Figure 3.3 shown earlier, which introduced the inversion coefficient, Figure 4.1 identifies the weak-, moderate-, and strong-inversion regions and subregions. As discussed earlier in

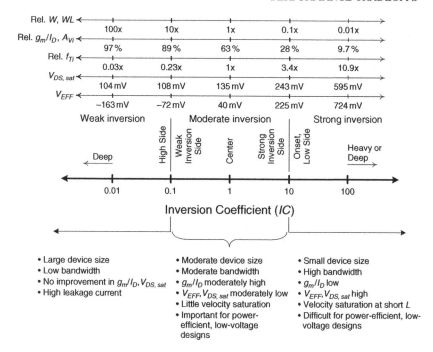

Figure 4.1 The inversion coefficient presented as a number line showing regions and subregions of MOS inversion with corresponding performance tradeoffs. Moderate inversion can provide optimal performance tradeoffs for low-voltage, power-efficient circuits because V_{EFF} and $V_{DS,sat}$ are low, while g_m/I_D, A_{Vi}, and f_{Ti} are moderately high. When velocity saturation is significant for short-channel devices at high IC, g_m/I_D, A_{Vi}, and f_{Ti} are lower, while V_{EFF} is higher than the values shown

Section 3.2.4, using the inversion coefficient in design permits an understanding and prediction of performance independent of the CMOS process. As a result, the values shown in Figure 4.1 apply to all bulk or partially depleted silicon-on-insulator (SOI) CMOS processes with only small changes resulting from the usual small variations in n.

In deep weak inversion at $IC = 0.01$, g_m/I_D and A_{Vi} are at their optimal maximum values (this assumes DIBL does not reduce r_{ds} and A_{Vi} for short-channel devices as discussed in Section 3.8.5), $V_{DS,sat}$ is at its optimal minimum value of 104 mV, but f_{Ti} is very low at 3 % of its $IC = 1$ value, and W and WL are very large at 100× their $IC = 1$ values. As IC increases to 0.1 at the weak-inversion side of moderate inversion, g_m/I_D and A_{Vi} decrease slightly to 89 % of their maximum values, $V_{DS,sat}$ increases slightly to 108 mV, but f_{Ti} increases significantly to 23 % of its $IC = 1$ value, and W and WL decrease significantly to 10× their $IC = 1$ values. This suggests that IC should rarely be set below 0.1, corresponding to operation in weak inversion, as there is little g_m/I_D, A_{Vi}, and $V_{DS,sat}$ advantage to counter the substantial loss of f_{Ti} and increase in W and WL. Additionally, weak-inversion operation results in increased leakage current because of the large W and WL required.

As IC increases to one at the center of moderate inversion, g_m/I_D and A_{Vi} decrease modestly to 63 % of their maximum values, $V_{DS,sat}$ increases modestly to 135 mV, V_{EFF} increases to a small positive value of 40 mV (corresponding to V_{GS} slightly above the threshold voltage), but f_{Ti} continues to increase significantly to 100 % of its $IC = 1$ value, and W and WL decrease significantly to 1× their $IC = 1$ values. This suggests operation near the center of moderate inversion for low-voltage design because of low $V_{DS,sat}$ and V_{EFF}, combined with moderately high g_m/I_D, A_{Vi}, and f_{Ti}. As IC increases to 10 at the onset of strong inversion, g_m/I_D and A_{Vi} decrease to 28 % of their maximum values,

$V_{DS,sat}$ increases to approximately 243 mV,[1] V_{EFF} increases to 225 mV, and f_{Ti} continues to increase, reaching 340 % of its $IC = 1$ value. Additionally, W and WL decrease to $0.1\times$ their $IC = 1$ values. This suggests operation at the onset of strong inversion or deeper in strong inversion for high-bandwidth applications, provided $V_{DS,sat}$ and V_{EFF} are sufficiently low for bias and signal compliance and reduced g_m/I_D does not require excessive supply current to provide a minimum value of g_m. A minimum value of g_m might be required, for example, to provide sufficiently low, gate-referred thermal-noise voltage.

Values listed on Figure 4.1 exclude the decrease in g_m/I_D, A_{Vi}, and f_{Ti} and increase in V_{EFF} for short-channel devices experiencing velocity saturation at high IC, usually in strong inversion. When evaluated in terms of IC, $V_{DS,sat}$ does not increase significantly due to velocity saturation effects as described earlier in Section 3.7.3.3. Operating at IC well below the critical value, IC_{CRIT}, described earlier in Section 3.8.2.2, ensures that velocity saturation effects are negligible. Increasing channel length significantly increases the value of IC_{CRIT} and the allowable value of IC. Selecting pMOS devices, which have higher values of critical, velocity saturation electric field, E_{CRIT}, also increases the value of IC_{CRIT} and the allowable value of IC. While having a much smaller effect than velocity saturation, the slight decrease in n with increasing IC, shown earlier in Figure 3.30, is also neglected for the values listed on Figure 4.1. Decreasing n causes a slight increase in g_m/I_D, A_{Vi}, and f_{Ti} and a slight decrease in V_{EFF} in strong inversion.

Modern, low-voltage applications suggest confining operation increasingly towards the moderate-inversion $(0.1 < IC < 10)$ portion of Figure 4.1 as anticipated over 25 years ago in 1982 by Tsividis [11]. Here, $V_{DS,sat}$ and V_{EFF} are sufficiently low to support low-voltage biasing, at least for the weak-inversion side of moderate inversion $(1 < IC < 3)$. Additionally, g_m/I_D is sufficiently high to support power-efficient designs operating at low values of drain current. f_{Ti}, while clearly below values available in strong inversion, is often adequate, especially for short-channel devices. Finally, operation in moderate inversion minimizes velocity saturation decreases in g_m/I_D, A_{Vi}, and f_{Ti} and increases in V_{EFF}. Moderate inversion will be used, especially for input devices, in the design examples described in Chapters 5 and 6.

Since traditional weak-inversion, exponential and strong-inversion, square-law drain current expressions cannot be used in moderate inversion, it is important to utilize expressions continuously valid in moderate inversion. Chapter 3 presented MOS performance expressions valid in weak, moderate, and strong inversion, inclusive of small-geometry effects like VFMR, velocity saturation, and DIBL.

4.3.3 Tradeoffs Common to All Devices

This section and Sections 4.3.4 and 4.3.5 that follow detail MOS performance tradeoffs using tables and graphs that build on the work of Chapter 3. This section details tradeoffs for all device applications, while Sections 4.3.4 and 4.3.5 detail tradeoffs specific to differential-pair and current-mirror applications. This leads to the development of figures of merit in Section 4.3.6 and the optimization of differential-pair and current-mirror devices in Section 4.4.

The performance tradeoffs considered are for native devices having no resistive source degeneration. Source degeneration lowers or degenerates transconductance, drain-referred thermal- and flicker-noise current, and drain mismatch current, while raising the drain output resistance. Additionally, source degeneration increases gate-referred thermal-noise voltage due to the presence of thermal noise in the source resistance and raises the drain voltage necessary for operation in saturation due to the voltage drop across the source resistance. Device performance predictions must then be modified when source degeneration is present. For example, device transconductance, drain-referred thermal- and flicker-noise current, and drain mismatch current are reduced by division by the factor $1 + g_{ms}r_s$,

---

[1] As discussed earlier in Section 3.7.3.2, $V_{DS,sat}$ is over-predicted at the onset of strong inversion above the expected, inversion charge pinch-off value of V_{EFF}/n. Deeper in strong inversion, $V_{DS,sat}$ approaches this expected value. In all cases, the value of $V_{DS,sat}$ used depends upon the physical or electrical definitions described earlier in Section 3.7.3.1.

where $g_{ms} = g_m + g_{mb} + g_{ds} \approx g_m + g_{mb}$ is the total source conductance and r_s is the external source degeneration resistance. Source degeneration is used to reduce the drain-referred noise current of non-input devices in the micropower, low-noise CMOS preamplifiers described later in Chapter 6.

Table 4.1 lists MOS geometry, capacitance, bias compliance voltage, small-signal parameter, intrinsic voltage gain, and intrinsic bandwidth expressions and performance tradeoffs common to all device applications. The table includes references to the relevant section and prediction table from Chapter 3 for each aspect of performance. The table also shows performance trends as the inversion

Table 4.1 MOS performance tradeoffs common to all devices versus the selected inversion coefficient, channel length, and drain current

Parameter	References			Trends		
	Section	Table	Figure	$IC \uparrow$ L, I_D fixed	$L \uparrow$ IC, I_D fixed	$I_D \uparrow$ IC, L fixed
Geometry: $W = \left(\dfrac{L}{IC}\right)\left(\dfrac{I_D}{I_0}\right)$	3.6.2	3.10–3.12	4.2, 4.3	↓	↑	↑
$WL = \left(\dfrac{L^2}{IC}\right)\left(\dfrac{I_D}{I_0}\right)$	3.6.3	3.10	4.2, 4.3, 3.8	↓	↑↑	↑
Capacitances:[a] $C_{gsi} + C_{gbi} = \left(\dfrac{L^2}{IC}\right)\left(\dfrac{I_D}{I_0}\right) C'_{OX} \cdot \left(\hat{C}_{gsi} + \hat{C}_{gbi}\right)$	3.9.2	3.23	4.4, 4.5, 3.53, 3.54	↓	↑↑	↑
$C_{DBO} \approx \left(\dfrac{L}{IC}\right)\left(\dfrac{I_D}{I_0}\right) \cdot \left(\dfrac{1}{2} W_{DIF} C_J + C_{JSW}\right)$	3.9.4	3.25	4.4, 4.5	↓	↑	↑
Bias compliance voltages:[b] $V_{EFF} = V_{GS} - V_T$ $= 2nU_T \ln\left(e^{\sqrt{IC}} - 1\right)$ $\approx 2nU_T\sqrt{IC}$ (SI)	3.7.2.4	3.14	4.6, 3.14, 3.15	↑, ↑, ↑	−, −, ↓	−
$V_{DS,sat} = 2U_T\sqrt{IC + 0.25} + 3U_T$	3.7.3.5	3.15	4.6, 3.16	−, ↑	−	−
Small-signal parameters:[c,d] $\dfrac{g_m}{I_D} = \dfrac{1}{nU_T\left(\sqrt{IC + 0.25} + 0.5\right)}$	3.8.2.4	3.17	4.7, 4.8, 3.26, 3.27	−, ↓, ↓	−, −, ↑	−
$V_A = I_D \cdot r_{ds} = I_D/g_{ds}$ $= V_{AL}(IC, L, V_{DS}) \cdot L$	3.8.4.6	3.21	4.7, 4.8, 3.43–3.47	−	↑	−
Intrinsic voltage gain and bandwidth:[d] $A_{Vi} = g_m \cdot r_{ds} = (g_m/I_D) V_A$ $= \dfrac{V_{AL}(IC, L, V_{DS}) \cdot L}{nU_T\left(\sqrt{IC + 0.25} + 0.5\right)}$	3.8.5	3.22	4.9, 4.10, 3.51, 3.52	−, ↓, ↓	↑, ↑, ↑↑	−

Table 4.1 (*continued*)

Parameter	References			Trends		
	Section	Table	Figure	$IC \uparrow$ L, I_D fixed	$L \uparrow$ IC, I_D fixed	$I_D \uparrow$ IC, L fixed
$f_{Ti} = \dfrac{g_m}{2\pi\left(C_{gsi}+C_{gbi}\right)}$ $= \left(\dfrac{IC}{\sqrt{IC+0.25}+0.5}\right)$ $\cdot\left(\dfrac{\mu_0 U_T}{\pi\left(\hat{C}_{gsi}+\hat{C}_{gbi}\right)L^2}\right)$	3.9.6	3.26	4.9, 4.10, 3.57, 3.58	$\uparrow, \uparrow, -$	$\downarrow\downarrow,$ $\downarrow\downarrow, \downarrow$	$-$

Trends include velocity saturation but exclude DIBL effects. Multiple trends are for WI, SI, and SI with full velocity saturation. Expressions and trends are for operation in saturation.

Trend key:

$-$	Generally little or no change.
$\uparrow\ \downarrow$	Generally increases sublinearly or decreases inversely as sublinear.
$\uparrow, \downarrow$	Generally increases linearly or decreases inversely.
$\uparrow\uparrow, \downarrow\downarrow$	Generally increases as the square or decreases as the inverse square.

Velocity saturation and VFMR effects may be included by replacing IC with $IC(1 + IC/IC_{CRIT})$ in $\sqrt{IC+0.25}$ terms and by replacing IC with $IC(1 + IC/(4IC_{CRIT}))$ in V_{EFF} expressions, where $IC_{CRIT} = [(LE_{CRIT})'/(4nU_T)]^2$ and $(LE_{CRIT})' = (1/\theta)||LE_{CRIT}$ as given in Table 3.17. When velocity saturation is significant, this results in channel length dependency on g_m/I_D and V_{EFF}, and additional channel length dependency on A_{Vi} and f_{Ti}.

[a] The trend of $C_{gsi} + C_{gbi}$ with IC, which tracks WL, is approximate because it neglects inversion-level dependency on $\hat{C}_{gsi} + \hat{C}_{gbi}$. C_{DBO} assumes an interdigitated drain where area and perimeter are minimized. Without interdigitation, C_{DBO} is nearly doubled from the value given.

[b] When evaluated in terms of IC, $V_{DS,sat}$ is nearly unaffected by velocity saturation, becoming a smaller fraction of V_{EFF} as this increases due to velocity saturation.

[c] When DIBL causes reduced V_A for short-channel devices, especially at low levels of inversion, V_A increases with increasing IC because decreasing g_m/I_D reduces the DIBL effect.

[d] When DIBL causes reduced V_A and A_{Vi} for short-channel devices, especially at low levels of inversion, these increase further as L increases because both DIBL and CLM effects are reduced.

coefficient, channel length, and drain current are increased. As noted, blank lines indicate little change in performance. Up and down arrows indicate performance that increases or decreases, with degree of the increase or decrease indicated by the size of the arrow. When multiple trends are included, the first is for weak inversion, the second is for strong inversion without small-geometry effects, and the third is for strong inversion with full velocity saturation.

The expressions and trends include velocity saturation and VFMR increases in V_{EFF} and decreases in g_m/I_D and intrinsic voltage gain and bandwidth predicted using the expression modifications listed at the bottom of the table that involve the critical inversion coefficient, IC_{CRIT}. IC_{CRIT}, which depends primarily on the channel length and critical, velocity saturation electric field, E_{CRIT}, corresponds to the inversion coefficient where g_m/I_D is down to approximately 70.7 % of the value present without velocity saturation and, secondarily, VFMR effects. All expressions and trends shown include velocity saturation effects, but exclude DIBL effects, which are noted at the bottom of the table. The inclusion of velocity saturation effects results in channel length dependency on V_{EFF} and g_m/I_D, and additional channel length dependency on the intrinsic voltage gain and bandwidth. As observed in the table, trends for figures of merit, for example A_{Vi}, comprising multiple aspects of performance can be found from the combination of trends for individual aspects of performance.

Table 4.1 also lists figures presented in this section and earlier in Chapter 3 that illustrate performance tradeoffs. The figures presented in this section show two complementary or competing aspects of performance compared to single aspects of performance shown earlier in Chapter 3. In addition to showing performance as a function of the inversion coefficient as done in Chapter 3, figures are included that show performance as a function of channel length.

The figures in this section are for nMOS devices in the example $0.18\,\mu m$ CMOS process described in Tables 3.2–3.4, with performance predicted or measured as described for the figures in Chapter 3. Performance is presented as a function of the inversion coefficient from deep weak inversion ($IC = 0.01$) to deep strong inversion ($IC = 100$) for drawn channel lengths of 0.18, 0.28, 0.38, 0.48, 1, 2, and $4\,\mu m$. The effective channel length is approximately $0.028\,\mu m$ less, corresponding to $DL = 0.028\,\mu m$ listed in Table 3.2. Additionally, performance is presented as a function of effective channel length from $0.152\,\mu m$ ($0.18\,\mu m - 0.028\,\mu m$) to $4\,\mu m$ for $IC = 0.1, 0.3, 1, 3, 10, 30$, and 100, corresponding to operation in weak ($IC = 0.1$), moderate ($IC = 0.3, 1$, and 3), and strong ($IC = 10, 30$, and 100) inversion. Finally, when performance is dependent on the drain current, a drain current of $100\,\mu A$ is used. As always unless otherwise specified, operation in saturation is assumed.

The figures in this section show the significant performance tradeoffs associated with inversion coefficient and channel length selections. Since the gate-oxide thickness is $4.1\,nm$ for the example $0.18\,\mu m$ CMOS process, gate leakage current and the resulting gate–source conductance, gate shot- and flicker-noise current, and increase in DC mismatch are negligible. These effects, described earlier in Section 3.12.2, may require a smaller channel length and gate area to balance the traditional improvements in gain, flicker noise, and local-area mismatch at increasing channel length and gate area with the deterioration of these associated with increasing gate leakage current caused by increasing gate area. Gate leakage current effects can become significant for gate-oxide thickness around $2\,nm$ and below.

4.3.3.1 Channel width and gate area

The geometry entries in Table 4.1 list MOS channel width, W, and gate area, WL. As mentioned, the table includes references to the relevant Chapter 3 prediction tables and Chapter 3 and 4 figures, and lists trends for the selected inversion coefficient, channel length, and drain current.

Figure 4.2 shows W and WL versus the inversion coefficient, IC, where both decrease inversely with IC because of the smaller shape factor, $S = W/L$, and W required for increasing inversion coefficient. Figure 4.3 shows W and WL versus the channel length, L, where W increases directly with L, while WL increases as the square of L. W must increase directly with L to maintain $S = W/L$ associated with the selected inversion coefficient. The increase of W and L together then results in gate area (WL) increasing as the square of L. As a result of the rapid increase of gate area with L, intrinsic gate capacitance, intrinsic bandwidth, flicker noise, and local-area mismatch are strong functions of L. Finally, W and WL scale directly with drain current, which is taken as $100\,\mu A$ in the figures, because $S = W/L$ and W must track drain current to maintain the selected inversion coefficient.

pMOS W and WL are approximately a factor of five larger than the nMOS values shown. This is because the technology current, $I_0 = 2n_0\mu_0 C'_{OX} U_T^2$, in the inversion coefficient definition, $IC = I_D/[I_0(W/L)]$, is approximately a factor of five smaller due to lower pMOS mobility. Table 3.10 summarized MOS sizing relationships, and Table 3.2 listed I_0 and the low-field mobility, μ_0, for the example $0.18\,\mu m$ CMOS process.

Figures 4.2 and 4.3 collectively show that W and WL are optimized or minimized for minimum layout area, gate area, and gate capacitance at high IC and short L. Unfortunately, intrinsic voltage gain is low here, and flicker noise and local-area mismatch are high.

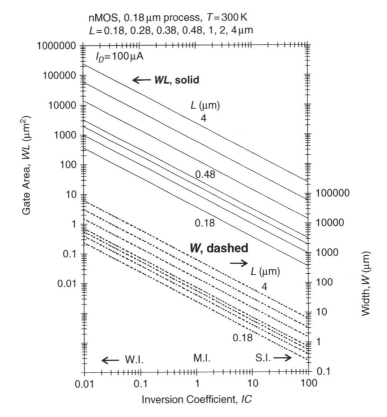

Figure 4.2   Channel width, W (dashed lines), and gate area, WL (solid lines), versus the inversion coefficient, IC, for $L = 0.18, 0.28, 0.38, 0.48, 1, 2,$ and $4\,\mu m$, nMOS devices in a $0.18\,\mu m$ CMOS process operating at a drain current of $100\,\mu A$. Width and gate area optimize (minimize) together at high IC. Both decrease inversely with IC and are large in weak inversion, resulting in large layout area and large device capacitances. Width and gate area scale directly with drain current and are nearly a factor of five larger for pMOS devices as required due to lower mobility

4.3.3.2 Intrinsic gate capacitance and drain–body capacitance

The capacitance entries in Table 4.1 list the intrinsic gate–source and gate–body capacitance, $C_{gsi} + C_{gbi}$, and the drain–body capacitance, C_{DBO}, which assumes the worst case, zero-bias, drain–body voltage of $V_{DB} = 0\,V$. These capacitances are closely related to W and WL given in Figures 4.2 and 4.3 since C_{DBO} is proportional to W and $C_{gsi} + C_{gbi}$ is nearly proportional to WL. $C_{gsi} + C_{gbi}$ has a complex inversion-level dependency through the normalized gate–source and gate–body capacitances, $\hat{C}_{gsi} = C_{gsi}/C_{GOX}$ and $\hat{C}_{gbi} = C_{gbi}/C_{GOX}$, as described earlier in Table 3.23. However, $C_{gsi} + C_{gbi}$ tracks WL in strong inversion where $\hat{C}_{gsi}$ is nearly constant at two-thirds and $\hat{C}_{gbi}$ is negligible, and tracks WL in weak inversion where $\hat{C}_{gsi}$ is negligible and $\hat{C}_{gbi}$ is nearly constant. $C_{gsi} + C_{gbi}$ represents the intrinsic gate input capacitance consisting of the intrinsic gate–source and gate–body capacitances, but excluding the intrinsic gate–drain capacitance that is negligible in saturation. C_{DBO} represents the drain output capacitance. W_{DIF}, C_J, and C_{JSW} are process values for the drain stripe width, area capacitance, and perimeter capacitance, respectively, where an even number of gate stripes or m is used with the drain

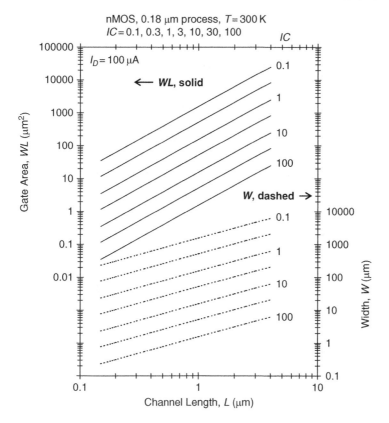

Figure 4.3 Channel width, W (dashed lines), and gate area, WL (solid lines), versus channel length, L, for nMOS devices in a 0.18 μm CMOS process operating at IC = 0.1, 0.3, 1, 3, 10, and 100 at a drain current of 100 μA. Width and gate area optimize (minimize) together at short L. Width increases directly with L, while gate area increases as the square of L. Width and gate area scale directly with drain current and are nearly a factor of five larger for pMOS devices as required due to lower mobility

area and perimeter minimized. As described earlier in Table 3.25, if drain interdigitation is not used (m = 1), C_{DBO} given in Table 4.2 and in the following figures is nearly doubled.

Figure 4.4 shows $C_{gsi} + C_{gbi}$ and C_{DBO} as a function of IC, where both capacitances decrease nearly inversely with IC, tracking WL and W, respectively. Figure 4.5 shows these capacitances as a function of L, where $C_{gsi} + C_{gbi}$ increases as the square of L, tracking WL, while C_{DBO} increases directly with L, tracking W. Both capacitances scale directly with drain current, which is taken as 100 μA in the figures, because WL and W must increase directly with the drain current.

pMOS $C_{gsi} + C_{gbi}$ and C_{DBO} are approximately a factor of five larger than the nMOS values shown because of the approximate factor-of-five increase in W and WL required because of lower pMOS mobility. The value of pMOS C_{DBO}, however, is increased further because of higher C_J listed in Table 3.4.

Figures 4.4 and 4.5 collectively show that operation at high IC and short L, where WL and W are minimum, minimizes the intrinsic gate capacitance ($C_{gsi} + C_{gbi}$) and drain–body capacitance (C_{DBO}). While not shown in the figures but discussed in Section 3.9.3, the extrinsic gate–source overlap capacitance, C_{GSO}, can add significantly to the intrinsic gate capacitance, especially for short-channel devices where gate area and intrinsic gate capacitances are small. Both C_{GSO} and the extrinsic

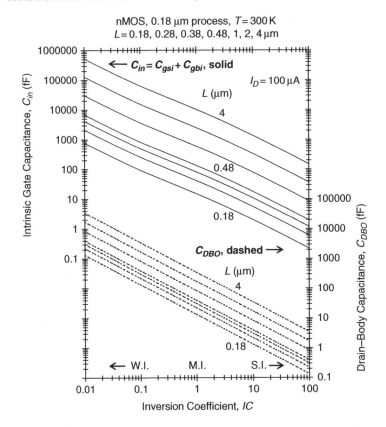

Figure 4.4 Predicted input, intrinsic gate capacitance, $C_{in} = C_{gsi} + C_{gbi}$ (solid lines), and output, drain–body capacitance, C_{DBO} (dashed lines), versus the inversion coefficient, IC, for $L = 0.18, 0.28, 0.38, 0.48, 1, 2,$ and $4\,\mu m$, nMOS devices in a $0.18\,\mu m$ CMOS process operating at a drain current of $100\,\mu A$. C_{in} and C_{DBO} optimize (minimize) together at high IC. Both decrease nearly inversely with IC, tracking gate area and channel width, respectively, and are large in weak inversion resulting in low circuit bandwidth. C_{in} and C_{DBO} scale directly with drain current and are nearly a factor of five larger for pMOS devices because of larger required channel width and gate area

gate–drain overlap capacitance, C_{GDO}, are proportional to W. The extrinsic gate–body overlap capacitance, C_{GBO}, which is often negligible, is proportional to L.

4.3.3.3 Effective gate–source voltage and drain–source saturation voltage

The bias compliance voltage entries in Table 4.1 list the effective gate–source voltage, $V_{EFF} = V_{GS} - V_T$, and the drain–source saturation voltage, $V_{DS,sat}$. These voltages govern bias compliance since the gate–source voltage, $V_{GS} = V_T + V_{EFF}$, is equal to the threshold voltage, V_T, summed with V_{EFF}. $V_{DS,sat}$ is the minimum drain–source voltage, V_{DS}, required for operation in saturation.

Figure 4.6 shows measured V_{EFF} and predicted $V_{DS,sat}$ as a function of IC. Both voltages are optimized or minimized at low IC, although they increase only modestly at the center of moderate inversion ($IC = 1$). Both voltages increase as the square root of IC in strong inversion, with V_{EFF} increasing directly with IC for short-channel devices at high IC experiencing velocity saturation. V_{EFF} and $V_{DS,sat}$

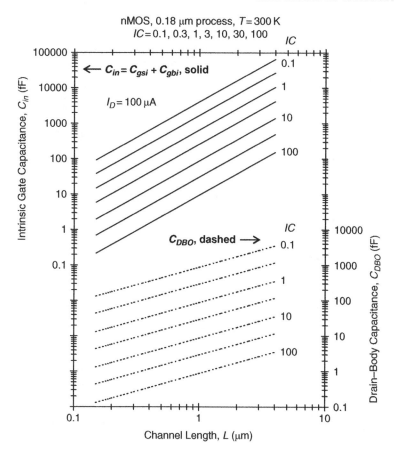

nMOS, 0.18 μm process, $T = 300$ K
$IC = 0.1, 0.3, 1, 3, 10, 30, 100$

Figure 4.5   Predicted input, intrinsic gate capacitance, $C_{in} = C_{gsi} + C_{gbi}$ (solid lines), and output, drain–body capacitance, C_{DBO} (dashed lines), versus channel length, L, for nMOS devices in a $0.18\,\mu$m CMOS process operating at $IC = 0.1$, 0.3, 1, 3, 10, 30, and 100 at a drain current of $100\,\mu$A. C_{in} and C_{DBO} optimize (minimize) together at short L. C_{in} increases as the square of L, tracking gate area, while C_{DBO} increases directly with L, tracking channel width. C_{in} and C_{DBO} scale directly with drain current and are nearly a factor of five larger for pMOS devices because of larger required channel width and gate area

are independent of L, except for the increase in V_{EFF} associated with velocity saturation for short-channel devices. Here, V_{EFF} decreases with increasing L because of the reduction of velocity saturation effects. $V_{DS,sat}$, when evaluated in terms of IC, is nearly insensitive to velocity saturation, becoming an increasingly smaller fraction of increasing V_{EFF} when velocity saturation is present. This was described earlier in Section 3.7.3.3. Both V_{EFF} and $V_{DS,sat}$ are independent of the drain current when evaluated in terms of IC. Finally, pMOS V_{EFF} and $V_{DS,sat}$ are similar to the nMOS values shown, excluding the smaller increase in V_{EFF} for short-channel devices at high IC associated with less pMOS velocity saturation.

Low-voltage design constrains V_{EFF} and $V_{DS,sat}$ to sufficiently low values to maintain gate–source and drain–source voltage compliance. Maintaining voltage compliance for operation in strong inversion ($IC > 10$) is increasingly difficult as supply voltage decreases in smaller-geometry processes. The weak-inversion side of moderate inversion ($1 < IC < 3$), where V_{EFF} and $V_{DS,sat}$ are low, and g_m/I_D, intrinsic voltage gain, and intrinsic bandwidth are moderately high, is increasingly important for

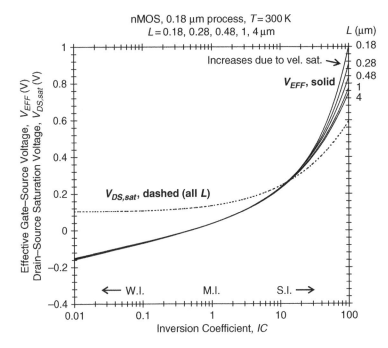

Figure 4.6 Measured effective gate–source voltage, $V_{EFF} = V_{GS} - V_T$ (solid lines), and predicted drain–source saturation voltage, $V_{DS,sat}$ (dashed line), versus the inversion coefficient, IC, for $L = 0.18, 0.28, 0.48, 1$, and $4\,\mu m$, nMOS devices in a $0.18\,\mu m$ CMOS process. V_{EFF} and $V_{DS,sat}$ optimize (minimize) together at low IC and increase as the square root of IC in strong inversion, with V_{EFF} increasing directly with IC for short-channel devices at high IC due to velocity saturation. V_{EFF} and $V_{DS,sat}$ are nearly independent of channel length, except for the increase in V_{EFF} associated with velocity saturation. V_{EFF} and $V_{DS,sat}$ are independent of the drain current and similar for pMOS devices, except for smaller pMOS V_{EFF} increases due to less velocity saturation

low-voltage design as described earlier for Figure 4.1. Moderate inversion is used, especially for input devices, in the design examples described later in Chapters 5 and 6.

4.3.3.4 Transconductance efficiency and Early voltage

The small-signal parameter entries in Table 4.1 list transconductance efficiency, g_m/I_D, and normalized, drain–source resistance or Early voltage, $V_A = I_D \cdot r_{ds} = I_D/g_{ds}$. Unlike W, WL, $C_{gsi} + C_{gbi}$, C_{DBO}, V_{EFF}, and $V_{DS,sat}$ shown in Figures 4.2–4.6 that are predicted except for V_{EFF}, g_m/I_D and V_A are measured in figures presented here. While g_m/I_D is well predicted, as shown earlier in Figures 3.26 and 3.27, V_A is difficult to predict given its complex inversion-level, channel length, and drain–source voltage dependencies. V_A is predicted using an Early voltage factor, V_{AL}, which is extracted near the operating inversion level, channel length, and drain–source voltage as shown earlier in Figure 3.48.

Figure 4.7 shows measured g_m/I_D and V_A as a function of IC for operation in saturation at constant $V_{DS} = 1\,V$. While g_m/I_D is relatively insensitive to V_{DS} for operation in saturation, V_{DS} must be specified for V_A measurements since, as shown in the measured data of Section 3.8.4.5, V_A is strongly dependent on the value of V_{DS}.

As seen in Figure 4.7, g_m/I_D is maximum in weak inversion, decreasing only modestly in moderate inversion before decreasing as the inverse square root of IC in strong inversion. For short-channel devices at high IC experiencing velocity saturation, g_m/I_D decreases inversely with IC. V_A behaves

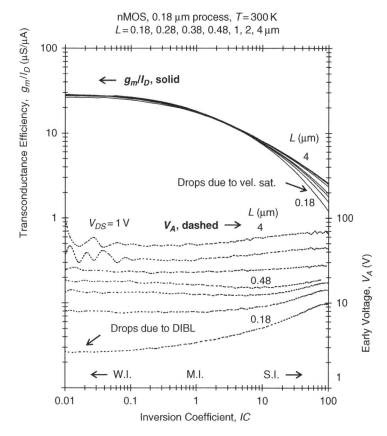

Figure 4.7 Measured transconductance efficiency, g_m/I_D (solid lines), and Early voltage, V_A (dashed lines), versus the inversion coefficient, IC, for $L = 0.18$, 0.28, 0.38, 0.48, 1, 2, and 4 μm, nMOS devices in a 0.18 μm CMOS process operating at $V_{DS} = 1$ V. g_m/I_D optimizes (maximizes) at low IC, while V_A is nearly independent of IC. In strong inversion, g_m/I_D decreases as the inverse square root of IC, before decreasing inversely with IC for short-channel devices at high IC due to velocity saturation. V_A generally increases slightly with IC, increasing more for short-channel devices due to reduced sensitivity to DIBL effects. g_m/I_D and V_A are independent of drain current and similar for pMOS devices, except for smaller pMOS g_m/I_D decreases due to less velocity saturation

very differently, generally increasing slightly as IC increases, except for short-channel devices where it increases more rapidly because of reduced sensitivity to DIBL effects. As IC increases and g_m/I_D decreases, there is less conversion of decreased threshold voltage with increasing V_{DS}, due to DIBL, into increased drain current and drain–source conductance, g_{ds}. This results in decreasing g_{ds} and increasing r_{ds} and V_A as IC increases as described earlier in Section 3.8.4.2.

Figure 4.8 shows g_m/I_D and V_A as a function of L. g_m/I_D is independent of L, excepting its decrease for short-channel devices at high IC due to velocity saturation. Here, g_m/I_D increases with L because of the reduction of velocity saturation effects. While g_m/I_D is nearly independent of L, V_A increases sublinearly to nearly linearly with L as shown in Figure 3.47, excepting its significant decrease for short-channel devices due to DIBL effects. Here, V_A increases more rapidly with L because of the reduction of DIBL effects. As described in Section 3.8.4.2, V_A decreases rapidly as L decreases to the process minimum because of rapidly increasing DIBL effects. The V_A decrease is less at high IC because, as mentioned, of reduced sensitivity to DIBL effects due to low g_m/I_D.

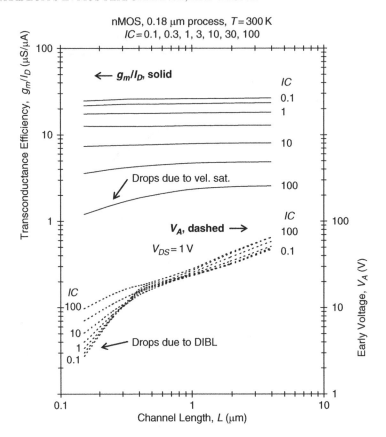

Figure 4.8 Measured transconductance efficiency, g_m/I_D (solid lines), and Early voltage, V_A (dashed lines), versus channel length, L, for nMOS devices in a $0.18\,\mu$m CMOS process operating at $IC = 0.1$, 0.3, 1, 3, 10, 30, and 100 at $V_{DS} = 1\,$V. V_A optimizes (maximizes) at long L, while g_m/I_D is independent of L, except for decreases for short-channel devices at high IC due to velocity saturation. V_A increases sublinearly with L, except for short-channel devices at low IC where DIBL effects are significant. g_m/I_D and V_A are independent of drain current and similar for pMOS devices, except for smaller pMOS g_m/I_D decreases due to less velocity saturation

g_m/I_D is maximized at low values of IC, while V_A is maximized at long channel lengths. As seen in Figure 4.7, there is no advantage of operation at IC below 0.1 in weak inversion because g_m/I_D ceases to increase here, while V_A remains nearly constant or decreases for short-channel devices due to DIBL effects.

Both g_m/I_D and V_A are independent of the drain current when evaluated in terms of IC. Additionally, pMOS g_m/I_D and V_A are similar to the nMOS values shown, excluding the smaller decrease in pMOS g_m/I_D for short-channel devices at high IC associated with less pMOS velocity saturation.

Maximizing g_m/I_D maximizes g_m for a given level of drain current, which favorably increases intrinsic voltage gain and decreases gate-referred thermal-noise voltage. High g_m/I_D corresponds to power-efficient production of g_m, where a given level of gate-referred thermal-noise voltage is obtained at minimum supply current. Section 6.3.1 describes this using a thermal-noise efficiency metric. Maximizing V_A maximizes r_{ds} for a given level of drain current, which increases the small-signal resistance appearing at the drain and raises the intrinsic voltage gain.

While g_m/I_D should be maximized for many applications, it should be minimized for applications where low g_m distortion is required. This is because the allowable input, 1 dB compression voltage, $V_{INDIF1dB}$, for a differential pair is inversely proportional to g_m/I_D as described later in Section 4.3.4.1. Additionally, as described for current-mirror devices later in Section 4.3.5.1, g_m/I_D should be minimized to minimize drain-referred thermal-noise current of non-input stages in low-noise circuits receiving voltage inputs through input transconductors.

4.3.3.5 Intrinsic voltage gain and bandwidth

The intrinsic voltage gain and bandwidth entries in Table 4.1 are intrinsic voltage gain, $A_{Vi} = g_m \cdot r_{ds} = g_m/g_{ds} = (g_m/I_D) \cdot V_A$, and intrinsic bandwidth, $f_{Ti} = g_m/[2\pi(C_{gsi} + C_{gbi})]$. A_{Vi} is the gate-to-drain voltage gain of a grounded-source device with its drain connected to an open-circuit current source load. f_{Ti} is the unity-gain frequency for the gate-to-drain current gain of a grounded-source device with its drain connected to a short-circuit load. A_{Vi} and f_{Ti} represent the optimal or unloaded voltage gain and bandwidth available for a single device having only intrinsic elements. μ_0 is the process value of low-field mobility where velocity saturation and VFMR effects are considered separately as described below Table 4.1 and in the sections and tables referenced.

Figure 4.9 shows measured A_{Vi} and predicted f_{Ti} as a function of IC. A_{Vi} is found by the product of measured g_m/I_D and V_A given in Figure 4.7 because of the earlier mentioned difficulty of accurately predicting V_A and, correspondingly, predicting A_{Vi}.

Since evaluated as a function of IC, A_{Vi} shown in Figure 4.9 generally tracks g_m/I_D versus IC shown in Figure 4.7 because V_A, also shown in Figure 4.7, is nearly constant with IC. A_{Vi} is maximum in weak inversion, decreasing only modestly in moderate inversion before decreasing as the inverse square root of IC in strong inversion. For short-channel devices at high IC experiencing velocity saturation reduction of g_m/I_D, A_{Vi} decreases inversely with IC. The shape of A_{Vi} versus IC is similar to the shape of g_m/I_D versus IC, although the value of A_{Vi} increases significantly with increasing channel length because of the increase in V_A.

Figure 4.9 shows the opposing tradeoffs of A_{Vi} and f_{Ti}, since f_{Ti} increases as IC in weak inversion before increasing as the square root of IC in strong inversion and leveling off for short-channel devices at high IC due to velocity saturation reduction of g_m. In strong inversion without significant velocity saturation, A_{Vi} decreases inversely as the square root of IC, while f_{Ti} increases as the square root of IC. Thus, the product of A_{Vi} and f_{Ti} remains nearly constant in strong inversion. This can be observed by visually observing the product or addition of the logarithmic presentation of A_{Vi} and f_{Ti} in Figure 4.9. When velocity saturation is significant for short-channel devices at high IC, A_{Vi} decreases more rapidly while f_{Ti} levels off, resulting in a decrease in the product of A_{Vi} and f_{Ti}. The product of A_{Vi} and f_{Ti} and other figures of merit for performance are discussed later in Section 4.3.6.

Figure 4.10 shows A_{Vi} and f_{Ti} as a function of channel length, L. Since evaluated as a function of L, A_{Vi} generally tracks V_A versus L shown in Figure 4.8 since g_m/I_D, also shown in Figure 4.8, is nearly constant with L. A_{Vi} increases sublinearly to nearly linearly with L, excepting the significant decrease for short-channel devices at low IC due to the DIBL reduction of V_A. Here, V_A and A_{Vi} increase more rapidly with L because of the reduction of DIBL effects. As mentioned, A_{Vi} decreases for short-channel devices at high IC due to velocity saturation reduction of g_m/I_D. Here, g_m/I_D increases with L and A_{Vi} increases more rapidly with L because of the reduction of velocity saturation effects. The shape of A_{Vi} versus L is similar to the shape of V_A versus L, although the value of A_{Vi} decreases significantly with increasing IC because of the decrease in g_m/I_D.

A_{Vi} and f_{Ti} move oppositely with L in Figure 4.10, just as they move oppositely with IC in Figure 4.9. f_{Ti} decreases as the inverse square of L, but decreases inversely with L for short-channel devices at high IC experiencing velocity saturation. Here, g_m increases with L because of the reduction of velocity saturation effects, resulting in less reduction of f_{Ti}. Unlike the product of A_{Vi} and f_{Ti} that is nearly constant with increasing IC in strong inversion, the product of A_{Vi} and f_{Ti} decreases with

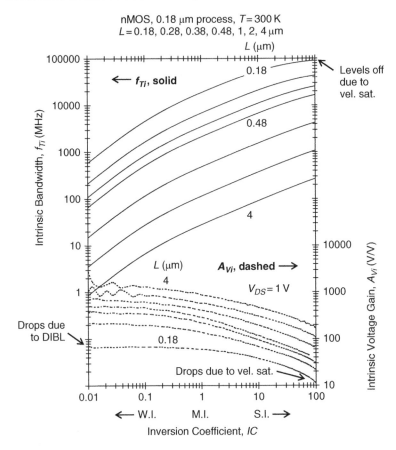

Figure 4.9 Measured intrinsic voltage gain, A_{Vi} (dashed lines), and predicted intrinsic bandwidth, f_{Ti} (solid lines), versus the inversion coefficient, IC, for $L = 0.18, 0.28, 0.38, 0.48, 1, 2,$ and $4\,\mu m$, nMOS devices in a $0.18\,\mu m$ CMOS process operating at $V_{DS} = 1\,V$. A_{Vi} and f_{Ti} optimize (maximize) oppositely from each other since A_{Vi} is maximum at low IC, while f_{Ti} is maximum at high IC. In strong inversion, A_{Vi} decreases inversely as the square root of IC, while f_{Ti} increases as the square root of IC, giving a constant product of A_{Vi} and f_{Ti}. When velocity saturation is significant for short-channel devices at high IC, A_{Vi} decreases inversely with IC, while f_{Ti} levels off. A_{Vi} and f_{Ti} are independent of drain current and similar for pMOS devices, except for smaller pMOS A_{Vi} and f_{Ti} decreases due to less velocity saturation and a nearly factor-of-five lower pMOS f_{Ti} due to larger intrinsic capacitances associated with larger required width and gate area

increasing L, assuming negligible velocity saturation and DIBL effects. This is because f_{Ti} decreases much more rapidly as the inverse square of L, assuming negligible velocity saturation, compared to the sublinear or near linear increase in A_{Vi} with L, assuming negligible DIBL effects.

Figures 4.9 and 4.10 show the opposing tradeoffs of gain and bandwidth. A_{Vi} is maximized at low IC and long channel lengths, while f_{Ti} is maximized under the opposite conditions of high IC and short channel lengths. As seen in Figure 4.9, there is no advantage of operation at IC below 0.1 in weak inversion because A_{Vi} ceases to increase here, while f_{Ti} continues decreasing. A balanced tradeoff of A_{Vi} and f_{Ti} might result from selecting IC on the strong-inversion side of moderate inversion, for example at $IC = 3$, and selecting L up modestly from the process minimum. This will be illustrated in the balanced optimization of differential-pair and current-mirror devices described later in Sections 4.4.4.4 and 4.4.4.5. Here, optimum DC performance, which includes A_{Vi}, is balanced with optimum AC performance, which includes f_{Ti}.

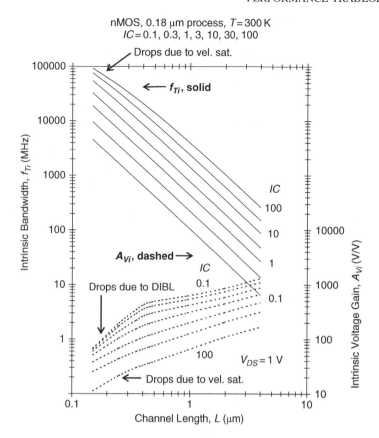

Figure 4.10   Measured intrinsic voltage gain, A_{Vi} (dashed lines), and predicted intrinsic bandwidth, f_{Ti} (solid lines), versus channel length, L, for nMOS devices in a $0.18\,\mu m$ CMOS process operating at $IC = 0.1, 0.3, 1, 3,$ 10, 30, and 100 at $V_{DS} = 1\,V$. A_{Vi} and f_{Ti} optimize (maximize) oppositely from each other since A_{Vi} is maximum at long L, while f_{Ti} is maximum at short L. Assuming velocity saturation and DIBL effects are negligible, A_{Vi} increases sublinearly with L, while f_{Ti} decreases rapidly as the inverse square of L. A_{Vi} and f_{Ti} are independent of drain current and similar for pMOS devices, except for smaller pMOS A_{Vi} and f_{Ti} decreases for short-channel devices at high IC due to less velocity saturation and a nearly factor-of-five lower pMOS f_{Ti} due to larger intrinsic capacitances associated with larger required width and gate area

Both A_{Vi} and f_{Ti} are independent of the drain current when evaluated in terms of IC. Additionally, pMOS A_{Vi} and f_{Ti} behave similarly to the nMOS values shown, excluding the smaller decrease in pMOS A_{Vi} and f_{Ti} for short-channel devices at high IC. This is due to the smaller decrease in g_m/I_D and g_m associated with less pMOS velocity saturation. However, pMOS f_{Ti} is approximately a factor-of-five smaller than the nMOS values shown because of the approximate factor-of-five increase in channel width, gate area, and intrinsic gate capacitance required because of lower pMOS mobility.

Extrinsic bandwidth, f_T, which includes the effects of extrinsic gate–source and gate–body overlap capacitances, C_{GSO} and C_{GBO}, and f_{diode}, which additionally includes the effect of the drain–body capacitance, C_{DB}, are lower than the intrinsic bandwidth, f_{Ti}, shown in Figures 4.9 and 4.10. As described in Section 3.9.7, f_T and f_{diode} can be well below f_{Ti} for short-channel devices because of small gate areas and small intrinsic gate capacitances. The trends with IC and L, however, are similar for each of the bandwidth measures, which are linked to the bandwidth of diode-connected devices, current mirrors, and other common circuits.

4.3.4 Tradeoffs Specific to Differential-Pair Devices

Section 4.3.3 presented MOS geometry, capacitance, bias compliance voltage, small-signal parameter, intrinsic voltage gain, and intrinsic bandwidth performance tradeoffs common to all device applications. This section presents transconductance distortion, thermal-noise, flicker noise, and local-area mismatch tradeoffs specific to differential-pair applications.

Differential-pair devices act as transconductors, converting the differential, gate input voltage to a differential, drain output current. When driven by differential, gate input voltages, the differential-pair source connection is at a virtual small-signal ground. The individual devices then act as grounded-source devices receiving opposite gate input voltages. As a result, the performance tradeoffs for a differential pair are similar to those for grounded-source devices receiving gate input voltages.

Like the performance tradeoffs presented in Section 4.3.3, tradeoffs are considered for native devices having no resistive source degeneration. This applies to differential-pair devices receiving differential, gate input voltages because the device sources are effectively at small-signal ground. Additionally, as before, tradeoffs are shown for nMOS devices in the example 0.18 μm CMOS process described in Tables 3.2–3.4. Finally, as before, tradeoffs exclude changes in performance associated with gate leakage current and the resulting gate–source conductance, gate shot- and flicker-noise current, and increase in DC mismatch described earlier in Section 3.12.2. Gate leakage current effects are negligible since the gate-oxide thickness is 4.1 nm for the example 0.18 μm CMOS process.

Table 4.2 lists transconductance distortion, thermal-noise, flicker noise, and local-area mismatch expressions and performance tradeoffs specific to differential-pair and grounded-source devices receiving gate input voltages. Like Table 4.1 that summarized tradeoffs earlier, Table 4.2 references the relevant section and prediction table in Chapter 3 for each aspect of performance. Also like Table 4.1, Table 4.2 shows performance trends as the inversion coefficient, channel length, and drain current are increased, using blank lines and up and down arrows of different sizes to indicate the direction and degree of trends. When two trends are included, the first is for weak inversion and the second is for strong inversion without small-geometry effects.

The abbreviated expressions and trends listed in the table exclude small-geometry effects like velocity saturation, VFMR, thermal-noise increases, and inversion-level or V_{EFF} increases in flicker noise. Full expressions and trends are included in the sections and tables referenced, with changes in trends resulting from small-geometry effects noted at the bottom of the table. Threshold-voltage mismatch is assumed to dominate the mismatch expressions and trends listed, while transconductance factor mismatch is included in the sections and tables referenced.

Like Table 4.1, Table 4.2 lists figures presented here and earlier in Chapter 3 that illustrate performance tradeoffs, inclusive of small-geometry effects. Again, two complementary or competing aspects of performance are shown here compared to single aspects of performance shown earlier in Chapter 3. Table 4.1, which shows performance tradeoffs common to all devices, and Table 4.2 collectively summarize performance tradeoffs for differential-pair and grounded-source devices receiving gate input voltages.

4.3.4.1 Transconductance distortion

The transconductance distortion entry in Table 4.2 lists the MOS differential-pair, input, 1 dB compression voltage, $V_{INDIF1dB}$. As mentioned, the table includes references to the relevant Chapter 3 prediction table and Chapter 3 and 4 figures, and lists trends for the selected inversion coefficient, channel length, and drain current. $V_{INDIF1dB}$ corresponds to the differential, gate input voltage where the differential, drain output current is compressed 1 dB (11 %) below its ideal, undistorted value. Higher values of $V_{INDIF1dB}$ correspond to higher allowable input voltages for a given level of distortion.

Figure 4.11 shows measured g_m/I_D and predicted $V_{INDIF1dB}$ versus the inversion coefficient, IC. As described earlier in Section 3.8.2.6, $V_{INDIF1dB}$ is nearly inversely proportional to g_m/I_D, having

Table 4.2 MOS performance tradeoffs specific to differential-pair and grounded-source devices versus the selected inversion coefficient, channel length, and drain current

Parameter	References			Trends		
	Section	Table	Figure	$IC \uparrow$ L, I_D fixed	$L \uparrow$ IC, I_D fixed	$I_D \uparrow$ IC, L fixed
Transcendance distortion:[a] $$V_{INDIF1dB} \propto \left(\frac{g_m}{I_D}\right)^{-1}$$ $\approx 1.22 \cdot nU_T$ (WI) $\approx 1.81 \cdot nU_T\sqrt{IC}$ (SI)	3.8.2.6	3.19	4.11, 3.28	$-, \uparrow$	$-$	$-$
Gate-referred thermal-noise voltage:[a,b] $$S_{VG}^{1/2} \text{ thermal} = \sqrt{\frac{4kT \cdot n\Gamma}{g_m}}$$	3.10.2.3	3.30	4.12, 4.13, 3.63	$-, \uparrow$	$-$	$\downarrow$
Gate-referred flicker-noise and local-area DC mismatch voltages:[c,d] $$S_{VG}^{1/2} \text{ flicker} = \sqrt{\frac{K_F'}{WLf^{AF}}}$$ $$= \left(\frac{\sqrt{IC}}{L}\sqrt{\frac{I_0}{I_D}}\right) \cdot \sqrt{\frac{K_F'}{f^{AF}}}$$	3.10.3.7	3.34	4.14, 4.15, 3.67	$\uparrow$	$\downarrow$	$\downarrow$
$$\Delta V_{GS}(1\sigma) = \frac{A_{VGS}}{\sqrt{WL}} \approx \frac{A_{VT}}{\sqrt{WL}}$$ $$\approx \left(\frac{\sqrt{IC}}{L}\sqrt{\frac{I_0}{I_D}}\right) \cdot A_{VT}$$	3.11.1.7	3.41	4.14, 4.15, 3.71	$\uparrow$	$\downarrow$	$\downarrow$

For simplicity, abbreviated expressions and trends are given with full expressions given in the tables referenced. Abbreviated expressions and trends exclude velocity saturation, VFMR, small-geometry thermal-noise increases, flicker-noise increases with increasing inversion or V_{EFF}, and increases in mismatch due to transcoductance factor mismatch. These effects are included in the full expressions with resulting trend effects described in the notes below. Expressions and trends are for operation in saturation.

[a] Velocity saturation, occurring for short-channel devices at high levels of inversion, causes $V_{INDIF1dB}$ and $S_{VG}^{1/2}$ thermal to increase further with increasing IC because of additional decreases in g_m/I_D and g_m. Under velocity saturation, $V_{INDIF1dB}$ and $S_{VG}^{1/2}$ thermal decrease with increasing L because of increases in g_m/I_D and g_m.

[b] When small-geometry effects cause increased $S_{VG}^{1/2}$ thermal, represented by increased Γ, the noise can decrease as L increases and Γ decreases to its normal value.

[c] $S_{VG}^{1/2}$ flicker may increase further with IC due to increases associated with increasing inversion or V_{EFF}.

[d] ΔV_{GS} may increase further with IC due to transconductance factor mismatch.

an approximate value of $1.22 \cdot (g_m/I_D)^{-1}$ in weak inversion and $1.81 \cdot (g_m/I_D)^{-1}$ in strong inversion. Correspondingly, $V_{INDIF1dB}$ is nearly directly proportional to the transconductance effective voltage, $V_{gm} = (g_m/I_D)^{-1}$, which is the reciprocal of g_m/I_D.

In weak inversion, g_m/I_D is high and $V_{INDIF1dB}$ is low, indicating that a small input voltage is required to steer differential-pair output current to the 1 dB compression point. Conversely, in strong inversion, g_m/I_D is low and $V_{INDIF1dB}$ is high, indicating that a large input voltage is required to steer the same proportion of output current. In strong inversion, g_m/I_D decreases inversely as the square root

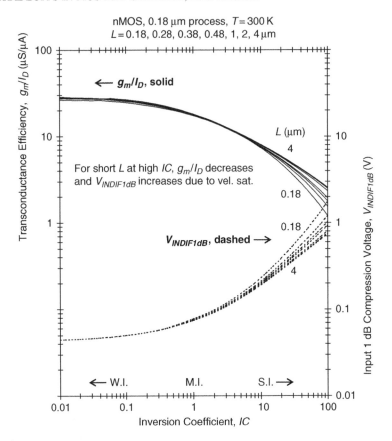

Figure 4.11 Measured transconductance efficiency, g_m/I_D (solid lines), and predicted differential-pair, input, 1 dB compression voltage, $V_{INDIF1dB}$ (dashed lines), versus the inversion coefficient, IC, for $L = 0.18, 0.28, 0.38,$ 0.48, 1, 2, and 4 μm, nMOS devices in a 0.18 μm CMOS process. g_m/I_D and $V_{INDIF1dB}$ optimize (maximize) oppositely from each other since g_m/I_D is maximum at low IC while $V_{INDIF1dB}$ is maximum at high IC. In strong inversion, g_m/I_D decreases as the inverse square root of IC, while $V_{INDIF1dB}$ increases as nearly the square root of IC. g_m/I_D decreases more rapidly and $V_{INDIF1dB}$ increases more rapidly for short-channel devices at high IC due to velocity saturation. g_m/I_D and $V_{INDIF1dB}$ are independent of drain current and similar for pMOS devices, except for smaller pMOS g_m/I_D decreases and smaller $V_{INDIF1dB}$ increases due to less velocity saturation

of IC, while $V_{INDIF1dB}$ increases nearly directly with the square root of IC. When velocity saturation is significant for short-channel devices at high IC, g_m/I_D decreases inversely with IC, while $V_{INDIF1dB}$ increases almost directly with IC. $V_{INDIF1dB}$ increases further due to linearization of the drain current associated with velocity saturation.

Figure 4.11 shows the opposing tradeoffs of g_m/I_D and $V_{INDIF1dB}$, where g_m/I_D is maximized at low IC and $V_{INDIF1dB}$ is maximized at high IC. This is an unfortunate result since the low g_m/I_D at high IC required for high $V_{INDIF1dB}$ and low distortion decreases g_m and increases the gate-referred thermal-noise voltage. g_m/I_D and $V_{INDIF1dB}$ are independent of L, except for their decrease and increase, respectively, for short-channel devices at high IC experiencing velocity saturation. Here, g_m/I_D increases and $V_{INDIF1dB}$ decreases with increasing L because of the reduction of velocity saturation effects. Both g_m/I_D and $V_{INDIF1dB}$ are independent of the drain current when evaluated in terms of IC. Finally, pMOS g_m/I_D and $V_{INDIF1dB}$ are similar to the nMOS values shown, excluding the smaller pMOS decreases and increases, respectively, associated with less pMOS velocity saturation.

4.3.4.2 Intrinsic gate capacitance and gate-referred thermal-noise voltage

The thermal-noise entry in Table 4.2 lists the gate-referred thermal-noise voltage density, $S_{VG}^{1/2}$ thermal, for a single device. Noise is referred to the gate because the gate receives a voltage input for differential-pair or grounded-source devices. The gate-referred thermal- and flicker-noise voltage density for a differential pair is 41 % above that of a single input device because of the addition of uncorrelated noise. Γ in the table is the thermal-noise factor relative to device transconductance. Γ has large-geometry values of one-half and two-thirds in weak and strong inversion, respectively.

Figure 4.12 shows $S_{VG}^{1/2}$ thermal and the input, intrinsic gate capacitance, $C_{gsi} + C_{gbi}$, as a function of IC. $C_{gsi} + C_{gbi}$, listed in Table 4.1 and described in Section 4.3.3.2, is shown to illustrate the tradeoffs

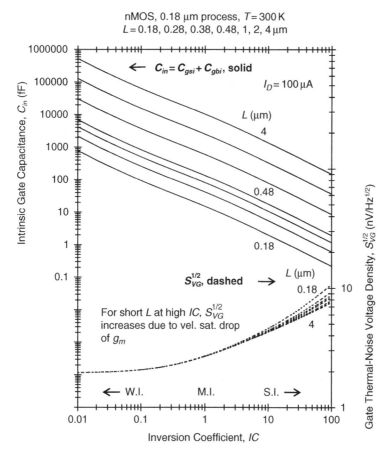

Figure 4.12 Predicted input, intrinsic gate capacitance, $C_{in} = C_{gsi} + C_{gbi}$ (solid lines), and gate-referred thermal-noise voltage density, $S_{VG}^{1/2}$ thermal (dashed lines), versus the inversion coefficient, IC, for $L = 0.18, 0.28, 0.38,$ 0.48, 1, 2, and 4 μm, nMOS devices in a 0.18 μm CMOS process operating at a drain current of 100 μA. C_{in} and $S_{VG}^{1/2}$ thermal optimize (minimize) oppositely from each other since C_{in} is minimum at high IC, while $S_{VG}^{1/2}$ thermal is minimum at low IC. C_{in} decreases nearly inversely with IC, tracking gate area. $S_{VG}^{1/2}$ thermal increases with IC, increasing further for short-channel devices at high IC due to velocity saturation reduction of g_m. Additionally, $S_{VG}^{1/2}$ thermal can increase beyond the values shown because of small-geometry noise increases. C_{in} increases directly with drain current, while $S_{VG}^{1/2}$ thermal decreases as the inverse square root of drain current. C_{in} is nearly a factor of five larger for pMOS devices because of larger required channel width and gate area, while $S_{VG}^{1/2}$ thermal is similar except for smaller pMOS increases due to less velocity saturation

between $S_{VG}^{1/2}$ thermal and $C_{gsi} + C_{gbi}$, which is the input, intrinsic gate capacitance loading for a single device. The corresponding differential input capacitance across two differential-pair devices is one-half the input capacitance of a single device. Increasing $C_{gsi} + C_{gbi}$ corresponds to decreasing bandwidth at the gate for a given source resistance.

Figure 4.12 shows the opposing tradeoffs of $S_{VG}^{1/2}$ thermal and $C_{gsi} + C_{gbi}$. $S_{VG}^{1/2}$ thermal increases as IC increases in moderate and strong inversion as a result of decreasing g_m/I_D (Figure 4.7) and, correspondingly, decreasing g_m. $C_{gsi} + C_{gbi}$, however, decreases nearly inversely with IC, nearly tracking the decrease of channel width and gate area. $S_{VG}^{1/2}$ thermal increases further for short-channel devices at high IC because of velocity saturation decreases in g_m. Additionally, as discussed earlier in Section 3.10.2.2, $S_{VG}^{1/2}$ thermal may increase beyond the values shown, especially for short-channel devices, due to small-geometry increases in thermal noise.

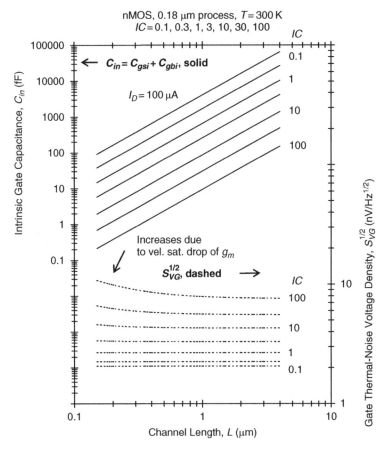

Figure 4.13 Predicted input, intrinsic gate capacitance, $C_{in} = C_{gsi} + C_{gbi}$ (solid lines), and gate-referred thermal-noise voltage density, $S_{VG}^{1/2}$ thermal (dashed lines), versus channel length, L, for nMOS devices in a 0.18 μm CMOS process operating at $IC = 0.1$, 0.3, 1, 3, 10, 30, and 100 at a drain current of 100 μA. C_{in} increases as the square of L, tracking gate area, while $S_{VG}^{1/2}$ thermal is independent of L, except for increases for short-channel devices at high IC due to velocity saturation reduction of g_m. Additionally, $S_{VG}^{1/2}$ thermal can increase beyond the values shown because of small-geometry noise increases. C_{in} increases directly with drain current, while $S_{VG}^{1/2}$ thermal decreases as the inverse square root of drain current. C_{in} is nearly a factor-of-five larger for pMOS devices because of larger required channel width and gate area, while $S_{VG}^{1/2}$ thermal is similar except for smaller pMOS increases due to less velocity saturation

Figure 4.13 shows $S_{VG}^{1/2}$ thermal and $C_{gsi} + C_{gbi}$ as a function of L. $S_{VG}^{1/2}$ thermal is independent of L, excepting its increase for short-channel devices at high IC when velocity saturation reduces g_m. Here, $S_{VG}^{1/2}$ thermal decreases with increasing L because of less velocity saturation reduction of g_m. Additionally, increasing L decreases $S_{VG}^{1/2}$ thermal when small-geometry noise increases are present. While $S_{VG}^{1/2}$ thermal is nearly independent of L, $C_{gsi} + C_{gbi}$ increases as the square of L, tracking the increase of gate area.

Figures 4.12 and 4.13 suggest that moderately low $S_{VG}^{1/2}$ thermal and moderately low $C_{gsi} + C_{gbi}$ (and moderately high intrinsic bandwidth) are obtained in moderate inversion for short-channel devices. Fortunately, velocity saturation is less likely in moderate inversion compared to strong inversion for short-channel devices, although channel length may need to be increased to avoid possible small-geometry noise increases. Additionally, the higher g_m/I_D in moderate inversion, again compared to strong inversion, results in power-efficient production of g_m, yielding lower $S_{VG}^{1/2}$ thermal for a given drain current. As seen in Figure 4.12 there is no advantage of operation at IC below 0.1 in weak inversion because $S_{VG}^{1/2}$ thermal ceases to decrease here, while $C_{gsi} + C_{gbi}$ continues increasing.

$S_{VG}^{1/2}$ thermal decreases as the inverse square root of drain current, taken as $100\,\mu A$ in the figures, because it is inversely proportional to the square root of g_m, which increases directly with the drain current (for a given value of IC). $C_{gsi} + C_{gbi}$, however, scales directly with the drain current. As a result, doubling the drain current decreases $S_{VG}^{1/2}$ thermal to 70.7 % ($1/\sqrt{2}$) of its original value, while doubling the required channel width, gate area, and $C_{gsi} + C_{gbi}$.

pMOS $S_{VG}^{1/2}$ thermal and $C_{gsi} + C_{gbi}$ behave similarly to the nMOS values shown, excluding the smaller increase in pMOS $S_{VG}^{1/2}$ thermal for short-channel devices at high IC due to less pMOS velocity saturation reduction of g_m. Additionally, pMOS $C_{gsi} + C_{gbi}$ is approximately a factor-of-five larger than the nMOS values shown because of the approximate factor-of-five increase in channel width and gate area required because of lower pMOS mobility.

Increasing $S_{VG}^{1/2}$ thermal at increasing IC, including the additional increase present for short-channel devices due to velocity saturation reduction of g_m, is problematic for power-efficient, low-noise, radio frequency (RF) preamplifiers. This is because operation at high IC at short channel lengths is required for low intrinsic gate capacitances and high intrinsic and operating bandwidths. However, drain current must increase significantly to counter the significant loss of g_m/I_D (Figure 4.7) and obtain the level of g_m required for a specified value of $S_{VG}^{1/2}$ thermal. Possible small-geometry increases in $S_{VG}^{1/2}$ thermal can further complicate the tradeoff of noise, bandwidth, and power.

4.3.4.3 Gate-referred flicker-noise voltage and gate–source mismatch voltage

The flicker noise and local-area mismatch entry in Table 4.2 lists the gate-referred flicker-noise voltage density, $S_{VG}^{1/2}$ flicker, for a single device and the local-area, gate–source mismatch voltage, ΔV_{GS}. ΔV_{GS} is the gate–source mismatch voltage between two identically sized and biased devices, typically configured in a differential pair. $S_{VG}^{1/2}$ flicker and ΔV_{GS} are considered together to illustrate their linkage through device gate area. K_F' is the flicker-noise factor in hand units of $(nV)^2 \cdot \mu m$, inclusive of inversion-level or V_{EFF} increases in the flicker noise, which is modeled in the section and table referenced. AF is the flicker-noise PSD slope factor, which is unity for $1/f$ noise, but is often slightly below unity. A_{VGS} and A_{VT} are the gate–source voltage and threshold-voltage mismatch factors, respectively, where A_{VGS} is equal to A_{VT} since transconductance factor mismatch is assumed negligible. Transconductance factor mismatch is included in the referenced section and table.

Figure 4.14 shows predicted $S_{VG}^{1/2}$ flicker at 1 Hz and ΔV_{GS} as a function of IC. Both increase as the square root of IC, tracking the inverse square root of gate area that decreases inversely with IC. However, when inversion-level or V_{EFF} increases in flicker noise are significant, $S_{VG}^{1/2}$ flicker increases further at high IC, especially for short-channel devices when V_{EFF} is increased due to velocity saturation. Additionally, when transconductance factor mismatch dominates threshold-voltage mismatch at high IC, ΔV_{GS} increases further at high IC, especially for short-channel devices when

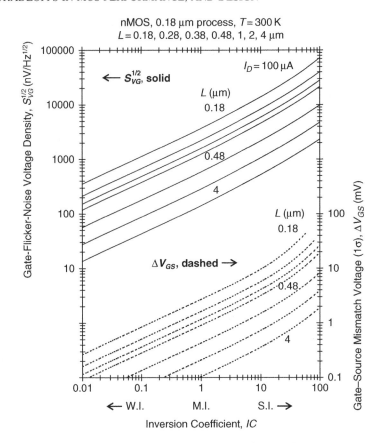

Figure 4.14 Predicted gate-referred flicker-noise voltage density at 1 Hz, $S_{VG}^{1/2}$ flicker (solid lines), and gate–source mismatch voltage, ΔV_{GS} (dashed lines), versus the inversion coefficient, IC, for $L = 0.18$, 0.28, 0.38, 0.48, 1, 2, and 4 μm, nMOS devices in a 0.18 μm CMOS process operating at a drain current of 100 μA. $S_{VG}^{1/2}$ flicker and ΔV_{GS} optimize (minimize) together at low IC where gate area and gate capacitance are large. Both increase as the square root of IC and decrease as the inverse square root of drain current, tracking the inverse square root of gate area. At high IC, $S_{VG}^{1/2}$ flicker and ΔV_{GS} can increase further due to inversion-level (V_{EFF}) increases in noise and transconductance factor mismatch, respectively. $S_{VG}^{1/2}$ flicker and ΔV_{GS} are nearly 45 % $(1/\sqrt{5})$ the nMOS values shown for pMOS devices having equal flicker-noise and mismatch factors because of larger required channel width and gate area. pMOS $S_{VG}^{1/2}$ flicker, however, often increases more rapidly at high IC due to higher inversion-level increases in noise

g_m/I_D is decreased due to velocity saturation. ΔV_{GS} increases further because drain current mismatch due to transconductance factor mismatch is divided by decreasing g_m to refer the mismatch to ΔV_{GS}.

Figure 4.15 shows $S_{VG}^{1/2}$ flicker at 1 Hz and ΔV_{GS} as a function of L. Both decrease inversely with L, tracking the inverse square root of gate area that increases as the square of L. When inversion-level or V_{EFF} increases in flicker noise are significant and V_{EFF} is increased for short-channel devices at high IC due to velocity saturation, $S_{VG}^{1/2}$ flicker decreases more rapidly with increasing L because of less velocity saturation increase in V_{EFF}. When transconductance factor mismatch dominates threshold voltage mismatch at high IC and g_m/I_D is decreased for short-channel devices at high IC due to velocity saturation, ΔV_{GS} decreases more rapidly with increasing L because of less velocity saturation decrease in g_m.

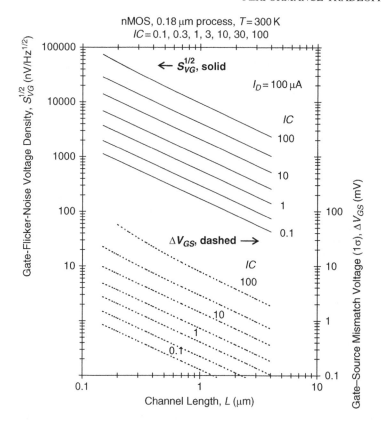

Figure 4.15   Predicted gate-referred flicker-noise voltage density at 1 Hz, $S_{VG}^{1/2}$ flicker (solid lines), and gate–source mismatch voltage, ΔV_{GS} (dashed lines), versus channel length, L, for nMOS devices in a 0.18 μm CMOS process operating at $IC = 0.1, 0.3, 1, 3, 10, 30$, and 100 at a drain current of 100 μA. $S_{VG}^{1/2}$ flicker and ΔV_{GS} optimize (minimize) together at long L where gate area and gate capacitance are large. Both decrease inversely with L and decrease as the inverse square root of drain current, tracking the inverse square root of gate area. At high IC, $S_{VG}^{1/2}$ flicker and ΔV_{GS} can increase further at short channel lengths due to inversion-level (V_{EFF}) increases in noise and transconductance factor mismatch, respectively. $S_{VG}^{1/2}$ flicker and ΔV_{GS} are nearly 45 % $(1/\sqrt{5})$ the nMOS values shown for pMOS devices having equal flicker-noise and mismatch factors because of larger required channel width and gate area. pMOS $S_{VG}^{1/2}$ flicker, however, often increases more rapidly at high IC due to higher inversion-level increases in noise

Figures 4.14 and 4.15 collectively show that $S_{VG}^{1/2}$ flicker and ΔV_{GS} are minimized together at low IC at long L where gate areas are large. However, increasing L has the greater effect on reducing $S_{VG}^{1/2}$ flicker and ΔV_{GS} since gate area increases as the square of L. Unfortunately, increasing L increases the input, intrinsic gate capacitance ($C_{gsi} + C_{gbi}$ described in Section 4.3.3.2) as the square of L, while decreasing intrinsic bandwidth (described in Section 4.3.3.5) as the inverse square of L, or as the inverse of L if full velocity saturation is present. This is an example of the opposing tradeoffs of minimum input capacitance or maximum intrinsic bandwidth desired for AC performance, and minimum flicker noise and DC mismatch desired for low-frequency and DC performance.

$S_{VG}^{1/2}$ flicker and ΔV_{GS} both decrease as the inverse square root of drain current, taken as 100 μA in the figures. This, again, is because $S_{VG}^{1/2}$ flicker and ΔV_{GS} decrease as the inverse square root of gate area, which directly tracks the drain current because of the required increase in channel width. As a result, doubling the drain current decreases $S_{VG}^{1/2}$ flicker and ΔV_{GS} to 70.7 % $(1/\sqrt{2})$ of their original values, while doubling the required channel width, gate area, and $C_{gsi} + C_{gbi}$.

As described earlier in Section 3.10.3.5, flicker-noise factors for weak or moderate inversion are often comparable for nMOS and pMOS devices in processes having minimum channel length below approximately 0.25 μm. For comparable flicker-noise factors, pMOS $S_{VG}^{1/2}$ flicker in weak and moderate inversion is lower than the nMOS values shown because of the approximate factor-of-five increase in pMOS channel width and gate area required due to lower pMOS mobility. However, as described earlier in Section 3.10.3.7, pMOS devices in the example 0.18 μm CMOS process have considerably greater inversion-level or V_{EFF} increases in flicker noise compared to nMOS devices. As a result, pMOS $S_{VG}^{1/2}$ flicker increases more rapidly with increasing IC from lower initial values before approaching or exceeding the nMOS values. This can be seen by comparing the pMOS $S_{VG}^{1/2}$ flicker in Figure 3.67 with the nMOS values shown in Figure 4.14.

For comparable values of threshold-voltage and transconductance factor mismatch factors, pMOS ΔV_{GS} is lower than the nMOS values shown because of the approximate factor-of-five increase in pMOS channel width and gate area. While non-zero V_{SB} is often present for differential-pair devices, $V_{SB} = 0$ V is assumed for the threshold-voltage mismatch factor of $A_{VTO} = 5$ mV $\cdot$ μm used here from Table 3.3. As described earlier in Section 3.11.1.5, the normally dominant, threshold-voltage component of mismatch is increased for non-zero V_{SB} because of body-effect mismatch. However, this increase is normally small or modest, especially since V_{SB} is usually necessarily limited to low values for low-voltage design.

Interestingly, as seen in Figures 4.14 and 4.15, $S_{VG}^{1/2}$ flicker and ΔV_{GS} optimize (minimize) together at low IC and long L. Flicker-noise results from carrier trapping and de-trapping near the Si–SiO$_2$ interface, while local-area, threshold-voltage mismatch results from mismatch in the number of active dopant atoms in the depletion region under the gate. Both decrease inversely with the square root of gate area.

As described earlier in Section 3.12.2, flicker-noise current and mismatch in gate leakage current may require the use of smaller channel lengths and gate areas to balance the traditional improvement in flicker noise and local-area mismatch at increasing gate area with the deterioration of these associated with gate leakage current at increasing gate area. While not a concern for the example 0.18 μm CMOS process considered here, gate leakage current can be important for gate-oxide thickness of 2 nm and below.

4.3.5 Tradeoffs Specific to Current-Mirror Devices

Section 4.3.3 presented MOS geometry, capacitance, bias compliance voltage, small-signal parameter, intrinsic voltage gain, and intrinsic bandwidth performance tradeoffs common to all device applications. Section 4.3.4 then presented transconductance distortion, thermal-noise, flicker noise, and local-area mismatch tradeoffs specific to differential-pair applications. This section presents thermal-noise, flicker noise, and local-area mismatch tradeoffs specific to current-mirror applications.

Like differential pairs, current mirrors are primary building blocks in analog MOS circuits. For simple current mirrors, performance is found directly from the devices since device sources are connected to small-signal ground through the power supply ground or rails. For cascode current mirrors, drain output current noise and mismatch are approximated from the bottom devices that have sources connected to small-signal ground, again, through the power supply ground or rails. Output current noise and mismatch are approximated from the bottom devices because noise and mismatch contributions from the top cascode devices are usually negligible due to the presence of high source degeneration resistances provided by drain–source resistances of the bottom devices. Frequency response at the input of simple current mirrors or low-voltage cascode mirrors can also be approximated from the bottom devices because the input connects to the gate of the bottom devices. As a result, tradeoffs in current-mirror performance can be evaluated directly from devices in simple current mirrors and often from bottom devices in cascode current mirrors. The drain output resistance of cascode current mirrors, however, must be found from the small-signal parameters of the top device in conjunction with the source degeneration resistance provided by the drain–source resistance of the bottom device. Small-signal resistance levels of MOS circuits were summarized earlier in Figure 3.22 and Table 3.16.

Like the performance tradeoffs presented in Sections 4.3.3 and 4.3.4, tradeoffs are considered for native devices having no resistive source degeneration. This applies to devices in simple current mirrors or the bottom devices in cascode current mirrors. Additionally, as before, tradeoffs are shown for nMOS devices in the example 0.18 μm CMOS process described in Tables 3.2–3.4. Finally, as before, tradeoffs exclude changes in performance associated with gate leakage current and the resulting gate–source conductance, gate shot- and flicker-noise current, and increase in DC mismatch described earlier in Section 3.12.2. Gate leakage current effects are negligible since the gate-oxide thickness is 4.1 nm for the example 0.18 μm CMOS process.

Table 4.3 lists thermal-noise, flicker-noise, and local-area mismatch expressions and performance tradeoffs specific to current-mirror devices. Like Tables 4.1 and 4.2 that summarized tradeoffs earlier, Table 4.3 references the relevant section and prediction table in Chapter 3 for each aspect of performance. Also like Tables 4.1 and 4.2, Table 4.3 shows performance trends as the inversion coefficient, channel length, and drain current are increased, using blank lines and up and down arrows of different sizes to indicate the direction and degree of trends. When two trends are included, the first is for weak inversion and the second is for strong inversion without small-geometry effects.

The abbreviated expressions and trends listed in the table exclude small-geometry effects like velocity saturation, VFMR, thermal-noise increases, and inversion-level or V_{EFF} increases in flicker noise. Full expressions and trends are included in the sections and tables referenced, with changes in trends resulting from small-geometry effects noted at the bottom of the table. Threshold-voltage mismatch is assumed to dominate the mismatch expressions and trends listed, while transconductance factor mismatch is included in the sections and tables referenced.

Like Tables 4.1 and 4.2, Table 4.3 lists figures presented here and earlier in Chapter 3 that illustrate performance tradeoffs, inclusive of small-geometry effects. Again, two complementary or competing aspects of performance are shown here compared to single aspects of performance shown earlier in Chapter 3. Table 4.1, which shows performance tradeoffs common to all devices, and Table 4.3 collectively summarize performance tradeoffs for current-mirror devices.

4.3.5.1 Intrinsic bandwidth and drain-referred thermal-noise current

The thermal-noise entry in Table 4.3 lists the drain-referred thermal-noise current density, $S_{ID}^{1/2}$ thermal, for a single device. Noise is referred to the drain because drain noise current from devices appears at the output of simple current mirrors. As mentioned, drain noise current from bottom or rail devices appears nearly directly at the output of cascode current mirrors. The output thermal- and flicker-noise current density for a single-output, 1:1 current mirror is 41 % above that of a single (bottom) device because of the addition of uncorrelated noise from the reference device. Γ in the table is the thermal-noise factor relative to device transconductance. Γ has large-geometry values of one-half and two-thirds in weak and strong inversion, respectively.

Figure 4.16 shows $S_{ID}^{1/2}$ thermal and the intrinsic bandwidth, f_{Ti}, as a function of IC. f_{Ti}, listed in Table 4.1 and described in Section 4.3.3.5, is shown to illustrate the tradeoffs between $S_{ID}^{1/2}$ thermal and f_{Ti}, which is related to the operating, current-mirror bandwidth. Operating bandwidth is well below f_{Ti} because the gate of the diode-connected, input device is paralleled with one or more gates of output devices. Thus, a single-output, 1:1 current mirror has an operating bandwidth immediately reduced below $f_{Ti}/2$ because of the paralleled capacitance of two gates. Another bandwidth measure, f_{diode}, described earlier in Section 3.9.7, better reflects the bandwidth associated with the diode-connected,

Table 4.3 MOS performance tradeoffs specific to current-mirror devices versus the selected inversion coefficient, channel length, and drain current

Parameter	References			Trends		
	Section	Table	Figure	$IC \uparrow$ L, I_D fixed	$L \uparrow$ IC, I_D fixed	$I_D \uparrow$ IC, L fixed
Drain-referred thermal-noise current:[a,b] $S_{ID}^{1/2}$ thermal $= \sqrt{4kT \cdot n\Gamma \cdot g_m}$	3.10.2.3	3.29	4.16, 4.17, 3.62	$-, \downarrow$	$-$	$\uparrow$
Drain-referred flicker-noise and local-area DC mismatch currents:[c−e] $S_{ID}^{1/2}$ flicker $= \sqrt{\dfrac{K_F'}{WLf^{AF}}} \cdot g_m$ $= \left(\dfrac{\sqrt{IC}}{L} \sqrt{\dfrac{I_0}{I_D}}\right) \cdot \sqrt{\dfrac{K_F'}{f^{AF}}} \cdot g_m$	3.10.3.7	3.34	4.18, 4.19, 3.68	$\uparrow, -$	$\downarrow$	$\uparrow$
$\dfrac{\Delta I_D}{I_D}(1\sigma) = \dfrac{A_{VGS} \cdot \dfrac{g_m}{I_D}}{\sqrt{WL}}$ $\approx \left(\dfrac{\sqrt{IC}}{L} \sqrt{\dfrac{I_0}{I_D}}\right) \cdot A_{VT} \cdot \dfrac{g_m}{I_D}$	3.11.1.7	3.41	4.18, 4.19, 3.72	$\uparrow, -$	$\downarrow$	$\downarrow$

For simplicity, abbreviated expressions and trends are given with full expressions given in the tables referenced. Abbreviated expressions and trends exclude velocity saturation, VFMR, small-geometry thermal noise increases, flicker-noise increases with increasing inversion or V_{EFF}, and increases in mismatch due to transconductance factor mismatch. These effects are included in the full expressions with resulting trend effects described in the notes below. Expressions and trends are for operation in saturation.
[a] Velocity saturation, occurring for short-channel devices at high levels of inversion, can cause $S_{ID}^{1/2}$ thermal to decrease further because of additional decreases in g_m. Under velocity saturation, $S_{ID}^{1/2}$ thermal can increase with increasing L because of increases in g_m.
[b] When small-geometry effects cause increased $S_{ID}^{1/2}$ thermal, represented by increased Γ, the noise can decrease as L increases and Γ decreases to its normal value.
[c] In strong inversion, $S_{ID}^{1/2}$ flicker and $\Delta I_D/I_D$ level off with increasing IC because the increase in $S_{VG}^{1/2}$ flicker and threshold-voltage mismatch caused by decreasing gate area is countered by decreasing g_m and g_m/I_D. Velocity saturation can cause a roll-over of $S_{ID}^{1/2}$ flicker and $\Delta I_D/I_D$ due to additional decreases in g_m and g_m/I_D.
[d] $S_{ID}^{1/2}$ flicker may increase further with IC due to increases associated with increasing inversion or V_{EFF}.
[e] $\Delta I_D/I_D$ may increase further with IC due to transconductance factor mismatch.

input device of a current mirror by including the extrinsic gate-overlap and drain–body capacitances along with the intrinsic gate capacitances. f_{diode} can be significantly below f_{Ti}, especially for short-channel devices that have small gate areas and small intrinsic capacitances. Still, f_{Ti} shown here illustrates trends of operating, current-mirror bandwidth, which generally tracks some fraction of f_{Ti}.

Figure 4.16 shows that $S_{ID}^{1/2}$ thermal favorably decreases as IC increases in moderate and strong inversion because of decreasing g_m/I_D (Figure 4.7) and, correspondingly, decreasing g_m. f_{Ti} favorably increases directly with IC in weak inversion and as the square root of IC in strong inversion, before leveling off for short-channel devices at high IC due to velocity saturation. The favorable increase in f_{Ti} as IC increases represents increased operating, current-mirror bandwidth even as $S_{ID}^{1/2}$ thermal favorably decreases. $S_{ID}^{1/2}$ thermal decreases further for short-channel devices at high IC because of velocity saturation decreases in g_m. However, like the gate-referred thermal-noise voltage density shown earlier in Figure 4.12, $S_{ID}^{1/2}$ thermal may increase beyond that shown, especially for short-channel devices, due to small-geometry increases in thermal noise that were discussed earlier in Section 3.10.2.2.

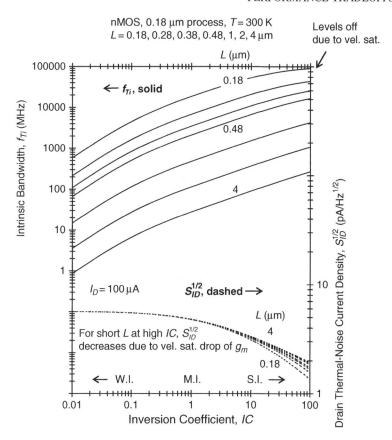

Figure 4.16 Predicted intrinsic bandwidth, f_{Ti} (solid lines), and drain-referred thermal-noise current density, $S_{ID}^{1/2}$ thermal (dashed lines), versus the inversion coefficient, IC, for $L = 0.18$, 0.28, 0.38, 0.48, 1, 2, and 4 μm, nMOS devices in a 0.18 μm CMOS process operating at a drain current of 100 μA. f_{Ti} and $S_{ID}^{1/2}$ thermal optimize together since f_{Ti} is maximum and $S_{ID}^{1/2}$ thermal minimum at high IC. In strong inversion, f_{Ti} increases as the square root of IC before leveling off for short-channel devices at high IC due to velocity saturation. $S_{ID}^{1/2}$ thermal decreases with IC, decreasing further when velocity saturation causes reduction of g_m. However, $S_{ID}^{1/2}$ thermal can increase beyond the values shown because of small-geometry noise increases. f_{Ti} is independent of drain current, while $S_{ID}^{1/2}$ thermal increases as the square root of drain current. f_{Ti} is nearly a factor-of-five lower for pMOS devices due to larger intrinsic capacitances associated with larger required width and gate area, while $S_{ID}^{1/2}$ thermal is similar. Both are less affected by velocity saturation for pMOS devices

Figure 4.17 shows $S_{ID}^{1/2}$ thermal and f_{Ti} as a function of L. $S_{ID}^{1/2}$ thermal is independent of L, excepting its decrease for short-channel devices at high IC when velocity saturation reduces g_m. Here, $S_{ID}^{1/2}$ thermal increases with L because of less velocity saturation reduction of g_m. However, increasing L decreases $S_{ID}^{1/2}$ thermal when small-geometry noise increases are present. While $S_{ID}^{1/2}$ thermal is nearly independent of L, f_{Ti} decreases rapidly as the inverse square of L, except for short-channel devices experiencing full velocity saturation where it decreases inversely with L.

Figures 4.16 and 4.17 show that $S_{ID}^{1/2}$ thermal is minimized and f_{Ti} is maximized at high IC, with f_{Ti} additionally maximized for short-channel devices. This is a fortunate result since many MOS performance trends oppose each other. However, as mentioned, $S_{ID}^{1/2}$ thermal may increase for short-channel devices, so the choice of high IC and short L may not be optimal. Additionally, as shown in Figure 4.6, $V_{DS,sat}$ is maximum at high IC, increasing the bias compliance voltages for current-mirror circuits. As a result, IC will usually be limited by $V_{DS,sat}$ requirements, which will raise $S_{ID}^{1/2}$ thermal

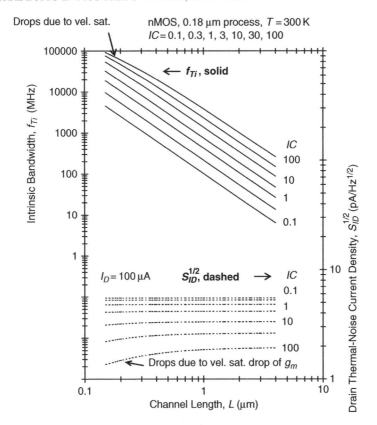

Figure 4.17   Predicted intrinsic bandwidth, f_{Ti} (solid lines), and drain-referred thermal-noise current density, $S_{ID}^{1/2}$ thermal (dashed lines), versus channel length, L, for nMOS devices in a 0.18 μm CMOS process operating at $IC = 0.1, 0.3, 1, 3, 10, 30,$ and 100 at a drain current of 100 μA. f_{Ti} decreases as the inverse square of L, except for short-channel devices at high IC experiencing velocity saturation where it decreases inversely with L. $S_{ID}^{1/2}$ thermal is independent of L unless velocity saturation effects decrease g_m and decrease $S_{ID}^{1/2}$ thermal. However, $S_{ID}^{1/2}$ thermal can increase beyond the values shown because of small-geometry noise increases. f_{Ti} is independent of drain current, while $S_{ID}^{1/2}$ thermal increases as the square root of drain current. f_{Ti} is nearly a factor-of-five lower for pMOS devices due to larger intrinsic capacitances associated with larger required width and gate area, while $S_{ID}^{1/2}$ thermal is similar. Both are less affected by velocity saturation for pMOS devices

and lower f_{Ti}. $S_{ID}^{1/2}$ thermal is evaluated in terms of $V_{DS,sat}$ later in Section 6.5.4.1 to illustrate the difficulty of minimizing $S_{ID}^{1/2}$ thermal for non-input devices in low-voltage applications where $V_{DS,sat}$ must be low. This is developed as part of the micropower, low-noise, CMOS preamplifier examples described later in Chapter 6.

$S_{ID}^{1/2}$ thermal increases as the square root of drain current, taken as 100 μA in the figures, because it is proportional to the square root of g_m, which increases directly with the drain current (for a given value of IC). f_{Ti}, however, is independent of the drain current. As a result, halving the drain current decreases $S_{ID}^{1/2}$ thermal to 70.7 % ($1/\sqrt{2}$) of its original value, while halving the transconductance and required gate area and capacitances, resulting in no change in f_{Ti}.

pMOS $S_{ID}^{1/2}$ thermal and f_{Ti} behave similarly to the nMOS values shown, excluding the smaller pMOS decreases in $S_{ID}^{1/2}$ thermal and f_{Ti} for short-channel devices at high IC due to less pMOS velocity saturation reduction of g_m. Additionally, pMOS f_{Ti} is approximately a factor-of-five smaller than the nMOS values shown because of the approximate factor-of-five increase in channel width and gate area required because of lower pMOS mobility.

4.3.5.2 Drain-referred flicker-noise current and drain mismatch current

The flicker noise and local-area mismatch entry in Table 4.3 lists the drain-referred flicker-noise current density, $S_{ID}^{1/2}$ flicker, for a single device and the local-area, drain mismatch current, $\Delta I_D/I_D$. $\Delta I_D/I_D$ is the relative drain mismatch current between two identically sized and biased devices, typically configured in a current mirror. $S_{ID}^{1/2}$ flicker and $\Delta I_D/I_D$ are considered together to illustrate their linkage through device gate area. K_F' is the flicker-noise factor in hand units of $(nV)^2 \cdot \mu m$, inclusive of inversion-level or V_{EFF} increases in the flicker noise, which is modeled in the section and table referenced. AF is the flicker-noise PSD slope factor, which is unity for $1/f$ noise, but is often slightly below unity. A_{VGS} and A_{VT} are the gate–source voltage and threshold-voltage mismatch factors, respectively, where A_{VGS} is equal to A_{VT} since transconductance factor mismatch is assumed to be negligible. Transconductance factor mismatch is included in the referenced section and table.

Figure 4.18 shows predicted $S_{ID}^{1/2}$ flicker at 1 Hz and $\Delta I_D/I_D$ as a function of IC. Both increase as the square root of IC in weak inversion, tracking the gate-referred flicker-noise voltage density, $S_{VG}^{1/2}$ flicker, and threshold-voltage mismatch that are multiplied by constant weak-inversion g_m to obtain drain noise and mismatch current. $S_{VG}^{1/2}$ flicker and threshold-voltage mismatch increase as the square root of IC, tracking the inverse square root of gate area that decreases inversely with IC. Interestingly, both $S_{ID}^{1/2}$ flicker and $\Delta I_D/I_D$ level off in strong inversion because the increasing $S_{VG}^{1/2}$ flicker and threshold-voltage mismatch associated with decreasing gate area are countered by decreasing g_m. For short-channel devices at high IC experiencing velocity saturation reduction of g_m, $S_{ID}^{1/2}$ flicker actually begins to roll down. $S_{ID}^{1/2}$ flicker then increases when inversion-level or V_{EFF} increases in flicker noise become significant. $\Delta I_D/I_D$ can also roll down for short-channel devices at high IC because of velocity saturation reduction of g_m. $\Delta I_D/I_D$ then increases when transconductance factor mismatch dominates threshold-voltage mismatch.

Figure 4.19 shows $S_{ID}^{1/2}$ flicker at 1 Hz and $\Delta I_D/I_D$ as a function of L. Both decrease inversely with L, tracking the inverse square root of gate area that increases as the square of L. $S_{ID}^{1/2}$ flicker and $\Delta I_D/I_D$ decrease somewhat for short-channel devices at high IC due to velocity saturation reduction of g_m, unless inversion-level or V_{EFF} increases in flicker noise are significant or transconductance factor mismatch dominates threshold-voltage mismatch.

Figures 4.18 and 4.19 collectively show that $S_{ID}^{1/2}$ flicker and $\Delta I_D/I_D$ are minimized together at low IC and long L where gate areas are large. These are minimized at low IC in weak inversion because the decrease in $S_{VG}^{1/2}$ flicker and threshold voltage mismatch associated with large gate area exceeds the effect of maximum and constant g_m. While $S_{ID}^{1/2}$ flicker and $\Delta I_D/I_D$ level off at increasing IC in strong inversion, increasing L consistently reduces them. Unfortunately, increasing L decreases f_{Ti} as the inverse square of L, or the inverse of L if full velocity saturation is present. This is an example of the opposing tradeoffs of maximum bandwidth desired for AC performance and minimum flicker noise and DC mismatch desired for low-frequency and DC performance.

$S_{ID}^{1/2}$ flicker increases as the square root of drain current, taken as $100\,\mu A$ in the figures, while $\Delta I_D/I_D$ decreases as the inverse square root of drain current. $S_{ID}^{1/2}$ flicker increases as the square root of drain current because g_m increases directly with the drain current (for constant IC), while $S_{VG}^{1/2}$ flicker decreases as the inverse square root of drain current because of the increase in required channel width and gate area. $\Delta I_D/I_D$ decreases as the inverse square root of drain current because of the increase in required channel width and gate area. As a result, halving the drain current decreases $S_{ID}^{1/2}$ flicker to 70.7 % $(1/\sqrt{2})$ of its original value and increases $\Delta I_D/I_D$ to 141 % $(\sqrt{2})$ of its original value. This also halves the transconductance and required gate area and capacitances, resulting in no change in f_{Ti}.

As mentioned in Section 4.3.4.3 for $S_{VG}^{1/2}$ flicker, flicker-noise factors for weak or moderate inversion are often comparable for nMOS and pMOS devices in processes having minimum process channel

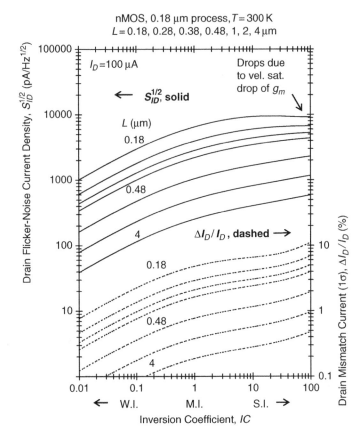

Figure 4.18 Predicted drain-referred flicker-noise current density at 1 Hz, $S_{ID}^{1/2}$ flicker (solid lines), and drain mismatch current, $\Delta I_D/I_D$ (dashed lines), versus the inversion coefficient, IC, for $L = 0.18, 0.28, 0.38, 0.48, 1,$ 2, and 4 μm, nMOS devices in a 0.18 μm CMOS process operating at a drain current of 100 μA. $S_{ID}^{1/2}$ flicker and $\Delta I_D/I_D$ optimize (minimize) together at low IC where gate area is large and bandwidth low. In weak inversion, $S_{ID}^{1/2}$ flicker and $\Delta I_D/I_D$ increase as the square root of IC, tracking the inverse square root of gate area. In strong inversion, both level off as described in the text. At high IC, $S_{ID}^{1/2}$ flicker and $\Delta I_D/I_D$ can decrease for short-channel devices due to velocity saturation reductions in g_m. However, $S_{ID}^{1/2}$ flicker can then increase due to inversion-level (V_{EFF}) increases in noise, and $\Delta I_D/I_D$ can increase due to transconductance factor mismatch. $S_{ID}^{1/2}$ flicker increases as the square root of drain current, while $\Delta I_D/I_D$ decreases as the inverse square root of drain current. $S_{ID}^{1/2}$ flicker and $\Delta I_D/I_D$ are nearly 45 % ($1/\sqrt{5}$) the nMOS values shown for pMOS devices having equal flicker-noise and mismatch factors because of larger required channel width and gate area. pMOS $S_{ID}^{1/2}$ flicker, however, often increases more rapidly at high IC due to higher inversion-level increases in noise

length below approximately 0.25 μm. In this case, pMOS $S_{ID}^{1/2}$ flicker in weak and moderate inversion is lower than the nMOS values shown because of the approximate factor-of-five increase in pMOS channel width and gate area required due to lower pMOS mobility. Again, as mentioned in Section 4.3.4.3, pMOS devices in the example 0.18 μm CMOS process have considerably greater inversion-level or V_{EFF} increases in flicker noise compared to nMOS devices. As a result, pMOS $S_{ID}^{1/2}$ flicker increases more rapidly with increasing IC from lower initial values before approaching or exceeding the nMOS values. This can be seen by comparing the pMOS $S_{ID}^{1/2}$ flicker in Figure 3.68 with the nMOS values shown in Figure 4.18.

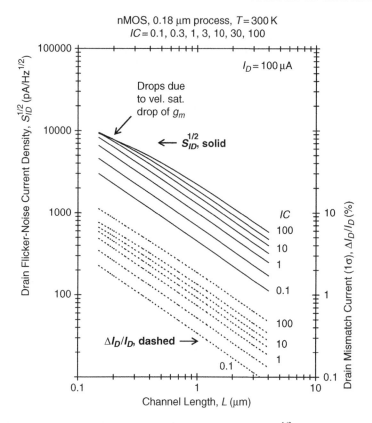

Figure 4.19 Predicted drain-referred flicker-noise current density at 1 Hz, $S_{ID}^{1/2}$ flicker (solid lines), and drain mismatch current, $\Delta I_D/I_D$ (dashed lines), versus channel length, L, for nMOS devices in a $0.18\,\mu m$ CMOS process operating at $IC = 0.1, 0.3, 1, 3, 10, 30,$ and 100 at a drain current of $100\,\mu A$. $S_{ID}^{1/2}$ flicker and $\Delta I_D/I_D$ optimize (minimize) together at long L where gate area is large and bandwidth low. Both decrease inversely with L, tracking the inverse square-root of gate area. At high IC, $S_{ID}^{1/2}$ flicker and $\Delta I_D/I_D$ can decrease for short-channel devices due to velocity saturation reductions in g_m, or these can increase due to inversion-level (V_{EFF}) increases in noise and transconductance factor mismatch. $S_{ID}^{1/2}$ flicker increases as the square root of drain current, while $\Delta I_D/I_D$ decreases as the inverse square root of drain current. $S_{ID}^{1/2}$ flicker and $\Delta I_D/I_D$ are nearly 45% $(1/\sqrt{5})$ the nMOS values shown for pMOS devices having equal flicker-noise and mismatch factors because of larger required channel width and gate area. pMOS $S_{ID}^{1/2}$ flicker, however, often increases more rapidly at high IC due to higher inversion-level increases in noise

Like the gate–source mismatch voltage, ΔV_{GS}, mentioned in Section 4.3.4.3, pMOS $\Delta I_D/I_D$ is lower compared to the nMOS values shown for comparable values of threshold-voltage and transconductance factor mismatch factors. This, again, is because of the approximate factor-of-five increase in pMOS channel width and gate area. $\Delta I_D/I_D$ is shown for zero V_{SB}, which is typical for devices in simple current mirrors or the bottom rail devices in cascode current mirrors. As mentioned in Section 4.3.4.3, mismatch increases for non-zero V_{SB} because of increased threshold-voltage mismatch resulting from body-effect mismatch.

Like $S_{VG}^{1/2}$ flicker and ΔV_{GS} described in Section 4.3.4.3 and shown in Figures 4.14 and 4.15, $S_{ID}^{1/2}$ flicker and $\Delta I_D/I_D$ optimize (minimize) together at low IC and long L as seen in Figures 4.18 and 4.19. Again, flicker-noise results from carrier trapping and de-trapping near the Si–SiO$_2$ interface, while local-area, threshold-voltage mismatch results from mismatch in the number of active dopant atoms in the depletion region under the gate. Both decrease inversely with the square root of gate area. $S_{ID}^{1/2}$ flicker and $\Delta I_D/I_D$, however, have more complex behavior with increasing IC compared to $S_{VG}^{1/2}$ flicker and ΔV_{GS}. This is because $S_{VG}^{1/2}$ flicker and ΔV_{GS} are multiplied by g_m and g_m/I_D, respectively, for conversion to $S_{ID}^{1/2}$ flicker and $\Delta I_D/I_D$. The most notable difference between $S_{VG}^{1/2}$ flicker and ΔV_{GS}, which are gate-referred voltages, and $S_{ID}^{1/2}$ flicker and $\Delta I_D/I_D$, which are drain-referred currents, is the leveling off of $S_{ID}^{1/2}$ flicker and $\Delta I_D/I_D$ in strong inversion where the increase in $S_{VG}^{1/2}$ flicker and ΔV_{GS} is countered by decreasing g_m and g_m/I_D, respectively.

Again, as described earlier in Section 3.12.2, flicker-noise current and mismatch in gate leakage current may require the use of smaller channel lengths and gate areas to balance the traditional improvement in flicker noise and local-area mismatch at increasing gate area with the deterioration of these associated with gate leakage current at increasing gate area. While not a concern for the example 0.18 μm CMOS process considered here, gate leakage current can be important for gate-oxide thickness of 2 nm and below.

4.3.6 Tradeoffs in Figures of Merit

Section 4.3.3 presented MOS performance tradeoffs common to all applications, while Sections 4.3.4 and 4.3.5 presented tradeoffs specific to differential-pair and current-mirror applications. This section presents tradeoffs in performance figures of merit. Figures of merit involve the combination of multiple aspects of performance and are useful for design intuition and optimization.

Table 4.4 lists expressions for figures of merit including intrinsic voltage gain and bandwidth, the product of intrinsic voltage gain and bandwidth, the product of transconductance efficiency and intrinsic bandwidth, the thermal-noise efficiency factor, and bandwidth–power–accuracy metrics involving both DC offset and thermal noise. The table includes references to tables, figures, and equations that

Table 4.4 MOS figure-of-merit tradeoffs versus the selected inversion coefficient and channel length. The figures of merit are independent of the drain current

Parameter	Reference	Trends	
		$IC \uparrow$ L fixed	$L \uparrow$ IC fixed
Intrinsic voltage gain and bandwidth:[a]			
$A_{Vi} = \dfrac{V_{AL}(IC, L, V_{DS}) \cdot L}{nU_T\left(\sqrt{IC+0.25}+0.5\right)}$	Table 4.1, Figures 4.9, 4.10	$-, \downarrow, \downarrow$	$\uparrow, \uparrow,$ $\uparrow\uparrow$
$f_{Ti} = \left(\dfrac{IC}{\sqrt{IC+0.25}+0.5}\right) \cdot \left(\dfrac{\mu_0 U_T}{\pi\left(\hat{C}_{gsi}+\hat{C}_{gbi}\right) L^2}\right)$	Table 4.1, Figures 4.9, 4.10	$\uparrow, \uparrow, -$	$\downarrow\downarrow,$ $\downarrow\downarrow, \downarrow$
Intrinsic gain-bandwidth product:[a]			
$A_{Vi} \cdot f_{Ti} = \dfrac{V_{AL}(IC, L, V_{DS}) \cdot \mu_0}{n\pi\left(\hat{C}_{gsi}+\hat{C}_{gbi}\right) L} \left(\dfrac{IC}{\left(\sqrt{IC+0.25}+0.5\right)^2}\right)$	Table 4.1, product of A_{Vi} and f_{Ti} in Figures 4.9, 4.10	$\uparrow, -, \downarrow$	$\downarrow, \downarrow, \uparrow$
Transconductance efficiency, intrinsic bandwidth product:			
$\left(\dfrac{g_m}{I_D}\right) f_{Ti} = \dfrac{\mu_0}{n\pi\left(\hat{C}_{gsi}+\hat{C}_{gbi}\right) L^2} \left(\dfrac{IC}{\left(\sqrt{IC+0.25}+0.5\right)^2}\right)$	Table 4.1, product of g_m/I_D and f_{Ti} in Figures 4.7–4.10	$\uparrow, -, \downarrow$	$\downarrow\downarrow, \downarrow\downarrow,$ $-$

Thermal-noise efficiency factor for single MOSFET:[b]

$$NEF' = \frac{S_{VG} \text{ thermal (MOS)}}{S_{VB} \text{ thermal (bipolar)}}$$

$$= 2n^2\Gamma\left(\sqrt{IC+0.25}+0.5\right)$$

Equation 6.3,
Figure 6.1

¬, ↑, ↑   ¬, ¬, ↓

Bandwidth–power–accuracy, DC offset:[b–d]

$$\frac{A_{CC,DC}^2 \cdot BW}{P} = \left(\frac{I_{IN}}{I_D}\right)^2 \frac{nU_T\left(\sqrt{IC+0.25}+0.5\right)}{48\pi C_{OX}' A_{VT}^2 V_{DD}}$$

(I)   Equation 3.166   ¬, ↑, ↑   ¬, ¬, ↓

$$\frac{A_{CC,DC}^2 \cdot BW}{P} = \frac{V_{IN}^2}{\dfrac{nU_T\left(\sqrt{IC+0.25}+0.5\right)}{12\pi C_{OX}' A_{VT}^2 V_{DD}}}$$

(V)   Equation 3.172   ¬, ↓, ↓   ¬, ¬, ↑

At design limit:

$$\frac{A_{CC,DC}^2 \cdot BW}{P} \propto \frac{1}{C_{OX}' A_{VT}^2} = \frac{1}{E_{matching}}$$

(both I and V) Equations 3.167,
3.173

–   –

Bandwidth–power–accuracy, thermal noise:[b,c]

$$\frac{A_{CC,AC}^2 \cdot BW}{P} = \left(\frac{I_{IN}}{I_D}\right)^2 \frac{U_T\left(\sqrt{IC+0.25}+0.5\right)}{16kT \cdot \Gamma \cdot V_{DD}}$$

(I)   Equation 4.3   ¬, ↑, ↑   ¬, ¬, ↓

$$\frac{A_{CC,AC}^2 \cdot BW}{P} = \frac{V_{IN}^2}{\dfrac{n^2 U_T\left(\sqrt{IC+0.25}+0.5\right)}{16kT \cdot \Gamma \cdot V_{DD}}}$$

(V)   Equation 4.6   ¬, ↓, ↓   ¬, ¬, ↑

At design limit:

$$\frac{A_{CC,AC}^2 \cdot BW}{P} \propto \frac{1}{kT} = \frac{1}{E_{thermal}}$$

(both I and V) Text   –   –

Trends include velocity saturation but exclude DIBL effects. Multiple trends are for WI, SI, and SI with full velocity saturation. See Table 4.1 for trend key. Expressions and trends are for operation in saturation.

Velocity saturation and VFMR effects may be included by replacing IC with $IC(1 + IC/IC_{CRIT})$ in $\sqrt{IC+0.25}$ terms, where $IC_{CRIT} = [(LE_{CRIT})'/(4nU_T)]^2$ and $(LE_{CRIT})' = (1/\theta)\|LE_{CRIT}$ as given in Table 3.17. When velocity saturation is significant, this results in channel length dependency on g_m/I_D, giving channel length dependency or additional channel length dependency on the figures of merit.

[a] When DIBL causes reduced V_A and A_{Vi} for short-channel devices, especially at low levels of inversion, A_{Vi} and $A_{Vi} \cdot f_{Ti}$ increase beyond the trends shown as L increases because DIBL effects are reduced.

[b] NEF' and current-mode, bandwidth–power–accuracy figures of merit are inversely proportional to g_m/I_D.

[c] Voltage-mode, bandwidth–power–accuracy figures of merit are proportional to g_m/I_D.

[d] Figures of merit involving DC offset assume threshold-voltage mismatch dominates transconductance factor mismatch. Full expressions and trends can be derived from individual measures of performance given in the tables referenced in Tables 4.1–4.3.

describe the figures of merit or their component parts. The table also shows figure-of-merit trends as the inversion coefficient and channel length are increased. As noted earlier in Table 4.1, blank lines indicate little change, while up and down arrows indicate increases or decreases, with the degree of the increase or decrease indicated by the size of the arrow. When multiple trends are included, the first is for weak inversion, the second is for strong inversion without small-geometry effects, and the third is for strong inversion with full velocity saturation.

The expressions and trends include velocity saturation and VFMR decreases in g_m/I_D predicted using the expression modifications listed at the bottom of the table that involve the critical inversion coefficient, IC_{CRIT}. IC_{CRIT}, which depends primarily on the channel length and critical, velocity saturation electric field, E_{CRIT}, corresponds to the inversion coefficient where g_m/I_D is down to approximately 70.7 % of the value present without velocity saturation and, secondarily, VFMR effects. All expressions and trends shown include velocity saturation effects, but exclude DIBL effects, which are noted at the bottom of the table. The inclusion of velocity saturation effects results in channel length dependency on g_m/I_D and, correspondingly, channel length dependency on the thermal-noise efficiency factor and bandwidth–power–accuracy metrics. Additionally, the inclusion of velocity saturation effects results in additional channel length dependency on intrinsic voltage gain and bandwidth, the product of intrinsic voltage gain and bandwidth, and the product of transconductance efficiency and intrinsic bandwidth. As observed in the table, trends for some figures of merit, for example, the product of intrinsic voltage gain and bandwidth, can be found from the combination of trends for individual component aspects of performance.

The figures of merit listed in Table 4.4 exclude gate leakage current and the resulting gate–source conductance, gate shot- and flicker-noise current, and increase in DC mismatch that was discussed earlier in Section 3.12.2. Gate leakage current effects, which can become significant for gate-oxide thickness around 2 nm and below, may require a smaller channel length and gate area than otherwise expected. This may be required to balance the traditional improvements in gain, flicker noise, and local-area mismatch at increasing channel length and gate area with the deterioration of these associated with increasing gate leakage current caused by increasing gate area. The modeling of gate leakage current and conductance described in Section 3.12.1 and the modeling of resulting effects on circuit performance described in Section 3.12.2 can be used to develop figures of merit that include gate leakage current effects.

4.3.6.1 Transconductance efficiency and Early voltage

While not shown separately in Table 4.4, the transconductance efficiency, g_m/I_D, and normalized, drain–source resistance or Early voltage, $V_A = I_D \cdot r_{ds} = I_D/g_{ds}$, are themselves figures of merit representing the production of transconductance, g_m, and drain–source resistance, r_{ds}, respectively, for a given level of drain current. g_m/I_D and V_A were included in Table 4.1 and shown in Figures 4.7 and 4.8, where g_m/I_D is maximized at low IC and V_A is maximized at long channel lengths. Maximizing g_m/I_D and V_A maximizes g_m and r_{ds}, respectively, for a given value of drain current.

4.3.6.2 Intrinsic voltage gain, bandwidth, and gain–bandwidth

The intrinsic voltage gain and bandwidth entries in Table 4.4 are intrinsic voltage gain, $A_{Vi} = g_m \cdot r_{ds} = g_m/g_{ds} = (g_m/I_D) \cdot V_A$, and intrinsic bandwidth, $f_{Ti} = g_m/[2\pi(C_{gsi} + C_{gbi})]$. These are figures of merit because they involve the combination of g_m and r_{ds}, and the combination of g_m and the intrinsic gate–source and gate–body capacitances, C_{gsi} and C_{gbi}, respectively. A_{Vi} and f_{Ti} were included in Table 4.1 and shown in Figures 4.9 and 4.10. A_{Vi} is maximized at low IC and long channel lengths, and f_{Ti} is maximized oppositely at high IC and short channel lengths. A_{Vi} and f_{Ti} are independent of the drain current.

Maximizing A_{Vi} and f_{Ti} maximizes the intrinsic voltage gain, described in Section 3.8.5, and intrinsic bandwidth, described in Section 3.9.6, for a single device. A_{Vi} and f_{Ti} are also proportional to the voltage gain and bandwidth across multiple devices having equal operating characteristics and can thus be used to estimate the performance of circuits larger than a single device.

The intrinsic gain–bandwidth product entry in Table 4.4 is the product of A_{Vi} and f_{Ti}, or $A_{Vi} \cdot f_{Ti}$. The behavior of $A_{Vi} \cdot f_{Ti}$ with increasing IC can be observed by visually observing the product or addition of the logarithmic presentation of A_{Vi} and f_{Ti} in Figure 4.9. In weak inversion, $A_{Vi} \cdot f_{Ti}$ increases directly with IC because A_{Vi} is constant while f_{Ti} increases directly with IC. In strong inversion without significant velocity saturation, $A_{Vi} \cdot f_{Ti}$ reaches a maximum and constant value because A_{Vi} decreases inversely as the square root of IC while f_{Ti} increases as the square root of IC. When velocity saturation effects cause A_{Vi} to decrease inversely with IC and f_{Ti} to level off, $A_{Vi} \cdot f_{Ti}$ decreases inversely with IC.

The behavior of $A_{Vi} \cdot f_{Ti}$ with increasing L can be observed by visually observing the product or addition of the logarithmic presentation of A_{Vi} and f_{Ti} in Figure 4.10. When velocity saturation and DIBL effects are not significant, $A_{Vi} \cdot f_{Ti}$ decreases nearly inversely with L because A_{Vi} increases nearly linearly with L while f_{Ti} decreases as the inverse square of L. When velocity saturation effects are significant for short-channel devices at high IC, $A_{Vi} \cdot f_{Ti}$ increases with L because A_{Vi} increases beyond its usual near-linear increase with L while f_{Ti} decreases inversely with L, both as a result of less velocity saturation reduction of g_m/I_D. The trends with L are approximate because V_A and, correspondingly, A_{Vi} increase sublinearly to nearly linearly with L as shown in Figure 3.47. V_A and A_{Vi} are predicted using the Early voltage factor, V_{AL}, described in Section 4.3.3.4. When DIBL effects are significant for short-channel devices at low IC, $A_{Vi} \cdot f_{Ti}$ increases with L because of less DIBL reduction of V_A, which significantly increases A_{Vi} beyond its usual near-linear increase with L.

Maximizing $A_{Vi} \cdot f_{Ti}$ at channel lengths near the process minimum near the onset of strong inversion, or at the strong-inversion side of moderate inversion if needed to minimize velocity saturation effects, corresponds to maximizing the product of intrinsic voltage gain and bandwidth for a single device or related devices having equal operating characteristics. $A_{Vi} \cdot f_{Ti}$, like its individual components of A_{Vi} and f_{Ti}, is independent of the drain current.

4.3.6.3 Transconductance efficiency and intrinsic bandwidth

The transconductance efficiency, intrinsic bandwidth product entry in Table 4.4 is the product of g_m/I_D and f_{Ti}, or $(g_m/I_D) \cdot f_{Ti}$. The behavior of $(g_m/I_D) \cdot f_{Ti}$ with increasing IC can be observed by visually observing the product or addition of the logarithmic presentation of g_m/I_D in Figure 4.7 and f_{Ti} in Figure 4.9. In weak inversion, $(g_m/I_D) \cdot f_{Ti}$ increases directly with IC because g_m/I_D is constant while f_{Ti} increases directly with IC. In strong inversion without significant velocity saturation, $(g_m/I_D) \cdot f_{Ti}$ reaches a maximum and constant value because g_m/I_D decreases inversely as the square root of IC while f_{Ti} increases as the square root of IC. When velocity saturation effects cause g_m/I_D to decrease inversely with IC and f_{Ti} to level off, $(g_m/I_D) \cdot f_{Ti}$ decreases inversely with IC.

The behavior of $(g_m/I_D) \cdot f_{Ti}$ with increasing IC is nearly identical to the behavior of $A_{Vi} \cdot f_{Ti}$ mentioned above because $A_{Vi} = (g_m/I_D) \cdot V_A$ where V_A is nearly constant with IC. However, the behavior of $(g_m/I_D) \cdot f_{Ti}$ with increasing L is different from the behavior of $A_{Vi} \cdot f_{Ti}$ because g_m/I_D is nearly constant with L compared to A_{Vi} that increases with L. The behavior of $(g_m/I_D) \cdot f_{Ti}$ with increasing L can be observed by visually observing the product or addition of the logarithmic presentation of g_m/I_D in Figure 4.8 and f_{Ti} in Figure 4.10. When velocity saturation is not significant, $(g_m/I_D) \cdot f_{Ti}$ decreases as the inverse square of L because g_m/I_D remains nearly constant with L while f_{Ti} decreases as the inverse square of L. When velocity saturation effects cause g_m/I_D to increase with L and f_{Ti} to decrease inversely with L, both due to less velocity saturation decreases in g_m/I_D, $(g_m/I_D) \cdot f_{Ti}$ is nearly constant with L.

Maximizing $(g_m/I_D) \cdot f_{Ti}$ at short channel lengths near the onset of strong inversion, or at the strong-inversion side of moderate inversion if needed to minimize velocity saturation effects, corresponds to maximizing the product of transconductance efficiency and intrinsic bandwidth for a single device or related devices having equal operating characteristics. Maximizing transconductance efficiency minimizes the gate-referred thermal-noise voltage for a given level of drain current as described in the next section, while maximizing intrinsic bandwidth maximizes the operating bandwidth. $(g_m/I_D) \cdot f_{Ti}$, like its individual components of g_m/I_D and f_{Ti}, is independent of the drain current.

In [12], $(g_m \cdot f_{Ti})/I_D$, which is equal to $(g_m/I_D) \cdot f_{Ti}$, is proposed as a figure of merit for optimizing the design of low-noise, low-power RF CMOS circuits. Here, it is observed that $(g_m/I_D) \cdot f_{Ti}$ is maximized for short channel lengths at the strong-inversion side of moderate inversion where velocity saturation reductions of g_m/I_D and f_{Ti} are modest. As mentioned, operating well into strong inversion where velocity saturation reduces g_m/I_D and f_{Ti} results in a declining value of $(g_m/I_D) \cdot f_{Ti}$.

4.3.6.4 Thermal-noise efficiency and flicker-noise area efficiency

The thermal-noise efficiency factor, NEF', for a single MOS device listed in Table 4.4 corresponds to the ratio of MOS, gate-referred thermal-noise voltage PSD, S_{VG} thermal, and bipolar transistor, base-referred thermal-noise voltage PSD, S_{VB} thermal, with both devices operating at equal bias currents. This figure of merit is derived and expanded later in Section 6.3.1 for evaluating the input-referred thermal-noise voltage of micropower, low-noise CMOS preamplifiers at different levels of supply current.

NEF' for a single MOS device is shown as a function of IC later in Figure 6.1. Being inversely proportional to g_m/I_D, it is minimum in weak inversion, increases as the square root of IC in strong inversion, and increases directly with IC when velocity saturation is significant. NEF' is independent of L, excepting its increase for short-channel devices at high IC due to velocity saturation reduction of g_m/I_D. Here, NEF' decreases with increasing L because g_m/I_D increases due to reduced velocity saturation effects. Additionally, when small-geometry thermal noise increases described in Section 3.10.2.2 are present, increasing L decreases the noise and NEF'. Because NEF' is inversely proportional to g_m/I_D, its trends are the opposite of those shown for g_m/I_D in Table 4.1.

Unlike the other figures of merit given in Table 4.4, NEF' should be minimized. This achieves the minimum gate-referred thermal-noise voltage for a given drain current. For a MOS device in a bulk CMOS process operating in weak inversion with $n = 1.4$ and a thermal-noise factor of $\Gamma = 0.5$, NEF' is equal to a minimum value of two. This indicates that the gate-referred thermal-noise voltage PSD is twice the base-referred value for a bipolar transistor operating at the same bias current. Alternatively, twice the bias current is required to obtain a specified level of MOS, gate-referred thermal-noise voltage compared to the bias current required for a bipolar transistor for the same level of noise. This assumes operation at the same IC by scaling MOS channel width directly with the drain current.

Operation in strong inversion where g_m/I_D is low increases the value of NEF' and, correspondingly, the drain current required for a specified value of gate-referred thermal-noise voltage. This is an especially large concern for low-noise RF amplifiers where operation at short channel lengths and high IC for high bandwidth results in low g_m/I_D, especially if velocity saturation is significant. This gives high NEF' and high required drain current for low input-referred thermal noise voltage.

While not included in Table 4.4, Equation 6.4 in Section 6.3.2 describes a flicker-noise area efficiency factor, NAF. This factor is equal to the circuit layout area divided by the layout area required for a single MOS device to achieve the same value of input-referred flicker-noise voltage. NAF should be minimized to minimize flicker noise for a given layout area. Correspondingly, minimizing NAF minimizes the layout area required to achieve a specified value of flicker noise.

4.3.6.5 Bandwidth, power, and accuracy with DC offset

The bandwidth–power–accuracy, DC offset figures of merit in Table 4.4 are intended to help maximize signal bandwidth, maximize accuracy or signal-to-noise ratio in the presence of DC offset, and minimize power consumption simultaneously in MOS circuits. This is important, for example, for data converter circuits where high bandwidth or speed, high accuracy, and low power consumption are desired. The inclusion of DC offset assumes circuits that are sensitive to DC offset where the offset is not removed by auto-zeroing or other techniques. The figures of merit, which were adapted from [13, 14], consider DC offsets resulting from local-area mismatch and were discussed in detail earlier in Sections 3.11.3.1 and 3.11.3.2.

The current-mode, bandwidth–power–accuracy, DC offset figure of merit in Table 4.4, which is applicable to current mirrors, considers the DC accuracy, $A_{CC,DC}$, given by Equation 3.162. $A_{CC,DC}$ is found by the input signal current, I_{IN}, expressed as a fraction of the bias current, I_D, divided by the 3σ value of input-referred DC offset current. $A_{CC,DC}$ is squared in the figure of merit and can be considered a signal-to-noise power ratio, where noise corresponds to the input-referred DC offset current. The figure of merit also considers the multiplication of signal bandwidth, BW, given by Equation 3.163, divided by the power consumption, $P = 2I_D \cdot V_{DD}$, where I_D is the bias current for each of two devices in a current mirror and V_{DD} is the supply voltage. The complete figure of merit from Equation 3.166 is listed in Table 4.4 where g_m/I_D is expressed in terms of IC from the expression in Table 4.1. Because the figure of merit is inversely proportional to g_m/I_D, its trends are opposite of those shown for g_m/I_D in Table 4.1. Its trends, but not values, can be observed by flipping vertically the logarithmic presentation of g_m/I_D versus IC in Figure 4.7.

The current-mode, bandwidth–power–accuracy figure of merit is proportional to $(I_{IN}/I_D)^2$ and is inversely proportional to g_m/I_D, V_{DD}, and the process metric, $C'_{OX} \cdot A_{VT}^2$. This indicates that current-mode circuits are optimized at high IC in strong inversion where g_m/I_D is low, combined with operation at low V_{DD}. However, as described in Section 3.11.3.1, while g_m/I_D decreases inversely as the square root of IC in strong inversion, $V_{DS,sat}$ increases directly as the square root of IC, increasing the required V_{DD}. As a result, the g_m/I_D and V_{DD} trends cancel assuming negligible velocity saturation, revealing that, at the design limit, the figure of merit is inversely proportional to the process value of $C'_{OX} \cdot A_{VT}^2$ alone [13, 14]. This process value will be discussed after discussing a voltage-mode figure of merit. Finally, the current-mode figure of merit is independent of the drain bias current, depending instead on the square of the input signal current relative to the bias current, $(I_{IN}/I_D)^2$. I_{IN} can remain a constant fraction of I_D regardless of the value of I_D, giving a value of I_{IN}/I_D sufficiently below unity as required to manage signal distortion.

The voltage-mode, bandwidth–power–accuracy, DC offset figure of merit in Table 4.4 is applicable, for example, to a buffer amplifier driving the voltage input of a comparator consisting of a differential pair. A similar figure of merit can also be developed for a simple differential pair in an operational amplifier using negative feedback [14]. $A_{CC,DC}$, given by Equation 3.169, is found by the voltage input signal, V_{IN}, divided by the 3σ value of input-referred DC offset voltage at the comparator input. $A_{CC,DC}$, again, is squared in the figure of merit and can be considered a signal-to-noise power ratio, where noise corresponds to the input-referred DC offset voltage. The figure of merit also considers the multiplication of signal bandwidth, BW, given by Equation 3.170, again, divided by the power consumption, $P = 2I_D \cdot V_{DD}$, where I_D is the bias current for each side of a differential signal path in the buffer amplifier and V_{DD} is the supply voltage. The complete figure of merit from Equation 3.172 is listed in Table 4.4 where g_m/I_D is expressed in terms of IC from the g_m/I_D expression in Table 4.1. Because the figure of merit is proportional to g_m/I_D for the buffer amplifier, its trends are the same as those shown for g_m/I_D in Table 4.1. Its trends, but not values, can be observed from g_m/I_D versus IC in Figure 4.7.

The voltage-mode, bandwidth–power–accuracy figure of merit is proportional to V_{IN}^2 and g_m/I_D for the buffer amplifier, and is inversely proportional to V_{DD} and $C'_{OX} \cdot A_{VT}^2$. This indicates that voltage-mode circuits are optimized at low IC for the buffer amplifier in weak or moderate inversion where g_m/I_D is high, combined with operation at low V_{DD}. As described in Section 3.11.3.2, when the buffer amplifier is an OTA having unity-gain negative feedback, high g_m/I_D for the transconductor (differential pair) devices gives high OTA g_m and low output resistance of $1/g_m$. This results in improved signal bandwidth at the comparator input. However, while g_m/I_D can be maximized for the buffer amplifier, V_{IN} is limited by the value of V_{DD}. As a result, at the design limit, the figure of merit is inversely proportional to $C'_{OX} \cdot A_{VT}^2$ alone [14]. This is the same result obtained for the current-mode figure of merit. Finally, like the current-mode figure of merit, the voltage-mode figure of merit is independent of the drain bias current.

Interestingly, both bandwidth–power–accuracy, DC offset figures of merit are independent of the gate area, because gate area cancels in the $A_{CC,DC}^2$ and BW terms as seen in the derivations of Sections 3.11.3.1 and 3.11.3.2. As a result, these figures of merit are independent of the channel length, except when velocity saturation effects cause reduced g_m/I_D for short-channel devices operating at high IC. Additionally, transconductance factor mismatch, neglected in the analysis, can be important at high IC and can contribute to DC offset and reduce the figures of merit, along with changes resulting from the reduction of g_m/I_D due to velocity saturation.

Operation at constant IC giving constant g_m/I_D results in constant bandwidth–power–accuracy figures of merit and, correspondingly, constant bandwidth-accuracy figures of merit for constant drain bias current and power consumption. This is seen in the continuation sheet of Figure 4.21 in Section 4.4.2 where the products of accuracy squared and bandwidth are constant with L. This is because accuracy, which is inversely proportional to local-area mismatch listed in Tables 4.2 and 4.3, increases as L while bandwidth listed in Table 4.1 decreases as $1/L^2$, both due to gate area that increases as L^2. The product of accuracy (not squared) and bandwidth, however, is not constant with L, but decreases as $1/L$. Bandwidth may actually decrease as $1/L$, changing the trends (Section 5.5.2.11). Finally, for constant IC and g_m/I_D, bandwidth-accuracy figures of merit (product of accuracy squared and bandwidth) increase directly with the bias current as seen in the continuation sheet of Figure 4.22 in Section 4.4.3.

When g_m/I_D and V_{DD} are set at their design limits along with I_{IN}/I_D and V_{IN} as required to manage signal distortion, the bandwidth–power–accuracy, DC offset figures of merit are largely set by $C'_{OX} \cdot A_{VT}^2$ [13, 14]. As a result, there is little the designer can do to improve the figures of merit for either current-mode or voltage-mode circuits [14]. $C'_{OX} \cdot A_{VT}^2$, given by Equation 3.168, is referred to as the process matching energy, $E_{matching}$ [14, 15]. It is equal to the product of the process gate-oxide capacitance per unit area, C'_{OX}, and the square of the threshold-voltage mismatch factor, A_{VT}^2. As described in Section 3.11.3.1, $C'_{OX} \cdot A_{VT}^2$ generally favorably decreases with decreasing t_{ox} in smaller-geometry processes because A_{VT}^2 decreases more rapidly than the increase in C'_{OX}. However, as mentioned, this technology trend may not continue if A_{VT} ceases to track the decreasing value of t_{ox} or begins to level off with decreasing t_{ox}.

The bandwidth–power–accuracy, DC offset figures of merit are lower than those associated with thermal noise as discussed later in Section 4.3.6.7. This is because MOS DC mismatch errors are more significant for circuits sensitive to DC offset compared to thermal noise errors for circuits sensitive only to thermal noise. This is well shown in the accuracy and error measures of Figure 4.21.

4.3.6.6 Bandwidth, power, and accuracy with thermal noise

The bandwidth–power–accuracy, DC offset figures of merit discussed in the previous section considered the loss of signal accuracy caused by DC offsets resulting from local-area mismatch. This section considers figures of merit where signal accuracy is affected by thermal noise. Unlike circuits that may be insensitive to DC offsets because of auto-zeroing or other techniques, thermal noise affects all circuits.

If flicker noise is significant compared to thermal noise considered below, it can be included in figures of merit similar to those developed for DC offsets. In this case, gate-area-dependent flicker noise would replace gate-area-dependent DC offset, yielding figures of merit having forms similar to those found for DC offsets.

The bandwidth–power–accuracy, thermal-noise figures of merit in Table 4.4 are intended to help maximize signal bandwidth, maximize accuracy or signal-to-noise ratio in the presence of thermal noise, and minimize power consumption simultaneously in MOS circuits. As mentioned, this can be important for data converter circuits where high bandwidth or speed, high accuracy, and low power consumption are desired. Unlike the figures of merit for the presence of DC offsets, the figures of merit for the presence of thermal noise were not previously described. These are developed here.

The current-mode, bandwidth–power–accuracy, thermal-noise figure of merit in Table 4.4, which is applicable to current mirrors, considers the AC accuracy, $A_{CC,AC}$. $A_{CC,AC}$ is given by

$$A_{CC,AC} = \frac{\left(\dfrac{I_{IN}}{I_D}\right) I_D}{\sqrt{2 \cdot 4kT \cdot n\Gamma \cdot g_m \cdot BW}} \tag{4.1}$$

where the numerator is the rms, input signal current, I_{IN}, expressed as a fraction of the bias current, I_D. The denominator is the rms, drain-referred thermal-noise current over the signal bandwidth, BW, for two current-mirror devices having transconductance of g_m each. This corresponds to the input-referred thermal-noise current at the current-mirror input. The noise is found from the drain-referred thermal-noise current density, $S_{ID}^{1/2}$ thermal, given in Table 4.3 for a single device, multiplied by the square root of two and by the square root of BW to give the rms noise current for two devices over the signal bandwidth. This assumes that the noise bandwidth is nearly equal to signal bandwidth.

Squaring Equation 4.1 gives

$$A_{CC,AC}^2 = \frac{\left(\frac{I_{IN}}{I_D}\right)^2 I_D^2}{2 \cdot 4kT \cdot n\Gamma \cdot g_m \cdot BW} = \frac{\left(\frac{I_{IN}}{I_D}\right)^2 I_D}{2 \cdot 4kT \cdot n\Gamma \cdot (g_m/I_D) \cdot BW} \tag{4.2}$$

where $A_{CC,AC}^2$ can be considered a signal-to-noise power ratio, where the noise corresponds to the input-referred thermal-noise current appearing at the current-mirror input. The power consumption is equal to $P = 2I_D \cdot V_{DD}$, where I_D is the bias current for each of two devices in the current mirror and V_{DD} is the supply voltage. The complete figure of merit is then given by

$$\frac{A_{CC,AC}^2 \cdot BW}{P} = \left[\frac{\left(\frac{I_{IN}}{I_D}\right)^2 I_D}{2 \cdot 4kT \cdot n\Gamma \cdot (g_m/I_D) \cdot BW} \right] \cdot BW \cdot \left(\frac{1}{2I_D \cdot V_{DD}} \right)$$

$$= \left(\frac{I_{IN}}{I_D}\right)^2 \frac{1}{16kT \cdot n\Gamma \cdot (g_m/I_D) \cdot V_{DD}} = \left(\frac{I_{IN}}{I_D}\right)^2 \frac{U_T \left(\sqrt{IC+0.25}+0.5\right)}{16kT \cdot \Gamma \cdot V_{DD}} \tag{4.3}$$

Conveniently, BW associated with the mean square, thermal-noise current in the denominator of $A_{CC,AC}^2$ cancels with BW appearing in the numerator of the figure of merit.

The complete figure of merit from Equation 4.3 is listed in Table 4.4 where g_m/I_D is expressed in terms of IC from the expression in Table 4.1. Because the figure of merit is inversely proportional to g_m/I_D, its trends are opposite of those shown for g_m/I_D in Table 4.1. Its trends, but not values, can be observed by flipping vertically the logarithmic presentation of g_m/I_D versus IC in Figure 4.7.

The current-mode, bandwidth–power–accuracy figure of merit is proportional to $(I_{IN}/I_D)^2$ and is inversely proportional to g_m/I_D, V_{DD}, and the thermal-noise energy, kT. This is nearly identical to the result for the current-mode, DC offset figure of merit described in the previous section and listed in Table 4.4, except the matching energy, $C_{OX}' \cdot A_{VT}^2$, is replaced with the thermal-noise energy, kT. Thus, current-mode circuits in the presence of either DC offset or thermal noise are optimized at high IC in strong inversion where g_m/I_D is low, combined with operation at low V_{DD}. However, as mentioned in the previous section, the g_m/I_D and V_{DD} trends cancel for current-mode circuits revealing that, at the design limit, the figures of merit are inversely proportional to the values of $C_{OX}' \cdot A_{VT}^2$ for DC offset and kT for thermal noise. Finally, the current-mode figures of merit for both DC offset and thermal noise are independent of the drain bias current, I_D, depending instead on the square of the input signal current relative to the bias current, $(I_{IN}/I_D)^2$. Again, I_{IN} can remain a constant fraction of I_D regardless of the value of I_D, giving a value of I_{IN}/I_D sufficiently below unity as required to manage signal distortion.

The voltage-mode, bandwidth–power–accuracy, thermal-noise figure of merit in Table 4.4 is applicable, for example, to the voltage input of a comparator consisting of a differential pair. Here, $A_{CC,AC}$ is given by

$$A_{CC,AC} = \frac{V_{IN}}{\sqrt{2 \cdot 4kT \cdot \frac{n\Gamma}{g_m} \cdot BW}} \tag{4.4}$$

where the rms, input voltage signal, V_{IN}, is divided by the rms, gate-referred thermal-noise voltage over the signal bandwidth, BW, for two differential-pair devices having transconductance of g_m each. The noise is found from the gate-referred thermal-noise voltage density, $S_{VG}^{1/2}$ thermal, given in Table 4.2 for a single device, multiplied by the square root of two and by the square root of BW to give the rms noise voltage for two devices over the signal bandwidth. As before, this assumes that the noise bandwidth is nearly equal to signal bandwidth.

Squaring Equation 4.4 gives

$$A_{CC,AC}^2 = \frac{V_{IN}^2}{2 \cdot 4kT \cdot \dfrac{n\Gamma}{g_m} \cdot BW} = \frac{V_{IN}^2 \cdot I_D \cdot (g_m/I_D)}{2 \cdot 4kT \cdot n\Gamma \cdot BW} \tag{4.5}$$

where $A_{CC,AC}^2$ can, again, be considered a signal-to-noise power ratio, where the noise corresponds to the input-referred thermal-noise voltage appearing at the input of the differential pair. The power consumption is, again, equal to $P = 2I_D \cdot V_{DD}$, where I_D is the bias current for each of two devices in the differential pair and V_{DD} is the supply voltage. The complete figure of merit is then given by

$$\frac{A_{CC,AC}^2 \cdot BW}{P} = \left(\frac{V_{IN}^2 \cdot I_D \cdot (g_m/I_D)}{2 \cdot 4kT \cdot n\Gamma \cdot BW} \right) \cdot BW \cdot \left(\frac{1}{2I_D \cdot V_{DD}} \right)$$

$$= \frac{V_{IN}^2 \cdot (g_m/I_D)}{16kT \cdot n\Gamma \cdot V_{DD}} = \frac{\dfrac{V_{IN}^2}{n^2 U_T \left(\sqrt{IC + 0.25} + 0.5 \right)}}{16kT \cdot \Gamma \cdot V_{DD}} \tag{4.6}$$

Again, conveniently, BW associated with the mean square, thermal-noise voltage in the denominator of $A_{CC,AC}^2$ cancels with BW appearing in the numerator of the figure of merit.

The complete figure of merit from Equation 4.6 is listed in Table 4.4 where g_m/I_D is expressed in terms of IC from the expression in Table 4.1. Because the figure of merit is proportional to g_m/I_D for the differential pair, its trends are the same as those shown for g_m/I_D in Table 4.1. Its trends, but not values, can be observed from g_m/I_D versus IC in Figure 4.7.

The voltage-mode, bandwidth–power–accuracy figure of merit is proportional to V_{IN}^2 and g_m/I_D, and is inversely proportional to V_{DD} and kT. This is nearly identical to the result for the voltage-mode, DC offset figure of merit described in the previous section and listed in Table 4.4, except that the matching energy, $C_{OX}' \cdot A_{VT}^2$, is replaced with the thermal-noise energy, kT. Additionally, g_m/I_D corresponds to buffer amplifier and differential pair devices for the DC offset and thermal-noise figures of merit, respectively. Thus, voltage-mode circuits in the presence of either DC offset or thermal noise are optimized at low IC in weak or moderate inversion where g_m/I_D is high, combined with operation at low V_{DD}. However, as mentioned in the previous section, beyond maximizing g_m/I_D, V_{IN} is limited by the value of V_{DD} for voltage-mode circuits revealing that, at the design limit, the figures of merit are ultimately inversely proportional to the values of $C_{OX}' \cdot A_{VT}^2$ for DC offset and kT for thermal noise. This is the same result obtained for both current-mode figures of merit. Finally, the voltage-mode figures of merit for both DC offset and thermal noise are independent of the drain bias current, I_D, as are both current-mode figures of merit.

Interestingly, current-mode and voltage-mode, bandwidth–power–accuracy figures of merit for both DC offset and thermal noise are independent of the gate area, because gate area cancels in the $A_{CC,DC}^2$, $A_{CC,AC}^2$, and BW terms. As a result, all figures of merit are independent of the channel length, except when velocity saturation effects cause reduced g_m/I_D for short-channel devices operating at high IC. Additionally, small-geometry increases in thermal noise (Section 3.10.2.2), considered by the value of the thermal-noise factor Γ, can reduce the thermal-noise figures of merit, along with changes resulting from the reduction of g_m/I_D due to velocity saturation.

Analogous to the DC figures of merit considered in the previous section, operation at constant IC giving constant g_m/I_D results in constant AC bandwidth–power–accuracy figures of merit and, correspondingly, constant bandwidth-accuracy figures of merit for constant drain bias current and power consumption. This is seen in the continuation sheet of Figure 4.21 in Section 4.4.2 where the products of accuracy squared and bandwidth are constant with L. This is because accuracy, which is inversely proportional to thermal noise over the square-root of bandwidth, increases as L while bandwidth listed in Table 4.1 decreases as $1/L^2$ due to gate area that increases as L^2. The product of accuracy (not squared) and bandwidth, however, is not constant with L, but decreases as $1/L$. Bandwidth may actually decrease as $1/L$, changing the trends (Section 5.5.2.11). Finally, for constant IC and g_m/I_D, bandwidth-accuracy figures of merit (products of accuracy squared and bandwidth) increase directly with the drain bias current as seen in the continuation sheet of Figure 4.22 in Section 4.4.3.

When g_m/I_D and V_{DD} are set at their design limits along with I_{IN}/I_D and V_{IN} as required to manage distortion, bandwidth–power–accuracy figures of merit for DC offset are largely set by the matching energy, $C'_{OX} \cdot A_{VT}^2$, while figures of merit for thermal noise are largely set by the thermal-noise energy, kT. As a result, there is little the designer can do to improve figures of merit for either current-mode or voltage-mode circuits [14].

4.3.6.7 Comparison of bandwidth, power, and accuracy for DC offset and thermal noise

In concluding the discussion on bandwidth–power–accuracy figures of merit, it is useful to compare the figures of merit given in Table 4.4 for both DC offset and thermal noise for equal operating conditions. Here, operation at the onset of strong inversion at $IC = 10$ will be considered for $V_{DD} = 1.8$ V. An input current signal of $I_{IN}/I_D = 0.5$ will modulate 50 % of the bias current for a DC input and 70.7 % of the bias current at the peak of an rms AC input, which should manage signal distortion. A voltage input signal of $V_{IN} = 0.1$ V for either a DC or rms AC input is below the input, 1 dB compression point of $V_{INDIF1dB} = 0.192$ V found from the expression in Table 4.2, which should also manage signal distortion. A strong inversion value of $n = 1.3$ and $\Gamma = 2/3$ will be assumed along with a thermal voltage of $U_T = 25.9$ mV at $T = 300$ K (room temperature). $C'_{OX} = 8.41$ fF/μm^2 and $A_{VT} = 5$ mV $\cdot \mu$m will be assumed for devices in the example 0.18 μm CMOS process described in Tables 3.2 and 3.3, giving a matching energy of $C'_{OX} \cdot A_{VT}^2 = 2.1 \times 10^{-19}$ J (W·s or W/Hz). Finally, a thermal-noise energy of $kT = 4.142 \times 10^{-21}$ J will be assumed as appropriate for $T = 300$ K.

The resulting current-mode, bandwidth–power–accuracy figures of merit are 5.5×10^{14} Hz/W and 3.0×10^{17} Hz/W for DC offset and thermal noise, respectively. The voltage-mode figures of merit are 5.6×10^{15} Hz/W and 7.8×10^{17} Hz/W for DC offset and thermal noise, respectively. The figures of merit for DC offset are over two orders of magnitude smaller than those for thermal noise. As discussed in [14], this is a result of the significantly higher matching energy of $C'_{OX} \cdot A_{VT}^2 = 2.1 \times 10^{-19}$ J, compared to the thermal noise energy of $kT = 4.142 \times 10^{-21}$ J. MOS circuits sensitive to DC offset caused by mismatch incur significantly lower accuracy and higher errors than circuits sensitive to thermal noise alone [14]. This is especially well shown in the accuracy and error measures in Figure 4.21 and is a motivation for using auto-zeroing or other techniques for minimizing DC offsets in MOS circuits.

The bandwidth–power–accuracy figures of merit estimated here are lower than those that could be obtained for separate, optimum operating conditions for current-mode and voltage-mode circuits and for V_{DD} below 1.8 V. However, these estimates show the significant decrease in figures of merit associated with DC offset compared to values associated with thermal noise. Estimates of figures of merit agree well with those shown in Figure 4.21 for the operating conditions considered.

4.3.6.8 Extensions

The listing of MOS figures of merit in Table 4.4 is not comprehensive, but does list selected figures of merit and illustrate their development. Additional figures of merit could include, for example, distortion by including the input, 1dB compression voltage, $V_{INDIF1dB}$. The development of figures of merit in terms of the inversion coefficient, channel length, and drain current, valid from weak through strong inversion, can be helpful in providing design intuition leading towards optimum design. Additionally, figures of merit can be helpful in understanding technology limitations and trends.

In practice, the designer will rarely maximize or minimize a single figure of merit. Instead, the designer will usually balance multiple design requirements using intuition gained from an understanding of tradeoffs in performance and tradeoffs in figures of merit that combine important aspects of performance.

In [16], performance–power relationships are reviewed that can be useful to understand and develop additional figures of merit. These relationships include the minimum power required for an analog circuit for a given load capacitance, signal voltage swing, signal frequency, signal-to-noise ratio, and overhead associated with the supply voltage and current. Additionally, relationships include the minimum power required for selected signal-to-noise and signal-to-noise-and-distortion ratios considering total integrated noise or noise over a selected bandwidth.

While not included in the performance–power relationships reviewed, [16] also discusses gate leakage current and its effects on circuit performance, which was discussed earlier in Sections 3.12.1 and 3.12.2. As mentioned at the beginning of Section 4.3.6, gate leakage current effects should be considered in the development of figures of merit for gate-oxide thickness around 2 nm and below where these effects can significantly affect circuit performance.

4.4 DESIGN OF DIFFERENTIAL PAIRS AND CURRENT MIRRORS USING THE *ANALOG CMOS DESIGN, TRADEOFFS AND OPTIMIZATION* SPREADSHEET

Section 4.2 described MOS performance trends for separate changes in the design choices of inversion coefficient, channel length, and drain current. Section 4.3 then presented graphical performance tradeoffs common to all applications, followed by tradeoffs specific to differential-pair and current-mirror applications. Additionally, Section 4.3 presented tradeoffs in various performance figures of merit. This final section presents numerical performance trends and tradeoffs that lead to the optimum design of differential pairs, current mirrors, and larger circuits.

The *Analog CMOS Design, Tradeoffs and Optimization* spreadsheet facilitates the design of differential pairs, current mirrors, and larger circuits by giving predicted performance for design inputs of MOS inversion coefficient, channel length, and drain current. This spreadsheet is described in the Appendix and is available on the book's web site listed on the cover. As considered in the previous discussions of performance tradeoffs, nMOS devices in the example 0.18 μm CMOS process described in Tables 3.2–3.4 are considered in the differential-pair and current-mirror design examples that follow.

The spreadsheet uses the 0.18 μm, nMOS process parameters from Tables 3.2–3.4 that are fixed or unbinned across device geometry and bias conditions, except for the Early voltage factor, $V_{AL} = V_A/L$. As described earlier in Section 3.8.4, no simple prediction of MOS Early voltage, $V_A = I_D \cdot r_{ds} = I_D/g_{ds}$, or the related drain–source resistance, r_{ds}, or conductance, $g_{ds} = 1/r_{ds}$, is available because of complex dependencies on the inversion level, channel length, and drain–source voltage. Instead, the process parameter $V_{AL} = V_A/L$ is extracted from V_A measurements near the operating inversion coefficient, channel length, and V_{DS}. V_{AL} is extracted from the V_A measurements given in Figure 3.45. These measurements are taken from weak through strong inversion for channel lengths ranging from $L = 0.18$ to 4 μm with V_{DS} held fixed at $V_{DS} = 0.5$ V, which is the value considered in the design examples.

The process values of V_{AL} used in the spreadsheet are similar to those shown in Figure 3.48, again, for nMOS devices in the example 0.18 μm CMOS process operating at $V_{DS} = 0.5$ V. However, the decrease in V_{AL} at short channel lengths caused by DIBL is removed because DIBL effects are included separately in the spreadsheet calculations. From Table 3.21, the total V_A is expressed as the parallel combination of V_A due to CLM and V_A due to DIBL. V_A values due to DIBL are calculated at 15.6 V for $L = 0.18\,\mu m$ and 71 V for $L = 0.28\,\mu m$ for operation at the onset of strong inversion at $IC = 10$ that is used for these channel lengths in the spreadsheet examples. The V_A values due to DIBL are found from Table 3.21 for $g_m/I_D = 8/V$ at $IC = 10$, $\delta V_T/\delta V_{DS} = DVTDIBL = -8\,mV/V$ for the minimum $L = 0.18\,\mu m$, and $DVTDIBLEXP = 3$, which describes the reduction of $\delta V_T/\delta V_{DS}$ as channel length increases. The process values of $DVTDIBL$ and $DVTDIBLEXP$ are taken from Table 3.2 along with $DL = 0.028\,\mu m$ used for finding effective channel length. Removing these values of V_A due to DIBL from the total V_A values measured in Figure 3.45 increases V_A from 3.1 V to 3.87 V for $L = 0.18\,\mu m$ and from 5.2 V to 5.6 V for $L = 0.28\,\mu m$. These values of V_A due to CLM alone are then divided by the effective channel length, giving $V_{AL} = V_A/L = 3.87/0.152\,\mu m = 25.5\,V/\mu m$ for $L = 0.18\,\mu m$ and $V_{AL} = 5.6\,V/0.252\,\mu m = 22.2\,V/\mu m$ for $L = 0.28\,\mu m$, again, for $IC = 10$ and $V_{DS} = 0.5\,V$. These V_{AL} values do not have the decreases shown in Figure 3.48 at short channel lengths caused by DIBL effects and are used as process values in the spreadsheet.

The removal of DIBL effects from the measured values of V_A in Figure 3.45 is not necessary for channel lengths of $0.48\,\mu m$ and longer because DIBL effects are not significant at these longer channel lengths. For these channel lengths, measured values of V_A are divided by effective channel length giving V_{AL} values similar to those shown in Figure 3.48.

The extraction of process V_{AL} values for the spreadsheet requires some effort, especially for a large circuit. However, V_{AL} only affects MOS $g_{ds} = 1/r_{ds}$ and the related $A_{Vi} = g_m \cdot r_{ds}$, so rough estimates for V_{AL} do not affect other aspects of performance. Linking the spreadsheet to a simulation MOS model that provides direct $g_{ds} = 1/r_{ds}$ modeling is anticipated in the future, so the reader is invited to check the book's web site for spreadsheet updates. This linkage would eliminate the need for extracting or estimating V_{AL}, although experimental verification is recommended given the complexity of predicting $g_{ds} = 1/r_{ds}$ and its long history of modeling errors.

Small-geometry velocity saturation and VFMR decreases in g_m/I_D and g_m, and the related increases in V_{EFF}, are included in the *Analog CMOS Design, Tradeoffs and Optimization* spreadsheet. Additionally, DIBL decreases in V_A and r_{ds} are included along with increases in gate-referred flicker-noise voltage with increasing inversion or V_{EFF}. However, gate leakage current and the resulting gate–source conductance, gate shot- and flicker-noise current, and increase in DC mismatch are negligible and not included in the spreadsheet results. These effects are negligible for the gate-oxide thickness of 4.1 nm in the example $0.18\,\mu m$ CMOS process, but can be significant for gate-oxide thickness around 2 nm and below. As described in Section 3.12.2, smaller channel length and gate area may be required to balance the traditional improvements in gain, flicker noise, and local-area mismatch at increasing channel length and gate area with the deterioration of these associated with increasing gate leakage current caused by increasing gate area. Later versions of the spreadsheet could include gate leakage current and its effects for processes where this is a concern.

Spreadsheet predictions of device performance are derived from the hand predictions of MOS performance in weak, moderate, and strong inversion that were described in Chapter 3. Sections A.2.2–A.2.4 in the Appendix reference the predictions used. The spreadsheet uses the slightly more accurate expression for g_m/I_D from Table 3.17 that involves two square-root terms compared to the g_m/I_D and related expressions that use the simpler one square-root term. All spreadsheet predictions of performance are approximate and intended only to provide design guidance leading towards optimum design. Simulations of a candidate design are then required using full, simulation MOS models for the actual production process. These simulations must include process, supply voltage, and temperature variations along with layout parasitics to validate a candidate design for production. Although well known to designers in industry, this is described further in the disclaimers in the spreadsheet and in Section 1.7.

The spreadsheet gives MOS geometry, W and WL, bias compliance voltages, V_{EFF} and $V_{DS,sat}$, small-signal parameters, g_m/I_D, g_m, V_A, and $g_{ds}(1/r_{ds})$, and intrinsic voltage gain, A_{Vi}. These are important for all applications and were displayed graphically in Section 4.3.3 and summarized in Table 4.1. The spreadsheet also gives capacitances, C_{GS}, C_{GB}, C_{GD}, and C_{DB}, and intrinsic bandwidth, f_{Ti}, which are also important for all applications. The spreadsheet gives $C_{GS} = C_{gsi} + C_{GSO}$, $C_{GB} = C_{gbi} + C_{GBO}$, and $C_{GD} = C_{GDO}$ (for operation in saturation), inclusive of the extrinsic gate overlap capacitances, C_{GSO}, C_{GBO}, and C_{GDO}. The intrinsic gate capacitances, C_{gsi} and C_{gbi}, along with C_{DB} and f_{Ti} were displayed graphically in Section 4.3.3 and summarized in Table 4.1.

The spreadsheet also gives $V_{INDIF1dB}$, $S_{VG}^{1/2}$ thermal, $S_{VG}^{1/2}$ flicker, and ΔV_{GS}, which describe the input, 1 dB compression voltage for a differential-pair and gate-referred noise and mismatch voltages. These are important for differential-pair applications and were displayed graphically in Section 4.3.4 and summarized in Table 4.2. Finally, the spreadsheet gives $S_{ID}^{1/2}$ thermal, $S_{ID}^{1/2}$ flicker, and $\Delta I_D/I_D$, which describe the drain-referred noise and mismatch currents. These are important for current-mirror applications and were displayed graphically in Section 4.3.5 and summarized in Table 4.3.

The spreadsheet gives additional performance measures, including figures of merit. These are described in design examples that follow and in Sections A.2.2–A.2.4.

4.4.1 Selecting Inversion Coefficient

Initially, differential-pair and current-mirror devices are designed using the *Analog CMOS Design, Tradeoffs and Optimization* spreadsheet for drain currents of $I_D = 100\,\mu A$, which are representative of millipower circuits. Holding constant drain current permits a "fair" study of performance tradeoffs at constant power consumption.

To illustrate the impact of selecting the inversion coefficient, channel length is held fixed at $L = 1\,\mu m$, while the inversion coefficient is varied. This permits a study of numerical performance tradeoffs leading to optimum inversion coefficient choices.

IC values are selected for the weak- and moderate-inversion boundary ($IC = 0.1$), the weak-inversion side of moderate inversion ($IC = 0.3$), the center of moderate inversion ($IC = 1$), the strong-inversion side of moderate inversion ($IC = 3$), the onset of strong inversion ($IC = 10$), and deep strong inversion ($IC = 30$). IC values below 0.1 deeper into weak inversion are not considered because there is little increase in transconductance efficiency and intrinsic voltage gain, while device geometries and capacitances continue increasing and bandwidth continues decreasing. Figure 4.1 presented the inversion coefficient as a number line and showed the regions and subregions of inversion along with a summary of key performance tradeoffs. Figure 3.1 presented the *MOSFET Operating Plane* that showed performance tradeoffs associated with inversion coefficient and channel length selections.

Actual IC values used are rounded slightly from the values given above as a result of trimming the reported drawn channel width to actual layout values. As discussed in Section A.2.3.6 in the Appendix, the spreadsheet gives trimmed values of IC for an optional number of gate stripes or interdigitations, m, for convenient values of drawn channel width for layout. Although such detail may not be needed for initial design guidance, this supports the usual design practice of interdigitating devices where the drain perimeter and area are minimized and corresponds to actual drawn layout width.

Figure 4.20 gives results from the *Analog CMOS Design, Tradeoffs and Optimization* spreadsheet for nMOS devices in the example 0.18 μm CMOS process for varying values of $IC = 0.1, 0.3, 1, 3,$ 10, and 30, and fixed values of $L = 1\,\mu m$, $I_D = 100\,\mu A$, $V_{DS} = 0.5\,V$, and $V_{SB} = 0\,V$. $V_{DS} = 0.5\,V$ is sufficiently high to ensure operation in saturation ($V_{DS} > V_{DS,sat}$), even at the maximum value of $IC = 30$. While the spreadsheet reports values for all IC selections, only values for $IC = 0.1, 1,$ and 10, corresponding to the onset of weak inversion, the center of moderate inversion, and the onset of strong inversion, respectively, are included in the following discussion for brevity. Trends for velocity saturation, present for short-channel devices at high IC in strong inversion, are mentioned with other trends, although velocity saturation is not significant for the $L = 1\,\mu m$ devices considered.

From Figure 4.20, the effective channel width, W, is 1500, 152, and 15 μm, and the effective gate area, WL, is 1460, 148, and 14.6 μm^2 for $IC = 0.1, 1,$ and 10, respectively. Both are maximum in weak inversion and decrease inversely with increasing IC. While not repeated here, the spreadsheet also includes the effective channel length, listed as L, which is equal to the input drawn channel length, listed as L_{DRAWN}, less the process lateral diffusion length, DL.

Analog CMOS Design, Tradeoffs and Optimization Spreadsheet -- MOSFETs Sheet						
For design guidance only as results are approximate and not for any particular CMOS process.						
© 2000–2007, David M. Binkley. See Disclaimer, Notes Sheet.						
---------- Optional User Design Information ----------						
Description		Illustration of changing IC at fixed $L = 1\,\mu m$ ($I_D = 100\,\mu A$)					
Device reference		M1	M2	M3	M4	M5	M6
Device notes		IC=0.1	IC=0.3	IC=1	IC=3	IC=10	IC=30
---------- Required User Design Inputs ----------						
Device model		nL1u	nL1u	nL1u	nL1u	nL1u	nIC30L1u
I_D	μA (+)	100	100	100	100	100	100
IC (fixed normalized)		0.101	0.296	0.998	3.011	10.117	30.351
L_{DRAWN}	μm	1	1	1	1	1	1
---------- Calculated Results ----------		OK	OK	OK	OK	OK	OK
Effective Width, Length, Gate Area							
W	μm	1504.00	512.00	152.00	50.40	15.00	5.00
L	μm	0.972	0.972	0.972	0.972	0.972	0.972
WL	μm²	1461.89	497.66	147.74	48.99	14.58	4.86
DC Bias Voltages							
$V_{EFF} = V_{GS} - V_T$	V	–0.064	–0.019	0.039	0.108	0.220	0.394
V_T (adjusted for V_{SB})	V	0.420	0.420	0.420	0.420	0.420	0.420
$V_{GS} = V_T + V_{EFF}$	V	0.356	0.401	0.459	0.528	0.640	0.814
$V_{DS,\,sat}$	V	0.108	0.116	0.137	0.173	0.250	0.375
Small Signal Parameters							
g_m/I_D	μS/μA	26.54	23.80	18.82	13.40	8.21	4.86
g_m	μS	2653.94	2379.96	1881.99	1340.28	820.60	485.56
V_A	V	12.70	12.71	12.74	12.76	12.79	10.19
g_{ds}	μS	7.8752	7.8668	7.8516	7.8350	7.8191	9.8130
$A_{Vi} = g_m/g_{ds}$	V/V	337.0	302.5	239.7	171.1	104.9	49.5
Transcond. Distortion (Diff. Pair)							
$V_{INDIF,1dB}$ (input 1–dB comp.)	V	0.050	0.057	0.075	0.108	0.196	0.377
Capacitances and Bandwidths							
$C_{GS} = C_{gsi} + C_{GSO}$	fF	2459.3	1259.3	564.9	238.9	82.4	29.4
$C_{GB} = C_{gbi} + C_{GBO}$	fF	2559.3	764.1	180.6	47.5	11.3	3.2
$C_{GD} = C_{GDO}$ (in saturation)	fF	1413.8	481.3	142.9	47.4	14.1	4.7
C_{DB} (at $V_{DB} = V_{DS} + V_{SB}$)	fF	742.6	253.6	75.8	25.3	7.7	2.8
$f_{Ti} = g_m/[2\pi(C_{gsi} + C_{gbi})]$	MHz	117.2	245.6	497.0	892.2	1640.6	2766.4
$f_T = g_m/[2\pi(C_{GS} + C_{GB})]$	MHz	84.2	187.2	401.8	744.7	1393.7	2368.0
$f_{DIODE} = g_m/[2\pi(C_{GS} + C_{GB} + C_{DB})]$,	MHz	73.3	166.4	364.7	684.2	1288.4	2183.7
Thermal and Flicker Noise							
$S_{VG}^{1/2}$ thermal	nV/Hz$^{1/2}$	2.04	2.20	2.57	3.14	4.08	5.31
$S_{VG}^{1/2}$ flicker at $f_{FLICKER}$	nV/Hz$^{1/2}$	175.23	300.34	572.82	1060.70	2141.50	4238.10
$S_{ID}^{1/2}$ thermal	pA/Hz$^{1/2}$	5.42	5.24	4.84	4.21	3.35	2.58
$S_{ID}^{1/2}$ flicker at $f_{FLICKER}$	pA/Hz$^{1/2}$	465.06	714.79	1078.04	1421.63	1757.31	2057.86
$f_{FLICKER}$, freq. for flicker noise	Hz	1	1	1	1	1	1
f_c, flicker-noise corner	MHz	0.0355	0.1056	0.3347	0.8888	2.5123	6.7377
DC Mismatch for Device Pair							
ΔV_{GS} (1σ)	mV	0.13	0.23	0.42	0.75	1.46	2.94
$\Delta I_D/I_D$ (1σ)	%	0.35	0.54	0.79	1.00	1.20	1.43

Figure 4.20 Predicted performance for nMOS devices in a $0.18\,\mu m$ CMOS process operating at $IC = 0.1$, 0.3, 1, 3, 10, and 30, $L = 1\,\mu m$, and $I_D = 100\,\mu A$ for $V_{DS} = 0.5\,V$ and $V_{SB} = 0\,V$. Values found from the *Analog CMOS Design, Tradeoffs and Optimization* spreadsheet illustrate tradeoffs in performance as inversion coefficient increases from weak through strong inversion at constant channel length and drain current, illustrating possible choices for differential-pair and current-mirror devices

*** MOSFETs, Sheet 2 ***							
Device reference		M1	M2	M3	M4	M5	M6
Device notes		IC = 0.1	IC = 0.3	IC = 1	IC = 3	IC = 10	IC = 30
Body Small Signal Parameters							
$\eta = g_{mb}/g_m$ (at V_{SB})		0.295	0.289	0.282	0.275	0.264	0.249
g_{mb} (at V_{SB})	µS	781.64	687.94	531.23	368.31	216.35	120.67
Source Parameters							
$g_{ms} = g_m + g_{mb} + g_{ds}$	µS	3443.46	3075.77	2421.07	1716.42	1044.77	616.05
C_{SB} (at V_{SB})	fF	888.1	318.1	103.9	40.6	15.8	5.6
Noise and Mismatch Parameters							
Γ, thermal noise factor (g_m)		0.516	0.539	0.585	0.627	0.653	0.662
($n\Gamma$), n multiplied by Γ		0.668	0.695	0.751	0.799	0.825	0.826
$K'_F{}^{1/2}$, flicker-noise factor	nV·µm	6700	6700	6963	7424	8177	9343
AF, flicker-noise slope		0.85	0.85	0.85	0.85	0.85	0.85
A_{VGS}, V_{GS} mismatch factor	V·µm	0.00506	0.00507	0.00511	0.00522	0.00556	0.00648
Velocity Sat. and VFMR Details							
Vel. sat. factor, $V_{EFF}/(LE_{CRIT})$		0.000	0.000	0.007	0.020	0.041	0.072
VFMR factor, $V_{EFF}\theta$		0.000	0.000	0.011	0.030	0.062	0.110
Critical IC, IC_{CRIT}		259.5	261.7	264.5	267.6	272.3	279.0
Gate Capacitance Details							
C_{GOX}	fF	12294.5	4185.4	1242.5	412.0	122.6	40.9
C_{gsi}	fF	1045.5	778.0	422.0	191.6	68.3	24.7
C_{gbi}	fF	2559.3	764.1	180.6	47.5	11.3	3.2
C_{GSO}	fF	1413.8	481.3	142.9	47.4	14.1	4.7
C_{GBO}	fF	0.0	0.0	0.0	0.0	0.0	0.0
C_{GDO}	fF	1413.8	481.3	142.9	47.4	14.1	4.7
Process and IC Details							
n (actual)		1.295	1.289	1.282	1.275	1.264	1.249
μ_0	cm²/Vs	422.00	422.00	422.00	422.00	422.00	422.00
$k_0 = \mu_0 C'_{OX}$	µA/V²	354.90	354.90	354.90	354.90	354.90	354.90
$U_T = kT/q$	V	0.02585	0.02585	0.02585	0.02585	0.02585	0.02585
$I_0 = 2n_0\mu_0 C'_{OX}U_T{}^2$	µA	0.641	0.641	0.641	0.641	0.641	0.641
IC (actual)		0.105	0.310	1.051	3.189	10.808	32.818
Layout Details, IC Trimming							
W_{DRAWN}, each; x m for total	µm	47.00	32.00	19.00	12.60	7.50	2.50
L_{DRAWN}	µm	1	1	1	1	1	1
AD, each; x m for total	µm²	14.10	9.60	5.70	3.78	2.25	0.75
PD, each; x m for total	µm	47.60	32.60	19.60	13.20	8.10	3.10
AS, each; x m for total	µm²	14.69	10.40	6.65	5.04	3.75	1.25
PS, each; x m for total	µm	50.56	36.65	24.45	19.70	16.00	6.00
IC (trimmed), may be used for IC input		0.101	0.296	0.998	3.011	10.117	30.351
---------- **Optional User Design Inputs** ----------							
	Default, units						
m, number of gate fingers	1	32	16	8	4	2	2
W_{DRAWN} (trimmed), each	none, µm	47	32	19	12.6	7.5	2.5
V_{DS} (for C_{DB})	V_{GS}, V (+)	0.5	0.5	0.5	0.5	0.5	0.5
V_{SB} (for g_{mb} and C_{SB})	0 V, V (+)						
$f_{FLICKER}$, freq. for flicker noise	1 Hz, Hz						
T, temperature	27 C, C						

Figure 4.20   (*continued*)

The effective gate–source voltage, V_{EFF}, is -0.064, 0.039, and $0.220\,\text{V}$, and the drain–source saturation voltage, $V_{DS,sat}$, is 0.108, 0.137, and $0.250\,\text{V}$ for $IC = 0.1$, 1, and 10, respectively. Both increase as IC increases, increasing with the square root of IC in strong inversion, illustrating the bias voltage compliance penalty associated with operation in strong inversion. The increase, however, is modest in moderate inversion. When velocity saturation is significant in strong inversion, V_{EEF} increases directly with IC. While not repeated here, the spreadsheet also includes the threshold voltage, V_T, and gate–source voltage, $V_{GS} = V_T + V_{EFF}$. These are useful for ensuring sufficiently low values of V_{GS} for bias voltage compliance.

Transconductance efficiency, g_m/I_D, is 26.5, 18.8, and $8.2\,\mu\text{S}/\mu\text{A}$ ($1/\text{V}$), and transconductance, g_m, is 2650, 1880, and $820\,\mu\text{S}$ for $IC = 0.1$, 1, and 10, respectively, illustrating their decrease as IC increases. The Early voltage, V_A, remains nearly constant at 12.7, 12.7, and $12.8\,\text{V}$ and is only slightly lower in weak inversion for moderate-length, $L = 1\,\mu\text{m}$ devices considered because of DIBL effects described earlier in Section 3.8.4.2. Drain–source conductance, g_{ds}, is also nearly constant at 7.9, 7.9, and $7.8\,\mu\text{S}$, inversely tracking V_A while I_D remains constant. Finally, the intrinsic voltage gain, A_{Vi}, is 337, 240, and $105\,\text{V/V}$, which, like g_m/I_D, is maximum in weak inversion. A_{Vi} decreases as IC increases due primarily to decreasing g_m/I_D, while V_A remains nearly constant. g_m/I_D, g_m, and A_{Vi} decrease inversely with the square root of IC in strong inversion or inversely with IC when velocity saturation is significant. As described earlier in Section 3.8.4.1, V_A, g_{ds}, and, correspondingly, A_{Vi} are strong functions of V_{DS}, which, again, is held fixed at $0.5\,\text{V}$.

The input, 1 dB compression voltage for a differential pair, $V_{INDIF1dB}$, is 0.050, 0.075, and $0.196\,\text{V}$ for $IC = 0.1$, 1, and 10, respectively, illustrating its increase as IC increases due to decreasing g_m/I_D. As described earlier in Section 3.8.2.6, $V_{INDIF1dB}$ is the differential input voltage for two equal devices in a differential pair where the differential output current is compressed by 1 dB (11 %) from its ideal, linear value.

The gate–source capacitance, C_{GS}, which includes both intrinsic, C_{gsi}, and extrinsic overlap, C_{GSO}, capacitances is 2460, 565, and $82\,\text{fF}$ for $IC = 0.1$, 1, and 10, respectively. The gate–body capacitance, C_{GB}, which includes both intrinsic, C_{gbi}, and extrinsic overlap, C_{GBO}, capacitances is 2560, 181, and $11\,\text{fF}$. C_{GB} is significant at $IC = 0.1$ in weak inversion where it actually exceeds C_{GS}. C_{GS} and C_{GB} decrease as IC increases due to decreasing gate area. The gate–drain capacitance, C_{GD}, which includes only the extrinsic overlap capacitance, C_{GDO}, for operation in saturation, is 1410, 143, and $14\,\text{fF}$. Finally, the drain–body capacitance, C_{DB}, is 743, 76, and $8\,\text{fF}$ at $V_{DB} = V_{DS} + V_{SB} = 0.5\,\text{V}$. C_{GD} and C_{DB} decrease inversely with increasing IC due to decreasing channel width.

Intrinsic bandwidth, f_{Ti}, is 117, 497, and $1640\,\text{MHz}$ for $IC = 0.1$, 1, and 10, respectively. Extrinsic bandwidth, f_T, which includes the extrinsic gate overlap capacitances, is 84, 402, and $1390\,\text{MHz}$. Finally, diode-connected bandwidth, f_{diode}, which includes the extrinsic gate overlap and drain–body capacitances, is 73, 365, and $1290\,\text{MHz}$. All bandwidths increase as IC increases, increasing with the square root of IC in strong inversion, because the decrease in capacitances exceeds the decrease in g_m. When velocity saturation is significant, the bandwidths level off with IC. f_T and f_{diode} are modestly lower than f_{Ti} for moderate-length, $L = 1\,\mu\text{m}$ devices considered, but are considerably lower than f_{Ti} for short-channel devices as illustrated in the next section.

The gate-referred thermal-noise voltage density, $S_{VG}^{1/2}$ thermal, is 2.0, 2.6, and $4.1\,\text{nV/Hz}^{1/2}$ for $IC = 0.1$, 1, and 10, respectively, illustrating its increase as IC increases due to decreasing g_m. Gate-referred flicker-noise voltage density at 1 Hz, $S_{VG}^{1/2}$ flicker, is 175, 573, and $2140\,\text{nV/Hz}^{1/2}$, illustrating its increase with the square root of IC as IC increases due to decreasing gate area. The increase can be greater at high IC due to inversion-level or V_{EFF} increases in flicker noise. The gate-referred noise voltage densities are 41 % larger for a differential pair containing two devices because of the addition of uncorrelated noise from two devices.

The drain-referred thermal-noise current density, $S_{ID}^{1/2}$ thermal, is 5.4, 4.8, and $3.4\,\text{pA/Hz}^{1/2}$ for $IC = 0.1$, 1, and 10, respectively, illustrating its decrease as IC increases due to decreasing g_m. Drain-referred flicker-noise current density at 1 Hz, $S_{ID}^{1/2}$ flicker, is 465, 1080, and $1760\,\text{pA/Hz}^{1/2}$, illustrating its increase as IC increases due initially to decreasing gate area. As described in Section 4.3.5.2, $S_{ID}^{1/2}$

flicker levels off in strong inversion or even decreases if velocity saturation is significant because the increase in $S_{VG}^{1/2}$ flicker due to decreasing gate area is countered by the decrease in g_m responsible for the conversion of $S_{VG}^{1/2}$ flicker to $S_{ID}^{1/2}$ flicker. However, the noise can then increase at high IC due to inversion-level or V_{EFF} increases in flicker noise. The drain-referred noise current densities are 41 % larger at the output of a 1:1 current mirror containing two devices because of the addition of uncorrelated noise from the reference device. Also, as described in Section 4.3.5, the output noise of a cascode current mirror can be estimated from noise of the bottom devices because of significant noise degeneration in the top devices.

The flicker-noise corner frequency, f_c, is 0.04, 0.33, and 2.5 MHz for $IC = 0.1$, 1, and 10, respectively, which is roughly a fixed fraction of f_{Ti} as described earlier in Section 3.10.3.8. f_c is an increasing fraction of f_{Ti} when inversion-level or V_{EFF} increases in flicker noise are significant. Flicker noise and f_c increase at high IC due to decreasing gate area, requiring the use of lower IC or longer channel lengths when flicker noise is important.

As discussed earlier in Section 3.10.2.2, thermal noise can be greater for short-channel devices than the values given because of possible small-geometry increases in the noise. The increase in flicker noise associated with increasing inversion-level or V_{EFF}, discussed earlier in Sections 3.10.3.3 and 3.10.3.6, however, is included in the spreadsheet results. This can be seen in the second sheet of the spreadsheet results where $K_F'^{1/2}$, which is the square root of the hand units flicker-noise factor K_F', increases from 6700 to 9343 nV · μm as IC increases from 0.1 to 30. This operating, normalized measure of flicker noise corresponds to $S_{VG}^{1/2}$ flicker at 1 Hz for a device having a gate area of 1 μm².

The 1σ, gate–source mismatch voltage, ΔV_{GS}, for a differential pair is 0.13, 0.42, and 1.5 mV for $IC = 0.1$, 1, and 10, respectively, illustrating its increase with the square root of IC as IC increases due to decreasing gate area. The increase can be greater at high IC due to transconductance factor mismatch. The 1σ, drain mismatch current, $\Delta I_D/I_D$, for a pair of current-mirror devices is 0.35, 0.79, and 1.2 %, illustrating its increase as IC increases due initially to decreasing gate area. As described in Section 4.3.5.2, $\Delta I_D/I_D$ levels off in strong inversion or even decreases if velocity saturation is significant because the increase in threshold-voltage mismatch associated with decreasing gate area is countered by the decrease in g_m/I_D responsible for the conversion of threshold-voltage mismatch to $\Delta I_D/I_D$. However, the mismatch can then increase at high IC due to transconductance factor mismatch. ΔV_{GS} and $\Delta I_D/I_D$ mismatch values are estimated 1σ values that would typically be multiplied by at least three for yield consideration in production design. As described in Section 4.3.5, the mismatch current of a cascode current mirror can be estimated from mismatch of the bottom devices because of significant mismatch degeneration in the top devices.

Increases in mismatch due to transconductance factor mismatch can be seen in the second sheet of spreadsheet results where the gate–source voltage mismatch factor, A_{VGS}, increases from 5 to 6.5 mV · μm as IC increases from 0.1 to 30. At low IC, A_{VGS} is equal to the threshold-voltage mismatch factor, A_{VT}, indicating negligible transconductance factor mismatch.

The continuation page of Figure 4.20 shows the second sheet of the spreadsheet results. These include the body-effect transconductance, g_{mb}, source transconductance, $g_{ms} = g_m + g_{mb} + g_{ds}$, source–body capacitance, C_{SB}, and thermal-noise, $n\Gamma$, flicker noise, K_F', and mismatch, A_{VGS}, parameters useful for circuit analysis. The sheet also includes velocity saturation and VFMR details, gate capacitance details, process and inversion coefficient details, layout details, and optional user inputs. Optional user inputs include the number of gate stripes or interdigitations, m, defaulted at unity, and a selected drawn channel width for each layout stripe that permits trimming the input value of IC to match the actual layout. Additionally, input values for V_{DS}, V_{SB}, the frequency for reported flicker noise, $f_{FLICKER}$, and temperature, T, may be included. Default values for these optional inputs are $V_{DS} = V_{GS}$, $V_{SB} = 0$ V, $f_{FLICKER} = 1$ Hz, and $T = 27$ °C. The second sheet of results and optional user inputs are described further in Section A.2.3 in the Appendix.

From the second sheet of results, the body-effect transconductance ratio, $\eta = g_{mb}/g_m$, is 0.30, 0.28, and 0.26 at $V_{SB} = 0$ V for $IC = 0.1$, 1, and 10, respectively. η decreases slightly as IC increases and decreases with increasing V_{SB}. Body-effect transconductance, g_{mb}, is 782, 531, and 216 μS, which decreases as IC

increases, primarily tracking the decrease in g_m. Finally, the source–body capacitance, C_{SB}, at $V_{SB} = 0\,\text{V}$ is 888, 104, and 16 fF, which decreases nearly inversely with IC as IC increases due to decreasing channel width.

Also from the second sheet of results, the velocity saturation factor, $V_{EFF}/(LE_{CRIT})$, is 0.000, 0.007, and 0.041, and the VFMR reduction factor, $V_{EFF} \cdot \theta$, is 0.000, 0.011, and 0.062 for $IC = 0.1$, 1, and 10, respectively. Values of 0.041 and 0.062 for $IC = 10$ at the onset of strong inversion indicate that drain current is down by approximately 4 and 6 %, respectively, due to velocity saturation and VFMR effects. The critical inversion coefficient, IC_{CRIT}, for velocity saturation and VFMR effects is 260, 265, and 272, which increases slightly as IC increases because of a slight decrease in the substrate factor, n. As described earlier in Section 3.8.2.2, IC_{CRIT} corresponds to the value of IC where g_m/I_D is down to approximately 70.7 % of the value present without velocity saturation and VFMR effects. IC_{CRIT} is lower for L shorter than 1 μm considered in this design example where velocity saturation effects are more significant.

A third sheet of spreadsheet results not shown here lists performance related to bandwidth–power–accuracy figures of merit. As IC increases, figures of merit involving the square of accuracy increase for current-mode circuits and decrease for voltage-mode circuits because these are inversely proportional to g_m/I_D and directly proportional to g_m/I_D, respectively. These figures of merit are included in spreadsheet results in the next section.

A glance across the columns in Figure 4.20 allows a rapid assessment of performance tradeoffs as IC is increased from $IC = 0.1$, 0.3, 1, 3, 10, and 30 from weak through strong inversion for fixed $L = 1\,\mu\text{m}$ and $I_D = 100\,\mu\text{A}$. This illustrates possible choices for differential-pair and current-mirror devices. Operation at low values of IC is a low-frequency and DC optimization because low-frequency gain measures of g_m/I_D, g_m, and A_{Vi} are large, while DC bias compliance voltage, low-frequency noise, and DC mismatch measures of V_{EFF}, $V_{DS,sat}$, $S_{VG}^{1/2}$ flicker, $S_{ID}^{1/2}$ flicker, f_c, ΔV_{GS}, and $\Delta I_D/I_D$ are small. This comes at the expense of geometry and capacitance measures of W, WL, C_{GS}, C_{GB}, C_{GD}, and C_{DB} that are large, and bandwidth and distortion measures of f_{Ti}, f_T, f_{diode}, and $V_{INDIF1dB}$ that are small. Operation at low values of IC also minimizes the thermal-noise measure of $S_{VG}^{1/2}$ thermal for voltage-mode devices receiving voltage signals at the gate, while maximizing $S_{ID}^{1/2}$ thermal. Finally, operation at low values of IC minimizes short-channel, velocity saturation reductions of g_m/I_D, g_m, and A_{Vi}, and minimizes $S_{VG}^{1/2}$ flicker increases associated with increases in inversion level or V_{EFF}.

Operation at high values of IC is an AC and distortion optimization because bandwidth and distortion measures of f_{Ti}, f_T, f_{diode}, and $V_{INDIF1dB}$ are large, while geometry and capacitance measures of W, WL, C_{GS}, C_{GB}, C_{GD}, and C_{DB} are small. This comes at the expense of DC bias compliance voltage, low-frequency noise, and DC mismatch measures of V_{EFF}, $V_{DS,sat}$, $S_{VG}^{1/2}$ flicker, $S_{ID}^{1/2}$ flicker, f_c, ΔV_{GS}, and $\Delta I_D/I_D$ that are large. Operation at high values of IC also minimizes the thermal-noise measure of $S_{ID}^{1/2}$ thermal for current-mode devices processing signal currents, while maximizing $S_{VG}^{1/2}$ thermal.

In addition to results given by the *Analog CMOS Design, Tradeoffs and Optimization* spreadsheet in Figure 4.20, Tables 4.1–4.3 summarized MOS performance tradeoffs for design selections of the inversion coefficient. Sections 4.3.3–4.3.5 showed these graphically, and Section 4.2.2 discussed performance trends.

4.4.2 Selecting Channel Length

Section 4.4.1 presented numerical results from the *Analog CMOS Design, Tradeoffs and Optimization* spreadsheet showing device performance for inversion coefficient selections from weak through strong inversion at constant $L = 1\,\mu\text{m}$ and $I_D = 100\,\mu\text{A}$. This permitted a study of numerical performance tradeoffs leading to optimum inversion coefficient selections. In this section, the inversion coefficient will be held fixed at $IC = 10$ at the onset of strong inversion, while channel length is decreased from $L = 4$, 2, 1, 0.48, 0.28, to 0.18 μm. This permits a study of numerical performance tradeoffs leading to optimum channel length selections. Drain current will continue to be held fixed at $I_D = 100\,\mu\text{A}$

to permit a "fair" study of performance tradeoffs at constant power consumption. nMOS devices in the example 0.18 μm CMOS process will again be considered for $V_{DS} = 0.5$ V and $V_{SB} = 0$ V. As described at the beginning of Section 4.4, process parameters described in Tables 3.2–3.4 are used, with V_{AL} values found from the measured V_A data of Figure 3.45.

As in the previous section, actual IC values are rounded slightly as a result of trimming the reported drawn channel width to actual layout values. Drawn channel length values are used as spreadsheet inputs where, for example, a drawn channel length of 0.18 μm corresponds to an effective channel length of 0.152 μm (0.18 – 0.028 μm) for the process lateral diffusion length of $DL = 0.028$ μm.

Figure 4.21 gives results from the *Analog CMOS Design, Tradeoffs and Optimization* spreadsheet for nMOS devices in the example 0.18 μm CMOS process for varying values of $L = 4, 2, 1, 0.48, 0.28,$ and 0.18 μm, and fixed values of $IC = 10$, $I_D = 100$ μA, $V_{DS} = 0.5$ V, and $V_{SB} = 0$ V. As in the previous section, $V_{DS} = 0.5$ V is sufficiently high to ensure operation in saturation. While the spreadsheet reports values for all L selections, only values for $L = 4, 1,$ and 0.18 μm are included in the following discussion for brevity. Spreadsheet results are not redefined as these were defined in the previous section.

From Figure 4.21, W (effective) is 62, 15, and 2.4 μm, and WL (effective) is 246, 14.6, and 0.4 μm² for $L = 4, 1,$ and 0.18 μm, respectively. Both are maximum for long-channel devices, with W decreasing directly with decreasing L, and WL decreasing very significantly with the square of decreasing L.

V_{EFF} is 0.218, 0.220, and 0.232 V and $V_{DS,sat}$ is nearly constant at 0.25 V for $L = 4, 1,$ and 0.18 μm, respectively. V_{EFF} increases slightly as L decreases due to slight velocity saturation. Velocity saturation is more significant for IC values greater than 10 or channel lengths shorter than the minimum of 0.18 μm considered here.

g_m/I_D is 8.3, 8.2, and 7.3 μS/μA (1/V), and g_m is 830, 820, and 730 μS for $L = 4, 1,$ and 0.18 μm, respectively. Both are nearly constant, but decrease modestly as L decreases due to slight velocity saturation. V_A is 29, 12.6, and 3.2 V, which is maximum for long-channel devices. It decreases significantly as L decreases, tracking L sublinearly, as a result of increased CLM effects for operation here at $IC = 10$ where DIBL effects are not significant. g_{ds} increases from 3.4, 7.9, to 31.7 μS, inversely tracking V_A while I_D remains constant. A_{Vi} is 241, 103, and 23.1 V/V, which, like V_A, is maximum for long-channel devices. A_{Vi} decreases significantly as L decreases due primarily to decreasing V_A and, correspondingly, increasing g_{ds}, while g_m/I_D remains nearly constant. As mentioned in the previous section, V_A, g_{ds}, and, correspondingly, A_{Vi} are strong functions of V_{DS}, which, again, is held fixed at 0.5 V. Finally, $V_{INDIF1dB}$ is 0.192, 0.196, and 0.230 V, illustrating its modest increase as L decreases due to the velocity saturation decrease in g_m/I_D.

C_{GS} is 1210, 82, and 4 fF, and C_{GB} is 192, 11, and 0.3 fF for $L = 4, 1,$ and 0.18 μm, respectively. Both decrease significantly as L decreases due to gate area that decreases with the square of decreasing L. C_{GD} is 58, 14, and 2.3 fF, and C_{DB} is 31, 8, and 1.5 fF at $V_{DB} = V_{DS} + V_{SB} = 0.5$ V. Both decrease directly with decreasing L due to decreasing channel width.

f_{Ti} is 99, 1640, and 58600 MHz, f_T is 94, 1400, and 27500 MHz, and f_{diode} is 92, 1290, and 20400 MHz for $L = 4, 1,$ and 0.18 μm, respectively. f_{Ti} increases significantly as L decreases, being inversely proportional to the square of L unless velocity saturation is significant where it is inversely proportional to L. For $L = 0.18$ μm, f_T is nearly one-half f_{Ti}, while f_{diode} is lower at nearly one-third f_{Ti}. As described earlier in Section 3.9.7, the reduction in f_T and f_{diode} compared to f_{Ti} is because gate overlap capacitances and the drain–body capacitance are significant for short-channel devices where gate area and intrinsic capacitances are small.

$S_{VG}^{1/2}$ thermal is 4.1, 4.1, and 4.3 nV/Hz$^{1/2}$ for $L = 4, 1,$ and 0.18 μm, respectively, increasing slightly as L decreases due to the velocity saturation decrease in g_m. $S_{VG}^{1/2}$ flicker at 1 Hz is 520, 2140, and 13700 nV/Hz$^{1/2}$, increasing inversely with decreasing channel length due to the significant decrease in gate area. As mentioned in the previous section, the gate-referred noise voltage densities are 41 % larger for a differential pair containing two devices because of the addition of uncorrelated noise from two devices.

Analog CMOS Design, Tradeoffs and Optimization Spreadsheet -- MOSFETs Sheet							
For design guidance only as results are approximate and not for any particular CMOS process.							
© 2000–2007, David M. Binkley. See Disclaimer, Notes Sheet.							
---------- Optional User Design Information ----------							
Description		Illustration of changing L at fixed $IC = 10$ ($I_D = 100\,\mu A$)					
Device reference		M1	M2	M3	M4	M5	M6
Device notes		L=4 um	L=2 um	L=1 um	L=0.48 um	L=0.28 um	L=0.18 um
---------- Required User Design Inputs ----------							
Device model		nSIL4u	nSIL2u	nSIL1u	nSIL05u	nSIL03u	nSIL02u
I_D	$\mu A\,(+)$	100	100	100	100	100	100
IC (fixed normalized)		10.002	9.996	10.117	10.081	9.836	9.888
L_{DRAWN}	μm	4	2	1	0.48	0.28	0.18
---------- Calculated Results ----------		OK	OK	OK	OK	OK	OK
Effective Width, Length, Gate Area							
W	μm	62.00	30.80	15.00	7.00	4.00	2.40
L	μm	3.972	1.972	0.972	0.452	0.252	0.152
WL	μm^2	246.26	60.74	14.58	3.16	1.01	0.36
DC Bias Voltages							
$V_{EFF} = V_{GS} - V_T$	V	0.218	0.218	0.220	0.223	0.224	0.232
V_T (adjusted for V_{SB})	V	0.420	0.420	0.420	0.420	0.420	0.420
$V_{GS} = V_T + V_{EFF}$	V	0.638	0.638	0.640	0.643	0.644	0.652
$V_{DS,\,sat}$	V	0.249	0.248	0.250	0.249	0.247	0.248
Small Signal Parameters							
g_m/I_D	$\mu S/\mu A$	8.32	8.30	8.21	8.07	7.88	7.32
g_m	μS	831.70	829.89	820.60	807.22	787.60	731.96
V_A	V	28.99	19.71	12.60	7.76	5.19	3.16
g_{ds}	μS	3.4492	5.0740	7.9390	12.8878	19.2577	31.6555
$A_{Vi} = g_m/g_{ds}$	V/V	241.1	163.6	103.4	62.6	40.9	23.1
Transcond. Distortion (Diff. Pair)							
$V_{INDIF,1dB}$ (input 1-dB comp.)	V	0.192	0.192	0.196	0.201	0.209	0.230
Capacitances and Bandwidths							
$C_{GS} = C_{gsi} + C_{GSO}$	fF	1210.0	313.0	82.4	21.4	8.5	4.0
$C_{GB} = C_{gbi} + C_{GBO}$	fF	192.0	47.3	11.3	2.5	0.8	0.3
$C_{GD} = C_{GDO}$ (in saturation)	fF	58.3	29.0	14.1	6.6	3.8	2.3
C_{DB} (at $V_{DB} = V_{DS} + V_{SB}$)	fF	31.0	15.4	7.7	3.7	2.3	1.5
$f_{Ti} = g_m/[2\pi(C_{gsi} + C_{gbi})]$	MHz	98.5	398.6	1640.6	7439.0	22822.7	58603.1
$f_T = g_m/[2\pi(C_{GS} + C_{GB})]$	MHz	94.4	366.6	1393.7	5386.7	13548.0	27450.3
$f_{DIODE} = g_m/[2\pi(C_{GS} + C_{GB} + C_{DB})]$, MHz		92.4	351.5	1288.4	4657.3	10885.2	20357.9
Thermal and Flicker Noise							
$S_{VG}^{1/2}$ thermal	$nV/Hz^{1/2}$	4.05	4.06	4.08	4.11	4.16	4.32
$S_{VG}^{1/2}$ flicker at $f_{FLICKER}$	$nV/Hz^{1/2}$	520.41	1048.12	2143.19	4608.50	8172.98	13676.95
$S_{ID}^{1/2}$ thermal	$pA/Hz^{1/2}$	3.37	3.37	3.35	3.32	3.28	3.16
$S_{ID}^{1/2}$ flicker at $f_{FLICKER}$	$pA/Hz^{1/2}$	432.82	869.82	1758.70	3720.05	6437.04	10010.99
$f_{FLICKER}$, freq. for flicker noise	Hz	1	1	1	1	1	1
f_c, flicker-noise corner	MHz	0.0915	0.4739	2.5169	14.9601	55.9916	172.6147
DC Mismatch for Device Pair							
ΔV_{GS} (1σ)	mV	0.35	0.71	1.46	3.14	5.59	9.43
$\Delta I_D/I_D$ (1σ)	%	0.29	0.59	1.20	2.53	4.40	6.91

Figure 4.21 Predicted performance for nMOS devices in a 0.18 μm CMOS process operating at $IC = 10$, $L = 4$, 2, 1, 0.48, 0.28, and 0.18 μm, and $I_D = 100\,\mu A$ for $V_{DS} = 0.5\,V$ and $V_{SB} = 0\,V$. Values found from the *Analog CMOS Design, Tradeoffs and Optimization* spreadsheet illustrate tradeoffs in performance from long to short channel length at constant inversion level and drain current, illustrating possible choices for differential-pair and current-mirror devices. The continuation page lists measures related to bandwidth–power–accuracy figures of merit

*** MOSFETs, Sheet 3 -- Performance Related to Bandwidth, Power, Accuracy Figures of Merit ***							
Device reference		M1	M2	M3	M4	M5	Approx.
Device notes		L=4um	L=2um	L=1um	L=0.48 um	L=0.28um	trend
I_D	μA	100	100	100	100	100	with L
IC (fixed normalized)		10.002	9.996	10.117	10.081	9.836	(excludes
L_{DRAWN}	μm	4	2	1	0.48	0.28	vel. sat.)
---------- Optional User Inputs for Figures of Merit ----------							
V_{DD} for P	V	1.8	1.8	1.8	1.8	1.8	
I_{IN}/I_D for I-mode FOMs	DC, rms	0.5	0.5	0.5	0.5	0.5	
V_{IN} for V-mode FOMs	Vdc, Vrms	0.1	0.1	0.1	0.1	0.1	
Current-Mode Figures of Merit, Power and Bandwidth							
Power, $P = 2I_D \cdot V_{DD}$	μW	360	360	360	360	360	NA
Bandwidth, $BW = f_{Ti}/2$	MHz	49.3	199.3	820.3	3719.5	11411.4	$\propto 1/L^2$
Current-Mode Figures of Merit, DC Offset (input current is relative to I_D)							
Input current, I_{IN}	μA	50	50	50	50	50	NA
Offset current, I_{OS} (3 σ)	μA	0.882	1.773	3.586	7.597	13.198	$\propto 1/L$
Accuracy, $A_{CC,DC} = I_{IN}/I_{OS}$ (3 σ)		56.7	28.2	13.9	6.6	3.8	$\propto L$
Error, $E_{RR,DC} = I_{OS}$ (3 σ)/I_{IN}	%	1.76	3.55	7.17	15.19	26.40	$\propto 1/L$
$A_{CC,DC} \cdot BW$	Hz	2.79E+09	5.62E+09	1.14E+10	2.45E+10	4.32E+10	$\propto 1/L$
$A_{CC,DC}^2 \cdot BW$	Hz	1.58E+11	1.58E+11	1.59E+11	1.61E+11	1.64E+11	$\approx$ const.
$A_{CC,DC}^2 \cdot BW/P$	Hz/W	4.40E+14	4.40E+14	4.43E+14	4.48E+14	4.55E+14	$\approx$ const.
Current-Mode Figures of Merit, Thermal Noise (input current is relative to I_D)							
Input current, I_{IN}	μA rms	50	50	50	50	50	NA
Noise current over BW, I_N	μA rms	0.0335	0.0672	0.1356	0.2864	0.4954	$\propto 1/L$
Accuracy, $A_{CC,AC} = I_{IN}/I_N$		1494	744	369	175	101	$\propto L$
Error, $E_{RR,AC} = I_N/I_{IN}$	%	0.0669	0.1344	0.2712	0.5728	0.9907	$\propto 1/L$
$A_{CC,AC} \cdot BW$	Hz	7.36E+10	1.48E+11	3.02E+11	6.49E+11	1.15E+12	$\propto 1/L$
$A_{CC,AC}^2 \cdot BW$	Hz	1.1E+14	1.1E+14	1.11E+14	1.13E+14	1.16E+14	$\approx$ const.
$A_{CC,AC}^2 \cdot BW/P$	Hz/W	3.06E+17	3.06E+17	3.1E+17	3.15E+17	3.23E+17	$\approx$ const.
Voltage-Mode Figures of Merit, Power and Bandwidth							
Power, $P = 2I_D \cdot V_{DD}$	μW	360	360	360	360	360	NA
Bandwidth, $BW = 2f_{Ti}$	MHz	197.0	797.2	3281.2	14878.0	45645.5	$\propto 1/L^2$
Voltage-Mode Figures of Merit, DC Offset (input voltage is fixed and given above)							
Offset voltage, V_{OS} (3 σ)	mV	1.061	2.137	4.370	9.411	16.757	$\propto 1/L$
Accuracy, $A_{CC,DC} = V_{IN}/V_{OS}$ (3 σ)		94.3	46.8	22.9	10.6	6.0	$\propto L$
Error, $E_{RR,DC} = V_{OS}$ (3 σ)/V_{IN}, %		1.06	2.14	4.37	9.41	16.76	$\propto 1/L$
$A_{CC,DC} \cdot BW$	Hz	1.86E+10	3.73E+10	7.51E+10	1.58E+11	2.72E+11	$\propto 1/L$
$A_{CC,DC}^2 \cdot BW$	Hz	1.75E+12	1.75E+12	1.72E+12	1.68E+12	1.63E+12	$\approx$ const.
$A_{CC,DC}^2 \cdot BW/P$	Hz/W	4.86E+15	4.85E+15	4.77E+15	4.67E+15	4.52E+15	$\approx$ const.
Voltage-Mode Figures of Merit, Thermal Noise (input voltage is fixed and given above)							
Noise voltage over BW, V_N	mV rms	0.0805	0.1620	0.3305	0.7096	1.2579	$\propto 1/L$
Accuracy, $A_{CC,AC} = V_{IN}/V_N$		1243	617	303	141	79	$\propto L$
Error, $E_{RR,AC} = V_N/V_{IN}$	%	0.0805	0.1620	0.3305	0.7096	1.2579	$\propto 1/L$
$A_{CC,AC} \cdot BW$	Hz	2.45E+11	4.92E+11	9.93E+11	2.10E+12	3.63E+12	$\propto 1/L$
$A_{CC,AC}^2 \cdot BW$	Hz	3.04E+14	3.04E+14	3E+14	2.95E+14	2.88E+14	$\approx$ const.
$A_{CC,AC}^2 \cdot BW/P$	Hz/W	8.46E+17	8.44E+17	8.34E+17	8.21E+17	8.01E+17	$\approx$ const.

Figure 4.21 (*continued*)

$S_{ID}^{1/2}$ thermal is 3.4, 3.4, and 3.2 pA/Hz$^{1/2}$ for $L = 4$, 1, and 0.18 μm, respectively, decreasing slightly as L decreases due to the velocity saturation decrease in g_m. $S_{ID}^{1/2}$ flicker at 1 Hz is 433, 1760, and 10000 pA/Hz$^{1/2}$, increasing inversely with decreasing channel length due to the significant decrease in gate area. As mentioned in the previous section, the drain-referred noise current densities are 41 % larger at the output of a 1:1 current mirror containing two devices because of the addition of uncorrelated noise from the reference device. Additionally, the output noise of a cascode current mirror can be estimated from noise of the bottom devices because of significant noise degeneration in the top devices.

Flicker noise f_c is 0.09, 2.5, and 173 MHz for $L = 4$, 1, and 0.18 μm, respectively, which is roughly a fixed fraction of f_{Ti} as mentioned in the previous section. Flicker noise and f_c increase significantly for short channel lengths due to the significant decrease in gate area, requiring the use of longer channel lengths when flicker noise is important.

As mentioned in the previous section, thermal noise can be greater for short-channel devices than the values given because of possible small-geometry increases in the noise. The increase in flicker noise associated with increasing inversion-level or V_{EFF}, however, is included in the spreadsheet results.

ΔV_{GS} (1σ) for a differential pair is 0.35, 1.5, and 9.4 mV, and $\Delta I_D/I_D$ (1σ) for a pair of current-mirror devices is 0.29, 1.2, and 6.9 % for $L = 4$, 1, and 0.18 μm, respectively. Both increase inversely with decreasing channel length due to the significant decrease in gate area. As mentioned in the previous section, increases in mismatch due to transconductance factor mismatch are included in the spreadsheet results. ΔV_{GS} and $\Delta I_D/I_D$ mismatch values are, again, estimated 1σ values that would typically be multiplied by at least three for yield consideration in production design. Like flicker noise, DC voltage and current mismatch increase significantly for short-channel devices due to the significant decrease in gate area, requiring the use of longer channel lengths when mismatch is important. As mentioned in the previous section, the mismatch current of a cascode current mirror can be estimated from mismatch of the bottom devices because of significant mismatch degeneration in the top devices.

From the second sheet of spreadsheet results, included in the previous section but not included here, $\eta = g_{mb}/g_m$ is nearly constant at 0.26 at $V_{SB} = 0$ V for $L = 4$, 1, and 0.18 μm, respectively. η is nearly constant with L, excepting its possible decrease for short-channel devices due to charge sharing, described earlier in Section 3.8.3.1. g_{mb} is 219, 216, and 192 μS, tracking the velocity saturation decrease in g_m. Finally, C_{SB} at $V_{SB} = 0$ V is 50, 16, and 3 fF, which decreases with decreasing L due to decreasing channel width.

Also from the second sheet of results, $V_{EFF}/(LE_{CRIT})$ is 0.010, 0.041, and 0.273, and $V_{EFF} \cdot \theta$ is 0.061, 0.062, and 0.065 for $L = 4$, 1, and 0.18 μm, respectively. Values of 0.273 and 0.065 for $L = 0.18$ μm at $IC = 10$ at the onset of strong inversion indicate that drain current is down by approximately 27.3 and 6.5 %, respectively, due to velocity saturation and VFMR effects. IC_{CRIT} is 554, 272, and 28, which decreases significantly as L decreases because of increased velocity saturation. For the minimum $L = 0.18$ μm, velocity saturation reduction of g_m/I_D is moderate because operating $IC = 10$ is moderately below the critical value of $IC_{CRIT} = 28$. Velocity saturation is more significant for IC values greater than 10 or, especially, for channel lengths shorter than the minimum of 0.18 μm considered here.

The continuation page of Figure 4.21 shows a third sheet of spreadsheet results giving performance related to bandwidth–power–accuracy figures of merit that were discussed in Sections 4.3.6.5–4.3.6.7 and listed in Table 4.4. Performance includes selected signal levels, DC offset and noise levels, and resulting errors, accuracies, and figures of merit. Performance is for $I_D = 100$ μA (for individual devices), $IC = 10$ at the onset of strong inversion, and varying channel lengths as considered on the main sheet, but also considers a supply voltage of $V_{DD} = 1.8$ V, a current-mirror input signal of $I_{IN} = 0.5 \cdot I_D = 50$ μA, and a differential-pair input signal of $V_{IN} = 0.1$ V. These operating conditions where discussed earlier in Section 4.3.6.7 for the calculation of figures of merit and correspond to constant power consumption of $P = 2I_D \cdot V_{DD} = 2 \cdot 100$ μA $\cdot 1.8$ V $= 360$ μW for a pair of current-mirror or differential-pair devices. Spreadsheet results will be discussed for current-mode circuits, for example a 1:1 current mirror, for $L = 4$ and 1 μm. Instead of a listing for $L = 0.18 \mu$m considered on the main sheet, the sheet includes trends versus the channel length.

Input-referred DC offset current is $I_{OS}(3\ \sigma) = 0.9$ and 3.6 μA, and input-referred AC noise current is $I_N = 0.034$ and 0.136 μArms for $L = 4$ and 1 μm, respectively. The offset and noise currents increase inversely with decreasing L because gate area for local-area, DC mismatch decreases as the square of decreasing L, while bandwidth for integrated thermal noise increases as the inverse square of decreasing L. The corresponding accuracy measures, which are equal to the input signals divided by the offset and noise currents, are $A_{CC,DC} = 56.7$ and 13.9 for DC offset, and $A_{CC,AC} = 1500$ and 370 for thermal noise. As observed, DC offset is much higher than noise resulting in much lower accuracy for DC offset compared to accuracy for thermal noise. Finally, the current-mirror bandwidth, approximated as $f_{Ti}/2$, is 49.2 and 820 MHz.

Operation at constant IC giving constant g_m/I_D results in constant bandwidth–power–accuracy figures of merit and, correspondingly, constant bandwidth-accuracy figures of merit for constant drain current and power consumption. This is seen where the products of DC accuracy squared and bandwidth are $A_{CC,DC}^2 \cdot BW = (56.7)^2 \cdot 49.3$ MHz and $(13.9)^2 \cdot 820$ MHz for $L = 4$ and 1 μm, respectively, which are both equal to 1.6×10^{11} Hz. The products of AC accuracy squared and bandwidth are $A_{CC,AC}^2 \cdot BW = (1500)^2 \cdot 49.3$ MHz and $(370)^2 \cdot 820$ MHz, which are both equal 1.1×10^{14} Hz. This clearly shows the significant tradeoffs available between DC or AC accuracy and bandwidth. For operation at constant IC and I_D, DC offset and thermal noise decrease inversely with L, resulting in DC and AC accuracy that increases directly with L. Accuracy squared then increases as the square of L, while bandwidth decreases inversely with the square of L, giving a constant product of accuracy squared and bandwidth. The product of accuracy (not squared) and bandwidth, however, decreases inversely with L as listed in the spreadsheet results. Bandwidth may actually decrease as $1/L$, changing the trends (Section 5.5.2.11).

Identical tradeoffs of accuracy and bandwidth versus channel length are also observed in the spreadsheet results for voltage-mode circuits. These circuits include, for example, a differential pair used as an operational amplifier with negative feedback.

A view across the columns in Figure 4.21 allows a rapid assessment of performance tradeoffs as L is decreased from $L = 4, 2, 1, 0.48, 0.28,$ to 0.18 μm for fixed $IC = 10$ at the onset of strong inversion and $I_D = 100 \mu$A. This illustrates possible choices for differential-pair and current-mirror devices. Operation at long L is a low-frequency and DC optimization because low-frequency, drain–source resistance and gain measures of V_A and A_{Vi} are large, while low-frequency noise and DC mismatch measures of $S_{VG}^{1/2}$ flicker, $S_{ID}^{1/2}$ flicker, f_c, ΔV_{GS}, and $\Delta I_D/I_D$ are small. This comes at the expense of geometry and capacitance measures of W, WL, C_{GS}, C_{GB}, C_{GD}, and C_{DB} that are large, and bandwidth measures of f_{Ti}, f_T, and f_{diode} that are small. Operation at long L also minimizes velocity saturation reductions of g_m/I_D, g_m, and A_{Vi}, and velocity saturation increases in V_{EFF} at high IC, while minimizing possible small-geometry increases in thermal noise. Operation at long L also minimizes DIBL reductions of V_A and A_{Vi}.

Operation at short L is an AC optimization because bandwidth measures of f_{Ti}, f_T, and f_{diode} are large, while geometry and capacitance measures of W, WL, C_{GS}, C_{GB}, C_{GD}, and C_{DB} are small. This comes at the expense of low-frequency, drain–source resistance and gain measures of V_A and A_{Vi} that are small, and low-frequency noise and DC mismatch measures of $S_{VG}^{1/2}$ flicker, $S_{ID}^{1/2}$ flicker, f_c, ΔV_{GS}, and $\Delta I_D/I_D$ that are large.

In addition to results given by the *Analog CMOS Design, Tradeoffs and Optimization* spreadsheet in Figure 4.21, Tables 4.1–4.3 summarized MOS performance tradeoffs for design selections of the channel length. Sections 4.3.3–4.3.5 showed these graphically, and Section 4.2.3 discussed performance trends. When velocity saturation is significant for short-channel devices at high IC, V_{EFF} increases and g_m/I_D decreases as channel length decreases. This results in additional channel-length dependencies beyond those described here for flicker noise and local-area mismatch. This is described in the Chapter 3 sections referenced in Tables 4.2 and 4.3.

4.4.3 Selecting Drain Current

Section 4.4.1 presented numerical results from the *Analog CMOS Design, Tradeoffs and Optimization* spreadsheet showing device performance tradeoffs for inversion coefficient selections from weak through strong inversion at constant $L = 1\,\mu m$ and $I_D = 100\,\mu A$. Section 4.4.2 then presented results showing performance tradeoffs for channel length selections from long to short lengths for constant $IC = 10$ at the onset of strong inversion and $I_D = 100\,\mu A$. Together, this permitted a study of performance tradeoffs leading to optimum inversion coefficient and channel length selections for a fixed value of drain current.

In this section, the inversion coefficient will be held fixed at $IC = 1$ at the center of moderate inversion and channel length will be held fixed at $L = 1\,\mu m$, while drain current is increased from $I_D = 1, 3, 10, 30,$ to $100\,\mu A$. This permits a study of performance tradeoffs leading to optimum drain current selection. As in the previous sections, actual IC values are rounded slightly as a result of trimming the reported drawn channel width to actual layout values. nMOS devices in the example $0.18\,\mu m$ CMOS process will again be considered for $V_{DS} = 0.5\,V$ and $V_{SB} = 0\,V$. As described earlier at the beginning of Section 4.4, process parameters described in Tables 3.2–3.4 are used, with V_{AL} values found from the measured V_A data of Figure 3.45.

Figure 4.22 gives results from the *Analog CMOS Design, Tradeoffs and Optimization* spreadsheet for nMOS devices in the example $0.18\,\mu m$ CMOS process for varying values of $I_D = 1, 3, 10, 30,$ and $100\,\mu A$, and fixed values of $IC = 1,$ $L = 1\,\mu m,$ $V_{DS} = 0.5\,V,$ and $V_{SB} = 0\,V$. As in the previous sections, $V_{DS} = 0.5\,V$ is sufficiently high to ensure operation in saturation. While the spreadsheet reports values for all I_D selections, only values for $I_D = 1, 10,$ and $100\,\mu A$ are included in the following discussion for brevity. Spreadsheet results are not redefined as these were defined in Section 4.4.1.

As observed in Figure 4.22, many aspects of MOS performance are independent of the drain current for the fixed selections of inversion coefficient and channel length. As summarized earlier in the sizing relationships of Table 3.10, drain current is changed by simply scaling width with the drain current, which maintains the selected inversion coefficient and channel length. For example, drain current is doubled by doubling the channel width or by paralleling two identical devices operating at the same level of inversion and channel length. Accurate current ratios are obtained by replicating device structures having identical channel widths, channel lengths, orientation, and external environment, which minimizes variations associated with lateral diffusion and other sources of deterministic device mismatch. This is often done using even ratios of even values of the interdigitation number, m, for indentical layout structures.

From Figure 4.22, bias compliance voltages are $V_{EFF} = 0.039\,V$ and $V_{DS,sat} = 0.137\,V$. Normalized small-signal parameters are $g_m/I_D = 18.8\,\mu S/\mu A,$ $V_A = 12.7\,V,$ and $\eta = 0.28,$ and the input, 1 dB compression voltage for a differential pair is $V_{INDIF1dB} = 0.075\,V$. η is from the second sheet of spreadsheet results not shown. Intrinsic voltage gain, intrinsic bandwidth, and extrinsic bandwidths are $A_{Vi} = 240\,V/V,$ $f_{Ti} = 497\,MHz,$ $f_T = 402\,MHz,$ and $f_{diode} = 364\,MHz$. Finally, the flicker-noise corner frequency is $f_c = 0.33\,MHz$. Each of these is independent of the drain current and equal to values shown for $IC = 1,$ $L = 1\,\mu m,$ and $I_D = 100\,\mu A$ in Figure 4.20 in Section 4.4.1. Additionally, the velocity saturation and VFMR measures of $V_{EFF}/(LE_{CRIT}),$ $V_{EFF} \cdot \theta,$ and IC_{CRIT} are independent of the drain current and equal to the values shown in Figure 4.20.

W (effective) is 1.5, 15.2, and $152\,\mu m,$ and WL (effective) is 1.5, 14.8, and $148\,\mu m^2$ for $I_D = 1, 10,$ and $100\,\mu A,$ respectively. C_{GS} is 5.6, 56, and 560 fF, C_{GB} is 1.8, 18, and 180 fF, and C_{GD} is 1.4, 14, and 140 fF. C_{DB} is 1.6, 7.8, and 76 fF at $V_{DB} = V_{DS} + V_{SB} = 0.5\,V,$ and C_{SB} is 1.8, 16, and 104 fF at $V_{SB} = 0\,V$. C_{SB} is from the second sheet of spreadsheet results not shown. The channel width, gate area, and capacitances increase directly with the drain current, except for C_{DB} and C_{SB} that increase less rapidly when device interdigitation is initially used, which reduces the drain and source areas and perimeters. Because interdigitation is not used at the lowest drain currents, the relative value of C_{DB} is higher giving lower f_{diode}. Table 3.25 lists expressions for the drain area and perimeter and resulting value of C_{DB}.

Analog CMOS Design, Tradeoffs and Optimization Spreadsheet -- MOSFETs Sheet						
For design guidance only as results are approximate and not for any particular CMOS process.						
© 2000–2007, David M. Binkley. See Disclaimer, Notes Sheet.						
---------- Optional User Design Information ----------						
Description		Illustration of changing I_D at fixed $IC=1$ and $L=1\,\mu m$				
Device reference		M1	M2	M3	M4	M5
Device notes		ID = 1 uA	ID = 3 uA	ID = 10 uA	ID = 30 uA	IC = 100 uA
---------- Required User Design Inputs ----------						
Device model		nL1u	nL1u	nL1u	nL1u	nL1u
I_D	μA (+)	1	3	10	30	100
IC (fixed normalized)		1.012	1.012	0.998	0.998	0.998
L_{DRAWN}	μm	1	1	1	1	1
---------- Calculated Results ----------		OK	OK	OK	OK	OK
Effective Width, Length, Gate Area						
W	μm	1.50	4.50	15.20	45.60	152.00
L	μm	0.972	0.972	0.972	0.972	0.972
WL	μm²	1.46	4.37	14.77	44.32	147.74
DC Bias Voltages						
$V_{EFF}=V_{GS}-V_T$	V	0.040	0.040	0.039	0.039	0.039
V_T (adjusted for V_{SB})	V	0.420	0.420	0.420	0.420	0.420
$V_{GS}=V_T+V_{EFF}$	V	0.460	0.460	0.459	0.459	0.459
$V_{DS,\,sat}$	V	0.137	0.137	0.137	0.137	0.137
Small Signal Parameters						
g_m/I_D	μS/μA	18.76	18.76	18.82	18.82	18.82
g_m	μS	18.76	56.27	188.20	564.60	1881.99
V_A	V	12.74	12.74	12.74	12.74	12.74
g_{ds}	μS	0.0785	0.2355	0.7852	2.3555	7.8516
$A_{Vi}=g_m/g_{ds}$	V/V	238.9	238.9	239.7	239.7	239.7
Transcond. Distortion (Diff. Pair)						
$V_{INDIF,1dB}$ (input 1-dB comp.)	V	0.075	0.075	0.075	0.075	0.075
Capacitances and Bandwidths						
$C_{GS}=C_{gsi}+C_{GSO}$	fF	5.6	16.8	56.5	169.5	564.9
$C_{GB}=C_{gbi}+C_{GBO}$	fF	1.8	5.3	18.1	54.2	180.6
$C_{GD}=C_{GDO}$ (in saturation)	fF	1.4	4.2	14.3	42.9	142.9
C_{DB} (at $V_{DB}=V_{DS}+V_{SB}$)	fF	1.6	4.3	7.8	23.0	75.8
$f_{Ti}=g_m/[2\pi(C_{gsi}+C_{gbi})]$	MHz	500.6	500.6	497.0	497.0	497.0
$f_T=g_m/[2\pi(C_{GS}+C_{GB})]$	MHz	404.9	404.9	401.8	401.8	401.8
$f_{DIODE}=g_m/[2\pi(C_{GS}+C_{GB}+C_{DB})]$,	MHz	332.6	338.9	363.9	364.3	364.7
Thermal and Flicker Noise						
$S_{VG}^{1/2}$ thermal	nV/Hz$^{1/2}$	25.76	14.87	8.13	4.69	2.57
$S_{VG}^{1/2}$ flicker at $f_{FLICKER}$	nV/Hz$^{1/2}$	5770.22	3331.44	1811.42	1045.82	572.82
$S_{ID}^{1/2}$ thermal	pA/Hz$^{1/2}$	0.48	0.84	1.53	2.65	4.84
$S_{ID}^{1/2}$ flicker at $f_{FLICKER}$	pA/Hz$^{1/2}$	108.23	187.46	340.91	590.47	1078.04
$f_{FLICKER}$, freq. for flicker noise	Hz	1	1	1	1	1
f_c, flicker-noise corner	MHz	0.3388	0.3388	0.3347	0.3347	0.3347
DC Mismatch for Device Pair						
ΔV_{GS} (1σ)	mV	4.23	2.44	1.33	0.77	0.42
$\Delta I_D/I_D$ (1σ)	%	7.94	4.59	2.50	1.44	0.79

Figure 4.22 Predicted performance for nMOS devices in a 0.18 μm CMOS process operating at $IC = 1$, $L = 1\,\mu m$, and $I_D = 1, 3, 10, 30,$ and $100\,\mu A$ for $V_{DS} = 0.5\,V$ and $V_{SB} = 0\,V$. Values found from the *Analog CMOS Design, Tradeoffs and Optimization* spreadsheet illustrate tradeoffs in performance as drain current increases at constant inversion level and channel length, illustrating possible choices for differential-pair and current-mirror devices. The continuation page lists measures related to bandwidth–power–accuracy figures of merit

*** MOSFETs, Sheet 3 -- Performance Related to Bandwidth, Power, Accuracy Figures of Merit ***							
Device reference		M1	M2	M3	M4	M5	Trend
Device notes		ID = 1uA	ID = 3uA	ID = 10uA	ID = 30uA	IC = 100uA	with I_D
I_D	μA	1	3	10	30	100	
IC (fixed normalized)		1.012	1.012	0.998	0.998	0.998	
L_{DRAWN}	μm	1	1	1	1	1	
---------- Optional User Inputs for Figures of Merit ----------							
V_{DD} for P	V	1.8	1.8	1.8	1.8	1.8	
I_{IN}/I_D for I-mode FOMs	DC, rms	0.5	0.5	0.5	0.5	0.5	
V_{IN} for V-mode FOMs	Vdc, Vrms	0.1	0.1	0.1	0.1	0.1	
Current-Mode Figures of Merit, Power and Bandwidth							
Power, $P = 2I_D \cdot V_{DD}$	μW	3.6	10.8	36	108	360	$\propto I_D$
Bandwidth, $BW = f_{Ti}/2$	MHz	250.3	250.3	248.5	248.5	248.5	≈ const.
Current-Mode Figures of Merit, DC Offset (input current is relative to I_D)							
Input current, I_{IN}	μA	0.5	1.5	5	15	50	$\propto I_D$
Offset current, I_{OS} (3 σ)	μA	0.238	0.413	0.751	1.300	2.374	$\propto \sqrt{I_D}$
Accuracy, $A_{CC,DC} = I_{IN}/I_{OS}$ (3 σ)		2.1	3.6	6.7	11.5	21.1	$\propto \sqrt{I_D}$
Error, $E_{RR,DC} = I_{OS}$ (3 σ)/I_{IN}	%	47.65	27.51	15.02	8.67	4.75	$\propto 1/\sqrt{I_D}$
$A_{CC,DC} \cdot BW$	Hz	5.25E+08	9.10E+08	1.65E+09	2.87E+09	5.23E+09	$\propto \sqrt{I_D}$
$A_{CC,DC}^2 \cdot BW$	Hz	1.10E+09	3.31E+09	1.10E+10	3.31E+10	1.10E+11	$\propto I_D$
$A_{CC,DC}^2 \cdot BW/P$	Hz/W	3.06E+14	3.06E+14	3.06E+14	3.06E+14	3.06E+14	≈ const.
Current-Mode Figures of Merit, Thermal Noise (input current is relative to I_D)							
Input current, I_{IN}	μA rms	0.5	1.5	5	15	50	$\propto I_D$
Noise current over BW, I_N	μA rms	0.0108	0.0187	0.0341	0.0591	0.1079	$\propto \sqrt{I_D}$
Accuracy, $A_{CC,AC} = I_{IN}/I_N$		46	80	147	254	464	$\propto \sqrt{I_D}$
Error, $E_{RR,AC} = I_N/I_{IN}$	%	2.1623	1.2484	0.6821	0.3938	0.2157	$\propto 1/\sqrt{I_D}$
$A_{CC,AC} \cdot BW$	Hz	1.16E+10	2.01E+10	3.64E+10	6.31E+10	1.15E+11	$\propto \sqrt{I_D}$
$A_{CC,AC}^2 \cdot BW$	Hz	5.35E+11	1.61E+12	5.34E+12	1.6E+13	5.34E+13	$\propto I_D$
$A_{CC,AC}^2 \cdot BW/P$	Hz/W	1.49E+17	1.49E+17	1.48E+17	1.48E+17	1.48E+17	≈ const.
Voltage-Mode Figures of Merit, Power and Bandwidth							
Power, $P = 2I_D \cdot V_{DD}$	μW	3.6	10.8	36	108	360	$\propto I_D$
Bandwidth, $BW = 2f_{Ti}$	MHz	1001.3	1001.3	994.0	994.0	994.0	≈ const.
Voltage-Mode Figures of Merit, DC Offset (input voltage is fixed and given above)							
Offset voltage, V_{OS} (3 σ)	mV	12.702	7.333	3.990	2.303	1.262	$\propto 1/\sqrt{I_D}$
Accuracy, $A_{CC,DC} = V_{IN}/V_{OS}$ (3 σ)		7.9	13.6	25.1	43.4	79.3	$\propto \sqrt{I_D}$
Error, $E_{RR,DC} = V_{OS}$ (3 σ)/V_{IN}, %		12.70	7.33	3.99	2.30	1.26	$\propto 1/\sqrt{I_D}$
$A_{CC,DC} \cdot BW$	Hz	7.88E+09	1.37E+10	2.49E+10	4.32E+10	7.88E+10	$\propto \sqrt{I_D}$
$A_{CC,DC}^2 \cdot BW$	Hz	6.21E+10	1.86E+11	6.25E+11	1.87E+12	6.25E+12	$\propto I_D$
$A_{CC,DC}^2 \cdot BW/P$	Hz/W	1.72E+16	1.72E+16	1.73E+16	1.73E+16	1.73E+16	≈ const.
Voltage-Mode Figures of Merit, Thermal Noise (input voltage is fixed and given above)							
Noise voltage over BW, V_N	mV rms	1.1528	0.6656	0.3625	0.2093	0.1146	$\propto 1/\sqrt{I_D}$
Accuracy, $A_{CC,AC} = V_{IN}/V_N$		87	150	276	478	872	$\propto \sqrt{I_D}$
Error, $E_{RR,AC} = V_N/V_{IN}$	%	1.1528	0.6656	0.3625	0.2093	0.1146	$\propto 1/\sqrt{I_D}$
$A_{CC,AC} \cdot BW$	Hz	8.69E+10	1.50E+11	2.74E+11	4.75E+11	8.67E+11	$\propto \sqrt{I_D}$
$A_{CC,AC}^2 \cdot BW$	Hz	7.53E+12	2.26E+13	7.57E+13	2.27E+14	7.57E+14	$\propto I_D$
$A_{CC,AC}^2 \cdot BW/P$	Hz/W	2.09E+18	2.09E+18	2.1E+18	2.1E+18	2.1E+18	≈ const.

Figure 4.22 *(continued)*

g_m is 18.8, 188, and 1880 μS, g_{ds} is 0.079, 0.79, and 7.9 μS, and g_{mb} is 5.3, 53, and 530 μS for $I_D = 1$, 10, and 100 μA, respectively. g_{mb} is from the second sheet of spreadsheet results not shown. The small-signal parameters increase directly with the drain current because their normalized measures of g_m/I_D, V_A, and η are constant for the fixed inversion coefficient and channel length.

$S_{VG}^{1/2}$ thermal is 25.8, 8.1, and 2.6 nV/Hz$^{1/2}$, and $S_{VG}^{1/2}$ flicker at 1 Hz is 5770, 1810, and 573 nV/Hz$^{1/2}$ for $I_D = 1$, 10, and 100 μA, respectively. $S_{VG}^{1/2}$ thermal decreases inversely with the square root of drain current because of the increase in g_m. $S_{VG}^{1/2}$ flicker decreases inversely with the square root of drain current because of the increase in gate area. Both noise PSDs, S_{VG} in units of (nV)^2/Hz, decrease inversely with the drain current. As mentioned in the previous sections, the gate-referred noise voltage densities are 41 % larger for a differential pair containing two devices because of the addition of uncorrelated noise from two devices.

$S_{ID}^{1/2}$ thermal is 0.48, 1.5, and 4.8 pA/Hz$^{1/2}$, and $S_{ID}^{1/2}$ flicker at 1 Hz is 108, 341, and 1080 pA/Hz$^{1/2}$ for $I_D = 1$, 10, and 100 μA, respectively. $S_{ID}^{1/2}$ thermal increases with the square root of drain current because of the increase in g_m. $S_{ID}^{1/2}$ flicker increases with the square root of drain current because the direct increase in g_m, which is responsible for the conversion of $S_{VG}^{1/2}$ flicker to $S_{ID}^{1/2}$ flicker, exceeds the decrease in $S_{VG}^{1/2}$ flicker with the inverse square root of drain current associated with increased gate area. Both noise PSDs, S_{ID} in units of (pA)^2/Hz, increase directly with the drain current. As mentioned in the previous sections, the drain-referred noise current densities are 41 % larger at the output of a 1:1 current mirror containing two devices because of the addition of uncorrelated noise from the reference device. Also, the output noise of a cascode current mirror can be estimated from noise of the bottom devices because of significant noise degeneration in the top devices.

ΔV_{GS} (1σ) for a differential pair is 4.2, 1.3, and 0.4 mV, and $\Delta I_D/I_D$ (1σ) for a pair of current-mirror devices is 7.9, 2.5, and 0.8 % for $I_D = 1$, 10, and 100 μA, respectively. Both decrease inversely with the square root of drain current because of the increase in gate area. ΔV_{GS} and $\Delta I_D/I_D$ mismatch values are, again, estimated 1σ values that would typically be multiplied by at least three for yield consideration in production design.

The continuation page of Figure 4.22, like the continuation page of Figure 4.21 described in the previous section, shows the third sheet of spreadsheet results giving performance related to bandwidth–power–accuracy figures of merit. These figures of merit, again, were discussed in Sections 4.3.6.5–4.3.6.7 and listed in Table 4.4. Performance, again, includes selected signal levels, DC offset and noise levels, and resulting errors, accuracies, and figures of merit. Performance is for $IC = 1$ at the center of moderate inversion, $L = 1$ μm, and varying drain currents (for individual devices) as considered on the main sheet, but also considers a supply voltage of $V_{DD} = 1.8$ V, a current-mirror input signal of $I_{IN} = 0.5 \cdot I_D$, and a differential-pair input signal of $V_{IN} = 0.1$ V. In addition to performance measures, the sheet shows trends versus the drain current that are described below.

Given that intrinsic voltage gain, A_{Vi}, bandwidths, f_{Ti}, f_T, and f_{diode}, and many other measures of performance are independent of the drain current for a selected inversion coefficient and channel length, the question arises: why increase the drain current from a low value giving low power consumption? Reasons for increasing the drain current include:

- To manage signal distortion where signal current must be sufficiently below the drain bias current.

- To obtain sufficiently large slew rate into an external load capacitance, where the slew rate is equal to the available drive current divided by the load capacitance.

- To obtain sufficiently large g_m needed to obtain sufficiently small small-signal resistance, related to $1/g_m$, associated with a diode-connected device, the source input resistance of a grounded-gate device, or the source output resistance of a source follower device.

- To obtain sufficiently high g_m needed for sufficiently low $S_{VG}^{1/2}$ thermal. Obtaining maximum g_m and, correspondingly, minimum $S_{VG}^{1/2}$ thermal for a given drain current is discussed later in Section 6.3.1 in Chapter 6, which describes micropower, low-noise, CMOS preamplifier design examples. Even in weak inversion where g_m/I_D is maximum, more bias current is required for MOS transistors compared to bipolar transistors to achieve equal values of g_m and input-referred thermal-noise voltage density. Significantly more MOS bias current is required in strong inversion where g_m/I_D is low.

- To obtain sufficiently small thermal noise errors proportional to the noise voltage density, $S_{VG}^{1/2}$ thermal, or the relative noise current density, $S_{ID}^{1/2}$ thermal$/I_D = (S_{VG}^{1/2}$ thermal$) \cdot g_m/I_D$. These thermal noise errors decrease inversely with the square root of drain current through the increase in g_m.

- To obtain sufficiently small flicker-noise errors proportional to the noise voltage density, $S_{VG}^{1/2}$ flicker, or the relative noise current density, $S_{ID}^{1/2}$ flicker $/ I_D = (S_{VG}^{1/2}$ flicker$) \cdot g_m/I_D$. These flicker-noise errors decrease inversely with the square root of drain current through the increase in required gate area.

- To obtain sufficiently small DC offset errors proportional to the mismatch voltage, ΔV_{GS}, or the relative mismatch current, $\Delta I_D/I_D = (\Delta V_{GS}) \cdot g_m/I_D$. These DC offset errors due to local-area mismatch decrease inversely with the square root of drain current through the increase in required gate area.

- To obtain sufficiently high accuracy that is inversely proportional to ΔV_{GS} or $\Delta I_D/I_D$ for DC offset, and inversely proportional to $S_{VG}^{1/2}$ thermal or $S_{ID}^{1/2}$ thermal $/ I_D$ for thermal noise. Accuracy increases with the square root of drain current because of the decrease in thermal noise and DC offset errors mentioned above. Accuracy squared then increases directly with the drain current.

- To obtain sufficiently large bandwidth-accuracy figures of merit for current- or voltage-mode circuits for DC offset or thermal noise. For constant bandwidth–power–accuracy figures of merit associated with constant IC and g_m/I_D, bandwidth-accuracy figures of merit, or products of accuracy squared and bandwidth, increase directly with the drain current. This can be seen in the continuation page of Figure 4.22 where bandwidth is independent of the drain current, while the accuracy squared increases directly with the drain current. Errors and accuracy for voltage-mode circuits depend on absolute values of signal, offset, and noise voltages, while errors and accuracy for current-mode circuits depend on values of signal, offset, and noise currents that are relative to the drain bias current.

- To obtain sufficiently large gate area needed for small flicker-noise and local-area mismatch errors simultaneously with sufficiently high IC and short L needed for high bandwidth measures of f_{Ti}, f_T, and f_{diode}. As discussed in the sizing relationships of Table 3.10, gate area is inversely proportional to the inversion coefficient, proportional to the square of channel length, and directly proportional to drain current. Although bandwidth is independent of the drain current, increased drain current allows the use of higher IC and, especially, shorter L for higher bandwidth for a given gate area. Equal gate-area optimizations for drain currents of 100 and 1 μA are discussed in the next section.

- To ensure device capacitances, which scale with the increasing channel width and gate area required at increasing drain current, sufficiently exceed layout or load capacitances that would otherwise reduce operating bandwidth.

- To ensure sufficiently large channel width for layout, especially for short-channel devices operating at high levels of inversion and low drain currents where the shape factor, $S = W/L$, is small.

In addition to results discussed here from the *Analog CMOS Design, Tradeoffs and Optimization* spreadsheet, Tables 4.1–4.3 summarized MOS performance tradeoffs for design selections of the drain current. Section 4.2.4 also discussed performance trends.

4.4.4 Optimizing for DC, Balanced, and AC Performance

Sections 4.4.1, 4.4.2, and 4.4.3 presented numerical results from the *Analog CMOS Design, Tradeoffs and Optimization* spreadsheet showing tradeoffs in device performance for separate inversion coefficient, channel length, and drain current selections. The observation of performance tradeoffs for these independent design selections leads to optimum selections for a given application. DC, balanced, and AC optimizations are discussed in this section, corresponding to performance tradeoffs shown at the upper left, center, and lower right portions of the *MOSFET Operating Plane* shown in Figure 3.1. These optimizations are considered for drain currents of 100 and 1 μA, representing millipower and micropower operation, and are applicable to devices in differential pairs, current mirrors, and other circuits.

Table 4.5 summarizes inversion coefficient and channel length selections for DC, balanced, and AC optimizations described below and describes relative values of MOS performance. Sections 4.2.2 and 4.2.3 overviewed the performance listed in the table. Sections 4.3.3–4.3.5 described the performance further using Tables 4.1–4.3 and corresponding graphs. Finally, Sections 4.4.1 and 4.4.2 illustrated the performance numerically using the *Analog CMOS Design, Tradeoffs and Optimization* spreadsheet.

Table 4.5 does not summarize performance in terms of the drain current. This was overviewed in Section 4.2.4 and described further in Sections 4.3.3–4.3.5 using Tables 4.1–4.3. This was also illustrated numerically in Section 4.4.3 using the spreadsheet, including a summary at the end.

Table 4.5 lists the total gate capacitances, $C_{GS} = C_{gsi} + C_{GSO}$, $C_{GB} = C_{gbi} + C_{GBO}$, and $C_{GD} = C_{GDO}$ (for operation in saturation), which include the intrinsic capacitances, C_{gsi} and C_{gbi}, and extrinsic overlap capacitances, C_{GSO}, C_{GBO}, and C_{GDO}. These total capacitances were considered in the spreadsheet results of Sections 4.4.1–4.4.3. The intrinsic gate capacitances, however, were considered separately in the other sections mentioned.

The DC, balanced, and AC optimizations described below track a path on the *Plane* from the upper left corner to the lower right corner. This corresponds to increasing inversion coefficients and simultaneously decreasing channel lengths across the DC, balanced, and AC optimizations. This gives the largest possible changes in performance. Different optimizations are also possible, for example for left-to-right or top-to-bottom paths on the *Plane*, corresponding to increasing inversion coefficients or decreasing channel lengths separately.

4.4.4.1 DC optimization

Section 4.4.1 described a DC optimization at low inversion coefficients towards weak inversion, while Section 4.4.2 described a DC optimization at long channel lengths. The combination of low inversion coefficients and long channel lengths, corresponding to operation at the upper left portion of the *Plane*, is then a DC optimization involving both inversion coefficient and channel length selections.

Table 4.5 summarizes performance tradeoffs for a DC optimization at low values of *IC* in weak inversion or the weak-inversion side of moderate inversion combined with long channel lengths. Here, geometry measures of W and WL are non-optimally large due to the combination of low inversion and long channel length. Bias compliance measures of V_{EFF} and $V_{DS,sat}$ are optimally small due to low inversion. Small-signal parameter measures of g_m/I_D and g_m are optimally large due to low inversion, while V_A and r_{ds} are optimally large due to long channel length. The combination of these gives especially large, optimal intrinsic voltage gain, A_{Vi}. The distortion measure of $V_{INDIF1dB}$ is non-optimally small due to high g_m/I_D. The combination of low inversion and, especially, long channel length gives especially large, non-optimal values of gate capacitances C_{GS} and C_{GB}, and especially small, non-optimal bandwidth measures of f_{Ti}, f_T, and f_{diode} due to large gate area. The drain–body and source–body capacitances, C_{DB} and C_{SB}, and gate–drain capacitance, C_{GD}, are non-optimally large due to large channel width. $S_{VG}^{1/2}$ thermal is optimally small, while $S_{ID}^{1/2}$ thermal is non-optimally large due to large g_m. Finally, the combination of low inversion and, especially, long channel length gives especially small, optimal values of flicker-noise and DC mismatch measures of $S_{VG}^{1/2}$ flicker, $S_{ID}^{1/2}$ flicker, flicker noise f_c, ΔV_{GS}, and $\Delta I_D/I_D$ due to large gate area.

V_{EFF}, $V_{DS,sat}$, g_m/I_D, V_A, $V_{INDIF1dB}$, A_{Vi}, f_{Ti}, f_T, f_{diode}, and f_c, which is a fraction of f_{Ti}, are independent of the drain current. However, W, WL, g_m, and $g_{ds} = 1/r_{ds}$ increase directly with the

Table 4.5 MOS inversion coefficient and channel length selections and resulting relative performance for DC, balanced, and AC optimization

Parameter	Relative value		
	DC optimize	Balanced optimize	AC optimize
Design selections			
IC	Small (WI/MI)	Medium (MI/SI)	Large (SI)
L		Long	Moderate	Short
Performance			
Geometry:			
W	Large	Medium	Small
WL	Large	Medium	Small
Bias compliance voltages:			
$V_{EFF} = V_{GS} - V_T$	Small	Medium	Large
$V_{DS,sat}$	Small	Medium	Large
Small-signal parameters:			
g_m/I_D	Large	Medium	Small
$V_A = I_D \cdot r_{ds} = I_D/g_{ds}$	Large	Medium	Small
$A_{Vi} = g_m \cdot r_{ds} = g_m/g_{ds}$	Large	Medium	Small
Transconductance distortion for differential pair:			
$V_{INDIF1dB}$ (related to linearity)	Small	Medium	Large
Capacitances and bandwidths:			
C_{GS}, C_{GB}, C_{GD}, C_{DB}, and C_{SB}	Large	Medium	Small
$f_{Ti} = g_m/[2\pi(C_{gsi} + C_{gbi})]$, f_T, and f_{diode}	Small	Medium	Large
Thermal and flicker noise:			
$S_{VG}^{1/2}$ thermal	Small	Medium	Large
$S_{ID}^{1/2}$ thermal	Large	Medium	Small
$S_{VG}^{1/2}$ flicker	Small	Medium	Large
$S_{ID}^{1/2}$ flicker	Small	Medium	Large
f_c flicker	Small	Medium	Large
DC mismatch (local area) for device pair:			
ΔV_{GS}	Small	Medium	Large
$\Delta I_D/I_D$	Small	Medium	Large

For micropower operation, e.g., drain currents of 1 µA, channel lengths above the process minimum and inversion coefficient selections below the onset of strong inversion ($IC < 10$) will likely be required to give sufficient channel width and gate area. This significantly reduces the bandwidth for the AC optimization.

drain current while $S_{VG}^{1/2}$ flicker, ΔV_{GS}, $\Delta I_D/I_D$, and $S_{VG}^{1/2}$ thermal decrease inversely with the square root of drain current, and $S_{ID}^{1/2}$ flicker and $S_{ID}^{1/2}$ thermal increase with the square root of drain current. As described later in Section 4.4.4.6, operation at micropower levels of drain current often requires increases in the channel length to obtain sufficient gate area to minimize $S_{VG}^{1/2}$ flicker, ΔV_{GS}, and

$\Delta I_D/I_D$. Finally, although device bandwidth measures are independent of the drain current, higher device capacitances for higher drain currents can minimize the loss of operating bandwidth associated with parasitic layout and load capacitances. The selection of drain current was described in Section 4.4.3.

Because small-signal parameter and gain measures apply to low-frequency signals, and low-frequency flicker noise is small, a DC optimization is also an optimization for low-frequency signals.

4.4.4.2 AC optimization

Section 4.4.1 described an AC optimization at high inversion coefficients in strong inversion, while Section 4.4.2 described an AC optimization at short channel lengths. The combination of high inversion coefficients and short channel lengths, corresponding to operation at the lower right portion of the *Plane*, is then an AC optimization involving both inversion coefficient and channel length selections.

Table 4.5 summarizes performance tradeoffs for an AC optimization at high values of *IC* in strong inversion and short channel lengths. Here, W and WL are optimally small due to the combination of high inversion and short channel length. V_{EFF} and $V_{DS,sat}$ are non-optimally large due to high inversion. g_m/I_D and g_m are non-optimally small due to high inversion, while V_A and r_{ds} are non-optimally small due to short channel length. The combination of these gives especially small, non-optimal A_{Vi}. $V_{INDIF1dB}$ is optimally high due to low g_m/I_D. The combination of high inversion and, especially, short channel length gives especially small, optimal values of C_{GS} and C_{GB}, and especially large, optimal values of f_{Ti}, f_T, and f_{diode} due to small gate area. C_{DB}, C_{SB}, and C_{GD} are optimally small due to small channel width. $S_{ID}^{1/2}$ thermal is optimally small, while $S_{VG}^{1/2}$ thermal is non-optimally large due to small g_m. Finally, the combination of high inversion and, especially, short channel length gives especially large, non-optimal values of $S_{VG}^{1/2}$ flicker, $S_{ID}^{1/2}$ flicker, f_c, ΔV_{GS}, and $\Delta I_D/I_D$ due to small gate area.

An AC optimization, as defined here, corresponds to high *IC* and short channel length, which maximizes signal bandwidth, while minimizing transconductance distortion, capacitances, and drain-referred thermal-noise current. This is an optimization for high-frequency voltage signals appearing at the gate with regard to bandwidth and distortion. However, this is not an optimization with regard to the signal-to-noise ratio for these signals because of the large gate-referred thermal-noise voltage. The AC optimization defined here is intended to illustrate primarily the optimization of signal bandwidth. Other optimizations involving different tradeoffs of signal bandwidth, distortion, and noise are possible as seen by different locations on the *Plane*.

As described for the DC optimization, drain current affects many aspects of performance and must be considered to obtain desired tradeoffs in performance.

4.4.4.3 Balanced optimization

Operation at low inversion coefficients and long channel lengths corresponds to a DC optimization while operation at high inversion coefficients and short channel lengths corresponds to an AC optimization. This suggests a combination of moderate inversion coefficients and moderate channel lengths, corresponding to the center portion of the *Plane*, for a balanced optimization of DC and AC performance.

Table 4.5 summarizes performance tradeoffs for a balanced optimization at moderate values of *IC*, most likely in moderate inversion, and moderate channel lengths. Here, performance is balanced between the DC and AC optimizations, where performance for DC and low-frequency signals and performance for high-frequency signals is simultaneously important. The inversion coefficient and channel length selections for a balanced optimization depend on the priorities of various aspects of performance and are complicated by the many aspects of performance and the varying dependencies of performance on the inversion coefficient and channel length. For example, bandwidth, flicker noise, and DC mismatch, which depend on gate area, are stronger functions of channel length compared to the inversion coefficient.

As described for the DC optimization, drain current affects many aspects of performance and must be considered to obtain desired tradeoffs in performance.

4.4.4.4 Optimizations at millipower operation

The *Analog CMOS Design, Tradeoffs and Optimization* spreadsheet will be used to illustrate DC, balanced, and AC optimizations that can be used for devices in differential-pair, current-mirror, and other circuits. As in Sections 4.4.1 and 4.4.2 that presented performance for separate inversion coefficient and channel length selections, drain current will continue to be held fixed at $I_D = 100\,\mu A$. This, again, permits a "fair" study of performance tradeoffs at constant, millipower power consumption. In the section that follows, drain current will be reduced to $I_D = 1\,\mu A$ to illustrate micropower optimizations.

nMOS devices in the example $0.18\,\mu m$ CMOS process will again be considered for $V_{DS} = 0.5\,V$ and $V_{SB} = 0\,V$. As described earlier at the beginning of Section 4.4, process parameters described in Tables 3.2–3.4 are used, with V_{AL} values found from the measured V_A data of Figure 3.45.

Operation at $IC = 1$ at the center of moderate inversion for a channel length of $L = 1\,\mu m$ is selected for the DC optimization. Operation at lower inversion or at longer channel length would provide better DC and low-frequency performance at the expense of AC performance. Operation at $IC = 10$ at the onset of strong inversion for the minimum channel length of $L = 0.18\,\mu m$ is selected for the AC optimization. Operation at higher levels of inversion would improve AC performance at the expense of DC performance. Finally, operation at $IC = 3$ at the strong inversion side of moderate inversion for a moderate channel length of $L = 0.48\,\mu m$ is selected for the balanced optimization. The balanced optimization choices are nearly geometrically centered between those used for the DC and AC optimizations and, as mentioned, do not correspond to any particular balance of DC and AC performance. As in Sections 4.4.1–4.4.3, actual IC values are rounded slightly from their nominal values as a result of trimming the reported drawn channel width to actual layout values.

Figure 4.23 gives results from the *Analog CMOS Design, Tradeoffs and Optimization* spreadsheet for nMOS devices in the example $0.18\,\mu m$ CMOS process for the three optimizations at fixed values of $I_D = 100\,\mu A$, $V_{DS} = 0.5\,V$, and $V_{SB} = 0\,V$. As in the previous sections, $V_{DS} = 0.5\,V$ is sufficiently high to ensure operation in saturation. Spreadsheet results are not redefined as these were defined in Section 4.4.1.

From Figure 4.23, W (effective) is 155, 23.6, and $2.4\,\mu m$, and WL (effective) is 151, 10.7, and $0.36\,\mu m^2$ for the DC, balanced, and AC optimization, respectively. Both are optimally small for the AC optimization due to small gate area.

V_{EFF} is 0.038, 0.108, and $0.232\,V$ and $V_{DS,sat}$ is 0.136, 0.173, and $0.248\,V$ for the DC, balanced, and AC optimization, respectively. Both are optimally small for the DC optimization due to low inversion.

g_m/I_D is 18.9, 13.4, and $7.3\,\mu S/\mu A$ (1/V), and g_m is 1890, 1340, and $730\,\mu S$ for the DC, balanced, and AC optimization, respectively. Both are optimally large for the DC optimization due to low inversion. V_A is 12.7, 8.0, and $3.2\,V$, and g_{ds} is 7.9, 12.4, and $31.7\,\mu S$. V_A is optimally large and g_{ds} is optimally small for the DC optimization due to long channel length. A_{Vi} is 241, 108, and $23.1\,V/V$, which is optimally large for the DC optimization because of high g_m and low g_{ds}. Differential-pair $V_{INDIF1dB}$ is 0.074, 0.109, and $0.230\,V$, which is optimally large for the AC optimization due to small g_m/I_D.

C_{GS} is 573, 64, and 4 fF and C_{GB} is 185, 10.4, and 0.3 fF for the DC, balanced, and AC optimization, respectively. Both are optimally small for the AC optimization due to small gate area. C_{GD} is 146, 22,

Analog CMOS Design, Tradeoffs and Optimization Spreadsheet -- MOSFETs Sheet							
For design guidance only as results are approximate and not for any particular CMOS process.							
© 2000–2007, David M. Binkley. See Disclaimer, Notes Sheet.							
---------- Optional User Design Information ----------							
Description		Illustration of DC, balanced, and AC optimization (I_D = 100 µA)					
Device reference		M1	M2	M3			
Device notes		DC	Balanced	AC			
---------- Required User Design Inputs ----------							
Device model		nMIL1u	nMIL05u	nSIL02u			
I_D	µA (+)	100	100	100			
IC (fixed normalized)		0.978	2.990	9.888			
L_{DRAWN}	µm	1	0.48	0.18			
---------- Calculated Results ----------		OK	OK	OK			
Effective Width, Length, Gate Area							
W	µm	155.20	23.60	2.40			
L	µm	0.972	0.452	0.152			
WL	µm^2	150.85	10.67	0.36			
DC Bias Voltages							
$V_{EFF} = V_{GS} - V_T$	V	0.038	0.108	0.232			
V_T (adjusted for V_{SB})	V	0.420	0.420	0.420			
$V_{GS} = V_T + V_{EFF}$	V	0.458	0.528	0.652			
$V_{DS, sat}$	V	0.136	0.173	0.248			
Small Signal Parameters							
g_m/I_D	µS/µA	18.92	13.37	7.32			
g_m	µS	1891.89	1337.40	731.96			
V_A	V	12.74	8.04	3.16			
g_{ds}	µS	7.8519	12.4307	31.6555			
$A_{Vi} = g_m/g_{ds}$	V/V	240.9	107.6	23.1			
Transcond. Distortion (Diff. Pair)							
$V_{INDIF.1dB}$ (input 1-dB comp.)	V	0.074	0.109	0.230			
Capacitances and Bandwidths							
$C_{GS} = C_{gsi} + C_{GSO}$	fF	573.4	63.8	4.0			
$C_{GB} = C_{gbi} + C_{GBO}$	fF	185.2	10.4	0.3			
$C_{GD} = C_{GDO}$ (in saturation)	fF	145.9	22.2	2.3			
C_{DB} (at $V_{DB} = V_{DS} + V_{SB}$)	fF	78.5	12.2	1.5			
$f_{Ti} = g_m/[2\pi(C_{gsi} + C_{gbi})]$	MHz	491.4	4092.5	58603.1			
$f_T = g_m/[2\pi(C_{GS} + C_{GB})]$	MHz	396.9	2868.8	27450.3			
$f_{DIODE} = g_m/[2\pi(C_{GS} + C_{GB} + C_{DB})]$,	MHz	359.7	2464.3	20357.9			
Thermal and Flicker Noise							
$S_{VG}^{1/2}$ thermal	nV/Hz$^{1/2}$	2.56	3.15	4.32			
$S_{VG}^{1/2}$ flicker at $f_{FLICKER}$	nV/Hz$^{1/2}$	566.28	2273.23	13666.19			
$S_{ID}^{1/2}$ thermal	pA/Hz$^{1/2}$	4.85	4.21	3.16			
$S_{ID}^{1/2}$ flicker at $f_{FLICKER}$	pA/Hz$^{1/2}$	1071.35	3040.22	10003.12			
$f_{FLICKER}$, freq. for flicker noise	Hz	1	1	1			
f_c, flicker-noise corner	MHz	0.3283	5.3313	172.2954			
DC Mismatch for Device Pair							
ΔV_{GS} (1σ)	mV	0.42	1.60	9.43			
$\Delta I_D/I_D$ (1σ)	%	0.79	2.14	6.91			

Figure 4.23 Predicted performance for nMOS devices in a 0.18 µm CMOS process with devices optimized for DC, balanced, and AC performance at a millipower drain current of 100 µA. Values found from the *Analog CMOS Design, Tradeoffs and Optimization* spreadsheet illustrate possible optimizations for differential-pair and current-mirror devices

and 2.3 fF, and C_{DB} is 79, 12.2, and 1.5 fF. Both are optimally small for the AC optimization due to small channel width.

f_{Ti} is 491, 4090, and 58600 MHz, f_T is 397, 2870, and 27500 MHz, and f_{diode} is 360, 2460, and 20400 MHz for the DC, balanced, and AC optimization, respectively. Bandwidths are optimally large for the AC optimization due to small capacitances.

$S_{VG}^{1/2}$ thermal is 2.6, 3.2, and 4.3 nV/Hz$^{1/2}$, and $S_{VG}^{1/2}$ flicker at 1 Hz is 566, 2270, and 13700 nV/Hz$^{1/2}$ for the DC, balanced, and AC optimization, respectively. $S_{VG}^{1/2}$ thermal is optimally small for the DC optimization due to large g_m, while $S_{VG}^{1/2}$ flicker is optimally small for the DC optimization due to large gate area. As mentioned in the previous sections, the gate-referred noise voltage densities are 41 % larger for a differential pair containing two devices because of the addition of uncorrelated noise from two devices.

$S_{ID}^{1/2}$ thermal is 4.9, 4.2, and 3.2 pA/Hz$^{1/2}$, and $S_{ID}^{1/2}$ flicker at 1 Hz is 1070, 3040, and 10000 pA/Hz$^{1/2}$ for the DC, balanced, and AC optimization, respectively. $S_{ID}^{1/2}$ thermal is optimally small for the AC optimization due to low g_m, while $S_{ID}^{1/2}$ flicker is optimally small for the DC optimization due to large gate area. As mentioned in the previous sections, the drain-referred noise current densities are 41 % larger at the output of a 1:1 current mirror containing two devices because of the addition of uncorrelated noise from the reference device. Additionally, the output noise of a cascode current mirror can be estimated from noise of the bottom devices because of significant noise degeneration in the top devices. f_c is 0.33, 5.3, and 172 MHz, which is optimally small for the DC optimization due to large gate area.

As discussed in the previous sections, thermal noise can be greater for short-channel devices than the values given because of possible small-geometry increases in the noise. The increase in flicker noise associated with increasing inversion-level or V_{EFF} however, is included in the spreadsheet results.

ΔV_{GS} (1σ) for a differential pair is 0.4, 1.6, and 9.4 mV, and $\Delta I_D/I_D$ (1σ) for a pair of current mirror devices is 0.8, 2.1, and 6.9 % for the DC, balanced, and AC optimization, respectively. These are estimated 1σ values that would typically be multiplied by at least three for yield consideration in production design. Like flicker noise, DC voltage and current mismatch are optimally small for the DC optimization due to large gate area.

The spreadsheet results given in Figure 4.23 along with relative performance summarized in Table 4.5 show that intrinsic voltage gain is maximized, and flicker noise and DC mismatch are minimized for the DC optimization, but bandwidth is very low. Intrinsic voltage gain is much smaller, and flicker noise and DC mismatch are much higher for the AC optimization, but bandwidth is maximized. Transconductance efficiency, transconductance distortion, and thermal noise, which depend only on the level of inversion unless velocity saturation is significant, vary less across the optimizations. The balanced optimization provides some balance of DC and AC performance. As observed, many performance tradeoffs are significant, requiring identification of important performance followed by careful inversion coefficient and channel length selection. The spreadsheet can facilitate optimum inversion coefficient, channel length, and drain current selections prior to launching time-consuming computer simulations of a candidate design. This will be illustrated later in the circuit design examples of Chapters 5 and 6.

4.4.4.5 *Optimizations at micropower operation*

The inversion coefficient and channel length selections for the DC, balanced, and AC optimizations shown in Figure 4.23 and Table 4.5 give the same relative performance tradeoffs, regardless of the operating drain current. However, operation at low drain currents for micropower circuits often requires lower inversion coefficients and longer channel lengths to achieve sufficient gate area to manage flicker noise and DC mismatch, or even to achieve sufficient channel width for layout.

This section considers DC, balanced, and AC optimizations at a drain current of $I_D = 1\,\mu A$ compared to $I_D = 100\,\mu A$ considered in the previous section. This corresponds to micropower operation compared to millipower operation for typical analog circuits.

nMOS devices in the example 0.18 µm CMOS process will again be considered for $V_{DS} = 0.5$ V and $V_{SB} = 0$ V. As described earlier at the beginning of Section 4.4, process parameters described in Tables 3.2–3.4 are used. V_{AL} values are found from the measured V_{AL} data of Figure 3.48. This permits an interpolation and extrapolation of values since some channel lengths used here are not equal to those in the V_A measurements shown in Figure 3.45.

Operation at $IC = 0.3$ at the weak-inversion side of moderate inversion for a long channel length of $L = 5.4$ µm is selected for the DC optimization. Operation at $IC = 3$ at the strong-inversion side of moderate inversion for a channel length of $L = 0.85$ µm is selected for the AC optimization. Finally, operation at $IC = 1$ at the center of moderate inversion for a channel length of $L = 2.6$ µm is selected for the balanced optimization. The balanced optimization choices are nearly geometrically centered between those used for the DC and AC optimizations and, as in the previous section, do not correspond to any particular balance of DC and AC performance. As in the previous section, actual IC values are rounded slightly from their nominal values as a result of trimming the reported drawn channel width to actual layout values.

It is important to note that lower inversion coefficients and longer channel lengths have been selected to give nearly equal gate areas as the devices described in Figure 4.23 for $I_D = 100$ µA in the previous section. This permits a comparison of performance at $I_D = 1$ µA, denoted as micropower optimizations, with performance at $I_D = 100$ µA, denoted as millipower optimizations, for equally large, moderate, and small gate areas associated with separate DC, balanced, and AC optimizations.

Figure 4.24 gives results from the *Analog CMOS Design, Tradeoffs and Optimization* spreadsheet for nMOS devices in the example 0.18 µm CMOS process for the three micropower optimizations at fixed values of $I_D = 1$ µA, $V_{DS} = 0.5$ V, and $V_{SB} = 0$ V. As in the previous section, $V_{DS} = 0.5$ V is sufficiently high to ensure operation in saturation. The tradeoffs in performance for the micropower, DC, balanced, and AC optimizations are identical to those of the millipower optimizations shown in Figure 4.23. Rather than discuss the individual results, the $I_D = 1$ µA, micropower optimizations shown in Figure 4.24 will be compared with the $I_D = 100$ µA, millipower optimizations shown in Figure 4.23. This will be done by comparing the two DC, two balanced, and two AC optimizations.

Gate areas (WL) are approximately 150, 10.5, and 0.35 µm² for the DC, balanced, and AC optimizations, respectively, for both the micropower and millipower optimizations. The gate area extends across a wide range, with the AC optimizations having gate areas well below 1 µm².

V_{EFF}, $V_{DS,sat}$, and $V_{INDIF1dB}$ are smaller and g_m/I_D is larger for the micropower optimizations due to lower levels of inversion. V_A is larger for micropower optimizations due to longer channel lengths. A_{Vi} is larger for the micropower optimizations due to the combination of lower inversion and longer channel lengths. g_m and g_{ds} are significantly smaller for the micropower optimizations due primarily to lower drain current.

C_{GS} and C_{GB} are similar for the micropower and millipower optimizations due to equal gate areas, but are somewhat different due to different levels of inversion. C_{GS} is larger for the millipower optimizations due to higher levels of inversion, while C_{GB} is larger for the micropower optimizations due to lower levels of inversion. C_{GD} and C_{DB} are smaller for the micropower optimizations due to smaller channel widths. f_{Ti}, f_T, f_{diode}, and f_c are significantly smaller for the micropower optimizations due to the combination of lower inversion coefficients and, especially, longer channel lengths.

$S_{VG}^{1/2}$ thermal is significantly larger and $S_{ID}^{1/2}$ thermal is significantly smaller for the micropower optimizations due to smaller values of g_m. However, $S_{VG}^{1/2}$ flicker is comparable for the micropower and millipower optimizations due to equal gate areas. $S_{VG}^{1/2}$ flicker is slightly higher for the millipower optimizations because of inversion level or V_{EFF} increases in the flicker noise associated with higher levels of inversion. While $S_{VG}^{1/2}$ flicker is comparable, $S_{ID}^{1/2}$ flicker is significantly smaller for the micropower optimizations due to smaller values of g_m responsible for the conversion of $S_{VG}^{1/2}$ flicker to $S_{ID}^{1/2}$ flicker.

Analog CMOS Design, Tradeoffs and Optimization Spreadsheet -- MOSFETs Sheet					
For design guidance only as results are approximate and not for any particular CMOS process.					
© 2000–2007, David M. Binkley. See Disclaimer, Notes Sheet.					
---------- Optional User Design Information ----------					
Description		Illustration of DC, balanced, and AC optimization ($I_D = 1\,\mu A$)			
Device reference		M1	M2	M3	
Device notes		DC	Balanced	AC	
---------- Required User Design Inputs ----------					
Device model		nMIL5u4	nMIL2u6	nMIL085u	
I_D	μA (+)	1	1	1	
IC (fixed normalized)		0.300	1.004	2.985	
L_{DRAWN}	μm	5.4	2.6	0.85	
---------- Calculated Results ----------		OK	OK	OK	
Effective Width, Length, Gate Area					
W	μm	28.00	4.00	0.43	
L	μm	5.372	2.572	0.822	
WL	μm²	150.42	10.29	0.35	
DC Bias Voltages					
$V_{EFF} = V_{GS} - V_T$	V	−0.019	0.039	0.108	
V_T (adjusted for V_{SB})	V	0.420	0.420	0.420	
$V_{GS} = V_T + V_{EFF}$	V	0.401	0.459	0.528	
$V_{DS,\,sat}$	V	0.116	0.137	0.173	
Small Signal Parameters					
g_m/I_D	μS/μA	23.77	18.80	13.44	
g_m	μS	23.77	18.80	13.44	
V_A	V	32.23	21.08	11.42	
g_{ds}	μS	0.0310	0.0474	0.0876	
$A_{Vi} = g_m/g_{ds}$	V/V	765.9	396.3	153.4	
Transcond. Distortion (Diff. Pair)					
$V_{INDIF,1dB}$ (input 1-dB comp.)	V	0.057	0.075	0.108	
Capacitances and Bandwidths					
$C_{GS} = C_{gsi} + C_{GSO}$	fF	263.0	33.2	1.8	
$C_{GB} = C_{gbi} + C_{GBO}$	fF	230.6	12.6	0.3	
$C_{GD} = C_{GDO}$ (in saturation)	fF	26.3	3.8	0.4	
C_{DB} (at $V_{DB} = V_{DS} + V_{SB}$)	fF	14.3	2.3	0.6	
$f_{Ti} = g_m / [2\pi(C_{gsi} + C_{gbi})]$	MHz	8.1	71.2	1241.2	
$f_T = g_m / [2\pi(C_{GS} + C_{GB})]$	MHz	7.7	65.4	1005.3	
$f_{DIODE} = g_m / [2\pi(C_{GS} + C_{GB} + C_{DB})]$, MHz		7.4	62.3	773.2	
Thermal and Flicker Noise					
$S_{VG}^{1/2}$ thermal	nV/Hz$^{1/2}$	22.02	25.72	31.38	
$S_{VG}^{1/2}$ flicker at $f_{FLICKER}$	nV/Hz$^{1/2}$	546.30	2171.25	12481.01	
$S_{ID}^{1/2}$ thermal	pA/Hz$^{1/2}$	0.52	0.48	0.42	
$S_{ID}^{1/2}$ flicker at $f_{FLICKER}$	pA/Hz$^{1/2}$	12.98	40.82	167.71	
$f_{FLICKER}$, freq. for flicker noise	Hz	1	1	1	
f_c, flicker-noise corner	MHz	0.001911	0.034092	1.307848	
DC Mismatch for Device Pair					
ΔV_{GS} (1σ)	mV	0.41	1.59	8.77	
$\Delta I_D/I_D$ (1σ)	%	0.98	3.00	11.79	

Figure 4.24 Predicted performance for nMOS devices in a 0.18 μm CMOS process with devices optimized for DC, balanced, and AC performance at a micropower drain current of 1 μA. Values found from the *Analog CMOS Design, Tradeoffs and Optimization* spreadsheet illustrate possible optimizations for differential-pair and current-mirror devices

Finally, ΔV_{GS} (1σ) for a differential pair is comparable for the micropower and millipower optimizations due to equal gate area, but is slightly higher for the millipower optimizations because transconductance factor mismatch contributes more to mismatch at the higher levels of inversion. Even though the gate areas are equal, $\Delta I_D/I_D$ (1σ) for a pair of current-mirror devices is modestly higher for the micropower optimizations. This is due to higher g_m/I_D responsible for the conversion of comparable threshold-voltage mismatch to $\Delta I_D/I_D$.

4.4.4.6 Summary of micropower performance considerations

Operation at low drain currents requires reduced inversion coefficients and increased channel lengths to obtain gate areas comparable to that at higher drain currents. While maintaining comparable gate area is not always a requirement, maintaining comparable or, at least, some minimum gate area is often required to manage low-frequency flicker noise and DC mismatch. Additionally, inversion coefficients may have to be reduced and channel lengths increased at low drain currents to achieve sufficient channel width for layout. In the $I_D = 100\,\mu A$, millipower and $I_D = 1\,\mu A$, micropower, DC, balanced, and AC optimizations described, gate area ranged from $150\,\mu m^2$ for the DC optimizations to a small value of $0.35\,\mu m^2$ for the AC optimizations.

While gate-referred flicker-noise voltage density, $S_{VG}^{1/2}$ flicker, gate–source mismatch voltage, ΔV_{GS}, and relative drain current mismatch, $\Delta I_D/I_D$, are comparable for the micropower and millipower optimizations because of equal gate areas, the bandwidth measures of f_{Ti}, f_T, and f_{diode} and the flicker-noise corner of f_c are much smaller for the micropower optimizations. This is not a result of lower drain current in the micropower optimizations, because these measures are independent of the drain current. Instead, this is a result of lower inversion coefficients and, especially, longer channel lengths required at low drain currents. Finally, $S_{VG}^{1/2}$ thermal is larger, and $S_{ID}^{1/2}$ thermal and $S_{ID}^{1/2}$ flicker are smaller at low drain currents due to smaller transconductance, g_m. Section 4.4.3 discussed performance tradeoffs related to the selection of drain current in further detail.

4.4.5 Summary Procedure for Device Optimization

The selection of inversion coefficient, channel length, and drain current for MOS device optimization involves many performance tradeoffs. The listing below summarizes a procedure for device optimization.

- Identify and prioritize performance including power consumption, bias compliance voltages (V_{EFF} and $V_{DS,sat}$), gain, bandwidth, distortion, layout area and related capacitances, DC accuracy and related local-area mismatch, and AC accuracy and related thermal and flicker noise. The *MOSFET Operating Plane* of Figure 3.1, the tables and figures of Sections 4.3.3–4.3.6, and Table 4.5 show performance tradeoffs. Additionally, the *Analog CMOS Design, Tradeoffs and Optimization* spreadsheet provides an immediate display of MOS performance for selections of the inversion coefficient, channel length, and drain current as illustrated in Sections 4.4.1–4.4.3. Section 4.4.4 illustrated specific examples of DC, balanced, and AC optimization for both millipower and micropower operation.

- Select the drain current using the listing at the end of Section 4.4.4 and performance listed in Figure 4.22. Increasing the drain current decreases signal current distortion, increases drive current and slew rate, lowers small-signal resistance levels, increases transconductance, and lowers gate-referred thermal-noise voltage. Increasing the drain current also increases channel width and gate area, increases device capacitances relative to layout and load capacitances, and increases power consumption. Finally, increasing the drain current increases DC accuracy associated with local-area mismatch, increases AC accuracy associated with thermal and flicker noise, and increases bandwidth-accuracy figures of merit.

- Select the inversion coefficient using the performance tradeoffs shown in Figures 3.1 and 4.1 and performance listed in Figure 4.20. As shown in Figure 4.1, low-voltage, power-efficient design restricts operation to moderate inversion where V_{EFF} and $V_{DS,sat}$ are low, g_m/I_D is high, gain and bandwidth are moderate, and immunity to velocity saturation effects is good. Increasing the inversion coefficient increases V_{EFF} and $V_{DS,sat}$, decreases g_m/I_D and related voltage gain, decreases transconductance distortion, increases bandwidth, and increases gate-referred thermal-noise voltage. Increasing the inversion coefficient also decreases channel width and gate area, decreases device capacitances, and increases local-area mismatch and flicker noise. Finally, increasing the inversion coefficient increases the bandwidth-accuracy and bandwidth–power–accuracy figures of merit for current-mode circuits, while decreasing these for voltage-mode circuits.

- Select the channel length using the tradeoffs shown in Figure 3.1 and performace listed in Figure 4.21. Increasing the channel length increases V_A and the related voltage gain, increases channel width, significantly increases gate area and gate capacitance, and significantly decreases bandwidth. Increasing the channel length primarily controls tradeoffs associated with constant bandwidth-accuracy figures of merit, where the product of accuracy squared and bandwidth remains nearly constant because of nearly constant g_m/I_D. Increasing the channel length directly increases the DC accuracy associated with DC offset and AC accuracy associated with integrated thermal noise, while decreasing the bandwidth as the inverse square of channel length (assuming negligible velocity saturation effects). DC accuracy is improved through the decrease in local-area mismatch associated with increased gate area, while AC accuracy is improved through the decrease in bandwidth.

- Iterate drain current, inversion coefficient, and channel length selections as required to meet desired performance tradeoffs. Optimization beyond individual devices, current mirrors, and differential pairs described here is developed in Chapters 5 and 6 where device performance is mapped to the performance of complete circuits.

REFERENCES[2]

[1] E. A. Vittoz, "Micropower techniques," in *Design of MOS VLSI Circuits for Telecommunications*, ed. J. Franca and Y. Tsividis, Prentice Hall, 1994.

[2] D. M. Binkley, "MOSFET modeling and circuit design: a methodology for transistor level analog CMOS design," Conference tutorial with D. Foty, "MOSFET modeling and circuit design: re-establishing a lost connection," *37th Annual Design Automation Conference (DAC)*, Los Angeles, June 2000.

[3] D. M. Binkley, "A methodology for analog CMOS design based on the EKV MOS Model," *2000 Fabless Semiconductor Association Design Modeling Workshop: SPICE Modeling*, Santa Clara, Oct. 2000.

[4] D. M. Binkley, "A methodology for analog CMOS design based on the EKV MOS model," Conference tutorial with D. Foty, "Re-connecting MOS modeling and circuit design: new methods for design quality," *2001 IEEE 2nd International Symposium on Quality Electronic Design (ISQED)*, San Jose, Mar. 2001.

[5] D. M. Binkley, "A methodology for analog CMOS design based on the EKV MOS model," Conference tutorial with D. Foty, "MOS modeling as a basis for design methodologies: new techniques for next-generation analog circuit design," *2001 15th European Conference on Circuit Theory and Design (ECCTD)*, Helsinki, Aug. 2001.

---

[2] The reader is referred to Chapter 3 for numerous references related to MOS operation and modeling for analog circuit design.

[6] D. M. Binkley, "A methodology for analog CMOS design based on the EKV MOS model," Conference tutorial with D. Foty, "MOS modeling as a basis for design methodologies: new techniques for modern analog design," *2002 IEEE International Symposium on Circuits and Systems (ISCAS)*, Scottsdale, May 2002.

[7] D. M. Binkley, C. E. Hopper, S. D. Tucker, B. C. Moss, J. M. Rochelle, and D. P. Foty, "A CAD methodology for optimizing transistor current and sizing in analog CMOS design," *IEEE Transactions on Computer-Aided Design of Integrated Circuits and Systems*, vol. 22, pp. 225–237, Feb. 2003.

[8] D. M. Binkley, M. Bucher, and D. Kazazis, "Modern analog CMOS design from weak through strong inversion," *Proceedings of the European Conference on Circuit Theory and Design (ECCTD)*, pp. I-8–I-13, Krakow, Aug. 2003.

[9] D. M. Binkley, B. J. Blalock, and J. M. Rochelle, "Optimizing drain current, inversion level, and channel length in analog CMOS design," *Journal of Analog Integrated Circuits and Signal Processing*, vol. 47, pp. 137–163, May 2006.

[10] C. Toumazou, G. Moschytz, and B. Gilbert, eds, *Trade-Offs in Analog Circuit Design: The Designer's Companion*, Kluwer Academic, 2002.

[11] Y. Tsividis, "Moderate inversion in MOS devices," *Solid-State Electronics*, vol. 25, pp. 1099–1104, Nov. 1982.

[12] A. Shameli and P. Heydari, "A novel power optimization technique for ultra-low power RFICs," *Proceedings of the 2006 International Symposium on Low Power Electronic Design (ISLPED)*, pp. 274–279, Oct. 2006.

[13] P. Kinget and M. Steyaert, "Impact of transistor mismatch on the speed-accuracy-power trade-off of analog CMOS circuits," *Proceedings of the IEEE 1996 Custom Integrated Circuits Conference (CICC)*, pp. 333–336, May 1996.

[14] P. Kinget, "Device mismatch and tradeoffs in the design of analog circuits," *IEEE Journal of Solid-State Circuits*, vol. 40, pp. 1212–1224, June 2005.

[15] M. Pelgrom, H. Tuinhout, and M. Vertregt, "Transistor matching in analog CMOS applications," *Technical Digest of the 1998 IEEE International Electron Devices Meeting (IEDM)*, pp. 915–918, Dec. 1998.

[16] A.-J. Annema, B. Nauta, R. van Langevelde, and H. Tuinhout, "Analog circuits in ultra-deep-submicron CMOS," *IEEE Journal of Solid-State Circuits*, vol. 40, pp. 132–143, Jan. 2005.

Part II
Circuit Design Examples Illustrating Optimization for Analog CMOS Design

Part II extends the MOS device performance, tradeoffs, and optimization developed in Part I to the optimum design of circuits. The contents of Part II were described earlier in Chapter 1 and are reviewed briefly here.

Chapter 5 presents the design of simple, $0.5\,\mu m$ and cascoded, $0.18\,\mu m$ CMOS operational transconductance amplifiers (OTAs) optimized for DC, balanced, and AC performance. Devices in a given version of the simple OTAs operate at equal drain currents, inversion coefficients, and channel lengths. Although not optimal for noise and mismatch performance because input devices do not dominate the noise and mismatch, these OTAs are representative of general-purpose designs and provide an introduction to the design methods described in this book. Devices in a given version of the cascoded OTAs, however, operate at different drain currents and inversion coefficients to ensure input devices dominate the thermal noise. Operating devices at low, moderate, and high inversion coefficients in the simple OTAs, combined with operating devices at long, moderate, and short channel lengths in both OTAs, provide DC, balanced, and AC optimized versions. The DC optimized OTAs have high output resistance and voltage gain, combined with low input-referred flicker-noise voltage and offset voltage due to local-area mismatch. The AC optimized OTAs have high transconductance bandwidth, combined with small input and output capacitances and small layout area. The balanced optimized OTAs have a balance of DC and AC performance. Additional performance measures of transconductance, transconductance distortion, input-referred thermal-noise voltage, and input and output voltage ranges are described for the OTAs. Because the measures of performance described apply to most analog circuits, the circuit analysis, performance trends, and optimization described can be extended to other analog CMOS circuits.

Chapter 6 presents the design of two micropower, low-noise, $0.35\,\mu m$ CMOS preamplifiers optimized for low thermal and flicker noise. Here, devices operate at different drain currents, inversion coefficients, and channel lengths to ensure input devices dominate both thermal and flicker noise. Expressions for noise are developed in terms of the bias current and bias compliance voltage, which is the minimum voltage required across a device or group of devices. These expressions show the difficulty of minimizing the drain noise current of non-input devices at low supply voltages.

Considerable attention is directed towards obtaining low preamplifier thermal noise at minimum supply current.

For the design examples in Part II, the *Analog CMOS Design, Tradeoffs and Optimization* spreadsheet is used to map the predicted performance of individual devices into overall circuit performance. The predicted circuit performance is compared to measured performance, validating the design methods. The spreadsheet is described further in the Appendix and is available on the book's web site listed on the cover.

5

Design of CMOS Operational Transconductance Amplifiers Optimized for DC, Balanced, and AC Performance

5.1 INTRODUCTION

Chapter 4 described tradeoffs in MOS device performance resulting from the design choices of inversion coefficient, channel length, and drain current. Tradeoffs involving geometry, intrinsic gate capacitances, drain–body capacitances, bias compliance voltages, small-signal parameters, and intrinsic voltage gain and bandwidth were described first as these are common to all applications. Then, tradeoffs specific to differential pairs or single, gate input, common source devices were described, including transconductance distortion using the input, 1 dB compression voltage, gate-referred thermal- and flicker-noise voltage, and gate–source voltage mismatch. Additionally, tradeoffs specific to current mirrors were described, including drain-referred thermal- and flicker-noise current, and drain current mismatch. Finally, tradeoffs in figures of merit were described that involve the combination of multiple aspects of performance. Chapter 4 built on the work of Chapter 3 that developed individual aspects of MOS device performance.

In addition to describing performance tradeoffs, Chapter 4 illustrated these numerically using the *Analog CMOS Design, Tradeoffs and Optimization* spreadsheet. This spreadsheet is described further in the Appendix and is available on the book's web site listed on the cover. The spreadsheet was used initially to explore the design of differential pairs and current mirrors for nMOS devices in a typical 0.18 μm CMOS process as inversion coefficient, channel length, and drain current were varied separately. Then, the spreadsheet was used for the design of differential pairs and current mirrors optimized for DC, balanced, and AC performance. The DC optimization where devices operate at low inversion coefficients and long channel lengths maximizes transconductance, drain–source resistance, and intrinsic voltage gain, while minimizing flicker noise and local-area mismatch. The AC optimization where devices operate at high inversion coefficients and short channel lengths maximizes intrinsic bandwidth, minimizes capacitances, and minimizes transconductance distortion at the expense of performance optimized by the DC optimization. The balanced optimization provides a compromise of DC and AC performance. Drain currents of 100 and 1 μA were considered for the device optimizations as representative of millipower and micropower circuits, respectively. The requirements of lower inversion coefficients and longer channel lengths for micropower circuits were discussed to achieve practical channel widths for layout and ensure sufficient gate area to manage flicker noise and local-area mismatch.

This chapter extends optimization beyond individual devices, differential pairs, and current mirrors described in Chapter 4 to complete circuits. Here, simple, 0.5 μm and cascoded, 0.18 μm CMOS operational transconductance amplifiers (OTAs) are optimized for DC, balanced, and AC performance. Devices in a given version of the simple OTAs operate at equal drain currents, inversion coefficients, and channel lengths. Although not optimal for noise or mismatch, these OTAs are representative of general-purpose circuits and provide an initial set of design examples. The DC optimized OTA uses long-channel devices operating in moderate inversion, resulting in large channel widths and gate areas. This OTA has high transconductance, output resistance, voltage gain, and input and output voltage ranges, combined with low input-referred flicker-noise voltage and offset voltage due to local-area mismatch and systematic offset. However, this comes at the expense of low transconductance bandwidth, high transconductance distortion, large input and output capacitances, and large layout area. The AC optimized OTA uses minimum channel length devices operating in strong inversion, resulting in small channel widths and gate areas. This OTA has high transconductance bandwidth, low transconductance distortion, small input and output capacitances, and small layout area, but at the expense of DC optimized performance. The balanced OTA provides a balance of DC and AC performance.

Unlike in the simple OTAs, devices in a given version of the cascoded OTAs operate at different inversion coefficients so that input devices dominate the thermal noise. Input devices operate at low inversion coefficients in moderate inversion for high transconductance, while critical non-input devices operate at high inversion coefficients in strong inversion for low transconductance and low drain noise current contributions. Additionally, some non-input devices operate at lower drain currents to further reduce their transconductance and noise contributions. Long, moderate, and short channel lengths provide DC, balanced, AC optimized versions for the cascoded OTAs, giving similar performance tradeoffs as those for the simple OTAs. However, the cascoded OTAs have equal transconductance, transconductance distortion, input-referred thermal-noise voltage, and input and output voltage ranges as a result of fixed inversion coefficients. Although input devices dominate thermal noise, input devices do not dominate flicker noise and mismatch because channel lengths and resulting gate areas are not sufficiently large for non-input devices. Chapter 6 describes two micropower, low-noise, 0.35 μm CMOS preamplifiers where input devices dominate both thermal and flicker noise.

This chapter first describes circuits and circuit analysis for the simple and cascoded OTAs. Then, OTA performance trends are related to device trends shown in Chapter 4. Following this, the optimization of device inversion coefficients, channel lengths, and drain currents is discussed, including for minimizing thermal noise, flicker noise, and local-area mismatch. The chapter then presents predicted performance for the OTAs using the *Analog CMOS Design, Tradeoffs and Optimization* spreadsheet. Predicted device performance in the *MOSFETs* sheet resulting from drain current, inversion coefficient, and channel length selections is mapped to the *Circuit Analysis* sheet. This sheet lists predicted OTA performance and compares this to measured performance. Predicted and measured performance is in good agreement and illustrates the wide range of performance available through the design choices of device inversion coefficient and channel length. Drain currents are not varied to illustrate performance optimization at constant power consumption.

5.2 CIRCUIT DESCRIPTION

This section describes circuits and circuit operation for the simple and cascoded OTAs.

5.2.1 Simple OTAs

Figure 5.1 shows a schematic diagram for the DC, balanced, and AC optimized, simple OTAs [1]. All signal-path bias currents are kept constant at $I_{BIAS} = 50\,\mu A$, and supply voltages are kept constant at $V_{DD} = +1.25\,V$ and $V_{SS} = -1.25\,V$. This gives a core current consumption of $200\,\mu A$ and power consumption of $500\,\mu W$, which excludes the external bias reference circuit. Maintaining constant bias currents and supply voltages permits optimization for the three OTAs at constant power consumption.

M_1 and M_2 comprise an input, nMOS differential pair with drain output currents mirrored by pMOS current mirrors consisting of M_5 and M_6, and M_7 and M_8. The M_6 drain output of the M_5 and M_6 current mirror connects directly to the OTA output, while the M_8 drain output of the M_7 and M_8 current mirror connects to an intermediate nMOS current mirror consisting of M_3 and M_4. The M_4 drain output of this current mirror connects to the OTA output, providing a complementary signal path. Finally, M_9 provides a bias current of $2I_{BIAS}$ to the M_1 and M_2 differential pair. The external bias reference circuit, not shown, is a diode-connected device forming a 1:1 current mirror with M_9.

The input, differential-pair devices, M_1 and M_2, operate at $I_{BIAS} = 50\,\mu A$ each with equal inversion coefficients, sizing, and transconductances. The nMOS current-mirror devices, M_3 and M_4, operate at I_{BIAS} each with equal inversion coefficients, sizing, and transconductances. The pMOS current-mirror devices, M_5–M_8, also operate at I_{BIAS} each with equal inversion coefficients, sizing, and transconductances. Although all pMOS devices have equal parameters, these will be assigned separately

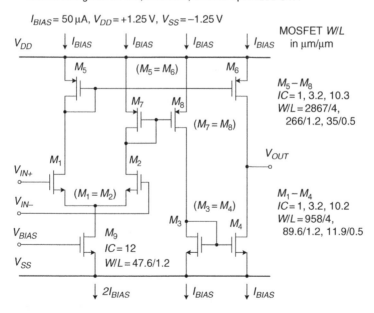

Figure 5.1 Schematic diagram of the simple OTAs optimized for DC, balanced, and AC performance. Figure 5.3 gives MOSFET drain current, inversion coefficient, and channel length selections with resulting device geometry and performance. Figure 5.4 gives predicted, simulated, and measured OTA performance

for the M_5 and M_6 mirror and M_7 and M_8 mirror. As seen later, considering individual differential pairs and current mirrors facilitates noise and local-area mismatch analysis. The OTAs presented here utilize unity-gain current mirrors, although the design optimization methods can be extended to different OTA configurations or other analog circuits.

As shown in Figure 5.1, MOSFET inversion coefficients, channel lengths, and drain currents are equal for all signal-path devices in a given OTA version. As described later, this does not result in optimal thermal noise, flicker noise, or local-area mismatch, but provides a simple, initial design illustration typical of general-purpose OTAs.

Input device operation at $IC = 1$ in the center of moderate inversion gives the DC optimized OTA high transconductance, low input-referred thermal-noise voltage, and a low input, 1 dB compression voltage. Long channel lengths at $L = 4\,\mu m$ give this OTA high output resistance and high voltage gain, while the resulting large channel widths and gate areas give large input and output capacitances, low transconductance bandwidth, and low input-referred flicker-noise voltage and offset voltage due to local-area mismatch. In contrast, input device operation at $IC = 10$ at the onset of strong inversion gives the AC optimized OTA low transconductance, high input-referred thermal-noise voltage, and a high input, 1 dB compression voltage. Short channel lengths at the process minimum of $L = 0.5\,\mu m$ give this OTA low output resistance and low voltage gain, while the resulting small channel widths and gate areas give small input and output capacitances, high transconductance bandwidth, and high input-referred flicker-noise voltage and offset voltage due to mismatch. Finally, device operation at the strong-inversion side of moderate inversion at $IC = 3.2$ and moderate channel lengths of $L = 1.2\,\mu m$ gives a balance of DC and AC performance for the balanced OTA. Inversion coefficient and channel length selections, and the resulting OTA performance, are described in detail in Section 5.4.

5.2.2 Cascoded OTAs

Figure 5.2 shows a schematic diagram for the DC, balanced, and AC optimized, cascoded OTAs [2, 3]. All signal-path bias currents are kept constant at the 50, 100, and 150 μA levels shown, and supply voltages are kept constant at $V_{DD} = +0.9\,V$ and $V_{SS} = -0.9\,V$. This gives a core current consumption of 300 μA and power consumption of 540 μW, which excludes the bias reference circuits. Maintaining constant bias currents and supply voltages permits optimization for the three OTAs at constant power consumption, as considered for the simple OTAs.

M_1 and M_2 comprise an input, nMOS differential pair with drain output currents received by pMOS folded-cascode devices, M_7 and M_8. The M_8 drain output connects directly to the OTA output, while the M_7 drain output connects to an nMOS, low-voltage, cascode current mirror. This current mirror consists of rail devices M_5 and M_6 and cascode devices M_9 and M_{10}. The M_{10} drain output of this current mirror connects to the OTA output, providing a complementary signal path. Finally, pMOS devices M_3 and M_4 are current sources supplying the drains of the M_1 and M_2 input pair devices and the sources of the M_7 and M_8 folded-cascode devices.

Rail devices M_{13} and M_{14}, and cascode devices M_{16} and M_{17} comprise an nMOS, low-voltage, cascode current mirror outside of the signal path that receives an external bias current for the OTA. M_{15} is mirrored off the rail devices and supplies current to the M_1 and M_2 input pair devices, while the M_{17} drain output supplies current to a pMOS, low-voltage, cascode current-mirror reference. This reference outside the signal path consists of rail device M_{11} and cascode device M_{12}, and provides a reference for the M_3 and M_4 current source devices and M_7 and M_8 folded-cascode devices. Reference and related signal-path devices have equal unit channel widths and lengths, and equal drain–source voltages to ensure accurate replication of bias currents.

The voltage at V_{INCM} is connected externally to ground and provides a low-voltage, cascode gate voltage for M_{16} and M_{17}. When the OTA inputs V_{IN+} and V_{IN-} are at a common-mode voltage equal to ground, the bias current in the M_{15} input pair current source is replicated from the current in the M_{13} bias reference because both devices have equal drain–source voltages.

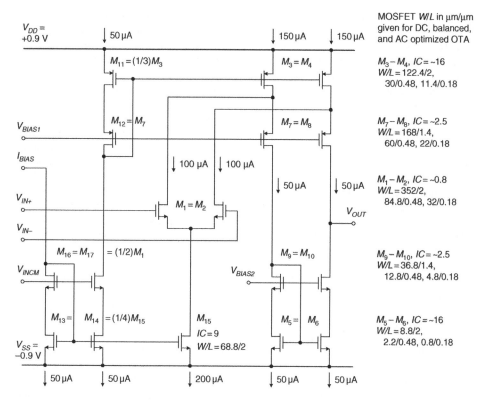

Figure 5.2 Schematic diagram of the cascoded OTAs optimized for DC, balanced, and AC performance. Figures 5.11–5.13 give MOSFET drain current, inversion coefficient, and channel length selections with resulting device geometry and performance. Figure 5.14 gives predicted, simulated, and measured OTA performance

The voltage at V_{BIAS1} provides a low-voltage, cascode gate voltage for M_7, M_8, and M_{12}, while the voltage at V_{BIAS2} provides this for M_9 and M_{10}. V_{BIAS1} connects to an external pMOS, diode-connected device connected to V_{DD}, and V_{BIAS2} connects to an external nMOS, diode-connected device connected to V_{SS}. Both of these diode-connected devices are biased and sized to establish a drain–source voltage of 0.4 V across the M_3–M_6 and M_{11} low-voltage, cascode rail devices. M_3–M_6 require sufficient drain–source voltages because they are operated well into strong inversion at high drain–source saturation voltages to minimize their drain thermal-noise current contributions. Higher drain–source voltages, unfortunately, reduce the OTA output voltage range. The external diode-connected devices are sized for $V_{EFF} = 0.6\,\text{V}$, where the $M_7 - M_{10}$ cascode devices have $V_{EFF} = 0.1\,\text{V}$ and a threshold voltage increase of 0.1 V for $V_{SB} = 0.4\,\text{V}$, which is also the desired voltage across $M_3 - M_6$.

As shown in Figure 5.2, the M_1 and M_2 input pair devices operate at $100\,\mu\text{A}$ each near the center of moderate inversion at $IC \approx 0.8$ for high g_m/I_D and g_m. This minimizes input pair, input-referred thermal-noise voltage and helps these devices dominate overall OTA noise. M_1 and M_2 have equal inversion coefficients, sizing, and transconductances.

The M_3 and M_4 current source devices operate at $150\,\mu\text{A}$ each to supply current to the M_1 and M_2 input pair and M_7 and M_8 cascode devices. M_3 and M_4, which can be significant noise contributors, operate well into strong inversion at $IC \approx 16$ for low g_m/I_D and g_m to minimize their drain thermal-noise current contributions. M_3 and M_4 have equal inversion coefficients, sizing, and transconductances. The M_{11} bias reference operates at the same inversion coefficient and channel length as M_3 and M_4, but operates at one-third the channel width and drain current.

The M_5 and M_6 current-mirror rail devices operate at $50\,\mu\text{A}$ each. Like M_3 and M_4, these operate well into strong inversion at $IC \approx 16$ to minimize their drain thermal-noise current contributions. M_5 and M_6 noise contributions are reduced further since these devices operate at $50\,\mu\text{A}$ compared to $100\,\mu\text{A}$ for the input devices. While this favorably lowers noise contributions, it unfavorably lowers the OTA slew rate. M_5 and M_6 have equal inversion coefficients, sizing, and transconductances.

The M_7 and M_8, and M_9 and M_{10} cascode devices operate at $50\,\mu\text{A}$ at the strong-inversion side of moderate inversion at $IC \approx 2.5$ to minimize their $V_{DS,sat}$, while maintaining moderately low device capacitances and moderately high intrinsic bandwidth. Low $V_{DS,sat}$ is required to minimize the loss of OTA output voltage range. These cascode devices contribute negligible noise and local-area mismatch because their effective transconductances are low due to high resistances appearing at their sources. However, they strongly influence OTA output resistance, voltage gain, and transconductance bandwidth. M_7 and M_8 have equal inversion coefficients, sizing, and transconductances and M_9 and M_{10} have equal inversion coefficients, sizing, and transconductances. Unlike the M_1–M_6 devices where parameters will be assigned to the lower-numbered device in a pair, parameters will be assigned to the higher-numbered devices for the M_7 and M_8, and M_9 and M_{10} pairs because these devices control the OTA output resistance. Finally, the M_{12} bias reference operates at the same inversion coefficient, channel length, channel width, and drain current as M_7 and M_8.

The inversion coefficient values given are nominal values because lateral diffusion results in variations in the effective channel length. The effective channel length is a smaller portion of the drawn channel length for short-channel devices used in the AC optimized OTA. As a result, the shape factors, $S = W/L$, are larger for this OTA resulting in smaller inversion coefficients. Variations in inversion coefficients exist because effective channel length was not considered in the original selection of inversion coefficients. These variations can be avoided in a new design by considering effective channel length as done in the *Analog CMOS Design, Tradeoffs and Optimization* spreadsheet.

Channel lengths for the input pair, current source, and current-mirror, rail devices M_1–M_6 are selected at 2, 0.48, and $0.18\,\mu\text{m}$ for the DC, balanced, and AC optimized OTAs, respectively. Channel lengths for the cascode devices M_7–M_{10} are selected at 1.4, 0.48, and $0.18\,\mu\text{m}$. A $1.4\,\mu\text{m}$ channel length is selected to improve transconductance bandwidth for the DC optimized OTA compared to that obtained using the same $2\,\mu\text{m}$ channel length as M_1–M_6. Long channel lengths for M_1–M_{10} give high output resistance, high voltage gain, and low transconductance bandwidth for the DC optimized OTA. As described in later noise and local-area mismatch analysis, M_1–M_6 are major contributors to thermal noise, flicker noise, and local-area mismatch. Long channel lengths for these devices in the DC optimized OTA result in large gate areas giving low input-referred flicker-noise voltage and offset voltage due to mismatch. Short channel lengths at the process minimum of $L = 0.18\,\mu\text{m}$ for the AC optimized OTA give low output resistance and voltage gain, and high transconductance bandwidth, while the resulting small gate areas give high input-referred flicker-noise voltage and offset voltage due to mismatch. Finally, the use of moderate channel lengths gives a balance of DC and AC performance for the balanced OTA. Inversion coefficient and channel length selections, and the resulting OTA performance, are described in more detail in Section 5.5.

5.3 CIRCUIT ANALYSIS AND PERFORMANCE OPTIMIZATION

Before exploring device drain currents, inversion coefficients, and channel lengths, it is necessary to analyze OTA performance in terms of the performance of individual devices. Following this, the *MOSFETs* sheet of the *Analog CMOS Design, Tradeoffs and Optimization* spreadsheet is used to select drain currents, inversion coefficients, and channel lengths and map the resulting device performance into complete OTA performance using the *Circuit Analysis* sheet.

This section describes circuit analysis, performance trends, and optimization for OTA transconductance, output resistance, voltage gain, transconductance frequency response, and input-referred thermal-noise voltage, flicker-noise voltage, and offset voltage due to both local-area mismatch and systematic offset. Additionally, this section describes analysis, trends, and optimization for input and output capacitances, slew rate, input and output voltage ranges, and the input, 1 dB compression voltage. Because of similarities in analysis, trends, and optimization, this section presents this for both the simple and cascoded OTAs. Following this, Sections 5.4 and 5.5 describe the optimization and resulting performance for the simple and cascoded OTAs separately.

This section discusses OTA performance trends and optimization in terms of the device design choices of inversion coefficient, channel length, and drain current. OTA performance trends are related to device performance trends shown earlier in summary tables and figures in Chapter 4. Although the device trends shown in Chapter 4 are for the 0.18 μm CMOS process used in the cascoded OTAs, the trends are nearly identical to those for the 0.5 μm CMOS process used in the simple OTAs. As discussed in Section 3.2.4, this is a result of the technology independence present in the inversion coefficient because of its normalization to process parameters. This section discusses the optimization of all aspects of OTA performance, including minimizing input-referred thermal-noise voltage, flicker-noise voltage, and offset voltage due to local-area mismatch through both input and non-input device design choices. Because the performance described applies to most analog circuits, the circuit analysis, performance trends, and optimization described in this section can be extended to other analog CMOS circuits.

The analysis and performance trends given in this section include velocity saturation and vertical field mobility reduction (VFMR) decreases in device g_m/I_D and g_m, and the related increases in $V_{EFF} = V_{GS} - V_T$. These effects are included by expression modifications described in the text. Velocity saturation effects are significant for short-channel devices operating at high levels of inversion, while smaller VFMR effects are present for all channel lengths at high levels of inversion. These effects can decrease OTA transconductance, voltage gain, and bandwidth, while increasing the input-referred thermal-noise voltage and the input, 1 dB compression voltage. These effects can also decrease the allowable input, common-mode voltage range as a result of increases in V_{EFF}. DIBL effects for short-channel devices operating at low levels of inversion can reduce MOS drain–source resistance and the related OTA output resistance and voltage gain. These effects are included in the Early voltage used for drain–source resistance predictions. Finally, increases in gate-referred flicker-noise voltage with increasing inversion or V_{EFF} can increase the OTA input-referred flicker-noise voltage. These increases, which can be especially significant for pMOS devices, are included through increases in the flicker noise-factor from its minimum value at low levels of inversion. The management of small-geometry effects to minimize the deterioration of OTA performance is discussed later in Section 5.3.13.

5.3.1 Transconductance

5.3.1.1 Simple OTAs

Transconductance for the simple OTAs shown in Figure 5.1 is nearly equal to the transconductance of the input pair because current mirrors route signal currents from the input pair drains to the OTA output with nearly unity current gain. OTA transconductance is then approximated by the transconductance of the input pair, which is equal to the transconductance of individual pair devices, M_1 and M_2, or

$$G_M = g_{m1} = \left(\frac{g_m}{I_D}\right)_1 (I_{D1} = I_{BIAS}) = \frac{I_{D1} = I_{BIAS}}{n_1 U_T \left(\sqrt{IC_1 + 0.25} + 0.5\right)} \qquad (5.1)$$

IC_1 is the inversion coefficient, n_1 is the substrate factor, and $I_{D1} = I_{BIAS}$ is the drain current for the M_1 and M_2 input pair devices. This expression excludes the slight loss of signal current at the current-mirror inputs due to loading by device drain–source resistances. The rightmost expression uses the g_m/I_D and g_m expressions summarized in Table 3.17, which exclude decreases due to velocity saturation and VFMR that can decrease G_M.

Velocity saturation effects, present for short-channel devices operating at high levels of inversion, and smaller VFMR effects, present for devices of all channel lengths operating at high levels of inversion, decrease g_m/I_D and g_m. As described in Table 3.17, these decreases in g_m/I_D and g_m can be included by replacing IC with $IC(1 + IC/IC_{CRIT})$ in the $\sqrt{IC + 0.25}$ term related to g_m/I_D, used here in Equation 5.1. As described in Section 3.8.2.2 and shown in Figures 3.24 and 3.25, IC_{CRIT} corresponds to a critical inversion coefficient where g_m/I_D is down to approximately 70.7 % of its value without velocity saturation and VFMR effects. $IC_{CRIT} = [(LE_{CRIT})'/(4nU_T)]^2$ depends primarily on the channel length and the critical, velocity saturation, electric field, E_{CRIT}. It also depends on the VFMR factor, θ, which affects g_m/I_D and g_m even for long-channel devices. $(LE_{CRIT})' = LE_{CRIT}||(1/\theta)$ is the equivalent velocity saturation voltage that includes both velocity saturation and VFMR effects.

5.3.1.2 Cascoded OTAs

The transconductance for the cascoded OTAs shown in Figure 5.2 is also nearly equal to the transconductance of the input pair. This is because the folded-cascode devices, M_7 and M_8, and low-voltage, cascode current mirror consisting of M_5, M_6, M_9, and M_{10} route signal currents from the input pair drains to the OTA output with nearly unity current gain. The OTA transconductance is then given by nearly the transconductance of the input pair or

$$G_M = g_{m1}\left(\frac{g_{ms7}}{g_{ms7} + g_{ds1} + g_{ds3}}\right) = \frac{I_{D1}}{n_1 U_T\left(\sqrt{IC_1 + 0.25} + 0.5\right)}\left(\frac{g_{ms7}}{g_{ms7} + g_{ds1} + g_{ds3}}\right)$$

$$\approx g_{m1} = \left(\frac{g_m}{I_D}\right)_1 I_{D1} = \frac{I_{D1}}{n_1 U_T\left(\sqrt{IC_1 + 0.25} + 0.5\right)} \tag{5.2}$$

IC_1 is the inversion coefficient, n_1 is the substrate factor, and I_{D1} is the drain current for the M_1 and M_2 input pair devices.

The first line of the expression includes the loss of signal current associated with current division at the source inputs of M_7 and M_8. This loss is included for the cascoded OTAs because operation at shorter channel lengths makes the loading of device drain–source resistances more significant compared to that in the simple OTAs. The signal current loss is found by the small-signal conductances appearing at the sources of M_7 and M_8, g_{ms7}, divided by the sum of this and the drain–source conductances of the M_1 and M_2 input pair, g_{ds1}, and M_3 and M_4 current source devices, g_{ds3}, that load the sources. $g_{ms7} = g_{m7} + g_{mb7} + g_{ds7}$ is the total source transconductance of M_7 and M_8. The small-signal conductance at the sources of M_7 and M_8 is found from Table 3.16 and is nearly equal to g_{ms7} because the drain of M_7 is connected to a low, small-signal resistance of nearly $1/g_{m5}$ of M_5, while the drain of M_8 is connected to ground when finding the OTA transconductance.

The second line of the G_M expression neglects the loss of signal current at the sources of M_7 and M_8 and is useful for design guidance. However, the full accuracy of the first line is used for predicting G_M by the *Analog CMOS Design, Tradeoffs and Optimization* spreadsheet. The g_m/I_D and g_m expressions used for G_M exclude velocity saturation and VFMR decreases that can decrease G_M. As described for the simple OTAs above, these effects are included by replacing IC with $IC(1 + IC/IC_{CRIT})$ in the $\sqrt{IC + 0.25}$ terms related to g_m/I_D.

5.3.1.3 Optimization

For both the simple and cascoded OTAs, G_M is maximized by maximizing g_m/I_D of the input pair devices by operating them at low inversion coefficients ($IC_1 = IC_2$) in weak or moderate inversion. G_M is constant and maximum for input pair devices in weak inversion, decreases as the inverse square root of IC_1 in strong inversion, and decreases inversely with IC_1 when velocity saturation is significant. Thus, strong-inversion operation requires increased bias current to achieve a given level of transconductance. In fact, G_M increases directly with input device drain current, which, as always, assumes that device width is scaled with drain current to maintain the selected inversion coefficient. Trends for G_M are the same as device g_m/I_D trends summarized in Table 4.1 and shown in Figures 4.7 and 4.8, where device trends are applied to the input devices.

5.3.2 Output Resistance

5.3.2.1 Simple OTAs

Output resistance for the simple OTAs shown in Figure 5.1 is equal to the parallel combination of drain–source resistances r_{ds4} and r_{ds6} for output devices M_4 and M_6. Output resistance is then given by

$$R_{OUT} = r_{ds4} \parallel r_{ds6} = \frac{V_{A4} \parallel V_{A6}}{I_{BIAS}} = \frac{(V_{AL4}L_4) \parallel (V_{AL6}L_6)}{I_{BIAS}} \qquad (5.3)$$

V_{A4} and V_{A6} are the Early voltages, V_{AL4} and V_{AL6} are the Early voltage factors, and L_4 and L_6 are the channel lengths for M_4 and M_6, respectively. I_{BIAS} is the drain current for each device. The rightmost terms express the drain–source resistance as $r_{ds} = V_A/I_D = (V_{AL} \cdot L)/I_D$. Here, V_A is found by $V_A = V_{AL} \cdot L$ from Table 3.21, where $V_{AL} = V_A/L$ is the Early voltage factor. As seen in the measured data of Figure 3.48, V_{AL} is not a fixed value for the process because of complex V_A dependency on IC, L, and V_{DS}.

5.3.2.2 Cascoded OTAs

Output resistance for the cascoded OTAs shown in Figure 5.2 is equal to the parallel combination of the output resistance r_{outD8} seen at the drain of cascode device M_8 and r_{outD10} seen at the drain of cascode device M_{10}. The output resistance at the drain of M_8 is given by

$$r_{outD8} = r_{ds8}\left[1 + g_{ms8}\left(r_{ds2} \parallel r_{ds4}\right)\right]$$
$$\approx r_{ds8} \cdot g_{ms8}\left(r_{ds2} \parallel r_{ds4}\right) = (nA_{Vi})_8 \left(r_{ds2} \parallel r_{ds4}\right) \qquad (5.4)$$

g_{ms8} is the total source transconductance of M_8, and r_{ds2}, r_{ds4}, and r_{ds8} are the drain–source resistances of M_2, M_4, and M_8, respectively. The first line of the expression shows that the drain output resistance of M_8 is equal to its drain–source resistance, r_{ds8}, increased by the source degeneration factor that appears inside the square brackets. The effective transconductance of M_8 is equal to its device transconductance, g_{m8}, divided by this same source degeneration factor. This makes the effective transconductance low and the related noise and local-area mismatch currents negligible for M_8 as described later in noise and mismatch analysis. The drain output resistance and effective transconductance for a MOSFET having external source degeneration resistance are summarized in Table 3.16.

When the source degeneration factor for M_8, $1 + g_{ms8} \cdot (r_{ds2} \parallel r_{ds4})$, is much greater than unity as is usually the case, the second line in Equation 5.4 shows that the drain output resistance of M_8 is nearly equal to the resistance connected to its source, $r_{ds2} \parallel r_{ds4}$, raised by n times its intrinsic voltage gain, A_{Vi8}, or $(nA_{Vi})_8$. This is found by substituting $g_{ms} \approx g_m + g_{mb} \approx n \cdot g_m$ using $g_{mb} \approx (n-1)g_m$ from the expression in Table 3.20 and by substituting $A_{Vi} = g_m \cdot r_{ds}$ from the expression in Table 3.22. The resistance connected to the source of M_8 is taken as r_{ds2} in parallel with r_{ds4}. A single value of r_{ds2} is used compared to the actual drain output resistance of $2r_{ds2}$ for M_2 because current flowing into the drain of M_2 flows out of the drain of M_1 and is mirrored into the OTA output. This doubles the loading effect from the drain output resistance of M_2 alone.

The output resistance at the drain of M_{10} is found similarly to that of M_8 and is given by

$$r_{outD10} = r_{ds10}\left(1 + g_{ms10} \cdot r_{ds6}\right)$$

$$\approx r_{ds10} \cdot g_{ms10} \cdot r_{ds6} = (nA_{Vi})_{10}\left(r_{ds6}\right) \tag{5.5}$$

g_{ms10} is the total source transconductance of M_{10}, and r_{ds6} and r_{ds10} are the drain–source resistances of M_6 and M_{10}, respectively. The first line of the expression shows that the drain output resistance of M_{10} is equal to its drain–source resistance, r_{ds10}, increased by the source degeneration factor that appears inside the brackets. As for M_8, the effective transconductance of M_{10} is equal to its device transconductance, g_{m10}, divided by its source degeneration factor. This makes the effective transconductance low and the related noise and local-area mismatch currents negligible as they are for M_8.

The second line in Equation 5.5 shows that the drain output resistance of M_{10} is equal to the resistance connected to its source, r_{ds6}, raised by n times its intrinsic voltage gain, A_{Vi10}, or $(nA_{Vi})_{10}$. This, again, assumes that the source generation factor for M_{10}, $1 + g_{ms10} \cdot r_{ds6}$, is much greater than unity as is usually the case.

Combining r_{outD8} from Equation 5.4 and r_{outD10} from Equation 5.5 gives the OTA output resistance as

$$R_{OUT} = r_{outD8} \parallel r_{outD10} \approx \{(nA_{Vi})_8\left(r_{ds2} \parallel r_{ds4}\right)\} \parallel \{(nA_{Vi})_{10}\, r_{ds6}\}$$

$$\approx (nA_{Vi})_{8,10}\left[r_{ds2} \parallel r_{ds4} \parallel r_{ds6}\right] \tag{5.6}$$

In the second line of the expression, the products of n and the intrinsic voltage gains for cascode devices M_8 and M_{10} are assumed equal and are denoted by $(nA_{Vi})_{8,10}$. These gains are assumed equal because M_8 and M_{10} operate at the same inversion coefficients, $IC_{8,10} = IC_8 = IC_{10}$, channel lengths, $L_{8,10} = L_8 = L_{10}$, and drain–source voltages, giving nearly equal values of $nA_{Vi} = ng_m r_{ds}$. This assumes that the Early voltage factors $V_{AL8,10} = V_{AL8} = V_{AL10}$ are equal giving equal drain–source resistances, although some differences can be expected between nMOS and pMOS devices. The second line of the expression shows that the OTA output resistance is nearly equal to the parallel combination of all resistances, r_{ds2}, r_{ds4}, and r_{ds6}, connected to the sources of M_8 and M_{10} multiplied by n times their intrinsic voltage gains, $(nA_{Vi})_{8,10}$. Although this approximation is useful for design guidance, the full accuracy of r_{outD8} and r_{outD10} given by Equations 5.4 and 5.5 is used for predicting R_{OUT} by the *Analog CMOS Design, Tradeoffs and Optimization* spreadsheet.

Equation 5.6 can be expressed using $r_{ds} = V_A/I_D = (V_{AL} \cdot L)/I_D$ as described for the simple OTAs above. This gives

$$R_{OUT} = r_{outD8} \parallel r_{outD10} \approx (nA_{Vi})_{8,10}\left[r_{ds2} \parallel r_{ds4} \parallel r_{ds6}\right]$$

$$\approx \frac{V_{AL8,10} \cdot L_{8,10}}{U_T\left(\sqrt{IC_{8,10} + 0.25} + 0.5\right)}\left[\frac{V_{AL2}L_2}{I_{D2}} \parallel \frac{V_{AL4}L_4}{I_{D4}} \parallel \frac{V_{AL6}L_6}{I_{D6}}\right] \tag{5.7}$$

Additionally, $(nA_{Vi})_{8,10}$ is expressed using $A_{Vi} = (g_m/I_D) \cdot V_A$ from the expression in Table 3.22. Velocity saturation and VFMR decreases in g_m/I_D that can decrease $(nA_{Vi})_{8,10}$ and R_{OUT} are excluded, but can be included by replacing IC with $IC(1 + IC/IC_{CRIT})$ in the $\sqrt{IC + 0.25}$ term related to g_m/I_D as described in Section 5.3.1.1. However, like the M_1 and M_2 input devices, cascode devices M_8 and M_{10} operate in moderate inversion and experience little velocity saturation. Although channel lengths of L_2, L_4, and L_6 for M_2, M_4, and M_6 are equal, these are considered separately because these devices operate at different drain currents of I_{D2}, I_{D4}, and I_{D6}, respectively. Finally, the Early voltage factors for these devices, V_{AL2}, V_{AL4}, and V_{AL6}, are also considered separately, although V_{AL4} is nearly equal to V_{AL6} because M_4 and M_6 operate at equal inversion coefficients, channel lengths, and drain–source voltages.

5.3.2.3 Optimization

R_{OUT} for the simple OTAs is maximized by operating the M_4 and M_6 output devices at long channel lengths ($L_4 = L_6$), which maximizes the V_A and r_{ds} values in Equation 5.3. R_{OUT} increases sublinearly with increasing output device channel length like V_A shown in Figure 3.47, unless DIBL effects are significant for short-channel devices where it increases more rapidly as these effects are reduced. As described in Section 3.8.4.2, DIBL effects lower V_A and r_{ds} for short-channel devices, especially for operation in weak or moderate inversion. R_{OUT} increases as output device drain current decreases since R_{OUT} is inversely proportional to the drain current (again, MOSFET width is scaled with drain current to maintain the selected inversion coefficient). Trends for R_{OUT} are similar to device V_A trends summarized in Table 4.1 and shown in Figures 4.7 and 4.8, where device trends are applied to the output devices.

R_{OUT} for the cascoded OTAs is maximized by operating M_2, M_4, M_6, M_8, and M_{10} at long channel lengths, which maximizes all V_A and r_{ds} values in Equation 5.7. Because these devices are paired with M_1, M_3, M_5, M_7, and M_9, this corresponds to operating all devices in the signal path or loading the signal path, M_1–M_{10}, at long channel lengths. Interestingly, if channel length is increased equally for M_1–M_{10}, R_{OUT} increases as nearly the square of the increasing device V_A. This is because R_{OUT} depends on the product of the drain–source resistance of cascode devices M_8 and M_{10}, through their intrinsic voltage gain, and the paralleled drain–source resistances of M_2, M_4, and M_6 connected to the M_8 and M_{10} sources. As a result, R_{OUT} is much higher for the DC optimized OTA that uses long-channel devices. The use of long-channel devices also ensures little decrease in V_A, r_{ds}, and, correspondingly, R_{OUT} due to DIBL effects. In addition to its dependency on channel length, R_{OUT} is proportional to g_m/I_D of M_8 and M_{10} through their intrinsic voltage gain. As a result, operating these devices in weak or moderate inversion ($IC_{8,10} \approx 2.5$) increases R_{OUT}, providing short-channel devices are not used where DIBL effects lower their intrinsic gain. Finally, R_{OUT} increases as device drain currents decrease since it is inversely proportional to the drain currents (again, MOSFET width is scaled with drain current to maintain the selected inversion coefficient).

Trends for R_{OUT} of the cascoded OTAs are similar to device V_A trends summarized in Table 4.1 and shown in Figures 4.7 and 4.8, where device trends are applied to M_1–M_{10} and are effectively squared. Although having a smaller effect, these trends are multiplied by trends of g_m/I_D for M_8 and M_{10} that are also shown in the table and figures.

5.3.3 Voltage Gain

5.3.3.1 Simple OTAs

Voltage gain for the simple OTAs shown in Figure 5.1 is equal to the product of G_M from Equation 5.1 and R_{OUT} from Equation 5.3 where I_{BIAS} cancels, giving

$$A_V = G_M \cdot R_{OUT} = g_{m1}(r_{ds4} \| r_{ds6}) = \left(\frac{g_m}{I_D}\right)_1 (V_{A4} \| V_{A6})$$

$$= \frac{(V_{AL4}L_4) \| (V_{AL6}L_6)}{n_1 U_T \left(\sqrt{IC_1 + 0.25} + 0.5\right)} \tag{5.8}$$

Like G_M given by Equation 5.1, Equation 5.8 excludes velocity saturation and VFMR decreases in g_m/I_D and g_m that can decrease A_V. As described in Section 5.3.1.1, these effects are included by replacing IC with $IC(1 + IC/IC_{CRIT})$ in the $\sqrt{IC + 0.25}$ term related to g_m/I_D.

5.3.3.2 Cascoded OTAs

Voltage gain for the cascoded OTAs shown in Figure 5.2 is equal to the product of G_M from Equation 5.2 and R_{OUT} from Equation 5.7 as given by

$$A_V = G_M \cdot R_{OUT} = g_{m1} \left(r_{outD8} \parallel r_{outD10} \right) = \left(\frac{g_m}{I_D} \right)_1 I_{D1} \left(r_{outD8} \parallel r_{outD10} \right)$$

$$\approx \frac{I_{D1}}{n_1 U_T \left(\sqrt{IC_1 + 0.25} + 0.5 \right)}$$

$$\cdot \frac{V_{AL8,10} \cdot L_{8,10}}{U_T \left(\sqrt{IC_{8,10} + 0.25} + 0.5 \right)} \left[\frac{V_{AL2} L_2}{I_{D2}} \parallel \frac{V_{AL4} L_4}{I_{D4}} \parallel \frac{V_{AL6} L_6}{I_{D6}} \right] \tag{5.9}$$

Like G_M given by Equation 5.2 and R_{OUT} given by Equation 5.7, Equation 5.9 excludes velocity saturation and VFMR decreases in g_m/I_D and g_m that can decrease A_V. These effects are included as described above for the simple OTAs. The approximation of A_V given by the second line of Equation 5.9 is useful for design guidance. However, the full accuracy of G_M given by Equation 5.2 and r_{outD8} and r_{outD10} given by Equations 5.4 and 5.5 is used for predicting A_V by the *Analog CMOS Design, Tradeoffs and Optimization* spreadsheet.

5.3.3.3 Optimization

A_V is maximized for the simple OTAs by operating input pair devices at low inversion coefficients ($IC_1 = IC_2$) in weak or moderate inversion for high g_m/I_D and g_m, combined with operating output devices at long channel lengths ($L_4 = L_6$) for high V_A and r_{ds}. A_V, which tracks g_m/I_D of the input devices, is constant and maximum for input devices in weak inversion, decreases as the inverse square root of IC_1 in strong inversion, and decreases inversely with IC_1 when velocity saturation is significant. A_V, which tracks V_A for the output devices, increases sublinearly as L_4 and L_6 increase, unless DIBL effects are significant for short-channel devices where it increases more rapidly as these effects are reduced. Finally, A_V is independent of the OTA bias current, I_{BIAS} (again, MOSFET width is scaled with drain current to maintain the selected inversion coefficient).

Trends for A_V of the simple OTAs are similar to trends for device intrinsic voltage gain, A_{Vi}, summarized in Table 4.1 and shown in Figures 4.9 and 4.10. Here, *IC* trends are applied to the input devices and *L* trends are applied to the output devices. The trends, however, are only directly applicable if the same *IC* and *L* values are used for both the input and output devices, which is the case for the simple OTAs.

A_V is maximized for the cascoded OTAs by operating the input pair devices in weak or moderate inversion ($IC_1 = IC_2 \approx 0.8$) for high g_m/I_D and g_m, combined with operating M_1–M_{10} at long channel lengths for high V_A and r_{ds}. A_V, which tracks g_m/I_D of the input devices, is constant and maximum for input devices in weak inversion, decreases as the inverse square root of IC_1 in strong inversion, and decreases inversely with IC_1 when velocity saturation is significant. A_V, which tracks the square of V_A for M_1–M_{10} assuming channel lengths increase equally, increases somewhat below the square of channel length, unless DIBL effects are significant for short-channel devices where it increases more rapidly as these effects are reduced. Additionally, A_V tracks g_m/I_D of the M_8 and M_{10} cascode devices through their intrinsic voltage gain and is maximum when these devices are operated in weak or moderate inversion ($IC_{8,10} \approx 2.5$). Finally, A_V is independent of the OTA bias currents if constant bias current ratios are maintained (again, MOSFET width is scaled with drain current to maintain the selected inversion coefficient).

Trends for A_V of the cascoded OTAs are similar to trends for g_m/I_D of the M_1 and M_2 input devices multiplied by trends for g_m/I_D of the M_8 and M_{10} cascode devices, where g_m/I_D is summarized in

Table 4.1 and shown in Figures 4.7 and 4.8. More importantly, these trends are then multiplied by the square of trends for V_A of M_1–M_{10}, shown in the same table and figures. As a result of the product of g_m/I_D for both input and cascode devices, combined with the product of V_A for both cascode devices and devices connected to the sources of the cascode devices, a very large design range is possible for A_V. As described later in Section 5.5.2.1, measured A_V is 490 V/V for the AC optimized OTA, increasing to 19100 V/V for the DC optimized OTA. Because device inversion coefficients and g_m/I_D are unchanged between the OTAs, the increase in A_V is due to the increase in R_{OUT} associated with increasing channel lengths.

Sections 5.4.2.12 and 5.5.2.11 summarize voltage gain tradeoffs for the simple and cascoded OTAs.

5.3.4 Frequency Response

5.3.4.1 Simple OTAs

Transconductance frequency response for the simple OTAs shown in Figure 5.1 is found by analyzing the frequency response of the current mirrors that route input pair, drain output current to the OTA output. Current-mirror frequency response is approximated by finding the pole frequency associated with the capacitance and small-signal resistance appearing at the diode-connected input device. This is easily found if the gate–drain capacitance, C_{GD}, and its Miller multiplication are neglected for the output device. This yields a usable approximation if C_{GD} is small compared to other circuit capacitances and the output device gate-to-drain voltage gain is low, which minimizes the Miller multiplication of C_{GD}. C_{GD} is well below the gate–source, C_{GS}, and drain–body, C_{DB}, capacitances for the 0.5 μm CMOS process considered for the simple OTAs. However, its value may not be negligible in smaller-geometry processes where the gate–drain overlap capacitance, C_{GDO}, which dominates the value of C_{GD} in saturation, can be more significant. In addition to a small value of C_{GD}, the gate-to-drain voltage gain of the current-mirror output devices is low because these devices drive the low, $1/g_m$ small-signal resistance of another current-mirror input or the short-circuit OTA load considered for calculating transconductance frequency response. C_{GD} for output devices will then be neglected in the frequency response.

Collecting capacitance and small-signal resistance at the diode-connected input device for the pMOS M_5 and M_6 current mirror gives a pole frequency of

$$f_{pPMOS} = \frac{g_{m5}}{2\pi\left(C_{DB1} + C_{DB5} + C_{GS5} + C_{GB5} + C_{GS6} + C_{GB6}\right)} \tag{5.10}$$

g_{m5} is the transconductance of the M_5 mirror input device, C_{DB1} is the drain–body capacitance of M_1 connected to M_5, and C_{DB5} is the drain–body capacitance of M_5. $C_{GS5} = C_{GS6}$ and $C_{GB5} = C_{GB6}$ are the gate–source and gate–body capacitances of M_5 and the output device M_6. The gate–source capacitance, $C_{GS} = C_{gsi} + C_{GSO}$, includes the intrinsic gate–source capacitance, C_{gsi}, and extrinsic gate–source overlap capacitance, C_{GSO}. The gate–body capacitance, $C_{GB} = C_{gbi} + C_{GBO}$, includes the intrinsic gate–body capacitance, C_{gbi}, and extrinsic gate–body overlap capacitance, C_{GBO}.

If C_{DB1} is assumed equal to C_{DB5}, f_{pPMOS} can be approximated as a function of the current mirror devices only, giving

$$\begin{aligned} f_{pPMOS} &\approx \frac{g_{m5}}{2\pi\left(2\,C_{DB5} + C_{GS5} + C_{GB5} + C_{GS6} + C_{GB6}\right)} \\ &\approx \frac{g_{m5}}{2\pi\left(2C_{DB5} + 2C_{GS5} + 2C_{GB5}\right)} \\ &\approx \frac{1}{2}\frac{g_{m5}}{2\pi\left(C_{DB5} + C_{GS5} + C_{GB5}\right)} = \tfrac{1}{2}f_{diode5} = \tfrac{1}{2}f_{diode6} \end{aligned} \tag{5.11}$$

This assumes a 1:1 current mirror and requires modification for current mirrors that are not 1:1 because of non-equal device capacitances. For example, a 1:2 current mirror would have nearly three units of capacitance giving an input pole frequency of approximately $\frac{1}{3}f_{diode}$. Because C_{DB1} is actually smaller than C_{DB5}, Equation 5.11 conservatively underpredicts f_{pPMOS} slightly. C_{DB1} is smaller than C_{DB5} because the channel width of nMOS device M_1 is three times smaller than the channel width of pMOS device M_5, where both devices operate at equal inversion coefficients, channel lengths, and drain currents. Channel width given by $W = (L/IC) \cdot (I_D/I_0)$ from Table 3.10 is three times smaller for nMOS devices having a technology current of $I_0 = 2n_0\mu_0 C'_{OX}U_T^2 = 0.21\,\mu A$ at $T = 300\,K$ (room temperature) compared to pMOS devices having $I_0 = 0.07\,\mu A$. This is a result of the higher mobility for nMOS devices as listed in the process parameters of Table 3.2.

f_{pPMOS} is approximately one-half f_{diode5}, where f_{diode} is the $-3\,dB$ frequency or pole frequency for a diode-connected MOSFET having small-signal resistance of $1/g_m$ (r_{ds} is generally negligible compared to $1/g_m$) shunted by its C_{GS}, C_{GB}, and C_{DB} capacitances. f_{diode}, as summarized in Table 3.27, is a lower and more conservative estimate of bandwidth compared to intrinsic bandwidth, f_{Ti}, because it includes the C_{GSO} and C_{GBO} extrinsic gate-overlap capacitances and C_{DB} extrinsic drain–body capacitance in addition to the C_{gsi} and C_{gbi} intrinsic gate capacitances included in f_{Ti}. As shown in Figure 3.59, f_{diode} can be less than one-half f_{Ti}, especially for short-channel devices having small gate area and small intrinsic capacitances.

f_{pPMOS}, which was derived for the M_5 and M_6 pMOS current mirror, is also the pole frequency associated with the identically sized and biased M_7 and M_8 pMOS current mirror. Because the two pMOS current mirrors appear in both sides of the OTA signal path, they collectively introduce a single pole having frequency f_{pPMOS}.

Collecting capacitance and small-signal resistance at the diode-connected input device for the nMOS M_3 and M_4 current mirror gives a pole frequency of

$$f_{pNMOS} = \frac{g_{m3}}{2\pi\left(C_{DB8} + C_{DB3} + C_{GS3} + C_{GB3} + C_{GS4} + C_{GB4}\right)} \tag{5.12}$$

g_{m3} is the transconductance of the M_3 mirror input device, C_{DB8} is the drain–body capacitance of M_8 connected to M_3, and C_{DB3} is the drain–body capacitance of M_3. $C_{GS3} = C_{GS4}$ and $C_{GB3} = C_{GB4}$ are the gate–source and gate–body capacitances of M_3 and the output device M_4.

If C_{DB8} is assumed equal to C_{DB3}, f_{pNMOS} can be approximated as a function of the current-mirror devices only, giving

$$f_{pNMOS} \approx \frac{1}{2}\frac{g_{m3}}{2\pi\left(C_{DB3} + C_{GS3} + C_{GB3}\right)} = \frac{1}{2}f_{diode3} = \frac{1}{2}f_{diode4} \tag{5.13}$$

The approximate pole frequency for the nMOS current mirror, like the approximate pole frequency for the pMOS current mirrors, is one-half f_{diode} of the individual current-mirror devices. This, again, applies to 1:1 current mirrors. Because C_{DB8} is actually larger than C_{DB3}, Equation 5.13 overpredicts f_{pNMOS} slightly. However, the accuracy of f_{pNMOS} is not critical since it will be shown that f_{pPMOS} dominates the overall OTA frequency response. C_{DB8} is larger than C_{DB3} because the channel width of pMOS device M_8 is three times larger than the channel width of nMOS device M_3, where both devices operate at equal inversion coefficients, channel lengths, and drain currents. This is a result of the factor-of-three lower technology current and mobility for pMOS devices mentioned earlier.

Finally, a zero is introduced at twice the pole frequency for the nMOS current mirror because one-half the OTA output signal is fed around the M_3 and M_4 nMOS current mirror by the M_5 and M_6 pMOS current mirror. The frequency associated with this zero is given by

$$f_z = 2f_{pNMOS} \tag{5.14}$$

As mentioned, equal inversion coefficients and channel lengths are used for both nMOS and pMOS current-mirror devices in the simple OTAs. As a result, the OTA transconductance, $-3\,dB$ frequency is approximately equal to f_{pPMOS} set by the pMOS current mirrors. This is because f_{pNMOS} set by the nMOS current mirror is approximately three times higher than f_{pPMOS} due to smaller areas, lower capacitances, and higher bandwidths for the nMOS current-mirror devices. The frequency associated with the zero, f_z, is then nearly six times higher than f_{pPMOS} since it is a factor of two higher than f_{pNMOS}. Both f_{pNMOS} and f_z will be neglected in predictions of the transconductance, $-3\,\mathrm{dB}$ frequency.

The OTA transconductance, $-3\,\mathrm{dB}$ frequency is then approximately $f_{pPMOS} \approx \frac{1}{2}f_{diode5}$ associated with the pMOS current mirrors. Expressing this as less than $\frac{1}{2}f_{Ti5}$ because of the extrinsic gate–source and gate–body overlap capacitances and drain–body capacitance not considered in f_{Ti5} gives

$$f_{-3dB} \approx f_{pPMOS} \approx \frac{1}{2}f_{diode5} < \frac{1}{2}f_{Ti5}$$

$$< \frac{1}{2}\left(\frac{IC_5}{\sqrt{IC_5 + 0.25} + 0.5}\right)\left(\frac{\mu_0 U_T}{\pi\left(\hat{C}_{gsi5} + \hat{C}_{gbi5}\right)L_5^2}\right) \tag{5.15}$$

where IC_5 and L_5 are the inversion coefficients and channel lengths for the pMOS current mirrors. This uses the f_{Ti} expression summarized in Table 3.26, which includes the normalized, intrinsic, gate–source, $\hat{C}_{gsi} = C_{gsi}/C_{GOX}$, and gate–body, $\hat{C}_{gbi} = C_{gbi}/C_{GOX}$, capacitances given in Table 3.23, where $C_{GOX} = WLC'_{OX}$ is the gate-oxide capacitance. $\hat{C}_{gsi}$ has values of nearly 0, $\frac{1}{3}$, and $\frac{2}{3}$, and $\hat{C}_{gbi}$ has values of nearly $(n-1)/n$, $\frac{2}{3}(n-1)/n$, and $\frac{1}{3}(n-1)/n$ in weak, moderate (taken at $IC = 1$), and strong inversion, respectively. The appearance of unmodified, low-field mobility, μ_0, in Equation 5.15 indicates that the equation excludes velocity saturation and VFMR decreases in g_m/I_D and g_m that can reduce the $-3\,\mathrm{dB}$ frequency. As mentioned in Section 5.3.1.1, these effects are included by replacing IC with $IC(1 + IC/IC_{CRIT})$ in the $\sqrt{IC + 0.25}$ term related to g_m/I_D.

Equation 5.15 provides an approximation for the OTA transconductance, $-3\,\mathrm{dB}$ frequency linked to f_{Ti} of the pMOS current-mirror devices. Although this is useful for design guidance, the improved accuracy of Equation 5.10 for f_{pPMOS} is used for the $-3\,\mathrm{dB}$ frequency prediction provided by the *Analog CMOS Design, Tradeoffs and Optimization* spreadsheet.

5.3.4.2 Cascoded OTAs

Transconductance frequency response for the cascoded OTAs shown in Figure 5.2 is found by analyzing the frequency response of devices that route input pair, drain output current to the OTA output. Unless there is significant interaction across circuit nodes, like Miller multiplication of gate–drain capacitance across a high inverting voltage gain between the gate and drain, the pole frequency associated with a circuit node can be approximated by collecting the small-signal resistances and capacitances at the node. This was done for finding the pole frequencies associated with current-mirror input devices for the simple OTAs. The pole frequency is approximated by $f_p \approx 1/(2\pi r_p C_p) = g_p/(2\pi C_p)$, where r_p, g_p, and C_p are the total small-signal resistance, conductance, and capacitance at the circuit node, respectively. Although a rough approximation, this estimation can provide useful design guidance.

The lowest-frequency pole for the cascoded OTAs appears at the sources of M_7 and M_8, which are pMOS cascode devices. A pole having approximately twice this frequency appears at the M_9 drain input of the nMOS, low-voltage, cascode current mirror. Also, a pole having approximately 10 times the frequency of the pole at the sources of M_7 and M_8 appears at the source of the M_{10} cascode device in the nMOS current mirror. Finally, a zero at twice the frequency of the pole at the M_9 drain input of the nMOS current mirror appears because one-half of the OTA output signal comes directly from M_8, bypassing this current mirror that provides the remaining one-half of the output signal. The frequency of this zero is then approximately four times the frequency of the pole at the sources of M_7 and M_8. For evaluating tradeoffs and providing design guidance, only the lowest-frequency pole at the sources of M_7 and M_8 is considered here. The OTA transconductance, $-3\,\mathrm{dB}$ bandwidth will be approximated as this pole frequency.

Collecting capacitance and small-signal resistance at the sources of M_7 and M_8 gives an approximate pole frequency and OTA transconductance, $-3\,\mathrm{dB}$ bandwidth of

$$f_{-3\mathrm{dB}} \approx f_{p,M7,M8} \approx \frac{g_{m7}}{2\pi\,(C_{GS7} + C_{SB7} + C_{DB1} + C_{GD1} + C_{DB3} + C_{GD3})} \tag{5.16}$$

g_{m7} is the transconductance for M_7 and M_8, C_{GS7} and C_{SB7} are the gate–source and source–body capacitances for M_7 and M_8, where $C_{GS} = C_{gsi} + C_{GSO}$ includes both the intrinsic gate–source capacitance and extrinsic gate–source overlap capacitance. C_{DB1} and C_{GD1} are the drain–body and gate–drain capacitances for the M_1 and M_2 input pair devices, where $C_{GD} = C_{GDO}$ is equal to the extrinsic gate–drain overlap capacitance for operation in saturation. C_{DB3} and C_{GD3} are the drain–body and gate–drain capacitances for current sources M_3 and M_4. The small-signal resistance at the sources of M_7 and M_8, used for finding G_M in Section 5.3.1.2, is nearly equal to $1/g_{ms7}$. The additional loading of drain–source resistances from M_1 and M_2, and M_3 and M_4 is neglected because these resistances are well above $1/g_{ms7}$. The small-signal resistance at the sources of M_7 and M_8 is taken at $1/g_{ms7}$, which is about 30 % higher than $1/g_{ms7}$ that includes the effects of g_{mb7}. This gives a lower, more conservative bandwidth estimate, which is intended since the pole at the M_9 drain input of the nMOS, low-voltage, cascode current mirror is neglected along with other poles and parasitic layout capacitances.

For OTA designs having differential outputs with common-mode feedback, the nMOS, low-voltage, cascode current mirror is replaced with low-voltage cascode current sources. This eliminates the pole associated with cascode current mirror and improves the transconductance frequency response.

In the previous section, transconductance bandwidth for the simple OTAs was conveniently linked to the diode-connected bandwidth of pMOS current-mirror devices, which permitted simple design guidance. If C_{DB1} and C_{GD1} from M_1 and M_2, and C_{DB3} and C_{GD3} from M_3 and M_4 are not significant compared to C_{GS7} and C_{SB7} from M_7 and M_8, it is possible to conveniently link the OTA bandwidth given by Equation 5.16 to the diode-connected bandwidth of the M_7 and M_8 devices. However, this is not possible because these capacitances are significant, especially for short-channel devices where extrinsic gate-overlap and drain–body capacitances are significant compared to the intrinsic gate capacitances. As a result, Equation 5.16 must be used for estimating OTA bandwidth.

5.3.4.3 Optimization

Transconductance bandwidth for the simple OTAs is maximized, like f_{diode5} and f_{Ti5}, by operating the pMOS mirrors in strong inversion at short channel lengths. Additionally, it is necessary to ensure that bandwidth remains dominated by the pMOS mirrors, which is ensured since all devices operate at equal inversion coefficients and channel lengths. The bandwidth, like f_{diode5} and f_{Ti5}, increases nearly directly with IC_5 ($IC_5 = IC_6 = IC_7 = IC_8$ for the pMOS mirrors) in weak inversion, increases as the square root of IC_5 in strong inversion, and levels off when velocity saturation effects are significant. The bandwidth decreases as $1/L_5^2$ ($L_5 = L_6 = L_7 = L_8$ for the pMOS mirrors) or as $1/L_5$ when velocity saturation effects are significant. These trends can be observed in Equation 5.15 that relates transconductance bandwidth to the intrinsic MOS bandwidth of the pMOS mirror devices, f_{Ti5}. These trends are approximate because extrinsic gate-overlap and drain–body capacitances decrease bandwidth, especially for short-channel devices having small gate areas and small intrinsic gate capacitances. Finally, like the voltage gain, bandwidth is independent of the bias current (again, MOSFET width is scaled with drain current to maintain the selected inversion coefficient). However, parasitic layout capacitances may reduce bandwidth more for operation at lower drain currents where device channel widths, gate areas, and capacitances are smaller.

Transconductance bandwidth for the cascoded OTAs is maximized by operating relevant devices in strong inversion at short channel lengths to minimize device capacitances. Additionally, it is necessary to ensure that bandwidth remains dominated by M_7 and M_8, and devices loading M_7 and M_8, which is ensured since drain currents and inversion coefficients remain fixed across the OTAs while only channel lengths are varied. Interestingly, as summarized in Section 5.5.2.11, bandwidth is nearly inversely proportional to channel length. This suggests width-dependent, extrinsic gate-overlap, drain–body, source–body, and layout capacitances dominate the capacitance, where width tracks channel length for fixed values of drain current and the inversion coefficient. Equation 5.16, in fact, shows that bandwidth is dominated by extrinsic capacitances because only the C_{GS7} term includes intrinsic gate capacitance. When gate-area-dependent, intrinsic gate capacitances dominate the capacitance, bandwidth depends inversely on the square of channel length, where gate area tracks the square of channel length. Both width-dependent ($1/L$) and gate-area-dependent ($1/L^2$) bandwidth trends assume negligible velocity saturation effects, where transconductance is independent of the channel length for a selected drain current and inversion coefficient. Finally, as for the simple OTAs, bandwidth is independent of the drain current, although parasitic layout capacitances may reduce bandwidth more for operation at lower drain currents.

Trends for transconductance bandwidth of the simple OTAs are similar to trends for device f_{Ti} summarized in Table 4.1 and shown in Figures 4.9 and 4.10. Here, device trends are applied to the pMOS current-mirror devices and, in fact, all devices to ensure the pMOS current-mirror devices dominate the bandwidth. Although bandwidth increases significantly at short channel lengths, the bandwidth does not increase as rapidly as expected due to the presence of increasingly significant extrinsic gate-overlap and drain–body capacitances. This can be seen in Figure 3.59 that compares device f_{diode}, which includes extrinsic and intrinsic capacitances, to device f_{Ti}, which includes only intrinsic capacitances. f_{diode} becomes a progressively smaller fraction of device f_{Ti} at short channel lengths where gate area and intrinsic gate capacitances are small.

Sections 5.4.2.12 and 5.5.2.11 summarize transconductance bandwidth tradeoffs for the simple and cascoded OTAs.

5.3.5 Thermal Noise

5.3.5.1 Simple OTAs

For the simple OTAs shown in Figure 5.1, thermal noise, like flicker noise and DC offset caused by local-area mismatch, is dominated by devices in the signal path, M_1–M_8. Noise from the input pair current source, M_9, and its external reference device is largely common-moded out or cancelled. This is because noise current from M_9 splits nearly equally into the M_1 and M_2 input pair devices and is later cancelled at the OTA output.

The OTA input-referred thermal-noise voltage, power spectral density (PSD) is given by

$$S_{VIN}(\text{thermal}) = 2 \cdot 4kT \left[\frac{(n\,\Gamma)_1}{g_{m1}} + \frac{(n\,\Gamma)_3\,g_{m3}}{g_{m1}^2} + \frac{(n\,\Gamma)_5\,g_{m5}}{g_{m1}^2} + \frac{(n\,\Gamma)_7\,g_{m7}}{g_{m1}^2} \right] \qquad (5.17)$$

g_{m1}, g_{m3}, g_{m5}, and g_{m7} are transconductances for the M_1 and M_2 input differential pair, and non-input M_3 and M_4, M_5 and M_6, and M_7 and M_8 current-mirror devices, respectively. $(n\Gamma)_1$, $(n\Gamma)_3$, $(n\Gamma)_5$, and $(n\Gamma)_7$ are the products of the substrate factor, n, and thermal-noise factor, Γ (Equation 3.106), for these devices. The "2" appears in the noise expression because pair devices are used throughout the signal path, doubling the noise PSD associated with single devices.

The first term in Equation 5.17 corresponds to the gate-referred thermal-noise voltage PSD of the input differential-pair devices from Table 3.30. The second, third, and fourth terms correspond to the drain-referred thermal-noise current PSD of the non-input, current-mirror devices from Table 3.29, with each term divided by the square of the input pair transconductance, g_{m1}^2. This refers the drain noise current PSD contributions to an OTA input-referred noise voltage PSD.

Factoring Equation 5.17 to show the noise PSD contributions of non-input devices relative to the noise contributions of input devices gives

$$S_{VIN}(\text{thermal}) = 2 \cdot \frac{4kT (n\Gamma)_1}{g_{m1}} \left[1 + \frac{(n\Gamma)_3}{(n\Gamma)_1} \frac{g_{m3}}{g_{m1}} + \frac{(n\Gamma)_5}{(n\Gamma)_1} \frac{g_{m5}}{g_{m1}} + \frac{(n\Gamma)_7}{(n\Gamma)_1} \frac{g_{m7}}{g_{m1}} \right] \qquad (5.18)$$

Substituting $g_m = I_D / \left[nU_T \left(\sqrt{IC + 0.25} + 0.5 \right) \right]$ from Table 3.17 into Equation 5.18 allows the noise to be expressed in terms of the inversion coefficient and drain current, giving

$$S_{VIN}(\text{thermal}) = 2 \cdot 4kT \left(n_1^2 \Gamma_1 \right) \left[\frac{U_T \left(\sqrt{IC_1 + 0.25} + 0.5 \right)}{I_{D1}} \right] \cdot$$

$$\left[1 + \left(\frac{\Gamma_3}{\Gamma_1} \right) \frac{\sqrt{IC_1 + 0.25} + 0.5}{\sqrt{IC_3 + 0.25} + 0.5} \left(\frac{I_{D3}}{I_{D1}} \right) + \left(\frac{\Gamma_5}{\Gamma_1} \right) \frac{\sqrt{IC_1 + 0.25} + 0.5}{\sqrt{IC_5 + 0.25} + 0.5} \left(\frac{I_{D5}}{I_{D1}} \right) \right.$$

$$\left. + \left(\frac{\Gamma_7}{\Gamma_1} \right) \frac{\sqrt{IC_1 + 0.25} + 0.5}{\sqrt{IC_7 + 0.25} + 0.5} \left(\frac{I_{D7}}{I_{D1}} \right) \right] \qquad (5.19)$$

IC_1 is the inversion coefficient for the nMOS, M_1 and M_2 differential-pair devices, IC_3 is the inversion coefficient for the nMOS, M_3 and M_4 mirror devices, and IC_5 and IC_7 are the inversion coefficients for the pMOS, M_5 and M_6, and M_7 and M_8, mirror devices, respectively. I_{D1}, I_{D3}, I_{D5}, and I_{D7} are the drain currents for these devices. Although inversion coefficients are equal for a given version of the simple OTAs and drain currents are equal at I_{BIAS}, these are expressed separately to show individual device noise contributions and provide a general analysis. Additionally, Γ values are nearly equal because of device operation at equal inversion coefficients, but these are expressed separately to provide a general analysis. Since Γ varies from $1/2$ in weak inversion to $2/3$ in strong inversion and may increase due to small-geometry noise increases described in Section 3.10.2.2, it can vary between input and output devices and affect the prediction of noise.

Equation 5.19 excludes g_m / I_D and g_m decreases associated with velocity saturation and VFMR effects. As described in Section 5.3.1.1, these effects are included by replacing IC with $IC(1 + IC/IC_{CRIT})$ in the $\sqrt{IC + 0.25}$ terms related to g_m / I_D.

In Equations 5.18 and 5.19, the unity terms correspond to the thermal-noise PSD contribution of the input differential pair. The second, third, and fourth terms correspond to the thermal-noise PSD contributions of the non-input, current-mirror pairs relative to the input differential pair. These terms will be referred to as relative noise powers, $RNPs$, where $RNP_1 = 1$, RNP_3, RNP_5, and RNP_7 are the relative noise powers for the M_1 and M_2 input differential pair, and non-input M_3 and M_4, M_5 and M_6, and M_7 and M_8 current-mirror devices, respectively. If RNP_3, RNP_5, and RNP_7 are sufficiently below unity, the input differential pair dominates the thermal noise. These terms are reduced below unity by reducing the transconductance of non-input devices below that of input devices.

5.3.5.2 Cascoded OTAs

The analysis of thermal noise for the cascoded OTAs shown in Figure 5.2 is similar to that described above for the simple OTAs. Unlike devices in the simple OTAs, devices in the cascoded OTAs operate at different drain currents and inversion coefficients so input devices dominate the thermal noise.

OTA thermal noise, flicker noise, and DC offset caused by local-area mismatch are dominated by the M_1 and M_2 input pair devices, the M_3 and M_4 current source devices, and the M_5 and M_6 rail devices in the low-voltage, cascode current mirror. While also in the signal path, the M_7–M_{10} cascode devices contribute negligible noise and mismatch because their sources are connected to the high drain–source resistances of other devices operating in saturation. This results in significant source degeneration for the cascode devices where their effective transconductance is equal to their device transconductance divided by the source degeneration factor, $1 + g_{ms} \cdot r_S$, as summarized earlier in Table 3.16. Here,

$g_{ms} = g_m + g_{mb} + g_{ds}$ is the cascode device, total source transconductance, and r_S is the external source resistance consisting of the drain–source resistances of devices connected to the source. Since the source degeneration factor is much greater than unity, the effective transconductance of the M_7–M_{10} cascode devices is low and their thermal-noise, flicker-noise, and mismatch currents are negligibly small.

The remaining devices in the circuit create noise that is largely common-moded out or cancelled. This includes noise from the M_{13} reference and M_{15} current source feeding the M_1 and M_2 input pair devices, and noise from the M_{11} reference for the M_3 and M_4 current sources. Additionally, this includes noise from the M_{12}, M_{16}, and M_{17} cascode reference devices that is already small because of resistive source degeneration present for these devices. Finally, this includes noise from the V_{INCM}, V_{BIAS1}, and V_{BIAS2} bias voltages that provide gate bias voltages for cascode devices in the circuit. Noise introduced by these devices or bias voltages results in noise currents that are later cancelled at the OTA output.

Thermal noise for the cascoded OTAs is identical to that given in Equations 5.17–5.19 for the simple OTAs, except the terms related to M_7 and M_8 are removed. For convenience, these equations are repeated with these terms removed.

The input-referred thermal-noise voltage PSD for the cascoded OTAs is given by

$$S_{VIN}(\text{thermal}) = 2 \cdot 4kT \left[\frac{(n\,\Gamma)_1}{g_{m1}} + \frac{(n\,\Gamma)_3\, g_{m3}}{g_{m1}^2} + \frac{(n\,\Gamma)_5\, g_{m5}}{g_{m1}^2} \right] \tag{5.20}$$

After factoring to show the noise PSD contributions of non-input devices relative to the noise contributions of input devices, this gives

$$S_{VIN}(\text{thermal}) = 2 \cdot \frac{4kT\,(n\,\Gamma)_1}{g_{m1}} \left[1 + \frac{(n\,\Gamma)_3}{(n\,\Gamma)_1} \frac{g_{m3}}{g_{m1}} + \frac{(n\,\Gamma)_5}{(n\,\Gamma)_1} \frac{g_{m5}}{g_{m1}} \right] \tag{5.21}$$

Finally, after expressing transconductance in terms of the inversion coefficient and drain current, this gives

$$S_{VIN}(\text{thermal}) = 2 \cdot 4kT \left(n_1^2\,\Gamma_1 \right) \left[\frac{U_T \left(\sqrt{IC_1 + 0.25} + 0.5 \right)}{I_{D1}} \right] \cdot$$

$$\left[1 + \left(\frac{\Gamma_3}{\Gamma_1} \right) \frac{\sqrt{IC_1 + 0.25} + 0.5}{\sqrt{IC_3 + 0.25} + 0.5} \left(\frac{I_{D3}}{I_{D1}} \right) + \left(\frac{\Gamma_5}{\Gamma_1} \right) \frac{\sqrt{IC_1 + 0.25} + 0.5}{\sqrt{IC_5 + 0.25} + 0.5} \left(\frac{I_{D5}}{I_{D1}} \right) \right] \tag{5.22}$$

g_{m1}, g_{m3}, and g_{m5} are the transconductances for the M_1 and M_2 input differential pair, and non-input M_3 and M_4 current source, and M_5 and M_6 current-mirror devices, respectively. $(n\Gamma)_1$, $(n\Gamma)_3$, and $(n\Gamma)_5$ are the products of the substrate factor and thermal-noise factor for these devices. Finally, IC_1, IC_3, and IC_5 are the inversion coefficients, and I_{D1}, I_{D3}, and I_{D5} are the drain currents for these devices.

Velocity saturation and VFMR decreases in transconductance are included in Equation 5.22 as described for Equation 5.19. Some velocity saturation effects are present in the AC optimized OTA for the M_5 and M_6 current-mirror devices because these nMOS devices operate at $IC_5 = IC_6 \approx 16$ well into strong inversion at minimum channel lengths of $L_5 = L_6 = 0.18\,\mu\text{m}$.

In Equations 5.21 and 5.22, the unity terms correspond to the thermal-noise PSD contribution of the input differential pair. The second and third terms correspond to the thermal-noise PSD contributions of the non-input devices relative to the input differential pair. As for the simple OTAs, these terms will be referred to as relative noise powers, where $RNP_1 = 1$, RNP_3, and RNP_5 are the relative noise powers for the M_1 and M_2 input differential pair, and non-input, M_3 and M_4 current source, and M_5 and M_6 current-mirror devices, respectively. If RNP_3 and RNP_5 are sufficiently below unity, the input differential pair dominates the thermal noise. These terms are reduced below unity by reducing the transconductance of non-input devices below that of input devices.

5.3.5.3 Optimization

Equations 5.18 and 5.19 for the simple OTAs and Equations 5.21 and 5.22 for the cascoded OTAs help the designer minimize the thermal noise of input devices and ensure these devices dominate overall OTA noise. For the simple OTAs, the non-unity terms in the bracketed portions of Equations 5.18 and 5.19 are the relative (compared to the input devices) thermal-noise power contributions of the non-input devices, RNP_3, RNP_5, and RNP_7. For the cascoded OTAs, the non-unity terms in the bracketed portions of Equations 5.21 and 5.22 are the relative thermal-noise power contributions of the non-input devices, RNP_3 and RNP_5. For both OTAs, the RNP values are nearly equal to the ratios of non-input to input device transconductances. As a result, reducing transconductances of non-input devices somewhat below that of the input devices ensures that non-input device noise contributions are small and input devices dominate the thermal noise.

For both OTAs, input-referred thermal-noise voltage is minimized by operating input devices at low inversion coefficients ($IC_1 = IC_2$) in weak or moderate inversion for high g_m/I_D and g_m, while operating non-input devices at high inversion coefficients ($IC_3 = IC_4$, $IC_5 = IC_6$, and, for the simple OTA only, $IC_7 = IC_8$) in strong inversion for low g_m/I_D and g_m. Non-input devices may also be operated at lower drain currents than input devices to further reduce their transconductance and noise contributions. Although drain current can be reduced for some non-input devices, some non-input devices must support the full drain current of the input devices. Increasing the drain current for input devices (again, MOSFET width is scaled with drain current to maintain the selected inversion coefficient) reduces the input-referred thermal-noise voltage by increasing the transconductance of the input devices. Increasing drain currents proportionally for all devices increases the transconductance equally for all devices, which holds the relative noise contributions of non-input devices constant.

In the simple OTAs, all devices operate at equal drain currents and inversion coefficients. As a result, all devices operate at equal transconductance and contribute equally to the thermal noise. This is shown in Equations 5.18 and 5.19 by the unity term, representing the noise power contribution of the input pair, combined with three terms, RNP_3, RNP_5, and RNP_7, also equal to unity, that represent the relative noise power contributions of the non-input pairs. RNP values will be shown later by the *Analog CMOS Design, Tradeoffs and Optimization* spreadsheet in Section 5.4.2.3.

In the cascoded OTAs, input devices operate at low inversion coefficients ($IC_1 = IC_2 \approx 0.8$) in moderate inversion for high g_m/I_D and g_m, while non-input devices operate at high inversion coefficients ($IC_3 = IC_4 \approx 16$ and $IC_5 = IC_6 \approx 16$) in strong inversion for low g_m/I_D and g_m. The drain currents of $I_{D3} = I_{D4} = 150\,\mu\text{A}$ for M_3 and M_4 are unfortunately greater than the drain currents of $I_{D1} = I_{D2} = 100\,\mu\text{A}$ for M_1 and M_2, as required for the folded-cascode circuit topology. This increases the noise contributions of M_3 and M_4. However, the drain currents of $I_{D5} = I_{D6} = 50\,\mu\text{A}$ for M_5 and M_6 are lower than those for M_1 and M_2, which decreases the noise contributions of M_5 and M_6. Because input devices operate in moderate inversion and non-input devices operate in strong inversion, input devices dominate the thermal noise. This is shown in Equations 5.21 and 5.22 by the unity term, representing the noise power contribution of the input pair, combined with two terms, RNP_3 and RNP_5, which represent the relative noise power contributions of the non-input devices. RNP_3 for the M_3 and M_4 devices is approximately 0.6, and RNP_5 for the M_5 and M_6 devices is approximately 0.2 as shown later by the spreadsheet in Section 5.5.2.3.

When input devices dominate the thermal noise, as they do for the cascoded OTAs, or when the relative noise contributions of non-input devices remain constant, as they do for both the simple and cascoded OTAs, the OTA input-referred thermal-noise voltage PSD tracks the gate-referred thermal-noise voltage PSD of the input devices. The OTA input-referred thermal-noise voltage PSD is then constant for input devices in weak inversion, increases as the square root of IC_1 in strong inversion, and increases as IC_1 when velocity saturation effects are significant. Small-geometry increases in the noise are possible for short-channel devices, which are modeled by increases in the value of the thermal-noise factor Γ as described in Section 3.10.2.2. OTA input-referred thermal-noise voltage density trends are the same as trends for device gate-referred thermal-noise voltage density (the square root of PSD) summarized in Table 4.2 and shown in Figures 4.12 and 4.13, where device trends are applied to the input devices. This again assumes constant relative noise contributions from non-input devices.

5.3.6 Flicker Noise

5.3.6.1 Simple OTAs

The input-referred flicker-noise voltage PSD for the simple OTAs shown in Figure 5.1 is given by

$$S_{VIN}(\text{flicker})$$

$$= 2\left[\frac{K'_{F1}}{(WL)_1 f^{AF1}} + \frac{K'_{F3}}{(WL)_3 f^{AF3}}\left(\frac{g_{m3}}{g_{m1}}\right)^2 + \frac{K'_{F5}}{(WL)_5 f^{AF5}}\left(\frac{g_{m5}}{g_{m1}}\right)^2 + \frac{K'_{F7}}{(WL)_7 f^{AF7}}\left(\frac{g_{m7}}{g_{m1}}\right)^2\right]$$

$$(5.23)$$

K'_{F1}, K'_{F3}, K'_{F5}, and K'_{F7} are flicker-noise factors in hand units of $(\text{nV})^2 \cdot \mu\text{m}^2$ for the M_1 and M_2 input differential pair, and non-input M_3 and M_4, M_5 and M_6, and M_7 and M_8 current-mirror devices, respectively. AF_1, AF_3, AF_5, and AF_7 are the flicker-noise slopes, and $(WL)_1$, $(WL)_3$, $(WL)_5$, and $(WL)_7$ are the gate areas in μm^2 for these devices. As in the preceding thermal noise analysis, "2" appears in the noise expression because pair devices are used throughout the signal path, doubling the noise PSD associated with single devices.

The first term in Equation 5.23 corresponds to the gate-referred flicker-noise voltage PSD of the input differential-pair devices from Tables 3.31 and 3.34. The second, third, and fourth terms correspond to the drain-referred flicker-noise current PSD of the non-input, current-mirror devices. This is the gate-referred flicker-noise voltage PSD of these devices multiplied by the square of their transconductances as shown in Table 3.34. Each non-input device term is divided by the square of the input pair transconductance, g^2_{m1}, which, as for the thermal noise, refers the drain noise current PSD contributions to an input-referred noise voltage PSD.

Increases in gate-referred flicker-noise voltage PSD with increasing inversion level or $V_{EFF} = V_{GS} - V_T$ are included in the values of K'_{F1}, K'_{F3}, K'_{F5}, and K'_{F7}. As described in Tables 3.31 and 3.34, and shown in Figure 3.64, flicker-noise increases are modeled by $K'_F = K'_{F0}(1 + V_{EFF}/V_{KF})^2$. K'_{F0} is the flicker-noise factor in weak inversion or the weak-inversion side of moderate inversion, and V_{KF} is a voltage describing the flicker-noise increase with V_{EFF}. At low levels of inversion where V_{EFF} is negative, it is clamped to zero in the K'_F expression. Although not significant for the simple, $0.5\,\mu\text{m}$ CMOS OTAs, flicker-noise increases are often significant in smaller-geometry processes, especially for pMOS devices. These increases are important for the cascoded, $0.18\,\mu\text{m}$ CMOS OTAs described in this chapter and for the micropower, low-noise, $0.35\,\mu\text{m}$ CMOS preamplifiers described in Chapter 6.

Factoring Equation 5.23 to show the noise PSD contributions of non-input devices relative to the contributions of input devices gives

$$S_{VIN}(\text{flicker}) = 2 \cdot \frac{K'_{F1}}{(WL)_1 f^{AF1}}\left[1 + \frac{K'_{F3}}{K'_{F1}}\frac{(WL)_1}{(WL)_3}\left(\frac{g_{m3}}{g_{m1}}\right)^2 f^{AF1-AF3}\right.$$

$$+ \frac{K'_{F5}}{K'_{F1}}\frac{(WL)_1}{(WL)_5}\left(\frac{g_{m5}}{g_{m1}}\right)^2 f^{AF1-AF5}$$

$$\left. + \frac{K'_{F7}}{K'_{F1}}\frac{(WL)_1}{(WL)_7}\left(\frac{g_{m7}}{g_{m1}}\right)^2 f^{AF1-AF7}\right]$$

$$(5.24)$$

Analogous to the thermal noise given by Equations 5.18 and 5.19, the unity term corresponds to the flicker-noise PSD contribution of the input differential pair. The second, third, and fourth terms correspond to the flicker-noise PSD contributions of the non-input, current-mirror pairs relative to the input differential pair.

Substituting $WL = (L^2/IC) \cdot (I_D/I_0)$ from Table 3.10 and $g_m = I_D/[nU_T(\sqrt{IC+0.25}+0.5)]$ from Table 3.17 into Equation 5.24 allows the noise to be expressed in terms of the inversion coefficient, channel length, and drain current, giving

$$S_{VIN}(\text{flicker}) = 2 \cdot \frac{K'_{F1}}{f^{AF1}} \left(\frac{IC_1}{L_1^2}\right)\left(\frac{I_{0,1}}{I_{D1}}\right) \cdot$$

$$\left\{ 1 + \frac{K'_{F3}}{K'_{F1}} \left[\left(\frac{L_1}{L_3}\right)^2 \left(\frac{IC_3}{IC_1}\right)\left(\frac{I_{0,3}}{I_{0,1}}\right)\left(\frac{I_{D1}}{I_{D3}}\right)\right]\left[\frac{n_1\left(\sqrt{IC_1+0.25}+0.5\right)}{n_3\left(\sqrt{IC_3+0.25}+0.5\right)}\left(\frac{I_{D3}}{I_{D1}}\right)\right]^2 \cdot f^{AF1-AF3} \right.$$

$$+ \frac{K'_{F5}}{K'_{F1}} \left[\left(\frac{L_1}{L_5}\right)^2 \left(\frac{IC_5}{IC_1}\right)\left(\frac{I_{0,5}}{I_{0,1}}\right)\left(\frac{I_{D1}}{I_{D5}}\right)\right]\left[\frac{n_1\left(\sqrt{IC_1+0.25}+0.5\right)}{n_5\left(\sqrt{IC_5+0.25}+0.5\right)}\left(\frac{I_{D5}}{I_{D1}}\right)\right]^2 \cdot f^{AF1-AF5}$$

$$\left. + \frac{K'_{F7}}{K'_{F1}} \left[\left(\frac{L_1}{L_7}\right)^2 \left(\frac{IC_7}{IC_1}\right)\left(\frac{I_{0,7}}{I_{0,1}}\right)\left(\frac{I_{D1}}{I_{D7}}\right)\right]\left[\frac{n_1\left(\sqrt{IC_1+0.25}+0.5\right)}{n_7\left(\sqrt{IC_7+0.25}+0.5\right)}\left(\frac{I_{D7}}{I_{D1}}\right)\right]^2 \cdot f^{AF1-AF7} \right\}$$

$$(5.25)$$

$I_{0,1}$, $I_{0,3}$, $I_{0,5}$, and $I_{0,7}$ are the technology currents, $I_0 = 2n_0\mu_0 C'_{OX}U_T^2$ from Table 3.6, and n_1, n_3, n_5, and n_7 are the substrate factors for the M_1 and M_2 input differential pair, and non-input M_3 and M_4, M_5 and M_6, and M_7 and M_8 current-mirror devices, respectively. Although inversion coefficients are equal for a given version of the simple OTAs and drain currents are equal at I_{BIAS}, these are expressed separately to show individual device noise contributions and provide a general analysis. Differences in AF values are considered because these can be important, especially when non-input devices contribute significantly to the noise. The ratio of n values is included in the transconductance ratio terms for completeness. Since n is a slight function of the inversion level (see Figure 3.30) and varies only slightly with process doping, the n values nearly cancel for most applications.

Equation 5.25 excludes g_m/I_D and g_m decreases associated with velocity saturation and VFMR effects. As mentioned in Section 5.3.1.1, these effects are included by replacing IC with $IC(1 + IC/IC_{CRIT})$ in the $\sqrt{IC+0.25}$ terms related to g_m/I_D. Velocity saturation can reduce the flicker-noise contributions of non-input devices by reducing the conversion of gate-referred flicker-noise voltage to drain-referred flicker-noise current. However, the increase in gate-referred flicker-noise voltage with increasing inversion or V_{EFF} can counter the reduction of drain-referred flicker-noise current caused by velocity saturation.

In Equations 5.24 and 5.25, the unity terms correspond to the flicker-noise PSD contribution of the input differential pair. The second, third, and fourth terms correspond to the flicker-noise PSD contributions of the non-input, current-mirror pairs relative to the input differential pair. These terms will, as for the thermal noise, be referred to as relative noise powers where $RNP_1 = 1$, RNP_3, RNP_5, and RNP_7 are the relative noise powers for the M_1 and M_2 input differential pair, and non-input M_3 and M_4, M_5 and M_6, and M_7 and M_8 current-mirror devices, respectively. If RNP_3, RNP_5, and RNP_7 are sufficiently below unity, the input differential pair dominates the flicker noise.

Each of the RNP_3, RNP_5, and RNP_7 terms is separated into a ratio of K'_F factors, a ratio of gate areas, a ratio of squared transconductances, and a ratio related to the frequency and differences in the flicker-noise slopes. Because the ratio of input to non-input device drain current appears in the gate-area ratio and the ratio of non-input to input device drain current appears squared in the transconductance ratio, this gives a net ratio of non-input to input device drain current. This same drain current ratio also appears in the thermal-noise contributions of Equation 5.19.

From Table 3.2 for $T = 300\,\mathrm{K}$, $I_{0,1} = I_{0,3} = 0.21\,\mu\mathrm{A}$ for nMOS devices, and $I_{0,5} = I_{0,7} = 0.07\,\mu\mathrm{A}$ for pMOS devices for the 0.5 μm CMOS process used for the simple OTAs. This gives a technology current ratio of $I_{0,3}/I_{0,1}$ equal to unity since the non-input M_3 and M_4 pair, and M_1 and M_2 input pair both use nMOS devices. The $I_{0,5}/I_{0,1}$ and $I_{07}/I_{0,1}$ ratios, however, are lower at one-third since the non-input M_5 and M_6, and M_7 and M_8 pairs use pMOS devices. This represents potentially lower flicker-noise contributions from pMOS devices due to their factor-of-three larger gate areas compared to nMOS devices for equal drain currents, inversion coefficients, and channel lengths. The pMOS flicker-noise contributions are lowered further since pMOS flicker-noise factors, listed in Table 3.3, are lower than those of nMOS devices in the 0.5 μm CMOS process. The technology current ratios mentioned can also be evaluated as ratios of low-field mobility, μ_0, or low-field transconductance factors, $k_0 = \mu_0 C'_{OX}$. These are also listed in the process information of Table 3.2.

5.3.6.2 Cascoded OTAs

The analysis of flicker noise for the cascoded OTAs shown in Figure 5.2 is similar to that described above for the simple OTAs. Unlike devices in the simple OTAs, devices in the cascoded OTAs operate at different drain currents and inversion coefficients. This affects their flicker-noise contributions.

Flicker noise for the cascoded OTAs is identical to that given in Equations 5.23–5.25 for the simple OTAs, except the terms related to M_7 and M_8 are removed. For convenience, these equations are repeated with these terms removed.

The input-referred flicker-noise voltage PSD for the cascoded OTAs is given by

$$S_{VIN}(\text{flicker}) = 2\left[\frac{K'_{F1}}{(WL)_1\,f^{AF1}} + \frac{K'_{F3}}{(WL)_3\,f^{AF3}}\left(\frac{g_{m3}}{g_{m1}}\right)^2 + \frac{K'_{F5}}{(WL)_5\,f^{AF5}}\left(\frac{g_{m5}}{g_{m1}}\right)^2\right] \qquad (5.26)$$

After factoring to show the noise PSD contributions of non-input devices relative to the noise contributions of input devices, this gives

$$S_{VIN}(\text{flicker}) = 2\cdot\frac{K'_{F1}}{(WL)_1\,f^{AF1}}\left[1 + \frac{K'_{F3}}{K'_{F1}}\frac{(WL)_1}{(WL)_3}\left(\frac{g_{m3}}{g_{m1}}\right)^2 f^{AF1-AF3} + \frac{K'_{F5}}{K'_{F1}}\frac{(WL)_1}{(WL)_5}\left(\frac{g_{m5}}{g_{m1}}\right)^2 f^{AF1-AF5}\right]$$

$$(5.27)$$

Finally, after expressing transconductance in terms of the inversion coefficient and drain current, and expressing gate area in terms of the inversion coefficient, channel length, and drain current, this gives

$$S_{VIN}(\text{flicker}) = 2\cdot\frac{K'_{F1}}{f^{AF1}}\left(\frac{IC_1}{L_1^2}\right)\left(\frac{I_{0,1}}{I_{D1}}\right)\cdot$$

$$\left\{1 + \frac{K'_{F3}}{K'_{F1}}\left[\left(\frac{L_1}{L_3}\right)^2\left(\frac{IC_3}{IC_1}\right)\left(\frac{I_{0,3}}{I_{0,1}}\right)\left(\frac{I_{D1}}{I_{D3}}\right)\right]\left[\frac{n_1\left(\sqrt{IC_1+0.25}+0.5\right)}{n_3\left(\sqrt{IC_3+0.25}+0.5\right)}\left(\frac{I_{D3}}{I_{D1}}\right)\right]^2\cdot f^{AF1-AF3}\right.$$

$$\left. + \frac{K'_{F5}}{K'_{F1}}\left[\left(\frac{L_1}{L_5}\right)^2\left(\frac{IC_5}{IC_1}\right)\left(\frac{I_{0,5}}{I_{0,1}}\right)\left(\frac{I_{D1}}{I_{D5}}\right)\right]\left[\frac{n_1\left(\sqrt{IC_1+0.25}+0.5\right)}{n_5\left(\sqrt{IC_5+0.25}+0.5\right)}\left(\frac{I_{D5}}{I_{D1}}\right)\right]^2\cdot f^{AF1-AF5}\right\}$$

$$(5.28)$$

K'_{F1}, K'_{F3}, and K'_{F5} are flicker-noise factors in hand units of $(\mathrm{nV})^2\cdot\mu\mathrm{m}^2$ for the M_1 and M_2 input differential pair, and non-input M_3 and M_4 current source, and M_5 and M_6 current-mirror devices, respectively. AF_1, AF_3, and AF_5 are the flicker-noise slopes, and $(WL)_1$, $(WL)_3$, and $(WL)_5$ are the gate areas in μm^2 for these devices. Finally, $I_{0,1}$, $I_{0,3}$, and $I_{0,5}$ are the technology currents, $I_0 = 2n_0\mu_0 C'_{OX} U_T^2$ from Table 3.6, and n_1, n_3, and n_5 are the substrate factors for these devices.

Velocity saturation and VFMR decreases in transconductance are included in Equation 5.28 as described for Equation 5.25. Additionally, increases in gate-referred flicker-noise voltage PSD with increasing inversion level or $V_{EFF} = V_{GS} - V_T$ are included in the values of K'_{F1}, K'_{F3}, and K'_{F5} as described for Equation 5.23. Flicker-noise increases are significant for pMOS devices in the $0.18\,\mu m$ CMOS process used for the cascoded OTAs as shown in Figures 3.65–3.66. In Section 5.5.2.4, flicker-noise increases will be shown to be significant for the pMOS, M_3 and M_4 current source devices that operate at $IC \approx 16$ well into strong inversion.

In Equations 5.27 and 5.28, the unity terms correspond to the flicker-noise PSD contribution of the input differential pair. The second and third terms correspond to the flicker-noise PSD contributions of the non-input devices relative to the input differential pair. These terms, as for the thermal noise, will be referred to as relative noise powers where $RNP_1 = 1$, RNP_3, and RNP_5 are the relative noise powers for the M_1 and M_2 input differential pair, and non-input M_3 and M_4 current source, and M_5 and M_6 current-mirror devices, respectively. If RNP_3 and RNP_5 are sufficiently below unity, the input differential pair dominates the flicker noise.

As described for the simple OTAs, each of the RNP_3 and RNP_5 terms is separated into a ratio of K'_F factors, a ratio of gate areas, a ratio of squared transconductances, and a ratio related to the frequency and differences in the flicker-noise slopes. Because the ratio of input to non-input device drain current appears in the gate-area ratio and the ratio of non-input to input device drain current appears squared in the transconductance ratio, this gives a net ratio of non-input to input device drain current. This same drain current ratio also appears in the thermal-noise contributions of Equation 5.22.

From Table 3.2 for $T = 300\,K$, $I_{0,1} = I_{0,5} = 0.64\,\mu A$ for nMOS devices, and $I_{0,3} = 0.135\,\mu A$ for pMOS devices for the $0.18\,\mu m$ CMOS process used in the cascoded OTAs. This gives a technology current ratio of $I_{0,5}/I_{0,1}$ equal to unity since the M_5 and M_6, non-input, current-mirror, and M_1 and M_2 input differential-pair devices are nMOS devices. The $I_{0,3}/I_{0,1}$ ratio, however, is lower at nearly one-fifth since the M_3 and M_4, non-input, current source devices are pMOS devices. This represents potentially lower flicker-noise contributions from pMOS devices due to their nearly factor-of-five larger gate areas compared to nMOS devices for equal drain currents, inversion coefficients, and channel lengths. However, for lower pMOS noise contributions, the flicker-noise factors, K'_F, must be comparable to those for nMOS devices, including the effects of increased noise with increasing inversion or V_{EFF}. As mentioned for the simple OTAs, technology current ratios can also be evaluated as ratios of low-field mobility, μ_0, or low-field transconductance factors, $k_0 = \mu_0 C'_{OX}$. These are also listed in the process information of Table 3.2.

5.3.6.3 Optimization

Equations 5.24 and 5.25 for the simple OTAs and Equations 5.27 and 5.28 for the cascoded OTAs help the designer minimize the flicker noise of input devices and ensure these devices dominate overall OTA noise. For the simple OTAs, the non-unity terms in the bracketed portions of Equations 5.24 and 5.25 are the relative (compared to the input devices) flicker-noise power contributions of the non-input devices, RNP_3, RNP_5, and RNP_7. For the cascoded OTAs, the non-unity terms in the bracketed portions of Equations 5.27 and 5.28 are the relative flicker-noise power contributions of the non-input devices, RNP_3 and RNP_5. For both OTAs, the RNP values are equal to the ratio of non-input to input device flicker-noise factors, multiplied by the ratio of input to non-input device gate areas, multiplied by the squared ratio of non-input to input device transconductances, finally multiplied by a factor that includes the frequency and differences in flicker-noise slopes. Minimizing non-input device transconductances has a squared effect on reducing the relative noise contributions of non-input devices, but ratios of flicker-noise factors and gate areas are also important.

For both OTAs, input-referred flicker-noise voltage is minimized by operating input devices at low inversion coefficients ($IC_1 = IC_2$) and long channel lengths ($L_1 = L_2$), which maximizes their gate area. Additionally, g_m/I_D and g_m are high at low inversion coefficients in weak or moderate inversion, helping input devices dominate the noise. Increasing the drain current for input devices (again, MOSFET width is scaled with drain current to maintain the selected inversion coefficient) reduces the input-referred flicker-noise voltage by increasing the gate area of the input devices. Increasing drain currents proportionally for all devices increases the transconductance and gate area equally for all devices, which holds the relative noise contributions of non-input devices constant.

If a preferred device type exists, input and non-input devices should be fabricated using this device type for applications where flicker noise is critical, although most designs require both nMOS and pMOS devices. pMOS devices have traditionally been favored for lower flicker noise, but, as mentioned in Section 3.10.3.5, this noise advantage may not be present in smaller-geometry processes. Additionally, as shown in Figures 3.64–3.66, pMOS devices in smaller-geometry processes may have significantly higher gate-referred flicker-noise voltage increases with increasing inversion or V_{EFF}.

In the simple OTAs, pMOS devices are preferred in the 0.5 μm CMOS process used as seen by comparing flicker noise-factors, K'_{F0}, in Table 3.3. Additionally, both pMOS and nMOS devices have small noise increases at increasing inversion or V_{EFF} such that K'_F remains nearly equal to its low inversion value of K'_{F0}.

In the cascoded OTAs, pMOS devices have slightly lower K'_{F0} in the 0.18 μm CMOS process used as seen by comparing flicker-noise factors in Table 3.3. However, the pMOS flicker-noise increases much more rapidly with increasing inversion or V_{EFF} and quickly exceeds that of nMOS devices in strong inversion. This can be seen in Figures 3.65 and 3.66 by comparing the measured gate-referred flicker-noise voltage of nMOS and pMOS devices having nearly equal gate area.

In addition to minimizing the flicker noise of input devices, minimizing overall OTA flicker noise requires minimizing the noise of non-input devices. The non-unity terms in Equations 5.25 and 5.28 for both OTAs indicate that non-input device flicker-noise contributions are minimized by operating these devices at low inversion coefficients in weak or moderate inversion. Here, even though g_m/I_D and g_m are maximum, maximizing the conversion of gate-referred flicker-noise voltage to drain-referred flicker-noise current, the large shape factors (W/L) and channel widths required result in large gate areas. This then lowers the gate-referred flicker-noise voltage and, correspondingly, the drain-referred flicker-noise current.

Although Equations 5.25 and 5.28 for both OTAs indicate that non-input devices should be operated at low inversion coefficients for large gate areas to minimize their flicker-noise contributions, Equations 5.19 and 5.22 indicate that non-input devices should be operated at high inversion coefficients in strong inversion for low g_m/I_D and g_m to minimize their thermal-noise contributions. This results in a design conflict for applications where both thermal and flicker noise are important. Fortunately, non-input device flicker-noise contributions do not necessarily continue increasing as these devices are operated at higher inversion coefficients ($IC_3 = IC_4$, $IC_5 = IC_6$, and, for the simple OTA only, $IC_7 = IC_8$) into strong inversion as required to minimize their thermal-noise contributions. This is observed in Equations 5.25 and 5.28 where the IC terms cancel in strong inversion ($IC > 10$) for the relative flicker-noise contributions of non-input devices. As IC increases in strong inversion, non-input device, drain-referred flicker-noise current levels off because the increasing gate-referred flicker-noise voltage associated with decreasing gate area is countered by decreasing g_m/I_D and g_m. However, the leveling off of drain-referred flicker-noise current occurs only if increases in gate-referred flicker-noise voltage with increasing inversion or V_{EFF} are small, which is indicated when K'_F remains near its low-inversion, K'_{F0} value.

The behavior of non-input device, drain-referred flicker-noise current is illustrated in Figures 4.18 and 3.68 for nMOS and pMOS devices in the 0.18 μm CMOS process used in the cascoded OTAs. The noise of both device types initially increases with increasing inversion coefficient in weak and moderate inversion because of the decrease in gate area. The noise of nMOS devices shown in Figure 4.18 then levels off in strong inversion as described above. However, the noise of pMOS devices shown in Figure 3.68 does not level off in strong inversion because of inversion-level or V_{EFF} increases in the gate-referred flicker-noise voltage. As a result, the pMOS noise can actually exceed the nMOS noise in strong inversion, even though the pMOS noise is initially lower at low levels of inversion due primarily to gate area that is nearly a factor of five larger. Although drain-referred flicker-noise current either levels off in strong inversion or increases with increasing inversion coefficient, the noise consistently decreases with increasing channel length because of the significant increase in gate area.

When flicker-noise increases are significant in strong inversion for non-input devices, the device relative noise contributions increase by IC instead of becoming constant as predicted by Equations 5.25 and 5.28 for constant values of $K'_F \approx K'_{F0}$. This is because $K'_F = K'_{F0}(1 + V_{EFF}/V_{KF})^2$ increases directly with IC for $V_{EFF} = 2nU_T\sqrt{IC}$ in strong inversion (assuming negligible velocity saturation) given in Table 3.14. These noise increases are included in the *Analog CMOS Design, Tradeoffs and Optimization* spreadsheet. Increases in flicker noise will be shown in Section 5.5.2.4 for non-input, pMOS, M_3 and M_4 current source devices in the cascoded OTAs that operate at $IC_3 = IC_4 \approx 16$ well into strong inversion. These noise increases will be shown in the spreadsheet through values of K'_F and through increases in the relative flicker-noise contributions of M_3 and M_4.

When input devices are operated at low inversion coefficients and non-input devices are operated at high inversion coefficients as required to minimize thermal noise, Equations 5.25 and 5.28 show that non-input device channel lengths ($L_3 = L_4$, $L_5 = L_6$, and, for the simple OTA only, $L_7 = L_8$) must be greater than input device channel lengths ($L_1 = L_2$) to achieve comparable device areas and ensure negligible non-input device flicker-noise contributions. Equations 5.24 and 5.27 also illustrate this where the product of flicker-noise factor and gate-area ratios must be in the vicinity of unity if non-input device flicker-noise contributions will be sufficiently reduced by the lower transconductance of these devices. Increasing non-input device channel lengths is required because operation at higher inversion coefficients requires smaller shape factors and smaller channel widths and gate areas for a given channel length. Increasing the channel length increases non-input device gate area as the square of channel length because channel width must increase directly with channel length to maintain the selected inversion coefficient. This significantly reduces the flicker-noise contributions, but at the expense of significantly reduced bandwidth (Tables 3.26 and 3.27) for devices in the signal path and significantly increased drain capacitive loading (Table 3.25) for devices providing bias to devices in the signal path.

The relative flicker-noise power contributions of non-input devices are proportional to the squared ratio of non-input to input device transconductances. As a result, the gate-area requirement for non-input devices is relaxed somewhat when non-input device transconductances are sufficiently below the transconductance of input devices as required to minimize thermal noise. However, minimizing flicker noise, unlike minimizing thermal noise, requires careful management of gate area for both input and non-input devices. This is even more important if inversion-level or V_{EFF} increases in flicker noise are present as in the cascoded, $0.18\,\mu m$ CMOS OTAs described in this chapter and in the micropower, low-noise, $0.35\,\mu m$ CMOS preamplifiers described in Chapter 6.

As for minimizing thermal noise, non-input devices may be operated at lower drain currents compared to input devices to reduce their transconductance and flicker-noise contributions. Although drain current can be reduced for some non-input devices, some non-input devices must support the full drain current of the input devices.

In the simple OTAs, all devices operate at equal drain currents, inversion coefficients, and channel lengths. As a result, all devices operate at equal transconductance and contribute equally to the thermal noise. Non-input nMOS devices contribute equal flicker noise as the nMOS input devices because these devices operate at equal transconductance and equal gate area. The larger required gate area and lower flicker-noise factor of non-input pMOS devices, however, significantly reduce their flicker-noise contributions. This is shown in Equations 5.24 and 5.25 by the unity term, representing the noise power contribution of the input pair, combined with three terms, RNP_3, RNP_5, and RNP_7, representing the relative noise power contributions of the non-input pairs. RNP_3 for the nMOS devices is unity, and RNP_5 and RNP_7 for the pMOS devices are much less than unity at 0.05 as shown later by the spreadsheet in Section 5.4.2.4.

In the cascoded OTAs, input devices operate at low inversion coefficients ($IC_1 = IC_2 \approx 0.8$) in moderate inversion for high g_m/I_D and g_m, while non-input devices operate at high inversion coefficients ($IC_3 = IC_4 \approx 16$ and $IC_5 = IC_6 \approx 16$) in strong inversion for low g_m/I_D and g_m. The drain currents of $I_{D3} = I_{D4} = 150\,\mu A$ for M_3 and M_4 are unfortunately greater than the drain currents of $I_{D1} = I_{D2} = 100\,\mu A$ for M_1 and M_2, as required for the folded-cascode circuit topology. This increases the flicker-noise contributions of M_3 and M_4. However, the drain currents of $I_{D5} = I_{D6} = 50\,\mu A$ for M_5 and M_6 are lower than those for M_1 and M_2, which decreases the noise contributions of M_5 and M_6. This is identical to the behavior described for the thermal noise.

Although input devices in the cascoded OTAs operate in moderate inversion and non-input devices operate in strong inversion such that input devices dominate the thermal noise, the input devices do not dominate the flicker noise. This is because the non-input, M_3–M_6 devices have equal channel lengths as the M_1 and M_2 input pair devices. This gives insufficient gate area for the non-input devices that operate at higher inversion coefficients, giving smaller shape factors and smaller channel widths and gate areas for a given channel length. This is shown in Equations 5.27 and 5.28 by the unity term, representing the noise power contribution of the input pair, combined with two terms, RNP_3 and RNP_5, representing the relative noise power contributions of the non-input devices. RNP_3 for the M_3 and M_4 devices and RNP_5 for the M_5 and M_6 devices are roughly equal to two as shown later by the spreadsheet in Section 5.5.2.4. As mentioned, increasing the channel length of non-input devices above that of the input devices reduces the flicker-noise power contributions of non-input devices, but significantly lowers the transconductance bandwidth.

In the micropower, low-noise, $0.35\,\mu m$ CMOS preamplifiers described in Chapter 6, input devices dominate the thermal noise as they do for the cascoded OTAs. However, unlike in the cascoded OTAs, longer channel lengths are used for non-input devices compared to input devices. This gives comparable gate area for input and non-input devices, giving nearly negligible non-input device flicker-noise contributions resulting from the combination of sufficient gate area and low transconductance.

When input devices dominate the flicker noise or when the relative noise contributions of non-input devices remain constant, as they do for both the simple and cascoded OTAs, the OTA input-referred flicker-noise voltage PSD tracks the gate-referred flicker-noise voltage PSD of the input devices. The OTA input-referred flicker-noise voltage PSD then increases as IC_1 and decreases as $1/L_1^2$, tracking the inverse of gate area for the input devices. When inversion-level or V_{EFF} increases in the noise are significant, the noise can increase further with IC as modeled by increases in K'_F. OTA input-referred flicker-noise voltage density trends are the same as trends for device gate-referred flicker-noise voltage density (the square root of PSD) summarized in Table 4.2 and shown in Figures 4.14 and 4.15, where device trends are applied to the input devices. This again assumes constant relative noise contributions from non-input devices.

Sections 5.4.2.12 and 5.5.2.11 summarize flicker-noise tradeoffs. Here, the input-referred flicker-noise voltage density is nearly inversely proportional to the square root of device gate areas for both the simple and cascoded OTAs.

5.3.7 Offset Voltage due to Local-Area Mismatch

5.3.7.1 Simple OTAs

Local-area mismatch in the zero-V_{SB} threshold voltage, ΔV_{TO}, and body-effect factor, $\Delta\gamma$, causes mismatch in the general, non-zero-V_{SB} threshold voltage, ΔV_T, between identically sized and biased devices. Additionally, mismatch in the transconductance factor, $\Delta K_P/K_P$, causes drain current mismatch between identically sized and biased devices having identical threshold voltages. As summarized in Table 3.41, threshold-voltage and transconductance factor mismatch together cause an overall drain current mismatch, ΔI_D, or an overall gate–source voltage mismatch, ΔV_{GS}, between two devices. Mismatch for multiple devices in a circuit then results in overall DC mismatch that, like thermal and flicker noise, can be referred to the input of the circuit.

Drain current mismatch for the M_1 and M_2 input differential pair, and non-input M_3 and M_4, M_5 and M_6, and M_7 and M_8 current-mirror pairs results in output offset current for the simple OTAs shown in Figure 5.1. Dividing the output offset current by the input pair transconductance refers the offset current to an input-referred offset voltage. The OTA mean square value or variance (square of the 1σ value) of the input-referred offset voltage is then given by

$$
\begin{aligned}
V_{INOFFSET}^2 \text{ (mismatch)} =\ & \frac{A_{VT1}^2}{(WL)_1} + \frac{A_{VT3}^2}{(WL)_3}\left(\frac{g_{m3}}{g_{m1}}\right)^2 + \frac{A_{VT5}^2}{(WL)_5}\left(\frac{g_{m5}}{g_{m1}}\right)^2 + \frac{A_{VT7}^2}{(WL)_7}\left(\frac{g_{m7}}{g_{m1}}\right)^2 \\
& + \frac{A_{KP1}^2}{(WL)_1}\left(\frac{I_{D1}}{g_{m1}}\right)^2 + \frac{A_{KP3}^2}{(WL)_3}\left(\frac{I_{D3}}{g_{m1}}\right)^2 + \frac{A_{KP5}^2}{(WL)_5}\left(\frac{I_{D5}}{g_{m1}}\right)^2 + \frac{A_{KP7}^2}{(WL)_7}\left(\frac{I_{D7}}{g_{m1}}\right)^2
\end{aligned}
$$

$$\tag{5.29}$$

A_{VT1}, A_{VT3}, A_{VT5}, and A_{VT7} are the threshold-voltage mismatch factors for the M_1 and M_2 input differential pair, and non-input M_3 and M_4, M_5 and M_6, and M_7 and M_8 current-mirror pair devices, respectively. A_{KP1}, A_{KP3}, A_{KP5}, and A_{KP7} are the transconductance factor mismatch factors, and I_{D1}, I_{D3}, I_{D5}, and I_{D7} are the drain currents for these devices.

The first row of terms in Equation 5.29 corresponds to the input-referred offset voltage due to device threshold-voltage mismatch, ΔV_T. The first term corresponds to the threshold-voltage mismatch for the M_1 and M_2 input differential pair, found by $\Delta V_{T1} = A_{VT1}/(WL)_1^{1/2}$ from Table 3.38. The second, third, and fourth terms correspond to drain current mismatch resulting from threshold-voltage mismatch for the non-input M_3 and M_4, M_5 and M_6, and M_7 and M_8 current-mirror pairs, respectively. This drain current mismatch is found, for example, for the M_3 and M_4 pair by $\Delta I_{D3} = \Delta V_{T3} \cdot g_{m3} = [A_{VT3}/(WL)_3^{1/2}] \cdot g_{m3}$, which corresponds to the threshold-voltage mismatch multiplied by the transconductance. The drain current mismatch of each non-input pair is then divided by the input pair transconductance, g_{m1}, to refer the current mismatch to an input-referred voltage mismatch.

The second row of terms in Equation 5.29 corresponds to the input-referred offset voltage due to device transconductance factor mismatch, $\Delta K_P/K_P$. This causes drain current mismatch that is found, for example, for the M_3 and M_4 pair by $\Delta I_{D3} = (\Delta K_{P3}/K_{P3}) \cdot I_{D3} = [A_{KP3}/(WL)_3^{1/2}] \cdot I_{D3}$, where $\Delta K_{P3}/K_{P3} = A_{KP3}/(WL)_3^{1/2}$ is found from Table 3.38. The drain current mismatch for each pair, including the input pair, is then divided by g_{m1} to refer the current mismatch to an input-referred voltage mismatch.

The addition of mismatch variance or mismatch power through the squaring of terms in Equation 5.29 assumes that the mismatch of each device pair is uncorrelated with the mismatch of other device pairs as expected for local-area mismatch. Additionally, threshold-voltage and transconductance factor mismatch are assumed to be uncorrelated for each device pair. As described in Section 3.11.1.3, some correlation can be expected, for example, through variations in the gate-oxide capacitance. However, as described below, transconductance factor mismatch is often negligible compared to threshold-voltage mismatch. In this case, the correlation between the two components of mismatch is unimportant.

Equation 5.29 can be conveniently expressed as

$$
V_{INOFFSET}^2 \text{ (mismatch)} = \frac{A_{VGS1}^2}{(WL)_1} + \frac{A_{VGS3}^2}{(WL)_3}\left(\frac{g_{m3}}{g_{m1}}\right)^2 + \frac{A_{VGS5}^2}{(WL)_5}\left(\frac{g_{m5}}{g_{m1}}\right)^2 + \frac{A_{VGS7}^2}{(WL)_7}\left(\frac{g_{m7}}{g_{m1}}\right)^2 \tag{5.30}
$$

where A_{VGS1}, A_{VGS3}, A_{VGS5}, and A_{VGS7} are gate–source voltage mismatch factors for the M_1 and M_2 input differential pair, and non-input M_3 and M_4, M_5 and M_6, and M_7 and M_8 current-mirror pair devices, respectively. A_{VGS}, which includes the effects of threshold-voltage, transconductance factor, and body-effect mismatch, is repeated from Equation 3.151 in Section 3.11.1.4 as

$$A_{VGS}^2 = \left[A_{VTO} \left(1 + \frac{V_{SB}}{2\phi_0} \right) \right]^2 + \left(\frac{A_{KP}}{g_m/I_D} \right)^2$$

$$= A_{VT}^2 + \left(\frac{A_{KP}}{g_m/I_D} \right)^2 \tag{5.31}$$

A_{VTO} and A_{KP} are the zero-V_{SB} threshold-voltage and transconductance factor mismatch factors, respectively. g_m/I_D is the operating transconductance efficiency, and V_{SB} is the source–body voltage. Finally, $\phi_0 \approx 2\phi_F + 4U_T$ is the approximate silicon surface potential taken in strong inversion, where $PHI = 2\phi_F$ is the process Fermi voltage (actually twice the Fermi voltage). As shown in Figure 3.9, V_{SB} is taken as zero or positively for both nMOS and pMOS devices, and $PHI = 2\phi_F$ is also taken positively for both device types.

As seen in Equation 5.31 and described in Section 3.11.1.4, the non-zero-V_{SB} threshold-voltage mismatch factor, A_{VT}, is increased from its zero-V_{SB} value, A_{VTO}, by $A_{VT} = A_{VTO}(1 + V_{SB}/(2\phi_0))$. Correspondingly, the non-zero-V_{SB} threshold-voltage mismatch, ΔV_T, is increased from its zero-V_{SB} value, ΔV_{TO}, by $\Delta V_T = \Delta V_{TO}(1 + V_{SB}/(2\phi_0))$. When $V_{SB} = 0\,\text{V}$ as typical for current-mirror devices having their sources connected to the power supply rails, $A_{VT} = A_{VTO}$ and $\Delta V_T = \Delta V_{TO}$. This applies to the OTA, non-input current-mirror devices, M_3–M_8. Additionally, since $V_{SB} \ll 2\phi_0$ is typically required for design in low-voltage processes, $A_{VT} \approx A_{VTO}$ and $\Delta V_T \approx \Delta V_{TO}$ for general devices having non-zero V_{SB} as shown in Figure 3.70. This applies to the input pair devices, M_1 and M_2.

In addition to usually assuming $A_{VT} = A_{VTO}$, the mismatch analysis is further simplified when threshold-voltage mismatch dominates transconductance factor mismatch. Threshold-voltage mismatch dominates for $g_m/I_D > A_{KP}/A_{VT}$ as observed in Equation 5.31 and described in Section 3.11.1.6. This condition is usually met unless devices are operated at unusually high levels of inversion where g_m/I_D is very low. Such high levels of inversion are not typical in low-voltage design because the resulting high values of V_{EFF} and $V_{DS,sat}$ are difficult to support at low supply voltages.

Assuming zero V_{SB} or sufficiently low values of V_{SB} where $A_{VT} = A_{VTO}$, and assuming threshold-voltage mismatch dominates, Equation 5.31 simply reduces to $A_{VGS}^2 = A_{VT}^2 = A_{VTO}^2$, revealing $A_{VGS} = A_{VT} = A_{VTO}$. As described later in Section 5.4.2.5, A_{VGS} is very nearly equal to the process value of A_{VTO}, indicating that threshold-voltage mismatch does indeed dominate for the OTAs illustrated here. As a result, Equation 5.30 and the equations derived below should be viewed with $A_{VGS} = A_{VT} = A_{VTO}$ to assist in their understanding. When high values of V_{SB} significantly increase threshold-voltage mismatch or when operation at high levels of inversion makes transconductance factor mismatch significant, the full expression of Equation 5.31 should be used for A_{VGS}. The full accuracy of Equation 5.31 is used for mismatch predictions provided by the *Analog CMOS Design, Tradeoffs and Optimization* spreadsheet.

Factoring Equation 5.30 to show the mismatch power contributions of non-input current-mirror pairs relative to the contribution input of the differential pair gives

$$V_{INOFFSET}^2 \text{ (mismatch)} = \frac{A_{VGS1}^2}{(WL)_1} \left[1 + \left(\frac{A_{VGS3}}{A_{VGS1}} \right)^2 \frac{(WL)_1}{(WL)_3} \left(\frac{g_{m3}}{g_{m1}} \right)^2 \right.$$

$$+ \left(\frac{A_{VGS5}}{A_{VGS1}} \right)^2 \frac{(WL)_1}{(WL)_5} \left(\frac{g_{m5}}{g_{m1}} \right)^2$$

$$\left. + \left(\frac{A_{VGS7}}{A_{VGS1}} \right)^2 \frac{(WL)_1}{(WL)_7} \left(\frac{g_{m7}}{g_{m1}} \right)^2 \right] \tag{5.32}$$

Analogous to the thermal- and flicker-noise analysis given earlier, the unity term corresponds to the mismatch power contribution of the input differential pair. The second, third, and fourth terms correspond to the mismatch power contributions of the non-input, current-mirror pairs relative to the input differential pair.

Substituting $WL = (L^2/IC) \cdot (I_D/I_0)$ from Table 3.10 and $g_m = I_D/\left[nU_T\left(\sqrt{IC+0.25}+0.5\right)\right]$ from Table 3.17 into Equation 5.32 allows the mismatch to be expressed in terms of the inversion coefficient, channel length, and drain current, giving

$$
V_{INOFFSET}^2 \text{ (mismatch)} = A_{VGS1}^2 \left(\frac{IC_1}{L_1^2}\right)\left(\frac{I_{0,1}}{I_{D1}}\right) \cdot
$$

$$
\left\{ 1 + \left(\frac{A_{VGS3}}{A_{VGS1}}\right)^2 \left[\left(\frac{L_1}{L_3}\right)^2 \left(\frac{IC_3}{IC_1}\right)\left(\frac{I_{0,3}}{I_{0,1}}\right)\left(\frac{I_{D1}}{I_{D3}}\right)\right]\left[\frac{n_1\left(\sqrt{IC_1+0.25}+0.5\right)}{n_3\left(\sqrt{IC_3+0.25}+0.5\right)}\left(\frac{I_{D3}}{I_{D1}}\right)\right]^2 \right.
$$

$$
+ \left(\frac{A_{VGS5}}{A_{VGS1}}\right)^2 \left[\left(\frac{L_1}{L_5}\right)^2 \left(\frac{IC_5}{IC_1}\right)\left(\frac{I_{0,5}}{I_{0,1}}\right)\left(\frac{I_{D1}}{I_{D5}}\right)\right]\left[\frac{n_1\left(\sqrt{IC_1+0.25}+0.5\right)}{n_5\left(\sqrt{IC_5+0.25}+0.5\right)}\left(\frac{I_{D5}}{I_{D1}}\right)\right]^2
$$

$$
\left. + \left(\frac{A_{VGS7}}{A_{VGS1}}\right)^2 \left[\left(\frac{L_1}{L_7}\right)^2 \left(\frac{IC_7}{IC_1}\right)\left(\frac{I_{0,7}}{I_{0,1}}\right)\left(\frac{I_{D1}}{I_{D7}}\right)\right]\left[\frac{n_1\left(\sqrt{IC_1+0.25}+0.5\right)}{n_7\left(\sqrt{IC_7+0.25}+0.5\right)}\left(\frac{I_{D7}}{I_{D1}}\right)\right]^2 \right\}
$$

$$
(5.33)
$$

Although inversion coefficients and channel lengths are equal for a given version of the simple OTAs and drain currents are equal at I_{BIAS}, these are expressed separately to show individual device mismatch contributions and provide a general analysis.

Equation 5.33 excludes g_m/I_D and g_m decreases associated with velocity saturation and VFMR effects. As mentioned in Section 5.3.1.1, these effects are included by replacing IC with $IC(1 + IC/IC_{CRIT})$ in the $\sqrt{IC+0.25}$ terms related to g_m/I_D. Velocity saturation effects can reduce the mismatch contributions of non-input pair devices by reducing the conversion of threshold-voltage mismatch to drain current mismatch. Although velocity saturation reduces the conversion of threshold-voltage mismatch to drain current mismatch, transconductance factor mismatch present at increasing inversion or V_{EFF} can counter the reduction of drain current mismatch.

In Equations 5.32 and 5.33, the unity terms correspond to the mismatch power contribution of the input differential pair. The second, third, and fourth terms correspond to the mismatch power contributions of the non-input, current-mirror pairs relative to the input differential pair. These terms will be referred to as relative mismatch powers, RMPs, where $RMP_1 = 1$, RMP_3, RMP_5, and RMP_7 are the relative mismatch powers for the M_1 and M_2 input differential pair, and non-input M_3 and M_4, M_5 and M_6, and M_7 and M_8 current-mirror pair devices, respectively. If RMP_3, RMP_5, and RMP_7 are sufficiently below unity, the input differential pair dominates the mismatch.

Each of the RMP_3, RMP_5, and RMP_7 terms is separated into a squared ratio of A_{VGS} factors, a ratio of gate areas, and a squared ratio of transconductances. Because the ratio of input to non-input device drain current appears in the gate-area ratio and the ratio of non-input to input device drain current appears squared in the transconductance ratio, this gives a net ratio of non-input to input device drain current. This same drain current ratio also appears in the thermal and flicker-noise contributions of Equations 5.19 and 5.25.

As described for the flicker noise given in Equation 5.25, the $I_{0,5}/I_{0,1}$ and $I_{0,7}/I_{0,1}$ ratios are equal to one-third since the non-input M_5 and M_6, and M_7 and M_8 pairs use pMOS devices compared to nMOS devices for the M_1 and M_2 input pair. This represents potentially lower mismatch contributions from pMOS devices due to their factor-of-three larger gate areas compared to nMOS devices for equal drain currents, inversion coefficients, and channel lengths. The pMOS mismatch contributions are lower since pMOS mismatch factors, listed in Table 3.3, are equal to those of nMOS devices for the $0.5\,\mu m$ CMOS process used for the simple OTAs.

Equations 5.30, 5.32, and 5.33 for input-referred offset voltage variance due to local-area mismatch are identical in form to Equations 5.23–5.25 for input-referred flicker-noise voltage PSD. In the offset voltage equations, device flicker-noise factors, K'_F, are replaced by the square of gate–source voltage mismatch factors, or A^2_{VGS}. As mentioned, A_{VGS} may usually be replaced with $A_{VGS} \approx A_{VTO}$, revealing that the terms consist of nearly the square of the threshold-voltage mismatch factors, or A^2_{VTO}. Also, in the offset voltage equations, the "2" terms are removed since mismatch is inherently between differential-pair and current-mirror pair devices. Finally, in the offset voltage equations, the frequency terms are removed.

5.3.7.2 Cascoded OTAs

The analysis of input-referred offset voltage due to local-area mismatch for the cascoded OTAs shown in Figure 5.2 is similar to that described above for the simple OTAs. Unlike devices in the simple OTAs, devices in the cascoded OTAs operate at different drain currents and inversion coefficients. This affects their mismatch contributions.

Mismatch for the cascoded OTAs is identical to that given in Equations 5.29, 5.30, 5.32, and 5.33 for the simple OTAs, except the terms related to M_7 and M_8 are removed. For convenience, these equations are repeated with these terms removed.

The mean square value or variance (square of the 1σ value) of the input-referred offset voltage for the cascoded OTAs is given by

$$V^2_{INOFFSET}\,(\text{mismatch}) = \frac{A^2_{VT1}}{(WL)_1} + \frac{A^2_{VT3}}{(WL)_3}\left(\frac{g_{m3}}{g_{m1}}\right)^2 + \frac{A^2_{VT5}}{(WL)_5}\left(\frac{g_{m5}}{g_{m1}}\right)^2$$
$$+ \frac{A^2_{KP1}}{(WL)_1}\left(\frac{I_{D1}}{g_{m1}}\right)^2 + \frac{A^2_{KP3}}{(WL)_3}\left(\frac{I_{D3}}{g_{m1}}\right)^2 + \frac{A^2_{KP5}}{(WL)_5}\left(\frac{I_{D5}}{g_{m1}}\right)^2 \qquad (5.34)$$

As shown for Equation 5.30, Equation 5.34 can be conveniently expressed as

$$V^2_{INOFFSET}\,(\text{mismatch}) = \frac{A^2_{VGS1}}{(WL)_1} + \frac{A^2_{VGS3}}{(WL)_3}\left(\frac{g_{m3}}{g_{m1}}\right)^2 + \frac{A^2_{VGS5}}{(WL)_5}\left(\frac{g_{m5}}{g_{m1}}\right)^2 \qquad (5.35)$$

After factoring to show the mismatch power contributions of non-input current-mirror pairs relative to the contribution of the input differential pair, this gives

$$V^2_{INOFFSET}\,(\text{mismatch}) = \frac{A^2_{VGS1}}{(WL)_1}\left[1 + \left(\frac{A_{VGS3}}{A_{VGS1}}\right)^2\frac{(WL)_1}{(WL)_3}\left(\frac{g_{m3}}{g_{m1}}\right)^2 + \left(\frac{A_{VGS5}}{A_{VGS1}}\right)^2\frac{(WL)_1}{(WL)_5}\left(\frac{g_{m5}}{g_{m1}}\right)^2\right]$$
$$(5.36)$$

Finally, after expressing transconductance in terms of the inversion coefficient and drain current, and expressing gate area in terms of the inversion coefficient, channel length, and drain current, this gives

$$V_{INOFFSET}^2 \text{ (mismatch)} = A_{VGS1}^2 \left(\frac{IC_1}{L_1^2}\right)\left(\frac{I_{0,1}}{I_{D1}}\right) \cdot$$

$$\left\{ 1 + \left(\frac{A_{VGS3}}{A_{VGS1}}\right)^2 \left[\left(\frac{L_1}{L_3}\right)^2 \left(\frac{IC_3}{IC_1}\right)\left(\frac{I_{0,3}}{I_{0,1}}\right)\left(\frac{I_{D1}}{I_{D3}}\right)\right]\left[\frac{n_1\left(\sqrt{IC_1 + 0.25} + 0.5\right)}{n_3\left(\sqrt{IC_3 + 0.25} + 0.5\right)}\left(\frac{I_{D3}}{I_{D1}}\right)\right]^2 \right.$$

$$\left. + \left(\frac{A_{VGS5}}{A_{VGS1}}\right)^2 \left[\left(\frac{L_1}{L_5}\right)^2 \left(\frac{IC_5}{IC_1}\right)\left(\frac{I_{0,5}}{I_{0,1}}\right)\left(\frac{I_{D1}}{I_{D5}}\right)\right]\left[\frac{n_1\left(\sqrt{IC_1 + 0.25} + 0.5\right)}{n_5\left(\sqrt{IC_5 + 0.25} + 0.5\right)}\left(\frac{I_{D5}}{I_{D1}}\right)\right]^2 \right\}$$

$$(5.37)$$

A_{VT1}, A_{VT3}, and A_{VT5} are the threshold-voltage mismatch factors for the M_1 and M_2 input differential pair, and non-input M_3 and M_4 current source, and M_5 and M_6 current-mirror pair devices, respectively. A_{KP1}, A_{KP3}, and A_{KP5} are the transconductance factor mismatch factors, and I_{D1}, I_{D3}, and I_{D5} are the drain currents for these devices. Finally, A_{VGS1}, A_{VGS3}, and A_{VGS5} are gate–source voltage mismatch factors for these devices, which are given by Equation 5.31.

As mentioned for the simple OTAs, assuming zero V_{SB} or sufficiently low values of V_{SB} where $A_{VT} = A_{VTO}$, and assuming threshold-voltage mismatch dominates, Equation 5.31 reveals $A_{VGS} = A_{VT} = A_{VTO}$. As described later in Section 5.5.2.5, A_{VGS} is nearly equal to the process value of A_{VTO}, indicating that threshold-voltage mismatch does indeed dominate for the OTAs illustrated here. As a result, Equations 5.34–5.37 should be viewed with $A_{VGS} = A_{VT} = A_{VTO}$ to assist in their understanding. The full accuracy of A_{VGS} given by Equation 5.31 is, again, used for mismatch predictions provided by the *Analog CMOS Design, Tradeoffs and Optimization* spreadsheet. Finally, velocity saturation and VFMR decreases in transconductance are included in Equation 5.37 as described for Equation 5.33.

In Equations 5.36 and 5.37, the unity terms correspond to the mismatch power contribution of the input differential pair. The second and third terms correspond to the mismatch power contributions of the non-input device pairs relative to the input differential pair. As described for the simple OTAs, these terms will be referred to as relative mismatch powers where $RMP_1 = 1$, RMP_3, and RMP_5 are the relative mismatch powers for the M_1 and M_2 input differential pair, and non-input M_3 and M_4 current source, and M_5 and M_6 current-mirror pair devices, respectively. If RMP_3 and RMP_5 are sufficiently below unity, the input differential pair dominates the mismatch.

As described for the simple OTAs, each of the RMP_3 and RMP_5 terms is separated into a squared ratio of A_{VGS} factors, a ratio of gate areas, and a squared ratio of transconductances. Because the ratio of input to non-input device drain current appears in the gate-area ratio and the ratio of non-input to input device drain current appears squared in the transconductance ratio, this gives a net ratio of non-input to input device drain current. This same drain current ratio also appears in the thermal and flicker-noise contributions of Equations 5.22 and 5.28.

As described for the flicker noise in Equation 5.28, the $I_{0,3}/I_{0,1}$ ratio is equal to nearly one-fifth since the non-input, M_3 and M_4 current source devices are pMOS devices compared to nMOS devices for the M_1 and M_2 input pair. This represents potentially lower mismatch contributions from pMOS devices due to their nearly factor-of-five larger gate areas compared to nMOS devices for equal drain currents, inversion coefficients, and channel lengths. The pMOS mismatch contributions are lower since pMOS mismatch factors, listed in Table 3.3, are equal to those of nMOS devices for the 0.18 μm CMOS process used for the cascoded OTAs.

Equations 5.35–5.37 for input-referred offset voltage variance due to local-area mismatch are identical in form to Equations 5.26–5.28 for input-referred flicker-noise voltage PSD. As mentioned for the simple OTAs, in the offset voltage equations, device flicker-noise factors, K_F', are replaced by the square of gate–source voltage mismatch factors, or A_{VGS}^2. Again, A_{VGS} may usually be replaced with $A_{VGS} \approx A_{VTO}$, revealing that the terms consist of nearly the square of the threshold-voltage mismatch factors, or A_{VTO}^2. Again, in the offset voltage equations, the "2" terms are removed since mismatch is inherently between differential-pair and current-mirror pair devices. Finally, in the offset voltage equations, the frequency terms are removed.

5.3.7.3 Optimization

Because the equations are identical in form, the OTA input-referred offset voltage due to local-area mismatch is minimized nearly identically as the input-referred flicker-noise voltage. This is especially true for operation at sufficiently low levels of inversion or V_{EFF} where transconductance factor mismatch contributions and gate-referred flicker-noise voltage increases are negligible. In this case, the gate–source voltage mismatch factor, A_{VGS}, is nearly equal to the process threshold-voltage mismatch factors, A_{VTO} or A_{VT}, and the flicker-noise factor, K'_F, remains near its low-inversion value of K'_{F0}. Under these conditions, the input-referred offset voltage variance due to local-area mismatch and the input-referred flicker-noise voltage PSD maintain simple and identical dependencies on the gate area and transconductance for devices in the OTAs. As a result, this discussion of offset voltage optimization closely parallels the earlier discussion for flicker noise.

Equations 5.32 and 5.33 for the simple OTAs and Equations 5.36 and 5.37 for the cascoded OTAs help the designer minimize the mismatch of input devices and ensure that these devices dominate overall OTA mismatch. For the simple OTAs, the non-unity terms in the bracketed portions of Equations 5.32 and 5.33 are the relative (compared to the input devices) mismatch power contributions of the non-input devices, RMP_3, RMP_5, and RMP_7. For the cascoded OTAs, the non-unity terms in the bracketed portions of Equations 5.36 and 5.37 are the relative mismatch power contributions of the non-input devices, RMP_3 and RMP_5. For both OTAs, the RMP values are equal to the squared ratio of non-input to input device gate–source voltage mismatch factors, multiplied by the ratio of input to non-input device gate areas, multiplied by the squared ratio of non-input to input device transconductances. Minimizing non-input device transconductances has a squared effect on reducing the relative mismatch contributions of non-input devices, but ratios of mismatch factors and gate areas are also important. Because threshold-voltage mismatch factors are usually comparable for nMOS and pMOS devices as seen in reported values in Table 3.39, RMP values are nearly equal to the ratio of input to non-input device gate areas, multiplied by the squared ratio of non-input to input device transconductances. This is true for typical operation where threshold-voltage mismatch dominates overall mismatch.

For both OTAs, input-referred offset voltage is minimized by operating input devices at low inversion coefficients ($IC_1 = IC_2$) and long channel lengths ($L_1 = L_2$), which maximizes their gate area. Additionally, g_m/I_D and g_m are high at low inversion coefficients in weak or moderate inversion, helping input devices dominate the mismatch. Increasing the drain current for input devices (again, MOSFET width is scaled with drain current to maintain the selected inversion coefficient) reduces the input-referred offset voltage by increasing the gate area of the input devices. Increasing drain currents proportionally for all devices increases the transconductance and gate area equally for all devices, which holds the relative mismatch contributions of non-input devices constant.

Unlike when minimizing flicker noise where a preferred device type may have lower flicker noise or lower increases in gate-referred flicker-noise voltage at increasing inversion or V_{EFF}, reported local-area, threshold-voltage mismatch factors suggest that neither nMOS nor pMOS devices are significantly superior for mismatch. For the $0.5\,\mu m$ CMOS process used in the simple OTAs, $A_{VTO} = 14\,mV \cdot \mu m$ as seen in mismatch process values listed in Table 3.3. This value decreases to $A_{VTO} = 5\,mV \cdot \mu m$ for the $0.18\,\mu m$ CMOS process used in the cascoded OTAs. As shown in reported values in Table 3.39 and discussed in Sections 3.11.1.1 and 3.11.1.2, A_{VTO} decreases with decreasing t_{ox}, but may not decrease as rapidly as t_{ox} decreases in smaller-geometry processes. Reported values for transconductance factor mismatch factors, A_{KP}, are also listed in Table 3.39. These show no clear technology trend. Values of $A_{KP} = 2\% \cdot \mu m$ from Table 3.3 are used for both OTAs. Because devices are not operated at unusually high levels of inversion, transconductance factor mismatch is not significant for the OTAs. As discussed in Sections 5.3.7.1 and 3.11.1.6, threshold-voltage mismatch dominates overall mismatch unless devices are operated at unusually high levels of inversion where transconductance factor mismatch can become significant.

In addition to minimizing the mismatch of input devices, minimizing overall OTA mismatch responsible for the input-referred offset voltage requires minimizing the mismatch of non-input devices. The non-unity terms in Equations 5.33 and 5.37 for both OTAs indicate that non-input device mismatch contributions are minimized by operating these devices at low inversion coefficients in weak or moderate inversion. Here, even though g_m/I_D and g_m are maximum, maximizing the conversion of threshold-voltage mismatch to drain current mismatch, the large shape factors (W/L) and channel widths required result in large gate areas. This then lowers the threshold-voltage mismatch and, correspondingly, the drain current mismatch.

Although Equations 5.33 and 5.37 for both OTAs indicate that non-input devices should be operated at low inversion coefficients for large gate areas to minimize their mismatch contributions, Equations 5.19 and 5.22 indicate that non-input devices should be operated at high inversion coefficients in strong inversion for low g_m/I_D and g_m to minimize their thermal-noise contributions. This results in a design conflict for applications where both thermal noise and mismatch are important. Fortunately, non-input device mismatch contributions do not necessarily continue increasing as these devices are operated at higher inversion coefficients ($IC_3 = IC_4$, $IC_5 = IC_6$, and, for the simple OTA only, $IC_7 = IC_8$) into strong inversion as required to minimize their thermal-noise contributions. This is observed in Equations 5.33 and 5.37 where the IC terms cancel in strong inversion ($IC > 10$) for the relative mismatch contributions of non-input devices. As IC increases in strong inversion, non-input device, drain mismatch current levels off because the increasing threshold-voltage mismatch associated with decreasing gate area is countered by decreasing g_m/I_D and g_m. However, the leveling off of drain mismatch current occurs only if transconductance factor mismatch remains negligible, which is indicated by $A_{VGS} \approx A_{VT}$ or $A_{VGS} \approx A_{VTO}$.

The behavior of non-input device, drain mismatch current is illustrated in Figure 3.72 for nMOS devices in the $0.18\,\mu m$ CMOS process used in the cascoded OTAs. The mismatch current initially increases with increasing inversion coefficient in weak and moderate inversion because of the decrease in gate area. The mismatch current then levels off in strong inversion as described above. However, when transconductance factor mismatch becomes significant, the mismatch current begins increasing again as shown by the solid lines in Figure 3.72. Although drain mismatch current either levels off in strong inversion or increases with increasing inversion coefficient, the mismatch current consistently decreases with increasing channel length because of the significant increase in gate area.

When transconductance factor mismatch is significant in deep strong inversion for non-input devices, the device relative mismatch contributions increase by IC instead of becoming constant as predicted by Equations 5.33 and 5.37 for constant values of $A_{VGS} \approx A_{VT}$ or $A_{VGS} \approx A_{VTO}$. This is because, when transconductance factor mismatch or A_{KP} dominates, $A_{VGS}^2 \approx A_{KP}^2/(g_m/I_D)^2$ from Equation 5.31 increases directly with IC for $g_m/I_D = 1/(nU_T\sqrt{IC})$ in strong inversion (assuming negligible velocity saturation) given in Table 3.17. These mismatch increases are included in the *Analog CMOS Design, Tradeoffs and Optimization* spreadsheet and will be shown to be nearly negligible for both OTAs in Sections 5.4.2.5 and 5.5.2.5 through reported A_{VGS} values.

When input devices are operated at low inversion coefficients and non-input devices are operated at high inversion coefficients as required to minimize thermal noise, Equations 5.33 and 5.37 show that non-input device channel lengths ($L_3 = L_4$, $L_5 = L_6$, and, for the simple OTA only, $L_7 = L_8$) must be greater than input device channel lengths ($L_1 = L_2$) to achieve comparable device areas and ensure negligible mismatch contributions. Equations 5.32 and 5.36 also illustrate this where the product of mismatch factor and gate-area ratios must be in the vicinity of unity if non-input device mismatch contributions are to be sufficiently reduced by the lower transconductance of these devices. Increasing non-input device channel lengths is required because operation at higher inversion coefficients requires smaller shape factors and smaller channel widths and gate areas for a given channel length. Increasing the channel length increases non-input device gate area as the square of channel length because channel width must increase directly with channel length to maintain the selected inversion coefficient. This significantly reduces the mismatch contributions, but at the expense of significantly reduced bandwidth (Tables 3.26 and 3.27) for devices in the signal path and significantly increased drain capacitive loading (Table 3.25) for devices providing bias to devices in the signal path.

The relative mismatch and flicker-noise power contributions of non-input devices are proportional to the squared ratio of non-input to input device transconductances. As a result, the gate area requirement for non-input devices is relaxed somewhat when non-input device transconductances are sufficiently below the transconductance of input devices as required to minimize thermal noise. However, minimizing local-area mismatch and flicker noise, unlike minimizing thermal noise, requires careful management of gate area for both input and non-input devices.

As for minimizing thermal and flicker noise, non-input devices may be operated at lower drain currents compared to input devices to reduce their transconductance and mismatch contributions. Although drain current can be reduced for some non-input devices, some non-input devices must support the full drain current of the input devices.

In the simple OTAs, all devices operate at equal drain currents, inversion coefficients, and channel lengths. As a result, all devices operate at equal transconductance and contribute equally to the thermal noise. Non-input nMOS devices contribute equal mismatch as the nMOS input devices because these devices operate at equal transconductance and equal gate area. The larger required gate areas of non-input pMOS devices, however, reduce their mismatch contributions. This is shown in Equations 5.32 and 5.33 by the unity term, representing the mismatch power contribution of the input pair, combined with three terms, RMP_3, RMP_5, and RMP_7, representing the relative mismatch power contributions of the non-input pairs. RMP_3 for the nMOS devices is unity, and RMP_5 and RMP_7 for the pMOS devices are approximately 0.4 as shown later by the spreadsheet in Section 5.4.2.5.

In the cascoded OTAs, input devices operate at low inversion coefficients ($IC_1 = IC_2 \approx 0.8$) in moderate inversion for high g_m/I_D and g_m, while non-input devices operate at high inversion coefficients ($IC_3 = IC_4 \approx 16$ and $IC_5 = IC_6 \approx 16$) in strong inversion for low g_m/I_D and g_m. The drain currents of $I_{D3} = I_{D4} = 150\,\mu A$ for M_3 and M_4 are unfortunately greater than the drain currents of $I_{D1} = I_{D2} = 100\,\mu A$ for M_1 and M_2, as required for the folded-cascode circuit topology. This increases the mismatch contributions of the M_3 and M_4 device pair. However, the drain currents of $I_{D5} = I_{D6} = 50\,\mu A$ for M_5 and M_6 are lower than those for M_1 and M_2, which decreases the mismatch contributions of the M_5 and M_6 device pair. This is identical to the behavior described for the thermal and flicker noise.

Although input devices in the cascoded OTAs operate in moderate inversion and non-input devices operate in strong inversion such that input devices dominate the thermal noise, the input devices do not dominate the mismatch. This is because the non-input M_3–M_6 devices have equal channel lengths as the M_1 and M_2 input pair devices. This gives insufficient gate area for the non-input devices that operate at higher inversion coefficients, giving smaller shape factors and smaller channel widths and gate areas for a given channel length. This is shown in Equations 5.36 and 5.37 by the unity term, representing the mismatch power contribution of the input pair, combined with two terms, RMP_3 and RMP_5, representing the relative mismatch power contributions of the non-input device pairs. RMP_3 for the M_3 and M_4 device pair and RMP_5 for the M_5 and M_6 device pair are roughly equal to unity as shown later by the spreadsheet in Section 5.5.2.5. As mentioned, increasing the channel length of non-input devices above that of the input devices reduces the mismatch power contributions of non-input devices, but significantly lowers the transconductance bandwidth.

When input devices dominate the mismatch or when the relative mismatch contributions of non-input devices remain constant, as they do for both the simple and cascoded OTAs, the OTA input-referred offset voltage variance due to local-area mismatch tracks the gate–source voltage variance of the input devices. The OTA input-referred offset voltage variance then increases as IC_1 and decreases as $1/L_1^2$, tracking the inverse of gate area for the input devices. When transconductance factor mismatch is significant, the offset can increase further as modeled by increases in A_{VGS}. The OTA input-referred offset voltage (the square root of the offset voltage variance) trends are the same as the trends for device gate–source voltage mismatch summarized in Table 4.2 and shown in Figures 4.14 and 4.15, where device trends are applied to the input devices. This again assumes constant relative mismatch contributions from non-input devices.

Sections 5.4.2.12 and 5.5.2.11 summarize offset voltage tradeoffs. Here, the input-referred offset voltage due to local-area mismatch is nearly inversely proportional to the square root of device gate areas for both the simple and cascoded OTAs.

5.3.8 Systematic Offset Voltage for Simple OTAs

Imbalances in drain–source bias voltages result in imbalances in drain bias currents due to finite drain–source resistances. When drain bias current imbalances appear in the signal path of circuits receiving voltage inputs, they give rise to an input, systematic offset voltage required to remove the current imbalances. In some designs, such as fully differential designs, systematic offset is not present because of inherent drain–source bias voltage balance. However, systematic offset can exceed the offset associated with local-area mismatch for the simple OTAs and is analyzed below. Systematic offset is not analyzed for the cascoded OTAs because it is not significant as a result of their high voltage gain.

For the simple OTAs shown in Figure 5.1, there is no drain current imbalance for the differential pair consisting of M_1 and M_2 because both device drains connect to equal bias voltages set by diode-connected devices M_5 and M_7. However, there is a drain current imbalance for the nMOS current mirror consisting of devices M_3 and M_4. The drain–source voltage of M_4 exceeds the drain–source voltage of M_3 by an amount equal to the output voltage, V_{OUT}, less the negative supply voltage, V_{SS}, less the drain–source voltage of M_3, which is equal to its gate–source voltage, V_{GS3}. This results in increased drain current for M_4 giving a negative (current sinking to V_{SS}) OTA output offset current expressed by

$$\Delta I_{OUT}|_{M3,M4} = -\frac{V_{OUT} - V_{SS} - V_{GS3}}{r_{ds3} = r_{ds4}} = -I_{BIAS} \cdot \frac{V_{OUT} - V_{SS} - V_{GS3}}{V_{A3} = V_{A4}} \qquad (5.38)$$

In the rightmost term, the drain–source resistance is expressed as $r_{ds} = V_A/I_D$, where $r_{ds3} = r_{ds4}$, $V_{A3} = V_{A4}$, and $I_{D3} = I_{D4} = I_{BIAS}$ are the drain–source resistance, Early voltage, and drain current, respectively, for M_3 and M_4.

The same drain voltage imbalance appearing between the drains of nMOS devices M_4 and M_3 also appears between the drains of pMOS devices M_6 and M_8. In this case, a higher voltage at the drain of M_4 compared to M_3 results in reduced drain–source voltage for M_6 compared to M_8, since M_6 and M_8 are referenced to the positive supply rail, V_{DD}. This causes increased drain current for M_8, which gives a negative OTA output offset current after mirroring by the M_3 and M_4 current mirror. The resulting OTA output offset current is then expressed by

$$\Delta I_{OUT}|_{M6,M8} = -\frac{V_{OUT} - V_{SS} - V_{GS3}}{r_{ds6} = r_{ds8}} = -I_{BIAS} \cdot \frac{V_{OUT} - V_{SS} - V_{GS3}}{V_{A6} = V_{A8}} \qquad (5.39)$$

In the rightmost term, the drain–source resistance is, again, expressed as $r_{ds} = V_A/I_D$, where $r_{ds6} = r_{ds8}$, $V_{A6} = V_{A8}$, and $I_{D6} = I_{D8} = I_{BIAS}$ are the drain–source resistance, Early voltage, and drain current, respectively, for M_6 and M_8.

The total output systematic offset current is the sum of Equations 5.38 and 5.39. Because the same $V_{OUT} - V_{SS} - V_{GS3}$ voltage imbalance appears in both equations, the sum of Equations 5.38 and 5.39 can be expressed by this voltage imbalance divided by the parallel combination of drain–source resistances responsible for the output systematic offset current. Dividing this then by the OTA transconductance refers the total output systematic offset current to an input systematic offset voltage given by

$$V_{INOFFSET} \text{ (systematic)} = \frac{1}{G_M} \left[\frac{-(V_{OUT} - V_{SS} - V_{GS3})}{(r_{ds3} = r_{ds4}) \parallel (r_{ds6} = r_{ds8})} \right]$$

$$= \frac{1}{\left(\frac{g_m}{I_D}\right)_1 \cdot I_{BIAS}} \left[\frac{-I_{BIAS} \cdot (V_{OUT} - V_{SS} - V_{GS3})}{(V_{A3} = V_{A4}) \parallel (V_{A6} = V_{A8})} \right]$$

$$= -\frac{1}{\left(\frac{g_m}{I_D}\right)_1} \left[\frac{V_{OUT} - V_{SS} - V_{GS3}}{(V_{A3} = V_{A4}) \parallel (V_{A6} = V_{A8})} \right]$$

(5.40)

The $r_{ds3} = r_{ds4}$, $V_{A3} = V_{A4}$, $r_{ds6} = r_{ds8}$, and $V_{A6} = V_{A8}$ terms emphasize that systematic offset results from drain–source voltage imbalances between nMOS devices M_3 and M_4 and pMOS devices M_6 and M_8.

For the simple OTAs, equal bias and sizing are used for the nMOS mirror devices, M_3 and M_4, giving unity current gain. Additionally, equal bias and sizing are used for the pMOS devices, M_6 and M_8, as well as their input current-mirror devices, M_5 and M_7, which have equal drain–source voltages and are not included in the systematic offset analysis. Equal biasing and sizing then also gives unity current gain for the pMOS current mirrors. A modified but similar circuit analysis is required if current mirror gain is used. Because all drain currents are equal to I_{BIAS}, the drain currents cancel in Equation 5.40.

An examination of Equation 5.40 reveals that the input systematic offset voltage is equal to the negated $V_{OUT} - V_{SS} - V_{GS3}$ voltage imbalance divided by the OTA voltage gain, A_V, given by Equation 5.8. This gives

$$V_{INOFFSET} \text{ (systematic)} = \frac{-(V_{OUT} - V_{SS} - V_{GS3})}{A_V}$$

$$= -\left[\frac{n_1 U_T \left(\sqrt{IC_1 + 0.25} + 0.5 \right)}{(V_{AL4} \cdot L_4) \parallel (V_{AL6} \cdot L_6)} \right] \cdot (V_{OUT} - V_{SS} - V_{GS3})$$

(5.41)

Like A_V given by Equation 5.8, Equation 5.41 excludes velocity saturation and VFMR decreases in g_m/I_D and g_m that can reduce A_V and increase the input systematic offset voltage. As mentioned in Section 5.3.1.1, these effects are included by replacing IC with $IC(1 + IC/IC_{CRIT})$ in the $\sqrt{IC + 0.25}$ term related to g_m/I_D.

The input systematic offset voltage is minimized when A_V is maximized by operating the input pair devices, M_1 and M_2, at low inversion coefficient ($IC_1 = IC_2$) in weak or moderate inversion for high g_m/I_D and g_m, and by selecting long channel lengths (L_4 and L_6) for the output devices, M_4 and M_6, and their related devices. The systematic offset voltage follows the inverse of the trends of A_V for the simple OTAs described in Section 5.3.3.3. Like A_V, the systematic offset voltage is independent of the OTA bias current (again, MOSFET width is scaled with drain current to maintain the selected inversion coefficient). Finally, $V_{OUT} = V_{SS} + V_{GS3}$ represents a balance point for the output where the systematic offset voltage is zero.

5.3.9 Input and Output Capacitances

5.3.9.1 Simple OTAs

The differential input capacitance for the simple OTAs shown in Figure 5.1 is found by considering a differential input voltage across the inputs. For a small-signal, differential input voltage, the voltage at the common source connection of the M_1 and M_2 input pair is static and can be considered a virtual ground. The resulting differential input capacitance is then one-half that seen at the gate of a single grounded-source device because the input gate capacitances appear in series for the two pair devices. The differential input capacitance is then given by

$$C_{INDIF} = \tfrac{1}{2} \left[C_{GS1} + C_{GB1} + C_{GD1} (1 + g_{m1}/g_{m5}) \right]$$

(5.42)

$C_{GS1} = C_{gsi1} + C_{GSO,1}$ and $C_{GB1} = C_{gbi1} + C_{GBO,1}$ are the gate–source and gate–body capacitances for the M_1 and M_2 input devices. C_{GS1} includes the intrinsic gate–source capacitance, C_{gsi1}, and extrinsic gate–source overlap capacitance, $C_{GSO,1}$. C_{GB1} includes the intrinsic gate–body capacitance, C_{gbi1}, and extrinsic gate–drain overlap capacitance, $C_{GBO,1}$. $C_{GD1} = C_{GDO,1}$ is the gate–drain capacitance for the input devices, which is essentially the extrinsic gate–drain overlap capacitance, $C_{GDO,1}$, alone for operation in saturation. These device capacitances are shown in Figure 3.21. C_{GD1} is Miller multiplied by $1 + g_{m1}/g_{m5}$, which is equal to $1 - A_{Vgd1}$. $A_{Vgd1} = -g_{m1}/g_{m5}$ is the gate-to-drain voltage gain for the input devices into the M_5 and M_7 diode-connected devices. This neglects the small reduction in gain caused by loading from drain–source resistances.

The evaluation of C_{INDIF} given by Equation 5.42 is complex because it involves the intrinsic gate capacitances given in Table 3.23, extrinsic gate-overlap capacitances given in Table 3.24, and the Miller multiplication, $1 + g_{m1}/g_{m5}$, which acts on C_{GD1}. Since C_{GD1} is usually small compared to the sum of C_{GS1} and C_{GB1}, and $1 + g_{m1}/g_{m5}$ is small for the input devices connected to diode-connected loads (it is only two for the OTAs considered here where $g_{m1} = g_{m5}$), the C_{GD1} term will be neglected, giving $C_{INDIF} \approx \frac{1}{2}(C_{GS1} + C_{GB1})$. For the purposes of design guidance, C_{INDIF} will be simplified further by considering only the intrinsic gate capacitances, giving $C_{INDIF} \approx \frac{1}{2}(C_{gsi1} + C_{gbi1})$. This provides a good estimate of capacitance for long-channel devices where gate-overlap capacitances are not significant, but under-predicts the capacitance for short-channel devices where the overlap capacitances are significant. This can be seen in Figure 3.55 by the ratio of gate–source overlap capacitance, C_{GSO}, to the gate-oxide capacitance, C_{GOX}. C_{GSO} becomes an increasingly larger fraction of C_{GOX} as channel length decreases, illustrating its increasing importance at short channel lengths where gate area, gate-oxide capacitance, and intrinsic gate capacitances are small.

Evaluating the approximation of $C_{INDIF} \approx \frac{1}{2}(C_{gsi1} + C_{gbi1})$ using C_{gsi1} and C_{gbi1} given by Table 3.23 and $WL = (L^2/IC) \cdot (I_D/I_0)$ given by Table 3.10 gives

$$C_{INDIF} \approx \frac{1}{2}\left(C_{gsi1} + C_{gbi1}\right) = \frac{1}{2}(WL)_1 C'_{OX}\left(\hat{C}_{gsi1} + \hat{C}_{gbi1}\right)$$

$$= \frac{1}{2}\left(\frac{L_1^2}{IC_1}\right)\left(\frac{I_{D1} = I_{BIAS}}{I_{0,1}}\right) C'_{OX}\left(\hat{C}_{gsi1} + \hat{C}_{gbi1}\right) \tag{5.43}$$

$\hat{C}_{gsi}$ and $\hat{C}_{gbi}$ are the normalized, intrinsic gate–source and gate–body capacitances with values described in Section 5.3.4.1 for the OTA bandwidth predicted by Equation 5.15. Equation 5.43 provides rough design guidance because, again, it excludes the extrinsic, gate-overlap capacitances that can be significant, especially for short-channel devices. Additionally, it excludes C_{GD1} and the Miller factor, $1 + g_{m1}/g_{m5}$. The full accuracy of Equation 5.42 is used for capacitance predictions provided later by the *Analog CMOS Design, Tradeoffs and Optimization* spreadsheet.

The output capacitance for the simple OTAs shown in Figure 5.1 is dominated by the drain–body capacitances, C_{DB4} and C_{DB6}, of the M_4 and M_6 output devices with a smaller contribution coming from their gate–drain capacitances, C_{GD4} and C_{GD6}. Output capacitance is then given by

$$C_{OUT} = C_{DB4} + C_{DB6} + C_{GD4} + C_{GD6} \tag{5.44}$$

For the purposes of design guidance, C_{OUT} can be roughly approximated using $C_{DB} \approx C_{DBO}$, where C_{DBO} is the maximum value of C_{DB} taken at $V_{DB} = 0\,\text{V}$. Additionally, the usually smaller value of C_{GD} can be neglected, partially compensating for the higher estimate of C_{DB}. This, then, gives the approximation of $C_{OUT} \approx C_{DBO,4} + C_{DBO,6}$.

As described in Table 3.25, the drain area and perimeter required for the calculation of C_{DB} or C_{DBO} can be estimated by assuming an even number of gate stripes. This provides a good estimate except for very small devices that do not use interdigitation. An even number of gate stripes corresponds to an even multiplier factor m, where the drain area and perimeter are minimized to minimize the value of C_{DB} or C_{DBO}. Substituting C_{DBO} from Table 3.25 into $C_{OUT} \approx C_{DBO,4} + C_{DBO,6}$ gives

$$C_{OUT} \approx C_{DBO,4} + C_{DBO,6}$$

$$\approx W_4 \left[\tfrac{1}{2} W_{DIF} C_{J4} + C_{JSW4} \right] + W_6 \left[\tfrac{1}{2} W_{DIF} C_{J6} + C_{JSW6} \right]$$

$$\approx \left(\frac{L_4}{IC_4} \right) \left(\frac{I_{BIAS}}{I_{0,4}} \right) \cdot \left[\tfrac{1}{2} W_{DIF} C_{J4} + C_{JSW4} \right]$$

$$+ \left(\frac{L_6}{IC_6} \right) \left(\frac{I_{BIAS}}{I_{0,6}} \right) \cdot \left[\tfrac{1}{2} W_{DIF} C_{J6} + C_{JSW6} \right] \tag{5.45}$$

This uses $W = (L/IC) \cdot (I_D/I_0)$ from Table 3.10 and assumes an even multiplier factor for the output devices. W_{DIF} is the process width of drain elements, C_J is the drain–body capacitance per unit area, C_{JSW} is the drain–body capacitance per unit perimeter length, and, again, $I_0 = 2n_0\mu_0 C'_{OX} U_T^2$ (Table 3.6) is the technology current. These process parameters are given in Tables 3.2 and 3.4 for the 0.5 μm CMOS process used in the simple OTAs. Equation 5.45, again, provides rough design guidance. The full accuracy of Equation 5.44 is used for capacitance predictions provided later by the *Analog CMOS Design, Tradeoffs and Optimization* spreadsheet.

Since $I_{0,6} = 0.07 \, \mu A$ for pMOS device M_6 compared to $I_{0,4} = 0.21 \, \mu A$ for nMOS device M_4, M_6 has three times the width of M_4 for operation at equal drain currents, inversion coefficients, and channel lengths. As a result, pMOS device M_6 dominates C_{OUT}.

5.3.9.2 Cascoded OTAs

The differential input capacitance for the cascoded OTAs shown in Figure 5.2 is given by

$$C_{INDIF} = \tfrac{1}{2} \left[C_{GS1} + C_{GB1} + C_{GD1} \left(1 + g_{m1}/g_{ms7} \right) \right] \tag{5.46}$$

where device capacitances apply to the M_1 and M_2 input devices.

This is identical to C_{INDIF} for the simple OTAs given by Equation 5.42, except for the Miller multiplication term, $1 + g_{m1}/g_{ms7} = 1 - A_{Vgd1}$. $A_{Vgd1} = -g_{m1}/g_{ms7}$ is the gate-to-drain voltage gain for the input devices into the source conductances of cascode devices M_7 and M_8, g_{ms7}. This source conductance, described earlier in Section 5.3.1.2, assumes that the M_7 and M_8 drain load resistances are sufficiently below their drain–source resistances as shown by the small-signal resistances summarized in Table 3.16. This condition is met for M_7 that connects to a low-voltage cascode, diode-connected load having a small-signal resistance of nearly $1/g_{m5}$, but assumes a sufficiently low impedance load for M_8 at the OTA output. As for Equation 5.42, the expression for A_{Vgd1} neglects the small reduction in gain caused by loading from drain–source resistances.

As described for the simple OTAs, neglecting C_{GD1} and its Miller multiplication gives $C_{INDIF} \approx \tfrac{1}{2}(C_{GS1} + C_{GB1})$. Again, this assumes that C_{GD1} is small compared to other capacitances combined with multiplication by $1 + g_{m1}/g_{ms7}$ that is small (approximately three for the cascoded OTAs). Additionally, as before, neglecting the gate–source and gate–body overlap capacitances gives $C_{INDIF} \approx \tfrac{1}{2}(C_{gsi1} + C_{gbi1})$ as expressed in Equation 5.43. Again, this is a rough estimate of C_{INDIF} for design guidance only since the overlap capacitances can be significant for short-channel devices having small gate areas and small intrinsic gate capacitances. The full accuracy of Equation 5.46 is used for capacitance predictions provided later by the *Analog CMOS Design, Tradeoffs and Optimization* spreadsheet.

The output capacitance for the cascoded OTAs shown in Figure 5.2 is dominated by the drain–body capacitances of the M_8 and M_{10} output devices, C_{DB8} and C_{DB10}, with a smaller contribution coming from their gate–drain capacitances, C_{GD8} and C_{GD10}. Output capacitance is then given by

$$C_{OUT} = C_{DB8} + C_{DB10} + C_{GD8} + C_{GD10} \tag{5.47}$$

As described for the simple OTAs, C_{OUT} can be approximated using the maximum drain–body capacitances, C_{DBO}, giving $C_{OUT} \approx C_{DBO,8} + C_{DBO,10}$. Expressing this as described for the simple OTAs in Equation 5.45 gives,

$$
\begin{aligned}
C_{OUT} &\approx C_{DBO,8} + C_{DBO,10} \\
&\approx W_8 \left[\tfrac{1}{2} W_{DIF} C_{J8} + C_{JSW8} \right] + W_{10} \left[\tfrac{1}{2} W_{DIF} C_{J10} + C_{JSW10} \right] \\
&\approx \left(\frac{L_8}{IC_8} \right) \left(\frac{I_{D8}}{I_{0,8}} \right) \cdot \left[\tfrac{1}{2} W_{DIF} C_{J8} + C_{JSW8} \right] \\
&\quad + \left(\frac{L_{10}}{IC_{10}} \right) \left(\frac{I_{D10}}{I_{0,10}} \right) \cdot \left[\tfrac{1}{2} W_{DIF} C_{J10} + C_{JSW10} \right]
\end{aligned}
\tag{5.48}
$$

which, again, assumes an even multiplier factor for the output devices.

The process parameters W_{DIF}, C_J, C_{JSW}, and $I_0 = 2n_0\mu_0 C'_{OX} U_T^2$ were described for the simple OTAs and are given in Tables 3.2 and 3.4 for the 0.18 μm CMOS process used in the cascoded OTAs. Equation 5.48, like Equation 5.45 for the simple OTAs, provides rough design guidance. The full accuracy of Equation 5.47 is used for capacitance predictions provided later by the *Analog CMOS Design, Tradeoffs and Optimization* spreadsheet.

Since $I_{0,8} = 0.135\,\mu A$ for pMOS device M_8 compared to $I_{0,10} = 0.64\,\mu A$ for nMOS device M_{10}, M_8 has nearly five times the width of M_{10} for operation at equal drain currents, inversion coefficients, and channel lengths. As a result, pMOS device M_8 dominates the output capacitance.

5.3.9.3 Optimization

For both OTAs, C_{INDIF} is minimized by operating input devices at high inversion coefficients ($IC_1 = IC_2$) in strong inversion at short channel lengths ($L_1 = L_2$), which minimizes their gate area. C_{INDIF} decreases nearly inversely with IC_1, tracking the gate area of the input devices, although the intrinsic gate capacitances have an additional complex dependency on IC as shown in Figure 3.54. C_{INDIF} increases significantly as nearly L_1^2 because both channel length and width must increase together to maintain the selected inversion coefficient, giving a squared increase in input device gate area. Finally, decreasing the input device drain currents decreases C_{INDIF} directly since channel width and, correspondingly, gate area scale directly with the drain current to maintain the selected inversion coefficient.

Trends for C_{INDIF} are similar to trends for device $C_{gsi} + C_{gbi}$ summarized in Table 4.1 and shown in Figures 4.4 and 4.5, where device trends are applied to the OTA input devices. At short channel lengths, C_{INDIF}, however, can be significantly larger than the value associated with intrinsic gate capacitances $C_{gsi} + C_{gbi}$ alone because of significant extrinsic gate-overlap capacitances.

For both OTAs, C_{OUT} is minimized by operating output devices at high inversion coefficients (IC_4 and IC_6 for the simple OTAs, and IC_8 and IC_{10} for the cascoded OTAs) in strong inversion at short channel lengths (L_4 and L_6 for the simple OTAs, and L_8 and L_{10} for the cascoded OTAs). This results in small channel width, giving small drain–body and gate–drain capacitances. C_{OUT} decreases inversely with IC for the output devices and increases directly with L for these devices, tracking device channel widths. Finally, decreasing the output device drain current decreases C_{OUT} directly since channel width scales directly with drain current to maintain the selected inversion coefficient.

Trends for C_{OUT} are similar to trends for device C_{DBO} summarized in Table 4.1 and shown in Figures 4.4 and 4.5, where device trends are applied to the OTA output devices.

5.3.10 Slew Rate

For both OTAs, the output slew rate is set by the maximum available output current divided by the sum of the output capacitance, C_{OUT}, and the external load capacitance, C_L.

5.3.10.1 Simple OTAs

For the simple OTAs shown in Figure 5.1, the maximum available output current is equal to the bias current from M_9 that feeds both input pair devices, or $2I_{D1} = 2I_{D2} = 2I_{BIAS} = 100\,\mu A$. This current is available for both sourcing and sinking, giving a symmetrical slew rate of

$$SR = \frac{\pm 2I_{BIAS}}{C_{OUT} + C_L} \tag{5.49}$$

5.3.10.2 Cascoded OTAs

For the cascoded OTAs shown in Figure 5.2, the maximum available output current is equal to twice the bias current of the M_8 and M_{10} output devices, or $2I_{D8} = 2I_{D10} = 2I_{D8,10} = 100\,\mu A$. This current is available for both sourcing and sinking, giving a symmetrical slew rate of

$$SR = \frac{\pm 2I_{D8,10}}{C_{OUT} + C_L} \tag{5.50}$$

The maximum output current is one-half the bias current of $2I_{D1} = 2I_{D2} = 200\,\mu A$ from M_{15} that feeds both input pair devices, which halves the available slew rate. This is a tradeoff resulting from reduced output device bias current used for reducing the input-referred thermal-noise voltage, flicker-noise voltage, and offset voltage due to local-area mismatch.

5.3.10.3 Optimization

For both OTAs, slew rate is maximized by maximizing the available output current ($2I_{BIAS}$ for the simple OTAs and $2I_{D8,10}$ for the cascoded OTAs) combined with minimizing C_{OUT} and C_L. As mentioned in the previous section, C_{OUT} is minimized by operating output devices at high inversion coefficients in strong inversion at short channel lengths.

5.3.11 Input and Output Voltage Ranges

5.3.11.1 Simple OTAs

The maximum common-mode input voltage for the simple OTAs shown in Figure 5.1 corresponds to the input voltage where the drain–source voltage, V_{DS1}, of the M_1 and M_2 input pair devices is equal to their drain–source saturation voltage, $V_{DS,sat1}$. An input voltage above this causes V_{DS1} to drop below $V_{DS,sat1}$ where M_1 and M_2 enter the ohmic, linear, or triode region causing their device transconductances and drain–source resistances to begin collapsing. As a result, operation above the maximum common-mode input voltage causes a deterioration of OTA performance.

The maximum common-mode input voltage is given by

$$V_{INCM+} = V_{DD} + V_{GS1} - V_{GS5} - V_{DS,sat1}$$

$$= V_{DD} + (V_{EFF1} - V_{EFF5}) + (V_{T1} - V_{T5}) - V_{DS,sat1} \tag{5.51}$$

V_{DD} is the positive supply voltage, V_{GS1} is the gate–source voltage of the M_1 and M_2 input pair, and V_{GS5} is the gate–source voltage of the M_5 and M_7 diode-connected loads. In the second line of the expression, the gate–source voltages are expressed as $V_{GS} = V_{EFF} + V_T$, where $V_{EFF} = V_{GS} - V_T$ is the effective gate–source voltage and V_T is the threshold voltage. As shown in Figure 3.9, all voltages are positive for nMOS devices and are also taken positively for pMOS devices. This includes V_{EFF} that is positive for operation above the threshold voltage for both device types.

Evaluating V_{INCM+} in terms of the inversion coefficient is complicated since V_{EFF} contains both exponential and square-root terms as seen in Table 3.14. However, in strong inversion ($IC > 10$), $V_{EFF} \approx 2nU_T\sqrt{IC}$, excluding velocity saturation increases for short-channel devices operating at high levels of inversion. V_{EFF} can also be approximated using $V_{EFF} \approx 2nU_T\sqrt{IC}$ in the strong-inversion side of moderate inversion ($1 < IC < 10$). This gives a maximum overprediction at the center of moderate inversion at $IC = 1$, where $V_{EFF} \approx 2nU_T\sqrt{IC}$ predicts V_{EFF} at 72 mV, which is 32 mV above its actual value of 40 mV, predicted in Table 3.14 for $n = 1.4$ and $U_T = 25.9$ mV at $T = 300$ K (room temperature). Although overpredicting the value slightly for $1 < IC < 10$ before providing a good prediction in strong inversion for $IC > 10$, $V_{EFF} \approx 2nU_T\sqrt{IC}$ is useful for design guidance, especially since small errors in V_{EFF} usually do not significantly affect the prediction of V_{INCM+} or the minimum common-mode input voltage, V_{INCM-}, described below.

Substituting $V_{EFF} \approx 2nU_T\sqrt{IC}$ for $IC > 1$ and $V_{DS,sat}$ from Table 3.15 into Equation 5.51 gives

$$V_{INCM+} \approx V_{DD} + 2U_T\left(n_1\sqrt{IC_1} - n_5\sqrt{IC_5}\right) + (V_{T1} - V_{T5}) - \left(2U_T\sqrt{IC_1 + 0.25} + 3U_T\right) \qquad (5.52)$$

This equation parallels the second line of Equation 5.51 where the right side contains V_{DD}, the difference of V_{EFF} values, the difference of V_T values, and $V_{DS,sat1}$. Increases in V_{EFF} associated with velocity saturation and VFMR are excluded, where increases in V_{EFF5} can result in an unfavorable decrease in V_{INCM+}. As described in Table 3.14, velocity saturation and VFMR effects are included by replacing IC with $IC(1 + IC/(4IC_{CRIT}))$ in the $\sqrt{IC}$ terms related to V_{EFF}. Equation 5.52 provides rough design guidance, while the full accuracy of Equation 5.51 is used for V_{INCM+} predictions provided later by the *Analog CMOS Design, Tradeoffs and Optimization* spreadsheet.

The minimum common-mode input voltage for the simple OTAs shown in Figure 5.1 corresponds to the input voltage where the drain–source voltage, V_{DS9}, of the M_9, input pair current source is equal to its drain–source saturation voltage, $V_{DS,sat9}$. An input voltage below this causes V_{DS9} to drop below $V_{DS,sat9}$ where M_9 enters the ohmic, linear, or triode region. This causes a collapse in its drain–source resistance and drain current that supplies the M_1 and M_2 input pair. As a result, operation below the minimum common-mode input voltage causes a deterioration of OTA performance.

The minimum common-mode input voltage is given by

$$V_{INCM-} = V_{SS} + V_{GS1} + V_{DS,sat9}$$
$$= V_{SS} + (V_{EFF1} + V_{T1}) + V_{DS,sat9} \qquad (5.53)$$

where V_{SS} is the negative supply voltage and V_{GS1} is, again, separated into its V_{EFF1} and V_{T1} components. Substituting $V_{EFF} \approx 2nU_T\sqrt{IC}$ for $IC > 1$ into Equation 5.53 gives

$$V_{INCM-} \approx V_{SS} + \left(2n_1 U_T\sqrt{IC_1} + V_{T1}\right) + V_{DS,sat9} \qquad (5.54)$$

Increases in V_{EFF1} associated with velocity saturation and VFMR are excluded, where increases in V_{EFF1} can result in an unfavorable increase in V_{INCM-}. As mentioned, velocity saturation and VFMR effects are included by replacing IC with $IC(1 + IC/(4IC_{CRIT}))$ in the $\sqrt{IC}$ term related to V_{EFF}. Equation 5.54 provides rough design guidance, while the full accuracy of Equation 5.53 is used for V_{INCM-} predictions provided later by the *Analog CMOS Design, Tradeoffs and Optimization* spreadsheet.

The maximum output voltage for the simple OTAs shown in Figure 5.1 corresponds to the output voltage where the drain–source voltage, V_{DS6}, of output device M_6 is equal to its drain–source saturation voltage, $V_{DS,sat6}$. An output voltage above this causes V_{DS6} to drop below $V_{DS,sat6}$ where M_6 enters the ohmic, linear, or triode region causing a collapse in its drain–source resistance. This results in a collapse of OTA output resistance and voltage gain.

The maximum output voltage is given by

$$V_{OUT+} = V_{DD} - V_{DS,sat6} \tag{5.55}$$

Although the unloaded output voltage will reach V_{DD}, V_{OUT+} is the maximum output voltage giving the expected OTA output resistance and voltage gain.

Substituting $V_{DS,sat}$ given by Table 3.15 into Equation 5.55 gives

$$V_{OUT+} = V_{DD} - \left(2U_T\sqrt{IC_6 + 0.25} + 3U_T\right) \tag{5.56}$$

This is a good prediction for V_{OUT+}, even if velocity saturation is present. This is because, when evaluated in terms of the inversion coefficient, $V_{DS,sat}$ does not increase significantly due to velocity saturation effects as described in Section 3.7.3.3.

The minimum output voltage for the simple OTAs shown in Figure 5.1 corresponds to the output voltage where the drain–source voltage, V_{DS4}, of output device M_4 is equal to its drain–source saturation voltage, $V_{DS,sat4}$. An output voltage below this causes V_{DS4} to drop below $V_{DS,sat4}$ where M_4 enters the ohmic, linear, or triode region causing a collapse in its drain–source resistance. This results in a collapse of OTA output resistance and voltage gain, just like operation above V_{OUT+}.

The minimum output voltage is given by

$$V_{OUT-} = V_{SS} + V_{DS,sat4} \tag{5.57}$$

Although the unloaded output voltage will reach V_{SS}, just as it will reach V_{DD}, V_{OUT-} is the minimum output voltage giving the expected OTA output resistance and voltage gain.

Substituting $V_{DS,sat}$ given by Table 3.15 into Equation 5.57 gives

$$V_{OUT-} = V_{SS} + \left(2U_T\sqrt{IC_4 + 0.25} + 3U_T\right) \tag{5.58}$$

As described for V_{OUT+}, this is a good prediction for V_{OUT-}, even if velocity saturation is present.

5.3.11.2 Cascoded OTAs

The maximum common-mode input voltage for the cascoded OTAs shown in Figure 5.2, like the maximum input voltage for the simple OTAs, corresponds to the input voltage where the drain–source voltage of the M_1 and M_2 input pair devices is equal to their drain–source saturation voltage. The maximum common-mode input voltage is then given by

$$V_{INCM+} = V_{DD} - V_{DS3} + V_{GS1} - V_{DS,sat1}$$
$$= V_{DD} - V_{DS3} + (V_{EFF1} + V_{T1}) - V_{DS,sat1} \tag{5.59}$$

As for the simple OTAs, V_{DD} is the positive supply voltage, V_{GS1} is the gate–source voltage of the M_1 and M_2 input pair, and $V_{DS,sat1}$ is the drain–source saturation voltage for M_1 and M_2. $V_{DS3} = V_{DS4}$ is the drain–source voltage across the M_3 and M_4 current source devices. In the second line of the expression, V_{GS1} is, again, expressed as $V_{GS1} = V_{EFF1} + V_{T1}$.

$V_{DS3} = V_{DS4}$ is set at 0.4 V by the external bias voltage, V_{BIAS1}, through an external, diode-connected MOSFET. $V_{DS3} = V_{DS4}$ is sufficiently above the drain–source saturation voltage of M_3 and M_4, which is approximately 0.3 V for an inversion coefficient of $IC_3 = IC_4 \approx 16$ well into strong inversion.

As mentioned for the simple OTAs, V_{EFF} can be approximated for $1 < IC < 10$ and well predicted in strong inversion for $IC > 10$ using $V_{EFF} \approx 2nU_T\sqrt{IC}$. Substituting this and $V_{DS,sat}$ given by Table 3.15 into Equation 5.59 gives

$$V_{INCM+} \approx V_{DD} - V_{DS3} + \left(2n_1 U_T\sqrt{IC_1} + V_{T1}\right) - \left(2U_T\sqrt{IC_1 + 0.25} + 3U_T\right) \tag{5.60}$$

As mentioned for the simple OTAs, velocity saturation and VFMR effects are included by replacing IC with $IC(1 + IC/(4IC_{CRIT}))$ in the $\sqrt{IC}$ term related to V_{EFF}. Equation 5.60 provides rough design guidance, while the full accuracy of Equation 5.59 is used for V_{INCM+} predictions provided later by the *Analog CMOS Design, Tradeoffs and Optimization* spreadsheet.

The minimum common-mode input voltage for the cascoded OTAs shown in Figure 5.2, like the minimum input voltage for the simple OTAs, corresponds to the input voltage where the drain–source voltage of the input pair current source is equal to its drain–source saturation voltage. The minimum common-mode input voltage is then given by

$$\begin{aligned} V_{INCM-} &= V_{SS} + V_{GS1} + V_{DS,sat15} \\ &= V_{SS} + (V_{EFF1} + V_{T1}) + V_{DS,sat15} \end{aligned} \tag{5.61}$$

As for the simple OTAs, V_{SS} is the negative supply voltage. $V_{DS,sat15}$ is the drain–source saturation voltage for the M_{15} input pair current source. Finally, in the second line of the expression, V_{GS1} is, again, expressed as $V_{GS1} = V_{EFF1} + V_{T1}$.

Substituting $V_{EFF} \approx 2nU_T\sqrt{IC}$ for $IC > 1$ into Equation 5.61 gives

$$V_{INCM-} \approx V_{SS} + \left(2n_1 U_T\sqrt{IC_1} + V_{T1}\right) + V_{DS,sat15} \tag{5.62}$$

Again, velocity saturation and VFMR effects are included by replacing IC with $IC(1 + IC/(4IC_{CRIT}))$ in the $\sqrt{IC}$ term related to V_{EFF}. Equation 5.62 provides rough design guidance, while the full accuracy of Equation 5.61 is used for V_{INCM-} predictions provided later by the *Analog CMOS Design, Tradeoffs and Optimization* spreadsheet.

The maximum output voltage for the cascoded OTAs shown in Figure 5.2, like the maximum output voltage for the simple OTAs, corresponds to the output voltage where the drain–source voltage of the positive-side output device is at its drain–source saturation value. The maximum OTA output voltage is then given by

$$V_{OUT+} = V_{DD} - V_{DS4} - V_{DS,sat8} \tag{5.63}$$

$V_{DS3} = V_{DS4}$ is the drain–source voltage across the M_3 and M_4 current source devices, which is equal to $0.4\,V$ as mentioned earlier. $V_{DS,sat8}$ is the drain–source saturation voltage for the M_8 output device. Although the unloaded output voltage will reach $V_{DD} - V_{DS4}$, V_{OUT+} is the maximum output voltage giving the expected OTA output resistance and voltage gain.

Substituting $V_{DS,sat}$ given by Table 3.15 into Equation 5.63 gives

$$\begin{aligned} V_{OUT+} &= V_{DD} - V_{DS4} - V_{DS,sat8} \\ &= V_{DD} - V_{DS4} - \left(2U_T\sqrt{IC_8 + 0.25} + 3U_T\right) \end{aligned} \tag{5.64}$$

As mentioned for the simple OTAs, this is a good prediction for V_{OUT+}, even if velocity saturation is present.

The minimum output voltage for the cascoded OTAs shown in Figure 5.2, like the minimum output voltage for the simple OTAs, corresponds to the output voltage where the drain–source voltage of the negative-side output device is at its drain–source saturation value. The minimum output voltage is then given by

$$V_{OUT-} = V_{SS} + V_{DS6} + V_{DS,sat10} \qquad (5.65)$$

$V_{DS5} = V_{DS6}$ is the drain–source voltage across the M_5 and M_6 current-mirror devices, and $V_{DS,sat10}$ is the drain–source saturation voltage for the M_{10} output device. Although the unloaded output voltage will reach $V_{SS} + V_{DS6}$, V_{OUT-} is the minimum output voltage giving the expected OTA output resistance and voltage gain.

$V_{DS5} = V_{DS6}$ is set at 0.4 V by the external bias voltage, V_{BIAS2}, through an external, diode-connected MOSFET. $V_{DS5} = V_{DS6}$ is sufficiently above the drain–source saturation voltage of M_5 and M_6, which is approximately 0.3 V for an inversion coefficient of $IC_5 = IC_6 \approx 16$ well into strong inversion. $V_{DS5} = V_{DS6}$ is set equal to $V_{DS3} = V_{DS4} = 0.4$ V because the M_3 and M_4 current source devices, and M_5 and M_6 current-mirror devices, operate at equal inversion coefficients and drain–source saturation voltages.

Substituting $V_{DS,sat}$ given by Table 3.15 into Equation 5.65 gives

$$\begin{aligned} V_{OUT-} &= V_{SS} + V_{DS6} + V_{DS,sat10} \\ &= V_{SS} + V_{DS6} + \left(2U_T\sqrt{IC_{10}+0.25} + 3U_T\right) \end{aligned} \qquad (5.66)$$

Again, this is a good prediction for V_{OUT-}, even if velocity saturation is present.

5.3.11.3 Optimization

For both OTAs, V_{INCM+} is nearly independent of the level of inversion for the nMOS M_1 and M_2 input pair devices because V_{INCM+} includes the term $V_{GS1} - V_{DS,sat1}$, where both V_{GS1} and $V_{DS,sat1}$ increase as IC_1 increases. For the simple OTAs, V_{INCM+} is favorably maximized by operating the M_5–M_8 pMOS current mirrors at low inversion coefficients ($IC_5 = IC_6 = IC_7 = IC_8$) in weak or moderate inversion, which minimizes V_{EFF5} and V_{GS5} that appear negated in the V_{INCM+} expressions. For the cascoded OTAs, V_{INCM+} is favorably maximized for low values of $V_{DS3} = V_{DS4}$ across the M_3 and M_4 pMOS current sources, which also appear negated in the V_{INCM+} expressions. However, for low noise applications, both the pMOS current mirrors in the simple OTAs and pMOS current sources in the cascoded OTAs must operate in strong inversion to minimize their g_m/I_D, g_m, and drain thermal-noise current contributions. This increases V_{EFF5} and V_{GS5} for the simple OTAs and increases the required value of $V_{DS3} = V_{DS4}$ for the cascoded OTAs, both of which lower the value of V_{INCM+}. This is an example of the opposing tradeoffs between bias compliance voltages and circuit noise. These tradeoffs are discussed in detail in Section 6.5 for the design of two micropower, low-noise CMOS preamplifiers.

For the simple OTAs, V_{INCM+} is equal to $V_{DD} - V_{DS,sat1}$ if the nMOS input pair and pMOS current-mirror devices have equal V_{GS} values resulting from equal values of V_{EFF} and V_T. Similarly, for the cascoded OTAs, V_{INCM+} is equal to $V_{DD} - V_{DS,sat1}$ if V_{GS} for the nMOS input pair and V_{DS} across the pMOS current sources are equal. If V_{GS} for the input pair exceeds these values, V_{INCM+} can reach or even exceed V_{DD}. This sometimes occurs when non-zero V_{SB} raises the threshold voltage for the input pair devices.

Trends for V_{INCM+} can be seen by observing the increasing V_{EFF} and $V_{DS,sat}$ with increasing IC shown in Figure 4.6. Here, IC corresponds to $IC_5 = IC_6 = IC_7 = IC_8$ for the pMOS current-mirror devices in the simple OTAs and $IC_3 = IC_4$ for the pMOS current source devices in the cascoded OTAs. The increasing value of V_{EFF} corresponds to a decreasing value of V_{INCM+} for the simple OTAs through the increase in V_{EFF5}, while the increasing value of $V_{DS,sat}$ corresponds to a decreasing value of V_{INCM+} for the cascoded OTAs through the required increase in $V_{DS3} = V_{DS4}$. A decreasing value of V_{INCM+} decreases the allowable window of OTA input voltages.

For both OTAs, V_{INCM-} is favorably minimized by operating the input pair devices at low inversion coefficients ($IC_1 = IC_2$) in weak or moderate inversion, which minimizes V_{EFF1} and V_{GS1}. Trends for V_{INCM-} can be seen by observing the increasing V_{EFF} with increasing IC shown in Figure 4.6, where IC corresponds to $IC_1 = IC_2$ for the input pair devices. The increasing value of V_{EFF1} corresponds to an increasing value of V_{INCM-}, which decreases the allowable window of OTA input voltages. Finally, V_{INCM-} is favorably minimized by reducing the drain–source saturation voltage of the input pair current sources. For the simple OTAs, $V_{DS,sat9}$ is approximately 0.25 V for $IC_9 = 12$ listed in the schematic of Figure 5.1 for $U_T = 25.9$ mV at $T = 300$ K (room temperature). For the cascoded OTAs, $V_{DS,sat15}$ is approximately 0.23 V for $IC_{15} = 9$ listed in the schematic of Figure 5.2. Operating both current source devices in weak inversion lowers their $V_{DS,sat}$ to approximately 0.1 V. This then lowers V_{INCM-} by approximately 0.15 V.

For both OTAs, V_{OUT+} is favorably maximized by operating the positive-side output devices at low inversion coefficients, IC_6 for M_6 in the simple OTAs and IC_8 for M_8 in the cascoded OTAs. This minimizes $V_{DS,sat}$ for the output devices, which appears negated in the V_{OUT+} expressions. Additionally, V_{OUT+} is favorably maximized for the cascoded OTAs for low values of $V_{DS3} = V_{DS4}$ across the M_3 and M_4 pMOS current sources, which also appear negated in the V_{OUT+} expressions. However, as mentioned for low noise applications, both the $M_5 - M_8$ pMOS current mirrors in the simple OTAs and M_3 and M_4 pMOS current sources in the cascoded OTAs must operate in strong inversion to minimize their g_m/I_D, g_m, and drain thermal-noise current contributions. This increases $V_{DS,sat6}$ for the simple OTAs and increases the required value of $V_{DS3} = V_{DS4}$ for the cascoded OTAs, both of which lower the value of V_{OUT+}. This, again, is an example of the opposing tradeoffs between bias compliance voltages and circuit noise.

Trends for V_{OUT+} can be seen by observing the increasing $V_{DS,sat}$ with increasing IC shown in Figure 4.6. Here, IC corresponds to $IC_5 = IC_6 = IC_7 = IC_8$ for pMOS current-mirror devices in the simple OTAs and $IC_3 = IC_4$ for pMOS current source devices in the cascoded OTAs, along with IC_8 for the pMOS output device. The increasing value of $V_{DS,sat}$ corresponds to a decreasing value of V_{OUT+} for the simple OTAs through the increase in $V_{DS,sat6}$. The increasing value of $V_{DS,sat}$ corresponds to a decreasing value of V_{OUT+} for the cascoded OTAs through the required increase in $V_{DS3} = V_{DS4}$, combined with the increase in $V_{DS,sat8}$. A decreasing value of V_{OUT+} decreases the allowable window of OTA output voltages.

For both OTAs, V_{OUT-} is favorably minimized by operating the negative-side output devices at low inversion coefficients, IC_4 for M_4 in the simple OTAs and IC_{10} for M_{10} in the cascoded OTAs. This minimizes $V_{DS,sat}$ for the output devices, which appears directly in the V_{OUT-} expressions. Additionally, V_{OUT-} is favorably minimized for the cascoded OTAs for low values of $V_{DS5} = V_{DS6}$ across the M_5 and M_6 nMOS current mirrors, which also appear directly in the V_{OUT-} expressions. However, for low noise applications, the M_3 and M_4 nMOS current mirrors in the simple OTAs and M_5 and M_6 nMOS current mirrors in the cascoded OTAs must operate in strong inversion to minimize their g_m/I_D, g_m, and drain thermal-noise current contributions. This increases $V_{DS,sat4}$ for the simple OTAs and increases the required value of $V_{DS5} = V_{DS6}$ for the cascoded OTAs, both of which increase the value of V_{OUT-}. This, again, is an example of the opposing tradeoffs between bias compliance voltages and circuit noise.

Trends for V_{OUT-} can be seen by observing the increasing $V_{DS,sat}$ with increasing IC shown in Figure 4.6. Here, IC corresponds to $IC_3 = IC_4$ for the nMOS current-mirror devices in the simple OTAs and $IC_5 = IC_6$ for the nMOS current-mirror devices in the cascoded OTAs, along with IC_{10} for the nMOS output device. The increasing value of $V_{DS,sat}$ corresponds to an increasing value of V_{OUT-} for the simple OTAs through the increase in $V_{DS,sat6}$. The increasing value of $V_{DS,sat}$ corresponds to an increasing value of V_{OUT-} for the cascoded OTAs through the required increase in $V_{DS5} = V_{DS6}$, combined with the increase in $V_{DS,sat10}$. An increasing value of V_{OUT-} decreases the allowable window of OTA output voltages.

5.3.12 Input, 1 dB Compression Voltage

As discussed in Section 3.8.2.6, the input, 1 dB compression voltage, $V_{INDIF1dB}$, for a differential pair is the differential input voltage where the differential output current is compressed 1 dB below its ideal value. The ideal value of output current is the input voltage multiplied by the bias point transconductance of the differential pair, which is also the bias point transconductance of the individual pair devices.

5.3.12.1 Simple OTAs

For the simple OTAs shown in Figure 5.1, $V_{INDIF1dB}$ for the OTAs is equal to $V_{INDIF1dB}$ for the input M_1 and M_2 differential pair because the differential-pair, output difference current appears directly at the OTA output. $V_{INDIF1dB}$ is given directly in the *Analog CMOS Design, Tradeoffs and Optimization* spreadsheet for differential-pair devices and thus requires no further calculation for finding OTA $V_{INDIF1dB}$.

OTA $V_{INDIF1dB}$ is inversely proportional to g_m/I_D of the M_1 and M_2 differential-pair devices with proportionality constants of 1.22 and 1.81 for operation in weak and strong inversion, respectively. It is given here for the M_1 and M_2 differential pair from Table 3.19 as

$$V_{INDIF1dB} = V_{INDIF1dB}(M_1, M_2)$$

$$= \frac{\left\{ \begin{matrix} 1.22,\ \text{WI} \\ 1.81,\ \text{SI} \end{matrix} \right\}}{\left(\dfrac{g_m}{I_D} \right)_1} = \left\{ \begin{matrix} 1.22,\ \text{WI} \\ 1.81,\ \text{SI} \end{matrix} \right\} \cdot n_1 U_T \left(\sqrt{IC_1 + 0.25} + 0.5 \right) \qquad (5.67)$$

Decreases in g_m/I_D associated with velocity saturation and VFMR are excluded. As discussed in Section 3.8.2.6, these can increase $V_{INDIF1dB}$ through its inverse proportionality with g_m/I_D, combined with an increase in the proportionality constant resulting from improved drain current linearity. As described in Section 5.3.1.1, velocity saturation and VFMR effects are included by replacing IC with $IC(1 + IC/IC_{CRIT})$ in the $\sqrt{IC + 0.25}$ term related to g_m/I_D. Additionally, the proportionality constant is increased as shown in Figure 3.29. Equation 5.67 provides design guidance, while the full accuracy of an iterative solution for $V_{INDIF1dB}$ is used for predictions provided later by the *Analog CMOS Design, Tradeoffs and Optimization* spreadsheet.

5.3.12.2 Cascoded OTAs

For the cascoded OTAs shown in Figure 5.2, $V_{INDIF1dB}$ for the OTAs is less than $V_{INDIF1dB}$ for the input M_1 and M_2 differential pair. This is because the OTA output device bias currents are lower at $I_{D8} = I_{D10} = 50\,\mu\text{A}$, compared to bias currents of $I_{D1} = I_{D2} = 100\,\mu\text{A}$ for the input pair devices. As a result, current compresses first at the output devices. The reduction of $V_{INDIF1dB}$ is a tradeoff resulting from lower output device bias currents used to lower the noise and mismatch contributions from the M_5 and M_6 current-mirror devices.

OTA $V_{INDIF1dB}$ is approximated by dividing the maximum output current, which is equal to twice the output device bias current, $2I_{D8,10}$, by the OTA transconductance, G_M, from Equation 5.2. This gives

$$V_{INDIF1dB} \approx \frac{2I_{D8,10}}{G_M}$$

$$\approx \frac{2I_{D8,10}}{\dfrac{I_{D1}}{n_1 U_T \left(\sqrt{IC_1 + 0.25} + 0.5 \right)}} = \frac{2I_{D8,10}}{I_{D1}} \cdot n_1 U_T \left(\sqrt{IC_1 + 0.25} + 0.5 \right) \qquad (5.68)$$

As for the simple OTAs, $IC_1 = IC_2$ is the inversion coefficient for the M_1 and M_2 input devices. Again, velocity saturation and VFMR effects are included by replacing IC with $IC(1 + IC/IC_{CRIT})$ in the $\sqrt{IC + 0.25}$ term related to g_m/I_D.

5.3.12.3 *Optimization*

For both OTAs, $V_{INDIF1dB}$ is favorably maximized by operating the M_1 and M_2 input devices at high inversion coefficients ($IC_1 = IC_2$) in strong inversion where g_m/I_D is minimum. However, the input-referred thermal-noise voltage is minimized by operating the input devices at low levels of inversion in weak or moderate inversion where g_m/I_D and g_m are maximum. Additionally, operating input devices at low levels of inversion helps the input devices dominate the thermal noise, flicker noise, and DC offset due to local-area mismatch. This illustrates the opposing tradeoffs between minimizing input-referred noise and mismatch and maximizing $V_{INDIF1dB}$, which minimizes signal distortion.

Trends for $V_{INDIF1dB}$ for the simple OTAs are the same as trends for differential-pair $V_{INDIF1dB}$ summarized in Table 4.2 and shown in Figure 4.11, where trends are applied to the OTA input devices. Trends for $V_{INDIF1dB}$ for the cascoded OTAs, however, are the same as the inverse of trends for device g_m/I_D, also summarized in Table 4.2 and shown in Figure 4.11. Again, trends are applied to the input devices. The inverse of trends for g_m/I_D can be seen by flipping vertically the logarithmic presentation of g_m/I_D in Figure 4.11. For both OTAs, $V_{INDIF1dB}$ is constant and minimum for input device operation in weak inversion, increasing as the square root of IC_1 in strong inversion and as IC_1 or more when velocity saturation reductions in g_m/I_D are significant. Although the $V_{INDIF1dB}$ trends are similar, $V_{INDIF1dB}$ is lower for the cascoded OTAs because signal compression occurs at the output devices prior to compression at the differential-pair devices. Finally, $V_{INDIF1dB}$ is independent of OTA bias currents (again, MOSFET width is scaled with drain current to maintain the selected inversion coefficient), assuming constant bias current ratios are maintained.

5.3.13 Management of Small-Geometry Effects

The preceding discussions of OTA circuit analysis and performance included small-geometry velocity saturation and VFMR decreases in g_m/I_D and g_m, and increases in V_{EFF} through the expression modifications mentioned. Additionally, DIBL decreases in drain–source resistance for short-channel devices and increases in gate-referred flicker-noise voltage with increasing inversion or V_{EFF} were included in the Early voltage factor, V_{AL}, and flicker-noise factor, K'_F, respectively. Finally, OTA performance trends were related to device performance trends shown in Chapter 4 that included small-geometry effects. In addition to their inclusion in the preceding discussions, small-geometry velocity saturation, VFMR, DIBL, and flicker-noise effects are included in the *Analog CMOS Design, Tradeoffs and Optimization* spreadsheet.

Velocity saturation decreases g_m/I_D and g_m by 18% in the AC optimized, cascoded OTA for the nMOS, M_5 and M_6 current-mirror devices that operate at $IC \approx 16$ in strong inversion at the minimum channel length of $L = 0.18\,\mu$m. DIBL decreases V_A by nearly 45% in the AC optimized, simple OTA for the M_4 and M_6 output devices that operate at $IC \approx 10$ in strong inversion at the minimum channel length of $L = 0.5\,\mu$m. Also, DIBL decreases V_A by nearly 30% in the AC optimized, cascoded OTA for the M_8 and M_{10} output devices that operate at $IC \approx 2.5$ in moderate inversion at the minimum channel length of $L = 0.18\,\mu$m. Finally, increases in flicker noise with increasing inversion or V_{EFF} are not significant for the simple OTAs, but these are significant in the cascoded OTAs for the pMOS, M_3 and M_4 current source devices that operate at $IC \approx 16$ in strong inversion. Here, K'_F increases nearly a factor of four from its K'_{F0} value. pMOS flicker-noise increases are also important for the micropower, low-noise, $0.35\,\mu$m CMOS preamplifiers described in Chapter 6.

The numerical values given for small-geometry effects are found by the spreadsheet and are included in predictions of OTA performance. Performance is compared to performance without velocity saturation, VFMR, and DIBL effects by setting process values of E_{CRIT} very high and setting Θ and *DVTDIBL* to zero.

Small-geometry effects can be managed by careful design. For example, velocity saturation decreases in g_m/I_D and g_m would typically be minimized for input pair devices to minimize the resulting decrease of OTA transconductance and voltage gain, and increase in input-referred thermal-noise voltage. As discussed in Sections 3.8.2.2 and 3.8.2.3, velocity saturation effects are minimized by avoiding the use of short-channel devices, particularly nMOS devices, at high levels of inversion. Input devices would typically not be operated at high levels of inversion because of the normal decrease in g_m/I_D and g_m. Velocity saturation increases in V_{EFF} would typically be minimized for devices where the increases would decrease the allowable, OTA input common-mode voltage range. DIBL decreases in drain–source resistance would typically be minimized for devices that influence OTA output resistance and, correspondingly, the voltage gain. As discussed in Sections 3.8.4.2 and 3.8.4.5, DIBL effects are minimized by avoiding the use of short-channel devices at low levels of inversion. Finally, a more difficult small-geometry effect to manage is the increase in gate-referred flicker-noise voltage with increasing inversion or V_{EFF} that can increase the OTA input-referred flicker-noise voltage. As discussed in Sections 3.10.3.3 and 3.10.3.6, flicker-noise increases are minimized by avoiding the use of pMOS devices at high levels of inversion.

Although not a concern for the OTAs illustrated here that have gate-oxide thickness of 4.1 nm or larger, gate leakage current can be significant for gate-oxide thickness around 2 nm and below as described in Section 3.12.1. Gate–source conductance, gate shot- and flicker-noise current, and mismatch due to gate leakage current can deteriorate circuit performance as described in Section 3.12.2. As a result, a smaller channel length and resulting gate area may be required to balance the traditional improvements in gain, flicker noise, and local-area mismatch at increasing channel length and gate area with the deterioration of these associated with increasing gate leakage current caused by increasing gate area. Although not included in the *Analog CMOS Design, Tradeoffs and Optimization* spreadsheet, gate leakage current effects could be included in later versions of the spreadsheet.

5.4 DESIGN OPTIMIZATION AND RESULTING PERFORMANCE FOR THE SIMPLE OTAS

To illustrate the OTA analysis and performance optimization discussed in Section 5.3, three versions of the simple, 0.5 μm CMOS OTAs shown in Figure 5.1 are optimized for DC, balanced, and AC performance [1]. The optimization of these OTAs parallels the optimization of differential pairs and current mirrors described earlier in Section 4.4.4. To investigate optimizations at constant power consumption, each OTA version maintains equal bias currents of 50 μA and constant supply voltages of ±1.25 V. This gives a core (excludes bias reference) current consumption of 200 μA and power consumption of 500 μW.

5.4.1 Selection of MOSFET Inversion Coefficients and Channel Lengths

The optimum design of analog CMOS circuits often requires operating devices at different inversion coefficients and channel lengths, corresponding to performance tradeoffs at different locations on the *MOSFET Operating Plane* shown in Figure 3.1. Additionally, devices may be operated at different drain currents to achieve optimal circuit performance. However, as shown in Figure 5.1, all signal-path devices operate at the same inversion coefficient, channel length, and drain current for a given version of the simple OTAs. As described in Sections 5.3.5–5.3.7, this does not optimize input-referred thermal-noise voltage, flicker-noise voltage, or offset voltage due to local-area mismatch because non-input devices contribute significantly to the noise and mismatch.

Although non-optimal for noise and mismatch performance, equally sized nMOS devices and equally sized pMOS devices, usually with increased channel width to give the same inversion coefficient or V_{EFF} as nMOS devices, are common in simple OTA circuits designed for general-purpose applications. This provides a starting point for design examples, while still demonstrating substantially different performance for the DC, balanced, and AC optimized OTA versions. Later, Section 5.5 presents the cascoded, CMOS OTAs where input devices dominate thermal noise. Following this, Chapter 6 presents micropower, low-noise, CMOS preamplifiers where input devices dominate both thermal and flicker noise.

Table 5.1 summarizes predictions of performance for the simple OTAs that were derived earlier in Section 5.3. The table also shows OTA performance trends as device inversion coefficient, channel length, and bias current are explored separately, where performance trends were described in detail in Section 5.3. As noted, blank lines indicate little change in performance. Up and down arrows indicate performance that increases or decreases, with the degree of the increase or decrease indicated by the size of the arrow. When multiple trends are included, the first is for weak inversion, and the second is for strong inversion without small-geometry effects. Table 5.1 is similar in format to Tables 4.1–4.3 that summarized performance trends common to all devices and devices used in differential pairs and current mirrors.

Although inversion coefficients, channel lengths, and drain currents are equal for devices in a given version of the simple OTAs, the trends shown in Table 5.1 are assigned to devices primarily responsible for the specified performance. Thermal-noise, flicker-noise, and local-area mismatch trends are assigned to the M_1 and M_2 input pair, but these also depend on the relative noise and mismatch contributions of non-input devices as reflected in the RNP and RMP values that were defined in Sections 5.3.5.1, 5.3.6.1, and 5.3.7.1. The trends shown assume that the relative noise and mismatch contributions of non-input devices remain constant, which they do since device inversion coefficients, channel lengths, and drain currents are equal for a given version of the simple OTAs.

The trends shown in Table 5.1 are simplified by excluding small-geometry velocity saturation and VFMR decreases in g_m/I_D and g_m, and related increases in V_{EFF}. Additionally, the trends shown exclude small-geometry DIBL decreases in V_A and r_{ds} for short-channel devices operating especially at low levels of inversion and exclude increases in the gate-referred flicker-noise voltage (included in the value of K_F') at increasing inversion or V_{EFF}. These effects were included in the analysis of Section 5.3 and, as mentioned in Section 5.3.13, are generally not significant for the simple OTAs. For completeness and application to general design, these effects are included through expression modifications summarized in the notes at the bottom of Table 5.1. As mentioned in Section 5.3.13, small-geometry effects that deteriorate OTA performance would typically be managed by careful design. Finally, trends shown in Table 5.1 exclude mismatch increases caused by transconductance factor mismatch at high levels of inversion. These increases are included in the value of A_{VGS}.

The DC, balanced, and AC optimizations described below for the simple OTAs track a diagonal path from the upper left corner to the lower right corner on the *MOSFET Operating Plane* shown in Figure 3.1. This corresponds to increasing inversion coefficients and simultaneously decreasing channel lengths across the DC, balanced, and AC optimizations. This gives the largest possible changes in performance and parallels the discussion of optimization for differential-pair and current-mirror devices described in Section 4.4.4. Different optimizations will be described in Section 5.5.1 for the cascoded OTAs, where the optimizations track a vertical path from top to bottom on the *Plane*. This corresponds to constant inversion coefficients, although different inversion coefficients for different devices, and decreasing channel lengths, giving modified DC, balanced, and AC optimizations. Here, many aspects of performance do not change while other aspects of performance change less compared to the simple OTAs.

Table 5.1 Summary of predicted performance and performance trends versus device inversion coefficient, channel length, and bias current for the simple OTAs. Trends are given for the devices primarily responsible for the performance identified where noise and mismatch trends assume constant relative noise and mismatch contributions from non-input devices

Parameter	Trends		
	$IC \uparrow$ L, I_{BIAS} fixed	$L \uparrow$ IC, I_{BIAS} fixed	$I_{BIAS} \uparrow$ IC, L fixed
Transconductance, output resistance, and voltage gain:			
$G_M = \dfrac{I_{BIAS}}{n_1 U_T \left(\sqrt{IC_1 + 0.25} + 0.5 \right)}$	$-, \downarrow M_{1,2}$	$-$	$\uparrow$
$R_{OUT} = \dfrac{(V_{AL4} \cdot L_4) \parallel (V_{AL6} \cdot L_6)}{I_{BIAS}}$	$-$	$\uparrow M_{4,6}$	$\downarrow$
$A_V = \dfrac{(V_{AL4} \cdot L_4) \parallel (V_{AL6} \cdot L_6)}{n_1 U_T \left(\sqrt{IC_1 + 0.25} + 0.5 \right)}$	$-, \downarrow M_{1,2}$	$\uparrow M_{4,6}$	$-$
Transconductance bandwidth:			
$f_{-3dB} <$ $\dfrac{1}{2} \left(\dfrac{IC_5}{\sqrt{IC_5 + 0.25} + 0.5} \right) \left(\dfrac{\mu_0 U_T}{\pi \left(\hat{C}_{gsi5} + \hat{C}_{gbi5} \right) L_5^2} \right)$	$\uparrow, \uparrow M_{5-8}$	$\downarrow\downarrow M_{5-8}$	$-$
Input-referred thermal-noise voltage PSD:			
$S_{VIN}(\text{thermal}) =$ $2 \cdot 4kT \left(n_1^2 \Gamma_1 \right) \left(\dfrac{U_T \left(\sqrt{IC_1 + 0.25} + 0.5 \right)}{I_{BIAS}} \right)$ $\cdot [1 + RNP_3 + RNP_5 + RNP_7]$	$-, \uparrow M_{1,2}$	$-$	$\downarrow$
Input-referred flicker-noise voltage PSD:			
$S_{VIN}(\text{flicker}) = 2 \cdot \dfrac{K'_{F1}}{f^{AF1}} \left(\dfrac{IC_1}{L_1^2} \right) \left(\dfrac{I_{0,1}}{I_{BIAS}} \right)$ $\cdot [1 + RNP_3 + RNP_5 + RNP_7]$	$\uparrow M_{1,2}$	$\downarrow\downarrow M_{1,2}$	$\downarrow$
Input-referred offset voltage variance due to local-area mismatch:			
$V_{INOFFSET}^2(\text{mismatch}) = A_{VGS1}^2 \left(\dfrac{IC_1}{L_1^2} \right) \left(\dfrac{I_{0,1}}{I_{BIAS}} \right)$ $\cdot [1 + RMP_3 + RMP_5 + RMP_7]$	$\uparrow M_{1,2}$	$\downarrow\downarrow M_{1,2}$	$\downarrow$
Input-referred offset voltage due to systematic offset:			
$V_{INOFFSET}(\text{systematic}) = \dfrac{-(V_{OUT} - V_{SS} - V_{GS3})}{A_V}$	$-, \uparrow M_{1,2}$	$\downarrow M_{4,6}$	$-$
Input and output capacitances:			
$C_{INDIF} \approx \dfrac{1}{2} \left(\dfrac{L_1^2}{IC_1} \right) \left(\dfrac{I_{BIAS}}{I_{0,1}} \right) C'_{OX} \left(\hat{C}_{gsi1} + \hat{C}_{gbi1} \right)$	$\downarrow M_{1,2}$	$\uparrow\uparrow M_{1,2}$	$\uparrow$

Table 5.1 *(continued)*

Parameter	Trends		
	$IC \uparrow$ L, I_{BIAS} fixed	$L \uparrow$ IC, I_{BIAS} fixed	$I_{BIAS} \uparrow$ IC, L fixed
$C_{OUT} \approx \left(\dfrac{L_4}{IC_4}\right)\left(\dfrac{I_{BIAS}}{I_{0,4}}\right) \cdot \left[\dfrac{1}{2} W_{DIF} C_{J4} + C_{JSW4}\right]$ $+ \left(\dfrac{L_6}{IC_6}\right)\left(\dfrac{I_{BIAS}}{I_{0,6}}\right) \cdot \left[\dfrac{1}{2} W_{DIF} C_{J6} + C_{JSW6}\right]$	$\downarrow M_{4,6}$	$\uparrow M_{4,6}$	$\uparrow$

Slew rate (trends assume $C_L \gg C_{OUT}$):

$SR = \dfrac{\pm 2 I_{BIAS}}{C_{OUT} + C_L}$	$-$	$-$	$\uparrow$

Input and output voltage ranges:

V_{INCM+} (Equation 5.52)	$-, \downarrow M_{5-8}$	$-$	$-$
$V_{INCM-} \approx V_{SS} + \left(2 n_1 U_T \sqrt{IC_1} + V_{T1}\right) + V_{DS,sat9}$	$-, \uparrow M_{1,2}$	$-$	$-$
$V_{OUT+} = V_{DD} - \left(2 U_T \sqrt{IC_6 + 0.25} + 3 U_T\right)$	$-, \downarrow M_6$	$-$	$-$
$V_{OUT-} = V_{SS} + \left(2 U_T \sqrt{IC_4 + 0.25} + 3 U_T\right)$	$-, \uparrow M_4$	$-$	$-$

Input, 1 dB compression voltage:

$V_{INDIF1dB} = \begin{Bmatrix} 1.22, & WI \\ 1.81, & SI \end{Bmatrix} \cdot n_1 U_T \left(\sqrt{IC_1 + 0.25} + 0.5\right)$	$-, \uparrow M_{1,2}$	$-$	$-$

Trend key (multiple trends are for WI, SI):

$-$	Generally little or no dependency.
$\uparrow, \downarrow$	Generally increases sublinearly or decreases inversely as sublinear.
$\uparrow, \downarrow$	Generally increases linearly or decreases inversely.
$\uparrow\uparrow, \downarrow\downarrow$	Generally increases as the square or decreases as the inverse square.

Trends and expressions exclude decreases in G_M, A_V, and f_{-3dB} and increases in S_{VIN} thermal, $V_{INOFFSET}$ systematic, and $V_{INDIF1dB}$ for short-channel devices at high inversion due to velocity saturation and VFMR decreases in g_m/I_D and g_m. These effects may be included by replacing IC with $IC(1 + IC/IC_{CRIT})$ in $\sqrt{IC + 0.25}$ terms, except for those terms in V_{OUT+} and V_{OUT-} expressions. As given in Table 3.17, $IC_{CRIT} = [(LE_{CRIT})'/(4 n U_T)]^2$ and $(LE_{CRIT})' = (1/\theta) \parallel LE_{CRIT}$.

Trends and expressions exclude decreases in V_{INCM+} and increases in V_{INCM-} due to velocity saturation and VFMR increases in V_{EFF}. These effects may be included by replacing IC with $IC(1 + IC/(4 IC_{CRIT}))$ in IC terms in V_{INCM+} and V_{INCM-} expressions.

Trends exclude decreases in R_{OUT} and A_V, and increases in $V_{INOFFSET}$ systematic, for short-channel devices at low inversion due to DIBL. CLM and DIBL effects are included in the value of V_{AL} (Figure 3.48) where $V_A = V_{AL} \cdot L$ (Figure 3.47) and, correspondingly, r_{ds} increase sublinearly with L, except for short-channel devices where the increase is greater as DIBL effects are reduced.

Trends exclude increases in gate-referred flicker-noise voltage PSD with increasing inversion or V_{EFF}. These effects are included in the value of K_F'.

Trends exclude increases in input-referred offset voltage variance due to transconductance factor mismatch at high levels of inversion. These effects are included in the value of A_{VGS}.

5.4.1.1 DC optimization

A DC optimization corresponds to operation at low inversion coefficients, in weak or moderate inversion, at long channel lengths. Performance tradeoffs associated with a DC optimization are shown in the upper left portion of the *Plane* in Figure 3.1. These are also listed in Table 4.5 and observed in Table 5.1 for the simple OTAs. A DC optimization for the simple OTAs parallels the DC optimization of differential-pair and current-mirror devices described in Section 4.4.4.1.

For devices in the DC optimized OTA, $IC = 1$ corresponding to the center of moderate inversion and long channel lengths of $4\,\mu m$ are selected. Device channel widths and gate areas are large because of low IC and long L. This results in large OTA layout area, and large input and output capacitances associated with input devices, M_1 and M_2, and output devices, M_4 and M_6. Additionally, device intrinsic bandwidths are small because of low IC and, especially, long L. This results in small OTA transconductance bandwidth associated with pMOS current-mirror devices, M_5–M_8. Device bias compliance voltage measures of V_{EFF} and $V_{DS,sat}$ are small because of low IC. This results in large OTA input and output voltage ranges. Device small-signal parameters of g_m/I_D and V_A are large because of low IC and long L, respectively. This results in large OTA transconductance, associated with high g_m/I_D of M_1 and M_2, and large output resistance, associated with high V_A of M_4 and M_6, where, as described in Section 5.3.2.3, V_A increases sublinearly with channel length, except for short-channel devices where the increase is greater as DIBL effects are reduced. The combination of high g_m/I_D and high V_A gives high voltage gain and, correspondingly, low input-referred offset voltage due to systematic offset. Additionally, device gate-referred thermal-noise voltage is small because of high g_m. This results in small OTA input-referred thermal-noise voltage through high g_m of M_1 and M_2, with, as mentioned, constant relative noise contributions from non-input devices, M_3–M_8. Device gate-referred flicker-noise voltage and gate–source mismatch voltage due to local-area mismatch are also small because of large gate area due to low IC and, especially, long L. This results in small OTA input-referred flicker-noise voltage and offset voltage due to mismatch due to large gate area of M_1 and M_2, with, as mentioned, constant relative noise and mismatch contributions from M_3–M_8. Finally, the input, 1 dB compression voltage for two devices in a differential pair is small, giving small OTA input, 1 dB compression voltage because of high g_m/I_D for M_1 and M_2.

A DC optimization maximizes OTA transconductance, output resistance, voltage gain, and input and output voltage ranges, while minimizing input-referred thermal-noise voltage, flicker-noise voltage, and offset voltage due to local-area mismatch and systematic offset. However, this comes at the expense of large layout area, large input and output capacitances, small transconductance bandwidth, and a small input, 1 dB compression voltage. Operating at IC below unity and channel length above $4\,\mu m$ selected for the DC optimized OTA improves the first aspects of performance mentioned, but at further expense of the second aspects of performance. This is especially true for increasing channel length, which requires an equal increase in channel width to maintain the selected inversion coefficient and drain current. This results in device areas and OTA input capacitances that increase as the square of channel length, and transconductance bandwidth that decreases nearly as the inverse square of channel length, or as the inverse of channel length if velocity saturation effects are significant.

In addition to listing OTA performance trends in terms of the inversion coefficient and channel length, Table 5.1 also lists these in terms of the bias or device drain current. As observed, many aspects of performance are independent of the drain current. However, transconductance increases directly with drain current through g_m of M_1 and M_2, and output resistance decreases inversely with drain current through r_{ds} of M_4 and M_6. Additionally, the input-referred thermal-noise voltage, flicker-noise voltage, and offset voltage due to local-area mismatch decrease as drain current increases through the increased g_m (affecting thermal noise) and gate area (affecting flicker noise and mismatch) of M_1 and M_2, again, with constant relative noise and mismatch contributions from M_3–M_8. Higher levels of drain current improve some aspects of DC performance, primarily as a result of increased M_1 and M_2 input pair transconductance and increased device gate areas. As always, channel width and, correspondingly, gate area scale directly with drain current to maintain the selected inversion coefficient.

High OTA transconductance, output resistance, and voltage gain benefit both DC and low-frequency signals. Additionally, low input-referred flicker-noise voltage at low frequencies benefits low-frequency signals. As a result, a DC optimization is also an optimization for low-frequency signals.

5.4.1.2 AC optimization

An AC optimization corresponds to operation at high inversion coefficients, in strong inversion, at short channel lengths. Performance tradeoffs associated with an AC optimization are shown in the lower right portion of the *Plane* in Figure 3.1. These are also listed in Table 4.5 and observed in Table 5.1 for the simple OTAs. An AC optimization for the simple OTAs parallels the AC optimization of differential-pair and current-mirror devices described in Section 4.4.4.2.

For devices in the AC optimized OTA, $IC = 10$ corresponding to the onset of strong inversion and minimum, process channel lengths of $0.5\,\mu\text{m}$ are selected. Device channel widths and gate areas are small because of high IC and short L. This results in small OTA layout area, and small input and output capacitances associated with input devices, M_1 and M_2, and output devices, M_4 and M_6. Additionally, device intrinsic bandwidths are large because of high IC and, especially, short L. This results in large OTA transconductance bandwidth associated with pMOS current mirror devices, M_5–M_8. Device bias compliance voltage measures of V_{EFF} and $V_{DS,sat}$ are high because of high IC. This results in small OTA input and output voltage ranges. Device small-signal parameters of g_m/I_D and V_A are small because of high IC and short L, respectively. This results in small OTA transconductance, associated with low g_m/I_D of M_1 and M_2, and low output resistance, associated with low V_A of M_4 and M_6. This gives low voltage gain and, correspondingly, high input-referred offset voltage due to systematic offset. Additionally, device gate-referred thermal-noise voltage is high because of small g_m. This results in large OTA input-referred thermal-noise voltage through small g_m of M_1 and M_2, again, with constant relative noise contributions from M_3–M_8. Device gate-referred flicker-noise voltage and gate–source mismatch voltage due to local-area mismatch are also large because of small gate area due to high IC and, especially, short L. This results in large OTA input-referred flicker-noise voltage and offset voltage due to mismatch due to small gate area of M_1 and M_2, again, with constant relative noise and mismatch contributions from M_3–M_8. Finally, the input, 1 dB compression voltage for two devices in a differential pair is large, giving large OTA input, 1 dB compression voltage because of small g_m/I_D for M_1 and M_2.

An AC optimization maximizes OTA transconductance bandwidth and the input, 1 dB compression voltage, while minimizing layout area, and input and output capacitances. However, this comes at the expense of small transconductance, output resistance, voltage gain, and input and output voltage ranges, and large input-referred thermal-noise voltage, flicker-noise voltage, and offset voltage due to local-area mismatch and systematic offset. Operating at IC above 10 selected for the AC optimized OTA improves the first aspects of performance mentioned, but at further expense of the second aspects of performance. Additionally, operating at increasing levels of inversion is increasingly problematic in low-voltage designs because of the increase in device V_{EFF} and $V_{DS,sat}$.

Again, as seen in Table 5.1, many aspects of OTA performance are independent of the bias or device drain current. However, OTA transconductance, input and output capacitances, and slew rate increase directly with the drain current. Input and output capacitances increase with the drain current because of the required gate-area increase of input devices, M_1 and M_2, and the required channel width increase of output devices, M_4 and M_6. Additionally, the input-referred thermal-noise voltage decreases as the drain current increases because of the increase in g_m of M_1 and M_2, again, with constant relative noise contributions from M_3–M_8. Higher levels of drain current improve some aspects of AC performance, primarily as a result of increased M_1 and M_2 input pair transconductance and increased slew rate, but at the expense of higher input and output capacitances. Again, as always, channel width and, correspondingly, gate area scale directly with drain current to maintain the selected inversion coefficient.

An AC optimization, as defined here for the simple OTAs and in Section 4.4.4.2 for differential-pair and current-mirror devices, corresponds to high IC and short channel lengths. This minimizes OTA input and output capacitances, maximizes transconductance bandwidth, and minimizes signal distortion through the large value of input, 1 dB compression voltage. This, then, is an optimization for high-frequency signals circuits with regard to bandwidth and distortion. However, the AC optimization is not an optimization for high-frequency signals in voltage-mode circuits with regard to signal-to-noise ratio because the high input-referred thermal-noise voltage decreases the signal-to-noise ratio. The AC optimization defined here primarily maximizes signal bandwidth and minimizes signal capacitances. Other AC optimizations involving different tradeoffs in signal bandwidth, distortion, and noise are possible as seen by different locations on the *Plane* shown in Figure 3.1.

5.4.1.3 Balanced optimization

A balanced optimization corresponds to operation at moderate inversion coefficients, most likely in moderate inversion, at moderate channel lengths. Performance tradeoffs associated with a balanced optimization are shown in the center portion of the *Plane* in Figure 3.1. These are also listed in Table 4.5 and observed in Table 5.1 for the simple OTAs. A balanced optimization for the simple OTAs parallels the balanced optimization of differential-pair and current-mirror devices described in Section 4.4.4.3 where performance is balanced between that given by DC and AC optimizations. This is required when performance for DC or low-frequency signals and high-frequency signals is important simultaneously.

For devices in the balanced OTA, $IC = 3.2$ corresponding to the strong-inversion side of moderate inversion and moderate channel lengths of 1.2 μm are selected. The selection of $IC = 3.2$ is geometrically centered between $IC = 1$ and $IC = 10$ used for the DC and AC optimizations. The selection of $L = 1.2$ μm is also nearly geometrically centered between $L = 4$ μm and $L = 0.5$ μm used for the DC and AC optimizations. The selections of inversion coefficient and channel length provide a general balance of DC and AC performance and were not chosen for any specific tradeoffs.

The inversion coefficient and channel length selections for a balanced optimization depend on the priorities of various aspects of performance. These selections are complicated by the many aspects of performance and the varying dependencies of performance on the inversion coefficient and channel length, as seen in Table 5.1. For example, OTA input-referred flicker-noise voltage and offset voltage due to local-area mismatch, both of which depend on gate area, are stronger functions of channel length compared to the inversion coefficient. Transconductance bandwidth is also a stronger function of channel length.

Again, as seen in Table 5.1, many aspects of OTA performance are independent of the bias or device drain current. However, OTA transconductance, output resistance, input and output capacitances, slew rate, and input-referred thermal-noise voltage, flicker-noise voltage, and offset voltage due to local-area mismatch depend on the drain current. As a result, the drain current can also be considered for a balanced optimization of performance. As mentioned, drain currents are held constant for the DC, balanced, and AC optimized OTAs.

5.4.2 Predicted and Measured Performance

The *Analog CMOS Design, Tradeoffs and Optimization* spreadsheet maps predicted device performance for selections of the inversion coefficient, channel length, and drain current into complete OTA circuit performance. The spreadsheet was described and illustrated in Section 4.4 for the optimization of differential-pair and current-mirror devices and is described further in the Appendix. The spreadsheet is available at the web site for this book listed on the cover.

For the simple OTAs, the spreadsheet uses the 0.5 μm CMOS process parameters from Tables 3.2–3.4 that are fixed or unbinned across device geometry and bias conditions, except for the Early voltage factor, $V_{AL} = V_A/L$. As described in Section 3.8.4, no simple prediction of MOS Early voltage, $V_A = I_D \cdot r_{ds} = I_D/g_{ds}$, or the related drain–source resistance, r_{ds}, or conductance, $g_{ds} = 1/r_{ds}$, is available because of complex dependencies on the inversion level, channel length, and V_{DS}. Instead, as described near the beginning of Section 4.4, the process parameter $V_{AL} = V_A/L$ is extracted from V_A measurements near the operating inversion coefficient, channel length, and V_{DS}.

For the 0.5 μm CMOS process used in the simple OTAs, V_{AL} values are found from the measured V_A values shown in Figure 3.50 for nMOS devices operating at $V_{DS} = V_{GS}$. Operation at $V_{DS} = V_{GS}$ corresponds to the operation of diode-connected devices in the simple OTAs and is similar to V_{DS} for other devices.

From Figure 3.50, $V_{AL} = V_A/L = 42\,\text{V}/4\,\mu\text{m} = 10\,\text{V}/\mu\text{m}$ is extracted for $L = 4\,\mu\text{m}$ at $IC = 1$, corresponding to operation for the DC optimized OTA. $V_{AL} = 18\,\text{V}/1.2\,\mu\text{m} = 15\,\text{V}/\mu\text{m}$ is extracted for $L = 1.2\,\mu\text{m}$ at $IC = 3$, corresponding to operation for the balanced optimized OTA. Finally, $V_{AL} = 8.3\,\text{V}/0.5\,\mu\text{m} = 16\,\text{V}/\mu\text{m}$ is extracted for $L = 0.5\,\mu\text{m}$ at $IC = 10$, corresponding to operation for the AC optimized OTA. V_{AL} values are rounded down slightly because pMOS values are generally slightly lower than nMOS values used here for both device types.

As described near the beginning of Section 4.4, process V_{AL} values used in the spreadsheet should include only CLM effects and not include the reduction associated with DIBL effects because this is considered separately. The measured value of $V_A = 4.7\,\text{V}$ for $L = 0.5\,\mu\text{m}$ at $IC = 10$ observed in Figure 3.50 corresponds to a calculated value of $V_A = 10.8\,\text{V}$ due to DIBL and an extracted value of $V_A = 8.3\,\text{V}$ due to CLM. This value due to CLM is used for the spreadsheet process value of V_{AL}. The separation of CLM and DIBL components of V_A was described near the beginning of Section 4.4 where, from Table 3.21, the measured V_A is expressed as the parallel combination of V_A due to CLM and V_A due to DIBL. The table also describes the prediction of V_A due to DIBL, which is predicted here using the process value of $\delta V_T/\delta V_{DS} = DVTDIBL = -12\,\text{mV/V}$ from Table 3.2 for the minimum, $L = 0.5\,\mu\text{m}$, nMOS device. The removal of DIBL effects from V_A and V_{AL} is not required for $L = 1.2$ and $4\,\mu\text{m}$ where these effects are negligible.

The extraction of process V_{AL} values for the spreadsheet requires some effort, especially for a larger circuit. However, V_{AL} only affects MOS $g_{ds} = 1/r_{ds}$ and resulting predictions of OTA output resistance and voltage gain, so rough estimates for V_{AL} do not affect other aspects of performance. Linking the spreadsheet to a simulation MOS model that provides direct $g_{ds} = 1/r_{ds}$ modeling is anticipated in the future, so the reader is invited to check the book's web site listed on the cover for spreadsheet updates. This linkage would eliminate the need for extracting or estimating V_{AL}, although experimental verification is recommended given the complexity of predicting $g_{ds} = 1/r_{ds}$ and its long history of modeling errors.

Figure 5.3 shows the *MOSFETs* sheet in the *Analog CMOS Design, Tradeoffs and Optimization* spreadsheet for the three simple OTAs shown in Figure 5.1. This sheet lists device drain current, inversion coefficient, and channel length selections for the DC, balanced, and AC optimized OTAs, and reports the resulting device geometry and performance. As mentioned, nMOS devices, $M_1–M_4$, and pMOS devices, $M_5–M_8$, are biased and sized equally within each OTA. Equally biased and sized devices are listed in a single column in the *MOSFETs* sheet. MOS geometry and performance given by this sheet was described earlier in Section 4.4 for the optimization of differential-pair and current-mirror devices.

The continuation page of Figure 5.3 shows a second page of the *MOSFETs* sheet. This sheet includes the body-effect transconductance, g_{mb}, source transconductance, $g_{ms} = g_m + g_{mb} + g_{ds}$, source–body capacitance, C_{SB}, and thermal noise, $n\Gamma$, flicker noise, K'_F, and mismatch, A_{VGS}, parameters useful for circuit analysis. The sheet also includes velocity saturation and VFMR details, gate capacitance details, process and inversion coefficient details, layout details, and optional user inputs. These optional user

Analog CMOS Design, Tradeoffs and Optimization Spreadsheet -- MOSFETs Sheet						
For design guidance only as results are approximate and not for any particular CMOS process.						
© 2000–2007, David M. Binkley. See Disclaimer, Notes Sheet.						
---------- Optional User Design Information ----------						
Description						
Device reference		M1–M4	M1–M4	M1–M4	M5–M8	M5–M8	M5–M8
Device notes		DC	BAL	AC	DC	BAL	AC
---------- Required User Design Inputs ----------						
Device model		nL4um	nL1_2um	nL0_5um	pL4um	pL1_2um	pL0_5um
I_D	μA (+)	50	50	50	50	50	50
IC (fixed normalized)		0.996	3.220	10.182	0.994	3.222	10.289
L_{DRAWN}	μm	4.1	1.3	0.6	4.1	1.3	0.6
---------- Calculated Results ----------		OK	OK	OK	OK	OK	OK
Effective Width, Length, Gate Area							
W	μm	956.80	88.80	11.70	2,865.60	265.20	34.60
L	μm	4.00	1.20	0.50	4.00	1.20	0.50
WL	μm²	3827.20	106.56	5.85	11462.40	318.24	17.30
DC Bias Voltages							
$V_{EFF} = V_{GS} - V_T$	V	0.040	0.117	0.231	0.038	0.113	0.221
V_T (adjusted for V_{SB})	V	0.700	0.700	0.700	0.950	0.950	0.950
$V_{GS} = V_T + V_{EFF}$	V	0.740	0.817	0.931	0.988	1.063	1.171
$V_{DS,sat}$	V	0.136	0.175	0.248	0.136	0.176	0.250
Small Signal Parameters							
g_m/I_D	μS/μA	17.75	12.40	7.63	18.61	12.98	8.12
g_m	μS	887.27	619.78	381.41	930.36	648.89	406.09
V_A	V	39.35	15.08	4.62	39.15	14.36	4.05
g_{ds}	μS	1.2708	3.3158	10.8270	1.2773	3.4819	12.3414
$A_{Vi} = g_m/g_{ds}$	V/V	698.2	186.9	35.2	728.4	186.4	32.9
Transcond. Distortion (Diff. Pair)							
$V_{INDIF,1dB}$ (input 1-dB comp.)	V	0.078	0.114	0.203	0.074	0.109	0.192
Capacitances and Bandwidths							
$C_{GS} = C_{gsi} + C_{GSO}$	fF	3486.5	146.6	10.8	10686.2	457.2	34.5
$C_{GB} = C_{gbi} + C_{GBO}$	fF	1772.7	39.8	2.3	4557.3	100.1	4.9
$C_{GD} = C_{GDO}$ (in saturation)	fF	201.1	18.8	2.5	802.8	74.5	9.8
C_{DB} (at $V_{DB} = V_{DS} + V_{SB}$)	fF	546.2	53.8	7.4	1956.0	181.2	24.5
$f_{Ti} = g_m/[2\pi(C_{gsi} + C_{gbi})]$	MHz	27.9	593.0	6048.9	10.3	214.5	2228.2
$f_T = g_m/[2\pi(C_{GS} + C_{GB})]$	MHz	26.9	529.1	4621.7	9.7	185.3	1640.1
$f_{DIODE} = g_m/[2\pi(C_{GS} + C_{GB} + C_{DB})]$,	MHz	24.3	410.6	2949.1	8.6	139.8	1011.7
Thermal and Flicker Noise							
$S_{VG}^{1/2}$ thermal	nV/Hz$^{1/2}$	3.87	4.78	6.18	3.69	4.56	5.85
$S_{VG}^{1/2}$ flicker at $f_{FLICKER}$	nV/Hz$^{1/2}$	33.52	200.89	857.50	4.57	27.44	117.70
$S_{ID}^{1/2}$ thermal	pA/Hz$^{1/2}$	3.43	2.96	2.36	3.43	2.96	2.37
$S_{ID}^{1/2}$ flicker at $f_{FLICKER}$	pA/Hz$^{1/2}$	29.74	124.51	327.06	4.25	17.81	47.80
$f_{FLICKER}$, freq. for flicker noise	Hz	100	100	100	100	100	100
f_c, flicker-noise corner	MHz	0.01936	0.91099	16.79802	0.00015	0.00362	0.04051
DC Mismatch for Device Pair							
ΔV_{GS} (1σ)	mV	0.23	1.37	5.89	0.13	0.79	3.42
$\Delta I_D/I_D$ (1σ)	%	0.40	1.69	4.49	0.24	1.02	2.78

Figure 5.3 *MOSFETs sheet in the Analog CMOS Design, Tradeoffs and Optimization spreadsheet for the simple OTAs optimized for DC, balanced, and AC performance. This sheet lists MOSFET drain current, inversion coefficient, and channel length selections with resulting device geometry and performance*

*** MOSFETs, Sheet 2 ***						
Device reference		M1–M4	M1–M4	M1–M4	M5–M8	M5–M8	M5–M8
Device notes		DC	BAL	AC	DC	BAL	AC
Body Small Signal Parameters							
$\eta = g_{mb}/g_m$ (at V_{SB})		0.373	0.362	0.347	0.305	0.296	0.284
g_{mb} (at V_{SB})	µS	330.82	224.31	132.45	284.17	192.36	115.53
Source Parameters							
$g_{ms} = g_m + g_{mb} + g_{ds}$	µS	1219.36	847.41	524.70	1215.80	844.74	533.96
C_{SB} (at V_{SB})	fF	796.5	79.1	16.6	2892.1	295.4	47.0
Noise and Mismatch Parameters							
Γ, thermal noise factor (g_m)		0.584	0.628	0.652	0.584	0.628	0.653
($n\Gamma$), n multiplied by Γ		0.802	0.855	0.879	0.763	0.815	0.838
$K_F'^{1/2}$, flicker-noise factor	nV·µm	13700	13701	13703	4895	4895	4896
AF, flicker-noise slope		0.82	0.82	0.82	1	1	1
A_{VGS}, V_{GS} mismatch factor	V·µm	0.01405	0.01409	0.01424	0.01404	0.01408	0.01421
Velocity Sat. and VFMR Details							
Vel. sat. factor, $V_{EFF}/(LE_{CRIT})$		0.002	0.024	0.115	0.001	0.009	0.042
VFMR factor, $V_{EFF}\theta$		0.006	0.016	0.032	0.007	0.019	0.038
Critical IC, IC_{CRIT}		1209.9	415.5	125.8	1460.7	894.6	436.1
Gate Capacitance Details							
C_{GOX}	fF	9797.6	272.8	15.0	29343.7	814.7	44.3
C_{gsi}	fF	3285.4	127.8	8.3	9883.4	382.7	24.7
C_{gbi}	fF	1768.6	38.5	1.7	4553.2	98.8	4.3
C_{GSO}	fF	201.1	18.8	2.5	802.8	74.5	9.8
C_{GBO}	fF	4.1	1.3	0.6	4.1	1.3	0.6
C_{GDO}	fF	201.1	18.8	2.5	802.8	74.5	9.8
Process and IC Details							
n (actual)		1.373	1.362	1.347	1.305	1.296	1.284
μ_0	cm²/Vs	438.00	438.00	438.00	152.00	152.00	152.00
$k_0 = \mu_0 C'_{OX}$	µA/V²	112.13	112.13	112.13	38.91	38.91	38.91
$U_T = kT/q$	V	0.02585	0.02585	0.02585	0.02585	0.02585	0.02585
$I_0 = 2n_0 \mu_0 C'_{OX} U_T^2$	µA	0.210	0.210	0.210	0.070	0.070	0.070
IC (actual)		1.016	3.310	10.580	1.028	3.355	10.814
Layout Details, IC Trimming							
W_{DRAWN}, each; x m for total	µm	119.70	11.20	5.95	179.20	33.25	8.75
L_{DRAWN}	µm	4.1	1.3	0.6	4.1	1.3	0.6
AD, each; x m for total	µm²	107.73	10.08	5.35	161.28	29.92	7.87
PD, each; x m for total	µm	121.50	13.00	7.75	181.00	35.05	10.55
AS, each; x m for total	µm²	125.69	11.76	8.92	174.72	34.91	10.50
PS, each; x m for total	µm	151.73	16.10	14.90	203.55	43.66	15.52
IC (trimmed), may be used for IC input		0.996	3.220	10.182	0.994	3.222	10.289
-------- Optional User Design Inputs ----------							
	Default, units						
m, number of gate fingers	1	8	8	8	16	8	4
W_{DRAWN} (trimmed), each	none, µm	119.7	11.2	5.95	179.2	33.25	8.75
V_{DS} (for C_{DB})	V_{GS}, V (+)						
V_{SB} (for g_{mb} and C_{SB})	0 V, V (+)						
$f_{FLICKER}$, freq. for flicker noise	1 Hz, Hz	100	100	100	100	100	100
T, temperature	27 C, C						

Figure 5.3   (*continued*)

inputs include the number of gate stripes or interdigitations, m, defaulted at unity, and a selected drawn channel width for each layout stripe that permits trimming the input value of IC to match the actual layout. Additionally, input values for V_{DS}, V_{SB}, the frequency for reported flicker noise, $f_{FLICKER}$, and temperature, T, may be included. Default values for these optional inputs are $V_{DS} = V_{GS}$, $V_{SB} = 0\,\text{V}$, $f_{FLICKER} = 1\,\text{Hz}$, and $T = 27\,^{\circ}\text{C}$. The second sheet of results and optional user inputs are described further in the Appendix.

Figure 5.4 shows the *Circuit Analysis* sheet in the spreadsheet for the three simple OTAs. This sheet gives predicted OTA performance from the circuit analysis equations derived earlier in Section 5.3 and calls device geometry and performance from the *MOSFETs* sheet of Figure 5.3. This permits automated updating of OTA performance as device drain current, inversion coefficient, and channel length selections are explored for optimization. Because the spreadsheet uses the slightly more accurate expression for g_m/I_D from Table 3.17 that involves two square-root terms, OTA performance related to this varies slightly from the design equations derived in Section 5.3 that use the simpler g_m/I_D expression that involves a single square-root term. Additionally, the spreadsheet uses a corrected value of the substrate factor, n, for the level of inversion, which also changes the results slightly. To permit comparisons with the predicted values, the *Circuit Analysis* sheet also lists simulated and measured performance for the OTAs. Simulations are from SPICE simulations using BSIM3V3 MOS models [4] for the $0.5\,\mu\text{m}$ CMOS process considered.

5.4.2.1 Transconductance, output resistance, and voltage gain

OTA transconductance, G_M, from Equation 5.1 is nearly equal to the transconductance, $g_{m1} = g_{m2}$, of the M_1 and M_2 input pair devices. Predicted G_M in Figure 5.4 is 887, 620, and $381\,\mu\text{S}$ for the DC, balanced, and AC optimized OTA, respectively, which is within 4.2 % of measured values of 912, 647, and $383\,\mu\text{S}$ shown later in Figure 5.9. Figure 5.5 shows simulated G_M of 842, 649, and $382\,\mu\text{S}$, which is within 7.7 % of measured values. Because the percentage errors given are calculated from unrounded performance values in the spreadsheet, these will vary slightly from errors calculated from the rounded performance values given. G_M is maximized for the DC optimized OTA because input devices operate at low inversion coefficients ($IC_1 = IC_2 = 1$) in the center of moderate inversion giving high g_m/I_D and g_m.

OTA output resistance, R_{OUT}, from Equation 5.3 is equal to the parallel combination of drain–source resistances, r_{ds4} and r_{ds6}, for the M_4 and M_6 output devices. Predicted R_{OUT} in Figure 5.4 is 0.392, 0.147, and $0.043\,\text{M}\Omega$ for the DC, balanced, and AC optimized OTA, respectively, which is within 13.5 % of measured values of 0.357, 0.170, and $0.044\,\text{M}\Omega$. Simulated R_{OUT} is 0.403, 0.195, and $0.047\,\text{M}\Omega$, which is within 14.7 % of measured values. Prediction errors in R_{OUT} result from prediction errors in r_{ds} as discussed later in Section 5.6. R_{OUT} is maximized for the DC optimized OTA because output devices operate at long channel lengths ($L_4 = L_6 = 4\,\mu\text{m}$). As discussed in Section 5.3.2.3, output device V_A and, correspondingly, device r_{ds}, OTA R_{OUT}, and OTA A_V increase sublinearly with channel length, except for short-channel devices where the increase is greater as DIBL effects are reduced.

OTA voltage gain, A_V, from Equation 5.8 is equal to the product of G_M and R_{OUT}. Predicted A_V in Figure 5.4 is 348, 91, and $16.5\,\text{V/V}$ for the DC, balanced, and AC optimized OTA, respectively, which is within 17.1 % of measured values of 326, 110, and $16.8\,\text{V/V}$. Figure 5.6 shows simulated A_V of 340, 127, and $18\,\text{V/V}$, which is within 15 % of measured values. Prediction errors in A_V, like errors in R_{OUT}, primarily result from prediction errors in r_{ds}.

A_V is maximized for the DC optimized OTA because of the combination of high g_m/I_D and g_m for input devices in moderate inversion and high V_A and r_{ds} for long-channel output devices. A_V is significantly lower for the AC optimized OTA because of the combination of low g_m/I_D and g_m for input devices in strong inversion and low V_A and r_{ds} for short-channel output devices. Section 5.4.2.12 summarizes OTA voltage gain, transconductance bandwidth, and the product of these.

Summary of Predicted, Simulated, and Measured Performance for Three, 0.5–µm OTAs					
MOSFET Notes					
$M1-M8$ operate at equal drain current ($I_D = 50$ µA) for all designs.					
nMOS $M1-M4$ are sized and biased equally; pMOS $M5-M8$ are sized and biased equally.					
Transconductance, G_M					
$G_M = g_{m1} = (g_m/I_D)_1 \cdot I_{BIAS}$					
			DC	Balanced	AC
		Predicted	887	620	381	µS
		Simulated	842	649	382	µS
		Measured	**912**	**647**	**383**	**µS**
Output Resistance, R_{OUT}					
$R_{OUT} = 1/(g_{ds4} + g_{ds6}) = (V_{A4}\|V_{A6})/I_{BIAS}$					
			DC	Balanced	AC
		Predicted	0.392	0.147	0.043	MΩ
		Simulated	0.403	0.195	0.047	MΩ
		Measured	**0.357**	**0.170**	**0.044**	**MΩ**
Voltage Gain, A_V					
$A_V = G_M \cdot R_{OUT} = g_{m1}/(g_{ds4} + g_{ds6}) = (g_m/I_D)_1 \cdot (V_{A4}\|V_{A6})$					
			DC	Balanced	AC
		Predicted	348.2	91.2	16.5	V/V
		Simulated	339.5	126.5	18.0	V/V
		Measured	**326**	**110**	**16.8**	**V/V**
Bandwidth, f_{-3dB}					
$f_{-3dB} \approx f_{p5} \approx g_{m5}/[2\pi(2C_{GS5} + 2C_{GBS} + C_{DB1} + C_{DBS})]$					
Pole associated with pMOS mirrors			DC	Balanced	AC
(approximately $f_{diode5}/2$)		Predicted	4.5	76.4	583.1	MHz
		Simulated	3.8	56.2	350.0	MHz
		Measured	**5**	**51**		**MHz**
Input-Referred Thermal Noise Voltage, $S_{VIN}^{1/2}$ (thermal)					
$S_{VIN}^{1/2} = \sqrt{[\{2 \cdot 4kT(n\Gamma)_1/g_{m1}\} \cdot (1 + RNP_3 + RNP_5 + RNP_7)]}$					
			DC	Balanced	AC
		Predicted	10.9	13.5	17.5	nV/Hz$^{1/2}$
		Simulated	12.5	14.4	19.4	nV/Hz$^{1/2}$
		Measured	**11.2**			**nV/Hz$^{1/2}$**
Relative Noise Power (RNP) -- relative to $M1$, $M2$					
			DC	Balanced	AC
$M1$, $M2$	1		1	1	1
$M3$, $M4$	$RNP_3 = [(n\Gamma)_3/(n\Gamma)_1] \cdot [g_{m3}/g_{m1}]$		1	1	1
$M5$, $M6$	$RNP_5 = [(n\Gamma)_5/(n\Gamma)_1] \cdot [g_{m5}/g_{m1}]$		1.00	1.00	1.02
$M7$, $M8$	$RNP_7 = [(n\Gamma)_7/(n\Gamma)_1] \cdot [g_{m7}/g_{m1}]$		1.00	1.00	1.02
Total:			4.00	3.99	4.03

Figure 5.4 *Circuit Analysis* sheet in the *Analog CMOS Design, Tradeoffs and Optimization* spreadsheet for the simple OTAs optimized for DC, balanced, and AC performance. This sheet lists predicted, simulated, and measured performance

Summary of Predicted, Simulated, and Measured Performance for Three, 0.5-μm OTAs

(second page)

Input-Referred Flicker-Noise Voltage, $S_{VIN}^{1/2}$ (flicker)

$$S_{VIN}^{1/2}(f) = \sqrt{[\{2 \cdot K'_{F1}/((WL)_1 f^{AF1})\} \cdot (1 + RNP_3 + RNP_5 + RNP_7)]}$$

				DC	Balanced	AC	(100 Hz)
			Predicted	69	411	1756	nV/Hz$^{1/2}$
			Simulated	69	408	1685	nV/Hz$^{1/2}$
			Measured	**80**	**450**	**2000**	**nV/Hz$^{1/2}$**

Relative Noise Power (RNP) -- relative to $M1$, $M2$ at $f = 1$ Hz

RNP_3 adjusted for frequency by (f^{AF1}/f^{AF3}), etc.

				DC	Balanced	AC	(1 Hz)
$M1, M2$	1			1	1	1	
$M3, M4$	$RNP_3 = [K'_{F3}/K'_{F1}] \cdot [(WL)_1/(WL)_3] \cdot [g_{m3}/g_{m1}]^2$			1	1	1	
$M5, M6$	$RNP_5 = [K'_{F5}/K'_{F1}] \cdot [(WL)_1/(WL)_5] \cdot [g_{m5}/g_{m1}]^2$			0.05	0.05	0.05	
$M7, M8$	$RNP_7 = [K'_{F7}/K'_{F1}] \cdot [(WL)_1/(WL)_7] \cdot [g_{m7}/g_{m1}]^2$			0.05	0.05	0.05	
Total:				2.09	2.09	2.10	

Input-Referred Offset Voltage Due to Local-Area Mismatch, $V_{INOFFSET}$ (mismatch)

$$V_{INOFFSET} = \sqrt{[\{A_{VGS1}^2/(WL)_1\} \cdot (1 + RMP_3 + RMP_5 + RMP_7)]}$$

				DC	Balanced	AC	(1σ)
			Predicted	0.4	2.3	9.8	mV
			Measured	**1.1**	**2.2**	**10.2**	**mV**

Relative Mismatch Power (RMP) -- relative to $M1$, $M2$

				DC	Balanced	AC	
$M1, M2$	1			1	1	1	
$M3, M4$	$[A_{VGS3}/A_{VGS1}]^2 \cdot [(WL)_1/(WL)_3] \cdot [g_{m3}/g_{m1}]^2$			1	1	1	
$M5, M6$	$[A_{VGS5}/A_{VGS1}]^2 \cdot [(WL)_1/(WL)_5] \cdot [g_{m5}/g_{m1}]^2$			0.37	0.37	0.38	
$M7, M8$	$[A_{VGS7}/A_{VGS1}]^2 \cdot [(WL)_1/(WL)_7] \cdot [g_{m7}/g_{m1}]^2$			0.37	0.37	0.38	
Total:				2.73	2.73	2.76	

Input-Referred Offset Voltage Due to Systematic Offset, $V_{INOFFSET}$ (systematic)

$$V_{INOFFSET} = \Delta I_{OUT}/g_{m1} = -[1/(g_m/I_D)_1] \cdot [(V_{OUT} - V_{SS} - V_{GS3})/(V_{A4} \| V_{A6})] = -(V_{OUT} - V_{SS} - V_{GS3})/A_V$$

$V_{OUT} = 0$ V and $V_{SS} = -1.25$ V.

				DC	Balanced	AC	
			Predicted	−1.5	−4.7	−19.4	mV
			Simulated	−1.6	−3.2	−14.9	mV
			Measured	**−1.1**	**−2.2**	**−16.3**	**mV**

Input and Output Capacitances and Voltage Ranges; Input, 1-dB Compression Voltage

All are predicted from MOS operating parameters. $V_{DD} = 1.25$ V, $V_{SS} = -1.25$ V, and $V_{DS,sat9} = 0.25$ V.

			DC	Balanced	AC	
$C_{INDIF} = (1/2)*(C_{GS1} + C_{GB1} + C_{GD1}(1 + g_{m1}/g_{m5}))$			2.826	0.112	0.009	pF
$C_{OUT} = C_{DB4} + C_{GD4} + C_{DB6} + C_{GD6}$			3.506	0.328	0.044	pF
$V_{INCM+} = V_{DD} - V_{GS5} + V_{GS1} - V_{DS,sat1}$			0.87	0.83	0.76	V
$V_{INCM-} = V_{SS} + V_{GS1} + V_{DS,sat9}$			−0.26	−0.18	−0.07	V
$V_{OUT+} = V_{DD} - V_{DS,sat6}$			1.11	1.07	1.00	V
$V_{OUT-} = V_{SS} + V_{DS,sat4}$			−1.11	−1.07	−1.00	V
$V_{INDIF1dB} = V_{INDIF1dB}$ ($M1, M2$)			0.078	0.114	0.203	V
		Measured	**0.078**	**0.115**	**0.218**	**V**

Figure 5.4 (*continued*)

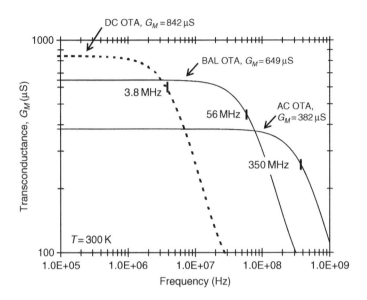

Figure 5.5 Simulated transconductance, G_M, for the simple OTAs. Simulated G_M is 842, 649, and 382 μS with −3 dB bandwidths of 3.8, 56, and 350 MHz for the DC, balanced, and AC optimized OTA, respectively

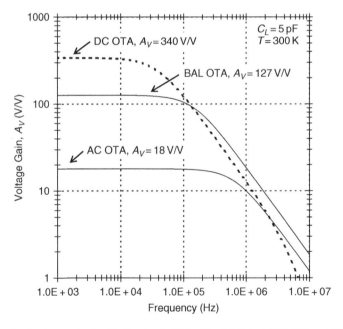

Figure 5.6 Simulated voltage gain, A_V, for the simple OTAs. Simulated A_V is 340, 127, and 18 V/V for the DC, balanced, and AC optimized OTA, respectively

5.4.2.2 Frequency response

OTA transconductance frequency response, f_{-3dB}, evaluated in the spreadsheet by Equation 5.10, is approximated as the input pole frequency associated with the pMOS, M_5 and M_6, and M_7 and M_8 current mirrors. Predicted f_{-3dB} in Figure 5.4 is 4.5, 76, and 583 MHz for the DC, balanced, and AC optimized OTA, respectively. This compares to measured values of 5 and 51 MHz for the DC and balanced OTAs, and simulated values shown in Figure 5.5 of 3.8, 56, and 350 MHz for all three OTAs. f_{-3dB} of the AC optimized OTA was not measured. While the predicted and measured f_{-3dB} are within 10 % for the DC optimized OTA, f_{-3dB} is overpredicted for the balanced and, especially, the AC optimized OTA. This is likely the result of parasitic layout capacitance, not considered in the prediction, which is more significant for the AC optimized OTA because of smaller device capacitances associated with smaller devices. The prediction of f_{-3dB}, while still providing useful design guidance, emphasizes the requirement of full circuit simulations that include the effects of layout parasitics.

As described by Equation 5.15, f_{-3dB} may be approximated by $f_{diode5}/2$ and very roughly as less than $f_{Ti5}/2$, where f_{diode5} and f_{Ti5} are the diode-connected and intrinsic bandwidths for the pMOS current mirror devices. From the *MOSFETs* sheet of Figure 5.3, this gives predicted $f_{-3dB} \approx f_{diode5}/2$ of 4.3, 70, and 500 MHz for the DC, balanced, and AC optimized OTA, respectively. This is slightly below the predicted values of f_{-3dB} mentioned above. While providing useful design guidance, the rough estimate of $f_{-3dB} \approx f_{Ti5}/2$ of 5.2, 107, and 1100 MHz is well above the predicted values of f_{-3dB} mentioned above, especially for the AC optimized OTA. As mentioned in Section 5.3.4.1, f_{Ti} significantly overpredicts bandwidth for short-channel devices present in the AC optimized OTA because it excludes extrinsic gate-overlap capacitances and the drain–body capacitance that are included in f_{diode}. All predictions of bandwidth exclude layout capacitances and significantly overpredict bandwidth for the AC optimized OTA.

f_{-3dB} is maximized in the AC optimized OTA because of the combination of high inversion coefficients and, especially, short channel lengths for the pMOS current-mirror devices. The combination of low inversion coefficients and, especially, long channel lengths significantly lowers f_{-3dB} for the DC optimized OTA. Section 5.4.2.12 summarizes OTA voltage gain, transconductance bandwidth, and the product of these.

5.4.2.3 Thermal noise

OTA thermal noise was analyzed in Section 5.3.5.1 in terms of input device, noise power contributions, and the relative noise power, *RNP*, contributions of non-input devices. As shown in Figure 5.4, RNP_3, RNP_5, and RNP_7 are unity, indicating that the M_3 and M_4, M_5 and M_6, and M_7 and M_8 current-mirror pairs contribute equal thermal-noise power as the M_1 and M_2 input pair. Clearly, input devices do not dominate thermal noise in these simple, general-purpose OTAs. Non-input device noise contributions can be reduced by operating these devices at high inversion coefficients for low g_m/I_D and g_m compared to the input devices. Additionally, non-input device noise contributions can be reduced by operating these devices at low drain currents for low g_m.

As seen in Figure 5.4, the total, relative, noise power contribution from the M_3–M_8 non-input devices is 3, resulting in a total OTA, input-referred thermal-noise voltage PSD of 4 $(1+3)$ times that of the input pair devices. This corresponds to a total input-referred thermal-noise voltage density of 200 % ($\sqrt{4}$) that of the input pair devices.

Predicted input-referred thermal-noise voltage density (the square root of PSD) in Figure 5.4 from Equation 5.18 is 10.9, 13.5, and 17.5 nV/Hz$^{1/2}$ for the DC, balanced, and AC optimized OTA, respectively. This compares to a measured value of 11.2 nV/Hz$^{1/2}$ for the DC optimized OTA shown in Figure 5.7, and simulated values of 12.5, 14.4, and 19.4 nV/Hz$^{1/2}$ for the three OTAs. Because of high flicker noise in the balanced and AC optimized OTAs, it was not possible to measure the thermal-noise

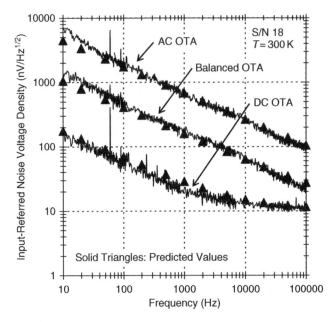

Figure 5.7 Measured (lines) and predicted (solid triangles) input-referred noise voltage density for the simple OTAs. Predicted noise for the DC, balanced, and AC optimized OTAs is from the *Circuit Analysis* sheet in the *Analog CMOS Design, Tradeoffs and Optimization* spreadsheet

floor for these. Predicted and measured thermal noise is within 2.3 % for the DC optimized OTA, and predicted and simulated thermal noise is within 14.7 % for all three OTAs.

Input-referred thermal-noise voltage is minimized for the DC optimized OTA because of the high, input device g_m/I_D and g_m resulting from operation at low inversion coefficients ($IC_1 = IC_2 = 1$) in moderate inversion. Section 5.4.3 describes the thermal-noise reduction available by operating the non-input M_3–M_8 devices at $IC = 16$ in strong inversion, while maintaining operation of the input M_1 and M_2 devices at $IC = 1$ in moderate inversion. This lowers the total, relative, noise power contribution from the M_3–M_8 non-input devices from 3 to 1.2. The predicted input-referred thermal-noise voltage density is then lower at $8.1\,\mathrm{nV/Hz^{1/2}}$ compared to $10.9\,\mathrm{nV/Hz^{1/2}}$ for the DC optimized OTA that operates with all devices at $IC = 1$ in moderate inversion.

5.4.2.4 Flicker noise

OTA flicker noise was analyzed in Section 5.3.6.1 by comparing the relative noise power contributions of non-input devices to that of input devices as described for thermal noise. As shown in Figure 5.4, RNP_3 is unity, and RNP_5 and RNP_7 are much smaller at 0.05 each. This indicates that the nMOS, M_3 and M_4 current-mirror pair contributes equal flicker-noise power as the M_1 and M_2 input pair, while the pMOS, M_5 and M_6, and M_7 and M_8 current-mirror pairs contribute only 5 % noise power each compared to the input pair. Although the pMOS current-mirror devices operate at the same transconductance as the input devices, their larger gate area and lower flicker-noise factor, K'_F, result in nearly negligible flicker-noise contributions. The *RNP* values listed are for a frequency of 1 Hz. These do not require adjustment for other frequencies because flicker noise is dominated by the nMOS M_1–M_4 devices, which have equal flicker-noise slopes, *AF*.

The total, relative, noise power contribution from the M_3–M_8 non-input devices is 1.1, resulting in a total OTA, input-referred flicker-noise voltage PSD of 2.1 $(1 + 1.1)$ times that of the input pair devices. This corresponds to a total input-referred flicker-noise voltage density of 145 % ($\sqrt{2.1}$) that of the input pair devices. As for the thermal noise, input devices do not dominate flicker noise in these simple, general-purpose OTAs. This is true for the nMOS M_3 and M_4 current-mirror pair that contributes equal flicker noise as the input pair.

Predicted input-referred flicker-noise voltage density in Figure 5.4 from Equation 5.24 is 69, 411, and 1750 nV/Hz$^{1/2}$ at 100 Hz for the DC, balanced, and AC optimized OTA, respectively. This is within 14.3 % of measured values of 80, 450, and 2000 nV/Hz$^{1/2}$ shown in Figure 5.7. Simulated noise is 69, 408, and 1690 nV/Hz$^{1/2}$, which is within 15.8 % of measured values. Figure 5.7 shows an overlay of noise predicted by the spreadsheet and measured noise. As observed, the predicted and measured noise nearly overlay.

Input-referred flicker-noise voltage is minimized for the DC optimized OTA because of the large device areas, particularly for nMOS devices M_1–M_4 that dominate the noise. Large device areas result from operation at low inversion coefficients and, especially, long channel lengths. Flicker noise is significantly higher for the AC optimized OTA because operation at high inversion coefficients and, especially, minimum channel lengths results in small device areas. As summarized in Section 5.4.2.12, the input-referred flicker-noise voltage density is inversely proportional to the square root of device gate areas for the OTAs.

5.4.2.5 Offset voltage due to local-area mismatch

OTA thermal and flicker noise were analyzed by comparing the relative noise power contributions of non-input devices to that of input devices. OTA offset voltage due to local-area mismatch was analyzed in a similar way in Section 5.3.7.1 by comparing the relative mismatch power, *RMP*, of non-input devices to that of input devices. As shown in Figure 5.4, RMP_3 is unity, and RMP_5 and RMP_7 are lower at 0.37 each. This indicates that the nMOS, M_3 and M_4 current-mirror pair contributes equal mismatch power as the M_1 and M_2 input pair, while the pMOS, M_5 and M_6, and M_7 and M_8 current-mirror pairs contribute 37 % mismatch power each compared to the input pair. Although the pMOS current-mirror devices operate at the same transconductance as the input devices, their larger gate area results in lower mismatch contributions.

The usual dominance of threshold-voltage mismatch over transconductance factor mismatch was discussed in Section 5.3.7.1 and applies to all OTA device pairs. This is seen by A_{VGS} values reported on the second page of the *MOSFETs* sheet in Figure 5.3 that are nearly equal to the $A_{VTO} = 14$ mV·μm process values listed in Table 3.3 for the 0.5 μm CMOS process considered.

The total, relative, mismatch power contribution from the M_3–M_8 non-input devices is 1.7, resulting in a total OTA, input-referred offset voltage variance of 2.7 $(1 + 1.7)$ times that of the input pair devices. This corresponds to a total, input-referred offset voltage of 164 % ($\sqrt{2.7}$) that of the input pair devices. This compares to a total, input-referred flicker-noise voltage density of 145 % that of the input pair devices, as mentioned in the previous section. Flicker noise and local-area mismatch optimize in the same way when threshold-voltage mismatch dominates and increases in gate-referred flicker-noise voltage with V_{EFF} are negligible as they are here. However, the lower flicker-noise factors and equal threshold-voltage mismatch factors for pMOS devices compared to nMOS devices lower the pMOS flicker-noise contributions more than their mismatch contributions. As for the thermal and flicker noise, input devices do not dominate local-area mismatch in these simple, general-purpose OTAs.

Predicted 1σ, input-referred offset voltage (the square root of the offset voltage variance) in Figure 5.4 from Equation 5.32 is 2.3 and 9.8 mV for the balanced and AC optimized OTAs. This is within 4 % of measured values of 2.2 and 10.2 mV. However, offset voltage is underpredicted at 0.4 mV for the DC optimized OTA compared to a measured value of 1.1 mV. This underprediction probably results because local-area mismatch is considered alone without distance mismatch. In Section 3.11.2.5, a critical spacing of $S = 131$ μm was calculated for the pMOS devices in the DC optimized OTA having gate areas of $WL = 2870$ μm · 4 μm $= 11500$ μm^2, distance threshold-voltage mismatch factors of $S_{VT} = S_{VTO} = 1\,\mu\text{V}/\mu\text{m}$, and local-area, threshold-voltage mismatch factors of $A_{VT} = A_{VTO} = 14\,\text{mV}\cdot\mu\text{m}$. However, as seen later in the photomicrograph of Figure 5.10, layout dimensions across pMOS device pairs in the DC optimized OTA exceed $S = 131\,\mu\text{m}$. This suggests that distance mismatch is greater than local-area mismatch for the DC optimized OTA. Given that offset voltage is well predicted for the balanced and AC optimized OTAs considering only local-area mismatch, distance mismatch appears to be a plausible explanation for the increase in measured offset voltage observed for the very large, DC optimized OTA.

Input-referred offset voltage due to local-area mismatch, like input-referred flicker-noise voltage density, is minimized for the DC optimized OTA because of the large device areas. As mentioned, large device areas result from operation at low inversion coefficients and, especially, long channel lengths. The offset voltage is significantly higher for the AC optimized OTA because, again, operation at high inversion coefficients and, especially, minimum channel lengths results in small device areas. As summarized in Section 5.4.2.12, with the exception of the DC optimized OTA, the input-referred offset voltage is inversely proportional to the square root of device gate areas for the OTAs.

5.4.2.6 Systematic offset voltage

OTA input-referred offset voltage due to systematic offset is equal to the drain–source voltage imbalance across M_6 and M_8, and M_3 and M_4 divided by A_V. Predicted offset voltage in Figure 5.4 from Equation 5.41 is -1.5, -4.7, and $-19.4\,\text{mV}$ for an output voltage of $0\,\text{V}$ for the DC, balanced, and AC optimized OTA, respectively. Predicted values are larger in magnitude than measured values of -1.1, -2.2, and $-16.3\,\text{mV}$. Simulated values are -1.6, -3.2, and $-14.9\,\text{mV}$. Prediction errors for systematic offset, like errors in R_{OUT} and A_V, result from prediction errors in MOS r_{ds}. Additionally, errors result from unknown levels of voltage imbalance.

The measured, input-referred systematic offset voltages of -1.1, -2.2, and $-16.3\,\text{mV}$ for the DC, balanced, and AC optimized OTA, respectively, are comparable to the measured, 1σ, input-referred offset voltages of 1.1, 2.2, and $10.2\,\text{mV}$ given in the previous section. Systematic offset is high for the simple OTAs because of the voltage imbalance mentioned above and low A_V.

Systematic offset voltage is minimized for the DC optimized OTA because of high A_V. It is significantly higher for the AC optimized OTA because of low A_V.

5.4.2.7 Input and output capacitances

OTA, differential input capacitance, C_{INDIF}, from Equation 5.42 is dominated by the gate–source and gate–body capacitances of the M_1 and M_2 input pair devices. Predicted C_{INDIF} in Figure 5.4 is 2.8, 0.11, and 0.009 pF for the DC, balanced, and AC optimized OTA, respectively. Output capacitance, C_{OUT}, from Equation 5.44 is dominated by the drain–body capacitances of the M_4 and M_6 output devices. Predicted C_{OUT} is 3.5, 0.33, and 0.044 pF for the three OTAs.

C_{INDIF} and C_{OUT} are minimized for the AC optimized OTA because operation at high inversion coefficients and minimum channel lengths results in small input device gate areas, minimizing C_{INDIF}, and small output device channel widths, minimizing C_{OUT}. Capacitances are significantly higher for the DC optimized OTA because operation at low inversion coefficients and long channel lengths results in large gate areas and large channel widths.

5.4.2.8 Slew rate

Slew rate is not listed in the predicted, simulated, and measured OTA performance summarized in Figure 5.4 because it depends strongly on the external load capacitance. From Equation 5.49, slew rate is equal to twice the bias current, $2I_{BIAS}$, of the individual M_1 and M_2 input devices divided by the total output capacitance, which is the sum of C_{OUT} and the external load capacitance. Since all three OTAs operate at the same input device bias currents of 50 μA, slew rate varies only with the total output capacitance. Slew rate would normally be maximized for the AC optimized OTA because its high transconductance bandwidth permits small capacitive loads in closed-loop configurations having high closed-loop bandwidths. Additionally, the AC optimized OTA has low C_{OUT}, reducing the total output capacitance and increasing the slew rate.

5.4.2.9 Input and output voltage ranges

The OTA, maximum, input common-mode voltage, V_{INCM+}, from Equation 5.51 is controlled by the positive supply voltage, the gate–source voltage of the M_5 and M_7 diode-connected devices, and the gate–source and drain–source saturation voltages of the M_1 and M_2 input devices. For a positive supply voltage of $+1.25$ V, predicted V_{INCM+} in Figure 5.4 is 0.87, 0.83, and 0.76 V for the DC, balanced, and AC optimized OTA, respectively. The minimum, input common-mode voltage, V_{INCM-}, from Equation 5.53 is controlled by the negative supply voltage, the gate–source voltage of the M_1 and M_2 input devices, and the drain–source saturation voltage of the M_9 current source. For a negative supply voltage of -1.25 V and a drain–source saturation voltage of 0.25 V for the M_9 current source, predicted V_{INCM-} is -0.26, -0.18, and -0.07 V for the three OTAs. Both V_{INCM+} and V_{INCM-} are underpredicted slightly because the increase in input device threshold voltage and gate–source voltage due to body effect is not considered in the spreadsheet where V_{SB} is set to its default value of zero. The input voltage range, which is the difference between V_{INCM+} and V_{INCM-}, is maximized for the DC optimized OTA. This is because operation at low inversion coefficients minimizes gate–source voltages of the input devices and related devices.

As described in Section 5.3.11.1, V_{INCM+} and V_{INCM-} are the maximum and minimum, input common-mode voltages where full OTA performance is maintained. Functionality, however, is maintained for inputs slightly outside these voltages.

The OTA, maximum output voltage, V_{OUT+}, from Equation 5.55 is controlled by the positive supply voltage and the drain–source saturation voltage of the M_6 output device. For a positive supply voltage of $+1.25$ V, predicted V_{OUT+} in Figure 5.4 is 1.11, 1.07, and 1.00 V for the DC, balanced, and AC optimized OTA, respectively. The minimum output voltage, V_{OUT-}, from Equation 5.57 is controlled by the negative supply voltage and the drain–source saturation voltage of the M_4 output device. For a complementary negative supply voltage of -1.25 V, predicted V_{OUT-} is -1.11, -1.07, and -1.00 V, which is complementary to the positive values for V_{OUT+}. The output voltage range, which is the difference between V_{OUT+} and V_{OUT-}, is maximized for the DC optimized OTA. This is because operation at low inversion coefficients minimizes drain–source saturation voltages of the output devices.

As described in Section 5.3.11.1, V_{OUT+} and V_{OUT-} are the maximum and minimum output voltages where full OTA output resistance and voltage gain are maintained. For a capacitive load, the output voltage will extend further to the supply voltages.

5.4.2.10 Input, 1 dB compression voltage

The last entry in Figure 5.4 is the OTA, differential input, 1 dB compression voltage, $V_{INDIF1dB}$. The application of a negative or positive input voltage at this value results in an output current that is 1 dB below its ideal, small-signal value. As described in Section 5.3.12.1, OTA $V_{INDIF1dB}$ is directly equal to $V_{INDIF1dB}$ for the M_1 and M_2 input differential pair. Predicted $V_{INDIF1dB}$ is 0.078, 0.114, and 0.203 V for the DC, balanced, and AC optimized OTA, respectively. This is within 6.7 % of measured values of 0.078, 0.115, and 0.218 V found in Figure 5.8 that shows measured short-circuit output current versus the input voltage. $V_{INDIF1dB}$ is maximized for the AC optimized OTA because the input

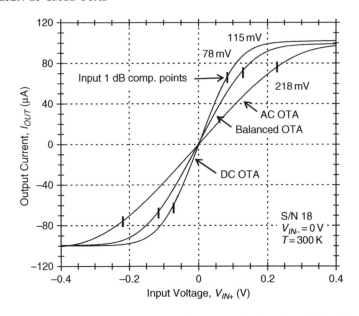

Figure 5.8 Measured short-circuit output current versus input voltage for the simple OTAs. The measured input, 1 dB compression voltage, $V_{INDIF1dB}$, is 78, 115, and 218 mV for the DC, balanced, and AC optimized OTA, respectively

differential pair is operated in strong inversion at low g_m/I_D. This requires a larger input voltage to steer an equal amount of current compared to operation at higher g_m/I_D present for the DC and balanced optimized OTAs.

Figure 5.9 shows measured G_M versus the input voltage for the three OTAs. G_M remains nearly constant for input voltages sufficiently below $V_{INDIF1dB}$, but drops significantly for voltages above this, indicating significant signal distortion. G_M remains nearly constant over a wider input voltage range for the AC optimized OTA compared to the DC optimized OTA because $V_{INDIF1dB}$ is higher at 0.218 V compared to 0.078 V.

5.4.2.11 Layout area

OTA layout area depends on the gate areas of individual devices. Gate areas and layout area are significantly higher for the DC optimized OTA because of operation at low inversion coefficients and, especially, long channel lengths. As described in Section 3.6.3, gate area is inversely proportional to the inversion coefficient, but is proportional to the square of channel length because channel width must increase directly with channel length to maintain the selected drain current and inversion coefficient. Gate area changes for the OTAs are summarized in Section 5.4.2.12.

Figure 5.10 shows a photomicrograph of the three, simple, 0.5 μm CMOS OTAs. Layout dimensions are approximately 545×463, 130×200, and $82 \times 89\,\mu m$ for the DC, balanced, and AC optimized OTA, respectively. This corresponds to approximate layout areas of 252000, 26000, and 7300 μm^2.

Tradeoffs in performance observed in Figure 5.4 are especially significant because both the inversion coefficient increases and the channel length decreases between the DC, balanced, and AC optimized OTAs. The high transconductance, output resistance, voltage gain, and input and output voltage ranges, combined with low input-referred thermal-noise, flicker-noise, and offset voltage due to local-area mismatch for the DC optimized OTA comes with the penalty of significantly larger layout area.

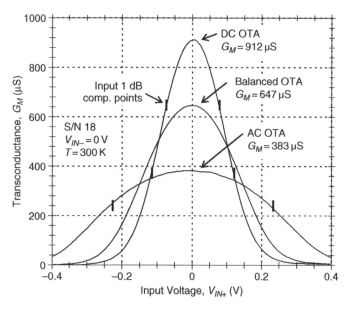

Figure 5.9 Measured transconductance, G_M, versus input voltage for the simple OTAs. G_M at an input voltage of zero is 912, 647, and 383 μS for the DC, balanced, and AC optimized OTA, respectively. G_M remains nearly constant for signals sufficiently below the input, 1 dB compression voltage, corresponding to low signal distortion. G_M remains nearly constant over a wider input voltage range for the AC optimized OTA because of the higher input, 1 dB compression voltage

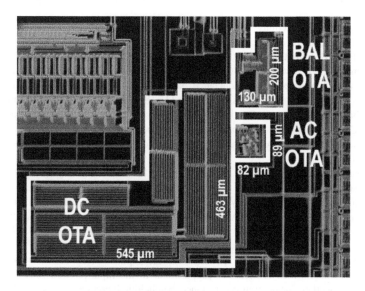

Figure 5.10 Photomicrograph of the simple OTAs with layout dimensions. Operation in moderate inversion combined with long channel lengths gives large gate areas and layout area for the DC optimized OTA. Conversely, operation in strong inversion combined with short channel lengths gives small gate areas and layout area for the AC optimized OTA. Reproduced by permission of Springer-Verlag GmbH © 2006, from [1]

5.4.2.12 Tradeoffs in DC accuracy, low-frequency AC accuracy, voltage gain, and transconductance bandwidth

The discussion below summarizes key performance tradeoffs in DC accuracy, low-frequency AC accuracy, voltage gain, and transconductance bandwidth for the simple OTAs. Performance discussed is from the *Circuit Analysis* sheet shown in Figure 5.4.

Because device drain currents remain constant across the OTA versions, device gate areas change as the inverse of inversion coefficient and the square of channel length, as seen in the sizing relationships of Table 3.10. Gate areas then decrease a factor of 36 ($(3.2/1)\cdot(4\ \mu m/1.2\ \mu m)^2$) between the DC and balanced optimized OTAs, and decrease a factor of 18 ($(10/3.2)\cdot(1.2\ \mu m/0.5\ \mu m)^2$) between the balanced and AC optimized OTAs. This is for $IC = 1$, 3.2, and 10, and $L = 4$, 1.2, and 0.5 μm for the DC, balanced, and AC optimized OTAs, respectively.

Because devices operate at equal inversion coefficients, channel lengths, and drain currents for a given OTA version, device relative flicker-noise and local-area mismatch contributions remain nearly constant across the OTA versions. As a result, both the input-referred flicker-noise voltage density and offset voltage due to local-area mismatch decrease inversely with the square root of device gate areas.

Measured input-referred flicker-noise voltage density at 100 Hz increases a factor of 5.6 from 80 to 450 nV/Hz$^{1/2}$ between the DC and balanced optimized OTAs, and increases a factor of 4.4 from 450 to 2000 nV/Hz$^{1/2}$ between the balanced and AC optimized OTAs. Measured 1σ, input-referred offset voltage increases a factor of 2 from 1.1 to 2.2 mV between the DC and balanced optimized OTAs, and increases a factor of 4.6 from 2.2 to 10.2 mV between the balanced and AC optimized OTAs. With the exception of offset voltage for the DC optimized OTA, the flicker noise and offset voltage closely follow the inverse square root of gate area predicted by increases of a factor of 6 ($\sqrt{36}$) between the DC and balanced optimized OTAs, and a factor of 4.2 ($\sqrt{18}$) between the balanced and AC OTAs. The offset voltage increases less rapidly between the DC and balanced optimized OTAs because the offset voltage is increased for the DC optimized OTA, most likely due to distance mismatch as described in Section 5.4.2.5.

As described earlier in Sections 3.11.3.2 and 4.3.6.5, DC accuracy, which is the reciprocal of DC signal error, is equal to a selected DC input signal voltage divided by the 3σ, input-referred offset voltage. For a DC input signal of 0.1V and measured 1σ, input-referred offset voltages of 1.1, 2.2, and 10.2 mV, the DC accuracy is 30.3, 15.2, and 3.3 for the DC, balanced, and AC optimized OTAs, respectively. Simulated transconductance bandwidth (-3 dB) is 3.8, 56.2, and 350 MHz, giving products of DC accuracy and bandwidth at 115, 850, and 1160 MHz. The product of DC accuracy and bandwidth is largest for the AC optimized OTA because bandwidth increases more rapidly than DC accuracy decreases across the DC, balanced, and AC optimized OTAs. Products of DC accuracy and bandwidth, however, are nearly constant for the cascoded OTAs as described later in Section 5.5.2.11.

Finally, measured voltage gain is 326, 110, and 16.8 V/V, giving products of voltage gain and transconductance bandwidth at 1240, 6180, and 5880 MHz for the DC, balanced, and AC optimized OTAs, respectively. The products of gain and bandwidth are largest for the balanced and AC optimized OTAs because bandwidth increases more rapidly than gain decreases across the DC and balanced optimized OTAs and across the DC and AC optimized OTAs.

DC accuracy, which is inversely proportional to the input-referred offset voltage, and low-frequency AC accuracy, which is inversely proportional to the input-referred flicker-noise voltage density, decrease together, while transconductance bandwidth increases across the DC, balanced, and AC optimized OTAs. Additionally, voltage gain decreases while transconductance bandwidth increases across the OTA versions. These are primary performance tradeoffs across the OTA versions.

5.4.3 Other Optimizations: Ensuring Input Devices Dominate Thermal Noise

The *Analog CMOS Design, Tradeoffs and Optimization* spreadsheet permits rapid exploration of alternate circuit optimizations, especially when configured to permit independent exploration of device design choices. In the *MOSFETs* sheet shown in Figure 5.3 for the simple OTAs, the nMOS, M_3 and M_4 current-mirror devices are assigned together with the nMOS, M_1 and M_2 input pair devices because these devices operate at equal drain currents, inversion coefficients, and channel lengths. To permit independent exploration of device choices, another version of the sheet was prepared where the nMOS, M_1 and M_2 input pair devices and nMOS, M_3 and M_4 current-mirror devices were placed in separate columns. No change was required for the columns associated with the pMOS, M_5–M_8 current-mirror devices because these were already separate.

As mentioned, input devices in the simple OTAs shown in Figure 5.1 do not dominate noise or mismatch because all devices operate at equal drain currents, inversion coefficients, and channel lengths in a given OTA version. New designs were then considered where input and non-input devices operate at different inversion coefficients. In the new OTAs, $IC_1 = IC_2 = 1$ is held fixed for the M_1 and M_2 input pair devices, while $IC_3 = IC_4 = 16$ and $IC_5 = IC_6 = IC_7 = IC_8 = 16$ are held fixed for the M_3 and M_4, and M_5–M_8, non-input, current-mirror devices. As described in Section 5.3.5.3, this should reduce the input-referred thermal-noise voltage by minimizing the gate-referred thermal-noise voltage of the input devices, while minimizing the drain noise current contributions of the non-input devices. To implement DC, balanced, and AC optimized versions, channel lengths are varied from 4, 1.2, and 0.5 μm as done for the original OTAs. As in the original OTAs, drain currents are maintained at 50 μA.

From the *Circuit Analysis* sheet (not shown), predicted OTA transconductance is 885 μS, input-referred thermal-noise voltage density is $8.1 \text{ nV/Hz}^{1/2}$, and the input, 1 dB compression voltage is 0.078 V for all three versions of the new OTAs. The transconductance and input, 1 dB compression voltage are the same as those for the original, DC optimized OTA because of equal $IC_1 = IC_2 = 1$. However, the input-referred thermal noise is lower than $10.9 \text{ nV/Hz}^{1/2}$ for the original, DC optimized OTA because non-input devices operate in strong inversion, lowering their thermal-noise contributions. In the new OTAs, *RNP* values are 0.4 for the M_3 and M_4, M_5 and M_6, and M_7 and M_8 non-input current mirrors. This gives a total *RNP* value of 1.2 for the non-input devices, indicating that these contribute nearly equal thermal-noise power as the input devices. Non-input thermal-noise contributions are higher in the original OTAs where the *RNP* values are unity for the non-input current mirrors. Here, non-input devices contribute three times the thermal-noise power as the input devices.

Output resistance and voltage gain are maximum, while input-referred flicker-noise voltage and offset voltage due to local-area mismatch are minimum for both the new and original, DC optimized OTAs. Additionally, transconductance bandwidth is maximum for both the new and original, AC optimized OTAs. However, performance changes less across the new, DC, balanced, and AC optimized OTAs because only channel lengths are varied, whereas both inversion coefficients and channel lengths are varied in the original OTAs. Transconductance, input, 1dB compression voltage, and thermal noise are constant across the new OTAs, whereas these vary across the original OTAs. This is because inversion coefficients do not change across the new OTAs, whereas these change across the orginal OTAs.

Operating input devices at low inversion coefficients while operating non-input devices at high inversion coefficients, as described for the new OTAs, is also done for the cascoded OTAs described in the next section. Many other optimizations are also possible. For example, as described in Sections 5.3.6.3 and 5.3.7.3, when non-input devices are operated at higher inversion coefficients than input devices to minimize their thermal-noise contributions, the channel lengths of these devices should be increased above those of the input devices. This reduces the flicker-noise and local-area mismatch contributions of non-input devices below those of the input devices by ensuring sufficient gate area for the non-input devices. This is done for the micropower, low-noise, CMOS preamplifiers described in Chapter 6, as summarized in the procedure of Section 6.10, where input devices dominate both thermal and flicker noise.

5.5 DESIGN OPTIMIZATION AND RESULTING PERFORMANCE FOR THE CASCODED OTAS

To further illustrate the OTA analysis and performance optimization discussed in Section 5.3, three versions of the cascoded, $0.18\,\mu m$ CMOS OTAs shown in Figure 5.2 are optimized for DC, balanced, and AC performance [2, 3]. However, unlike the simple OTAs described in Section 5.4, different drain currents and inversion coefficients are used for devices in the cascoded OTAs so input devices dominate the thermal noise. Additionally, only the channel lengths are changed between the DC, balanced, and AC optimized versions, compared to changing both inversion coefficients and channel lengths in the simple OTAs. Again, to investigate optimizations at constant power consumption, each cascoded OTA version maintains equal bias currents and constant supply voltages. This gives a core (excluding bias references) current consumption of $300\,\mu A$ and power consumption of $540\,\mu W$ for supply voltages of $\pm 0.9\,V$.

5.5.1 Selection of MOSFET Inversion Coefficients and Channel Lengths

As mentioned in Section 5.4.1 for the simple OTAs, the optimum design of analog CMOS circuits often requires operating devices at different inversion coefficients and channel lengths, corresponding to performance tradeoffs at different locations on the *MOSFET Operating Plane* shown in Figure 3.1. Additionally, devices may be operated at different drain currents to achieve optimal circuit performance.

Table 5.2, like Table 5.1 for the simple OTAs, summarizes predictions of performance for the cascoded OTAs that were derived earlier in Section 5.3. Table 5.2 also shows OTA performance trends as device inversion coefficient, channel length, and drain current are explored separately, where performance trends were described in detail in Section 5.3. Again, as noted, blank lines indicate little change in performance. Up and down arrows indicate performance that increases or decreases, with the degree of the increase or decrease indicated by the size of the arrow. When multiple trends are included, the first is for weak inversion, and the second is for strong inversion without small-geometry effects.

The trends shown in Table 5.2, like those shown in Table 5.1, are assigned to devices primarily responsible for the specified performance. Thermal-noise, flicker-noise, and local-area mismatch trends are assigned to the M_1 and M_2 input pair, but these also depend on the relative noise and mismatch contributions of non-input devices as reflected in the *RNP* and *RMP* values that were defined in Sections 5.3.5.2, 5.3.6.2, and 5.3.7.2. Again, the trends shown assume that the relative noise and mismatch contributions of non-input devices remain constant. These contributions remain constant because device inversion coefficients and drain currents remain constant across different versions of the OTAs, while channel lengths change equally for the different versions. As described in Sections 5.3.2.3 and 5.3.3.3, output resistance and, correspondingly, voltage gain depend on the channel lengths of M_2, M_4, M_6, M_8, and M_{10}, or, when paired with other devices, effectively the channel lengths of all devices. Trends associated with output resistance and voltage gain are thus assigned to all devices. Finally, as described in Section 5.3.4.2, transconductance frequency response is related to the diode-connected bandwidth, f_{diode}, of M_7 and M_8, along with additional capacitive loading from other devices as noted.

The trends shown in Table 5.2, like those shown in Table 5.1, are simplified by excluding small-geometry velocity saturation and VFMR decreases in g_m/I_D and g_m, and related increases in V_{EFF}. Additionally, the trends shown exclude small-geometry DIBL decreases in V_A and r_{ds} for short-channel

Table 5.2 Summary of predicted performance and performance trends versus device inversion coefficient, channel length, and drain current for the cascoded OTAs. Trends are given for the devices primarily responsible for the performance identified where noise and mismatch trends assume constant relative noise and mismatch contributions from non-input devices

Parameter	Trends		
	$IC \uparrow$ L, I_D fixed	$L \uparrow$ IC, I_D fixed	$I_D \uparrow$ IC, L fixed
Transconductance, output resistance, and voltage gain:			
$G_M = \dfrac{I_{D1}}{n_1 U_T \left(\sqrt{IC_1 + 0.25} + 0.5\right)}$	$-, \downarrow M_{1,2}$	$-$	$\uparrow M_{1,2}$
$R_{OUT} \approx \dfrac{V_{AL8,10} \cdot L_{8,10}}{U_T \left(\sqrt{IC_{8,10} + 0.25} + 0.5\right)} \left[\dfrac{V_{AL2}L_2}{I_{D2}} \left\| \dfrac{V_{AL4}L_4}{I_{D4}} \right\| \dfrac{V_{AL6}L_6}{I_{D6}}\right]$	$-, \downarrow M_{8,10}$	$\uparrow\uparrow$ all	$\downarrow$ all
$A_V \approx \dfrac{I_{D1}}{n_1 U_T \left(\sqrt{IC_1 + 0.25} + 0.5\right)}$ $\cdot \dfrac{V_{AL8,10} \cdot L_{8,10}}{U_T \left(\sqrt{IC_{8,10} + 0.25} + 0.5\right)} \left[\dfrac{V_{AL2}L_2}{I_{D2}} \left\| \dfrac{V_{AL4}L_4}{I_{D4}} \right\| \dfrac{V_{AL6}L_6}{I_{D6}}\right]$	$-, \downarrow M_{1,2}$ and $-, \downarrow M_{8,10}$	$\uparrow\uparrow$ all	$-$ all
Transconductance bandwidth:			
$f_{-3dB} \approx \dfrac{g_{m7}}{2\pi \left(C_{GS7} + C_{SB7} + C_{DB1} + C_{GD1} + C_{DB3} + C_{GD3}\right)}$	$\uparrow, \uparrow M_{7,8}$ and others	$\downarrow M_{7,8}$ and others	$-M_{7,8}$ and others
Input-referred thermal-noise voltage PSD:			
$S_{VIN}(\text{thermal}) = 2 \cdot 4kT \left(n_1^2 \Gamma_1\right) \left(\dfrac{U_T \left(\sqrt{IC_1 + 0.25} + 0.5\right)}{I_{D1}}\right)$ $\cdot [1 + RNP_3 + RNP_5]$	$-, \uparrow M_{1,2}$	$-$	$\downarrow M_{1,2}$
Input-referred flicker-noise voltage PSD:			
$S_{VIN}(\text{flicker}) = 2 \cdot \dfrac{K'_{F1}}{f^{AF1}} \left(\dfrac{IC_1}{L_1^2}\right) \left(\dfrac{I_{0,1}}{I_{D1}}\right) \cdot [1 + RNP_3 + RNP_5]$	$\uparrow M_{1,2}$	$\downarrow\downarrow M_{1,2}$	$\downarrow M_{1,2}$
Input-referred offset voltage variance due to local-area mismatch:			
$V_{INOFFSET}^2(\text{mismatch}) = A_{VGS1}^2 \left(\dfrac{IC_1}{L_1^2}\right) \left(\dfrac{I_{0,1}}{I_{D1}}\right)$ $\cdot [1 + RMP_3 + RMP_5]$	$\uparrow M_{1,2}$	$\downarrow\downarrow M_{1,2}$	$\downarrow M_{1,2}$
Input and output capacitances:			
$C_{INDIF} \approx \dfrac{1}{2} \left(\dfrac{L_1^2}{IC_1}\right) \left(\dfrac{I_{D1}}{I_{0,1}}\right) C'_{OX} \left(\hat{C}_{gsi1} + \hat{C}_{gbi1}\right)$	$\downarrow M_{1,2}$	$\uparrow\uparrow M_{1,2}$	$\uparrow M_{1,2}$
$C_{OUT} \approx \left(\dfrac{L_8}{IC_8}\right) \left(\dfrac{I_{D8}}{I_{0,8}}\right) \cdot \left[\dfrac{1}{2} W_{DIF} C_{J8} + C_{JSW8}\right]$ $+ \left(\dfrac{L_{10}}{IC_{10}}\right) \left(\dfrac{I_{D10}}{I_{0,10}}\right) \cdot \left[\dfrac{1}{2} W_{DIF} C_{J10} + C_{JSW10}\right]$	$\downarrow M_{8,10}$	$\uparrow M_{8,10}$	$\uparrow M_{8,10}$

Table 5.2 (*continued*)

Parameter	Trends		
	$IC \uparrow$ L, I_D fixed	$L \uparrow$ IC, I_D fixed	$I_D \uparrow$ IC, L fixed
Slew rate (trends assume $C_L \gg C_{OUT}$):			
$SR = \dfrac{\pm 2I_{D8,10}}{C_{OUT} + C_L}$	–	–	$\uparrow M_{8,10}$
Input and output voltage ranges:			
V_{INCM+} (Equation 5.60); decreases as $V_{DS3} = V_{DS4}$ increases	–	–	–
$V_{INCM-} \approx V_{SS} + \left(2n_1 U_T \sqrt{IC_1} + V_{T1}\right) + V_{DS,sat15}$	$-, \uparrow M_{1,2}$	–	–
$V_{OUT+} = V_{DD} - V_{DS4} - \left(2U_T \sqrt{IC_8 + 0.25} + 3U_T\right)$; decreases as $V_{DS3} = V_{DS4}$ increases	$-, \downarrow M_8$	–	–
$V_{OUT-} = V_{SS} + V_{DS6} + \left(2U_T \sqrt{IC_{10} + 0.25} + 3U_T\right)$; increases as $V_{DS5} = V_{DS6}$ increases	$-, \uparrow M_{10}$	–	–
Input, 1 dB compression voltage:			
$V_{INDIF1dB} \approx \dfrac{2I_{D8,10}}{I_{D1}} \cdot n_1 U_T \left(\sqrt{IC_1 + 0.25} + 0.5\right)$	$-, \uparrow M_{1,2}$	–	–

Trend key (multiple trends are for WI, SI):

- $-$ Generally little or no dependency.
- $\uparrow, \downarrow$ Generally increases sublinearly or decreases inversely as sublinear.
- $\uparrow, \downarrow$ Generally increases linearly or decreases inversely.
- $\uparrow\uparrow, \downarrow\downarrow$ Generally increases as the square or decreases as the inverse square.

Trends exclude small-geometry effects described in the notes of Table 5.1. These effects are included by the expression modifications described in the notes of Table 5.1.

devices operating especially at low levels of inversion and exclude increases in the gate-referred flicker-noise voltage (included in the value of K_F') at increasing inversion or V_{EFF}. These effects were included in the analysis of Section 5.3 and are included through expression modifications summarized in the notes at the bottom of Table 5.1. As mentioned in Section 5.3.13, small-geometry effects that deteriorate OTA performance would typically be managed by careful design. However, as shown later in Section 5.5.2.4, flicker-noise increases are significant for non-input pMOS devices in the cascoded OTAs. Finally, trends shown in Table 5.2 exclude mismatch increases caused by transconductance factor mismatch at high levels of inversion. These increases are included in the value of A_{VGS}.

As observed in Table 5.2, cascoded OTA transconductance and voltage gain are maximized and input-referred thermal-noise voltage minimized by operating the M_1 and M_2 input devices at low inversion coefficients for high g_m/I_D and g_m. The input, 1 dB compression voltage, however, is unfavorably minimized for operation here. As discussed in Section 5.2.2 and shown in Figure 5.2, the M_1 and M_2 input pair devices operate in moderate inversion at $IC \approx 0.8$ so these devices dominate thermal noise. These devices operate at drain currents of $100\,\mu A$. The pMOS, M_3 and M_4 current source devices and nMOS, M_5 and M_6 current-mirror devices operate well into strong inversion at $IC \approx 16$ for low g_m/I_D and g_m so these devices contribute reduced drain thermal noise current. M_3 and M_4 operate at drain currents of $150\,\mu A$, above $100\,\mu A$ for M_1 and M_2, which unfortunately increases their noise and mismatch contributions somewhat. M_5 and M_6 operate at lower drain currents of $50\,\mu A$, which reduces their noise and mismatch contributions. As mentioned in Section 5.2.2, the inversion coefficient values given are nominal values because of lateral diffusion variations in effective channel length.

OTA output resistance and, correspondingly, voltage gain are maximized by operating the M_8 and M_{10} cascode devices at low inversion coefficients, which increases their intrinsic voltage gain, resulting in increased drain output resistance. However, transconductance bandwidth depends strongly on the pMOS, M_7 and M_8 devices and is maximized by operating these at high inversion coefficients in strong inversion. The pMOS, M_7 and M_8, and nMOS, M_9 and M_{10} cascode devices operate in moderate inversion at $IC \approx 2.5$, which provides a compromise between high OTA output resistance and voltage gain, combined with moderate transconductance bandwidth. The M_7–M_{10} cascode devices operate at drain currents of $50\,\mu A$. As discussed in Section 5.3.5.2, the M_7–M_{10} cascode devices are negligible contributors to noise and mismatch because their effective transconductance is low due to considerable resistive source degeneration.

OTA input and output voltage ranges are maximized by operating devices at low inversion coefficients because this minimizes their V_{EFF} and $V_{DS,sat}$, and minimizes the required V_{DS} across the M_3 and M_4 current source and M_5 and M_6 current-mirror devices as described in Section 5.3.11.3. Because M_3–M_6 operate at high inversion coefficients in strong inversion to minimize their thermal-noise contributions, input and output voltage ranges are not maximized due to high $V_{DS,sat}$ for these devices.

Because device inversion coefficients and drain currents remain constant across the DC, balanced, and AC optimized versions of the cascoded OTAs, the transconductance, input-referred thermal-noise voltage and input, 1 dB compression voltage remain constant. Additionally, the input and output voltage ranges remain constant. This is in contrast to the simple OTAs, where inversion coefficients increased across the DC, balanced, and AC optimized versions resulting in changes for each of these aspects of performance.

Channel lengths for the M_1–M_6 devices are equal for a given version of the cascoded OTAs. This, as mentioned in Sections 5.3.6.3 and 5.3.7.3, does not minimize input-referred flicker-noise voltage and offset voltage due to local-area mismatch because gate areas for the non-input M_3–M_6 devices are insufficient compared to gate areas of the M_1 and M_2 input devices. However, as described later in Section 5.5.2.5, the measured 3σ offset voltage is only 0.72 mV for the DC optimized OTA, indicating very good offset performance. Channel lengths for the M_7–M_{10} devices are equal to channel lengths for the M_1–M_6 devices for a given version of the OTAs, except for the DC optimized OTA where the M_7–M_{10} channel lengths are reduced slightly to improve transconductance bandwidth.

The DC, balanced, and AC optimizations for the simple OTAs tracked a diagonal path from the upper left corner to the lower right corner on the *MOSFET Operating Plane* shown in Figure 3.1. This corresponded to increasing inversion coefficients and simultaneously decreasing channel lengths across the DC, balanced, and AC optimizations, which gives the largest changes in performance. DC, balanced, and AC optimizations for the cascoded OTAs, however, track a vertical path from top to bottom on the *Plane*. This corresponds to constant inversion coefficients, although different inversion coefficients for different devices, and decreasing channel lengths across the DC, balanced, and AC optimizations. Here, as mentioned, many aspects of performance do not change. Other aspects of performance change less compared to the simple OTAs.

5.5.1.1 DC optimization

In the DC optimized, cascoded OTA, channel lengths of $2\,\mu m$ are selected for the M_1 and M_2 input devices and M_3–M_6 non-input devices, which are all critical for noise and mismatch performance. Channel lengths are selected slightly lower at $1.4\,\mu m$ for the M_7–M_{10} cascode devices to improve transconductance bandwidth. As observed in Table 5.2, output resistance and voltage gain are maximized for long channel lengths in the DC optimized OTA. As described in Sections 5.3.2.3

and 5.3.3.3, output resistance and voltage gain increase as nearly the square of device V_A, which itself increases sublinearly with channel length, except for short-channel devices where the increase is greater as DIBL effects are reduced. The squared dependency on V_A is because the cascode output resistance depends on the product of the intrinsic voltage gain for the M_8 and M_{10} cascode devices, which increases with their channel lengths, and drain–source resistances that connect to their sources, which also increase with channel lengths. Unfortunately, transconductance bandwidth decreases nearly inversely with channel length for M_7 and M_8 and related devices due to dominant width-dependent, extrinsic gate-overlap, drain–body, and source–body capacitances as described in Section 5.3.4.3. Input-referred flicker-noise voltage and offset voltage due to local-area mismatch are minimized in the DC optimized OTA because of large gate areas for the M_1 and M_2 input devices, with, as mentioned, constant relative noise and mismatch contributions from non-input devices, M_3–M_6. Input capacitance is maximized, increasing as the square of channel length for the M_1 and M_2 input devices, tracking their gate area. Output capacitance is also maximized, increasing directly with channel length for the M_8 and M_{10} output devices, tracking their channel width. Gate area is proportional to the square of channel length because channel width must increase directly with length to maintain the selected inversion coefficient. Operating devices at longer channel lengths than those selected for the DC optimized OTA increases the output resistance and voltage gain, and reduces the flicker noise and local-area mismatch. However, this comes at the expense of increased device gate areas and input and output capacitances, and decreased transconductance bandwidth.

In addition to listing OTA performance trends in terms of the inversion coefficient and channel length, Table 5.2 also lists these in terms of the device drain current. As observed, many aspects of performance are independent of the drain current. However, transconductance increases directly with the M_1 and M_2 drain current through increased g_m, and output resistance decreases inversely with device drain currents, assuming constant drain current ratios are maintained. Additionally, the input-referred thermal-noise voltage, flicker-noise voltage, and offset voltage due to local-area mismatch decrease with increasing M_1 and M_2 drain current because of increased g_m (affecting thermal noise) and gate area (affecting flicker noise and mismatch), again, with constant relative noise and mismatch contributions from M_3–M_6. Higher levels of drain current improve some aspects of DC performance, primarily as a result of increased M_1 and M_2 input pair transconductance and increased device gate areas. As always, channel width and, correspondingly, gate area scale directly with drain current to maintain the selected inversion coefficient.

High OTA output resistance and voltage gain benefit both DC and low-frequency signals. Additionally, low input-referred flicker-noise voltage at low frequencies benefits low-frequency signals. As a result, the DC optimization is also an optimization for low-frequency signals

5.5.1.2 AC optimization

In the AC optimized, cascoded OTA, minimum process channel lengths of $0.18\,\mu m$ are selected for the M_1–M_{10} devices. As observed in Table 5.2, transconductance bandwidth is maximized, and input and output capacitances are minimized resulting from small gate areas and channel widths associated with short channel lengths. However, this comes at the expense of low output resistance and voltage gain because of low device V_A and r_{ds}. Additionally, this comes at the expense of high input-referred flicker-noise voltage and offset voltage due to local-area mismatch because of the significant decrease in gate area for M_1 and M_2, again, for constant relative noise and mismatch contributions from M_3–M_6.

Again, as seen in Table 5.2, many aspects of OTA performance are independent of the device drain current. However, OTA transconductance, input and output capacitances, and slew rate increase directly with the drain current. Input and output capacitances increase with the drain current because of the required gate-area increase for input devices, M_1 and M_2, and the required channel width increase for output devices, M_8 and M_{10}. Additionally, the input-referred thermal-noise voltage decreases as the drain current increases because of the increase in g_m for M_1 and M_2, again, for constant relative noise contributions from M_3–M_6. Higher levels of drain current improve some aspects of AC performance, primarily as a result of increased M_1 and M_2 input pair transconductance and increased slew rate, but at the expense of higher input and output capacitances. Again, as always, channel width and, correspondingly, gate area scale directly with drain current to maintain the selected inversion coefficient.

An AC optimization as defined here primarily maximizes signal bandwidth and minimizes signal capacitances, as mentioned also in Section 5.4.1.2 for the simple OTAs. Maximum signal bandwidth combined with low input-referred thermal-noise voltage maximizes the signal-to-noise ratio for the cascoded, AC optimized OTA. However, maximum signal bandwidth combined with high input, 1 dB compression voltage maximizes the signal-to-distortion ratio for the simple, AC optimized OTA. The differences in optimization are a result of operating input devices in moderate inversion for low input-referred thermal-noise voltage in the cascoded OTAs, and operating input devices in strong inversion for high input, 1 dB compression voltage in the simple, AC optimized OTA. Other AC optimizations involving different tradeoffs in signal bandwidth, distortion, and noise are possible as seen by different locations on the *Plane* shown in Figure 3.1.

5.5.1.3 Balanced optimization

In the balanced optimized, cascoded OTA, moderate channel lengths of 0.48 μm are selected for the M_1–M_{10} devices. This channel length selection is nearly geometrically centered between 2 μm used for most devices in the DC optimized OTA and 0.18 μm used in the AC optimized OTA. The effective channel lengths are lower at 0.152 and 0.129 μm for nMOS and pMOS devices in the AC optimized OTA for $DL = 0.028$ and 0.051 μm listed in Table 3.2, moving the geometric center close to 0.48 μm for the balanced OTA. This selection of channel length provides a general balance of DC and AC performance and was not chosen for any specific tradeoffs. A balanced optimization is required when performance for DC or low-frequency signals and high-frequency signals is important simultaneously.

As mentioned for the simple OTAs in Section 5.4.1.3, the inversion coefficient and channel length selections for a balanced optimization depend on the priorities of various aspects of performance. These selections are complicated by the many aspects of performance and the varying dependencies of performance on the inversion coefficient and channel length, as seen in Table 5.2. For example, OTA input-referred flicker-noise voltage and offset voltage due to local-area mismatch, both of which depend on gate area, are stronger functions of channel length compared to the inversion coefficient. Transconductance bandwidth is also a stronger function of channel length.

Again, as seen in Table 5.2, many aspects of OTA performance are independent of the device drain current. However, OTA transconductance, output resistance, input and output capacitances, slew rate, and input-referred thermal-noise voltage, flicker-noise voltage, and offset voltage due to local-area mismatch depend on the drain current. As a result, the drain current can also be considered for a balanced optimization of performance. As mentioned, drain currents are not varied across the DC, balanced, and AC optimized OTAs.

5.5.2 Predicted and Measured Performance

As described in Section 5.4.2 for the simple OTAs, the *Analog CMOS Design, Tradeoffs and Optimization* spreadsheet maps predicted device performance for selections of the inversion coefficient, channel length, and drain current into complete OTA circuit performance.

For the cascoded OTAs, the spreadsheet uses the 0.18 μm CMOS process parameters from Tables 3.2–3.4 that are fixed or unbinned across device geometry and bias conditions, except for the Early voltage factor, $V_{AL} = V_A/L$. As mentioned for the simple OTAs, no simple prediction of MOS Early voltage, $V_A = I_D \cdot r_{ds} = I_D/g_{ds}$, or the related drain–source resistance, r_{ds}, or conductance,

$g_{ds} = 1/r_{ds}$, is available because of complex dependencies on the inversion level, channel length, and V_{DS}. Instead, as described near the beginning of Section 4.4, the process parameter $V_{AL} = V_A/L$ is extracted from V_A measurements near the operating inversion coefficient, channel length, and V_{DS}.

For the 0.18 μm CMOS process used in the cascoded OTAs, V_{AL} values are found from the measured V_A values shown in Figure 3.45 for nMOS devices operating at $V_{DS} = 0.5$ V and from additional data not shown for pMOS devices. However, V_A and, correspondingly, V_{AL} values for the pMOS, M_3 and M_4 current source devices are reduced to 40 % of their values at $V_{DS} = 0.5$ V because these devices operate at $V_{DS} = 0.4$ V, which is close to their $V_{DS,sat} = 0.3$ V at $IC \approx 16$. Additionally, V_A and V_{AL} are reduced to 50 % of their values at $V_{DS} = 0.5$ V for the nMOS, M_5 and M_6 current-mirror devices because of the same operating conditions where V_{DS} is near $V_{DS,sat}$. V_{DS} is sufficiently above the lower $V_{DS,sat}$ of the M_1 and M_2, and M_7–M_{10} devices, so these devices require no adjustment in V_A and V_{AL} at their operating V_{DS} near 0.5 V.

The increase in g_{ds}, corresponding to a decrease in r_{ds} and V_A, for V_{DS} approaching $V_{DS,sat}$ was discussed in Section 3.8.4.4 and shown in Figures 3.36 and 3.38. This can also be seen in Figure 3.45 by the decrease in V_A at $IC > 30$ where $V_{DS,sat}$ approaches the operating value of $V_{DS} = 0.5$ V. In the 0.18 μm CMOS process used, the decrease in r_{ds} and V_A is greater for pMOS devices operating at V_{DS} near $V_{DS,sat}$ compared to nMOS devices. This can be seen in the measured data shown in Figures 3.43 and 3.44 for operation at $IC = 100$ at $V_{DS} = 1$ V where $V_{DS,sat} = 0.6$ V. At this operating condition, nMOS V_A is almost unaffected in Figure 3.43, while pMOS V_A is beginning to roll off in Figure 3.44.

As mentioned for the simple OTAs, the process value of $V_{AL} = V_A/L$ used in the spreadsheet should not include the reduction associated with DIBL effects for short-channel devices because DIBL effects are separately included in the spreadsheet. As described near the beginning of Section 4.4, the DIBL component of V_{AL} is removed for the minimum channel, $L = 0.18$ μm devices, leaving only the CLM component required for the spreadsheet. The removal of the DIBL component is not required for the other channel lengths, which are 0.48 μm or larger where DIBL effects are negligible.

The final process values of V_{AL} used in the spreadsheet for the M_1 and M_2, and M_7–M_{10} devices are similar to those shown in Figure 3.48 for devices in the 0.18 μm CMOS process operating at $V_{DS} = 0.5$ V. The values for the M_3–M_6 devices are reduced as described because they operate at a lower $V_{DS} = 0.4$ V, near their $V_{DS,sat}$. Because of the removal of DIBL effects, V_{AL} values for the $L = 0.18$ μm devices do not decrease as shown in Figure 3.48, but continue increasing slightly as channel length decreases.

As mentioned for the simple OTAs, the extraction of process V_{AL} values for the spreadsheet requires some effort, especially for a larger circuit. However, V_{AL} only affects MOS $g_{ds} = 1/r_{ds}$ and resulting predictions of OTA output resistance and voltage gain, so rough estimates for V_{AL} do not affect other aspects of performance. Linking the spreadsheet to a simulation MOS model that provides direct $g_{ds} = 1/r_{ds}$ modeling is anticipated in the future, so the reader is invited to check the book's web site listed on the cover for spreadsheet updates. This linkage would eliminate the need for extracting or estimating V_{AL}, although experimental verification is recommended given the complexity of predicting $g_{ds} = 1/r_{ds}$ and its long history of modeling errors.

Figure 5.11 shows the *MOSFETs* sheet in the spreadsheet for the nMOS, M_1 and M_2 input devices in the three cascoded OTAs shown in Figure 5.2. This sheet lists device drain current, inversion coefficient, and channel length selections for the DC, balanced, and AC optimized OTAs, and reports the resulting device geometry and performance. Figure 5.12 shows this for the pMOS, M_3 and M_4 current source devices and the nMOS, M_5 and M_6 current-mirror devices. Finally, Figure 5.13 shows this for the pMOS, M_7 and M_8, and nMOS, M_9 and M_{10} cascode devices. As for the simple OTAs, equally biased and sized devices are listed in a single column in the *MOSFETs* sheet. MOS geometry and performance given by this sheet was described earlier in Section 4.4 for the optimization of differential-pair and current-mirror devices.

For brevity, the second sheets of the *MOSFETs* sheets are not shown. A second-page sheet was shown in Figure 5.3 for the simple OTAs. As mentioned, this sheet includes the body-effect transconductance, g_{mb}, source transconductance, $g_{ms} = g_m + g_{mb} + g_{ds}$, source–body capacitance, C_{SB},

Analog CMOS Design, Tradeoffs and Optimization Spreadsheet -- MOSFETs Sheet						
For design guidance only as results are approximate and not for any particular CMOS process.						
© 2000–2007, David M. Binkley. See Disclaimer, Notes Sheet.						
---------- Optional User Design Information ----------						
Description		------------ Input pair ------------				
Device reference		M1, M2	M1, M2	M1, M2		
Device notes		DC	BAL	AC		
---------- Required User Design Inputs ----------						
Device model		nMIL2	nMIL05	nMIL02		
I_D	μA (+)	100	100	100		
IC (fixed normalized)		0.866	0.832	0.742		
L_{DRAWN}	μm	1.98	0.48	0.18		
---------- Calculated Results ----------		OK	OK	OK		
Effective Width, Length, Gate Area						
W	μm	352.00	84.80	32.00		
L	μm	1.952	0.452	0.152		
WL	μm²	687.10	38.33	4.86		
DC Bias Voltages						
$V_{EFF} = V_{GS} - V_T$	V	0.032	0.030	0.024		
V_T (adjusted for V_{SB})	V	0.420	0.420	0.420		
$V_{GS} = V_T + V_{EFF}$	V	0.452	0.450	0.444		
$V_{DS,sat}$	V	0.133	0.132	0.130		
Small Signal Parameters						
g_m/I_D	μS/μA	19.50	19.66	20.08		
g_m	μS	1949.59	1965.63	2008.14		
V_A	V	17.93	8.13	2.27		
g_{ds}	μS	5.5758	12.3038	44.0606		
$A_{Vi} = g_m/g_{ds}$	V/V	349.7	159.8	45.6		
Transcond. Distortion (Diff. Pair)						
$V_{INDIF,1dB}$ (input 1-dB comp.)	V	0.072	0.071	0.071		
Capacitances and Bandwidths						
$C_{GS} = C_{gsi} + C_{GSO}$	fF	2188.1	181.7	42.4		
$C_{GB} = C_{gbi} + C_{GBO}$	fF	865.3	48.7	6.3		
$C_{GD} = C_{GDO}$ (in saturation)	fF	330.9	79.7	30.1		
C_{DB} (at $V_{DB} = V_{DS} + V_{SB}$)	fF	175.1	42.8	16.3		
$f_{Ti} = g_m/[2\pi(C_{gsi} + C_{gbi})]$	MHz	114.0	2077.1	17139.3		
$f_T = g_m/[2\pi(C_{GS} + C_{GB})]$	MHz	101.6	1358.2	6559.0		
$f_{DIODE} = g_m/[2\pi(C_{GS} + C_{GB} + C_{DB})]$, MHz		96.1	1145.4	4915.0		
Thermal and Flicker Noise						
$S_{VG}^{1/2}$ thermal	nV/Hz$^{1/2}$	2.51	2.50	2.46		
$S_{VG}^{1/2}$ flicker at $f_{FLICKER}$	nV/Hz$^{1/2}$	37.24	157.39	439.44		
$S_{ID}^{1/2}$ thermal	pA/Hz$^{1/2}$	4.90	4.91	4.95		
$S_{ID}^{1/2}$ flicker at $f_{FLICKER}$	pA/Hz$^{1/2}$	72.61	309.38	882.45		
$f_{FLICKER}$, freq. for flicker noise	Hz	100	100	100		
f_c, flicker-noise corner	MHz	0.0569	1.7102	19.8233		
DC Mismatch for Device Pair						
ΔV_{GS} (1σ)	mV	0.19	0.82	2.31		
$\Delta I_D/I_D$ (1σ)	%	0.38	1.62	4.64		

Figure 5.11 *MOSFETs sheet in the Analog CMOS Design, Tradeoffs and Optimization spreadsheet listing M_1 and M_2 input pair devices in the cascoded OTAs optimized for DC, balanced, and AC performance. This sheet lists MOSFET drain current, inversion coefficient, and channel length selections with resulting device geometry and performance*

Analog CMOS Design, Tradeoffs and Optimization Spreadsheet -- MOSFETs Sheet							
For design guidance only as results are approximate and not for any particular CMOS process.							
© 2000–2007, David M. Binkley. See Disclaimer, Notes Sheet.							
---------- Optional User Design Information ----------							
Description		--- pMOS current source ---			---------- nMOS mirror ----------		
Device reference		M3, M4	M3, M4	M3, M4	M5, M6	M5, M6	M5, M6
Device notes		DC	BAL	AC	DC	BAL	AC
---------- Required User Design Inputs ----------							
Device model		pSIL2	pSIL05	pSIL02	nSIL2	nSIL05	nSIL02
I_D	μA (+)	150	150	150	50	50	50
IC (fixed normalized)		17.461	15.843	12.537	17.316	16.038	14.832
L_{DRAWN}	μm	1.98	0.48	0.18	1.98	0.48	0.18
---------- Calculated Results ----------		OK	OK	OK	OK	OK	OK
Effective Width, Length, Gate Area							
W	μm	122.40	30.00	11.40	8.80	2.20	0.80
L	μm	1.929	0.429	0.129	1.952	0.452	0.152
WL	μm²	236.11	12.87	1.47	17.18	0.99	0.12
DC Bias Voltages							
$V_{EFF} = V_{GS} - V_T$	V	0.298	0.285	0.258	0.292	0.286	0.293
V_T (adjusted for V_{SB})	V	0.420	0.420	0.420	0.420	0.420	0.420
$V_{GS} = V_T + V_{EFF}$	V	0.718	0.705	0.678	0.712	0.706	0.713
$V_{DS,sat}$	V	0.301	0.290	0.267	0.302	0.294	0.286
Small Signal Parameters							
g_m/I_D	μS/μA	6.37	6.57	6.97	6.47	6.45	5.76
g_m	μS	955.51	985.33	1045.98	323.57	322.69	288.17
V_A	V	7.91	3.20	1.50	11.71	4.57	2.16
g_{ds}	μS	18.9688	46.8880	99.9052	4.2703	10.9432	23.1248
$A_{Vi} = g_m/g_{ds}$	V/V	50.4	21.0	10.5	75.8	29.5	12.5
Transcond. Distortion (Diff. Pair)							
$V_{INDIF,1dB}$ (input 1-dB comp.)	V	0.271	0.261	0.244	0.263	0.267	0.308
Capacitances and Bandwidths							
$C_{GS} = C_{gsi} + C_{GSO}$	fF	1236.3	81.8	14.3	92.6	6.9	1.3
$C_{GB} = C_{gbi} + C_{GBO}$	fF	181.1	10.0	1.2	12.3	0.7	0.1
$C_{GD} = C_{GDO}$ (in saturation)	fF	78.3	19.2	7.3	8.3	2.1	0.8
C_{DB} (at $V_{DB} = V_{DS} + V_{SB}$)	fF	63.0	15.8	6.5	4.9	1.4	1.0
$f_{Ti} = g_m/[2\pi(C_{gsi} + C_{gbi})]$	MHz	113.6	2158.4	20299.9	533.0	9217.8	67620.8
$f_T = g_m/[2\pi(C_{GS} + C_{GB})]$	MHz	107.3	1707.2	10742.5	491.0	6722.6	32066.8
$f_{DIODE} = g_m/[2\pi(C_{GS} + C_{GB} + C_{DB})]$, MHz		102.7	1456.3	7577.9	469.0	5693.8	19091.9
Thermal and Flicker Noise							
$S_{VG}^{1/2}$ thermal	nV/Hz$^{1/2}$	3.82	3.76	3.65	6.51	6.52	6.89
$S_{VG}^{1/2}$ flicker at $f_{FLICKER}$	nV/Hz$^{1/2}$	73.77	308.42	865.93	295.01	1220.40	3510.29
$S_{ID}^{1/2}$ thermal	pA/Hz$^{1/2}$	3.65	3.71	3.82	2.11	2.10	1.99
$S_{ID}^{1/2}$ flicker at $f_{FLICKER}$	pA/Hz$^{1/2}$	70.49	303.89	905.75	95.45	393.80	1011.56
$f_{FLICKER}$, freq. for flicker noise	Hz	100	100	100	100	100	100
f_c, flicker-noise corner	MHz	0.0281	0.4415	3.3442	0.7893	22.2382	234.2794
DC Mismatch for Device Pair							
ΔV_{GS} (1σ)	mV	0.38	1.63	4.75	1.42	5.90	17.45
$\Delta I_D/I_D$ (1σ)	%	0.24	1.07	3.31	0.92	3.81	10.06

Figure 5.12 *MOSFETs sheet in the Analog CMOS Design, Tradeoffs and Optimization spreadsheet listing* M_3 *and* M_4 *current source and* M_5 *and* M_6 *current-mirror devices in the cascoded OTAs optimized for DC, balanced, and AC performance*

Analog CMOS Design, Tradeoffs and Optimization Spreadsheet -- MOSFETs Sheet							
For design guidance only as results are approximate and not for any particular CMOS process.							
© 2000–2007, David M. Binkley. See Disclaimer, Notes Sheet.							
---------- Optional User Design Information ----------							
Description		-------- pMOS cascode --------			-------- nMOS cascode --------		
Device reference		M7, M8	M7, M8	M7, M8	M9, M10	M9, M10	M9, M10
Device notes		DC	BAL	AC	DC	BAL	AC
---------- Required User Design Inputs ----------							
Device model		pMIL1_4	pMIL05	pMIL02	nMIL1_4	nMIL05	nMIL02
I_D	µA (+)	50	50	50	50	50	50
IC (fixed normalized)		2.922	2.641	2.166	2.868	2.757	2.472
L_{DRAWN}	µm	1.38	0.48	0.18	1.38	0.48	0.18
---------- Calculated Results ----------		OK	OK	OK	OK	OK	OK
Effective Width, Length, Gate Area							
W	µm	168.00	60.00	22.00	36.80	12.80	4.80
L	µm	1.329	0.429	0.129	1.352	0.452	0.152
WL	µm²	223.27	25.74	2.84	49.75	5.79	0.73
DC Bias Voltages							
$V_{EFF}=V_{GS}-V_T$	V	0.107	0.100	0.087	0.104	0.102	0.097
V_T (adjusted for V_{SB})	V	0.420	0.420	0.420	0.420	0.420	0.420
$V_{GS}=V_T+V_{EFF}$	V	0.527	0.520	0.507	0.524	0.522	0.517
$V_{DS,sat}$	V	0.171	0.167	0.159	0.171	0.170	0.165
Small Signal Parameters							
g_m/I_D	µS/µA	13.39	13.85	14.70	13.65	13.77	13.96
g_m	µS	669.71	692.55	735.23	682.45	688.59	698.07
V_A	V	12.74	6.20	2.31	15.24	8.25	2.55
g_{ds}	µS	3.9251	8.0633	21.6022	3.2805	6.0624	19.5823
$A_{Vi}=g_m/g_{ds}$	V/V	170.6	85.9	34.0	208.0	113.6	35.6
Transcond. Distortion (Diff. Pair)							
$V_{INDIF,1dB}$ (input 1–dB comp.)	V	0.108	0.104	0.099	0.106	0.105	0.107
Capacitances and Bandwidths							
$C_{GS}=C_{gsi}+C_{GSO}$	fF	971.6	135.8	24.3	227.2	34.2	7.2
$C_{GB}=C_{gbi}+C_{GBO}$	fF	233.7	27.5	3.2	48.7	5.7	0.7
$C_{GD}=C_{GDO}$ (in saturation)	fF	107.5	38.4	14.1	34.6	12.0	4.5
C_{DB} (at $V_{DB}=V_{DS}+V_{SB}$)	fF	85.5	30.7	11.3	19.3	6.9	2.7
$f_{Ti}=g_m/[2\pi(C_{gsi}+C_{gbi})]$	MHz	97.1	882.2	8728.6	450.1	3925.8	32030.6
$f_T=g_m/[2\pi(C_{GS}+C_{GB})]$	MHz	88.4	674.8	4257.3	393.7	2743.4	13921.4
$f_{DIODE}=g_m/[2\pi(C_{GS}+C_{GB}+C_{DB})]$, MHz		82.6	568.1	3016.2	368.0	2340.3	10445.5
Thermal and Flicker Noise							
$S_{VG}^{1/2}$ thermal	nV/Hz$^{1/2}$	4.48	4.40	4.25	4.40	4.38	4.34
$S_{VG}^{1/2}$ flicker at $f_{FLICKER}$	nV/Hz$^{1/2}$	49.38	142.55	413.36	148.18	433.71	1215.36
$S_{ID}^{1/2}$ thermal	pA/Hz$^{1/2}$	3.00	3.05	3.12	3.00	3.01	3.03
$S_{ID}^{1/2}$ flicker at $f_{FLICKER}$	pA/Hz$^{1/2}$	33.07	98.72	303.92	101.13	298.65	848.41
$f_{FLICKER}$, freq. for flicker noise	Hz	100	100	100	100	100	100
f_c, flicker-noise corner	MHz	0.0096	0.0754	0.6120	0.3924	4.9736	57.4524
DC Mismatch for Device Pair							
ΔV_{GS} (1σ)	mV	0.35	1.03	3.08	0.74	2.16	6.09
$\Delta I_D/I_D$ (1σ)	%	0.47	1.42	4.52	1.01	2.98	8.50

Figure 5.13 *MOSFETs* sheet in the *Analog CMOS Design, Tradeoffs and Optimization* spreadsheet listing M_7–M_{10} cascode devices in the cascoded OTAs optimized for DC, balanced, and AC performance

and thermal noise, $n\Gamma$, flicker-noise, K'_F, and mismatch, A_{VGS}, parameters useful for circuit analysis. The sheet also includes velocity saturation and VFMR details, gate capacitance details, process and inversion coefficient details, layout details, and optional user inputs. The optional user inputs include the number of gate stripes or interdigitations, m, defaulted at unity, and a selected drawn channel width for each layout stripe that permits trimming the input value of IC to match the actual layout. Additionally, input values for V_{DS}, V_{SB}, the frequency for reported flicker noise, $f_{FLICKER}$, and temperature, T, may be included. Default values for these optional inputs are $V_{DS} = V_{GS}$, $V_{SB} = 0\,\text{V}$, $f_{FLICKER} = 1\,\text{Hz}$, and $T = 27\,°\text{C}$. The second sheet of results and optional user inputs are described further in the Appendix.

Figure 5.14 shows the *Circuit Analysis* sheet in the spreadsheet for the three cascoded OTAs. This sheet gives predicted OTA performance from the circuit analysis equations derived earlier in Section 5.3 and calls device geometry and performance from the *MOSFETs* sheets of Figures 5.11–5.13. As for the simple OTAs, this permits automated updating of OTA performance as device drain current, inversion coefficient, and channel length selections are explored for optimization. Again, because the spreadsheet uses the slightly more accurate expression for g_m/I_D from Table 3.17 that involves two square-root terms, OTA performance related to this varies slightly from the design equations derived in Section 5.3 that use the simpler g_m/I_D expression that involves a single square-root term. Additionally, the spreadsheet uses a corrected value of the substrate factor, n, for the level of inversion, which also changes the results slightly. To permit comparisons with the predicted values, the *Circuit Analysis* sheet also lists simulated and measured performance for the OTAs. Simulations are from SPICE simulations using BSIM3V3 MOS models [4] for the $0.18\,\mu\text{m}$ CMOS process considered.

5.5.2.1 *Transconductance, output resistance, and voltage gain*

OTA transconductance, G_M, from Equation 5.2 is nearly equal to the transconductance, $g_{m1} = g_{m2}$, of the M_1 and M_2 input pair devices. Predicted G_M in Figure 5.14 is 1930, 1920, and $1900\,\mu\text{S}$ for the DC, balanced, and AC optimized OTA, respectively, which is within 0.7, 0.8, and 5.8 % of measured values of 1920, 1940, and $1800\,\mu\text{S}$ shown later in Figure 5.22. Figure 5.15 shows simulated G_M of 1960, 1940, and $1830\,\mu\text{S}$, which is within 2.4 % of measured values. Because the percentage errors given are calculated from unrounded performance values in the spreadsheet, these will vary slightly from errors calculated from the rounded performance values given. G_M is equal and maximized for all OTAs because input devices operate at low inversion coefficients ($IC_1 = IC_2 \approx 0.8$) near the center of moderate inversion giving high g_m/I_D and g_m.

OTA output resistance, R_{OUT}, from Equation 5.6 is equal to the parallel combination of drain output resistances for the M_8 and M_{10} cascoded output devices. Predicted R_{OUT} in Figure 5.14 is 9.0, 1.9, and $0.35\,\text{M}\Omega$ for the DC, balanced, and AC optimized OTA, respectively, which is within 9.7, 15, and 28 % of measured values of 10.0, 2.3, and $0.27\,\text{M}\Omega$. Simulated R_{OUT} is 11.0, 2.7, and $0.36\,\text{M}\Omega$, which is within 33 % of measured values. The largest prediction and simulation errors are present for the AC optimized OTA that uses the minimum process channel length of $L = 0.18\,\mu\text{m}$. As discussed in Section 5.3.2.3, because of cascode action, R_{OUT} tracks the square of device V_A, which itself increases sublinearly with channel length except for short channel devices where the increase is greater as DIBL effects are reduced. Prediction errors in R_{OUT}, discussed later in Section 5.6, then result from prediction errors in r_{ds} that are effectively squared. R_{OUT} is maximized for the DC optimized OTA because devices operate at long channel lengths (L_1–$L_6 = 2\,\mu\text{m}$ and L_7–$L_{10} = 1.4\,\mu\text{m}$).

OTA voltage gain, A_V, from Equation 5.9 is equal to the product of G_M and R_{OUT}. Predicted A_V in Figure 5.14 is 17400, 3700, and 660 V/V for the DC, balanced, and AC optimized OTA, respectively, which is within 9.1, 16, and 35 % of measured values of 19100, 4400, and 490 V/V. Figure 5.16 shows simulated A_V of 21600, 5200, and 660 V/V, which is within 35 % of measured values. Prediction errors in A_V, like errors in R_{OUT}, primarily result from the square of prediction errors in r_{ds}. These errors are maximum for the AC optimized OTA because of its error in R_{OUT}.

Summary of Predicted, Simulated, and Measured Performance for Three, 0.18–μm OTAs							
Transconductance, G_M							
$G_M = g_{m1} \cdot [g_{ms7}/(g_{ms7} + g_{ds1} + g_{ds3})]$							
					DC	Balanced	AC
				Predicted	1,931	1,921	1,904	μS
				Simulated	1,964	1,935	1,833	μS
				Measured	**1,918**	**1,937**	**1,800**	μS
Output Resistance, R_{OUT}							
$R_{OUT} = r_{outD8} \| r_{outD10}$							
					DC	Balanced	AC
				Predicted	9.0	1.93	0.35	MΩ
				Simulated	11.0	2.70	0.36	MΩ
				Measured	**10.0**	**2.27**	**0.27**	MΩ
$r_{outD8} = r_{ds8} \cdot [1 + g_{ms8}/(g_{ds2} + g_{ds4})]$					10.5	2.25	0.42	MΩ
$r_{outD10} = r_{ds10} \cdot (1 + g_{ms10}/g_{ds6})$					62.7	13.49	2.06	MΩ
Voltage Gain, A_V							
$A_V = G_M \cdot R_{OUT} = g_{m1} \cdot [g_{ms7}/(g_{ms7} + g_{ds1} + g_{ds3})] \cdot (r_{outD8} \| r_{outD10})$							
					DC	Balanced	AC
				Predicted	17,353	3,702	661	V/V
				Simulated	21,600	5,230	662	V/V
				Measured	**19,100**	**4,400**	**490**	V/V
Bandwidth, f_{-3dB}							
$f_{-3dB} \approx f_{p7} \approx g_{m7}/[2\pi(C_{GS7} + C_{SB7} + C_{DB1} + C_{GD1} + C_{DB3} + C_{GD3})]$							
Pole at source of pMOS cascode devices, $M7$ and $M8$					DC	Balanced	AC
				Predicted	61	320	1075	MHz
				Simulated	76	286	860	MHz
Input-Referred Thermal Noise Voltage, $S_{VIN}^{1/2}$ (thermal)							
$S_{VIN}^{1/2} = \sqrt{[\{2 \cdot 4kT(n\Gamma)_1/g_{m1}\} \cdot (1 + RNP_3 + RNP_5)]}$							
					DC	Balanced	AC
				Predicted	4.7	4.7	4.6	nV/Hz$^{1/2}$
				Simulated				nV/Hz$^{1/2}$
				Measured	**5**	**5**		nV/Hz$^{1/2}$
Relative Noise Power (RNP) -- relative to $M1$, $M2$							
					DC	Balanced	AC
$M1, M2$	1				1	1	1
$M3, M4$	$RNP_3 = [(n\Gamma)_3/(n\Gamma)_1] \cdot [g_{m3}/g_{m1}]$					0.56	0.57	0.60
$M5, M6$	$RNP_5 = [(n\Gamma)_5/(n\Gamma)_1] \cdot [g_{m5}/g_{m1}]$					0.18	0.18	0.16
Total:						1.74	1.75	1.76

Figure 5.14 *Circuit Analysis* sheet in the *Analog CMOS Design, Tradeoffs and Optimization* spreadsheet for the cascoded OTAs optimized for DC, balanced, and AC performance. This sheet lists predicted, simulated, and measured performance

Summary of Predicted, Simulated, and Measured Performance for Three, 0.18–μm OTAs							
						(second page)	
Input-Referred Flicker-Noise Voltage, $S_{VIN}^{1/2}$ (flicker)							
$S_{VIN}^{1/2}(f) = \sqrt{[\{2 \cdot K'_{F1}/((WL)_1 f^{AF1})\} \cdot (1+RNP_3+RNP_5)]}$							
					DC	Balanced	AC	(100 Hz)
				Predicted	101	421	1140	nV/Hz$^{1/2}$
				Simulated				nV/Hz$^{1/2}$
				Measured	**96**	**420**	**1700**	**nV/Hz$^{1/2}$**
Relative Noise Power (RNP) -- relative to $M1$, $M2$ at $f = 1$ Hz							
RNP_3 adjusted for frequency by (f^{AF1}/f^{AF3}), etc.					DC	Balanced	AC	(1 Hz)
$M1$, $M2$	1				1	1	1
$M3$, $M4$	$RNP_3 = [K'_{F3}/K'_{F1}] \cdot [(WL)_1/(WL)_3] \cdot [g_{m3}/g_{m1}]^2$				2.37	2.42	2.65
$M5$, $M6$	$RNP_5 = [K'_{F5}/K'_{F1}] \cdot [(WL)_1/(WL)_5] \cdot [g_{m5}/g_{m1}]^2$				1.73	1.62	1.31
Total:					5.10	5.04	4.96
Input-Referred Offset Voltage Due to Local-Area Mismatch, $V_{INOFFSET}$ (mismatch)							
$V_{INOFFSET} = \sqrt{[\{A_{VGS1}^2/(WL)_1\} \cdot (1+RMP_3+RMP_5)]}$							
					DC	Balanced	AC	(1σ)
				Predicted	0.36	1.51	4.21	mV
				Measured	**0.24**	**1.10**	**3.20**	**mV**
Relative Mismatch Power (RMP) -- relative to $M1$, $M2$							
					DC	Balanced	AC	
$M1$, $M2$	1				1	1	1
$M3$, $M4$	$[A_{VGS3}/A_{VGS1}]^2 \cdot [(WL)_1/(WL)_3] \cdot [g_{m3}/g_{m1}]^2$				0.94	0.99	1.15
$M5$, $M6$	$[A_{VGS5}/A_{VGS1}]^2 \cdot [(WL)_1/(WL)_5] \cdot [g_{m5}/g_{m1}]^2$				1.46	1.38	1.17
Total:					3.40	3.37	3.32
Input and Output Capacitances and Voltage Ranges; Input, 1-dB Compression Voltage							
Unless specified, all are predicted from MOS operating parameters.							
Design inputs							
V_{DD}			0.9	V	Positive supply voltage		
V_{SS}			−0.9	V	Negative supply voltage		
$V_{DS3}=V_{DS4}=V_{DS5}=V_{DS6}$			0.4	V	V_{DS} set by low-voltage cascode bias		
$V_{DS,sat15}$			0.23	V	$V_{DS,sat}$ of pair current source		
					DC	Balanced	AC	
$C_{INDIF} = (1/2)*[C_{GS1}+C_{GB1}+C_{GD1}(1+g_{m1}/g_{ms7})]$					2.061	0.241	0.070	pF
$C_{OUT} = C_{DB8}+C_{GD8}+C_{DB10}+C_{GD10}$					0.247	0.088	0.033	pF
$V_{INCM+} = V_{DD}-V_{DS3}+V_{GS1}-V_{DS,sat1}$					0.82	0.82	0.81	V
$V_{INCM-} = V_{SS}+V_{DS,sat15}+V_{GS1}$					−0.22	−0.22	−0.23	V
$V_{OUT+} = V_{DD}-V_{DS4}-V_{DS,sat8}$					0.33	0.33	0.34	V
$V_{OUT-} = V_{SS}+V_{DS6}+V_{DS,sat10}$					−0.33	−0.33	−0.33	V
$V_{INDIF1dB} = (2I_{D8} = 2I_{D10})/G_M$					0.052	0.052	0.053	V
				Measured	**0.055**	**0.055**	**0.060**	**V**

Figure 5.14   (*continued*)

A_V is maximized for the DC optimized OTA because devices operate at long channel lengths where, as mentioned, R_{OUT} is increased by nearly the square of device V_A values. The high, measured A_V of 19100 V/V for the DC optimized OTA required careful laboratory measurement to ensure that the output remained within its full-gain, linear range. The voltage gain was measured in a closed-loop configuration by measuring the loop gain response to an injected error signal. Section 5.5.2.11 summarizes OTA voltage gain, transconductance bandwidth, and the product of these.

5.5.2.2 Frequency response

OTA transconductance frequency response, f_{-3dB}, evaluated in the spreadsheet by Equation 5.16, is approximated as the pole frequency associated with the sources of the pMOS, M_7 and M_8 cascode devices. Predicted f_{-3dB} in Figure 5.14 is 61, 320, and 1080 MHz for the DC, balanced, and AC optimized OTA, respectively, which is within 20, 12, and 25 % of simulated values of 76, 290, and 860 MHz. Transconductance frequency response was not measured, but Figure 5.15 shows simulated frequency response along with identified values of f_{-3dB}. f_{-3dB} is overpredicted by 25 % compared to the simulated value for the AC optimized OTA, where some error is likely the result of parasitic layout capacitance, not considered in the prediction. Parasitic layout capacitance is more significant for the AC optimized OTA because of smaller device capacitances associated with smaller devices.

f_{-3dB} is maximized in the AC optimized OTA because of short channel lengths for the pMOS cascode devices and related devices included in Equation 5.16. Long channel lengths lower f_{-3dB} for the DC optimized OTA since f_{-3dB} is nearly inversely proportional to the channel length as summarized in Section 5.5.2.11. As described in Section 5.3.4.3, this is because capacitance is dominated by width-dependent, extrinsic gate-overlap, drain–body, and source–body capacitances compared to intrinsic gate capacitances that depend on gate area.

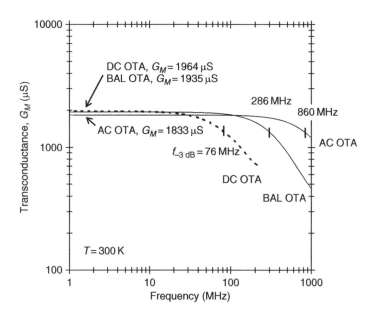

Figure 5.15   Simulated transconductance, G_M, for the cascoded OTAs. Simulated G_M is 1964, 1935, and 1833 μS with −3 dB bandwidths of 76, 286, and 860 MHz for the DC, balanced, and AC optimized OTA, respectively

5.5.2.3 Thermal noise

OTA thermal noise was analyzed in Section 5.3.5.2 in terms of input device, noise power contributions and the relative noise power, *RNP*, contributions of non-input devices. As shown in Figure 5.14, RNP_3 is approximately 0.57, indicating that the M_3 and M_4 current source devices contribute 57 % of thermal-noise power compared to the M_1 and M_2 input pair. This noise contribution is reasonably low, especially considering that M_3 and M_4 operate at drain currents of 150 μA compared to 100 μA for the input devices. RNP_5 is lower at approximately 0.18, indicating that the M_5 and M_6 current-mirror devices contribute 18 % of thermal-noise power compared to the input pair. This noise contribution is one-third that of M_3 and M_4 because M_5 and M_6 operate at drain currents of 50 μA, one-third that of M_3 and M_4, while the devices operate at the same inversion coefficients. All critical non-input devices, M_3–M_6, operate at inversion coefficients of approximately 16 well into strong inversion for low g_m/I_D and g_m compared to input devices that operate at inversion coefficients of approximately 0.8 in moderate inversion for high g_m/I_D and g_m. Operation in strong inversion ensures reduced drain thermal-noise contributions from non-input devices, while operation in moderate inversion ensures low input-referred thermal-noise voltage from the input devices.

As seen in Figure 5.14, the total, relative, noise power contribution from the M_3–M_6 non-input devices is approximately 0.75, resulting in a total OTA, input-referred thermal-noise voltage PSD of 1.75 $(1 + 0.75)$ times that of the input pair devices. This corresponds to a total input-referred thermal-noise voltage density of 132 % $(\sqrt{1.75})$ that of the input pair devices, where input devices nearly dominate the thermal noise.

Predicted input-referred thermal-noise voltage density (the square root of PSD) in Figure 5.14 from Equation 5.21 is approximately $4.7\,\text{nV}/\text{Hz}^{1/2}$ for the three OTAs. This is within 6.4 % of measured values of $5\,\text{nV}/\text{Hz}^{1/2}$ for the DC and balanced optimized OTAs shown in Figure 5.17. Because of high flicker noise, it was not possible to measure the thermal-noise floor for the AC optimized OTA. Input-referred thermal-noise voltage is constant for the three OTAs because device drain currents and inversion coefficients do not change across the OTA versions.

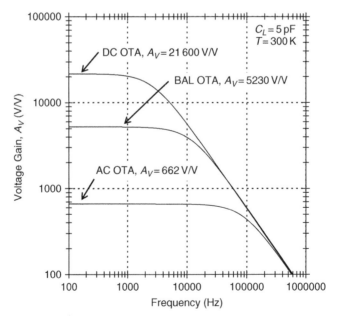

Figure 5.16 Simulated voltage gain, A_V, for the cascoded OTAs. Simulated A_V is 21600, 5230, and 662 V/V for the DC, balanced, and AC optimized OTA, respectively

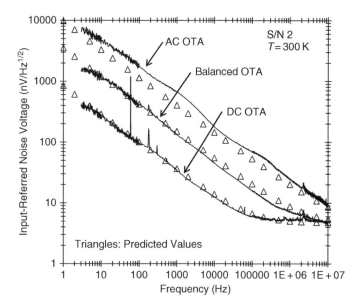

Figure 5.17 Measured (lines) and predicted (triangles) input-referred noise voltage density for the cascoded OTAs. Predicted noise for the DC, balanced, and AC optimized OTAs is from the *Circuit Analysis* sheet in the *Analog CMOS Design, Tradeoffs and Optimization* spreadsheet

5.5.2.4 Flicker noise

OTA flicker noise was analyzed in Section 5.3.6.2 by comparing the relative noise power contributions of non-input devices to that of input devices as described for thermal noise. As shown in Figure 5.14, RNP_3 is approximately 2.4, indicating that the pMOS, M_3 and M_4 current source devices contribute approximately 240 % of flicker-noise power compared to the M_1 and M_2 input pair. Interestingly, this noise contribution would be much lower at approximately 0.52 or 52 % if the flicker-noise factor for M_3 and M_4 did not increase to approximately $K'_F = (12\,400\,\mathrm{nV})^2 \cdot \mu\mathrm{m}^2$ or 460 % of its low inversion value of $K'_{F0} = (5800\,\mathrm{nV})^2 \cdot \mu\mathrm{m}^2$ listed in the process parameters of Table 3.3. The operating value of K'_F is listed on the second page of the *MOSFETs* sheet of Figure 5.12, which is not shown here but was shown in Figure 5.3 for the simple OTAs. As shown in Figure 3.64 for one 0.18 μm CMOS process and in Figures 3.65 and 3.66 for the 0.18 μm CMOS process used, pMOS devices experience a significant increase in gate-referred flicker-noise voltage for operation at high values of V_{EFF} in strong inversion. As seen in the figures, nMOS devices experience a much smaller noise increase. As described in Section 3.10.3.3, flicker-noise increases are modeled by $K'_F = K'_{F0}(1 + V_{EFF}/V_{KF})^2$, where V_{KF} is a voltage that models the increase in flicker noise with V_{EFF}. For the 0.18 μm CMOS process used, V_{KF} is 1 V and 0.25 V for nMOS and pMOS devices, respectively, as listed in Table 3.3.

    RNP_5 is smaller than RNP_3 at approximately 1.6, indicating that the nMOS, M_5 and M_6 current-mirror devices contribute approximately 160 % of flicker-noise power compared to the input pair. This noise contribution would be modestly lower at 0.97 or 97 % if the flicker-noise factor for M_5 and M_6 did not increase to approximately $K'_F = (8600\,\mathrm{nV})^2 \cdot \mu\mathrm{m}^2$ or 165 % of its low inversion value of $K'_{F0} = (6700\,\mathrm{nV})^2 \cdot \mu\mathrm{m}^2$. Although also operating at $IC \approx 16$ in strong inversion, the increase in flicker noise for these nMOS devices is much less than that of the pMOS, M_3 and M_4 devices.

    The *RNP* values listed in Figure 5.14 are for a frequency of 1 Hz. These require the adjustment noted for other frequencies because the flicker-noise slopes, AF, are different between nMOS and pMOS devices. Variations in the *RNP* values across different OTA versions, especially for the AC optimized version, result from variations in inversion coefficients and gate areas due to variations in effective channel length. Section 5.2.2 described variations in effective channel length and the resulting variations in inversion coefficients. These variations are present because effective channel length was not considered in the original selections of device inversion coefficients. These variations in inversion coefficients can be avoided in a new design by considering effective channel length as done in the spreadsheet.

The total, relative, noise power contribution from the M_3–M_6 non-input devices is 4, resulting in a total OTA, input-referred flicker-noise voltage PSD of 5 $(1+4)$ times that of the input pair devices. This corresponds to a total input-referred flicker-noise voltage density of 224 % $(\sqrt{5})$ that of the input pair devices, where input devices do not dominate the flicker noise. As described later in Section 5.5.4, the flicker-noise contributions of non-input devices can be lowered by increasing the channel lengths of these devices above that of the input devices. This increases the gate area of the non-input devices, lowering their flicker-noise contributions.

Predicted input-referred flicker-noise voltage density in Figure 5.14 from Equation 5.27 is 101, 421, and 1140 nV/Hz$^{1/2}$ at 100 Hz for the DC, balanced, and AC optimized OTA, respectively. This compares to measured values of 96, 420, and 1700 nV/Hz$^{1/2}$ shown in Figure 5.17, which also shows predicted noise from the spreadsheet. Predicted noise for the DC and balanced optimized OTAs overlays measured noise within 5.2 %, while noise for the AC optimized OTA is underpredicted by 33 %. The underprediction of flicker noise for the AC optimized OTA likely results from increases in K_F' for the minimum channel, $L = 0.18\,\mu m$ devices. As described in Section 3.10.3.5, the trap density near the Si–SiO$_2$ interface under the gate may be higher for short-channel devices because of increased damage caused by the close proximity of the implanted source and drains. Additionally, as discussed in Section 3.10.3.6, there is uncertainty in the flicker noise of small-area devices because of humps in the spectrum associated with discrete trapping locations. This is observed in the measured noise spectrum for the AC optimized OTA shown in Figure 5.17. Unlike the larger-area, DC and balanced optimized OTAs, the spectrum for the AC optimized OTA contains noticeable humps.

Input-referred flicker-noise voltage is minimized for the DC optimized OTA because of the large device areas resulting from long channel lengths. Flicker noise is substantially higher for the AC optimized OTA because operation at minimum channel lengths results in small device areas. As summarized in Section 5.5.2.11, with the exception of the AC optimized OTA, input-referred flicker-noise voltage density is inversely proportional to the square root of gate areas and, correspondingly, inversely proportional to the channel lengths of M_1–M_6.

5.5.2.5 Offset voltage due to local-area mismatch

OTA thermal and flicker noise were analyzed by comparing the relative noise power contributions of non-input devices to that of input devices. OTA offset voltage due to local-area mismatch was analyzed in a similar way in Section 5.3.7.2 by comparing the relative mismatch power, RMP, of non-input devices to that of input devices. As shown in Figure 5.14, RMP_3 is approximately unity, indicating that the M_3 and M_4 current source devices contribute approximately 100 % of mismatch power compared to the M_1 and M_2 input pair. RMP_5 is somewhat higher at approximately 1.4, indicating that the M_5 and M_6 current-mirror devices contribute approximately 140 % of mismatch power compared to the input pair. The mismatch contributions from the pMOS, M_3 and M_4 devices that operate at drain currents of 150 μA are actually less than contributions from the nMOS, M_5 and M_6 devices that operate at lower drain currents of 50 μA, even though the devices operate at equal inversion coefficients and channel lengths. This is a result of equal threshold-voltage mismatch factors, but a nearly factor-of-five increase in channel width and gate area for pMOS devices compared to nMOS devices for operation at the same inversion coefficient, channel length, and drain current. This can be seen by $W = (L/IC)(I_D/I_0)$ given in Table 3.10 for the technology current, I_0, values given in Table 3.2 that track the low-field mobility, μ_0.

The operating value of the gate–source voltage mismatch factor, $A_{VGS} = 5.9\,mV \cdot \mu m$, is slightly higher for the M_3–M_6 non-input devices than the threshold-voltage mismatch factor, $A_{VTO} = 5\,mV \cdot \mu m$, listed in Table 3.3. This indicates, as described in Section 5.3.7.1, that threshold-voltage mismatch nearly dominates transconductance factor mismatch for the M_3–M_6 devices, even though these operate well into strong inversion at $IC \approx 16$. The increased value of A_{VGS} above A_{VTO}, however, indicates the presence of transconductance factor mismatch. The value of $A_{VGS} = 5.1\,\text{mV}\cdot\mu\text{m}$ is very nearly equal to $A_{VTO} = 5\,\text{mV}\cdot\mu\text{m}$ for the M_1 and M_2 input devices that operate at $IC \approx 0.8$, indicating that threshold-voltage mismatch dominates for these devices also. When V_{SB} is assigned a non-zero value in the spreadsheet, A_{VGS} increases consistent with A_{VT} increasing above A_{VTO}. The second page of the *MOSFETs* sheet of Figure 5.12, not shown here but shown in Figure 5.3 for the simple OTAs, lists values for A_{VGS} that are used in mismatch circuit analysis.

Like variations in flicker-noise contributions, variations in the mismatch contributions across different OTA versions, especially for the AC optimized version, result from variations in inversion coefficients and gate areas resulting from variations in effective channel length. Additionally, RMP_5 is reduced for the AC optimized OTA because some velocity saturation reduction of g_m/I_D and g_m is present that lowers the conversion of threshold-voltage mismatch to drain current mismatch.

The total, relative, mismatch power contribution from the M_3–M_6 non-input devices is approximately 2.4, resulting in a total OTA, input-referred offset voltage variance of 3.4 $(1 + 2.4)$ times that of the input pair devices. This corresponds to a total input-referred offset voltage of 184 % ($\sqrt{3.4}$) that of the input pair devices. This compares to a total, input-referred flicker-noise voltage of 224 % that of the input pair devices, as mentioned in the previous section. Flicker noise and local-area mismatch optimize in the same way when threshold-voltage mismatch dominates and increases in gate-referred flicker-noise voltage with V_{EFF}, corresponding to increases in K_F', are negligible. However, increases in K_F' for the pMOS, M_3 and M_4 devices increase their flicker-noise contributions above their mismatch contributions.

As for the flicker noise, input devices do not dominate the local-area mismatch in the cascoded OTAs. As described later in Section 5.5.3, the mismatch contributions of non-input devices can be lowered by increasing the channel lengths of these devices above that of the input devices. This increases the gate area of the non-input devices, lowering their mismatch contributions.

Predicted 1σ, input-referred offset voltage (the square root of the offset voltage variance) in Figure 5.14 from Equation 5.36 is 0.36, 1.5, and 4.2 mV for the DC, balanced, and AC optimized OTA, respectively. The measured offset voltage, however, is more than 30 % lower at 0.24, 1.1, and 3.2 mV as shown in the measured offset histograms of Figures 5.18, 5.19, and 5.20. Predicted offset voltage is conservatively high, but closely matches the trends of measured offset across the OTA versions. Predicted offset voltage drops to 0.3, 1.3, and 3.6 mV, which is closer to the measured offset, if device A_{VTO} is lowered to $4\,\text{mV}\cdot\mu\text{m}$ from the process value of $5\,\text{mV}\cdot\mu\text{m}$ used from Table 3.3 for the $0.18\,\mu\text{m}$ CMOS process.

The measured 1σ offset voltage of 0.24 mV shown in Figure 5.18 for the DC optimized OTA corresponds to a 3σ offset voltage of 0.72 mV achieved, as mentioned, without input device dominance of the mismatch. This is an unusually low value of offset voltage for a CMOS circuit and is suggestive of a bipolar transistor circuit.

Input-referred offset voltage due to local-area mismatch, like input-referred flicker-noise voltage density, is minimized for the DC optimized OTA because of large device areas associated with long channel lengths. The offset voltage is substantially higher for the AC optimized OTA because of small device areas associated with short channel lengths. As summarized in Section 5.5.2.11, the input-referred offset voltage is inversely proportional to the square root of gate areas and, correspondingly, inversely proportional to the channel lengths of M_1–M_6.

In addition to listing the measured 1σ, input-referred offset voltages denoted by standard deviation values, Figures 5.18, 5.19, and 5.20 also list the measured, input-referred systematic offset voltages denoted by mean values. The systematic offset voltage is 0.05, -0.32, and 0.35 mV for the DC, balanced, and AC optimized OTA, respectively, which is significantly below their offset voltages of 0.24, 1.1, and 3.2 mV. Systematic offset was not analyzed because it is low due to high voltage gain for the cascoded OTAs.

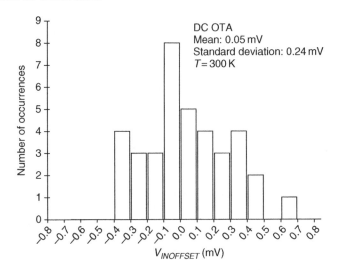

Figure 5.18 Measured input-referred offset voltage for the cascoded, DC optimized OTA. The measured 1σ offset voltage of 0.24 mV, which corresponds to a 3σ offset voltage of 0.72 mV, results from long MOS channel lengths and large gate areas. The offset voltage is unusually low for analog CMOS circuits and is more typical of bipolar transistor circuits

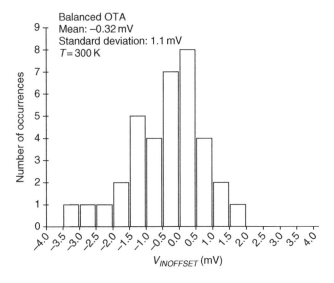

Figure 5.19 Measured input-referred offset voltage for the cascoded, balanced optimized OTA. The measured 1σ offset voltage of 1.1 mV, which corresponds to a 3σ offset voltage of 3.3 mV, is representative of analog CMOS circuits having moderate channel lengths and moderate gate areas

5.5.2.6 Input and output capacitances

OTA, differential input capacitance, C_{INDIF}, from Equation 5.46 is dominated by the gate–source and gate–body capacitances of the M_1 and M_2 input pair devices. Predicted C_{INDIF} in Figure 5.14 is 2.1, 0.24, and 0.07 pF for the DC, balanced, and AC optimized OTA, respectively. Output capacitance,

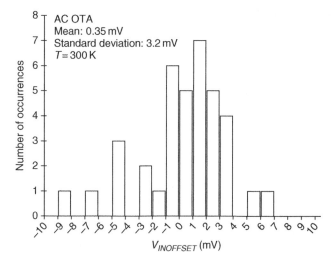

Figure 5.20 Measured input-referred offset voltage for the cascoded, AC optimized OTA. The measured 1σ offset voltage of 3.2 mV, which corresponds to a 3σ offset voltage of 9.6 mV, is high and representative of analog CMOS circuits having short channel lengths and small gate areas

C_{OUT}, from Equation 5.47 is dominated by the drain–body capacitances of the M_8 and M_{10} output devices. Predicted C_{OUT} is 0.25, 0.09, and 0.03 pF for the three OTAs.

C_{INDIF} and C_{OUT} are minimized for the AC optimized OTA because operation at minimum channel lengths results in small input device gate areas, minimizing C_{INDIF}, and small output device channel widths, minimizing C_{OUT}. Capacitances are significantly higher for the DC optimized OTA because operation at long channel lengths results in large channel widths and large gate areas.

5.5.2.7 Slew rate

Slew rate is not listed in the predicted, simulated, and measured OTA performance summarized in Figure 5.14 because it depends strongly on the external load capacitance. From Equation 5.50, slew rate is equal to twice the bias current, $2I_{D8} = 2I_{D10}$, of the M_8 and M_{10} output devices divided by the total output capacitance, which is the sum of C_{OUT} and the external load capacitance. Since all three OTAs operate at the same output device bias currents of 50 μA, slew rate varies only with the total output capacitance. Slew rate would normally be maximized for the AC optimized OTA because its high transconductance bandwidth permits small capacitive loads in closed-loop configurations having high closed-loop bandwidths. Additionally, the AC optimized OTA has low C_{OUT}, reducing the total output capacitance and increasing the slew rate.

5.5.2.8 Input and output voltage ranges

The OTA, maximum, input common-mode voltage, V_{INCM+}, from Equation 5.59 is controlled by the positive supply voltage, the operating drain–source voltage of the M_3 and M_4 current source devices, and the gate–source and drain–source saturation voltages of the M_1 and M_2 input devices. For a positive supply voltage of +0.9 V, and M_3 and M_4 drain–source voltage of 0.4 V (described in Section 5.2.2), predicted V_{INCM+} in Figure 5.14 is 0.82 V for the three OTAs. The minimum, input common-mode voltage, V_{INCM-}, from Equation 5.61 is controlled by the negative supply voltage, the gate–source voltage of the M_1 and M_2 input devices, and the drain–source saturation voltage of the M_{15} current source. For a negative supply voltage of $-0.9\,V$ and a drain–source saturation voltage of $0.23\,V$ for the M_{15} current source, predicted V_{INCM-} is $-0.22\,V$ for the three OTAs. Both V_{INCM+} and V_{INCM-} are underpredicted slightly because the increase in input device threshold voltage and gate–source voltage due to body effect is not considered in the spreadsheet where V_{SB} is set to its default value of zero. The input voltage range, which is the difference between V_{INCM+} and V_{INCM-}, is constant for all OTAs because devices operate at the same inversion coefficients across the OTA versions.

As described in Section 5.3.11.2, V_{INCM+} and V_{INCM-} are the maximum and minimum, input common-mode voltages where full OTA performance is maintained. Functionality, however, is maintained for inputs slightly outside these voltages.

The OTA, maximum, output voltage, V_{OUT+}, from Equation 5.63 is controlled by the positive supply voltage, the operating drain–source voltage of the M_3 and M_4 current source devices, and the drain–source saturation voltage of the M_8 output device. For a positive supply voltage of $+0.9\,V$, and M_3 and M_4 drain–source voltage of $0.4\,V$, predicted V_{OUT+} in Figure 5.14 is $0.33\,V$ for the three OTAs. The minimum output voltage, V_{OUT-}, from Equation 5.65 is controlled by the negative supply voltage, the operating drain–source voltage of the M_5 and M_6 current-mirror devices, and the drain–source saturation voltage of the M_{10} output device. For a complementary negative supply voltage of $-0.9\,V$, and M_5 and M_6 drain–source voltage of $0.4\,V$ (described in Section 5.2.2), predicted V_{OUT-} is $-0.33\,V$ for the three OTAs, which is complementary to the positive value for V_{OUT+}. The output voltage range, which is the difference between V_{OUT+} and V_{OUT-}, is constant for all OTAs because devices operate at the same inversion coefficients across the OTA versions.

As described in Section 5.3.11.2, V_{OUT+} and V_{OUT-} are the maximum and minimum output voltages where full OTA output resistance and voltage gain are maintained. For a capacitive load, the output voltage will extend further to the supply voltages less the operating drain–source voltages for the M_3 and M_4 current source, and M_5 and M_6 current-mirror devices.

5.5.2.9 Input, 1 dB compression voltage

The last entry in Figure 5.14 is the OTA, differential input, 1 dB compression voltage, $V_{INDIF1dB}$. The application of a negative or positive input voltage at this value results in an output current that is 1 dB below its ideal, small-signal value. As described in Section 5.3.12.2 and given in Equation 5.68, $V_{INDIF1dB}$ is approximately equal to twice the bias current, $2I_{D8} = 2I_{D10}$, of the M_8 and M_{10} output devices divided by G_M for the OTA. For the output device bias current of $50\,\mu A$, predicted $V_{INDIF1dB}$ is $0.052, 0.052$, and $0.053\,V$ for the DC, balanced, and AC optimized OTA, respectively, which is within $5.8, 5.3$, and $12.4\,\%$ of measured values of $0.055, 0.055$, and $0.060\,V$. Measured $V_{INDIF1dB}$ is found from Figure 5.21 that shows measured short-circuit output current versus the input voltage. $V_{INDIF1dB}$ is nearly constant for the three OTAs because input devices operate at the same inversion coefficients resulting in nearly constant G_M.

Figure 5.22 shows measured G_M versus the input voltage for the three OTAs. G_M remains nearly constant for input voltages sufficiently below $V_{INDIF1dB}$, but drops significantly for voltages above this, indicating significant signal distortion.

5.5.2.10 Layout area

OTA layout area depends on the gate areas of individual devices. Gate areas and layout area are largest for the DC optimized OTA because of operation at long channel lengths. As described in Section 3.6.3, gate area is proportional to the square of channel length because channel width must increase directly

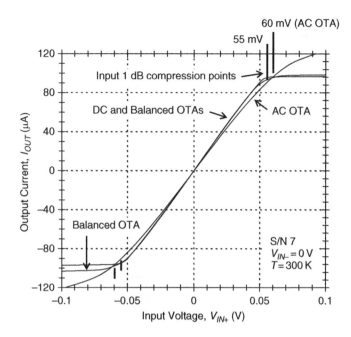

Figure 5.21 Measured short-circuit output current versus input voltage for the cascoded OTAs. The measured input, 1 dB compression voltage, $V_{INDIF1dB}$, is 55 mV for the DC and balanced optimized OTAs and 60 mV for the AC optimized OTA. OTA output current limits abruptly at nearly 100 μA due to limited output stage bias current

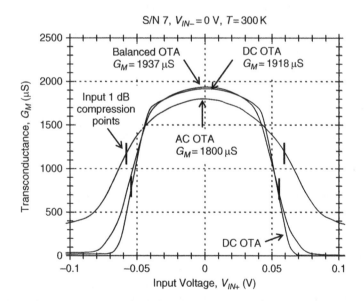

Figure 5.22 Measured transconductance, G_M, versus input voltage for the cascoded OTAs. G_M at an input voltage of zero is 1918, 1937, and 1800 μS for the DC, balanced, and AC optimized OTA, respectively. G_M remains nearly constant for signals sufficiently below the input, 1 dB compression voltage corresponding to low signal distortion. For input signals near the input, 1 dB compression voltage, G_M drops suddenly due to limited output stage bias current

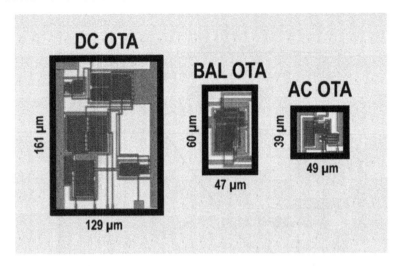

Figure 5.23 Photomicrograph of cascoded OTAs with layout dimensions. Operation at long channel lengths gives large gate areas and layout area for the DC optimized OTA, while operation at short channel lengths gives small gate areas and layout area for the AC optimized OTA. Because the level of inversion does not change for the different OTAs, the area differences are less than those shown in Figure 5.10 for the simple OTAs. Layout area for the DC optimized OTA can be reduced by better placement of devices. Photomicrograph was made after an SF$_6$ reactive ion etch removed the top passivation layer and intermetal dielectric

with channel length to maintain the selected drain current and inversion coefficient. Gate area changes for the OTAs are summarized in Section 5.5.2.11.

Figure 5.23 shows a photomicrograph of the three cascoded, 0.18 µm OTAs. Layout dimensions are approximately 129 × 161, 47 × 60, and 49 × 39 µm for the DC, balanced, and AC optimized OTA, respectively. This corresponds to approximate layout areas of 21 000, 2800, and 1900 µm^2 for the three OTAs. The layout area for the DC optimized OTA can be reduced somewhat because of unused area within the layout, but is still considerably larger than that of the other OTAs.

Tradeoffs in performance summarized in Figure 5.14 result from channel length decreases between the DC, balanced, and AC optimized OTAs. The high output resistance and voltage gain for the DC optimized OTA, combined with low input-referred flicker-noise voltage and offset voltage due to local-area mismatch, come with the penalty of significantly larger layout area.

5.5.2.11 Tradeoffs in DC accuracy, low-frequency AC accuracy, voltage gain, and transconductance bandwidth

The discussion below summarizes key performance tradeoffs in DC accuracy, low-frequency AC accuracy, voltage gain, and transconductance bandwidth for the cascoded OTAs. Performance discussed is from the *Circuit Analysis* sheet shown in Figure 5.14.

Because device inversion coefficients and drain currents remain constant (although at different values for different devices) across the DC, balanced, and AC optimized versions of the cascoded OTAs, device gate areas change as the square of channel length, as seen in the sizing relationships of Table 3.10. Channel lengths decrease a factor of 4.4 (2 µm/0.45 µm) between the DC and balanced optimized OTAs, and decrease a factor of 3 (0.45 µm/0.15 µm) between the balanced and AC optimized OTAs. Gate areas then decrease a factor of 20 ((2 µm/0.45 µm)^2) between the DC and balanced optimized OTAs, and decrease a factor of 9 ((0.45 µm/0.15 µm)^2) between the balanced and

AC optimized OTAs. Effective channel lengths used in the channel-length ratios are equal to 2 μm, 0.45 μm (0.48 – 0.028 μm), and 0.15 μm (0.18 – 0.028 μm) for the DC, balanced, and AC optimized OTAs, respectively, where the lateral diffusion length is taken at $DL = 0.028$ μm for nMOS devices from Table 3.2. These channel lengths, channel-length ratios, and gate-area ratios apply to all primary OTA devices (M_1–M_{10}), except for M_7–M_{10} in the DC optimized OTA where channel length is reduced slightly from 2 to 1.4 μm.

Because device inversion coefficients and drain currents remain constant across the OTA versions, and channel lengths for M_1–M_6 are constant for a given OTA version, device relative flicker noise and local-area mismatch contributions remain nearly constant across the OTA versions. As a result, both the input-referred flicker-noise voltage density and offset voltage due to local-area mismatch decrease inversely with the square root of gate areas for M_1–M_6 that control the noise and mismatch. The flicker-noise and offset voltage then decrease inversely with channel length for these devices since gate area is proportional to the square of channel length.

Measured input-referred flicker-noise voltage density at 100 Hz increases a factor of 4.4 from 96 to 420 nV/Hz$^{1/2}$ between the DC and balanced optimized OTAs, and increases a factor of 4 from 420 to 1700 nV/Hz$^{1/2}$ between the balanced and AC optimized OTAs. Measured 1σ, input-referred offset voltage increases a factor of 4.6 from 0.24 to 1.1 mV between the DC and balanced optimized OTAs, and increases a factor of 2.9 from 1.1 to 3.2 mV between the balanced and AC optimized OTAs. With the exception of flicker noise for the AC optimized OTA, the flicker-noise and offset voltage closely follow the inverse of channel length predicted by increases of a factor of 4.4 (2 μm/0.45 μm) between the DC and balanced optimized OTAs, and a factor of 3 (0.45 μm/0.15 μm) between the balanced and AC OTAs. Flicker-noise increases more rapidly between the balanced and AC optimized OTAs because the flicker noise is increased for the AC optimized OTA, most likely due to increases in the flicker-noise factor for short-channel devices as described in Section 5.5.2.5.

As described earlier in Sections 3.11.3.2 and 4.3.6.5, DC accuracy, which is the reciprocal of DC signal error, is equal to a selected DC input signal voltage divided by the 3σ, input-referred offset voltage. OTA DC accuracy then increases directly with the channel length for M_1–M_6 because the input-referred offset voltage is inversely proportional to the channel length of these devices. For a DC input signal of 0.1 V and measured 1σ, input-referred offset voltages of 0.24, 1.1, and 3.2 mV, the DC accuracy is 139, 30.3, and 10.4 for the DC, balanced, and AC optimized OTAs, respectively. Simulated transconductance bandwidth (−3 dB) is nearly inversely proportional to channel length at values of 76, 286, and 860 MHz, giving nearly constant products of DC accuracy and bandwidth at 10,600, 8,700, and 8,900 MHz since DC accuracy is proportional to channel length. The product of DC accuracy and bandwidth is modestly higher for the DC optimized OTA because channel lengths are reduced slightly from 2 to 1.4 μm for M_7–M_{10}, increasing the expected bandwidth.

Had transconductance bandwidth been inversely proportional to the square of channel length, as expected through the gate area, the products of DC accuracy *squared* and bandwidth would have been nearly constant across the OTA versions. This is analogous to constant bandwidth-accuracy figures of merit described in Sections 3.11.3.2 and 4.3.6.5 for voltage-mode circuits operating at constant g_m/I_D and I_D with varying channel lengths (and gate areas). Devices across the OTA versions also operate at constant g_m/I_D (through constant inversion coefficients) and I_D with varying channel lengths (and gate areas).

Simulated OTA bandwidth (−3 dB) increases a factor of 3.8 from 76 to 286 MHz between the DC and balanced optimized OTAs, and increases a factor of 3 from 286 to 860 MHz between the balanced and AC optimized OTAs. With the exception of the DC optimized OTA, the bandwidth closely follows the inverse of channel length predicted by increases of a factor of 4.4 (2 μm/0.45 μm) between the DC and balanced optimized OTAs, and a factor of 3 (0.45 μm/0.15 μm) between the balanced and AC OTAs. The bandwidth increases less rapidly between the DC and balanced optimized OTAs because the bandwidth, again, is increased for the DC optimized OTA because channel lengths are reduced slightly from 2 to 1.4 μm for M_7–M_{10}. OTA transconductance bandwidth is not inversely proportional to the square of channel length as expected when intrinsic gate capacitances dominate the bandwidth through the gate area. As described in Section 5.3.4.3, transconductance bandwidth is nearly inversely proportional to channel length because width-dependent, extrinsic gate-overlap, drain–body, and source–body, and layout capacitances dominate the capacitance and bandwidth. Device channel width tracks the selected channel length for operation at constant inversion coefficients and drain currents.

Finally, measured voltage gain is 19100, 4400, and 490 V/V, giving products of voltage gain and transconductance bandwidth at 1.45, 1.26, and 0.42 x 10^{12} Hz for the DC, balanced, and AC optimized OTAs, respectively. The product of gain and bandwidth is largest for the DC optimized OTA because gain decreases more rapidly than bandwidth increases across the DC, balanced, and AC optimized OTAs.

DC accuracy, which is inversely proportional to the input-referred offset voltage, and low-frequency AC accuracy, which is inversely proportional to the input-referred flicker-noise voltage density, decrease together directly with decreasing channel length, while transconductance bandwidth increases nearly inversely with decreasing channel length across the DC, balanced, and AC optimized OTAs. Additionally, voltage gain decreases while transconductance bandwidth increases across the OTA versions. These are primary performance tradeoffs across the OTA versions.

5.5.2.12 *Comparison of performance tradeoffs with those of simple OTAs*

Tradeoffs in performance for the simple OTAs summarized earlier in Figure 5.4 are even greater than those for the cascoded OTAs summarized in Figure 5.14. This is because, in addition to decreasing channel length, the inversion coefficient increases between the simple DC, balanced, and AC optimized OTAs. Changes in the inversion coefficient result in changes in the transconductance, input-referred thermal-noise voltage, input 1 dB compression voltage, and input and output voltage ranges not present for the cascoded OTAs. This also results in greater changes in the voltage gain, transconductance bandwidth, input-referred flicker-noise voltage and offset voltage due to local-area mismatch, and layout area.

5.5.3 Other Optimizations: Ensuring Input Devices Dominate Flicker Noise and Local-Area Mismatch

As mentioned in Section 5.4.3 for the simple OTAs, the spreadsheet permits rapid exploration of alternative circuit optimizations. The cascoded, 0.18 μm CMOS OTAs described here are designed for minimum thermal noise and various tradeoffs in performance across the DC, balanced, and AC optimized versions. Although input devices operate in moderate inversion and non-input devices operate in strong inversion so input devices dominate the thermal noise, input devices do not dominate the flicker noise or local-area mismatch. This is because channel lengths are equal for input and non-input devices, resulting in insufficient gate area for non-input devices that have small shape factors and small channel widths because of operation at high inversion coefficients.

Alternative optimizations could then involve increasing the channel lengths and, correspondingly, the channel widths of non-input devices so input devices dominate the flicker noise and local-area mismatch, as well as the thermal noise. This will significantly increase the gate area and capacitance for non-input devices and reduce the transconductance bandwidth. Increased channel lengths are used for non-input devices in the micropower, low-noise, CMOS preamplifiers described in Chapter 6 where input devices dominate both the thermal and flicker noise. This design method is summarized in the procedure of Section 6.10.

5.5.4 Other Optimizations: Complementing the Design

Complementing the existing OTA design would make interesting alternative optimizations where input devices would continue to dominate the thermal noise. Here, nMOS and pMOS devices are interchanged while maintaining the same drain currents, inversion coefficients, and channel lengths. This can be observed in the schematic of Figure 5.2 where the M_1 and M_2 input pair devices are switched from nMOS to pMOS devices, the M_3 and M_4 current source devices are switched from pMOS to nMOS devices, and the M_5 and M_6 current-mirror devices are switched from nMOS to pMOS devices. Of course, all other devices must be switched, but these do not contribute significantly to noise and mismatch. By complementing the design, flicker noise and local area mismatch might be decreased as a result of the larger gate area required for pMOS input devices. Additionally, the increase in the flicker-noise factor, K'_F, for nMOS, M_3 and M_4 devices that operate well into strong inversion would be reduced compared to that of pMOS devices in the original design that have higher increases in gate-referred flicker-noise voltage at high levels of inversion.

Evaluation of performance for a complemented design is easily done by changing only the model names in the *MOSFETs* sheet of the spreadsheet shown in Figures 5.11–5.13. This allows the spreadsheet to call MOS models for the switched device types. This was done for the complemented design, while maintaining the existing drain current, inversion coefficient, and channel length selections present in the original design.

The *Circuit Analysis* sheet shown in Figure 5.14 for the original design requires no modification for the complemented design because predictions of OTA circuit performance are automatically updated to reflect changes in device geometries and performance. The OTA transconductance, input-referred thermal-noise voltage, input 1 dB compression voltage, and input and output voltage ranges are nearly unchanged from the DC, balanced, and AC optimized versions of the original design because inversion coefficients are unchanged. The output resistance and voltage gain are also similar because inversion coefficients and channel lengths are unchanged. The input capacitance is increased by nearly a factor of four because of the increase in channel width and gate area required for pMOS input devices in the complemented design, while the output capacitance is nearly unchanged. Finally, the transconductance bandwidth is decreased by approximately 25 % in the complemented design.

Although most aspects of performance are unchanged or similar, flicker-noise analysis reveals significant differences in performance between the original and complemented designs. Predicted input-referred flicker-noise voltage density is modestly higher at 126 nV/Hz$^{1/2}$ at 100 Hz for the DC optimized, complemented design compared to 101 nV/Hz$^{1/2}$ for the DC optimized, original design. However, the complemented design has much higher relative flicker-noise power contributions at 1 Hz of $RNP_3 = 28$ for the M_3 and M_4 current source devices, and $RNP_5 = 4.2$ for the M_5 and M_6 current-mirror devices, compared to values of 2.4 and 1.7 for the original design. As a result of pMOS, M_1 and M_2 input devices that have a nearly factor-of-five increase in gate area compared to the original nMOS devices, and nMOS, M_3 and M_4 current source devices that have a nearly factor-of-five decrease in gate area compared to the original pMOS devices, flicker noise is totally dominated by the much smaller M_3 and M_4 devices in the complemented design. This can be seen by the $(WL)_1/(WL)_3$ gate area ratio in the RNP_3 term in Equation 5.27, also listed in Figure 5.14, which is combined with ratios of flicker-noise factors and transconductances squared, and a frequency correction term associated with variations in flicker-noise slopes. pMOS channel widths and, correspondingly, gate areas are approximately a factor-of-five larger than those of nMOS devices for equal drain currents, inversion coefficients, and channel lengths because of the lower technology current, I_0, and low-field mobility, μ_0, listed in Table 3.2 for the 0.18 μm CMOS process. Finally, the frequency correction term of $f^{AF1-AF3}$ reveals that RNP_3 increases a factor of 2.5 from 28 at 1 Hz to 70 at 100 Hz for the complemented design. This is a result of the shallower $AF_3 = 0.85$ slope for the non-input, nMOS, M_3 and M_4 devices compared to $AF_1 = 1.05$ for the input, pMOS, M_1 and M_2 devices, where the noise contribution from the non-input devices increase further above that of the input devices as frequency increases. This emphasizes the importance of considering flicker-noise slopes, listed in Table 3.3 for the 0.18 μm CMOS process, in flicker-noise analysis.

The substantial change in flicker-noise contributions from both input and non-input devices between the complemented and original OTA designs illustrates the complexity of low-flicker-noise design. As observed in the complemented design, the use of large-area input devices does not guarantee optimally low flicker noise as the noise contributions of non-input stages can be dominant. Flicker-noise analysis is complicated by ratios of flicker-noise factors, gate areas, transconductances squared, and frequency corrections associated with variations in flicker-noise slopes. Additionally, increases in gate-referred flicker-noise voltage at high levels of inversion, included in the operating value of K'_F, complicate flicker-noise analysis.

Although overall mismatch is similar, the relative mismatch power, RMP_3, for M_3 and M_4 is very large at 21 in the complemented design compared to nearly unity in the original design shown in the *Circuit Analysis* sheet of Figure 5.14. As they do for flicker noise, M_3 and M_4 dominate the input-referred offset voltage due to local-area mismatch in the complemented design because of their small gate areas. The substantial change in mismatch contributions between the complemented and original design also illustrates the complexity of low-mismatch design. As for flicker noise, the use of large-area input devices alone does not guarantee low mismatch as mismatch contributions of non-input devices can be dominant.

The spreadsheet is useful for rapid predictions of thermal noise, flicker noise, and local-area mismatch, including evaluation of the relative noise and mismatch contributions of non-input devices compared to the contributions of input devices. An interesting optimization problem is minimizing flicker noise or local-area mismatch for a given layout area. This may require careful balances of input and non-input noise and mismatch contributions and could be considered in the spreadsheet by displaying the total gate area of devices through a simple sum. Evaluating performance over the independent choices of device inversion coefficient, channel length, and drain current provides an effective methodology for optimizing noise, mismatch, and all aspects of circuit performance.

5.6 PREDICTION ACCURACY FOR DESIGN GUIDANCE AND OPTIMIZATION

Predictions of device and circuit performance using the *Analog CMOS Design, Tradeoffs and Optimization* spreadsheet are derived from hand predictions of MOS performance in weak, moderate, and strong inversion that were described in Chapter 3. These predictions are generally not as accurate as predictions found by computer simulations using full, simulation MOS models.[1] However, as shown in the *Circuit Analysis* sheets of Figures 5.4 and 5.14 for the simple and cascoded OTAs, the spreadsheet provides immediate performance predictions as device drain currents, inversion coefficients, and channel lengths are explored. Additionally, the predictions include the effects of local-area mismatch that are often excluded in computer simulations. Finally, the spreadsheet provides useful design details such as the relative noise and mismatch contributions of non-input devices.

The spreadsheet predictions illustrate tradeoffs, for example in OTA voltage gain, output resistance, transconductance bandwidth, and input-referred flicker-noise voltage and offset voltage due to local-area mismatch, that span one or more decades. Here, precise prediction of performance is not required to manage performance tradeoffs during initial design. The spreadsheet can minimize or eliminate time-consuming, trial-and-error computer simulations during the initial design process, while providing design intuition.

---

[1] The *Analog CMOS Design, Tradeoffs and Optimization* spreadsheet may ultimately be linked to computer simulation MOS models where prediction accuracy would approach that obtained in computer simulations. The reader is invited to check the book's web site listed on the cover for updates to the spreadsheet.

Prediction errors are given in the discussion of predicted and measured performance in Sections 5.4.2 and 5.5.2 for the simple and cascoded OTAs. Additionally, the *Circuit Analysis* sheets of Figures 5.4 and 5.14 show adjacent predicted, simulated, and measured circuit performance for the OTAs.

Prediction errors for OTA transconductance and the input, 1 dB compression voltage are well below 10 %, except for the prediction of compression voltage for the cascoded, AC optimized OTA. Prediction errors for the output resistance and related voltage gain, however, are higher. This is due to the complexity of predicting MOS drain–source resistance, r_{ds}, or conductance, $g_{ds} = 1/r_{ds}$. As described in Section 3.8.4 and shown in Section 3.8.4.5 through measurements of the Early voltage, $V_A = r_{ds} \cdot I_D = I_D/g_{ds}$, the drain–source resistance is a complex function of the inversion level and, especially, the channel length and drain–source voltage.

As described in Section 3.8.4.4, errors in the drain–source resistance for the prediction of circuit small-signal resistance and the resulting voltage gain appear directly in simple circuits, squared in cascode circuits, and cubed in regulated cascode circuits. As a result, significant prediction errors are possible for circuits using cascode or regulated cascode topologies to boost resistances and gains. These errors increase even more as V_{DS} approaches $V_{DS,sat}$ where the drain–source resistance drops rapidly. Unfortunately, V_{DS} is necessarily confined close to $V_{DS,sat}$ in low-voltage circuits, increasing the uncertainty of circuit resistances and gains.

For the cascoded OTAs, the maximum prediction error for output resistance and voltage gain is 35 %. Interestingly, simulation errors from full, simulation MOS models are equally as high. However, output resistance and voltage gain span over a range of nearly 40 across the DC, balanced, and AC optimized OTAs, indicating that the prediction is still useful for initial design guidance leading towards optimum design. For the simple OTAs, the maximum prediction error for output resistance and voltage gain is one half that of the cascoded OTAs at 17%. Fortunately, uncertainties in open-loop gains associated with errors in MOS drain–source resistance are often mitigated by closed-loop, negative feedback gain regulation. This requires only a sufficient level of open-loop gain and, correspondingly, loop gain to regulate the closed-loop gain.

Prediction errors for thermal and flicker noise are less than 15 % and are often well below this, except for the prediction of flicker noise for the cascoded, AC optimized OTA that uses minimum, channel length, $L = 0.18\,\mu\text{m}$ devices. Transconductance bandwidth is overpredicted by a maximum of 67 % for the simple, AC optimized OTA and 25 % for the cascoded, AC optimized OTA, due in part to the exclusion of parasitic layout capacitances that are unavailable during initial design. The predictions, however, are useful because bandwidth spans nearly two decades for the simple OTAs and over one decade for the cascoded OTAs. Finally, prediction errors for the input-referred offset voltage due to local-area mismatch are less than 4 % for the simple OTAs, except for the DC optimized version where distance mismatch likely contributes to the mismatch. Mismatch is conservatively overpredicted for the cascoded OTAs.

As noted in Footnote 1, the spreadsheet may ultimately be linked to a full, simulation MOS model. If linked to an accurate MOS model, the prediction accuracies described above could be improved. Additionally, this would eliminate the need for extracting values of the Early voltage factor, V_{AL}, near the device geometry and operating conditions as described near the beginning of Sections 5.4.2 and 5.5.2.

As mentioned in Section 5.3.13, small-geometry effects of velocity saturation, VFMR, DIBL, and flicker-noise increases with inversion or V_{EFF} are included in the spreadsheet. However, as mentioned, gate leakage current and the resulting gate–source conductance, gate shot- and flicker-noise current, and increase in DC mismatch are negligible and not included for the simple and cascoded OTAs. These effects could be included in later versions of the spreadsheet for processes having gate-oxide thickness less than 2 nm.

Predictions of device and circuit performance provided by the spreadsheet are approximate and intended only to provide design guidance leading towards optimum design. Simulations of a candidate design are then required using full, simulation MOS models for the actual production process.

These simulations must include process, supply voltage, and temperature variations along with layout parasitics in order to validate a candidate design for production. Although well known to industry designers, this is described further in the disclaimers in the spreadsheet and in Section 1.7.

REFERENCES[2]

[1] D. M. Binkley, B. J. Blalock, and J. M. Rochelle, "Optimizing drain current, inversion level, and channel length in analog CMOS design," *Analog Integrated Circuits and Signal Processing*, vol. 47, pp. 137–163, May 2006.

[2] D. M. Binkley, "Tradeoffs and optimization in analog CMOS design," *Proceedings of the 14th International Conference Mixed Design of Integrated Circuits and Systems* (*MIXDES*), pp. 47–60, Ciechocinek, Poland, June 2007.

[3] D. M. Binkley, "Tradeoffs and optimization in analog CMOS design," Conference tutorial with H. Graeb, G. G. E. Gielen, and J. Roychowdhury, "From Transistor to PLL – Analog Design and EDA Methods," *2008 Design, Automation, and Test in Europe Conference (DATE)*, Munich, March, 2008.

[4] University of California at Berkeley BSIM3 and BSIM4 MOS models, available on-line at http://www-device.eecs.Berkeley.edu/bsim3.

---

[2] The reader is referred to references in Chapter 2 for previously reported analog CMOS optimization methods and to references in Chapter 3 for MOS device performance.

6

Design of Micropower CMOS Preamplifiers Optimized for Low Thermal and Flicker Noise

6.1 INTRODUCTION

Chapter 5 presented the design of simple, $0.5\,\mu m$ and cascoded, $0.18\,\mu m$ CMOS operational transconductance amplifiers (OTAs) optimized for DC, balanced, and AC performance. Devices in a given version of the simple OTAs operated at equal drain currents, inversion coefficients, and channel lengths. This is representative of general-purpose designs where input devices do not dominate the noise and mismatch, and provided an introduction to the design methods described in this book. Devices in a given version of the cascoded OTAs, however, operated at different drain currents and inversion coefficients to ensure that input devices dominate the thermal noise. Here, input devices operated at low inversion coefficients in moderate inversion for high g_m/I_D and g_m to minimize input-referred thermal-noise voltage. Non-input devices operated at high inversion coefficients in strong inversion for low g_m/I_D and g_m to minimize their drain thermal-noise current contributions.

Operating devices at low, moderate, and high inversion coefficients in the simple OTAs, combined with operating devices at long, moderate, and short channel lengths in both OTAs, provided DC, balanced, and AC optimized versions. The DC optimized OTAs had high output resistance and voltage gain, combined with low input-referred flicker-noise voltage and offset voltage due to local-area mismatch. The AC optimized OTAs had high transconductance bandwidth, combined with small input and output capacitances and small layout area. The balanced optimized OTAs had a balance of DC and AC performance. Additional performance measures of transconductance, transconductance distortion, input-referred thermal-noise voltage, and input and output voltage ranges were also described for the OTAs. Predictions and trends of OTA performance were linked to device inversion coefficients, channel lengths, and drain currents and can be extended to other analog CMOS circuit topologies.

Rather than examining tradeoffs and optimization for the multiple aspects of circuit performance described in Chapter 5 for OTAs, this chapter emphasizes the minimization of input-referred thermal- and flicker-noise voltage. This is illustrated through the design of two micropower, $0.35\,\mu m$, silicon-on-insulator (SOI) CMOS preamplifiers. A thermal-noise efficiency and flicker-noise area efficiency factor are first developed to evaluate designs for a given power consumption and layout area. Space and biomedical preamplifier applications are then discussed, and the thermal noise, flicker noise, circuit

Tradeoffs and Optimization in Analog CMOS Design   David M. Binkley
© 2008 John Wiley & Sons, Ltd

topology, and power efficiency of reported micropower, CMOS preamplifiers are compared to the preamplifiers presented in this chapter.

Considerable attention is directed towards minimizing non-input device, drain thermal- and flicker-noise current that appears as a preamplifier input noise voltage after division by the transconductance of input devices. In order to minimize drain noise current for a given bias compliance voltage, transconductance is linked to the drain–source saturation voltage. This allows thermal and flicker noise to be expressed in terms of the DC bias current and bias compliance voltage, which is the minimum voltage required across a device or group of devices. The analysis is then extended to include resistive source degeneration. This shows that for a given bias compliance voltage, resistive degeneration provides significant reduction of flicker noise while having little effect on thermal noise because of the added thermal noise of the degeneration resistance. With or without the presence of resistive degeneration, drain thermal- and flicker-noise current decreases as the bias compliance voltage increases. Unfortunately, bias compliance voltage is limited in modern, low-voltage processes, making the management of non-input, drain noise current difficult.

For the preamplifiers presented in this chapter, input devices operate at low inversion coefficients in moderate inversion for high g_m/I_D and g_m to minimize input-referred thermal-noise voltage and help ensure overall noise dominance. Non-input devices operate at high inversion coefficients in strong inversion for low g_m/I_D and g_m to minimize their drain thermal-noise current contributions. Additionally, unlike what was done for the OTAs described in Chapter 5, the channel length of non-input devices is increased above that of the input devices to ensure sufficient gate area and negligible flicker-noise contributions. Non-input nMOS devices utilize resistive degeneration provided by deep ohmic MOS devices to further reduce their thermal and flicker noise. The noise contributions of non-input devices are studied in detail and are related to the DC bias current and bias compliance voltage. The significant increase in gate-referred flicker-noise voltage for non-input pMOS devices operating in strong inversion is also included in the analysis.

As for the OTAs described in Chapter 5, drain currents, inversion coefficients, and channel lengths for the preamplifiers are selected using the *MOSFETs* sheet in the *Analog CMOS Design, Tradeoffs and Optimization* spreadsheet. The resulting device geometry and performance are then mapped to the *Circuit Analysis* sheet, which includes the effects of resistive source degeneration for non-input nMOS devices. This sheet lists predictions of preamplifier voltage gain, unity-gain bandwidth, and input-referred thermal- and flicker-noise voltage and compares these to computer-simulated and measured performance. Measured thermal and flicker noise closely overlays the predicted noise, validating the design methods.

6.2 USING THE LATERAL BIPOLAR TRANSISTOR FOR LOW-FLICKER-NOISE APPLICATIONS

When low-flicker-noise, low DC offset, or exponential current characteristics are required, the lateral bipolar transistor inherent in bulk CMOS processes can be used [1]. In [1], this transistor is used in current-mirror, OTA, and bandgap reference circuits. In [2], this transistor is used in a variable gain circuit that uses the Gilbert multiplier. This multiplier circuit utilizes the exponential current characteristics of bipolar transistors and the resulting linear relationship between the transconductance and bias current. In [3], the lateral bipolar transistor is used to generate exponential control currents for a CMOS variable gain circuit. In [4], the lateral bipolar transistor is used in the input stage of a low-noise, variable-gain amplifier. This transistor is also used in [5] in the second stage of a low-noise preamplifier and in [6] in the input stage of a general-purpose, low-noise operational amplifier. Finally, in [7], this transistor is used in the input and base-band stages of a very low-frequency (VLF), micropower, direct-conversion receiver.

Although the lateral bipolar transistor can be useful for low-flicker-noise and low-DC-offset circuits along with variable gain circuits, circuits containing only MOS devices are often desired for layout simplicity. Additionally, all-MOS circuits do not experience the bias current and small-signal resistance loading present at the bipolar transistor base input, unless the MOS gate-oxide thickness is smaller than approximately 2 nm where gate leakage current is significant. Unfortunately, all-MOS circuits, including the preamplifiers described in this chapter, require a larger layout area compared to bipolar transistor circuits to achieve comparable flicker noise.

6.3 MEASURES OF PREAMPLIFIER NOISE PERFORMANCE

6.3.1 Thermal-Noise Efficiency Factor

Low levels of input-referred thermal-noise voltage density are available at arbitrarily high levels of drain current where the transconductance of MOS input devices is high. For noise and supply current optimization, it is useful then to evaluate noise while considering the level of supply current.

The noise efficiency factor, *NEF*, introduced in [8], describes the ratio of preamplifier, input-referred, root-mean-square (rms) noise voltage, $V_{rms,in}$, to that of a single bipolar transistor operating over the same $-3\,$dB bandwidth, *BW*, and supply current, I_{SUP}. The *NEF* is given by [8]

$$NEF = \frac{V_{rms,in}}{\propto V_{rms,in}(\text{bipolar, thermal})} = \frac{V_{rms,in}}{\sqrt{\dfrac{4kT \cdot (BW \cdot \pi/2)}{g_m(\text{bipolar})}}} = \frac{V_{rms,in}}{\sqrt{\dfrac{4kT \cdot (BW \cdot \pi/2)}{I_{SUP}/U_T}}}$$

$$= V_{rms,in}\sqrt{\frac{2I_{SUP}}{\pi \cdot U_T \cdot 4kT \cdot BW}} \tag{6.1}$$

The numerator in the expressions in the top line is the preamplifier $V_{rms,in}$ over the noise bandwidth, $BW \cdot (\pi/2)$. The denominator in the expressions in the top line is proportional to the input-referred, rms, thermal-noise voltage of a single bipolar transistor operating over the same noise bandwidth. The bipolar transistor noise is for a transconductance of g_m (bipolar) and a collector current of I_{SUP}, and excludes flicker noise, base resistance noise, and base shot noise. If the noise bandwidth is $BW \cdot (\pi/2)$ as expected for a $-3\,$dB, single-pole bandwidth of *BW*, a $\frac{1}{2}$ term should appear in the denominator terms since the input-referred thermal-noise voltage power spectral density (PSD) for a bipolar transistor is actually $4kT(\frac{1}{2})/(g_m \text{ (bipolar)})$. However, the *NEF* is defined as in [8] and is a widely used noise and power–efficiency measure for preamplifiers.

The *NEF* includes preamplifier flicker noise and requires knowledge of the signal bandwidth. Increasing values of *NEF* correspond to increasing levels of preamplifier, input-referred, rms noise voltage compared to that of a single bipolar transistor, again, operating over the same bandwidth at the preamplifier supply current, but having no flicker noise, base resistance noise, or base shot noise.

Here, a modified factor called the thermal-noise efficiency factor, NEF', is introduced that considers only thermal noise and requires no knowledge of the signal bandwidth. For a given level of input-referred thermal-noise voltage PSD or density, the NEF' describes the ratio of preamplifier supply current to that of a single bipolar transistor. The NEF' provides an effective measure of the power supply current efficiency for low-noise circuits and is useful for comparing performance of reported preamplifiers as described in Section 6.4. When using the NEF', flicker noise can be considered separately by evaluating the flicker-noise corner frequency relative to the signal bandwidth of interest.

The NEF' is defined by

$$NEF' = \frac{S_{VIN} (\text{thermal})}{\left(\dfrac{4kT\left(\frac{1}{2}\right)}{g_m(\text{bipolar})}\right)} = \frac{S_{VIN} (\text{thermal})}{\left(\dfrac{4kT\left(\frac{1}{2}\right)}{I_{SUP}/U_T}\right)} = \frac{S_{VIN} (\text{thermal})}{4kT}\left(\frac{2I_{SUP}}{U_T}\right)$$

$$= \frac{S_{VIN} (\text{thermal})}{(14.63 \text{ nV})^2/\text{Hz}}\left(\frac{I_{SUP}}{1\mu\text{A}}\right) \quad \text{at } T = 300 \text{ K} \tag{6.2}$$

S_{VIN} (thermal) is the preamplifier input-referred thermal-noise voltage PSD, and I_{SUP}, again, is the preamplifier supply current. The denominator terms in the first line involving g_m (bipolar) are the input-referred thermal-noise voltage PSD of a bipolar transistor operating at I_{SUP}. Finally, the second line, motivated by expressions in [9, p. 51], conveniently gives the NEF' at $T = 300\,\text{K}$ (room temperature) for $4kT = 1.657 \times 10^{-20}\,\text{VA} \cdot \text{s}$.

Substituting the MOS gate-referred thermal-noise voltage PSD from Table 3.30 for S_{VIN} (thermal) and MOS transconductance from Table 3.17 for g_m (MOS) gives the NEF' for a single MOS device of

$$NEF' \text{ (single MOSFET)} = \frac{\left(\dfrac{4kT\,(n\Gamma)}{g_m(\text{MOS})}\right)}{\left(\dfrac{4kT\left(\frac{1}{2}\right)}{g_m(\text{bipolar})}\right)} = 2\,(n\Gamma)\,\frac{g_m(\text{bipolar})}{g_m(\text{MOS})}$$

$$= 2\,(n\Gamma)\,\frac{\dfrac{I_{SUP}}{U_T}}{I_{SUP}\left(\dfrac{g_m}{I_D}\right)_{MOS}} = \frac{2\,(n\Gamma)}{U_T\left(\dfrac{g_m}{I_D}\right)_{MOS}} = 2n^2\Gamma\left(\sqrt{IC+0.25}+0.5\right)$$

(6.3)

IC is the inversion coefficient, n is the substrate factor, and Γ is the thermal-noise factor for the MOS device. The rightmost expression in the second line expresses g_m/I_D (MOS) without velocity saturation and VFMR decreases that can increase the NEF'. The inclusion of these effects will be described later.

In weak inversion ($IC < 0.1$) for Γ given by Equation 3.106 of nearly $\frac{1}{2}$ and $n = 1.4$ for a typical bulk CMOS process, the NEF' is equal to n^2 or two. In weak inversion where g_m/I_D and g_m are maximum and the NEF' is minimum, the NEF' is two because n reduces MOS g_m/I_D and g_m by a factor of 1.4 compared to the bipolar transistor, while simultaneously increasing the thermal-noise PSD by this same factor. At the center of moderate inversion ($IC = 1$) for $\Gamma = 0.583$ and $n = 1.35$, the NEF' increases to 3.4. Finally, at the onset of strong inversion ($IC = 10$) for Γ of nearly $\frac{2}{3}$ and $n = 1.33$, the NEF' increases significantly to 8.7. Thus a MOS device operating in weak, in the center of moderate, and at the onset of strong inversion requires nearly 2, 3.4, and 8.7 times the supply current of a bipolar transistor to achieve the same level of input-referred thermal-noise voltage PSD or density.

Velocity saturation effects, present for short-channel devices operating at high levels of inversion, and smaller VFMR effects, present for devices of all channel lengths operating at high levels of inversion, decrease g_m/I_D and g_m. This increases the MOS gate-referred thermal-noise voltage and, correspondingly, increases the NEF'. As described in Table 3.17, velocity saturation and VFMR decreases in g_m/I_D and g_m can be included by replacing IC with $IC(1 + IC/IC_{CRIT})$ in the $\sqrt{IC+0.25}$ term related to g_m/I_D, seen above in Equation 6.3. As described in Section 3.8.2.2 and shown in Figures 3.24 and 3.25, IC_{CRIT} corresponds to a critical inversion coefficient where g_m/I_D is down to approximately 70.7 % of its value without velocity saturation and VFMR effects. $IC_{CRIT} = [(LE_{CRIT})'/(4nU_T)]^2$ depends primarily on the channel length and the critical, velocity saturation, electric field, E_{CRIT}. It also depends on the VFMR factor, θ, which affects g_m/I_D and g_m even for long-channel devices. $(LE_{CRIT})' = LE_{CRIT}||(1/\theta)$ is the equivalent velocity saturation voltage that includes both velocity saturation and VFMR effects.

Figure 6.1 shows the predicted NEF' for $IC = 0.01$–100, corresponding to operation from deep weak inversion to deep strong inversion, for $L = 0.18$, 0.28, 0.48, and $1\,\mu\text{m}$, nMOS devices in the $0.18\,\mu\text{m}$ CMOS process described in Table 3.2. The NEF' is found from Equation 6.3 with the modifications described above for velocity saturation and VFMR effects. Again, Γ is found from Equation 3.106, which is shown in Figure 3.61. $U_T = 25.9\,\text{mV}$ ($T = 300\,\text{K}$) and $n = 1.33$, which is held fixed and equal to the value used for g_m/I_D shown in Figure 3.26, are used along with process values of $E_{CRIT} = 5.6\,\text{V}/\mu\text{m}$ and $\theta = 0.28/\text{V}$ from Table 3.2. Channel length is reduced by $DL = 0.028\,\mu\text{m}$ (also given in Table 3.2) from

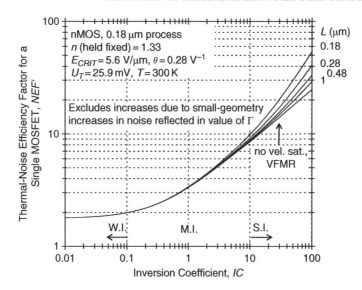

Figure 6.1   Predicted thermal-noise efficiency factor, NEF', versus the inversion coefficient, IC, for $L = 0.18$, 0.28, 0.48, and $1\,\mu m$, nMOS devices in a $0.18\,\mu m$ CMOS process. The NEF' corresponds to the MOS drain current compared to the bipolar transistor collector current for equal input-referred thermal-noise voltage PSD or density. The NEF', which tracks the reciprocal of g_m/I_D, increases as the square root of IC in strong inversion and increases directly with IC when velocity saturation is significant. As a result, substantially more MOS drain current is required in strong inversion to achieve a given level of gate-referred thermal-noise voltage. The NEF' can increase beyond the values shown because of small-geometry increases in MOS thermal noise, which can be included in the value of the thermal-noise factor, Γ

the values given to include lateral diffusion effects. In addition to the channel lengths shown, the NEF' is also shown assuming velocity saturation and VFMR reductions of g_m/I_D and g_m are not present.

The NEF' shown in Figure 6.1 is the ratio of MOS drain current to bipolar transistor collector current for equal values of input-referred thermal-noise voltage PSD or density. The NEF' is constant and minimum in weak inversion, increases as the square root of IC in strong inversion for long-channel devices, and increases nearly directly with IC for short-channel devices experiencing significant velocity saturation. Since the NEF' is inversely proportional to g_m/I_D, its increase is primarily the result of decreasing g_m/I_D. As observed, the NEF' can exceed 10 in strong inversion, especially for short-channel devices experiencing velocity saturation. This indicates that MOS devices can require over 10 times the bias current of bipolar transistor devices to obtain equal input-referred thermal noise.

If small-geometry effects discussed in Section 3.10.2.2 increase thermal noise, modeled through an increase in Γ, the NEF' becomes larger than the values shown in Figure 6.1. These effects can result in an even higher NEF' and, correspondingly, even lower power supply current efficiency for radio frequency (RF) amplifiers that require operation in strong inversion at short channel lengths to achieve sufficient bandwidth. Input devices for the preamplifiers presented later in this chapter operate near the center of moderate inversion to achieve high g_m/I_D, low NEF', and, correspondingly, good power supply current efficiency. Additionally, the channel lengths are long, minimizing possible small-geometry noise increases.

The NEF' for pMOS devices is nearly equal to that of nMOS devices shown in Figure 6.1 if pMOS channel length is decreased by a factor of 2.5 $((14\,V/\mu m)/(5.6\,V/\mu m))$, the ratio of pMOS and nMOS, E_{CRIT}. This is a result of nearly equal g_m/I_D for $L = 0.18\,\mu m$, pMOS devices shown in Figure 3.27 compared to $L = 0.48\,\mu m$, nMOS devices shown in Figure 3.26. For long-channel devices operating at high IC, pMOS NEF' increases slightly more than that of nMOS devices because g_m/I_D decreases more due to higher VFMR, represented by a higher value of θ.

Because the NEF' is directly proportional to Γ and inversely proportional to g_m/I_D, it is largely independent of process technology when evaluated in terms of the inversion coefficient, except for its dependency on the square of n. Like g_m/I_D, the NEF' is independent of channel length when small-geometry effects are excluded, but depends on the channel length and E_{CRIT} if velocity saturation decreases in g_m/I_D and g_m are present. It also depends on θ when smaller VFMR decreases in g_m/I_D and g_m are present. The NEF' can also have additional dependency on the channel length, inversion level, and process if small-geometry increases in thermal noise, included in the value of Γ, are present. Since n varies slightly between 1.3 and 1.4 for typical bulk CMOS processes, decreasingly slightly with increasing inversion as shown in Figure 3.30, the NEF' varies little over these processes. It can, however, decrease for fully depleted (FD) SOI CMOS processes having lower n between 1.1 and 1.2.

The NEF' increases by a factor of four for a differential-pair input stage compared to an equivalently designed single-device input stage where devices operate at the same IC and g_m/I_D. This is because one-half the gate-referred thermal-noise voltage PSD is required at twice the drain current and transconductance for each differential pair device to obtain the same, overall input-referred thermal noise as a single device. Thus, to achieve equal input-referred thermal-noise voltage PSD or density, a differential-pair input stage requires a factor-of-four higher supply current compared to the equivalently designed single-device input stage. A minimum value of the NEF' indicates that minimum supply current is obtained using a single input device in weak inversion with no additional current consumed for subsequent devices. The NEF' is 38 for the differential input preamplifier and 11.6 for the single-ended preamplifier presented in Sections 6.7 and 6.8, as summarized later in Table 6.1.

6.3.2 Flicker-Noise Area Efficiency Factor

Since low levels of gate-referred thermal-noise voltage density are available at arbitrarily high supply currents, the NEF' was suggested as a measure of supply current efficiency for achieving a specified level of thermal noise. Low levels of gate-referred flicker-noise voltage density are available for arbitrarily large MOS gate areas, suggesting the need for some measure of area efficiency for achieving a specified level of flicker noise.

A flicker-noise area factor, NAF, is defined here as the ratio of preamplifier layout area to the gate area of a single MOS device of the preferred device type that gives the same level of input-referred flicker-noise voltage PSD at 1 Hz. The NAF is given by

$$NAF = \frac{S_{VIN}(\text{flicker, 1 Hz})}{\left(\dfrac{K'_{F0}(\text{preferred})}{AREA \cdot 1\,\text{Hz}}\right)} = S_{VIN}(\text{flicker, 1 Hz}) \cdot \left(\frac{AREA \cdot 1\,\text{Hz}}{K'_{F0}(\text{preferred})}\right) \tag{6.4}$$

S_{VIN} (flicker, 1 Hz) is the preamplifier input-referred flicker-noise voltage PSD at 1 Hz, $AREA$ is the preamplifier layout area, and K'_{F0} (preferred) is the low-inversion value of the flicker-noise factor for the preferred device type offering lower flicker noise in the process. The denominator of the first expression is the gate-referred flicker-noise voltage PSD at 1 Hz from Table 3.31 for a MOS device of the preferred type having gate area equal to the preamplifier layout area. Equation 6.4 expresses the NAF using K'_{F0} in convenient units of $(nV)^2 \cdot \mu m^2$. K'_{F0} can be found from K_{F0} in semiconductor units of C^2/cm^2 or $K_{FSPICE0}$ in SPICE units of $V^2 \cdot F$ using the unit conversions given in Table 3.32.

The NAF given by Equation 6.4 considers in the value of preamplifier S_{VIN} (flicker, 1 Hz) possible increases in flicker noise caused by increases in gate-referred flicker-noise voltage at increasing inversion or V_{EFF}. As described in Tables 3.31 and 3.34, and shown in Figure 3.64, flicker-noise increases are modeled by $K'_F = K'_{F0}(1 + V_{EFF}/V_{KF})^2$, where K'_{F0} is the flicker-noise factor in weak inversion or the weak-inversion side of moderate inversion, and V_{KF} is a voltage describing the flicker-noise increase with V_{EFF}. At low levels of inversion where V_{EFF} is negative, it is clamped to zero. When increases in flicker noise are significant, additional device gate area and, correspondingly, preamplifier layout area are required. This increases the value of the *NAF*.

Just as the *NEF'* is four times larger for differential-pair input stages compared to equivalently designed single-device input stages, the *NAF* is four times larger for differential-pair input stages. This is because one-half the gate-referred flicker-noise voltage PSD is required at twice the gate area for each differential-pair device to obtain the same, overall input-referred flicker noise as a single device. Thus, differential-pair input stages require four times the layout area to achieve the same level of input-referred flicker-noise voltage PSD or density of equivalently designed, single-device input stages.

A minimum value of the *NAF* indicates that minimum layout area is obtained using a single input device of the preferred type at low levels of inversion or V_{EFF} where flicker-noise increases are small. Additionally, non-input devices should contribute negligible flicker noise and layout area. As seen later in Sections 6.7.3 and 6.8.3 for the preamplifiers presented here, non-input devices do not have negligible gate area when these devices contribute negligible flicker noise.

Given the difficulty of assessing total layout area, the *NAF* is not reported for preamplifiers reviewed in Section 6.4, including the two preamplifiers presented here. However, it is presented here to assist in area-efficient, low-flicker-noise design. The total gate area of preamplifier devices could easily be included in the *Circuit Analysis* sheet of the *Analog CMOS Design, Tradeoffs and Optimization* spreadsheet. Additionally, an estimation of the *NAF* could be included that would lead to the optimization of the *NAF* for a specified level of input-referred flicker-noise voltage PSD or density.

6.4 REPORTED MICROPOWER, LOW-NOISE CMOS PREAMPLIFIERS

Micropower, low-noise CMOS preamplifiers are used in biomedical applications to amplify signals from neural, nerve, cardiac, and other biological probes. Additionally, with sufficient radiation hardening, these preamplifiers are used in space applications to amplify signals from gyroscopes, accelerometers, and other mission sensors. Both application areas require micropower operation because of battery operation or limited power supply capacity. Additionally, thermal noise must often be managed, while flicker-noise management can be even more important for signals having important low-frequency content. Even for radio frequency (RF) applications where signal frequencies are high, flicker noise may need to be managed to minimize phase noise in oscillators resulting from upconverted noise near the carrier frequency.

For micropower preamplifiers, "low noise" refers to low, input-referred noise voltage for micropower power consumption versus an absolute measure of noise where "low noise" often refers to input-referred noise voltage densities below $1 \, nV/Hz^{1/2}$. Such low levels of noise are available only with high input device transconductance associated with millipower power consumption.

Table 6.1 lists reported micropower, low-noise CMOS preamplifiers having power consumption below $550 \, \mu W$, including the differential and single-ended input, $0.35 \, \mu m$ SOI CMOS preamplifiers described in Sections 6.7 and 6.8. Listed in the table, when available, are the process, input-referred flicker-noise voltage density at 1 Hz, input-referred thermal-noise voltage density, flicker-noise corner frequency, thermal-noise efficiency factor, power consumption, and circuit topology for the preamplifiers. Flicker noise when not reported at 1 Hz is extrapolated to 1 Hz assuming a flicker-noise power slope of unity.

Most of the preamplifiers reported in Table 6.1 are for recording neural [15–18], nerve [10, 12], cardiac [14], and electroencephalogram (EEG) [16] biomedical signals. For brevity, earlier neural preamplifiers [19–21] in $2 \, \mu m$ or larger CMOS processes are not included in the table. Another neural preamplifier [22] is not included in the table, but is similar to the differential input preamplifier

Table 6.1 Noise measures, power consumption, and circuit topologies of reported micropower CMOS preamplifiers

Process	Input ref. flicker noise at 1 Hz $(\text{nV}/\text{Hz}^{1/2})$	Input ref. thermal noise $(\text{nV}/\text{Hz}^{1/2})$	Flicker-noise corner (Hz)	Thermal-noise efficiency factor (NEF')	Power (μW)	Circuit topology (input diff. pair unless specified)	Ref.
0.35 μm SOI	240	63.8 at 1 kHz	12	38	6.6 at 3.3 V	Folded-cascode	Sect. 6.7
0.35 μm SOI	160	35.3 at 1 kHz	19	11.6	6.6 at 3.3 V	Single-ended with reg. cascode	Sect. 6.8
0.35 μm		7, freq. not given		31	245 at 1.8 V	Folded-cascode	[10]
0.35 μm	~1600 (500 at 10 Hz)				5.2 at 1.2 V	Two-stage (Miller)	[11]
0.5 μm		4.8 average over 400 Hz–4 kHz			275 at ±1.5 V	Cascode with ohmic MOS load	[12]
0.7 μm	80				210 at ±2 V	Single-ended with folded-cascode	[13]
0.8 μm		400 average over 300 Hz BW			2.7 at 2.7 V	Two-stage (Miller)	[14]
1.5 μm	~2200 (220 at 100 Hz)				115 at 3 V	Two-stage (Miller)	[15]
1.5 μm	500	21 at 1 kHz	100	33	80 at ±2.5 V	(Diff. pair) with cascode mirror	[16]
1.5 μm	400	200 at 10 Hz	2.3	34	0.9 at ±2.5 V	(Diff. pair) with cascode mirror	[16]
1.5 μm	~2500 (800 at 10 Hz)				0.8 at 1.5 V	(Diff. pair) with current mirror	[17]
1.5 μm	105 average over 10 Hz–7.3 kHz				52 at 3.3 V	Folded-cascode	[18]

described in Section 6.7. Also included in Table 6.1 are preamplifiers for general biomedical applications [11] and for a pyroelectric sensor for infrared imaging [13]. For brevity, other preamplifiers in 2 μm or larger CMOS processes, including a general-purpose amplifier [8] and hearing aid amplifier [23], are not included in the table.

The differential and single-ended input preamplifiers described in Sections 6.7 and 6.8 use SOI CMOS technology for the extreme temperature and radiation environment of space applications [24]. The differential input preamplifier has been evaluated from 77 to 400 K [25] for proton irradiation to 1 Mrad (Si) [26].

The differential and single-ended input preamplifiers described in Sections 6.7 and 6.8 have lower flicker noise compared to other preamplifiers in Table 6.1, except for the preamplifier reported in [13]. This single-ended input preamplifier has a measured, 1 Hz flicker noise of $80\,\mathrm{nV/Hz}^{1/2}$ compared to $160\,\mathrm{nV/Hz}^{1/2}$ for the single-ended input preamplifier described in Section 6.8. As mentioned in Section 6.3.2, single-ended designs have lower flicker noise compared to equal-area differential designs. While the preamplifiers described in Sections 6.7 and 6.8 compare favorably with others reported in the table, flicker noise can be made arbitrarily low if layout area is made arbitrarily large. As mentioned in Section 6.3.2, the *NAF* considers the efficiency of layout area for a given level of flicker noise. *NAF* values are not compared in Table 6.1 because layout areas are often not reported.

Thermal noise is not reported for many of the preamplifiers listed in Table 6.1 because the thermal-noise floor was not reached over the measurement bandwidth. For these applications, flicker noise is clearly important. The preamplifiers with reported thermal noise have *NEF'* values found from Equation 6.2 that are well above unity. These preamplifiers require a supply current of *NEF'* times that required for a single bipolar transistor having the same input-referred thermal-noise voltage density.

Interestingly, the differential input preamplifiers reported in [10, 16] and in Section 6.7 have comparable values of the *NEF'* between 31 and 38 as listed in Table 6.1. The single-ended input preamplifier described in Section 6.8, however, has the lowest *NEF'* at 11.6 of any of the reported preamplifiers. As mentioned in Section 6.3.1, single-ended designs have a *NEF'* of a factor-of-four below that of comparable differential designs. The *NEF'* values listed in the table can be compared because the preamplifier supply current does not include the supply current of additional circuitry, such as output buffers for driving large off-chip loads. When output buffers, analog-to-digital converters, or other circuits are included in the power supply current, the value of the *NEF'* does not reflect the power supply efficiency of the preamplifier circuit alone and cannot be directly compared.

The single-ended input, 0.8 μm CMOS preamplifier reported in [27], which includes a buffer for off-chip loads, has a measured input-referred thermal-noise voltage density of $0.65\,\mathrm{nV/Hz}^{1/2}$ for a supply current of 3 mA. From Equation 6.2, this corresponds to a *NEF'* of 5.9 that is well below the values reported in Table 6.1. Although this preamplifier is a millipower design not reported in the table, its noise and power efficiency can be directly compared to micropower designs in the table using the *NEF'*. In this millipower preamplifier, a single, $W = 4800\,\mu\mathrm{m}/L = 0.8\,\mu\mathrm{m}$, nMOS input device operates at a drain current of 2 mA and an inversion coefficient of 1.67, giving a measured transconductance of 29.2 mS and gate-referred thermal-noise voltage density of $0.6\,\mathrm{nV/Hz}^{1/2}$ [27]. The input device inversion coefficient for the single-ended input, micropower preamplifier described in Section 6.8 is nearly the same at 1.9. Its drain current is a factor of 2000 lower at 1 μA for operation at nearly the same inversion coefficient, giving proportionally lower g_m of approximately $14.6\,\mu\mathrm{S} = 29.2\,\mathrm{mS}/2000$ and proportionally higher gate-referred thermal-noise voltage PSD of approximately $(27\,\mathrm{nV})^2/\mathrm{Hz} = (0.6\,\mathrm{nV})^2/\mathrm{Hz} \cdot 2000$ compared to the input device in the millipower preamplifier.

The millipower preamplifier reported in [27] uses an input device inversion coefficient of 1.67, while the differential and single-ended input, micropower preamplifiers described in Table 6.1 and Sections 6.7 and 6.8 use input device inversion coefficients of 0.95 and 1.9, respectively. Input device inversion coefficients are selected at approximately 0.7 for the nerve preamplifier reported in [10] and 0.43 for the neural preamplifier reported in [16]. Each of these input devices operates in moderate inversion for high g_m/I_D and good power supply efficiency, while avoiding the significant loss of bandwidth associated with operation in weak inversion. Finally, an input device inversion coefficient of 0.0034, deep in weak inversion, is selected for the EEG preamplifier reported in [16]. The use of the inversion coefficient in recent reported research suggests that the inversion coefficient is being increasingly used to optimize analog CMOS circuits. Design using the inversion coefficient [28–34] is, of course, a major subject of this book.

6.5 MOS NOISE VERSUS THE BIAS COMPLIANCE VOLTAGE

Input transconductor devices in circuits that receive voltage inputs should operate at low inversion coefficients in weak or moderate inversion where g_m/I_D and g_m are high. As described in Section 3.10.2.3, this minimizes the gate-referred thermal-noise voltage. Additionally, high input device g_m minimizes the drain noise current contributions of non-input devices because these are divided by input device g_m to refer them to a noise voltage at the input. Fortunately, $V_{DS,sat}$ is low for input devices operating at low inversion coefficients, giving low bias compliance voltages. The bias compliance voltage is defined here as the minimum voltage required across a device or group of devices for normal operation in saturation. Unlike input devices, non-input devices should operate at high inversion coefficients in strong inversion where g_m/I_D and g_m are low, minimizing the drain-referred thermal-noise current. Unfortunately, $V_{DS,sat}$ is high for non-input devices operating at high inversion coefficients, resulting in high bias compliance voltages. This complicates low-noise, CMOS design because high bias compliance voltages are difficult to support for circuits in modern processes that operate at low supply voltages.

The following discussions consider minimizing gate-referred thermal- and flicker-noise voltage of input devices, and minimizing drain-referred thermal- and flicker-noise current of non-input devices, both against the constraint of the bias compliance voltage. Source resistive degeneration of drain-referred noise current for non-input devices is also considered. Here, the bias compliance voltage increases above the device $V_{DS,sat}$ by the voltage drop across the degeneration resistance. While having only a modest effect on the thermal noise for a given bias compliance voltage, resistive degeneration significantly reduces the flicker noise.

6.5.1 Transconductance in Saturation

In order to evaluate MOS noise against the constraint of the bias compliance voltage, it is necessary to express transconductance in terms of $V_{DS,sat}$. $V_{DS,sat}$ is the bias compliance voltage for a single MOS device operating in saturation since this is the minimum allowable voltage (V_{DS}) across the device. $V_{DS,sat}$ is given from Table 3.15 by

$$V_{DS,sat} = 2U_T\left[\left(\sqrt{IC+0.25}+0.5\right)+1\right] = 2U_T\sqrt{IC+0.25}+3U_T \tag{6.5}$$

g_m/I_D for operation in saturation ($V_{DS} > V_{DS,sat}$) is given from Table 3.17 by

$$\frac{g_m}{I_D} = \frac{1}{V_{gm}} = \frac{1}{nU_T\left(\sqrt{IC+0.25}+0.5\right)} = \frac{2}{n\left(V_{DS,sat}-2U_T\right)} \tag{6.6}$$

In the rightmost term, g_m/I_D is linked to $V_{DS,sat}$ through Equations 6.5 and 6.6. V_{gm} is the transconductance effective voltage [28, 31–34], which is the reciprocal of g_m/I_D given by

$$V_{gm} = \left(\frac{g_m}{I_D}\right)^{-1} = nU_T\left(\sqrt{IC+0.25}+0.5\right) = \frac{n}{2}\left(V_{DS,sat}-2U_T\right) \tag{6.7}$$

g_m is found by I_D/V_{gm} in a similar way that drain–source conductance, g_{ds}, is found by I_D/V_A, where V_A is an Early voltage described near the beginning of Section 3.8.4 that depends on the operating IC, L, and V_{DS}. Finally, g_m is the product of I_D and g_m/I_D, or the quotient of I_D divided by V_{gm}, given by

$$g_m = \left(\frac{g_m}{I_D}\right) I_D = \frac{I_D}{V_{gm}} = \frac{I_D}{nU_T\left(\sqrt{IC+0.25}+0.5\right)} = \frac{2I_D}{n\left(V_{DS,sat}-2U_T\right)} \tag{6.8}$$

The g_m/I_D, V_{gm}, and g_m expressions given by Equations 6.6–6.8 are linked to $V_{DS,sat}$ given by Equation 6.5. This linkage enables noise optimization in terms of $V_{DS,sat}$, which is the bias compliance voltage for a single device and part of the bias compliance voltage for a device having resistive source degeneration. Velocity saturation effects, present for short-channel devices operating at high levels of inversion, and smaller VFMR effects, present for devices of all channel lengths operating at high levels of inversion, decrease g_m/I_D and g_m, and increase $V_{EFF} = V_{GS} - V_T$. When these effects are significant, g_m/I_D and g_m become increasingly decoupled from their simple inverse relationship with $V_{DS,sat}$. This is because g_m/I_D and g_m decrease more rapidly than $V_{DS,sat}$ increases when these effects are significant. As discussed in Section 3.7.3.3, when $V_{DS,sat}$ is evaluated in terms of the inversion coefficient, it is relatively unaffected by velocity saturation and VFMR effects because it becomes an increasing smaller fraction of V_{EFF} as this increases due to these effects. Velocity saturation and VFMR decreases in g_m/I_D and g_m are not significant for the micropower, low-noise preamplifiers described in Sections 6.7 and 6.8 that use long-channel devices for low flicker noise. These effects, however, may be included by the expression modification described for Equation 6.3 in Section 6.3.1, which is also described for g_m/I_D given in Table 3.17.

The first expression in Table 6.2 summarizes MOS transconductance from Equation 6.8 in terms of both the inversion coefficient and $V_{DS,sat}$. Additionally, the table summarizes the drain–source resistance and transconductance of MOS devices operating in the deep ohmic region as resistors. This will be described in the next section for devices acting as source degeneration resistors.

Figure 6.2 shows measured g_m/I_D from deep weak inversion ($IC = 0.01$) to deep strong inversion ($IC = 100$) for $L = 1.2\,\mu\mathrm{m}$, nMOS and pMOS devices in the 0.35 μm SOI CMOS process used for the preamplifiers described in Sections 6.7 and 6.8. The inversion coefficient for measured data is found

Table 6.2 MOS transconductance in saturation, and transconductance and drain–source resistance in the deep ohmic region, expressed in terms of the inversion coefficient and DC bias conditions. DC bias expressions permit noise analysis in terms of the bias compliance voltage

Parameter	Expression
Transconductance in saturation[a] ($V_{DS} > V_{DS,sat}$)	$g_m = \left(\dfrac{g_m}{I_D}\right) I_D = \dfrac{I_D}{V_{gm}} = \dfrac{I_D}{nU_T\left(\sqrt{IC+0.25}+0.5\right)}$ $= \dfrac{2I_D}{n\left(V_{DS,sat}-2U_T\right)}$
Transconductance in deep ohmic region[b] ($V_{DS} \ll V_{DS,sat}$)	$g_m\,(\mathrm{ohmic}) = \left(\dfrac{g_m}{I_D}\right)_{ohmic} \cdot I_D(\mathrm{ohmic}) = \dfrac{I_D(\mathrm{ohmic})}{V_{GS}-V_T} = \dfrac{I_D(\mathrm{ohmic})}{V_{EFF}}$
Drain–source resistance in deep ohmic region[b] ($V_{DS} \ll V_{DS,sat}$)	$g_{ds}(\mathrm{ohmic}) = \dfrac{1}{r_{ds}(\mathrm{ohmic})} = \dfrac{I_D(\mathrm{ohmic})}{V_{DS}}$

[a] The decrease of g_m due to velocity saturation and VFMR effects is excluded. These effects are included in the expression for g_m given in Table 3.17.

[b] The inversion coefficient is not used here for operation in the ohmic region. Ohmic region expressions are for operation in strong inversion.

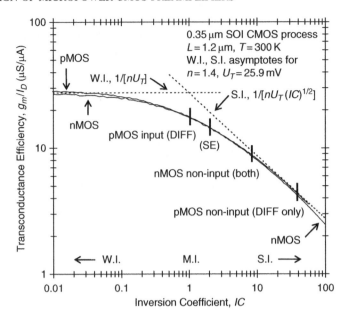

Figure 6.2 Measured transconductance efficiency, g_m/I_D, versus the inversion coefficient, IC, with operating points for devices in a 0.35 μm SOI CMOS process used for the preamplifiers described in Sections 6.7 and 6.8. Input devices operate in moderate inversion for high g_m/I_D and g_m, giving low gate-referred thermal-noise voltage. Non-input devices operate in strong inversion for low g_m/I_D and g_m, giving low drain-referred thermal-noise current. Additionally, non-input, nMOS devices utilize resistive source degeneration to lower their drain-referred thermal- and flicker-noise current

by $IC = I_D/[I_0(W/L)]$ from Table 3.6 using the technology current, $I_0 = 2n_0k_0U_T^2 = 2n_0\mu_0C'_{OX}U_T^2$. From the process data of Table 3.2, $I_0 = 0.3$ and 0.105 μA for nMOS and pMOS devices, respectively.

As shown in the figure, g_m/I_D is constant at $1/[nU_T]$ in weak inversion and is equal to $1/[nU_TIC^{1/2}]$ in strong inversion, decreasing as the inverse square root of IC. While measurements were made on $L = 1.2$ μm devices, g_m/I_D is nearly constant for all channel lengths, excepting the decrease at high inversion for short-channel devices where velocity saturation is significant. As mentioned, short-channel devices are not used in the preamplifiers. The characteristic and technology-independent shape of the g_m/I_D curve, discussed in Section 3.8.2.5, is observed in Figure 6.2 for 0.35 μm SOI CMOS devices, in Figure 3.24 for 0.5 μm CMOS devices, and in Figures 3.25–3.27 for 0.18 μm CMOS devices.

Figure 6.2 shows operating points on the g_m/I_D curves for input and non-input devices in the preamplifiers. Devices used in the differential input preamplifier described in Section 6.7 are denoted as *DIFF*, while those used in the single-ended input preamplifier described in Section 6.8 are denoted as *SE*. pMOS input devices operate in moderate inversion for high g_m/I_D and g_m, which minimizes their gate-referred thermal-noise voltage and helps ensure noise dominance for both thermal and flicker noise. Non-input pMOS devices operate deep into strong inversion for low g_m/I_D and g_m, which minimizes their drain-referred thermal-noise current. Finally, non-input nMOS devices operate near the onset of strong inversion for moderately low g_m/I_D and g_m. Additionally, these devices use resistive source degeneration to reduce their drain-referred thermal- and flicker-noise current.

Figure 6.3 shows predicted $V_{DS,sat}$ and $V_{gm} = (g_m/I_D)^{-1}$ from the boundary of weak and moderate inversion ($IC = 0.1$) to deep strong inversion ($IC = 100$) as calculated from Equations 6.5 and 6.7. $V_{DS,sat}$ and V_{gm} are calculated assuming $U_T = 25.9$ mV at room temperature ($T = 300$ K),

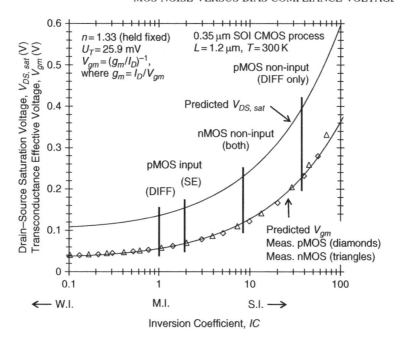

Figure 6.3   Drain–source saturation voltage, $V_{DS,sat}$, and transconductance effective voltage, $V_{gm} = (g_m/I_D)^{-1}$, versus the inversion coefficient, IC, with operating points for devices in a 0.35 μm SOI CMOS process used for the preamplifiers described in Sections 6.7 and 6.8. In strong inversion, increasing V_{gm}, corresponding to decreasing g_m/I_D, decreases non-input device g_m and the drain-referred thermal-noise current, while increasing $V_{DS,sat}$ increases the bias compliance voltage. Values of $V_{DS,sat}$ and V_{gm} are linearly linked as shown in the figure

and V_{gm} is calculated assuming n is held fixed at 1.33. n actually decreases slightly with increasing inversion as illustrated in Figure 3.30 for a 0.18 μm CMOS process, but is assumed fixed for design predictions. Figure 6.3 also shows measured V_{gm}, taken as the reciprocal of measured g_m/I_D in Figure 6.2 for the preamplifier process. Predicted and measured V_{gm} nearly overlay, excepting the slight increase in measured nMOS V_{gm} at high levels of inversion. Velocity saturation effects cause a slight decrease of g_m/I_D and a correspondingly slight increase in V_{gm} for the relatively long channel length of 1.2 μm evaluated.

As shown in Equation 6.7 and Figure 6.3, $V_{DS,sat}$ and V_{gm} are tightly linked. Both are constant and minimum in weak inversion, increasing modestly in moderate inversion before increasing as the square root of IC in strong inversion. Figure 6.3 shows $V_{DS,sat}$ and V_{gm} for preamplifier input and non-input devices, including the g_m/I_D operating points shown in Figure 6.2. In strong inversion, increasing V_{gm}, corresponding to decreasing g_m/I_D, favorably decreases g_m and the drain-referred thermal-noise current of non-input devices. Unfortunately, this is accompanied by increasing $V_{DS,sat}$, which unfavorably increases the bias compliance voltage.

6.5.2   Drain–Source Resistance and Transconductance in the Deep Ohmic Region

The previous section considered transconductance for the usual case where MOS devices operate in saturation where $V_{DS} > V_{DS,sat}$. MOS devices, however, may be operated as drain–source resistors in the ohmic, linear, or triode region where $V_{DS} < V_{DS,sat}$. The micropower, low-noise preamplifiers described in Sections 6.7 and 6.8 utilize MOS resistors to provide resistive source degeneration of drain flicker-noise current for non-input nMOS devices. These resistor devices operate in the deep ohmic region ($V_{DS} \ll V_{DS,sat}$), which provides good resistor linearity combined with low layout area and capacitance.

Strong-inversion drain current in the ohmic region ($V_{DS} < V_{DS,sat}$) is given by

$$I_D(\text{ohmic}) = \mu\, C'_{OX}\left(\frac{W}{L}\right)\left[(V_{GS} - V_T) - \frac{V_{DS}}{2}\right]V_{DS}$$

$$= \mu\, C'_{OX}\left(\frac{W}{L}\right)(V_{GS} - V_T)\, V_{DS}, \quad V_{DS} \ll V_{DS,sat} \tag{6.9}$$

where the second expression considers drain current in the deep ohmic region. Here, $V_{DS} \ll (V_{DS,sat} \approx V_{GS} - V_T = V_{EFF})$, where $V_{DS,sat}$ is approximately equal to V_{EFF} in strong inversion. The inversion charge, pinch-off definition for $V_{DS,sat}$ is actually closer to V_{EFF}/n as described in Section 3.7.3.1.

Differentiating the drain current in Equation 6.9 with respect to V_{DS} gives the ohmic drain–source conductance, which is the reciprocal of the ohmic drain–source resistance, as

$$g_{ds}(\text{ohmic}) = \frac{1}{r_{ds}(\text{ohmic})} = \mu\, C'_{OX}\left(\frac{W}{L}\right)[(V_{GS} - V_T) - V_{DS}]$$

$$= \mu\, C'_{OX}\left(\frac{W}{L}\right)(V_{GS} - V_T), \quad V_{DS} \ll V_{DS,sat} \tag{6.10}$$

The second expression gives g_{ds} (ohmic) in the deep ohmic region where its value is nearly independent of the drain–source voltage. Expressing g_{ds} (ohmic) using drain current in the deep ohmic region from Equation 6.9 shows that

$$g_{ds}(\text{ohmic}) = \frac{1}{r_{ds}(\text{ohmic})} = \frac{I_D(\text{ohmic})}{V_{DS}}, \quad V_{DS} \ll V_{DS,sat} \tag{6.11}$$

This is the expected value from the Ohm's law ratio of current and voltage. g_{ds} (ohmic) given by Equation 6.11 will be used later in Section 6.5.5 for predicting the thermal and flicker noise of a MOS device having resistive source degeneration provided by another MOS device operating in the deep ohmic region.

The transconductance of ohmic MOS devices is needed to evaluate the conversion of gate-referred flicker-noise voltage to drain-referred flicker-noise current. Differentiating the drain current in Equation 6.9 with respect to V_{GS} gives the ohmic transconductance as

$$g_m(\text{ohmic}) = \mu\, C'_{OX}\left(\frac{W}{L}\right)V_{DS} \tag{6.12}$$

Interestingly, g_m (ohmic) collapses to zero as V_{DS} (and I_D) collapses to zero. This resulted in serious underprediction of thermal noise in the deep ohmic region for early SPICE MOS models where noise was erroneously linked to transconductance, as appropriate for operation in saturation, instead of being linked to the ohmic drain–source conductance, as appropriate for operation in the deep ohmic region. MOS noise is correctly predicted continuously across the ohmic and saturation regions by using expressions like Equation 3.104 that consider transconductance, body-effect transconductance, and drain–source conductance.

Dividing g_m (ohmic) by the deep ohmic drain current given by the second expression in Equation 6.9 gives $(g_m/I_D)_{ohmic}$ in the deep ohmic region as

$$\left(\frac{g_m}{I_D}\right)_{ohmic} = \frac{1}{V_{GS} - V_T} = \frac{1}{V_{EFF}}, \quad V_{DS} \ll V_{DS,sat} \tag{6.13}$$

Multiplying this by the drain current gives g_m (ohmic) in the deep ohmic region as

$$g_m(\text{ohmic}) = \left(\frac{g_m}{I_D}\right)_{ohmic} \cdot I_D(\text{ohmic}) = \frac{I_D(\text{ohmic})}{V_{GS} - V_T} = \frac{I_D(\text{ohmic})}{V_{EFF}}, \quad V_{DS} \ll V_{DS,sat} \qquad (6.14)$$

$(g_m/I_D)_{ohmic}$ and g_m (ohmic) are the expected values for a linear-law device where deep ohmic drain current increases linearly with $V_{EFF} = V_{GS} - V_T$. Interestingly, a fully velocity-saturated device operating in saturation is also a linear-law device, having the same values for g_m/I_D and g_m given by Equations 6.13 and 6.14.

$(g_m/I_D)_{ohmic}$ and g_m (ohmic) given by Equations 6.13 and 6.14 apply only for operation in strong inversion in the deep ohmic region where $V_{DS} \ll (V_{DS,sat} \approx V_{GS} - V_T = V_{EFF})$. As V_{DS} increases, $(g_m/I_D)_{ohmic}$ and g_m (ohmic) increase, ultimately doubling from the equations given as expected for MOS square-law operation in saturation (this assumes strong-inversion operation with negligible velocity saturation). Operating at high V_{EFF} keeps $(g_m/I_D)_{ohmic}$ inefficiently low, giving a low value for g_m (ohmic). This minimizes the conversion of gate-referred flicker-noise voltage to drain-referred flicker-noise current and helps ensure deep ohmic operation where the device acts as a simple, fixed conductance or resistance between the drain and source.

g_m in saturation given by the rightmost expression in Equation 6.8 depends on the DC bias conditions of I_D and $V_{DS,sat}$, and is independent of process parameters except for n. Similarly, g_{ds} (ohmic) and g_m (ohmic) in the deep ohmic region given by Equations 6.11 and 6.14 depend on the DC bias conditions of I_D, V_{DS}, and V_{EFF}. Mobility, μ, is significantly degraded from its low-field value, μ_0, when high values of V_{EFF} are used to ensure deep ohmic MOS operation. Although this must be carefully considered to size devices for a given value of g_{ds} (ohmic) using Equation 6.10, or the expression at the bottom of Table 3.28 that includes a simple model for mobility reduction, it does not have to be considered for predicting g_{ds} (ohmic) and g_m (ohmic) from the DC bias conditions.

Table 6.2, which earlier summarized the g_m expressions of Equation 6.8, also summarizes g_{ds} (ohmic) and g_m (ohmic) expressions given by Equations 6.11 and 6.14. This table will be used later in noise analysis developed in terms of DC bias conditions. This noise analysis will ultimately be expressed in terms of the bias compliance voltage, which is necessarily limited in modern, low-voltage processes.

6.5.3 Gate Noise Voltage

MOS gate-referred noise voltage should be minimized for input transconductors such as differential pairs or single devices that receive voltage signals at the gate. In this case, gate-referred noise voltage appears directly in series with the signal.

In the following discussions, gate-referred thermal-noise voltage is linked to both the inversion coefficient and $V_{DS,sat}$, while gate-referred flicker-noise voltage is linked to the inversion coefficient. Linking noise to $V_{DS,sat}$ permits an analysis of noise in terms of the bias compliance voltage, which is useful for minimizing the noise of MOS circuits operating at low supply voltages.

6.5.3.1 Thermal noise

The first expression in Table 6.3 summarizes MOS gate-referred thermal-noise voltage PSD in terms of both the inversion coefficient and $V_{DS,sat}$. The noise is expressed in terms of the inversion coefficient from Table 3.30 and is expressed in terms of $V_{DS,sat}$ using g_m expressed from $V_{DS,sat}$ from Table 6.2.

The expression for gate-referred thermal-noise voltage PSD in Table 6.3 excludes velocity saturation and VFMR effects. These effects can be included by modifying the expression for g_m described in Table 3.17 and using this in the noise expression in Table 6.3. Gate-referred thermal-noise voltage increases if g_m is reduced by velocity saturation and VFMR effects. However, operation at low levels of

Table 6.3 MOS thermal and flicker noise expressed in terms of the inversion coefficient and DC bias conditions. DC bias expressions permit the minimization of noise in terms of the bias compliance voltage

Parameter	Expression
Gate-referred thermal-noise voltage PSD	$S_{VG}(\text{thermal}) = 4kT\left[\dfrac{n\Gamma}{g_m}\right]$
	$= 4kT\left[\dfrac{n^2\Gamma\,U_T\left(\sqrt{IC+0.25}+0.5\right)}{I_D}\right]$
	$= 4kT\left[\dfrac{n^2\,\Gamma\left(V_{DS,sat}-2U_T\right)}{2I_D}\right]$
Gate-referred flicker-noise voltage PSD	$S_{VG}(\text{flicker}) = \dfrac{K'_F}{WLf^{AF}} = \left(\dfrac{K'_F}{f^{AF}}\right)\left(\dfrac{IC}{L^2}\right)\left(\dfrac{I_0}{I_D}\right)$
Drain-referred thermal-noise current PSD	$S_{ID}(\text{thermal}) = 4kT\left[n\,\Gamma\,g_m\right]$
	$= 4kT\left[\dfrac{\Gamma I_D}{U_T\left(\sqrt{IC+0.25}+0.5\right)}\right]$
	$= 4kT\left[\dfrac{2\Gamma I_D}{V_{DS,sat}-2U_T}\right]$
Drain-referred flicker-noise voltage PSD	$S_{ID}(\text{flicker}) = \dfrac{K'_F}{WLf^{AF}}\cdot g_m^2$
	$= \left(\dfrac{2K'_F}{f^{AF}}\right)\left(\dfrac{k_0}{n}\right)\left(\dfrac{I_D}{L^2}\right)\dfrac{IC}{\left(\sqrt{IC+0.25}+0.5\right)^2}$
	$\approx \dfrac{2K'_F}{f^{AF}}\left(\dfrac{k_0}{n}\right)\left(\dfrac{I_D}{L^2}\right)$, in strong inversion
	$= \dfrac{K'_F}{WLf^{AF}}\left(\dfrac{2I_D}{n\left(V_{DS,sat}-2U_T\right)}\right)^2$

Expressions are for operation in saturation where $V_{DS} > V_{DS,sat}$.

Expressions exclude increases in gate-referred thermal-noise voltage and decreases in drain-referred thermal- and flicker-noise current due to g_m decreases caused by velocity saturation and VFMR effects. These effects are included by using the g_m expression from Table 3.17 that includes these effects.

Γ is the thermal-noise factor given in Equation 3.106. Excluding possible small-geometry increases, Γ varies from $1/2$ to $2/3$ from weak to strong inversion.

K'_F is the flicker-noise factor in units of $(\text{nV})^2\cdot\mu\text{m}^2$. As described in Tables 3.31 and 3.34, inversion-level or V_{EFF} increases in flicker noise are included by $K'_F = K'_{F0}(1+V_{EFF}/V_{KF})^2$, where K'_{F0} is the flicker-noise factor in weak inversion or the weak-inversion side of moderate inversion, and V_{KF} is a voltage describing the flicker-noise increase with V_{EFF}. V_{EFF} is clamped to zero when it is negative in weak inversion or the weak-inversion side of moderate inversion.

$k_0 = \mu_0 C'_{OX}$ is the transconductance factor, and $I_0 = 2n_0\mu_0 C'_{OX}U_T^2$ is the technology current.

inversion for high g_m/I_D and g_m and low gate-referred thermal-noise voltage minimizes these effects. Noise may also increase due to small-geometry effects described in Section 3.10.2.2 where the thermal-noise factor Γ may increase from its expected value of $1/2$ to $2/3$ from weak through strong inversion. Using long-channel input devices to maximize gate area and minimize flicker noise minimizes small-geometry thermal-noise increases for the micropower, low-noise preamplifiers described in Sections 6.7 and 6.8.

The expression for gate-referred thermal-noise voltage PSD in Table 6.3 reveals that the noise inversely tracks drain current, while in strong inversion tracking the square root of the inversion coefficient and tracking $V_{DS,sat}$ directly. When velocity saturation is significant, the noise increases nearly directly with the inversion coefficient, increasing more potentially if noise increases in Γ are present. For a given drain current, the noise is minimized at low inversion coefficients in weak or moderate inversion where g_m/I_D and g_m are high and $V_{DS,sat}$ is low. Low $V_{DS,sat}$ easily permits low-voltage design for voltage input stages where gate-referred thermal-noise voltage should be minimized. As shown by Equation 6.3 and Figure 6.1, operation in weak or moderate inversion also results in an optimally low thermal-noise efficiency factor, NEF'. This corresponds to minimum required drain current for a given level of gate-referred thermal-noise voltage.

The behavior of gate-referred thermal-noise voltage density (the square root of PSD) versus the inversion coefficient is shown in Figure 3.63. This is shown for nMOS devices in the $0.18\,\mu m$ CMOS process used for the cascoded OTAs described in Chapter 5. Because it is expressed in terms of the inversion coefficient, the noise is nearly the same for both device types in all processes having similar substrate factors, n, excluding differences associated with velocity saturation and possible small-geometry noise increases in Γ.

6.5.3.2 Flicker noise

The second expression in Table 6.3 summarizes MOS gate-referred flicker-noise voltage PSD in terms of the inversion coefficient. The noise is expressed from Tables 3.31 and 3.34 where K'_F is the flicker-noise factor in convenient units of $(nV)^2 \cdot \mu m^2$ with unit conversions from other flicker-noise factors detailed in Table 3.32. AF is the flicker-noise PSD slope factor. As noted in Table 6.3, increases in the gate-referred flicker-noise voltage with increasing inversion or V_{EFF} are considered through the increase in K'_F from its value, K'_{F0}, in weak inversion or the weak-inversion side of moderate inversion. As described in Sections 3.10.3.5–3.10.3.7 and shown in Figures 3.64–3.66, increases in gate-referred flicker-noise voltage can be significant, especially for pMOS devices. These noise increases, however, are avoided for pMOS input devices in the micropower, low-noise preamplifiers described in Sections 6.7 and 6.8 because these devices operate in moderate inversion. Process values of K'_{F0}, V_{KF}, and AF are given in Table 3.3.

The expression for gate-referred flicker-noise voltage PSD in Table 6.3 uses gate area expressed in terms of the inversion coefficient as $WL = (L^2/IC) \cdot (I_D/I_0)$ from Table 3.10, where $I_0 = 2n_0k_0U_T^2 = 2n_0\mu_0C'_{OX}U_T^2$ is the technology current from Table 3.6. Process values for I_0 are given in Table 3.2. Because of the complex mathematical relationship between gate area and $V_{DS,sat}$, flicker noise is not expressed in terms of $V_{DS,sat}$.

The expression for gate-referred flicker-noise voltage PSD in Table 6.3 reveals that the noise increases with the inversion coefficient, decreases as the inverse square of channel length, and decreases inversely with drain current, tracking the inverse of gate area. When inversion-level or V_{EFF} increases in the noise are present, the noise increases more rapidly with the inversion coefficient. For a given drain current, the noise is minimized at low inversion coefficients in weak or moderate inversion where channel width and gate area are large and $V_{DS,sat}$ is low, again, facilitating low-voltage design. Increasing channel length has the greatest effect on reducing flicker noise because of the large increase in gate area. Gate area increases as the square of channel length because width must increase with channel length to maintain a given inversion coefficient and drain current.

The behavior of gate-referred flicker-noise voltage density versus the inversion coefficient is shown in Figures 4.14 and 3.67 for nMOS and pMOS devices in the $0.18\,\mu m$ CMOS process used for the cascoded OTAs described in Chapter 5. As observed in the figures, inversion-level or V_{EFF} increases in the noise are significant for pMOS devices in the process. These noise increases are also significant for non-input pMOS devices in the preamplifier described in Section 6.7.

6.5.4 Drain Noise Current

MOS drain-referred noise current should be minimized for non-input devices when input transconductors such as differential pairs or single devices receive voltage signals at the gate. In this case, drain noise current from non-input devices is divided by the input transconductance, which refers the noise current to an input noise voltage.

In the following discussions, drain-referred thermal-noise and flicker-noise current is linked to both the inversion coefficient and $V_{DS,sat}$. As described for the gate-referred thermal-noise voltage, linking drain-referred noise current to $V_{DS,sat}$ permits an analysis of noise in terms of the bias compliance voltage. This is useful for minimizing the noise of MOS circuits operating at low supply voltages.

6.5.4.1 Thermal noise

The third expression in Table 6.3 summarizes MOS drain-referred thermal-noise current PSD in terms of both the inversion coefficient and $V_{DS,sat}$. The noise is expressed in terms of the inversion coefficient from Table 3.29 and is expressed in terms of $V_{DS,sat}$ using g_m expressed from $V_{DS,sat}$ from Table 6.2. Like the gate-referred thermal-noise voltage, the drain-referred thermal-noise current given in Table 6.3 excludes velocity saturation and VFMR effects. These effects, again, can be included by modifying the expression for g_m described in Table 3.17 and using this in the noise expression in Table 6.3. Non-input devices are often operated well into strong inversion to minimize their g_m and drain-referred thermal-noise current. As a result, velocity saturation can reduce g_m significantly for short-channel devices, while VFMR can reduce g_m modestly for devices of all channel lengths. Fortunately, drain-referred thermal-noise current is reduced if g_m is reduced by these effects. However, as for gate-referred thermal-noise voltage, the drain-referred thermal-noise current may be increased due to small-geometry increases in the noise modeled by increases in Γ.

The expression for drain-referred thermal-noise current PSD in Table 6.3 reveals that the noise directly tracks drain current, while in strong inversion inversely tracking the square root of inversion coefficient and inversely tracking $V_{DS,sat}$. When velocity saturation is significant, the noise decreases nearly inversely with the inversion coefficient, assuming no noise increases in Γ. For a given drain current, the noise is minimized at high inversion coefficients deep into strong inversion where g_m/I_D and g_m are low and $V_{DS,sat}$ high. Minimizing drain-referred thermal-noise current then poses a significant challenge for low-voltage circuits where the bias compliance voltage for a single device, $V_{DS,sat}$, must necessarily be limited. In Section 6.5.5.2, drain-referred thermal-noise current will be evaluated as a function of the bias compliance voltage.

The behavior of drain-referred thermal-noise current density versus the inversion coefficient is shown in Figure 3.62. This is shown for nMOS devices in the $0.18\,\mu m$ CMOS process used for the cascoded OTAs described in Chapter 5. Because it is expressed in terms of the inversion coefficient, the noise is nearly the same for both device types in all processes, excluding differences associated with velocity saturation and possible small-geometry noise increases in Γ.

6.5.4.2 Flicker noise

MOS drain-referred flicker-noise current PSD is equal to the gate-referred flicker-noise voltage PSD multiplied by g_m^2, which converts the gate-referred noise voltage PSD to a drain-referred noise current PSD. The fourth and final expression in Table 6.3 summarizes drain-referred flicker-noise current PSD in terms of both the inversion coefficient and $V_{DS,sat}$. The inversion coefficient term is from Table 3.34, where gate area is found in terms of the inversion coefficient and technology current as described in Section 6.5.3.2 for the gate-referred flicker-noise voltage. The technology current,

$I_0 = 2n_0 k_0 U_T^2 = 2n_0 \mu_0 C'_{OX} U_T^2$, is then expressed in terms of the transconductance factor, $k_0 = \mu_0 C'_{OX}$, which results in the cancellation of some terms. The drain-referred flicker-noise current is also expressed in terms of $V_{DS,sat}$ using g_m expressed from $V_{DS,sat}$ from Table 6.2. Like the gate-referred and drain-referred thermal noise given in Table 6.3, the drain-referred flicker-noise current excludes velocity saturation and VFMR effects. These effects, again, can be included by modifying the expression for g_m described in Table 3.17 and using this in the noise expression in Table 6.3. Decreases in g_m in strong inversion caused by velocity saturation and VFMR effects decrease the drain-referred flicker-noise current, provided the gate-referred flicker-noise voltage does not increase significantly with increasing inversion or V_{EFF}.

The expression for drain-referred flicker-noise current PSD in Table 6.3 reveals that the noise is minimized at low inversion coefficients in weak inversion, even though g_m is constant and maximum here, maximizing the conversion of gate-referred flicker-noise voltage to drain-referred flicker-noise current. The drain-referred flicker-noise current is minimized in weak inversion because the gate-referred flicker-noise voltage is minimized here due to the large gate area resulting from the large required shape factor (W/L) and channel width. However, as mentioned, non-input devices are usually operated in strong inversion to minimize their drain-referred thermal-noise current.

Interestingly, as seen by the expression in Table 6.3, the drain-referred flicker-noise current PSD ceases to continue increasing in strong inversion ($IC > 10$) as the inversion coefficient increases. Instead, the noise becomes almost independent of the inversion coefficient and, correspondingly, $V_{DS,sat}$. As always, this assumes that only one parameter, the inversion coefficient in this case, is changed while the others, the channel length and drain current in this case, are held fixed. Drain-referred flicker-noise current becomes constant as the inversion coefficient increases in strong inversion because the increasing gate-referred flicker-noise voltage associated with smaller gate area resulting from the smaller required shape factor and width is countered by decreasing g_m/I_D and g_m. However, the leveling off of drain-referred flicker-noise current occurs only if increases in gate-referred flicker-noise voltage with increasing inversion or V_{EFF} are small, which is indicated when K'_F remains near its low-inversion, K'_{F0} value.

The expression for drain-referred flicker-noise current PSD in Table 6.3 reveals that for a fixed gate area and gate-referred flicker-noise voltage, the drain-referred flicker-noise current PSD decreases inversely as the square of $V_{DS,sat}$ due to decreasing g_m/I_D and g_m. Unfortunately, increasing $V_{DS,sat}$ increases the bias compliance voltage, again, equal to $V_{DS,sat}$ for a single device. In Section 6.5.5.3, drain-referred flicker-noise current will be evaluated as a function of the bias compliance voltage.

While non-input devices should be operated in strong inversion for low g_m/I_D and g_m to minimize their drain-referred thermal-noise current, drain-referred flicker-noise current, as mentioned, is nearly constant with the inversion coefficient in strong inversion, or increases due to inversion-level or V_{EFF} increases in the noise. Fortunately, increasing channel length significantly reduces flicker noise in all regions of operation because gate area increases as the square of channel length. Decreasing drain current also lowers drain-referred flicker-noise current because g_m decreases more rapidly than the increase in gate-referred flicker-noise voltage associated with the smaller required channel width and gate area at lower drain currents.

The behavior of drain-referred flicker-noise current density is illustrated in Figures 4.18 and 3.68 for nMOS and pMOS devices in the 0.18 μm CMOS process used in the cascoded OTAs described in Chapter 5. The noise of both device types initially increases with increasing inversion coefficient in weak and moderate inversion because of the decrease in gate area. The noise of nMOS devices shown in Figure 4.18 then levels off in strong inversion as described above. However, the noise of pMOS devices shown in Figure 3.68 does not level off in strong inversion because of inversion-level or V_{EFF} increases in the gate-referred flicker-noise voltage. For the preamplifier described in Section 6.7, a similar increase in flicker noise is present for non-input, pMOS devices that operate deep into strong inversion to minimize their drain-referred thermal-noise current.

6.5.5 Drain Noise Current with Resistive Source Degeneration

Resistive source degeneration raises the MOS drain output resistance while degenerating or reducing the transconductance and drain-referred noise current. Figure 6.4 shows a MOS current source consisting of device M_1 with resistive source degeneration. The degeneration resistance, R, can be fabricated by a process resistor or by another MOS device, M_2, operating in the deep ohmic region. The following discussions consider the bias compliance voltage, thermal noise, and flicker noise for a resistively degenerated MOS current source.

6.5.5.1 Bias compliance voltage

The bias compliance voltage, V_{COMP}, for a resistively degenerated MOS current source shown in Figure 6.4 is the minimum voltage across the current source required to keep M_1 in saturation. V_{COMP} is given by

$$V_{COMP} = V_{DS,sat} + V_R \qquad (6.15)$$

$V_{DS,sat}$ is the drain–source saturation voltage of M_1, and V_R is the DC bias voltage across R.
  The value of R is set by the ratio of its DC bias voltage, V_R, and current, I_D, as

$$R = \frac{V_R}{I_D} \qquad (6.16)$$

Like MOS drain-referred noise current without resistive degeneration developed in Section 6.5.4, noise of a resistively degenerated MOS current source will be developed in terms of DC bias conditions to permit minimization for a given value of V_{COMP}.

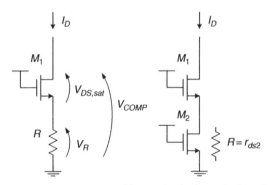

M_2 operates in deep-ohmic region
where $(V_{DS} = V_R) \ll V_{DS,sat}$

V_{COMP} is the bias compliance voltage. $V_{COMP} = V_{DS,sat} + V_R$

Figure 6.4 Schematic diagram of a MOS current source with resistive source degeneration showing components of the bias compliance voltage. $V_{DS,sat}$ is the MOS drain–source saturation voltage, and V_R is the bias voltage drop across the degeneration resistance. $V_{COMP} = V_{DS,sat} + V_R$ is the bias compliance voltage corresponding to the minimum allowable voltage across the current source to maintain MOS saturation operation. In the right side of the figure, a MOS device operating in the deep ohmic region provides the degeneration resistance. This device generates similar thermal noise as a resistor and generally has negligible flicker noise because of low transconductance in the deep ohmic region

When a MOS device in the deep ohmic region, shown in the right side of Figure 6.4, provides the degeneration resistance, R, the resistor thermal-noise current PSD is taken as that of a regular resistor having value of R. This approximation is good for deep ohmic MOS resistors where $V_{DS} \ll V_{DS,sat}$. For the micropower, low-noise preamplifiers described in Sections 6.7 and 6.8, this condition is established by connecting the supply voltage, $V_{DD} = 3.3$ V, to the gate of nMOS devices used as degeneration resistors and ensuring sufficiently small voltage across these devices.

6.5.5.2 Thermal noise

The drain-referred thermal-noise current PSD for a resistively degenerated MOS current source shown in Figure 6.4 is given by

$$S_{ID}(\text{thermal}) = 4kT \left[\frac{n\,\Gamma\,g_m}{(1+g_{ms}R)^2} + \frac{1}{R}\left(\frac{g_{ms}R}{1+g_{ms}R}\right)^2 \right] \tag{6.17}$$

For the saturated MOS device, M_1, g_m is the transconductance, g_{ms} is the source transconductance, and Γ is the thermal-noise factor. g_{ms} is the sum of g_m, the body-effect transconductance, g_{mb}, and the drain–source conductance, g_{ds}, as given by

$$g_{ms} = g_m + g_{mb} + g_{ds}$$

$$\approx g_m + g_{mb}$$

$$\approx n\,g_m \tag{6.18}$$

In the approximation above, g_{ds} is neglected since it is usually much smaller than g_m, and g_{mb} is approximated as $g_{mb} \approx (n-1) \cdot g_m$ from Table 3.20. This gives $g_{ms} \approx n \cdot g_m$ for the usual case when the local MOS body is not driven by the source. When the local body is driven by the source, the g_{mb} term disappears leaving $g_{ms} \approx g_m$.

The left term in Equation 6.17 corresponds to the native drain-referred thermal-noise current density (the square root of PSD) of M_1 divided by the denominator term, $1+g_{ms}R$, which describes the degree of noise current reduction due to resistive source degeneration. The degenerated MOS transconductance, G_M listed in Table 3.16, is equal to the native device transconductance, g_m, divided by this same denominator term, where R is equal to r_s, the external source resistance shown in the small-signal Norton circuit model of Figure 3.22.

Expressing the $1+g_{ms}R$ denominator term in Equation 6.17 using $g_{ms} \approx n \cdot g_m$ from Equation 6.18, g_m from Equation 6.8, and R from Equation 6.16 gives

$$1+g_{ms}R = 1+n\,g_m R = 1+n \cdot \frac{2I_D}{n\left(V_{DS,sat}-2U_T\right)} \cdot \frac{V_R}{I_D} = 1+\frac{2V_R}{V_{DS,sat}-2U_T} \tag{6.19}$$

Equation 6.19 is independent of the drain current, depending only on the voltage drop across the degeneration resistance, the drain–source saturation voltage of M_1, and the thermal voltage. Maximizing the voltage across the degeneration resistance, V_R, relative to $V_{DS,sat}-2U_T$, maximizes this equation and the degree of resistive noise degeneration. While the native drain noise current of M_1 decreases by division by this equation, the drain output resistance of M_1, $r_{out1} = r_{ds1}(1+g_{ms}R)$, increases by multiplication by this same equation. Like the degenerated transconductance listed in Table 3.16, the drain output resistance is listed in Table 3.16 for the small-signal Norton circuit model of Figure 3.22.

Corresponding to the left term in Equation 6.17, the drain-referred thermal-noise current of M_1 is multiplied by the reciprocal of the term given by Equation 6.19. The reciprocal of Equation 6.19 is given by

$$\frac{1}{1+g_{ms}R} = \frac{1}{1 + \dfrac{2\,V_R}{V_{DS,sat} - 2U_T}} = \frac{V_{DS,sat} - 2U_T}{2\,V_R + V_{DS,sat} - 2U_T} \tag{6.20}$$

This is a degeneration factor that is multiplied by the native MOS transconductance and drain-referred noise current to describe their reduction due to resistive source degeneration. Increasing V_R relative to $V_{DS,sat} - 2U_T$ minimizes Equation 6.20, which minimizes the left term in Equation 6.17. This corresponds to minimizing the MOS drain-referred thermal-noise current.

The right term in Equation 6.17 corresponds to the thermal-noise current density of the degeneration resistance R reduced by the effects of current division into the source of M_1, where the source input resistance is equal to $1/g_{ms}$ for a short-circuit drain. Like the noise degeneration factor given by Equation 6.20, a current-division factor can also be expressed in terms of V_R, $V_{DS,sat}$, and U_T, giving

$$\frac{g_{ms}R}{1+g_{ms}R} = \frac{\dfrac{2\,V_R}{V_{DS,sat} - 2U_T}}{1 + \dfrac{2\,V_R}{V_{DS,sat} - 2U_T}} = \frac{2\,V_R}{2\,V_R + V_{DS,sat} - 2U_T} \tag{6.21}$$

Although increasing V_R relative to $V_{DS,sat} - 2U_T$ increases the degree of resistive degeneration and reduces the drain-referred thermal-noise current from M_1 as described by decreasing values of Equation 6.20, the current-division factor of Equation 6.21 approaches its maximum unity value for this condition. This permits the full thermal-noise current of degeneration resistance R to enter the source of M_1 and appear at its drain.

Expressing the drain-referred thermal-noise current PSD in Equation 6.17 using g_m from Equation 6.8, g_{ms} from Equation 6.18, R from Equation 6.16, the noise degeneration factor from Equation 6.20, and the current-division factor from Equation 6.21 gives

$$S_{ID}(\text{thermal}) = 4kT\left[\left(\frac{2\Gamma\,I_D}{V_{DS,sat} - 2U_T}\right)\left(\frac{V_{DS,sat} - 2U_T}{2\,V_R + V_{DS,sat} - 2U_T}\right)^2 + \left(\frac{I_D}{V_R}\right)\left(\frac{2\,V_R}{2\,V_R + V_{DS,sat} - 2U_T}\right)^2\right] \tag{6.22}$$

This expression gives the noise in terms of the DC operating and bias conditions, $V_{DS,sat}$, V_R, I_D, and effectively the bias compliance voltage, $V_{COMP} = V_{DS,sat} + V_R$. This permits minimizing the noise under the constraint of limited bias compliance voltages in low-voltage processes.

Since $V_{COMP} = V_{DS,sat} + V_R$, the voltage across the degeneration resistance is $V_R = V_{COMP} - V_{DS,sat}$. V_R is then maximized for M_1 operating in weak inversion where $V_{DS,sat} = V_{DS,sat}(\text{WI}) = 4U_T \approx 0.1\,\text{V}$ (at room temperature) is at its minimum value. Here, the maximum value of V_R corresponds to the maximum value of degeneration resistance, $R = V_R/I_D$. Conversely, the degeneration resistance is equal to zero when V_R is equal to zero. This corresponds to $V_{DS,sat}$ being directly equal to V_{COMP} where M_1 is operated at its maximum available level of inversion for the allowable V_{COMP}.

When no resistive degeneration is used ($V_R = 0$ V and $R = 0\,\Omega$), the noise degeneration factor given by Equation 6.20 reduces to unity and Equations 6.17 and 6.22 reduce to the drain-referred thermal-noise current PSD of a MOS device without degeneration, as summarized in Table 6.3. Under this condition, the thermal-noise current PSD is inversely proportional to $V_{DS,sat} - 2U_T$ and the noise is minimized by maximizing $V_{DS,sat}$, which corresponds to minimizing g_m. This unfortunately maximizes the bias compliance voltage.

When full resistive degeneration is used ($V_R \gg (V_{DS,sat} - 2U_T)$ and $R \gg 1/g_{ms}$), Equations 6.17 and 6.22 reduce to nearly $4kT/R$, the thermal-noise current PSD of the degeneration resistance R alone. Here, the current-division factor given by Equation 6.21 approaches unity, indicating that nearly all of the thermal noise current from R enters the source of M_1 and appears at its drain. Under conditions of full resistive degeneration, the drain-referred thermal-noise current PSD is nearly inversely proportional to V_R and is minimized by maximizing V_R, which corresponds to maximizing the degeneration resistance. This unfortunately maximizes the bias compliance voltage.

Figure 6.5 shows the drain-referred thermal-noise current density of a resistively degenerated MOS current source (Figure 6.4) operating at a drain current of $1\,\mu A$. This is found from the square root of the drain-referred thermal-noise current PSD given by Equations 6.17 and 6.22. Although having a nearly negligible effect, the thermal-noise factor Γ is varied from $1/2$ to $2/3$ for MOS operation from weak through strong inversion using Equation 3.106, where the inversion coefficient is calculated from $V_{DS,sat}$ given by Equation 6.5. The substrate factor is held fixed at $n = 1.33$, and $U_T = 25.9$ mV for $T = 300$ K. Finally, the bias compliance voltage, V_{COMP}, is held constant at 0.25 and 0.5 V for different levels of resistive degeneration denoted by values of V_R, corresponding to different voltages across the degeneration resistance.

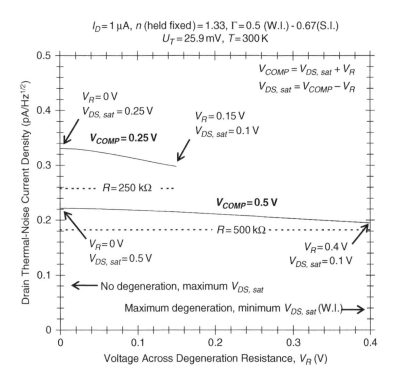

Figure 6.5 Drain-referred thermal-noise current density for a resistively degenerated MOS current source versus the degeneration resistance voltage, V_R. The noise is shown for constant bias compliance voltages, $V_{COMP} = V_{DS,sat} + V_R$, and a drain current of $1\,\mu A$. For constant V_{COMP}, the thermal-noise current decreases only slightly with increasing resistive degeneration (increasing V_R) because of thermal-noise contributions from the degeneration resistance. However, the thermal-noise current decreases significantly as V_{COMP} increases. Drain-referred thermal-noise current density should be minimized for non-input devices in circuits having input transconductors that receive voltage inputs

As seen in Figure 6.5, for a given value of V_{COMP}, the drain-referred thermal-noise current density is reduced, but only slightly, by maximizing V_R and, correspondingly, the level of resistive degeneration. Although increasing the level of resistive degeneration significantly reduces the drain-referred thermal-noise current of M_1, by the amount shown in Equation 6.20, the degeneration resistance itself contributes thermal noise.

Figure 6.5 also shows the thermal-noise current density for resistors alone operating at the same current of 1 μA and values of V_{COMP} considered. This corresponds to resistor values of 250 and 500 kΩ (V_R/I_D) for $V_R = V_{COMP}$ equal to 0.25 and 0.5 V, respectively. The resistor thermal-noise current density is slightly less than the drain-referred thermal-noise current density of a MOS current source without resistive source degeneration operating in strong inversion at the same current and equal $V_R = V_{DS,sat} = V_{COMP}$. For a resistor operating at current I_D with a voltage drop of $V_R = V_{COMP}$, the thermal-noise current PSD is $4kT[1/R] = 4kT[I_D/V_{COMP}]$. For a MOS device operating at I_D with $V_{DS,sat} = V_{COMP}$, the drain-referred thermal-noise current PSD is $4kT[2\Gamma I_D/(V_{COMP} - 2U_T)]$ from Table 6.3. Assuming operation in strong inversion where $\Gamma = \frac{2}{3}$ and $(V_{DS,sat} = V_{COMP}) \gg 2U_T$, the drain-referred thermal-noise current PSD is approximately $4kT[(4/3)I_D/(V_{COMP})]$, which is a factor of 4/3 above that of the resistor alone. This corresponds to a slight increase of 15 % in noise current density, but the increase is more if $V_{DS,sat}$ is not significantly greater than $2U_T$. In fact, the MOS noise current density is 41% above that of a resistor alone for operation in weak inversion at $V_R = V_{DS,sat} = 4U_T = V_{COMP}$. These estimations and the predictions shown in Figure 6.5 exclude possible decreases in MOS g_m and drain-referred thermal-noise current due to velocity saturation and VFMR effects at high levels of inversion. Additionally, possible small-geometry increases in MOS thermal noise are excluded. These effects were described in Section 6.5.4.1.

As seen in Equation 6.22, the drain-referred thermal-noise current PSD of a resistively degenerated MOS current source decreases nearly inversely with increasing V_{COMP}, which, again, is the sum of MOS $V_{DS,sat}$ and degeneration resistance voltage drop, V_R. Thermal-noise current density, then, decreases nearly inversely with the square root of increasing V_{COMP} as shown in Figure 6.5. As a result, minimizing the drain-referred thermal-noise current poses special problems for low-voltage design where V_{COMP} cannot be increased much. The noise can also be reduced by decreasing the drain current. As seen in Equation 6.22, drain-referred thermal-noise current PSD increases directly with drain current and the noise current density increases as the square root of drain current.

As shown in Figure 6.5, for a given value of V_{COMP}, the presence or degree of resistive source degeneration has little effect on the drain-referred thermal-noise current of a MOS current source because of added thermal noise from the degeneration resistance. However, resistive source degeneration can significantly reduce MOS drain-referred flicker-noise current as described in the next section.

6.5.5.3 Flicker noise

The drain-referred flicker-noise current PSD for a MOS device without resistive source degeneration is equal to the gate-referred flicker-noise voltage PSD, S_{VG} (flicker), multiplied by g_m^2 as seen in Table 6.3. The drain-referred flicker-noise current PSD for a resistively degenerated MOS current source shown in Figure 6.4 is then equal to this multiplied by the square of the degeneration factor given in Equation 6.20, giving

$$S_{ID}(\text{flicker}) = S_{VG}(\text{flicker}) \cdot \left(\frac{g_m}{1 + g_{ms}R} \right)^2 \tag{6.23}$$

The right term before squaring, $g_m/(1 + g_{ms}R)$, is the degenerated MOS transconductance for M_1 as listed in Table 3.16 for the small-signal Norton circuit model of Figure 3.22. As mentioned in the preceding section for thermal noise, g_m is the native device transconductance, R is the source degeneration resistance, and $1 + g_{ms}R$ is a denominator term describing the degree of resistive source degeneration.

The degeneration resistance R contributes virtually no flicker noise unless it is implemented by a MOSFET, M_2, operating in the deep ohmic region as shown on the right side of Figure 6.4. The flicker noise of this MOS device, which is often negligible because of its collapsed transconductance, will be considered later.

Expressing the drain-referred flicker-noise current PSD in Equation 6.23 using g_m from Equation 6.8, g_{ms} from Equation 6.18, R from Equation 6.16, and the noise degeneration factor from Equation 6.20 gives

$$S_{ID}(\text{flicker}) = S_{VG}(\text{flicker}) \cdot \left(\frac{g_m}{1 + g_{ms}R} \right)^2 = S_{VG}(\text{flicker}) \cdot \left[\frac{\dfrac{2 I_D}{n\left(V_{DS,sat} - 2 U_T\right)}}{1 + \dfrac{2 V_R}{V_{DS,sat} - 2 U_T}} \right]^2$$

$$= S_{VG}(\text{flicker}) \cdot \left[\frac{2 I_D}{n\left(2 V_R + V_{DS,sat} - 2 U_T\right)} \right]^2 \tag{6.24}$$

Like the thermal-noise expression given by Equation 6.22, this expression gives the flicker noise in terms of the DC operating and bias conditions, $V_{DS,sat}$, V_R, I_D, and effectively the bias compliance voltage, $V_{COMP} = V_{DS,sat} + V_R$. Again, this permits minimizing the noise under the constraint of limited bias compliance voltages in low-voltage processes.

For a given value of gate-referred flicker-noise voltage PSD corresponding to a fixed MOS gate area, Equation 6.24 shows that the drain-referred flicker-noise current PSD decreases as $V_{DS,sat}$ and the voltage across the degeneration resistance, V_R, increase, corresponding to increasing bias compliance voltage. When no resistive degeneration is used ($V_R = 0\,V$ and $R = 0\,\Omega$), Equations 6.23 and 6.24 reduce to the drain-referred flicker-noise current PSD of a MOS device having no degeneration, as summarized in Table 6.3. Under this condition, the flicker-noise current PSD is inversely proportional to the square of $V_{DS,sat} - 2 U_T$ and the noise is minimized by maximizing $V_{DS,sat}$, which corresponds to minimizing g_m. When resistive degeneration is used, the flicker-noise current PSD is inversely proportional to the square of $2 V_R + V_{DS,sat} - 2 U_T$ and the noise is minimized by maximizing V_R over $V_{DS,sat}$. This corresponds to maximizing the level of resistive degeneration.

Figure 6.6 shows the drain-referred flicker-noise current density of a resistively degenerated MOS current source (Figure 6.4) operating at a drain current of $1\,\mu A$. The noise is shown for a constant gate-referred flicker-noise voltage density of $500\,nV/Hz^{1/2}$ at 1 Hz, corresponding to constant M_1 gate area. The noise is found from the square root of the drain-referred flicker-noise current PSD given by Equations 6.23 and 6.24 using $n = 1.33$ and $U_T = 25.9\,mV$ for $T = 300\,K$. As done in Figure 6.5 for the thermal noise, the bias compliance voltage, V_{COMP}, is held constant at 0.25 and 0.5 V for different levels of resistive degeneration denoted by values of V_R, corresponding to different voltages across the degeneration resistance.

As seen in Figure 6.6, for a given value of V_{COMP}, the drain-referred flicker-noise current density is reduced significantly by maximizing the level of resistive degeneration, indicated by the maximum value of V_R. For maximum degeneration, flicker-noise current density approaches the factor-of-two reduction expected for a factor-of-four reduction in flicker-noise current PSD predicted by Equation 6.24. The decrease in thermal-noise current density, shown in Figure 6.5, however, was considerably less because, as mentioned, the degeneration resistance itself contributes thermal noise. As mentioned in the preceding section for thermal noise, maximum degeneration occurs when M_1 is operated in weak inversion, giving minimum $V_{DS,sat} = V_{DS,sat}(\text{WI}) = 4 U_T \approx 0.1\,V$ (at room temperature) and maximum $V_R = V_{COMP} - V_{DS,sat}(\text{WI})$ and $R = V_R/I_D$. Full degeneration ($R \gg 1/g_{ms}$) occurs when $V_R \gg (V_{DS,sat} - 2 U_T)$ and nearly the full bias compliance voltage is dropped across the degeneration resistance ($V_R \approx V_{COMP}$).

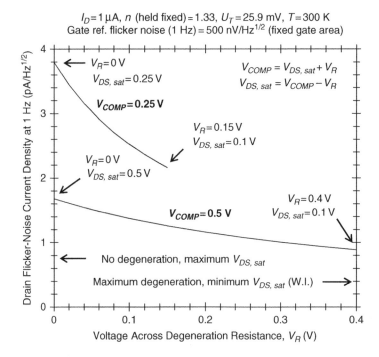

Figure 6.6   Drain-referred flicker-noise current density at 1 Hz for a resistively degenerated MOS current source versus the degeneration resistance voltage, V_R. The noise is shown for constant gate-referred flicker-noise voltage density of 500 nV/Hz$^{1/2}$, corresponding to constant gate area, and is shown for constant bias compliance voltages, $V_{COMP} = V_{DS,sat} + V_R$, and a drain current of 1 μA. For constant V_{COMP}, the flicker-noise current decreases significantly with increasing resistive degeneration (increasing V_R) and approaches one-half that of a device without resistive degeneration. The flicker-noise current also decreases significantly as V_{COMP} increases. Drain-referred flicker-noise current density should be minimized for non-input devices in circuits having input transconductors that receive voltage inputs

Under conditions of full degeneration, Equation 6.24 shows that the drain-referred flicker-noise current PSD is nearly inversely proportional to the square of $2V_R \approx 2V_{COMP}$, compared to being nearly inversely proportional to the square of $V_{DS,sat} = V_{COMP}$ (for operation in strong inversion) when no resistive degeneration is used. As a result, the flicker-noise current PSD is nearly one-fourth and the flicker-noise current density one-half that of a non-degenerated device having $V_{DS,sat}$ equal to the same bias compliance voltage, V_{COMP}. This assumes operation in strong inversion for the non-degenerated device where $V_{DS,sat} \gg 2U_T$. This reduction in noise is equivalent to that provided by a MOS device having a factor-of-four smaller gate-referred flicker-noise voltage PSD, corresponding to a factor-of-four larger gate area. As a result, resistive source degeneration can be advantageous to minimize non-input-device, drain-referred flicker-noise current for a given layout area.

The reduction of drain-referred flicker-noise current resulting from resistive source degeneration occurs because of reduced MOS effective transconductance that converts gate-referred flicker-noise voltage to drain-referred flicker-noise current. In the case of full degeneration where $V_R \approx V_{COMP}$, the degenerated transconductance is $g_m \approx 1/(nR) = I_D/(nV_R) = I_D/(nV_{COMP})$. However, the transconductance associated with a MOS device operating in strong inversion without degeneration where $V_{DS,sat} = V_{COMP}$ is $g_m = 2I_D/V_{EFF} = 2I_D/(nV_{DS,sat}) = 2I_D/(nV_{COMP})$, which is a factor of two higher for the same value of V_{COMP}. As a result, MOS gate-referred flicker-noise voltage is less efficiently converted to drain-referred flicker-noise current when resistive source degeneration is used.

Full degeneration requires that V_{COMP} significantly exceeds $V_{DS,sat}$, which can pose special problems for low-voltage design where V_{COMP}, the sum of $V_{DS,sat}$ and V_R, must necessarily be limited.

If full velocity saturation occurs for a MOS device operating in strong inversion, the device g_m is reduced to I_D/V_{EFF}, one-half that of a MOS device operating in strong inversion without velocity saturation or VFMR effects. This potentially reduces the drain-referred flicker-noise current because of the lower conversion of gate-referred flicker-noise voltage to drain-referred flicker-noise current, providing a similar effect to resistive source degeneration. However, full velocity saturation requires high levels of inversion and short channel lengths, both resulting in small channel width and gate area, and, correspondingly, high flicker noise. Additionally, the gate-referred flicker-noise voltage can increase significantly at high levels of inversion or V_{EFF}, especially for pMOS devices, as described in Sections 3.10.3.5–3.10.3.7 and shown in Figures 3.64–3.66. Reduced drain-referred flicker-noise current for devices experiencing velocity saturation requires that the gate-referred flicker-noise voltage does not increase significantly with the inversion level or V_{EFF}.

As seen in Equation 6.24, the drain-referred flicker-noise current PSD of a resistively degenerated MOS current source decreases nearly inversely with the square of increasing V_{COMP}, which, again, is the sum of MOS $V_{DS,sat}$ and the voltage across the degeneration resistance, V_R. Flicker-noise current density, then, decreases nearly inversely with increasing V_{COMP} as shown in Figure 6.6. As a result, minimizing the drain-referred flicker-noise current poses special problems for low-voltage design where V_{COMP} cannot be increased much. The noise can also be reduced by decreasing the drain current. As seen in Equation 6.24, drain-referred flicker-noise current PSD increases with the square of drain current and the noise current density increases directly with the drain current. Decreasing the drain current reduces the flicker-noise current because of the reduction of g_m and increase in R. Again, a constant gate-referred flicker-noise voltage, corresponding to constant MOS gate area, is assumed for these trends, as well as for the previously described trends.

A general noise analysis including variable MOS gate area is available using the drain-referred flicker-noise current PSD given in Table 6.3 expressed in terms of the inversion coefficient, channel length, and drain current. This can then be multiplied by the square of Equation 6.20 to show the decrease of noise PSD associated with resistive source degeneration, with $g_{ms} = n \cdot g_m$ found from Table 6.2. General noise analyses are always used in the *Analog CMOS Design, Tradeoffs and Optimization* spreadsheet, and this is illustrated for the micropower, low-noise preamplifiers described in Sections 6.7 and 6.8.

The drain-referred flicker-noise current density predicted by the square root of Equation 6.24 and shown in Figure 6.6 assumes a degeneration resistance having no flicker noise. Often, as shown in the right side of Figure 6.4, a MOS device (M_2) operating in the deep ohmic region ($V_{DS} \ll V_{DS,sat}$) provides the degeneration resistance. This is especially advantageous for micropower designs where resistor values are large, requiring large layout area when fabricated from process resistors.

The additional drain-referred flicker-noise current PSD resulting from a deep ohmic MOS device used as the degeneration resistance is given by

$$S_{ID}(\text{flicker, ohmic}) = S_{VG}(\text{flicker, ohmic}) \cdot g_m^2(\text{ohmic}) \cdot \left(\frac{g_{ms}R}{1+g_{ms}R} \right)^2 \qquad (6.25)$$

S_{VG} (flicker, ohmic) is the gate-referred flicker-noise voltage PSD for the ohmic MOS device, M_2, g_m (ohmic) is the transconductance for this device, and the rightmost term describes the loss of noise current due to current division at the source of saturated MOS device, M_1. This current division term or factor was given earlier by Equation 6.21 for evaluating the loss of thermal-noise current from the degeneration resistance, R. Expressing g_m (ohmic) from Equation 6.14 and the current division factor from Equation 6.21 gives

$$S_{ID}(\text{flicker, ohmic}) = S_{VG}(\text{flicker, ohmic}) \cdot \left(\frac{I_D}{V_{EFF}(\text{ohmic})} \right)^2 \left(\frac{2V_R}{2V_R + V_{DS,sat} - 2U_T} \right)^2 \qquad (6.26)$$

V_{EFF} (ohmic) is the effective gate–source voltage ($V_{EFF} = V_{GS} - V_T$) for the deep ohmic (M_2) device acting as the degeneration resistance. As before, V_R is the voltage drop across the degeneration resistance, and $V_{DS,sat}$ is the drain–source saturation voltage for the saturated (M_1) device. When significant resistive degeneration is used ($V_R \gg V_{DS,sat}$ and $R \gg 1/g_{ms}$), the rightmost current division factor approaches unity as nearly all the flicker-noise current from the deep ohmic device enters the source of the saturated device. The flicker-noise PSD contribution from the deep ohmic device given by Equation 6.26 is added to the flicker noise PSD from the saturated device given by Equation 6.24.

Flicker-noise current contributions from deep ohmic MOS devices used for degeneration resistances will be shown to be negligible for the micropower, low-noise preamplifiers described in Sections 6.7 and 6.8. This is because V_{EFF} (ohmic) is high since the full supply voltage, $V_{DD} = 3.3\,\text{V}$, is impressed across V_{GS} of the ohmic devices. This ensures deep ohmic operation where $V_{DS} \ll (V_{DS,sat} \approx V_{EFF}$ (ohmic)) and results in small g_m(ohmic) $= I_D/(V_{EFF}(\text{ohmic}))$. Small g_m (ohmic) limits the conversion of gate-referred flicker-noise voltage to drain-referred flicker-noise current.

6.6 EXTRACTION OF MOS FLICKER-NOISE PARAMETERS

Section 3.10.3.5 listed reported flicker-noise factors for multiple CMOS processes. Although reported factors provide insight on technology trends, MOS flicker noise is process dependent and should be evaluated for the process of interest. The following discussion describes flicker-noise measurements for the 0.35 μm SOI CMOS process used for the micropower, low-noise preamplifiers described in Sections 6.7 and 6.8. A similar flicker-noise study was done for the 0.18 μm CMOS process used for the cascoded, operational transconductance amplifiers (OTAs) described in Chapter 5. This study was described in Section 3.10.3.6 and shown in Figures 3.65 and 3.66. The noise measurements shown in this current section were taken using the noise test set and methods described in Section 3.10.3.6.

As described in Sections 3.10.3.5–3.10.3.7 and shown in Figures 3.64–3.66, increases in gate-referred flicker-noise voltage with increasing inversion or V_{EFF} can be significant, especially for pMOS devices. These noise increases will also be shown to be significant for non-input pMOS devices in the differential input preamplifier described in Section 6.7. As noted in Table 6.3 and earlier in Tables 3.31 and 3.34, increases in the gate-referred flicker-noise voltage with increasing inversion or V_{EFF} are considered in the operating value of the flicker-noise factor given by $K_F' = K_{F0}'(1 + V_{EFF}/V_{KF})^2$. When V_{EFF} is negative at low levels of inversion, it is clamped to zero in the expression. K_{F0}' is the flicker-noise factor in weak inversion or the weak-inversion side of moderate inversion, and V_{KF} is a voltage describing the flicker-noise increase with V_{EFF}. Noise measurements should then provide for an extraction of K_{F0}' as well as V_{KF} when flicker-noise increases are significant.

6.6.1 Preamplifier Input Devices

Figure 6.7 shows the measured gate-referred flicker-noise voltage density for pMOS input devices used in the differential input preamplifier described, again, in Section 6.7. The figure also shows flicker noise for non-input devices that will be described in the following section. The pMOS input devices (M_1 and M_2 shown later in Figure 6.8) are sized at $W = 100\,\mu\text{m}$ and $L = 20\,\mu\text{m}$, and operate at a drain current of 0.5 μA. This gives an inversion coefficient of $IC = I_D/[(W/L)I_0] = 0.95$ from Table 3.6 for the pMOS technology current of $I_0 = 2n_0\mu_0 C_{OX}' U_T^2 = 0.105\,\mu\text{A}$ listed in the process parameters of Table 3.2. This inversion coefficient corresponds to nearly the center of moderate inversion, which is intended to minimize gate-referred thermal- and flicker-noise voltage.

Figure 6.7 also shows flicker-noise lines corresponding to extracted flicker-noise factors and slopes. The flicker-noise lines represent the gate-referred flicker-noise voltage density found from the square root of the gate-referred flicker-noise voltage PSD, $S_{VG}(f) = K_F'/(WLf^{AF})$, given in Table 6.3 and

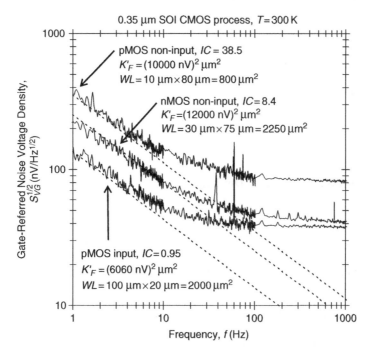

0.35 μm SOI CMOS process, $T = 300$ K

pMOS non-input, $IC = 38.5$
$K'_F = (10000\ \text{nV})^2\ \mu\text{m}^2$
$WL = 10\ \mu\text{m} \times 80\ \mu\text{m} = 800\ \mu\text{m}^2$

nMOS non-input, $IC = 8.4$
$K'_F = (12000\ \text{nV})^2\ \mu\text{m}^2$
$WL = 30\ \mu\text{m} \times 75\ \mu\text{m} = 2250\ \mu\text{m}^2$

pMOS input, $IC = 0.95$
$K'_F = (6060\ \text{nV})^2\ \mu\text{m}^2$
$WL = 100\ \mu\text{m} \times 20\ \mu\text{m} = 2000\ \mu\text{m}^2$

Gate-Referred Noise Voltage Density, $S_{VG}^{1/2}$ (nV/Hz$^{1/2}$)

Frequency, f (Hz)

Figure 6.7 Measured gate-referred flicker-noise voltage density for input and non-input devices in a 0.35 μm SOI CMOS process used in the differential input preamplifier. Devices are sized and biased as operated in the preamplifier. Flicker-noise extractions of $K'_F = K'_{F0}(1 + V_{EFF}/V_{KF})^2$ show that $K'_{F0} = (5700\ \text{nV})^2 \cdot \mu\text{m}^2$ and $V_{KF} = 0.55$ V for pMOS devices, where $V_{EFF} = 0.035$ and 0.418 V for operation at $IC = 0.95$ and 38.5, respectively. Flicker-noise extractions show that $K'_F = K'_{F0} = (12000\ \text{nV})^2 \cdot \mu\text{m}^2$ for nMOS devices where inversion-level or V_{EFF} increases in the noise are much less for the process and are not modeled. Because of unequal gate areas, flicker noise cannot be directly compared across the devices. However, their extracted K'_F values permit a comparison of the noise. The flicker-noise corner frequency is in the vicinity of 10 Hz for these long-channel, large-gate-area devices

earlier in Tables 3.31 and 3.34. The extracted flicker-noise factor for the pMOS input devices is $K'_F = (6060\ \text{nV})^2 \cdot \mu\text{m}^2 = S_{VG}(f) \cdot (f^{AF}) \cdot WL$. This is extracted as described in Table 3.32, where $S_{VG}(f)$ is the gate-referred flicker-noise voltage PSD taken here at $f = 1$ Hz. Taking the extraction at $f = 1$ Hz avoids errors associated with uncertainties in the slope AF and helps ensure that flicker noise dominates thermal noise in the measurement. The flicker-noise factor my also be expressed as K_F in units of C^2/cm^2 or K_{FSPICE} in units of V^2/F from the conversions detailed in the table, assuming $C'_{OX} = 4.31$ fF/μm^2 from the process parameters given in Table 3.2. K'_F in convenient units of (nV)$^2 \cdot \mu$m^2 is used here since gate-referred flicker-noise voltage PSD is easily found by dividing K'_F by the gate area in μm^2 and the frequency in Hz, which is raised to the AF power. Taking the square root of this then gives the noise density in design units of nV/Hz$^{1/2}$. Finally, as seen in Figure 6.7, assuming a flicker-noise slope of unity for the pMOS input devices gives a good fit with the measured noise.

Flicker-noise measurements were also made for the input pMOS device (M_1 shown later in Figure 6.13) in the single-ended input preamplifier described in Section 6.8. This device is equally sized as input devices in the differential input preamplifier, but operates at twice the drain current at 1 μA. This corresponds to $IC = 1.9$ deeper in moderate inversion compared to 0.95 for input devices in the differential input preamplifier. The extracted flicker-noise factor is $K'_F = (6500\ \text{nV})^2 \cdot \mu\text{m}^2$, which is up slightly at the higher inversion level compared to $K'_F = (6060\,\text{nV})^2 \cdot \mu\text{m}^2$ for input devices in the differential input preamplifier. The extracted flicker-noise slope is again nearly unity.

As noted in the caption of Figure 6.7, flicker-noise extractions using $K'_F = K'_{F0}(1 + V_{EFF}/V_{KF})^2$ show that $K'_{F0} = (5700\,\text{nV})^2 \cdot \mu\text{m}^2$ and $V_{KF} = 0.55\,\text{V}$ for input pMOS devices operating at $V_{EFF} = 0.035$ at $IC = 0.95$. The value of V_{EFF} is found from Table 3.14 for $n = 1.35$ and $U_T = 25.9\,\text{mV}$ at $T = 300\,\text{K}$ (room temperature). The extraction requires at least one other noise measurement, preferably at a considerably different inversion level or V_{EFF}. This will be shown below for non-input pMOS devices that operate deep into strong inversion.

6.6.2 Preamplifier Non-Input Devices

Figure 6.7 also shows the measured gate-referred flicker-noise voltage density for non-input nMOS devices (M_3 and M_4 shown later in Figures 6.8 and 6.13) used in both the differential and single-ended input preamplifiers described in Sections 6.7 and 6.8. These devices are sized at $W = 30\,\mu\text{m}$ and $L = 75\,\mu\text{m}$, and operate at a drain current of $1\,\mu\text{A}$. This gives an inversion coefficient of $IC = 8.4$ for the nMOS technology current of $I_0 = 0.30\,\mu\text{A}$ listed in the process parameters of Table 3.2, where the calculation of inversion coefficient was described in the previous section. This corresponds to operation near the onset of strong inversion. As described later in Sections 6.7.1 and 6.8.1, resistive source degeneration is used to lower drain thermal-noise and, especially, flicker-noise current contributions of these devices. This reduction in noise was described earlier in Section 6.5.5.

The extracted flicker-noise factor for the non-input nMOS devices is $K'_F = (12000\,\text{nV})^2 \cdot \mu\text{m}^2$, which is a factor of four larger than $K'_F = (6060\,\text{nV})^2 \cdot \mu\text{m}^2$ found for pMOS input devices in the differential input preamplifier. This indicates some pMOS flicker-noise advantage, but pMOS gate-referred flicker-noise voltage PSD is only one-fourth and density one-half that of equal gate area nMOS devices. As described below, increasing pMOS flicker noise at increasing levels of inversion or V_{EFF} reduces the pMOS flicker-noise advantage. The extracted nMOS flicker-noise slope is taken at unity, although a careful examination of Figure 6.7 shows the slope is likely less than unity. As seen in Table 3.33, reported nMOS flicker-noise slopes often vary from slightly less than unity to unity.

Finally, Figure 6.7 shows the gate-referred flicker-noise voltage density for non-input pMOS devices (M_7 and M_8 shown later in Figure 6.8) in the differential input preamplifier. These devices are sized at $W = 10\,\mu\text{m}$ and $L = 80\,\mu\text{m}$, and operate at a drain current of $0.5\,\mu\text{A}$. This gives an inversion coefficient of $IC = 38.5$ for the pMOS technology current of $I_0 = 0.105\,\mu\text{A}$. This corresponds to operation deep into strong inversion, intended to minimize drain thermal-noise current contributions without the use of resistive source degeneration. The extracted flicker-noise factor is $K'_F = (10000\,\text{nV})^2 \cdot \mu\text{m}^2$, which is up by nearly a factor of three from $K'_F = (6060\,\text{nV})^2 \cdot \mu\text{m}^2$ for the pMOS input devices in the differential input preamplifier that operate in moderate inversion. The pMOS flicker-noise factor in deep strong inversion begins to approach $K'_F = (12000\,\text{nV})^2 \cdot \mu\text{m}^2$ for nMOS devices, nearly eliminating the pMOS noise advantage. The extracted pMOS flicker-noise slope is again observed at nearly unity.

Flicker-noise extractions using $K'_F = K'_{F0}(1 + V_{EFF}/V_{KF})^2$ show that $K'_{F0} = (5700\,\text{nV})^2 \cdot \mu\text{m}^2$ and $V_{KF} = 0.55\,\text{V}$ for pMOS devices, found for $V_{EFF} = 0.418\,\text{V}$ at $IC = 38.5$ for non-input devices and $V_{EFF} = 0.035\text{V}$ at $IC = 0.95$ for input devices described in the previous section. The value of V_{EFF} for operation at $IC = 38.5$ is found from Table 3.14 for $n = 1.3$ and $U_T = 25.9\,\text{mV}$ at $T = 300\,\text{K}$.

An extracted value of $K'_{F0} = (12000\,\text{nV})^2 \cdot \mu\text{m}^2$ is found for the non-input nMOS devices, which is directly equal to K'_F as shown in Figure 6.7. Flicker-noise increases with inversion level or V_{EFF} are much less for nMOS devices compared to pMOS devices in the 0.35 μm SOI CMOS process considered, which has a thick gate-oxide thickness of $t_{ox} = 8\,\text{nm}$. As a result, flicker-noise increases are neglected for nMOS devices. However, for the 0.18 μm CMOS process used in the cascoded OTAs described in Chapter 5, nMOS flicker-noise increases are not negligible. As shown in the process values of Table 3.3, V_{KF} is $1\,\text{V}$ for nMOS devices and $0.25\,\text{V}$ for pMOS for this 0.18 μm CMOS process compared to a value of $0.55\,\text{V}$ for pMOS devices alone in the 0.35 μm SOI CMOS process.

A lower value of V_{KF} corresponds to greater flicker-noise increases with inversion level or V_{EFF}. Possible mechanisms for the flicker-noise increases were described in Section 3.10.3.3.

6.6.3 Comparisons of Flicker Noise

As noted in the caption of Figure 6.7, flicker-noise measurements cannot be directly compared across the devices shown because of different gate areas. However, the extracted K'_F values permit a comparison of flicker noise because these values are normalized to the gate area. As labeled on the figure, the minimum value of K'_F is for input pMOS devices operating in moderate inversion. This value increases significantly for non-input pMOS devices operating deep into strong inversion, approaching the value for non-input nMOS devices.

It is interesting to compare the flicker-noise measurements of Figure 6.7 for the 0.35 μm SOI CMOS process used in the micropower, low-noise preamplifiers described in this chapter with the measurements of Figures 3.65 and 3.66 for the 0.18 μm CMOS process used in the cascoded OTAs described in Chapter 5. In Figures 3.65 and 3.66 equal gate areas are used, so the modest increase in nMOS (Figure 3.65) and significant increase in pMOS (Figure 3.66) gate-referred flicker-noise voltage density at increasing inversion or V_{EFF} is directly observed for the 0.18 μm CMOS process. As mentioned, flicker-noise increases are significant for pMOS devices, but not significant for nMOS devices for the 0.35 μm SOI CMOS process.

The flicker noise is significantly lower in Figure 6.7 for the large-area, preamplifier, 0.35 μm SOI CMOS devices compared to the smaller-area, OTA, 0.18 μm CMOS devices shown in Figures 3.65 and 3.66. The flicker-noise corner frequency is in the vicinity of 10 Hz for the 0.35 μm SOI CMOS devices, but exceeds the maximum measurement frequency of 10 kHz for the 0.18 μm CMOS devices. As discussed in Section 3.10.3.8, the flicker-noise corner frequency, f_c, is linked to the intrinsic device bandwidth, f_{Ti}. Both are much lower for the long-channel, preamplifier, 0.35 μm SOI CMOS devices shown in Figure 6.7 compared to the shorter-channel, OTA, 0.18 μm CMOS devices shown in Figures 3.65 and 3.66.

6.7 DIFFERENTIAL INPUT PREAMPLIFIER

This section and Section 6.8 that follows illustrate methods for minimizing thermal and flicker noise in analog CMOS circuits through the design of two micropower, 0.35 μm SOI CMOS preamplifiers. The first preamplifier, described in this section, is a differential input preamplifier that permits high input impedance, voltage gain using non-inverting operational amplifier configurations. Here, the input signal appears at the non-inverting input, while a fed-back signal appears at the inverting input through external resistive feedback connected to the output signal. Later, Section 6.8 describes a single-ended input preamplifier that permits transimpedance and inverting amplifier configurations. Both preamplifiers are designed to amplify low-frequency signals from gyros and other sensors used in space missions and require minimization of thermal and flicker noise at micropower operation [24–26]. An SOI CMOS process is used to permit operation over extreme temperature and radiation environments. Noise performance for the differential input preamplifier is described for temperatures from 77 to 400 K in [25] and for 1 Mrad (Si) proton radiation in [26].

6.7.1 Description

Figure 6.8 shows a schematic diagram of the differential input preamplifier. M_1 and M_2 are pMOS, input differential-pair devices. pMOS devices are used because, as described in Sections 6.6.1 and 6.6.2, these have lower flicker-noise factors compared to nMOS devices in the 0.35 μm SOI CMOS

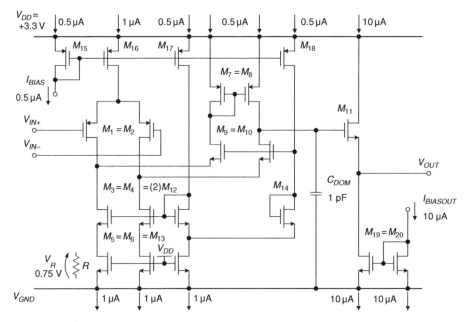

MOSFET *W/L* in μm/μm

M_1, M_2, $IC=0.95$	M_5, M_6, M_{13}	M_9, M_{10}, $IC=6.7$	M_{14}, $IC=83$	$M_{12}=(1/2)M_3$
W/L=100/20	*W/L*=1/300	*W/L*=5/20	*W/L*=1/50	
M_3, M_4, $IC=8.4$	M_7, M_8, $IC=38.5$	M_{11}, M_{19}, M_{20}, $IC=8.3$	M_{15}, M_{17}, M_{18}, $IC=9.5$	$M_{16}=(2)M_{15}$
W/L=30/75	*W/L*=10/80	*W/L*=8/2	*W/L*=5/10	

Figure 6.8 Schematic diagram of the differential input preamplifier. Figure 6.9 gives MOSFET drain current, inversion coefficient, and channel length selections with resulting device geometry and performance. Figure 6.11 gives predicted and measured performance for the preamplifier

process used. This is especially true for operation in moderate inversion before pMOS flicker noise increases significantly with increasing inversion or V_{EFF}. However, as mentioned in Section 3.10.3.5, even for operation at low levels of inversion, pMOS devices may not have lower flicker noise in smaller-geometry processes and thus be the preferred devices in these processes.

The M_1 and M_2 input pair connects to nMOS, M_9 and M_{10} folded-cascode devices, which then connect to a current-mirror load consisting of pMOS, M_7 and M_8 devices. Each of these devices operates at a drain current of 0.5 μA. The nMOS, M_{11} source follower device buffers the high-impedance, dominant-pole signal appearing at the drain of M_8 across the dominant-pole compensation capacitor C_{DOM}. This permits the output signal to drive external feedback resistors. The M_{11} source follower, which operates at a high drain current of 10 μA, may be omitted to save power when capacitive loads and feedback networks are driven.

nMOS, M_3 and M_4 current source devices, with resistive source degeneration provided by nMOS, M_5 and M_6 deep ohmic devices, provide bias currents for the M_1 and M_2 input pair, and M_9 and M_{10} folded-cascode devices. As discussed in Section 6.5.5.3, resistive source degeneration reduces the drain flicker-noise current of M_3 and M_4. Although not in the signal path, the M_3 and M_4 current sources can contribute significant thermal and flicker noise, especially since they operate at 1 μA, twice the drain current of the M_1 and M_2 input pair devices. Additionally, M_5 and M_6 that provide source degeneration resistances for M_3 and M_4 can contribute significant thermal noise. The noise contributions of M_3 and M_4, and M_5 and M_6 will be given considerable attention in later noise analysis. Finally, nMOS device M_{12} and deep-ohmic device M_{13} set the gate voltage for M_3 and M_4, and, correspondingly, the drain–source voltage for M_5 and M_6.

pMOS device M_{15} provides a reference for M_{16} that provides bias current for the M_1 and M_2 input pair. M_{15} also provides a reference for M_{17} and M_{18}. M_{17} provides bias current for the M_{12} and M_{13} reference devices, and M_{18} provides bias current for nMOS device M_{14} that provides level shifting to set the gate voltage for the M_9 and M_{10} folded-cascoded devices. Finally, nMOS, M_{19} and M_{20} devices comprise a current mirror to provide bias current for the M_{11} source follower.

The preamplifier supply voltage is $V_{DD} = 3.3\,\text{V}$. The core current consumption is $2\,\mu\text{A}$ for M_1–M_{10}, giving a core power consumption of $6.6\,\mu\text{W}$. This excludes the reference devices and the output source follower, which is required only to drive resistive loads.

6.7.2 Circuit Analysis, Performance Optimization, and Predicted Performance

This section describes circuit analysis, performance optimization, and predicted performance for the differential input preamplifier. The circuit analysis equations developed in this section are included later in Section 6.7.3 in the *Circuit Analysis* sheet of the *Analog CMOS Design, Tradeoffs and Optimization* spreadsheet. This sheet calls device geometry and performance predicted by the *MOSFETs* sheet of the spreadsheet for selected device drain currents, inversion coefficients, and channel lengths, and gives predicted preamplifier performance. Predicted performance is described in this section as part of discussions on performance optimization. Predicted performance is summarized and compared to measured performance later in Section 6.7.3.

Parasitic pole frequencies, input-referred offset voltage due to local-area mismatch, input and output voltage ranges, slew rate, the input, 1 dB compression voltage, and input and output capacitances are not developed for the preamplifier because, as described below, these are generally not critical for the application. These aspects of circuit performance were developed in Section 5.3 for OTA design examples.

A high preamplifier, closed-loop voltage gain of 51 V/V, set by external feedback resistors for the test application, helps ensure reduced loop gain and good phase margin in combination with a sufficiently large choice of C_{DOM}. As a result, parasitic pole frequencies are not developed. The input-referred offset voltage due to local-area mismatch is not developed because the application requires subsequent AC coupling, although the low-flicker-noise performance suggests the offset voltage will also be low. Additionally, input and output voltages are small, so slew rate, input and output voltage ranges, and the input, 1 dB compression voltage are not developed. Finally, the output capacitance appearing at the M_{11} source follower buffer is small for this small device and is not developed.

The preamplifier input capacitance, however, can be significant for the application. The open-loop, differential input capacitance can be approximated as one-half the sum of the gate–source, C_{GS1}, gate–body, C_{GB1}, and gate–drain, C_{GD1}, capacitances associated with the M_1 and M_2 input pair devices. This assumes little Miller multiplication of C_{GD1} resulting from the low gate-to-drain voltage gains of M_1 and M_2 into the sources of the M_3 and M_4 folded-cascode devices. The differential input capacitance is one-half the sum of gate capacitances of individual input devices because the common source connection of M_1 and M_2 is centered between differential input signals at a virtual ground. The closed-loop, input capacitance at the positive input is smaller than the differential input capacitance because the C_{GS1} component is nearly bootstrapped out by the fed-back signal appearing at the negative input. The C_{GB1} component is also nearly bootstrapped out if the local bodies of M_1 and M_2 are driven by their common source connection. This is generally only possible for SOI CMOS processes, like the one used for the preamplifier, or for pMOS devices that reside in bodies separate from the main p-type substrate in bulk CMOS processes.

6.7.2.1 Voltage gain

The preamplifier open-loop voltage gain is equal to the product of differential pair transconductance, $g_{m1} = g_{m2}$, and dominant-pole resistance, r_{DOM}, appearing at C_{DOM}, multiplied by the voltage gain of the M_{11} source follower. The differential-pair transconductance is directly equal to the transconductance, $g_{m1} = g_{m2}$, of the individual M_1 and M_2 input pair devices. The unloaded gate-to-source voltage gain of M_{11} is $A_{VGS11} = g_{m11}/(g_{m11} + g_{mb11} + g_{ds11})$, as found from the summary of small-signal analysis given in Table 3.16. g_{m11}, g_{mb11}, and g_{ds11} are the transconductance, body-effect transconductance, and drain–source conductance for M_{11}. As typical for open-loop voltage gain calculations, the decrease in source follower voltage gain from its ideal value of unity is neglected here. For a typical value of $\eta = g_{mb}/g_m = 0.3$ for the partially depleted (PD) SOI CMOS process used here or for common bulk CMOS processes, the open-circuit, source follower voltage gain is approximately 0.77 V/V when the body is not locally driven by the source. The gain becomes near unity when the body is driven by the source.

The dominant-pole resistance is equal to the drain output resistance of M_{10} in parallel with the drain–source resistance of mirror device M_8, r_{ds8}. Since M_{10} is a cascode device having high external source degeneration resistance, its drain output resistance is raised significantly above its drain–source resistance, r_{ds10}, and r_{ds8}. As a result, the dominant-pole resistance can be approximated by r_{ds8} alone, giving an open-loop voltage gain of

$$A_{VOL} = g_{m1} \cdot r_{DOM} \approx g_{m1} \cdot r_{ds8} = \left(\frac{g_m}{I_D}\right)_1 \cdot V_{A8}$$

$$\approx \frac{V_{AL8} \cdot L_8}{n_1 U_T \left(\sqrt{IC_1 + 0.25} + 0.5\right)}$$

$$\approx \frac{2 \cdot V_{AL8} \cdot L_8}{n_1 \left(V_{DS,sat1} - 2U_T\right)} \tag{6.27}$$

The first line in Equation 6.27 gives the voltage gain as the transconductance efficiency, $(g_m/I_D)_1$, for the M_1 and M_2 input pair devices, having $g_{m1} = g_{m2} = (g_m/I_D)_1 \cdot I_{D1}$, multiplied by the Early voltage, V_{A8}, for the M_8 mirror output device, having drain–source resistance of $r_{ds8} = V_{A8}/I_{D8}$. This expression assumes equal M_1, M_2, (M_7), and M_8 drain current as present for the preamplifier. The second line gives $(g_m/I_D)_1$ from Table 6.2 in terms of the inversion coefficient, IC_1, for M_1 and M_2, and also gives the Early voltage, $V_{A8} = V_{AL8} \cdot L_8$, for M_8, where V_{AL8} is the Early voltage factor and L_8 is the channel length. The prediction of drain–source resistance using the Early voltage factor was described in Table 3.21, where the Early voltage factor is a complex function of IC, L, and V_{DS} and is not fixed for the process. Finally, the third line gives $(g_m/I_D)_1$ from Table 6.2 in terms of the bias compliance voltage or drain–source saturation voltage, $V_{DS,sat1}$, for M_1 and M_2. In the second and third lines, n_1 is the substrate factor for M_1 and M_2.

Velocity saturation effects, present for short-channel devices operating at high levels of inversion, and smaller VFMR effects, present for devices of all channel lengths operating at high levels of inversion, decrease g_m/I_D and g_m. The inclusion of these effects is not considered for the preamplifier because it uses long-channel devices to give large gate areas and low flicker noise. However, as described in Table 3.17, velocity saturation and VFMR decreases in g_m/I_D and g_m can be included by replacing IC with $IC(1 + IC/IC_{CRIT})$ in the $\sqrt{IC + 0.25}$ term related to g_m/I_D and g_m, used here in Equation 6.27. As described in Section 3.8.2.2 and shown in Figures 3.24 and 3.25, IC_{CRIT} corresponds to a critical inversion coefficient where g_m/I_D is down to approximately 70.7 % of its value without velocity saturation and VFMR effects.

The voltage gain A_{VOL} is maximized by operating the M_1 and M_2 input pair devices in weak or moderate inversion for high $(g_m/I_D)_1$ and low $V_{DS,sat1}$, which also minimizes their gate-referred thermal-noise voltage. Operating these devices in weak or moderate inversion at long channel lengths also minimizes their flicker noise by maximizing their gate area. These devices are operated at $IC_1 = 0.95$ at the center of moderate inversion, giving $(g_m/I_D)_1 = 18.8\,\mu S/\mu A$ and $V_{DS,sat1} = 0.135\,V$. Additionally, A_{VOL} is maximized by operating the M_7 and M_8, output, current-mirror devices at long channel lengths for high V_{A8} and r_{ds8}, which also minimizes their flicker-noise contributions by maximizing their gate area. These devices are operated at $L_8 = 80\,\mu m$, giving an unusually high $V_{A8} = 800\,V$ for $V_{AL8} = 10\,V/\mu m$, with the value of V_{AL8} described later in Section 6.7.3.

The predicted A_{VOL} is nearly 15000 V/V from the product of $(g_m/I_D)_1 = 18.8\,\mu S/\mu A$ and $V_{A8} = 800\,V$, where n_1 is approximately 1.3. MOSFET operating parameters, predicted preamplifier performance, and experimental measurements are summarized later in Section 6.7.3 using the spreadsheet.

6.7.2.2 Frequency response

The preamplifier open-loop, unity-gain frequency is approximated by

$$f_T \approx \frac{g_{m1}}{2\pi\,(C_{DOM} + C_{DB8} + C_{GD8})} \tag{6.28}$$

This assumes that the M_8 drain–body capacitance, C_{DB8}, and gate–drain capacitance, C_{GD8}, sufficiently exceed capacitances for the smaller M_{10} folded-cascode device, where M_8 is made larger to manage its flicker noise. Additionally, this assumes that the M_8 capacitances sufficiently exceed the input capacitance of the small M_{11} source follower device.

f_T is controlled by the input pair transconductance, g_{m1}, and the dominant-pole capacitance, C_{DOM}, selected at 1000 fF, which dominates the M_8 capacitances. f_T is set at approximately 1.5 MHz to limit the closed-loop bandwidth to approximately 30 kHz (1.5 MHz/51) for a non-inverting, closed-loop voltage gain of 51 V/V.

6.7.2.3 Thermal noise

The input-referred thermal-noise voltage PSD for the preamplifier is given by

$$S_{VIN}\,(\text{thermal}) = 2\cdot 4kT\cdot\left[\frac{(n\,\Gamma)_1}{g_{m1}} + \frac{(n\,\Gamma)_3\,g_{m3}}{(1+g_{ms3}R)^2\,g_{m1}^2} + \left(\frac{1}{R}\right)\left(\frac{g_{ms3}R}{1+g_{ms3}R}\right)^2\left(\frac{1}{g_{m1}^2}\right) + \frac{(n\,\Gamma)_7\,g_{m7}}{g_{m1}^2}\right]$$

$$\tag{6.29}$$

g_{m1}, g_{m3}, and g_{m7} are the transconductances for the M_1 and M_2 input pair, M_3 and M_4 degenerated current source, and M_7 and M_8 current-mirror devices, respectively. $(n\Gamma)_1$, $(n\Gamma)_3$, and $(n\Gamma)_7$ are the product of the substrate factors, n, and thermal-noise factors, Γ (Equation 3.106), for these devices. g_{ms3} is the source transconductance for M_3 and M_4, where $g_{ms} \approx g_m + g_{mb} \approx n\cdot g_m$ from Equation 6.18. Finally, R is the value of degeneration resistances placed at the sources of M_3 and M_4. These degeneration resistances are developed by the M_5 and M_6 deep ohmic devices. The "2" appears in the noise expression because pair devices are used throughout the signal path, doubling the noise PSD associated with single devices.

The first term in Equation 6.29, from Table 6.3, corresponds to the gate-referred thermal-noise voltage PSD of the M_1 and M_2 input pair. The second and third terms, from Equation 6.17, correspond to the drain-referred thermal-noise current PSD of the M_3 and M_4, degenerated current source devices, including noise from their degeneration resistances, R, developed by M_5 and M_6. The fourth term, from Table 6.3, corresponds to the drain-referred thermal-noise current PSD of the M_7 and M_8 current-mirror devices. The second, third, and fourth terms are divided by the input pair transconductance squared, g_{m1}^2, which refers the drain noise current PSD contributions to a preamplifier input-referred noise voltage PSD.

The noise analysis given by Equation 6.29 includes only noise contributions from M_1–M_8, where the drain–source resistance of the M_5 and M_6 deep ohmic devices is considered by the degeneration resistance R. Noise of the M_9 and M_{10} folded-cascode devices is neglected because of significant resistive source degeneration. The significant reduction of effective MOS transconductance and drain-referred thermal- and flicker-noise current for a cascode device having significant resistive source degeneration was discussed in Section 5.3.5.2. The noise of the M_{11} source follower and the M_{19} and M_{20} current mirror that provides its bias current is neglected because the voltage gain from the preamplifier input to the M_{11} gate is high, being essentially A_{VOL}. The noise current from the M_{16}, input pair current source and its M_{15} reference creates a common-mode noise current in the input pair that is largely cancelled by the M_7 and M_8 current mirror. Finally, the noise voltage from the M_{12} and M_{13} bias voltage reference creates a common-mode noise current in the M_3 and M_4, degenerated current source devices, and the noise from the M_{13} and M_{14} bias voltage reference creates a very small common-mode noise current in the M_9 and M_{10} cascode devices. These common-mode noise currents are also cancelled by the M_7 and M_8 current mirror.

Equation 6.29 may be factored to show the thermal-noise PSD contributions of non-input devices relative to that of the input pair devices. This gives

$$S_{VIN} \text{ (thermal)} = 2 \cdot \frac{4kT\,(n\Gamma)_1}{g_{m1}} \cdot \left[1 + \left(\frac{(n\Gamma)_3}{(n\Gamma)_1}\right)\left(\frac{g_{m3}}{g_{m1}}\right)\left(\frac{1}{1+g_{ms3}R}\right)^2 \right.$$
$$\left. + \frac{1}{(n\Gamma)_1\, g_{m1}R}\left(\frac{g_{ms3}R}{1+g_{ms3}R}\right)^2 + \left(\frac{(n\Gamma)_7}{(n\Gamma)_1}\right)\frac{g_{m7}}{g_{m1}} \right] \qquad (6.30)$$

The term to the left of the square bracket is the input-referred thermal-noise voltage PSD of the input pair devices, and the terms inside the square brackets correspond to the relative noise PSD contributions of the input pair and non-input devices. The unity term corresponds to the noise PSD contribution of the input pair devices. The second, third, and fourth terms correspond to the additional noise PSD contributions from the non-input devices. In the *Circuit Analysis* sheet of the spreadsheet described later in Section 6.7.3, these terms will be referred to as relative noise powers (*RNPs*). RNP_3, RNP_5, and RNP_7 denote the thermal-noise PSD contributions of the M_3 and M_4, degenerated current source devices, the M_5 and M_6, deep ohmic degeneration resistances denoted again by R, and the M_7 and M_8 current-mirror devices, with all noise contributions relative to the input pair. If RNP_3, RNP_5, and RNP_7 are sufficiently below unity, the input devices dominate the thermal noise as desired for low-noise design.

Although Γ is often assumed fixed at $2/3$ and cancelled in thermal-noise analysis, Γ varies from $1/2$ in weak inversion to $2/3$ in strong inversion and may increase further due to small-geometry noise increases described in Section 3.10.2.2. As a result, Γ is included in the analysis because it can vary between input and output devices and affect the prediction of thermal noise.

6.7.2.4 Thermal noise expressed from DC bias conditions

Equation 6.30 helps the designer minimize thermal noise from input devices and ensure these devices dominate the overall preamplifier noise by ensuring small non-input device noise contributions. This equation is in terms of device transconductances and degeneration resistance values. Section 6.5, however, developed MOS noise in terms of the DC bias conditions, including the bias compliance voltage, V_{COMP}, which is the minimum voltage allowed across a MOS device or a device and its external source degeneration resistance. In Section 6.5.5, it was observed that the drain noise current of non-input devices is reduced at higher values of V_{COMP}, and that for a given value of V_{COMP}, flicker noise is reduced through the use of resistive source degeneration. It is useful then to develop preamplifier noise in terms of DC bias conditions and bias compliance voltages. This permits the minimization of noise under the constraint of limited bias compliance voltages resulting from low supply voltages in small-geometry processes.

Table 6.4 shows the relative thermal-noise PSD contributions of devices in the differential input preamplifier. Additionally, the table shows the relative noise contributions of devices in the single-ended input preamplifier described later in Section 6.8.

For thermal-noise PSD contributions, Table 6.4 shows traditional expressions from Equation 6.30, expressions in terms of the DC bias conditions and bias compliance voltages, and expressions in terms of the inversion coefficient. The last two expressions permit noise minimization through either the choice of DC bias conditions and bias compliance voltages or the inversion coefficient. Expressions in terms of the DC bias conditions and bias compliance voltages use I_D, $V_{DS,sat}$, and the DC bias voltage across the degeneration resistances, V_R. In these expressions, g_m, the degeneration resistance, R, the degeneration factor, $1/(1 + g_{ms3}R)$, and the current-splitting factor, $g_{ms3}R/(1 + g_{ms3}R)$, are expressed from Equations 6.8, 6.16, 6.20, and 6.21, respectively. In the expressions in terms of the inversion coefficient, g_m and g_{ms} are expressed using Equations 6.8 and 6.18. Finally, in the expression for overall preamplifier thermal noise noted below the table, the gate-referred thermal-noise voltage PSD of the input devices is expressed from Table 6.3.

All expressions assume negligible g_m/I_D and g_m decreases due to velocity saturation and VFMR effects. The inclusion of these effects is, again, not considered for the preamplifier because it uses long-channel devices to give large gate areas and low flicker noise. As mentioned in Section 6.7.2.1, these effects can be included by simple modifications to the g_m/I_D and g_m terms expressed in terms of IC.

The first row in Table 6.4 shows that the thermal-noise PSD contribution of the M_1 and M_2 input pair devices is 100 % relative to their own noise contribution, which would ideally be the only preamplifier noise contribution. The input-referred thermal-noise voltage PSD for the input pair, noted below the table, is minimized for low $V_{DS,sat}$ and low IC in weak or moderate inversion because of high g_m/I_D and g_m. Additionally, operation at high g_m helps ensure that the input pair devices dominate the overall preamplifier thermal and flicker noise. Increasing input pair drain current, which increases g_m, also lowers the thermal-noise voltage and helps ensure that input pair devices dominate the overall preamplifier noise.

As noted below Table 6.4, the input-referred thermal-noise voltage density associated with the input pair devices is 51.7 nV/Hz$^{1/2}$ for the selected $I_{D1} = 0.5\,\mu A$, $V_{DS,sat1} = 0.135\,V$, and $I_{C1} = 0.95$ (corresponding to near the center of moderate inversion). MOSFET operating parameters, predicted preamplifier performance including thermal-noise contributions, and experimental measurements are summarized later in Section 6.7.3 using the spreadsheet. Because the spreadsheet uses slightly more accurate predictions of MOS performance, as described in Section 6.7.3, spreadsheet predictions of performance will vary slightly from the design equations given. Additionally, the spreadsheet includes velocity saturation and VFMR effects not considered in the design equations.

The second row in Table 6.4 shows the thermal-noise PSD contribution of the M_3 and M_4, degenerated current source devices relative to the noise contribution of the input pair devices. The relative thermal-noise PSD contribution from M_3 and M_4 is only 1.1 % for the selected $I_{D3} = 1\,\mu A$, $V_{DS,sat3} = 0.232\,V$, $IC_3 = 8.4$ (near the onset of strong inversion), and $V_R = 0.75\,V$. This noise contribution is low even though these devices operate at twice the drain current of the input pair devices in the folded-cascode, preamplifier circuit topology. Substantial thermal and flicker-noise degeneration occurs for the selected value of $V_R = 0.75\,V$, where the degeneration factor, $1/(1 + g_{ms3}R)$, given by Equation 6.20 has a value of 0.1.

The native thermal- and flicker-noise PSD contributions of M_3 and M_4 are multiplied by the square of the degeneration factor or 0.01. Without source degeneration, their thermal-noise PSD contribution would have increased from 1.1 % to 110 %. Alternatively, these devices could have been

Table 6.4   Thermal-noise PSD contributions of devices compared to input devices for the differential (denoted DIFF) and single-ended (denoted SE) input preamplifiers. Traditional expressions, expressions in terms of DC bias conditions, and expressions in terms of the inversion coefficient enable design where input devices dominate the noise. The note at the bottom of the table summarizes total preamplifier, input-referred thermal-noise voltage PSD

Devices	Noise PSD	Expressions = traditional expression = DC bias expression = inversion coefficient expression
M_1 (M_2)	100 % BOTH	$=1$
M_3, M_4 (with R degen.)	1.1 % DIFF 2.7 % SE	$= \left(\dfrac{(n\Gamma)_3}{(n\Gamma)_1} \right) \left(\dfrac{g_{m3}}{g_{m1}} \right) \left(\dfrac{1}{1+g_{ms3}R} \right)^2$ $= \left(\dfrac{\Gamma_3}{\Gamma_1} \cdot \dfrac{I_{D3}}{I_{D1}} \cdot \dfrac{V_{DS,sat1}-2U_T}{V_{DS,sat3}-2U_T} \right) \left(\dfrac{V_{DS,sat3}-2U_T}{2V_R+V_{DS,sat3}-2U_T} \right)^2$ $= \left(\dfrac{\Gamma_3}{\Gamma_1} \cdot \dfrac{I_{D3}}{I_{D1}} \cdot \dfrac{\sqrt{IC_1+0.25}+0.5}{\sqrt{IC_3+0.25}+0.5} \right) \left[\dfrac{U_T\left(\sqrt{IC_3+0.25}+0.5\right)}{V_R+U_T\left(\sqrt{IC_3+0.25}+0.5\right)} \right]^2$ For the SE preamplifier, multiply expressions by "2"
M_5, M_6 (acting as R)	15.0 % DIFF 23.6 % SE	$= \dfrac{1}{(n\Gamma)_1\,g_{m1}R} \cdot \left(\dfrac{g_{ms3}R}{1+g_{ms3}R} \right)^2$ $= \left(\dfrac{1}{\Gamma_1} \cdot \dfrac{I_{D5}}{I_{D1}} \cdot \dfrac{V_{DS,sat1}-2U_T}{2V_R} \right) \left(\dfrac{2V_R}{2V_R+V_{DS,sat3}-2U_T} \right)^2$ $= \left[\dfrac{1}{\Gamma_1} \cdot \dfrac{U_T}{V_R} \cdot \dfrac{I_{D5}}{I_{D1}} \cdot \left(\sqrt{IC_1+0.25}+0.5\right) \right] \left[\dfrac{V_R}{V_R+U_T\left(\sqrt{IC_3+0.25}+0.5\right)} \right]^2$ For the SE preamplifier, multiply expressions by "2"
M_7, M_8 (mirror pair)	26.1 % DIFF NA SE	$= \dfrac{(n\Gamma)_7}{(n\Gamma)_1} \cdot \dfrac{g_{m7}}{g_{m1}}$ $= \dfrac{\Gamma_7}{\Gamma_1} \cdot \dfrac{I_{D7}}{I_{D1}} \cdot \dfrac{V_{DS,sat1}-2U_T}{V_{DS,sat7}-2U_T}$ $= \dfrac{\Gamma_7}{\Gamma_1} \cdot \dfrac{I_{D7}}{I_{D1}} \cdot \dfrac{\sqrt{IC_1+0.25}+0.5}{\sqrt{IC_7+0.25}+0.5}$ For the SE preamplifier, eliminate these expressions
Total	142.2 % DIFF 126.3 % SE	Sum of above expressions

Note: Preamplifier input-referred thermal-noise voltage PSD can be found from the expression below giving input device noise multiplied by the total noise PSD contributions given above:

$$S_{VIN}\,(\text{thermal}) = 2 \cdot 4kT \cdot (n\Gamma)_1$$

$$\cdot \left[\frac{1}{g_{m1}} = \frac{n_1\left(V_{DS,sat1}-2U_T\right)}{2I_{D1}} = \frac{n_1 U_T\left(\sqrt{IC_1+0.25}+0.5\right)}{I_{D1}} \right] \cdot [\text{Total noise PSD contrib.}]$$

For the SE preamplifier, remove the "2" before the $4kT$ term.
$S_{VIN}\,(\text{thermal}) = (51.7\,\text{nV})^2/\text{Hz} \cdot 1.422 = (61.6\,\text{nV})^2/\text{Hz},\ \text{DIFF}$
$S_{VIN}\,(\text{thermal}) = (29.1\,\text{nV})^2/\text{Hz} \cdot 1.263 = (32.7\,\text{nV})^2/\text{Hz},\ \text{SE}$

operated without degeneration deep into strong inversion at high $V_{DS,sat3}$ to reduce their thermal-noise contribution.

The thermal-noise PSD contribution of M_3 and M_4 is primarily minimized by maximizing the value of V_R, the voltage across the source degeneration resistance, which maximizes the value of this resistance, R. However, this comes at the expense of higher bias compliance voltage for M_3 and M_4, which is $V_{COMP} = V_{DS,sat3} + V_R$, where V_R adds directly to the value of $V_{DS,sat3}$. For a given value of V_{COMP}, as described in Section 6.5.5.2, resistive source degeneration does not significantly lower the overall drain thermal-noise current at the output of M_3 and M_4 because of the additional noise of the degeneration resistances, considered next in the noise analysis. However, for a given value of V_{COMP}, as described in Section 6.5.5.3, resistive source degeneration significantly reduces drain flicker-noise current. Lowering the drain currents for M_3 and M_4, which increases the value of R, decreases both their thermal- and flicker-noise contributions.

The third row in Table 6.4 shows the thermal-noise PSD contribution of the M_5 and M_6, deep ohmic degeneration resistances relative to the noise contribution of the input pair devices. The relative thermal-noise PSD contribution from the M_5 and M_6 degeneration resistances is 15 % for the selected $I_{D5} = 1\,\mu A$ and $V_R = 0.75\,V$, which gives $R = 0.75\,V/1\,\mu A = 750\,k\Omega$. This noise contribution would be lower for lower drain currents, but is modest even though the M_5 and M_6 drain currents, like the M_3 and M_4 drain currents, are twice those of the input pair devices, which necessitates lower value degeneration resistances. As for M_3 and M_4, the noise contribution of the M_5 and M_6, deep ohmic degeneration resistances is primarily minimized by maximizing V_R, which maximizes the value of degeneration resistance at the expense of higher bias compliance voltage. The current-division factor, $g_{ms3}R/(1 + g_{ms3}R)$, which describes the portion of degeneration resistance noise current that enters the sources of M_3 and M_4, is equal to unity less the noise degeneration factor, $1/(1 + g_{ms3}R)$. The current-division factor has a value of 0.9 for the degeneration factor of 0.1 mentioned above. When significant degeneration is used, the current-division factor approaches unity.

The fourth row in Table 6.4 shows the thermal-noise PSD contribution of the M_7 and M_8 current-mirror devices relative to the noise contribution of the input pair devices. The relative thermal-noise PSD contribution from M_7 and M_8 is 26.1 % for the selected $V_{DS,sat7} = 0.41\,V$, $IC_7 = 38.5$ (deep into strong inversion), and drain current ($I_{D7} = I_{D1} = 0.5\,\mu A$) equal to that of the input pair devices. This noise contribution is minimized by minimizing the g_{m7}/g_{m1} ratio, done by minimizing the drain current and maximizing $V_{DS,sat7}$ and IC_7 for M_7 and M_8 in strong inversion. This lowers their g_m relative to the input pair devices that operate in moderate inversion. Interestingly, the DC bias expression shows that the g_{m7}/g_{m1} ratio is equal to the drain current ratio, I_{D7}/I_{D1}, multiplied by nearly the ratio of $V_{DS,sat1}/V_{DS,sat7}$. Thus, the M_7 and M_8 noise contribution decreases nearly inversely with $V_{DS,sat7}$. Just as increasing V_R minimizes the noise contributions of M_3 and M_4 and the degeneration resistances provided by M_5 and M_6, increasing $V_{DS,sat7}$ minimizes the M_7 and M_8 noise contribution. Increasing V_R and $V_{DS,sat7}$ both come at the expense of increasing the bias compliance voltage, which poses special challenges in low-voltage design. $V_{DS,sat7}$ is the bias compliance voltage for M_7 and M_8 directly since these devices do not use resistive source degeneration.

Finally, Table 6.4 shows the sum of all thermal-noise PSD contributions relative to the noise contribution of the input pair devices. This sum is 142.2 %, where non-input devices contribute 42.2 % additional noise PSD compared to noise from the input pair devices. This corresponds to a modest 19.2 % increase in preamplifier input-referred thermal-noise voltage density compared to that provided by the input pair devices alone. As noted below the table, the total predicted preamplifier, input-referred thermal-noise voltage density is $61.6\,nV/Hz^{1/2}$ for the input pair device noise of $51.7\,nV/Hz^{1/2}$. The input pair devices have low gate-referred thermal-noise voltage for micropower drain currents of $0.5\,\mu A$, resulting from high g_m/I_D and g_m in moderate inversion. Additionally, their high g_m assists them in dominating overall preamplifier noise, although careful management of non-input device noise is also required. As described later in Section 6.7.4, the very small noise contributions from M_3 and M_4 indicate over-design, where less source degeneration could have been used for these devices.

6.7.2.5 Flicker noise

The input-referred flicker-noise voltage PSD for the preamplifier is given by

$$S_{VIN} \text{(flicker)} = 2 \cdot$$

$$\left[\frac{K'_{F1}}{(WL)_1 f^{AF1}} + \frac{K'_{F3}}{(WL)_3 f^{AF3}} \left(\frac{g_{m3}}{1 + g_{ms3}R} \right)^2 \left(\frac{1}{g_{m1}^2} \right) \right.$$

$$\left. + \frac{K'_{F5}}{(WL)_5 f^{AF5}} \cdot g_{m5}^2 \text{(ohmic)} \cdot \left(\frac{g_{ms3}R}{1 + g_{ms3}R} \right)^2 \left(\frac{1}{g_{m1}^2} \right) + \frac{K'_{F7}}{(WL)_7 f^{AF7}} \cdot g_{m7}^2 \left(\frac{1}{g_{m1}^2} \right) \right]$$

$$(6.31)$$

K'_{F1}, K'_{F3}, K'_{F5}, and K'_{F7} are the flicker-noise factors in units of $(nV)^2 \cdot \mu m^2$ for the M_1 and M_2 input pair devices, the M_3 and M_4, degenerated current source devices, the M_5 and M_6 deep ohmic devices providing degeneration resistances, and the M_7 and M_8 current-mirror devices, respectively. $AF1$, $AF3$, $AF5$, and $AF7$ are the flicker-noise slopes, and $(WL)_1$, $(WL)_3$, $(WL)_5$, and $(WL)_7$ are the gate areas in μm^2 for these devices. g_{m5} (ohmic) is the transconductance for the M_5 and M_6 deep ohmic devices. As in Equation 6.29 in the previous thermal-noise analysis, the "2" appears because pair devices are used throughout the signal path, doubling the noise PSD associated with single devices.

The first term in Equation 6.31, from Table 6.3, corresponds to the gate-referred flicker-noise voltage PSD of the M_1 and M_2 input pair. The second term, from Equation 6.23, corresponds to the drain-referred flicker-noise current PSD of the M_3 and M_4, degenerated current source devices. The third term, from Equation 6.25, corresponds to the drain-referred flicker-noise current PSD of the M_5 and M_6 deep ohmic devices acting as source degeneration resistors for M_3 and M_4. This term includes current division occurring at the sources of M_3 and M_4. In the second and third terms, the needed gate-referred flicker-noise voltage PSD is found from Table 6.3. Finally, the fourth term, from Table 6.3, corresponds to the drain-referred flicker-noise current PSD of the M_7 and M_8 current-mirror devices. As in the previous thermal-noise analysis, the second, third, and fourth terms are divided by g_{m1}^2 to refer the drain noise current PSD contributions to a preamplifier input-referred noise voltage PSD. Also as in the previous thermal-noise analysis, only noise contributions from M_1–M_8 are considered as noise contributions from the other devices are negligible.

Analogous to the previous thermal-noise analysis, Equation 6.31 can be factored to show the flicker-noise PSD contributions of non-input devices relative to that of the input pair devices. This gives

$$S_{VIN} \text{(flicker)} = 2 \cdot \frac{K'_{F1}}{(WL)_1 f^{AF1}} \cdot \left[1 + f^{(AF1-AF3)} \left(\frac{K'_{F3}}{K'_{F1}} \right) \left(\frac{(WL)_1}{(WL)_3} \right) \left(\frac{g_{m3}}{g_{m1}} \right)^2 \left(\frac{1}{1 + g_{ms3}R} \right)^2 \right.$$

$$+ f^{(AF1-AF5)} \left(\frac{K'_{F5}}{K'_{F1}} \right) \left(\frac{(WL)_1}{(WL)_5} \right) \left(\frac{g_{m5}\text{(ohmic)}}{g_{m1}} \right)^2 \left(\frac{g_{ms3}R}{1 + g_{ms3}R} \right)^2$$

$$\left. + f^{(AF1-AF7)} \left(\frac{K'_{F7}}{K'_{F1}} \right) \left(\frac{(WL)_1}{(WL)_7} \right) \left(\frac{g_{m7}}{g_{m1}} \right)^2 \right]$$

$$(6.32)$$

The term to the left of the square bracket is the input-referred flicker-noise voltage PSD of the input pair devices, and the terms inside the square brackets correspond to the relative noise PSD contributions of the input pair and non-input devices. As in the previous thermal-noise analysis, the unity term corresponds to the noise PSD contribution of the input pair devices. The second, third, and fourth terms correspond to the additional noise PSD contributions from the non-input devices. As in the previous thermal-noise analysis, these terms will be referred to as relative noise powers, where RNP_3, RNP_5, and RNP_7 denote the flicker-noise PSD contributions of the M_3 and M_4, degenerated current source devices, the M_5 and M_6 deep ohmic devices, and the M_7 and M_8 current-mirror devices, with all noise contributions relative to the input pair. If RNP_3, RNP_5, and RNP_7 are sufficiently below unity, the input devices dominate the flicker noise as desired for low-noise design. The $RNPs$ for the flicker noise are included in the *Circuit Analysis* sheet of the spreadsheet described later in Section 6.7.3.

The frequency correction terms that appear in front of the non-input device noise contributions are required for frequencies away from $f = 1$ Hz when input and non-input device flicker-noise slopes are unequal and non-input device noise contributions are significant. These terms are not important for the preamplifier described here, where flicker-noise slopes are nearly equal and input devices dominate the noise. However, the correction terms, for example, can be very important when non-input, nMOS devices have shallower flicker-noise slope and contribute significant noise at frequencies above 1 Hz compared to input, pMOS devices. This situation was described in Section 5.5.4 for cascoded, OTA design examples.

6.7.2.6 Flicker noise expressed from DC bias conditions

Table 6.5, analogous to Table 6.4 given for the thermal noise, shows the relative flicker-noise PSD contributions of devices in the differential input preamplifier. Additionally, the table shows the relative noise contributions of devices in the single-ended input preamplifier described later in Section 6.8.

For flicker-noise PSD contributions, Table 6.5 shows traditional expressions from Equation 6.32, expressions in terms of the DC bias conditions and bias compliance voltages, and expressions in terms of the inversion coefficient. The last two expressions permit noise minimization through either the choice of DC bias conditions and bias compliance voltages or the inversion coefficient. Expressions in terms of the DC bias conditions and bias compliance voltages use I_D, $V_{DS,sat}$, and the DC bias voltage across the degeneration resistances, V_R, and are found using the substitutions described in Section 6.7.2.4 for the thermal noise. Additionally, the deep ohmic transconductance of M_5 and M_6 is expressed by $g_{m5}(\text{ohmic}) = I_{D5}/V_{EFF5}$ from Table 6.2. Expressions in terms of the inversion coefficient are also found using the substitutions described for the thermal noise. Additionally, gate area is expressed from Table 3.10 as $WL = (L^2/IC) \cdot (I_D/I_0)$, where $I_0 = 2n_0\mu_0 C'_{OX} U_T^2$ is the technology current given in the process information of Table 3.2. Finally, in the expression for overall preamplifier flicker noise noted below the table, the gate-referred flicker-noise voltage PSD of the input devices is expressed from Table 6.3.

As for the thermal noise, the expressions assume negligible g_m/I_D and g_m decreases due to velocity saturation and VFMR effects. The inclusion of these effects is, again, not considered for the preamplifier because it uses long-channel devices to give sufficient gate areas and minimize flicker noise. As mentioned in Section 6.7.2.1, these effects can be included by simple modifications to the g_m/I_D and g_m terms expressed in terms of IC. Finally, the ratio of substrate factor values, n, is included in the transconductance ratio terms for completeness. Since n is a slight function of the inversion level (see Figure 3.30) and process doping, the n values nearly cancel for most applications.

The first row in Table 6.5 shows that the flicker-noise PSD contribution of the M_1 and M_2 input pair devices is 100 % relative to their own noise contribution, which would ideally be the only preamplifier noise contribution. The input-referred flicker-noise voltage PSD for the input pair, noted below the table, is minimized at low $V_{DS,sat}$ and low IC in weak or moderate inversion because channel width and gate area are large here. Operation at low IC also minimizes the thermal-noise voltage, noted below Table 6.4, because of high g_m/I_D and g_m, and helps ensure that the input pair devices dominate both thermal and flicker noise. Although operation at low IC decreases flicker noise, increasing channel length decreases the noise more because gate area increases as the square of channel length. This is because channel width must increase equally with channel length to maintain the selected inversion coefficient. Finally, increasing input pair drain current, which requires proportionally larger channel width and gate area, decreases the flicker-noise voltage and increases g_m. The increase in g_m again lowers the thermal-noise voltage and helps ensure that input pair devices dominate overall preamplifier noise.

Table 6.5 Flicker-noise PSD contributions of devices compared to input devices for the differential (denoted DIFF) and single-ended (denoted SE) input preamplifiers. Traditional expressions, expressions in terms of DC bias conditions, and expressions in terms of the inversion coefficient enable design where input devices dominate the noise. The note at the bottom of the table summarizes total preamplifier, input-referred flicker-noise voltage PSD

Devices	Noise PSD (at 1 Hz)	Expressions = traditional expression = DC bias expression = inversion coefficient expression
M_1 (M_2)	100 % BOTH	$= 1$

M_3, M_4 (with R degen.) — 3.1 % DIFF, 4.0 % SE

$$= \left(\frac{K'_{F3}}{K'_{F1}}\right)\left(\frac{(WL)_1}{(WL)_3}\right)\left(\frac{g_{m3}}{g_{m1}}\right)^2\left(\frac{1}{1+g_{ms3}R}\right)^2$$

$$= \left(\frac{K'_{F3}}{K'_{F1}}\right)\left(\frac{(WL)_1}{(WL)_3}\right)\left[\frac{I_{D3}}{I_{D1}}\cdot\frac{n_1\left(V_{DS,sat1}-2U_T\right)}{n_3\left(V_{DS,sat3}-2U_T\right)}\right]^2\left(\frac{V_{DS,sat3}-2U_T}{2V_R+V_{DS,sat3}-2U_T}\right)^2$$

$$= \left(\frac{K'_{F3}}{K'_{F1}}\right)\left(\frac{L_1}{L_3}\right)^2\left(\frac{IC_3}{IC_1}\right)\left(\frac{I_{0,3}}{I_{0,1}}\right)\left(\frac{I_{D3}}{I_{D1}}\right)\left[\frac{n_1\left(\sqrt{IC_1+0.25}+0.5\right)}{n_3\left(\sqrt{IC_3+0.25}+0.5\right)}\right]^2$$

$$\cdot\left[\frac{U_T\left(\sqrt{IC_3+0.25}+0.5\right)}{V_R+U_T\left(\sqrt{IC_3+0.25}+0.5\right)}\right]^2$$

For $f \neq 1$ Hz, multiply each term by $f^{AF1-AF3}$
For the SE preamplifier, multiply expressions by "2"

M_5, M_6 (acting as R) — 3.3 % DIFF, 2.9 % SE

$$= \left(\frac{K'_{F5}}{K'_{F1}}\right)\left(\frac{(WL)_1}{(WL)_5}\right)\left(\frac{g_{m5}(\text{ohmic})}{g_{m1}}\right)^2\left(\frac{g_{ms3}R}{1+g_{ms3}R}\right)^2$$

$$= \left(\frac{K'_{F5}}{K'_{F1}}\right)\left(\frac{(WL)_1}{(WL)_5}\right)\left[\frac{I_{D5}}{I_{D1}}\cdot\frac{n_1\left(V_{DS,sat1}-2U_T\right)}{2V_{EFF5}}\right]^2\left(\frac{2V_R}{2V_R+V_{DS,sat3}-2U_T}\right)^2$$

$$= \left(\frac{K'_{F5}}{K'_{F1}}\right)\left(\frac{1}{IC_1}\right)\left(\frac{L_1^2}{(WL)_5}\right)\left(\frac{n_1U_T}{V_{EFF5}}\right)^2\left(\frac{I_{D5}^2}{I_{D1}I_{0,1}}\right)\left(\sqrt{IC_1+0.25}+0.5\right)^2$$

$$\cdot\left[\frac{V_R}{V_R+U_T\left(\sqrt{IC_3+0.25}+0.5\right)}\right]^2$$

For $f \neq 1$ Hz, multiply each term by $f^{AF1-AF5}$
For the SE preamplifier, multiply expressions by "2"

M_7, M_8 (mirror pair) — 40.3 % DIFF, NA SE

$$= \left(\frac{K'_{F7}}{K'_{F1}}\right)\left(\frac{(WL)_1}{(WL)_7}\right)\left(\frac{g_{m7}}{g_{m1}}\right)^2$$

$$= \left(\frac{K'_{F7}}{K'_{F1}}\right)\left(\frac{(WL)_1}{(WL)_7}\right)\left[\frac{I_{D7}}{I_{D1}}\cdot\frac{n_1\left(V_{DS,sat1}-2U_T\right)}{n_7\left(V_{DS,sat7}-2U_T\right)}\right]^2$$

$$= \left(\frac{K'_{F7}}{K'_{F1}}\right)\left(\frac{L_1}{L_7}\right)^2\left(\frac{IC_7}{IC_1}\right)\left(\frac{I_{0,7}}{I_{0,1}}\right)\left(\frac{I_{D7}}{I_{D1}}\right)\left[\frac{n_1\left(\sqrt{IC_1+0.25}+0.5\right)}{n_7\left(\sqrt{IC_7+0.25}+0.5\right)}\right]^2$$

For $f \neq 1$ Hz, multiply each term by $f^{AF1-AF7}$
For the SE preamplifier, eliminate these expressions

Total	146.8 % DIFF 106.9 % SE	Sum of above expressions

Note: Preamplifier input-referred flicker-noise voltage PSD can be found from the expression below giving input device noise multiplied by the total noise PSD contributions given above:

$$S_{VIN}(\text{flicker}) = 2\cdot\left[\frac{K'_{F1}}{(WL)_1 f^{AF1}} = \left(\frac{K'_{F1}}{f^{AF1}}\right)\left(\frac{IC_1}{L_1^2}\right)\left(\frac{I_{0,1}}{I_{D1}}\right)\right]\cdot[\text{Total noise PSD contrib.}]$$

For the SE preamplifier, remove the "2."
S_{VIN} (flicker) at 1 Hz $= (192.8\,\text{nV})^2/\text{Hz}\cdot 1.468 = (233.5\,\text{nV})^2/\text{Hz}$, DIFF
S_{VIN} (flicker) at 1 Hz $= (145.7\,\text{nV})^2/\text{Hz}\cdot 1.069 = (150.7\,\text{nV})^2/\text{Hz}$, SE

As noted below Table 6.5, the input-referred flicker-noise voltage density associated with the input pair devices is 192.8 nV/Hz$^{1/2}$ at $f = 1$ Hz for the selected $I_{D1} = 0.5\,\mu A$, $V_{DS,sat1} = 0.135$ V, $IC_1 = 0.95$ (corresponding to near the center of moderate inversion), and $L = 20\,\mu m$. This results in a required $W = 100\,\mu m$ found by $W = (L/IC)\cdot(I_D/I_0)$ from Table 3.10. As mentioned for the thermal noise, MOSFET operating parameters, predicted preamplifier performance including flicker-noise contributions, and experimental measurements are summarized later in Section 6.7.3 using the spreadsheet. Again, spreadsheet predictions of performance will vary slightly from the design equations given because the spreadsheet uses slightly more accurate predictions and includes velocity saturations and VFMR effects.

The second row in Table 6.5 shows the flicker-noise PSD contribution of the M_3 and M_4, degenerated current source devices relative to the noise contribution of the input pair devices. The relative flicker-noise PSD contribution from M_3 and M_4 is only 3.1 % for the selected $I_{D3} = 1\,\mu A$, $V_{DS,sat3} = 0.232$ V, $IC_3 = 8.4$ (near the onset of strong inversion), $L = 75\,\mu m$ (resulting in a required $W = 30\,\mu m$), and $V_R = 0.75$ V. This noise contribution is low even though these nMOS devices have higher flicker-noise factors and operate at twice the drain current compared to the pMOS input pair devices. As for the thermal noise, the flicker-noise PSD contribution of M_3 and M_4 is primarily minimized by maximizing the value of V_R, the voltage across the source degeneration resistance, which maximizes the value of this resistance, R. Again, this comes at the expense of higher bias compliance voltage for M_3 and M_4, which is $V_{COMP} = V_{DS,sat3} + V_R$, where V_R adds directly to the value of $V_{DS,sat3}$. Lowering the drain currents for M_3 and M_4, which increases the value of R, also decreases their flicker-noise contributions.

As mentioned for the thermal noise, substantial thermal- and flicker-noise degeneration occurs for the selected value of $V_R = 0.75$ V, where the degeneration factor, $1/(1 + g_{ms3}R)$, given by Equation 6.20 has a value of 0.1. When the degeneration factor of 0.1 is squared, this indicates that the native flicker-noise PSD contribution of M_3 and M_4 is multiplied by 0.01, or reduced by a factor of 100. Thus, without source degeneration, the flicker-noise PSD contribution of M_3 and M_4 would have been 310 % and these devices would have dominated the preamplifier flicker noise. These devices could have been operated (with sufficient gate area) deep into strong inversion without source degeneration to lower their flicker noise, but, as described in Section 6.5.5.3, degeneration reduces the flicker noise for a given value of V_{COMP}. Although degeneration reduces the flicker-noise contribution of M_3 and M_4, their gate area must also be sufficiently large. Minimizing the $[K'_{F3}/K'_{F1}]\cdot[(WL)_1/(WL)_3]$ terms in Table 6.5 also helps minimize the flicker-noise contribution. In the micropower preamplifier described here, unusually long channel lengths, which result in long channel widths as well, are used to ensure sufficient gate area and low flicker noise.

The third row in Table 6.5 shows the flicker-noise PSD contribution of the M_5 and M_6, deep ohmic degeneration devices relative to the noise contribution of the input pair devices. These devices operate in the deep ohmic region providing degeneration resistance, but introduce flicker noise as described earlier in Section 6.5.5.3. The relative flicker-noise PSD contribution from the M_5 and M_6 deep ohmic devices is only 3.3 % for the selected $I_{D5} = 1\,\mu A$, $V_{EFF5} = 2.65$ V, $W = 300\,\mu m$, and $L = 1\,\mu m$, required for the value of $R = 0.75$ V/1$\mu A = 750$ kΩ. This noise contribution would be lower for lower drain current, but is low even though these nMOS devices have higher flicker-noise factors and operate at twice the drain current compared to the pMOS input pair devices. The noise contribution is primarily minimized by minimizing the g_{m5}(ohmic)/g_{m1} ratio, which is done by maximizing V_{EFF5} relative to $V_{DS,sat1}$. Small or collapsed g_{m5} (ohmic) results from high $V_{EFF5} = V_{GS} - V_T = 3.3$ V $- 0.65$ V $= 2.65$ V, which is equal to the supply voltage less the device threshold voltage. In addition to minimizing the g_{m5}(ohmic)/g_{m1} ratio, M_5 and M_6 must also have sufficient gate area to minimize their flicker-noise contribution. Minimizing the $[K'_{F5}/K'_{F1}]\cdot[(WL)_1/(WL)_5]$ terms in Table 6.5 also helps minimize the flicker-noise contribution. As for the drain thermal-noise current associated with the degeneration resistances provided by M_5 and M_6, the $g_{ms3}R/(1 + g_{ms3}R)$ current-division factor describes the portion of drain flicker-noise current that enters the sources of M_3 and M_4. As mentioned, the current-division factor has a value of 0.9 for the degeneration factor, $1/(1 + g_{ms3}R)$, equal to 0.1.

The fourth row in Table 6.5 shows the flicker-noise PSD contribution of the M_7 and M_8 current-mirror devices relative to the noise contribution of the input pair devices. The relative flicker-noise PSD contribution from M_7 and M_8 is 40.3 % for the selected $V_{DS,sat7} = 0.41$ V, $IC_7 = 38.5$ (deep into strong inversion), $L = 80\,\mu m$ (resulting in a required $W = 10\,\mu m$), and drain current ($I_{D7} = I_{D1} = 0.5\,\mu A$) equal to that of the input pair devices. This noise contribution is minimized by minimizing the square of the g_{m7}/g_{m1} ratio, done by minimizing the drain current and maximizing $V_{DS,sat7}$ and IC_7 for M_7 and M_8 in strong inversion. This lowers their g_m relative to the input pair devices that operate in moderate inversion. The DC bias expression in Table 6.5 shows that the g_{m7}/g_{m1} ratio is equal to the drain current ratio, I_{D7}/I_{D1}, multiplied by nearly the ratio of $V_{DS,sat1}/V_{DS,sat7}$. Thus, the M_7 and M_8 noise contribution decreases nearly inversely with the square of $V_{DS,sat7}$, which, as mentioned for the thermal noise, comes at the expense of increased bias compliance voltage. In addition to minimizing the g_{m7}/g_{m1} ratio, M_7 and M_8 must also have sufficient gate area to minimize the flicker-noise contribution. Minimizing the $[K'_{F7}/K'_{F1}] \cdot [(WL)_1/(WL)_7]$ terms in Table 6.5 also helps minimize the flicker-noise contribution.

The M_7 and M_8 flicker-noise PSD contribution of 40.3 % is larger than expected and significantly exceeds the noise contributions of 3.1 % and 3.3 % for the M_3 and M_4, and M_5 and M_6 devices, respectively. The increased noise contribution results from the increase in pMOS $K'_{F7} = (10250)^2\,(nV)^2 \cdot \mu m^2$ for M_7 and M_8 operating in deep strong inversion at $IC = 38.5$, compared to the pMOS $K'_{F1} = (6077)^2\,(nV)^2 \cdot \mu m^2$ for the input pair devices operating in moderate inversion at $IC = 0.95$. These flicker-noise factors are summarized later in Section 6.7.3 in the *MOSFETs* sheet of the spreadsheet. Had their pMOS flicker-noise factor not increased by a factor of 2.84 above that of the input devices, the M_7 and M_8 flicker-noise contribution would have been divided by this factor and equal to 14.2 %. This illustrates the importance of measuring gate-referred flicker-noise voltage and extracting the low-inversion, flicker-noise factor, K'_{F0}, and voltage, V_{KF}, describing the flicker-noise increase with inversion or V_{EFF}, especially for pMOS devices operating at high levels of inversion. Figure 6.7 illustrated this measurement and extraction for devices used in the preamplifier. As mentioned, in the 0.35 μm SOI CMOS process used for the preamplifier, the pMOS flicker-noise advantage is almost gone for pMOS devices operating in deep strong inversion. Here, the pMOS K'_F begins to approach the nMOS $K'_F = (12000)^2\,(nV)^2 \cdot \mu m^2$, which has only a small inversion-level or V_{EFF} increase in the process.

Finally, Table 6.5 shows the sum of all flicker-noise PSD contributions relative to the noise contribution of the input pair devices. This sum is 146.8 %, where non-input devices contribute 46.8 % additional noise PSD compared to noise from the input pair devices. This corresponds to a modest 21 % increase in preamplifier input-referred flicker-noise voltage density compared to that provided by the input pair devices alone. As noted below the table, the total predicted preamplifier, input-referred flicker-noise voltage density is 233.5 nV/Hz$^{1/2}$ at $f = 1$ Hz for the input pair device noise of 192.8 nV/Hz$^{1/2}$. The input pair devices have low gate-referred flicker-noise voltage resulting from large gate area, primarily resulting from the selection of long channel length. Additionally, their high g_m in moderate inversion assists them in dominating both thermal and flicker noise, although careful management of non-input device noise is also required. As described later in Section 6.7.4, the very small noise contributions from M_3 and M_4 indicate over-design, where less source degeneration and gate area could have been used for these devices.

6.7.3 Summary of Predicted and Measured Performance

Section 6.7.2 presented circuit analysis, performance optimization, and predicted performance for voltage gain, bandwidth, thermal noise, and flicker noise for the differential input preamplifier. This section summarizes predicted and measured performance for the preamplifier using the *Analog CMOS Design, Tradeoffs and Optimization* spreadsheet.

The spreadsheet maps predicted device performance for selections of the inversion coefficient, channel length, and drain current into complete circuit performance. Section 4.4 described and illustrated the spreadsheet for the optimization of differential-pair and current-mirror devices. Additionally, Sections 5.4.2 and 5.5.2 illustrated the spreadsheet for the optimization of simple and cascoded OTAs. The spreadsheet is described further in the Appendix and is available at the web site for this book listed on the cover.

For the differential input preamplifier described here and the single-ended input preamplifier described later in Section 6.8, the spreadsheet uses the 0.35 μm SOI CMOS process parameters from Tables 3.2–3.4 that are fixed or unbinned across device geometry and bias conditions, except for the Early voltage factor, $V_{AL} = V_A/L$. As described in Section 3.8.4, no simple prediction of MOS Early voltage, $V_A = I_D \cdot r_{ds} = I_D/g_{ds}$, or the related drain–source resistance, r_{ds}, or conductance, $g_{ds} = 1/r_{ds}$, is available because of complex dependencies on the inversion level, channel length, and V_{DS}. Instead, as described near the beginning of Sections 4.4, 5.4.2, and 5.5.2, the process parameter $V_{AL} = V_A/L$ is extracted from V_A measurements near the operating inversion coefficient, channel length, and V_{DS}.

The drain–source resistance for the 0.18 μm CMOS process described in Tables 3.2–3.4 was extensively studied through V_A measurements shown in Figures 3.43–3.49. For this process used for device measurements in Chapter 3, optimization of differential-pair and current-mirror devices in Chapter 4, and cascoded OTA design examples in Chapter 5, V_{AL} varies from 3 to 34 V/μm as summarized in Figure 3.48. The drain–source resistance, however, was not studied in detail for the 0.35 μm SOI CMOS process used for the preamplifiers. Instead, limited measurements on long-channel devices similar to those used in the preamplifiers indicate that V_{AL} is in the vicinity of 10 V/μm, and this value is used for the spreadsheet. The removal of DIBL effects, which are considered separately in the spreadsheet, is not required for this value of V_{AL} because DIBL effects are negligible for the long-channel devices used. The removal of DIBL effects from V_{AL} for short-channel devices was described near the beginning of Sections 4.4, 5.4.2, and 5.5.2. The fixed value of $V_{AL} = 10$ V/μm compared to variable values reflecting different operating conditions and channel lengths affects only the prediction accuracy of preamplifier open-loop voltage gain. As described later in Section 6.9, the spreadsheet is intended only for design guidance where computer simulation of a candidate design is required using production MOS models.

Because the spreadsheet uses the slightly more accurate expression for g_m/I_D from Table 3.17 that involves two square-root terms, spreadsheet predictions of performance will vary slightly from the design equations given in Section 6.7.2 that use the simpler g_m/I_D expressions that involves a single square-root term. Additionally, the spreadsheet includes velocity saturation and VFMR effects not considered in the design equations. These effects, which are nearly negligible for the long-channel devices used, can be included in the design equations by modifications described in Section 6.7.2.1. Finally, the spreadsheet uses a corrected value of the substrate factor, n, for the level of inversion, which also changes the results slightly.

6.7.3.1 MOSFET design selections

Figure 6.9 shows the *MOSFETs* sheet in the spreadsheet for the differential input preamplifier (schematic shown in Figure 6.8) and the single-ended input preamplifier (schematic shown in Figure 6.13) described later in Section 6.8. Device performance given by the *MOSFETs* sheet is mapped into *Circuit Analysis* sheets for the two preamplifiers as described later in Sections 6.7.3.2 and 6.8.3.2. MOS device performance given by the *MOSFETs* sheet was described earlier in Section 4.4 for the optimization of differential-pair and current-mirror devices.

The M_1, M_2 column in Figure 6.9 for the differential input preamplifier describes the pMOS input pair devices. The M_3, M_4 column describes the nMOS current source devices that use external resistive source degeneration. Finally, the M_7, M_8 column describes the pMOS current-mirror devices.

Analog CMOS Design, Tradeoffs and Optimization Spreadsheet -- MOSFETs Sheet						
For design guidance only as results are approximate and not for any particular CMOS process.						
© 2000–2007, David M. Binkley. See Disclaimer, Notes Sheet.						
---------- Optional User Design Information ----------						
Description		**Differential Preamplifier**				**Single End. Preamp.**
Device reference		M1, M2	M3, M4	M7, M8		M1	M3, M4
Device notes		100/20	30/75	10/80		100/20	30/75
---------- Required User Design Inputs ----------						
Device model		pMIL20	nSIL75	pSIL80		pMIL20	nSIL75
I_D	μA (+)	0.5	1	0.5		1	1
IC (fixed normalized)		0.954	8.382	38.453		1.907	8.382
L_{DRAWN}	μm	20	75	80		20	75
---------- Calculated Results ----------		OK	OK	OK		OK	OK
Effective Width, Length, Gate Area						
W	μm	99.60	29.80	9.90		99.60	29.80
L	μm	19.95	74.95	79.95		19.95	74.95
WL	μm²	1987.02	2233.51	791.51		1987.02	2233.51
DC Bias Voltages						
$V_{EFF} = V_{GS} - V_T$	V	0.036	0.206	0.439		0.077	0.206
V_T (adjusted for V_{SB})	V	0.900	0.650	0.900		0.900	0.650
$V_{GS} = V_T + V_{EFF}$	V	0.936	0.856	1.339		0.977	0.856
$V_{DS,\ sat}$	V	0.135	0.232	0.411		0.155	0.232
Small Signal Parameters						
g_m/I_D	μS/μA	18.84	8.56	4.48		15.52	8.56
g_m	μS	9.42	8.56	2.24		15.52	8.56
V_A	V	199.5	749.5	799.5		199.5	749.5
g_{ds}	μS	0.0025	0.0013	0.0006		0.0050	0.0013
$A_{Vi} = g_m/g_{ds}$	V/V	3758	6414	3580		3096	6414
Transcond. Distortion (Diff. Pair)						
$V_{INDIF,1dB}$ (input 1-dB comp.)	V	0.074	0.178	0.409		0.091	0.178
Capacitances and Bandwidths						
$C_{GS} = C_{gsi} + C_{GSO}$	fF	2843	5236	2087		3562	5236
$C_{GB} = C_{gbi} + C_{GBO}$	fF	1326	1133	270		1143	1133
$C_{GD} = C_{GDO}$ (in saturation)	fF	0.0	0.0	0.0		0.0	0.0
C_{DB} (at $V_{DB} = V_{DS} + V_{SB}$)	fF	29.4	14.5	3.3		29.4	14.5
$f_{Ti} = g_m/[2\pi(C_{gsi} + C_{gbi})]$	MHz	0.360	0.214	0.151		0.525	0.214
$f_T = g_m/[2\pi(C_{GS} + C_{GB})]$	MHz	0.360	0.214	0.151		0.525	0.214
$f_{DIODE} = g_m/[2\pi(C_{GS} + C_{GB} + C_{DB})]$, MHz		0.357	0.213	0.151		0.522	0.213
Thermal and Flicker Noise						
$S_{VG}^{1/2}$ thermal	nV/Hz$^{1/2}$	36.53	41.17	78.49		29.07	41.17
$S_{VG}^{1/2}$ flicker at $f_{FLICKER}$	nV/Hz$^{1/2}$	136.32	253.97	364.29		145.74	253.97
$S_{ID}^{1/2}$ thermal	pA/Hz$^{1/2}$	0.34	0.35	0.18		0.45	0.35
$S_{ID}^{1/2}$ flicker at $f_{FLICKER}$	pA/Hz$^{1/2}$	1.28	2.17	0.82		2.26	2.17
$f_{FLICKER}$, freq. for flicker noise	Hz	1	1	1		1	1
f_c, flicker-noise corner	MHz	1.39E-05	3.81E-05	2.15E-05		2.51E-05	3.81E-05
DC Mismatch for Device Pair						
ΔV_{GS} (1σ)	mV	0.20	0.20	0.36		0.20	0.20
$\Delta I_D/I_D$ (1σ)	%	0.38	0.17	0.16		0.32	0.17

Figure 6.9 *MOSFETs* sheet in the *Analog CMOS Design, Tradeoffs and Optimization* spreadsheet for the differential and single-ended input preamplifiers. This sheet lists MOSFET drain current, inversion coefficient, and channel length selections with resulting geometry and performance

*** MOSFETs, Sheet 2 ***		M1, M2	M3, M4	M7, M8		M1	M3, M4
Device reference		M1, M2	M3, M4	M7, M8		M1	M3, M4
Device notes		100/20	30/75	10/80		100/20	30/75
Body Small Signal Parameters							
$\eta = g_{mb}/g_m$ (at V_{SB})		0.302	0.348	0.256		0.296	0.348
g_{mb} (at V_{SB})	μS	2.84	2.97	0.57		4.60	2.97
Source Parameters							
$g_{ms} = g_m + g_{mb} + g_{ds}$	μS	12.26	11.53	2.81		20.12	11.53
C_{SB} (at V_{SB})	fF	47.9	28.1	8.2		47.9	28.1
Noise and Mismatch Parameters							
Γ, thermal noise factor (g_m)		0.583	0.650	0.663		0.611	0.650
($n\Gamma$), n multiplied by Γ		0.759	0.875	0.833		0.792	0.875
$K'_F{}^{1/2}$, flicker-noise factor	nV·μm	6077	12003	10249		6496	12003
AF, flicker-noise slope		1	1	1		1	1
A_{VGS}, V_{GS} mismatch factor	V·μm	0.00906	0.00930	0.01005		0.00909	0.00930
Velocity Sat. and VFMR Details							
Vel. sat. factor, $V_{EFF}/(LE_{CRIT})$		0.000	0.001	0.001		0.000	0.001
VFMR factor, $V_{EFF}\theta$		0.009	0.041	0.105		0.018	0.041
Critical IC, IC_{CRIT}		921	1245	1019		929	1245
Gate Capacitance Details							
C_{GOX}	fF	8564	9626	3411		8564	9626
C_{gsi}	fF	2843	5236	2087		3562	5236
C_{gbi}	fF	1326	1133	270		1143	1133
C_{GSO}	fF	0.0	0.0	0.0		0.0	0.0
C_{GBO}	fF	0.0	0.0	0.0		0.0	0.0
C_{GDO}	fF	0.0	0.0	0.0		0.0	0.0
Process and IC Details							
n (actual)		1.302	1.348	1.256		1.296	1.348
μ_0	cm^2/Vs	135.00	372.00	135.00		135.00	372.00
$k_0 = \mu_0 C'_{OX}$	μA/V^2	58.19	160.33	58.19		58.19	160.33
$U_T = kT/q$	V	0.02585	0.02585	0.02585		0.02585	0.02585
$I_0 = 2n_0\mu_0 C'_{OX}U_T^2$	μA	0.105	0.300	0.105		0.105	0.300
IC (actual)		0.989	8.707	41.326		1.987	8.707
Layout Details, IC Trimming							
W_{DRAWN}, each; x m for total	μm	12.50	7.50	5.00		12.50	7.50
L_{DRAWN}	μm	20	75	80		20	75
AD, each; x m for total	μm^2	7.50	4.50	3.00		7.50	4.50
PD, each; x m for total	μm	13.70	8.70	6.20		13.70	8.70
AS, each; x m for total	μm^2	8.75	6.00	5.00		8.75	6.00
PS, each; x m for total	μm	17.03	12.85	12.00		17.02	12.85
IC (trimmed), may be used for IC input		0.954	8.382	38.453		1.907	8.382
---------- **Optional User Design Inputs** ----------							
	Default, units						
m, number of gate fingers	1	8	4	2		8	4
W_{DRAWN} (trimmed), each	none, μm	12.5	7.5	5		12.5	7.5
V_{DS} (for C_{DB})	V_{GS}, V (+)	0.8	0.8	0.8		0.8	0.8
V_{SB} (for g_{mb} and C_{SB})	0 V, V (+)						
$f_{FLICKER}$, freq. for flicker noise	1 Hz, Hz	1	1	1		1	1
T, temperature	27 C, C						

Figure 6.9   *(continued)*

Degeneration Details -- R, degen. factors, and deep-ohmic $M5 = M6$ details						
		Preamplifier Version			Notes	
		Diff. Input	Single-Ended Input			
Degeneration Resistance						
V_R	(Des. input)	0.75	0.5	V	Bias voltage across R	
I_{D5}	(Des. input)	1	1	μA	Drain current (equal for MOSFETs and Rs)	
$R = V_R / I_{D5} = r_{ds5}$		0.75	0.5	MΩ	Degeneration resistance value	
Degeneration Factors						
$1 + g_{ms3}R$		9.650	6.767		r_{ds3} (r_{ds4}) raised by this factor	
$1/(1 + g_{ms3}R)$		0.104	0.148		$M3$ $(M4)$ drain noise reduced by this factor	
Current division factor						
$g_{ms3}R/(1 + g_{ms3}R)$		0.896	0.852		Current division into source of $M3$ $(M4)$	
Deep Ohmic Transconductance of $M5 = M6$						
V_{EFF5}	(Des. input)	2.65	2.65	V	$V_{EFF} = V_{GS}$ (3.3 V) $-V_{TO}$ (0.65 V)	
$g_{m5} = I_{D5} / V_{EFF5}$		0.377	0.377	μS	Collapsed g_m of deep ohmic devices	
Gate Area, Flicker-Noise Factor for $M5 = M6$						
$K'^{1/2}_F$	(Des. input)	12000	12000	nVμm	Flicker-noise factor, nMOS value	
$(WL)_5$	(Des. input)	300	200	μm²	Gate area	
W_5	(Des. input)	300	200	μm	Channel width	
L_5	(Des. input)	1	1	μm	Channel length	

Figure 6.10 Portion of *Circuit Analysis* sheet in the *Analog CMOS Design, Tradeoffs and Optimization* spreadsheet giving resistive source degeneration details for the differential and single-ended input preamplifiers. This sheet lists degeneration voltages, degeneration resistances, degeneration and current division factors, and transconductances and gate areas of deep ohmic MOSFETs acting as degeneration resistances. This information is used in the gain and noise analysis given in Figures 6.11 and 6.14

Because the *MOSFETs* sheet considers only devices in saturation, the M_5 and M_6 devices, which operate in the deep ohmic region to provide degeneration resistances for M_3 and M_4, are considered separately in Figure 6.10 as part of the *Circuit Analysis* sheet in the spreadsheet. This figure summarizes the degeneration voltage, V_R, degeneration resistance, R, degeneration factor, $1/(1 + g_{ms3}R)$, current-division factor, $g_{ms3}R/(1 + g_{ms3}R)$, ohmic transconductance, g_{m5} (ohmic), gate area, $(WL)_5$, and flicker-noise factor, K'_{F5}, associated with M_5 and M_6. These parameters were discussed earlier in Sections 6.7.2.3–6.7.2.6.

As described in Section 6.7.2 and summarized in Figure 6.9, the M_1 and M_2, pMOS, input pair devices operate at inversion coefficients of 0.95 near the center of moderate inversion for high g_m/I_D and g_m. This minimizes their gate-referred thermal-noise voltage and helps ensure that they dominate both preamplifier thermal and flicker noise. Additionally, this helps maximize the preamplifier open-loop voltage gain. Channel lengths of 20 μm are selected for M_1 and M_2. This results in required widths of 100 μm found from $W = (L/IC) \cdot (I_D/I_0)$ from Table 3.10 for $I_D = 0.5$ μA, assuming a pMOS process $I_0 = 0.105$ μA from Table 3.2. In addition to minimizing gate-referred thermal-noise voltage for operation at low inversion coefficients, this combined with operation at long channel lengths results in large channel widths and gate areas, minimizing the gate-referred flicker-noise voltage.

The M_3 and M_4, nMOS, current source devices operate at inversion coefficients of 8.4 near the onset of strong inversion. This gives them lower g_m/I_D compared to the input pair devices, but comparable g_m because they operate at twice the drain current of the input pair devices. Their effective g_m, however, is significantly reduced by multiplication with the degeneration factor given by Equation 6.20 and summarized in Figure 6.10, which has a value of 0.1. This significantly reduces their drain thermal- and flicker-noise current. As described in Sections 6.7.2.4 and 6.7.2.6, the thermal- and flicker-noise contributions of M_3 and M_4 would have exceeded the input device noise for the preamplifier if source degeneration were not used. However, as mentioned, these devices could have been operated without degeneration deep into strong inversion at high $V_{DS,sat3}$ to reduce their noise contributions, but degeneration reduces the flicker-noise contribution for a given bias compliance voltage as described in Section 6.5.5.3. Channel lengths of $75\,\mu m$ are selected for M_3 and M_4, which results in required widths of $30\,\mu m$ for $I_D = 1\,\mu A$, assuming an nMOS process $I_0 = 0.3\,\mu A$ from Table 3.2. The resulting large gate area also minimizes the flicker-noise contributions of M_3 and M_4. Because the flicker-noise contributions of these devices are small, their channel length and corresponding width and gate area could have been reduced.

The M_7 and M_8, pMOS mirror devices operate at the same $0.5\,\mu A$ drain currents as the input pair devices, but at inversion coefficients of 38.5 deep into strong inversion, which gives them considerably lower g_m/I_D and g_m. This is required to minimize their thermal-noise contributions. Channel lengths of $80\,\mu m$ are selected for M_7 and M_8, which results in required channel widths of $10\,\mu m$ for the pMOS process $I_0 = 0.105\,\mu A$. The resulting large gate area is required to minimize flicker-noise contributions, while the long channel length maximizes the M_8 drain–source resistance, r_{ds8}, which maximizes the preamplifier voltage gain.

The channel length of $80\,\mu m$ for the M_7 and M_8, pMOS mirror devices exceeds the channel length of $20\,\mu m$ for the M_1 and M_2, pMOS, input pair devices. When input devices operate at low inversion coefficients in weak or moderate inversion for high g_m/I_D and g_m, and non-input devices operate at high inversion coefficients in strong inversion for low g_m/I_D and g_m, it is necessary to increase the channel length of non-input devices above that of the input devices to manage flicker noise. As mentioned in Section 5.5.3 for the OTA design examples of Chapter 5 and mentioned later in Section 6.10, this is required because the smaller shape factor $(S = W/L)$ of non-input devices operating at high levels of inversion requires an increase in channel length and, correspondingly, channel width to achieve sufficient gate area to reduce flicker-noise contributions.

The M_5 and M_6, nMOS, deep ohmic devices summarized in Figure 6.10 operate with drain–source voltage drops of $V_R = 0.75\,V$ at $I_D = 1\,\mu A$, giving degeneration resistance values of $R = 0.75\,V/1\,\mu A = 750\,k\Omega$. These devices are sized at $W = 1\,\mu m$ and $L = 300\,\mu m$ with their gate–source voltages connected to the full supply voltage of $3.3\,V$ to ensure deep ohmic operation and low ohmic transconductance, g_{m5} (ohmic). As described in Section 6.5.5.3, low g_{m5} (ohmic) minimizes the drain-referred flicker-noise current of M_5 and M_6, which are intended to operate as resistors having only thermal noise.

As described in Section 6.5.2, only the DC bias conditions are required to predict the deep ohmic, drain–source resistance and transconductance. However, the substantial loss of mobility due to VFMR at the high value of $V_{EFF} = V_{GS} - V_T$ must be carefully considered to properly size devices when solving Equation 6.10 for the desired deep ohmic, drain–source resistance. Since the *MOSFETs* sheet in the spreadsheet considers only MOSFETs operating in saturation, M_5 and M_6 are sized from SPICE simulations using BSIM3V3 MOS models [35].

6.7.3.2 Resulting preamplifier performance

Figure 6.11 shows the *Circuit Analysis* sheet in the *Analog CMOS Design, Tradeoffs and Optimization* spreadsheet for the differential input preamplifier. This sheet gives predicted preamplifier open-loop voltage gain, unity-gain bandwidth, thermal noise, and flicker noise from the circuit analysis equations developed earlier in Section 6.7.2. The *Circuit Analysis* sheet calls geometries and performance from the *MOSFETs* sheet of Figure 6.9 for the M_1 and M_2 input pair, M_3 and M_4, degenerated current source, and M_7 and M_8 current-mirror devices. Additionally, the *Circuit Analysis* sheet calls the portion of the sheet given in Figure 6.10 that lists details for the M_5 and M_6, deep ohmic degeneration

Summary of Predicted and Measured Performance for 0.35–um SOI CMOS Micropower Preamp.

*** Differential Input Version ***

MOSFET Notes

The following devices are biased and sized equally: $M1 = M2$ (diff. pair), $M3 = M4$ (current source), $M5 = M6$ (deep ohmic degeneration resistors), and $M7 = M8$ (mirror). Other devices have little effect on performance and are not included in circuit analysis.

Open Loop Voltage Gain, A_V

$A_V = g_{m\,1} \cdot r_{ds\,8} = (g_m/I_D)_1 \cdot V_{A8}$; Drain currents are equal. Cascoded $R_{outD\,10}$ neglected.

	Predicted	Measured	
	15063	**15500**	V/V

Unity-Gain Bandwidth Product, f_T

$f_T = g_{m1} / [2\pi(C_{DOM} + C_{DB8} + C_{GD8})]$; $C_{DB\,10}$ and $C_{GD\,10}$ are neglected.

C_{DOM}	1000	fF		Predicted	Measured	
				1.49	**1.53**	MHz

Input-Referred Thermal Noise Voltage, $S_{VIN}^{1/2}$ (thermal)

$S_{VIN}^{1/2} = \sqrt{[\{2 \cdot 4kT(n\Gamma)_1/g_{m1}\} \cdot (1 + RNP_3 + RNP_5 + RNP_7)]}$

	Predicted	Measured	
	61.6	**63.8**	nV/Hz$^{1/2}$

Relative Noise Power (RNP) -- relative to $M1$, $M2$					Predicted	
$M1, M2$	1				1	
$M3, M4$	$RNP_3 = [(n\Gamma)_3/(n\Gamma)_1] \cdot [g_{m3}/g_{m1}] \cdot [1/(1 + g_{ms\,3}R)]^2$				0.011	
$M5, M6$	$RNP_5 = \{1/[(n\Gamma)_1 g_{m\,1}R]\} \cdot [g_{ms\,3}R/(1 + g_{ms\,3}R)]^2$				0.150	
$M7, M8$	$RNP_7 = [(n\Gamma)_7/(n\Gamma)_1] \cdot [g_{m\,7}/g_{m\,1}]$				0.261	
Total:					1.422	

Input-Referred Flicker-Noise Voltage, $S_{VIN}^{1/2}$ (flicker)

$S_{VIN}^{1/2}(f) = \sqrt{[\{2 \cdot K'_{F\,1}/((WL)_1 f^{AF1})\} \cdot (1 + RNP_3 + RNP_5 + RNP_7)]}$

	Predicted	Measured	(1 Hz)
	233.5	**231.0**	nV/Hz$^{1/2}$

Relative Noise Power (RNP) -- relative to $M1$, $M2$ at $f = 1$ Hz						
RNP_3 adjusted for frequency by (f^{AF1}/f^{AF3}), etc.					Predicted	(1 Hz)
$M1, M2$	1				1	
$M3, M4$	$RNP_3 = [K'_{F3}/K'_{F1}] \cdot [(WL)_1/(WL)_3] \cdot [g_{m\,3}/g_{m\,1}]^2 \cdot [1/(1 + g_{ms\,3}R)]^2$				0.031	
$M5, M6$	$RNP_5 = [K'_{F5}/K'_{F1}] \cdot [(WL)_1/(WL)_5] \cdot [g_{m5}/g_{m1}]^2 \cdot [g_{ms\,3}R/(1 + g_{ms\,3}R)]^2$				0.033	
$M7, M8$	$RNP_7 = [K'_{F7}/K'_{F1}] \cdot [(WL)_1/(WL)_7] \cdot [g_{m7}/g_{m1}]^2$				0.403	
Total:					1.468	

Figure 6.11 *Circuit Analysis* sheet in the *Analog CMOS Design, Tradeoffs and Optimization* spreadsheet for the differential input preamplifier. This sheet lists predicted and measured performance for the preamplifier

devices and the degeneration and current-division factors required for circuit analysis. The *Circuit Analysis* sheet provides automatic updating of predicted preamplifier performance as MOSFET drain current, inversion coefficient, and channel length selections, and degeneration selections are explored for optimization. To permit comparisons with the predicted values, the *Circuit Analysis* sheet also lists measured performance.

As described in Equation 6.27 of Section 6.7.2.1 and summarized in the *Circuit Analysis* sheet of Figure 6.11, the preamplifier open-loop voltage gain, A_{VOL}, is approximated by the transconductance of the M_1 and M_2 input pair devices, $g_{m1} = 9.42\,\mu S$, multiplied by the drain–source resistance of the M_8 mirror device, $r_{ds8} = 1600\,M\Omega$ ($1/g_{ds8}$ given in Figure 6.9). This gives a predicted gain of 15063 V/V, which agrees well with the measured gain of 15500 V/V, also listed in Figure 6.11.

For the measured open-loop voltage gain of 15500 V/V, the negative feedback loop gain is 304 V/V (15500/51) for a preamplifier, non-inverting, closed-loop voltage gain of 51 V/V intended for the application. This loop gain is sufficiently high to ensure feedback regulation of the closed-loop voltage gain, even for significant variations in the open-loop gain. Interestingly, computer simulations using BSIM3V3 MOS models [35] significantly underpredicted preamplifier r_{ds8} and resulting open-loop voltage gain, compared to the simple predictions provided by the spreadsheet where predicted gain agreed closely with the measured gain. As discussed later in Section 6.9, the largest prediction errors are usually for the prediction of drain–source resistances and the resulting circuit gains.

The measurement of high open-loop voltage gain requires careful laboratory methods to ensure the output remains within its full-gain, linear range. Like measurements for the cascoded OTAs described in Section 5.5.2.1, the preamplifier open-loop voltage gain was measured in a closed-loop configuration by measuring the loop gain response to an injected error signal.

As described in Equation 6.28 of Section 6.7.2.2 and summarized in Figure 6.11, the preamplifier unity-gain bandwidth, f_T, is approximated by $g_{m1} = 9.42\,\mu S$ divided by 2π and the sum of the dominant-pole compensation capacitor, C_{DOM}, and the drain–body and drain–gate capacitances of M_8, C_{DB8} and C_{GD8}. With a value of 1000 fF, C_{DOM} dominates, giving a predicted unity-gain bandwidth of 1.49 MHz. This is in close agreement with a measured value of 1.53 MHz ($30\,kHz \cdot 51$) found from the measured closed-loop, $-3\,dB$ frequency of 30 kHz for a non-inverting voltage gain of 51 V/V. Some overprediction of the unity-gain bandwidth is expected because the prediction excludes the capacitive loading from the cascode device, M_{10}, the source follower, M_{11}, and the layout.

Preamplifier thermal noise was described in detail in Sections 6.7.2.3 and 6.7.2.4 and was summarized in Table 6.4 in terms of noise PSD contributions from the M_3 and M_4, degenerated current source, M_5 and M_6, deep ohmic degeneration resistance, and M_7 and M_8 current-mirror devices. These noise PSD contributions were given relative to the noise of the M_1 and M_2 input devices and were referred to as relative noise powers, *RNPs*. Figure 6.11 summarizes *RNP* values from Table 6.4 using traditional equations based on device transconductance and degeneration resistance values. To facilitate the minimization of noise against the constraint of limited bias compliance voltage, Table 6.4 additionally includes equations based on the DC bias conditions and the bias compliance voltage.

RNP_3, RNP_5, and RNP_7 values for the thermal noise are 1.1 %, 15.0 %, and 26.1 %, corresponding to relative noise PSD contributions from the M_3 and M_4, M_5 and M_6, and M_7 and M_8 non-input devices, respectively. The total relative noise PSD contribution is 142.2 % or 1.422 as an absolute factor, which includes the 100 % noise contribution from the M_1 and M_2 input pair devices. The total preamplifier, input-referred thermal-noise voltage density, also summarized in Figure 6.11, includes the gate-referred thermal-noise voltage density of $36.5\,nV/Hz^{1/2}$ from Figure 6.9 for the individual M_1 and M_2 input pair devices multiplied by $\sqrt{2} = 1.414$ to consider the noise density of two devices. This gives an input pair noise of $51.7\,nV/Hz^{1/2}$. This noise is then multiplied further by $\sqrt{1.422} = 1.192$ to consider the noise density contributions from the non-input devices, giving a total preamplifier, input-referred thermal-noise voltage density of $61.6\,nV/Hz^{1/2}$. Predicted and measured noise values will be compared later.

Preamplifier flicker noise was described in detail in Sections 6.7.2.5 and 6.7.2.6 and was summarized in Table 6.5 in terms of noise PSD contributions from the M_3 and M_4, M_5 and M_6, and M_7 and M_8 devices relative to the noise of the M_1 and M_2 input pair devices. As for the thermal noise, these relative noise contributions were referred to as RNP_3, RNP_5, and RNP_7. Figure 6.11 summarizes *RNP* values for flicker noise from Table 6.5 using traditional equations based on device transconductance, gate area, and degeneration resistance values. Like Table 6.4 given for the thermal noise, Table 6.5 additionally includes equations based on the DC bias conditions and the bias compliance voltage.

RNP_3, RNP_5, and RNP_7 values for the flicker noise are 3.1 %, 3.3 %, and 40.3 %, corresponding to relative noise PSD contributions from the M_3 and M_4, M_5 and M_6, and M_7 and M_8 non-input devices, respectively. The total relative noise PSD contribution is 146.8 % or 1.468 as an absolute factor, which includes the 100 % noise contribution from the M_1 and M_2 input pair devices. The total preamplifier, input-referred flicker-noise voltage density, also summarized in Figure 6.11, includes the gate-referred flicker-noise voltage density of 136.3 nV/Hz$^{1/2}$ at 1 Hz from Figure 6.9 for the individual M_1 and M_2 input pair devices multiplied by $\sqrt{2} = 1.414$ to consider the noise density of two devices. This gives an input pair noise of 192.8 nV/Hz$^{1/2}$ at 1 Hz. This noise is then multiplied further by $\sqrt{1.468} = 1.212$ to consider the noise density contributions from the non-input devices. This gives a total preamplifier, input-referred flicker-noise voltage density of 233.5 nV/Hz$^{1/2}$ at 1 Hz.

The total preamplifier noise is the sum of the thermal- and flicker-noise PSD or the square root of the sum of the squared noise densities. For the predicted input-referred thermal-noise voltage density of 61.6 nV/Hz$^{1/2}$ and flicker-noise voltage density of 233.5 nV/Hz$^{1/2}$ at 1 Hz, the total predicted noise voltage density is increased slightly from the flicker noise alone to 241.5 nV/Hz$^{1/2}$ at 1 Hz.

Figure 6.12 shows the predicted and measured, input-referred noise voltage density for the differential input preamplifier. The preamplifier was configured for a non-inverting, closed-loop voltage gain of 51 V/V, set by external $R_F = 1\,M\Omega$ and $R_G = 20\,k\Omega$ feedback resistors. The measured $-3\,dB$ frequency was 30 kHz, which closely matches the predicted closed-loop frequency of 29.2 kHz (($f_T = 1.49\,MHz$)/51). No frequency peaking was observed, indicating sufficient phase margin. As discussed at the beginning of Section 6.7.2, no analysis of preamplifier parasitic poles was presented here as the dominant-pole compensation and high closed-loop gain ensure good negative feedback stability.

The measured input-referred noise voltage density is 240 nV/Hz$^{1/2}$ at 1 Hz, compared to the predicted noise of 241.5 nV/Hz$^{1/2}$ shown by the triangles in Figure 6.12. The average measured thermal-noise floor is 63.8 nV/Hz$^{1/2}$, which is 3.6 % above the predicted thermal noise of 61.6 nV/Hz$^{1/2}$. Removing the measured thermal noise of 63.8 nV/Hz$^{1/2}$ from the measured noise of 240 nV/Hz$^{1/2}$

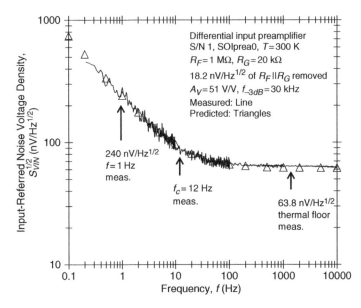

Figure 6.12 Measured (line) and predicted (triangles) input-referred noise voltage density for the differential input preamplifier. Predicted noise is from the *Circuit Analysis* sheet in the *Analog CMOS Design, Tradeoffs and Optimization* spreadsheet shown in Figure 6.11

at 1 Hz gives a measured flicker noise of $231.4\,\mathrm{nV/Hz^{1/2}}$ at 1 Hz, compared to the predicted flicker noise of $233.5\,\mathrm{nV/Hz^{1/2}}$ at 1 Hz. The measured and predicted flicker-noise overlay in the figure, being within the measurement uncertainty. Finally, the measured flicker-noise corner frequency, found by the intersection of flicker-noise and thermal-noise asymptotes, is very low for MOS circuits at 12 Hz.

As mentioned in the description of Section 6.7.1 and observed in the schematic diagram of Figure 6.8, the differential input preamplifier operates at a core current consumption of $2\,\mu\mathrm{A}$, giving a core power consumption of $6.6\,\mu\mathrm{W}$ for $V_{DD} = 3.3\,\mathrm{V}$. This excludes the power consumption of bias reference devices, which can be common for multiple channels, and the source follower, which can be omitted for on-chip capacitive loads. The core power consumption of $6.6\,\mu\mathrm{W}$ will be used for predicting the preamplifier thermal-noise efficiency factor.

The thermal-noise efficiency factor for the preamplifier is $NEF' = 38$, found from Equation 6.2. This compares favorably with other preamplifiers reported in Table 6.1. As discussed in Section 6.3.1, the NEF' for a differential design, like this preamplifier, is a factor of four higher than that for an equivalent single-ended design. The single-ended input preamplifier described in Section 6.8 has an NEF' of nearly a factor of four lower at 11.6.

Both the differential and single-ended input preamplifiers use an input stage current of $1\,\mu\mathrm{A}$. However, the differential input preamplifier uses an additional current of $1\,\mu\mathrm{A}$ for non-input devices, while the single-ended input preamplifier uses an additional current of $1\,\mu\mathrm{A}$ for devices providing regulated cascode action. As described in Sections 6.7.4 and 6.8.4, operation at the input stage current only would halve the thermal-noise efficiency factor for each preamplifier.

Following the discussion of the single-ended input preamplifier in Section 6.8, Figure 6.16 shows the layout with dimensions for both the differential and single-ended input preamplifiers.

6.7.4 Design Improvements

The differential input preamplifier has a measured input-referred thermal-noise voltage density of $63.8\,\mathrm{nV/Hz^{1/2}}$, resulting primarily from input pair devices operating at $IC = 0.95$ near the center of moderate inversion at micropower drain currents of $0.5\,\mu\mathrm{A}$. The noise could be lowered modestly by operation closer to weak inversion. For example, operating at $IC = 0.1$ at the boundary of weak inversion increases input device g_m by nearly 40 % and decreases the thermal-noise voltage density by nearly 20 %. Operation here, however, comes at the expense of a nearly a factor-of-ten increase in device shape factor, $S = W/L$, and potentially significantly increased layout area and input capacitance. As a result, operation near the center of moderate inversion can provide a good compromise between thermal noise, layout area, and input capacitance.

The measured input-referred noise voltage density is $240\,\mathrm{nV/Hz^{1/2}}$ at 1 Hz with a flicker-noise corner frequency of 12 Hz, resulting primarily from flicker noise from large-area input devices. Increasing input device area would lower the flicker noise, but at the expense of increased layout area and input capacitance.

The preamplifier, input-referred thermal- and flicker-noise voltage density decreases as the inverse square root of the input device drain current if channel width is increased with the drain current to maintain the selected inversion coefficient. This, of course, assumes constant noise PSD contributions from non-input devices, which can be arranged by also scaling non-input device channel width with increased drain current. At higher drain currents, preamplifier channel lengths would normally be decreased from the large values used in this micropower design.

Noise predictions indicate that non-input devices contribute only 19.2 % and 21.2 % additional noise density for thermal and flicker noise, respectively, above that generated by the input pair devices. This indicates that input devices dominate the noise, while non-input devices contribute only a small amount of noise.

Design improvements, however, can be made, especially since thermal- and flicker-noise contributions from the M_3 and M_4, degenerated current source devices are very low, suggesting a substantial over-design of too much resistive source degeneration and too much gate area for these devices. The degree of degeneration could be decreased from the level provided by $V_R = 0.75\,\text{V}$, which would lower the degeneration resistances provided by M_5 and M_6, and favorably decrease the bias compliance voltage. Alternatively, as mentioned in Sections 6.7.2.4 and 6.7.2.6, M_3 and M_4 could be operated without source degeneration deep into strong inversion at high $V_{DS,sat}$ to minimize their thermal-noise contributions. However, as mentioned in Section 6.5.5.3, source degeneration reduces the drain flicker-noise current for a given bias compliance voltage, lowering the gate-area requirement.

Additionally, the M_7 and M_8 current-mirror devices could be operated at drain currents below those of the input pair devices to lower their g_m and relative thermal- and flicker-noise contributions without increasing their level of inversion and bias compliance voltage from $IC = 38.5$ and $V_{DS,sat} = 0.41\,\text{V}$. As described in Section 6.7.2.6, these pMOS devices contribute most of the non-input device flicker noise because their flicker-noise factors increase by a factor of 2.84 above that of the input pair devices. This is a result of the significant increase in gate-referred flicker-noise voltage for pMOS devices with increasing inversion or V_{EFF}, which was described in the device measurements of Section 6.6. Such noise increases are lower at lower levels of inversion generally required in low-voltage design.

Lowering non-input device drain current lowers g_m and thermal- and flicker-noise contributions without the increase of bias compliance voltage associated with operation deeper in strong inversion or the use of resistive source degeneration. However, circuit topologies require devices that capture the drain currents of input devices, as in the folded-cascode topology of the differential input preamplifier shown in Figure 6.8. Here, certain devices necessarily operate at drain currents equal to or above that of the input devices, unless input device drain currents are captured by non-MOS devices such as resistors where flicker noise could be potentially reduced.

In addition to lowering their noise contributions, lowering the drain current for the M_7 and M_8 current-mirror devices lowers the core preamplifier supply current because these devices operate at a supply current that is separate from the supply current of $1\,\mu\text{A}$ for the input pair devices. Reducing the M_7 and M_8 supply current from 1 to $0.25\,\mu\text{A}$ would reduce the preamplifier core supply current from 2 to $1.25\,\mu\text{A}$. Even without the effect of lower noise contributions from M_7 and M_8, the lower supply current would lower the thermal-noise efficiency factor from $NEF' = 38$, described in Section 6.7.3.2, to $NEF' = 24$. Operation at the same level of noise using only the supply current of the input devices would halve the $NEF' = 38$ to $NEF' = 19$. Supply current for a specified level of input-referred thermal-noise voltage, specified by the NEF', is minimized by operating input devices in weak or moderate inversion for high g_m/I_D, ensuring small non-input device noise contributions, and minimizing supply current beyond that required for input devices.

The minimization of thermal noise for a given power consumption and the minimization of flicker noise for a given layout area require a careful balance of input and non-input device noise contributions where both over- and under-design are avoided. This design optimization can be done using the *Analog CMOS Design, Tradeoffs and Optimization* spreadsheet where the thermal-noise efficiency factor, NEF', described in Section 6.3.1, and the flicker-noise area factor, NAF, described in Section 6.3.2, are included on the *Circuit Analysis* sheet. This is done easily by using Equation 6.2 for the NEF' and Equation 6.4 for the NAF. The NAF requires the inclusion of gate area for all devices, but this is readily available from the *MOSFETs* sheet of the spreadsheet.

The differential input preamplifier shown in Figure 6.8 is one design example for illustrating the minimization of thermal and flicker noise. Here, MOS operation in moderate and strong inversion and the use of external resistive source degeneration are considered, providing a general design method. The method, however, can also be applied to other circuit topologies, including a fully differential circuit where the M_9 and M_{10} folded-cascode and M_7 and M_8 current-mirror devices are removed, halving the core current consumption. Here, a differential output would be taken directly at the drains of input devices M_1 and M_2 or at the drains of added telescopic cascode devices. The required common-mode feedback could be applied at the gates of the M_5 and M_6, deep ohmic degeneration resistance devices. The noise analysis for this fully differential circuit is identical to that described for the differential input preamplifier, requiring only the removal of terms associated with M_7 and M_8.

6.8 SINGLE-ENDED INPUT PREAMPLIFIER

Section 6.7 presented the circuit description, circuit analysis, performance optimization, and predicted and measured performance for a micropower, differential input preamplifier. This section describes this for a single-ended input preamplifier intended for transimpedance and inverting amplifier applications. Here, external resistive feedback is connected between the output signal and the single, inverting input terminal. Like the differential input preamplifier, the single-ended input preamplifier is designed to amplify low-frequency signals from gyros and other sensors used in space missions that require minimization of thermal and flicker noise at micropower operation [24]. Both preamplifiers use the same $0.35\,\mu\text{m}$ SOI CMOS process to permit operation over extreme temperature and radiation environments [24–26]. Being a single-ended design with single devices in the signal path, the single-ended input preamplifier has lower input-referred thermal- and flicker-noise voltage compared to the differential input preamplifier.

6.8.1 Description

Figure 6.13 shows a schematic diagram of the single-ended input preamplifier. M_1 is a single input device having the same geometry as one of the input pair devices in the differential input preamplifier shown earlier in Figure 6.8. Like the input pair devices in the differential input preamplifier, M_1 is a pMOS device because these have lower flicker noise in the process. However, as mentioned in Section 6.7.1, pMOS devices may not be the preferred device in smaller-geometry processes and can have significant increases in flicker noise with increasing inversion or V_{EFF}. M_1 operates at a drain current of $1\,\mu\text{A}$, which is the current of both input devices in the differential input preamplifier.

The M_1 input device connects to the pMOS, M_2 cascode device, with the pMOS, M_7 device providing regulated cascode action. This raises the already high, cascoded, drain output resistance of M_2 by the factor $1 + A_{VGD7}$, where A_{VGD7} is the gate-to-drain voltage gain of M_7 taken positively. The regulated cascode action also simultaneously reduces the input resistance at the source of M_2 by division by this same factor. M_1, M_2, and M_7 each operate at a drain current of $1\,\mu\text{A}$. The pMOS, M_{10} source follower device buffers the high-impedance, dominant-pole signal appearing at the drain of M_2 across the dominant-pole compensation capacitor C_{DOM}. This permits the output signal to drive external feedback resistors. The M_{10} source follower, which operates at a high drain current of $10\,\mu\text{A}$, may be omitted to save power when capacitive loads and feedback networks are driven.

The nMOS, M_3 current source device, with resistive source degeneration provided by the nMOS, M_5 deep ohmic device, provides bias current for the M_2 cascode device. Resistive source degeneration raises the M_3 drain output resistance, raising the dominant-pole resistance level and, correspondingly, raising the preamplifier open-loop voltage gain. Also, as discussed in Section 6.5.5.3, resistive source degeneration reduces the drain flicker-noise current of M_3. Diode-connected device M_4, with resistive source degeneration provided by deep ohmic device M_6, is a bias reference, providing a unity-gain current mirror with M_3 and M_5. Although not in the signal path, the M_3 current source can contribute significant thermal and flicker noise. Additionally, M_5 that provides source degeneration resistance for M_3 can contribute significant thermal noise. The noise contributions of M_3 and M_5, including noise contributions coming from their M_4 and M_6 bias references, will be given considerable attention in later noise analysis. $M_3 - M_6$ each operate at a drain current of $1\,\mu\text{A}$.

Devices M_8 and M_9 are mirrored off the M_4 and M_6 bias references, providing a bias current of $1\,\mu\text{A}$ for M_7. Capacitor C_1 provides compensation for the local feedback loop associated with the regulated

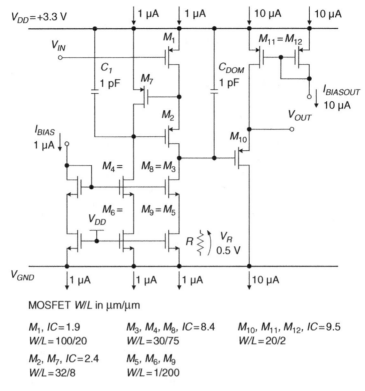

MOSFET *W/L* in µm/µm

M_1, IC=1.9 M_3, M_4, M_8, IC=8.4 M_{10}, M_{11}, M_{12}, IC=9.5
W/L=100/20   W/L=30/75   W/L=20/2

M_2, M_7, IC=2.4 M_5, M_6, M_9
W/L=32/8   W/L=1/200

Figure 6.13 Schematic diagram of single-ended input preamplifier. Figure 6.9 gives MOSFET drain current, inversion coefficient, and channel length selections with resulting device geometry and performance. Figure 6.14 gives predicted and measured performance for the preamplifier

cascode circuit of M_1, M_2, and M_7. The compensation of this internal feedback loop can be tested in circuit simulations by observing the response to an impulse of current. Finally, pMOS, M_{11} and M_{12} devices comprise a current mirror to provide bias current for the M_{10} source follower.

The preamplifier supply voltage is $V_{DD} = 3.3$ V. The core current consumption is 2 µA for devices M_1, M_2, M_3, M_5, and M_7–M_9, giving a core power consumption of 6.6 µW. This excludes the reference devices and the output source follower, which is required only to drive resistive loads. Because the input signal is referenced to the V_{DD} terminal, this terminal can be connected to circuit ground with the V_{GND} terminal used as a negative V_{SS} power supply terminal.

6.8.2 Circuit Analysis, Performance Optimization, and Predicted Performance

This section describes circuit analysis, performance optimization, and predicted performance for the single-ended input preamplifier, paralleling Section 6.7.2 that described this for the differential input preamplifier. The circuit analysis equations developed in this section are included later in Section 6.8.3 in the *Circuit Analysis* sheet of the *Analog CMOS Design, Tradeoffs and Optimization* spreadsheet. Again, this sheet calls device geometry and performance predicted by the *MOSFETs* sheet of the spreadsheet for selected device drain currents, inversion coefficients, and channel lengths, and gives predicted preamplifier performance. Predicted performance is described in this section as part of discussions on performance optimization. Predicted performance is summarized and compared to measured performance later in Section 6.8.3.

As for the differential input preamplifier, parasitic pole frequencies, input-referred offset voltage due to local-area mismatch, input and output voltage ranges, slew rate, the input, 1 dB compression voltage, and input and output capacitances are not developed for the single-ended input preamplifier. Again, these aspects of circuit performance were developed in Section 5.3 for OTA design examples.

A high preamplifier, inverting, closed-loop voltage gain of 50 V/V, set by external feedback resistors for the test application, helps ensure reduced loop gain and good phase margin in combination with a sufficiently large choice of C_{DOM}. As a result, parasitic pole frequencies are not developed. Transimpedance applications having unity high-frequency feedback from the output to inverting input, however, may require an increase in the value of C_{DOM} to ensure sufficient phase margin. The input-referred offset voltage due to local-area mismatch is not developed because the application requires subsequent AC coupling. Additionally, input and output voltages are small, so slew rate, input and output voltage ranges, and the input, 1 dB compression voltage are not developed. Finally, the output capacitance appearing at the M_{10} source follower buffer is small for this small device and is not developed.

As for the differential input preamplifier, the input capacitance, however, can be significant for the application. The input capacitance for the single-ended input preamplifier can be easily approximated as the sum of the gate–source, gate–body, and gate–drain capacitances, C_{GS1}, C_{GB1}, and C_{GD1}, associated with the M_1 input device. This assumes little Miller multiplication of C_{GD1}, resulting from the low M_1 gate-to-drain voltage gain into the low source resistance of M_2. As mentioned, the input resistance at the source of M_2 is low because regulated cascode action lowers this resistance.

6.8.2.1 Voltage gain

The open-loop voltage gain for the single-ended input preamplifier is equal to the product of input device transconductance, g_{m1}, and dominant-pole resistance, r_{DOM}, appearing at C_{DOM}, multiplied by the voltage gain of the M_{10} source follower. The voltage gain of the source follower was described in Section 6.7.2.1 for the differential input preamplifier and will again be approximated at unity, as typical for open-loop voltage gain calculations.

The dominant-pole resistance for the single-ended input preamplifier is equal to the drain output resistance of M_2 in parallel with the drain output resistance of M_3. Since the drain of M_2 is the output of a regulated cascode stage, the M_2 drain output resistance is raised by the factor of $1 + A_{VGD7}$, mentioned in Section 6.8.1, above its already high cascode output resistance. As a result, the dominant-pole resistance can be approximated by the output resistance of the M_3, degenerated current source alone, giving an open-loop voltage gain of

$$A_{VOL} = g_{m1} \cdot r_{DOM} \approx g_{m1} \cdot r_{ds3} \left(1 + g_{ms3}R\right) = \left(\frac{g_m}{I_D}\right)_1 \cdot V_{A3} \cdot \left(1 + g_{ms3}R\right)$$

$$\approx \frac{V_{AL3} \cdot L_3}{n_1 U_T \left(\sqrt{IC_1 + 0.25} + 0.5\right)} \cdot \left(1 + g_{ms3}R\right)$$

$$\approx \frac{2 \cdot V_{AL3} \cdot L_3}{n_1 \left(V_{DS,sat1} - 2U_T\right)} \cdot \frac{2V_R + V_{DS,sat3} - 2U_T}{V_{DS,sat3} - 2U_T} \tag{6.33}$$

The first line in Equation 6.33 gives the voltage gain as the transconductance efficiency, $(g_m/I_D)_1$, for the M_1 input device, having $g_{m1} = (g_m/I_D)_1 \cdot I_{D1}$, multiplied by the Early voltage, V_{A3}, for the M_3, degenerated current source device, having drain–source resistance of $r_{ds3} = V_{A3}/I_{D3}$. This expression assumes equal M_1 and M_3 drain current as present for the preamplifier. The voltage gain is additionally multiplied by the factor, $1 + g_{ms3}R$, which describes the increase in the drain output resistance of M_3 and, correspondingly, the increase in preamplifier gain due to resistive source degeneration. $g_{ms3} \approx g_{m3} + g_{mb3}$ from Equation 6.18 is the source transconductance of M_3, and R is the source degeneration resistance provided by the M_5 deep ohmic device. The second line gives $(g_m/I_D)_1$ from Table 6.2 in terms of the inversion coefficient, IC_1, for M_1, and also gives the Early voltage, $V_{A3} = V_{AL3} \cdot L_3$, for M_3, where V_{AL3} is the Early voltage factor and L_3 is the channel length. The prediction of drain–source resistance using the Early voltage factor was described in Table 3.21, where the Early voltage factor is a complex function of IC, L, and V_{DS} and is not fixed for the process. Finally, the third line gives $(g_m/I_D)_1$ from Table 6.2 in terms of the bias compliance voltage or drain–source saturation voltage, $V_{DS,sat1}$, for M_1, and also gives the factor, $1 + g_{ms3}R$, from Equation 6.19 in terms of the voltage across the degeneration resistance, V_R, and the drain–source saturation voltage, $V_{DS,sat3}$, of M_3. In the second and third lines, n_1 is the substrate factor for M_1.

As for the differential input preamplifier, the expressions assume negligible $(g_m/I_D)_1$ decreases due to velocity saturation and VFMR effects. The inclusion of these effects is not considered for the preamplifier because it uses long-channel devices to give large gate area and low flicker noise. However, these effects can be included as described earlier in Section 6.7.2.1.

The voltage gain A_{VOL} is maximized by operating the M_1 input device in weak or moderate inversion for high $(g_m/I_D)_1$ and low $V_{DS,sat1}$, which also minimizes its gate-referred thermal-noise voltage. Operating this device in weak or moderate inversion at long channel lengths also minimizes its flicker noise by maximizing its gate area. This device is operated at $IC_1 = 1.9$ near the center of moderate inversion, giving $(g_m/I_D)_1 = 15.5\,\mu S/\mu A$ and $V_{DS,sat1} = 0.155\,V$. Additionally, A_{VOL} is maximized by operating the M_3, degenerated current source device at long channel lengths for high V_{A3} and r_{ds3}, which also minimizes its flicker-noise contributions by maximizing its gate area. This device is operated at $L_3 = 75\,\mu m$, giving an unusually high $V_{A3} = 750\,V$ for $V_{AL3} = 10\,V/\mu m$, with the value of V_{AL3} described in Section 6.7.3. Finally, A_{VOL} is maximized by maximizing the degeneration resistance voltage, V_R, which maximizes the degeneration resistance, R. Although maximizing V_R maximizes R and A_{VOL}, this increases the bias compliance voltage for M_3 and reduces the preamplifier output voltage range. V_R is set at $V_R = 0.5\,V$, giving $R = V_R/I_D = 0.5V/1\,\mu A = 500\,k\Omega$.

The predicted A_{VOL} is nearly 78700 V/V from the product of $(g_m/I_D)_1 = 15.5\,\mu S/\mu A$, $V_{A3} = 750\,V$, and $1 + g_{ms3}R = 6.8$, where n_1 is approximately 1.3. MOSFET operating parameters, predicted preamplifier performance, and experimental measurements are summarized later in Section 6.8.3 using the spreadsheet.

The predicted single-ended input preamplifier voltage gain of 78700 V/V is over five times greater than 15000 V/V predicted in Section 6.7.2.2 for the differential input preamplifier. The gain increase is due to increased drain output resistance of M_3 at r_{DOM} caused by resistive source degeneration compared to the output resistance of M_8 in the differential input preamplifier that does not utilize source degeneration. The gain increase can be seen by the presence of the factor, $1 + g_{ms3}R$, in Equation 6.33 for the single-ended input preamplifier, which does not appear in Equation 6.27 for the differential input preamplifier.

6.8.2.2 Frequency response

The open-loop, unity-gain frequency for the single-ended input preamplifier is approximated by

$$f_T \approx \frac{g_{m1}}{2\pi\,(C_{DOM} + C_{DB3} + C_{GD3})} \tag{6.34}$$

This assumes that the M_3 drain–body capacitance, C_{DB3}, and gate–drain capacitance, C_{GD3}, sufficiently exceed capacitances for the smaller M_2 cascode device, where M_3 is made larger to manage its flicker noise. Additionally, this assumes that the M_3 capacitances sufficiently exceed the input capacitance of the small M_{10} source follower device.

f_T is controlled by the input device transconductance, g_{m1}, and the dominant-pole capacitance, C_{DOM}, selected at 1000 fF for both the single-ended and differential input preamplifiers. f_T is higher at approximately 2.4 MHz for the single-ended input preamplifier compared to 1.5 MHz for the differential input preamplifier because of higher input device transconductance. Input device transconductance is higher for the single-ended input preamplifier because its single input device operates at 1 μA, compared to 0.5 μA for each input device in the differential input preamplifier, where the input pair transconductance is equal to the transconductance of one device. Input devices in both preamplifiers operate near the center of moderate inversion for high g_m/I_D and g_m.

6.8.2.3 Thermal noise

Thermal-noise analysis for the single-ended input preamplifier is similar to analysis given earlier in Sections 6.7.2.3 and 6.7.2.4 for the differential input preamplifier. Device numbers have been assigned to permit similar noise expressions. g_{m1} is the transconductance and $(n\Gamma)_1$ is the product of substrate and thermal-noise factors for the M_1 input device in the single-ended input preamplifier (schematic shown in Figure 6.13). g_{m1} and $(n\Gamma)_1$ also correspond to the M_1 and M_2 input pair devices in the differential input preamplifier (schematic shown in Figure 6.8). g_{m3} is the transconductance and $(n\Gamma)_3$ is the product of substrate and thermal-noise factors for the M_3 and M_4, degenerated current source devices in both preamplifiers. g_{ms3} is the source transconductance for M_3 and M_4 in both preamplifiers. Finally, R is the degeneration resistance provided by the M_5 and M_6 deep ohmic devices in both preamplifiers.

The input-referred thermal-noise voltage PSD for the single-ended input preamplifier is given by

$$S_{VIN} \text{ (thermal)} = 4kT \cdot \left[\frac{(n\Gamma)_1}{g_{m1}} + 2\frac{(n\Gamma)_3 \, g_{m3}}{(1+g_{ms3}R)^2 \, g_{m1}^2} + 2\left(\frac{1}{R}\right)\left(\frac{g_{ms3}R}{1+g_{ms3}R}\right)^2 \left(\frac{1}{g_{m1}^2}\right) \right] \quad (6.35)$$

The first term, from Table 6.3, corresponds to the gate-referred thermal-noise voltage PSD of the M_1 input device. The second and third terms, from Equation 6.17, correspond to the drain-referred thermal-noise current PSD of the M_3 and M_4, degenerated current source devices, including noise from their degeneration resistances, R, developed by M_5 and M_6. The second and third terms are divided by the input device transconductance squared, g_{m1}^2, which refers the drain noise current PSD contributions to a preamplifier input-referred noise voltage PSD.

The thermal noise given by Equation 6.35 is similar to that given by Equation 6.29 for the differential input preamplifier. However, no overall "2" appears in front of Equation 6.35 because the single-ended input preamplifier uses single devices in the signal path. Additionally, only two non-input device terms appear because the single-ended input preamplifier does not have the additional M_7 and M_8 current-mirror stage. Finally, a "2" appears in front of the non-input device terms for the single-ended input preamplifier. This is because of noise current PSD contributions from the M_4 and M_6 bias references that form a unity-gain current mirror with M_3 and M_5.

Although single-ended preamplifiers offer lower noise because of single devices in the signal path, they may introduce bias reference device noise and power supply noise that is normally cancelled or common-moded out in differential designs. Creating a step-down current mirror by increasing the drain currents and channel widths for the M_4 and M_6 bias references compared to their mirrored devices, M_3 and M_5, reduces the noise contributions of M_4 and M_6 at the expense of additional power consumption. For example, creating a 4:1 step-down mirror lowers the "2" terms in front of the non-input device terms in Equation 6.35 to 1.25. Such a design choice might be good for a single reference system for multiple preamplifier channels where the additional power consumption of the bias reference is shared over multiple channels.

The noise analysis given by Equation 6.35 includes only noise contributions from M_1, M_3 and its bias reference M_4, and M_5 and its bias reference M_6, where, again, M_5 and M_6 provide degeneration resistances having values of R. Noise of the M_2 cascode device is neglected because of significant noise degeneration due to high external resistance at its source. As mentioned for the differential input preamplifier, the significant reduction of effective MOS transconductance and drain-referred thermal- and flicker-noise current for a cascode device having significant resistive source degeneration was discussed in Section 5.3.5.2. Additionally, the noise of M_7, which provides regulated cascode action for M_2, is neglected because the M_7, gate-referred noise voltage is converted to a small drain noise current at M_1 by division by the high value of M_1 drain–source resistance, r_{ds1}. Finally, the noise of the M_{10} source follower is neglected because the voltage gain from the preamplifier input to its gate is high, being essentially A_{VOL}.

Factoring Equation 6.35 to show the thermal-noise PSD contributions of non-input devices relative to that of the input device gives

$$S_{VIN} \text{ (thermal)} = \frac{4kT \, (n\Gamma)_1}{g_{m1}} \cdot \left[1 + 2 \left(\frac{(n\Gamma)_3}{(n\Gamma)_1} \right) \left(\frac{g_{m3}}{g_{m1}} \right) \left(\frac{1}{1 + g_{ms3}R} \right)^2 + \frac{2}{(n\Gamma)_1 \, g_{m1} R} \left(\frac{g_{ms3}R}{1 + g_{ms3}R} \right)^2 \right]$$

$$(6.36)$$

This expression is similar to thermal noise given by Equation 6.30 for the differential input preamplifier, where the term to the left of the square bracket is the input-referred thermal-noise voltage PSD of the input device, and the terms inside the square brackets correspond to the relative noise PSD contributions of the input and non-input devices. The unity term corresponds to the noise PSD contribution of the input device, while the second and third terms correspond to the additional noise PSD contributions from the non-input devices. In the *Circuit Analysis* sheet of the spreadsheet described later in Section 6.8.3, these terms will be referred to as relative noise powers as done for the differential input preamplifier. RNP_3 and RNP_5 denote the thermal-noise PSD contributions of the M_3 and M_4, degenerated current source devices, and the M_5 and M_6, deep ohmic degeneration resistances denoted by R, with all noise contributions relative to the input device. If RNP_3 and RNP_5 are sufficiently below unity, the input device dominates the thermal noise as desired for low-noise design. As mentioned for the differential input preamplifier, the thermal-noise factor Γ is considered separately for each device because it depends on the inversion level and can include small-geometry noise increases.

6.8.2.4 Thermal noise expressed from DC bias conditions

Table 6.4 showed the relative thermal-noise PSD contributions of devices in the differential input preamplifier that were discussed in Section 6.7.2.4. This table also shows the relative thermal-noise PSD contributions of devices in the single-ended input preamplifier that are described in this section. The table lists traditional expressions from Equation 6.36, expressions in terms of the DC bias conditions and bias compliance voltages, and expressions in terms of the inversion coefficient, where the last two expressions are developed as described for the differential input preamplifier. Again, expressions in terms of the DC bias conditions and bias compliance voltages permit noise minimization under the constraint of limited bias compliance voltages in low-voltage processes. Again, the expressions assume negligible g_m/I_D and g_m decreases due to velocity saturation and VFMR effects. The inclusion of these effects is not considered for the differential and single-ended input preamplifiers because these use long-channel devices to give large gate areas and low flicker noise. As mentioned in Section 6.7.2.1, velocity saturation and VFMR effects can be included by simple modifications to the g_m/I_D and g_m terms expressed in terms of IC.

The first row in Table 6.4 shows that the thermal-noise PSD contribution of the M_1 input device is 100 % relative to its own noise contribution, which would ideally be the only preamplifier noise contribution. The input-referred thermal-noise voltage PSD for the input device, noted below the table, is minimized for low $V_{DS,sat}$ and low IC in weak or moderate inversion because of high g_m/I_D and g_m. Additionally, operation at high g_m helps ensure that the input device dominates the overall preamplifier thermal and flicker noise. Increasing input device drain current, which increases g_m, also lowers the thermal-noise voltage and helps ensure that the input device dominates the overall preamplifier noise.

As noted below Table 6.4, the input-referred thermal-noise voltage density associated with the input device is $29.1 \, \text{nV}/\text{Hz}^{1/2}$ for the selected $I_{D1} = 1 \, \mu\text{A}$, $V_{DS,sat1} = 0.155 \, \text{V}$, and $IC_1 = 1.9$ (corresponding to near the center of moderate inversion). This is well below $51.7 \, \text{nV}/\text{Hz}^{1/2}$ given for both the input pair devices in the differential input preamplifier that operate at the same total supply current. MOSFET operating parameters, predicted preamplifier performance including thermal-noise contributions, and experimental measurements are summarized later in Section 6.8.3 using the spreadsheet. As mentioned in Section 6.7.2.4 for the differential input preamplifier, the spreadsheet predictions of performance will vary slightly from the design equations given because the spreadsheet uses slightly more accurate predictions. Also, the spreadsheet includes velocity saturation and VFMR effects not considered in the design equations.

The second row in Table 6.4 shows the thermal-noise PSD contribution of the M_3 and M_4, degenerated current source devices relative to the noise contribution of the input device. As noted, expressions are multiplied by "2" for the single-ended input preamplifier, corresponding to the second term inside the square brackets in Equation 6.36. The relative thermal-noise PSD contribution from M_3 and M_4 is only 2.7 % for the selected $I_{D3} = 1 \, \mu\text{A}$, $V_{DS,sat3} = 0.232 \, \text{V}$, $IC_3 = 8.4$ (near the onset of strong inversion), and $V_R = 0.5 \, \text{V}$. M_3 and M_4 are biased and sized as in the differential input preamplifier, but V_R is lower at 0.5 V compared to 0.75 V for the differential input preamplifier. Still, substantial thermal- and flicker-noise degeneration occurs for the selected value of $V_R = 0.5 \, \text{V}$, where the degeneration factor, $1/(1 + g_{ms3}R)$, given by Equation 6.20 has a value of 0.15.

As described for the differential input preamplifier, the native thermal- and flicker-noise PSD contributions of M_3 and M_4 are multiplied by the square of the degeneration factor, or 0.0225 for the single-ended input preamplifier. Without source degeneration, their thermal-noise PSD contribution would have increased from 2.7 % to 120 %. Alternatively, these devices could have been operated without degeneration deep into strong inversion at high $V_{DS,sat3}$ to reduce their thermal-noise contribution.

As described for the differential input preamplifier, the thermal-noise PSD contribution of M_3 and M_4 is primarily minimized by maximizing the value of V_R, the voltage across the source degeneration resistance, which maximizes the value of this resistance, R. Again, this comes at the expense of higher bias compliance voltage for M_3 and M_4, which is $V_{COMP} = V_{DS,sat3} + V_R$, where V_R adds directly to the value of $V_{DS,sat3}$. Again, for a given value of V_{COMP}, as described in Section 6.5.5.2, resistive source degeneration does not significantly lower the overall drain thermal-noise current at the output of M_3 and M_4 because of the additional noise of the degeneration resistances. However, for a given value of V_{COMP}, as described in Section 6.5.5.3, resistive source degeneration significantly reduces drain flicker-noise current. Lowering the drain currents for M_3 and M_4, which increases the value of R, decreases both their thermal- and flicker-noise contributions.

The third row in Table 6.4 shows the thermal-noise PSD contribution of the M_5 and M_6, deep ohmic degeneration resistances relative to the noise contribution of the input device. As noted, expressions are multiplied by "2" for the single-ended input preamplifier, corresponding to the third term inside the square brackets in Equation 6.36. The relative thermal-noise PSD contribution from the M_5 and M_6 degeneration resistances is 23.6 % for the selected $I_{D5} = 1 \, \mu\text{A}$ and $V_R = 0.5 \, \text{V}$, which gives $R = 0.5 \, \text{V}/1 \, \mu\text{A} = 500 \, \text{k}\Omega$. As for M_3 and M_4, the noise contribution of the M_5 and M_6, deep ohmic degeneration resistances is primarily minimized by maximizing V_R, which maximizes the value of degeneration resistance at the expense of higher bias compliance voltage. Lowering the drain current, which raises R, also lowers the noise contribution.

Finally, Table 6.4 shows the sum of all thermal-noise PSD contributions relative to the noise contribution of the input device. This sum is 126.3 %, where non-input devices contribute 26.3 % additional noise PSD compared to noise from the input device. This corresponds to a slight 12.4 % increase in preamplifier input-referred thermal-noise voltage density compared to that provided by the input device alone. As noted below the table, the total predicted preamplifier, input-referred thermal-noise voltage density is $32.7\,\text{nV}/\text{Hz}^{1/2}$ for the input device noise of $29.1\,\text{nV}/\text{Hz}^{1/2}$. This is well below $61.6\,\text{nV}/\text{Hz}^{1/2}$ given for the differential input preamplifier that operates at the same core supply current. For the single-ended and differential input preamplifiers, input devices have low gate-referred thermal-noise voltage for micropower drain currents of 1 and $0.5\,\mu\text{A}$, resulting from high g_m/I_D and g_m in moderate inversion. Additionally, their high g_m assists them in dominating both thermal and flicker noise, although careful management of non-input device noise is also required. As for the differential input preamplifier, the very small noise contributions from M_3 and M_4 indicate over-design, where less source degeneration could have been used for these devices. This is described later in Section 6.8.4.

6.8.2.5 Flicker noise

Flicker-noise analysis for the single-ended input preamplifier is similar to analysis given earlier in Sections 6.7.2.5 and 6.7.2.6 for the differential input preamplifier. As mentioned in the thermal-noise discussion of Section 6.8.2.3 for the single-ended input preamplifier, device numbers have been assigned to permit similar noise expressions for the two preamplifiers.

The input-referred flicker-noise voltage PSD for the single-ended input preamplifier is given by

$$S_{VIN}\,(\text{flicker}) = \frac{K'_{F1}}{(WL)_1\,f^{AF1}} + \frac{2K'_{F3}}{(WL)_3\,f^{AF3}}\left(\frac{g_{m3}}{1+g_{ms3}R}\right)^2\left(\frac{1}{g_{m1}^2}\right)$$

$$+ \frac{2K'_{F5}}{(WL)_5\,f^{AF5}} \cdot g_{m5}^2(\text{ohmic})\left(\frac{g_{ms3}R}{1+g_{ms3}R}\right)^2\left(\frac{1}{g_{m1}^2}\right) \tag{6.37}$$

For the single-ended input preamplifier (schematic shown in Figure 6.13), as for the differential input preamplifier (schematic shown in Figure 6.8), K'_{F1}, K'_{F3}, and K'_{F5} are the flicker-noise factors in units of $(\text{nV})^2 \cdot \mu\text{m}^2$ for the M_1 input device, M_3 and M_4, degenerated current source devices, and M_5 and M_6 deep ohmic devices providing degeneration resistances. $AF1$, $AF3$, and $AF5$ are the flicker-noise slopes, and $(WL)_1$, $(WL)_3$, and $(WL)_5$ are the gate areas in μm^2 for these devices. g_{m5} (ohmic) is the transconductance for the M_5 and M_6 deep ohmic devices.

The first term in Equation 6.37, from Table 6.3, corresponds to the gate-referred flicker-noise voltage PSD of the M_1 input device. The second term, from Equation 6.23, corresponds to the drain-referred flicker-noise current PSD of the M_3 and M_4, degenerated current source devices. The third term, from Equation 6.25, corresponds to the drain-referred flicker-noise current PSD of the M_5 and M_6 deep ohmic devices acting as source degeneration resistors for M_3 and M_4. In the second and third terms, the needed gate-referred flicker-noise voltage PSD is found from Table 6.3. As in the previous thermal-noise analysis, the second and third terms are divided by g_{m1}^2 to refer the drain noise current PSD contributions to a preamplifier input-referred noise voltage PSD. Also, as in the previous thermal-noise analysis, only noise contributions from M_1, M_3, M_5, and bias references M_4 and M_6 are considered as noise contributions from the other devices are negligible.

The flicker noise given by Equation 6.37 is similar to that given in Equation 6.31 for the differential input preamplifier. However, as for the thermal noise, no overall "2" appears in front of Equation 6.37 because the single-ended input preamplifier uses single devices in the signal path. Additionally, as for the thermal noise, only two non-input device terms appear, and a "2" appears in front of these terms because of noise current PSD contributions from the M_4 and M_6 bias references, which form a unity-gain current mirror with M_3 and M_5. As mentioned for the thermal noise, creating a step-down current mirror by increasing the currents and widths of the M_4 and M_6 bias references compared to their mirror devices, M_3 and M_5, reduces the noise contributions of the M_4 and M_6 bias references. This, again, comes at the expense of increased power consumption.

Analogous to the previous thermal-noise analysis, Equation 6.37 can be factored to show the flicker-noise PSD contributions of non-input devices relative to that of the input device. This gives

$$S_{VIN} \text{ (flicker)} = \frac{K'_{F1}}{(WL)_1 f^{AF1}} \cdot$$

$$\left[1 + 2f^{(AF1-AF3)} \left(\frac{K'_{F3}}{K'_{F1}} \right) \left(\frac{(WL)_1}{(WL)_3} \right) \left(\frac{g_{m3}}{g_{m1}} \right)^2 \left(\frac{1}{1+g_{ms3}R} \right)^2 \right.$$

$$\left. + 2f^{(AF1-AF5)} \left(\frac{K'_{F5}}{K'_{F1}} \right) \left(\frac{(WL)_1}{(WL)_5} \right) \left(\frac{g_{m5}(\text{ohmic})}{g_{m1}} \right)^2 \left(\frac{g_{ms3}R}{1+g_{ms3}R} \right)^2 \right]$$

$$(6.38)$$

This expression is similar to flicker noise given by Equation 6.32 for the differential input preamplifier, where the term to the left of the square bracket is the input-referred thermal-noise voltage PSD of the input device, and the terms inside the square brackets correspond to the relative noise PSD contributions of the input and non-input devices. The unity term corresponds to the noise PSD contribution of the input device, while the second and third terms correspond to the additional noise PSD contributions from the non-input devices. In the *Circuit Analysis* sheet of the spreadsheet described later in Section 6.8.3, these terms will again be referred to as relative noise powers, where RNP_3 and RNP_5 denote the flicker-noise PSD contributions of the M_3 and M_4, degenerated current source devices, and the M_5 and M_6 deep ohmic devices, with all noise contributions relative to the input device. If RNP_3 and RNP_5 are sufficiently below unity, the input device dominates the flicker noise as desired for low-noise design. As mentioned for the differential input preamplifier, the frequency correction terms that appear in front of the non-input device noise contributions are required for frequencies away from $f = 1\,Hz$ when input and non-input device flicker-noise slopes are unequal and non-input device noise contributions are significant.

6.8.2.6 Flicker noise expressed from DC bias conditions

Table 6.5 showed the relative flicker-noise PSD contributions of devices in the differential input preamplifier that were discussed in Section 6.7.2.6. This table also shows the relative flicker-noise PSD contributions of devices in the single-ended input preamplifier that are described in this section. The table lists traditional expressions from Equation 6.38, expressions in terms of the DC bias conditions and bias compliance voltages, and expressions in terms of the inversion coefficient, where the last two expressions are developed as described for the differential input preamplifier. Again, expressions in terms of the DC bias conditions and bias compliance voltages permit noise minimization under the constraint of limited bias compliance voltages in low-voltage processes. Again, the expressions assume negligible g_m/I_D and g_m decreases due to velocity saturation and VFMR effects. The inclusion of these effects is not considered for the differential and single-ended input preamplifiers because these use long-channel devices to give sufficient gate areas and minimize flicker noise. As mentioned in Section 6.7.2.1, velocity saturation and VFMR effects can be included by simple modifications to the g_m/I_D and g_m terms expressed in terms of IC.

The first row in Table 6.5 shows that the flicker-noise PSD contribution of the M_1 input device is 100 % relative to its own noise contribution, which would ideally be the only preamplifier noise contribution. The input-referred flicker-noise voltage PSD for the input device, noted below the table, is minimized at low $V_{DS,sat}$ and low IC in weak or moderate inversion because channel width and gate area are large here. Operation at low IC also minimizes the thermal-noise voltage, noted below Table 6.4, because of high g_m/I_D and g_m, and helps ensure that the input device dominates both thermal and flicker noise. Although operation at low IC decreases flicker noise, increasing channel length decreases the noise more because gate area increases as the square of channel length. As always, this assumes drain current and inversion coefficient are held fixed as channel length is varied. Finally, increasing input device drain current, which requires proportionally larger channel width and gate area, decreases the flicker noise and increases g_m. The increase in g_m again lowers thermal noise and helps ensure that the input device dominates overall preamplifier noise.

As noted below Table 6.5, the input-referred flicker-noise voltage density associated with the input device is 145.7 nV/Hz$^{1/2}$ at $f = 1$ Hz for the selected $I_{D1} = 1\,\mu$A, $V_{DS,sat1} = 0.155$ V, $IC_1 = 1.9$ (corresponding to near the center of moderate inversion), and $L = 20\,\mu$m. This results in a required $W = 100\,\mu$m found by $W = (L/IC) \cdot (I_D/I_0)$ from Table 3.10, where I_0 is the technology current given in the process information of Table 3.2. The input device noise is below 192.8 nV/Hz$^{1/2}$ given for both input pair devices in the differential input preamplifier that operate at the same total supply current (gate-referred flicker-noise voltage depends only on the drain current through the required channel width and resulting gate area, and potentially on the inversion level). As mentioned for the thermal noise, MOSFET operating parameters, predicted preamplifier performance including flicker-noise contributions, and experimental measurements are summarized later in Section 6.8.3 using the spreadsheet. Again, spreadsheet predictions of performance will vary slightly from the design equations given because the spreadsheet uses slightly more accurate predictions and includes velocity saturation and VFMR effects.

The second row in Table 6.5 shows the flicker-noise PSD contribution of the M_3 and M_4, degenerated current source devices relative to the noise contribution of the input device. As noted, expressions are multiplied by "2" for the single-ended input preamplifier, corresponding to the second term inside the square brackets in Equation 6.38. The relative flicker-noise PSD contribution from M_3 and M_4 is only 4.0 % for the selected $I_{D3} = 1\,\mu$A, $V_{DS,sat3} = 0.232$ V, $IC_3 = 8.4$ (near the onset of strong inversion), $L = 75\,\mu$m (resulting in a required $W = 30\,\mu$m), and $V_R = 0.5$ V. As for the thermal noise, the flicker-noise PSD contribution of M_3 and M_4 is primarily minimized by maximizing the value of V_R, the voltage across the source degeneration resistance, which maximizes the value of this resistance, R. Again, this comes at the expense of higher bias compliance voltage for M_3 and M_4, which is $V_{COMP} = V_{DS,sat3} + V_R$, where V_R adds directly to the value of $V_{DS,sat3}$. Lowering the drain currents for M_3 and M_4, which also increases for value of R, also decreases their flicker-noise contributions.

As mentioned for the thermal noise, substantial thermal- and flicker-noise degeneration occurs for the selected value of $V_R = 0.5$ V, where the degeneration factor, $1/(1 + g_{ms3}R)$, given by Equation 6.20 has a value of 0.15. When the degeneration factor of 0.15 is squared, this indicates that the native flicker-noise PSD contribution of M_3 and M_4 is multiplied by 0.0225, or reduced by a factor of 44. Thus, without source degeneration, the flicker-noise PSD contribution of M_3 and M_4 would have been 176 % and these devices would have nearly dominated the preamplifier flicker noise. As mentioned for the differential input preamplifier, these devices could have been operated (with sufficient gate area) deep into strong inversion without source degeneration to lower their flicker noise, but, as described in Section 6.5.5.3, degeneration reduces the flicker noise for a given value of V_{COMP}. Although resistive source degeneration reduces the flicker-noise contribution of M_3 and M_4, their gate area must also be sufficiently large. Minimizing the $[K'_{F3}/K'_{F1}] \cdot [(WL)_1/(WL)_3]$ terms in Table 6.5 also helps minimize the flicker-noise contribution. In both the single-ended and differential input preamplifiers, unusually long channel lengths, which result in long channel widths, are used to ensure sufficient gate area and low flicker noise.

The third row in Table 6.5 shows the flicker-noise PSD contribution of the M_5 and M_6 deep ohmic devices relative to the noise contribution of the input device. These devices operate in the deep ohmic region providing degeneration resistance, but introduce flicker noise as described earlier in Section 6.5.5.3. As noted, expressions are multiplied by "2" for the single-ended input preamplifier, corresponding to the third term inside the square brackets in Equation 6.38. The relative flicker-noise PSD contribution from the M_5 and M_6 deep ohmic devices is only 2.9 % for the selected $I_{D5} = 1\,\mu$A, $V_{EFF5} = 2.65$ V, $W = 200\,\mu$m, and $L = 1\,\mu$m, required for the value of $R = 0.5$V/1 μA $= 500$ kΩ As mentioned for the differential input preamplifier, this noise contribution is minimized by maximizing V_{EFF5} relative to $V_{DS,sat1}$, which minimizes the g_{m5}(ohmic)$/g_{m1}$ ratio. The noise contribution is also reduced by lowering the drain currents in M_5 and M_6. Although the collapse of device transconductance resulting from operation in the deep ohmic region significantly reduces the flicker-noise contribution of M_5 and M_6, their gate area must also be sufficiently large. Minimizing the $[K'_{F5}/K'_{F5}] \cdot [(WL)_1/(WL)_3]$ terms in Table 6.5 also helps minimize the flicker-noise contribution.

Finally, Table 6.5 shows the sum of all flicker-noise PSD contributions relative to the noise contribution of the input device. This sum is 106.9%, where non-input devices contribute 6.9% additional noise PSD compared to noise from the input device. This corresponds to a slight 3.4% increase in preamplifier input-referred flicker-noise voltage density compared to that provided by the input device alone. As noted below the table, the total predicted preamplifier, input-referred flicker-noise voltage density is 150.7 nV/Hz$^{1/2}$ at $f = 1$ Hz for the input device noise of 145.7 nV/Hz$^{1/2}$. This is well below 233.5 nV/Hz$^{1/2}$ given for the differential input preamplifier that operates at the same core supply current. Like input devices in the differential input preamplifier, the input device has low gate-referred flicker-noise voltage resulting from large gate area, primarily resulting from the selection of long channel length. Additionally, the high g_m in moderate inversion assists the input device in dominating both thermal and flicker noise, although careful management of non-input device noise is also required. As for the differential input preamplifier, the very small noise contributions from M_3 and M_4 indicate over-design, where less source degeneration and gate area could have been used for these devices. This is described later in Section 6.8.4.

6.8.3 Summary of Predicted and Measured Performance

Section 6.8.2 presented circuit analysis, performance optimization, and predicted performance for voltage gain, bandwidth, thermal noise, and flicker noise for the single-ended input preamplifier. This section summarizes predicted and measured performance for the preamplifier using the *Analog CMOS Design, Tradeoffs and Optimization* spreadsheet.

The spreadsheet maps predicted device performance for selections of the inversion coefficient, channel length, and drain current into complete circuit performance. The spreadsheet and the 0.35 μm SOI CMOS process parameters used for both the single-ended and differential input preamplifiers were described in Section 6.7.3. As mentioned, the spreadsheet is described further in the Appendix and is available at the web site for this book listed on the cover.

Because the spreadsheet uses slightly more accurate predictions of MOS performance, as described in Section 6.7.3, spreadsheet predictions of performance will vary slightly from the design equations given in Section 6.8.2. Additionally, the spreadsheet includes velocity saturation and VFMR effects not considered in the design equations. These effects, which are nearly negligible for the long-channel devices used, can be included in the design equations by modifications described in Section 6.7.2.1.

6.8.3.1 MOSFET design selections

Figure 6.9 showed the *MOSFETs* sheet in the spreadsheet for the differential input preamplifier (schematic shown in Figure 6.8) described earlier in Section 6.7. This sheet also includes devices for the single-ended input preamplifier (schematic shown in Figure 6.13) described here. Device performance given by the *MOSFETs* sheet was mapped into the *Circuit Analysis* sheet of Figure 6.11 in Section 6.7.3.2 to give predicted performance for the differential input preamplifier. Device performance is mapped in the next section to give predicted performance for the single-ended input preamplifier.

The M_1 column in Figure 6.9 for the single-ended input preamplifier describes the pMOS input device. The M_3, M_4 column describes the nMOS current source devices that use external resistive source degeneration.

As mentioned in Section 6.7.3.1 for the differential input preamplifier, the *MOSFETs* sheet considers only devices in saturation. As a result, the M_5 and M_6 devices, which operate in the deep ohmic region to provide degeneration resistances for M_3 and M_4, are considered separately in Figure 6.10 as part of the *Circuit Analysis* sheet in the spreadsheet. This figure was shown earlier for M_5 and M_6 used in the differential input preamplifier and includes information for M_5 and M_6 used in the single-ended input preamplifier. The figure summarizes the degeneration voltage, V_R, degeneration resistance, R, degeneration factor, $1/(1 + g_{ms3}R)$, current-division factor, $g_{ms3}R/(1 + g_{ms3}R)$, ohmic transconductance, g_{ms5} (ohmic), gate area, $(WL)_5$, and flicker-noise factor, K'_{F5}, associated with M_5 and M_6. These parameters were discussed earlier in Section 6.8.2.3–6.8.2.6.

As described in Section 6.8.2 and summarized in Figure 6.9, the M_1, pMOS input device operates at an inversion coefficient of 1.9 near the center of moderate inversion for high g_m/I_D and g_m. This minimizes its gate-referred thermal-noise voltage and helps ensure that it dominates both preamplifier thermal and flicker noise. Additionally, this helps maximize preamplifier open-loop voltage gain. A channel length of $20\,\mu m$ is selected for M_1. This results in a required width of $100\,\mu m$ found from $W = (L/IC) \cdot (I_D/I_0)$ from Table 3.10 for $I_D = 1\,\mu A$, assuming a pMOS process $I_0 = 0.105\,\mu A$ from Table 3.2. In addition to minimizing gate-referred thermal-noise voltage for operation at a low inversion coefficient, this combined with operation at a long channel length results in large channel width and gate area, minimizing the gate-referred flicker-noise voltage.

If M_1 were operated at an inversion coefficient of 0.95, like the input devices in the differential input preamplifier, its required channel width would be $200\,\mu m$. This would have resulted in slightly lower gate-referred thermal-noise voltage because of slightly higher g_m/I_D and g_m at the lower inversion coefficient. However, this would have resulted in a more significant reduction in the gate-referred flicker-noise voltage because of the doubling of gate area. This design for M_1 would have been preferable to give a nearly equivalent single-ended version of the differential input preamplifier, where the input-referred thermal- and flicker-noise voltage densities would have been nearly one-half those of the differential input preamplifier. This is described further in the suggested design improvements given in Section 6.8.4.

As summarized in Figure 6.9, the M_3 and M_4, nMOS, current source devices for both the single-ended and differential input preamplifiers operate at inversion coefficients of 8.4 near the onset of strong inversion, which gives lower g_m/I_D compared to the input devices. Their effective g_m, however, is significantly reduced by multiplication with the degeneration factor given by Equation 6.20 and summarized in Figure 6.10. This has a value of 0.15 for the single-ended and 0.1 for the differential input preamplifier for degeneration voltages, V_R, of 0.5 V and 0.75 V, respectively. As described in Sections 6.7.2.4, 6.7.2.6, 6.8.2.4, and 6.8.2.6, the thermal- and flicker-noise contributions of M_3 and M_4 would have exceeded the input device noise for both preamplifiers if source degeneration were not used. However, as mentioned, these devices could have been operated without degeneration deep into strong inversion at high $V_{DS,sat3}$ to reduce their noise contributions, but degeneration reduces the flicker-noise contribution for a given bias compliance voltage as described in Section 6.5.5.3. Channel lengths of $75\,\mu m$ are selected for M_3 and M_4, which results in required widths of $30\,\mu m$ for $I_D = 1\,\mu A$, assuming an nMOS process $I_0 = 0.3\,\mu A$ from Table 3.2. The resulting large gate area also minimizes the flicker-noise contributions of M_3 and M_4. Because the flicker-noise contributions of these devices are small, their channel length and corresponding width and gate area could have been reduced.

The M_5 and M_6, nMOS, deep ohmic devices summarized in Figure 6.10 operate with drain–source voltage drops of $V_R = 0.5\,V$ at $I_D = 1\,\mu A$, giving degeneration resistance values of $R = 0.5\,V/1\,\mu A = 500\,k\Omega$. These devices are sized at $W = 1\,\mu m$ and $L = 200\,\mu m$ with their gate–source voltages connected to the full supply voltage of 3.3 V to ensure deep ohmic operation and low ohmic transconductance, g_{ms5} (ohmic). As described in Section 6.5.5.3, low g_{ms5} (ohmic) minimizes the drain-referred flicker-noise current of M_5 and M_6, which are intended to operate as resistors having only thermal noise. As described in Section 6.5.2, only the DC bias conditions are required to predict the deep ohmic, drain–source resistance and transconductance of M_5 and M_6, but these must be carefully sized by considering the substantial loss of mobility due to VFMR at the high $V_{EFF} = V_{GS} - V_T$.

6.8.3.2 Resulting preamplifier performance

Figure 6.14 shows the *Circuit Analysis* sheet in the spreadsheet for the single-ended input preamplifier. Like the sheet shown in Figure 6.11 for the differential input preamplifier, this sheet gives predicted open-loop voltage gain, unity-gain bandwidth, thermal noise, and flicker noise from the circuit analysis equations developed earlier in Section 6.8.2. The *Circuit Analysis* sheet calls geometries and performance from the *MOSFETs* sheet of Figure 6.9 for the M_1 input device and M_3 and M_4, degenerated current source devices. Additionally, the *Circuit Analysis* sheet calls the portion of the sheet given in Figure 6.10 that lists details for the M_5 and M_6, deep ohmic degeneration devices and the degeneration and current-division factors required for circuit analysis. As described for the differential input preamplifier, the *Circuit Analysis* sheet provides automatic updating of predicted preamplifier performance as MOSFET drain current, inversion coefficient, and channel length selections, and degeneration selections are explored for optimization. To permit comparisons with the predicted values, the *Circuit Analysis* sheet also lists measured performance.

As described in Equation 6.33 of Section 6.8.2.1 and summarized in the *Circuit Analysis* sheet of Figure 6.14, the preamplifier open-loop voltage gain, A_{VOL}, is approximated by the transconductance of the M_1 input device, $g_{m1} = 15.5\,\mu$S, multiplied by the drain–source resistance of the M_3, degenerated current source device, $r_{ds3} = 750\,M\Omega$ ($1/g_{ds3}$ given in Figure 6.9), followed by multiplication by the degeneration factor, $1 + g_{ms3}R = 6.8$ (given in Figure 6.10). This gives a predicted gain of 78723 V/V, which is 55 % above the measured gain of 50900 V/V, also listed in Figure 6.14.

For the measured open-loop voltage gain of 50900 V/V, the negative feedback loop gain is 998 V/V(50900/(50 + 1)) for a preamplifier, inverting, closed-loop voltage gain of 50 V/V intended for the application. This loop gain is sufficiently high to ensure feedback regulation of the closed-loop voltage gain, even for significant variations in the open-loop gain. Although the gain is overpredicted by 55% in the spreadsheet, the gain was underpredicted by a greater amount in computer simulations using BSIM3V3 MOS models [35]. As discussed later in Section 6.9, the largest circuit prediction errors are usually for the prediction of drain–source resistances and the resulting circuit gains. As described in Section 3.8.4.4, this is especially significant for cascode and regulated cascode circuits where errors in drain–source resistance are effectively squared or cubed.

The measured open-loop voltage gain of 50900 V/V for the single-ended input preamplifier is over three times greater than the measured gain of 15500 V/V for the differential input preamplifier. This is primarily the result of resistive source degeneration in M_3, which raises its drain output resistance and r_{DOM} well above the native drain–source resistance of M_3. In the differential input preamplifier, r_{DOM} is set by the native drain–source resistance of M_8 because resistive source degeneration is not used for this device.

As mentioned for the differential input preamplifier, the measurement of high open-loop voltage gain requires careful laboratory methods to ensure that the output remains within its full-gain, linear range. Again, the preamplifier open-loop voltage gain was measured in a closed-loop configuration by measuring the loop gain response to an injected error signal.

As described in Equation 6.34 of Section 6.8.2.2 and summarized in Figure 6.14, the preamplifier unity-gain bandwidth, f_T, is approximated by $g_{m1} = 15.5\,\mu$S divided by 2π and the sum of the dominant-pole compensation capacitor, C_{DOM}, and the drain–body and drain–gate capacitances of M_3, C_{DB3} and C_{GD3}. With a value of 1000 fF, C_{DOM} dominates, giving a predicted unity-gain bandwidth of 2.4 MHz. The measured bandwidth is 1.8 MHz (36 kHz · (50 + 1)) found from the measured closed-loop, −3 dB frequency of 36 kHz for an inverting voltage gain of 50 V/V. The unity-gain bandwidth is likely overpredicted because the prediction excludes the capacitive loading from the cascode device, M_2, the source follower, M_{10}, and the layout.

Preamplifier thermal noise was described in detail in Sections 6.8.2.3 and 6.8.2.4 and was summarized in Table 6.4 in terms of noise PSD contributions from the M_3 and M_4, degenerated current source and M_5 and M_6, deep-ohmic degeneration resistance devices. These noise PSD contributions were given relative to the noise of the M_1 input device and, as for the differential input preamplifier,

Summary of Predicted and Measured Performance for 0.35–um SOI CMOS Micropower Preamp.							
*** Single-Ended Input Version ***							
MOSFET Notes							
*M*1 is input device. The following devices are biased and sized equally: *M*3 = *M*4 (current source and bias reference) and *M*5 = *M*6 (deep ohmic degeneration resistors). Other devices have little effect on performance and are not included in circuit analysis.							
Open Loop Voltage Gain, A_V							
$A_V = g_{m1} \cdot r_{ds3}(1+g_{ms\,3}R) = (g_m/I_D)_1 \cdot V_{A\,3}(1+g_{ms\,3}R);$ Drain currents are equal.							
Regulated cascoded $R_{outD\,2}$ neglected.							
					Predicted	Measured	
					78723	**50900**	V/V
Unity-Gain Bandwidth Product, f_T							
$f_T = g_{m1} / [2\pi(C_{DOM}+C_{DB3}+C_{GD3})];$ $C_{DB\,2}$ and $C_{GD\,2}$ are neglected.							
	C_{DOM}	1000	fF		Predicted	Measured	
					2.43	**1.80**	MHz
Input-Referred Thermal-Noise Voltage, $S_{VIN}^{1/2}$ (thermal)							
$S_{VIN}^{1/2} = \sqrt{[\{4kT(n\Gamma)_1/g_{m1}\}\cdot(1+RNP_3+RNP_5)]}$							
					Predicted	Measured	
					32.7	**35.3**	nV/Hz$^{1/2}$
Relative Noise Power (*RNP*) -- relative to *M*1						Predicted	
*M*1	1					1	
*M*3, *M*4	$RNP_3 = 2\cdot[(n\Gamma)_3/(n\Gamma)_1]\cdot[g_{m3}/g_{m1}]\cdot[1/(1+g_{ms\,3}R)]^2$					0.027	
*M*5, *M*6	$RNP_5 = 2\cdot\{1/[(n\Gamma)_1 g_{m1}R]\}\cdot[g_{ms\,3}R/(1+g_{ms\,3}R)]^2$					0.236	
Total:						1.263	
Input-Referred Flicker-Noise Voltage, $S_{VIN}^{1/2}$ (flicker)							
$S_{VIN}^{1/2}(f) = \sqrt{[\{K'_{F1}/((WL)_1 f^{AF1})\}\cdot(1+RNP_3+RNP_5)]}$							
					Predicted	**Measured**	(1 Hz)
					150.7	**156.0**	nV/Hz$^{1/2}$
Relative Noise Power (*RNP*) -- relative to *M*1 at *f* = 1 Hz							
RNP_3 adjusted for frequency by (f^{AF1}/f^{AF3}), etc.						Predicted	(1 Hz)
*M*1	1					1	
*M*3, *M*4	$RNP_3 = 2\cdot[K'_{F3}/K'_{F1}]\cdot[(WL)_1/(WL)_3]\cdot[g_{m\,3}/g_{m1}]^2\cdot[1/(1+g_{ms\,3}R)]^2$					0.040	
*M*5, *M*6	$RNP_5 = 2\cdot[K'_{F5}/K'_{F1}]\cdot[(WL)_1/(WL)_5]\cdot[g_{ms\,3}/g_{m1}]^2\cdot[g_{ms\,3}R/(1+g_{ms\,3}R)]^2$					0.029	
Total:						1.069	

Figure 6.14 *Circuit Analysis* sheet in the *Analog CMOS Design, Tradeoffs and Optimization* spreadsheet for the single-ended input preamplifier. This sheet lists predicted and measured performance for the preamplifier

were referred to as relative noise powers, *RNPs*. Figure 6.14 summarizes *RNP* values from Table 6.4 using traditional equations based on device transconductance and degeneration resistance values. As mentioned for the differential input preamplifier, Table 6.4 additionally includes equations based on the DC bias conditions and the bias compliance voltage. These facilitate the minimization of noise against the constraint of limited bias compliance voltage.

RNP_3 and RNP_5 values for the thermal noise are 2.7% and 23.6%, corresponding to relative noise PSD contributions from the M_3 and M_4, and M_5 and M_6, non-input devices, respectively. The total relative noise PSD contribution is 126.3% or 1.263 as an absolute factor, which includes the 100% noise contribution from the M_1 input device. The total preamplifier, input-referred thermal-noise voltage density, also summarized in Figure 6.14, is equal to the gate-referred thermal-noise voltage density of 29.1 nV/Hz$^{1/2}$ from Figure 6.9 for the M_1 input device multiplied by $\sqrt{1.263} = 1.124$ to consider the noise density contributions from the non-input devices. This gives a total preamplifier, input-referred thermal-noise voltage density of 32.7 nV/Hz$^{1/2}$, which is 53% or nearly one-half of 61.6 nV/Hz$^{1/2}$ predicted for the differential input preamplifier. Reduced thermal noise for the single-ended input preamplifier is a result of a single input device operating at twice the drain current of the individual pair devices in the differential input preamplifier. Predicted and measured noise values will be compared later.

Preamplifier flicker noise was described in detail in Sections 6.8.2.5 and 6.8.2.6 and was summarized in Table 6.5 in terms of noise PSD contributions from the M_3 and M_4, and M_5 and M_6 devices relative to the noise of the M_1 input device. As for the thermal noise, these relative noise contributions were referred to as RNP_3 and RNP_5. Figure 6.14 summarizes RNP values for flicker noise from Table 6.5 using traditional equations based on device transconductance, gate area, and degeneration resistance values. Like Table 6.4 given for the thermal noise, Table 6.5 additionally includes equations based on the DC bias conditions and the bias compliance voltage.

RNP_3 and RNP_5 values for the flicker noise are 4.0% and 2.9%, corresponding to relative noise PSD contributions from the M_3 and M_4, and M_5 and M_6, non-input devices, respectively. The total relative noise PSD contribution is 106.9% or 1.069 as an absolute factor, which includes the 100% noise contribution from the M_1 input device. The total preamplifier, input-referred flicker-noise voltage density, also summarized in Figure 6.14, is equal to the gate-referred flicker-noise voltage density of 145.7 nV/Hz$^{1/2}$ at 1 Hz from Figure 6.9 for the M_1 input device multiplied by $\sqrt{1.069} = 1.034$ to consider the noise density contributions of all devices. This gives a total preamplifier, input-referred flicker-noise voltage density of 150.7 nV/Hz$^{1/2}$ at 1 Hz. This is 65% of 233.5 nV/Hz$^{1/2}$ at 1 Hz predicted for the differential input preamplifier.

The total preamplifier noise is the sum of the thermal- and flicker-noise PSD or the square root of the sum of the squared noise densities. For the predicted input-referred thermal-noise voltage density of 32.7 nV/Hz$^{1/2}$ and flicker-noise voltage density of 150.7 nV/Hz$^{1/2}$ at 1 Hz, the total predicted noise voltage density is increased slightly from the flicker noise alone to 152.4 nV/Hz$^{1/2}$ at 1 Hz. This is 63% of 241.5 nV/Hz$^{1/2}$ at 1 Hz predicted for the differential input preamplifier.

Figure 6.15 shows the predicted and measured, input-referred noise voltage density for the single-ended input preamplifier. The preamplifier was configured for an inverting, closed-loop voltage gain of 50 V/V, set by external $R_F = 255$ kΩ and $R_G = 5.1$ kΩ feedback resistors. The measured -3 dB frequency was 36 kHz, which is below the predicted closed-loop frequency of 47 kHz $((f_T = 2.4 \text{ MHz})/(50 + 1))$. No frequency peaking was observed, indicating sufficient phase margin. As discussed at the beginning of Section 6.8.2, no analysis of preamplifier parasitic poles was presented here as the dominant-pole compensation and high closed-loop gain ensure good negative feedback stability.

The measured input-referred noise voltage density is 160 nV/Hz$^{1/2}$ at 1 Hz, compared to the predicted noise of 152.4 nV/Hz$^{1/2}$ shown by the triangles in Figure 6.15. The average measured thermal-noise floor is 35.3 nV/Hz$^{1/2}$, which is 8% above the predicted thermal noise of 32.7 nV/Hz$^{1/2}$. Removing the measured thermal noise of 35.3 nV/Hz$^{1/2}$ from the measured noise of 160 nV/Hz$^{1/2}$ at 1 Hz gives a measured flicker noise of 156 nV/Hz$^{1/2}$ at 1 Hz, compared to the predicted flicker noise of 150.7 nV/Hz$^{1/2}$ at 1 Hz. The measured and predicted flicker-noise overlay in the figure, being within the measurement uncertainty. Finally, the measured flicker-noise corner frequency, found by the intersection of flicker-noise and thermal-noise asymptotes, is very low for MOS circuits at 19 Hz.

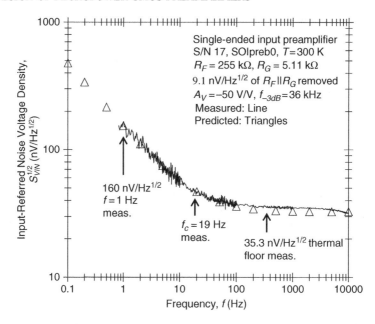

Figure 6.15 Measured (line) and predicted (triangles) input-referred noise voltage density for the single-ended input preamplifier. Predicted noise is from the *Circuit Analysis* sheet in the *Analog CMOS Design, Tradeoffs and Optimization* spreadsheet shown in Figure 6.14

As mentioned in the description of Section 6.8.1 and observed in the schematic diagram of Figure 6.13, the single-ended input preamplifier operates at a core current consumption of $2\,\mu$A, giving a core power consumption of $6.6\,\mu$W for $V_{DD} = 3.3\,$V. This is equal to the core power consumption of the differential input preamplifier shown in Figure 6.8. As for the differential input preamplifier, the core power consumption excludes the power consumption of bias reference devices, which can be common for multiple channels, and the source follower, which can be omitted for on-chip capacitive loads. The core power consumption of $6.6\,\mu$W will be used for predicting the preamplifier thermal-noise efficiency factor.

The thermal-noise efficiency factor for the singled-ended input preamplifier is $NEF' = 11.6$, found from Equation 6.2. This is the lowest value for the preamplifiers reported in Table 6.1. The NEF' of 11.6 is nearly a factor-of-four lower than 38 found for the differential input preamplifier, as described in Section 6.7.3.2. As discussed in Section 6.3.1, the NEF' for a single-ended design is a factor of four lower than that for an equivalent differential design. As a result, potentially four times less supply current is required for a single-ended design compared to a differential design to obtain equivalent thermal-noise performance.

Both the differential and single-ended input preamplifiers use an input stage current of 1 μA. However, the differential input preamplifier uses an additional current of 1 μA for non-input devices, while the single-ended input preamplifier uses an additional current of 1 μA for devices providing regulated cascode action. As described in Sections 6.7.4 and 6.8.4, operation at the input stage current only would halve the thermal-noise efficiency factor for each preamplifier.

Figure 6.16 shows a layout plot for both the differential and single-ended input preamplifiers. Although photomicrographs were given in Figures 5.10 and 5.23 for the OTA design examples described in Chapter 5, the thick oxide layers associated with the SOI CMOS process could not be successfully etched away to permit photomicrographs of the preamplifiers.

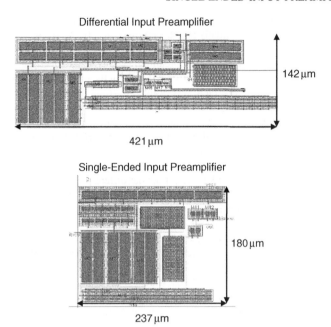

Differential Input Preamplifier

142 μm

421 μm

Single-Ended Input Preamplifier

180 μm

237 μm

Figure 6.16 Layout plot of the differential and single-ended input preamplifiers with layout dimensions. A photomicrograph was unattainable because it was not possible to successfully etch the thick oxide layers present in the SOI CMOS process

The layout dimensions are approximately 421×142 and $237 \times 180\,\mu\text{m}$ for the differential and single-ended input preamplifiers, respectively. This corresponds to approximate layout areas of 59800 and $42700\,\mu\text{m}^2$. Because of open areas in the layouts, the layout areas cannot be directly compared. However, the layout area is less, as expected, for the single-ended input preamplifier that uses fewer devices compared to the differential input preamplifier.

6.8.4 Design Improvements

The single-ended input preamplifier has a measured input-referred thermal-noise voltage density of $35.3\,\text{nV}/\text{Hz}^{1/2}$, resulting primarily from a single input device operating at $IC = 1.9$ near the center of moderate inversion at a micropower drain current of $1\,\mu\text{A}$. As estimated mathematically in Section 6.7.4 for the differential input preamplifier, the noise could be lowered modestly by operation closer to weak inversion, which gives higher g_m/I_D and g_m. As mentioned, this comes at the expense of potentially significantly increased layout area and input capacitance.

The measured input-referred noise voltage density for the single-ended input preamplifier is $160\,\text{nV}/\text{Hz}^{1/2}$ at 1 Hz with a flicker-noise corner frequency of 19 Hz, resulting primarily from flicker noise from a large-area input device. Increasing input device area would lower the flicker-noise, but at the expense of increased layout area and input capacitance.

As mentioned for the differential input preamplifier, increasing input device drain current while increasing channel width to maintain the selected inversion coefficient decreases both the input-referred thermal- and flicker-noise voltage density by the inverse square root of the drain current. Again, this assumes constant non-input device noise PSD contributions, which can be arranged by also scaling non-input device channel width with increased drain current. Again, at higher drain currents, preamplifier channel lengths would normally be reduced from the large values used in this micropower design.

The input device for the single-ended input preamplifier was sized equally at $W = 100\,\mu\mathrm{m}$ and $L = 20\,\mu\mathrm{m}$ as the input pair devices operating at $I_D = 0.5\,\mu\mathrm{A}$ in the differential input preamplifier. For the higher single device drain current of $1\,\mu\mathrm{A}$, this resulted in $IC = 1.9$ versus $IC = 0.95$ for input pair devices in the differential input preamplifier. Doubling the channel width of the input device in the single-ended input preamplifier would have resulted in $IC = 0.95$, giving slightly higher g_m/I_D and g_m, and slightly lower gate-referred thermal-noise voltage density. However, this would have reduced the gate-referred flicker-noise voltage density by $1/\sqrt{2}$ through the doubling of device gate area.

Doubling the channel width of the input device in the single-ended input preamplifier would have resulted in a nearly equivalent single-ended version of the differential input preamplifier where both thermal- and flicker-noise voltage density would be approximately one-half that of the differential input preamplifier. Here, all input devices would operate at $IC = 0.95$, but the single-ended input preamplifier device operates at twice the drain current with now twice the gate area of the individual pair devices in the differential input preamplifier. Since the measured input-referred thermal-noise voltage density for the single-ended input preamplifier is $35.3\,\mathrm{nV/Hz^{1/2}}$ or $55\,\%$ of $63.8\,\mathrm{nV/Hz^{1/2}}$ measured for the differential input preamplifier, only a slight decrease in thermal noise would be expected. However, the measured noise at $1\,\mathrm{Hz}$, which is dominated by the flicker noise, for the single-ended input preamplifier is $160\,\mathrm{nV/Hz^{1/2}}$ or $67\,\%$ of $240\,\mathrm{nV/Hz^{1/2}}$ measured for the differential input preamplifier. This would decrease in the vicinity of $120\,\mathrm{nV/Hz^{1/2}}$ if the input device channel width were doubled in the single-ended input preamplifier.

Noise predictions for the single-ended input preamplifier indicate that non-input devices contribute only $12.4\,\%$ and $3.4\,\%$ additional thermal- and flicker-noise voltage density, respectively, above that generated by the input device. These noise contributions are lower than those for the differential input preamplifier, primarily because no pMOS, M_7 and M_8 current-mirror devices are used. In addition to generating thermal noise, the flicker noise of these devices increased significantly because of their operation deep into strong inversion.

As observed also for the differential input preamplifier, the essentially negligible noise contributions from the M_3 and M_4, degenerated current source devices suggest a substantial over-design of too much resistive source degeneration and too much gate area for these devices. The degree of degeneration could be decreased from the level provided by $V_R = 0.5\,\mathrm{V}$, which would lower the degeneration resistances provided by M_5 and M_6, and favorably decrease the bias compliance voltage. Alternatively, as mentioned in Sections 6.8.2.4 and 6.8.2.6, M_3 and M_4 could be operated without source degeneration deep into strong inversion at high $V_{DS,sat}$ to minimize their thermal-noise contributions. However, as mentioned in Section 6.5.5.3, source degeneration reduces the drain flicker-noise current for a given bias compliance voltage, lowering the gate area requirement.

As mentioned for the differential input preamplifier, lowering non-input device drain current lowers g_m and thermal- and flicker-noise contributions. However, as discussed, certain devices must capture the input device current, unless this current is captured by non-MOS devices such as resistors where flicker noise could be potentially reduced.

Because the M_7 device that provides regulated cascode action uses a supply current that is separate from the supply current of $1\,\mu\mathrm{A}$ for the input device, reducing the M_7 supply current from 1 to $0.25\,\mu\mathrm{A}$ would reduce the preamplifier core supply current from 2 to $1.25\,\mu\mathrm{A}$. The lower supply current would lower the thermal-noise efficiency factor from $NEF' = 11.6$, described in Section 6.8.3.2, to $NEF' = 7.25$. Operation at the same level of noise using only the supply current of the input device would halve the $NEF' = 11.6$ to $NEF' = 5.8$. Supply current for a specified level of input-referred thermal-noise voltage, specified by the NEF', is minimized by operating the input device in weak or moderate inversion for high g_m/I_D, ensuring small non-input device noise contributions, and minimizing supply current beyond that required for the input device.

As described in Section 6.7.4 for the differential input preamplifier, thermal noise and flicker noise can be minimized for a given power consumption and layout area by including the thermal-noise efficiency factor, *NEF'*, and flicker-noise area factor, *NAF*, in the circuit analysis. As mentioned, this can be included in the *Circuit Analysis* sheet of the *Analog CMOS Design, Tradeoffs and Optimization* spreadsheet. This would permit a careful balance of input and non-input device noise contributions where both over- and under-design are avoided. Finally, as mentioned, the noise minimization methods can be extended to other circuit topologies for operation from weak through strong inversion, with or without the presence of external source degeneration resistance.

6.9 PREDICTION ACCURACY FOR DESIGN GUIDANCE AND OPTIMIZATION

As described in Section 5.6 for the OTA design examples of Chapter 5, predictions of device and circuit performance using the *Analog CMOS Design, Tradeoffs and Optimization* spreadsheet are derived from hand predictions of MOS performance in weak, moderate, and strong inversion that were described in Chapter 3. Additionally, separate predictions for preamplifier devices operating in the deep ohmic region as degeneration resistors were derived from approximations described in Section 6.5.2. As a result, the spreadsheet predictions are generally not as accurate as predictions found by computer simulations using full simulation MOS models.[1] However, as shown in the *Circuit Analysis* sheets of Figures 6.11 and 6.14 for the single-ended and differential input preamplifiers, the spreadsheet provides immediate performance predictions as device drain currents, inversion coefficients, and channel lengths are explored. Additionally, the spreadsheet permits the exploration of resistive source degeneration for reducing the drain-referred noise current of non-input devices and includes useful design details such as the relative noise contributions of non-input devices. Finally, as mentioned in Section 6.7.4, the spreadsheet can be easily enhanced to include optimization metrics like the thermal-noise efficiency factor, *NEF'*, and the flicker-noise area factor, *NAF*.

As seen in Figures 6.12 and 6.15, flicker noise predicted by the spreadsheet overlays measured noise for the differential and single-ended input preamplifiers. Additionally, the average value of measured thermal noise is 3.6 % above the predicted value for the differential input preamplifier and 8 % above the predicted value for the single-ended input preamplifier. This is clearly adequate for the purpose of the spreadsheet, which is to provide design intuition and guidance leading to the rapid design of optimum analog CMOS circuits.

The largest spreadsheet errors, and often simulation errors, occur for the prediction of MOS drain–source resistance and the resulting circuit small-signal resistances and gains. This is due to the complexity of predicting MOS drain–source resistance, r_{ds}, or conductance, $g_{ds} = 1/r_{ds}$. As described in Section 3.8.4 and shown in Section 3.8.4.5 through measurements of the Early voltage, $V_A = r_{ds} \cdot I_D = I_D/g_{ds}$, the drain–source resistance is a complex function of the inversion level and, especially, the channel length and V_{DS}.

As described in Section 3.8.4.4, errors in the drain–source resistance appear directly in simple circuits, squared in cascode circuits, and cubed in regulated cascode circuits for the prediction of circuit small-signal resistance and the resulting voltage gain. As a result, significant prediction errors are possible for circuits using cascode or regulated cascode topologies to boost resistances and gains. These errors increase even more as V_{DS} approaches $V_{DS,sat}$ where the drain–source resistance begins dropping rapidly. Unfortunately, V_{DS} is necessarily confined close to $V_{DS,sat}$ in low-voltage circuits, increasing the uncertainty of circuit resistances and gains.

Fortunately, uncertainties in open-loop gains associated with errors in MOS drain–source resistance are often mitigated by closed-loop, negative feedback gain regulation. This requires only a sufficient level of open-loop gain and the resulting negative feedback loop gain to regulate the closed-loop gain.

---

[1] The *Analog CMOS Design, Tradeoffs and Optimization* spreadsheet may ultimately be linked to computer simulation MOS models where prediction accuracy would approach that obtained in computer simulations. The reader is invited to check the book's web site listed on the cover for updates to the spreadsheet.

As described in Sections 6.7.3.2 and 6.8.3.2, measured values of open-loop voltage gains give negative feedback loop gains of 304 and 998 V/V for the differential and single-ended input preamplifiers, respectively, as applied for their intended applications. As a result of the high loop gains, wide variations in open-loop voltage gains will have only a small effect on the closed-loop voltage gain.

The largest spreadsheet prediction error is for the open-loop voltage gain for the single-ended input preamplifier, which utilized resistive source degeneration to develop high small-signal resistance and gain. Here, the predicted gain of 78700 V/V is 55 % above the measured gain of 50900 V/V. Approximately one half the gain overprediction is likely caused by the source degeneration term, $1 + g_{ms3}R$, that appears in Equation 6.33 for the voltage gain. Although this term is correctly used in the spreadsheet, the actual SOI CMOS preamplifiers were fabricated with their local MOS bodies tied to their sources. This lowers the g_{ms} expression given by Equation 6.18 by approximately 25% through the removal of the g_{mb} component. The remaining gain overprediction is likely caused by errors in $V_A = V_{AL} \cdot L$ used for predicting MOS drain–source resistance. As described near the beginning of Section 6.7.3, extractions of the MOS Early voltage factor, V_{AL}, were not taken for different device channel lengths and operating conditions as done for the OTA design examples of Chapter 5. Instead, a single value of $V_{AL} = 10\,\text{V}/\mu\text{m}$ was used for all preamplifier device predictions, which resulted in a good prediction of gain for the differential input preamplifier.

As summarized in Figure 3.48, V_{AL} varies from 3 to $34\,\text{V}/\mu\text{m}$ for devices in a $0.18\,\mu\text{m}$ CMOS process used for the cascoded OTA design examples in Chapter 5. This variation was observed from weak through strong inversion, across a wide range of channel lengths, and across values of V_{DS} sufficiently above $V_{DS,sat}$. This emphasizes the importance of extracting V_{AL} near the device channel length and operating conditions or, ideally, using a MOS model that accurately predicts drain–source resistance. As noted in Footnote 1, the spreadsheet may ultimately be linked to a full simulation MOS model. This could potentially provide an accurate prediction of drain–source resistance without the need for V_{AL} extractions, although experimental verification is recommended given the long history of drain–source resistance modeling errors.

Small-geometry effects of velocity saturation, VFMR, DIBL, and flicker-noise increases with inversion or V_{EFF} are included in the spreadsheet. However, gate leakage current and the resulting gate–source conductance and gate shot- and flicker-noise current are negligible and not included for the preamplifiers. These effects, described in Section 3.12.2, could be included in later versions of the spreadsheet for processes having gate-oxide thickness less than 2 nm.

Predictions of device and circuit performance provided by the spreadsheet are approximate and intended only to provide design guidance leading towards optimum design. Simulations of a candidate design are then required using full, simulation MOS models for the actual production process. These simulations must include process, supply voltage, and temperature variations along with layout parasitics in order to validate a candidate design for production. Although well known to industry designers, this is described further in the disclaimers in the spreadsheet and in Section 1.7.

Of particular importance for low-noise designs is verifying the modeling of MOS noise from weak through strong inversion in both the saturation and the ohmic regions. Additionally, it is important to verify the modeling of flicker-noise increases with the inversion level or V_{EFF}, discussed in Section 3.10.3.6, as well as verifying the modeling of possible small-geometry increases in thermal noise, discussed in Section 3.10.2.2.

6.10 SUMMARY OF LOW-NOISE DESIGN METHODS AND RESULTING CHALLENGES IN LOW-VOLTAGE PROCESSES

The discussion below considers MOS circuits having input transconductor devices that receive voltage inputs. These input devices can be either differential-pair or single devices as illustrated by the differential input preamplifier described in Section 6.7 or the single-ended input preamplifier described in Section 6.8.

The input-referred thermal-noise voltage is minimized by:

- Operating input devices at low inversion coefficients in weak or moderate inversion for high g_m/I_D and g_m. This minimizes the gate-referred thermal-noise voltage of input devices and helps ensure that these devices dominate the overall circuit noise. Operation at low inversion coefficients corresponds to low $V_{DS,sat}$, which is compatible with low-voltage design in small-geometry processes. Input devices should be operated at increasing levels of drain current as needed to obtain sufficiently low gate-referred thermal-noise voltage.

- Operating non-input devices at high inversion coefficients in strong inversion for low g_m/I_D and g_m. This minimizes the drain-referred thermal-noise current of non-input devices that is divided by the transconductance of input devices and referred to the input as a noise voltage. Unfortunately, operation at high inversion coefficients corresponds to high $V_{DS,sat}$, which is not compatible with low-voltage design in small-geometry processes. When possible, non-input devices should be operated at lower drain currents than input devices to further reduce their g_m and drain-referred thermal-noise current.

Additionally, the input-referred flicker-noise voltage is minimized by:

- Operating input devices with sufficiently large gate area to minimize their gate-referred flicker-noise voltage.

- Operating non-input devices with sufficiently large gate area to minimize their drain-referred flicker-noise current. Because input devices operate at low inversion coefficients and non-input devices operate at high inversion coefficients to minimize the overall input-referred thermal-noise voltage, it is necessary to increase the channel length of non-input devices above that of the input devices. As mentioned in Section 5.3.6.3 for the OTA design examples of Chapter 5, this is required because the smaller shape factor $(S = W/L)$ of non-input devices operating at high levels of inversion requires an increase in channel length (with an equal increase in channel width to maintain the inversion coefficient) to achieve sufficient gate area for small flicker-noise contributions. Unfortunately, as described in Section 3.9.6, this significantly decreases the intrinsic bandwidth of non-input devices, which is inversely proportional to the square of channel length (or inversely proportional to channel length if velocity saturation is significant). Finally, as described in Section 6.5.5.3, resistive source degeneration provides a reduction of drain-referred flicker-noise current for non-input devices for a given bias compliance voltage.

Operating devices for low $V_{DS,sat}$ in moderate inversion or perhaps up to the onset of strong inversion at $IC = 10$ becomes increasingly necessary in low-voltage design where the bias compliance voltage is limited. Unfortunately, operation here at moderately high g_m/I_D and g_m increases the drain-referred thermal-noise current contributions of non-input devices. Additionally, using resistive source degeneration to reduce drain-referred flicker-noise current becomes increasingly difficult due to the increase in bias compliance voltage. As described in Section 6.5.5, resistive source degeneration lowers drain-referred flicker-noise current for a given bias compliance voltage while having little effect on drain-referred thermal-noise current because of added thermal noise from the degeneration resistance.

Interestingly, velocity-saturated devices have lower $g_m/I_D = 1/V_{EFF}$ compared to $g_m/I_D = 2/V_{EFF}$ in strong inversion without velocity saturation, giving low g_m for their bias compliance voltage of $V_{DS,sat}$. As described in Section 3.7.3.3, $V_{DS,sat}$ is relatively insensitive to velocity saturation effects when evaluated in terms of the inversion coefficient because it becomes a smaller fraction of $V_{EFF} = V_{GS} - V_T$ as this increases due to velocity saturation. The low $V_{DS,sat}$ and low g_m/I_D of velocity-saturated devices suggest that these might be candidates for non-input devices having low drain-referred thermal-noise current. However, velocity saturation requires short-channel devices (typically nMOS devices because of lower E_{CRIT}) operating in or near strong inversion. This results in small-area devices having high flicker noise and also increases the risks of inversion-level or V_{EFF} increases in flicker noise, described in Section 3.10.3.6, and small-geometry increases in thermal noise, described in Section 3.10.2.2.

Clearly, the management of non-input device, drain-referred thermal-noise current becomes an increasingly difficult problem in low-voltage processes that necessarily limit MOS operation to moderate inversion or perhaps the onset of strong inversion. Thicker gate-oxide options are helpful because these allow operation at higher supply voltage and higher levels of inversion for non-input devices. Additionally, a thick gate oxide minimizes gate leakage current and its contributions to gate–source conductance and gate shot- and flicker-noise current. These effects were discussed in Section 3.12.2 and can become significant for gate-oxide thicknesses below approximately 2 nm.

REFERENCES[2]

[1] E. Vittoz, "MOS transistors operated in the lateral bipolar mode and their application in CMOS," *IEEE Journal of Solid-State Circuits*, vol. SC-18, pp. 273–279, June 1983.

[2] T. W. Pan and A. A. Abidi, "50-dB variable gain amplifier using parasitic bipolar transistors in CMOS," *IEEE Journal of Solid-State Circuits*, vol. 24, pp. 951–961, Aug. 1989.

[3] R. Gomez and A. A. Abidi, "A 50-MHz CMOS variable gain amplifier for magnetic data storage systems," *IEEE Journal of Solid-State Circuits*, vol. 27, pp. 935–939, June 1992.

[4] J. Hauptmann, F. Dielacher, R. Steiner, C. C. Enz, and F. Krummenacher, "A low-noise amplifier with automatic gain control and anticlipping control in CMOS technology," *IEEE Journal of Solid-State Circuits*, vol. 27, pp. 974–981, July 1992.

[5] B. Stefanelli, J.-P. Bardyn, A. Kaiser, and D. Billet, "A very low-noise CMOS preamplifier for capacitive sensors," *IEEE Journal of Solid-State Circuits*, vol. 28, pp. 971–978, Sept. 1993.

[6] W. T. Holman and J. A. Connelly, "A compact low noise operational amplifier for a 1.2 μm digital CMOS technology," *IEEE Journal of Solid-State Circuits*, vol. 30, pp. 710–714, June 1995.

[7] D. M. Binkley, J. M. Rochelle, B. K. Swann, L. G. Clonts, and R. N. Goble, "A micropower CMOS, direct-conversion, VLF receiver chip for magnetic-field wireless applications," *IEEE Journal of Solid-State Circuits*, vol. 33, pp. 344–358, Mar. 1998.

[8] M. S. J. Steyaert, W. M. C. Sansen, and C. Zhongyuan, "A micropower low-noise monolithic instrumentation amplifier for medical purposes," *IEEE Journal of Solid-State Circuits*, vol. SC-22, pp. 1163–1168, Dec. 1987.

[9] Y. Tsividis, *Mixed Analog-Digital VLSI Devices and Technology: An Introduction*, McGraw-Hill, 1996.

[10] J. H. Nielsen and E. Bruun, "A CMOS low-noise instrumentation amplifier using chopper modulation," *Analog Integrated Circuits and Signal Processing*, vol. 42, pp. 65–76, Jan. 2005.

[11] E. Raisanen-Routsalainen, K. Lasanen, and J. Kostamovaara, "A 1.2 V micropower CMOS op amp with floating-gate input transistors," *Proceedings of the 43rd IEEE Midwest Symposium on Circuits and Systems*, pp. 794–797, Aug. 2000.

[12] J. H. Nielsen and T. Lehmann, "An implantable CMOS amplifier for nerve signals," *Analog Integrated Circuits and Signal Processing*, vol. 36, pp. 153–164, July 2003.

[13] H. J. Weller, D. Setiadi, and T. D. Binnie, "Low-noise charge sensitive readout for pyroelectric sensor arrays using PVDF thin film," *Sensors and Actuators*, vol. 85, pp. 267–274, Aug. 2000.

[14] L. Lentola, A. Mozzi, A. Neviani, and A. Baschirotto, "A 1-μA front end for pacemaker atrial sensing channels with early sensing capability," *IEEE Transactions on Circuits and Systems II*, vol. 50, pp. 397–403, Aug. 2003.

[15] P. Mohseni and K. Najafi, "A fully integrated neural recording amplifier with DC input stabilization," *IEEE Transactions on Biomedical Engineering*, vol. 51, pp. 832–837, May 2004.

[16] R. R. Harrison and C. Charles, "A low-power low-noise CMOS amplifier for neural recording applications," *IEEE Journal of Solid-State Circuits*, vol. 38, pp. 958–965, June 2003.

[17] T. Horiuchi, T. Swindell, D. Sander, and P. Abshire, "A low-power CMOS neural amplifier with amplitude measurements for spike sorting," *Proceedings of the IEEE International Symposium on Circuits and Systems (ISCAS)*, vol. 4, pp. 29–32, May 2004.

---

[2] The reader is referred to Chapter 2 for references on previously reported analog CMOS optimization methods and to Chapter 3 for references on MOS analog modeling and performance.

[18] W. R. Patterson, Y.-K. Song, C. W. Bull, I. Ozden, A. P. Deangellis, C. Lay, J. L. McKay, A. V. Nurmikko, J. D. Donoghue, and B. W. Connors, "A microelectrode/microelectronic hybrid device for brain implantable neuroprosthesis applications," *IEEE Transactions on Biomedical Engineering*, vol. 51, pp. 1845–1853, Oct. 2004.

[19] D. T. Kewley, M. D. Hills, D. A. Borkholder, I. E. Opris, N. I. Maluf, C. W. Storment, J. M. Bower, and G. T. A. Kovacs, "Plasma etched neural probes," *Sensors and Activators*, vol. 58, pp. 27–35, Jan. 1997.

[20] R. H. Olsson, D. L. Buhl, A. M. Sirota, G. Buzsaki, and K. D. Wise, "Band-tunable and multiplexed integrated circuits for simultaneous recording and stimulation with microelectrode arrays," *IEEE Transactions on Biomedical Engineering*, vol. 52, pp. 1303–1311, July 2005.

[21] K. Najafi and K. D. Wise, "An implantable multielectrode array with on-chip signal processing," *IEEE Journal of Solid-State Circuits*, vol. SC-21, pp. 1035–1044, Dec. 1986.

[22] M. Mojarradi, D. M. Binkley, B. J. Blalock, R. Andersen, N. Ulshoefer, T. Johnson, and L. Del Castillo, "A miniaturized, neuroprosthesis suitable for implantation into the brain," *IEEE Transactions on Neural Systems and Rehabilitation Engineering*, vol. 11, pp. 38–42, Mar. 2003.

[23] G. Torelli, E. Chioffi, and F. Maloberti, "A CMOS micropower input amplifier for hearing aids," *Microelectronics Journal*, vol. 27, pp. 477–484, Sept. 1996.

[24] D. M. Binkley, D. H. Ihme, B. J. Blalock, and M. M. Mojarradi, "Micropower, 0.35-μm partially depleted SOI CMOS preamplifiers having low white and flicker noise," *2003 IEEE International SOI Conference*, Newport Beach, California, pp. 85–86, Sept. 2003.

[25] D. M. Binkley, C. E. Hopper, B. J. Blalock, M. M. Mojarradi, J. D. Cressler, and L. K. Yong, "Noise performance of 0.35-μm SOI CMOS devices and micropower preamplifier from 77–400 K," *2004 IEEE Aerospace Conference*, Big Sky, Montana, pp. 2495–2506, Mar. 2004.

[26] D. M. Binkley, C. E. Hopper, J. D. Cressler, M. M. Mojarradi, and B. J. Blalock, "Noise performance of 0.35-μm SOI CMOS devices and micropower preamplifier following 63-MeV, 1-Mrad (Si) proton irradiation," *IEEE Transactions on Nuclear Science*, vol. 51, pp. 3788–3794, Dec. 2004.

[27] D. M. Binkley, M. E. Casey, B. S. Puckett, R. Lecomte, and A. Saoudi, "A power efficient, low noise, wideband, integrated CMOS preamplifier for LSO/APD PET systems," *IEEE Transactions on Nuclear Science*, vol. 47, pp. 810–817, June 2000.

[28] D. M. Binkley, "MOSFET modeling and circuit design: a methodology for transistor level analog CMOS design," Conference tutorial with D. Foty, "MOSFET modeling and circuit design: re-establishing a lost connection," *37th Annual Design Automation Conference (DAC)*, Los Angeles, June 2000.

[29] D. M. Binkley, "A methodology for analog CMOS design based on the EKV MOS model," Conference tutorial with D. Foty, "MOS modeling as a basis for design methodologies: new techniques for modern analog design," *2002 IEEE International Symposium on Circuits and Systems (ISCAS)*, Scottsdale, May 2002.

[30] D. M. Binkley, M. Bucher, and D. Kazazis, "Modern analog CMOS design from weak through strong inversion," *Proceedings of the European Conference on Circuit Theory and Design (ECCTD) 2003*, Krakow, pp. I-8–I-13, Aug. 2003.

[31] D. M. Binkley, C. E. Hopper, S. D. Tucker, B. C. Moss, J. M. Rochelle, and D. P. Foty, "A CAD methodology for optimizing transistor current and sizing in analog CMOS design," *IEEE Transactions on Computer-Aided Design of Integrated Circuits and Systems*, vol. 22, pp. 225–237, Feb. 2003.

[32] D. M. Binkley, B. J. Blalock, and J. M. Rochelle, "Optimizing drain current, inversion level, and channel length in analog CMOS design," *Analog Integrated Circuits and Signal Processing*, vol. 47, pp. 137–163, May 2006.

[33] D. M. Binkley, "Tradeoffs and optimization in analog CMOS design," *Proceedings of the 14th International Conference Mixed Design of Integrated Circuits and Systems (MIXDES)*, pp. 47–60, Ciechocinek, Poland, June 2007.

[34] D. M. Binkley, "Tradeoffs and optimization in analog CMOS design," Conference tutorial with H. Graeb, G. G. E. Gielen, and J. Roychowdhury, "From transistor to PLL – Analog design and EDA methods," *2008 Design, Automation, and Test in Europe Conference (DATE)*, Munich, Mar. 2008.

[35] University of California at Berkeley BSIM3 and BSIM4 MOS models, available on-line at http://www-device.eecs.Berkeley.edu/bsim3.

7

Extending Optimization Methods to Smaller-Geometry CMOS Processes and Future Technologies

7.1 INTRODUCTION

Chapters 3 and 4 described predictions and measurements of MOS device performance, mostly for a typical, $0.18\,\mu$m, bulk CMOS process having a gate-oxide thickness of 4.1 nm. Additionally, Chapter 5 described operational transconductance amplifier (OTA) design examples in this $0.18\,\mu$m process, along with design examples in a $0.5\,\mu$m bulk CMOS process. Finally, Chapter 6 described micropower, low-noise preamplifier design examples in a $0.35\,\mu$m, partially depleted (PD), silicon-on-insulator (SOI) CMOS process.

Although at publication time some designers in large companies are working in 0.09 and $0.065\,\mu$m CMOS processes, $0.18\,\mu$m and even larger feature-size processes are still in widespread use. Additionally, the $0.18\ \mu$m process, and even the 0.35 and 0.5 μm processes, illustrated in this book exhibited important effects seen in smaller-geometry processes. These small-geometry effects, which were included in predictions and measurements of device performance, include velocity saturation, vertical field mobility reduction (VFMR), drain-induced barrier lowering (DIBL), and inversion-level or $V_{EFF} = V_{GS} - V_T$ increases in gate-referred flicker-noise voltage.

The inclusion of small-geometry effects and the technology independence enabled by the inversion coefficient permit ready extension of the optimization methods presented in this book to different CMOS processes, including smaller-geometry processes. Additionally, the optimization methods can be potentially extended to emerging, non-CMOS technologies. This concluding chapter describes extension of the optimization methods.

Tradeoffs and Optimization in Analog CMOS Design   David M. Binkley
© 2008 John Wiley & Sons, Ltd

7.2 USING THE INVERSION COEFFICIENT FOR CMOS PROCESS INDEPENDENCE AND FOR EXTENSION TO SMALLER-GEOMETRY PROCESSES

7.2.1 Universal g_m/I_D, V_{EFF}, and $V_{DS,sat}$ Characteristics Across CMOS Processes

As described earlier in Section 3.4.2, the inversion coefficient provides a numerical measure of MOS inversion. Inversion coefficient values of less than 0.1 correspond to weak inversion, values between 0.1 and 10 correspond to moderate inversion with unity corresponding to the center of moderate inversion, and values greater than 10 correspond to strong inversion. These inversion coefficient boundaries are independent of the MOS process.

Additionally, as described earlier in Section 3.8.2.5, the structure of the transconductance efficiency, g_m/I_D, versus the inversion coefficient is independent of the process, nMOS or pMOS device type, and channel length, except for short-channel devices. This structure is illustrated by Figures 3.24–3.27, where g_m/I_D remains constant and maximum in weak inversion, decreases modestly in moderate inversion, and decreases inversely with the square root of the inversion coefficient in strong inversion before decreasing inversely with the inversion coefficient for short-channel devices experiencing significant velocity saturation. As seen in the expressions given in Table 3.17, even the values of g_m/I_D are nearly the same across processes, with values varying only slightly with the inverse of the substrate factor, n, which typically ranges from 1.4 to 1.3 from weak through strong inversion for bulk CMOS processes. The g_m/I_D decrease for short-channel devices at high levels of inversion due to velocity saturation and the smaller decrease for all devices due to VFMR are also similar across processes, depending primarily on the critical, horizontal, velocity saturation electric field, E_{CRIT}, and the VFMR reduction factor, θ. The process-independent structure of g_m/I_D versus the inversion coefficient is the result of the process normalization present in the inversion coefficient definition that includes the process mobility and gate-oxide capacitance.

In addition to g_m/I_D, the structures of the effective gate–source voltage, $V_{EFF} = V_{GS} - V_T$, and drain–source saturation voltage, $V_{DS,sat}$, versus the inversion coefficient are independent of the process, nMOS or pMOS device type, and channel length, except, again, for short-channel devices. The structure of V_{EFF} is illustrated by Figures 3.13–3.15, where V_{EFF} remains logarithmically proportional to the inversion coefficient in weak inversion, increases modestly in moderate inversion, and increases as the square root of the inversion coefficient in strong inversion before increasing directly with the inversion coefficient for short-channel devices experiencing significant velocity saturation. The structure of $V_{DS,sat}$ is shown in Figure 3.16, where $V_{DS,sat}$ remains constant and minimum in weak inversion, increases modestly in moderate inversion, and increases as the square root of the inversion coefficient in strong inversion with little dependency on velocity saturation effects as described in Section 3.7.3.3. As seen in the expressions given in Tables 3.14 and 3.15, even the values of V_{EFF} and $V_{DS,sat}$ are nearly the same across processes, with V_{EFF} varying only slightly, tracking n. The V_{EFF} increase for short-channel devices at high levels of inversion due to velocity saturation and the smaller increase for all devices due to VFMR are also similar across processes, depending primarily on E_{CRIT} and θ.

7.2.2 Other Nearly Universal Performance Characteristics Across CMOS Processes

MOS intrinsic gate capacitances, gate-referred flicker-noise voltage, and local-area threshold-voltage mismatch are linked to the gate area, which is inversely proportional to the inversion coefficient as shown in the expressions in Table 3.10. As a result, the structures of these versus the inversion coefficient are similar across CMOS processes. Additionally, MOS thermal noise is linked to the transconductance, g_m, while drain-referred flicker-noise current and local-area mismatch current are linked to the gate area and g_m. Finally, the intrinsic voltage gain, A_{Vi}, and intrinsic bandwidth, f_{Ti}, are linked to g_m (additionally these are linked to the drain–source conductance and intrinsic gate capacitances, respectively) while the input, 1 dB compression voltage, $V_{INDIF1dB}$, for a differential pair is linked to the reciprocal of g_m/I_D. As a result of the universal structures of g_m/I_D, V_{EFF}, and $V_{DS,sat}$ mentioned in the previous section, universal gate area trends, and various linkages in performance, most measures of MOS performance for both nMOS and pMOS devices retain a nearly universal behavior across processes when evaluated in terms of the inversion coefficient.

Table 4.1 summarizes the universal MOS performance characteristics and tradeoffs for the channel width, gate area, intrinsic gate–source and gate–body capacitances, V_{EFF}, $V_{DS,sat}$, g_m/I_D, normalized drain–source conductance, g_{ds}, expressed as $V_A = I_D/g_{ds}$, A_{Vi}, and f_{Ti}. Of these, V_A, which increases primarily with the channel length and drain–source voltage, and A_{Vi}, which is linked to V_A, exhibit considerably more variation across processes as described in Section 3.8.4.5. Table 4.2 summarizes the universal MOS performance characteristics and tradeoffs for $V_{INDIF1dB}$ and the gate-referred thermal-noise, flicker-noise, and local-area mismatch voltage. Finally, Table 4.3 summarizes the universal MOS performance characteristics and tradeoffs for the drain-referred thermal-noise, flicker-noise, and local-area mismatch current. Tables 4.1–4.3 include simplified performance expressions, references to book sections, tables, and figures that provide further details, and arrows showing the direction and degree of performance dependence on the inversion coefficient, channel length, and drain current. Figure 3.1 shows the *MOSFET Operating Plane*, which illustrates tradeoffs in performance for selected choices of the inversion coefficient and channel length. Figure 4.1 shows performance tradeoffs for selected choices of the inversion coefficient.

When MOS performance is evaluated in terms of the inversion coefficient, performance trends are also similar across CMOS processes as the channel length and drain current are varied. This is shown in Tables 4.1–4.3 and is a result, for example, of g_m directly tracking the drain current for fixed values of the inversion coefficient and channel length (as mentioned in the previous section, g_m is independent of the channel length unless velocity saturation effects are present). Similar performance trends across processes are also a result of the channel width and gate-area relationships shown in Table 3.10 for design choices of the inversion coefficient, channel length, and drain current.

7.2.3 Porting Designs Across CMOS Processes

Using the inversion coefficient in design facilitates the porting of designs across CMOS processes. As mentioned in the previous sections, trends and, in some cases, actual values of MOS performance are independent of the CMOS process when evaluated in terms of the inversion coefficient, channel length, and drain current. As a result, a design might be ported to a new CMOS process by first maintaining the selected MOS inversion coefficients, channel lengths, and drain currents. Additionally, a design might be "complemented" by maintaining the selected inversion coefficients, channel lengths, and drain currents and interchanging nMOS and pMOS devices.

As an example, nearly equal g_m/I_D and, correspondingly, nearly equal g_m and gate-referred thermal-noise voltage, can be realized for both nMOS and pMOS devices in all CMOS processes for a selected inversion coefficient and drain current. Here, g_m/I_D, g_m, and the thermal noise are also independent of the channel length unless velocity saturation decreases in g_m or small-geometry increases in thermal noise are present. Maintaining a selected inversion coefficient, channel length, and drain current in a new process involves only resizing the MOS channel width required for layout, using the expression given in Table 3.10. Channel widths for the new process will depend on the process and the nMOS or pMOS device type through an inverse relationship with the technology current, $I_0 = 2n_0\mu_0 C'_{OX}U_T^2$, which depends on the mobility, gate-oxide capacitance, and temperature. The substrate factor, n, usually has a small effect since it is nearly constant across typical bulk CMOS processes, having values of approximately 1.4 to 1.3 from weak through strong inversion.

As mentioned in Section 7.2.1, MOS g_m/I_D, V_{EFF}, and $V_{DS,sat}$ will be nearly unchanged for a design ported to a new CMOS process for equal MOS inversion coefficients as the original design, although velocity saturation effects, if present, will depend additionally on the channel length and process value of E_{CRIT}. Additionally, as mentioned, MOS g_m and thermal noise will be nearly unchanged for a design ported with equal MOS inversion coefficients and drain currents as the original design, although velocity saturation effects and possible small-geometry increases in the noise, if present, will depend additionally on the channel length and process.

Interestingly, as described in Section 3.9.1, the gate-oxide capacitance and, correspondingly, intrinsic gate capacitances and bandwidths, are nearly independent of the process gate-oxide capacitance, C'_{OX}, for a fixed value of the inversion coefficient, channel length, and drain current because the required channel width and gate area decrease as C'_{OX} increases. However, the capacitances are inversely proportional to the mobility and bandwidths are proportional to the mobility, which is why capacitances are greater and bandwidths lower for pMOS devices compared to nMOS devices operating at the same inversion coefficient, channel length, and drain current. Although intrinsic gate capacitances and bandwidths depend on the mobility, the required channel width and gate area depend inversely on the mobility and C'_{OX} and are smaller in smaller-geometry processes, primarily because of the increase in C'_{OX}. The smaller gate areas may then result in increased values of flicker noise and local-area mismatch.

As mentioned in the previous sections, the universal structures of g_m/I_D, V_{EFF}, and $V_{DS,sat}$, universal gate-area trends, and various linkages in performance result in most aspects of MOS performance retaining a nearly universal behavior when evaluated in terms of the inversion coefficient. This suggests that designs based on the inversion coefficient, channel length, and drain current can be ported systematically to different CMOS processes by simply adjusting channel widths to maintain the selected inversion coefficients, channel lengths, and drain currents. A detailed analysis of process scaling could be developed for designs based on the inversion coefficient, channel length, and drain current. This could, for example, suggest increases in channel length for smaller-geometry processes having large C'_{OX}, if needed to maintain sufficient gate areas when flicker noise and local-area mismatch are important.

The listing below summarizes changes in MOS performance for designs ported to different CMOS processes. The listing includes references to tables and figures in Chapters 3 and 4 that describe MOS performance and the often universal characteristics when performance is evaluated in terms of the inversion coefficient, channel length, and drain current. When MOS inversion coefficients, channel lengths, and drain currents are held constant for a design ported to a different CMOS process:

- **The required shape factor, $S = W/L$, channel width, W, and gate area, WL,** change inversely with process low-field mobility, μ_0, gate-oxide capacitance, C'_{OX}, and average substrate factor, n_0, which are included in the technology current, $I_0 = 2n_0\mu_0 C'_{OX} U_T^2$. If channel width and gate area become too small in smaller-geometry processes having high C'_{OX} caused by small gate-oxide thickness, t_{ox}, channel length can be increased (with the corresponding required increase in channel width) to increase channel width and gate area. This may be required to manage flicker noise and local-area mismatch in smaller-geometry processes. See Tables 3.10 and 4.1, and Figures 3.7, 3.8, 4.2, and 4.3.

- **The gate-oxide capacitance, $C_{GOX} = WLC'_{OX}$, and intrinsic gate capacitances, C_{gsi} and C_{gbi},** change inversely with μ_0 and nearly inversely with n_0, but do not depend on C'_{OX}. This is because as C'_{OX} increases, the required channel width and gate area decrease, holding the gate-oxide capacitance constant. See Tables 3.23 and 4.1, and Figures 3.53, 3.54, 4.4, and 4.5.

- **The gate–source, gate–body, and gate–drain overlap capacitances, C_{GSO}, C_{GBO}, and C_{GDO},** change inversely with μ_0, C'_{OX}, and n_0, tracking the required channel width. These also track the process gate–source, gate–body, and gate–drain overlap capacitance parameters, $CGSO$, $CGBO$, and $CGDO$, with additional complex behavior and bias dependency associated with the source and drain extensions. See Table 3.24 and Figure 3.55.

- **The drain–body and source–body capacitances, C_{DB} and C_{SB},** change inversely with μ_0, C'_{OX}, and n_0, tracking the required channel width. These also track the process drain/source area and perimeter capacitance parameters, C_J and C_{JSW}. See Tables 3.25 and 4.1, and Figures 3.55, 3.56, 4.4, and 4.5.

- **The DC bias measure of the effective gate–source voltage, $V_{EFF} = V_{GS} - V_T$,** changes only slightly, directly with the substrate factor, n. Velocity saturation increases in V_{EFF} for short-channel devices operating at high levels of inversion depend on the channel length and critical, horizontal, velocity saturation electric field, E_{CRIT}. Smaller VFMR increases in V_{EFF} for all devices operating at high levels of inversion depend on the VFMR reduction factor, θ. See Tables 3.14 and 4.1, and Figures 3.13–3.15 and 4.6.

- **The DC bias measure of the drain–source saturation voltage, $V_{DS,sat}$,** is nearly independent of the process. When expressed in terms of the (fixed-normalized) inversion coefficient used here, $V_{DS,sat}$ is nearly insensitive to velocity saturation effects because it becomes a smaller fraction of the increasing V_{EFF}. See Tables 3.15 and 4.1, and Figures 3.16–3.20 and 4.6.

- **The small-signal parameter measures of transconductance efficiency, g_m/I_D, and transconductance, g_m,** change only slightly with the inverse of n. Velocity saturation decreases in g_m/I_D and g_m for short-channel devices operating at high levels of inversion depend on the channel length and E_{CRIT}. Smaller VFMR decreases in g_m/I_D and g_m for all devices operating at high levels of inversion depend on θ. See Tables 3.17 and 4.1, and Figures 3.24–3.27, 4.7, and 4.8.

- **The small-signal parameter measures of drain–source conductance, $g_{ds} = I_D/V_A$, or drain–source resistance, $r_{ds} = 1/g_{ds} = V_A/I_D$, and their normalized representation using the Early voltage, $V_A = I_D/g_{ds} = I_D \cdot r_{ds}$,** do not have a fundamental characteristic across CMOS process. However, their trends are similar across CMOS processes. See Tables 3.21 and 4.1, and Figures 3.43–3.48, 4.7, and 4.8.

- **The intrinsic voltage gain, $A_{Vi} = g_m/g_{ds} = g_m \cdot r_{ds} = (g_m/I_D) \cdot V_A$,** does not have a fundamental characteristic across CMOS processes because of the non-fundamental characteristic of V_A. However, its trends are similar across CMOS processes. See Tables 3.22 and 4.1, and Figures 3.51, 3.52, 4.9, and 4.10.

- **The intrinsic bandwidth, $f_{Ti} = g_m/[2\pi(C_{gsi} + C_{gbi})]$,** changes directly with μ_0, has a slight dependency on n, and is independent of C'_{OX}. Velocity saturation decreases in f_{Ti}, caused by decreases in g_m/I_D, depend on the channel length and E_{CRIT}. Smaller VFMR decreases in f_{Ti}, caused by smaller decreases in g_m/I_D, depend on θ. See Tables 3.26 and 4.1, and Figures 3.57, 3.58, 4.9, and 4.10.

- **The input, 1-dB compression voltage for a differential pair, $V_{INDIF1dB}$,** is nearly inversely proportional to g_m/I_D and changes only slightly, directly with n. Velocity saturation increases in $V_{INDIF1dB}$, caused by decreases in g_m/I_D, depend on the channel length and E_{CRIT}. Smaller VFMR increases in $V_{INDIF1dB}$, caused by smaller decreases in g_m/I_D, depend on θ. See Tables 3.19 and 4.2, and Figures 3.28, 3.29, and 4.11.

- **The gate-referred thermal-noise voltage density** changes slightly, directly with n, while the **drain-referred thermal-noise current density** is nearly independent of the process. Because the gate-referred noise and drain-referred noise are linked to the inverse square root of g_m and the square root of g_m directly, the noise is nearly independent of the process. However, increases in the gate-referred noise and decreases in the drain-referred noise, caused by velocity saturation decreases in g_m, depend on the channel length and E_{CRIT}. Additionally, possible small-geometry increases in the thermal noise depend on the channel length and the process, as modeled by the thermal noise factor, Γ. See Tables 3.29, 3.30, 4.2, and 4.3, and Figures 3.61–3.63, 4.12, 4.13, 4.16, and 4.17.

- **The gate-referred flicker-noise voltage density** is linked to the inverse square root of gate area, resulting in similarities in trends across processes. Additionally, the **drain-referred flicker-noise voltage density** is linked to the inverse square root of gate area and the transconductance directly, resulting in similarities across processes. However, the value of flicker noise depends strongly on the process through the flicker-noise factor, K'_{F0} (or K'_F that includes the noise increase with inversion), which varies widely across processes as described in Section 3.10.3.5. Additionally, increases in gate-referred flicker-noise voltage with increasing inversion or V_{EFF}, and corresponding increases in drain-referred flicker-noise current, depend on the process through the voltage, V_{KF}, which describes the increase in flicker noise with V_{EFF}. See Tables 3.31, 3.34, 4.2, and 4.3, and Figures 3.67, 3.68, 4.14, 4.15, 4.18, and 4.19.

- **The local-area gate-source mismatch voltage** is linked to the inverse square root of gate area, resulting in similarities in trends across processes. Additionally, the **local-area drain mismatch current** is linked to the inverse square root of gate area and the transconductance directly, resulting in similarities across processes. However, the value of mismatch depends strongly on the process through the local-area threshold voltage mismatch factor, A_{VTO}, which nearly tracks t_{ox} as described in Section 3.11.1.2. Additionally, increases in mismatch at high levels of inversion due to transconductance factor mismatch depend on the process through the transconductance mismatch factor, A_{KP}. See Tables 3.41, 4.2, and 4.3, and Figures 3.71, 3.72, 4.14, 4.15, 4.18, and 4.19.

7.2.4 Extending Design Methods to Smaller-Geometry Processes

The similar and often universal characteristics and tradeoffs of MOS performance versus the inversion coefficient, channel length, and drain current are a primary advantage of using the inversion coefficient, channel length, and drain current as the three, independent degrees of design freedom for optimizing analog CMOS design. As mentioned in Section 3.2, design using the inversion coefficient, channel length, and drain current permits the optimization of these separately, design in moderate and all regions of inversion, design inclusive of small-geometry effects, design with process independence, design optimization with a minimum of trial-and-error simulations, observation of performance trends and tradeoffs, and cross-checking with simulation MOS models. These advantages also facilitate the extension of design methods to smaller-geometry CMOS processes.

Interestingly, because of the lower supply voltage requirements for smaller-geometry CMOS processes, MOS operation must be constrained to inversion coefficients below approximately 3 near the center of moderate inversion to give sufficiently low values of V_{EFF} and $V_{DS,sat}$ to support the reduced supply voltage. Fortunately, there is little decrease in g_m/I_D (Figures 3.24–3.27) and increases in V_{EFF} (Figures 3.14 and 3.15) and $V_{DS,sat}$ (Figure 3.16) for short-channel devices here. This is a result of significantly reduced velocity saturation effects for operation near the center of moderate inversion compared to that present in strong inversion for inversion coefficients greater than 10, although $V_{DS,sat}$ itself is relatively unaffected by velocity saturation effects when evaluated in terms of the inversion coefficient as described in Section 3.7.3.3. Thus, there is some inherent management of velocity saturation effects in smaller-geometry, low-voltage CMOS processes because of restrictions on the maximum allowable values of the inversion coefficient. However, gate leakage current effects can be significant for smaller-geometry CMOS processes having gate-oxide thickness below 2 nm as discussed in the next section.

7.3 ENHANCING OPTIMIZATION METHODS BY INCLUDING GATE LEAKAGE CURRENT EFFECTS

Velocity saturation and VFMR decreases in g_m/I_D and g_m, and related increases in V_{EFF}, were included in design methods described in this book and in predictions of MOS performance provided in the *Analog CMOS Design, Tradeoffs and Optimization* spreadsheet described in the Appendix. Additionally, DIBL decreases in V_A, and the related increases in g_{ds}, were included for short-channel devices operating at low levels of inversion along with increases in gate-referred flicker-noise voltage for devices operating at high levels of inversion or V_{EFF}. Although these important small-geometry effects were included, gate leakage current effects were generally not included. Gate leakage current effects were not present for device measurements in the example 0.18 μm CMOS process or for design examples in the example 0.5, 0.35, and 0.18 μm CMOS processes because the gate-oxide thickness is well above 2 nm.

Gate leakage current can be significant for gate-oxide thickness around 2 nm and below as described in Section 3.12.1 and should be included when significant. Additionally, gate–source conductance, gate shot- and flicker-noise current, and mismatch due to gate leakage current can deteriorate circuit performance as described in Section 3.12.2 and should also be included when significant.

Gate leakage current and the resulting gate–source conductance, gate shot- and flicker-noise current, and mismatch due to gate leakage current can be included for design methods developed in terms of the inversion coefficient, channel length, and drain current. As an example, Figure 3.74 shows predicted gate–source leakage current versus the inversion coefficient and channel length for nMOS devices operating at drain currents of 100 μA in a process having a gate-oxide thickness of 2 nm. As described in Section 3.12.1, gate leakage current increases significantly above that shown as gate-oxide thickness decreases below 2 nm.

The inclusion of gate leakage current effects, when these are significant, will show that smaller channel length and gate area are required to balance the traditional improvements in gain, flicker noise, and local-area mismatch at increasing channel length and gate area with the deterioration of these associated with increasing gate leakage current caused by increasing gate area. As a result, new optimizations and trends will be present that will necessarily restrict the gate area. Gate leakage current effects may be included later in the *Analog CMOS Design, Tradeoffs and Optimization* spreadsheet as mentioned in Section A.6. The reader is invited to check the book's web site listed on the cover for possible enhancements to the spreadsheet, which is a "work-in-progress."

Gate leakage current and the resulting gate–source conductance, gate shot- and flicker-noise current, and mismatch due to gate leakage current can be avoided in small-geometry CMOS processes having thick gate-oxide options. Clearly, if available, these options are desirable for analog CMOS design.

7.4 USING AN INVERSION COEFFICIENT MEASURE FOR NON-CMOS TECHNOLOGIES

As described in Section 3.4.2.2, the MOS, fixed-normalized inversion coefficient is a normalized drain-current, geometry, and technology measure given by $IC = I_D/[I_0(W/L)]$. Here, I_D is the drain current, W is the effective channel width, L is the effective channel length, and $I_0 = 2n_0\mu_0 C'_{OX}U_T^2$ is a technology current found for an average value of the substrate factor, n_0, the low-field mobility, μ_0, the gate-oxide capacitance per unit area, C'_{OX}, and the thermal voltage, U_T. As described in Section 3.4.2.3, decreases in mobility with increasing inversion or $V_{EFF} = V_{GS} - V_T$ caused by velocity saturation and VFMR are considered externally from the fixed-normalized inversion coefficient definition.

A similar inversion coefficient measure can also be considered for organic or amorphous silicon, thin-film, field-effect transistors and carbon nanotube, field-effect transistors, which both have channel width, channel length, drain current, and gate–source voltage measures analogous to MOS devices. An inversion coefficient measure could be useful for these devices, which have exponential- and square-law (at least for the thin-film devices) drain-current versus gate–source voltage for operation sufficiently below and above the threshold voltage, respectively. These distinct regions of drain current and, correspondingly, g_m/I_D operation are analogous to weak- and strong-inversion operation in MOS devices. Additionally, the thin-film and carbon nanotube devices have a transitional drain-current and g_m/I_D region, analogous to moderate inversion in MOS devices. However, the width of this transitional region may not equal the two-decade width of the inversion coefficient ($0.1 < IC < 10$) or drain current (for a device having fixed geometry) present for MOS devices operating in moderate inversion. Still, normalization could be arranged where an inversion coefficient of unity corresponds to the transition of g_m/I_D for the exponential- and square-law drain-current regions. Figures 3.24–3.27 illustrate this for MOS devices.

Organic or amorphous silicon, thin-film, field-effect transistors are candidates for large-area, flexible, low-cost electronics. These devices have carrier mobility much below that of MOS devices at approximately 0.01–1 cm^2/Vs and generally have much higher subthreshold swings that can exceed 1 V/decade [1–9]. Carbon nanotube, field-effect transistors are candidates for nanoscale circuits not realizable using existing MOS devices. These devices have potentially higher carrier mobility and current densities than MOS devices, and subthreshold swings that are comparable to approximately 90 mV/decade present for bulk MOS devices [10–14]. The lower subthreshold swing, given for room temperature at $T = 300$ K, corresponds to lower off-state current for digital circuits, while the much smaller geometries and higher mobility correspond to much higher analog bandwidths and digital circuit speeds for the carbon nanotube devices compared to the thin-film devices.

Evaluating measures of performance analogous or equivalent to MOS effective gate–source voltage, drain–source saturation voltage, transconductance efficiency, normalized drain–source conductance, intrinsic voltage gain, capacitances, intrinsic bandwidth, thermal noise, flicker noise, mismatch, and leakage can permit an extension of the design methods described in this book to non-CMOS devices. As for MOS devices, performance tradeoffs and optimization could be potentially considered for analogous or equivalent design choices of device bias current, inversion coefficient, and channel length.

REFERENCES

[1] O. Shchekin, R. Wenz, R. Rotzoll, M. Grigas, J. Barad, K. Dimmler, and A. Dodabalapur, "Pentacene organic field-effect transistors for flexible electronics," *Proceedings of SPIE - The International Society for Optical Engineering*, vol. 5522, Organic Field-Effect Transistors III, pp. 17–21, Denver, Aug. 2004.

[2] D. M. Binkley, N. Verma, R. L. Crawford, E. Brandon, and T. N. Jackson, "Design of an auto-zeroed, differential, organic thin-film field-effect transistor amplifier for sensor applications," *Proceedings of SPIE - The International Society for Optical Engineering*, vol. 5522, Organic Field-Effect Transistors III, pp. 41–52, Denver, Aug. 2004.

[3] E. Calvetti, L. Colalongo, and Zs. M. Kovacs-Vajna, "Organic thin film transistors: a DC/dynamic analytical model," *Solid-State Electronics*, vol. 49, pp. 567–577, Apr. 2005.

[4] A. Sazonov, D. Striakhilev, C.-H. Lee, and A. Nathan, "Low-temperature materials and thin-film transistors for flexible electronics," *Proceedings of the IEEE*, vol. 93, pp. 1420–1428, Aug. 2005.

[5] E. Calvetti, A. Savio, Zs. M. Kovacs-Vajna, and L. Colalongo, "Analytical model for organic thin-film transistors operating in the subthreshold region," *Applied Physics Letters*, vol. 87, no. 22, pp.223506–1–3, 2005.

[6] C. Rolin, S. Steudel, K. Myny, D. Cheyns, S. Verlaak, J. Genoe, and P. Heremans, "Pentacene devices and logic gates fabricated by organic vapor phase deposition," *Applied Physics Letters*, vol. 89, no. 20, pp. 203502–1–3, 2006.

[7] J. G. Park, R. Vasic, J. S. Brooks, and J. E. Anthony, "Characterization of functionalized pentacene field-effect transistors and its logic gate application," *Journal of Applied Physics*, vol. 100, no. 4, pp. 044511–1–6, 2006.

[8] D. Oberhoff, K. P. Pernstich, D. J. Gundlach, and B. Batlogg, "Arbitrary density of states in an organic thin-film field-effect transistor model and application to pentacene devices," *IEEE Transactions on Electron Devices*, vol. 54, pp. 17–25, Jan. 2007.

[9] B. Iniguez, R. Picos, M. Estrada, A. Cerdeira, T. A. Ytterdal, W. Jackson, A. Koudymov, D. Veksler, and M. S. Shur, "Modelling of thin film transistors for circuit simulation," *Proceedings of the 14th International Conference on Mixed Design of Integrated Circuits and Systems* (*MIXDES*), Ciechocinek, Poland, pp. 35–40, June 2007.

[10] P. Avouris, J. Appenzeller, R. Martel, and S. J. Wind, "Carbon nanotube electronics," *Proceedings of the IEEE*, vol. 91, pp. 1772–1784, Nov. 2003.

[11] J.-M. Bethoux, H. Happy, A. Siligaris, G. Dambrine, J. Borghetti, V. Derycke, and J.-P. Bourgoin, "Active properties of carbon nanotube field-effect transistors deduced from s parameters measurements," *IEEE Transactions on Nanotechnology*, vol. 5, pp. 335–342, July 2006.

[12] A. Raychowdhury, A. Keshavarzi, J. Kurtin, V. De, and K. Roy, "Carbon nanotube field-effect transistors for high-performance digital circuits – DC analysis and modeling toward optimum transistor structure," *IEEE Transactions on Electron Devices*, vol. 53, pp. 2711–1717, Nov. 2006.

[13] L. Latessa, A. Pecchia, and A. Di Carlo, "DFT modeling of bulk-modulated carbon nanotube field-effect transistors," *IEEE Transactions on Nanotechnology*, vol. 6, pp. 13–21, Jan. 2007.

[14] A. Hazeghi, T. Krishnamohan, and H.-S. Phillip Wong, "Schottky-barrier carbon nanotube field-effect transistor modeling," *IEEE Transactions on Electron Devices*, vol. 54, pp. 439–445, Mar. 2007.

Appendix
The *Analog CMOS Design, Tradeoffs and Optimization* Spreadsheet

A.1 OVERVIEW

The *Analog CMOS Design, Tradeoffs and Optimization* spreadsheet facilitates rapid initial optimization of analog CMOS design prior to time-consuming simulations of circuit performance required for production release. Using the spreadsheet, the designer explores MOS device drain current, inversion coefficient, and channel length and observes the resulting performance of devices and complete circuits.

The spreadsheet, available from the book's web site listed on the cover, contains separate sheets entitled *MOSFETs, Circuit Analysis, Process,* and *Disclaimer, Notes* that are described below.

A.2 THE *MOSFETS* SHEET

The *MOSFETs* sheet contains columns for individual MOSFETs or for collections of related MOSFETs, such as equally sized and biased devices used in differential pairs and current mirrors. The *MOSFETs* sheet provides predictions of MOS device performance as the user explores device drain current, inversion coefficient, and channel length for circuit optimization. Columns for additional MOSFETs can be copied from existing columns.

The *MOSFETs* sheet was illustrated in Figures 4.20–4.24 in Section 4.4 for the optimum design of differential pairs and current mirrors. Additionally, the *MOSFETs* sheet was illustrated in Figure 5.3 in Section 5.4.2 and in Figures 5.11–5.13 in Section 5.5.2 for 0.5 μm, simple and 0.18 μm, cascoded CMOS operational transconductance amplifiers (OTAs) that were optimized for DC, balanced, and AC performance. Finally, the *MOSFETs* sheet was illustrated in Figure 6.9 in Sections 6.7.3 and 6.8.3 for 0.35 μm, differential and single-ended input, micropower CMOS preamplifiers that were optimized for minimum thermal and flicker noise. The *MOSFETs* sheet illustrated here is taken from Figure 5.11 that was used for the design of input differential-pair devices for the 0.18 μm, cascoded CMOS OTAs.

Tradeoffs and Optimization in Analog CMOS Design   David M. Binkley
© 2008 John Wiley & Sons, Ltd

A.2.1 User Inputs

Figure A.1 shows optional user design information and required user design inputs located at the top of the *MOSFETs* sheet. Although not included on the actual *MOSFETs* sheet, the figure includes references to book sections and figures that provide further details on the required user inputs.

A.2.1.1 Optional User Design Information

The *Optional User Design Information* shown in Figure A.1 contains a *Description*, which can include text describing device applications in input differential pairs, current sources, or other subcircuits. The *Device reference* can include text describing device references from a circuit schematic. Finally, the *Device notes* can include text denoting different optimizations or process or environmental conditions that the designer wishes to explore. In the example shown, the optional user design information describes an input pair consisting of devices M_1 and M_2 that is optimized for DC, balanced, and AC performance. As mentioned, this is part of the 0.18 μm, cascoded CMOS OTAs described in Section 5.5.2.

A.2.1.2 Required User Design Inputs

The *Required User Design Inputs* shown in Figure A.1 contain a *Device model*, which is a text name that must match a model name on the *Process* sheet. In the example shown, the model names specify nMOS or pMOS devices, the region of device operation in weak, moderate, or strong inversion, and the channel length. Channel length is specified in the model names because the process Early voltage, channel length modulation parameter V_{AL} depends on the channel length. The user, however, may use any text convention for naming MOS models. Required user design inputs also include the drain current, given positively for both nMOS and pMOS devices, the fixed–normalized inversion coefficient, and the drawn channel length. As described in Section 3.2, these three, independent degrees of design freedom are used in this book to evaluate tradeoffs in performance and optimize designs.

The spreadsheet returns error messages if the device model is not found on the *Process* sheet. Additionally, the spreadsheet returns error messages if the selected drain current, inversion coefficient, channel length, or optional interdigitation multiplier described below results in a channel width that is below the process minimum. The error messages contain information suggesting changes in the user

Analog CMOS Design, Tradeoffs and Optimization Spreadsheet -- MOSFETs Sheet					
For design guidance only as results are approximate and not for any particular CMOS process.					
© 2000 – 2007, David M. Binkley. See Disclaimer, Notes Sheet.					
---------- Optional User Design Information ----------					
Description		------------ Input pair ------------			
Device reference		M1, M2	M1, M2	M1, M2	
Device notes		DC	BAL	AC	
---------- Required User Design Inputs ----------				**Section**	
Device model		nMIL2	nMIL05	nMIL02	**Reference**
I_D	μA (+)	100	100	100	3.2.1, Figure 3.9
IC (fixed normalized)		0.866	0.832	0.742	3.2.1, 3.4.2.2
L_{DRAWN}	μm	1.98	0.48	0.18	3.2.1, 3.3.2

Figure A.1   Optional user design information and required user design inputs located at the top of the *MOSFETs* sheet in the *Analog CMOS Design, Tradeoffs and Optimization* spreadsheet. Book section and figure references have been added on the right

inputs to obtain sufficient channel width. These include increasing the drain current, decreasing the inversion coefficient, increasing the channel length, or decreasing the interdigitation multiplier. The spreadsheet also returns error messages if the selected channel length is below the process minimum.

A.2.1.3 Optional User Design Inputs

Figure A.2 shows optional user design inputs located at the bottom of the *MOSFETs* sheet. Like Figure A.1, the figure includes references to book sections and figures that provide further details on the user inputs.

The *Optional User Design Inputs* include m, which is the interdigitation multiplier. This is the number of device interdigitations or individual gate fingers. When unspecified, m is assigned a default value of unity as listed. W_{DRAWN} *(trimmed), each* is the desired drawn drain and source finger width for the specified value of m. The optional specification of m and W_{DRAWN} *(trimmed), each* results in a displayed value of *IC (trimmed)* shown later in Figure A.4, which is the value of the inversion coefficient required for the specified values of m and W_{DRAWN} *(trimmed), each*. The value of *IC (trimmed)* can then be pasted, using the spreadsheet paste value and not paste equation feature, into the user input of *IC* shown in Figure A.1. This will then result in a "trimmed" device having the specified values of m and W_{DRAWN} *(trimmed), each* that are convenient and rounded for layout. In design practice, m is usually set to an even number when not set to unity.

Although device "trimming" can be useful to reflect actual device layout and accurately predict drain and source areas and perimeters used in the calculation of drain–body and source–body capacitances, the user should initially explore MOS device performance by leaving entries m and W_{DRAWN} *(trimmed), each* blank. Then, after the desired MOS drain current, inversion coefficient, and channel length are determined, the designer can optionally "trim" the device for practical values for m and W_{DRAWN} *(trimmed), each* desired for device layout. The spreadsheet returns error messages if W_{DRAWN} *(trimmed), each* is below the process minimum and, as mentioned, gives suggestions for adjusting user inputs of drain current, inversion coefficient, channel length, or m.

The *Optional User Design Inputs* also include the drain–source voltage, V_{DS}, which has the default value listed of V_{GS} corresponding to a diode-connected device. If specified, V_{DS} is given positively for both nMOS and pMOS devices, just as V_{GS} is given positively in this book for both device types. The source–body voltage, V_{SB}, has the default value of 0 V listed. If specified, V_{SB} is also given positively for both nMOS and pMOS devices. As for m and W_{DRAWN} *(trimmed), each* mentioned above, the user could initially explore MOS device performance by leaving the V_{DS} and V_{SB} entries blank. Then, values for these could be optionally entered if desired. The spreadsheet returns error messages if the values of the drain–body voltage, $V_{DB} = V_{DS} + V_{SB}$, or V_{SB} exceed those permitted by the maximum supply voltage for the process.

The *Optional User Design Inputs* also include the selected frequency, $f_{FLICKER}$, for evaluating flicker noise, which has the default value listed of 1 Hz. The final optional input is temperature, which has the default value listed of 27 °C. As mentioned in Section 3.5.3, the inversion coefficient

---------- Optional User Design Inputs ----------					Section	
	Default, units				Reference	
m, number of gate fingers	1	16	8	4	3.9.4	
W_{DRAWN} (trimmed), each	none, μm	22	10.6	8	3.9.4	
V_{DS} (for C_{DB})	V_{GS}, V (+)	0.5	0.5	0.5	Figure 3.9	
V_{SB} (for g_{mb} and C_{SB})	0 V, V (+)				Figure 3.9	
$f_{FLICKER}$, freq. for flicker noise	1 Hz, Hz	100	100	100	3.10.3.7	
T, temperature	27 C, C				3.5	

Figure A.2 Optional user design inputs located at the bottom of the *MOSFETs* sheet in the *Analog CMOS Design, Tradeoffs and Optimization* spreadsheet. Book section and figure references have been added on the right

changes with temperature for a fixed-geometry device because of changes in the technology current, $I_0 = 2n_0\mu_0 C'_{OX} U_T^2$. However, as mentioned in Section 3.5.5, the inversion coefficient can be adjusted or trimmed to maintain a constant layout channel width as different temperatures are studied. This can be done using different columns in the *MOSFETs* sheet for different temperatures. Alternatively, a design can be optimized at the nominal temperature and then verified, as always required, by simulations over process variations and temperature.

A.2.2 Reported MOS Performance, Primary

Figure A.3 shows the primary portion of the *MOSFETs* sheet. This includes at the top the primary user inputs shown in Figure A.1, followed by primary measures of calculated MOS performance. Although not included on the actual *MOSFETs* sheet, the figure includes references to book sections and figures that describe the aspects of MOS performance. The figure additionally includes references to tables and equations used for the calculation of performance. The calculation references, however, are intended for guidance only as details of the actual calculations are contained in the spreadsheet software code and its comments, which are accessible to the user. Additionally, as described later in Section A.6, the spreadsheet calculations are "works in progress" that could be improved in the future by including, for example, linkage to full simulation MOS models. The reader is invited to periodically check the book's web site listed on the cover for possible improvements to the spreadsheet.

The primary portion of the *MOSFETs* sheet shown in Figure A.3 is designed to summarize MOS device design choices and performance on a single sheet of paper. Calculated MOS performance begins below *Calculated Results*, which displays *OK* or *ERROR* following successful or unsuccessful calculation of MOS performance. If *ERROR* is reported, the spreadsheet will usually provide an error message describing errors in user inputs or MOS model parameters.

A.2.2.1 Effective Width, Length, and Gate Area

The first group of calculated MOS results in Figure A.3 contains the effective channel width, W, channel length, L, and gate area, WL. These indicate device sizing and are used implicitly in other calculations of MOS performance. Additionally, the gate area is used for calculations of circuit flicker noise and local-area mismatch in the *Circuit Analysis* sheet. Channel width and gate area are total values inclusive of interdigitation.

A.2.2.2 DC Bias Voltages

The second group of calculated MOS results contains the effective gate–source voltage, $V_{EFF} = V_{GS} - V_T$, threshold voltage, V_T, gate–source voltage, $V_{GS} = V_T + V_{EFF}$, and drain–source saturation voltage, $V_{DS,sat}$. Collectively, these voltages are referred to as bias voltages. $V_{DS,sat}$ and V_T, however, are not bias voltages directly. $V_{DS,sat}$ represents the minimum V_{DS} required for operation in saturation, which is assumed in the spreadsheet. V_T directly influences the value of V_{GS}.

V_{EFF} is taken positively for both nMOS and pMOS devices when the magnitude of V_{GS} is above the magnitude of V_T, corresponding to operation above the threshold voltage. V_{EFF} is taken negatively for both device types when the magnitude of V_{GS} is below the magnitude of V_T, corresponding to operation below the threshold voltage. Consistent with this book, V_T, V_{GS}, and $V_{DS,sat}$ are taken positively for both nMOS and pMOS devices. V_{GS} and $V_{DS,sat}$ are used for calculations of circuit, input and output voltage ranges in the *Circuit Analysis* sheet.

Although Table 3.14 provides simple expressions for V_{EFF}, this is found in the initial version of the spreadsheet by an iterative solution of the enhanced drain current equation given in Table 3.13. Both

Analog CMOS Design, Tradeoffs and Optimization Spreadsheet -- MOSFETs Sheet					
For design guidance only as results are approximate and not for any particular CMOS process.					
© 2000 – 2007, David M. Binkley. See Disclaimer, Notes Sheet.					
---------- Optional User Design Information ----------					
Description		------------ **Input pair** ------------			
Device reference		M1, M2	M1, M2	M1, M2	
Device notes		DC	BAL	AC	
---------- **Required User Design Inputs** ----------					Section
Device model		nMIL2	nMIL05	nMIL02	Reference
I_D	μA (+)	100	100	100	3.2.1, Figure 3.9
IC (fixed normalized)		0.866	0.832	0.742	3.2.1, 3.4.2.2
L_{DRAWN}	μm	1.98	0.48	0.18	3.2.1, 3.3.2
					Calculation
---------- **Calculated Results** ----------		OK	OK	OK	Section	Table
Effective Width, Length, Gate Area					Reference	Reference
W	μm	352.00	84.80	32.00	3.6.2	3.10
L	μm	1.952	0.452	0.152	3.3.2	L_{DRAWN}-DL
WL	μm²	687.10	38.33	4.86	3.6.3	3.10
DC Bias Voltages						
$V_{EFF}=V_{GS}-V_T$	V	0.032	0.030	0.024	3.7.2	See text.
V_T (adjusted for V_{SB})	V	0.420	0.420	0.420	3.7.1.7	3.7, 3.20
$V_{GS}=V_T+V_{EFF}$	V	0.452	0.450	0.444	Figure 3.9	
$V_{DS, sat}$	V	0.133	0.132	0.130	3.7.3	3.15
Small Signal Parameters						
g_m/I_D	μS/μA	19.50	19.66	20.08	3.8.2	3.17
g_m	μS	1949.59	1965.63	2008.14	3.8.2	$(g_m/I_D)I_D$
V_A	V	17.93	8.13	2.27	3.8.4	See text.
g_{ds}	μS	5.5758	12.3038	44.0606	3.8.4	I_D/V_A
$A_{Vi}=g_m/g_{ds}$	V/V	349.7	159.8	45.6	3.8.5	
Transcond. Distortion (Diff. Pair)						
$V_{INDIF,1dB}$ (input 1-dB comp.)	V	0.072	0.071	0.071	3.8.2.6	See text.
Capacitances and Bandwidths						
$C_{GS}=C_{gsi}+C_{GSO}$	fF	2188.1	181.7	42.4	Figure 3.21	
$C_{GB}=C_{gbi}+C_{GBO}$	fF	865.3	48.7	6.3	Figure 3.21	
$C_{GD}=C_{GDO}$ (in saturation)	fF	330.9	79.7	30.1	Figure 3.21	
C_{DB} (at $V_{DB}=V_{DS}+V_{SB}$)	fF	175.1	42.8	16.3	3.9.4	3.25
$f_{Ti}=g_m/[2\pi(C_{gsi}+C_{gbi})]$	MHz	114.0	2077.1	17139.3	3.9.6	
$f_T=g_m/[2\pi(C_{GS}+C_{GB})]$	MHz	101.6	1358.2	6559.0	3.9.7	
$f_{DIODE}=g_m/[2\pi(C_{GS}+C_{GB}+C_{DB})]$,MHz		96.1	1145.4	4915.0	3.9.7	
Thermal and Flicker Noise						
$S_{VG}^{1/2}$ thermal	nV/Hz$^{1/2}$	2.51	2.50	2.46	3.10.2.3	3.30
$S_{VG}^{1/2}$ flicker at $f_{FLICKER}$	nV/Hz$^{1/2}$	37.24	157.39	439.44	3.10.3.7	3.34
$S_{ID}^{1/2}$ thermal	pA/Hz$^{1/2}$	4.90	4.91	4.95	3.10.2.3	3.29
$S_{ID}^{1/2}$ flicker at $f_{FLICKER}$	pA/Hz$^{1/2}$	72.61	309.38	882.45	3.10.3.7	3.34
$f_{FLICKER}$, freq. for flicker noise	Hz	100	100	100	3.10.3.7	3.34
f_c, flicker-noise corner	MHz	0.0569	1.7102	19.8233	3.10.3.8	3.35
DC Mismatch for Device Pair						
ΔV_{GS} (1σ)	mV	0.19	0.82	2.31	3.11.1.7	3.41
$\Delta I_D/I_D$ (1σ)	%	0.38	1.62	4.64	3.11.1.7	3.41

Figure A.3 Primary portion of the *MOSFETs* sheet in the *Analog CMOS Design, Tradeoffs and Optimization* spreadsheet listing the primary user inputs shown in Figure A.1 and primary measures of calculated MOS performance. Book section and figure references, and table and equation references for calculations, have been added on the right

the expressions and iterative solution for finding V_{EFF} include velocity saturation and vertical field mobility reduction (VFMR) effects, which increase the V_{EFF} required to support a given drain current.

A.2.2.3 Small-Signal Parameters

The third group of calculated MOS results contains the small-signal parameters of transconductance efficiency, g_m/I_D, transconductance, $g_m = (g_m/I_D) \cdot I_D$, Early voltage, $V_A = I_D/g_{ds}$, drain–source conductance, $g_{ds} = I_D/V_A$, and intrinsic voltage gain, $A_{Vi} = g_m/g_{ds}$. g_m is considered implicitly in other MOS calculations and is used for calculations of circuit transconductance, noise, and local-area mismatch in the *Circuit Analysis* sheet. g_m and g_{ds} are used for calculations of small-signal resistance, gain, and bandwidth. g_m/I_D, V_A, and A_{Vi} are useful for evaluating the performance and trends for individual devices.

g_m/I_D is found in the initial version of the spreadsheet from the double square-root expression contained in Table 3.17. V_A is found by the parallel combination of V_A (CLM) and V_A (DIBL) as described in Table 3.21. V_A (DIBL) is found as described in the table from process parameters *DVTDIBL* and *DVTDIBLEXP*, while V_A (CLM) is found from the process, channel length modulation parameter, V_{AL}, by V_A (CLM) $= V_{AL} \cdot L$. As described at the beginning of Sections 4.4, 5.4.2, and 5.5.2, the process value of V_{AL} must be extracted near the channel length and V_{DS} used to permit accurate prediction of V_A and, correspondingly, g_{ds}. This is because, as shown in Figure 3.48, V_{AL} is not constant for a process, but is a strong function of the channel length and V_{DS}. As mentioned in Sections 4.4, 5.4.2, and 5.5.2 and later in Section A.6, linking the spreadsheet to a full simulation MOS model could permit prediction of V_A and g_{ds}, which are often predicted with significant errors, without requiring the extraction of V_{AL}. In the initial version of the spreadsheet, "binned" extraction (near the device channel length and V_{DS} used) is required only for the process value of V_{AL}.

A.2.2.4 Transconductance Distortion for a Differential Pair

The fourth group of calculated MOS results contains the input, 1 dB compression voltage, $V_{INDIF,1dB}$, for two equal devices operating in a differential pair. This is used for calculations of circuit distortion in the *Circuit Analysis* sheet.

Although Table 3.19 contains approximations for $V_{INDIF,1dB}$, which is nearly inversely proportional to g_m/I_D, $V_{INDIF,1dB}$ is found by an iterative solution of the enhanced drain current expression given in Table 3.13, applied for a differential pair. As mentioned, this drain current expression includes velocity saturation and VFMR effects.

A.2.2.5 Capacitances and Bandwidths

The fifth group of calculated MOS results contains device capacitance and bandwidth measures. C_{GS}, C_{GB}, and C_{GD} are the total gate–source, gate–body, and gate–drain capacitances, inclusive of intrinsic gate capacitances, C_{gsi}, C_{gbi}, and C_{gdi}, and extrinsic, gate-overlap capacitances, C_{GSO}, C_{GBO}, and C_{GDO}. C_{DB} is the drain–body capacitance associated with the drain–body diode junction. References for the calculation of intrinsic gate and extrinsic gate-overlap capacitances are included in the secondary, reported MOS results shown later in Figure A.4.

$f_{Ti} = g_m/[2\pi(C_{gsi} + C_{gbi})]$ is the intrinsic bandwidth associated with the intrinsic gate–source and gate–body capacitances. $f_T = g_m/[2\pi(C_{GS} + C_{GB})]$ is the extrinsic bandwidth associated with the total gate–source and gate–body capacitances, which include the extrinsic gate-overlap capacitances.

Finally, $f_{DIODE} = g_m/[2\pi(C_{GS} + C_{GB} + C_{DB})]$ is a diode-connected bandwidth measure that includes the total gate–source and gate–body capacitances, along with the drain–body capacitance.

The capacitances are used for calculations of circuit capacitance and related bandwidth in the *Circuit Analysis* sheet. The bandwidth measures are useful for evaluating the performance and trends of individual devices. As described in Section 3.9.7, the extrinsic and diode-connected bandwidths can be significantly below the intrinsic bandwidth, especially for short-channel devices having small gate area and small intrinsic capacitances.

A.2.2.6 Thermal and Flicker Noise

The sixth group of calculated MOS results contains device noise measures. $S_{VG}^{1/2}$ thermal is the gate-referred thermal-noise voltage density. $S_{VG}^{1/2}$ flicker is the gate-referred flicker-noise voltage density at the user-selected frequency of $f_{FLICKER}$, mentioned in Section A.2.1.3, which is repeated in the calculated results for convenience. $S_{ID}^{1/2}$ thermal is the drain-referred thermal-noise current density, and $S_{ID}^{1/2}$ flicker is the drain-referred flicker-noise current density at $f_{FLICKER}$. Finally, f_c is the flicker-noise corner frequency.

Gate-referred and drain-referred noise are used for calculations of circuit thermal and flicker noise in the *Circuit Analysis* sheet. f_c is useful for evaluating the flicker-noise performance of individual devices.

A.2.2.7 DC Mismatch for Device Pair

The seventh and final group of calculated MOS results shown in Figure A.3 contains measures of local-area DC mismatch for a pair of equally sized and biased devices. ΔV_{GS} is the gate–source mismatch voltage for two devices in a differential pair. $\Delta I_D/I_D$ is the relative drain current mismatch for two devices in a current mirror. Device-pair mismatch voltage and current are used for calculations of circuit mismatch in the *Circuit Analysis* sheet. These are also useful for evaluating the mismatch of individual differential pairs and current mirrors.

A.2.3 Reported MOS Performance, Secondary

Figure A.4 shows the secondary portion of the *MOSFETs* sheet, noted as *Sheet 2*. For convenience, this contains at the top a repeat of the *Device reference* and *Device notes* optional user design information shown earlier in Figure A.1. Additionally, this contains at the bottom the optional user design inputs shown earlier in Figure A.2. Located between the top and bottom rows are additional measures of MOS performance not included on the primary portion of the *MOSFETs* sheet shown in Figure A.3. The secondary portion of the *MOSFETs* sheet shown in Figure A.4 is intended to be presented on a second sheet of paper to provide additional details of MOS performance.

Like Figure A.3, Figure A.4 includes references to book sections and figures that describe the aspects of MOS performance. Figure A.4 also includes references to tables and equations used for the calculation of performance. The calculation references, as for Figure A.3, are intended for guidance only as details of the actual calculations are contained in the spreadsheet software code and its comments. Again, the spreadsheet calculations are "works in progress". The reader is invited to periodically check the book's web site listed on the cover for possible improvements to the spreadsheet.

*** MOSFETs, Sheet 2 ***							**Calculation**
Device reference		M1, M2	M1, M2	M1, M2		**Section**	**Table**
Device notes		DC	BAL	AC		**Reference**	**Reference**
Body Small Signal Parameters							
$\eta = g_{mb}/g_m$ (at V_{SB})		0.283	0.283	0.284		3.8.3	3.20
g_{mb} (at V_{SB})	μS	551.99	556.96	570.28		3.8.3	ηg_m
Source Parameters							
$g_{ms} = g_m + g_{mb} + g_{ds}$	μS	2507.16	2534.89	2622.48		3.8.1	
C_{SB} (at V_{SB})	fF	219.6	58.6	26.1		3.9.4	3.25
Noise and Mismatch Parameters							
Γ, thermal noise factor (g_m)		0.579	0.578	0.573		3.10.2	Eq. 3.106
$(n\Gamma)$, n multiplied by Γ		0.744	0.742	0.736		3.10.2	
$K'_F{}^{1/2}$, flicker-noise factor	nV·μm	6912	6899	6861		3.10.3.4	3.31
AF, flicker-noise slope		0.85	0.85	0.85		3.10.3.4	3.3
A_{VGS}, V_{GS} mismatch factor	V·μm	0.00510	0.00510	0.00510		3.11.1.4	Eq. 3.151
Velocity Sat. and VFMR Details							
Vel. sat. factor, $V_{EFF}/(LE_{CRIT})$		0.003	0.012	0.028		3.7.1.2	
VFMR factor, $V_{EFF}\theta$		0.009	0.008	0.007		3.7.1.3	
Critical IC, IC_{CRIT}		411.5	124.6	26.8		3.8.2.2	3.17
Gate Capacitance Details							
C_{GOX}	fF	5778.5	322.4	40.9		3.9.1	3.23
C_{gsi}	fF	1857.2	102.0	12.3		3.9.2	3.23
C_{gbi}	fF	865.3	48.7	6.3		3.9.2	3.23
C_{GSO}	fF	330.9	79.7	30.1		3.9.3	3.24
C_{GBO}	fF	0.0	0.0	0.0		3.9.3	3.24
C_{GDO}	fF	330.9	79.7	30.1		3.9.3	3.24
Process and IC Details							
n (actual)		1.283	1.283	1.284		3.4.1	3.6
μ_0	cm²/Vs	422.00	422.00	422.00		3.3.2	3.7
$k_0 = \mu_0 C'_{OX}$	μA/V²	354.90	354.90	354.90		3.3.2	
$U_T = kT/q$	V	0.02585	0.02585	0.02585		3.3.1	3.7
$I_0 = 2n_0\mu_0 C'_{OX}U_T{}^2$	μA	0.641	0.641	0.641		3.4.2.2	3.6
IC (actual)		0.911	0.875	0.780		See text.	See text.
Layout Details, IC Trimming							
W_{DRAWN}, each; x m for total	μm	22.00	10.60	8.00		3.9.4	See text.
L_{DRAWN}	μm	1.98	0.48	0.18		3.2.1, 3.3.2	User input
AD, each; x m for total	μm²	6.60	3.18	2.40		3.9.4	See text.
PD, each; x m for total	μm	22.60	11.20	8.60		3.9.4	See text.
AS, each; x m for total	μm²	7.15	3.71	3.20		3.9.4	See text.
PS, each; x m for total	μm	25.40	13.95	12.80		3.9.4	See text.
IC (trimmed), may be used for IC input		0.866	0.832	0.742		See text.	See text.
---------- Optional User Design Inputs ----------						**Section**	
	Default, units					**Reference**	
m, number of gate fingers	1	16	8	4		3.9.4	
W_{DRAWN} (trimmed), each	none, μm	22	10.6	8		3.9.4	
V_{DS} (for C_{DB})	V_{GS}, V (+)	0.5	0.5	0.5		Figure 3.9	
V_{SB} (for g_{mb} and C_{SB})	0 V, V (+)					Figure 3.9	
$f_{FLICKER}$, freq. for flicker noise	1 Hz, Hz	100	100	100		3.10.3.7	
T, temperature	27 C, C					3.5	

Figure A.4 Secondary portion of the *MOSFETs* sheet in the *Analog CMOS Design, Tradeoffs and Optimization* spreadsheet listing secondary measures of calculated MOS performance and the optional user design inputs shown in Figure A.2. Book section and figure references, and table and equation references for calculations, have been added on the right

A.2.3.1 Body Small-Signal Parameters

The first group of calculated MOS results in Figure A.4 contains the body-effect transconductance ratio, $\eta = g_{mb}/g_m$, and body-effect transconductance, g_{mb}. g_{mb} is used for calculations of circuit small-signal resistance and gain in the *Circuit Analysis* sheet, especially for source follower stages. η is useful for evaluating the body-effect performance and trends for individual devices.

η is calculated from $\eta = g_{mb}/g_m = n - 1$, listed in Table 3.20, where n is the substrate factor. As mentioned in Section 3.8.3.3, this simple prediction can provide better accuracy than the more complex expressions given in the table. Both η and n decrease slightly with the inversion level or V_{EFF} and decrease slightly with increasing V_{SB}.

A.2.3.2 Source Parameters

The second group of calculated MOS results contains the total source small-signal conductance, $g_{ms} = g_m + g_{mb} + g_{ds}$, and the source–body capacitance, C_{SB}, associated with the source–body diode junction. g_{ms} is used for calculations of circuit small-signal resistance, gain, and bandwidth in the *Circuit Analysis* sheet. C_{SB} is used for calculations of circuit capacitance and related bandwidth.

A.2.3.3 Noise and Mismatch Parameters

The third group of calculated MOS results contains noise and mismatch parameters. Γ is the thermal-noise factor relative to the transconductance in saturation, and $n\Gamma$ is the product of n and Γ. $K_F'^{1/2}$ is the square root of the operating value of the flicker-noise factor, K_F'. $K_F'^{1/2}$ includes possible inversion-level or V_{EFF} increases in the gate-referred flicker-noise voltage above that predicted in weak inversion or the weak-inversion side of moderate inversion by the MOS process parameter K_{F0}' contained in the *Process* sheet. AF is the flicker-noise slope, repeated for convenience from the process parameter contained in the *Process* sheet. Finally, A_{VGS} is the operating value of the gate–source voltage mismatch factor, which includes both threshold-voltage and transconductance factor mismatch components. A_{VGS} includes increases in gate–source mismatch voltage caused by increases in the threshold-voltage mismatch due to non-zero V_{SB} and includes mismatch increases caused by transconductance factor mismatch.

The noise and mismatch parameters are used for calculations of circuit noise and local-area mismatch in the *Circuit Analysis* sheet. These parameters are also useful for understanding possible thermal-noise, flicker-noise, and local-area mismatch increases for individual devices or device pairs. Comparing $K_F'^{1/2}$ to the fixed process value of $K_{F0}'^{1/2}$ illustrates the degree of inversion-level or V_{EFF} increases in gate-referred flicker-noise voltage density. Comparing A_{VGS} to the fixed process value of A_{VTO} illustrates the degree that local-area mismatch is increased due to non-zero V_{SB} threshold-voltage mismatch and transconductance factor mismatch.

A.2.3.4 Velocity Saturation and VFMR Details

The fourth group of calculated MOS results contains details on the level of velocity saturation and vertical field mobility reduction (VFMR) present for devices. The velocity saturation factor, $V_{EFF}/(LE_{CRIT})$, is a measure of the degree of velocity saturation, which can be significant for short-channel devices operating at high levels of inversion. A value well below unity indicates negligible velocity saturation. A value of 0.1, for example, indicates that the drain current is reduced by approximately 10 % below the value present without velocity saturation. Finally, a value above unity indicates nearly full velocity saturation.

The VFMR reduction factor, $V_{EFF} \cdot \theta$, provides a similar measure for the degree of VFMR, which can be significant for long-channel devices operating at high levels of inversion with little velocity saturation. A value well below unity indicates negligible VFMR. A value of 0.1, for example, indicates that the drain current is reduced by approximately 10 % below the value present without VFMR.

IC_{CRIT} corresponds to the value of the (fixed–normalized) inversion coefficient, IC, where g_m/I_D is down to approximately 70.7% of its value without velocity saturation or VFMR effects. Operation at selected IC values well below IC_{CRIT} results in nearly negligible velocity saturation or VFMR effects, whereas operation at IC values above IC_{CRIT} results in significant velocity saturation or VFMR effects.

The velocity saturation and VFMR factors and IC_{CRIT} are useful for observing the degree of velocity saturation or VFMR present for devices. Decreasing the level of inversion or IC decreases both effects, while increasing the channel length significantly decreases velocity saturation.

A.2.3.5 Gate Capacitance Details

The fifth group of calculated MOS results contains gate capacitance details. $C_{GOX} = C'_{OX}WL$ is the total gate-oxide capacitance. C_{gsi} and C_{gbi} are the intrinsic gate–source and gate–body capacitances, which can be observed as a fraction of C_{GOX}. The intrinsic gate–drain capacitance, C_{gdi}, is not included because it is assumed negligible in saturation. Finally, C_{GSO}, C_{GBO}, and C_{GDO} are the extrinsic gate–source, gate–body, and gate–drain overlap capacitances.

The gate capacitance details are not used explicitly in calculations of circuit performance in the *Circuit Analysis* sheet because the composite values of C_{GS}, C_{GB}, and C_{GD} are used from Figure A.3. However, gate capacitance details are useful to assess the contributions of different capacitance components. The extrinsic gate-overlap capacitances described in Section 3.9.3, along with the drain–body capacitance described in Section 3.9.4, can be significant for short-channel devices having small gate area and small intrinsic capacitances.

A.2.3.6 Process and IC Details

The sixth group of calculated MOS results contains process and inversion coefficient details. n (actual) is the operating value of the substrate factor for the operating value of V_{EFF} and V_{SB}. n (actual) is used for calculations of MOS performance and is slightly less than n_0 when operation is in strong inversion or at non-zero V_{SB}. n_0 is the average process value of n in moderate inversion at zero V_{SB}, given in the *Process* sheet. μ_0 is the low-field mobility, and $k_0 = \mu_0 C'_{OX}$ is the low-field transconductance factor. $U_T = kT/q$ is the thermal voltage, and $I_0 = 2n_0\mu_0 C'_{OX}U_T^2$ is the technology current.

IC (actual) $= I_D/[2n\mu_0 C'_{OX}U_T^2(W/L)]$ is the operating value of the inversion coefficient for n (actual). IC (actual) is used for calculations of MOS performance compared to the user input of IC, which is used to find the effective channel width and gate area from I_0 that includes n_0. IC (actual) will be slightly above the user input of IC for the usual case where n (actual) is slightly below the process value of n_0.

As used throughout this book and described in Sections 3.4.2.2 and 3.4.2.3, IC is taken as the fixed–normalized inversion coefficient where the mobility is assumed fixed at its low-field value of μ_0. The loss of mobility, drain current, and transconductance associated with velocity saturation and VFMR is considered outside of the inversion coefficient definition. This permits linear relationships between the drain current, inversion coefficient, channel length, channel width, and gate area as described in Table 3.10.

All process parameters are corrected for the operating temperature as described in Section 3.5. As mentioned in Section A.2.1.3, the operating temperature is an optional user input having a listed default value of $T = 27\,°C$ (300 K).

A.2.3.7 *Layout Details and IC Trimming*

The seventh and final group of calculated MOS results contains details on layout and the trimming of *IC* required for desired layout dimensions. W_{DRAWN}, *each* is the drawn width for each drain and source finger. As noted, this is multiplied by the interdigitation multiplier, m, to give the total drawn channel width. L_{DRAWN} is the drawn channel length, repeated for convenience from the user input shown in Figure A.1. *AD, each, PD, each, AS, each,* and *PS, each* are the drain area, drain perimeter, source area, and source perimeter for each drain and source finger. As noted, these are multiplied by m to give the total values. The formatting of W_{DRAWN}, *each*, L_{DRAWN}, *AD, each, PD, each, AS, each,* and *PS, each* permits their use directly in simulation statements of MOS geometry using the multiplier value of m given in the *Optional User Design Inputs*, which is located below these calculated geometry values.

Although Table 3.25 gives values for the drain and source areas and perimeters assuming m is unity or even, these are calculated in the spreadsheet for all values of m, even though odd values are normally avoided when replicating or ratioing devices. The areas and perimeters are calculated using the values of *WDIFINT* and *WDIFEXT* from the *Process* sheet, which describe the drain and source finger, process dimensions for interior and exterior fingers. The overall area and perimeter are minimized for the drain.

IC (trimmed) reports the required trimmed input value for *IC*, described earlier in Section A.2.1.3, for optional user inputs of m and W_{DRAWN} *(trimmed), each*. As mentioned, the value of *IC (trimmed)* can be copied using the copy value (not copy formula) feature to the input value of *IC* shown in Figure A.1 or at the top of Figure A.3. This will result in a value of W_{DRAWN}, *each* that matches the desired value of W_{DRAWN} *(trimmed), each*. Again, as mentioned in Section A.2.1.3, this permits the optional trimming of the input *IC* to reflect desired, rounded layout dimensions, with the use of device interdigitation. As mentioned, it is recommended that the optional inputs of m and W_{DRAWN} *(trimmed), each* shown in Figure A.2 or at the bottom of Figure A.4 be left blank during initial design exploration. After this initial design exploration, inputs may be entered to reflect actual, rounded layout dimensions.

A.2.4 Reported MOS Performance: Bandwidth, Power, and Accuracy Figures of Merit

The third and final portion of the *MOSFETs* sheet, noted as *Sheet 3 – Performance Related to Bandwidth, Power, Accuracy Figures of Merit*, contains bandwidth, power, and accuracy figures of merit that were discussed in Sections 4.3.6.5–4.3.6.7 and summarized in Table 4.4. The reader is referred to the continuation pages of Figures 4.21 and 4.22 for illustrations of figures of merit contained in the *MOSFETs* sheet, which are described in the accompanying discussions of Sections 4.4.2 and 4.4.3. The trends listed in the right-most columns of Figures 4.21 and 4.22 are not part of the spreadsheet calculations, but were added to summarize performance trends for separate explorations of the channel length and drain current.

The first group of rows related to figures of merit in the *MOSFETs* sheet, shown in the continuation pages of Figures 4.21 and 4.22, contains a repeat of the *Device reference* and *Device notes* optional user design information from Figure A.1 and the top of Figure A.3. Additionally, the rows contain a repeat of the drain current, inversion coefficient, and channel length from the required user design inputs. This information is repeated so performance related to figures of merit can be displayed or printed on a separate page.

The second group of rows, entitled *Optional User Inputs for Figures of Merit*, contains user inputs for the supply voltage, V_{DD}, the input current relative to the drain bias current, I_{IN}/I_D, and the input voltage, V_{IN}. These user inputs are optional for calculations of MOS performance, but are required for calculations of figures of merit.

The third group of rows, entitled *Current-Mode Figures of Merit, Power and Bandwidth*, lists the power, $P = 2I_D \cdot V_{DD}$, and intrinsic bandwidth, $BW = f_{Ti}/2$, for a current-mode circuit consisting of a current mirror having two identical devices.

The fourth group of rows, entitled *Current-Mode Figures of Merit, DC Offset*, lists the input DC signal current, $I_{IN} = (I_{IN}/I_D) \cdot I_D$, and 3 σ input-referred offset current, I_{OS} (3 σ), for the current-mode circuit. The offset current is found from I_D multiplied by three times the 1 σ, relative drain current mismatch, $\Delta I_D/I_D$ (1 σ), displayed on the primary portion of the *MOSFETs* sheet shown in Figure A.3. The remaining entries are the DC accuracy, $A_{CC,DC} = I_{IN}/I_{OS}$ (3 σ), DC error, $E_{RR,DC} = I_{OS}$ (3 σ)/I_{IN}, and the bandwidth-accuracy and bandwidth-power-accuracy figures of merit, $A_{CC,DC} \cdot BW$, $A_{CC,DC}^2 \cdot BW$, and $A_{CC,DC}^2 \cdot BW/P$.

The fifth group of rows, entitled *Current-Mode Figures of Merit, Thermal Noise*, lists the input AC, rms signal current, $I_{IN} = (I_{IN}/I_D) \cdot I_D$, and input-referred, rms noise current, I_N, for the current-mode circuit. The noise current is found from the drain-referred thermal-noise current density, $S_{ID}^{1/2}$ thermal, displayed on the primary portion of the *MOSFETs* sheet, multiplied by the square root of two for two devices, finally multiplied by the square root of the bandwidth, BW, which is assumed to be the noise bandwidth. The remaining entries are the AC accuracy, $A_{CC,AC} = I_{IN}/I_N$, AC error, $E_{RR,AC} = I_N/I_{IN}$, and the bandwidth-accuracy and bandwidth-power-accuracy figures of merit, $A_{CC,AC} \cdot BW$, $A_{CC,AC}^2 \cdot BW$, and $A_{CC,AC}^2 \cdot BW/P$.

The sixth group of rows, entitled *Voltage-Mode Figures of Merit, Power and Bandwidth*, lists the power, $P = 2I_D \cdot V_{DD}$, and intrinsic bandwidth, $BW = 2f_{Ti}$, for a voltage-mode circuit consisting, for example, of a differential pair driven from a source impedance equal to $1/g_m$, where g_m is also the transconductance for each differential pair device.

The seventh group of rows, entitled *Voltage-Mode Figures of Merit, DC Offset*, lists the input DC signal voltage, V_{IN}, and 3 σ input-referred offset voltage, V_{OS} (3 σ), for the voltage-mode circuit. The offset voltage is found from three times the 1 σ, gate–source mismatch voltage, ΔV_{GS} (1 σ), displayed on the primary portion of the *MOSFETs* sheet. The remaining entries are the DC accuracy, $A_{CC,DC} = V_{IN}/V_{OS}$ (3 σ), DC error, $E_{RR,DC} = V_{OS}$ (3 σ)/V_{IN}, and the bandwidth-accuracy and bandwidth-power-accuracy figures of merit, $A_{CC,DC} \cdot BW$, $A_{CC,DC}^2 \cdot BW$, and $A_{CC,DC}^2 \cdot BW/P$.

The final group of rows, entitled *Voltage-Mode Figures of Merit, Thermal Noise*, lists the input AC, rms signal voltage, V_{IN}, and input-referred, rms noise voltage, V_N, for the voltage-mode circuit. The noise voltage is found from the gate-referred thermal-noise voltage density, $S_{VG}^{1/2}$ thermal, displayed on the primary portion of the *MOSFETs* sheet, multiplied by the square root of two for two devices, finally multiplied by the square root of the bandwidth, BW, which is assumed to be the noise bandwidth. The remaining entries are the AC accuracy, $A_{CC,AC} = V_{IN}/V_N$, AC error, $E_{RR,AC} = V_N/V_{IN}$, and the bandwidth-accuracy and bandwidth-power-accuracy figures of merit, $A_{CC,AC} \cdot BW$, $A_{CC,AC}^2 \cdot BW$, and $A_{CC,AC}^2 \cdot BW/P$.

As mentioned, the bandwidth, power, and accuracy figures of merit were discussed in Sections 4.3.6.5–4.3.6.7, summarized in Table 4.4, and illustrated in the continuation pages of Figures 4.21 and 4.22 with accompanying discussions given in Sections 4.4.2 and 4.4.3. It is important to note that bandwidth measures consider the intrinsic device bandwidth. As described in Section 3.9.7, extrinsic and diode-connected bandwidth measures, which include extrinsic gate-overlap and drain–body capacitances, can be significantly lower than the intrinsic bandwidth, especially for short-channel devices having small gate areas and small intrinsic capacitances. Correspondingly, the operating bandwidth for current-mode or voltage-mode circuits can be significantly below the predictions given that are based on intrinsic device bandwidth. This can also affect operating bandwidth trends where bandwidth increases less rapidly than the expected inverse square of decreasing channel length, which assumes negligible velocity saturation effects.

A.3 THE *CIRCUIT ANALYSIS* SHEET

The *Circuit Analysis* sheet in the *Analog CMOS Design, Tradeoffs and Optimization* spreadsheet is an optional, user-defined sheet that maps MOS device performance from the *MOSFETs* sheet into predictions of complete circuit performance. This sheet requires user-defined circuit analysis for the circuit topology of interest and "calls" device performance from the *MOSFETs* sheet. The *Circuit Analysis* sheet provides an immediate display of circuit performance as the designer explores MOS device design choices of drain current, inversion coefficient, and channel length. This leads towards rapid, optimum design.

The reader is referred to the *Circuit Analysis* sheets illustrated in Figures 5.4 and 5.14 in Sections 5.4.2 and 5.5.2. These sheets are for 0.5 μm, simple and 0.18 μm, cascoded CMOS OTAs that were optimized for DC, balanced, and AC performance. Additionally, the reader is referred to Figures 6.11 and 6.14 in Sections 6.7.3 and 6.8.3. These sheets are for 0.35 μm, differential and single-ended input, micropower CMOS preamplifiers that were optimized for minimum thermal and flicker noise.

Figures 5.4 and 5.14 include analysis of OTA transconductance, output resistance, voltage gain, transconductance bandwidth, and input-referred thermal-noise voltage, flicker-noise voltage, and offset voltage due to local-area mismatch. Additionally, these figures include analysis of input and output capacitances, input and output voltage ranges, and the input, 1 dB compression voltage. Figures 6.11 and 6.14 include analysis of preamplifier voltage gain, unity-gain bandwidth, and input-referred thermal- and flicker-noise voltage. Every analysis of thermal-noise, flicker-noise, and offset voltage includes an analysis of the noise and mismatch power contributions of non-input devices relative to the input devices. This helps facilitate low-noise or low-offset design by ensuring that input devices dominate the noise or mismatch.

Additional measures of circuit performance, including approximate layout area and user-defined figures of merit such as the thermal-noise efficiency factor described in Section 6.3.1, could be added to the *Circuit Analysis* sheets. As shown in Figures 5.4, 5.14, 6.11, and 6.14, the spreadsheet format facilitates calculations of predicted performance from user-defined circuit analysis. Also included are manual entries of simulated and measured performance that permit a comparison with spreadsheet predicted performance.

A.4 THE *PROCESS* SHEET

The *Process* sheet in the *Analog CMOS Design, Tradeoffs and Optimization* spreadsheet contains columns for individual MOSFET models. Each model column contains a model name followed by model parameters used for the calculation of MOS performance shown in the *MOSFETs* sheet of Figures A.3 and A.4.

Figure A.5 shows an excerpt of the *Process* sheet for the 0.18 μm, cascoded CMOS OTAs that were optimized for DC, balanced, and AC performance in Section 5.5.2. The *Process* sheet contains text notes, followed by columns of names, descriptions, units, and values for MOS model parameters. Since model parameter descriptions are included in the figure, these will not be re-described here. Instead, the model parameters will be linked to process parameters used in this book. The model parameters will also be linked to various aspects of MOS performance given in the *MOSFETs* sheet.

A.4.1 Description of Process Parameters

The first process parameter listed in Figure A.5 is *NAME*. This process model name must match the model name in the required user design inputs in the *MOSFETs* sheet, where these user inputs are shown in Figure A.1 and at the top of Figure A.3. As noted in Figure A.5, process parameters *TYPE* and *LEVEL* are not used, but are included to permit model enhancements. Similarly, process parameters *TINT* and *CINTEX* are not used, but are included for model enhancements involving low-temperature increases in interface state capacitance and the substrate factor that were mentioned earlier in Section 3.5.1.

Analog CMOS Design, Tradeoffs and Optimization Spreadsheet -- Process Parameters Sheet				
Notes:				
1. Model parameters appear in a vertical column for each MOS model.				
2. Some model parameter names are the same as those used in the EKV 2.6 MOS model, but units may be different.				
3. Model parameter polarities, e.g., DVTDIBL, are taken the same for both nMOS and pMOS devices.				
4. TYPE, LEVEL, TINT, and CINT are not used, but permit future enhancements.				
5. Model parameters of first model, starting with model name, must begin at cell D12. Up to 32 models are supported.				
6. **Changes to model parameters require recalculation, which can be done by "clicking" on design inputs.**				

	(Model values for TNOM)			
Name	**Description**	**Units**		
NAME	Model name	text	**nMIL2**	**nMIL1_4**
TYPE	Model type (not used)			
LEVEL	Model level (not used)			
U0	Low field mobility	cm^2/Vs	**422**	**422**
COX	Gate oxide capacitance	fF/μm^2	**8.41**	**8.41**
GAMMA	Body effect factor	V$^{1/2}$	**0.56**	**0.56**
PHI	(Twice) Fermi potential of channel ($2\phi_F$), positive N and P	V	**0.85**	**0.85**
N0	Average value of substrate factor for IC definition		**1.35**	**1.35**
VTO	Large geometry, zero-V_{SB} threshold voltage, positive N and P	V	**0.42**	**0.42**
ECRIT	Velocity saturation critical electric field (horizontal field)	V/μm	**5.6**	**5.6**
VSATEXP	Velocity saturation transition exponent		**1.3**	**1.3**
THETA	Mobility reduction factor (vertical field)	V^{-1}	**0.28**	**0.28**
DW	Lateral diffusion width, $W_{EFF} = W_{DRAWN} - DW$	μm	**0**	**0**
DL	Lateral diffusion length, $L_{EFF} = L_{DRAWN} - DL$	μm	**0.028**	**0.028**
VAL	Early voltage factor for channel length modulation	V/μm	**9.2**	**11.3**
DVTDIBL	DIBL drop in $|V_{TO}|$ with V_{DS} at min. L, negative N and P	V/V	**−0.008**	**−0.008**
DVTDIBLEX	Channel length exponent for DIBL drop in threshold voltage		**3**	**3**
KF0	Flicker-noise factor in weak inversion	C^2/cm^2	**3.17E−31**	**3.17E−31**
VKF	Voltage describing flicker-noise increase with $V_{EFF} = V_{GS} - V_T$	V	**1**	**1**
AF	Flicker-noise slope		**0.85**	**0.85**
AVTO	Local area threshold voltage mismatch factor, 1σ	V*μm	**0.005**	**0.005**
AKP	Local area transconductance factor mismatch factor, 1σ	μm	**0.02**	**0.02**
BEX	Mobility temperature exponent		**−1.5**	**−1.5**
UCEX	Velocity saturation critical electric field temperature exponent		**0.8**	**0.8**
TCV	Temperature coefficient for $|V_{TO}|$, negative N and P	V/K	**−0.001**	**−0.001**
TNOM	Temperature for model parameters	K	**300**	**300**
TINT	Temp. corner for interface state cap. increase (not used)	K		
CINTEX	Temp. exponent for interface state cap. increase (not used)			
CGDO	Gate–drain overlap capacitance	fF/μm	**0.94**	**0.94**
CGSO	Gate–souce overlap capacitance	fF/μm	**0.94**	**0.94**
CGBO	Gate–body overlap capacitance	fF/μm	**0**	**0**
CJ	Drain, source junction area capacitance	fF/μm^2	**0.96**	**0.96**
PB	Drain, source junction area capacitance built-in potential	V	**0.8**	**0.8**
MJ	Drain, source junction area capacitance voltage exponent		**0.38**	**0.38**
CJSW	Drain, source junction perimeter capacitance	fF/μm	**0.27**	**0.27**
PBSW	Drain, source junction perimeter capacitance built-in potential	V	**0.8**	**0.8**
MJSW	Drain, source junction perimeter capacitance voltage exponent		**0.15**	**0.15**
WDIFEXT	Drain, source exterior junction process width	μm	**0.5**	**0.5**
WDIFINT	Drain, source interior junction process width	μm	**0.6**	**0.6**
WDRAWNMIN	Minimum drawn channel width	μm	**0.3**	**0.3**
LDRAWNMIN	Minimum drawn channel length	μm	**0.18**	**0.18**
VDDMAX	Maximum V_{DD} allowed for process	V	**1.8**	**1.8**

Figure A.5   The *Process* sheet in the *Analog CMOS Design, Tradeoffs and Optimization* spreadsheet listing MOS process parameters and their descriptions

U0 (μ_0) and *COX* (C'_{OX}) pertain to calculations of the inversion coefficient, drain current, transconductance, effective gate–source voltage, intrinsic gate capacitances, and related aspects of performance. *N0* (n_0) additionally pertains to relationships between the inversion coefficient, drain current, channel width, and channel length. *GAMMA* (γ) and *PHI* pertain to calculations of the operating value of the substrate factor or n (actual) and the operating value of the inversion coefficient or *IC* (actual), which depends on n (actual). *GAMMA* and *PHI* also pertain to the body-effect increase in threshold voltage and local-area threshold-voltage mismatch for non-zero V_{SB}, and pertain to the body-effect transconductance. *VTO* (V_{TO}) pertains to the calculation of the threshold voltage. *ECRIT* (E_{CRIT}), *VSATEXP* (α), and *THETA* (θ) pertain to velocity saturation and VFMR effects on the drain current, transconductance, and effective gate–source voltage. *DW* and *DL* pertain to lateral diffusion adjustments for effective channel width and length. Finally, *VAL* (V_{AL}), *DVTDIBL*, and *DVTDIBLEXP* pertain to calculations of the drain–source conductance. These basic process parameters are related to the drain current, small-signal parameters, and intrinsic gate capacitances and are listed in Table 3.2 for the example 0.5, 0.35, and 0.18 μm CMOS processes. These parameters are described in Section 3.3.2 and referenced in relevant calculations of MOS performance used in the *MOSFETs* sheet shown in Figures A.3 and A.4.

KF0 (K_{F0}), *VKF* (V_{KF}), and *AF* pertain to the prediction of flicker noise, where *VKF* models the increase in gate-referred flicker-noise voltage with increasing inversion or effective gate–source voltage, $V_{EFF} = V_{GS} - V_T$. *AVTO* (A_{VTO}) and *AKP* (A_{KP}) pertain to the prediction of local-area threshold-voltage and transconductance factor mismatch. These noise and local-area mismatch process parameters are listed in Table 3.3 for the example 0.5, 0.35, and 0.18 μm CMOS processes. These parameters are described in Section 3.3.3 and referenced in relevant calculations of MOS performance used in the *MOSFETs* sheet.

CGDO, *CGSO*, and *CGBO* pertain to the prediction of gate-overlap capacitances. *CJ* (C_J), *PB* (P_B), and *MJ* (M_J) pertain to the prediction of the area component of the drain–body and source–body junction capacitances. *CJSW* (C_{JSW}), *PBSW* (P_{BSW}), and *MJSW* (M_{JSW}) pertain to the prediction of the perimeter component of these capacitances. *WDIFEXT* and *WDIFINT* are the process widths for exterior and interior, drain and source fingers that are used for the calculation of drain and source areas and perimeters, described earlier in Section A.2.3.7. These areas and perimeters are needed for calculation of the drain–body and source–body capacitances. The extrinsic, gate-overlap, drain–body, and source–body capacitance process parameters are listed in Table 3.4 for the example 0.5, 0.35, and 0.18 μm CMOS processes. These parameters are described in Section 3.3.4 and referenced in the relevant calculations of MOS performance used in the *MOSFETs* sheet. In Table 3.4 and for simplified calculations shown in Table 3.25, a single dimension of W_{DIF} is used instead of separate values of *WDIFEXT* and *WDIFINT*. As mentioned in Sections 3.9.3 and 3.9.4, the gate-overlap, drain–body, and source–body capacitance parameters can include estimated capacitance components from the source and drain extensions that complicate the prediction of capacitances.

BEX, UCEX, and *TCV* pertain to temperature effects on the mobility, critical horizontal, velocity saturation electric field, and threshold voltage. *TNOM* is the temperature used for the extraction of model parameters. These temperature process parameters are listed in Table 3.5 for the example 0.5, 0.35, and 0.18 μm CMOS processes. These parameters are described in Section 3.3.5 with calculations of temperature effects detailed in Section 3.5.

Finally, *WDRAWNMIN, LDRAWNMIN*, and *VDDMAX* are the minimum drawn channel width (for each drain or source finger when interdigitation is used), minimum drawn channel length, and maximum supply voltage for the process. These are not included in the process parameters for the example 0.5, 0.35, and 0.18 μm CMOS processes described in Tables 3.2–3.5, but are used for error checking in the *MOSFETs* sheet. As mentioned in Sections A.2.1.2 and A.2.1.3, the spreadsheet returns error messages if the drawn channel width or drawn channel length is less than the process minimum, or if MOS bias voltages are greater than the maximum process supply voltage.

A.4.2 Binning of V_{AL} Process Parameter

The *Analog CMOS Design, Tradeoffs and Optimization* spreadsheet uses process parameters, like those shown in the *Process* sheet of Figure A.5, that are fixed or unbinned across device geometry and bias conditions, except for the channel length modulation, Early voltage factor, $V_{AL} = V_A/L$. As described in Section 3.8.4, no simple prediction of MOS Early voltage, $V_A = I_D/g_{ds}$, or the related drain–source conductance, $g_{ds} = I_D/V_A$, is available because of complex dependencies on the inversion level, channel length, and drain–source voltage. Instead, the process parameter $V_{AL} = V_A/L$ is extracted from V_A measurements near the operating inversion coefficient, channel length, and drain–source voltage.

The extraction of $V_{AL} = V_A/L$ from V_A measurements was described near the beginning of Section 4.4 for the design of differential pairs and current mirrors and near the beginning of Sections 5.4.2 and 5.5.2 for the design of OTAs. For example, for the design of differential pairs and current mirrors in the example 0.18 μm CMOS process, V_A measurements were taken from Figure 3.45. These measurements were taken from weak through strong inversion for channel lengths ranging from $L = 0.18$ to 4 μm with the drain–source voltage held fixed at $V_{DS} = 0.5$ V, which is the value considered for the differential pairs and current mirrors. The corresponding process values of $V_{AL} = V_A/L$ used in the *Process* sheet for the differential pairs and current mirrors are similar to those shown in Figure 3.48 for nMOS devices in the example 0.18 μm CMOS process operating at $V_{DS} = 0.5$ V. However, the decrease in V_{AL} at short channel lengths caused by DIBL is removed because DIBL effects are included separately in the spreadsheet calculations. The removal of DIBL effects for short-channel devices was described near the beginning of Sections 4.4, 5.4.2, and 5.5.2 such that process $V_{AL} = V_A/L$ extractions consider channel length modulation only.

The extraction of process V_{AL} values for the *Process* sheet requires some effort, especially for a circuit having a wide range of MOS channel lengths or drain–source voltages. However, V_{AL} only affects g_{ds} and related small-signal resistances and signal gains, so rough estimates of V_{AL} do not affect other aspects of performance. Linking the *Analog CMOS Design, Tradeoffs and Optimization* spreadsheet to a full simulation MOS model that provides good g_{ds} modeling could eliminate binned model parameter extraction of V_{AL}. However, experimental evaluation is still recommended given the complexity of predicting g_{ds} and its long history of modeling errors. Section A.6 describes possible enhancements to the spreadsheet.

A.4.3 Forcing Recalculation Whenever Process Parameters Are Changed

Contained in bold near the top of the *Process* sheet (Figure A.5) in the *Analog CMOS Design, Tradeoffs and Optimization* spreadsheet is the important note: **Changes to model parameters require recalculation, which can be done by "clicking" on design inputs**. Recalculation can be invoked by activating or changing any of the required or optional user inputs shown in Figures A.1 and A.2. This is required whenever process parameters on the *Process* sheet are changed because changes on this sheet do not invoke spreadsheet recalculation. This is because process parameters on the *Process* sheet are called from array functions on the *MOSFETs* sheet. Further discussion of the spreadsheet software architecture and operation is contained in the spreadsheet code.

A.5 THE *DISCLAIMER, NOTES* SHEET

A.5.1 Overview

The *MOSFETs* sheet in the *Analog CMOS Design, Tradeoffs and Optimization* spreadsheet contains the disclaimer, "For design guidance only as results are approximate and not for any particular CMOS process." The *Disclaimer, Notes* sheet contains a more detailed disclaimer given in Section A.5.2 below. Additionally, the *Disclaimer, Notes* sheet contains notes for using the spreadsheet.

As mentioned in Section 1.7 and at various places in this book, and well known to experienced designers, all candidate designs must be thoroughly verified by computer simulations using MOS models with parameters appropriate for the production process. This normally involves extensive simulations, inclusive of layout parasitics, over nominal and corner process conditions, temperature, and supply voltage. Portions of the disclaimer below appear at various places in this book and in the *Disclaimer, Notes* sheet as a reminder.

A.5.2 Disclaimer Statement

The design tradeoff and optimization methods, predictions, examples, and measurements given in this book or its associated spreadsheet software are intended for design guidance only, not for actual design, and do not correspond to any particular CMOS fabrication process. The designer must independently validate designs using MOS models and parameters appropriate for the actual fabrication process used.

Use of information in this book or its associated spreadsheet software expressly indicates the assumption of risk that this information should only be used for design guidance and should not be used for actual design or the validation of actual design. Additionally, use of information in this book or its associated spreadsheet software expressly indicates acknowledgement of responsibility for independently validating designs using MOS models and parameters appropriate for the actual fabrication process used.

The information in this book and associated spreadsheet software is provided without express or implied warranties that the information is accurate or reliable, and there are no warranties as to fitness for any particular purpose. Neither the author nor John Wiley & Sons, Ltd accept any responsibility or liability for loss or damage occasioned to any person or property through using the material, instructions, methods, or ideas contained herein, or acting or refraining from acting as a result of such use. The author and Publisher expressly disclaim all implied warranties, including merchantability of fitness for any particular purpose. There will be no duty on the author or Publisher to correct any errors or defects in this book or its associated spreadsheet software.

A.6 FUTURE ENHANCEMENTS

Velocity saturation and VFMR decreases in transconductance efficiency and transconductance, and the related increases in V_{EFF}, are included in predictions of MOS performance in the *MOSFETs* sheet in the *Analog CMOS Design, Tradeoffs and Optimization* spreadsheet. Additionally, DIBL decreases in V_A, and the related increases in g_{ds}, are included along with increases in gate-referred flicker-noise voltage with increasing inversion or V_{EFF}.

Although these important small-geometry effects are included, gate leakage current and the resulting gate–source conductance, gate shot- and flicker-noise current, and increase in DC mismatch are not included in the spreadsheet calculations. These effects are negligible for the gate-oxide thickness of 4.1 nm or larger for the example 0.5, 0.35, and 0.18 μm CMOS processes, but can be significant for smaller-geometry processes having gate-oxide thickness below 2 nm. As described in Section 3.12.2, smaller channel length and gate area may be required to balance the traditional improvements in gain, flicker noise, and local-area mismatch at increasing channel length and gate area with the deterioration of these associated with increasing gate leakage current caused by increasing gate area. Enhanced versions of the spreadsheet could then include gate leakage current effects.

As mentioned in Section A.4.2, the channel length modulation, Early voltage factor, V_{AL}, requires binning near the actual inversion coefficient, channel length, and drain–source voltage used for accurate predictions of V_A and, correspondingly, g_{ds}. Linking the spreadsheet to full simulation MOS models like the EKV (Enz, Krummenacher, Vittoz) inversion charge or PSP (Philips, Pennsylvania State University) surface-potential models could potentially eliminate the need for V_{AL} binning. Although the spreadsheet is not intended to provide simulation accuracy or, as mentioned in Section A.5, to replace simulations, linking the spreadsheet to simulation MOS models could result in accuracy approaching that of simulations. Additionally, gate leakage current effects could be included for processes having gate-oxide thickness below 2 nm. One potential disadvantage of linking the spreadsheet to simulation MOS models could be the large number of process parameters required and the availability of these parameters for typical processes. In this regard, the EKV models have fewer model parameters, leading also to simpler predictions of MOS performance. This, in fact, inspired the initial spreadsheet development followed by the writing of this book.

The reader is invited to periodically check the book's web site listed on the cover for possible enhancements to the *Analog CMOS Design, Tradeoffs and Optimization* spreadsheet. As described in Chapter 7, the technology and process normalization inherent in the inversion coefficient facilitates design optimization in smaller-geometry CMOS processes and even potentially in emerging, non-CMOS technologies. The evaluation of performance tradeoffs and the optimization of design are certainly not limited to the example processes used for device and circuit design examples in this book.

Index

Printed and bound in the UK by
CPI Antony Rowe, Eastbourne

Printed and bound by CPI Group (UK) Ltd, Croydon, CR0 4YY

16/04/2025

14658472-0001